Bubbles, Drops, and Particles in Non-Newtonian Fluids

The third edition of *Bubbles, Drops, and Particles in Non-Newtonian Fluids* provides comprehensive coverage of the scientific foundations and the latest advances in particle motion in non-Newtonian media.

Thoroughly updating and expanding its best-selling predecessor, this edition addresses numerical and experimental developments in non-Newtonian particulate systems. It includes a new chapter on heat transfer in non-Newtonian fluids in the free and mixed convection regimes and thus covers forced convection regimes separately in this edition.

Salient Features:

- Demonstrates how dynamic behavior of single particles can yield useful information for modeling transport processes in complex multiphase flows.
- Addresses heat transfer in *Generalized Newtonian Fluids* (GNF), visco-plastic and visco-elastic fluids throughout the book and outlines potential strategies for heat transfer enhancement.
- Provides a new detailed section on the effect of confinement on heat transfer from bluff-bodies in non-Newtonian fluids.

Written in a clear and concise manner, this book remains an excellent handbook and reference. It is essential reading for students and researchers interested in exploring particle motion in different types of non-Newtonian systems encountered in disciplines across engineering and the sciences.

Chemical Industries

Founding Editor: Heinz Heinemann
Series Editor:
James G. Speight

The **Chemical Industries** Series offers in-depth texts related to all aspects of the chemical industries from experts and leaders in academia and industry. The titles explore recent developments and best practices that facilitate successful process control and commercialization of industrial processes and products to help meet changing market demands and match the stringent emission standards. The series focuses on technologies, process development improvements, and new applications to ensure proper performance in industrial units and evaluation of novel process designs that will result in production of valuable products from efficient and economical processes.

Refining Used Lubricating Oils
James Speight and Douglas I. Exall

Petroleum and Gas Field Processing, Second Edition
Hussein K. Abdel-Aal, Mohamed A. Aggour, and Mohamed A. Fahim

Handbook of Refinery Desulfurization
Nour Shafik El-Gendy and James G. Speight

Handbook of Petroleum Refining
James G. Speight

Advances in Refining Catalysis
Deniz Uner

Lubricant Additives: Chemistry and Applications, Third Edition
Leslie R. Rudnick

Handbook of Petrochemical Processes
James G. Speight

Transport Phenomena Fundamentals, Fourth Edition
Joel L. Plawsky

Synthetics, Mineral Oils, and Bio-Based Lubricants: Chemistry and Technology, Third Edition
Leslie R. Rudnick

Transport Phenomena Fundamentals, Fourth Edition
Joel Plawsky

Introduction to Process Control, Third Edition
Jose A. Romagnoli and Ahmet Palazoglu

Bubbles, Drops, and Particles in Non-Newtonian Fluids, Third Edition
Raj P. Chhabra and Swati A. Patel

For more information about this series, please visit: https://www.crcpress.com/Chemical-Industries/book-series/CRCCHEMINDUS

Bubbles, Drops, and Particles in Non-Newtonian Fluids

Third Edition

Raj P. Chhabra
Department of Chemical Engineering
Shiv Nadar University
Delhi NCR Campus, Dadri

AND

Swati A. Patel
Department of Chemical Engineering
Indian Institute of Technology Ropar
Rupnagar

CRC Press is an imprint of the
Taylor & Francis Group, an **informa** business

Third edition published 2024
by CRC Press
6000 Broken Sound Parkway NW, Suite 300, Boca Raton, FL 33487-2742

and by CRC Press
4 Park Square, Milton Park, Abingdon, Oxon, OX14 4RN

CRC Press is an imprint of Taylor & Francis Group, LLC

© 2024 Raj P. Chhabra and Swati A. Patel

First edition published by CRC Press 1992
Second edition published by CRC Press 2007

Reasonable efforts have been made to publish reliable data and information, but the author and publisher cannot assume responsibility for the validity of all materials or the consequences of their use. The authors and publishers have attempted to trace the copyright holders of all material reproduced in this publication and apologize to copyright holders if permission to publish in this form has not been obtained. If any copyright material has not been acknowledged please write and let us know so we may rectify in any future reprint.

Except as permitted under U.S. Copyright Law, no part of this book may be reprinted, reproduced, transmitted, or utilized in any form by any electronic, mechanical, or other means, now known or hereafter invented, including photocopying, microfilming, and recording, or in any information storage or retrieval system, without written permission from the publishers.

For permission to photocopy or use material electronically from this work, access www.copyright.com or contact the Copyright Clearance Center, Inc. (CCC), 222 Rosewood Drive, Danvers, MA 01923, 978-750-8400. For works that are not available on CCC please contact mpkbookspermissions@tandf.co.uk

Trademark notice: Product or corporate names may be trademarks or registered trademarks and are used only for identification and explanation without intent to infringe.

ISBN: 9780367203023 (hbk)
ISBN: 9781032417554 (pbk)
ISBN: 9780429260759 (ebk)

DOI: 10.1201/9780429260759

Typeset in Times
by codeMantra

Contents

Preface to the Third Edition xiii
Preface to the Second Edition xv
Preface to the First Edition xvii
Acknowledgments xix
Authors xxi

Introduction 1

1. Non-Newtonian Fluid Behavior 7
    1.1   Introduction 7
    1.2   Definition of a Newtonian Fluid 7
    1.3   Non-Newtonian Fluids 11
        1.3.1   Time-Independent Fluid Behavior 11
                1.3.1.1   Shear-Thinning or Pseudoplastic Fluids 12
                1.3.1.2   Visco-Plastic Fluids 17
                1.3.1.3   Shear-Thickening Fluids 21
        1.3.2   Time-Dependent Behavior 24
                1.3.2.1   Thixotropy 24
                1.3.2.2   Rheopexy or Negative Thixotropy 26
        1.3.3   Visco-elastic Behavior 28
                1.3.3.1   Normal-Stress Effects in Steady Shearing Flows 29
                1.3.3.2   Elongational Flow 32
                1.3.3.3   Small-Amplitude Oscillatory Shearing Motion 34
                1.3.3.4   Mathematical Models for Visco-Elastic Behavior 37
    1.4   Dimensional Considerations in the Fluid Mechanics of Visco-elastic Fluids 39
    1.5   Experimental Techniques: Rheometry 41
    1.6   Concluding Remarks 42
    Nomenclature 42
    Greek Symbols 43
    Subscripts 44
    Superscripts 44

2. Rigid Particles in Time-Independent Liquids without a Yield Stress 45
    2.1   Introduction 45
    2.2   Governing Equations for a Sphere 45
    2.3   Spherical Particles in Newtonian Fluids 48
        2.3.1   Drag Force 49
        2.3.2   Free-Fall Velocity 50
        2.3.3   Flow Regimes 51
        2.3.4   Unsteady Motion 54
    2.4   Spheres in Shear-Thinning Liquids 55
        2.4.1   Drag Force 55
                2.4.1.1   Theoretical Developments in Creeping Flow Region 55
                2.4.1.2   Experimental Results 66

		2.4.1.3 Drag Force at High Reynolds Numbers	71

2.4.2 Free-Fall Velocity 77
2.4.3 Flow Field and Flow Regime 78
2.4.4 Unsteady Motion 82
2.4.5 Effect of Imposed Fluid Motion 82
2.5 Spheres in Shear-Thickening Liquids 83
2.6 Drag on Light Spheres Rising in Pseudoplastic Media 84
2.7 Pressure Drop Due to a Settling Sphere 85
2.8 Nonspherical Particles 86
   2.8.1 Introduction 86
   2.8.2 Drag Force 87
       2.8.2.1 Newtonian Fluids 87
       2.8.2.2 Shear-Thinning Liquids 90
2.9 Conclusions 101
Nomenclature 102
Greek Symbols 103
Superscript 103

3. Rigid Particles in Visco-Plastic Liquids 105
3.1 Introduction 105
3.2 Spheres in Visco-Plastic Liquids 107
   3.2.1 Static Equilibrium 107
   3.2.2 Flow Field 112
   3.2.3 Drag Force 118
       3.2.3.1 Theoretical Developments 118
       3.2.3.2 Experimental Drag Correlations 122
   3.2.4 Values of Yield Stress Used in Correlations 128
   3.2.5 Time Dependence of Velocity in Visco-Plastic Fluids 129
3.3 Flow Past a Circular Cylinder 131
3.4 Flow Normal to a Plate 137
3.5 Nonspherical Particles 138
3.6 Conclusions 139
Nomenclature 139
Greek Symbols 140
Subscripts 140

4. Rigid Particles in Visco-Elastic Fluids 141
4.1 Introduction 141
4.2 Flow Over a Sphere 146
   4.2.1 Theoretical Developments 147
       4.2.1.1 Drag Force on an Unbounded ($\beta = 0$) Sphere in Creeping Region (Re $\rightarrow$ 0) 147
       4.2.1.2 Drag Force on a Sphere for $\beta = 0.5$ and Re $\rightarrow$ 0: The Benchmark Problem 149
       4.2.1.3 Wake Phenomenon 151
   4.2.2 Experimental Results 153
       4.2.2.1 Shear-Thinning Visco-Elastic Liquids 153
       4.2.2.2 Nonshear-Thinning Visco-Elastic Liquids (Boger Fluids) 156
   4.2.3 The Time Effect 159
   4.2.4 Velocity Overshoot 160
   4.2.5 Drag-Reducing Fluids 161

		4.2.6	Sphere in Mixed Flows	162
4.3	Flow Over a Long Circular Cylinder			164
4.4	Interaction between Visco-Elasticity, Particle Shape, Multiple Particles, Confining Boundaries, and Imposed Fluid Motion			168
4.5	Conclusions			172
Nomenclature				173
Greek Symbols				174

5. Fluid Particles in Non-Newtonian Media 175
- 5.1 Introduction 175
- 5.2 Formation of Fluid Particles 176
  - 5.2.1 Bubbles 176
    - 5.2.1.1 Davidson–Schuler Model 177
    - 5.2.1.2 Kumar–Kuloor Model 178
  - 5.2.2 Drops 182
    - 5.2.2.1 Criterion I: Low-Viscosity Systems 183
    - 5.2.2.2 Criterion II: High-Viscosity Systems 183
  - 5.2.3 Disintegration (or Breakup) of Jets and Sheet 185
  - 5.2.4 Growth or Collapse of Bubbles 186
- 5.3 Shapes of Bubbles and Drops in Free Rise or Fall 188
  - 5.3.1 Newtonian Continuous Media 188
  - 5.3.2 Non-Newtonian Continuous Media 191
- 5.4 Terminal Velocity–Volume Behavior in Free Motion 206
- 5.5 Drag Behavior of Single Particles 213
  - 5.5.1 Theoretical Developments 213
    - 5.5.1.1 Newtonian Continuous Phase 214
    - 5.5.1.2 Shear-Thinning Continuous Phase 216
    - 5.5.1.3 Visco-Elastic Continuous Phase 222
    - 5.5.1.4 Non-Newtonian Drops 223
  - 5.5.2 Experimental Results 225
- 5.6 Bubble and Drop Ensembles in Free Motion 228
- 5.7 Coalescence of Bubbles and Drops 230
  - 5.7.1 Bubble Coalescence 231
  - 5.7.2 Drop Coalescence 233
- 5.8 Breakage of Drops 234
- 5.9 Motion and Deformation of Bubbles and Drops in Confined Flows ... 235
- 5.10 Conclusions 236
- Nomenclature 236
- Greek Symbols 238
- Subscripts 238

6. Non-Newtonian Fluid Flow in Porous Media and Packed Beds 239
- 6.1 Introduction 239
- 6.2 Porous Medium 240
  - 6.2.1 Definition of a Porous Medium, Its Classification, and Examples 240
  - 6.2.2 Description of a Porous Medium 241
- 6.3 Newtonian Liquids 244
  - 6.3.1 Flow Regimes 244
  - 6.3.2 Pressure Loss–Throughput Relationship 246
    - 6.3.2.1 Dimensionless Empirical Correlations 247
    - 6.3.2.2 The Conduit or Capillary Models 249

		6.3.2.3	The Submerged Objects Models or Drag Theories	253
		6.3.2.4	Use of the Field Equations for Flow through a Porous Medium	257
		6.3.2.5	Flow in Periodically Constricted Tubes (PCTs)	257
		6.3.2.6	Volume Averaging of the Navier–Stokes Equations	259
	6.3.3	Wall Effects		260
	6.3.4	Effects of Particle Shape, Particle Roughness, and Size Distribution		264
	6.3.5	Fibrous Porous Media		266
	6.3.6	Theoretical Treatments		269
		6.3.6.1	Flow Parallel to an Array of Rods	270
		6.3.6.2	Transverse Flow over an Array of Rods	271
		6.3.6.3	Creeping Flow Region	271
		6.3.6.4	Inertial Effects	274
6.4	Non-Newtonian Fluids			275
	6.4.1	Flow Regimes		288
	6.4.2	Pressure Loss for Generalized Newtonian Fluids		288
		6.4.2.1	The Capillary Model	289
		6.4.2.2	Submerged Object Models or Drag Theories	299
		6.4.2.3	Volume Averaging of Equations	303
		6.4.2.4	Other Methods	303
	6.4.3	Visco-Elastic Effects in Porous Media		304
	6.4.4	Dilute/Semidilute Drag Reducing Polymer Solutions		308
	6.4.5	Wall Effects		311
	6.4.6	Effect of Particle Shape and Size Distribution		312
	6.4.7	Flow in Fibrous Media		312
		6.4.7.1	Generalized Newtonian fluids	312
		6.4.7.2	Visco-Elastic Fluids	319
	6.4.8	Mixing in Packed Beds		320
6.5	Miscellaneous Effects			320
	6.5.1	Polymer Retention in Porous Media		321
	6.5.2	Slip Effects		323
	6.5.3	Flow-Induced Mechanical Degradation of Flexible Molecules in Solutions		324
6.6	Two-Phase Gas/Liquid Flow			326
6.7	Conclusions			327
Nomenclature				327
Greek Symbols				329
Subscripts				330
Superscript				330

7. Fluidization and Hindered Settling 331

7.1	Introduction			331
7.2	Two-Phase Fluidization			332
	7.2.1	Minimum Fluidization Velocity		332
		7.2.1.1	Definition	332
		7.2.1.2	Prediction of V_{mf}	333
		7.2.1.3	Non-Newtonian Systems	334
	7.2.2	Bed Expansion Behavior		337
		7.2.2.1	Inelastic Non-Newtonian Systems	341
	7.2.3	Effect of Visco-elasticity		350
7.3	Three-Phase Fluidized Beds			352
	7.3.1	Introduction		352
	7.3.2	Minimum Fluidization Velocity		352
	7.3.3	Bed Expansion Behavior		352

	7.3.4 Gas Holdup 354

- 7.4 Sedimentation or Hindered Settling 357
  - 7.4.1 Non-Newtonian Studies 358
- 7.5 Conclusions 360
- Nomenclature 360
- Greek Symbols 361
- Subscripts 361

8. Heat and Mass Transfer in Particulate Systems: Forced Convection 363
- 8.1 Introduction 363
- 8.2 Boundary Layer Flows 370
  - 8.2.1 Plates 370
  - 8.2.2 Cylinders 375
  - 8.2.3 Spheres 384
- 8.3 Visco-elastic Effects in Boundary Layers 394
- 8.4 Bubbles 398
  - 8.4.1 Large Peclet Number (Pe $\gg$ 1) 398
  - 8.4.2 Small Peclet Number (Pe $\ll$ 1) 400
- 8.5 Drops 401
- 8.6 Ensembles of Bubbles and Drops 403
- 8.7 Fixed Beds 405
- 8.8 Liquid–Solid Fluidized Beds 409
- 8.9 Three-Phase Fluidized Bed Systems 410
- 8.10 Heat Transfer from Tube Bundles 411
- 8.11 Conclusions 412
- Nomenclature 413
- Greek Symbols 415
- Subscripts 415

9. Heat and Mass Transfer in Particulate Systems: Free and Mixed Convection 417
- 9.1 Introduction 417
- 9.2 Governing Equations 417
- 9.3 Vertical Plate 424
  - 9.3.1 Free Convection 424
    - 9.3.1.1 Newtonian Fluids 424
    - 9.3.1.2 Power-Law Fluids 425
    - 9.3.1.3 Bingham Plastic Fluids 430
  - 9.3.2 Mixed Convection 430
    - 9.3.2.1 Newtonian Fluids 432
    - 9.3.2.2 Power-Law Fluids 433
    - 9.3.2.3 Visco-plastic Fluids 434
- 9.4 Horizontal Cylinders 434
  - 9.4.1 Free Convection 434
    - 9.4.1.1 Newtonian Fluids 434
    - 9.4.1.2 Power-Law Fluids 436
    - 9.4.1.3 Bingham Plastic Fluids 440
  - 9.4.2 Mixed Convection 443
    - 9.4.2.1 Newtonian Fluids 443
    - 9.4.2.2 Power-Law Fluids 444
    - 9.4.2.3 Bingham Plastic Fluids 448
- 9.5 Spheres 448

	9.5.1	Free Convection	448
		9.5.1.1 Newtonian Fluids	448
		9.5.1.2 Power-Law Fluids	450
		9.5.1.3 Bingham Plastic Fluids	457
	9.5.2	Mixed Convection	461
		9.5.2.1 Newtonian Fluids	461
		9.5.2.2 Power-Law Fluids	463
		9.5.2.3 Bingham Plastic Fluids	465
9.6	Visco-elastic Effects in Boundary Layers		466
9.7	Conclusions		467
Nomenclature			467
Greek Symbols			469
Subscripts			469

10. Wall Effects 471

- 10.1 Introduction 471
- 10.2 Definition 472
- 10.3 Rigid Spheres 472
  - 10.3.1 Newtonian Fluids 472
    - 10.3.1.1 Theoretical Treatments 472
    - 10.3.1.2 Experimental Results and Correlations 475
  - 10.3.2 Inelastic Non-Newtonian Liquids 482
    - 10.3.2.1 Theoretical and Numerical Treatments 482
    - 10.3.2.2 Experimental Studies 484
  - 10.3.3 Visco-plastic Liquids 491
  - 10.3.4 Visco-elastic Liquids 493
    - 10.3.4.1 Boger Fluids 494
- 10.4 Nonspherical Rigid Particles 496
  - 10.4.1 Newtonian Liquids 497
  - 10.4.2 Inelastic Non-Newtonian Liquids 499
- 10.5 Effect of Blockage on Heat Transfer from a Sphere 500
- 10.6 Drops and Bubbles 502
  - 10.6.1 Newtonian Continuous Phase 504
    - 10.6.1.1 Low Reynolds Number Regime 504
    - 10.6.1.2 High Reynolds Number Regime 504
  - 10.6.2 Non-Newtonian Continuous Phase 505
- 10.7 Conclusions 506
- Nomenclature 506
- Greek Symbols 507
- Subscripts 508

11. Falling Object Rheometry 509

- 11.1 Introduction 509
- 11.2 Falling Ball Method 509
  - 11.2.1 Newtonian Fluids 509
  - 11.2.2 Non-Newtonian Fluids 510
    - 11.2.2.1 Zero-Shear Viscosity 511
    - 11.2.2.2 Shear-Dependent Viscosity 517
    - 11.2.2.3 Yield Stress 519
    - 11.2.2.4 Characteristic Time for Visco-elastic Fluids 521
- 11.3 Rolling Ball Method 521

		11.3.1	Newtonian Fluids 521
		11.3.2	Non-Newtonian Fluids (Shear-Dependent Viscosity) 522
		11.3.3	Yield Stress 523
	11.4	Rotating Sphere Viscometer 523
	11.5	Falling Cylinder Viscometer 524
		11.5.1	Newtonian Fluids 524
		11.5.2	Non-Newtonian Fluids 527
			11.5.2.1 Shear-Dependent Viscosity 527
			11.5.2.2 Yield Stress 528
	11.6	Concluding Summary 528
	Nomenclature 529
	Greek Symbols 530
	Subscripts 530

References 531

Author Index 659

Subject Index 701

Preface to the Third Edition

The second edition of this title appeared more than 15 years ago so we thought it is time to prepare the next edition. The fact that this title has never gone out of print is a testimony to the timelessness of its contents as well as the continuing interest in the topics covered in this book. In the fast-changing scientific landscape, 15 years is a long time and indeed significant advances have occurred directly relevant to the subject matter of this book. Therefore, a new edition is required for this title to remain relevant and competitive.

The objectives of this edition remain the same as that of the previous editions, namely, to provide a reference text book for senior undergraduate and postgraduate students, researchers and industrial practitioners who must deal with such complex fluids on a daily basis at their work place. Similarly, we have retained the approach to the organization of the material and style of presentation of the previous editions. Each chapter has been thoroughly reviewed and updated as deemed appropriate to reflect the current state of the art. One major change in this edition is the splitting of the chapter on interphase heat and mass transfer into two separate chapters, namely, the forced convection regime is treated in Chapter 8, whereas the free- and mixed convection regimes are dealt with in Chapter 9.

Lastly, we thank the numerous readers who not only spotted typos and mistakes in the previous editions but also made helpful and constructive suggestions for further improvement. We have tried our best to incorporate such suggestions in this edition. Finally, we sincerely hope that this edition would be received as warmly as its predecessors.

Raj P. Chhabra and Swati A. Patel
Dadri and Rupnagar

Preface to the Second Edition

Since the publication of its first edition in 1993, significant advances have occurred in this field that impinge on almost all topics covered in this book. This, coupled with the fact that the first edition has been out of print for a while now, prompted me to prepare a new edition of this book. The goals and the structure of this edition are the same as in the first edition: to provide a reference text for graduate students and researchers by presenting a comprehensive and critical evaluation of the available extensive literature relating to the non-Newtonian effects in multiphase flows encompassing single particles at one extreme and concentrated systems such as porous media at the other extreme, and to provide reliable estimation methods required by practicing professionals to perform day-to-day routine design calculations relating to the process technology of non-Newtonian fluids, such as the prediction of the settling velocity of individual particles and concentrated suspensions or the frictional pressure drop in porous media flows.

Bearing in mind the independent reviews of the first edition and the suggestions made by the readers of the first edition, the entire book has been reviewed. Where the need was recognized, the presentation has been improved by reorganizing the material for easier understanding, new material to facilitate comprehension has been added, and the most current viewpoints and research results have been incorporated in this revised edition. Apart from the general updating of all chapters, the specific changes made from the first edition are summarized as follows.

The introductory material in Chapter 1 has been completely overhauled and greatly expanded to illustrate both the pervasive and the ubiquitous nature of non-Newtonian fluid behavior as encountered in everyday life and in technology. In Chapter 2, a short section on extensional flow has been added due to its direct relevance in porous media flows and in the growth of gaseous inclusions in polymer melts and other non-Newtonian media encountered in food-processing and geological engineering applications. New research results relating to the flow of time-independent fluids (without a yield stress) past axisymmetric particles like a cylinder or a circular disk or spheroidal particles have been incorporated into Chapter 3. Likewise, the corresponding body of experimental results for the free settling behavior of a range of nonspherical shapes of particles has been treated in detail in this chapter. Also, the interesting behavior exhibited by ascending light spheres in quiescent polymer solutions has now been introduced in this chapter. For added emphasis, the hydrodynamics of rigid particles in visco-plastic media has now been treated separately in Chapter 4. Since the first edition, the significant advances made in the field of particle motion in visco-elastic fluids, especially for the two benchmark problems of a sphere in a tube and a cylinder in a planar slit, are reflected in the new look of Chapter 5. The treatment of the dynamics of bubbles and drops and their ensembles in a wide variety of stagnant and moving non-Newtonian media has been completely revised in Chapter 6 to take into account the new developments since the publication of the first edition. In particular, the measurements of the detailed velocity fields close to the rising bubbles have been included here. An extensive section on the flow of both Newtonian and non-Newtonian liquids in fibrous media has been added to Chapter 7. The other changes to this chapter include a slight expansion of the section on the use of the volume averaging methods for modeling porous media flow and a new section on the two-phase flow of a gas and a non-Newtonian liquid in packed beds. The treatment of the role of visco-elasticity in liquid–solid fluidization and in the sedimentation of concentrated suspensions of noninteracting particles has been somewhat sharpened in Chapter 8, especially in view of the new evidence showing the formation of preferred flow passages in these systems. The new Chapter 9 presents an overview of the extensive literature on the non-Newtonian effects in boundary layer flows and on the interphase heat/mass transfer from particles immersed in a variety of non-Newtonian fluids. To highlight their significance, the additional effects arising from the presence of confining walls on the sedimentation of particles have now been treated on their own in a separate Chapter 10. Chapter 11 has been considerably expanded by adding a whole range of viscometers including the falling cylinder (needle) and the rolling ball devices to evaluate a variety of non-Newtonian characteristics, all of which are routinely used for monitoring the quality of a range of products during the course of their manufacture. Finally, in this edition, the list of references cited has been placed at the end of the text, for this format not only avoids duplication between chapters, but it is also thought to be more convenient for the reader.

And lastly, I would like to thank the readers who have made such helpful suggestions and have drawn my attention to the errors in the first edition, many of which I would have never spotted myself. Please keep up the good work and let me know if, no, not if, but when, you find errors in this edition.

Raj P. Chhabra
Kanpur, India

Preface to the First Edition

For the past 300 years or so, the simple Newtonian fluid model has been accepted as the standard fluid behavior. Though most gases and low-molecular-weight substances do exhibit this kind of fluid behavior, in recent years, there has been an increasing recognition of the importance of non-Newtonian flow characteristics displayed by most materials encountered in everyday life, both in nature (gums, proteins, biological fluids such as blood, synovial fluid, etc.) and in technology (polymers and plastics, emulsions, slurries, etc.). Indeed, so widespread is the non-Newtonian behavior that it would be no exaggeration to say that the Newtonian fluid behavior is an exception rather than the rule. Consequently, the last three to four decades have witnessed a remarkable upsurge of interest and research activity in the mechanics of non-Newtonian fluids. An unprecedented increase in the number of research papers, books, and conference proceedings testifies to the rapid growth of this field. There are even specialized serial publications (*Journal of Rheology*, *Rheologica Acta*, and *Journal of Non-Newtonian Fluid Mechanics*) devoted largely to the publication of new results in this area.

This book is neither about the non-Newtonian fluid behavior per se nor does it purport even to touch on, let alone cover, all aspects of the non-Newtonian fluid mechanics. Rather, it deals with a narrow but important class of problems involving the motion of small, rigid, and deformable particles (and their ensembles) in a viscous medium. Although many years of research have been devoted to the behavior of particles in Newtonian media, many challenging problems in both theory and applications remain unresolved. But even more challenging is the subject of particle motion in non-Newtonian media, such as the motion of oil drops in polymer solutions through a porous medium during enhanced oil recovery, the dissolution of gas bubbles during fermentation, the key role played by the bubbles encapsulated in the production of foam plastics, etc. Equally important are the numerous applications involving significant inter-particle interactions such as those encountered in the fluid flow in packed and fluidized beds, bubble columns, slurry reactors, and in hindered settling of concentrated suspensions. In spite of such overwhelming pragmatic and fundamental significance, none of the existing books on non-Newtonian fluid mechanics deals with these topics adequately. This book is an effort to fulfill this need. In particular, it has been written to accomplish the following two specific goals:

1. To provide a reference text for research workers by way of presenting a comprehensive and critical evaluation of the voluminous literature available on the particle motion in non-Newtonian media and the related multiparticle process applications, as mentioned in the foregoing.
2. To provide useful design information often required by practicing scientists and engineers in day-to-day routine calculations such as the estimation of the minimum fluidization velocity or the prediction of pressure drop for flow through a fixed bed of particles, etc.

Aside from these two primary objectives, parts of the material presented herein have been used at the Indian Institute of Technology in Kanpur in a first-year graduate course on Engineering Applications of Rheology for the past 5 years. Thus, depending upon one's interest and taste, I believe that this material can form a part of an elective course on Process Applications of Rheology for senior undergraduate or first-year graduate students.

The limitations of this book should also be mentioned. The most important of all is that it concentrates primarily on the free motion of spherical particles under the influence of gravity. The scant work relating to the behavior of nonspherical particles is included only by way of providing the pertinent references. Likewise, no additional effects arising from the presence of electrical (electro-rheological fluids) or magnetic fields, the compressibility of the fluids, etc. are included unless specifically mentioned otherwise.

Raj P. Chhabra
Kanpur, India

Acknowledgments

Every book is in part the product of the knowledge and enthusiasm of the authors' friends. This one is no exception. It is a pleasure to acknowledge the assistance we have received from many individuals during the preparation of this book. While it is impossible to acknowledge all of them, RPC is particularly indebted to his Ph.D. advisor, Dr. Peter Uhlherr (retired professor from Monash University, Melbourne) for introducing him to the fascinating world of non-Newtonian fluids, and to his postdoc mentor, late Professor Jack Richardson (University of Wales, Swansea) for further sharpening this interest in the mechanics of such complex fluids.

Over the past 40 years or so, we have greatly benefited from our numerous collaborators from all over the world: Professor Tam Sridhar (Professor Emeritus, Monash University), Professor Pierre Carreau (Professor Emeritus, Ecole Polytechnique de Montreal), Professor Daniel De Kee (Professor Emeritus, University of Toronto), Professor P. N. Kaloni (Professor Emeritus, University of Windsor, Windsor), Professor Rakesh Gupta (West Virginia University), Professor R. Shankar Subramanian (Clarkson University), Professor Jacques Comiti (GEPEA, St. Nazaire), late Professor Jose Ferreira (Universidade de Tras-os-Montes e Alto Douro, Vila Real, Portugal), Professors Evelyne Mauret and Maurice Renaud (both of EFPG, INPG, Grenoble), Professor Nori Kawase (Toyo University, Japan), Professor Rob Poole (University of Liverpool), Professor T. Sundararajan (Indian Institute of Technology, Chennai), and Professor V. Eswaran (Indian Institute of Technology, Hyderabad). Many others have been gracious and generous in responding to our requests for their experimental results and for some of the photographs that appear in this work. Thoughtful suggestions have consistently come from our past and present students via their critical comments and alternate points of view on many issues. It is impossible to mention everyone, but we would like to single out Ram Prakash, Anoop Gupta, Neelkanth Nirmalkar, Chandi Sasmal, Preeti Suri, Sri Mourya, Abhishek Srivastava, Anshuman Verma, Khyati Aherwar, Jyoti Yadav and Dr. N. Dinesh Babu for their extensive help with the preparation of this edition.

We would also like to thank Barbara Knott, Alexis O'Brien, Shelly Thomas, Julia Tanner, Solomon Pace-McCarrick, and Tiffany Cameron, the wonderful team at Taylor & Francis for their patience and help at every stage of this project. It has been a pleasure to work with these outstanding individuals who went way beyond their call of duty to bring this project to completion. Finally, we are also grateful to the team at Code Mantra, especially Karthik Orukaimani for creating the final product you are looking at now!

Finally, it is not a Hindu custom to express gratitude to your spouse in public, but there is no doubt that this book would have been never completed without the unlimited and unconditional love and support of Neeru and of Dinesh.

Authors

Raj P. Chhabra has been a Professor of Chemical Engineering at the Indian Institutes of Technology in Kanpur and in Ropar for the past 38 years. After receiving his B.S. and M.S. in Chemical Engineering from the University of Roorkee (now the Indian Institute of Technology, Roorkee) and the Indian Institute of Science, Bangalore, respectively, he obtained his Ph.D. from Monash University in Melbourne (Australia). He has been a Visiting Professor at several universities, including the University of New South Wales, Sydney; Clarkson University, Potsdam (NY); The State University of New York at Buffalo; Ecole Polytechnique de Montreal; Universite de Nantes, Nantes; Technology University of Lodz, Lodz (Poland); China University of Petroleum-Beijing and the Cape Peninsula University of Technology, Cape Town. Currently, he is a Professor of Chemical Engineering at Shiv Nadar University.

His current teaching and research interests are in the general areas of non-Newtonian fluid mechanics, multiphase flows, and research methods and skills. He has 10 books, 35 invited book chapters and reviews, and over 400 technical papers to his credit. He is a recipient of the Amar Dye Chem (1988) and the Herdillia (1996) awards (of the Indian Institute of Chemical Engineers) for excellence in research. He is a Fellow of the Indian National Academy of Engineering and of the Indian National Science Academy.

Swati A. Patel is an Assistant Professor of Chemical Engineering at the Indian Institute of Technology at Ropar (Punjab). After earning her doctorate in Chemical Engineering from the Indian Institute of Technology in Kanpur, she worked as a post-doc researcher at the New Jersey Institute of Technology before joining the Indian Institute of Technology Ropar in May 2017. Her current research interests are in the broad areas of Computational Fluid Dynamics (CFD) of non-Newtonian multiphase flows and particulate product engineering. She has published numerous technical papers in these fields.

Introduction

The presence and motion of bubbles, drops and particles in fluids impinge on our everyday life to such an extent that it would be no exaggeration to say that the phenomenon is ubiquitous. The water we drink has suspended particles and dissolved gases in it, some contributing to our well-being, while others have harmful effects; the air we breathe has minute solid particles in it; the fizzy drinks (soda, alcoholic beverages and ales, liqueurs) and aerated foods (beating of an egg to entrain as much air as possible to create a fluffy omelet!) have gases dissolved in the form of small gas bubbles. Think about bubbles next time you enjoy a glass of beer! (Donald, 1994; Lohse and Zijm, 2003; Benilov et al., 2013; Lee et al., 2018; Zenit and Rodriguez-Rodriguez, 2018; Yotsumoto and Shiona, 2019; Watamura et al., 2019, 2021, etc.). Blood, a vital element of life, immediately comes to mind with its red and white blood cells suspended in plasma. One can think of numerous other examples involving interaction between particles and fluids encountered literally everywhere in everyday life and technology. Thus, to understand the behavior of particles in a fluid has been a challenge throughout the history of mankind.

From a technological standpoint, numerous operations in chemical and processing industries involve fluid-particle systems. Fluidization technology relies almost solely on fluid-particle (solids and bubbles) interactions which lead to the enhanced levels of interphase heat and mass transfer, mixing and chemical reactions. Hydraulic and pneumatic transportation of particulate materials involves hydrodynamic interactions between the conveying medium (liquid/gas) and the particulate material to be conveyed from one location to another. Other examples of solid-fluid interactions include the filtration of polymer melts, sewage sludges and paper coating suspensions, suspension of debris in drilling muds in oil recovery processes, sedimentation, dewatering and thickening of a range of slurries, degassing of polymer melts, disposal of waste streams from mineral industries, interpretation of the rheological behavior of suspensions, operation of trickle bed, fixed bed and slurry reactors, and the other multiphase contacting devices employed in chemical and process engineering applications. Less-appreciated applications of fluid-particle systems include the motion of red blood cells in capillary flow, chromatographic separations, electrophoresis, separation of macromolecules according to their sizes, etc. The most common method to affect gas/liquid contacting is to introduce the gas through a multi-hole distributor in the form of tiny gas bubbles. Naturally, the rates of transfer processes, and of chemical reactions, are essentially governed by the flow field established around the ascending gas bubbles. Introduction of an inert gas into a pool of liquid to improve mixing is a commonly used process in metallurgical and chemical industries. Dispersion of a liquid in another immiscible liquid, resulting in an emulsion, is indeed a basis for the manufacture of a class of polymers, cosmetics, toiletries, foodstuffs, metallic foams, alcoholic beverages, etc. In recent years, there has been an increasing interest in studying the behavior of nanobubbles, nanosuspensions, and behavior of fluids in nanofluidic and microfluidic devices, especially in the context of health-care-related point-of-care applications as well as a process intensification tool. Thus, there is no dearth of examples of industrial relevance involving interactions between bubbles, drops and particles and a flowing or stagnant continuous fluid phase.

Ever since the discovery of the laws of motion enunciated by Newton more than three centuries ago, it has been a customary practice to accept the Newtonian fluid model as the standard fluid behavior. Indeed, the fluid mechanics of Newtonian fluid-particle systems has witnessed remarkable years of progress in recent times, especially with the advent of elegant numerical techniques and due to an enormous improvement in our computational abilities. While it has reached a respectable level of maturity, many complex multiphase flows (turbulent flows, granular flows, large-scale atmospheric and oceanic flows, for instance) continue to pose challenges and remain unresolved to a large extent. Yet it is a simple task to generate experiments that could never be explained (even qualitatively) by the standard Newtonian fluid model (Bingham, 1932; Boger and Walters, 1992). Indeed, the Newtonian fluid behavior seems to be an exception rather than the rule.

Over the past five to six decades, there has been an increasing recognition of the fact that most materials of practical and industrial interest do not conform to the simple Newtonian fluid behavior and are accordingly known as "rheologically complex" or "non-Newtonian fluids", "complex fluids", "soft matter", among others (albeit the earliest reference to non-Newtonian bahavior dates back to 700 BC (Slattery, 1972). Indeed, many interesting historical accounts (Markovitz, 1968, 1977, 1985; Scott Blair, 1942, 1982; Joseph, 1986; Litt, 1989; Doraiswamy, 2002) show that not only the field of rheology (in some form or the other) is an ancient science but also has an incredibly rich history (Tanner and Walters, 1998). The development and growth of this discipline has been touched upon by many pioneering individuals, including Newton, Maxwell, Lord Kelvin, Boltzmann, Bingham, Reiner, Oldroyd, and Einstein (Physics Noble Prize, 1921). Flory (Chemistry Nobel Prize, 1974), de Gennes (Physics Nobel Prize, 1991), and Chu (Physics Nobel Prize, 1997) to mention a few.

The simplest and possibly the commonest type of departure from the Newtonian flow behavior is shear-thinning (or pseudoplasticity) wherein the apparent viscosity (shear stress divided by shear rate) of a fluid decreases with the increasing shear rate. Besides, some of these materials behave like a viscous fluid in long-time experiments, while the initial (short-time) response to applied external stress is like that of an elastic solid. This phenomenon can easily be illustrated by considering the response of a ball made of silly putty (polymeric material). Such a ball when dropped from height bounces like an elastic rubber ball (short-time response like an elastic solid) whereas given sufficient time, it will level out under the influence of gravity (long-time response like a fluid). One can qualitatively explain this type of behavior by postulating that these visco-elastic materials possess a sort of "memory". Yet another example of non-Newtonian behavior is the so-called time dependence wherein the value of the stress generated in a fluid by an imposed velocity gradient varies with the duration of experiment. Finally, it is not uncommon to encounter all these complexities in a single material and/or application under appropriate circumstances. While the early developments in this field were almost exclusively motivated by the increasing use of rubbers, polymers and plastics as potential substitutes for the traditional materials of construction, namely glass, wood, concrete, and metals, it is no longer so (White, 1990). The non-Newtonian fluid behavior is much more widespread than is generally perceived. For instance, as one walks down the aisles of a modern supermarket and department store, they pass a wide range of processed food products (both manufactured and natural) which are either in the form of a powder, or an emulsion (or suspension) or have been in an emulsified form at some stage during their production. Flavored milks, creams, precooked meals, cereals, salad dressings, dips, fruit yogurts, mayonnaise, cheese spreads, nibbles and snacks, for instance, immediately come to mind! Similarly, many pharmaceutical products, including suspensions, laxatives, gels and creams, are all multiphase mixtures. Indeed, frogs combine the visco-elasticity of their tongue with the non-Newtonian properties of their saliva to catch prey (Noel et al., 2016). Similarly, the effort required in chewing, swallowing different types of food, mouth feel, etc., are all strongly influenced by the interaction between the rheological behavior of the food and the oral processing (Stokes et al., 2013). Table 0.1 provides a representative summary of diverse industrial materials and settings where rheologically complex behavior is encountered at some stage of processing. It is virtually impossible to imagine modern life without emulsions or suspensions (pastes) or dispersions. Thus, over the years, due to the ever-increasing needs and demands of consumers coupled with the advancements in technology, chemical and process engineering applications extensively use a wide range of rheologically complex materials including polymeric melts and solutions, multiphase mixtures (foams, emulsions, dispersions).

In many of the applications involving bubbles/drops/particles – fluid systems cited in the foregoing, the liquid phase displays complex non-Newtonian flow behavior on its own, or due to the presence of multiple entities contained therein. Typical examples include the use of dilute polymer solutions in enhanced oil recovery operations, in drag reduction applications, in pipeline transportation of coarse grains (Barnes, 1993; Maciejewski et al., 1997; Darby and Chhabra, 2017; Wilson et al., 2006; Chhabra and Richardson, 2008; Barabati et al., 2016, etc.), as thickening agents in food processing (Holdsworth, 1992), as vehicles in pharmaceutical formulations (Berney and Deasy, 1979; Miller and Drabik, 1984, etc.), the use of drilling muds to provide lubrication as well as to keep the cuttings in suspension, the handling and processing of fermentation broths, emulsion polymerization, processing of filled polymers, downstream processing in nuclear, agricultural, dairy and biotechnological processes, hydraulic

TABLE 0.1

Diversity of Systems Exhibiting Non-Newtonian Fluid Behavior

Systems	Reference
Dairy products and waste slurries	Hart et al. (1966), Staley et al. (1973), Prentice (1992), Camacho et al. (2005)
Poultry waste slurries	Hashimoto and Chen (1976), Chen and Hashimoto (1976), Bjerkholt et al. (2005a,b,c)
Sewage sludge	Eshtiaghi et al. (2013), Baudez et al. (2011)
Polymeric materials used as thickening agents in pharmaceutical applications, creams, gels, foods, etc.	Berney and Deasy (1979), Miller and Drabik (1984), Korhonen et al. (2000, 2002), Chang et al. (2002)
Printing inks and paper coating colors	Arzate et al. (2004), Higgins (1997), Park et al. (2010), Phair et al. (2009)
Pharmaceutical products	Allahham et al. (2005)
Food stuffs (fruit yogurts, cheese and cheese products, butter, egg albumen, jams, jellies, marmalades, fruit juices, salad dressing, ice cream and cake toppings, cake mixes, ice creams, soups, spreads, etc.)	Sherman (1970), Tung et al. (1971), Faridi (1989), Ramaswamy and Basak (1991), Borwankar and Shoemaker (1992), Rao (1999), Rao et al. (2005), Mc Clements (2004), Bertsch et al. (2019)
Food processing	Steffe (1996), Holdsworth (1992), Heldman and Lund (1992), Stokes et al. (2013)
Polymers and polymeric solutions	Bird et al. (1987a,b), Larson (1998), Gupta (2000), Carreau et al. (2021)
Cosmetics and toiletries, personal care products (lipsticks, nail polish, creams and body lotions, shampoos, conditioners, etc.)	Laba (1993), Garcia-Morales et al. (2004), Brummer (2006), Ardakani et al. (2011)
Thick pastes, cement pastes and fresh concrete, bitumen	Yaron and Ish-Shalom (1975), Tatersall (1983), Banfill (1991), Yahia and Khayat (2001), Brower and Ferraris (2003), Murali Krishnan and Rajagopal (2003), Coussot (1997,2007), Hanehara and Yamada (2008), Toussaint et al. (2009)
Biological fluids (blood, synovial fluid, saliva, semen, mucous)	Gabelnick and Litt (1973), Dunn and Picologlou (1977a,b), Briedis et al. (1980), Chmiel and Walitza (1980), Rosentrater and Flores (1997), Shi et al. (2004), Haward et al. (2011), Noel et al. (2016)
Industrial polysaccharides	Lapasin and Pricl (1995)
Molten lava, magmas, rocks and soils	Cristescu (1988), Hailemariam and Mulugeta (1998), Saar et al. (2001), Petford (2003)
Mineral slurries and mine tailings; Agricultural chemicals, fly ash slurries; nuclear waste slurries; pulp fiber suspensions	Chang and Smith (1996), Schramm (2014), Usui et al. (2001), Hou et al. (2005), Derakhshandeh et al. (2011,2012), Pullum et al. (2018)
Mud and debris flows; sea ice	Coussot (1997, 2005), Feltham (2008), Maciel et al. (2009), Skadsem et al. (2019)
Lubricants	Davenport (1973), Jean (1989)
Paints and pigments; coatings and inks	Patton (1979)

conveying of large lumps of materials in a carrier liquid containing fine particles, etc. Similarly, the dynamics of bubbles in rheologically complex systems directly impinges on the degassing of polymeric melts and molten glass, in the performance of freshly prepared concrete as well as on the process of aeration used extensively in food processing applications and in treatment of sludges (Gauglitz et al., 2003). Similarly, the bubble mobility in muds and volcanic matter is also of considerable interest (Tran et al., 2015). Litt (1989) provided an exhaustive list of industrial examples where non-Newtonian behavior is of central importance in determining the efficiency of a process as well as in achieving the desired product quality. Similarly, it is often required to cool or heat a non-Newtonian process stream such as in the thermal treatment of food particles in non-Newtonian carriers, or thermal treatment of other foodstuffs to achieve the desired product quality in terms of their lethality, in the melting of polymer pellets prior to their liquid phase processing, gas-liquid mass transfer in fermentation processes, sterilization of milk and other foodstuffs to extend their shelf life, etc. Due to the spatial distribution of the rate of deformation (shear rate for instance), viscosity stratification occurs in complex flow situations which, in turn, influences the rate of convective heat and mass transport even from as simple shapes as a sphere or a cylinder or a plane surface which act as sources or sinks. Suffice it to add here that owing to their generally high viscosity, much of the processing is carried out at low Reynolds numbers and therefore the corresponding convective coefficients not only tend to be quite low but also the vast body of knowledge obtained for Newtonian fluids is of little help in this regard.

There is no question that in real-life applications, one often encounters swarms or clusters of particles rather than an isolated single particle. Notwithstanding the complexity of multiparticle systems, experience has shown that an understanding of the behavior of single particles not only provides useful insights into the underlying physical processes but also often serves as a launching pad for analyzing and modeling the more realistic multiparticle situations. In addition to this, single particle studies are also of considerable theoretical interest in their own right dating back to the times of Galileo, Newton, Stokes and Basset in the context of their motion in air and water. Indeed, significant differences are observed in terms of their hydrodynamics in non-Newtonian media as compared to the behavior of single particles in linear (Newtonian) fluids, some of which may have far reaching implications including aggregation in suspensions under flow conditions. For instance, the shapes of bubbles and drops in free fall/rise observed in a non-Newtonian medium are qualitatively and quantitatively different from those encountered in a Newtonian liquid phase under nominally identical conditions. Undoubtedly, such differences in shape directly influence their coalescence behavior, interphase heat and mass transfer and chemical reactions to varying extents. Additional motivation to study the flow of a fluid past single objects (sphere, cylinder, bubble, drop, for instance) stems from the fact that these highly idealized shapes are used to benchmark the suitability of a constitutive relation and/or of numerical solution methodologies prior to using them for complex flow problems. Conversely, the flow around a sphere or a cylinder is also used to explore the extent of molecular orientation (Haward and Odell, 2004), or to ascertain other rheological properties (de Bruyn, 2004, 2006; Dollet et al., 2005a,b, 2019; Chakrabarti and Chaudhury, 2013a,b; Spratt et al., 2018), or have direct relevance to their use as probe particles in polymer solutions (Ye et al., 1998). These results also find applications in the design of systems to separate magnetic particles from high viscosity fluids by employing permanent magnets (Garraud et al., 2016) or using a hydro-cyclone, in employing the MRI technique for cardiovascular studies (Mathieu et al., 2006). Similarly, the study of convective heat and mass transport from idealized shaped objects in model non-Newtonian fluids leads to useful insights. Another interesting example of non-Newtonian effects is the phenomenon of drag reduction in external flows such as over a sphere or a cylinder or more realistic shapes e.g., ship hulls, submarines, etc. Small doses of macromolecules added to water can reduce the skin friction by 30%–40%, which directly influences the power requirements for ships, submarines, etc. This book aims to provide an up-to-date and critical account of the progress made in this evolving field of multiphase flows over the past 50–60 years. At each stage, considerable effort has been made to present the most reliable information, and generally accepted methods for performing process engineering calculations.

0.1 Scope and Organization

Though this work is primarily concerned with the behavior of rigid and fluid particles in rheologically complex systems, a terse background material on the corresponding flow in Newtonian fluids is included in each chapter. This not only facilitates the subsequent treatment for non-Newtonian fluids but also serves as a base case for the purpose of making qualitative comparisons between different types of fluids. It is also important to mention here the chief limitations of this work: the only body force considered herein is the one due to the earth's gravitational field. Similarly, the fluids are assumed to be incompressible and isotropic. This is a good approximation for most of non-Newtonian fluids referred to in this work, except for foams, dispersions and emulsions which at times can exhibit some compressibility and anisotropic effects. Similarly, highly structured liquid crystalline polymers, fiber reinforced polymers

Introduction

and plastics do exhibit a degree of anisotropy. Likewise, with the notable exception of Chapters 8 and 9, no temperature or concentration effects are considered in any other chapter. Finally, rheological behavior associated with the presence of electric or magnetic fields is also not considered herein. Each chapter is concluded by a short summary of the current state of the art and by outlining the areas meriting further systematic study.

In Chapter 1, a brief exposition to the non-Newtonian flow characteristics is provided. Different types of non-Newtonian fluid behavior are described. A selection of the widely used rheological equations along with their merits and demerits is presented. By way of examples, typical values of model parameters are cited to give the reader a feel for numbers for wide-ranging materials of industrial significance. This chapter is closed by outlining the role of dimensional analysis in analyzing the fluid mechanics of visco-elastic fluids, especially about the definition of Deborah or Weissenberg number, as used in the interpretation and correlation of experimental and theoretical results for such fluids.

The steady motion of rigid particles (sphere, cylinder and of other nonspherical shapes) in purely viscous fluids without a yield stress is treated in Chapter 2. Currently available theoretical and experimental results have been critically examined for a variety of generalized Newtonian fluid models. In particular, the influence of shear rate-dependent viscosity on the prediction of drag coefficient, the terminal falling velocity and on the detailed flow field for a sphere is elucidated. In the context of nonspherical particles, the difficulties of quantifying their size, shape and orientation during their free fall in fluids are highlighted, and these are further accentuated in the case of non-Newtonian liquids.

The static and dynamic behavior of rigid particles (spherical and nonspherical) in visco-plastic media is considered in Chapter 3. Notwithstanding the long and inconclusive debate about the existence of a true yield stress, the capacity of such fluids to support particles is used extensively in mineral, processed foods, pharmaceutical and personal care product industries. Starting with the issue of static equilibrium, attention is given to the size and shape of fluid-like zones created by a moving particle, its drag force and terminal settling behavior of spherical and nonspherical particles undergoing steady translation motion in such fluids. The role of regularization schemes used in numerical simulations and the difficulties of an unambiguous measurement of yield stress in such experiments are highlighted. Due to the improved numerics and experiments, the gap between the predictions and observations has been bridged to some extent. The available scant information pertaining to nonspherical particles is also included here.

Chapter 4 addresses the analogous problem of a sphere and a cylinder undergoing steady translation motion in visco-elastic media. A detailed progress report on the numerical simulations and experimental developments on the steady sedimentation of a sphere in a cylindrical tube is presented. Next, the corresponding developments relating to the two-dimensional flow of visco-elastic liquids past a confined and an unconfined long circular cylinder are considered. Many other interesting effects not reported hitherto are also discussed briefly. Combined effects of the fluid visco-elasticity and shear-thinning viscosity are also treated in this chapter. The unusual phenomena such as migration, alignment, aggregation and cluster formation, development of shear-induced structures, etc. observed during the motion of single- and multiple particles in visco-elastic media in a variety of flow fields are also discussed here.

The behavior of fluid particles with mobile interface-bubbles and drops is treated in Chapter 5. Starting with the available phenomenological models for their formation from a single orifice or a nozzle are presented and discussed. The other methods of producing drops such as atomization and breakup of liquid jets and sheets are also mentioned, especially with regard to the role of non-Newtonian fluid properties. Quantitatively different shapes of bubbles and drops are observed in rheologically complex media as compared to those encountered in Newtonian fluids. Terminal falling or rising velocity-volume behavior and drag coefficient of single particles and their ensembles in the free fall conditions are analyzed in detail. Finally, consideration is given to the problem of coalescence and breakage in non-Newtonian continuous phase under various flow conditions.

The hydrodynamics of non-Newtonian fluid flow on its own and together with a gas in unconsolidated porous media and packed beds is dealt with in Chapter 6. An overview of the wide variety of approaches employed to model fluid flow in porous media is presented. The two commonly used frameworks are described in detail: the capillary bundle approach and the submerged objects approach. First, the flow of purely viscous fluids is considered, followed by that of Visco-plastic and visco-elastic systems. The main emphasis here is on the establishment of pressure loss-throughout relationships for a given system, followed by the effect of particle shape and of the containing walls on the frictional pressure drop across the bed. The flow of visco-elastic and of the so-called drag-reducing polymer solutions is also considered briefly. Included here is also the hydrodynamics of Newtonian and non-Newtonian fluids in fibrous media as that encountered during the flow in tumors, and in polymer processing applications aimed at the manufacturing of fiber composites. Consideration is given here to the prediction of permeability of the fibrous medium. The chapter is concluded by presenting a brief discussion on the range of anomalous effects observed in packed beds such as wall slip, polymer adsorption, gel formation, mechanical degradation of polymers, pore wall-molecule interactions, etc. each of which contributes to the differences observed between the bulk and in situ rheological characteristics of a polymer solution. The scant literature dealing with the simultaneous flow of a gas and a non-Newtonian fluid through granular packed beds is also summarized in this chapter.

Hydrodynamically similar problems of fluidization and sedimentation (hindered settling) are treated in Chapter 7. In particular, the influence of non-Newtonian characteristics on the bed expansion dynamics, minimum fluidization velocity and the rate of sedimentation of concentrated suspensions in the initial constant concentration region are examined in this chapter. These areas thus far remain virtually unexplored. Even less studied are the three-phase fluidized bed systems involving non-Newtonian liquid phase. Based on the available limited information, an attempt has been made to develop suitable predictive expressions for the macroscopic process parameters. The scant literature pertaining to the sedimentation behavior of fibrous suspensions (such as pulp fibers) is also included here.

Boundary-layer flows are considered in Chapters 8 and 9. In the absence of free convection effects, starting with the three commonly used idealized geometries, namely a plate, or a sphere or a cylinder, the laminar momentum thermal and concentration boundary layers for power law fluids are first considered in the forced convection regime in Chapter 8. The effects of buoyancy-induced flow on heat and mass transfer in the pure free and mixed convection regimes are taken up in Chapter 9. An attempt has been made here to delineate the limits of the applicability of the boundary-layer analysis in terms of the modified Rayleigh number and the relevant rheological parameters. Next, the role of visco-elasticity on heat transfer is presented. Included in this chapter are also the scant numerical and experimental correlations which allow the prediction of heat and mass transfer coefficients for single particles at one end and for packed and fluidized beds, tube bundles, at the other end of the spectrum of applications.

The confining walls are known to influence the hydrodynamic behavior of rigid and fluid articles, both by exerting extra retardation force and by altering the shape of fluid particles. The resulting modifications to the detailed velocity field established in the vicinity of the submerged particle also influence the interphase heat and mass transfer. Chapter 10 examines such wall effects in detail. Starting with the extent of wall effects in Newtonian fluids, the available scant body of knowledge for non-Newtonian fluids is reviewed here for rigid and fluid particles. While the chapter is mainly limited to the sedimentation of spherical and nonspherical particles in cylindrical domains, some key references are provided for the other shapes of confining boundaries. Similarly, the wall effects on heat transfer in non-Newtonian fluids are mainly limited to spheres and spheroids.

In the last Chapter 11, attention is focused on the applicability of the falling object rheometry for Newtonian and non-Newtonian systems. Consideration has been given here to the methods based on the application of the falling ball, the rolling ball, the rotating sphere and the falling cylinder configurations. While all these methods have proved to be of immense value (particularly at high pressures and temperatures) for both qualitative and quantitative measurements of viscosity and of product quality for Newtonian fluids, their utility for non-Newtonian fluids is severely limited. This is simply due to the complexity of the flow field prevailing in these devices. However, under appropriate circumstances, it is possible to infer some useful rheological information, especially about the zero-shear viscosity, yield stress, etc. from a knowledge of their settling velocity and geometrical configuration. Despite these limitations and the lack of sound theoretical justification, many of these configurations are extensively used in scores of industries (food, paints, pharmaceuticals) for routine quality control purposes.

1

Non-Newtonian Fluid Behavior

1.1 Introduction

Three distinct stages can be discerned in the evolution of fluid dynamics. The first stage of development deals with the study of imaginary perfect or ideal fluids, that is, those without viscosity or elasticity and incompressible in nature. Evidently, the shearing motion will not give rise to any shearing forces in such fluids, and hence, the flow is said to be frictionless. Exact analyses for various physical situations involving ideal fluids have been developed over the years, some of which have proved to be useful approximations for the performance of real fluids under certain special conditions. The seminal works of Pascal, Bernoulli, Euler, etc. belong to this period.

The concept of boundary layer, introduced by Prandtl (1904), marked the beginning of the second stage of the development of classical fluid dynamics. Indeed, without the notion of a boundary layer, the solutions developed for ideal fluids would have been of little practical utility. Prandtl (1904) simply postulated that in flows over a solid surface, the frictional effects are confined to a relatively thin layer, known as the boundary layer, which exists adjacent to the solid surface. Thus, the flow domain can be divided into two regions: the flow outside the boundary layer, which is adequately modeled by assuming the ideal fluid behavior, and the fluid friction is limited to the part of the fluid within the boundary layer (the so called inner region). The latter has led to the development, although in less detail, of a dynamical theory for the simplest class of real fluids commonly referred to as Newtonian fluids.

Finally, the third stage of fluid dynamical theory is still currently being developed and is in its infancy (Doraiswamy, 2002). This development was prompted by the increasing significance of a range of materials encountered in a large variety of commercial applications cutting across a number of industrial settings. The flow behavior of this new class of materials does not conform to the Newtonian postulate, and accordingly, such materials are known as non-Newtonian fluids. Such materials are also variously known as "soft matter", "complex," "nonlinear" fluids, etc. Typical examples of materials exhibiting non-Newtonian flow characteristics include multiphase mixtures (slurries, emulsions, and gas–liquid dispersions), polymer melts and solutions, soap solutions, personal care products, including cosmetics and toiletries, food products (jams, jellies, cheese, butter, mayonnaise, meat extract, soups, yogurt, etc.), biological fluids (blood, synovial fluid, saliva, semen, tissue, etc.), building materials, natural products (gums, protein solutions, extracts, etc.), agricultural and dairy wastes, magmas, lava (Griffiths, 2000; Petford, 2003), etc., as discussed in Chapter 1. Indeed, non-Newtonian fluid behavior is so widespread that it would be no exaggeration to say that simple Newtonian fluid behavior is an exception rather than the rule! In Section 1.2, we begin with the definition of a Newtonian fluid that, in turn, sets the stage for the subsequent treatment of non-Newtonian substances.

1.2 Definition of a Newtonian Fluid

Consider a thin layer of fluid confined between two wide parallel plates separated from each other by a distance dy, as shown in Figure 1.1. Now if a constant shearing force F is applied tangentially to the top plate, at a steady state, it will be balanced by an internal frictional force in the fluid arising from its viscosity; the resulting steady-state linear velocity profile is also sketched in Figure 1.1. For a Newtonian fluid in a streamline flow, the shear stress is proportional to the shear rate, that is,

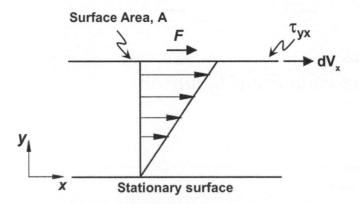

FIGURE 1.1 Schematic representation of unidirectional shearing flow.

$$\frac{F}{A} = \tau_{yx} = \mu\left(-\frac{dV_x}{dy}\right) = \mu\dot{\gamma}_{yx} \tag{1.1}$$

The minus sign on the right-hand side of Equation 1.1 suggests that the shear stress is a resisting force. Alternately, one can also interpret it, by analogy with the corresponding laws for diffusive heat and mass transfer (i.e., Fourier's and Fick's laws), to imply that the momentum transfer takes place in the direction of the decreasing momentum (velocity), i.e., in the negative y-direction.

The constant of proportionality, μ, a characteristic property of each substance, is variously known as the Newtonian viscosity, shear viscosity, dynamic viscosity, or simply viscosity. Sometimes the reciprocal of viscosity, known as fluidity, is also used to indicate the mobility of a substance. By the definition of a Newtonian fluid, the value of μ is independent of the shear rate, and it depends only on temperature and pressure for a given substance. A graph of shear stress (τ_{yx}) vs. shear rate $(\dot{\gamma}_{yx})$, the so-called flow curve or rheogram, for a Newtonian fluid is therefore a straight line of slope μ that passes through the origin. Thus, the single constant completely characterizes the flow behavior of a Newtonian fluid at a fixed temperature and pressure. All gases, low-molecular-weight (<5000 or so) liquids and their solutions, solutions of inorganic salts, molten salts, and liquid metals behave as Newtonian fluids. Typical rheograms for a cooking oil and a corn syrup are shown in Figure 1.2, while Table 1.1 lists viscosity values for some familiar systems to give a feel for the numbers.

Equation 1.1 describes the simplest case, where the velocity vector has only one component, namely, in the x-direction, which varies in the y-direction. This type of flow is known as the simple shear flow. For the case of a 3-D flow, there will be additional shearing and normal stresses (as shown in Figure 1.3), and it is therefore necessary to write expressions similar to Equation 1.1 for the other components of the stress tensor. The more general case for a Newtonian fluid may be expressed for the y-plane (i.e., the plane oriented normal to the y-direction) as

$$\tau_{yx} = -\mu\left(\frac{\partial V_x}{\partial y} + \frac{\partial V_y}{\partial x}\right) \tag{1.2}$$

$$\tau_{yy} = -2\mu\frac{\partial V_y}{\partial y} + \frac{2}{3}\mu(\nabla \cdot V) \tag{1.3}$$

$$\tau_{yz} = -\mu\left(\frac{\partial V_z}{\partial y} + \frac{\partial V_y}{\partial z}\right) \tag{1.4}$$

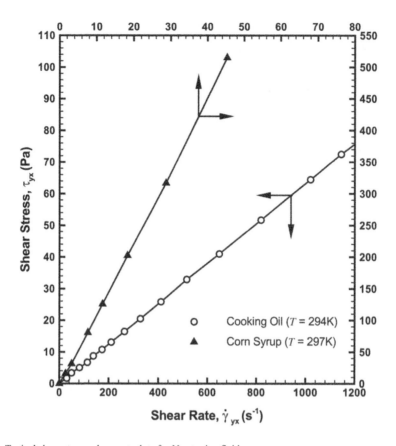

FIGURE 1.2 Typical shear stress–shear rate data for Newtonian fluids.

TABLE 1.1

Typical Viscosity Values of Common Substances at Room Temperature

Substance	μ (mPa s)
Air	10^{-2}
Water	1
Sodium chloride (1173 K)	1.01
Mercury	1.55
Molten lead (673 K)	2.33
Olive oil	100
Castor oil	600
100% Glycerine	1500
Honey	10^4
Corn syrup	10^5
Molten polymers	10^6
Bitumen	10^{11}
Molten glass	10^{15}
Earth mantle	10^{25}
Glass	10^{43}

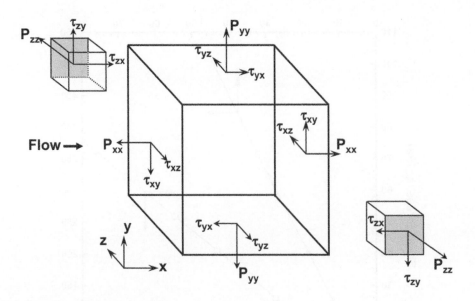

FIGURE 1.3 Stress components in 3-D flow.

Similar expressions can be written for the *x*- and *z*-planes, respectively. Altogether, there are nine stress components that completely describe the stress field in a Newtonian fluid under flow conditions. The normal stresses can be viewed as made up of two constituents: isotropic pressure *p* and a component resulting from fluid motion, that is,

$$P_{xx} = -p + \tau_{xx}$$
$$P_{yy} = -p + \tau_{yy} \qquad (1.5)$$
$$P_{zz} = -p + \tau_{zz}$$

where τ_{xx}, τ_{yy}, and τ_{zz}, known as deviatoric normal stresses, are the contributions arising from the shearing motion. By definition, the isotropic pressure *p* is given by:

$$p = -\frac{1}{3}\left(P_{xx} + P_{yy} + P_{zz}\right) \qquad (1.6)$$

Combining Equations 1.5 and 1.6, one obtains:

$$\tau_{xx} + \tau_{yy} + \tau_{zz} = 0 \qquad (1.7)$$

Furthermore, for an incompressible Newtonian fluid in simple shear, deviatoric normal stresses are identically zero, that is,

$$\tau_{xx} = \tau_{yy} = \tau_{zz} = 0 \qquad (1.8)$$

Thus, the complete definition of a Newtonian fluid is that it not only has a constant viscosity, but it also conforms to Equation 1.8 in simple shear, or simply, it satisfies the complete Navier–Stokes equations for Newtonian fluids (Bird et al., 1987a, 1987b, 2002). During the last 30–40 years or so, a new class of synthetic fluids (Boger, 1977a; Choplin et al., 1983; James, 2009; Prilutski et al., 1983; Tam et al., 1989) has emerged that in steady shear displays nearly a constant viscosity over a limited but finite range of shear rate and nonequal normal stresses. Obviously, such a liquid will not satisfy the complete Navier–Stokes equations and, thus, despite its constant viscosity, must be classed as a non-Newtonian fluid. Indeed, as will be seen later, these so-called *Boger fluids* have proved to be of considerable interest in both experimental and analytical/numerical modeling of a range of complex flows (James, 2009). Before leaving this section, thus, nine components of stresses (six shearing and three normal) are needed to describe the stress field in a given flow. However, the constraint of zero trace and the assumption of the symmetry of the stress tensor, i.e., $\tau_{ij} = \tau_{ji}$ where i≠j, allow the stress field to be expressed in terms of three shearing components, two normal-stress differences and the pressure developed in the fluid.

1.3 Non-Newtonian Fluids

A substance is termed non-Newtonian when its flow curve is nonlinear or it is linear, but it does not pass through the origin, that is, when its viscosity (shear stress divided by shear rate) is not constant at a given temperature and pressure and it exhibits nonequal normal stresses in a simple shearing flow. Instead, the value of the viscosity depends upon flow conditions, such as flow geometry, shear rate (or stress) developed within the fluid, time of shearing, and kinematic history of the sample. Most materials, under appropriate conditions, exhibit a blend of solid- and fluid-like responses to varying extents under appropriate conditions. Consider a ball made of silly putty (Cross, 2012) (silicon polymers) when released from a height bounces upon hitting the floor (like a rubber ball) and/or breaks with a sharp blow, yet it finds its own level in a container when left for a long time (like a fluid). Thus, this substance behaves like an elastic solid in "short" timescale processes (bouncing and breaking) and like a viscous fluid in long time slow flow (Edgeworth et al., 1984). Also, visit the pitch drop experiment at www.smp.uq.edu.au/pitch-drop-experiment for such dual behavior recorded over a period of almost 100 years. Thus, the distinction between a solid and a fluid is not as sharp as we think! Notwithstanding such difficulties, it is customary to classify, though somewhat arbitrarily, the non-Newtonian fluid behavior into three general categories as (Boger et al., 1980)

1. Substances for which the rate of shear is dependent only on the current value of the shear stress or vice versa; this class of materials is variously known as purely viscous, time-independent, or generalized Newtonian fluids (GNFs).
2. More complex materials for which the relation between the shear stress and the shear rate also depends upon the duration of shearing, previous kinematic history, etc.; these are known as time-dependent systems.
3. Materials exhibiting mixed characteristics of both an elastic solid and a viscous fluid and showing partial elastic and recoil recovery after deformation, the so-called visco-elastic fluids.

This classification is quite arbitrary and probably unscientific as well in that most real materials often display a combination of two or even all the three types of non-Newtonian characteristics under appropriate conditions. It is also possible for the same substance to behave as an elastic or viscous material depending upon the circumstances as seen above in the context of the pitch drop experiment and the bouncing putty. In most cases, however, it is possible to identify the dominating non-Newtonian feature and to use it as the basis for subsequent process engineering calculations (Chhabra, 2010, 2016; Patel and Chhabra, 2020). We now discuss each type of non-Newtonian fluid behavior in some detail.

1.3.1 Time-Independent Fluid Behavior

Fluids of the time-independent type whose flow properties are independent of the duration of shearing and past history may be described in simple shear by a rheological equation of the form

$$\tau_{yx} = f\left(\dot{\gamma}_{yx}\right) \tag{1.9a}$$

or its inverse form

$$\dot{\gamma}_{yx} = f^{-1}\left(\tau_{yx}\right) \tag{1.9b}$$

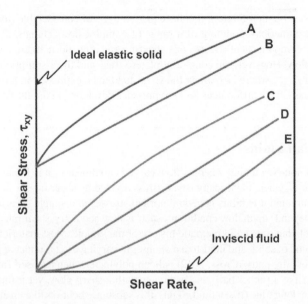

FIGURE 1.4 Qualitative flow curves for different types of non-Newtonian fluids: (a) Herschel–Bulkley (shear-thinning and yield stress), (b) Bingham, (c) pseudoplastic (shear-thinning), (d) Newtonian, and (e) dilatant fluid (shear-thickening).

Equations 1.9a and 1.9b imply that the rate of shear at any point within the sheared fluid is determined solely by the current value of the shear stress at that point, or vice versa. Depending upon the form of Equations 1.9a and 1.9b, these fluids may be further subdivided into the following three different types:

1. Shear-thinning or pseudoplastics,
2. Visco-plastics, and
3. Shear-thickening or dilatant.

Figure 1.4 shows qualitative flow curves for these three types of fluid behaviors; the linear relation typical of Newtonian fluids is also included as are the inviscid fluid (x-axis)- and elastic solid (y-axis)-limiting behaviors.

1.3.1.1 Shear-Thinning or Pseudoplastic Fluids

There is no question that shear-thinning is the most commonly encountered type of time-independent fluid behavior. A shear-thinning or pseudoplastic substance is characterized by an apparent viscosity (shear stress divided by shear rate) that decreases with the increasing shear rate. Figure 1.5 clearly illustrates the shear-thinning behavior observed in a range of aqueous polymer solutions. Evidently, the rate of decrease of the apparent viscosity is not the same for each fluid. Furthermore, if these measurements are extended over a sufficiently wide range of shear rates, most polymeric solutions seem to exhibit regions of constant viscosity both at very low and at very high shear rates, that is,

$$\lim_{\dot{\gamma}_{yx} \to 0} \frac{\tau_{yx}}{\dot{\gamma}_{yx}} = \mu_0 \quad \text{(zero shear viscosity)} \tag{1.10}$$

$$\lim_{\dot{\gamma}_{yx} \to \infty} \frac{\tau_{yx}}{\dot{\gamma}_{yx}} = \mu_\infty \quad \text{(infinite shear viscosity)} \tag{1.11}$$

Hence, the apparent viscosity of a shear-thinning substance decreases from μ_0 to μ_∞ with shear rate. Data demonstrating the so-called upper and lower Newtonian regions are difficult to obtain and are scarce in the literature. One such set of data (Boger, 1977b) for an aqueous solution of polyacrylamide is replotted in Figure 1.6. Note that three different instruments have been used to encompass seven orders of

Non-Newtonian Fluid Behavior

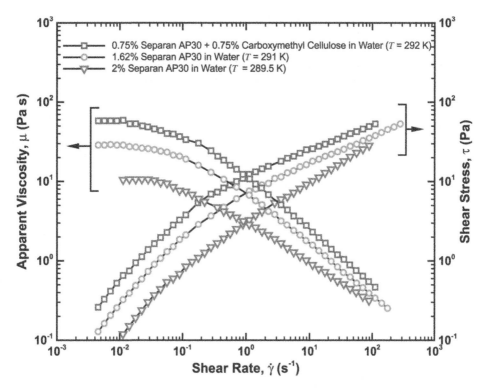

FIGURE 1.5 Representative shear stress and apparent viscosity plots for three pseudoplastic polymer solutions (Chhabra, 1980).

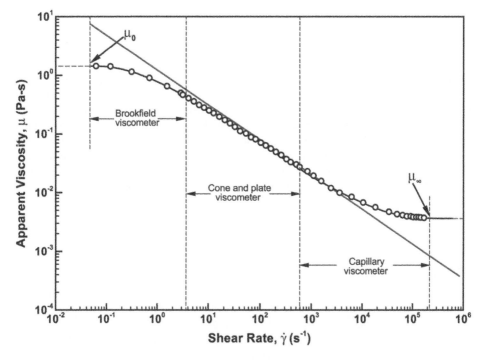

FIGURE 1.6 Demonstration of zero-shear and infinite-shear viscosities for a polymer solution.

magnitudes of shear rates. Also, note that the apparent viscosity of this polymer solution drops from 1400 to 4.2 mPa s. The values of the shear rate at which the two limiting forms of behavior are observed, and the rate of decrease of viscosity in the intermediate shear-thinning region depend upon the physicochemical factors, such as the type and concentration of polymer, its molecular weight distribution, and the type of solvent. Graessley (1974, 2004), Larson (1988, 1998), Morrison (2001), and Witten and Pincus (2004) have dealt with these aspects, thereby shedding some light on the bulk rheological characteristics and the micro-structural aspects of commonly encountered polymeric systems. Generally, the range of shear rate over which the apparent viscosity is constant (in the zero-shear region) increases as the molecular weight of the polymer falls, as its molecular weight distribution becomes narrower, and as the polymer concentration (in solution) drops. Almost all non-Newtonian fluids display shear-thinning behavior under appropriate circumstances. However, while most polymer solutions and melts will eventually reach the zero-shear viscosity behavior, only polymeric solutions seem to display the infinite-shear viscosity region.

1.3.1.1.1 Mathematical Models for Pseudoplastic Behavior

Numerous mathematical expressions of varying forms and complexities are available in the literature to model shear-thinning characteristics; some represent the straightforward attempts at curve fitting of the shear stress–shear rate data, while others have some theoretical basis in the statistical mechanics—as an extension of the kinetic theory as applied to the liquid state, and the theory of rate processes. Only a selection of the more widely used (and referred to in the subsequent chapters) is presented here. More complete and detailed descriptions of such models are available in the literature (Bird, 1976; Bird et al., 1987a; Carreau et al., 2021; Chhabra and Richardson, 2008; Holdsworth, 1992, 1993).

1.3.1.1.2 The Power Law or Ostwald-de Waele Model

The relationship between shear stress and shear rate, plotted on log–log coordinates, for pseudoplastic fluids can often be approximated by a straight line over a limited range of shear rate, and hence, this part of the flow curve can be described by the following power-law expression:

$$\tau_{yx} = m(\dot{\gamma}_{yx})^n \tag{1.12a}$$

or

$$\mu = m|\dot{\gamma}_{yx}|^{n-1} \tag{1.12b}$$

where n and m, known as the power-law index and the fluid consistency coefficient, are the two model parameters; for a Newtonian fluid, $n=1$, and for a pseudoplastic substance, $n<1$. The lower the value of the index n, the greater the degree of shear-thinning. Admittedly, Equation 1.12 provides the simplest description of shear-thinning behavior, but it also has a number of limitations. Generally, the applicability of Equation 1.12 is limited to a narrow range of shear rates; thus, the values of n and m are somewhat dependent on the shear rate range. Furthermore, it predicts unrealistically infinite and zero values of the apparent viscosity in the limits of very low and high shear rates, respectively. In spite of these deficiencies, this is perhaps the most criticized, most maligned, and yet most widely used equation in all of rheology (Schowalter, 1978). Some typical values of n and m are listed in Table 1.2 for a few aqueous solutions and suspensions (Chhabra, 1980).

TABLE 1.2

Typical Values of n and m

Liquid	Temperature (K)	n (–)	m (Pa s^n)	Shear Rate Range (s^{-1})
0.77% Carboxymethyl cellulose	294	0.95	0.044	44–560
0.10% Separan MG-500	294	0.55	0.205	0.56–883
32% Kaolin in water	303	0.103	19.5	1–50

Non-Newtonian Fluid Behavior

Considerable confusion exists in the literature regarding the effect of temperature on the values of power-law constants. While the flow behavior index is only weakly dependent on temperature (over a moderate temperature interval), the consistency coefficient m decreases with increasing temperature, though there exists some evidence that even the power-law index can also decrease with the increasing temperature (Alvarez et al., 2018; Lomba et al., 2002). In most process engineering applications, it is sufficient to neglect the temperature dependence of the flow behavior index and to use the usual Arrhenius-type expression to approximate the temperature dependence of the consistency coefficient (Chhabra, 1999b). The effect of pressure on the rheology of polybutene and other melts has been studied by Bair (2001) and Kadijk and van den Brule (1994).

1.3.1.1.2.1 The Eyring Model and Its Modifications Based on the theory of rate processes, Ree and Eyring (1965) developed the following constitutive relation for shear-thinning behavior:

$$\mu = R_1 \theta_E \left(\frac{\text{arc sinh}(\theta_E \dot{\gamma}_{yx})}{\theta_E \dot{\gamma}_{yx}} \right) \tag{1.13}$$

where R_1 and θ_E are the two model parameters. In a sense, Equation 1.13 represents the first attempt to obtain a rough molecular explanation for shear-thinning behavior. Only recently have statistical mechanical theories been able to describe shear-thinning characteristics for dilute polymer solutions. Sutterby (1966) modified Equation 1.13 to achieve more flexibility in fitting shear stress–shear rate data for polymer solutions by introducing another parameter and slightly rearranging it as

$$\mu = \mu_0 \left(\frac{\text{arc sinh}(\theta_E \dot{\gamma}_{yx})}{\theta_E \dot{\gamma}_{yx}} \right)^{A_0} \tag{1.14}$$

As expected, in the limit of zero-shear rate, Equation 1.14 predicts a constant viscosity (μ_0). Generally, a single set of values of the model parameters (μ_0, θ_E, A_0) can be used to approximate the flow curve over a reasonable range of shear rate, including the zero-shear viscosity region. Some representative values of the model parameters for a series of aqueous solutions of Natrosol polymer are listed in Table 1.3 (Sutterby, 1966).

Another useful empiricism (Salt et al., 1951) is obtained by adding a constant term (zero-shear viscosity) on the right-hand side of Equation 1.13, which results in the so-called Powell–Eyring model as

$$\mu = \mu_0 + R_1 \theta_E \left(\frac{\text{arc sinh}(\theta_E \dot{\gamma}_{yx})}{\theta_E \dot{\gamma}_{yx}} \right) \tag{1.15}$$

Obviously, as $\dot{\gamma}_{yx} \to 0$, the apparent viscosity approaches the zero-shear viscosity value μ_0.

1.3.1.1.1.3 The Cross Model Based on the assumption that the shear-thinning behavior is caused by the formation and breakdown of "structural linkages or units" present in a complex fluid, Cross (1965) put forward an equation. For 1-D steady shearing flow, this three-constant equation can be written as

$$\frac{\mu - \mu_\infty}{\mu_0 - \mu_\infty} = \frac{1}{1 + (\lambda \dot{\gamma}_{yx})^{2/3}} \tag{1.16}$$

TABLE 1.3

Typical Values of Sutterby Model Parameters at 298 K

Concentration (wt%)	μ_0 (Pa s)	θ_E (s)	A_0 (–)	Range of $\dot{\gamma}$ (s^{-1})
0.3	0.0245	0.014	0.5	<20,000
0.5	0.126	0.068	0.60	<20,000
0.7	0.510	0.225	0.66	<20,000

TABLE 1.4

Typical Values of Cross Model Parameters ($T = 298$ K)

System	μ_0 (Pa s)	μ_∞ (Pa s)	λ (ms)
3% Sodium carboxymethyl cellulose	0.8	~0	4.02
7% Ammonium polymethacrylate	7.52	~0	44.2
Polyvinyl acetate (thickened)	11.10	0.2	12.8

TABLE 1.5

Typical Values of Carreau Model Parameters

Material	μ_0 (Pa s)	μ_∞ (Pa s)	λ (s)	n (–)
2% PIB in Primol 355	923	0.15	191	0.36
7% Aluminum soap in decalin/m-cresol	89.6	0.01[a]	1.41	0.20
High-density polyethylene	8920	0[b]	1.58	0.50

Source: Bird, R.B., Armstrong, R.C., and Hassager, O., *Dynamics of Polymeric Liquids vol. 1: Fluid Dynamics*, 2nd ed., Wiley, New York (1987a).

[a] Assumed equal to solvent viscosity.
[b] Assumed.

where μ_0 and μ_∞, respectively, are the zero- and infinite-shear viscosities; λ is a constant with units of time. Cross (1965) reported satisfactory fits to shear stress–shear rate data for a wide variety of pseudoplastic systems. Typical values of the Cross model constants for a few aqueous systems are presented in Table 1.4 (Cross, 1965).

Many workers have reported significant improvement in fitting viscosity/shear rate data by replacing the index (2/3) in Equation 1.16 with a fourth parameter, e.g., see Barnes et al. (1989) and Struble and Ji (2000).

1.3.1.1.1.4 The Carreau Viscosity Equation This model, which has its origin in molecular network theories (Carreau, 1972; Carreau et al., 2021), accounts for all the features displayed by a pseudoplastic fluid in simple shear; it is written as:

$$\frac{\mu - \mu_\infty}{\mu_0 - \mu_\infty} = \left[1 + \left(\lambda \dot{\gamma}_{yx}\right)^2\right]^{(n-1)/2} \tag{1.17}$$

where μ_0 and μ_∞ are the zero-shear and infinite-shear viscosities, respectively, and n (<1) and λ are two disposable curve-fitting parameters. Typical values of the model parameters for some systems are listed in Table 1.5.

The Carreau viscosity equation, Equation 1.17, has been further modified by introducing yet another disposable parameter to improve the degree of fit (Yasuda, 1979). The modified form purports to improve the prediction of the onset of the power-law region, albeit at the expense of an additional parameter.

1.3.1.1.1.5 The Ellis Fluid Model The four models presented so far are examples of the form $\tau_{yx} = f(\dot{\gamma}_{yx})$. The three-parameter Ellis fluid model is an illustration of the inverse form and it is written as

$$\mu = \frac{\mu_0}{1 + \left(\tau_{yx}/\tau_{1/2}\right)^{\alpha-1}} \tag{1.18}$$

where μ_0 is the zero-shear viscosity; $\tau_{1/2}$ is a model parameter denoting the value of the shear stress at which the apparent viscosity has dropped to a value of $\mu_0/2$, and α (>1) is a measure of the extent of shear-thinning behavior. This form of the equation is advantageous in that it permits an easy calculation

TABLE 1.6

Typical Values of Ellis Model Parameters

Material	T (K)	μ_0 (Pa s)	$\tau_{1/2}$ (Pa)	α (–)	Range of $\dot{\gamma}$ (s^{-1})
0.85% Separan AP-30	292	8.2	1.64	2.70	0.005–113
1.46% Methocel HG-90	295.5	3.3	28.8	2.38	0.5–896
2% Separan AP-30	289.5	59	6.49	3.13	0.004–113

Source: Chhabra, R.P., Tiu, C., and Uhlherr, P.H.T., *Rheol. Acta*, **20**, 346 (1981a).

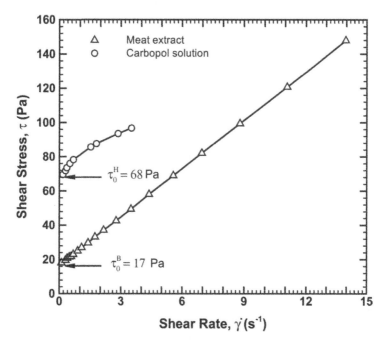

FIGURE 1.7 Shear stress–shear rate data for a meat extract and a carbopol solution displaying Bingham plastic and visco-plastic behaviors.

of the velocity profile from a known stress distribution but renders the reverse operation cumbersome. Representative values of μ_0, α, and $\tau_{1/2}$ for a few aqueous polymer solutions are given in Table 1.6.

1.3.1.2 Visco-Plastic Fluids

This class of materials is characterized by the existence of a yield stress (τ_0) that must be exceeded before the fluid will deform or flow like a viscous fluid. Once the applied stress exceeds the yield stress, the flow curve may be linear (curve B in Figure 1.4) or nonlinear (curve A in Figure 1.4). It should be noted that such a substance is not really a fluid according to the strict physical definition. One can, however, explain this type of behavior by postulating that the substance at rest consists of a 3-D structure of sufficient rigidity or strength to resist any external stress less than τ_0. For stress levels greater than τ_0, the structure disintegrates and the material behaves like a viscous fluid.

A fluid with a linear flow curve for $|\tau_{yx}| > |\tau_0|$ is called the Bingham plastic fluid and is obviously characterized by a constant value of plastic viscosity. Figure 1.7 illustrates this type of flow behavior for a meat extract; the resulting values of the yield stress and plastic viscosity are $\tau_0^B = 17$ Pa and $\mu_B = 9.40$ Pa s, respectively. In contrast, a substance possessing a yield stress and a nonlinear flow curve (for $|\tau_{yx}| > |\tau_0|$) is simply known as a visco-plastic material, as shown schematically in Figure 1.4 (curve A), and for an aqueous carbopol solution in Figure 1.7. Strictly speaking, it is very difficult to ascertain whether any real

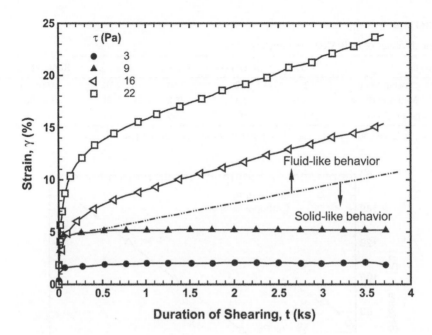

FIGURE 1.8 Creep behavior in a 1% bleached softwood kraft fiber suspension at 296 K. (Data courtesy of Dr. Derakhshandeh and from Derakhshandeh et al., 2011.)

material has a true yield stress (Houwink and de Decker, 1971; Barnes and Walters, 1985; Cheng, 1986; Hartnett and Hu, 1989; Astarita, 1990; Schurz, 1990; Evans, 1992; Barnes, 1992,1999, 2001; Balan, 1999; Balmforth et al. 2014; Bonn et al., 2017). Furthermore, such a substance will behave like an elastic solid (or will flow *en masse* like a rigid body) when the externally applied stress is less than the yield stress (τ_0). Also, it stands to reason that, in the absence of surface tension effects, such a material will not level out under gravity to form a flat free surface.

From another vantage point, such material will respond to an external stress by attaining a constant strain (zero rate of strain) in line with Hooke's law of elasticity. Figure 1.8 shows this type of behavior (creep) in a 1% bleached softwood kraft fiber suspension. The regions of fluid-like and solid-like behaviors are separated by a broken line corresponding to the yield stress (9 Pa $< \tau_0 <$ 16 Pa) for this suspension. In contrast, many researchers (O'Donovan and Tanner, 1984; Derakhshandeh et al., 2011; Balmforth et al., 2014) treat such a fluid by displaying vastly different viscosity levels depending upon the magnitude of the externally applied stress, as can be clearly seen in Figure 1.9 for a printing ink (Valverde et al., 2004), wherein the apparent viscosity is seen to drop by several orders of magnitudes over an extremely narrow interval of shear stress. Additional difficulties arise when the material also exhibits time-dependent effects (Dinkgreve et al., 2017). Notwithstanding these difficulties, the concept of a yield stress is convenient in practice because some fluids of great industrial significance closely approximate this type of flow behavior. The answer to the question of whether a fluid possesses a true yield stress depends upon the choice of an appropriate time scale of observation that will clearly vary from one application to another, as discussed by Cheng (1986), Astarita (1990), and Barnes (2001) and subsequently by others (Denn and Bonn, 2011; Coussot, 2014, 2017; Bonn et al., 2017). In view of this uncertainty, some authors have advocated the use of the term "apparent yield stress," but in this work, the term "yield stress" will be used. Common examples of materials displaying visco-plastic behavior are greases, foams, drilling muds, paints, concentrated slurries, blood, processed foodstuffs, diamond mine tailings, mucus, molten lava, filled polymers, etc. (Bird et al., 1983; Barnes, 1999; Balmforth et al., 2014; Coussot, 2014; Bonn et al., 2017; De kee, 2021). In recent years, there has been a spurt in the rheological behavior of the so-called electro-rheological fluids (Parthasarathy and Klingenberg, 1996; Rankin et al., 1998, 1999; See and Brian, 2005) and magneto-rheological fluids (Dang et al., 2000).

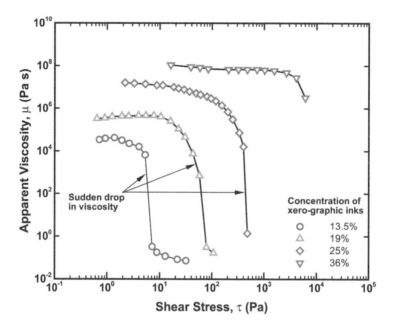

FIGURE 1.9 Large drop in apparent viscosity of a model ink (Valverde et al., 2004).

1.3.1.2.1 Mathematical Models for Visco-Plastic Behavior

Over the years, numerous empirical expressions have been proposed in the literature; a model based on sound theory is yet to emerge for visco-plastic behavior. Some simple and widely used expressions are given here, whereas thorough and critical accounts of the developments in this area have been presented by, among others, Bird et al. (1983), Balmforth et al. (2014), and Bonn et al. (2017).

1.3.1.2.1.1 The Bingham Plastic Model This is the simplest and perhaps the most widely used fluid model to approximate the visco-plastic flow behavior. For a simple shearing flow, it is written as:

$$\tau_{yx} = \tau_0^B + \mu_B \dot{\gamma}_{yx} \qquad \text{for } \tau_{yx} > \tau_0^B$$

$$\dot{\gamma}_{yx} = 0 \qquad \text{for } \tau_{yx} \leq \tau_0^B \tag{1.19}$$

where τ_0^B is the Bingham yield stress and μ_B is the plastic viscosity. Typical comparisons of this model with typical experimental data for two aqueous silica suspensions and two aqueous polymer solutions are shown in Figure 1.10; Equation 1.19 is seen to provide a satisfactory fit to the data only for $\dot{\gamma}_{yx} > 40 s^{-1}$.

The available limited information suggests that both the Bingham plastic viscosity and yield stress are much more sensitive to temperature than to pressure (Briscoe et al., 1994). However, the available data regarding the influence of temperature, especially on the yield stress, are far from being conclusive (Vryzas et al., 2016; Alvarez et al., 2018; Poole, 2018). Coussot et al. (2017) have presented an account of the historical development of the Bingham model.

1.3.1.2.1.2 The Casson Model This is another two-parameter model that for a simple shearing motion is written as

$$\left(\tau_{yx}\right)^{1/2} = \left(\tau_0\right)^{1/2} + \left(\mu_C \dot{\gamma}_{yx}\right)^{1/2} \qquad \text{for } \tau_{yx} > \tau_0^C$$

$$\dot{\gamma}_{yx} = 0 \qquad \text{for } \tau_{yx} \leq \tau_0^C \tag{1.20}$$

The model parameters are the Casson yield stress, τ_0^C and a consistency parameter, μ_C. Experimental data for the same polymer solutions and suspensions as those used in Figure 1.10 are compared with the predictions of Equation 1.20 in Figure 1.11 where the agreement is seen to be moderate, especially at the low shear rate end.

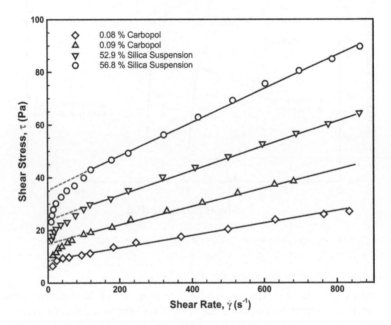

FIGURE 1.10 Typical comparison between data and predictions of Equation 1.19 for two polymer solutions and two silica suspensions. (From Chhabra and Uhlherr, 1988a.)

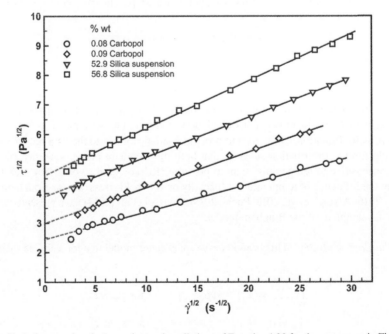

FIGURE 1.11 Typical comparison between data and predictions of Equation 1.20 for the systems as in Figure 1.10.

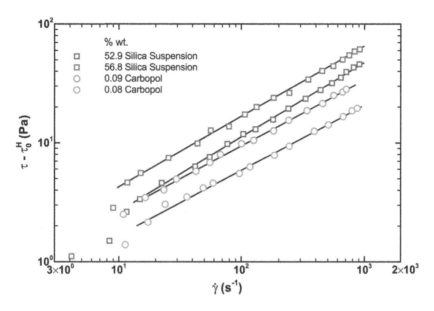

FIGURE 1.12 Typical comparison between data and prediction of Equation 1.21 for the same fluid as in Figure 1.10.

1.3.1.2.1.3 The Herschel–Bulkley Model This is a generalization of the simple Bingham model in which the linear shear rate dependence has been replaced by a power-law behavior. It is thus a three-parameter model, written for a simple shear flow as

$$\tau_{yx} = \tau_0^H + m(\dot{\gamma}_{yx})^n \qquad \text{for } \tau_{yx} > \tau_0^H$$
$$\dot{\gamma}_{yx} = 0 \qquad \text{for } \tau_{yx} \leq \tau_0^H$$
(1.21)

This model is more flexible than the two previous ones and generally fits experimental data over a somewhat wider range of conditions and/or slightly better than Equations 1.19 and 1.20. Typical comparisons with the experimental data, as that used in Figures 1.10 and 1.11, are shown in Figure 1.12. Limited efforts have also been made to establish the nature of dependences of the fluid-model parameters on composition, temperature, etc. (Briscoe et al., 1994; de Larrard et al., 1998).

Finally, it is appropriate to add here that for a given set of shear stress–shear rate data, it is not uncommon that Equations 1.19–1.21 will yield different values of τ_0^B, τ_0^C, and τ_0^H, and these as such must not be confused with the true value of the yield stress (if any!) of the material. It also needs to be borne in mind that many visco-plastic materials also exhibit time-dependent effects (thixotropy discussed in Section 1.3.2.1) and none of the viscosity models presented here account for this complexity. Nor do these capture the elastic behavior for the stress levels below the fluid yield stress. Some attempts have been made at incorporating these features into constitutive equations, which, however, tend to be far more complex than the simple models included here. Malkin et al. (2017) have recently revisited the notion of yield stress. On the other hand, Thompson et al. (2018) have presented the idea of so called yield stress tensor for 3D flows.

1.3.1.3 Shear-Thickening Fluids

These materials, also known as dilatant fluids, are similar to shear-thinning materials in that they show no yield stress, but their apparent viscosity increases with the increasing shear rate. This type of flow behavior is encountered in concentrated suspensions of solids and can be qualitatively explained as follows: when a suspension is at rest, its voidage is minimum and the liquid present is just sufficient to fill the void spaces. At low shear rates, the liquid lubricates the motion of one particle past another, and the resulting stresses are consequently low. At high shear rates, in contrast, the dense packing of solids breaks down and the material expands or dilates slightly causing an increase in the voidage, and thus the amount of liquid available is no longer sufficient to lubricate the solid motion of one particle past another and the resulting solid–solid friction causes the stresses to increase rapidly, which, in turn, causes an increase in the apparent viscosity. Evidently, one will observe such effects only in highly concentrated suspensions and that too at reasonably high shear rates. A similar mechanism for the dilatant behavior encountered in rice-starch suspensions has been described by Andrade and Fox (1949). The term *dilatant* is also used for all other fluids that exhibit an increase in apparent viscosity with shear rate. Many of such materials, such as starch pastes, are neither true suspensions nor do they show any dilation on shearing. The above explanation therefore is irrelevant, but nevertheless, the term dilatant is still used to describe the rheological behavior of such materials.

Of the time-independent fluids, this type has received very scant attention; consequently, few reliable data illustrating dilatant behavior are available in the literature. This is partly so due to the fact that, until recently, the dilatant behavior was considered to be much less widespread in chemical and processing industries. However, in recent years, with the growing importance of the processing of highly loaded systems in a broad spectrum of applications, including armors, sports equipment, and medical implants, there has been a renewed interest in studying the dilatant behavior (Barnes, 1989; Boersma et al., 1990; Goddard and Bashir, 1990; Brown et al., 2010; Jiang et al., 2010; Cheng et al., 2011; Galindo-Rosales et al., 2011a,b; Van Hecke, 2012; Waitukaitis and Jaeger, 2012; Zarei and Aalaie, 2020). Typical examples of systems exhibiting dilatant behavior include TiO_2/water suspensions (Metzner and Whitlock, 1958) (shown in Figure 1.13), concentrated china-clay suspensions (Bullivant and Jones, 1981), cornflour-in-water suspensions (Griskey et al., 1985; Dail and Steffe, 1990), solutions of polyethylene oxide and sodium dodecyl sulfate (Rivero et al., 2012), carbon fiber-reinforced polymers (Pinto and Meo, 2017), etc. Furthermore, under appropriate conditions, solutions of certain polymers (such as partially hydrolyzed polyacrylamide) in glycerol–water mixtures have been shown to exhibit pronounced shear-thickening behavior at high shear rates (e.g., see Wang et al. 1996, Briscoe et al. 1999, Mewis and Wagner 2012). Some attempts have also been made at explaining the shear-thickening behavior of dilute polymer solutions via Brownian dynamics calculations (Hatzikiriakos and Vlassopoulos, 1996). A survey (Bagley and Dintzis, 1999) provides an overview of the activity in this field with special reference to food stuffs and biopolymers. Also, there has been a growing interest in the dilatant characteristics of the dilute solutions of wormlike micelles, which is generally explained via the formation of shear-induced

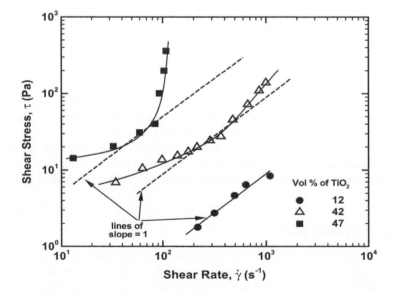

FIGURE 1.13 Typical shear stress–shear rate data for TiO_2 suspensions displaying shear-thickening behavior. (Adapted from Metzner and Whitlock, 1958.)

structures (Barentin and Liu, 2001) or via dynamic simulations (Dratler et al., 1997). As noted earlier, it is not uncommon for the same suspension to exhibit Newtonian and/or shear-thinning behavior at low shear rates and shear-thickening at high shear rates followed by another shear-thinning region. For instance, Figure 1.14 displays various rheological regimes in a suspension of fumed silica spheres in polyethylene glycol (PEG) (Galindo-Rosales et al., 2009). Based on phenomenological considerations and the Stokesian dynamics simulations, it is suggested that particles do not form 3-D structures or networks under rest conditions, at least in non-flocculating systems, though aggregates can be formed. Thus, at low-stress levels (shear rates), the formation and destruction of such aggregates are responsible for the initial Newtonian and/or shear-thinning behavior, seen in Figures 1.13 and 1.14. This behavior will continue as long as the interparticle interactions and the hydrodynamic forces cancel each other out. As the shear rate (or shear stress) is gradually increased, the hydrodynamic forces rise rapidly and eventually far exceed the interparticle forces, thereby leading to the formation of "hydro-clusters" or the so-called "jammed" structures. Though these clusters are of transient nature, this phenomenon, in turn, manifests in the form of a steep rise in the apparent viscosity of the suspension, as seen in Figures 1.13 and 1.14. The familiar Peclet number, defined as $Pe = \dot{\gamma} l^2 / D_m$, denotes the ratio of the hydrodynamic (convective or shear) to diffusive (Brownian motion) forces. Thus, $Pe < 1$ regime is dominated by Brownian or particle diffusion effects whereas $Pe > 1$ corresponds to strong hydrodynamic effects. However, it is not yet possible to predict the onset of each of the viscosity regimes, as seen in Figures 1.13 and 1.14 for a given system. Therefore, no single viscosity equation is available to capture the shear stress–shear rate behavior of such systems (Galindo-Rosales et al., 2011a, b). Much more complex shear-thickening behavior of corn starch-in-water/polypropylene glycol (PPG) mixtures has been demonstrated by Taylor (2013). Indeed, depending upon the composition of the fluid in terms of the relative proportions of corn starch, water, and PPG, it is possible to encounter shear-thinning, or shear-thickening, or visco-plastic behavior under appropriate conditions. Over the years, shear-thickening behavior has been reported in wide-ranging systems, including nanosize fumed silica in PEG(Gürgen, 2023; Gürgen et al., 2016a, b; Han et al., 2018; Zhang et al., 2008c) and semi-dilute entangled polymer solutions (Xu et al., 2011). Morris (2020) has recently reviewed the available body of information on shear-thickening behavior, particularly in the context of concentrated suspensions. Therefore, the viscosity models that account for all these aspects of shear-thickening materials tend to be rather complex and contain many additional

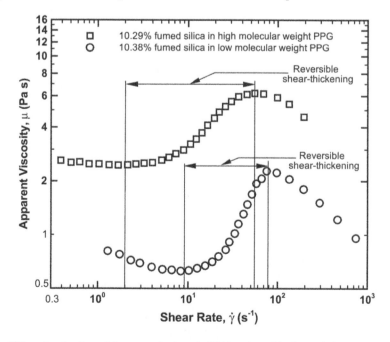

FIGURE 1.14 Effect of molecular weight on steady shear (exhibiting shear-thinning and shear-thickening) behavior of silica in polypropylene glycol (PPG). (Replotted from Galindo-Rosales et al., 2009.)

disposable parameters, which must be evaluated using a range of experimental tools (Galindo-Rosales et al., 2011a,b). However, from an engineering perspective, the power-law model (Equation 1.12) may be used with $n > 1$ in this case over the range of shear rate over which shear-thickening behavior is observed. This simple approach, however, is severely limited in its utility because it is virtually impossible to guesstimate the expected range of shear rate in a given application. Notwithstanding this intrinsic difficulty, this empirical approach is believed to be useful in providing crude estimates for process engineering design calculation purposes.

1.3.2 Time-Dependent Behavior

For many industrially important materials, the shear flow properties depend on both the rate of shear and the time of shearing and the past kinematic history of the substance. For instance, when the aqueous suspensions of red mud and bentonite, crude oils, building materials, toothpaste, hand lotions, and certain foodstuffs, etc., are sheared at a constant rate of shear following a period of rest, their apparent viscosity gradually decreases as their internal "structure" is progressively broken down. As the number of "structural linkages" available for breaking down decreases, the rate of change of viscosity with time also drops to zero. In contrast, the rate at which the linkages can reform increases, and eventually, a state of dynamic equilibrium is reached when the rates of buildup and breakdown of linkages become equal. This type of fluid behavior may be further divided into two categories: thixotropy and rheopexy or negative thixotropy.

1.3.2.1 Thixotropy

A material is said to exhibit thixotropy if its apparent viscosity (or shear stress) decreases with time when sheared at a constant rate of shear. If the flow curve is measured in a single experiment in which the shear rate is steadily increased at a constant rate from zero to a maximum value and then decreased at the same rate to zero again, a hysteresis loop, as shown schematically in Figure 1.15, is obtained. The

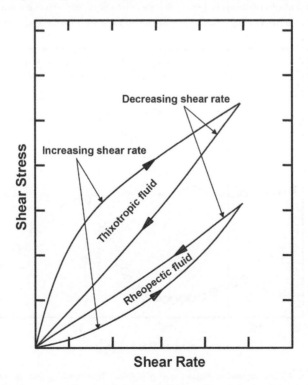

FIGURE 1.15 Qualitative shear stress–shear rate behavior for thixotropic and rheopectic materials.

Non-Newtonian Fluid Behavior

height, shape, and the enclosed area of the loop depend upon the kinematic parameters, such as the duration and rate of shearing and the past deformation history of the sample. Figure 1.16 shows the hysteresis effects in a cement paste (Struble and Ji, 2000), and Figure 1.17 shows the thixotropic behavior in commercial yogurts (Ramaswamy and Basak, 1991). Similarly, the data plotted in Figure 1.18 exemplify the thixotropic behavior of a red mud suspension (Nguyen and Uhlherr, 1983) wherein the shear stress decreases with the duration of shearing and eventually approaches an equilibrium value corresponding

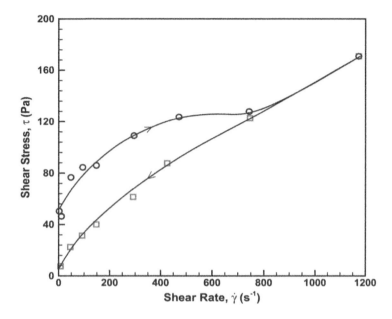

FIGURE 1.16 Thixotropy in a cement paste.

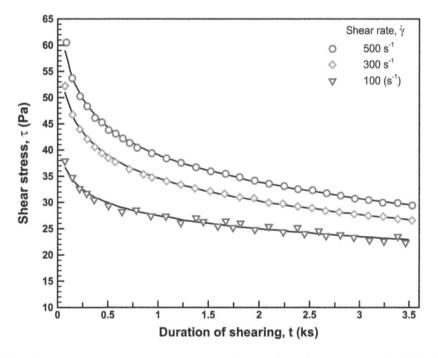

FIGURE 1.17 Stress decay curves of stirred yogurts over a 60 min shear at selected shear rates ($T=10°C$). (Replotted from Ramaswamy and Basak, 1991.)

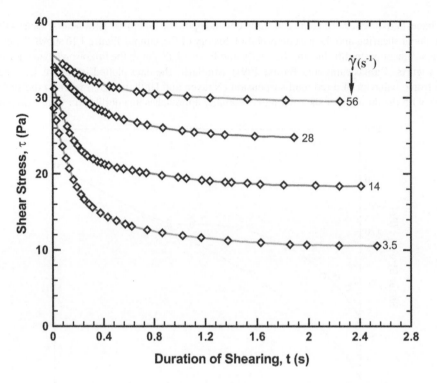

FIGURE 1.18 Typical experimental data showing thixotropic behavior in a 59% (by weight) red mud suspension (Nguyen and Uhlherr, 1983).

to the applied shear rate. The term false body has been used to describe the thixotropic behavior of visco-plastic materials. Though the thixotropic behavior is associated with the breakdown of structure, visco-plastic materials do not completely lose their solid-like characteristics, such as yield stress, which may well be less than its initial value; the latter is regained (if at all) only after a long recovery period. Similarly, Dolz-Planas et al. (1991) have studied the thixotropic behavior of Carboxy Methyl Cellulose (CMC) solutions whereas Gilchrist and Chandler (1996) have reported on thixotropy in polymer-flocculated clay suspensions. Mewis (1979), Mujumdar et al. (2002), and Mewis and Wagner (2009) have studied some transient effects in such systems.

1.3.2.2 Rheopexy or Negative Thixotropy

The relatively few systems for which the apparent viscosity increases with the duration of shearing are said to display rheopexy or negative thixotropy. Again, hysteresis effects are observed in the flow curve, but in this case, it is inverted as compared to that for a thixotropic material (see Figure 1.15). By analogy with thixotropy, rheopexy is associated with a gradual buildup of "structure" as the fluid is sheared, though it is not certain whether an equilibrium will ever be reached. Thus, in a rheopectic material, the structure builds up by shear and it breaks down when the material is at rest. For instance, Freundlich and Juliusburger (1935), using a 42% aqueous gypsum paste, found that, after shaking, this material resolidified in 40 min if at rest, but in only 20 s if the container was rolled gently in the palms of hands. Thus, the structure formation is facilitated by a gentle shearing motion (rolling), but a more intense motion (shaking) destroys the structure. This suggests that there is a critical amount of shear beyond which re-formation of the structure is not induced but breakdown occurs. They attributed this behavior to the anisometric shape of gypsum particles. It is not uncommon for the same suspension/dispersion to display both thixotropy and rheopexy under appropriate conditions of shear rate and the solid contents. Figure 1.19 illustrates the gradual onset of rheopexy in a saturated polyester at 333 K (Steg and Katz,

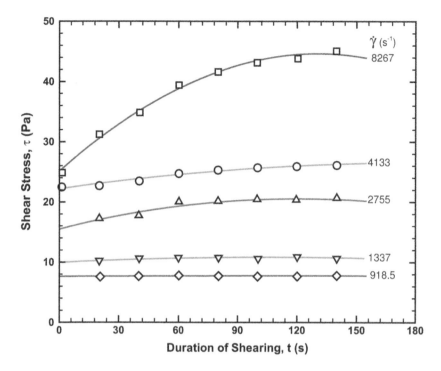

FIGURE 1.19 Rheopectic behavior in a saturated polyester (Steg and Katz, 1965).

1965). Qualitatively, similar behavior is reported to occur in protein solutions (Pradipasena and Rha, 1977), coal–water slurries (Keller and Keller Jr., 1990), suspensions of ammonium oleate, colloidal dispersions of vanadium pentaoxide at moderate shear rates (Tanner, 1982, 2000), and highly concentrated water-in-oil emulsions (Malkin et al., 2004; Masalova et al., 2006). The available mathematical equations to describe the thixotropic and rheopectic behavior are much more complex in form than those mentioned in the preceding section for the time-independent fluid behavior, and it is usually necessary to make measurements over the range of conditions of interest. Besides, most of these constitutive models have been custom built to describe the behavior of a specific material in a specific application. Cheng (1979, 1987, 2003), Mewis (1979), Govier and Aziz (1982), Barnes (1997), and Warson (2003) have written state-of-the-art informative accounts in the field of thixotropy.

Suffice it to add here that though much confusion exists in the literature regarding thixotropy (Garlaschelli et al., 1994; Atkinson, 2001), there has been a spurt in research activity in this field (e.g., see the recent work of Dullaert and Mewis (2005, 2006), Mewis and Wagner (2009), de Souza Mendes and Thompson (2012), Larson and co-workers (Mahaut et al., 2008; Larson, 2015; Larson and Wei, 2019), Madraki et al. (2017), Ewoldt and McKinley (2017), N'Gouamba et al. (2019), Varchanis et al. (2019), and Ramya et al. (2020)). Depending upon the nature of the fluid, it is not uncommon to encounter elasticity, yield stress, and thixotopy simultaneously, and therefore, the constituitive equations tend to be rather complex. However, the basic ideas underpinning these developments are summarized here. Broadly, three distinct approaches can be discerned in the literature: continuum, micro-structural, and structural kinetics. Within the framework of the continuum modeling, depending upon the equilibrium rheological characteristics of the substance, existing viscosity models (such as Bingham model, Casson model, or any other such equation) are modified by splitting the viscosity, yield stress, etc., into two parts corresponding to its equilibrium value and a time-dependent component. Obviously, this approach takes no cognizance of the micro-structure and variations thereof with shearing and time of shearing. Consequently, it is not possible to establish links between the model parameters and the molecular properties and physical processes responsible for the structural changes under shear and for the subsequent building of structure upon the removal of shear. The micro-structural approach, in contrast, necessitates detailed information of interparticle forces and potentials that are rarely available for the systems of industrial relevance, thereby severely impeding the developments in this direction. Lastly, the thixotropy models based on the structural-kinetics idea hinge on the validity of a single scalar parameter, ξ, which "somehow" indicates the state of the "structure" in a system. Typically, this parameter takes on values between 0 (no structure or structureless limit) and 1 (complete buildup of structure or no more buildup is possible). Thus, this approach relies on the applicability of two coupled equations which, in the case of steady shear, are $\tau - \dot{\gamma}$ relationship for a fixed value of ξ and ξ–t relationship, similar to the rate expression for a chemical reaction that describes the evolution of ξ with time depending upon the shearing motion. This idea goes back almost 60 years to the work of Moore (1959b) and subsequently utilized by Houska (1981). For a Herschel–Bulkley material, the thixotropic effects are incorporated by rewriting Equation 1.21 as follows:

$$\tau_{yx} = \left(\tau_0^H + \tau_{0_1}^H\right) + \left(m_0 + m_t \xi\right)\dot{\gamma}^n \tag{1.22}$$

and

$$\frac{d\xi}{dt} = a_o\left(1 - \xi\right) - b_o \xi \dot{\gamma}^\beta \tag{1.23}$$

In Equation 1.22, τ_0^H and m_0 are the so-called permanent or equilibrium values; $\tau_{0_1}^H$ and m_t are the corresponding time-dependent contribution. Thus, Equation 1.22 is written for a constant value of ξ evolving with time in accordance with Equation 1.23. The first term on the right-hand side of Equation 1.23 describes the rate of buildup of structure (assumed to occur under rest and is thus independent of $\dot{\gamma}$), whereas the second term describes the rate of breakdown of structure that is a function of ξ and $\dot{\gamma}$. Thus, altogether this model contains eight material constants in simple shear: five rheological $\left(\tau_0^H, \tau_{0_1}^H, m_o, m_t, n\right)$ and three kinetic (a_o, b_o, β) constants. Clearly, their evaluation necessitates experimental protocols that are far more challenging than that required for the rheological characterization of time-independent and visco-elastic fluids. However, some useful tips in this regard are available in the literature (Mewis and Wagner, 2009, 2012; Larson and Wei, 2019).

1.3.3 Visco-elastic Behavior

In the classical theory of elasticity, the stress in a sheared body is directly proportional to the strain. For tension, Hooke's law applies, and the coefficient of proportionality, G, is called Young's modulus:

$$\tau_{yx} = G\frac{dx}{dy} \tag{1.24}$$

When a solid deforms within the elastic limit, it regains its original form on the removal of the stress. However, if the applied stress exceeds the characteristic yield point of the material, complete recovery will not occur and creep will occur. In other words, the solid will have *flowed*. Table 1.7 lists representative values of Young's modulus G for a range of materials, including metals, concrete, stones, plastics, polymer and colloidal solutions, and foodstuffs. These values as such provide the basis to identify some of these substances as "soft" solids or "soft" matter because their values of G are orders of magnitudes lower than that of real solids, such as steel and concrete. At the other extreme is the Newtonian fluid for which the shearing stress is proportional to the rate of shear (Equation 1.1). Many substances show both elastic and viscous effects under appropriate circumstances. In the absence of time-dependent effects mentioned in Section 1.3.2, such materials are known as visco-elastic fluids. Perfectly viscous flow and elastic deformation are indeed the two limiting cases of visco-elastic behavior. For some materials, it is only these limiting conditions that are observed in practice. The viscosity of ice and the elasticity of water may generally pass unnoticed! The behavior of such materials depends not only on their nature, flow field, etc., but also on their past kinematic history. Thus, the distinctions between a solid and a fluid and between elastic and viscous deformations are to some extent arbitrary. Many materials of practical interest, such as polymer melts and solutions, emulsions, micellar systems, synovial fluid, soap solutions,

TABLE 1.7

Representative (Approximate) Values of Young's Modulus, G, for a Range of Substances

Material	Value of G
Polymer and colloidal solutions	1–100 Pa
Gelatin gel	0.2 MPa
Banana	0.8–3 MPa
Rubber	0.1–5 MPa
Potatoes	6–14 MPa
Raw apples	6–14 MPa
Peaches	2–20 MPa
Pears	10–30 MPa
Carrots	20–40 MPa
Leather	1–100 MPa
Dry spaghetti	3 GPa
Ice	10 GPa
Wood	1–10 GPa
Concrete	10–20 GPa
Engineering plastics	5–20 GPa
Stones	40–60 GPa
Glass	70 GPa
Aluminum, copper, and alloys	100 GPa
Steel	200 GPa
High modulus-oriented fibers	>300 GPa

Source: R.P. Chhabra, *Non-Newtonian Fluids: An Introduction, in Rheology of Complex Fluids*, (Eds. A.P. Deshpande, J. Murali Krishnan, P.B. Sunil Kumar), Springer, Berlin (2010); J.F. Steffe, *Rheological Methods in Food Process Engineering*, Freeman, East Lansing, MI (1996); A.Y. Malkin, A.I. Isayev, *Rheology: Concepts, Methods and Applications*, Chem Tec, Toronto (2006).

gels, and foams, exhibit visco-elastic behavior; they have some ability to store energy and thus show partial recovery upon the removal of stress. These fluids also show memory effects in so far that they can remember the "events" that occurred in the past.

Numerous other unusual phenomena associated with visco-elastic behavior include die swell, rod climbing, tubeless siphon, "soup-bowl" effect, the development of secondary flows, aggregation of particles, migration of particles across streamlines, unexpected shapes of bubbles and drops, negative wake, etc. These and other similar phenomena have been adequately dealt with by Bird et al. (1987a), Barnes et al. (1989), and Boger and Walters (1992) in their excellent books. Obviously, exhaustive coverage of visco-elastic effects is beyond the scope of this work, and interested readers are referred to several excellent sources of information available on this subject (Bird et al., 1987a; Lapasin and Pricl, 1995; Carreau et al., 2021; Larson, 1998; Lakes, 1999; Gupta, 2000; Tanner, 2000; Morrison, 2001). Here, we shall only touch upon the so-called primary and secondary normal-stress differences observed in steady shearing, loss and storage module in small-amplitude oscillatory shearing, and the Trouton ratio in uniaxial elongation flows that are used both to label a fluid as visco-elastic or visco-inelastic and to ascertain the importance of visco-elastic effects for a given material in an envisaged application.

1.3.3.1 Normal-Stress Effects in Steady Shearing Flows

Let us consider the steady 1-D shearing flow of a fluid; the stresses so developed are shown in Figure 1.20 where a cubical element of material sheared between the two planes is depicted. Obviously, by nature of the unidirectional shearing flow, the y- and z-components of the velocity vector are zero. In addition to

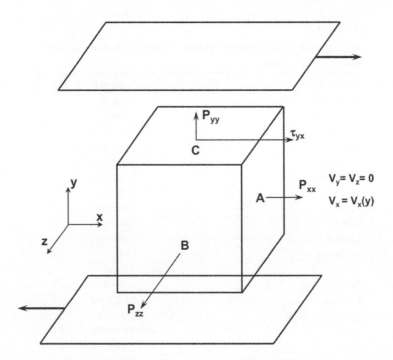

FIGURE 1.20 Nonzero components of stress tensor in unidirectional shearing motion of a visco-elastic fluid.

the shear stress τ_{yx}, there are also three normal stresses denoted by P_{xx}, P_{yy}, and P_{zz}, which are given in Equation 1.5. Weissenberg (1947) was the first to notice that the shearing motion of a visco-elastic fluid results in unequal normal stresses, that is,

$$P_{xx} \neq P_{yy} \neq P_{zz} \tag{1.25}$$

The pressure in a fluid can be determined in a straightforward manner, and Equation 1.6 is still applicable so that the three individual normal stresses can be found from two independent measurements in a unidirectional steady shearing flow. The differences $(P_{xx} - P_{yy})$ $(= N_1)$ and $(P_{yy} - P_{zz})$ $(= N_2)$ are easier to measure than the individual stresses, and it is customary to use N_1 and N_2 together with τ_{yx} as functions of shear rate to describe the rheological behavior of a visco-elastic fluid in a simple shearing flow. Sometimes the coefficients of two normal-stress differences are used, which are defined as

$$\text{Primary normal stress coefficient: } \psi_1 = \frac{N_1}{(\dot{\gamma}_{yx})^2} \tag{1.26a}$$

$$\text{Second normal stress coefficient: } \psi_2 = \frac{N_2}{(\dot{\gamma}_{yx})^2} \tag{1.26b}$$

Typical dependence of N_1 on shear rate is shown in Figure 1.21 for a series of polystyrene-in-toluene solutions. Generally, the rate of decrease of ψ_1 with shear rate is higher than that of apparent viscosity. At very low shear rates, the first normal-stress difference is expected to be proportional to the square of shear rate, that is, ψ_1 tends to a constant value; this limit is seen to be reached by some of the data shown in Figure 1.21.

There is, however, no evidence of ψ_1 reaching another constant value in the limit of very high shear rates. Overall, the first normal-stress difference has not been studied as extensively as the shear stress.

The second normal-stress difference (N_2) has received even less attention than that given to N_1. The most important points to note about the second normal-stress difference are that it is about one-tenth of

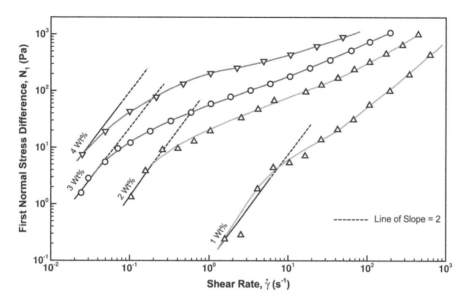

FIGURE 1.21 Typical first normal-stress difference data for polystyrene-in-toluene solutions (Kulicke and Wallbaum, 1985).

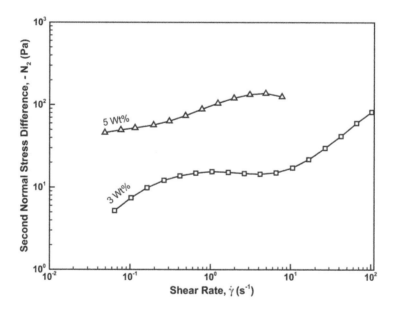

FIGURE 1.22 Typical second normal-stress difference data for polystyrene-in-toluene solutions (Kulicke and Wallbaum, 1985).

N_1 and that it is negative. Until recently, the second normal-stress difference was assumed to be zero—the so-called Weissenberg hypothesis is no longer thought to be correct. Some literature data seem to suggest that N_2 may even change sign and become positive at high shear rates (Maklad and Poole, 2021). Figure 1.22 shows the typical dependence of the (magnitude of) second normal difference on shear rate for the same polymer solutions as that shown in Figure 1.21. The two normal-stress differences are characteristic of a substance and as such are used to classify a fluid as visco-inelastic ($N_1 \sim 0$) or visco-elastic ($N_1 \napprox 0$): the magnitude of N_1 is taken to be a measure of the level of the fluid visco-elasticity. For a given fluid, the ratio (N_1/τ) provides a crude but simple measure of the severity of visco-elastic behavior, specifically ($N_1/2\tau$) is called the recoverable shear and values greater than 0.5 are not uncommon for highly visco-elastic fluids. However, there is no guarantee that a substance that exhibits a significant first normal-stress difference in steady shear will also show large values of the storage modulus in an oscillatory experiment (see Section 1.3.3.3). Thus, whether a fluid behaves like a weakly or strongly visco-elastic fluid is somewhat dependent upon the test it is subjected to (Barnes et al., 1989). It is useful to mention here that the so-called time-independent fluids introduced in Section 1.3.1 when subjected to simple steady shear exhibit $N_1=0$ and $N_2=0$, and hence, these are also known as inelastic, or viscous, or GNFs.

1.3.3.2 Elongational Flow

Flows that result in fluids being subjected to stretching in one or more directions also occur in several engineering processes, fiber spinning and polymer film blowing being only two of the most common examples. Likewise, when two bubbles coalesce, a very similar stretching of the liquid film between them occurs that finally ruptures. Additional examples of extensional flows are found with the flow of polymer solutions in porous media, as encountered in the enhanced oil recovery process and inkjet printing, spraying, etc. It is customary to categorize the elongational flow into three types, namely, uniaxial, biaxial, and planar, as shown schematically in Figure 1.23. Naturally, the mode of extension affects the way the fluid resists deformation and, although this resistance can be loosely quantified in terms of an elongational or extensional viscosity (which depends upon the type of elongation, that is, uniaxial, biaxial, or planar), this parameter is, in general, not necessarily constant. For an incompressible liquid being stretched at a constant rate $\dot{\varepsilon}$ in the x-direction (Figure 1.24), the element must contract in both the y- and z-directions at the rate of $\dot{\varepsilon}/2$, if the system is symmetrical in those directions. The normal stresses P_{yy} and P_{zz} will then be equal. Under these conditions, the velocity components are written as

$$V_x = \dot{\varepsilon}x, V_y = -(\dot{\varepsilon}/2)y \text{ and } V_z = -(\dot{\varepsilon}/2)z \quad (1.27)$$

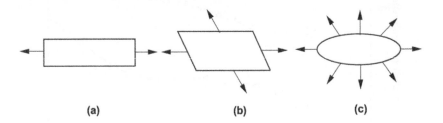

FIGURE 1.23 Schematic representation of uniaxial (a), biaxial (b), and planar (c) extension.

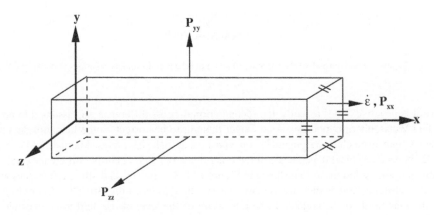

FIGURE 1.24 Schematics of uniaxial extensional flow.

Non-Newtonian Fluid Behavior

and the rate of elongation in the x-direction is given by

$$\dot{\varepsilon} = \frac{\partial V_x}{\partial x} \quad (1.28)$$

In uniaxial extension, the elongational viscosity, μ_E, is then defined as:

$$\mu_E = \frac{P_{xx} - P_{yy}}{\dot{\varepsilon}} = \frac{\tau_{xx} - \tau_{yy}}{\dot{\varepsilon}} \quad (1.29)$$

Owing to the symmetry of the fluid element in the y- and z-directions, P_{yy} and τ_{yy} can be replaced by P_{zz} and τ_{zz}, respectively, in Equation 1.29.

The earliest determinations of elongational viscosity were made for the uniaxial stretching of a filament of Newtonian liquids. Trouton (1906) and many later investigators found that, at low elongation rates, the elongational viscosity was three times the corresponding shear viscosity μ (Barnes et al., 1989). The ratio (μ_E/μ) is referred to as the Trouton ratio, Tr, that is,

$$\mathrm{Tr} = \frac{\mu_E}{\mu} \quad (1.30)$$

The value of 3 for the Trouton ratio for an incompressible Newtonian liquid applies to all values of shear and elongation rates. By analogy, one may define the Trouton ratio for a non-Newtonian fluid as

$$\mathrm{Tr} = \frac{\mu_E(\dot{\varepsilon})}{\mu(\dot{\gamma})} \quad (1.31)$$

Since the Trouton ratio given by Equation 1.31 depends on both $\dot{\varepsilon}$ and $\dot{\gamma}$, some convention must be adopted to relate the strain rates in extension and shear. To obviate this difficulty and at the same time to provide a convenient estimate of behavior in extension, Jones et al. (1987) proposed the following definition of Tr:

$$\mathrm{Tr} = \frac{\mu_E(\dot{\varepsilon})}{\mu(\dot{\gamma} = \dot{\varepsilon}\sqrt{3})} \quad (1.32)$$

that is, in the denominator, the shear viscosity is evaluated at $\dot{\gamma} = \dot{\varepsilon}\sqrt{3}$. They also suggested that for inelastic isotropic liquids, the Trouton ratio is equal to 3 for all values of $\dot{\varepsilon}$ and $\dot{\gamma}$, and any departure from the value of 3 can be ascribed unambiguously to the visco-elasticity of the liquid. In other words, Equation 1.32 implies that for an inelastic shear-thinning fluid, the extensional viscosity must also decrease with the increasing rate of extension, resulting in the so-called tension thinning. Obviously, a shear-thinning visco-elastic liquid (for which Tr>3) will thus have an extensional viscosity that increases with the rate of extension, the so-called strain hardening. Many materials including polymer melts and solutions and protein solutions thus exhibit shear-thinning in simple shear and strain hardening in uniaxial extension. Except in the limit of very small rates of deformation, there does not appear to be any simple relationship between the elongational viscosity and the other rheological properties and, thus to date, its determination rests entirely on experiments that themselves are often constrained by the difficulty of establishing and maintaining a well-defined elongational flow field long enough for the steady state to be reached (e.g., see James and Walters 1994 and Gupta and Sridhar 1998). Measurements made on the same fluid using different methods seldom show quantitative agreement, especially for low-viscosity systems. Figure 1.25 shows representative results on the Trouton ratio for a 0.31% Polyiso Butylene (PIB) solution in a mixture of polybutene and tetradecane at 292.5 K (Tirtaatmadja and Sridhar, 1993); there is some evidence of the Trouton ratio starting at a value of 3 for small values of time and $\dot{\varepsilon}$, whereas the solution shows enormous strain hardening at high values of $\dot{\varepsilon}$. Suffice it to add here that the limiting

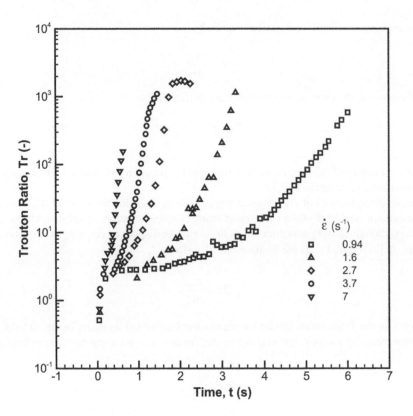

FIGURE 1.25 Extensional behavior of a PIB solution.

Trouton ratios for biaxial and planar extensions at low strain rates have values of 6 and 9, respectively, for all inelastic fluids, and at all values of $\dot{\varepsilon}$ and $\dot{\gamma}$ for Newtonian fluids.

1.3.3.3 Small-Amplitude Oscillatory Shearing Motion

This is another common form of motion that is used to characterize the flow behavior of visco-elastic fluids by analyzing its response to a small-amplitude oscillatory shearing motion. It is useful to consider here the response of a viscous Newtonian fluid and that of a Hookean solid to an oscillatory shear strain, which varies sinusoidally with time as (also see Figure 1.26)

$$\gamma = \gamma_m \sin \omega t \tag{1.33}$$

where γ_m is the amplitude and ω is the frequency of the applied strain.

For an elastic solid obeying Hooke's law, the stress is related linearly to strain, i.e.,

$$\tau = G\gamma = G\gamma_m \sin \omega t \tag{1.34}$$

Thus, there is no phase shift between the stress and applied strain in this case, as shown schematically (by broken curve) in Figure 1.26. In contrast, for a viscous Newtonian fluid, the shear stress is related to the rate of shear strain, i.e.,

$$\dot{\gamma} = \frac{d\gamma}{dt} = \gamma_m \omega \cos \omega t = \gamma_m \omega \sin\left(\frac{\pi}{2} + \omega t\right) \tag{1.35}$$

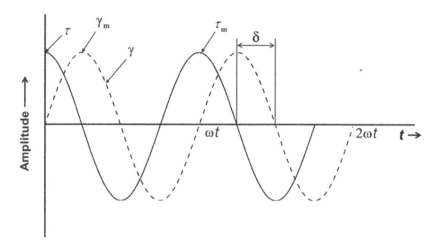

FIGURE 1.26 Schematic representation of response of a viscous fluid (solid line) and elastic solid (broken line) to oscillatory shear strain.

and hence,

$$\tau = \mu\dot{\gamma} = \mu\gamma_m\omega\sin\left(\frac{\pi}{2}+\omega t\right) = \tau_m\sin\left(\frac{\pi}{2}+\omega t\right) \quad (1.36)$$

Obviously, in this case, the resulting shear stress leads the applied strain by a phase difference of $(\pi/2)$. Thus, the measurement of the phase shift, δ, which can vary between zero (purely elastic response, in accordance with Equation 1.34) and $\pi/2$ (purely viscous response, in accordance with Equation 1.36), provides a convenient means of quantifying the level of visco-elasticity of a substance. Needless to add here that small values of δ represent predominantly elastic behavior, whereas large values of δ correspond to viscous fluid-like behavior. Thus, for a visco-elastic substance, one can re-write Equation 1.36, generalized for a phase difference of δ, as:

$$\tau = \tau_m\sin(\omega t + \delta) = (\tau_m\cos\delta)\sin\omega t + (\tau_m\sin\delta)\cos\omega t \quad (1.37)$$

On dividing Equation 1.37 by the applied strain γ_m, one can get the modulus as:

$$\frac{\tau}{\gamma_m} = G = G'(\omega)\sin\omega t + G''(\omega)\cos\omega t \quad (1.38)$$

where

$$G'(\omega) = \frac{\tau_m\cos\delta}{\gamma_m} \quad \text{and} \quad G''(\omega) = \frac{\tau_m\sin\delta}{\gamma_m} \quad (1.39)$$

The in-phase component $G'(\omega)$ is appropriately called the storage modulus (elastic) and the out-of-phase component $G''(\omega)$ is called the loss modulus (viscous).

For the linear visco-elastic region, one can also define the so-called complex viscosity μ^* as follows:

$$\mu^* = \mu' - i\mu'' \quad (1.40)$$

where the real and imaginary parts, μ' and μ'', in turn, are related to the storage (G') and loss (G'') modulii as:

$$\mu'' = \frac{G'}{\omega} \quad \text{and} \quad \mu' = \frac{G''}{\omega} \tag{1.41}$$

It is readily seen from Equation 1.39 that

$$\tan\delta = \frac{G''}{G'} \tag{1.42}$$

Naturally, G', G'', and δ are functions of the frequency of oscillation, ω, of the imposed strain in the linear visco-elastic range. Such oscillatory shear results are also used to classify the fluid behavior as being "weakly" ($G' \ll G''$) or "strongly" elastic ($G' \gg G''$) fluids. Figure 1.27 shows representative data on the variation in G' and G'' with frequency, ω, for a 40% silver nanosol suspension in solvents with different values of the molar ratio, R, of $[COO^-]/[Ag^+]$, which are representatives of inkjet printing applications (Park et al., 2010). In all cases, the storage modulus is seen to be larger than the corresponding loss modulus, thereby testifying to their increasingly elastic behavior with the increasing concentration. Figure 1.28 shows the corresponding phase shift data, where $\tan\delta \leq \sim 1$, i.e., $\delta \leq 45°$, which suggests significant elastic effects (Park et al., 2010).

Many commercially available instruments are equipped for performing such oscillating shear tests (Macosko, 1994), albeit the interpretation of experimental data requires extra care especially to ensure that the amplitude and frequency used are within the linear visco-elastic range, which will naturally vary from one substance to another.

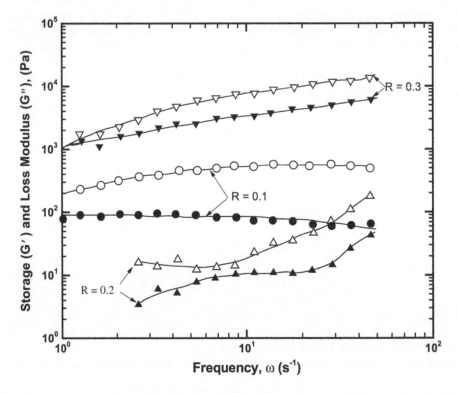

FIGURE 1.27 Effect of molar ratio (R) of [COO⁻]/[Ag⁺] ions on the storage (G', hollow symbols) and loss (G'', filled symbols) modulii for a 40% silver nanosol inkjet printing ink. (Replotted from Park et al., 2010.)

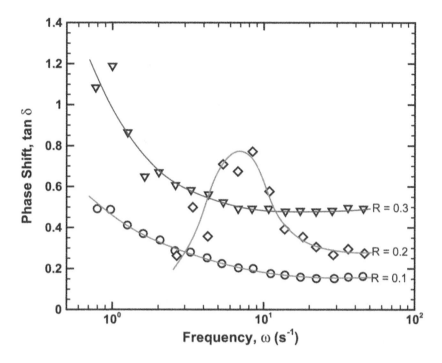

FIGURE 1.28 Dependence of tan θ on the frequency for the inks shown in Figure 1.27 (Park at al., 2010).

1.3.3.4 Mathematical Models for Visco-Elastic Behavior

Though the results of experiments with oscillatory, or steady shear, or elongational flow, or many other transient experiments (Carreau et al., 2021) may be used to calculate viscous and elastic parameters for a fluid, in general, the equations, however, need to be quite elaborate to describe a real fluid behavior adequately for this class of materials. Certainly, the most striking feature connected with the deformation of a visco-elastic material is its simultaneous display of fluid-like and solid-like characteristics. It is therefore not at all surprising that the early attempts at the quantitative description of visco-elastic behavior hinged on the notion of a linear combination of viscous and elastic properties. The Maxwell model represents the cornerstone of the so-called linear visco-elastic models. This model, which is synthesized by a series combination of viscous (dashpot) and elastic (spring) elements (Figure 1.29a), is written as

$$\tau + \lambda_0 \dot{\tau} = \mu_0 \dot{\gamma} \tag{1.43}$$

The feature that distinguishes it from a Newtonian fluid is the term reflecting an influence of the rate of change of stress whose importance is weighed by a relaxation time, λ_0. An important characteristic of the Maxwell fluid is its predominant fluid-like behavior. Figure 1.30 shows the response of a Maxwell fluid when subject to a constant value of strain γ (i.e., $\dot{\gamma} = 0$), including the limiting cases of a purely viscous substance ($\lambda_0 = 0$) and an elastic substance ($\lambda_0 = \infty$). Thus, the rate of stress relaxation is directly influenced by the value of the relaxation time, λ_0, of the fluid. A more "solid-like" response is obtained by considering the so-called Kelvin–Voigt model for a visco-elastic material (Figure 1.29b). Likewise, the Maxwell model in series with the Kelvin–Voigt model gives rise to the so-called Burgers model (Figure 1.29c). One of the main virtues of the linear visco-elastic models is that they can be conveniently superposed by introducing a spectrum of relaxation times, as exhibited by polymeric materials in practice, or by including higher time derivatives.

Alternatively, using the idea of superposition, one can assume the stress to be due to a summation of a number of small partial stresses, each pertaining to a partial strain, and each stress relaxing according to some relaxation law. This approach yields what are known as "integral models." In addition to

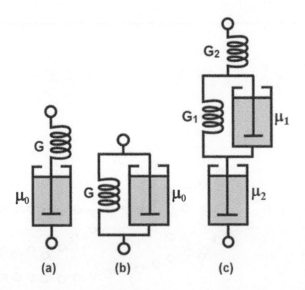

FIGURE 1.29 Schematic representation of (a) the Maxwell model, (b) the Kelvin–Voigt model, and (c) the Burgers model.

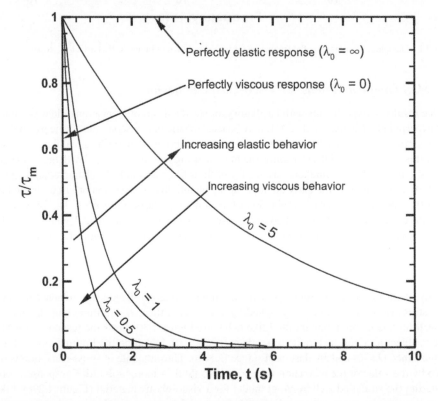

FIGURE 1.30 Schematic representation of stress decay in a Maxwell fluid.

these ideas, numerous other approaches have been used to construct elementary visco-elastic models, including the dumbbell model, the bead–spring models of Rouse and Zimm, network theories, molecular models (Marrucci and Ianniruberto, 2000), reptation models, and Pom-Pom model. All such attempts entail varying degrees of idealizations and empiricisms, their most notable limitation being "small" strain and "low" strain rate, or in the so-called linear ranges. Furthermore, in laminar shear flows, these models predict the viscosity to be independent of shear rate that is contrary to experimental results for most polymer solutions and melts under conditions of practical interest. In spite of the aforementioned deficiencies, linear visco-elastic theories have proved to be useful in predicting certain time-dependent features of fluid behavior and in providing useful qualitative insights into the nature of flow. The next stage of the development in this subject is marked by the formulation of the so-called nonlinear models relaxing the restriction of small deformation and deformation rates. These developments are based on a number of ideas including those by Coleman and Noll (1960, 1961), Green and Rivlin (1957, 1960), Green et al. (1959), Oldroyd (1958), Bernstein et al. (1963), de Gennes (1984), Doi and Edwards (1986), Gutierrez-Lemini (2014), and Doi (1997), for instance. All such attempts involve varying levels of empiricism, and each equation obviously predicts fluid response in certain flow configurations better than that in the others. Excellent critical accounts of the merits and demerits of various constitutive equations are available in the literature (Astarita and Marrucci, 1974; Doi and Edwards, 1986; Bird et al., 1987a,b; Larson, 1988,1998; Malkin, 1995; Bird and Wiest, 1995; Carreau et al., 2021; Gupta, 2000; Tanner, 2000; Morrison, 2001; Kroger, 2004, Mewis and Wagner, 2012). A book by Barnes (2000) is an excellent source of information (including an exhaustive list of the books available on rheology and non-Newtonian fluid mechanics) both for an uninitiated reader and an advanced reader.

When one sees such a bewildering variety of constitutive equations of varying complexity, one may well inquire which one is the most useful. Because of the wide variety of flows and the fact that each non-Newtonian fluid is unique in itself, it is not at all possible to recommend a single constitutive equation as "the equation" for use in all calculations. Therefore, considerable intuition is (and will continue to be) required to choose an appropriate constitutive relation for an envisaged application. Some guidelines in this regard have been provided by Tanner (2000) and Morrison (2001). However, it should be borne in mind that the constitutive relation chosen for an application must be ultimately combined with the conservation equations (mass, momentum, energy) to solve an engineering flow problem. Therefore, a compromise between the choice of a rheological equation of state and the likelihood of achieving a solution to the governing equations is inevitable. This difficulty is also encountered in the interpretation and correlation of experimental data. A constitutive equation with numerous material parameters yields correspondingly a rather large number of dimensionless groups. Therefore, there exists a delicate balance between the efficacy of a constitutive relation to describe the behavior of a fluid in well-defined flows and the utility of the fluid-model parameters so evaluated in interpreting the behavior of the same fluid in a complex flow problem.

1.4 Dimensional Considerations in the Fluid Mechanics of Visco-elastic Fluids

From a practical standpoint, it is a common practice to describe the visco-elastic fluid behavior in steady unidirectional shear in terms of a shear stress (τ_{yx}) and the first normal-stress difference (N_1) as functions of shear rate $(\dot{\gamma}_{yx})$ or the Trouton ratio if elongational effects are believed to be important. Generally, a fluid relaxation or characteristic time (or a spectrum) is defined to quantify the importance of the visco-elastic effects in a flow problem. There are several ways of defining a characteristic time by combining the shear stress and the first normal-stress difference as

$$\theta_f = \frac{N_1}{2\tau_{yx}(\dot{\gamma}_{yx})} \tag{1.44}$$

This is also known as the Maxwellian relaxation time, and evidently, it is shear rate dependent for a real material; it approaches a constant value of ($\psi_1/2\mu_0$) in the limit of zero-shear rate. Although Equation 1.44 defines a fluid characteristic time as a function of shear rate, its practical utility is severely limited by the fact that in most applications, the characteristic shear rate is not known a priori. Leider and Bird (1974) and Grimm (1978), among others, have obviated this difficulty empirically by re-defining θ_f as

$$\theta_f = \left(\frac{m_1}{2m}\right)^{\frac{1}{s-n}} \tag{1.45}$$

Equation 1.45 makes use of the fact that both τ_{yx} and N_1 can be approximated as power-law functions of shear rate over the range of conditions of interest, that is, $\tau_{yx} = m(\dot{\gamma}_{yx})^n$ and $N_1 = m_1(\dot{\gamma}_{yx})^s$. Thus, there is no need to extend the rheological measurements to the zero-shear region. Note that in the limit of zero shear, Equations 1.44 and 1.45 coincide as $s \rightarrow 2$ and $n \rightarrow 1$.

The state of the flow of Newtonian fluids in a given geometry can be described by Reynolds number (Re), Froude number (Fr), or whatever, but for a visco-elastic fluid, it is necessary to introduce at least one additional group involving a measure of elastic forces, that is, N_1, recoverable shear, etc. The Reynolds number is a ratio of the inertial to viscous forces, and thus, it is reasonable to expect that a ratio involving elastic and inertial forces would be useful. Unfortunately, attempts to achieve meaningful correlations have not been very successful, frequently being defeated by the complexity of natural situations and real materials. One simple parameter that has found widespread use is the ratio of a characteristic time of deformation to a natural time for the fluid, akin to relaxation or characteristic time. The precise definitions for these two-time scales are a matter for argument, but it is evident that at least for processes involving very slow deformation of the fluid elements, it is conceivable for the elastic forces to be released by the normal forces of relaxation. In operations that are carried out rapidly, the extent of viscous flow will be minimal and the deformation will be followed by recovery when the stress is removed (think about the bouncing ball example in Section 1.3). Let us illustrate this point by considering the flow of a 1% aqueous solution of polyacrylamide, which is known to have a relaxation time of about 10 ms. For the flow of this fluid through a packed bed of glass beads of 25 mm diameter at a superficial velocity of 250 mm s^{-1}, intuitively one would not expect the elastic effects to be significant in this situation. However, in the case of a packed bed consisting of 0.25 mm size beads, some visco-elastic effects for the same fluid and velocity will begin to manifest. The ratio known as the Deborah number has been defined by Metzner et al. (1966) in these terms:

$$\text{De} = \frac{\text{Characteristic time of fluid}}{\text{Characteristic time of process}} \tag{1.46}$$

In the example considered above, the values of De for the two sets of conditions, respectively, are 0.1 and 10, using the characteristic time of process being given by the time required to traverse a distance of one particle diameter, that is, $25/250 \approx 0.1$ s in the first case and $0.25/250 \approx 1$ ms in the second case. The larger the value of De, the more significant the visco-elastic effects. Thus, small values of the Deborah number ($\ll 1$) correspond to viscous fluid-like behavior, and large values ($\gg 1$) denote elastic solid-like behavior. Therefore, the distinction between an elastic solid and a viscous fluid is not as sharp as one might imagine. Unfortunately, the Deborah number depends on the assignment of a single characteristic time to the fluid. While this is better than no description at all, it appears to be inadequate for many visco-elastic materials that show different relaxation behaviors under differing conditions. Reiner (1964), Huilgol (1975), Lauga (2009) and Poole (2012) have written interesting articles on the usefulness and limitations of the Deborah number, whereas Astarita (1974, 1997) and Zlokarnik (2001) have alluded to some additional difficulties encountered in applying the dimensional considerations to the flow of visco-elastic liquids.

Many other non-dimensional groups have also been introduced to account for visco-elastic effects. One such description is the so-called Weissenberg number (We), which is defined as

$$\text{We} = \frac{\theta_f V}{l} \tag{1.47}$$

while some authors (e.g., see Broer, 1957) have introduced the so-called elasticity number, El, as

$$\text{El} = \frac{\mu \theta_f}{\rho l^2} \tag{1.48}$$

that is really the ratio of elastic to inertial forces, where θ_f is some characteristic time associated with the fluid. Poole (2012) has revisited the conditions when the Deborah number and the Weissenberg number coincide, albeit the two really quantify different effects. Rajagopal (2012) has examined the origins of the Deborah number.

In 1965, Bird made an interesting suggestion that a fluid characteristic time derived from shear viscosity data (such as θ_E in Equation 1.14, λ in Equations 1.16 or 1.17, and $(\mu_0/\tau_{1/2})$ in Equation 1.18) can also be used to ascertain the importance of visco-elastic effects in a flow problem. This approach was indeed used by Sadowski and Bird (1965) to explain the visco-elastic effects in flow through porous media. However, Astarita (1966a) has severely criticized this idea while Slattery (1968) provided further support to Bird's assertion. This idea is explored further in Chapter 4.

Before closing this section, it is appropriate to mention here that the notion of the so-called elastic turbulence has been introduced recently to account for the nonlinear phenomena observed in various flow configurations with visco-elastic liquids in low Reynolds number situations (Groisman and Steinberg, 2000; Larson, 2000). Since such nonlinear effects (including the distortion of a jet and a large increase in flow resistance) are observed at vanishingly small Reynolds numbers, the traditional notion of turbulence as used in low-viscosity Newtonian fluids is therefore irrelevant here. Suffice it to add here that such nonlinear effects are direct manifestations of fluid visco-elasticity even at such low Reynolds numbers.

1.5 Experimental Techniques: Rheometry

Determination of the rheological characteristics of a fluid requires an apparatus that enables the measurement of shear stress and shear rate at the same point in the fluid, and if visco-elastic properties are needed, also the measurement of normal stresses in steady shear, elongational viscosity in uniaxial flow, loss and storage moduli in oscillatory shear, creep behavior, etc. It is not the purpose of this section to undertake a detailed discussion of various measurement techniques available but to give the reader only an idea of what is involved in performing such measurements. All rheometers are developed and operated in such a manner that the resulting flow field is known (or at least approximately in limiting conditions) from the equations of continuity and momentum without a priori choice of a specific fluid model. Usually, this means that there is only one nonzero component of velocity that varies only in one direction. Under these conditions, the kinematics of the flow is known exactly, and the shear stress–shear rate can be calculated from the easily measured quantities. By far the techniques for the measurement of steady shear stress–shear rate data in unidirectional flow are the most advanced, though extra care is required when handling multiphase mixtures, such as foams, emulsions, and suspensions, owing to the possible slip effects, loss of sample, stratification of phases, etc. (Barnes, 1995; Coussot, 2005). This is not so in the case of the first normal-stress difference, especially at high shear rates as encountered in polymer-processing applications. The measurement of the second normal-stress difference has received only very sparse attention, partly due to its very small magnitude. Aside from the steady shearing motions, the response of visco-elastic substances in various steady (elongational) and unsteady (stress growth/decay, small- and large-amplitude oscillatory) motions is also used to derive useful quantitative information regarding their rheological behavior and to evaluate the model parameters. While different geometrical configurations developed for the measurement of shear stress in steady shear motions have been dealt with, among others, by van Wazer et al. (1963), Walters (1975), Whorlow (1992), Macosko (1994), Barnes et al. (1999), etc., Kulicke and Classen (2004) have given an excellent description of the techniques available for the measurement of N_1, N_2, and the response of visco-elastic substances in transient experiments. Similarly, James and Walters (1994), Gupta and Sridhar (1998), and McKinley and Sridhar (2002) have addressed the problems associated with the measurement of elongational viscosity, especially for low-viscosity mobile systems. Ferraris (1999) and Brower and Ferraris (2003) have similarly reviewed the rheometers used for concrete mixtures. Experimental methods for rheological characteristics of multiphase systems, including foams, emulsions, and suspensions, have also been described by Coussot (2005).

The measurement of yield stress for visco-plastic materials has also generated considerable interest; consequently, various methods (De Kee et al., 1980, 1990b; Keentok, 1982; Nguyen and Boger, 1983, 1985, 1992; Uhlherr et al., 1984, 2002; Park et al., 1988; De Kee et al. 1990b; Zhu et al., 2001, 2002) have been developed to measure this property.

Detailed evaluations highlighting the merits and demerits of each method have been reported by Yoshimura et al. (1987), Nguyen and Boger (1992), and De Kee (2021). The Vane method of Nguyen and Boger (1983, 1992) for measuring the true yield stress seems to have gained wide acceptance (Liddell and Boger, 1996; Barnes and Nguyen, 2001). Similarly, the simple slump test is also gradually gaining acceptance as a convenient method to measure the yield stress of mineral slurries and mine tailings (Pashias et al., 1996; Coussot and Boyer, 1995; Roussel and Coussot, 2005). A critical evaluation of the available experimental methods and their relative merits and demerits has been reported, among others, by Nguyen et al. (2006) and Sun and Gunasekaran (2009).

1.6 Concluding Remarks

From the preceding discussion, it is abundantly clear that each non-Newtonian fluid is unique in its characteristics, and reliable information about its rheology comes only from experimental tests. Also, considerable intuition is required to identify the dominant non-Newtonian characteristics depending upon the envisaged application. In steady pipe flow applications, shear-dependent viscosity dominates the pressure loss characteristics under laminar flow conditions, whereas at high Reynolds numbers, visco-elastic effects cannot be always neglected. Thus, the complexity of the real material behavior combined with the nature of flow often dictates the choice of a reasonable rheological equation of state. Provided that there are sufficient data points, interpolation usually poses no difficulty, and extrapolation should, however, be avoided as far as possible as it can frequently lead to erroneous results. Certainly, the fitting of an empirical rheological model should not be used as a justification for such extrapolations. Similarly, it is always possible to fit a number of models to a given set of data equally well, and the choice is primarily governed by convenience. Often, it is not possible to ascertain whether a true yield stress is present or not, and in the absence of an independent test, the constants (τ_0^B, τ_0^C, τ_0^H) in the Bingham plastic, Casson, and Herschel–Bulkley models, respectively, must be regarded as the operational parameters only. For instance, it is not uncommon that these constants will have different values for the same material, and none of these may bear any direct relationship with the true yield stress if at all it exists for a given fluid. Needless to say that extreme caution is needed in analyzing, interpreting, and using such rheological data. The characterization of visco-elastic materials is much more difficult than that of time-independent fluids.

We close this chapter by reiterating that the presentation here has been fairly elementary and far from being rigorous. In most cases, discussion has been limited only to the so-called unidirectional steady shearing or elongational flows. Only elementary fluid models that are used extensively in the experimental branch of non-Newtonian fluid mechanics are presented, especially in the type of flows addressed in the following chapters in this book. Likewise, only those aspects of dimensional analysis that are of relevance in the subsequent chapters are included here. For more rigorous and thorough treatments, interested readers are referred to the numerous books available in this continually growing field (Ferry, 1980; Barnes et al., 1989; Gupta, 2000; Tanner, 2000; Morrison, 2001; Graessley, 2004; Malkin and Isayev, 2006; Pal, 2006; Chhabra and Richardson, 2008; Mewis and Wagner, 2012; Metzger, 2011; Schramm, 2014; Carreau et al., 2021).

Nomenclature

A	Area, (m^2)	
A_0	constant in Equation 1.14 (-)	
a_o	constant in Equation 1.23 (-)	
b_o	constant in Equation 1.23 (-)	
De	Deborah number, Equation 1.46 (-)	
D_m	molecular diffusivity (m^{2}s^{-1})	
El	elasticity number, Equation 1.48 (-)	
F	shearing force (N)	
Fr	Froude number (-)	
G	elasticity modulus (Pa)	

G'	storage modulus, Equation 1.39 (Pa)
G''	Loss modulus, Equation 1.39 (Pa)
l	Characteristic linear dimension (m)
m	power-law consistency coefficient for shear stress (Pa s^n)
m_o, m_t	constants, Equation 1.22 (Pa s^n)
m_1	power-law consistency coefficient for N_1, Equation 1.45 (Pa s^s)
n	power-law index for shear stress (-)
N_1, N_2	first and second normal-stress differences, respectively (Pa)
p	isotropic pressure (Pa)
P	total normal stress (Pa)
Pe	Peclet number (-)
Re	Reynolds number (-)
R_1	fluid parameter, Equation 1.13 (-)
s	power-law index for primary normal-stress difference (-)
Tr	Trouton ratio, Equation 1.31 (-)
V	characteristic velocity (m/s)
We	Weissenberg number, Equation 1.47 (-)
x, y, z	cartesian coordinates (m)

Greek Symbols

α	Ellis model parameter, Equation 1.18 (-)
β	constant in Equation 1.23 (-)
γ	Strain, Equation 1.33 (-)
γ_m	Amplitude of strain, Equation 1.33 (-)
$\dot{\gamma}$	shear rate or velocity gradient (s^{-1})
δ	Phase angle, Equation 1.37 (-)
$\dot{\varepsilon}$	rate of extension, Equation 1.28 (s^{-1})
ξ	structure parameter, Equation 1.22 (-)
θ_E	model parameter, Equation 1.13 (s)
θ_f	fluid characteristic time, Equation 1.47 or 1.48 (s)
λ	Carreau or Cross model parameter (s)
λ_0	Maxwell model time parameter (s)
μ	apparent or Newtonian viscosity (Pa s)
μ^*	complex viscosity, Equation 1.40 (Pa s)
μ'	real part of μ^*, Equation 1.40 (Pa s)
μ''	imaginary part of μ^*, Equation 1.40 (Pa s)
ρ	fluid density (kg/m^3)
τ	extra stress (Pa)
τ_m	amplitude of stress in oscillatory shear, Equation 1.36 (Pa)
τ_0	yield stress (Pa)
$\tau_{1/2}$	Ellis model parameter, Equation 1.18 (Pa)
ψ_1, ψ_2	first and second normal-stress coefficients, respectively (Pa s^2)
ω	Frequency (Hz)

Subscripts

B	Bingham values
C	Casson model values
E	extensional
0	zero shear
∞	infinite shear
x, y, z	components
xx, yy, zz	normal stresses
yx, zy, xz	shear stresses

Superscripts

B	Bingham model parameter, Equation 1.19
C	Casson equation parameter, Equation 1.20
H	Herschel–Bulkley model parameter, Equation 1.21
$*$	complex
$'$	in-phase component
$''$	out-of-phase component

2

Rigid Particles in Time-Independent Liquids without a Yield Stress

2.1 Introduction

A knowledge of the terminal falling velocity of particles in stationary and moving fluid streams is frequently needed in a wide spectrum of process engineering applications, including liquid-solid separations, fluidization and transportation of solids, falling ball viscometry, drilling applications, stability of suspensions, disposal of mine tailing, mineral slurries, medical diagnostics (Gavignet and Sobey, 1989; Machado and Aragao, 1990; Darby, 2000; Li and Kuru, 2003; Couto et al., 2004; Bonadonna et al., 2005; Harris et al., 2009; Kulkarni et al., 2012, 2016a, 2016b; Arnipally and Kuru, 2018), etc. The terminal falling velocity of a particle depends upon a rather large number of variables, including the size, shape, and density of particles, its orientation, properties of the liquid medium (density, rheology), size and shape of the fall vessels, and whether the liquid is stationary or moving. The discussion presented in this chapter is mainly concerned with the motion of particles falling freely in quiescent fluids, albeit some of the results apply equally well when the particle is held stationary in a stream of moving fluid. Perhaps the time-independent fluid behavior represents the most commonly encountered type of fluid behavior. In this chapter, consideration is therefore given to the influence of fluid characteristics on global quantities such as drag coefficient and sedimentation velocity as well as on the detailed structure (wake length, angle of separation, for instance) of the flow field for the steady motion of rigid spherical and nonspherical particles. Within the framework of the time-independent fluid behavior, attention is given to the particle motion in shear-thinning and shear-thickening liquids (without a yield stress) in this chapter and the analogous treatment for viscoplastic liquids is presented in Chapter 3. Likewise, the effect of confining boundaries on the hydrodynamic behavior of particles is considered in Chapter 10. A terse discussion of the significant results on particle motion in incompressible Newtonian fluids is also included here, not only because it is a special case of the time-independent fluid behavior but it also lays the foundation for the subsequent treatment for non-Newtonian fluids. It is convenient to begin with the motion of a spherical particle in a Newtonian fluid medium.

2.2 Governing Equations for a Sphere

Consider a rigid spherical particle of radius R (or diameter d) moving relative to an incompressible fluid of infinite extent with a steady velocity V, as shown schematically in Figure 2.1. From the symmetry considerations, the flow is two-dimensional, the φ-component of the velocity vector, V, being zero, and the flow variables do not vary with φ; hence, one can write

$$V_r = V_r(r,\theta) \qquad (2.1)$$

$$V_\theta = V_\theta(r,\theta) \qquad (2.2)$$

$$V_\varphi = 0 \qquad (2.3)$$

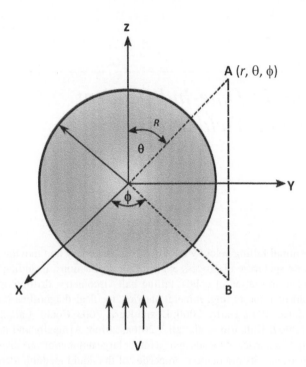

FIGURE 2.1 Schematic representation of flow around a sphere.

It is customary to introduce a stream function, ψ, defined as

$$V_r = -\frac{1}{r^2 \sin\theta} \frac{\partial \psi}{\partial \theta} \tag{2.4}$$

$$V_\theta = \frac{1}{r \sin\theta} \frac{\partial \psi}{\partial r} \tag{2.5}$$

The fundamental physical laws governing the motion of a sphere in an incompressible fluid under isothermal conditions are the conservation of mass, and the Newton's second law of motion, that is, the Cauchy's momentum equation. Application of these laws to an infinitesimal control volume of an incompressible fluid leads to the so-called continuity and momentum equations (Bird et al., 2002), respectively, written as

$$\nabla \cdot V = 0 \tag{2.6}$$

$$\rho\left(\frac{\partial V}{\partial t} + (V \cdot \nabla)V\right) = -\nabla p + \nabla \cdot \left\{\mu\left(\nabla V + (\nabla V)^T\right)\right\} \tag{2.7}$$

where p is the nongravitational pressure. For steady-state, $(\partial V/\partial t) = 0$ in Equation 2.7.

The appropriate boundary conditions for this flow problem are that of no-slip at the sphere surface, and the free stream velocity far away from the sphere. Taking a reference frame fixed to the particle with the origin at its center, these boundary conditions in the spherical coordinate system can be written as

At $r = R$

$$V_r = 0 \tag{2.8a}$$

$$V_\theta = 0 \tag{2.8b}$$

At $r \to \infty$

$$V_r = V\cos\theta; \quad V_\theta = -V\sin\theta \tag{2.9}$$

In addition to the field equations and the boundary conditions (Equations 2.6–2.9), a rheological equation of state relating the components of the extra stress tensor to that of the rate of deformation tensor for the fluid is also needed to define the problem completely. This will enable the viscosity term appearing in the momentum in Equation 2.7 to be expressed in terms of the relevant velocity components and their gradients. However, some progress can be made without choosing an equation of state at this stage.

The velocity and pressure fields (V and p) are the two unknowns here. In principle, therefore, Equations 2.6 and 2.7 together with the boundary conditions (Equations 2.8 and 2.9) are sufficient to solve for these two unknowns. In practice, however, a general solution has proved to be a formidable task even for Newtonian fluids. This is so mainly due to the nonlinear inertial terms present in the momentum equation. Additional complications arise in the case of non-Newtonian fluids on account of the nonlinear rheological equation of state and/or when the flow occurs in a bounded domain.

Once the pressure and velocity fields are known, the drag force acting on a moving particle can be evaluated as

$$F_D = \int_0^{2\pi}\int_0^\pi \left(-p_{r=R}\cos\theta\right)R^2\sin\theta\,d\theta\,d\varphi + \int_0^{2\pi}\int_0^\pi \left(\tau_{r\theta}|_{r=R}\sin\theta\right)R^2\sin\theta\,d\theta\,d\varphi \tag{2.10}$$

The two components on the right-hand side of Equation 2.10 are known as the form or pressure and the friction drag, respectively.

It is convenient to introduce the following dimensionless variables:

$$V^* = \frac{V}{V}; \; p^* = \frac{p}{\rho V^2}; \; r^* = \frac{r}{R}; \; \nabla^* = \nabla R; \; \mu^* = \frac{\mu}{\mu_{\text{ref}}}; \; \psi^* = \frac{\psi}{VR^2} \tag{2.11}$$

Using Equation 2.11, Equations 2.6–2.9 can be rewritten as

$$\nabla^* \cdot V^* = 0 \tag{2.12}$$

$$V^* \cdot \nabla^* V^* = -\nabla^* p^* + \left(\frac{2}{\text{Re}}\right)\nabla^* \cdot \left[\mu^*\left(\nabla^* V^* + \left(\nabla^* V^*\right)^T\right)\right] \tag{2.13}$$

At $r^* = 1$

$$V_r^* = 0; \quad V_\theta^* = 0 \tag{2.14}$$

At $r^* \to \infty$

$$V_r^* = \cos\theta; \quad V_\theta^* = -\sin\theta \tag{2.15}$$

where Re is the usual Reynolds number defined as

$$\text{Re} = \frac{\rho V d}{\mu_{\text{ref}}} \tag{2.16}$$

A reference viscosity, μ_{ref}, has been introduced here to develop a general formulation applicable to the flow of time-independent fluids. Finally, the drag force is made dimensionless by introducing a drag coefficient C_D as

$$C_D = \frac{8F_D}{\pi d^2 \rho V^2} \tag{2.17}$$

Thus, Equation 2.10 now becomes,

$$C_D = 4\int_{-1}^{1}\left(-p^*\big|_{r^*=1} z\right)dz + \left(\frac{8}{\text{Re}}\right)\int_{-1}^{1}\left(\tau^*_{r\theta}\big|_{r^*=1}\right)\left(1-z^2\right)^{1/2} dz \tag{2.18}$$

where $z = \cos\theta$; C_{DP} and C_{DF} are known as the pressure and friction drag coefficients, respectively.

The treatment and equations presented so far are general and thus are applicable for the incompressible flow of purely viscous non-Newtonian fluids, of which the Newtonian fluid behavior is a special case. It is thus instructive to begin with the case of Newtonian fluids; this then sets the stage for the subsequent treatment for the time-independent non-Newtonian fluids.

2.3 Spherical Particles in Newtonian Fluids

The Newtonian fluid represents the simplest realistic class of materials and, as such, the hydrodynamic behavior of rigid spheres in Newtonian media has received considerable attention since the pioneering work of Stokes (1851). The rheological equation of state for an incompressible Newtonian fluid is given by:

$$\tau_{ij} = 2\mu\varepsilon_{ij}, \quad i,j = r,\theta,\phi \tag{2.19}$$

where ε_{ij} represents the components of the rate of deformation tensor and these are related to the two nonzero components of the velocity vector (V_r, V_θ) in a spherical coordinate system as

$$\varepsilon_{rr} = \frac{\partial V_r}{\partial r}; \quad \varepsilon_{\theta\theta} = \left(\frac{1}{r}\frac{\partial V_\theta}{\partial \theta} + \frac{V_r}{r}\right); \quad \varepsilon_{\varphi\varphi} = \left(\frac{V_r + V_\theta \cot\theta}{r}\right);$$

$$\varepsilon_{r\theta} = \varepsilon_{\theta r} = \left(\frac{1}{2}\right)\left\{r\frac{\partial}{\partial r}\left(\frac{V_\theta}{r}\right) + \frac{1}{r}\frac{\partial V_r}{\partial \theta}\right\}; \quad \varepsilon_{r\varphi} = \varepsilon_{\varphi r} = \varepsilon_{\theta\varphi} = \varepsilon_{\varphi\theta} = 0 \tag{2.20}$$

Using Equations 2.19 and 2.20, one can now rewrite the momentum balance equation, Equation 2.13, in terms of ψ^* as

$$E^4\psi^* = \frac{\text{Re}}{2}\left[\frac{\partial \psi^*}{\partial \theta}\frac{\partial}{\partial r^*}\left(\frac{E^2\psi^*}{r^{*2}\sin^2\theta}\right) - \frac{\partial \psi^*}{\partial r^*}\frac{\partial}{\partial \theta}\left(\frac{E^2\psi^*}{r^{*2}\sin^2\theta}\right)\right]\sin\theta \tag{2.21}$$

where

$$E^2 \equiv \frac{\partial^2}{\partial r^{*2}} + \frac{\sin\theta}{r^{*2}}\left(\frac{1}{\sin\theta}\frac{\partial}{\partial \theta}\right)$$

In this case, the reference viscosity, μ_{ref}, turns out to be equal to μ, the Newtonian viscosity and therefore $\mu^* = 1$, and this leads to the usual definition of the Reynolds number. Now, we turn our attention to the progress made in obtaining solutions to Equation 2.21 subject to the boundary conditions given in Equations 2.14 and 2.15 eventually yielding expressions for drag coefficient of a sphere in an unconfined expanse of an incompressible Newtonian fluid.

2.3.1 Drag Force

The highly nonlinear form of Equation 2.21 has precluded the possibility of general solutions and hence only approximate solutions are available. For instance, the creeping flow approximation (Re = 0) allows the nonlinear inertial terms to be dropped, thereby reducing Equation 2.21 to the form

$$E^4 \psi^* = 0 \tag{2.22}$$

This fourth-order partial differential equation was solved by Stokes (1851), and the resulting expression for the so-called Stokes drag is given by

$$F_D = 6\pi\mu R V \tag{2.23}$$

The individual contributions arising from pressure and friction components are given by $2\pi\mu RV$ and $4\pi\mu RV$, respectively. Equation 2.23 can be rewritten in a more familiar form as

$$C_D = \frac{24}{\text{Re}} \tag{2.24}$$

It is important to recall that Equation 2.23 or 2.24 is applicable only when the inertial effects are negligible. Experiments have shown that Equation 2.24 is useful for Reynolds numbers up to about 0.1, and beyond this value, the deviation between experiments and theory increases. Oseen (1927) extended the range of applicability of the Stokes law by partially taking into account the fluid inertia, and presented the expression

$$C_D = \frac{24}{\text{Re}}\left(1 + \frac{3}{16}\text{Re}\right) \tag{2.25}$$

This equation predicts the value of drag coefficient for Reynolds numbers up to about 1 with a maximum error of 1%. It was further improved upon by Proudman and Pearson (1957), and subsequently Ockendon and Evans (1972) have obtained higher order terms in the series expansion as

$$C_D = \frac{24}{\text{Re}}\left\{1 + \frac{3}{16}\text{Re} + \frac{9}{160}\text{Re}^2 \log\frac{\text{Re}}{2} + \frac{0.1879}{4}\text{Re}^2 + ...\right\} \tag{2.26}$$

The drag formula due to Proudman and Pearson is obtained by omitting the last term in Equation 2.26. Numerous other extensions of the creeping flow solutions are available in the literature (Goldstein, 1929; Chester and Breach, 1969; Liao, 2002), including one with 24 terms of the series (Van Dyke, 1970).

As the value of the Reynolds number increases (>1), the inertial terms become increasingly significant in the momentum equation and no analytical solutions are possible. Therefore, numerical solutions must be sought for finite values of the Reynolds numbers. Jenson (1959) employed a finite difference scheme to obtain numerical results up to Re = 40. LeClair et al. (1970) solved the complete Navier–Stokes equations and reported accurate values of drag coefficient for a sphere up to Re = 400 or so. Subsequently, Fornberg (1988) has documented detailed results on the wake characteristics and drag coefficient of single spheres up to the Reynolds number of 5000. In recent years, there has been an upsurge in exploring different aspects of the flow past a sphere. Thus, for instance, Weisenborn and Ten Bosch (1993) have employed the method of induced forces to evaluate the Oseen drag coefficient in the limit Re → ∞. Similarly, while Johnson and Patel (1999) and Cliffe et al. (2000) have studied both numerically and experimentally different flow regimes for a sphere including steady and unsteady laminar flow at Reynolds numbers up to 300, Mittal (1999) has delineated the range of conditions for the flow over a sphere to gradually change from axisymmetric and steady, to nonaxisymmetric and steady, to finally nonaxisymmetric and nonsteady regime. Almost identical values of drag coefficient up to Re = 200 have been reported earlier by Tabata and Itakura (1996). Most of these and other developments in this field have been reviewed by O'Neill (1981), Zapryanov and Tabakova (1999), and by Michaelides (1997, 2002, 2003). The range of the

TABLE 2.1

Recommended Drag Correlation (Clift et al., 1978)

Range	Correlation
Re < 0.01	$C_D = \dfrac{24}{\text{Re}}\left(1+\dfrac{3}{16}\text{Re}\right)$
0.01 < Re < 20	$C_D = \dfrac{24}{\text{Re}}\left(1+0.1315\text{Re}^{(0.82-0.05w)}\right)$
20 < Re < 260	$C_D = \dfrac{24}{\text{Re}}\left(1+0.1935\text{Re}^{0.6305}\right)$
260 < Re < 1500	$\log C_D = 1.6435 - 1.1242\omega + 0.1558\omega^2$
1500 < Re < 1.2×10⁴	$\log C_D = -2.4571 + 2.5558\omega - 0.9295\omega^2 + 0.1049\omega^3$
1.2×10⁴ < Re < 4.4×10⁴	$\log C_D = -1.9181 + 0.637\omega - 0.0636\omega^2$
4.4×10⁴ < Re < 3.38×10⁵	$\log C_D = -4.339 + 1.5809\omega - 0.1546\omega^2$
3.38×10⁵ < Re < 4×10⁵	$C_D = 29.78 - 5.3\omega$
4×10⁵ < Re < 10⁶	$C_D = 0.1\omega - 0.49$
10⁶ < Re	$C_D = 0.19 - \left(\dfrac{8\times 10^4}{\text{Re}}\right)$

Source: Clift, R., Grace, J.R., and Weber, M.E., *Bubbles, Drops and Particles*, Academic, New York (1978).
Note: $\omega = \log \text{Re}$.

numerical results has been extended up to Re ~ 10⁵ including the so-called "drag crisis", characterized by a sudden drop in the value of the drag coefficient by Fornberg (1988), Deshpande et al. (2017), and Tiwari et al. (2020a,b). Owing to the computational difficulties stemming from the three-dimensional and time-dependent nature of the flow, in general, most works reported for Re >5000 have been in the form of experimental correlations. Most of these have been critically reviewed by Clift et al. (1978), Khan and Richardson (1987), and recently by Kalman and Matana (2022). The recommendations of Clift et al. (1978) for calculating the value of drag coefficient for a fixed value of the Reynolds number embracing the complete standard drag curve are given in Table 2.1 whereas Khan and Richardson (1987) presented the following equation that purports to predict the value of drag coefficient with an average uncertainty of less than 5% in the range of $10^{-2} < \text{Re} < 3 \times 10^5$:

$$C_D = \left[2.25\text{Re}^{-0.31} + 0.36\text{Re}^{0.06}\right]^{3.45} \qquad (2.27)$$

Numerous other expressions of varying forms and complexity are available in the literature with their precision comparable to that of Equation 2.27 as well as the multi-interval expressions summarized in Table 2.1 (Haider and Levenspiel, 1989; Duan et al., 2015; Song et al., 2017; Goossens, 2019, 2020; Kalman and Matana, 2022, etc.).

2.3.2 Free-Fall Velocity

For a given value of the Reynolds number, it is straightforward to calculate the value of the drag coefficient. However, for the reverse operation, that is, for the calculation of the free-fall velocity for a given sphere/fluid combination, the forms of correlations presented in Section 2.3.1 are not convenient as the unknown velocity appears in both the dimensionless groups, that is, the Reynolds number and the drag coefficient. This difficulty can be obviated by introducing a new dimensionless group variously known as *Archimedes*, *Galileo*, or *Best* number (Best, 1950) which is defined as

$$\mathrm{Ar} = C_D \, \mathrm{Re}^2 = \frac{4}{3}\left(\frac{g\rho(\rho_p - \rho)\, d^3}{\mu^2}\right) \tag{2.28}$$

Evidently, for a given sphere–fluid combination, the Archimedes number, Ar, is independent of the unknown velocity V. Several attempts have been made to establish the functional relationship between the Archimedes number and the Reynolds number (Clift et al., 1978; Khan and Richardson, 1987; Goossens, 2020). Based on a large body of experimental data, Khan and Richardson (1987) proposed an equation that has been corrected subsequently by Yu and Liu (2003) as

$$\mathrm{Re} = \left(1.47 k_D^{-0.14} + 0.11 k_D^{-0.4}\right)^{3.56} \tag{2.29}$$

where $k_D = (3C_D/4\,\mathrm{Re})$ and Equation 2.29 encompasses the range $10^{-6} < \mathrm{Re} < 10^6$. Clearly, the new factor k_D and the Archimedes number Ar are related as $\mathrm{Ar} = (4/3) k_D \, \mathrm{Re}^3$. This problem continues to receive attention in the literature (Liao, 2002; Brown and Lawler, 2003; Yang and Mao, 2005; Majumder, 2007; Cheng, 2009; Song et al., 2017; Goossens, 2019, 2020; Ganguli and Lele, 2019).

Other aspects of this flow configuration such as the effects of surface roughness and of the intensity of turbulence, etc. on drag coefficient as well as the vast literature concerning the detailed flow field around a sphere have been reviewed by Torobin and Gauvin (1959, 1960), Clift et al. (1978), Kim and Karrila (1991), Le Roux (1992), and Zapryanov and Tabakova (1999). Some PIV data for a single sphere settling under gravity have recently been reported by Ten Cate et al. (2002) and Tiwari et al. (2020a, b).

2.3.3 Flow Regimes

In the simplest case involving the unconfined flow of an incompressible fluid over a sphere, the sole influencing dimensionless parameter is the Reynolds number. The value of the Reynolds number denotes the ratio of the inertial and viscous forces induced in the fluid by the relative motion between the fluid and the sphere. The viscous forces suppress minor disturbances in the flow whereas the inertial forces tend to amplify them eventually leading to instabilities. It thus stands to reason that, similar to the case of the two-dimensional flow over a long cylinder, the flow past a sphere also exhibits numerous flow regimes depending upon the value of the Reynolds number. In spite of the perfect symmetry of the shape, the three-dimensional flow over a sphere can exhibit more complicated kinematics and vortical features in the flow whence these have not been explored as extensively for a sphere as that in the case of a cylinder (Taneda, 1956b; Zdravkovich, 1997, 2003; Sumer and Fredsoe, 2006; Riberio et al., 2012). A good state of art summary is available in the literature (Taneda, 1956a; Johnson and Patel, 1999; Reddy et al., 2010; Tiwari et al., 2020b).

At the front stagnation point (point A, Figure 2.2), fluid comes to a complete stop and it gradually accelerates attaining its maximum velocity at point B or C (Figure 2.2) and the pressure is minimum at this point. Due to the diverging area, the fluid slows down between points B and D (Figure 2.2) accompanied by a concomitant increase in pressure, that is, flow occurs against the increasing pressure. Such an adverse pressure gradient further slows down the slow-moving fluid near the surface, bringing it to rest, and eventually reversing the flow direction at the solid boundary. The point of separation S thus corresponds to the zero-wall shear stress. The location of the point of separation is denoted by the angle θ_s (also shown in Figure 2.2). For a sphere, this is the first flow regime transition which occurs at about $\mathrm{Re} \sim 20$–21, though as low a value as $\mathrm{Re} = 7$–8 has also been reported in the literature for this transition (Masliyah, 1972; Nakamura, 1976). Beyond this value of the Reynolds number, a recirculating region of toroidal shape is formed (shown schematically in Figure 2.2). The flow is axisymmetric and steady under these conditions. The gross features are captured in terms of the separation angle θ_s, the length of the recirculating wake region L_w (usually nondimensionalized using the sphere diameter), and the so-called "eye" of the vortex where the fluid velocity is zero in the sphere's reference frame. The fluid separated at θ_s rejoins at a distance L_w (from the rear of the sphere) on the axis of the flow to form a closed wake. In the range $\sim 20 \leq \mathrm{Re} \leq \sim 200$–$210$, the topology of the flow remains the same (axisymmetric and steady) except that the values of the θ_s, L_w, and the position of the eye of the toroidal vortex vary with the value of

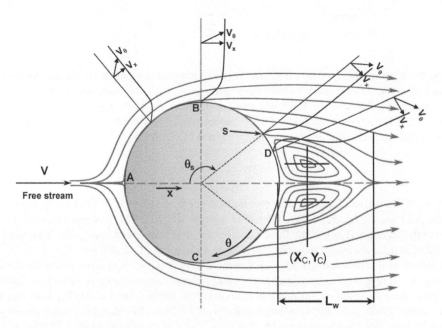

FIGURE 2.2 Schematics of the recirculation length (L_w), flow separation angle (θ_s), and eye of the vortex (X_c, Y_c).

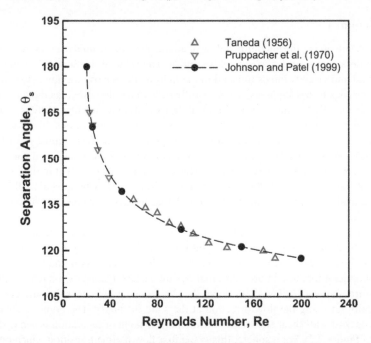

FIGURE 2.3 Dependence of the separation angle of sphere on Reynolds number in Newtonian fluids.

the Reynolds number. Figures 2.3–2.5 show typical results culled from various sources. The agreement between different results (both numerical and experimental) is seen to be fair.

Based on their experimental results, Kalra and Uhlherr (1973) put forward the following predictive expressions for θ_s over the range ~ $30 \leq \text{Re} \leq 750$,

$$\theta_s = 83 + 262 \, \text{Re}^{-0.372} \tag{2.30}$$

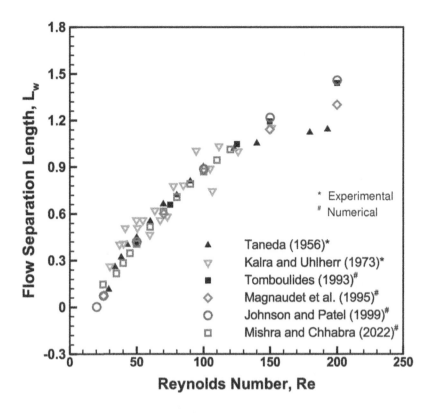

FIGURE 2.4 Variation of the flow separation length for sphere with Reynolds number in Newtonian fluids.

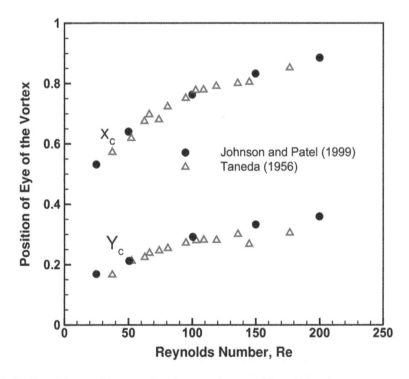

FIGURE 2.5 Position of the eye of the vortex for sphere over the range of Reynolds number.

At about Re > ~211, though the flow continues to be steady, but it is no longer axisymmetric. This loss of axisymmetric flow results in the net lift force on the sphere in the lateral direction and indeed the value of the lift coefficient (defined in the same manner as the drag coefficient with the drag force replaced by the lift force in Equation 2.17) is monitored to locate this transition. Naturally, the lift forces are identically zero in the axisymmetric flow regime. For instance, the lift coefficient jumped by a factor of ~1000 as the Reynolds number was incremented from Re = 211 to Re = 212 thereby pinpointing this transition. This flow regime persists up to about Re ~ 270 (Magarvey and Bishop, 1961; Nakamura, 1976; Tomboulides et al., 1993). At about Re ~ 270–300, the flow loses its steady character as well and all flow variables exhibit temporal variation and vortex shedding is observed in this regime. More detailed discussions on this subject are available in the literature (Johnson and Patel, 1999; Fornberg, 1988; Deshpande et al., 2017; Tiwari et al., 2020a, b).

2.3.4 Unsteady Motion

The analysis and prediction of the fluid motion and drag is much more difficult when the motion is unsteady. Even simple dimensional considerations reveal that the unsteady motion of a particle is governed by at least four dimensionless groups, namely, Reynolds number, drag coefficient, density ratio, and dimensionless distance ($x^* = x/d$) as opposed to the only first two required to describe the steady motion. Additional complications arise due to the nonspherical shape of particles, wall effects, and the imposed motion of the fluid itself as opposed to the free-settling motion of the sphere (Clift et al., 1978; Michaelides, 1997, 2006; Michaelides and Roig, 2011; Zapryanov and Tabakova, 1999). It is convenient to distinguish between two cases at this stage; the first kind of motion is characterized by a rapid change of the Reynolds number with x^* and under these conditions, the instantaneous drag may differ significantly from the corresponding steady-state value. This situation is encountered when a particle settles in a liquid. The second type of motion is characterized by a slow change of the Reynolds number with x^* and in this case, the instantaneous drag is similar to the steady-state value of drag relevant to the current value of Re. This situation arises when the density ratio (ρ_p/ρ) is high, such as for particle motion in gases. In their pioneering contributions, the so-called Boussinesq–Basset expression (Boussinesq, 1885; Basset, 1888) for the transient hydrodynamic force, F, exerted by an infinite, quiescent fluid on a sphere of radius R, initially at rest, is given as

$$F(t) = 6\pi\mu R V(t) + \frac{m_f}{2}\frac{dV}{dt} + 6R^2\sqrt{\pi\rho\mu}\int_0^t \frac{(dV/dt)}{\sqrt{t-\tau}}d\tau \quad (2.31)$$

where $V(t)$ is the arbitrary velocity of the sphere. The three terms appearing on the right-hand side are the so-called steady-state drag, the added mass effect, and the Basset or the history integral force. Strictly, Equation 2.31 is applicable in the limit of zero Reynolds number. By combining Equation 2.31 or the modifications thereof with the macroscopic force balance on a moving sphere, an integral–differential equation is obtained which relates to the distance or velocity with time. It is appropriate to remark here that it is a common practice to neglect the awkward Basset history force in Equation 2.31 and to retain the added mass force term. Though this is not at all justifiable, this practice continues in the literature (Clift et al., 1978). Therefore, a range of empirical modifications to various terms in Equation 2.31 have been introduced over the years to account for finite values of the Reynolds number as encountered in most practical applications (Michaelides, 1997, 2006; Maris, 2019). From a practical standpoint, it is frequently required to estimate the distance traveled and the time required for a sphere to attain its terminal falling velocity in a quiescent fluid. This type of information is required while designing falling ball viscometers or measuring terminal falling velocity to estimate the required contact time during the design of solid separation equipment. The results available to date on these aspects of the unsteady motion for rigid spheres have been compiled by many investigators (Clift et al., 1978; Michaelides, 1997, 2006; Kim et al., 1998; Zapryanov and Tabakova, 1999). Broadly speaking, under the creeping flow conditions, the time required to attain a velocity within 95% of the terminal falling velocity increases rapidly as the value of (ρ_p/ρ) increases. Analogous but highly approximate results at high Reynolds numbers have been reported among others by Ferreira and Chhabra

(1998) and Ferreira et al. (1998). For example, a 3.96 mm diameter steel ball attains a value of settling velocity which is within 95% of the terminal velocity value in an oil ($\rho = 960$ kgm^{-3}, $\mu = 910$ mPa s) in 23.3 ms after falling a distance of 1 mm only. On the other hand, the corresponding values for another liquid ($\rho = 1178$ kgm^3, $\mu = 18.2$ mPa s) are 165 ms and 74 mm, respectively. Suffice it to add here that all such predictions beyond the creeping flow region are based on empirical modifications and simplified forms of Equation 2.31 and hence must be treated with reserve. Loewenberg (1993) has reported preliminary results on the effects of added mass and Basset forces for finite-size cylinders.

2.4 Spheres in Shear-Thinning Liquids

At the outset, it is instructive to recall that a time-independent fluid resembles a Newtonian fluid as far as the normal stresses are concerned (i.e., in simple shear, both N_1 and N_2 are zero for time-independent fluids), but the two differ in a significant manner, as this class of purely viscous fluids is characterized by a shear-rate-dependent viscosity. In other words, no memory effects are exhibited by these fluids, that is, the Deborah number is identically zero here. It can readily be shown that the apparent viscosity of a purely viscous fluid is a function of the three invariants of the rate of deformation tensor, that is,

$$\mu = \mu(I_1, I_2, I_3) \tag{2.32}$$

For an incompressible fluid, the first invariant is identically zero. Strictly, the third invariant is zero only for viscometric flows, but the available experimental evidence seems to suggest that it is of little consequence in most flow configurations of practical interest (Slattery and Bird, 1961; Tanner, 1966; Bird et al., 1987a). Hence, Equation 2.32 simplifies to

$$\mu = \mu(I_2) \tag{2.33}$$

In simple shear, the second invariant is related to the rate of shear as

$$\dot{\gamma} = \sqrt{\frac{I_2}{2}} \tag{2.34}$$

In the present case, the second invariant I_2 is related to the individual components of the rate of deformation tensor as

$$I_2 = \varepsilon_{rr}^2 + \varepsilon_{\theta\theta}^2 + \varepsilon_{\phi\phi}^2 + 2\varepsilon_{\theta r}^2 \tag{2.35}$$

where the various terms of Equation 2.35 can be expressed in terms of the two nonzero velocity components, as given by Equation 2.20.

2.4.1 Drag Force

2.4.1.1 Theoretical Developments in Creeping Flow Region

The starting point for a creeping flow analysis is the continuity equation, Equation 2.12, and the momentum, Equation 2.13, in which the inertial terms on the left-hand side are altogether neglected. However, the resulting simplified version is still highly nonlinear owing to the shear-dependent viscosity. Thus, the extension of the Stokes solution to a shear-thinning fluid is a nontrivial task and indeed various types of approximations have been invoked to realize the solutions of the governing equations. Early theoretical developments in this field hinge on the application of either the variational principles or on the linearization of the viscous term in Equation 2.13 and perturbation methods, or the numerical schemes such as the extended moment, finite volume, finite difference, boundary elements, finite element methods, etc. Among the three approaches, the variational principles have been used most extensively in the early years, followed by the linearization/perturbation approach whereas the numerical simulations of this problem are of relatively recent vintage (Crochet et al., 1984; Gu and Tanner, 1985; Graham and Jones, 1994; Tripathi et al., 1994; Dhole et al., 2006a; Song et al., 2009; Suri et al., 2023a). It seems to be instructive and desirable to recapitulate the salient features of these methods particularly as applied to the steady sphere motion in a purely viscous fluid in the absence of inertial effects (Re → 0). Detailed descriptions of the extremum principles as well as their adaptations to applications involving purely viscous fluids are available in the literature (Hill and Power, 1956; Bird, 1960; Johnson, 1960, 1961; Schechter, 1962; Haddow and Luming, 1965, 1966; Yoshioka and Adachi, 1971, 1973, 1974; Slattery, 1972; Astarita, 1977, 1983; Leonov, 1988).

At the outset, it must be recognized that the extremum principles are applicable only to the steady flow of incompressible shear-thinning fluids under the creeping flow conditions. In essence, this method involves the choice of a trial velocity field, with some unknown parameters, which satisfies the equation of continuity as well as the boundary conditions on velocity. This trial velocity field is used to evaluate an energy functional whose minimum value corresponds to the upper bound on the drag force. This is known as the velocity variational principle. Likewise, one can choose a trial stress profile, with some unknown parameters, which satisfies the momentum equation together with the explicit boundary conditions on stress. In this case, one maximizes the so-called complementary energy functional. This leads to the lower bound on drag force, and is known as the stress variational principle. Strictly speaking, this approach yields rigorous upper- and lower bounds on drag force only for those fluid models for which the two energy functionals referred to above are homogeneous functions. It can be shown that only the Newtonian and power-law model fluids fulfill this condition. In all other cases, the resulting bounds are less accurate. Evidently, in this approach, the equations of continuity and momentum are not satisfied simultaneously, and in the absence of any definitive information regarding the exact location of the true solution between the two bounds, the arithmetic average of the two bounds has been generally used in practice (Wasserman and Slattery, 1964; Hopke and Slattery, 1970a, b), without any theoretical justification.

The second class of approximate solutions is characterized by the fact that the viscosity term appearing in the momentum equation is evaluated by using the Newtonian flow field, and the resulting partial differential equation is solved for the unknown velocity and pressure. The early works of Hirose and Moo-Young (1969) and of Acharya et al. (1976) are illustrative of this approach. Some closed-form analytical results have also been obtained by employing the standard perturbation method. In this scheme, the flow variables are expanded in the form of a series with the coefficients involving non-Newtonian fluid parameters.

Thus, for the flow of power-law fluids, the series expansions are of the following general form

$$\psi = \psi_0 + \varepsilon\psi_1 + \varepsilon^2\psi_2 + \cdots \tag{2.36a}$$

$$p = p_0 + \varepsilon\, p_1 + \varepsilon^2 p_2 + \cdots \tag{2.36b}$$

where ψ_0 and p_0, etc. are the zeroth-order solutions corresponding to the Newtonian fluid behavior and for a power-law fluid, $\varepsilon = (n - 1)/2$. To date, only first-order corrections, that is, ψ_1 and p_1, etc. have been evaluated. The works of Koizumi (1974), Kawase and Moo-Young (1985, 1986), and of Rodrigue et al. (1996b, c) exemplify the usefulness of this approach. Owing to the nature of approximations involved, this method yields results which are restricted to a small degree of shear-thinning behavior, that is, when the value of the power-law index is not too different from unity.

Since 1980s, there has been a spurt in the use of numerical methods to solve the field equations describing the steady motion of a sphere in shear-thinning and shear-thickening fluids (Bush and Phan-Thien, 1984; Crochet et al., 1984; Gu and Tanner, 1985; Carew and Townsend, 1988; Graham and Jones, 1994; Tripathi et al., 1994; Tripathi and Chhabra, 1995; Whitney and Rodin, 2001; Fortin et al., 2004; Song et al., 2009). Undoubtedly, a majority of the research efforts has been directed at obtaining what might be called a non-Newtonian equivalent of the standard drag curve for power-law fluids. At the outset, it should be recognized that all such attempts and the resulting drag relations are going to be fluid model-dependent. Indeed, the steady creeping motion of a sphere has been studied in a wide variety of generalized Newtonian fluids model. In the following sections, only representative results obtained with some of the more widely used fluid models are presented. However, an extensive compilation of the pertinent literature is given in Table 2.2.

TABLE 2.2

Summary of Theoretical Investigations of Sphere Motion in Shear-Thinning Liquids (Creeping Flow Regime)

Investigator	Fluid Model	Type of Approach	Remarks
Tomita (1959)	Power-law	Velocity variational principle	Upper bound on Y in range $0.5 < n < 2$. Results have been corrected by Wallick et al. (1962).
Slattery (1961), Foster and Slattery (1962)	Reiner-Rivlin[a]	Velocity variational principle	Upper bound on Y was calculated. However, the assumption of the constant cross viscosity is questionable (Leigh, 1962).
Ziegenhagen et al. (1961)	Truncated power law	Velocity variational principle	Upper bound on Y
Slattery (1962)	Power-law and Sisko[b] models	Velocity variational principle	Two trial velocity fields were used to get the estimates of upper bound on Y.
Rathna (1962)	Reiner–Rivlin	Perturbation scheme	First-order correction to the Stokes drag formula.
Wasserman and Slattery (1964)	Power law	Velocity and stress variational principles	Upper and lower bounds on Y in the range $1 > n > 0.1$ were computed. Lower bound was corrected by Yoshioka and Adachi (1973) and by Mohan (1974a, b).
Ziegenhagen (1964)	Powell-Eyring	Velocity variational principle	Approximate upper-bound results
Yoshioka and Nakamura (1966)	GNF model with 4 constants[c]	Perturbation method	Very little deviation from the Stokes drag is predicted.
Hopke and Slattery (1970a)	Ellis model	Velocity and stress variational principles	Upper and lower bounds on drag force are given in the range: $1 < \alpha < 3$; $0 < El < 10$. These ranges have been extended (Chhabra et al., 1984).
Mitsuishi et al. (1971)	Sutterby model	Velocity variational principle	Approximate upper bound on Y.
Adachi et al. (1973)	Power-law	Numerical solution	Drag coefficient and other flow variables are reported for $0.8 < n < 1$ and $Re_{PL} = 60$
Koizumi (1974)	Power-law	Perturbation method	Approximate closed from expression for drag coefficient is obtained.
Acharya et al. (1976)	Power-law	Linearized momentum equations	Approximate closed from expression for drag coefficient was reported which was subsequently corrected by Lockyer et al. (1980) and by Kawase and Ulbrecht (1981a).
Shmakov and Shamakova (1977)	Power-law	Perturbation method	Evaluated drag on a sphere in shear field.
Adachi et al. (1977/78)	Extended Williamson model[d]	Numerical solution	Drag force and other flow details are reported for $0.1 < Re < 60$.
Chhabra et al. (1980a) and Chhabra and Uhlherr (1980a)	Power-law and Carreau Viscosity model	Velocity variational principle	A comparative study of upper bounds on drag for power-law fluids and new results for the Carreau model fluids.
Cho and Hartnett (1983a, b)	Power-law	Velocity and stress variational principles	Extended the results of Wasserman and Slattery (1964).
Kawase and Ulbrecht (1983c)	Power-law	Linearization of field equations	Influence of power-law parameters on wall effects and drag coefficient is examined.
Crochet et al. (1984)	Power-law	Finite element method	Detailed flow field, and drag results ($1 > n > 0.1$)
Gu and Tanner (1985)	Power-law	Finite element solution	Drag and wall effects are reported for $1 > n > 0.1$.
Bush and Phan-Thien (1984)	Carreau viscosity equation	Numerical solution	Their results are in excellent agreement with the approximate upper bound (Chhabra and Uhlherr, 1980a)
Kawase and Moo-Young (1986)	Power-law	Perturbation method	Closed-form formula for drag.

(Continued)

TABLE 2.2 (*Continued*)

Summary of Theoretical Investigations of Sphere Motion in Shear-Thinning Liquids (Creeping Flow Regime)

Investigator	Fluid Model	Type of Approach	Remarks
Leonov (1988)	Power-law	Variational principles	Improved upper and lower bounds on drag force.
Chhabra (1990a)	Ellis model; Carreau viscosity Eq.; Allen and Uhlherr viscosity (1986)	Velocity variational principle	Comparative study of three models is reported.
Tripathi et al. (1994) and Tripathi and Chhabra (1995)	Power-law	Finite element method	Values of drag for spheres and spheroids for $1.8 > n > 0.4$ and $Re_{PL} < 100$.
Graham and Jones (1994)	Power-law	Numerical solution	Drag on spheres up to $Re_{PL} < 130$.
Rodrigue et al. (1996b)	Carreau model	Perturbation method	Approximate expression for Y.
Ceylan et al. (1999)	Power-law	Mixed ideas	Approximate expression for Y.
Whitney and Rodin (2001)	Power-law	Numerical method	Results for Y for a sphere and a cylinder.

a Reiner–Rivlin fluid model: $\tau_{ij} = 2\mu(I_2,I_3)\varepsilon_{ij} + 2\mu_c(I_2,I_3)\varepsilon_i^k \varepsilon_{kj}$.

b Sisko fluid model: $\tau_{yx} = A(\dot{\gamma}_{yx}) + B(\dot{\gamma}_{yx})^n$, A, B and n are model parameters.

c GNF model used by Yoshioka and Nakamura (1966): $\mu = \dfrac{\mu_1}{1+\varepsilon_1 I_1} + \dfrac{\mu_2}{1+\varepsilon_2 I_2}$.

d Extended Willamson model: $(\mu-\mu_\infty)/(\mu_0-\mu_\infty) = \left[1+(\alpha I_2/2)^\beta\right]^{-1}$, μ_0, μ_∞, α, β are model parameters.

2.4.1.1.1 Power-Law Fluids

An examination of Table 2.2 reveals the preponderance of the available theoretical results based on the usual two-parameter power-law model. This is so presumably due to the fact that the single parameter, namely n, can be used to gauge the importance of shear-thinning (or shear-thickening) effects in a flow problem. Theoretical results for the drag on a sphere are usually expressed in the form of a drag correction factor, Y, to be applied to the Stokes formula defined as

$$F_D = 3\pi\mu dVY \qquad (2.37)$$

where the viscosity, μ, is evaluated by assuming the characteristic shear rate around a sphere to be equal to (V/d). This leads to $\mu^* = (I_2^*/2)^{(n-1)/2}$ in Equation 2.13. Equation 2.37 can be rearranged to yield

$$Y = \dfrac{C_D \, Re_{PL}}{24} \qquad (2.38)$$

where the Reynolds number for a sphere falling with its terminal velocity, V, in a power-law medium is defined as

$$Re_{PL} = \dfrac{\rho V^{2-n} d^n}{m} \qquad (2.39)$$

Using dimensional arguments, it can readily be shown that in the limit of creeping flow, the drag correction factor Y would be a function of the power-law index, n, alone. The functional dependence of Y on the power-law index, n, as reported by different investigators is shown in Figure 2.6 where it is clearly seen that the results reported by different investigators differ widely from each other.

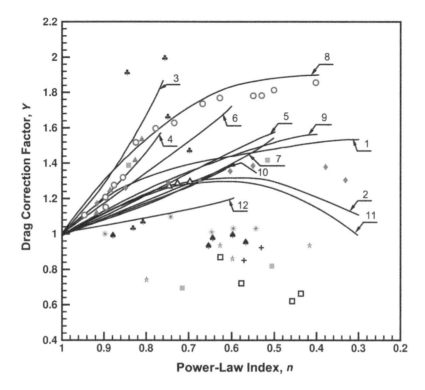

FIGURE 2.6 Comparisons between predictions and experiments on sphere sedimenting in unconfined power-law fluids.

Curve number	References	Symbol	Reference
1,2	Cho and Hartnett (1983a)	△	Uhlherr et al. (1976)
3,4	Slattery (1962)	□	Dallon (1976)
5	Nakano and Tien (1968)	○	Chhabra et al. (1980a)
6	Acharya et al. (1976) and Kawase and Ulbrecht (1981a)	▲	Slattery (1959)
7	Chhabra et al. (1980a)	♠	Kato et al. (1972)
8	Tomita (1959); Wallick et al. (1962)	∗	Yoshioka and Adachi (1973)
10	Kawase and Moo-Young (1986)	■	Acharya et al. (1976)
9,11	Wasserman and Slattery (1964); Mohan (1974a,b)	+	Prakash (1976)
12	Mohan (1974b)	☆	Peden and Luo (1987)
		♦	Cho and Hartnett (1983b)
		♣	Reynolds and Jones (1989)

The principal reason for such a poor correspondence is the fact that all upper-bound calculations are based on an arbitrary choice of the velocity field which does not necessarily satisfy the momentum equations. For instance, Slattery (1962) and Ziegenhagen (1964) have used the Newtonian flow field as a first approximation which has been subsequently improved upon by including an additional term in the stream function with one or more unknown parameters. This approach presupposes that the flow field for a power-law fluid to be expressible as a sum of two components or in a form which reduces to the Stokes stream function in the limit of $n = 1$: the Newtonian stream function as a base solution and the non-Newtonian effects are included by adding another term to it. The variety of stream functions which have been used for calculating the upper bound on drag is also really diverse; a compilation is presented in Table 2.3. Even more surprising is the fact that, though all stream functions listed here yield the correct value of Y in the limiting case of $n = 1$, some of these do not even reduce to the Stokes' stream function. In view of the uniqueness of the Stokes solution for $n = 1$, not all stream functions will therefore satisfy the momentum equation. In contrast to this, there is only one lower-bound result available, which is based on the trial stress

TABLE 2.3

Stream Functions Used for Calculation of Creeping Sphere Motion

Reference	Expression for ψ^*	Observations
Stokes (1851)	$\sin^2\theta\left[-\dfrac{1}{4\xi}+\dfrac{3\xi}{4}-\dfrac{\xi^2}{2}\right]$	Theoretically derived
Tomita (1959)	$\sin^2\theta\left[\dfrac{\xi}{\sqrt{2}}-\dfrac{\xi}{\sqrt{2}}\dfrac{n-1}{n}\right]^2$	Assumed and does not reduce to the Stokes stream function for $n = 1$.
Slattery (1962) Zeroth approximation	$\sin^2\theta\left[-\dfrac{1}{4\xi}+\dfrac{3\xi}{4}-\dfrac{\xi^2}{2}\right]$	C_1 is an unknown parameter which is zero for $n = 1$.
First approximation	$\sin^2\theta\left[-\dfrac{1}{4\xi}+\dfrac{3\xi}{4}-\dfrac{\xi^2}{2}-\dfrac{3C_1}{4}\left\{\xi-\dfrac{2}{\xi}+\dfrac{1}{\xi^3}\right\}\right]$	
Ziegenhagen (1964)	$\sin^2\theta\left[-\dfrac{1}{4\xi}+\dfrac{3\xi}{4}-\dfrac{\xi^2}{2}\right.$ $\left.+C_1\left\{\dfrac{1}{\xi B_3}+\dfrac{B_1-B_3}{B_2-B_1}\xi^{B_2}+\dfrac{B_3-B_2}{B_2-B_1}\xi^{A_2}\right\}\right]$	A_1, B_1, B_2, B_3, C_1 etc. are unknown parameters
Wasserman and Slattery (1964) and Hopke and Slattery (1970a)	$\sin^2\theta\left[\dfrac{\xi}{\sqrt{2}}-\left(\dfrac{1}{\sqrt{2}\xi}\right)^{C_1}\right]^2$	Assumed profile but does not reduce to the Stokes stream function for $n = 1$. C_1 is an unknown constant.
Nakano and Tien (1968)	$\sin^2\theta\left[-\dfrac{C_1-2}{2(C_1+1)\xi}-\dfrac{\xi^2}{2}+\dfrac{3\xi^{C_1}}{2(C_1+1)}\right]$	Assumed profile but does reduce to the Stokes stream function for $n = 1$. C_1 is an unknown parameter.
Acharya et al (1976)	$\sin^2\theta\left[\dfrac{1}{4\xi}-\dfrac{3\xi}{4}+\dfrac{9n(n-1)}{(2n+1)}\left(\xi-\dfrac{1}{\xi}-2\xi\ln\xi\right)\right]$	Obtained by solving the linearized from of the momentum equation. Note that it does not reduce to the Stokes stream function for $n = 1$. Results corrected by Kawase and Ulbrecht (1981a).
Kawase and Moo-Young (1986)	$\sin^2\theta\left[-\dfrac{1}{4\xi}+\dfrac{3\xi}{4}-\dfrac{\xi^2}{2}+\dfrac{n-1}{2}\left\{-\dfrac{9\xi}{5}+\dfrac{9}{5\xi}+3\xi\ln\xi+\dfrac{3}{5\xi}\ln\xi\right\}\right]$	Obtained by solving the field equations by perturbation method.
Ceylan et al. (1999)	$\sin^2\theta\left(\dfrac{\xi^2}{2}-\dfrac{3}{4}\xi^{2n-1/n}+\dfrac{1}{4}\xi^{2n+3/n}\right)$	Assumed form which satisfies the boundary conditions.

profile introduced originally by Wasserman and Slattery (1964), though these results have been corrected subsequently for numerical errors (Yoshioka and Adachi, 1973; Mohan, 1974a, b). It has been argued that the lowest of the upper bounds and the highest of lower bounds would be closest to the exact solution.

In addition to these upper- and lower-bound solutions, some analytical, though approximate, results are also available for this problem. Using a linearized form of the momentum equation and/or using the Stokes stream function to approximate the second invariant I_2, Hirose and Moo-Young (1969) and Acharya et al. (1976) presented the following closed-form expression (as corrected by Kawase and Ulbrecht, 1981a):

$$Y = 3^{\left(\frac{3n-3}{2}\right)}\left\{\frac{-22n^2+29n+2}{n(n+2)(2n+1)}\right\} \qquad (2.40)$$

Rigid Particles in Time-Independent Liquids without a Yield Stress 61

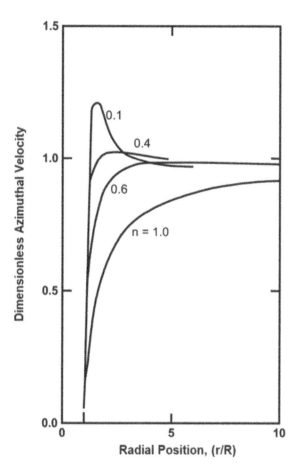

FIGURE 2.7 Predicted velocity profiles around a sphere moving slowly in power-law fluids. (Replotted from Crochet, M.J., Davies, A.R., and Walters, K., *Numerical simulation of Non-Newtonian flow*. Elsevier, Amsterdam, Chapter 9, 1984.)

which predicts $Y(n = 1) = 1$. The predictions of Equation 2.40 are also included in Figure 2.6 and the corresponding stream function is listed in Table 2.3 where the first term is seen to be the Stokes stream function. Subsequently, Kawase and Moo-Young (1986) have reinvestigated this problem and arrived at the following expression for the drag correction factor:

$$Y = 3^{\left(\frac{3n-3}{2}\right)} \left\{ \frac{-7n^2 - 4n + 26}{5n(n+2)} \right\} \tag{2.41}$$

It was asserted to be applicable only for mild shear-thinning behavior ($1 > n > \sim 0.75$). Equation 2.41 also reduces to the expected value of $Y(n = 1) = 1$. The corresponding stream function, corrected to the first order only, is included in Table 2.3 whereas the predictions of Equation 2.41 are shown in Figure 2.6.

Since mid-1980s, this problem has been tackled using numerical methods. Crochet et al. (1984) and Gu and Tanner (1985) solved the governing equations for the creeping sphere motion in power-law fluids. The study of Crochet et al. (1984) assumed the sphere to be fixed at the axis of a cylindrical tube which was moving with a constant velocity whereas Gu and Tanner (1985) considered both, namely sphere-in-sphere and sphere-in-tube configurations to assess the importance of wall effects on the drag force. The resulting numerical values of the drag correction factor Y are presented in Table 2.4. The agreement between the different numerical values of the drag correction factor, Y, is seen to be excellent (±3%). Suffice it to add here that the other numerical predictions of Y (Tripathi et al., 1994; Graham and Jones, 1994; Whitney and Rodin, 2001; Ahmed, 2002; Song et al., 2009) show a similar degree of correspondence thereby lending further credibility to the results of Crochet et al. (1984) and of Gu and

TABLE 2.4
Numerical Values of Drag Correction Factor, Y

n	Crochet et al. (1984)	Gu and Tanner (1985)[b]	Tripathi et al. (1994)[a]	Song et al. (2009)
1	1.02	1.002	1.003	1.0183
0.9	1.18	1.140	1.141	–
0.8	1.27	1.240	1.23	1.258
0.7	1.35	1.320	1.316	–
0.6	1.44	1.382	1.381	1.416
0.5	1.47	1.420	1.42	–
0.4	1.51	1.442	1.44	1.50
0.3	1.48	1.458	1.46	1.485
0.2	1.46	1.413	1.398	1.434
0.1	1.39	1.354	1.36	1.343

[a] Far-away boundary at 50 R.
[b] Extrapolated to *unbounded conditions*.

Tanner (1985). Tanner (1990) has also alluded to the possible interplay between the numerical solution procedure and the domain and mesh size, and their effect on the accuracy of the results. On the other hand, not only the upper- and lower-bound predictions differ from the numerical values of Y increasingly as the value of the index n decreases but also the two bounds themselves diverge appreciably for markedly shear-thinning conditions (Figure 2.6). Despite the considerable scatter present in Figure 2.6, all analyses seem to suggest that the shear-thinning behavior causes an increase in the drag force above its Newtonian value, with a maximum occurring ca. $n \sim 0.35$ or so.

In addition to the calculation of drag coefficient, Crochet et al. (1984) also presented some details about the velocity field. Their results reveal that as the value of the power-law index decreases, the disturbance in the flow field caused by a sedimenting sphere is felt over shorter and shorter distances, that is, the resulting boundary layer is thinner than that in Newtonian fluids otherwise under identical conditions. A typical variation of the azimuthal velocity with radial position is shown in Figure 2.7, where it is clearly seen that the fluid at a distance of ~2R moves like a rigid body. Subsequently, similar results have been reported by Dhole et al. (2006a) and Song et al. (2009). As will be seen in Chapter 3, similar behavior is observed for a sphere moving in viscoplastic fluids and during the agitation of highly shear-thinning fluids in mixing tanks (Chhabra, 2003). Furthermore, with reference to Figure 2.7, for instance, for $n = 0.4$, the presence of the sphere is felt only up to $z/R = 7.7$ and $r/R = 1.7$, indicating thereby that a sphere attains its terminal velocity much quicker in a power-law fluid than that in a Newtonian fluid otherwise under identical conditions. This observation is qualitatively consistent with the observations of Sigli and Coutanceau (1977) and Funatsu et al. (1986), and is also reaffirmed by the subsequent study of Whitney and Rodin (2001).

It is useful to recall here that the flow field created by a moving sphere is nonviscometric and the shear rate varies from zero to a maximum value on the sphere surface. Therefore, a part of the sphere, howsoever small, does come in contact with a fluid of viscosity equal to its zero-shear viscosity, whereas the power-law fluid model does not predict the transition to the zero-shear viscosity region. Several authors (Wasserman and Slattery, 1964; Chhabra and Uhlherr, 1980a) have therefore argued that a fluid model containing a zero-shear viscosity should be preferred over the power-law model in describing such slow flows involving stagnation points and vanishingly small shear rates. Therefore, limited results based on the application of the variational principles that tend to be inherently less accurate (Astarita, 1983) are available on sphere motion in a few other generalized Newtonian fluid models containing the zero-shear viscosity, as one of the model parameter and some of these are described in Sections 2.4.1.1.2–2.4.1.1.4.

2.4.1.1.2 Ellis Model Fluids

The only theoretical results on the creeping motion of spheres in Ellis model fluids are those of Hopke and Slattery (1970a), which have been extended subsequently to encompass wide ranges of non-Newtonian fluid behavior and kinematic conditions (Chhabra et al., 1984). Hopke and Slattery (1970a) employed the velocity and stress variational principles to obtain upper- and lower bounds on drag force. As mentioned earlier, these bounds are less accurate than those for power-law fluids (Slattery, 1972; Astarita, 1983). Dimensional considerations suggest that the use of this fluid model gives rise to an additional dimensionless group called the Ellis number, El, which is defined as

$$\text{El} = \frac{\sqrt{2}\mu_0 V}{d\tau_{1/2}} \qquad (2.42a)$$

The relevant Reynolds number is now defined as

$$\text{Re}_0 = \frac{\rho V d}{\mu_0} \qquad (2.42b)$$

This is tantamount to choosing $\mu_{\text{ref}} = \mu_0$ which, in turn, results in the following expression for μ^* to be used in Equation 2.13:

$$\mu^* = \left\{ 1 + \left(\text{El}^2 I_{2,\tau}^*\right)^{\frac{\alpha-1}{2}} \right\}^{-1} \qquad (2.43)$$

The drag correction factor (Y), still defined by Equation 2.38 with Re_{PL} replaced by Re_0, is obviously now a function of the Ellis number and the fluid parameter α. The upper- and lower bounds on the drag correction factor, Y, are plotted in Figures 2.8 and 2.9 for wide ranges of conditions. Note that this model approaches the correct Newtonian limit only when $\tau_{1/2} \to \infty$ (i.e., when El tends to zero), and this expectation is indeed borne out by the results shown in Figures 2.8 and 2.9. Apart from this limiting behavior, the two bounds diverge increasingly with the increasing extent of the non-Newtonian fluid behavior as

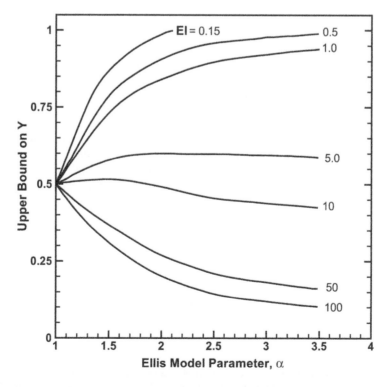

FIGURE 2.8 Upper bound on sphere drag in Ellis model fluids. (Replotted from Chhabra, R.P., Machac, I. and Uhlherr, P.H.T., *Rheol. Acta*, **23**, 457, 1984.)

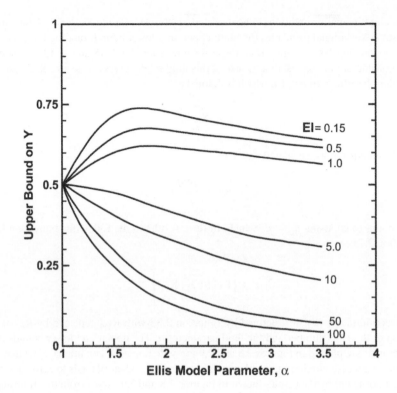

FIGURE 2.9 Lower bound on sphere drag in Ellis model fluids. (Replotted from Chhabra R.P., Machac, I. and Uhlherr, P.H.T., *Rheol. Acta*, **23**, 457, 1984.)

reflected by the increasing values of the Ellis number, El and/or α. Again, Hopke and Slattery (1970a, b) recommended the use of the arithmetic average of the two bounds.

For $\alpha = 1$, both upper- and lower bounds converge to $Y = 0.5$. It needs to be emphasized here that it is a direct consequence of the fact that the Ellis fluid model predicts $\mu = \mu_0/2$ in the limit of $\alpha = 1$.

2.4.1.1.3 Carreau Model Fluids

An approximate upper bound for the drag on a sphere moving slowly ($Re_0 = 0$) through a Carreau model fluid was obtained by Chhabra and Uhlherr (1980a). By noting that the shear rates at which a fluid is likely to display the infinite shear viscosity are unlikely to be reached in the creeping region, one can drop the infinite shear viscosity, $\mu\infty$, from Equation 1.17 and thus this four-parameter model simplifies to the following dimensionless form:

$$\mu^* = \left\{1 + \left(\frac{\Lambda^2 I_2^*}{2}\right)\right\}^{\frac{n-1}{2}} \quad (2.44)$$

The definition of the relevant Reynolds number, Re_0, is still given by Equation 2.42b and the new dimensionless group Λ (Carreau number), akin to the Ellis number, is defined as:

$$\Lambda = \frac{2\lambda V}{d} \quad (2.45)$$

In this case, the drag correction factor is a function of Λ and n, as shown in Figure 2.10. Note that, in this case, the Newtonian result is obtained as $\Lambda \to 0$ or $n \to 1$ or both. More rigorous (but still approximate) upper- and lower bounds and numerical results are also available for this fluid model (Bush and

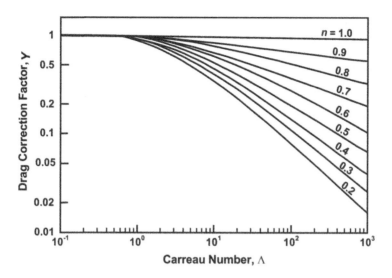

FIGURE 2.10 Approximate drag correction factor for a sphere falling in Carreau model fluids. (Replotted from Chhabra, R.P., and Uhlherr, P.H.T., *Rheol. Acta*, **19**, 187, 1980. With permission.)

Phan-Thien, 1984; Chhabra and Dhingra, 1986). Suffice it to add here that the numerical results of Bush and Phan-Thien (1984) are in excellent agreement with the approximate upper-bound calculations (Chhabra and Uhlherr, 1980a). In the limit of vanishingly small values of the Carreau number and for values of the index, n not too different from unity, a few closed-form expressions for the drag correction factor are also available in the literature (Kawase and Moo-Young, 1985; Rodrigue et al., 1996b, c). As expected, under these conditions, these analyses predict very weak influence of the Carreau number and the flow behavior index. The role of the ratio (μ_0/μ_∞) on the drag of a sphere has subsequently been investigated by Dolecek et al. (2004).

2.4.1.1.4 Sutterby Model Fluids

An analogous development for the creeping sphere motion in Sutterby model fluids (Equation 1.14) has been presented by Mitsuishi et al. (1971). The additional dimensionless group, the Sutterby number, Su, is of the same form as Λ (Equation 2.45), that is,

$$\text{Su} = \frac{\theta_E V}{d} \tag{2.46}$$

The functional dependence of the approximate upper bound on the drag correction factor Y on the model parameters A and Su is similar to that for the Carreau model fluids as shown in Figure 2.10 and hence is not shown here.

A common feature of all such predictions based on the non-Newtonian viscosity models containing the zero-shear viscosity is that the drag on a sphere is seen to be reduced below its Newtonian value, as implied by the values of Y smaller than unity. This inference is at variance with the conclusions reached when one uses the power-law fluid model. It should be emphasized here that there is no contradiction or inconsistency as to whether the value of Y is smaller or larger than unity, for it is simply a matter of the choice of the reference viscosity used in the definition of the Reynolds number. Finally, it is worthwhile to note here that both the Ellis fluid model and the Carreau viscosity as well as the Sutterby model equations contain the power-law model as a special case. For instance, for $(\lambda \dot{\gamma})^2 \gg 1$, Equation 1.17 reduces to

$$\mu = \left(\mu_0 \lambda^{n-1}\right) \dot{\gamma}^{n-1} \tag{2.47}$$

which is identical to the power-law fluid model with $m = \mu_0 \lambda^{n-1}$. Thus, intuitively, one would expect the results shown in Figure 2.10 to approach those for power-law fluids for large values of the Carreau number. This expectation has been confirmed by the studies of Bush and Phan-Thien (1984) and of Chhabra and Dhingra (1986). However, it is not yet possible to specify the values of n and Λ a priori beyond which the power-law model analysis can be used. Similar equivalence between the Ellis model and the power-law fluid can also be established for large values of the ratio $(\tau/\tau_{1/2}) \gg 1$. Chung and Vaidya (2010) have considered the general case of a sphere falling in non-constant viscosity fluids.

2.4.1.2 Experimental Results

A large number of investigators have measured drag coefficients of spheres falling freely under the influence of gravity in shear-thinning fluids (Table 2.5). Evidently, there is a preponderance of studies based on the use of the simple power-law model. It is thus possible to make comparisons between experimental and theoretical results for this problem. Prior to undertaking such an exercise, it is appropriate to add here that considerable confusion exists in the literature regarding the limiting value of the Reynolds number for the so-called creeping flow regime. For instance, both Uhlherr et al. (1976) and Acharya et al. (1976) concluded that the creeping flow persists up to about $Re_{PL} = 10$ whereas Hopke and Slattery (1970b) suggested a value of $Re_0 = 0.1$ as the limit of the creeping flow regime. The evidence in both cases, however, is indirect and tenuous. In this work, by analogy with the behavior in Newtonian fluids, the critical value of the Reynolds number is arbitrarily taken to be $Re_{PL} = 0.1$, albeit there is some experimental evidence (Hopke and Slattery, 1970b; Peden and Luo, 1987; Koziol and Glowacki, 1988) suggesting that the inertial effects are not negligible even at Reynolds numbers as low as 0.005. Data culled from various sources are plotted in Figure 2.6 together with the available theoretical analyses. The wild scatter seen in this figure is indeed disturbing and extremely discomforting. Until now one was confronted with the widely divergent theoretical predictions, but one is now faced with an even more complicated scenario due to the poor correspondence between experimental results reported by different investigators. It is not at all surprising that conflicting conclusions can be (and have been) drawn regarding the influence of the power-law index on drag correction factor. For instance, the results of Slattery (1959), Uhlherr et al. (1976), Chhabra et al. (1980a), and Cho and Hartnett (1983b) suggest that the drag force is augmented above its Newtonian value due to the shear-thinning behavior approximated by the power-law model whereas the works of Turian (1964, 1967), Kato et al. (1972), Yoshioka and Adachi (1973, 1974), Acharya et al. (1976), and Lali et al. (1989) though exhibiting considerable spread (especially see the results of Reynolds and Jones (1989) and Peden and Luo (1987)) show no such drag enhancement. In fact, some studies even suggest the drag to reduce below its Newtonian value, for example, see Leonov and Isayev (1989). Generally, this lack of agreement between theory and experiments as well as the poor reproducibility of the latter have been attributed primarily to the possible viscoelastic effects, the inadequacy of power-law model, and to the uncertainty of wall effects. For instance, in early studies (Slattery and Bird, 1961; Turian, 1967), the polymer solutions used were seldom checked for possible viscoelastic effects. Likewise, wall effects were either assumed to be negligible or the Newtonian wall correction was applied (Turian 1964, 1967; Acharya et al., 1976). Neither of these procedures are generally applicable or justifiable (Chhabra et al., 1977; Chhabra and Uhlherr, 1980c). Unfortunately, even when experimental results are free from these uncertainties, these lie almost completely outside the rigorous upper- and lower bounds (Chhabra, 1983). One is therefore left with no choice but to question the suitability of the power-law fluid model for describing the creeping sphere motion with stagnation points. As noted earlier, it is important to recognize that one must evaluate the power-law constants (n, m) using the viscometric data in the same shear rate range as that encountered in a falling sphere test. However, the characteristic shear rate associated with a falling sphere cannot be calculated a priori and in fact it is a function of the rheological parameters of the fluid itself (Gottlieb, 1979; Cho et al., 1984; Suri et al., 2023a). Often, the corresponding surface averaged value for Newtonian fluids, that is, $(2V/d)$ has been used as a general guide (Slattery and Bird, 1961; Dallon, 1967; Chhabra and Uhlherr, 1981), albeit there is some experimental evidence suggesting it to be smaller than $(2V/d)$ in shear-thinning fluids (Sato et al., 1966). This uncertainty coupled with the fact that the power-law fluid model does not predict the transition from the

TABLE 2.5

Experimental Studies on Free Settling of Spheres in Shear-Thinning Liquids

Investigator	Test Liquids	Fluid Model	Remarks
Chase (1955)	Solutions of CMC	–	No rheological parameters measured.
Slattery (1959), Slattery and Bird (1961), Wasserman and Slattery (1964)	Solutions of three different grades of CMC	Power-law and Ellis fluid models	Experimental values showed better agreement with power-law than the Ellis model theory. Wall effects were reported to be negligible.
Sato et al. (1966)	Solutions of CMC and Natrosol	Power-law	Wall effects are less significant in non-Newtonian fluids.
Turian (1964, 1967)	Solutions of HEC and PEO	Ellis fluid model	The main thrust of the work was on the estimation of zero-shear viscosity and wall effects. A correlation for drag coefficient in the creeping flow range was presented.
Dallon (1967)	Solutions of CMC, HEC and PEO	Ellis fluid model	Correlation for drag coefficient.
Mitsuishi et al. (1971)	Solutions of CMC, HEC and PEO	Sutterby fluid model	Correlation for drag coefficient.
Kato et al. (1972)	Solutions of PEO and CMC	Power-law	In creeping flow, experimental results showed poor agreement with upper and lower bounds whereas drag reduction was observed at high Reynolds numbers.
Adachi (1973), Yoshioka and Adachi (1974)	Solutions of CMC	Power-law	Comparison with upper and lower bounds showed poor agreement.
Uhlherr et al. (1976)	Solutions of cellulose gum and Methocel	Power-law	Drag results showed good agreement with the analysis of Tomita (1959) and Slattery (1962).
Acharya et al. (1976)	Solutions of CMC, HEC, PAA, Carbopol and PEO	Power-law	Extensive results on drag coefficients embracing a sixfold variation in Re_{PL}.
Prakash (1976, 1983, 1986)	Solutions of CMC	Power-law	Empirical correlations for drag coefficient.
Chhabra (1980), Chhabra and Uhlherr (1980a,b,c), Chhabra et al. (1981a,b, 1984)	Solutions of Natrosol, PEO, HEC, and PAA	Power-law, Carreau model and Ellis fluid model	A comprehensive study on wall effects and drag coefficient. Detailed comparisons with theoretical analyses.
Shah (1982, 1986)	Solutions of HEC and HPG	Power-law	Empirical correlation for the estimation of free-fall velocity of spheres.
Peev and Mateeva (1982)	Solutions of CMC and PAA	Power-law	New experimental results on drag.
Subramaniam and Zuritz (1990)	Solutions of CMC	Power-law	Correlation for drag coefficient of suspended spheres.
Cho and Hartnett (1983b)	Solutions of CMC, PAA and Carbopol	Power-law	New results on drag coefficients and comparison with new upper and lower bounds.
Dunand et al. (1984), Dunand and Soucemarianadin (1985)	Solutions of HPG	Ellis fluid	New experimental results on drag coefficient are reported.
Roodhart (1985)	HEC & HPG solutions	Truncated power-law	Settling velocities under static and dynamic conditions.
Dolecek et al. (1983), Machac et al. (1987, 1995)	Solutions of PAA, PEO, and Natrosol, and Clay suspensions	Power-law, and Carreau model and Ellis model	Extensive results on wall effects and drag coefficient at low and high Reynolds numbers.

(Continued)

TABLE 2.5 (*Continued*)
Experimental Studies on Free Settling of Spheres in Shear-Thinning Liquids

Investigator	Test Liquids	Fluid Model	Remarks
Peden and Luo (1987)	Solutions of CMC, XC and HEC	Power-law	New predictive correlation for spherical and nonspherical particles are presented.
Koziol and Glowacki (1988)	Solutions of CMC	Power-law	New experimental results, and a graph is presented for estimating the free-fall velocity of particles.
Lali et al. (1989)	Solutions of CMC	Power-law	In the creeping flow regime, drag correction factor seems to be a function of Reynolds number. The high Reynolds number data conform to the Newtonian standard drag curve.
Reynolds and Jones (1989)	Solutions of CMC and HEC	Power-law	New experimental data without any analysis.
Briscoe et al. (1993)	Bentonite suspensions	Power-law and Bingham plastic	Wall effects and drag correlation.
Jin and Chenevert (1994)	HEC, Xanthan gum, and PHPA Solutions	No fluid model	Effective viscosity used at $\dot{\gamma} = (V/R)$ to correlate drag results.
Ataide et al. (1998, 1999)	Solutions of CMC	Power-law	Results on wall effects and drag coefficients
Matijasic and Glasnovic (2001)	Solutions of CMC	Power-law	Data on drag coefficient $0.54 < n < 1$ and $Re_{PL} < 1000$.

Note: CMC, Carboxymethyl cellulose; HEC, Hydroxyethyl cellulose; PEO, Polyethylene Oxide; PAA, Polyacrylamide; XC, Cellulose gum; PHPA, Partially hydrolyzed polyacrylamide; HPG, Hydroxypropyl gum.

zero-shear viscosity region to shear-thinning has led to a sort of general consensus that the power-law fluid model is not adequate for this flow configuration, and one must therefore resort to the use of a fluid model containing the zero-shear viscosity as one of its parameters (Adachi et al., 1977/78; Chhabra and Uhlherr, 1981; Clark et al., 1985; Chhabra, 1986) in order to improve the degree of drag prediction.

In a series of papers, Chhabra (1980), Chhabra and Uhlherr (1980a), and Chhabra et al. (1984) reported extensive experimental data on drag coefficients of spheres in aqueous solutions of a wide variety of chemically different polymers. The rheological behavior of test fluids was approximated either by the three-parameter Ellis model or by the Carreau viscosity equation. A typical comparison between their experiments and predictions (Chhabra and Uhlherr, 1980a; Bush and Phan-Thien, 1984) is shown in Figure 2.11 for the Carreau viscosity equation. The agreement is seen to be excellent. Other experimental data available in the literature (Dallon, 1967; Dolecek et al., 1983) also conform (within ±10%) to theoretical predictions. Figure 2.12 demonstrates that the theoretical estimates of the drag correction factor based on the Carreau viscosity equation and power-law models approach each other for large values of the Carreau number (Cho and Hartnett, 1983b). By introducing the notion of a reference viscosity coupled with simple dimensional considerations, Bush and Phan-Thien (1984) approximated the theoretical dependence of Y on the index n and Λ by the following expression:

$$Y = \left(1 + k^2 \Lambda^2\right)^{\frac{n-1}{2}} \tag{2.48}$$

Based on extensive experimental data encompassing wide ranges of conditions ($1 \geq n \geq 0.4$; $\Lambda < \sim 400$), Bush and Phan-Thien (1984) suggested $k = 0.275$. This value seems to correlate most of the literature data with an average error not exceeding 10%. The predictions of Equation 2.48 are also plotted in Figure 2.12. Figure 2.13 shows a similar comparison between experiments (Chhabra, 1980) and theory (Hopke and Slattery, 1970a; Chhabra et al., 1984) for the Ellis model fluids. The two bounds, albeit approximate, seem to enclose most of the experimental results with a slight propensity of the data points being close to the upper bound at low values of the Ellis number and gradually shifting toward the lower bound with the increasing value of the Ellis number. Based on the extensive results available in

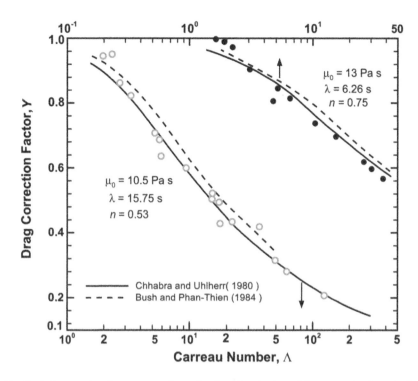

FIGURE 2.11 Typical comparison between experimental and predicted values of drag correction factor in Carreau model fluids.

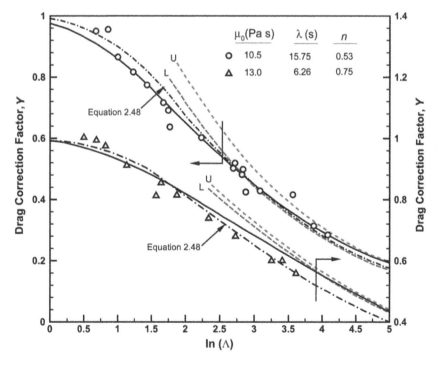

FIGURE 2.12 Comparison between experiments and predictions of Equation 2.48. This figure also shows that for large values of Λ, the predictions of Carreau viscosity equation (shown as solid line) approach those for power-law fluid (shown as U [upper] and L [lower] bounds).

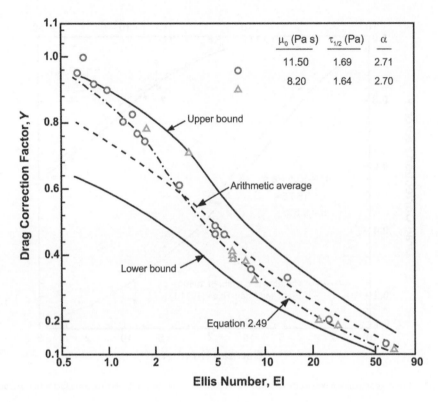

FIGURE 2.13 Typical comparison between experiments (Chhabra, 1980) and predictions (Chhabra et al., 1984) for Ellis model fluids.

the literature (Dolecek et al., 1983; Dunand et al., 1984; Dunand and Soucemarianadin, 1985), Chhabra et al. (1981a) found that the following empirical expression correlated the values of drag correction factor reasonably well (±15%), as seen in Figure 2.13:

$$Y = \left[1 + 0.50 \text{El}^{1.65} (\alpha - 1)^{0.38}\right]^{-0.35} \tag{2.49}$$

Equation 2.49 encompasses the following ranges of conditions: $1 < \alpha < 3.22$ and $0.10 < \text{El} < 141$. Based on this fluid model, numerous other (Slattery and Bird, 1961; Dallon, 1967; Turian, 1967) empirical formulae are also available, however all of these are either too tentative or restricted to narrow ranges of conditions to be included here.

The aforementioned detailed comparisons clearly establish that the use of a generalized Newtonian fluid model containing the zero-shear viscosity leads to much better predictions of drag coefficient than the power-law fluid model, though at the expense of an extra dimensionless parameter like Ellis number, Carreau number, or the Sutterby number. Besides, the experimental results reported by different workers are also consistent with each other within the limits of experimental uncertainty. This improvement is solely attributable to the inclusion of the zero-shear viscosity, both in theoretical analyses as well as in the interpretation of experimental data. Therefore, once more the caveat: as far as possible the use of power-law fluid model should be avoided in calculating the drag on a sphere in the creeping flow region.

Another interesting and significant observation is that even though the experimental liquids employed by Mitsuishi et al. (1971), Chhabra and Uhlherr (1980a), Dolecek et al. (1983), and Dunand et al. (1984) exhibited (or were of sufficiently high polymeric concentration to display) viscoelastic behavior (i.e., nonzero primary normal stress difference in simple shear), yet the drag results are in line with the theoretical results based on the purely viscous models, for example, Carreau model, Ellis fluid models, etc. This agreement is believed to be realistic and hence it does lend support to Bird's intuitive assertion

(Bird, 1965), but as will be seen in Chapter 5, the viscoelastic effects are only of minor significance in this flow configuration, at least at low Weissenberg and Deborah numbers. Conversely, one can argue that the drag on a sphere is primarily determined by the shear-thinning viscosity of the fluid which also overshadows the viscoelastic effects if any at such low Reynolds numbers.

2.4.1.3 Drag Force at High Reynolds Numbers

In contrast to the voluminous literature in the creeping flow region, little numerical work on sphere motion in shear-thinning fluids is available at intermediate Reynolds numbers. Adachi et al. (1973, 1977/78) solved the governing equations numerically for the steady flow of power-law and of the extended Williamson fluid models for the steady flow over a sphere. Owing to the unrealistic restrictions imposed on the material constants, their results for the extended Williamson fluid model are of little practical interest. For the case of power-law fluid, their results are limited to a single value of $Re_{PL} = 60$ and for minor shear-thinning behavior $1 \leq n \leq 0.8$. Within this narrow range of conditions, both the kinematics of the flow (streamline and isovorticity contours, surface pressure profiles) and the values of the drag coefficient deviate very little from the corresponding Newtonian results. Qualitatively similar results have been reported by Hua and Ishii (1981), albeit their numerical results appear to be in error as their values of drag coefficient for $n = 1$ deviate from the expected values by as much as 100%. Indeed, there have been only a few numerical studies for the flow of power-law fluids past a sphere beyond the creeping flow regime. Tripathi et al. (1994) reported the values of the pressure, friction, and total drag coefficients for a sphere up to $Re_{PL} = 100$. Figure 2.14 shows the variation of the ratio of the pressure to friction drag components with the flow behavior index and the Reynolds number, Re_{PL}. While the ratio (C_{DP}/C_{DF}) is independent of the Reynolds number below 0.1 (the limiting value for the creeping flow), it increases with the increasing degree of shear-thinning behavior. Clearly, the pressure drag rises much more steeply with the increasing value of Re_{PL} and decreasing value of n than the frictional drag. In general, the non-Newtonian effects are more prominent at low Reynolds numbers and these progressively

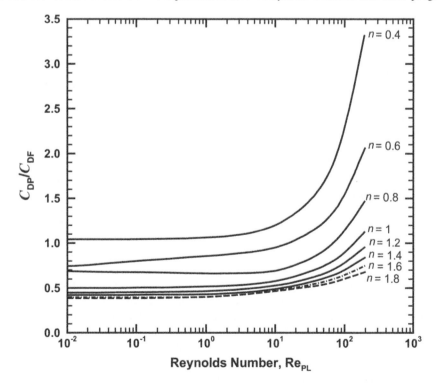

FIGURE 2.14 Dependence of (C_{DP}/C_{DF}) on Reynolds number and power-law index for spheres. (From Tripathi, A., Chhabra, R.P., and Sundararajan, T., *Ind. Eng. Chem. Res.*, **33**, 403, 1994.)

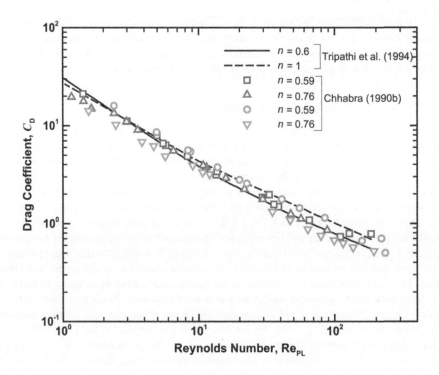

FIGURE 2.15 Typical comparison between numerically predicted (Tripathi et al., 1994) and experimental values (Chhabra, 1980) of drag coefficient in power-law fluids ($Re_{PL} > 1$).

diminish at high Reynolds numbers. The independent study of Graham and Jones (1994) is almost in complete agreement with that of Tripathi et al. (1994). Figure 2.15 shows a representative comparison between the numerical and experimental results up to $Re_{PL} = 100$ and for $1 \geq n \geq 0.6$. The correspondence is seen to be fair, as the experimental values are well enclosed by the standard drag curve for $n = 1$ and by the corresponding line for $n = 0.6$. Some attempts (Chhabra, 1995a; Darby, 1996; Darby and Chhabra, 2017; Renaud et al., 2004) have also been made to correlate these numerical results via simple analytical expressions. Subsequent results reported by Dhole et al. (2006a), Song et al. (2009), Daunais et al. (2023) and others (Shah et al., 2007) are also in line with these observations.

In contrast to the limited numerical developments, several investigators have attempted to establish the drag curve for non-Newtonian fluids via experimental results. Consequently, there is no scarcity of empirical expressions available in the literature. A thorough review of the pertinent literature (Chhabra, 1990b, c, 2002a; El Fadili, 2005; Agwu et al., 2018; Sun et al., 2020b) suggests that much of the available experimental data on spheres in power-law fluids at high Reynolds numbers is well correlated by the so-called standard drag curve (Clift et al., 1978) for Newtonian fluids about as accurately (±30%) as can be expected in this type of work. This viewpoint is shared, among others, by Kato et al. (1972), Lali et al. (1989), Briscoe et al. (1993), and Machac et al. (1995). Figure 2.16, showing data culled from a range of sources together with the standard drag curve, supports this notion. Clearly, there are no discernable trends with respect to the power-law index. In the range of conditions, $0.535 < n < 1$ and $1 < Re_{PL} < 1000$, the resulting average deviation from the standard Newtonian drag curve is 18% for 460 data points, albeit the maximum error is of the order of 70% in a few cases. Subsequently reported scant data (Chien, 1994; Ford et al., 1994; Matijasic and Glasnovic, 2001; Miura et al., 2001a; Pinelli and Magelli, 2001; Kelessidis and Mpandelis, 2004; Kelessidis, 2004) also conforms to this expectation. For engineering design calculations, this level of accuracy is quite acceptable. However, in spite of this, several empirical correlations are available in the literature which purport to offer improved predictions of sphere drag in power-law fluids at high Reynolds numbers. Thus, for instance, Acharya et al. (1976) put forward the following correlation:

Rigid Particles in Time-Independent Liquids without a Yield Stress

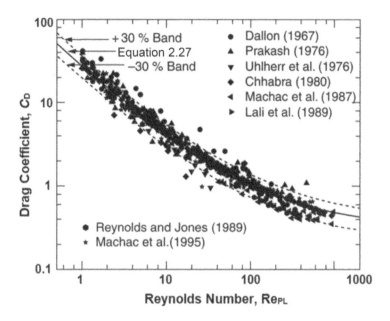

FIGURE 2.16 Comparison of experimental values of drag coefficient in power-law fluids with the standard drag curve for Newtonian media, Equation 2.27 ($Re_{PL} > 1$).

$$C_D = \frac{24}{Re_{PL}} Y + (10.5n - 3.5) Re_{PL}^{-(0.32n + 0.13)} \tag{2.50}$$

where the value of Y is given by Equation 2.40.

Equation 2.50 was stated to be applicable in the range $0.5 < n < 1$, and $Re_{PL} < 1000$. A typical comparison between the predictions of Equation 2.50 and experimental results is displayed in Figure 2.17. Subsequently, Ceylan et al. (1999) have presented the following correlation which is also applicable up to $Re_{PL} \leq 1000$:

$$C_D = \left(\frac{24}{Re_{PL}}\right)\left[3^{2n-3}\left\{\frac{n^2 - n + 3}{n^{3n}}\right\} + \frac{4n^4}{24(Re_{PL})^{(n-3)/3}}\right] \tag{2.51}$$

In Equation 2.51, while the first term on the right-hand side has been derived using the stream function listed in Table 2.3, they used the experimental data of Lali et al. (1989) to evaluate the empirical constants appearing in the second term. The predictions of Equations 2.27 and of 2.51 are also included in Figure 2.17 where it is clearly seen that the standard Newtonian drag curve correlates these results nearly as well as the correlations of Acharya et al. (1976) or Ceylan et al. (1999). Thus, it is safe to conclude that the standard drag curve for Newtonian fluids also provides an adequate (±30%) representation of the available data on drag of spheres in power-law fluids in the range $0.5 \leq n \leq 1$ and $1 \leq Re_{PL} \leq 1000$.

Numerous other correlations have been developed but none of these have been tested using independent data or seem to offer any significant improvements over Equation 2.27, for example, see Prakash (1983, 1986), Ren (1991), Matijasic and Glasnovic (2001), Wilson et al. (2003), and Kelessidis (2004).

Subsequently, Renaud et al. (2004) have extended their previous work (Mauret and Renaud, 1997) to develop the following general framework relating the drag coefficient for spheres and cylinders in power-law fluids:

$$C_D = C_{D0} + \chi C_{D\infty} C_{D0}^{2A_0} B_0 \left\{\frac{6Yb_o}{6Yb_0 + C_{D0}}\right\}^{A_o} + C_{D\infty}\left\{\frac{6Yb_0}{6Yb_0 + 128 C_{D0}}\right\}^{11/12} \tag{2.52}$$

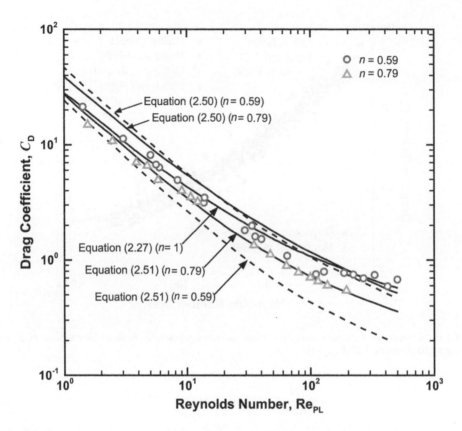

FIGURE 2.17 Comparison between the predictions of Equations 2.27, 2.50, 2.51, and experiments.

where C_{D0} and $C_{D\infty}$ are the values of the drag coefficient in the Stokes and Newton regimes, respectively, and χ is the ratio of the surface area to the projected area of the particle. The remaining three constants, namely, A_0, B_0, and b, were evaluated using the numerical predictions of Tripathi et al. (1994) in the ranges $1 \geq n \geq 0.4$ and $Re_{PL} \leq 100$. The constants may be expressed as

$$b_0 = \exp\{3(C - \ln 6)\} \tag{2.53a}$$

$$B_0 = \frac{C_0 - C}{2C_0 C} \exp\left[3\left\{\frac{C_0 - C}{2C_0 C}\right\} \ln 3\right] \tag{2.53b}$$

$$A_0 = \frac{11}{48}\sqrt{6}\left[1 - \exp\left\{\left(\frac{C_0 - C}{C(C_0 - 1)}\right)^2 \ln \frac{\sqrt{6}-1}{\sqrt{6}}\right\}\right] \tag{2.53c}$$

The remaining two functions Y, defined by Equation 2.38, and C are related to each other as follows. The numerical values of Y due to Tripathi et al. (1994) reported in Table 2.4 can be approximated by the following simple form (Renaud et al., 2004):

$$Y = 6^{\frac{n-1}{2}}\left\{\frac{3}{n^2 + n + 1}\right\}^{n+1} \tag{2.54}$$

The factor C represents the correction for the average shear rate and is related to Y as follows:

$$C = \left\{ 6^{\frac{1-n}{2}} Y \right\}^{\frac{1}{n+1}} \tag{2.55}$$

Finally, C_0 denotes the value of C for $n = 0$, that is, $C_0 = 3$. It is worthwhile to point out here that in the limit of $n = 1$, Equation 2.52 reduces correctly to the limiting form of drag for spheres in Newtonian fluids (Mauret and Renaud, 1997). At this juncture, it is important to make some comments about the value of $C_{D\infty}$. In Newtonian fluids, $C_{D\infty} \sim 0.44$ for a sphere and $C_{D\infty} \sim 1$ for an infinitely long cylinder-oriented normal to the flow. Unfortunately, currently available data for drag in power-law liquids are limited to the maximum Reynolds number of 1000 even for spheres, let alone for a cylinder or any other shape. At this stage, the value of $C_{D\infty} = 0.44$ has been retained in Equation 2.52 as a first approximation. For a spherical particle, evidently the area ratio factor $\chi = 4$, whereas it will take on different values depending upon the shape and orientation of nonspherical particles during the course of its sedimentation.

In order to demonstrate the applicability of Equation 2.52, experimental data for spheres have been culled from as many sources as possible (Table 2.6). Figures 2.18 and 2.19 contrast the predictions of Equation 2.52 with the literature data for $Re_{PL} < 1$ and $Re_{PL} > 1$, respectively, that is, in the so-called creeping and intermediate flow regions. It needs to be emphasized here that the constants A_0, B_0, and, b appearing in Equation 2.52 have been evaluated using the numerical predictions of Tripathi et al. (1994) over the range of $Re_{PL} \leq 100$ and $1 \geq n \geq 0.4$, and therefore the comparisons shown in Figures 2.18 and 2.19 must be seen as an independent validation for the reliability of Equation 2.52. While the mean deviation is only ~26%, a few data points show deviations greater than 100%. Bearing in mind that some of the values of n and Re_{PL} are outside the ranges, the degree of correspondence seen in these figures is certainly acceptable for the purpose of engineering design calculations.

Likewise, for spheres settling in Ellis model fluids, Dallon (1967) developed the following empirical correlation which is valid in the range of conditions as $0.2 < Re^* < 300$; $1 \leq \alpha < 2.22$; $0.00212 \leq El \leq 4.24$ and $0.2 \leq Re_0 \leq 550$,

$$C_D = \frac{23.2}{Re^*} \left(1 + a Re^{*b} \right) \tag{2.56}$$

TABLE 2.6

Summary of Experimental Data for Spheres Filling in Power-Law Fluids

Source	Range of n	Range of Re_{PL}
Dallon (1967)	0.63–0.94	7.5×10^{-3} – 740
Uhlherr et al. (1976)	0.71–0.92	6.65×10^{-4} – 34
Prakash (1976)	0.54–0.85	7.6×10^{-3} – 465
Chhabra (1980)	0.50–0.95	2.9×10^{-4} – 940
Shah (1982, 1986)	0.281–0.762	5×10^{-3} – 67
Machac et al. (1987)	0.58–0.72	0.15–980
Lali et al. (1989)	0.56–0.85	1.54×10^{-3} – 157
Reynolds and Jones (1989)	0.50–0.71	3.6×10^{-6} – 10.5
Ford et al. (1994)	0.06–0.29	0.9 – 65
Machac et al. (1995)	0.39–0.87	1.4×10^{-3} – 549
Miura et al. (2001a)	0.60–0.75	0.45 – 770
Pinelli and Magelli (2001)	0.73	1.25 – 28
Kelessidis (2004)	0.75–0.92	0.11–64
Kelessidis and Mpandelis (2004)	0.75–0.91	0.48–39.5

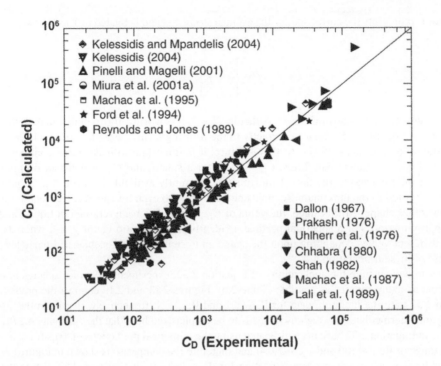

FIGURE 2.18 Comparison between the predictions of Equation 2.52 and experimental data for spheres ($\chi = 4$) in power-law fluids for $Re_{PL} < 1$. (Modified from Renaud, M., Mauret, E., and Chhabra, R.P., *Can, J. Chem. Eng.*, **82**, 1066, 2004.)

where

$a = 0.198 + 0.023\,(\alpha - 1)$
$b = 0.634 - 0.044\,(\alpha - 1) + 0.007\,(\alpha - 1)^2$

$$Re^* = \frac{\rho V d}{\bar{\mu}_0}; \quad \bar{\mu}_0 = \frac{\mu_0 + k_0 \mu_m}{1 + k_0}$$

$$k_0 = 0.812 \alpha^{1.1} El^{0.6} Re_0^{0.12}$$

and the minimum viscosity, μ_m, is calculated at $\dot{\gamma}_{max} = (3V/d)Re_0^{0.36}$.

Though Equation 2.56 purports to correlate the experimental results of Dallon (1967) satisfactorily, subsequent comparisons (Chhabra, 1990b) of Equation 2.56 with independent data sets have been less satisfactory, thereby highlighting its deficiencies. However, when this set of data is recast in the form of power-law fluids, it correlates just as well with the Newtonian standard drag curve (Figure 2.11) or with Equation 2.52, Figures 2.18 and 2.19, as with Equation 2.56.

Similarly, based on the use of the Carreau viscosity equation, Chhabra and Uhlherr (1980b) proposed the following modification of the well-known Schiller and Naumann (1933) drag formula:

$$C_D = \frac{24}{Re_0}\left(1 + 0.15 Re_0^{0.687}\right)\left(1 + 0.65(n-1)\Lambda^{0.20}\right) \tag{2.57}$$

Equation 2.57 predicts 300 individual data points with a maximum error of $\pm 14\%$ in the following ranges of conditions: $0.032 < \Lambda < 630$, $0.032 < Re_0 < 400$, and $0.52 < n < 1$. Also, in the limit of $n = 1$ and/or $\Lambda = 0$, it reverts to the original form for Newtonian fluids. Recent numerical predictions (Daunais et al., 2023) are consistent with Equation 2.57

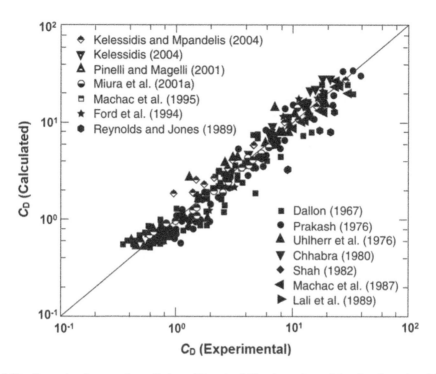

FIGURE 2.19 Comparison between the predictions of Equation 2.52 and experimental data for spheres ($\chi = 4$) in power-law fluids in the range $1 \leq \text{Re}_{PL} \leq \sim 1000$. (Modified from Renaud, M., Mauret, E., and Chhabra, R.P., *Can. J. Chem. Eng.*, 82, 1066, 2004.)

2.4.2 Free-Fall Velocity

In process design calculations, a knowledge of the free-fall velocity of a sphere is frequently required for the given rheological and physical properties of the liquid, density, and size of the sphere. For a sphere falling under its own weight in a stationary power-law fluid, Equation 2.37 can be rewritten as:

$$V = \left\{ \frac{gd^{n+1}(\rho_p - \rho)}{18mY} \right\}^{\frac{1}{n}} \tag{2.58}$$

Note that Equation 2.58 reduces to the familiar Stokes equation for $n = 1$. One can thus estimate the free-settling velocity V for a given sphere/power-law liquid combination as long as the value of the Reynolds number is less than 0.1. However, a difficulty arises outside the creeping flow regime as the unknown velocity, V, appears in both the dimensionless groups, that is, the Reynolds number and drag coefficient, and therefore, an iterative procedure is required to solve for the velocity. To overcome this difficulty, by analogy with the Newtonian fluids (Equation 2.28), one can define a modified Archimedes number as:

$$\text{Ar} = C_D \text{Re}_{PL}^{2/2-n} = \frac{4}{3} g d^{\frac{2+n}{2-n}} (\rho_P - \rho) \rho^{n/2-n} m^{2/n-2} \tag{2.59}$$

For a given sphere/power-law liquid combination, the right-hand side of Equation 2.59 is thus known, and now the Archimedes number is expected to be a function of the Reynolds number, Re_{PL} and the power-law index, n. It is appropriate to add here that other definitions of the Archimedes number are also used, but these are all interrelated through a function of n and arbitrary constants.

Several attempts have been made to establish the functional dependence of the Archimedes number on the Reynolds number and the flow behavior index (Novotny, 1977; Daneshy, 1978; Harrington et al., 1979; Clark and Quadir, 1981; Shah, 1982, 1986; Clark and Guler, 1983; Prakash, 1983, 1986; Roodhart, 1985; Clark et al., 1985; Acharya, 1986, 1988; Peden and Luo, 1987; Koziol and Glowacki, 1988; Chhabra, 1990b, 1995a, 2002a; Briens, 1991; Chhabra and Peri, 1991; Jin and Penny, 1995; Darby, 1996; Darby and Chhabra, 2017; Prakash et al., 2002; Shah et al., 2007; Sun et al., 2020b). Unfortunately, most of these have proved to be either of limited utility or too restrictive in their range of application. Furthermore, in view of the fact that the values of drag coefficient of spheres falling freely in power-law fluids at high Reynolds numbers ($Re_{PL} > 1$) are in line with the standard Newtonian drag curve, it is reasonable to presume that this would be so also for the reverse calculation, namely for the estimation of the free-fall velocity of a sphere. Based on this premise, it has been suggested (Chhabra, 1990b, 2002b) that in the first instance, Equation 2.29 can also be used for power-law fluids with appropriate definitions of the Reynolds and Archimedes numbers. This method seems to yield the values of the free-fall velocity with an average error of ±31% and unfortunately, none of the custom-built correlations developed for the estimation of the free-fall velocity in power-law liquids seem to offer any significant improvement over Equation 2.29 (Chhabra, 2002a).

Additional dimensionless parameters appear even under creeping flow conditions when the formulation of similar expressions is attempted for the other fluid models containing three parameters such as the Ellis and Carreau model fluids. Thus, for instance, Machac et al. (2000) used their approximate calculations and experimental data to develop the following scheme for calculating the terminal settling velocity of a sphere in Carreau model fluids at low Reynolds numbers ($Re_0 < 0.1$):

for $0 \leq \Lambda \leq 10$

$$Y = \left[1 + \left\{\left(1.25 - 0.821n + 0.256n^2\right)\Lambda\right\}^{0.814}\right]^{(n-1)/2} \quad (2.60a)$$

and for $10 \leq \Lambda \leq 1500$

$$Y = \left[1 + \left\{\left(0.525 - 0.635n + 0.308n^2\right)\Lambda\right\}^{1.817}\right]^{(n-1)/2} \quad (2.60b)$$

Equation 2.60 in turn is combined with the Stokes formula and solved iteratively to extract the value of V for a given fluid (i.e., known λ, n, μ_0, ρ) and sphere (ρ_p, d) provided the Reynolds number is small ($Re_0 < 0.1$). Equation 2.60 was shown to predict the value of V with an average error of 15%–20% in the range of conditions as: $0.33 \leq n \leq 0.9$, $0.6 \leq \Lambda \leq 375$ and $Re_0(1 + 0.25\Lambda^2)^{(1-n)/2} \leq 0.27$. Obviously, the accuracy of these predictions is only marginally better than that of Equation 2.29. Subsequently, this approach has been extended to the settling in the transition regime (Siska et al., 2005). Similarly, Wilson et al. (2003) have outlined a graphical procedure to estimate the free-falling velocity of a sphere in quiescent generalized Newtonian fluids. This approach obviates the necessity of choosing a rheological model, and really hinges on the evaluation of a factor which is the ratio of the areas under the flow curve for the non-Newtonian fluid to that one under the flow curve for an equivalent Newtonian medium, and this forces the results for non-Newtonian liquids to collapse on to the curve (or equation) for Newtonian media. While there is a degree of empiricism involved in this scheme, it circumvents the use of a viscosity model. The resulting mean errors are of the order of ~20% in this method.

2.4.3 Flow Field and Flow Regime

In the preceding sections, consideration has been mainly limited to the prediction of drag coefficient and the free-falling velocity. The differences observed between these parameters for a Newtonian and power-law or other types of GNF must also be reflected in the detailed kinematics of flow. Unfortunately, not many investigators have provided details about the flow field. Thus, for instance, Crochet et al. (1984) observed the flow field to decay much faster in power-law shear-thinning fluids than that anticipated from the Stokes stream function for a sphere settling in a quiescent power-law medium at small Reynolds numbers (as seen in Figure 2.7). Similarly, the value of the pressure on the surface of a sphere is believed to progressively decrease with the decreasing value of n for a fixed value of Re_{PL}, albeit the form of the pressure distribution is qualitatively similar to that in Newtonian fluids (Adachi, 1973; Adachi et al., 1973, 1977/78; Tripathi et al., 1994; Tripathi and Chhabra, 1995; Dhole et al., 2006a; Song et al., 2009). Representative results are shown in Figure 2.20 for three values of $n = 0.6$, 1, and 2. At low Reynolds numbers, the power-law index influences the surface pressure over the entire sphere, albeit the role of power-law index flips over at the point of flow separation. As the Reynolds number is gradually increased, the effect of n is seen only in the wake region. Likewise, the shear-thinning behavior seems to increase the value of the surface vorticity above that in a Newtonian fluid up to the stagnation point which itself keeps moving forward with the increasing degree of pseudoplasticity. Another more detailed study (Whitney and Rodin, 2001) shows similar features of the flow pattern. By comparing the constant shear rate contours for $n = 1$ and $n = 0.11$, they reported the flow to be confined to a very thin layer near the surface of the sphere as opposed to several radii away from the sphere for $n = 1$. For power-law fluids, the flow field displays the complete fore and aft symmetry up to $Re_{PL} \approx 1$, thereby suggesting the end of the viscous regime to occur somewhere in this region. A detailed investigation to pinpoint the critical value of Re_{PL}, which is likely to be a function of n, is not yet available and it is therefore suggested that $Re_{PL} \approx 1$ can be taken to denote the end of the creeping flow regime in shear-thinning liquids. Figure 2.21 shows representative streamline plots for $Re_{PL} = 5$ and $Re_{PL} = 200$ for shear-thinning ($n = 0.5$), Newtonian ($n = 1$) and shear-thickening ($n = 2$) fluids, respectively. Broadly speaking, in shear-thinning fluids, not only the wake formation is delayed but the wakes also tend to be shorter than that in Newtonian fluids (Figure 2.22). As expected, shear-thickening fluids display exactly the opposite type of behavior. This counterintuitive result is due to the scaling variables used in this work. Thus, for instance, while (V/d) is probably a good approximation for the shear rate in the vicinity of the sphere, it cannot be justified far away from the sphere. Figure 2.23 shows a typical comparison between various predictions and observations of wake length for a sphere in Newtonian fluids.

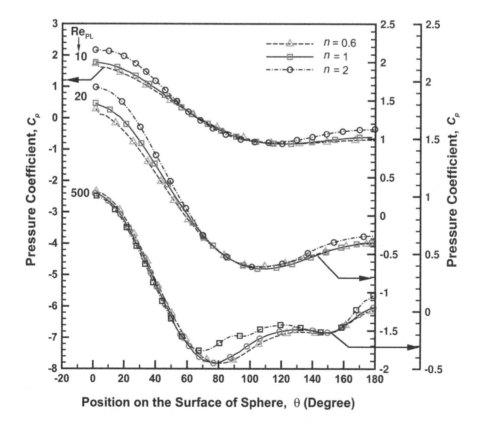

FIGURE 2.20 Surface pressure profiles for a power-law fluid flow past a sphere. (Adapted from Dhole, S. D., Chhabra, R.P. and Eswaran, V., *Ind. Eng. Chem. Res.*, **45**, 4773, 2006.)

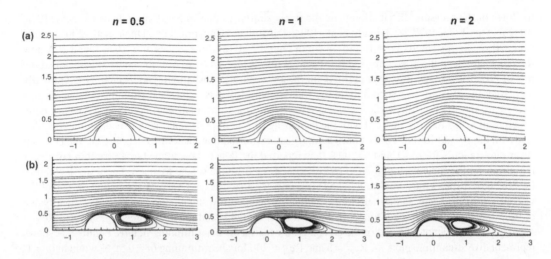

FIGURE 2.21 Streamline patterns for a sphere moving in power-law fluids at (a) $Re_{PL} = 5$ and (b) $Re_{PL} = 200$ (flow is from left to right). (After Dhole, S. D., Chhabra, R.P. and Eswaran, V., *Ind. Eng. Chem. Res.*, **45**, 4773, 2006.)

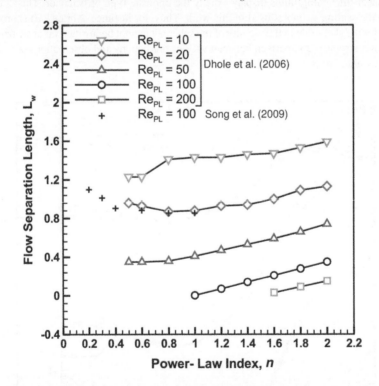

FIGURE 2.22 Dependence of the flow separation length on Reynolds number and power-law index for a sphere. (After Dhole, S. D., Chhabra, R.P. and Eswaran, V., *Ind. Eng. Chem. Res.*, **45**, 4773, 2006 and Song, D., Gupta, R.K. and Chhabra, R.P., *Ind. Eng. Chem. Res.*, **48**, 5845, 2009.)

By analogy with the flow regime transitions seen in Newtonian fluids, it is expected that the flow of power-law fluids over a sphere will also undergo such transitions (Mishra and Chhabra, 2022). For instance, the critical Reynolds number indicating the onset of flow separation shows a strong dependence on the power-law index (Figure 2.24). As postulated above, flow separation occurs at lower values of the Reynolds number in shear-thickening fluids whereas the flow remains attached up to $Re_{PL} \geq 23$, though the peak value of $Re_{PL} = 28$ is seen at $n \sim 0.5$. Similarly, Vasukiran et al. (2018) have studied the next

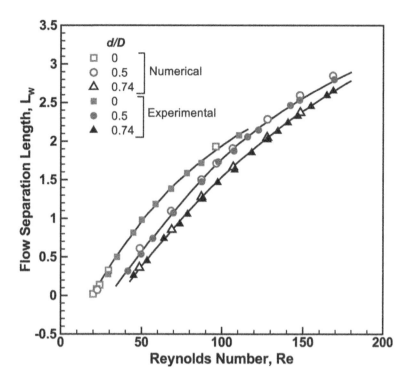

FIGURE 2.23 Comparison between predictions and observations for flow separation length of a sphere in Newtonian fluids. (Oh, J. H. and Lee, S.J., *Korean J. of Chem. Eng.*, **5**, 190, 1988.)

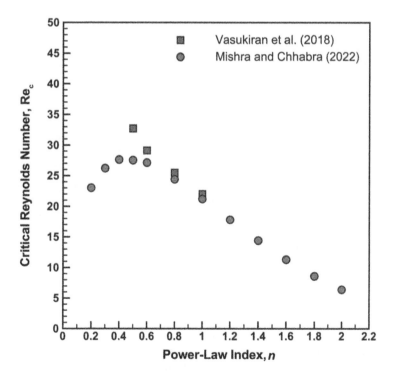

FIGURE 2.24 Critical Reynolds number for a sphere for the onset of flow separation in power-law fluids. (Based on Mishra, G. and Chhabra, R.P., *J. Chem. Eng. Japan*, **55**, 132, 2022.)

transition denoting the loss of axisymmetry of flow. They reported the critical values of the Reynolds number of 193, 163, and 158 for $n = 0.85$, 0.6, and 0.54, respectively, thereby indicating an early loss of the axisymmetric flow region for a sphere.

2.4.4 Unsteady Motion

In contrast to the voluminous literature available on the unsteady motion of a sphere in Newtonian fluids, very little is known about the analogous problem in power-law fluids and in the other types of GNFs. Only in the studies of Bagchi and Chhabra (1991a) and Chhabra et al. (1998), the accelerating motion of a sphere (initially at rest) in power-law fluids has been studied by neglecting the Basset history term. While Bagchi and Chhabra (1991a) approximated the instantaneous drag on the sphere by the corresponding Newtonian value, Chhabra et al. (1998) used the numerical results of Tripathi et al. (1994) (as fitted by Darby, 1996), thereby limiting their results to a maximum value of the Reynolds number of 100. In order to attain the sphere settling velocity within 1% of the ultimate terminal velocity, both dimensionless time elapsed and distance traveled by the particle increase with the increasing Reynolds number as well as with the decreasing value of the flow behavior index. Thus, for instance, a steel sphere ($d = 3.96$ mm, $\rho_p = 8200$ kg/m^3) settling in a polymer solution ($\rho = 1000$ kg/m^3; $n = 0.62$ and $m = 4$ Pa s$^{0.62}$) travels 290 μm and for 17 ms before attaining the velocity which is within 1% of the ultimate terminal velocity. The corresponding values are 43 mm and 162 ms if the same sphere falls in a liquid with $\rho = 1000$ kg/m^3, $n = 0.86$, and $m = 0.16$ Pa s$^{0.863}$. Qualitatively, these trends are similar to that observed in Newtonian fluids under such conditions. In the absence of more rigorous theoretical results, one can use these results for the purpose of designing falling ball viscometers and to measure the terminal falling velocity of particles in shear-thinning media.

2.4.5 Effect of Imposed Fluid Motion

Thus far, consideration has been given to the case of a sphere settling in a quiescent fluid and a uniform flow of fluid imposed on a fixed sphere. Many interesting effects (which have far-reaching implications on the rheology and processing of suspensions and pastes, e.g., see Segre and Silberberg (1963) and Mewis and Wagner (2009, 2012)) can occur when sedimentation occurs in shear and extensional flow fields even in incompressible Newtonian fluids. Perhaps one of the most striking phenomenon involves the radial migration of particles in Poiseuille flow. Early analyses (Jeffrey, 1922) indicate that particles tend to accumulate near the axis of the tube. The experimental study of Segre and Silberberg (1961) shows that neutrally buoyant particles migrated away both from the axis and the wall, attaining an equilibrium position at about 0.6 times the tube radius from the axis, irrespective of their initial positions. On the other hand, Goldsmith and Mason (1962) found that in Poiseuille flow at low Reynolds numbers, the radial position of a single sphere remained constant over prolonged periods of time. The apparent discrepancy between the observations of Segre and Silberberg (1961) and earlier works (Jeffrey, 1922) is thought to be due to the slip between the fluid and the particle. Broadly speaking, migration occurs toward the wall when particles travel faster than the fluid, and the reverse occurs when the fluid moves faster than the particles (Oliver, 1962; Repetti and Leonard, 1964). Thus, for instance, Oliver (1962) observed that neutrally buoyant particles near the wall migrated away from the wall, while those near the axis of the pipe move toward the wall thereby attaining an equilibrium position ranging from 0.5 to 0.7 times the tube radius. Other studies on radial migration under Poiseuille and Couette flow conditions (Karnis et al., 1967) further corroborate the earlier findings of Goldsmith and Mason (1962). Most of the pertinent literature has been critically reviewed, among others, by Leal (1979, 1980), Sastry and Zuritz (1987), and by Lareo et al. (1997). As pointed out by Leal (1980), indeed many other interesting effects including preferred orientation, aggregation of particles, Jeffrey orbits, have been predicted and observed experimentally. However, all such effects are ascribed to either nonspherical particle shape, or to the deformability of particles (bubbles/drops), or to inertial effects or to non-Newtonian effects. Leal (1980), D'Avino et al. (2017), Wang et al. (2018), and Zenit and Feng (2018) have collated and critically evaluated much of the available literature. In most cases, the non-Newtonian effects refer to the viscoelasticity of fluids and this body of information as such is included in Chapter 5.

Similarly, scant experimental results are available on the effect of shear flow (in a Couette system, or in a moving belt parallel plate arrangement, etc.) on the sedimentation velocity of single spheres in inelastic power-law and Ellis model fluids (Novotny, 1977; Hannah and Harrington, 1981; Harrington et al., 1979; Shah, 1982, 1986; Clark et al., 1985; Acharya, 1986, 1988; Roodhart, 1985; McMechan and Shah, 1991; Briscoe et al., 1993, etc.) and in tube flow (Subramaniam and Zuritz, 1990; Subramaniam et al., 1991; Song et al., 2011; Busch and Johansen, 2019). Qualitatively, the sedimentation velocity of a sphere is higher in a fluid subject to shearing than that under quiescent conditions otherwise under identical conditions. This is perhaps due to the shear-thinning behavior of the fluids. However, not only considerable confusion exists in the literature regarding the role of shear rate on settling but also the interpretation of such experiments is further complicated by anomalous effects such as clustering, migration, wall effects, etc. (Barree and Conway, 1995). Furthermore, some evidence exists that the use of an effective or apparent viscosity alone is inadequate to correlate these results (Acharya, 1986, 1988). In spite of all these complexities, some investigators (de Kruijf et al., 1993; Goel et al., 2002) have attempted to link the molecular parameters of fracturing fluids (used in oil industry) to their bulk rheological parameters and their capacity to carry proppant particles. Obviously, these tend to be highly system-specific and extrapolations to other systems are fraught with danger. However, little is available in terms of predictive equations. By way of example, Subramaniam et al. (1991) put forward the following expression for the drag coefficient of a sphere fixed at the axis of a tube with the laminar flow of a power-law fluid:

$$C_D = \frac{35(1-\beta)^{-1.4}}{(\mathrm{Re}_{PL})^{0.7/n}} \tag{2.61}$$

Equation 2.61 is obtained from data embracing rather narrow range of conditions as follows: $0.7 \leq n \leq 0.78$; Re_{PL} (based on the local fluid velocity) $\leq \sim 10$, and for $\beta = 0.21$, 0.47, and 0.67. These conditions are typical of that encountered in aseptic processing of food products (Lareo et al., 1997; Lareo and Fryer, 1998) and the predictions of Equation 2.61 are qualitatively consistent with the numerical predictions of Song et al. (2011).

In many food processing applications, particle distributions, and residence time distribution and holding time of particles in moving fluids in tubes is an important consideration. Consequently, this aspect has received some attention in horizontal, vertical, and inclined flows, for example, see Tucker and Withers (1994), Pordesimo et al. (1994), Sandeep and Zuritz (1995), Lareo et al. (1997), Sandeep et al. (1997), Fairhurst et al. (1999), Fairhurst and Pain (1999), Prokunin et al. (1992), and Barigou et al. (2003). In all these studies, the behavior of nearly neutrally buoyant spherical particles (at moderate to high concentrations) in non-Newtonian polymer solutions, frequently approximated as power-law or Ellis model fluids, has been investigated. Depending upon the flow conditions, particle size, and concentration, different types of passage time distributions are obtained. The salient observations can be summarized as follows: under certain circumstances, an annular region (about one-particle thick) of slow-moving particles close to the tube wall is present, especially for small particles and viscous carrier fluids in vertical flow. For instance, Fairhurst et al. (1999) observed this phenomenon with 5 mm particles in a 0.5% CMC solution, but not with 10 mm particles. It is likely that tube diameter also plays a role in it. A normalized passage time correlates rather well with the concentration of solids for a given fluid. However, the role of liquid rheology is less clear. The passage time distributions can vary from the one with a single peak with a tail, to a bimodal distribution, to one with a wide single peak, and finally, to one with a single narrow peak. These transformations suggest the possibility of radial migration of particles from their initial position under appropriate conditions. Excellent accounts of the developments in this field are available in the literature (Lareo et al., 1997; Lareo and Fryer, 1998; Fairhurst et al., 1999; Eesa and Barigou, 2008, 2009).

2.5 Spheres in Shear-Thickening Liquids

There have been a very few investigations relating to the motion of spheres in dilatant fluids. This is so partly due to the fact that dilatant fluids were thought to be encountered rarely in process applications, albeit this is no longer so (see Chapter 1). Tomita (1959) and Wallick et al. (1962) reported preliminary

TABLE 2.7

Values of C_D for Spheres in Dilatant Fluids (Tripathi and Chhabra, 1995; Dhole et al., 2006a)

$Re_{PL} \to n\downarrow$	0.01	1	10	50	100	200	500
1	2400.19	27.15	4.28	1.55	1.062	0.74	0.486
1.2	1984.55	24.16	4.52	1.78	1.26	0.90	0.584
1.4	1366.58	21.27	4.62	1.97	1.41	1.04	0.685
1.6	936.76	19.43	4.72	2.13	1.57	1.16	0.786
1.8	627.57	17.08	4.85	2.28	1.69	1.28	0.893
2	–	–	4.913	2.415	1.835	1.408	0.996

results on the creeping motion of spheres in power-law fluids with flow behavior index greater than unity. Notwithstanding the fact that the variational principles are not applicable to dilatant fluids, and that Tomita merely calculated the rate of energy dissipation based on an arbitrary flow field, his results must therefore be treated with reserve. Tripathi and Chhabra (1995) numerically solved the governing equations for power-law fluids flowing over a sphere up to $Re_{PL} \leq 100$ and $1 \leq n \leq 1.8$. However, their results (especially at high Reynolds numbers and for large values of the flow behavior index) seem to be rather inaccurate as revealed by subsequent numerical simulations (Dhole et al., 2006a; Tian et al., 2018). At low Reynolds number ($Re_{PL} \leq \sim 0.1$ or so), the flow behavior index exerts a strong influence on the value of the drag correction factor Y whereas at high Reynolds numbers, the role of n progressively diminishes (Table 2.7). The predictions of Tian et al. (2018) are generally within 5% of those of Dhole et al. (2006a) in the range $1 \leq n \leq 1.8$ and $0.01 \leq Re_{PL} \leq 100$. Based on a detailed examination of the flow field and the drag behavior, it is perhaps reasonable to assume that $Re_{PL} \approx 0.1$ marks the end of the creeping flow in these fluids, albeit only slight errors are incurred in using the values up to $Re_{PL} \sim 1$. The resulting values of the drag correction factor Y, defined by Equation 2.37, for $n = 1.2, 1.4, 1.6$, and 1.8, respectively, are 0.83, 0.57, 0.39, and 0.26. This is in stark contrast to the behavior observed in shear-thinning fluids for which $Y > 1$ for all values of $n < 1$. However, as the value of the Reynolds number increases, the drag is strongly influenced by the inertial forces. The ratio (C_{DP}/C_{DF}) deviates a little from its value of 0.5, even for a highly shear-thickening fluid with $n = 1.8$ as the value of the Reynolds number is increased from 0.001 to 100. Table 2.7 summarizes their results on drag coefficient. Qualitatively, the streamline patterns appear to be similar for shear-thickening as that shown in Figure 2.21.

Prakash (1976) measured drag coefficients of a few spheres falling in a shear-thickening starch solution ($n = 1.2$); however, his all data points pertain to the values of Reynolds number greater than 1000 and hence cannot be used to validate the aforementioned numerical predictions. This remains an area for future research activity, especially with the growing trends in the processing of paste-like materials which invariably display dilatant behavior under appropriate conditions of shearing. In a recent study Wu and Mohammadigoushki (2019) have studied the flow of a micellar shear-thickening solutions over a sphere. At low Reynolds numbers the drag was similar to the Stokes value whereas significant drag reduction is observed at Reynolds number value of 10. Also, the wake extended up to 80 times the diameter of the sphere.

2.6 Drag on Light Spheres Rising in Pseudoplastic Media

It has been generally assumed that the downward motion of a heavy sphere ($\rho_p > \rho$) and the upward motion of a light sphere ($\rho_p < \rho$) in a quiescent liquid are hydrodynamically similar. However, in recent years, some experimental results in Newtonian fluids (Karamanev and Nikolov, 1992; Karamanev et al., 1996; Karamanev, 2001) cast doubt on the validity of this assumption, not only in terms of differences in drag coefficient-Reynolds number relationships for falling and rising spheres but also in terms of the trajectory of the particles. For instance, the aforementioned experimental studies clearly show that the drag coefficient of a rising sphere follows the standard drag curve up to about Re = 135 and thereafter, the drag coefficient becomes constant at $C_D = 0.95$. Clearly this limiting value is not only more than twice

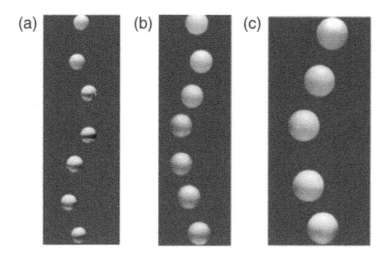

FIGURE 2.25 Trajectories of light spheres rising in shear-thinning liquids (a) $d = 49$ mm, $Re_{PL} = 4800$ (b) $d = 60$ mm, $Re_{PL} = 6240$ (c) $d = 75$ mm, $Re_{PL} = 9010$. (After Dewsbury et al., *AIChE J*, **46**, 46, 2000.)

the value expected in the Newton's region for Newtonian fluids, but this transition also seems to occur at much smaller value of the Reynolds number as opposed to the usual value of Re > 1000 for falling spheres. Furthermore, the trajectory of a rising sphere in Newtonian media (for Re > 135) seems to be a spiral as opposed to a rectilinear one for a falling sphere. The angle between the velocity vector (the tangent to the spiral) and the horizontal plane is always found to be close to 61°. Subsequent studies with gas bubbles and light spheres rising in power-law liquids show qualitatively similar behavior which is obviously at variance from that of falling spheres (Dewsbury et al., 1999, 2000, 2002a). In fact, what is even more puzzling is the fact that even in the creeping flow region ($Re_{PL} < \sim 0.1$), the drag follows the standard Newtonian drag curve thereby suggesting that shear-thinning behavior exerts no influence in determining the drag on a sphere. This intriguing effect awaits a theoretical justification, for the numerical simulations do not distinguish between the two cases of $(\rho_p < \rho)$ and $(\rho_p > \rho)$. Again, in this case, the drag coefficient approaches a constant value of 0.95 at $Re_{PL} \geq \sim 135$, and the trajectory is also seen to be a spiral as shown in Figure 2.25 for spheres rising in CMC solutions. While the exact reasons for these anomalous findings are not immediately obvious, it has been attributed to an imbalance of the nonvertical forces caused by wake shedding that is known to occur at about the same value (ca. 130) of the Reynolds number (Torobin and Gauvin, 1959). Such an imbalance coupled with the low inertia of light spheres leads to a spiral trajectory that, in turn, alters the drag behavior. Qualitatively similar results have also been reported for gas bubbles rising through pseudoplastic fluids and through gases (Dewsbury et al., 1999; Karamanev, 1994, 2001). In a subsequent study, a new drag correlation for light spheres rising in power-law liquids has been developed that applies up to $Re_{PL} = 55,000$ (Dewsbury et al., 2002a). Similarly, Jenny et al. (2003, 2004), Veldhuis et al. (2009), and Horowitz and Williamson (2010) have postulated that a free-sphere falling or rising even in a Newtonian fluid undergoes a regular, symmetry breaking bifurcation making the sphere trajectory deviate from the vertical direction. This is thus consistent with the behavior observed for rising spheres and probably this reasoning is also relevant to the behavior in shear-thinning fluids.

2.7 Pressure Drop Due to a Settling Sphere

For a sphere settling in a Newtonian fluid, Brenner (1962) showed that at low Reynolds number flow past a particle in a tube, the hydrodynamic forces acting on the duct walls are not always negligible as compared to the drag force on the particle. Thus, for instance, in the case of a single particle settling in an incompressible Newtonian liquid in the Oseen regime, the forces on the wall are comparable to the drag experienced by the particle. Based on the consideration of the rate of dissipation of mechanical energy coupled with the particle being viewed as a perturbation to the flow, he derived the following expression for the so-called additional pressure drop (Δp^+) for a Newtonian fluid:

$$\frac{(\Delta p^+)A}{F_D} = \frac{V^0}{<V>} \qquad (2.62)$$

where F_D is the viscous drag on the particle, A is the cross-section area of the duct, V^0 is the velocity of the undisturbed (without particle) flow at the center of mass of the particle in the disturbed (with particle) flow, and (V) is the mean velocity which is maintained constant in both the undisturbed and the disturbed flow conditions. In this result, the right-hand side of Equation 2.62 not only entails only the values relating to the undisturbed flow but also the fluid properties do not appear explicitly. It can be shown in a straightforward manner that the additional force (F_W^+) exerted on the walls by the fluid is given by

$$F_W^+ = \left(\frac{V^0}{<V>} - 1\right) F_D \qquad (2.63)$$

Thus, for a particle settling through a quiescent Newtonian fluid at the center of a cylindrical tube, one can invoke the Poiseuille flow assumption to infer

$$\frac{V^0}{<V>} = 2, \quad F_W^+ = F_D \quad \text{and} \quad \frac{(\Delta p^+)A}{F_D} = 2 \qquad (2.64)$$

Experimental verification of these results has been provided by Pliskin and Brenner (1963), Feldman and Brenner (1968) for spherical particles and by Langins et al. (1971) for cones and composite particles. As anticipated, the particle shape is unimportant, but the wall effects can be significant (Langins et al., 1971). However, the value of $[(\Delta p^+A)/F_D]$ begins to drop from a value of 2 at about the sphere Reynolds number ca. 30 and beyond the sphere Reynolds number >1000, this quantity approaches unity. This analysis has recently been extended to spheres (Ribeiro et al., 1994) and to cylindrical particles (Pereira, 2000) settling in power-law fluids. In the laminar flow conditions, for a particle settling at the center of a tube filled with a power-law fluid, we have

$$\frac{V^0}{<V>} = \frac{3n+1}{n+1} \qquad (2.65)$$

and therefore,

$$\frac{(\Delta p^+)A}{F_D} = \frac{3n+1}{n+1} \qquad (2.66)$$

For a polymer solution ($n = 0.874$), Ribeiro et al. (1994) reported good correspondence between their experiments and the predictions of Equation 2.66 provided the sphere-to-tube diameter ratio is smaller than 0.1, that is, $\beta < 0.1$. However, additional dimensionless groups are required to correlate their results in viscoelastic fluids. Subsequently, the effects of power-law index, Reynolds number, and β on the value of $[(\Delta p^+A)/F_D]$ have been studied numerically in detail, supported by limited experimental results (Pereira, 2000).

2.8 Nonspherical Particles

2.8.1 Introduction

It is readily conceded that one encounters nonspherical particles in process engineering applications much more frequently than spherical particles. Considerable research effort has thus been expended in exploring the hydrodynamics of isolated particles of various shapes in Newtonian liquids. Consequently, a wealth of information is now available on various aspects of particle motion, albeit mainly limited to isometric shapes of particles. Notwithstanding the importance of the detailed kinematics of flow, most of the research activity in this area endeavors to develop simple and reliable predictive expressions for parameters like drag coefficient, terminal falling velocity, wall effects, etc. Therefore, the ensuing discussion is primarily concerned with the prediction of these quantities, albeit reference will frequently be made to the other related aspects also. We begin with the prediction of drag on nonspherical particles in Newtonian fluids, followed by the analogous developments in shear-thinning and shear-thickening fluids.

2.8.2 Drag Force

2.8.2.1 Newtonian Fluids

Ever since the pioneering work of Stokes in 1851, a considerable body of knowledge has accrued on the drag force experienced by a range of shapes of particles settling in quiescent media or held stationary in moving fluids. A spherical particle is unique in that it presents the same projected area to the oncoming fluid irrespective of its orientation. For nonspherical particles, on the other hand, the orientation must be known before their terminal settling velocity or the drag force acting on them can be calculated. Conversely, under appropriate circumstances, nonspherical particles have a propensity to attain a preferred or most stable orientation irrespective of their initial orientation. All these phenomena are strongly influenced not only by the shape of the particle, its size, and density, fluid properties but also by the shape and size of confining boundaries and the imposed fluid motion, etc. In this section, our main concern is the prediction of drag in the case of unconfined flow, whereas the wall effects per se are dealt with in a later chapter (Chapter 10). The vast literature, although not as extensive and rich as that for spherical particles, available on the regular and irregular-shaped nonspherical particles in incompressible Newtonian media has been reviewed, among others, by Happel and Brenner (1965), Clift et al. (1978), Kim and Karrila (1991), Loth (2008a), and Michaelides and Feng (2023) whereas the corresponding aerodynamic literature has been summarized by Hoerner (1965). Hence, only the significant developments are reviewed here.

As mentioned previously, the terminal falling velocity (and hence drag coefficient) of a particle is strongly influenced by its size, shape, and orientation in addition to the viscosity of the medium and the densities of the particle and of the fluid medium. Indeed, the lack of an unambiguous measure of shape, size, and orientation during settling is really the main impediment in developing universally applicable predictive expressions. Theoretical and numerical solutions are possible only for axisymmetric shapes and there has been only limited progress beyond what is available in the treatises of Clift et al. (1978) and of Kim and Karrila (1991). Most progress in this area therefore has been made via experimental results aided by dimensional considerations. Currently available schemes to handle this problem fall into two distinct categories. The first approach endeavors to develop drag expressions for a particle of fixed shape and orientation via numerical solutions of the field equations and experimental results. The works of Bowen and Masliyah (1973), Dwyer and Dandy (1990), and Tripathi et al. (1994) for oblates and prolates, of Huner and Hussey (1977), Ui et al. (1984), Allan and Brown (1986), and Liu et al. (2004) for finite cylinders in axial motion, of Michael (1966), Shail and Norton (1969), Davis (1990), Wang (1996a), Pulley et al. (1996), Field et al. (1997), Munshi et al. (1999), Nitin and Chhabra (2005a,b), Shenoy and Kleinstreuer (2008), Mishra and Chhabra (2022) and Suri and Patel (2023) for thin circular disks (falling broad-face wise) and of Lasso and Weidman (1986) for hollow cylinders, cross-flow past square bars (Dhiman et al., 2005), cubes (Saha, 2004), plates (Saha, 2007), for instance, demonstrate the utility of this approach. Limited results are also available for cones, hemispheres, and sectors of a sphere in Newtonian fluids (Kim and Choi, 2003; Mishra and Chhabra, 2022). Undoubtedly, this class of expressions/correlations work rather well for the selected shapes and orientations, but extrapolations to other shapes and orientations are clearly not possible. In the second approach, on the other hand, efforts are directed at the formulation of a single correlation for all shapes and orientations of nonspherical particles. Obviously, this approach tends to be less accurate than the former, but appears to be more appealing from an engineering applications perspective. This approach thus relies heavily on experimental results and indeed does afford interpolations/extrapolations for the missing shapes. The attempts of Heiss and Coull (1952),

Haider and Levenspiel (1989), Thompson and Clark (1991), Ganser (1993), Chien (1994), Venumadhav and Chhabra (1995), Hartman et al. (1994), Tsakalakis and Stamboltzis (2001), Göğüs et al. (2001), Tang et al. (2004), She et al. (2005), Kelbaliyev (2011), Breakey et al. (2018), Knoll et al. (2019), etc. exemplify this category of methods. Though many descriptions of the shape and size are available for nonspherical particles, for example, see Yow et al. (2005), Taylor (2002), Podczeck and Newton (1994, 1995), Podczeck (1997), Brown et al. (2005), Rajitha et al. (2006), Aschenbrenner (1956), Williams (1965a), Kasper (1982), Clark (1987), Young et al. (1990), Illenberger (1992), Benn and Ballantyne (1992), Le Roux (1996, 2002, 2004), and Lin and Miller (2005), most of the aforementioned studies of drag have used the so-called equal volume sphere diameter (d_s) as the representative particle size and the sphericity (ψ) to quantify the shape of the particle. Thus, most expressions are of the general form

$$f(\text{Re}, C_D, \psi) = 0 \tag{2.67}$$

Note that the functional dependence expressed in Equation 2.67 does not account for the orientation of the freely settling particle.

Numerous expressions of varying forms and complexity of this functional relationship are available in the literature (Clift et al., 1978; Chhabra et al., 1999; She et al., 2005; Yow et al., 2005; Rajitha et al., 2006; Loth, 2008a; Breakey et al., 2018), and their comparative performance in predicting the value of drag coefficient for scores of particle shapes has been evaluated thoroughly using nearly 2000 individual data points relating to a variety of particle shapes (Chhabra et al., 1999). Based on an extensive data base in the range $10^{-4} \leq \text{Re} \leq 10^5$ and $0.1 \leq \psi \leq 1$ thereby covering scores of particle shapes, the following three equations are believed to be most reliable. The first one, due to Ganser (1993), is written as

$$C_D = \left(\frac{24}{\text{Re}_1}\right)\left\{1 + 0.1118(k_2 \text{Re}_1)^{0.657}\right\} + \frac{0.4305 k_2}{1 + (3305/k_2 \text{Re}_1)} \tag{2.68}$$

where the modified Reynolds number, $\text{Re}_1 = k_1 \text{Re}$; k_1 and k_2 are the two shape factors which are, in turn, related to ψ and d_n as follows:

$$(1/k_1) = \left[0.33 + 0.67(\psi)^{-0.5}\right] \tag{2.69a}$$

for isometric shapes, and

$$(1/k_1) = (d_n/3 d_s) + (2/3)\psi^{-0.5} \tag{2.69b}$$

for nonisometric shapes. The shape factor k_2 is given by,

$$\log k_2 = 1.815(-\log \psi)^{0.574} \tag{2.69c}$$

Equations 2.68 and 2.69 reproduce bulk of the literature data with an overall average error of 16%, albeit the maximum error can be as high as 180%.

The second most reliable method is due to Haider and Levenspiel (1989):

$$C_D = \left(\frac{24}{\text{Re}}\right)\left[1 + 8.172 \exp(-4.066\psi) \text{Re}^{0.0964 + 0.557\psi}\right] + \frac{73.69 \text{Re} \ \exp(-5.075\psi)}{\text{Re} + 5.378 \exp(6.212\psi)} \tag{2.70}$$

The overall mean error for this correlation is 21.5% which could rise to a maximum of 275%.

At this juncture, it is appropriate to return to Equation 2.52. The composite parameter χ can really be rewritten as

$$\chi = \frac{4}{\psi}\left(\frac{d_s}{d_n}\right)^2 \tag{2.71}$$

Thus, it combines the three features of a nonspherical particle, that is, its size (d_s), orientation (d_n), and shape (ψ). In principle, therefore, Equation 2.52 can be used to predict drag coefficient of nonspherical particles in Newtonian fluids with $n = 1$. Figure 2.26 shows the influence of χ on drag coefficient over a range of the particle Reynolds number as predicted by Equation 2.52, and Figure 2.27 shows a typical comparison between the predictions of Equation 2.52 and the experimental (and limited numerical) values for a variety of particle shapes including prolates, oblates, needles, cylinders, cones, cubes and rectangles, and prisms encompassing the ranges of the particle sphericity ($0.33 \leq \psi \leq 0.98$) and of the particle Reynolds number as $10^{-6} \leq \text{Re} \leq 400$. Out of nearly 1000 data points, 66% of the data are predicted with the average error of 32.3% that rises to a maximum of 100%. On the other hand, about 86% of the data are predicted with an average error of 58% that rises to a maximum of 100%. Overall, about 12% of the population shows errors larger than 100%. Considering the wide variety of particle shapes included in Figure 2.27, it is probably not too bad a prediction using Equation 2.52. Subsequent data obtained with cubes (Agarwal and Chhabra, 2007) is also in line with the prediction of Equation 2.52.

Finally, a much less accurate but particularly simple expression (and hence convenient) due to Chien (1994) is given as follows

$$C_D = \frac{30}{\text{Re}} + 67.289\exp(-5.03\psi) \tag{2.72}$$

Evidently, Equation 2.72 overpredicts the drag on a sphere in the creeping flow region by 16%. The mean and average errors in using Equation 2.72 are 23% and 152%, respectively. Generally, the smaller the value of ψ, poorer is the prediction of drag. Bearing in mind the simplicity of these expressions, coupled with the fact that no measure of orientation is needed (at least in Equations 2.70 and 2.72), this is about as good an accuracy as can be expected for regular shaped nonspherical particles. Goharzadeh et al. (2012) have reported experimental findings on flow visualization around a hemisphere. Additional complications arise in case of irregular-shaped particles owing to the inherent problems in measuring their size and surface area (to evaluate d_n and ψ).

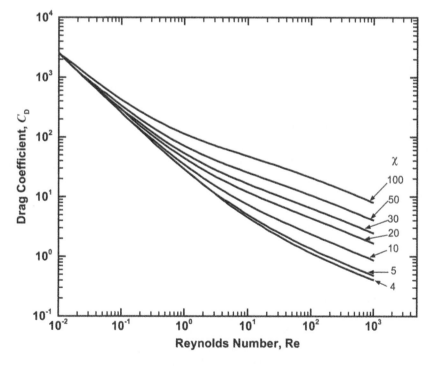

FIGURE 2.26 Drag coefficient - Reynolds number relationship for nonspherical particles in Newtonian media according to Equation 2.52 with $n = 1$. (Modified from Rajitha, P., Chhabra, R.P., Sabiri, N.E., and Comiti, J., *Int. J. Min. Process.*, **78**, 110, 2006.)

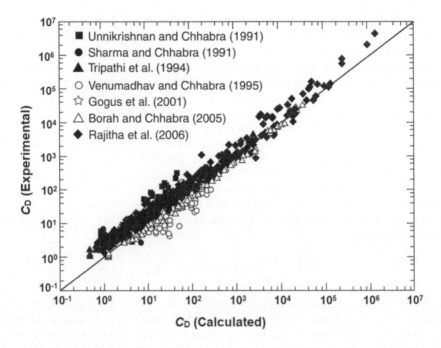

FIGURE 2.27 Comparison between the predictions of Equation 2.52 and experimental data for a range of particle shapes in Newtonian media. (Modified from Rajitha, P., Chhabra, R.P., Sabiri, N.E., and Comiti, J., *Int. J. Min. Process.*, **78**, 110, 2006.)

Aside from the aforementioned studies for regular shapes, many other unusual and interesting shapes including particles with jagged edges (Huilgol et al., 1995), hexagonal flakes (Maul et al., 1994), particle aggregates (Chhabra et al., 1995; Yaremko et al., 1997), dendrite fragments (Zakhem et al., 1992), natural particles (Dietrich, 1982), pebbles (Komar and Reimers, 1978), irregular-shaped particles (Losenno and Easson, 2002; Tran-Cong et al., 2022), Davaadorj et al., 2013), chains of spheres (Kasper et al., 1985; Chhabra et al., 1995; Verma et al., 2022), etc., have also been studied due to their wide occurrence in photographic, material processing, and geological engineering applications.

2.8.2.2 Shear-Thinning Liquids

In contrast to the voluminous literature for drag on nonspherical particles in Newtonian fluids, our understanding of the effects of shear-thinning and shear-thickening viscosity on the hydrodynamic drag of nonspherical particles is still in its embryonic stage (Chhabra, 1996a). Table 2.8 provides a concise summary of the research activity in this field.

An examination of this table clearly shows the paucity of theoretical investigations, even for simple axisymmetric shapes. Thus, for instance, Tripathi et al. (1994) and Tripathi and Chhabra (1995) have numerically solved the governing equations of motion for the flow of power-law fluids ($0.4 \leq n \leq 1.8$) over prolates and oblates covering Reynolds numbers up to 100 and aspect ratio $0.2 \leq E \leq 5$, albeit their results for dilatant fluids seem to be less reliable than that for shear-thinning fluids (Dhole et al., 2006a). The dependence of drag coefficient on the flow behavior index is qualitatively similar to that for spherical particles. In the low Reynolds number regime ($Re_{PL} < 1$) the drag is higher in pseudoplastic liquids ($n < 1$) and lower in shear-thickening ($n > 1$) fluids as compared to that in an equivalent Newtonian fluid. The limiting value of the Reynolds number for viscous flow is somewhat dependent on the values of the aspect ratio, E and the power-law index, n. For instance, it seems to lie somewhere in the range of $Re_{PL} \sim 2$–5 for shear-thinning fluids and $Re_{PL} \sim 0.1$–0.5 for shear-thickening fluids. The overall trends can be summarized as follows:

TABLE 2.8

Nonspherical Particles in Newtonian, Shear-Thinning and Shear-Thickening Liquids

Particle Shape	Fluid Model	Reference
Cylinders (cross-flow & axial)	Power-law	Unnikrishnan and Chhabra (1990, 1991); Chhabra, 1992; Chhabra et al. (2001a); Tanner (1993); Rodrigue et al. (1994); D'Alessio and Pascal (1996); Whitney and Rodin (2001); Chhabra et al. (2004); Soares et al. (2005a); Xu et al. (2019b)
Thin rods and wires	Power-law	Manero et al. (1987); Chiba et al. (1986); Cho et al. (1991); Venumadhav and Chhabra (1994, 1995); Rajitha et al. (2006)
Thin discs, plates, Chips, etc.	Power-law	Reynolds and Jones (1989); Maul et al. (1994); Chhabra et al. (1996c); Rami et al. (2000); Nitin and Chhabra (2006); Saha (2007); Shenoy and Kleinstreuer (2008); Zhong et al. (2011); Zhong and Lee (2012); Chrust et al. (2013); Bobinski et al. (2014); Wang et al. (2016); Lau et al. (2019)
Prisms, rectangles, cubes	Power-law	Rodrigue et al. (1994); Venumadhav and Chhabra (1994); Agarwal and Chhabra (2007); Seyed-Ahmadi and Wachs (2019); Xu et al. (2019b)
Ellipsoids and discs	Power-law	Peden and Luo (1987); Taamneh (2011); Xu et al. (2019b)
Oblates and prolates	Power-law	Tripathi et al. (1994); Tripathi and Chhabra (1995); Verrelli (2014); Wang et al. (2009)
Porous sphere	Power-law	Kawase and Ulbrecht (1981f)
Cones	Power-law	Sharma and Chhabra (1991); Borah and Chhabra (2005); Mishra et al. (2019a,b); Yaginuma and Ito (2008); Yu et al. (2014)
Irregular-shaped particles	Power-law	Clark and Guler (1983); Kirkby and Rockefeller (1985); Torrest (1983); Subramanayam and Chhabra (1990)
Chains of spheres & conglomerates of spheres	Power-law	Jefri et al. (1985); Chhabra et al. (1995)

- In the creeping flow regime, the drag of prolate-shaped particles is less sensitive to the value of n (<1) than that of an oblate; the reverse is, however, true in shear-thickening fluids.
- The role of the power-law index progressively diminishes as the value of the Reynolds number is gradually increased, akin to that in the case of a sphere. For oblates in shear-thinning fluids, and for a fixed value of the aspect ratio, the ratio (C_{DF}/C_{DP}) decreases both with the decreasing value of the power-law index and the increasing value of the Reynolds number (Figure 2.28). The opposite type of behavior is observed for prolates. In contrast, the flow behavior index, n, exerts a little influence in shear-thickening fluids.
- The surface pressure profiles are qualitatively similar to that observed in Newtonian fluids, except that the shear-thinning behavior lowers the value of the surface pressure whereas the shear-thickening augments the corresponding value. Similarly, the streamline patterns are also more sensitive to the aspect ratio than the flow behavior index.

The only other analytical study on the creeping flow over spheroidal particles is that of Sigli (1971) who derived an approximate stream function for the Reiner–Rivlin fluid model and predicted very little departure from the Newtonian fluid kinematics. This finding is qualitatively consistent with the numerical results of Tripathi et al. (1994) and the experimental observations of Maalouf and Sigli (1984). Subsequently, Hsu et al. (2005a, 2006) have numerically studied the steady translation of spheroidal and cylindrical particles in Carreau model fluids, with and without significant wall effects, up to the particle Reynolds number of 40 and $\Lambda \le 1$.

Likewise, as far as known to us, in contrast to the vast literature for Newtonian fluids, there have been a few two-dimensional numerical studies relating to the cross-flow of power-law fluids over a long circular cylinder. At the outset, it is worthwhile to note here that the so-called Stokes paradox is not relevant for the creeping flow of power-law fluids past a cylinder for shear-thinning fluids ($n < 1$) (Tanner, 1993; Morusic-Paloka, 2001). For the creeping flow, Tanner (1993) thus presented approximate numerical

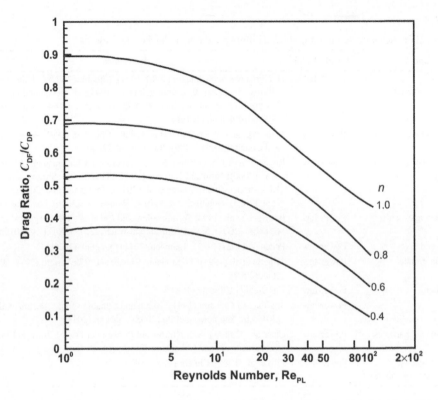

FIGURE 2.28 Dependence of (C_{DF}/C_{DP}) on Reynolds number and power-law index for $E = 0.5$. (From Chhabra, R.P., *Handbook of Applied Polymer Processing Technology*, N.P. Cheremisinoff and P.N. Cheremisinoff, (Eds.), Chapter 1, Marcel Dekker, New York, 1996a.)

predictions of drag as a function of the power-law index. It is convenient to introduce a dimensionless drag force F_D^* defined as follows:

$$F_D^* = \frac{2F_D}{md(2V/d)^n} \qquad (2.73)$$

where F_D is the drag force per unit length of cylinder. Equation 2.73 can readily be rearranged in terms of the usual drag coefficient-Reynolds number (based on diameter, d) relationship as

$$C_D = \frac{2^n F_D^*}{Re_{PL}} \qquad (2.74)$$

The numerical results of Tanner (1993) are summarized in Table 2.9.

He asserted that the accuracy of these results improved with the decreasing value of the flow behavior index, n. Qualitatively F_D^* displays similar dependence on n as that seen for a sphere, that is, F_D^* attains a maximum value at $n \sim 0.4$. Wall effects are also believed to be less severe in shear-thinning fluids than in a Newtonian fluid which is in line with the observations for spheres (Chhabra and Uhlherr, 1980c).

Subsequently, Whitney and Rodin (2001) have analyzed the low Reynolds number cross-flow of power-law fluids past circular cylinders of finite and infinite length-to-diameter ratio. While their results embrace a much wider range of values of the flow behavior index ($0.092 \leq n \leq 0.9$), the results of Tanner (1993) appear to be more reliable. D'Alessio and Pascal (1996) reported the numerical values of drag coefficient, angle of separation and the wake length for the flow of power-law fluids over a long cylinder for $Re_{PL} = 5$, 20 and 40 and for $0.65 \leq n \leq 1.2$ thereby covering moderately shear-thinning and mildly shear-thickening conditions. However, their results appear to be in error due to the inadvertent omission

TABLE 2.9

Values of F_D^* in Equation 2.74

n	F_D^*	
	Approximate	Numerical
0.1	3.31	–
0.2	7.85	–
0.3	11.36	–
0.4	12.67	15.48
0.5	11.92	12.15
0.6	9.80	9.10
0.7	7.05	6.70
0.8	4.22	4.88
0.9	1.76	3.55

TABLE 2.10

Values of Drag Coefficient for a Circular Cylinder in Cross-Flow of Power-Law Fluids

n	Reynolds number, Re_{PL}						
	1	20	40	50	100	120	140
0.2	16.95	1.38	0.63	–	–	–	–
0.4	20.80	1.82	1.10	1.12	1.135	1.181	1.23
0.6	16.93	1.98	1.39	1.23	1.18	1.20	1.22
0.8	12.86	2.00	1.48	–	–	–	–
1.0	10.43	1.99	1.48	1.435	1.340	1.331	1.328
1.2	9.05	2.03	1.53	–	–	–	–
1.4	8.12	2.09	1.62	1.61	1.5	1.46	1.46

of a factor in one of their equations (D'Alessio, 2003; D'Alessio and Finlay, 2005). These results have been subsequently corrected by Chhabra et al. (2004) and Soares et al. (2005a). Table 2.10 summarizes the numerical values of drag coefficient for the cross-flow of power-law fluids past an unconfined circular cylinder in the steady and vortex shedding (time-averaged values) regime for a cylinder (Patnana et al., 2009). Vortex shedding characteristics and pressure profiles for the flow of inelastic and visco-elastic polymer solutions have been studied by Coelho and Pinho (2003).

Likewise scant results are also available for the two-dimensional flow of power-law fluids over a square cylinder, with and without the wall effects (Gupta et al., 2003; Paliwal et al., 2003; Dhiman et al., 2006, 2008a; Sahu et al., 2009a, 2010a; Rao et al., 2011b, c) and for a long bar of rectangular cross section (Nitin and Chhabra, 2005b) cylinders of elliptical cross section (Sivakumar et al., 2007; Koteswara Rao et al., 2010; Patel and Chhabra, 2013), etc. Analogous results for the steady flow of a Carreau model fluid over a circular cylinder are also available in the literature (Pantokratoras, 2016). For a square cylinder, in the range $5 \leq Re_{PL} \leq 40$ and $0.5 \leq n \leq 1.4$, the drag correction factor Y hovers around unity, being slightly above unity in shear-thinning fluids and slightly below unity in shear-thickening fluids (see Figure 2.29 for a square obstacle). Figure 2.29 clearly reveals that the drag coefficient normalized with respect to the corresponding value in an equivalent Newtonian fluid is governed by an intricate interplay between the power-law index and the Reynolds number. However, in the range of conditions shown here, the drag is seen to be higher than the corresponding Newtonian value by up to ~15% for $n < 1$ and it can drop below the Newtonian value by up to about ~10% for $n > 1$ under appropriate circumstances. Once again, the effect of power-law rheology is seen to diminish as the value of the Reynolds number is progressively increased. Figure 2.30 shows the variation of the wake length with n and Re_{PL} in the steady flow regime for a circular cylinder where the wakes are seen to be shorter in shear-thinning fluids and longer in shear-thickening fluids than that in Newtonian fluids.

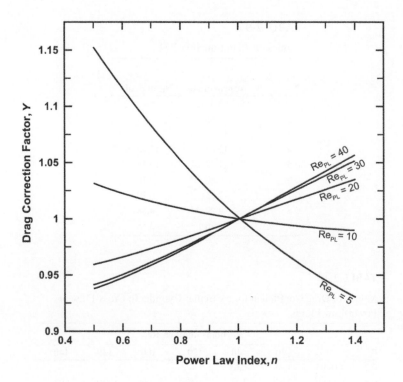

FIGURE 2.29 Drag correction factor for a long square obstacle. (Modified from Gupta, A.K., Sharma, A., Chhabra, R.P., and Eswaran, V., *Ind. Eng. Chem. Res.*, **42**, 5674, 2003.)

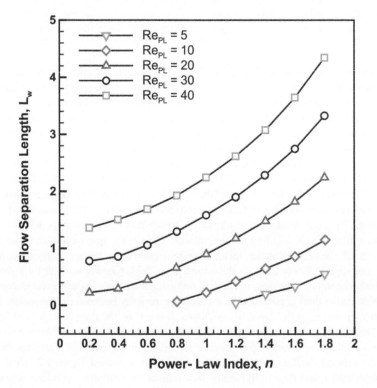

FIGURE 2.30 Dependence of recirculation (L_w) on Reynolds number (Re_{PL}) and power-law index (n) for a circular cylinder. (Based on Sivakumar et al., *Chem. Eng. Sci.*, **60**, 6035, 2006.)

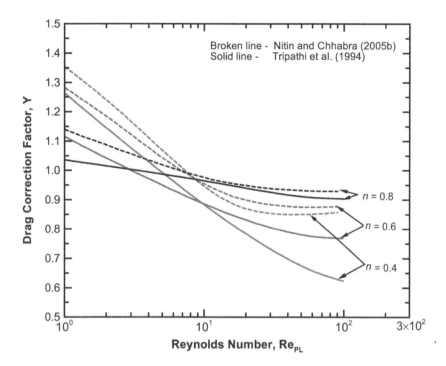

FIGURE 2.31 Drag correction factor for a disk (broken lines) and for a sphere (solid lines) in power-law fluids. (Modified from Nitin, S., and Chhabra, R.P., *J. Colloid Interface Sci.*, **295**, 520, 2006.)

Limited numerical results for the steady two-dimensional flow of power-law fluids past a circular disk (oriented) normal to the direction of flow are available in the range $1 \geq n \geq 0.4$ and $1 \leq Re_{PL} \leq 100$ (Nitin and Chhabra, 2006). The drag values normalized with respect to the corresponding Newtonian values are shown in Figure 2.31. The results do not seem to be influenced by the value of the power-law index beyond $Re_{PL} \sim 20$. At low Reynolds numbers, the drag is seen to increase above the Newtonian value in shear-thinning fluids, but it drops rather rapidly below its Newtonian value as the value of the Reynolds number is increased gradually. Included in this figure are also the results for a sphere (Tripathi et al., 1994). The effect of power-law rheology is seen to be much stronger in the case of a sphere than that for a circular disk. At low Reynolds numbers, the values of the drag correction factor for a sphere and a disk are very similar, but at high Reynolds numbers a sphere experiences much lower drag than a disk otherwise under identical conditions.

Finally, before leaving this section, it is appropriate to mention here that the drag coefficient (based on the projected area) for disks in power-law fluids is adequately described by the following correlation initially developed for Newtonian fluids (Clift et al., 1978):

$$C_D = \left(\frac{64}{\pi Re_{PL}}\right)(1 + 10^x) \quad \text{for } Re_{PL} < 1.5 \tag{2.75a}$$

where

$$x = -0.883 + 0.906 \log(Re_{PL}) - 0.025 (\log Re_{PL})^2$$

$$C_D = \left(\frac{64}{\pi Re_{PL}}\right)(1 + 0.138 Re_{PL}^{0.792}) \quad \text{for } 1.5 < Re_{PL} < 133 \tag{2.75b}$$

Here, the Reynolds number, Re_{PL}, is based on the diameter of the disk. Thus, Equation 2.75 works well for thin circular disks and square plates sedimenting in both Newtonian and power-law liquids. Figure 2.32 shows a comparison between numerical and experimental wake lengths for a disk in Newtonian fluids.

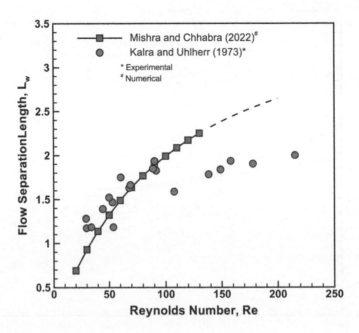

FIGURE 2.32 Comparison between the predicted and experimental flow separation length for a circular disk in Newtonian fluids (Modified after Mishra, G and Chhabra, R.P., *J. Chem. Eng. Japan*, 55, 132, 2022.)

TABLE 2.11

Critical Reynolds Number (Re_{c1}) for the Onset of Wake Formation in Power-Law Fluids

n	Square (Rao et al., 2011c)		Circular (Sivakumar et al., 2006)	
	No Wake	Wake	No Wake	Wake
0.2	1.2	1.3	–	–
0.4	3.2	3.3	10.5	11
0.6	3.1	3.2	9.5	10
0.8	2.2	2.3	7.5	8
1	1.2	1.3	6	6.5
1.2	0.55	0.6	3.5	4
1.4	0.15	0.2	2.5	3
1.6	–	–	1.5	2
1.8	–	–	0.5	1

The agreement is seen to be good up to about Re ≈ 100. Suffice it to add here that the flow ceases to be steady at about Re ~ 110–120 and perhaps the meaning of the wake length is less clear in the vortex shedding regime.

The preceding numerical results for the two-dimensional cylinders of various cross-sections are limited to the steady flow regime in which the flow either remains attached to the surface of the cylinder and/or the wake remains attached. Scant results are available in the laminar vortex shedding regime in power-law fluids for a circular cylinder (Patnana et al., 2009; Bailoor et al., 2019), for a square bar (Rao et al., 2011a). The vortex shedding tends to be somewhat sluggish in these fluids due to viscosity stratification in the flow field.

By analogy with the variety of flow regimes encountered with the flow of Newtonian fluids past a circular cylinder, Sivakumar et al. (2006) delineated the values of the critical Reynolds number for the onset of flow detachment, Re_{c1} (Table 2.11) and for the loss of steadiness of the flow (Re_{c2}) for a circular

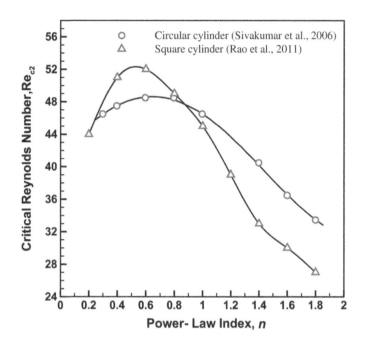

FIGURE 2.33 Critical Reynolds number for the loss of the steady power-law fluid flow over a circular cylinder. (Based on P. Sivakumar, R.P. Bharti. R.P. Chhabra, *Chem. Eng. Sci.*, **60**, 6035, 2006 and P.K. Rao, A.K. Sahu, R. P. Chhabra, *Int. J. Heat Mass Transfer*, **54**, 390, 2011.)

cylinder, Figure 2.33. Evidently, the flow remains attached almost up to Re ~ 10–11 in highly shear-thinning fluids whereas it detaches itself at as low a value of the Reynolds number as Re ~ 0.5 in highly shear-thickening fluids (Table 2.11). Included in this table are all the corresponding values for a square bar (Rao et al., 2011c). While qualitatively similar trends with respect to the power-law index are seen in both cases, the flow separates at much lower Reynolds number for a square cylinder. Also, these two geometries differ in another manner in this regard. For a square cylinder, the flow detachment initiates at the corners of the square cylinder whereas this point moves forward on the surface of a circular cylinder with the increasing Reynolds number. Furthermore, for a square cylinder, the first signature of the flow separation is seen in the rear of the bluff body and beyond a critical Reynolds number, the fluid also detaches itself along the top and bottom surfaces. The results shown in Table 2.11 correspond to the primary separation in the rear of the square cylinder. Now turning to Figure 2.33 for the onset of the laminar vortex shedding regime for a square and circular cylinder, the dependence of the critical Reynolds number on the power-law index is similar in the two cases and this transition is seen to be advanced to lower values of the Reynolds number in highly shear-thickening fluids. On the other hand, the critical Reynolds number is seen to exhibit a peak at about $n \approx 0.5 - 0.6$ in both cases. Subsequently, Lashgari et al. (2012) have studied the onset of vortex shedding from a circular cylinder in Carreau model fluids. The corresponding experimental results have been reported in a series of papers by Coelho and Pinho (2003). Figures 2.34 and 2.35 show typical comparisons between the experiments and predictions of the wake length and surface pressure for a cylinder in Newtonian fluids; the good correspondence seen here inspires confidence in the reliability and precision of numerical simulations in this field. Limited numerical results for cones and hemispheres for their steady translation in power-law fluids have also been reported in recent years (Sasmal et al., 2013; Mishra et al., 2019a). Some interesting irregular shapes like a hemisphere, triangular, trapezoidal have been investigated by Yang and Mao (2003, 2005) and they also reported extensive numerical results on streamline patterns.

Similarly, many investigators (see Table 2.8) have developed empirical correlations for the prediction of drag on freely falling nonspherical particles. Indeed, a cursory inspection shows that a diverse variety of shapes including cylinders (cross- and axial orientation), needles, cones, cubes and rectangles,

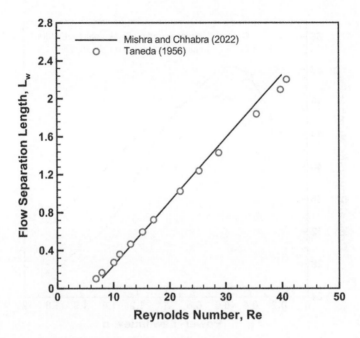

FIGURE 2.34 Comparison between predicted and experimental flow separation lengths for a cylinder in Newtonian fluids. (Based on Mishra, G and Chhabra, R.P., *J. Chem. Eng. Japan*, **55**, 132, 2022.)

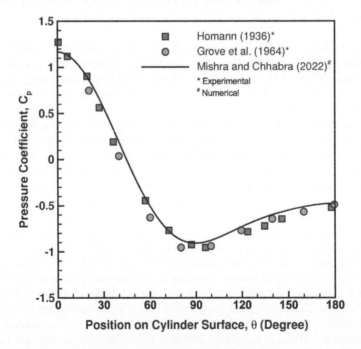

FIGURE 2.35 Comparison between predicted and experimental surface pressure for a cylinder in Newtonian fluids at Re = 40. (Based on Mishra, G and Chhabra, R.P., *J. Chem. Eng. Japan*, **55**, 132, 2022.)

chains of spheres, thin plates, and disks and irregularly shaped particles have been employed. In most cases, the correlations so developed relate to only one or two specific shapes and none of these have been tested using independent experimental data. Hence, many of these are too tentative and restrictive to be included here. In a comprehensive study, Chhabra et al. (2001a) measured the terminal falling velocity of scores of cylinders (length-to-diameter ratio ≤10) in viscoelastic shear-thinning polymer solutions in the creeping flow regime. They postulated that it was possible to isolate the effects of shape and of non-Newtonian fluid behavior. Thus, in the creeping flow conditions, they related the velocity ratio, K (velocity of a non-spherical particle divided by that of an equal volume sphere) in the following fashion:

$$\frac{K_{NN}}{K_N} = 1 + 0.317\{(1-n)\Lambda\}^{0.69} \tag{2.76}$$

where for a cylinder in axial motion,

$$\log K_N = \left\{ -\frac{0.27(\varphi-1)}{\psi^{0.5}\varphi^{0.345}} + \log(\varphi\psi^{0.5}) \right\} \tag{2.77}$$

and for a cylinder in cross-flow orientation

$$\log K_N = -0.25(\psi\varphi)^{0.5}(\varphi-1) + \log(\varphi\psi^{0.5}) \tag{2.78}$$

with, $\varphi = d_s/d_n$

Equations 2.77 and 2.78 are due to the pioneering study of Heiss and Coull (1952). Equation 2.76 correlates 248 data points embracing the ranges of kinematics conditions as: $0.31 \le \Lambda \le 201$; $0.69 \le \psi \le 0.87$ and $0.6 \le n \le 1$. The overall correlation is shown in Figure 2.36 where a satisfactorily fit is evident.

Finally, in view of the fact that Equation 2.52 successfully correlates the extensive literature data on spheres falling in power-law fluids (Figures 2.18 and 2.19) and on nonspherical particles in Newtonian liquids (Figure 2.27), it is worthwhile to examine its applicability to the settling of nonspherical particles in inelastic power-law liquids. However, prior to embarking upon the comparison with the pertinent experimental results, it is instructive to delineate the interplay between the flow behavior index and the composite shape factor χ. Figure 2.37 shows the predictions of Equation 2.52 for a range of values of the power-law index and the shape factor χ. As the value of the power-law index is decreased, the drag coefficient also reduces below its Newtonian value under otherwise identical conditions of the Reynolds number and the shape factor. Figure 2.38 shows a comparison between the experimental results and the predictions of Equation 2.52 for a variety of particle shapes and over wide ranges of conditions of the particle Reynolds number and particle sphericity, as summarized in Table 2.12. Once again, the correspondence seen in Figure 2.38 is about as good as can be expected in this type of work. The mean

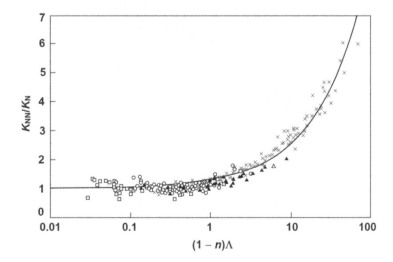

FIGURE 2.36 Overall drag correlation for cylinders showing the fit of Equations 2.76. (Data from Chhabra, R.P., Rami, K., Uhlherr, P.H.T., *Chem. Eng. Sci.*, 56, 2221, 2001.)

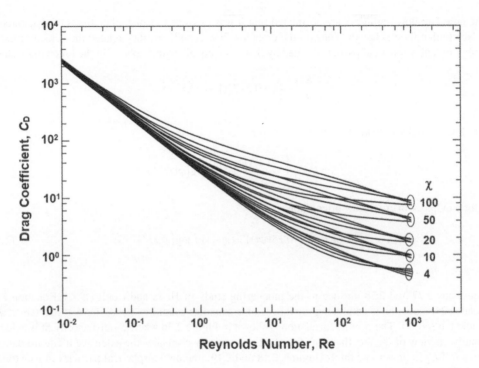

FIGURE 2.37 Predictions of Equation 2.52 for nonspherical particles in power-law fluids. For each value of χ, the four lines relate from top to $n = 1$, 0.8, 0.6, and 0.4, respectively.

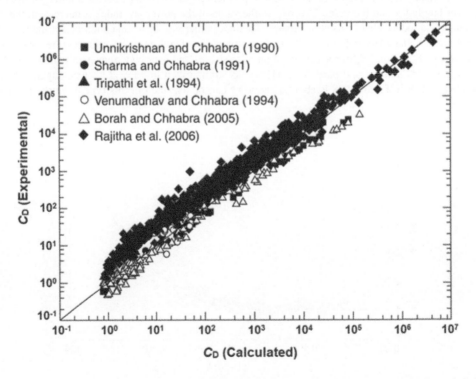

FIGURE 2.38 Comparison between the predictions of Equation 2.52 and experimental results for nonspherical particles falling in power-law fluids for the conditions summarized in Table 2.12.

TABLE 2.12

Sources of Experimental Data Shown in Figure 2.38

Source	Particle shape	ψ	n	Re_{PL}
Unnikrishnan and Chhabra (1990)	Vertically falling cylinders	0.6–0.95	0.48–0.6	0.01–1.7
Sharma and Chhabra (1991)	Cones	0.64–0.80	0.3–0.85	0.01–36
Venumadhav and Chhabra (1994)	Cylinders, prisms, rectangles, cubes	0.33–0.98	0.77–0.96	0.1–140
Tripathi et al. (1994) (numerical study)	Oblates and prolates	0.62–0.93	0.4–1	0.01–100
Borah and Chhabra (2005)	Cones	0.66–0.94	0.4–0.72	10–113
Rajitha et al. (2006)	Cylinders (Both orientations)	0.63–0.87	0.31–0.86	10^{-5}–270

Source: Modified after Rajitha et al. (2006).

deviation is of the order of ~30% which rises to a maximum of about ~80%, both of which are slightly greater than that seen in Figure 2.27 for Newtonian fluids.

In summary, it is thus possible to estimate the drag on spherical and nonspherical particles in both Newtonian and power-law fluids with a reasonable level of reliability using the single expression given by Equation 2.52. This necessitates a knowledge of d_s, d_n, and ψ in addition to the physical and rheological properties of the particle and the liquid medium. Scant results available for irregular-shaped particles also seem not to be inconsistent with these predictions (Clark and Guler, 1983; Kirkby and Rockefeller, 1985). However, if the orientation of the particle is not known, one can use either Equation 2.70 or 2.72 for nonspherical particles falling in Newtonian liquids. No such simple expression, however, is available for power-law fluids.

Furthermore, an examination of the overall data suggests that for a given particle shape and orientation, the role of non-Newtonian properties gradually diminishes as the inertial effects become more important outside the creeping flow conditions. This has been shown convincingly in the case of thin disks and plates (Chhabra et al., 1996c; Rami et al., 2000; Nitin and Chhabra, 2006). Other pertinent studies dealing with the hydrodynamical behavior of nonspherical particles in non-Newtonian fluids have been reviewed elsewhere (Chhabra, 1996a; Machac et al., 2002). The vortex shedding characteristics and the transition from one flow regime to another for the flow of aqueous polymer solutions past an unconfined circular cylinder at high Reynolds numbers ($50 \leq Re \leq 9000$) have been recently studied experimentally (Coelho and Pinho, 2003).

Suffice it to add here that experimental results for cubes (Agarwal and Chhabra, 2007) and numerical predictions for cones and hemisphere (Sasmal et al., 2013; Mishra et al., 2019a) are also well represented by Equation 2.52. The aforementioned treatment is primarily restricted to the regular shaped particles, and it is thus possible to evaluate their size, shape and orientation in terms of d_s, ψ and d_n. On the other hand, it is not at all simple to evaluate these parameters for irregularly shaped particles and therefore some other measures (such as fractal dimensions) are required to characterize their settling behavior (Brown et al., 2005; Tang et al., 2004). Furthermore, the foregoing treatment for nonspherical particles also excludes the variety of time-dependent phenomena such as fluttering, tumbling, etc. observed during the free-fall of particles in Newtonian and non-Newtonian media (Craze, 1977; Field et al., 1997; Mittal et al., 2004; Yaginuma and Ito, 2008; Yu et al., 2014; Chrust et al., 2013; Wang et al., 2016) and in the rolling motion (Prokunin et al., 1992).

2.9 Conclusions

Evidently, a considerable body of information on the drag of rigid spherical particles falling freely in incompressible Newtonian media is now available. Consequently, satisfactory methods for estimating the values of drag coefficient and terminal falling velocity under most conditions of interest have evolved over the years. During the past 30–40 years, results have also accrued on the detailed structure (velocity profiles and wake size and structure, etc.) of the flow field around a sphere. Likewise, the motion of spheres in time-independent fluids (without a yield stress) has been studied extensively, albeit the majority of efforts have been directed at elucidating the behavior in the creeping flow region and for shear-thinning fluids only. Combined together, accurate and reliable numerical values of drag on spheres in power-law fluids are now available up to Reynolds numbers of 200 or so. Fortunately, this gap in the existing literature is not very serious, as non-Newtonian effects manifest much more prominently in the low Reynolds number region than that at high Reynolds numbers and, in fact, one can use the standard Newtonian drag curve between $\sim 1 \leq Re_{PL} \leq 1000$ with an accuracy of ±30%, albeit the maximum errors of up to about ~100% can also occur. The available literature on the estimation of drag on nonspherical-regular and -irregular particles in time-independent fluids (without a yield stress) is less extensive and less coherent, as compared to the analogous developments in Newtonian fluids. Suffice it to add here that the current status of the prediction of drag on particles in time-independent fluids is nowhere near as mature and healthy as that in Newtonian fluids. The new correlation, Equation 2.52, reconciles most of the literature data (experimental and numerical) on spherical and nonspherical particles falling in Newtonian and power-law liquids. Notwithstanding the aforementioned limitations, appropriate theoretical/empirical predictive expressions for drag coefficient for spheres and nonspherical particles moving in a variety of GNFs have been presented. Every effort has been made to present the expressions which have been tested adequately for their reliability. However, extrapolation beyond the range of their validity must be treated with reserve. In recent years, there has been a spurt in numerical activity in this field. Consequently, limited results on the flow transitions, wake properties in power-law fluids for a sphere and a cylinder are now available.

Nomenclature

A	Cross-sectional area (m²)
A_0, B_0	Constants, Equation 2.52
Ar	Archimedes number, Equations 2.28 and 2.59 (–)
b_o	Constant, Equation 2.53a (–)
C_D	Drag coefficient (–)
C_{DF}	Friction drag coefficient (–)
C_{DN}	Drag coefficient in Newtonian media (–)
C_{D0}	Drag coefficient in the low Reynolds number region (–)
C_{DP}	Pressure drag coefficient (–)
$C_{D\infty}$	Drag coefficient in Newton's regime (–)
D	Sphere (or cylinder) diameter (m)
d_n	Diameter of a circle with area equal to the projected area of a particle (m)
d_s	Equal volume sphere diameter (m)
D	Container or fall tube diameter (m)
El	Ellis number, Equation 2.42a (–)
F_D	Drag force (N)
F_D^*	Dimensionless drag force, Equation 2.73 (–)
F_W^+	Additional drag on duct walls, Equation 2.63 (N)
g	Acceleration due to gravity (ms⁻²)
I_1, I_2, I_3	Three invariants of the rate of deformation tensor (s⁻²)
k_1, k_2	Shape factors, Equation 2.69 (–)
k_D	Parameter, Equation 2.29 (–)
K_N	Velocity ratio in a Newtonian fluid, Equation 2.76 (–)
K_{NN}	Velocity ratio in a non-Newtonian fluid, (Equation 2.76) (–)
K	Constant in Equation 2.48 (–)
L_w	Recirculation (or wake) length, (–)
m	Consistency index in power-law model (Pa.s^n)
n	Power-law index (–)

P	Pressure (Pa)
Δp^+	Pressure drop induced by a settling particle, Equation 2.62 (Pa)
R	Radius of sphere or cylinder (m)
Re	Reynolds number for a Newtonian fluid (–)
Re_1	Modified Reynolds number, Equation 2.68 (–)
Re_0	Reynolds number based on zero-shear viscosity (–)
Re_{PL}	Reynolds number for power-law fluids, Equation 2.39 (–)
Re^*	Modified Reynolds number, Equation 2.56 (–)
r	Spherical coordinate (m)
Su	Sutterby number, Equation 2.46 (–)
V	Free-settling velocity in infinite medium (m/s)
V	Velocity vector (m/s)
V_r, V_θ	$r-, \theta$ - components of V
V^0	Local velocity, Equation 2.62 (m/s)
$<V>$	Area averaged velocity, Equation 2.62 (m/s)
X_c, Y_c	Position of the eye of the vortex, (–)
Y	Drag correction factor, Equation 2.37 (–)

Greek Symbols

α	Ellis model parameter (–)
β	Sphere-to-tube diameter ratio (–)
$\dot{\gamma}$	Shear rate (s^{-1})
ε_{ij}	Components of the rate of deformation tensor (s^{-1})
ξ	Dimensionless radial coordinate, Table 2.3, ($= r/R$) (–)
θ	Spherical coordinate (–)
θ_E	Sutterby model parameter (s)
θ_s	Angle of flow separation (measured from the front stagnation point), (degree)
λ	Carreau model parameter (s)
Λ	Carreau number, Equation 2.45(–)
μ:	Viscosity (Pa.s)
μ_0	Zero-shear viscosity (Pa.s)
μ_{ref}	Reference viscosity (Pa.s)
ρ	Fluid density (kg/m^3)
ρ_p	Particle density (kg/m^3)
τ_{ij}	Component of the extra stress tensor (Pa)
$\tau_{1/2}$	Ellis model parameter (Pa)
ϕ	Spherical coordinate (–)
χ	Area ratio, Equation 2.52 (–)
ψ	Stream function or sphericity (m^2/s) or (–)
∇	Del operator (m^{-1})

Superscript

* Nondimensional variable

3

Rigid Particles in Visco-Plastic Liquids

3.1 Introduction

By virtue of its yield stress, a visco-plastic medium in an unsheared condition has the capacity to support the weight of an embedded particle for an indefinite (or sufficiently long) period of time. Figure 3.1 shows 25 mm gravel particles (of density 2700 kg/m^3) floating in a Kimberlite slime slurry of mass density ~1300 kg/m^3, and likewise Figure 3.2 shows static air bubbles trapped in a Carbopol solution having a yield stress of about 8 Pa. Over the years, this phenomenon has been utilized successfully in the design of long-distance slurry pipelines for conveying coarse solids in dense media (Thomas, 1977; Traynis, 1977; Chien and Wan, 1983; Duckworth et al., 1986; Hill and Shook, 1998; Talmon and Huisman, 2005; Talmon, 2010; Talmon and Mastbergen, 2004; Talmon et al., 2013, Tokpavi and Magnin, 2016; Wachs and Frigaard, 2016). Similarly, the sedimentation of suspended particles in many processed foodstuffs and household and pharmaceutical products is undesirable and is often prevented by inducing a small value of yield stress by adding suitable thickening agents to the suspending medium (Berney and Deasy, 1979; Miller and Drabik, 1984) or in the trapping of dense colloids in gels (Laxton and Berg, 2005). Additional examples are found in the context of proppant transport in drilling muds in oil recovery, segregation of coarse particles in self-compacting cements, and movement of solid objects in highly stabilized foams (Cox et al., 2006). On the other hand, the existence of a yield stress is highly detrimental in effecting liquid–solid separation and/or during the pouring of such products into or out of containers and bottles. Evidently, depending upon the value of the yield stress and the size and density of particle, a particle may or may not settle in a visco-plastic medium under its own weight. Hence, before one can talk about the hydrodynamic aspects of particle motion in such media, it is important to establish the criterion for the movement of a particle in a visco-plastic medium of known density and yield stress. In literature, this aspect is usually referred to as static equilibrium. In contrast to the voluminous literature on the motion of particles in viscous fluids without a yield stress (Newtonian, generalized Newtonian fluids, and visco-elastic fluids), there have been only a few investigations on the equilibrium and motion of particles in visco-plastic media. Furthermore, the bulk of the literature deals with spherical particles and circular cylinders. Table 3.1 provides a listing of the pertinent works in this field. An examination of this table shows that most of the research effort has been directed at the elucidation of three aspects of the particle behavior in visco-plastic media. They are

1. Static equilibrium, and the development of a criterion for the initiation or cessation of motion under the influence of gravity.
2. Qualitative flow visualization studies aimed at identifying the regions of flow; some results on wall effects are also available.
3. Drag measurements, both at terminal fall conditions under gravity and in towing tanks, to establish what might be called the standard drag curve for variously shaped particles in visco-plastic media.

Each of these aspects is now considered in detail in the ensuing sections. Following the organizational structure of Chapter 2, we begin with the sedimentation behavior of spherical particles in quiescent visco-plastic media followed by the scant work for nonspherical particles.

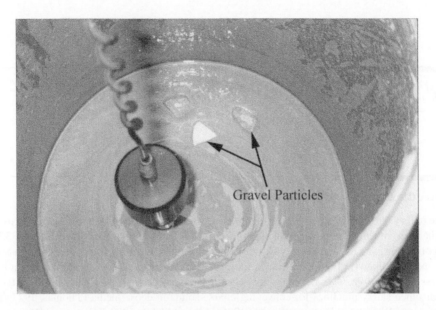

FIGURE 3.1 25 mm gravel particles (ρ_p=2700 kg/m^3) floating in a Kimberlite slime slurry of density about 1300 kg/m^3. (Photograph courtesy Dr. Angus Paterson, Paterson and Cooke Consulting Engineers, Cape Town, South Africa.)

FIGURE 3.2 Air bubbles trapped in a Carbopol solution with a yield stress of 8 Pa. (Photograph courtesy Professor I. Frigaard, University of British Columbia, Vancouver, Canada.)

3.2 Spheres in Visco-Plastic Liquids

3.2.1 Static Equilibrium

While studying the behavior of spheres under gravity (as is the case with most studies referred to herein), it is convenient to introduce an additional dimensionless group that is a measure of the relative magnitudes of the forces due to the yield stress ($\sim \tau_0 d^2$) and the gravitational ($\sim d^3 g(\rho_p-\rho)$) effects. Neglecting all arbitrary constants, the simplest form of a yield-gravity parameter is given as

$$Y_G = \frac{\tau_0}{gd(\rho_p - \rho)} \tag{3.1}$$

It thus stands to reason that for a particle to fall under its own weight, the value of Y_G must be smaller than unity. Some authors prefer to rewrite Equation 3.1 in terms of a critical diameter of particle to capture this transition (Petrou et al., 2000; Saak et al., 2001; Bethmont et al., 2003; Ramge et al., 2010; Tarlet et al., 2019). The question of whether a sphere would or would not move in an unsheared visco-plastic medium has received considerable attention in the literature, as can be seen in Table 3.1. Based on simple heuristics (Andres, 1961; Uhlherr, 1986; Schurz, 1990), it has been assumed that the buoyant weight of a sphere is supported by the vertical component of the force due to the yield stress acting over the sphere surface. In this framework, this component turns out to be ($\pi^2 d^2 \tau_0/4$). This component when equated to the buoyant weight leads to the criterion for the initiation of motion as $Y_G=0.212$. Similarly, some others (Johnson, 1970; Zeppenfeld, 1988) have erroneously set the net buoyant weight of the sphere equal to $\pi d^2 \tau_0$, thereby leading to $Y_G=0.167$ as the criterion for the commencement of particle fall. However, it must be recognized that once the motion ceases, neither the shear stress nor the normal stress distribution is known over the sphere surface (Boardman and Whitmore, 1961; Ansley and Smith, 1967); thus, the shear stress acting on the sphere surface may not be equal to the yield stress everywhere. Besides, the pressure may not be hydrostatic, thereby casting some doubt on the relevance of the buoyant weight of the sphere (Mitishita et al., 2018). In spite of this lack of understanding, the yield-gravity group Y_G may still be used to express the results of experiments and theoretical analyses in a coherent manner, without ascribing any physical interpretation to its definition. The critical values of Y_G signifying the conditions for the initiation of sphere motion as reported in the literature are compiled in Table 3.2. Clearly, not only the reported values of Y_G show a five-fold variation, but more importantly, most of the results can be broadly categorized into two groups. One group with $Y_G \sim 0.040-0.080$ contains the numerical solutions of the equations of motion (Yoshioka et al., 1971; Beris et al., 1985; Beaulne and Mitsoulis, 1997; Blackery and Mitsoulis, 1997; Chen et al., 2016; Di Vaira, 2016), experimental results on the observation of motion/no motion in free fall, and the measurement of the residual force after cessation of the fluid motion (Jossic and Magnin, 2001; Kiljanski, 2004; Merkak et al., 2006) and the extrapolation of force–velocity data to zero velocity (Chafe and de Bruyn, 2005). The second group, with $Y_G \sim 0.2$, consists of the original postulates of Andres (1961) and the measurements of static equilibrium of a tethered sphere in an unsheared medium, though the experimental results of Tabuteau et al. (2007) also point to the value of $Y_G \sim 0.047$ or so. Overall, the experimental results in this category seem to range from ~ 0.047 to 0.13. Such a broad variability of these results has been attributed to strong wall effects, aging, viscoelastic, and time-dependent effects. Comparisons between the bulk and microscale flow behavior also suggest the potential difficulties in making cross comparisons across the fluids as well as across the experimental results (Oppong and de Bruyn, 2007; Bonn and Denn, 2009, Emady et al., 2013, etc.). The similarity of the methods used by Uhlherr (1986) and by Boardman and Whitmore (1961), and the close correspondence of their results, together with the large discrepancy between these results and those of all others, seem to suggest a fundamental difference in the underlying mechanisms of the two approaches. In other words, there seems to be an inherent difference whether the yield point is approached from above (as in the pendulum experiment of Uhlherr 1986) or from below (as in tow-tank experiments or in tests where the sphere is made progressively heavier). Also, another possible reason for such divergent values of Y_G is the different values of yield stress itself obtained using various methods that is discussed in great detail in a later section. In recent years, wall slip has also been invoked to explain such differences in the results

TABLE 3.1

Investigations of Sphere Motion in Visco-plastic Media

Investigator	Test Fluids or Fluid Model	Comments and Results
Khomikovskii and Shilov (1946)	Clay suspension	Static equilibrium of sand in clay suspension; recognized a critical size of sand grains remaining in suspension; critical diameter increased 10–30 fold with a 2–4 fold increase in yield stress.
Tyabin (1949)	Bingham plastic	Integrated the simplified equation of visco-plastic motion about a sphere but the resulting flow field (velocity distribution) is identical to that of a viscous medium.
Volarovich and Gutkin (1953) (and reply by Tyabin 1953)	Bingham plastic	First to note that the region of flow of the medium caused by a falling sphere is bounded. Outside this region, there is a zone of no shear but elastic deformation occurs. The boundary isolating the two zones is a surface of revolution and moves along with the sphere.
Andres (1961)	Not clear $0 < \tau_0^B < 4$ Pa	Proposed a stability criterion for a sphere and presented a correlation for the settling velocity of a sphere.
Boardman and Whitmore (1960, 1961); Whitmore and Boardman (1962)	Clay suspensions $\tau_0^B < 4$ Pa	Static measurements of yield stress using immersed bodies including a sphere, which gave a value comparable to that obtained from the extrapolation of viscometric data. Other shapes yielded unrealistic values.
Rae (1962)	-----	Argued that bodies immersed in flowing visco-plastic media have unsheared material attached to them, and that the yield stress acts over the entire surface.
Bulina et al. (1967)	Bingham plastic Clay suspensions	Experimentally verified the existence of stagnant regions of visco-plastic medium at the front of 2-D blunt bodies when towed through a stationary medium.
Valentik and Whitmore (1965)	Bingham plastic Clay suspensions $7.8 < \tau_0^B < 59$ Pa	Postulated the existence of a concentric sphere of unsheared fluid surrounding the moving sphere, and correlated their data on fall velocity. The diameter of the unsheared shell was estimated by assuming that the Newtonian standard drag curve is applicable.
du Plessis and Ansley (1967)	Kaolin suspensions $1 < \tau_0^B < 3.4$ Pa	Presented a correlation for drag coefficient of spheres and sand grains using Bingham (Bi) and Reynolds number (Re_B >> 1).
Ansley and Smith (1967)	Tomato sauce $\tau_0^B = 29$ Pa	Correlated drag coefficient of spheres with a modified Reynolds number. Suggested the existence of an envelope of sheared medium of toroidal shape and of diameter $d\sqrt{2}$. Also, presented a criterion for static equilibrium supported by limited experimental evidence.
Brookes and Whitmore (1968, 1969); Whitmore (1969)	Clay suspensions	Based on the measurement of the residual force on bodies flowing in visco-plastic medium upon the cessation of flow, postulated a criterion for static equilibrium. Also, attempted to correlate the drag coefficient of sphere by introducing the notion of an effective viscosity in such a manner that the points fall on the Newtonian standard drag curve. Also observed small stagnant region at the front and rear of 2-D bodies which raise some doubt about their earlier ideas (Valentik and Whitmore, 1965).
Ito and Kajiuchi (1969)	Clay suspensions $0.47 < \tau_0^B < 2.83$ Pa	Correlated the results on free fall of spheres using a modified Reynolds number to collapse data on to the Newtonian drag curve.
Traynis (1977)	Bingham plastic Coal suspensions	Suggested an expression for the static equilibrium of sphere under gravity and claimed it to be consistent with experimental results without giving any details of experimental work.
Yoshioka et al. (1971, 1975); Adachi and Yoshioka (1973)	Bingham plastic	Used variational principles to obtain upper and lower bounds on the drag coefficient of a sphere and a cylinder in creeping motion. Also showed the extent of sheared zone surrounding the sphere along with two small stagnant caps attached to the sphere in the front and rear. First to highlight the large discrepancies in the published results.
Pazwash and Robertson (1971, 1975)	Bingham plastic Clay suspensions	Reported drag force measurement on spheres and discs in a visco-plastic medium, and correlated the results in terms of deviations from the corresponding Newtonian values.

(Continued)

TABLE 3.1 (*Continued*)

Investigations of Sphere Motion in Visco-plastic Media

Investigator	Test Fluids or Fluid Model	Comments and Results
Kenchington (1976)	No details	Developed a correlation for the settling velocity of spheres in Bingham plastic media.
Thomas (1977)	No details	General discussion on static equilibrium; discrepancies in the values of the yield stress obtained by different methods.
Wan (1982, 1985)	Bingham plastic Bentonite and kaolin suspensions	Allowed plastic beads and sand to be carried up by or to settle through a uniform upward flow of visco-plastic medium. Correlated drag results using the method of Ansley and Smith (1967). Also, examined the static equilibrium.
Xu and Wu (1983)	Bingham plastic Clay suspensions	Correlated the drag results using the method of Ansley and Smith (1967).
Hanks and Sen (1983), Sen (1984)	Herschel–Bulkley Laponite solutions $0.04 < \tau_0^H < 4.5$ Pa $0.57 < n < 0.85$	Modified the approach of Ansley and Smith to correlate their drag results.
Beris et al. (1985)	Bingham plastic	Numerical solution for the creeping sphere motion. Velocity distribution, shape, and dimensions of sheared zone and unsheared caps. Extensive comparisons with prior theoretical and experimental studies.
Uhlherr (1986)	Carbopol solutions $7.3 < \tau_0 < 73$ Pa	Experimental results on static equilibrium of a simple pendulum.
Atapattu et al. (1986, 1988, 1990, 1995), Atapattu and Uhlherr (1988), Atapattu (1989)	Carbopol solutions $3.3 < \tau_0 < 25.2$ Pa	Extensive experimental results on wall effects, drag coefficient, static equilibrium, and velocity distribution around spheres in Herschel–Bulkley fluids in the low Reynolds number (< 1) region.
Dedegil (1987)	–	Reanalysis of the data of Valentik and Whitmore (1965).
Hartnett and Hu (1989)	Carbopol solutions	Evaluated yield stress by observing motion/no motion of a sphere. Also, see the criticism by Astarita (1990).
Schurz (1990)	Kaolin suspension	Used rolling motion of a sphere to evaluate yield stress.
Wunsch (1990)	Carbopol solutions	Used falling sphere method to evaluate yield stress.
Saha et al. (1992)	Bingham plastic	Empirical correlation for the settling velocity based on the data of Valentik and Whitmore (1965).
Briscoe et al. (1992a, 1993)	Bentonite suspensions (Bingham plastic model)	Correlation for drag under static and dynamic conditions. Possible thixotropic effects?
Tran et al. (1993)	Carbopol solutions Clay suspensions	Correlations for drag of a sphere.
Machac et al. (1995)	Kaolin and TiO_2 suspensions (power-law, Bingham plastic, and Herschel–Bulkley models)	Correlations for drag coefficient of a sphere.
Blackery and Mitsoulis (1997), Beaulne and Mitsoulis (1997)	Bingham plastic and Herschel–Bulkley models	Numerical predictions of drag in creeping flow regime, static equilibrium, wall effects, and regions of flow.
Song and Chiew (1997)	Clay/water suspensions (Bingham plastic model)	Criterion of static equilibrium.
Hariharaputhiran et al. (1998)	Carbopol solutions (Bingham plastic model)	Time dependence of settling velocity of spheres.
He et al. (2001)	Magnetic suspensions (Bingham and Casson models)	Some data on settling velocity of particles.

(*Continued*)

TABLE 3.1 (*Continued*)

Investigations of Sphere Motion in Visco-plastic Media

Investigator	Test Fluids or Fluid Model	Comments and Results
Jossic and Magnin (2001)	Carbopol 940 in water (Herschel–Bulkley model)	Drag and static equilibrium of spheres, disc, cylinders, cubes, and cones.
Liu et al. (2002, 2003)	Bingham plastic model	Numerical simulation of creeping flow.
Ferroir et al. (2004)	Laponite suspensions (Thixotropic)	Sedimentation behavior of spheres in quiescent and vibrated fluids.
de Bruyn (2004), Chafe and de Bruyn (2005)	Foams and bentonite clay suspensions (Herschel–Bulkley model and Bingham model)	Extraction of yield stress of foams from velocity–force data on a sphere and to infer some information about relaxation.
Merkak et al. (2006)	Carbopol solutions (Herschel–Bulkley model)	Interaction between two in-line and side-by-side spheres and their static equilibrium.
Gueslin et al. (2006)	Aging laponite suspension	Falling velocity decreases with aging in an exponential manner.
de Bruyn (2006)	Aqueous foams	Effect of aging on the drag force exerted on a sphere (low Reynolds number regime).
Yu and Wachs (2007)	Bingham fluid	Results are consistent with Blackery and Mitsoulis (1997).
Putz et al. (2008)	Carbopol solution	Detailed PIV data reveal the asymmetry of flow field in the front and rear of the sphere.
Ireland and Jameson (2012)	Foams	Effect of foam bubble size and wettability on sphere drag.
Nirmalkar et al. (2013a, b)	Bingham and Herschel–Bulkley models	Extensive numerical results on yield surfaces and drag coefficient over the range: $1 \le Re \le 100$; $0 \le Bi \le 10^4$; $0.2 \le n \le 1$.
Emady et al. (2013)	Carbopol solution and a colloidal gel of micro-fibrous cellulose	Static equilibrium.
Ahonguio et al. (2014)	Carbopol solution	Effect of surface properties and elasticity of the fluid on the stability criterion for a sphere.
Gumulya et al. (2014)	Shear-thinning with thixotropy	Different settling regimes depending upon structure parameter.
Chen et al. (2016)	Numerical; Bingham fluid model	Drag coefficient for range of value of Bi and static equilibrium.
Di Vaira (2016)	Numerical; Bingham fluid model	Drag coefficient as a function of Bi and critical value of Y_G.
Fraggedakis et al. (2016)	Elasto-visco-plastic fluids	Effects of yield stress, elasticity, and slip on drag of a sphere.
Maleki-Jirsaraei et al. (2018)	Thixotropy and yield stress	Varying microstructure close to the sphere.
Freund et al. (2018)	Thixotropy and yield stress	Varying microstructure close to the sphere.
Kim and Park (2020)	Thixotropy and yield stress	Complex settling behavior.
Mirzaagha et al. (2017)	Colloidal fibers in a surfactant micellar solution (Herschel–Bulkley model)	Rising sphere and bubbles in low yield stress fluids and static equilibrium.
Sgreva et al. (2020)	Superabsorbent polymers in water	Different regimes of sphere settling, depending upon the relative sizes of gel grains and test spheres.

TABLE 3.2

Limiting Values of Y_G for Incipient Motion of a Sphere under Gravity

Investigator	Y_G	Technique	Observation
Andres (1961)	0.212	Postulated (based on τ_0^B)	Good summary of the Russian experimental work embracing $0.056 < Y_G < 0.59$.
Boardman and Whitmore (1961)	0.20	Experimental (based on τ_0^B)	Spheres suspended using a cantilever beam.
Ansley and Smith (1967)	~0.068–0.084	Theory and experimental (based on τ_0^B)	Model based on slip-line theory of soil mechanics.
Brookes and Whitmore (1968)	0.04	Experimental (based on τ_0^B)	Obtained by the direct measurement of the residual horizontal force.
Traynis (1977)	0.083–0.10 for coal particles and 0.167 for spheres	Postulated and experimental (based on τ_0^B)	No details presented.
Wan (1982, 1985)	0.056–0.067	Postulated and experimental (based on τ_0^B)	Based on their experiments on motion/no motion of spheres in bentonite suspensions, and the work of Ansley and Smith (1967).
Beris et al. (1985), Blackery and Mitsoulis (1997)	0.048	Theoretical (based on τ_0^B)	Extrapolation of numerical results.
Uhlherr (1986)	0.181–0.206	Experimental (based on τ_0^B)	Equilibrium of a simple pendulum.
Atapattu et al. (1986)	0.095–0.111	Experimental (based on τ_0^B)	Observation of motion/no motion of sphere under gravity.
Zeppenfeld (1988)	0.167	-	Simply equated the motive force to the yield stress force.
Beaulne and Mitsoulis (1997)	0.048	Numerical (based on τ_0^H)	Extrapolation of numerical results.
Song and Chiew (1997)	0.083	Based on τ_0^B	No details available; probably based on experimental results.
Jossic and Magnin (2001), Merkak et al. (2006)	~0.062–0.088	Based on τ_0^B	Force measurement on spheres being pulled at a constant velocity.
Chafe and de Bruyn (2005)	0.048	Based on τ_0^B	Extrapolation of force to zero velocity of spheres in bentonite suspensions.
Chen et al. (2016)	0.046	Numerical solution	Using multiple relaxation time (MRT) Lattice Boltzmann method (LBM).
Di Vaira (2016)	0.049, 0.062	Numerical solution	Used implicit regularization of Bingham model along with MRT–LBM.

(Ahonguio et al., 2014). The wall slip is believed to increase the limiting value of Y_G slightly. Suffice it to add here that Rankin et al. (1998, 1999) have noted similar difficulties in the suspensions of iron particles in magneto-rheological fluids.

Thus, not only does considerable confusion exist regarding the critical value of Y_G, but more importantly, it also acts as a constant reminder about the complexity of visco-plastic materials. The detailed study of Emady et al. (2013) makes this point very well. They have elucidated the role of microstructure in terms of its response to the onset of particle sedimentation in a visco-plastic fluid. For this purpose, they classified the fluids exhibiting visco-plastic behavior into two types; first, glasses (highly loaded systems) which tend to be elastic, and the so called gels (rather dilute systems) with significant inter-particle interactions that lead to the formation of elastic networks. These two types of microstructures are a subset of the state diagram put forward by Trappe and Sandkühler (2004). On this basis, widely used aqueous Carbopol solutions in experimental studies as model test fluids belong to the colloidal glass category and usually do not exhibit any time-dependent (thixotropic) effects. On the other hand, colloidal gels (much lower viscosity, yield stress, and concentration of particles than glass) are exemplified by clays (kaolin, bentonite, etc.), which exhibit thixotropic effects due to their aging (Bonn and Denn, 2009). Thus, the two fluids with identical yield stress values would behave very differently depending upon whether it is a "glass" or a "gel". Emady et al. (2013) used two fluids, one from each category, namely aqueous Carbopol solution and micro-fibrous cellulose, as model fluids. They reported the values of Y_G ranging from ~0.001 to ~0.026 for the micro-fibrous cellulose systems and from ~0.01 to ~0.08 for the Carbopol solution. This has been attributed to the way the two systems yield, in line with the suggestion of Putz et al. (2008) as well as that of Oppong et al. (2006) and Oppong and de Bruyn (2007) about the micro structural and bulk rheological characteristics. Thus, in general a visco-plastic fluid behavior may be accompanied either by elasticity or by thixotropy depending upon its microstructure. This aspect has often been overlooked in early experimental and numerical studies of visco-plastic flow past variously shaped objects.

The complementary situation of thixotropy effects on the steady translation of a sphere has been numerically investigated by Gumulya et al. (2014), Maleki-Jirsaraei et al. (2018), Freund et al. (2018), Kim and Park (2020), etc. for a sphere and by Bui and Ho (2019, 2020) and by Rossi et al. (2022) for a cylinder. At small (nondimensional) settling velocity (<1), Brownian effects are negligible, and the resulting hydrodynamics deviates little from the corresponding Newtonian response. With the gradual increase in the velocity, shear-induced structure field is established. The microstructure of the fluid (hence its rheology) in the proximity of the sphere assumes an important role. The falling velocity is determined by three factors: Brownian effects, shear-induced microstructure breakdown, and convection effects. At very high velocity, Brownian effects become negligible and the other two factors determine the overall behavior. These predictions are based on the four-parameter Moore model for thixotropy. Some of these trends are borne out, at least qualitatively, by the experimental results of Sgreva et al. (2020). They used aqueous superabsorbent polymers (SAP) wherein the SAP powder grains swell in water significantly attaining particle sizes of 1–8 mm. This gives rise to the EVP (elastic, viscous, and plastic)-type bulk rheological behavior. Sgreva et al. (2020) used test spheres of 3–30 mm diameter ranging in density from 2200 to ~15,000 kg/m³. They reported the five possible settling regimes depending upon the size ratio of the swollen powder grain (d_g) to the settling sphere (d) as follows: linear regime in which the sphere settles in a straight path attaining its terminal falling velocity. In the second regime, the falling velocity of the sphere varies around a mean value without any regular pattern. The third regime is characterized by the combination of periods of "no motion" and "irregular settling". The fourth regime is characterized by a slow fall and the sphere's velocity drops in a logarithmic fashion, and finally "no-motion" regime wherein the buoyant weight of the sphere is not able to overcome the yield stress. The limited experimental data pertaining to the first three regimes could be explained using the Herschel–Bulkley model representation of the fluid wherein both the yield stress $\left(\tau_0^H\right)$ and consistency index (m) bear a positive dependence on the grain size of SAP. Furthermore, the critical value of Y_G decreases for (d_g/d) < 2.

3.2.2 Flow Field

At the outset, it is useful to recall here that the flow field created by a falling sphere extends to several sphere radii in Newtonian media and in fluids without any yield stress (theoretically up to ∞). In contrast, however, in the case of a visco-plastic liquid, it will not be so, since as soon as the prevailing stress levels drop below the value of the yield stress, the substance will no longer shear (or flow) and it will behave like an elastic solid. The shape and size of such a cavity will obviously depend upon the value of the yield stress, the size and the density of sphere, and the relative velocity between the sphere and the medium. Unfortunately, the exact shape and size of such a cavity cannot be predicted a priori, and on the contrary it adds to the complexity of the problem from a theoretical standpoint. Indeed, Volarovich and Gutkin (1953), together with Tyabin (1953), were the first to postulate the existence of a restricted zone of flow for the motion of a sphere in a visco-plastic medium. In their important contribution, Volarovich and Gutkin wrote,

> The solution is complicated by the difficulty in finding the boundaries of the region within which flow of the medium due to motion of the sphere occurs. Beyond the limits of this region, shear is not transmitted and only a zone of elastic deformation exists. This boundary, in any event, is a surface of rotation and moves together with the sphere.

They went on to add further and proposed that a first-order approximation may be reached by *solving the differential equation of motion for a viscous liquid taking into account that the flow takes place within a certain region*. This approach was used by Lipscomb and Denn (1984) for solving a number of flow problems involving visco-plastic media. Rae (1962) intuitively asserted that, within the region of plastic deformation surrounding a body, there would be unsheared material adhering to certain parts of the sphere. Evidently, the yield stress does not necessarily act over the surface of the sphere only, but rather over the sphere surface as well as that of the unsheared material. The existence of such unsheared material that moves with the sphere has been observed experimentally, albeit without any sharp boundaries (Whitmore and Boardman, 1962; Bulina et al., 1967; Atapattu, 1989; Atapattu et al., 1995). Numerical solutions (Beris et al., 1985; Beaulne and Mitsoulis, 1997; Blackery and Mitsoulis, 1997; Liu et al., 2002; Deglo de Besses et al., 2004 and subsequent studies listed in Table 3.1) also confirm, at least qualitatively, the existence of such zones of fluid-like and solid-like behavior of a visco-plastic material close to a moving particle.

Numerous models of varying shapes and complexity, as shown in Figure 3.3, have been postulated for the unsheared region surrounding a moving sphere in visco-plastic media. The simplest of all is a spherical shell concentric with the solid sphere (Valentik and Whitmore, 1965), though not shown in Figure 3.3. Subsequently, as expected, this grossly over-simplified picture was proved to be unrealistic, as in some cases the diameter of the fictitious shell turned out to be larger than the cylindrical fall tubes in which the experiments were performed (Whitmore, 1969; Brookes and Whitmore, 1969). For instance, this model resulted in the following ratio of diameters for the unsheared zone to that of the solid sphere: 1–1.8 (Valentik and Whitmore, 1965) and 8.2–85 (Brookes and Whitmore, 1969). Based on the elegant slip-line theory of solid mechanics, the shape of the unsheared material as proposed by Ansley and Smith (1967) is also shown in Figure 3.3. It is perhaps appropriate to quote Ansley and Smith:

> The stress distribution imposed on the plastic material by the motive force on the sphere causes the material to become fluid in an envelope surrounding the sphere. Within the envelope, the motions of the sphere and the displaced fluid are steady as the sphere-envelope system moves through the plastic material causing instantaneous, localized transformation between the plastic and fluid states.

The shape of the envelope (Figure 3.3a) is a kind of truncated toroid with its section centered on the surface of the sphere with a diameter of $d\sqrt{2}$.

The problem of creeping sphere motion in a Bingham plastic fluid has been solved approximately using the velocity and stress variational principles by Yoshioka and Adachi (1971), and using numerical techniques by Beris et al. (1985) and Blackery and Mitsoulis (1997). The predicted shapes of the unsheared material are also included in Figure 3.3. Surprisingly, the shape predicted by numerical solutions is in good agreement with that postulated by Ansley and Smith, but differs significantly from that of Yoshioka and Adachi (1971). Subsequently, Beaulne and Mitsoulis (1997) and Deglo de Besses et al. (2004) have reported extensive numerical results for the creeping sphere motion in Herschel–Bulkley fluids. Qualitatively similar zones of deformation/no deformation have been reported by them,

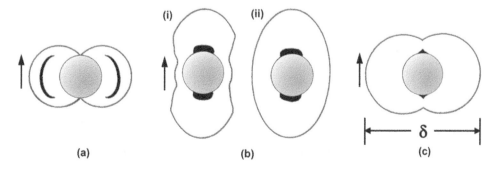

FIGURE 3.3 Shape of the sheared envelope surrounding a sphere in creeping motion in visco-plastic fluids: (a) Ansley and Smith (1967); (b) Yoshioka et al. (1971) (i) maximum principle (ii) minimum principle; (c) Beris et al. (1985).

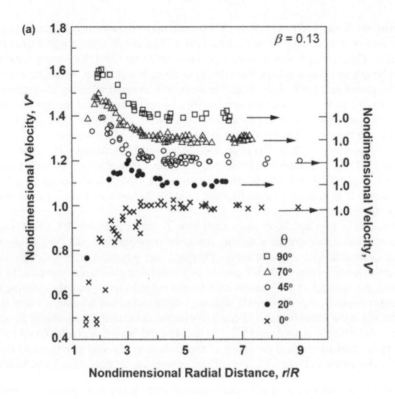

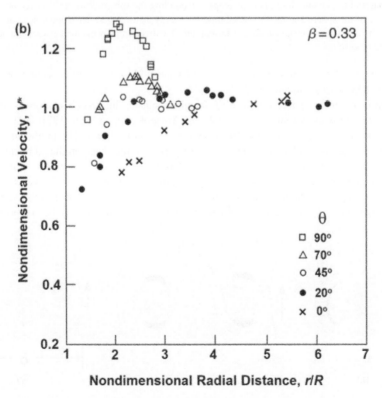

FIGURE 3.4 Velocity profiles for creeping visco-plastic flow past a sphere in a Carbopol solution for two values of diameter ratio (a) $\beta=0.13$ (b) $\beta=0.33$. (After Atapattu, D. D., Ph.D. dissertation, Monash University, Melbourne, Australia, 1989.)

though depending upon the value of the sphere-to-tube diameter ratio, β, and the Bingham number, some solid-like regions may also be present close to the wall of the tube. Using an elegant optical technique, Atapattu (1989) and Atapattu et al. (1995) have carried out quantitative flow visualization experiments. Notwithstanding the additional effects arising from the presence of cylindrical walls, typical velocity profiles at different angular positions are shown in Figure 3.4 for two values of the sphere-to-tube diameter ratio ($\beta=d/D$). In addition to providing an insight into the physics of the flow field, such measurements also facilitate the delineation of the boundaries of the shear zone. To put these profiles in perspective, the corresponding profiles for a Newtonian fluid together with theoretical predictions are shown in Figure 3.5. Note the similarity between the profiles for visco-plastic fluids and those for pseudoplastic fluids shown in Figure 2.7. A representative comparison between the numerical results (Beaulne and Mitsoulis, 1997) and the experimental results (Atapattu et al., 1995) for one experimental fluid is shown in Figure 3.6 where the match is seen to be as good as can be expected in this type of work, especially for large values of β. Note the difference in the slopes of the velocity profile before and after the maxima. The numerical simulations also seem to capture well the leveling-off of the velocity profiles as the distance from the surface of the sphere increases. The experimental results for a sphere falling in a visco-plastic medium clearly show the existence of unsheared zones, as has been postulated by numerous workers. The extent of the shear zone can easily be measured by locating the radial distance at which the dimensionless velocity becomes unity. The typical shapes of the sheared zone so obtained are displayed in Figure 3.7 and are compared with those predicted by Beaulne and Mitsoulis (1997) in Figure 3.8. In addition, these figures also elucidate the effect of Bingham number (Bi) and diameter ratio (β) on the shape of the sheared zone. Overall, good correspondence is seen to exist in this figure. Qualitatively similar match is obtained for the other fluids used by Atapattu et al. (1995). The actual shape of the sheared zone also seems to resemble qualitatively with the postulate of Ansley and Smith

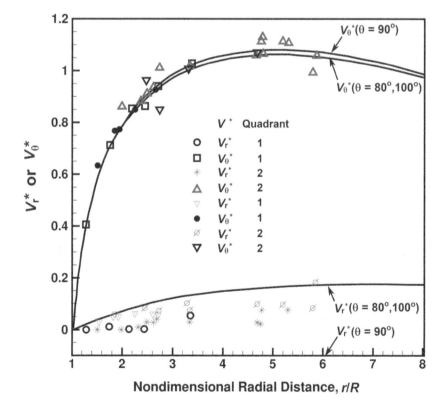

FIGURE 3.5 Comparison between experimental and predicted (Haberman and Sayre, 1958) velocity profiles for the creeping Newtonian flow past a sphere. (After Atapattu, D.D., Ph.D. dissertation, Monash University, Melbourne, Australia, 1989.)

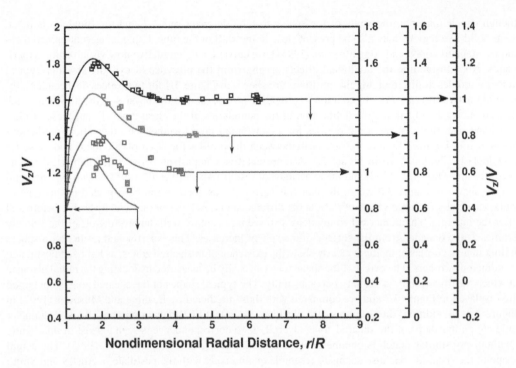

FIGURE 3.6 Comparison between the experimental and predicted axial velocities for the creeping sphere motion in Hershel–Bulkley model fluids $\tau_0^H = 46.5$ Pa; $m = 23.89$ Pa·s^n; $n = 0.50$. (From Beaulne, M. and Mitsoulis, E., *J. Non-Newt. Fluid Mech.*, **72**, 55, 1997.)

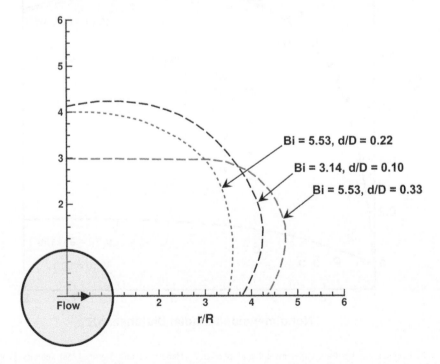

FIGURE 3.7 Effect of Bingham number and diameter ratio on the size of the fluid-like regions. (Based on the results of Atapattu, D.D., Ph.D. dissertation, Monash University, Melbourne, Australia, 1989.)

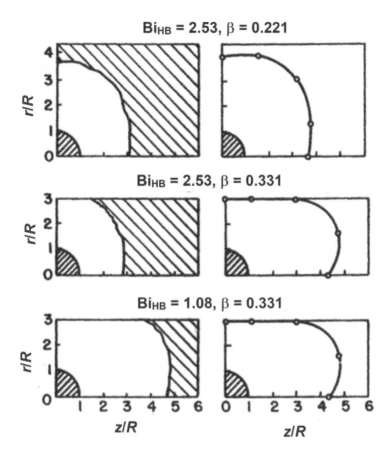

FIGURE 3.8 Comparison between the observed (Atapattu, 1989, shown on right) and predicted (Beaulne and Mitsoulis, 1997, shown on left) shapes of fluid-like regions around a sphere. (Modified after Beaulne, M. and Mitsoulis, E., *J. Non-Newt. Fluid Mech.*, **72**, 55, 1997.)

(1967). However, the possibility of major changes in the shape of the sheared zone with the increasing value of Bi and for the other values of β cannot be ruled out at this stage, as has been predicted by the numerical studies of Blackery and Mitsoulis (1997) and Beaulne and Mitsoulis (1997).

The subsequent work of de Bruyn (2004) and Chafe and de Bruyn (2005) on the towing of a sphere through visco-plastic foams and bentonite clay suspensions suggests the fluid-like zone to be about $2R$ in size in the presence of significant wall effects. While the force–velocity was governed initially by a single value of time constant (inversely proportional to the imposed shear rate), the stoppage of the flow seems to be influenced by distinct processes that, in turn, are linked to the behavior of individual bubbles in the foam.

In summary, within the range of conditions studied thus far, the deformation cavity seems to extend up to about 4–5 sphere radii in both axial and lateral directions when the wall effects are negligible. This is in stark contrast to the much larger values of the distances up to which the disturbance caused by a sphere is felt in Newtonian and power-law fluids, as detailed in Chapter 2.

Detailed particle image velocimetry (PIV) measurement in the proximity of a sphere translating slowly in visco-plastic fluids reveals the differences in the velocity field in the front and rear of the sphere even when the inertial effects are negligibly small (Putz et al., 2008; Putz and Frigaard, 2010). Qualitatively similar results have also been reported for a circular cylinder (Tokpavi et al., 2008, 2009). This asymmetry has been attributed to the elasticity of the fluid in the unyielded state (Fraggedakis et al., 2016). Therefore, the assumption of the fore-and-aft symmetry in the experiments of Atapattu et al. (1995) and in some of the numerical studies is open to criticism on this count.

3.2.3 Drag Force

3.2.3.1 Theoretical Developments

Only limited attention has been devoted to the solution of field equations describing the sphere motion in visco-plastic media, even in the creeping flow region. Yoshioka et al. (1971) used the stress and velocity variational principles to obtain approximate upper and lower bounds on drag coefficient for the creeping motion of a sphere in a Bingham fluid medium. As noted in Chapter 2, this approach necessitates assumptions regarding the plausible forms of the stream function and stress distribution subject to the boundary conditions. Bearing in mind the unrealistic shapes of the sheared zone surrounding a sphere, as discussed in the preceding section, their results must be treated with reserve. Their upper bound calculation, however, is in good agreement with the numerical results of Beris et al. (1985). This is surprising, in view of the major differences in the shapes of sheared zones predicted by these two analyses. As mentioned previously, the most reliable values of drag coefficient in Bingham plastic fluids for creeping sphere motion are those of Beris et al. (1985) for the unbounded case and those of Blackery and Mitsoulis (1997) and Deglo de Besses et al. (2004) with significant wall effects. In the range $0 \leq Bi \leq 1000$, Blackery and Mitsoulis (1997) have synthesized their numerical results as

$$Y - Y_N = a(\text{Bi})^b \tag{3.2}$$

where the drag correction factor Y, still defined as $(C_D \text{Re}_B/24)$, now also includes the contribution of wall effects. For $\beta = 0$, obviously $Y_N = (1/f_0) = 1$. Table 3.3 lists the values of Y_N, a, b for a range of values of β. These results have been subsequently substantiated by Prashant and Derksen (2011) who used a direct numerical approach.

Furthermore, based on the criterion of static equilibrium of $Y_G = 0.048$, Blackery and Mitsoulis (1997) demonstrated that in the limit of $Bi \rightarrow \infty$, $Y = 1.17 Bi$. Indeed, this limiting behavior is obtained for $Bi \geq 1000$. Subsequently, Beaulne and Mitsoulis (1997) have reported the values of drag coefficients for two test fluids and diameter ratio β as used by Atapattu et al. (1995), and they reported excellent agreement between their predictions and the experimental data of Atapattu et al. (1995) for Herschel–Bulkley model fluids.

In an interesting study, Deglo de Besses et al. (2004) studied numerically the creeping motion of a sphere in a tube filled with a Herschel–Bulkley fluid. In particular, they examined the role of total slip/no-slip condition at the bounding wall by prescribing zero-shear stress or a constant translational velocity at the wall. For a fluid without any yield stress, the drag coefficient is found to be always lower in the presence of the total slip than that with the no-slip condition otherwise under identical conditions. Indeed, this effect gets accentuated in shear-thinning fluids with a yield stress. Similarly, the size of the fluid-like zones is more sensitive to the type of the boundary condition than to the value of the flow behavior index.

Before leaving this topic, it is appropriate to mention here some other aspects of the numerical simulations of visco-plastic fluid flows in general and of this problem in particular. In view of the inherently discontinuous form of the visco-plastic constitutive equations (Bingham, Herschel–Bulkley, Casson), it is not possible to incorporate them directly into a numerical solution methodology. Therefore, over the years, a few regularization models have evolved which convert the abrupt transition between the unyielded- and fluid-like regions into a gradual one. The simple shearing (unidirectional) flows of

TABLE 3.3
Value of a, b, and Y_N in Equation 3.2

β	Y_N	a	b
0	1	2.93	0.83
0.02	1.05	2.59	0.86
0.10	1.26	2.33	0.88
0.125	1.35	2.28	0.89
0.25	1.98	1.92	0.92
0.50	5.94	1.63	0.95

visco-plastic fluids (like in a circular tube, thin slit and planar films, etc.) are amenable to analytical solutions and it is thus possible to locate the true yield surfaces separating the "fluid-like" and "solid-like" regions established in the flow domain. Thus, no regularization approximation is needed in these situations. In contrast, it is not possible to guess a priori even approximately the location of the yield surfaces which need not be contiguous in complex flows, such as that the flow past bluff bodies (sphere or cylinder for instance). Even, for power-law fluids, numerical solutions are needed in these cases as seen in Chapter 2.

The three regularization approaches that have gained wide acceptance in the literature in the context of visco-plastic fluids flow over a sphere (and other axisymmetric, or 2-D shaped objects) are briefly introduced here. In the first approach by O'Donovan and Tanner (1984), the unyielded part of the fluid is modeled as a highly viscous fluid with its viscosity (μ_Y) being orders of magnitude higher than the Bingham model viscosity (μ_B), i.e., the so-called yielding viscosity $\mu_Y \gg \mu_B$ and hence the term "biviscous" model. This approach reflects the experimental observation seen in Figure 1.9 wherein the viscosity drops steeply over a very narrow range of shear stress (or shear rate). Figure 3.9b here shows this idealization, and the ideal Bingham viscosity equation can now be rewritten as:

$$\eta_{Bv} = \frac{\mu_y}{\mu_B} \quad \text{if} \quad |\tau| \leq \text{Bi} \tag{3.3a}$$

$$\eta_{Bv} = 1 + \frac{\text{Bi}}{|\dot{\gamma}|}\left(1 - \frac{1}{\mu_y/\mu_B}\right) \quad \text{if} \quad |\tau| > \text{Bi} \tag{3.3b}$$

In the second approach, by Papanastasiou (1987), shown here in Figure 3.9a, the Bingham fluid model is rewritten as:

$$\eta_P = 1 + \frac{\text{Bi}}{|\dot{\gamma}|}\left(1 - e^{(-M|\dot{\gamma}|)}\right) \tag{3.4}$$

Naturally, in the limit of $\mu_Y \to \infty$ or $M \to \infty$, both Equations 3.3 and 3.4 reduce to the ideal Bingham behavior, as shown in Figure 3.9. Therefore, as large as possible values of μ_Y and M are required in numerical solutions to make reliable predictions. However, excessively large values can also lead to convergence and other numerical difficulties (Glowinski, 2003; Glowinski and Wachs, 2011).

In the third approach, the "solid-like" region admits extremely small but a finite strain rate as opposed to the expected zero strain rate in an ideal elastic solid medium. This approach leads to a family of

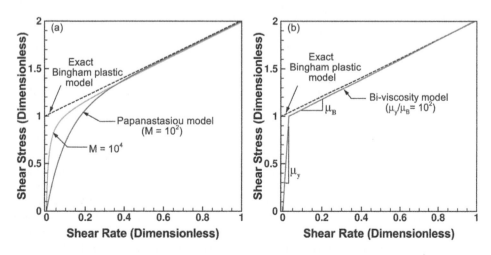

FIGURE 3.9 Schematics of the regularization models (a) Papanastasiou model (b) Bi-viscosity model.

regularization models (also known as simple regularization models (Frigaard and Nouar, 2005)). The following specific form by Bercovier and Engelman (1980) is popular:

$$\eta_{BE} = 1 + \text{Bi}\left(\frac{1}{|\dot{\gamma}| + \delta}\right) \tag{3.5}$$

Clearly, in this case, the true Bingham behavior is approached in the limit of $\delta \to 0$. The relative merits and demerits of these and other approximations proposed in the literature have been presented, among others, by Dean et al. (2007), Glowinski and Wachs (2011), Saramito and Wachs (2017), Mitsoulis and Tsamopoulos (2017), etc. It is generally believed that the exponential regularization given by Equation 3.4 is probably the best, provided a suitable value of M is used which may vary from one flow to another. It is, however, fair to add here that new algorithms such as the damped Newton method (Saramito, 2016), operator-splitting approach (Huilgol and Kefayati, 2015; Huilgol, 2015), augmented Lagrangian method (Saramito and Wachs, 2017), etc. do not necessitate the regularized visco-plastic viscosity models, but these tend to be far more computationally intensive. Furthermore, for the few cases where results are available with and without invoking the regularization method, the two predictions of integral quantities such as drag and heat transfer coefficient are in close agreement, thereby inspiring confidence in the validity of the regularized forms of the Bingham model (Mitsoulis and Tsamopoulos, 2017). Also, it needs to be added here that none of the regularization schemes resolve the stress field entirely satisfactorily and therefore the resulting "yield surfaces", delineated using the common von Mises yielding criterion (Macosko, 1994), must be seen as constant stress surfaces within the prechosen tolerance level of the fluid yield stress. This seems to be reasonable for steady flows, whereas significant errors can be incurred in transient flows (Liu et al., 2002; Frigaard and Nouar, 2005; Glowinski and Wachs, 2011; Saramito and Wachs, 2017; Mitsoulis and Tsamopoulos, 2017; Ahmadi and Karimfazli, 2021). In recent years, there has been a shift from the finite element-based methods to the lattice Boltzmann methods with two or multiple relaxation times as well as implicitly regularized form of the constitutive equations (Chen et al., 2014, 2016; Di Vaira, 2016).

Returning to the pioneering study, Beris et al. (1985) solved a free-boundary problem and therefore the yield surfaces are treated as unknown boundaries. In other words, their approach requires calculations of the flow field only in the fluid-like zone. While from a numerical standpoint, this approach is efficient and fast, it requires a priori knowledge of the general location of the yield surfaces. On the other hand, Blackery and Mitsoulis (1997) solved the governing equations for both the fluid- and solid-like regions. This strategy obviously allows the boundaries of the yielded/unyielded regions to be determined as a part of the solution. While the two predictions of drag differ only slightly, the location of the fluid- and solid-like domains shows significant differences. Subsequent drag results of Liu et al. (2002) are also consistent with that of Beris et al. (1985), but deviate up to about 10%–12% from that of Blackery and Mitsoulis (1997). Liu et al. (2002) attributed these minor differences to the meshes used in the two studies. Subsequently, this work has been extended to the behavior of two spheres falling along their lines of centers in a Bingham plastic medium numerically by Liu et al. (2003) and Jie and Ke-Qin (2006), and experimentally by Merkak et al. (2006). The numerical errors precluded the use of regularization parameters to determine the location of the yield surfaces from the yield condition. However, for initial separations larger than $6R$, the yield surfaces were almost identical to that for a single sphere thereby suggesting no interactions between the two spheres which is in line with subsequent experimental findings (Merkak et al., 2006). Both Blackery and Mitsoulis (1997) and Beaulne and Mitsoulis (1997) have approximated the yield stress term (in the Bingham and Herschel–Bulkley models) by an exponential term of the form $\tau_0^B\left(1 - e^{-M\dot{\gamma}}\right)$ where ε is the stress growth exponent (Papanastasiou, 1987), i.e., Equation 3.4 here. Clearly, the larger the value of M, the better the approximation. There is, however, no way of choosing an appropriate value of this parameter a priori; indeed, this will vary from one fluid to another and from one flow to another. For the problem of falling spheres that involves relatively small values of shear rate, a value of $M > 1000$ seems satisfactory. While this approach has been found to be convenient to evaluate the macroscopic parameters like drag coefficient, it is open to criticism when used to delineate the shape of yielded/unyielded regions. This weakness stems from the fact that this model does predict a finite deformation (albeit extremely small) for all values of *M*, no matter how large a value is used. Hence, the demarcation between the so-called yielded and unyielded region is arbitrary and not as sharp as one might imagine which further depends upon the choice of the yielding criterion (Liu et al., 2016; Tanner, 2022). Therefore, it is preferable to use the term apparently unyielded as opposed to unyielded regions in such situations. Frigaard and Nouar (2005) have also pointed out some further weaknesses of such viscosity regularization approximations used to simulate the flow of visco-plastic fluids. There is another subtle difference between the strategies used by Beris et al. (1985) and Blackery and Mitsoulis (1997) on one hand and that of Beaulne and Mitsoulis (1997) on the other. The former realized the range of Bingham numbers by varying the value of the yield stress while keeping the velocity constant as opposed to Beaulne and Mitsoulis (1997) who maintained the value of the yield stress (same fluid) constant but varied the velocity. This explains why the results of Beaulne and Mitsoulis (1997), Chen et al. (2016), Di Vaira (2016), Tokpavi et al. (2008, 2009), and Jossic and Magnin (2009) approach the so-called no-motion condition far better than the others. In other words, the simulations of Beris et al. (1985) and Blackery and Mitsoulis (1997) represent the case of pulling a sphere at a constant velocity whereas that of Beaulne and Mitsoulis (1997) relate to the situation in which the sphere is progressively made smaller/lighter so that it is surrounded by unyielded material.

Only Yu and Wachs (2007), Tokpavi et al. (2008, 2009), Jossic and Magnin (2009), Nirmalkar et al. (2013a, b) have reported the drag results at finite Reynolds numbers ($Re_B \leq 100$) in Bingham and Herschel–Bulkley Model fluids. Their results are based on the exponential regularization model, Equation 3.4. However, limited results obtained using the other two models, Equations 3.3 and 3.5, were found to be in perfect agreement with these results. The aforementioned numerical predictions of drag on a sphere in Bingham plastic fluids have also been corroborated by Lovato et al. (2022).

Figure 3.10 shows the (apparent) yield surfaces for scores of values of the Reynolds and Bingham numbers and the power-law index including the limiting cases of Bingham plastic fluid ($n = 1$). Evidently, there are two regions of unyielded material: polar caps and the faraway region wherein the material is in piston-type motion. The small island at the front stagnation point disappears quickly with the increasing Reynolds number whereas that at the rear stagnation point grows spatially. The fluid-like cavity

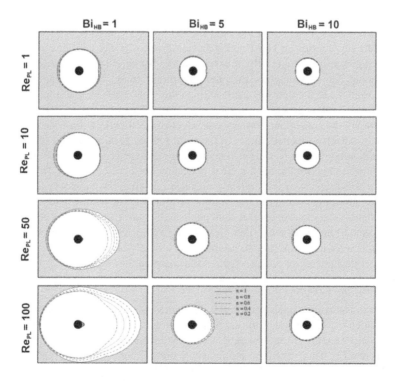

FIGURE 3.10 Effect of Reynolds number, Re_{PL}, Bingham number, Bi_{HB}, and power-law index, *n*, on the apparent yield surfaces for flow past a sphere (Nirmalkar et al., 2013a, b).

encapsulating the sphere expands a little; this tendency, however, is countered by the increasing Bingham number. Also, fluid yielding was seen to be somewhat suppressed by the shear-thinning behavior. With the increasing Bingham number, the role of power-law index gradually disappears, as seen in Figure 3.10. While the yield surfaces approach a limiting shape beyond certain Bingham number, the drag continues to increase with Bingham number. Also, for fixed values of the Reynolds number and power-law index, there exists a limiting value of the Bingham number beyond which the flow remains attached even at high Reynolds number such as $Re_{HB} = 100$. This value increases with the increasing Reynolds number.

In recent years, there has been further recognition of the fact that a visco-plastic fluid may also exhibit elasticity and/or thixotropy. Thus, Fraggedakis et al. (2016) have studied the creeping motion of a sphere in elasto-visco-plastic fluids (Saramito, 2007; Cheddadi et al., 2012). This fluid behaves like an elastic solid below the yield stress and like a visco-elastic fluid above the yield stress. They also incorporated the wall slip into their analysis via the Navier linear slip relation. With suitable choices of the influencing parameters, their predictions are consistent with the available numerical and experimental results for a single sphere. They explained the negative wake in terms of the differences in the way the shear and normal stresses decay spatially downstream from the sphere. Furthermore, the yield surfaces delineated using the von Mises criterion and the PIV data (Holenberg et al., 2012) did not superimpose. Similarly, the existence of wall slip can not only lower the drag exerted on a sphere but also lead to significant modifications to the resulting yield surfaces. They correlated their drag results in terms of the Bingham and Deborah numbers, and a slip parameter in the creeping flow regime.

3.2.3.2 Experimental Drag Correlations

The literature is inundated with various empirical formulate developed to relate the drag coefficient with the other pertinent dimensionless groups for spheres in visco-plastic fluids. In the absence of wall effects, the drag coefficient of a sphere in a visco-plastic medium is a function of different dimensionless parameters depending upon the choice of a fluid model. For instance, the relevant parameters are the Reynolds number (Re_B) and the Bingham number (Bi) in the case of the Bingham fluid model, whereas another parameter, namely n, emerges if the Herschel–Bulkley fluid model is used. Additional parameters are also needed to account for elastic and/or thixotropic effects. Irrespective of the choice of the fluid model, any attempt at constructing the drag curve (Drag coefficient–Reynolds number relationship) inevitably results in a family of curves, unlike the single standard drag curve for spheres falling in Newtonian fluids. Numerous attempts (Bhavaraju et al., 1978; Atapattu, 1989; Atapattu et al., 1995) have been made at collapsing the resulting family of curves on to a single curve by combining the relevant dimensionless groups in various ways. The task is made more difficult by the desire to incorporate accurately the Newtonian limit in such a generalized correlation. The value of drag coefficient is extremely sensitive to small values of the yield-gravity parameter, that is, near the yielding/no yielding or motion/no-motion transition. This indeed makes it very difficult to include accurately the Newtonian limit in the correlations for visco-plastic media as the transition from the fluid-like to solid-like (or vice versa) behavior is frequently modeled as being abrupt, rather than being gradual. Additional complications arise from the uncertainty surrounding the determination of the yield stress of experimental fluids. The experimental results published so far also testify to both these problems. However, before embarking upon a detailed presentation and discussion of some of the more widely used empirical equations, it is important to establish the critical value of the Reynolds number (which itself is fluid model-dependent) marking the end of the creeping flow region.

3.2.3.2.1 Criterion for Creeping Flow Regime

For sphere motion in Newtonian fluids, Bi = He = 0, the critical value of the Reynolds number is generally taken to be 0.1, which marks the end of the creeping flow regime. It is useful to recall here that, in general, one of the distinct features of the creeping flow is the inverse relation between the drag coefficient and the Reynolds number. Thus, it will be assumed here that the behavior of the type $C_D \cdot Re = k_1 f_1(Bi)$ or any variation thereof can be interpreted as the signature of the creeping flow regime. Hence, for a constant value of the Bingham (Bi) or Hedstrom (He = $Re_B \cdot Bi$) number, the slope of C_D-Re plots (on log-log coordinates) of −1 would imply the creeping flow conditions. Conversely, the point of departure from

Rigid Particles in Visco-Plastic Liquids

such an inverse relation can be seen as the end of the creeping flow region. Based on the limited experimental evidence (~2.5 ≥ Bi ≥ ~200), Chhabra and Uhlherr (1988a) proposed the following approximate criterion for the maximum value of the Reynolds number marking the end of the creeping flow regime for spheres falling in Bingham plastic fluids:

$$\mathrm{Re}_{B,\max} \sim 100 \mathrm{Bi}^{0.4} \quad (3.6)$$

Unfortunately, Equation 3.6 does not approach the expected limiting behavior as Bi→0. Broadly, the larger the value of the Bingham number, the larger is the value of the critical Reynolds number up to which the creeping flow conditions can be realized. This inference is in line with the currently available numerical results (Nirmalkar et al., 2013a, b) and the general notion that the fluid yield stress imparts stability to the flow.

3.2.3.2.2 Drag Expressions

Most attempts at developing universal drag curves for spheres in visco-plastic fluids fall into two distinct categories. In the first category, these are based on the use of the rheological model parameters evaluated from viscometric data, and these formulae do not necessarily include the Newtonian result in the limit of Bi=He=0, for example, see Andres (1961) and du Plessis and Ansley (1967). In the second category, the definition of either the Reynolds number or the drag coefficient is modified with the objective of forcing the results for visco-plastic media to coincide with the standard drag curve for Newtonian fluids, for example, see Atapattu et al. (1995) and Chafe and de Bruyn (2005). Any discussion on drag correlations for visco-plastic systems must inevitably begin with the pioneering work of Andres (1961) whose original (somewhat awkward) drag formula can be rearranged in the following form without incurring any loss in its accuracy:

$$C_D \mathrm{Re}_B^{1.8} = 10^{(12.26 - 0.78 \mathrm{He})} \quad (1 \leq \mathrm{He} \leq 27) \quad (3.7)$$

However, Andres presented no details regarding the range of experimental conditions associated with Equation 3.7, spheres or fluids used, etc. Pazwash and Robertson (1971, 1975), on the other hand, argued that the difference between the values of drag coefficient for a Bingham fluid and that for a Newtonian fluid, at the same value of the Reynolds number, must be a function of the Bingham or Hedstorm number, as also noted by Blackery and Mitsoulis (1997) via Equation 3.2. On this basis, Pazwash and Robertson (1975) proposed the correlation for drag as

$$C_D - C_{\mathrm{DN}} = 36 \left(\frac{\mathrm{He}}{\mathrm{Re}_B^2} \right) \quad (3.8)$$

Their data embrace the following ranges of conditions: $920 \leq \mathrm{He} \leq 3600$ and $60 \leq \mathrm{Re}_B \leq 2000$. Furthermore, Pazwash and Robertson (1971, 1975) observed that for small values of Re_B and when $C_{\mathrm{DN}} \propto 1/\mathrm{Re}_B$ (hence, C_{DN} will be much smaller than C_D) and therefore $C_D \sim 36 \mathrm{He}/\mathrm{Re}_B^2$, and this is consistent with the form of Equation 3.2; on the other hand, for large values of Re_B, Equation 3.8 predicts a constant value of the drag coefficient, that is, $C_D = C_{\mathrm{DN}}$. The predictions of Equation 3.8 together with the data of Pazwash and Robertson (1971, 1975), Valentik and Whitmore (1965) and Ansley and Smith (1967) are shown in Figure 3.11. Except for the original data of Pazwash and Robertson (1971, 1975), Equation 3.8 is not particularly successful in correlating the independent sets of data available in the literature.

Similarly, Chen et al. (2016) also expressed the total drag force as the sum of the Newtonian and Bingham model contribution, though their numerical constant is ~29 (20% lower) as opposed to the value of 36 in Equation 3.8.

Based on simple intuitive considerations, du Plessis and Ansley (1967) postulated that the drag on a sphere in a visco-plastic medium consists of two contributions, namely, the dynamic and the yield stress component acting over the sphere surface. Thus, they analyzed their results in terms of a dynamic parameter Q, which is in line with the recent suggestion of Thompson and Soares (2016), defined as

$$Q = \frac{\mathrm{Re}_B}{1 + \mathrm{Bi}} \quad (3.9)$$

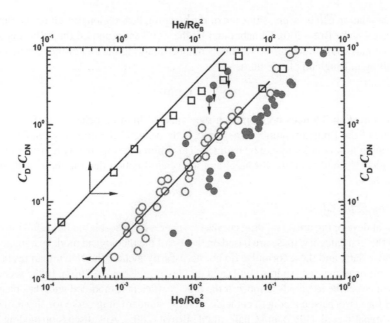

FIGURE 3.11 Literature data plotted according to the method of Pazwash and Robertson (1971, 1975). — Prediction of Equation 3.8. ●: Valentik and Whitmore (1965); □: Ansley and Smith (1967); ○: Pazwash and Robertson (1971, 1975).

For spheres, du Plessis and Ansley (1967) proposed the following expression for drag (in the ranges $0.7 \leq \text{Re}_B \leq 1200$; $2.9 \leq \text{Bi} \leq 16.5$):

$$C_D = 5Q^{-0.49} \tag{3.10}$$

Unfortunately, Equation 3.10 does not approach the Newtonian behavior as Bi → 0, but it is based on experimental results obtained with clusters of particles rather than single spheres. This work, however, laid the foundation for the subsequent model of Ansley and Smith (1967) who, using the slip-line theory of soil mechanics, modified the definition of the dynamic parameter Q as

$$Q_{AS} = \frac{\text{Re}_B}{1+(7\pi/24)\text{Bi}} \tag{3.11}$$

to correlate their results on sphere drag measurements as

$$C_D = \frac{34}{Q_{AS}} \quad \text{for } Q_{AS} < 20 \tag{3.12a}$$

$$C_D = 0.4 \quad \text{for } Q_{AS} > 200 \tag{3.12b}$$

Unfortunately, the two expressions are not quite additive in the intermediate range, viz, $20 \leq Q_{AS} \leq 200$. These equations also do not include the Newtonian limit. Besides, as $(7\pi/24) \approx 1$, the two definitions of the dynamic parameter Q and Q_{AS} are virtually identical. Figure 3.12 shows a comparison between the predictions of Equation 3.12 and some of the experimental results available in the literature. The data of Pazwash and Robertson (1971, 1975) are seen to be in rather poor agreement with Equation 3.12 whereas those of Valentik and Whitmore (1965) are closer to the Newtonian line up to about $Q_{AS} \leq \sim 20$. Subsequently, Hanks and Sen (1983) and Sen (1984), while working with bentonite suspensions, extended the approach of Ansley and Smith (1967) to Herschel–Bulkley model fluids by redefining the dynamic parameter Q_{HS} as

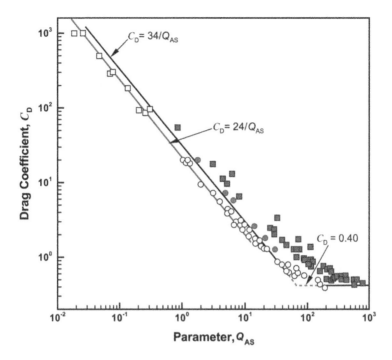

FIGURE 3.12 Comparison between the predictions of Equation 3.12 and the literature data. -----Prediction of Equation 3.12a and b. ■: Pazwash and Robertson (1971, 1975); □: Atapattu (1989); ○: Valentik and Whitmore (1965); re-calculated results of ●: Valentik and Whitmore (1965).

$$Q_{HS} = \frac{Re_{HS}}{1+(7\pi/24)Bi_{HS}} \quad (3.13a)$$

where

$$Re_{HS} = (4n/(3n+1))^n 8^{1-n} Re_{PL} \quad (3.13b)$$

and

$$Bi_{HS} = \frac{\rho d^2}{\tau_0^H}(\tau_0^H/m)^{2/n} \quad (3.13c)$$

They simply stated that the expression $C_D = 100/Q_{HS}$ correlates their results satisfactorily (without giving any details) over the ranges of conditions as $0.66 \le n \le 0.85$; $0.04 \le \tau_0^H \le 4.5$ Pa and $0.2 \le Q_{HS} \le 10^4$. Irrespective of the reliability and accuracy of this expression, it certainly does not incorporate any limiting behaviors such as the power-law behavior $(\tau_0^H = 0)$ or Bingham plastic model ($n=1$) or the Newtonian limit ($\tau_0^H = 0$ and $n=1$). Furthermore, the Newtonian drag curve itself is plotted incorrectly in their original paper that casts some doubt on the reliability of this study.

In a comprehensive experimental study, Atapattu (1989) and Atapattu et al. (1995) have not only presented a critical appraisal of the available results in this field, but also developed a new predictive expression for drag on a sphere taking into account their own measurements as well as the literature data. For the Herschel–Bulkley model fluids, they defined the Bingham number (Bi_{HB}) as

$$Bi_{HB} = \frac{\tau_0^H}{m(V/d)^n} \quad (3.14)$$

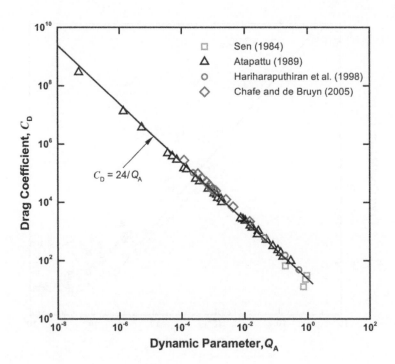

FIGURE 3.13 Drag correlation of Atapattu et al. (1995).

whereas the Reynolds number, Re_{PL}, is still defined by Equation 2.39. These definitions, in turn, can be combined to yield the modified dynamic parameter Q_A as

$$Q_A = \frac{Re_{PL}}{1 + k Bi_{HB}} \quad (3.15)$$

Figure 3.13, replotted from Atapattu et al. (1995), clearly shows that the behavior conforms to $C_D = 24/Q_A$ thereby including the Newtonian result (as $Bi_{HB} \to 0$, $n=1$), with the resulting deviations ranging from −22% to 30%. The data of Atapattu et al. (1995) extended over wide ranges of conditions as $9.6 \times 10^{-5} \le Re_{PL} \le 0.36$, $0.25 \le Bi_{HB} \le 280$, and $0.43 \le n \le 0.84$. At this juncture, it is worthwhile to make two observations. First, the value of $k = (7\pi/24) = 0.916$ as postulated by Ansley and Smith (1967) and subsequently used by others (Hanks and Sen, 1983; Atapattu et al. 1995; Chafe and de Bruyn, 2005) is not only fairly close to that inferred by Beaulne and Mitsoulis (1997) but also appears to be independent of β. Thus, for instance, for the rheology of the S-10 fluid of Atapattu et al. (1995), and in the range $0.10 \le \beta \le 0.33$, the value of k ranges from 0.82 to 0.89 which is well within 10% of the value of 0.916 postulated by Ansley and Smith (1967). The second point concerns the form of the dependence of the drag coefficient C_D on the dynamic parameter Q_A for Herschel–Bulkley fluids. Dimensional considerations suggest the drag coefficient to be a function of the dynamic parameter Q_A and the flow behavior index n. Thus, in the creeping flow regime, the simplest form to capture this relationship is expressed as

$$C_D = 24Y/Q_A \quad (3.16)$$

The advantage of Equation 3.16 is that it incorporates all desirable limits. Unfortunately, the experimental results of Atapattu (1989), Hanks and Sen (1983), and others (Hariharaputhiran et al., 1998) correlate better if $Y = 1$ is used in Equation 3.16 than that when the appropriate value of Y is used corresponding to the value of n, albeit in the case of one fluid, Equation 3.16 seems to work better with the relevant value of Y (Beaulne and Mitsoulis, 1997). This is somewhat surprising that in the range $0.43 \le n \le 0.84$, the value of Y for power-law fluids ranges from ~1.20 to ~1.41 (Table 2.4), but perhaps this effect is overshadowed

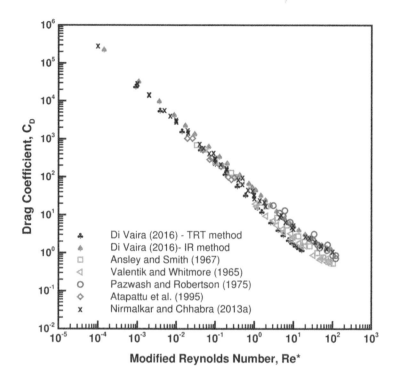

FIGURE 3.14 Comparison of drag coefficient for a sphere in Bingham plastic fluids over the range of finite Reynolds number.

by the rather large errors associated with Equation 3.16. Also, under these conditions, the drag force is heavily weighted toward the yield stress contribution.

Finally, Figure 3.14 contrasts the numerical predictions of drag coefficient (Nirmalkar et al., 2013a, b) with the available experimental results at finite Reynolds number. Following the proposal of Thompson and Soares (2016), the modified Reynolds number used in Figure 3.14 is defined as:

$$\text{Re}^* = \frac{\rho V^{2-n} d^n}{m\{1+\text{Bi}_{HB}\}} = \frac{\text{Re}_{PL}}{1+\text{Bi}_{HB}} \quad (3.17)$$

This definition thus includes the limiting cases of the Bingham fluid behavior ($n=1$) coinciding with the form of Equation 3.11, as well as the power-law fluid behavior $(\tau_0^H = 0)$, and of course the Newtonian limit is recovered by settling $\tau_0^H = 0$ and $n=1$. The agreement is seen to be fair in Figure 3.14. Over the range of condition as $0.2 \leq n \leq 1$, $1 \leq \text{Re}_{PL} \leq 100$, $10^{-3} \leq \text{Bi}_{HB} \leq 10$, and $1 \leq \text{Re}^* \leq 100$, the numerical drag results of Nirmalkar and Chhabra (2013a, b) can be correlated as:

$$C_D = \frac{a_0}{\text{Re}^*}\left(1 + b_0 \text{Re}^{*c_0}\right) \quad (3.18)$$

The values of a_0, b_0, and c_0 are influenced by the value of power-law index as shown in Figure 3.15. For a fixed value of Re*, the drag coefficient is lower in Herschel–Bulkley fluids than that in a Bingham fluid. Also, the value of (C_D.Re*) begins to deviate at gradually increasing value of Re* with the decreasing value of n. Included here are the numerical predictions of Di Vaira (2016) based on the implicitly regularized (IR) and two-relaxation time lattice Boltzmann model. These are seen to be in good agreement with Nirmalkar et al. (2013a).

In addition to these methods, some investigators (Ito and Kajiuchi, 1969) have presented graphs that can be used to estimate the value of drag coefficient in a given application, whereas others have

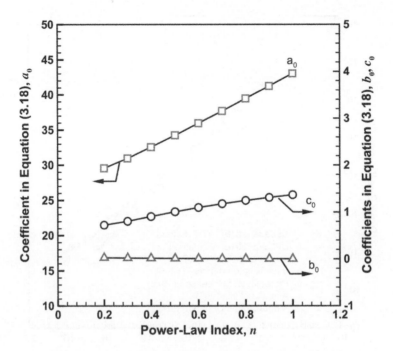

FIGURE 3.15 Values of coefficients in Equation 3.18 over the range of power-law index.

outlined methods that rely on the applicability of the Newtonian drag curve (Whitmore, 1969; Brookes and Whitmore, 1968, 1969; Briscoe et al., 1993; Machac et al., 1995). Indeed, the experimental data of Valentik and Whitmore (1965) have formed the basis of many empirical correlations reported in the literature (Dedegil, 1987; Saha et al., 1992; Wilson et al., 2003).

3.2.4 Values of Yield Stress Used in Correlations

Undoubtedly, the value of the yield stress plays a central role in the interpretation and correlation of experimental results on sphere motion in visco-plastic media. Indeed, much of the disagreement among authors can be attributed to the uncertainties in the values of yield stress used, though other mechanisms such as slip (Deglo de Besses et al., 2004), elasticity and time dependence of yield stress (Cheng, 1979) have also been identified as possible causes of this discrepancy. In view of the recent developments, it is not always safe to assume model experimental fluids as being purely visco-plastic fluids, for these can also concurrently exhibit elastic and thixotropic effects. The commonly used Carbopol solution can exhibit various other features in addition to yield stress (Younes et al., 2020). Each of these contribute to the lack of correspondence between different studies. The difficulty of ascribing a value to the yield stress is illustrated in Table 3.4 where different methods are shown to yield wide ranging values in line with the inter-laboratory study of Nguyen et al. (2006). In most cases, shear stress–shear rate data (or equivalently, force–velocity data in tow tanks) are extrapolated to zero-shear rate, and the intercept on the ordinate is taken to be the yield stress. Unfortunately, the values of the yield stress so obtained are extremely sensitive to the range (particularly the lowest value of shear rate) of the data being extrapolated corresponding to zero-shear rate or zero velocity. This approach intrinsically assumes that the same type of relationship between the shear stress and the shear rate (or force and velocity) will continue all the way down to zero values. This practice is thus open to criticism and as such has no theoretical justification. The approximate minimum values of the shear rate reached in viscometric measurements of various experimental studies are summarized in Table 3.5.

For a sphere moving through a fluid in creeping motion, the surface-averaged shear rate is of the order of (V/d) and this quantity is generally smaller than $100\,s^{-1}$ and possibly as small as $2\text{--}3\,s^{-1}$ or even smaller

TABLE 3.4

Comparison of Yield Stress Values Obtained Using Different Methods

Measurement Technique	Method	Yield Stress (Pa)			
		Carbopol Solutions		Silica Suspensions	
		0.08%	0.09%	53%	56.8%
Haake rotational viscometer	Extrapolation of flow curve (low shear rate)	2.6	6.5	13.1	18.8
	Bingham model	15.0	18.8	26.5	48.9
	Herschel–Bulkley model	4.2	8.8	17.5	20.0
	Casson model	4.8	9.0	16.9	25.0
	Stress relaxation model (Keentok, 1982)	2.4	5.8	12.7	13.2
	Vane torsion (Nguyen and Boger, 1983)	4.0	6.2	17.5	21.8
Steady flow on an inclined plane	Extrapolation of flow curve (small shear rate)	2.0	5.8	12.5	–
	Bingham model	2.0	5.9	12.5	–
	Stability of flow on a plane (Uhlherr et al., 1984)	2.0	5.4	11.5	14.3

Sources: Park, K.H., Ph.D. Dissertation, Monash University, Melbourne, Australia (1986).
Chhabra, R.P. and Uhlherr, P.H.T., *Encyclopedia of Fluid Mech.*, **7**, 611 (1988a).

TABLE 3.5

Lowest Shear Rate Reached by Various Investigators

Investigators	Minimum shear rate (s^{-1})
Valentik and Whitmore (1965)	150
du Plessis and Ansley (1967)	10
Ansley and Smith (1967)	~24
Ito and Kajiuchi (1969)	2000
Pazwash and Robertson (1971, 1975)	5
Hanks and Sen (1983); Sen (1984)	50
Atapattu (1989)	~0.1

near the yielding point. Hence, the fluid characterization of Ito and Kajiuchi (1969) is of little relevance to their falling sphere results. In this regard, a comment about the work of Ansley and Smith (1967) is also in order. They evaluated the plastic viscosity and yield stress as the slope and intercept of a tangent to the flow curve at a point where $\dot{\gamma} = (V/d)$. This approach has an inherent weakness that the yield stress is no longer a constant material parameter. Moreover, this approach generates artificial values of τ_0^B and, μ_B even for simple pseudoplastic fluids without any real yield stress. Thus, intuitively, one would expect the correlation of Ansley and Smith to be applicable to pseudoplastic materials also and indeed this was shown by Atapattu (1989). Another study of Wilson et al. (2003) also echoes similar views. It is abundantly clear that an unambiguous evaluation of the yield stress is of crucial importance for a realistic analysis of flow problems involving visco-plastic media.

3.2.5 Time Dependence of Velocity in Visco-Plastic Fluids

Unlike in the case of Newtonian and purely shear-thinning and shear-thickening fluids, the reproducibility of the terminal settling velocity of spheres in visco-plastic fluids is known to be poor (Valentik and Whitmore, 1965; Ansley and Smith, 1967; Sgreva et al., 2020). Perhaps Atapattu et al. (1995) were the first to document this effect, who found that the reproducible results of the terminal falling velocities could only be obtained after 4–10 individual spheres had been dropped in the fluid consecutively at intervals of a few minutes. While the drag results presented in the preceding section are based on the asymptotic values of the terminal velocity, this time effect itself is intriguing. Prompted by the preliminary findings of Atapattu et al. (1995), Hariharaputhiran et al. (1998) conducted a systematic study to understand this phenomenon. They worked with one Carbopol solution, similar to one of the fluids used by Atapattu et al. (1995) with $n=0.55$, $m=7.06$ Pa s^n, and $\tau_0^H = 8.25$ Pa and used several spheres ($6.35 \leq d \leq 9.54$ mm; $3208 \leq \rho_p \leq 8539$ kg m^{-3}). They used a different experimental protocol as follows. The sets of four spheres were dropped at different intervals of time. The "time interval" is defined as the elapsed time between the instant when the last sphere in a set is released and the instant when the first sphere in the next set is released, that is, the time interval is a crude measure of the period during which the fluid is left undisturbed. Also, the terminal falling velocity of the first sphere in each set is taken to be the measure of the recovery/healing of the network from the disturbance caused by the passage of the previous sets of spheres. Figure 3.16 shows the typical time dependence for a 6.35 mm stainless-steel sphere; since similar results were obtained with other spheres, these are not shown here. Clearly, within each set of four spheres, the terminal velocity of successive spheres increases and attains an asymptotic value after three spheres, that is, the velocity of the fourth sphere in each set is nearly the same irrespective of the time interval. This observation is in line with that of Atapattu et al. (1995). Also, perhaps the most important observation here is that the velocity of the first sphere in each set decreases with increasing time interval between each set. Finally, a time interval of the order of about 2 h was found adequate to eliminate the effects of the previous spheres. It is reasonable to postulate that this value (healing time) will vary from one polymer to another as well as with polymer concentration and the sphere characteristics. Qualitatively similar observations have also been made by Cho et al. (1984) and Gheissary and van den Brule (1996) who ascribed this effect to the selective sweeping of polymer molecules by a falling sphere and then the polymer molecules return, by diffusion, to their initial position. However, Ambeskar and Mashelkar (1990) were unable to measure any differences in the concentration of polymer in the solution in the front and rear of the sphere. Likewise, the time scale of diffusion is much longer than the healing time of 3–20 min as reported by Cho et al. (1984). On the other hand, Hariharaputhiran et al. (1998) argued that the Carbopol-type resins are polymers of acrylic acid cross-linked with poly

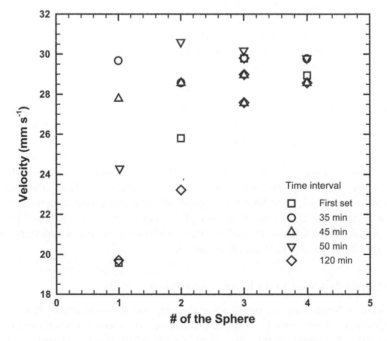

FIGURE 3.16 Time-dependent terminal falling velocity of 6.35 mm steel spheres in a Carbopol solution. (Replotted from Hariharaputhiran et al., *J. Non- Newt. Fluid Mech.*, **79**, 87, 1998.)

alkenyl ethers or divinyl glycol and contain some linear polymer (linear chain) impurities. In an aqueous solution, the individual molecules take up water and swell, and the linear chains present induce the formation of polymer networks involving swollen polymer molecules. With the passage of a sphere, the linear chains may become disentangled and in due course of time, these diffuse back to again act as linkages between the swollen molecules. Based on realistic values of molecular diffusivity and the size of linear chains, the estimated healing time of about 2 h compares favorably with the terminal velocity data. Undoubtedly, this seems like a plausible mechanism for healing, but clearly a more detailed study is needed to put this matter on a sound footing. Furthermore, this effect is expected to vary from one polymer system to another and may even vary with the weight of spheres.

This section is concluded by mentioning two interesting studies. Chafe and de Bruyn (2005) have measured the force required to pull a sphere through a 6% bentonite clay suspension. By choosing the experimental conditions carefully, the values of the force are only weakly dependent on the speed of the sphere, that is, the required force is mainly due to the presence of the yield stress. Therefore, not only were they able to extract the values of the yield stress by extrapolating force–velocity data to zero velocity, but were also able to follow the kinetics of gelling using the force–time data. However, the values of the yield stress evaluated using the residual force were found to be smaller than those obtained by extrapolating force–velocity data. Similarly, in another interesting study, Ferroir et al. (2004) have studied the sedimentation of spheres in laponite suspensions with and without the externally imposed vibrations in the direction normal to the sedimentation of the sphere. The laponite suspensions were modeled as thixotropic fluids with a yield stress. Depending upon the experimental protocols, different types of settling behaviors were recorded.

For sedimentation in the suspensions at rest, for short times of rest, the terminal velocity was found to be approximately constant. Following long periods of rest, the sphere either did not move or stopped after moving for a short period. While the vibrations had no significant influence on the settling velocity in a viscous Newtonian fluid, the falling velocity was greatly enhanced due to vibrations in the laponite suspensions. Some of these features resemble that observed by Hariharaputhiran et al. (1998) for spheres falling in a Carbopol solution and that observed by Sgreva et al. (2020) who used superabsorbent polymer grains. Merkak et al. (2008, 2009) and Jossic and Magnin (2004) have studied structuring, migration, and sedimentation of particles in shearing flow of visco-plastic fluids in a pipe. Interactions between two spheres falling in Bingham plastic fluids have been experimentally examined by Fahs et al. (2018).

3.3 Flow Past a Circular Cylinder

The creeping flow of visco-plastic liquids past a confined and unconfined circular cylinder has received very little attention. Table 3.6 provides a bird's eye view of the activity in this field for 2-D cylinders of various cross-sections. In a key paper, Adachi and Yoshioka (1973) studied the 2-D creeping flow of a Bingham plastic fluid past an unconfined circular cylinder. They considered the case of a cylinder moving with a constant velocity. They argued that since the deformation induced by the cylinder is restricted to a localized region near the cylinder, the inertial effects in the far-flow field can be neglected and thus the so-called Stokes paradox is irrelevant here, similar to the flow of pseudoplastic power-law fluids past a cylinder (Tanner, 1993). The size of this fluid region, in general, decreases with the increasing value of the Bingham number, $\text{Bi} = \left(\tau_0^B d/\mu_B V\right)$. Their upper and lower bounds on the dimensionless drag force $F_D^* = (F_D/6\mu_B dV)$ are shown in Figure 3.17 where the two bounds are seen to differ appreciably at small values of the Bingham number, though the two bounds are pretty close for $\text{Bi} > \sim 50$. Based on the idea of slip lines, Adachi and Yoshioka (1973) also extended the analysis of Ansley and Smith (1967) to the flow past a cylinder to deduce the approximate formula for F_D^* as

$$F_D^* = \frac{(\pi+2)\text{Bi}}{6} \tag{3.19}$$

The predictions of Equation 3.19 are also included in Figure 3.17. Finally, Adachi and Yoshioka (1973) recommended the use of the lower bound for $\text{Bi} < 100$ and Equation 3.19 for $\text{Bi} > 100$. No justification, however, was provided for this suggestion. Subsequently, Yoshioka et al. (1975) reported an experimental

TABLE 3.6

Investigations of Visco-plastic Fluid Flow over Circular and Noncircular Cylinders

Investigators	Test Fluids or Fluid Model	Comment and Results
Patel and Chhabra (2013)	Elliptical cylinder in Bingham fluids	Extensive numerical results on yield surfaces and drag coefficient.
Nirmalkar and Chhabra (2014)	Circular cylinder in Bingham fluids	Extensive numerical results on yield surfaces and drag coefficient.
Tiwari and Chhabra (2015b)	Semi-circular cylinder in Bingham fluids	Extensive numerical results on yield surfaces and drag coefficient.
Bose et al. (2014)	Cylinder of equilateral triangular cross-section in Bingham plastic fluids	Extensive numerical results on yield surfaces and drag coefficient.
Nirmalkar et al. (2012, 2013c)	Cylinder of square cross-section in Bingham plastic fluid	Extensive numerical results on yield surfaces and drag coefficient.
Tokpavi et al. (2008, 2009)	Circular cylinder in Carbopol solution	Asymmetric flow field in the front and rear of the cylinder observed using PIV results in the zero Reynold number limit.
Jossic and Magnin (2009)	Circular cylinder in Carbopol solution	Drag on single cylinder and interaction between two cylinders.
Mossaz et al. (2010, 2012a, b)	Circular cylinder in Carbopol solution	Effect of Reynold number on drag and flow separation in the steady flow regime. Onset of unsteady flow regime.
Ozogul et al. (2015)	Circular cylinder in Carbopol solution	Effect of wall slip on the hydrodynamics.
Mahmood et al. (2018)	Circular cylinder in a Bingham fluid	Used the exponential regularization up to $Bi = 50$ and $Re_B \leq 75$; asymmetrical confined cylinder in a slit.
Kanaris et al. (2015)	3-D, direct numerical simulation for Bingham fluid flow over a cylinder in a slit ($\beta = 0.2$)	Established the conditions for the onset of 3-D effect and/or turbulence.
Chen et al. (2014)	$Bi \leq 5$; $150 \leq Re_B \leq 600$ Numerical study in a Bingham fluid	Drag coefficient as a function of Bi and Re_B for a circular cylinder.
Syrakos et al. (2015)	Thixotropic flow over a cylinder at $Re = 45$	In the absence of thixotropy, multiple unyielded regions inside the wake can be observed. Vortex shredding is suppressed.
Fonseca et al. (2013)	Thixotropic flow over a confined cylinder in a slit ($\beta = 0.125$)	Role of viscous effects is modulated by the thixotropic behavior.

study, but no drag measurements were made to substantiate/refute these predictions. In the intervening 25 years, there has been a spurt in research related to the visco-plastic flow over a long cylinder. Thus, Deglo de Besses et al. (2003) have reported analogous numerical simulations for the unconfined flow of Herschel–Bulkley model fluids of which the Bingham plastic is a limiting case, with $n = 1$. Their predictions are also included in Figure 3.17 where the agreement is seen to be fair. This numerical study also shows the lower bound predictions of Adachi and Yoshioka (1973) to be reasonable. For the zero-Reynolds number flow of Herschel–Bulkley model fluids, the 31 numerical data points encompassing the ranges $0.26 \leq n \leq 1$ and $Bi_{HB} < 100$ are well approximated by the equation

$$C_D = \frac{2}{Re_{PL}}(A + B Bi_{HB}) \quad (3.20)$$

where

$$A(n) = 19.39 \sin\{1.022(1-n)\} \quad (3.21a)$$

$$B(n, Bi_{HB}) = 11.63\left[1 + 1.845(n - 0.160) Bi_{HB}^{-0.375(n+0.73)}\right] \quad (3.21b)$$

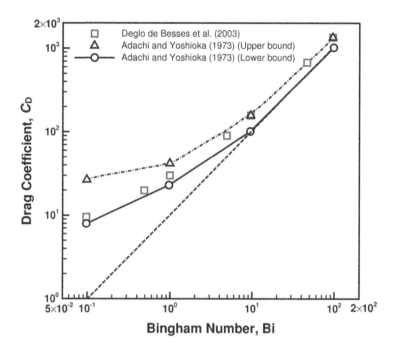

FIGURE 3.17 Predicted drag coefficients for the unconfined creeping cross-flow of Bingham fluids past a cylinder; - - - - Slip-line approximation, Equation 3.19 (Adachi and Yoshioka, 1973). (Replotted from Deglo de Besses, et al., *J. Non-Newt. Fluid Mech.*, **115**, 27, 2003.)

In qualitative terms, the effect of shear-thinning (value of n) diminishes gradually with the increasing value of the Bingham number, Bi_{HB}. Thus, one can expect $B(n, \infty) = B_\infty = 11.63$, and this value compares rather well with the value of 10.28 predicted by the slip-line analysis of Adachi and Yoshioka (1973) and the subsequent numerical results in the fully plastic flow limit.

The corresponding situation involving the creeping flow of Bingham plastic fluids past a circular cylinder confined symmetrically between two parallel plates has been numerically studied by Zisis and Mitsoulis (2002). They considered the Poiseuille flow case and used the standard exponential modification (Papanastasiou model) for the yield stress term in the Bingham model equation. Their drag results in the range $Bi \leq (1000\,\beta)$ are well correlated by the equation

$$Y = \frac{C_D}{C_{DN}} = \left\{1 + a\left(\frac{Bi}{\beta}\right)\right\}^b \tag{3.22}$$

where β (< 1) is the blockage ratio defined as the cylinder diameter/slit width. The drag coefficient in a Newtonian fluid, C_{DN}, is given by

$$C_{DN} = \frac{2F_N^*}{Re} \tag{3.23}$$

The values of a, b, and F_N^* for different values of β are provided in Table 3.7.

Figure 3.18 shows these results graphically in terms of a dimensionless drag force F_D^* ($= F_D/\mu_B V L$), or the drag correction factor, Y, as a function of the Bingham number for scores of values of β. The solid lines correspond to the predictions of Equation 3.22. At large values of Bi ($>1000\beta$), the drag force seems to vary linearly with the Bingham number. This finding is consistent with that observed for spheres and with Equation 3.19.

Subsequently, these results have been supplemented by studying the case of a stationary cylinder confined between the walls moving with a constant velocity (Mitsoulis, 2004) equivalent to that of the freely

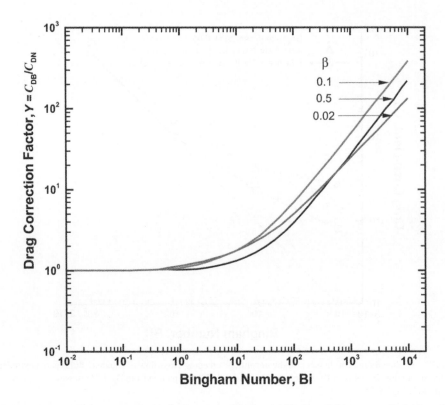

FIGURE 3.18 Drag coefficient values for the laminar Poiseuille flow of Bingham fluids past a cylinder in a planar slit. (After Zisis, Th. and Mitsoulis, E., *J. Non-Newt. Fluid Mech.*, **105**, 1, 2002.)

TABLE 3.7
Value of a, b, and F_N^* in Equations 3.22 and 3.23

β	F_N^*	a	b
0.02	6.29	0.0732	0.746
0.05	9.04	0.0755	0.840
0.10	13.37	0.0747	0.901
0.125	15.72	0.0726	0.916
0.250	31.95	0.057	0.959
0.50	132.31	0.031	0.944

Source: Zisis, Th. and Mitsoulis, E., *J. Non-Newt. Fluid Mech.*, **105**, 1 (2002).

falling cylinder. While qualitatively similar behavior is observed for this case, the resulting values of the drag force are slightly different. Mitsoulis (2004) correlated his numerical results on drag coefficient for this case using the slightly different form of Equation 3.22 as

$$Y = \frac{C_D}{C_{DN}} = (1 + a_1 \text{Bi})^{b_1} \qquad (3.24)$$

The drag coefficient in a Newtonian fluid, C_{DN}, is still given by Equation 3.23, but with slightly different values of F_N^*. The values of a_1, b_1, and F_N^* for this case are summarized in Table 3.8 for a range of values of the blockage ratio β in the range $0 \leq \text{Bi} \leq 1000$.

Although some experimental results on cylinder drag in visco-plastic fluids are available in the literature (Jossic and Magnin, 2001), those relate to the towing of cylinders of finite length-to-diameter ratio

TABLE 3.8
Value of a_1, b_1, and F_N^* for use in Equation 3.24

β	F_N^*	a_1	b_1
0.02	4.20	5.16	0.913
0.05	6.03	3.39	0.916
0.10	8.96	2.14	0.921
0.25	21.97	0.72	0.943
0.50	99.32	0.15	0.971

Source: Mitsoulis, E., *Chem. Engng. Sci.*, **59**, 789, (2004).

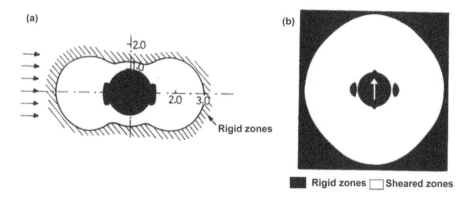

FIGURE 3.19 Shape and size of fluid-like regions for the creeping Bingham fluid flow across a circular cylinder according to (a) Adachi and Yoshioka (1973) and (b) Deglo de Besses et al. (2003) for $Bi_{HB}=10$ and $n=0.26$.

at constant velocity under unconfined conditions and, therefore, cannot be compared directly with the numerical predictions of Zisis and Mitsoulis (2002), or of Mitsoulis (2004).

In addition to the drag results, Adachi and Yoshioka (1973) also provided approximate size and shape of the fluid-like and solid-like zones close to the cylinder, as shown schematically in Figure 3.19a. While the numerical calculations of Deglo de Besses et al. (2003) and Roquet and Saramito (2003) reveal qualitatively similar shapes (Figure 3.19b) of the yielded and unyielded regions, the analysis of Adachi and Yoshioka (1973) seems to underestimate the size of the deformed regions. Notwithstanding the additional complications arising from the wall effects, the simulations of Zisis and Mitsoulis (2002) also corroborate these findings on the shape and size of these regions. Broadly speaking, the size of the outer (far away) boundary between the yielded/unyielded regions is little influenced by the power-law index for $Bi_{HB} > \sim 5$. The polar caps of the unyielded material grow with the increasing value of Bi_{HB}, but shear-thinning has an opposite effect on the size of these regions. Finally, there are unyielded regions present on the sides of the cylinder. These pockets of unyielded material have been found to undergo rigid-body rotation and translation. Depending upon whether the cylinder is ascending or falling, the rotation could be clockwise for one and anticlockwise for the other. Moreover, these regions also grow in size with the increasing Bingham number and decreasing degree of shear-thinning. Their relative positioning is also strongly dependent on the values of n and Bi_{HB}, being close to the cylinder at small values of the power-law index and high values of Bi_{HB}. However, to date, no flow visualization experimental results are available to provide confirmation of these predictions.

Subsequent more detailed numerical studies (Nirmalkar and Chhabra, 2013a, b; Chen et al., 2014; Di Vaira, 2016) reveal that the unyielded material present at $\theta = \pi/2$ and $3\pi/2$ (Figure 3.19b) is also undergoing rigid-body rotation whereas the polar caps at $\theta = 0$ and $\theta = \pi$ are static in nature in the creeping flow region. Figure 3.20 shows the combined effect of the Bingham number and Reynolds number on the yield surfaces delineating the "fluid-like" and "solid-like" regions. The "fluid-like" cavity encapsulating the cylinder expands with the increasing strength of the flow (Reynold number) whereas the size of the polar caps diminishes with the increasing Reynolds number. The unyielded material present at the front stagnation point completely disappears with the increasing Reynolds number and the unyielded material at the rear stagnation point grows a little. At very high Bingham number (Bi $\geq 10^4$), the fluid inertia is completely outweighed by the yield stress effects and thus the yield surface approached a limiting size and shape commensurate with the fully plastic flow condition. Broadly, the currently available various numerical predictions for a confined and unconfined cylinder with no-slip condition on the surface of the cylinder are mutually consistent, as shown in Figure 3.21.

Similar to the flow regime transitions for a cylinder in Newtonian fluids, Mossaz et al. (2010, 2012a) have delineated the conditions for the onset of flow separation and for the loss of steady-state flow regime, angle of separation, and wake properties in Bingham plastic and Herschel–Bulkley model fluids. Some of these predictions are in qualitative agreement with the scant experimental results. Broadly, the fluid yield stress tends to stabilize the flow and thus the first two instabilities (flow detachment and vortex shedding) occur at higher Reynolds numbers in yield stress fluids than that in Newtonian fluids. Limited results, both numerical and experimental, on the role of wall slip (on the surface of cylinder) in the low Reynolds number regime are also available in the literature (Ozogul et al., 2015; Tokpavi et al., 2008, 2009).

Generally, the hydrodynamic drag on the cylinder is reduced due to the wall slip and the fluid-like regions formed are generally smaller in this case than that for the no-slip condition. The numerical predictions based on the Herschel–Bulkley model fluid are consistent with the asymptotic analyses and scant experiment data available in the literature (Tokpavi et al., 2008, 2009; Jossic and Magnin, 2009). In the fully plastic limit, the drag on a cylinder with no-slip condition can be ~30% higher than that with

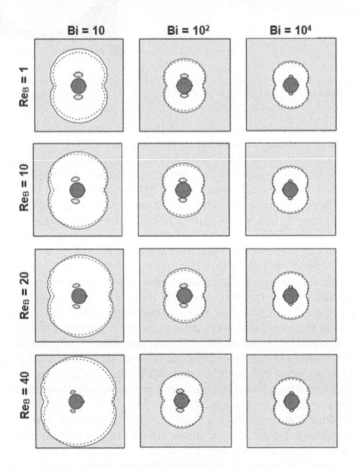

FIGURE 3.20 Effect of Reynolds number, Re_B, and Bingham number, Bi, on the apparent yield surfaces for flow over a cylinder. (From Nirmalkar, N. and Chhabra, R.P., *Int. J. Heat Mass Transf.*, **70**, 564, 2014.)

Rigid Particles in Visco-Plastic Liquids

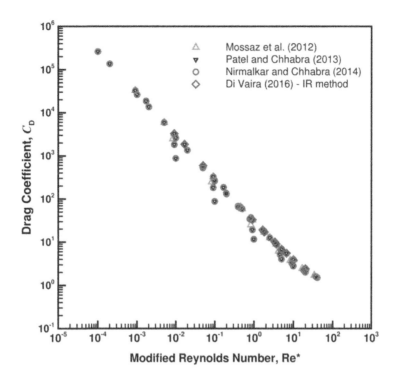

FIGURE 3.21 Comparison of drag coefficient for a cylinder in Bingham plastic fluids over the range of finite Reynolds number.

fully slip condition prescribed on the surface of the cylinder. Similarly, Ouattara et al. (2018) have studied the drag on a cylinder in a visco-plastic fluid moving parallel to a plane wall. Ozogul et al. (2016), on the other hand, have examined the hydrodynamics of a cylinder in a Poiseuille flow in a slit.

In another study, Kanaris et al. (2015) have investigated the condition when the signature of 3-D effects begins to manifest for Bingham fluid flow over a cylinder confined in a slit. For $\beta=0.2$, they carried out direct numerical simulation over the range $150 \leq Re_B \leq 600$ and $Bi \leq 5$ and concluded that the critical value of the Reynolds number marking the transition from 2-D to 3-D flow increased linearly with the Bingham number. Also, they identified two possible modes for this transition.

The numerical drag results at finite Reynolds numbers for a 2-D cylinder (Nirmalkar and Chhabra, 2014; Chen et al., 2014; Di Vaira, 2016) are in good agreement up to about $Re^* \leq 50$. Nirmalkar and Chhabra (2014) put forward the following expression for drag valid over the range of conditions as $1 \leq Re_B \leq 40$; $Bi \leq 10^4$:

$$C_D = \frac{24.75}{Re^*} \quad (3.25)$$

On the other hand, based on their results ($0 \leq Bi \leq 41$ and $0.2 \leq Re_B \leq \sim 100$), Chen et al. (2014) consolidated their numerical results by an equation of the form of Equation 3.8 with the numerical constant of two on the right-hand side. Figure 3.21 contrasts various drag predictions which seem to be internally consistent with each other.

3.4 Flow Normal to a Plate

In an extensive study, Savreux et al. (2005) have studied the steady flow of Bingham plastic fluids normal to a plate. They reported extensive results on drag and on the shape of yielded/unyielded regions as functions of the Reynolds number and Bingham number, both defined using the height of the plate oriented normal to the direction of flow. Over the range of conditions of $0.01 \leq Re_B \leq 10$ and that of the Bingham number $0.001 \leq Bi \leq 100$, they approximated their numerical results by analytical expressions such as the following.

For $Re_B = 0.01$

$$C_D = \frac{1}{Re_B}\left[1.05 + \left(5.77 + 6.37 Bi^{-0.57}\right) Bi\right] \quad (3.26a)$$

For $Re_B = 1$

$$C_D = \frac{1}{Re_B}\left[2.34 + \left(5.51 + 5.57 Bi^{-0.46}\right) Bi\right] \quad (3.26b)$$

For $Re_B = 10$

$$C_D = \frac{1}{Re_B}\left[6.8 + \left(5.82 + 3.47 Bi^{-0.5}\right) Bi\right] \quad (3.26c)$$

An interesting feature of this work is that for large values of the Bingham number (> ~2–5), Equations 3.26a–c all merge into one curve, thereby suggesting that the value of ($C_D \cdot Re_B$) is a unique function of the Bingham number only. As the value of the Reynolds number is gradually increased, a vortex is formed downstream of the plate. The formation and the size of the vortex are strongly influenced by the values of the Reynolds number and Bingham number. Generally, the vortex in visco-plastic fluids is smaller than that in Newtonian fluids at a fixed Reynolds number.

3.5 Nonspherical Particles

In addition to the aforementioned body of information on the flow of visco-plastic liquids past a sphere, a cylinder, and a plate, scant work has also been reported on the objects of various shapes including discs, hemisphere, cones, and plates, prisms, cubes, thin rods and wires, and ellipsoids and a terse summary of such studies is given in Table 3.9, and these have been reviewed elsewhere (Chhabra, 1996a).

The early studies of Brookes and Whitmore (1968, 1969) and Pazwash and Robertson (1971, 1975) primarily reported on drag coefficient behavior in free settling conditions and in tow tanks and presented empirical predictive correlations. However, the inadequate rheological characterization of the test fluids used in these studies raises some doubts about the reliability of these correlations. Subsequent study of Jossic and Magnin (2001), on the other hand, presented extensive results on the criterion of static equilibrium and on drag coefficient. The role of possible slip was also examined by using roughened test particles. Generally, there was very little influence of this parameter on drag coefficient, except in the case of a sphere and a cylinder oriented vertically. The values of Y_G vary from ~ 0.02 for short vertical ($L/d < 0.14$) cylinders to about

TABLE 3.9

Nonspherical Particles in Visco-Plastic Liquids

Particle Shape	Fluids	Reference
Thin rods and wires	Clay suspensions	Brookes and Whitmore (1968, 1969)
Prisms, rectangles, and cubes	Clay suspensions and Carbopol solutions	Brookes and Whitmore (1968, 1969), Jossic and Magnin (2001)
Ellipsoids and discs	Clay suspensions	Pazwash and Robertson (1971, 1975)
Ellipsoids	Foams	Dollet et al. (2006)
Spheroids	Bingham plastics	Gupta and Chhabra (2014)
Cones	Bingham plastics	Ahonguio et al. (2014), Mishra et al. (2019b)
Plate	Bingham plastics	Raja et al. (2015), Ahonguio et al. (2016a, b), Ouattara et al. (2019)
Hemisphere	Bingham plastics	Patel et al. (2015)
Thin disc	Bingham plastics	Gupta et al. (2017b)

~ 0.15 for long ($L/d \sim 5$) vertical cylinders. Similarly, the dimensionless drag force (scaled with the yield stress multiplied by the projected area) ranges from ~ 6 to ~32 for smooth and roughened particles of the shapes studied by them. Aside from the aforementioned studies dealing with the measurement of terminal falling velocity and drag, de Bruyn (2004) and Dollet et al. (2005a, b) have reported interesting results on the drag experienced by variously shaped obstacles being towed at constant velocities through visco-plastic foams. By gradually reducing the velocity to vanishing small values, they were able to delineate the contributions of drag due to the yield stress and viscous effects. Such detailed studies indeed provide useful insights into the nature of dissipative processes that can be linked to the dynamics of bubbles making up the foam and/or to study the extent of molecular orientation (Haward and Odell, 2004).

In recent years, as summarized in Table 3.9, there has been a renewed interest in numerically studying the steady flow of Bingham plastic fluids past spheroids (Gupta and Chhabra, 2014), cones of different apex angles (Mishra et al., 2019b), plate (Raja et al., 2015; Ahonguio et al., 2014), hemisphere (Patel et al., 2015), thin circular discs (Gupta et al., 2017b), etc. In most cases, drag coefficient has been correlated with the modified Reynolds number, similar to the one defined by Equation 3.17 and reference must be made to the original publications for further details.

3.6 Conclusions

The available body of knowledge on the flow of visco-plastic media past variously shaped objects has been thoroughly reviewed here. As usual, the steady creeping flow past a sphere occupies the center stage, and excellent numerical predictions of drag and yielded/unyielded regions are now available for both the Bingham plastic and Herschel–Bulkley fluid models. These show fair to moderate correspondence with the corresponding experimental results, provided the yield stress is evaluated appropriately. The large body of experimental data relating to the conditions beyond the creeping flow has been correlated using empirical expressions; however, none of these has proved to be completely satisfactory. However, these seem to be consistent with the limited numerical predictions for an unconfined sphere up to $Re_B = 100$. The cross-flow past a confined and unconfined circular cylinder has also received much attention, and excellent numerical results on drag and flow field are now available up to the Reynolds number value of 40–50 in the steady flow regime. Significant differences are observed in term of the wake properties, morphology of yielded/unyielded subdomains. Drag and fluid yielding are reduced due to the wall slip. Scant results are also available now for objects of scores of 2-D and axisymmetric shapes. At the same time, there has been a growing recognition of the fact that visco-plasticity is often accompanied either by elasticity or thixotropy. Some results are also available on the additional effects arising from these aspects on the detailed kinematics and gross parameters of bluff body flows which point to even more complex dynamics than that predicted in the absence of elasticity and thixotropy. However, these have not yet been contrasted with the experimental results. Little is known about the drag or static equilibrium criterion for nonspherical particles of other shapes.

Nomenclature

a	constant (-), Equation 3.2
a_0	constant (-), Equation 3.18
a_1	constant (-), Equation 3.24
b	constant (-), Equation 3.2
b_0	constant (-), Equation 3.18
b_1	constant (-), Equation 3.24
Bi	Bingham number (-)
Bi_{HB}	Bingham number for Herschel–Bulkley fluids (-)
c_0	constant (-), Equation 3.18
C_D	drag coefficient (–)
C_{DN}	drag coefficient in an equivalent Newtonian fluid (–)

d	diameter of sphere or cylinder (m)
D	diameter of tube (m)
F_D	drag force on cylinder per unit length (N/m)
F_D^*	dimensionless drag force (-)
F_N^*	dimensionless drag force in a Newtonian medium (-)
f_0	wall correction factor in the creeping region (-)
g	acceleration due to gravity (m/s^2)
$He = (Re \times Bi)$	Hedstrom number (-)
L	length of cylinder (m)
m	consistency coefficient in Herschel–Bulkley fluid model (Pa.s^n)
M	parameter, Equation 3.4 (-)
n	flow behavior index (-)
Q	dynamic parameter, Equation 3.9 (-)
Q_A	modified dynamic parameter, Equation 3.15 (-)
Q_{AS}	modified dynamic parameter, Equation 3.11 (-)
$Q_{H\text{-}S}$	modified dynamic parameter, Equation 3.13a (-)
Re_B	Reynolds number based on Bingham plastic viscosity (-)
Re_{PL}	power-law Reynolds number (-)
Re^*	modified Reynolds number, Equation 3.18 (-)
V	terminal velocity of sphere or cylinder or faraway uniform flow velocity (m/s)
Y	drag correction factor (-)
Y_G	dimensionless yield parameter, Equation 3.1 (-)
Y_N	drag correction factor for a sphere or a cylinder in Newtonian fluids (-)

Greek Symbols

β	blockage ratio, (d/D) for spheres or cylinder diameter/slit height, (-)
$\dot{\gamma}$	shear rate (s^{-1})
δ	parameter, Equation 3.5 (-)
η	apparent viscosity (-)
μ	Newtonian viscosity (Pa.s)
μ_B	Bingham plastic viscosity (Pa.s)
μ_y	yielding viscosity, Equation 3.3 (-)
ρ	fluid density (kg/m^3)
ρ_p	particle density (kg/m^3)
τ_0	true yield stress (Pa)
τ_0^B	Bingham model parameter (Pa)
τ_0^H	Herschel–Bulkley model parameter (Pa)

Subscripts

BE	Bercovier Engelman model
BV	Biviscous model
P	Papanastasiou model

4

Rigid Particles in Visco-Elastic Fluids

4.1 Introduction

During the past 50–60 years, the field of particle motion in visco-elastic liquids has witnessed unprecedented growth and, consequently, significant advances have been made to our understanding of the underlying fluid dynamical phenomena. Such studies have been motivated by two distinct but interrelated perspectives. First, an adequate knowledge of gross fluid mechanical aspects such as drag force and torque acting on a particle moving in a stationary or sheared visco-elastic medium is frequently needed in a range of process engineering applications such as sedimentation of muds and suspensions, handling of fluid systems, falling-ball viscometry, micro-rheometry in diffusing wave spectroscopy (Levine and Lubensky, 2001), formation of weld lines in polymer processing (Nguyen–Chung et al., 1998), in assembly process for anisotropic conductive joints (Ogunjimi et al., 1995), and in biological processes (Holzwarth et al., 2002; Foo et al., 2004). Second, the flow over a sphere or a cylinder yields a complex (but free from geometric singularities) nonviscometric flow and it thus affords a fairly stringent test for establishing the validity of visco-elastic fluid models whose parameters are invariably evaluated from data obtained in well-defined viscometric flows. These two objectives coupled with the simplicity of the geometry have also made such flows very attractive for the validation of numerical solution procedures. In addition, there has been a renewed interest in studying mixed flows such as that induced by a translating and rotating sphere in a stationary visco-elastic fluid, or the steady settling of a sphere in a fluid medium subjected to shearing flow. Additional potential applications are found in the rheometry of suspensions (in visco-elastic suspending media) which can be challenging due to the migration of particles across streamlines (D'Avino et al., 2017; Wang et al., 2018) and in the emerging area of micro-rheology (He et al. 2020). In view of all these features, the creeping motion of a sphere in a visco-elastic fluid filled in a cylindrical tube (sphere-to-tube diameter ratio of 0.5) and the uniform flow over a long cylinder in a planar slit (cylinder diameter-to-slit width ratio of 0.5) have been used extensively for the benchmarking of a diverse variety of numerical solution procedures developed and adapted for computing visco-elastic flows (Brown et al., 1993; Szady et al., 1995; Saramito, 1995; Owens and Phillips, 2002, Dou and Phan-Thien, 2001, 2007; Xiong et al., 2017). Concentrated efforts have led to the development of highly refined and very successful numerical algorithms for computing steady and unsteady visco-elastic flows, which are continually being improved to overcome the so-called high Weissenberg number syndrome (Keunings, 2000; Reddy and Gartling, 2001; McKinley, 2002; Owens and Phillips, 2002; Petera, 2002). The exponential growth in the numerical activity in this field has also been matched by the development of elegant and improved experimental techniques for resolving the spatial and temporal features of the flow field around submerged objects.

These advances, when combined with good rheological characterization of the test liquids, have led to a greater understanding of the motion of particles in complex fluids in general and of spheres and cylinders in particular, albeit the bulk of the activity in this field relates to the benchmark sphere or cylinder problem. In this chapter, consideration is given to the roles of rheology (visco-elasticity with and without shear-thinning), geometry (shape of particles, interactions with walls and other particles) and transient effects. Excellent reviews of the pertinent voluminous body of knowledge are available in the literature (Walters and Tanner, 1992; McKinley, 2002; Caswell et al., 2004; Zenit and Feng, 2018).

Table 4.1 provides a succinct summary of the theoretical and numerical research activity in this field. Evidently, the bulk of the research effort has been directed at the elucidation of the role of fluid rheology

TABLE 4.1

Summary of Theoretical Investigations of Visco-Elastic Flow past Spheres and Cylinders (Re ≪1)

Investigator	Constitutive Equation	Main Results and Observations
Leslie and Tanner (1961)	Oldroyd fluid	Small reduction in drag from the Stokes value and a slight shift of streamlines (results corrected by Giesekus, 1963).
Giesekus (1963)	Third-order Coleman–Noll fluid	Main emphasis on the flow field for a rotating and translating sphere.
Caswell and Schwarz (1962)	Rivlin–Ericksen fluid	Slight reduction in drag.
Clegg and Power (1964)	Oldroyd fluid	Actually, the flow between two concentric spheres is studied. The case of a single sphere is recovered as a special case.
Srivastava and Maiti (1966)	Second-order fluid	Flow around a cylinder has been investigated analytically.
Thomas and Walters (1966)	Walters fluid	The fluid elasticity does not seem to influence the time required for a sphere to attain its terminal velocity.
Rajvanshi (1969)	Oldroyd fluid	Drag expression is derived for a porous sphere, of which the solid sphere is a special case with zero porosity. Their result coincides with that of Leslie and Tanner (1961).
Gilligan and Jones (1970)	Walters fluid	Unsteady flow past a circular cylinder is studied.
Caswell (1970, 1972)	Rivlin–Ericksen fluid	The wall effects are less severe in visco-elastic fluids than in Newtonian fluids. Also, examined the stability of a particle near a plane wall and predicted radial migration in cylindrical tubes which is consistent with observations (Tanner, 1964).
Ultman and Denn (1971)	Convected Maxwell fluid	This analysis is equivalent to that of Oseen for Newtonian fluids. A reduction in drag is predicted.
Verma and Rajvanshi (1971)	Second-order fluid	This study parallels that of Clegg and Power (1964), and the case of a single sphere is recovered by letting the radius of the outer sphere go to ∞.
King and Waters (1972)	Walters fluid model	Up and down bouncing of a sphere is predicted which has been observed experimentally (Walters and Tanner, 1992).
Lai (1973, 1974, 1975)	Maxwell fluid	Drag on accelerating sphere and discs.
Mena and Caswell (1974)	Oldroyd fluid	Very little upstream shift in streamlines is predicted.
Verma and Sacheti (1975)	Second-order fluid	Oseen-type approximation is used to take into account the inertial effects.
Gupta (1976)	Walters fluid	Drag for the unsteady motion of a spheroid moving from rest has been studied.
Brunn (1977, 1979, 1980)	Second-order fluid	Dynamics of spherical and slightly deformed particles is examined, and the radial migration is predicted due to fluid elasticity.
Pilate and Crochet (1977)	Second-order fluid	The elasticity reduces drag on circular and elliptic cylinders whereas the reverse effect is observed at high Reynolds numbers. These are consistent with experimental results.
Sigli and Coutanceau (1977)	Convected Maxwell fluid model	The fluid elasticity causes a drag reduction and the confining walls further enhance the visco-elastic effects on sphere motion.
Townsend (1980)	Four-constant Oldroyd fluid	Drag and lift on a translating and rotating circular cylinder are calculated for Reynolds number values between 5 and 40.
Crochet (1982)	Maxwell fluid model	Slight reduction in a drag of a sphere and downstream shift in streamlines for We ≤ 2.
Tiefenbruck and Leal (1982)	Oldroyd fluid	Similar conclusions as those reached by Crochet (1982).
Hassager and Bisgaard (1982, 1983)	Maxwell fluid model	Combined effect of walls and visco-elasticity on drag of a sphere are investigated.
Marchal et al. (1984)	Maxwell and Oldroyd-B models	Drag results for $\beta = 0.2$. Slight drag reduction is predicted.
Daireneih and McHugh (1985)	Third-order fluid	Results for spherical and nonspherical particles are reported.

(Continued)

TABLE 4.1 (*Continued*)

Summary of Theoretical Investigations of Visco-Elastic Flow past Spheres and Cylinders (Re $\ll$ 1)

Investigator	Constitutive Equation	Main Results and Observations
Sugeng and Tanner (1986)	Phan-Thien–Tanner (PTT) fluid model	Slight downstream shift in streamlines is predicted. The resulting drag values are in qualitative agreement with the experimental results of Chhabra et al. (1980b).
Luo and Tanner (1986)	Maxwell type differential model	The drag of a sphere translating in a tube ($\beta = 0.5$) shows good agreement with the literature values.
El Kayloubi et al. (1987)	Generalized Maxwell fluid model	Detailed information on the components of the rate of deformation tensor for weakly elastic flows for a sphere-in-sphere and a sphere-in-cylinder configuration. The fluid elasticity seems to suppress the wall effects.
Carew and Townsend (1988)	PTT model	Simulations for $\beta = 0.5$ are in qualitative agreement with experiments for a sphere.
Chilcott and Rallison (1988)	FENE dumbbell model	Finite difference solution for flow around spheres and cylinders.
Crochet (1988)	Oldroyd-B fluid model	Numerical predictions of drag on a sphere falling in a cylindrical tube for $\beta = 0.5$.
Debbaut and Crochet (1988)	Maxwell fluid with shear-rate-dependent viscosity	The importance of extensional effect is examined on sphere motion in a cylindrical tube for $\beta = 0.5$.
Marchal and Crochet (1988)	Oldroyd-B model	Sphere drag reduces up to about We ~ 1 and beyond this value drag increases.
Lunsmann et al. (1989, 1993)	UCM (upper-convected Maxwell), Oldroyd-B and dumbbell model of Chilcott and Rallison	Good summary of UCM-based calculations. Both drag reduction and enhancement predicted for suitable value of β and rheological parameters.
Ramkissoon (1990)	Oldroyd fluid model	Perturbation analysis for a slightly deformed sphere.
Zheng et al. (1990a)	Oldroyd-B and UCM model	Predicted a reduction in drag and also a limiting value of Weissenberg number.
Harlen (1990); Harlen et al. (1990)	FENE model	Numerical drag prediction ($\beta = 0$) which suggests the possibility of drag enhancement beyond a critical Weissenberg number.
Phan-Thien et al. (1991)	Maxwell, Oldroyd-B, and PTT models	Limiting We only for the bounded sphere case and it seems to be an artifact of numerics.
Zheng et al. (1991)	Newtonian, Carreau, and PTT models	Studied effects of inertia, shear-thinning, and visco-elasticity on flow field for a sphere ($\beta = 0.5$).
Jin et al. (1991)	UCM and PTT models	Limiting We depends upon mesh, and drag reduction is mainly due to shear-thinning.
Gervang et al. (1992)	Oldroyd-B model	For $\beta = 0$, no change in drag up to We $\approx$ 2. Slight reduction and enhancement at large We.
Georgiou and Crochet (1993)	UCM	Wall effects diminish with increasing We and decreasing β.
Mitsoulis et al. (1993)	UCM and K-BKZ (Kaye-Bernstein Kearsley Zapas) models	For UCM and $\beta = 0.5$, drag reduction is predicted and for M1 fluid ($\beta = 0.02$), no change in drag from its Newtonian value.
Bush (1993)	Boger fluids	For $\beta = 0.5$, visco-elasticity extends the wake region and causes a slight upstream shift close to the sphere. Wall effects exert a significant influence on flow field.
Barakos and Mitsoulis (1993)	K-BKZ model for M1 fluid	For $\beta = 0.02$ and up to We ≤ 0.3, no significant change in drag of a sphere.
Sun and Tanner (1994)	UCM and PTT models	For $\beta = 0.5$, only slight reduction in drag for UCM but dramatic drag reduction for a sphere for PTT model at We ~ 3 to 4.
Bodart and Crochet (1994)	Oldroyd-B model	Visco-elastic effects diminish as $\beta \to 0$. The overshoot in sphere velocity also decreases with increasing value of We.

(*Continued*)

TABLE 4.1 (*Continued*)
Summary of Theoretical Investigations of Visco-Elastic Flow past Spheres and Cylinders (Re $\ll$ 1)

Investigator	Constitutive Equation	Main Results and Observations
Satrape and Crochet (1994)	UCM, Oldroyd-B, and FENE models	For $\beta = 0.5$, various levels of drag reduction predicted for a sphere. Qualitative agreement with the results of Chhabra et al. (1980b).
Ianniruberto and Marrucci (1994)	Two-fluid model	Migration of macro-molecules induced by a falling sphere is limited close to the sphere.
Baaijens (1994)	UCM	For $\beta = 0.5$ and in the range $0 \leq$ We ≤ 4, drag reduction and upturn in drag after reaching a minimum value.
Arigo et al. (1995); Arigo and McKinley (1994, 1997, 1998)	Experiments and simulations (UCM, PTT, FENE-CR)	Qualitative match between experiments and predictions.
Joseph and Feng (1995)	Second-order fluid	No negative wake is predicted.
Becker et al. (1996)	Modified FENE-CR model	3-D creeping flow near a plane wall and weak non-Newtonian effects up to second order are calculated. All the three parameters, that is elasticity, N_2, and shear-thinning enhance drag reduction.
Feng et al. (1996)	Oldroyd-B model	Interaction between a cylinder and a wall and between two spheres resulting in repulsion and attraction under appropriate circumstances.
Rasmussen and Hassager (1996)	Rivlin–Sawyers type fluid	Simulated the experimental results of Becker et al. (1994) for $\beta = 0.243$ and good match between theory and experiments.
Warichet and Legat (1997)	UCM	For $\beta = 0.5$ and We ≤ 2.5, following a region of drag reduction, it levels off.
Baaijens et al. (1994, 1995, 1997)	PTT and Giesekus models	While for sphere ($\beta = 0.5$), their drag results are consistent with others, new results for cylinder are presented.
Mitsoulis et al. (1998a)	K-BKZ model to mimic W1 and S1 fluids	For $\beta = 0.88$, up to 80% drag reduction is predicted in S1, but only 20% drag reduction in nonshear-thinning W1 fluid.
Mutlu et al. (1996)	Oldroyd-B model	For $\beta = 0.5$ and Re =1, main thrust is on numerical aspects.
Owens and Phillips (1996a, b)	UCM and Oldroyd-B models	For UCM, results for a sphere are consistent with those of Lunsmann et al. (1993) for $\beta = 0.5$. The maximum attainable value of We does not increase with mesh refinement.
Luo (1996)	UCM model	For $\beta = 0.5$, good match between the drag values for seven different algorithms for a sphere.
Rameshwaran et al. (1998)	Shear-thinning and extension-hardening model (Debbaut and Crochet, 1988)	Combined effects of walls, rotation, and inertia on drag of sphere. Strain-hardening exerts little influence on drag.
Luo (1998)	UCM model	Predicts 33% drag reduction, followed by a drag increase at We > 2.2.
Hu and Joseph (1999)	Second-order fluid	Calculated lift on a sphere close to a plane wall.
Yang and Khomami (1999)	FENE and multimode Giesekus models	Simulated the experiments of Arigo et al. (1995) for $\beta = 0.121, 0.243$. None of the model predicts drag behavior quantitatively.
Peters et al. (2000)	UCM and Rivlin–Sawyers model	For UCM, results are in good agreement with previous studies.
Harlen (2000, 2002)	FENE and Giesekus models	The wake characteristics are primarily governed by the extensional behavior. Depending upon the relative magnitudes of elastic recoil of shear stress and high extensional stress, there may or may not be a negative wake.
Yurun (2003a, b), Yurun and Crochet (1995); Yurun et al. (1999)	UCM and Oldroyd-B models	Numerical simulation of the limiting behavior up to which fully converged results are available for a sphere using various numerical tools.

(*Continued*)

TABLE 4.1 (*Continued*)

Summary of Theoretical Investigations of Visco-Elastic Flow past Spheres and Cylinders (Re $\ll$ 1)

Investigator	Constitutive Equation	Main Results and Observations
Dou and Phan-Thien (2001, 2007)	Oldroyd-B model	Cylinder confined in a slit ($\beta = 0.5$). In the limit of Re = 0, numerical difficulties encountered beyond a critical point are attributed to the inadequate resolution of the extensional flow in the rear and/or shearing flow in the annular gap. This can also be explained via the inflectional velocity profile near the cylinder and the convection of oscillating shear-layer downstream.
Ardekani et al. (2008)	Second-order fluid	Extends the results of Stimson and Jeffery for two spheres (Re = 0).
Richter et al. (2010, 2012)	FENE-P	Combined effects of polymer extensibility and inertia on 3-D flow past a cylinder are studied. Both the wake length and drag (with reference to the values in Newtonian fluids) increase with the rising degree of polymer extensibility parameter (L^2). Also, the B-mode instability observed in Newtonian fluid is suppressed.
D'Avino et al. (2012, 2019)	Worm-like micellar surfactant solutions (Boger fluids)	Migration of a sphere (in a Couette cell) occurs in the direction of decreasing shear-rate gradient (toward the outer cylinder) unless the sphere is initially close to the inner cylinder. The migration velocity is reduced in a shear-thinning fluid. Subsequently, this work has been extended to the case of a spheroid in visco-elastic fluids flowing in a slit. In most cases, the body migrates to the center plane of the channel.
Abedijaberi and Khomami (2012)	Multi-segment bead–spring chain model	In the creeping flow limit, numerical predictions match the drag results of Arigo et al. (1995) for $\beta = 0.121$ and $\beta = 0.243$. It is possible to achieve higher Weissenberg number results than that with the FENE-family of fluid models.
Lee and Zhang (2012)	Johnson and Segalman fluid	Careful choice of the value of the rheological model can lead to the self-sustaining regular or irregular oscillations in the falling velocity of a sphere (Re = 0.0325). The falling velocity gradually decreases and the sphere suddenly accelerates.
Garduno et al. (2016)	Combined White-Metzner and FENE-CR model	Numerical predictions reproduced the experimental results of Jones et al. (1994) for a translating sphere. The method of increasing the value of the Weissenberg number by increasing falling velocity seems to be less reliable for drag prediction.
Housiadas and Tanner (2016)	PTT model	In the creeping flow, higher order perturbation approximation is obtained. The drag reduction occurs at low Deborah numbers followed by significant enhancement in line with the experimental observations. Though no negative wake is predicted, flow field decays slowly with the increasing De.
Xiong et al. (2017)	Oldroyd-B fluid model	Inertial effects (Re = 500) on the flow over a cylinder confined in a planar slit ($\beta = 0.5$) are studied numerically. At low Reynolds numbers, drag is modified beyond a critical value of We and the polymer viscosity ratio exerts an influence at low Re and We. Both drag reduction and drag enhancement are possible depending upon whether vortex shedding is suppressed or not.
Castillo et al. (2019)	Oldroyd-B (Boger) fluid	For a translating sphere, drag is increased beyond the Stokes value as predicted by the numerical and asymptotic regular perturbation results.
Sahin and Atalik (2019)	Cross flow of power-law, Oldroyd-B, Giesekus, and FENE-P fluids past a cylinder	Extensive results on drag, lift, wake size and wake angle, onset of vortex shedding, etc. are presented in the range 80 $\leq$ Re $\leq$ 300; 0.6 $\leq n \leq$ 1.2 and We < 1.2.

(*Continued*)

TABLE 4.1 (*Continued*)

Summary of Theoretical Investigations of Visco-Elastic Flow past Spheres and Cylinders (Re ≪1)

Investigator	Constitutive Equation	Main Results and Observations
Faroughi et al. (2020)	Oldroyd-B model	3-D Direct Numerical Simulation (DNS) study for the flow over a sphere. In line with the previous studies, in the creeping flow, drag reduction occurs for De < 1 and drag enhancement for De > 1. At finite Reynolds numbers, fluid elasticity always augments drag above its Newtonian value. This is ascribed to the asymmetry of the flow field.
Gkormpatsis et al. (2020)	UCM, Oldroyd-B, Giesekus, exponential PTT models	In the creeping flow regime, the effect of Navier-type slip prescribed on the sphere surface is investigated. For fixed values of We, drag increases above the Stokes drag with slip, though the extent of increase is model-dependent. In Newtonian fluids (We = 0), sphere drag decreases with slip.
He et al. (2020)	–	Analysis of visco-elastic flow induced by a small amplitude oscillation motion of a colloidal sphere.
Su et al. (2022)	Giesekus and Oldroyd models	In the creeping regime, drag on a confined and unconfined sphere; negative wake is predicted in Giesekus fluid but not in an Oldroyd fluid.

and wall effects on the drag of sphere and the wake behind a sphere, velocity overshoot, followed by the flow over a circular cylinder confined between two plane walls. Some results are also available on ellipsoidal and rod-like particles. However, we begin with the simplest case, namely, that of the flow over an unconfined sphere.

4.2 Flow Over a Sphere

The flow problem of interest is shown schematically in Figure 2.1 ($\beta = 0$). A solid sphere of diameter d (or radius R) and density ρ_p sediments under gravity in an infinite expanse of a visco-elastic liquid of density ρ, a zero-shear viscosity μ_0, and characteristic relaxation time λ. Additional parameters such as the solvent viscosity (μ_s) and shear-thinning characteristics are also frequently used to describe the drag of a sphere depending upon the choice of a suitable constitutive equation. Finally, it is a common practice to perform such experimental work in cylindrical tubes of radius (R/β). In a typical experiment, the terminal falling velocity, V, of a sphere is measured experimentally using the direct observation method (Chhabra et al., 1980b), a photographic method (Jones et al., 1994; Malhotra and Sharma, 2014), a digital video-imaging technique (Becker et al., 1994), or an ultrasound method (Watanabe et al., 1998). The scaling of the field equations or a simple dimensional analysis of the problem yields the following functional relationship

$$\frac{\lambda V}{R} = f\left(\frac{\rho_p}{\rho}, \frac{\mu(\dot{\gamma})}{\mu_0}, \frac{\mu_s}{\mu_0}, \beta, \text{Re}\right) \quad (4.1)$$

In Equation 4.1, all dimensionless groups have their usual meanings. At times, the so-called elasticity number, El ($\equiv$ De/Re) is also used. Similarly, for a sphere falling under its own weight, one can replace the density ratio (ρ_p/ρ) by a ratio of the buoyant weight to the ($\mu_0 d^2/\lambda$) thus yielding the dimensionless group ($d\Delta\rho g\lambda/\mu_0$). Clearly, the actual form and the number of dimensionless groups are somewhat dependent upon the choice of a visco-elastic fluid model to portray the shear-rate dependence of viscosity and the other visco-elastic parameters. At low Reynolds numbers, it is reasonable to assume (V/R) to be the representative rate of deformation for a sedimenting sphere and therefore the ratio of two-time scales

Rigid Particles in Visco-Elastic Fluids

($\lambda V/R$) is appropriately called the Deborah number, De. In the rheological literature, the term Weissenberg number $\left(\text{We} = \lambda \dot{\gamma}_s\right)$ is frequently used interchangeably; evidently, this is correct only in the limit of $\beta \to 0$, because for moderate values of β, the shear rate in the annular region is given by $\dot{\gamma}_s \sim O\left(V/R(1-\beta^2)\right)$; therefore, the numerical values of De and We will be significantly different (Degand and Walters, 1995; McKinley, 2002; Poole, 2012). On the other hand, in numerical simulations, a steady Eulerian reference frame (moving downward with a velocity V) is selected and hence the sphere is stationary, and the walls of the container move upward with a constant velocity V. Therefore, extreme caution must be exercised in making quantitative comparisons between experiments and predictions as noted by Mena et al. (1987), Arigo et al. (1995), and Garduno et al. (2016).

It is worthwhile to add here that it is virtually impossible to consider the effects of all parameters appearing in Equation 4.1 in experiments or in numerical simulations, and it is thus customary to first consider the limiting cases of this flow problem by assigning appropriate limits of zero or infinity to various terms. Furthermore, it is convenient to present the current state of the art of the visco-elastic fluid flow past a sphere in the following sections: drag coefficient for creeping flow past an unbounded sphere, the benchmark problem, inertial effects, wake structure, and time effects. Each of these is discussed in detail in the ensuing sections.

4.2.1 Theoretical Developments

4.2.1.1 Drag Force on an Unbounded ($\beta = 0$) Sphere in Creeping Region (Re → 0)

The field equations and the boundary conditions, together with the choice of a suitable visco-elastic fluid model, presented in Chapter 2 are still valid. As such, there are no general guidelines available for the selection of a particular model for an envisaged application and therefore considerable intuition is frequently required in making this choice (Tanner, 2000; Carreau et al., 2021). The final choice invariably reflects a judicious compromise between an accurate portrayal of the fluid behavior on the one hand and the amenability of the resulting equations to numerical solution on the other. An examination of Table 4.1 reveals that most studies up to about 1970 (and a few since then) are based on the perturbation types of solutions that involve fairly lengthy and tedious algebra and are generally prone to errors (Caswell and Schwarz, 1962; Rathna, 1962; Caswell, 1962, 1970; Rajvanshi, 1979; Sharma, 1979). In this approach, it is assumed a priori that the flow field for a visco-elastic fluid in the creeping flow regime is expressible in terms of a series solution in which the first term represents the Newtonian solution (zero-order approximation) and the coefficient of the subsequent terms involve material parameters. Thus, the flow variables are written as

$$\psi = \psi_0 + \text{We}\,\psi_1 + \text{We}^2 \psi_2 + \cdots \tag{4.2}$$

$$p = p_0 + \text{We}\, p_1 + \text{We}^2 p_2 + \cdots \tag{4.3}$$

and so on.

The investigations of Leslie and Tanner (1961), Caswell and Schwarz (1962), Rajvanshi (1969), etc. are illustrative of this approach. By way of example, the final drag expression obtained by Leslie and Tanner (corrected by Giesekus, 1963) for a sphere undergoing steady translation in an eight-constant Oldroyd model is given as

$$Y = \frac{C_D \text{Re}_0}{24} = \left[1 - \frac{1}{3}\left(\frac{V}{R}\right)^2 \left\{0.016(\lambda_1 - \lambda_2)(3\lambda_1 - \lambda_2) - 0.618(\sigma_2 - \sigma_1)\right\}\right] \tag{4.4}$$

In addition to the creeping flow assumption $\text{Re}_0 \ll 1$, Equation 4.4 is subject to the following constraints on the fluid behavior:

$$\sigma_1 > \sigma_2 \geq \frac{\sigma_1}{9}; \qquad \sigma_1 = \lambda_1 \mu_0 + (\lambda_1 - 1.5\mu_0)\gamma_1$$

$$\sigma_2 = \lambda_2 \mu_0 + (\lambda_1 - 1.5\mu_0)\gamma_2; \qquad (\lambda_1 - 1.5\mu_0)(\lambda_1 \gamma_2 - \lambda_2 \gamma_1) \geq 0$$

(4.5)

where λ_1, λ_2, σ_1, etc. are the Oldroyd fluid parameters. All perturbation analyses, besides predicting a slight decrease ($\propto V^2$) in the value of drag coefficient below its Newtonian value, also seem to suggest a small downstream shift in the streamlines around a sphere. Most such analyses are not only valid for the creeping flow region but are also limited to the so-called weak flow (We $\ll$ 1), albeit some results that supposedly take into account inertial (Verma and Sacheti, 1975) and finite elastic effects (Ultman and Denn, 1971) are also available. Overall, within the range of validity, these latter analyses also predict very little difference between the kinematics of Newtonian and visco-elastic flows around a sphere or a cylinder. Owing to the near equivalence of all fluid models to the second-order behavior at low deformation rates, it is sufficient to treat all perturbation problems in terms of the second-order fluid to elucidate qualitatively the role of visco-elasticity on the flow field, especially for slowly varying flows that are quite suitable for both using the second-order fluid behavior and employing the series expansion to obtain qualitatively useful results for nonviscometric flow problems. In essence, all such analyses elucidate the influence of visco-elasticity on drag coefficient and on the detailed flow field in the absence of shear-rate-dependent viscosity effects. Furthermore, the standard perturbation procedure involves expansions about Reynolds number, which represents the ratio of the inertial to viscous forces. Such an expansion is thus tantamount to, to zeroth order, neglecting all nonlinear effects irrespective of whether these arise from inertial effects or from non-Newtonian fluid behavior, and thereby leaving the zero Reynolds number Newtonian flow as the zeroth approximation. The fact that the first-order terms (in We) add "equal amounts" of inertial and visco-elastic effects to the solutions appears to be at odds with experimental results that clearly show that the effects of shear-rate-dependent viscosity begin to manifest at much lower values of the dynamic parameters than do the nonlinear effects of fluid inertia (Hill, 1969). Subsequently, the limitation of the weak visco-elastic effects has been partially relaxed (Ultman and Denn, 1971). This analysis also suggests a fundamental change in the mechanism of flow (reflected by a change in the type of field equations) at $Re_0We = 1$. Though such a change is unlikely to occur for zero Reynolds number flows (Walters and Tanner, 1992), it has also been suggested subsequently by others (see Joseph, 1990). For a Maxwell model fluid, the drag on a sphere is given by

$$Y = 1 - 0.425 \, We \, Re_0 \qquad (4.6)$$

Equation 4.6 is applicable in the limits of $Re_0 \to 0$ and $Re_0 We \ll 0.05$. Clearly, within these ranges of conditions, Equation 4.6 predicts very small drag reduction as long as the fluid velocity is below the shear wave velocity (Ultman and Denn, 1970; Denn and Porteous, 1971). Ultman and Denn (1971) predicted and experimentally observed an appreciable upstream shift of the streamlines for large values of We; however, subsequent studies have failed to reveal such a shift (Broadbent and Mena, 1974; Zana et al., 1975; Manero and Mena, 1981), albeit such shifting of streamline patterns with respect to Weissenberg number is in line with the general hypothesis of Walters and Barnes (1980). However, the Oseen-type of linearization of the visco-elastic terms used by Ultman and Denn (1971) has also come under severe criticism (Broadbent and Mena, 1974; Zana et al., 1975).

Since 1970s, a range of numerical methods has been (and continue to be) used extensively for calculating non-Newtonian flows in general and visco-elastic flows in particular (Crochet and Walters, 1983, 1993; Crochet et al., 1984; Crochet, 1988, 1989; Baaijens, 1994, 1998; Owens and Phillips, 2002; Petera, 2002, Abedijaberi and Khomami, 2012; Alves et al, 2021). Consequently, unprecedented computational efforts have been expended in solving the creeping visco-elastic flow around a sphere by using various numerical solution methods such as finite difference, finite element, boundary elements, elastic-viscous stress splitting, spectral methods, and modifications thereof. Detailed discussions of their relative merits and demerits and of the associated numerical aspects such as type and fineness of mesh, discretization errors, error estimates, and convergence issues are available in the literature (see Crochet et al., 1984; Dupret et al., 1985; Crochet et al., 1990; Luo, 1996, 1998; Owens and Phillips, 1996a, b; Baaijens, 1998;

Matallah et al., 1998; Owens, 1998; Chauviere and Owens, 2000; Reddy and Gartling, 2001). One possible reason for this upsurge in the research activity in this area is the formulation of a new class of model visco-elastic liquids, the so-called *Boger* fluids (Boger, 1977a, 1984, 1995; James, 2009), which facilitate a direct comparison between predictions and observations (more on this in a later section), at least in a limited range of conditions.

The state-of-the-art reviews of Walters and Tanner (1992), McKinley (2002), and Zenit and Feng (2018) on visco-elastic flow past a sphere provide excellent progress reports on the numerical developments in this field, and Table 4.1 testifies to the degree of attention accorded to this flow. An examination of Table 4.1 shows that indeed most of the numerical activity has been directed at the benchmark problem of $\beta = 0.5$ and only limited results are available for the unconfined flow, that is, $\beta \rightarrow 0$. This is partly due to the fact that the faraway boundary conditions are required to be imposed at a distance of $\geq 50R$ thereby making computations prohibitively expensive in terms of CPU time and memory requirements. It is convenient to consolidate these results in the form of a drag correction factor Y defined as $(C_D \text{Re}_0/24)$. Evidently for a Newtonian fluid $Y = 1$ and in the limits of $\text{Re}_0 \rightarrow 0$ and $\beta \rightarrow 0$, it is expected to be a function of Weissenberg number, We, alone. Under these conditions, the few numerical results for the drag on an unconfined sphere in a visco-elastic fluid predict a small reduction in drag, that is, the value of $Y \sim 1$; for example, see Crochet (1982) for Maxwell model fluids. Similarly, for a Finitely Extensible Nonlinear Elastic (FENE) model, the analysis of Chilcott and Rallison (1988) predicts a monotonic decrease (very small, however) in the value of Y for extensibility parameter $L < 10$, whereas a small decrease followed by a large increase in drag was predicted only for $L \geq 10$. Unfortunately, such large values of L are believed to be rather unrealistic (McKinley, 2002). Gervang et al. (1992) used a spectral method to compute the value of Y for an Oldroyd-B model (used to model Boger fluids, Mackay and Boger, 1987) for a range of values of (μ_s/μ_0). These calculations predict a reduction in drag of the order of 0.2% at We ~ 0.5, followed by a 3% increase in drag before the loss of convergence at We ~ 2. On the other hand, Housiadas (2021) has identified a singularity at a finite Weissenberg number when the Navier slip condition is prescribed on the surface of a sphere undergoing steady translation in UCM (upper-convected Maxwell) as well as in Oldroyd-B model fluids.

Thus, in summary, virtually no change in sphere drag for unconfined flow is predicted due to visco-elasticity at low to moderate values of the Weissenberg number whereas the drag may slightly increase above the Newtonian value ($Y > 1$) as the Weissenberg number is progressively increased. It appears that the limiting value of the Weissenberg number, ~8 (beyond which fully converged results are not yet available), is weakly dependent on the numerical solution procedure including the details of the mesh, and possibly on the choice of the fluid model. These small changes in drag are reflected by only minor changes in the structure of the flow field. At low Weissenberg (or Deborah) numbers, nonlinear elastic effects due to normal stress differences or extensional effects are negligible and, therefore, the drag deviates a little (maximum by a factor of μ_s/μ_0) from the Stokes value. However, as the value of We is gradually incremented, extensional effects in the wake region become increasingly more important that can result in an increase in drag due to strain hardening of the fluid. Additional complications also arise from the formation of the negative wake under some conditions. All in all, the drag on a sphere is determined by a complex interplay between the thermodynamic quality of the solvent, molecular weight of polymers, shear and extensional rheology of solutions (Solomon and Muller, 1996; Doi, 1997; McKinley, 2002).

4.2.1.2 Drag Force on a Sphere for $\beta = 0.5$ and Re → 0: The Benchmark Problem

A cursory glance over Table 4.1 shows that much more research effort has been directed at the evaluation of drag on a sphere in a visco-elastic fluid for $\beta = 0.5$ than that for the case of an unbounded sphere. Furthermore, a diverse range of numerical solution procedures coupled with a selection of rheological models have been used, but most studies have endeavored to elucidate the role of visco-elasticity on drag in the absence of shear-thinning, and only limited results on the combined effects of elasticity, shear-thinning, and inertia on drag are available in the literature (e.g., see Jin et al., 1991; Mendoza-Fuentes et al., 2009; Kaur et al., 2019). Under these conditions, the drag correction factor Y denotes the ratio of the drag force in the visco-elastic fluid for $\beta = 0.5$ to the Stokes drag force corresponding to $\beta = 0.5$, and hence Y now is a function of both β and We. As noted above, though the computations of Y (We, β

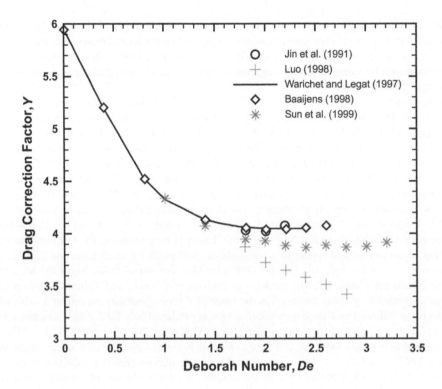

FIGURE 4.1 Summary of numerical predictions of drag in a UCM (upper-convected Maxwell) fluid ($\beta = 0.5$).

= 0.5) for the creeping flow of a UCM fluid past a sphere in a tube was identified as a benchmark flow problem in 1988 (Hassager, 1988; Caswell, 1996), it remains a challenge even today, primarily due to the numerical difficulties in coping with the stress boundary layers near the rigid walls of the tube and near the sphere (Lunsmann et al., 1989, 1993; Warichet and Legat, 1997; Baaijens, 1998; McKinley, 2002; Yurun, 2003a, b). The most recent as well as most reliable data are shown in Figure 4.1 where it is clearly seen that the value of Y decreases from its Newtonian value of 5.947 gradually (Haberman and Sayre, 1958) as the value of the Deborah (or Weissenberg) number is progressively increased. There seems to be a good agreement between various predictions up to about De or We ~ 1.5 beyond which some simulations suggest a monotonic decrease in the value of Y while others point to the possibility of a slight increase in drag, before the loss of convergence in the range De = 2.5–3.5. Generally, it has been found that once the value of β exceeds ~ 0.25, numerical results reveal that the highest stresses occur in the nip region between the sphere and the tube wall and not in the wake region (Georgiou and Crochet, 1993; McKinley, 2002) and, therefore, in hindsight the choice of $\beta = 0.5$ for the benchmark problem may not have been the optimal choice. Subsequently, the values of $\beta = 0.125$ and 0.25 have been suggested (Brown and McKinley, 1994), but less extensive results are available for these values of β (Arigo et al., 1995; Abedijaberi and Khomami, 2012).

Undoubtedly, such benchmark problems are of great value in establishing that fully converged and internally consistent results can be obtained using various numerical methods; however, the efforts for extending the domain of convergence to higher value of the Deborah number, De, have been less successful, presumably due to the sharp gradients in the polymeric stresses, as mentioned above (Crochet, 1988; Zheng et al., 1990a, b; Phan-Thien et al., 1991; Yurun and Crochet, 1995; Owens and Phillips, 1996a, b; Yurun et al., 1999; Chauviere and Owens, 2000). On the other hand, a boundary layer approach for both the extensional stress in the wake region and the polymeric stresses near the sphere may be more suitable for analyzing the flow at very large values of the Deborah number (Harlen, 1990; Renardy, 2000a, b). Similarly, limited time-dependent 2-D and 3-D simulations (Brown et al., 1993; Smith et al., 2000) and linear stability analysis (Oztekin et al., 1997) point to the possible loss of stability at high values of

Deborah number. Such a transition for the case of a cylinder in Boger fluids is well documented in the literature (McKinley et al., 1993).

Analogous studies demonstrating complex interplay between the flow kinematics and the other visco-elastic constitutive models can be found in the studies of Lunsmann et al. (1993), Satrape and Crochet (1994), Arigo et al. (1995), Rasmussen and Hassager (1996), and Yang and Khomami (1999). Though multimode FENE dumbbell models, with an appropriate choice of the model parameter values, do predict the initial decrease and subsequent increase in the drag correction factor, none of these, however, is able to match the experimental results over wide ranges of β for which experimental results are now available except for the recent results of Abedijaberi and Khomami (2012). This failure is ascribed to the inadequate modeling of the rapid transient molecular extension in the fluid and the resulting stress-conformation hysteresis. Similar modifications are needed for bead–spring chain models with internal configurational degrees of freedom to resolve the complicated looped/kinked structures seen in recent transient elongational flow tests (Smith and Chu, 1998; Li et al., 2000).

This section is concluded by noting that all the aforementioned analyses predict the visco-elasticity to cause a drag reduction at low values of a suitably defined Weissenberg number whereas some results also point to the possibility of drag enhancement for moderate to large values of the Weissenberg number (Chilcott and Rallison, 1988; Crochet, 1988; Debbaut and Crochet, 1988; Harlen, 1990; Harlen et al., 1990). This type of "switch-over" in the manifestation of visco-elastic effects has also been observed for visco-elastic fluid flow across circular and elliptic cylinders in relation to the value of Reynolds number (Pilate and Crochet, 1977). The drag enhancement is due to the existence of a thin region of highly extended polymer molecules on the downstream side of the sphere. Therefore, the velocity in the wake region decays much more slowly in visco-elastic fluids than that in a Newtonian fluid. This also suggests that longer tubes are needed to minimize/eliminate the end effects on sphere motion in visco-elastic fluids than those for purely viscous media. To date, most of the simulation work has been carried out at a sphere-to-tube diameter ratio of 0.5, and it is not at all obvious whether similar trends would apply, even qualitatively, for the other values of the blockage ratio including the unbounded conditions or not (Jones et al., 1994).

Very little is known about the combined effects of fluid inertia and visco-elasticity on sphere motion. Utilizing semi-theoretical arguments, El Kayloubi et al. (1987) and Sigli and Kaddioui (1988) noted that though an increase in the value of Reynolds number did not cause any shift in the streamlines, the rate of deformation and stress fields was significantly influenced. Several investigators have considered boundary layer flow of visco-elastic fluid around submerged objects, and it is now readily accepted that the nature and extent of manifestation of visco-elastic effect is strongly dependent on the shape of the object (Maalouf and Sigli, 1984). Similarly, the role of inertial effects on the sphere drag in visco-elastic fluids has been investigated by Mutlu et al. (1996) and Matallah et al. (1998), whereas the combined influence of walls, fluid rotation, and fluid inertia (Reynolds number) on sphere drag in shear-thinning, strain-hardening fluids has been studied by Rameshwaran et al. (1998). Finally, some simulations even suggest that a rheological model found suitable for a specific value of β may be altogether inappropriate for the other values of β (Yang and Khomami, 1999).

Excellent review articles summarizing the other related theoretical developments in the general field of particle motion in visco-elastic media are available in the literature (Caswell, 1977; Giesekus, 1978; Leal, 1979; Brunn, 1980; Dairenieh and McHugh, 1985; Kaloni and Stastna, 1992; McKinley, 2002; Caswell et al., 2004; Zenit and Feng, 2018).

4.2.1.3 Wake Phenomenon

It is readily conceded that an integral parameter such as drag is relatively insensitive to the detailed structure of the flow field. On the other hand, numerical simulations over the years have provided useful information about the detailed kinematics of the flow that sheds some light on the overall trends (Harlen, 1990; Chilcott and Rallison, 1988). The first generation of (perturbation-type) solutions was limited to the prediction of shift in streamline with reference to that in the Stokes flow. The findings tend to be somewhat dependent upon the possibilities of inertial effects and changes in the type of the field equations (Delvaux and Crochet, 1990). Subsequently, considerable attention has been given to the importance of extensional flow in the wake region, downstream of the rear stagnation point. Some of these studies have also been motivated by the fact that excellent experimental results are now available on the wake of spheres in Boger fluids (Arigo et al., 1995; Baaijens et al., 1995; Pakdel and McKinley, 1997; Fabris et al., 1999). Broadly speaking, the available experimental results suggest that the length of the downstream wake region increases significantly with the increasing levels of fluid visco-elasticity and the velocity in the wake region also decays over a distance of up to 30 sphere radii, as seen in Figure 4.2. While the biaxial flow field in front of the sphere is hardly influenced by the fluid elasticity, severe extension and high tensile stresses in the wake region retard the decay of the velocity field under these conditions. This also results in a high degree of molecular orientations in the wake region. Qualitatively, similar trends are present in the 3-D wake behind a sphere settling in a rectangular channel (Harrison et al., 2001; Lawson et al., 2004) and near a wall (Tatum et al., 2005).

While some of the numerical simulations (Bush, 1993, 1994; Lunsmann et al., 1993; Yang and Khomami, 1999) do capture some of these aspects, the modest changes predicted by these simulations are nowhere near the dramatic effects observed experimentally. Such a rather large discrepancy between predictions and observations is believed to be due to the stress-conformation hysteresis associated with fast transient extensional flow as has been observed in some of the recent Brownian dynamics calculations; for example, see McKinley (2002) for more details. While an analysis based on the second-order fluid does predict a negative wake (Joseph and Feng, 1995), the so-called negative wake has been observed in visco-elastic shear-thinning fluids thereby suggesting another recirculating region downstream of the sphere in which the fluid velocity is in the opposite direction to that of the sphere and the vorticity changes sign. Similar negative wake has also been observed behind the rising bubbles in visco-elastic liquids (Bisgaard and Hassager, 1982; Bisgaard, 1983; Belmonte, 2000). The available theoretical

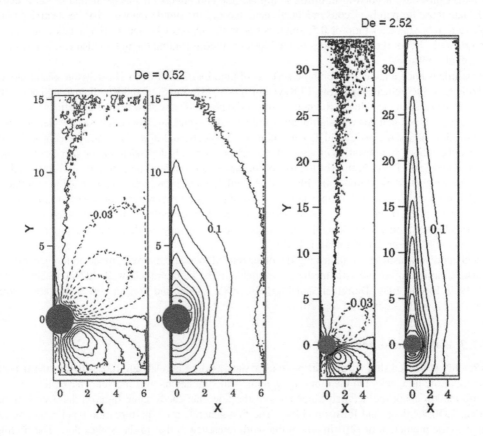

FIGURE 4.2 PIV measurements of the spatial structure in the wake of a 6.35 mm sphere falling in a polystyrene Boger fluid in a tube ($\beta = 0.083$) showing contours of dimensionless radial and axial velocity for De = 0.52 (left) and De = 2.52 (right). (Re-drawn from Fabris, D., Muller, S.J., and Liepmann, D., *Phys. Fluids*, **11**, 3599, 1999.)

analyses (Jin et al., 1991; Zheng et al., 1991; Bush, 1994) suggest that both visco-elasticity and shear-thinning are necessary for the formation of negative wakes and the detailed structure of the wake also seems to be strongly influenced by these characteristics. These findings are in qualitative agreement with the available experimental results of Oh and Lee (1992) and Arigo and McKinley (1998). The early observations also exhibited fluctuations in the wake velocity (Bisgaard, 1983); however, such fluctuations have not been seen in later experimental studies. The complex interactions between the geometry of the obstacle and the roles of elastic and shear-thinning properties on the detailed flow field have been experimentally studied by Maalouf and Sigli (1984) also. Some of the other phenomena documented in the literature include oscillatory shear-layer downstream of a cylinder (Dou and Phan-Thien, 2001, 2007), self-sustaining oscillations in the settling velocity of a sphere (Lee and Zhang, 2012), etc.

4.2.2 Experimental Results

The theoretical research effort expended on the falling sphere (and to lesser extent in the crossflow over a cylinder) problem has been matched by the growing experimental activity (see Table 4.2). Most studies have dealt with the measurement of the drag force on a sphere falling under gravity and, only in a very few of these, additional information on the wall effects and the detailed flow field around a sphere has been provided. Depending upon the type of experimental fluids used, one can easily classify the available body of experimental results into three types: those involving the combined effects of shear-thinning and visco-elasticity (quantified in terms of the primary normal stress difference in steady shear); those obtained using nonshear-thinning but highly elastic fluids (Boger fluids); and finally, using the so-called drag-reducing fluids. The latter studies are included here simply because the phenomenon of drag reduction observed under turbulent conditions in pipes and external flows has often been ascribed to the visco-elastic characteristics of the dilute solutions, though seldom any rheological property other than the shear viscosity has been measured to substantiate this assertion. It is thus convenient to discuss the hydrodynamics of a falling sphere separately in each of these types of fluids.

4.2.2.1 Shear-Thinning Visco-Elastic Liquids

Early studies (Leslie and Tanner, 1961; Ultman and Denn, 1971; Kato et al., 1972; Acharya et al., 1976; Yamanaka et al., 1976a; Sigli and Coutanceau, 1977; Kanchanalakshana and Ghajar, 1986; Malhotra and Sharma, 2014; Mendoza-Fuentes, 2009; Mrokowska and Krzton-Maziopa, 2019; Kaur et al., 2019, etc.) that purport to elucidate the role of fluid visco-elasticity on sphere motion used polymer solutions that displayed both shear-thinning and visco-elastic characteristics. Indeed, in some cases (Leslie and Tanner, 1961; Kato et al., 1972; Cho and Hartnett, 1979; Cho et al., 1980; Kanchanalakshana and Ghajar, 1986; Navez and Walters, 1996), only shear-rate-dependent viscosity was measured, and the test fluids were simply asserted to be visco-elastic based on indirect and intuitive tests including the bouncing of a sphere, stickiness/tackiness of the solution, etc. It is therefore not at all surprising that the early literature abounds with conflicting conclusions. For instance, in the creeping flow regime, Broadbent and Mena (1974) and others (Yamanaka et al., 1976a; Sigli and Coutanceau, 1977) reported the visco-elasticity to cause drag reduction that is in stark contrast with the works of Ultman and Denn (1971) and many others (Kato et al., 1972; Chhabra and Uhlherr, 1980b, 1988b; Chhabra et al., 1980b, 1981a; Bush and Phan-Thien, 1984) who concluded that the shear-thinning completely overshadows the minor visco-elastic effects, if any. Indeed, it was shown in Chapter 2 that the measured values of drag coefficient in highly visco-elastic media were in excellent agreement with purely viscous theories (Chhabra and Uhlherr, 1980a: Bush and Phan-Thien, 1984). However, at high Reynolds numbers (>1), the fluid visco-elasticity appears to exert appreciable influence on the value of drag coefficient as is evidenced by the following empirical equation proposed by Acharya et al. (1976):

$$C_{De} = C_D \left[1 - 0.18 (Re_{PL} We)^{0.19} \right] \tag{4.7}$$

where C_{De} is the drag coefficient in elastic fluids; C_D is the value of the drag coefficient in power-law fluids at the same value of Reynolds number and is given by Equation 2.50; and We is the Weissenberg

TABLE 4.2

Experimental Studies of Visco-Elastic Flow Past Spheres and Cylinders

Investigator	Experimental Fluids	Fluid Model	Remarks
Leslie and Tanner (1961)	Polyisobutylene in carbon tetrachloride	—	Results on sphere drag agree qualitatively with their analysis.
Ultman and Denn (1971)	Aqueous solutions of CMC (Carboxy Methyl Cellulose)	—	Slight shift in streamlines is observed which is in line with their theory for the crossflow over a cylinder.
Kato et al. (1972)	Aqueous solutions of CMC and PEO	Power-law model	The role of visco-elasticity on drag is completely overshadowed by the variable viscosity.
Broadmont and Mena (1974)	Polyacrylamide in glycerol/water mixtures	—	No change in streamline for a sphere and a cylinder was observed.
Zana et al. (1975)	0.5% Polyacrylamide in glycerol/water mixtures	—	No shift in streamlines is reported for spheres.
Acharya et al. (1976)	Aqueous solutions of Separan, PEO, and HEC	Power-law for both μ and N_1	Proposed a correlation for sphere drag involving power-law constants.
Yamanaka et al. (1976a)	Aqueous solutions of sodium polyacrylate and methyl cellulose	Sutterby model for viscosity and Spriggs model for visco-elastic behavior	Empirical correlation for sphere drag.
Sigli and Coutanceau (1977) Riddle et al. (1977)	Aqueous solutions of PEO solutions of HEC and Separan	Power-law viscosity	The combined effects of elasticity and walls are examined on sphere motion. The two spheres falling in line may diverge/converge depending upon the rheological properties and initial separation.
Cho and Hartnett (1979); Cho et al. (1980)	Aqueous solutions of CMC and Separan	—	Visco-elasticity reduces wall effects. Also, presented a method for evaluating a relaxation time for visco-elastic fluids.
Chhabra et al. (1980b, 1981b); Chhabra and Uhlherr (1988b)	Boger fluids; aqueous solutions of polyacrylamide and CMC	Carreau viscosity equation; Maxwellian relaxation time	Extensive experimental results on wall effects and drag coefficients.
Bisgaard (1982, 1983)	1% polyacrylamide solution in glycerol	—	Negative wake undergoing oscillations.
Manero et al. (1986, 1987)	Aqueous solutions of Separan, Carbopol in ethylene glycol, and Boger fluids	—	Drag on confined spheres in visco-elastic fluids.
Kanchanalakshana and Ghajar (1986)	Aqueous solutions of Separan	Power-law model	Falling-ball viscometry for visco-elastic fluids.
Chmielewski et al. (1990a)	Boger fluids	—	Drag coefficient of a sphere in various Boger fluids show good agreement, but there is a possibility of drag increase for We > 0.3.
Tirtaatmadja et al. (1990)	Boger fluids	—	Extensive results on wall effects and sphere drag showing drag reduction and enhancement.
Ambeskar and Mashelkar (1990)	Aqueous solutions of polyacrylamide	—	Time effects and their influence on terminal velocities of spheres.
Oh and Lee (1992)	Aqueous solutions of PEO and Separan	Carreau model	Wake size may increase or decrease depending upon the rheology. Wall effects are suppressed due to elasticity.

(Continued)

TABLE 4.2 (Continued)
Experimental Studies of Visco-Elastic Flow Past Spheres and Cylinders

Investigator	Experimental Fluids	Fluid Model	Remarks
Bush (1993)	Boger fluids	—	Wake region increases with increasing We.
Van den Brule and Gheissary (1993)	Boger fluids	—	Migration occurs for $\beta > 0.2$. Significant overshoot in velocity is observed.
Jones et al. (1994)	Boger fluids	—	At moderate values of We, drag enhancement is predicted. The Y–We relationship is strongly dependent on the value of β.
Degand and Walters (1995)	Boger fluids	—	For $\beta \sim 0.9$, drag reduction occurs.
Navez and Walters (1996)	Polyacrylamide and S1 solutions	—	At low We, drag is dominated by shear-thinning.
Solomon and Muller (1996)	Polystyrene-in-Di Octyl Phthalate (DOP) solutions	—	Drag behavior is strongly influenced by solvent quality and molecular weight.
Arigo et al. (1995); Rajagopalan et al. (1996)	PIB–PB-type Boger fluids	Phan-Thien–Tanner model	Effects of elasticity and walls on drag and flow field under steady-state and transient conditions.
Arigo and McKinley (1997, 1998)	Polyacrylamide in glycerine/water mixtures	—	Transient behavior and negative wake in shear-thinning visco-elastic fluids.
Fabris et al. (1999)	Same fluids as used by Solomon and Muller (1996)	—	Detailed flow visualization, wall effects, and drag results for a sphere.
Jayaraman and Belmonte (2003)	Worm-like micellar solutions	—	Oscillatory setting behavior of a sphere.
Chen and Rothstein (2004)	Worm-like micellar solutions	—	At a critical Deborah number, unstable flow regime develops about a sphere.
Weidman et al. (2012)	HPG gelling agent-based fluids	—	Various types of instabilities recorded for a falling sphere.
Mendoza-Fuentes et al. (2009)	Spheres falling in associative polymer solutions (Acrysol TT-935)	—	PIV data show that the negative wake formation is influenced by the blockage effects, in line with the prediction of Dou and Phan-Thien (2003, 2004).
Malhotra and Sharma (2014)	Sphere in surfactant and visco-elastic solutions	—	Settling velocity results in confined fluids. In the absence of wall effects, drag reduction is observed.
Freire et al. (2019)	Aqueous solutions of polyacrylamide	—	Interaction between two inline spheres. Depending upon the rheological properties, two settling regimes are obtained. In regime I, two spheres drift away from each other for initial separations above a critical value. In regime II, the two spheres may come together for an initial separation below the critical separation, whereas no repulsion is present when the initial separations is greater than the critical value. These trends are qualitatively consistent with the studies of Ardekani et al. (2008) and Riddle et al. (1977).
Kaur et al. (2019)	Aqueous carboxymethyl cellulose and polyacrylamide solutions	—	Drag (spheres and short cylinders, $1 \leq L/d \leq 2.4$) enhancement is predicted for We > 60, but the definition of fluid relaxation time is not clear.
Amaratunga et al. (2020)	Oscillatory motion of a sphere in aqueous solutions of poly-anionic cellulose and carboxymethyl cellulose	—	Detailed PIV measurements indicate overall reduction in the effective viscosity of the fluid as seen by the oscillating sphere. Minor visco-elastic effects are observed.

number using the fluid characteristic time defined by Equation 1.45. Equation 4.7 was stated to be applicable in the following range of conditions: $1 \leq Re_{PL} \leq 10^4$; $1 \leq We \leq 100$; $0.5 \leq n \leq 1.0$; $0.8 \leq s \leq 1.7$. In addition to presenting this correlation, Acharya et al. (1976) also carried out flow visualization experiments revealing three different types of wakes including a dual wake behind a sphere depending upon the relative magnitudes of the viscous, inertial, and elastic forces prevailing in the flow field close to the surface of the sphere.

From a theoretical standpoint, however, what is more important is to recognize the fact that none of the aforementioned studies permit a direct and quantitative comparison with the theoretical developments presented in Section 4.2.1 even in the creeping flow region. This difficulty stems simply from the fact that most theoretical developments elucidate the effect of visco-elasticity on the sphere drag in the absence of shear-thinning, whereas the experimental results relate to the conditions wherein both the viscosity and relaxation time show strong shear-rate dependences. This gap between theory and experiments is encountered in almost all disciplines, but is indeed acute in rheology that is well illustrated by an anonymous remark quoted by Willets (1967): "Rheologists can be divided into two classes: the practical rheologists who observe things that cannot be explained, and the theoretical rheologists who explain things that cannot be observed." Walters (1979) described this gulf between theory and experiments by calling them the "left" and "right" wings of rheology. There are two possible ways to bring the left and right wings closer: either the existing theoretical treatments must be improved to elucidate the combined effects of shear-rate-dependent viscosity and visco-elasticity, or experiments must be performed in model visco-elastic media that display varying levels of visco-elasticity but no shear-thinning effects. Admittedly, considerable progress has been made in bridging this gap by the currently available numerical simulations of the falling-ball problem, but unfortunately, to date it has not been possible to incorporate a realistic description of shear-rate-dependent viscosity together with the fluid visco-elasticity as encountered in such experimental studies. The second approach, as will be seen in the next section, has had some success in narrowing the existing gap between theory and experiments in this field. However, it needs to be emphasized here that it is not possible to split the combined nonlinear effects into two individual nonlinear contributions, i.e., that of shear-thinning and nonlinear visco-elasticity.

4.2.2.2 Nonshear-Thinning Visco-Elastic Liquids (Boger Fluids)

In 1977, Boger (1977a) reported the steady shear characteristics (namely, shear stress and primary normal stress difference) of a new class of materials formulated by dissolving small amounts of polyacrylamide in highly viscous Newtonian corn syrups diluted with small amounts of water. Over a narrow but finite shear-rate range, these fluids exhibit almost a constant shear viscosity, but varying levels of primary normal stress difference (visco-elastic effects) including the expected quadratic region at low shear rates; the extent of shear-thinning increases with the increasing proportion of water in the solution. Typical shear stress and primary normal stress difference data for such a fluid are shown in Figure 4.3. This indeed proved to be a turning point in the mechanics of visco-elastic fluids. Walters (1979) greeted the emergence of Boger fluids by noting:

> The discovery of the Boger constant viscosity highly elastic liquids has recently given renewed hope... at least for a small sub-set of real elastic liquids with the side effect that some of the early analysis for simple fluid models may not be as obsolete as their dating would suggest.

In 1983, Prilutski et al. described the formulation of another type of constant viscosity elastic fluids prepared by dissolving polyisobutylene (PIB) in polybutene (PB). The preparation of these fluids is facilitated by using a small amount of kerosene, akin to the role of water in the corn syrup-based fluids. Detailed descriptions of the recipes of preparing these fluids and the associated difficulties are described in a number of publications (Boger, 1977a; Nguyen and Boger, 1978; Choplin et al., 1983; Binnington and Boger, 1985, 1986; Gupta et al., 1986). Typical shear stress and primary normal stress difference data (Boger, 1984) for a PIB/PB Boger fluid are also included in Figure 4.3. Evidently, this fluid exhibits a constant shear viscosity and quadratic behavior (with respect to N1) over a wider range of shear rates than that observed for the corn syrup-based fluids. Aside from the steady shear stress and N1 data,

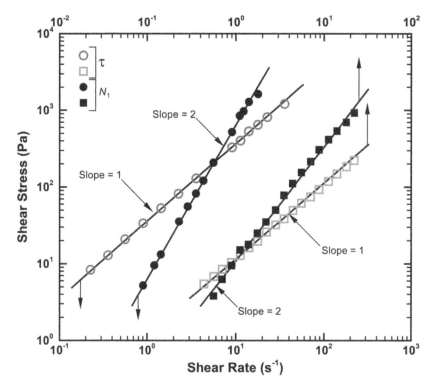

FIGURE 4.3 Steady shear rheological data for two Boger fluids. (□■) Polyacrylamide in corn syrup; (○●) PIB in polybutene system.

numerous investigators have reported several other rheological characteristics of these systems including the second normal stress difference (Keentok et al., 1980), dynamic oscillatory measurements (Sigli and Maalouf, 1981; Jackson et al., 1984; Phan-Thien et al., 1985), normal stress relaxation and recovery of strain after the cessation of steady shear (Magda and Larson, 1988), uniaxial drawing (Sridhar et al., 1986), etc. Based on the available information, it is now generally agreed that the Oldroyd-B model provides an adequate representation of the rheological behavior of Boger fluids. Much of this discussion about the family of Boger fluids has been summarized by James (2009).

Most of the experimental results on falling balls in such fluids available to date are compiled in Figure 4.4; since all measurements relate to Re < 0.01, the creeping flow conditions can be assumed to exist in such studies. It is clearly seen that there is very little quantitative agreement among different workers. Though the general picture is quite provocative and confusing, 30 years ago, Walters and Tanner (1992) postulated the general Y–We map in the absence of wall effects ($\beta = 0$) that is shown schematically in Figure 4.5. The horizontal portion (A–B) of the map is to be expected from continuum mechanics requirements, and its presence has been confirmed experimentally (Chhabra et al., 1980b; Mena et al., 1987; Chmielewski et al., 1990a; Tirtaatmadjta et al., 1990) up to We ~ 1, followed by a region of drag reduction (B–C) with Chhabra et al. (1980a) and Mena et al. (1987) demonstrating a drop of about ~25%, the point C being located at We ~ 1.0. On the other hand, Chmielewski et al. (1990a) and Tirtaatmadja et al. (1990) reported a drag reduction of about ~10% with corn syrup-based Boger fluids. Virtually, no drag reduction has been obtained with PIB/PB Boger fluids. The difference in the responses of the corn syrup- and PIB-based Boger fluids has been attributed to the differences in solvent–polymer interactions and it demands further study. The presence of the plateau region (C–D) has been confirmed by Mena et al. (1987) and Chhabra et al. (1980b). Walters and Tanner (1992) conjectured that the plateau region C–D may or may not always be present, but what is certain is the region of substantial drag enhancement, denoted here schematically by D–E. Such drag enhancement has been observed experimentally by Tirtaatmadja et al. (1990) and several others (Chmielewski et al., 1990a; Jones et al.,

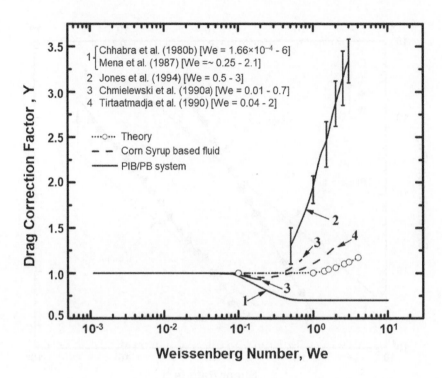

FIGURE 4.4 Experimental results on drag for an unconfined sphere ($\beta = 0$) in Boger fluids in the creeping flow regime.

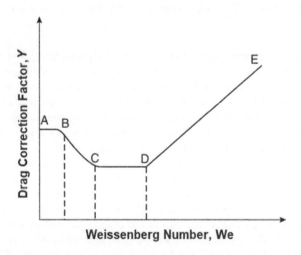

FIGURE 4.5 Postulated overall drag–Weissenberg number map. (After Walters, K. and Tanner, R.I., *Transport Processes in Bubbles, Drops, and Particles*, Chhabra, R.P. and De Kee, D., Eds., Hemisphere, New York, Chapter 3, 1992.)

1994; Arigo et al., 1995; Solomon and Muller, 1996; Fabris et al., 1999). Indeed, the values of the drag correction factor, Y, as large as 8 (corresponding to We ~ 15) have been reported in the literature (Fabris et al., 1999). As seen in Section 4.2.1, the analyses of Chilcott and Rallison (1988), Marchal and Crochet (1988), and Harlen (1990) as well as the numerical studies (Castillo et al., 2019; Housiadas and Tanner, 2016; Faroughi et al., 2020, etc.) for instance do allude to such an increase in drag. Over the years, there has been a growing realization of the fact that steady shear properties alone (namely viscosity and first normal stress difference) are not sufficient to reconcile the drag results even in seemingly similar fluids (McKinley, 2002). Owing to the presence of a strong extensional flow in the wake region, extensional

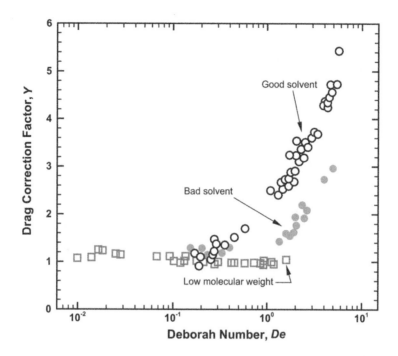

FIGURE 4.6 Effect of polymer–solvent interactions on sphere drag in polystyrene-based Boger fluids. (○) high molecular weight (2×10^7 g mol^{-1}) in a good solvent; (●) high molecular weight in a poor solvent; (□) low molecular weight polymer (2×10^6 g mol^{-1}). (Replotted from Solomon, M.J. and Muller, S.J., *J. Non-Newt. Fluid Mech.*, 62, 81, 1996).

behavior of the test fluids plays an important role in determining the values of Y, especially at high values of We. Similarly, the elegant study of Solomon and Muller (1996) elucidates the effects of solvent quality, molecular weight, extensibility, etc. on drag as shown in Figure 4.6. The value of the drag correction factor Y hardly deviates from its Newtonian value of unity in a fluid involving a low molecular weight and less extensible polymer molecules. Thus, all in all, the drag on a sphere in purely elastic fluids is determined by a complex interplay between the shear and extensional rheological characteristics that in turn are strongly influenced by the detailed molecular architecture, quality of solvent, polymer conformation and hysteresis, etc. (McKinley, 2002; Zenit and Feng, 2018). Therefore, unless all these features are built into numerical simulations, the correspondence between predictions and experiments is unlikely to improve.

4.2.3 The Time Effect

A few investigators (Cho and Hartnett, 1979; Bisgaard and Hassager, 1982; Bisgaard, 1983; Cho et al., 1984; Manero et al., 1986; Ambeskar and Mashelkar, 1990; Yoo et al., 1994) have demonstrated another unexpected experimental difficulty with the deceptively simple sphere drop experiment. In highly viscoelastic polymer solutions, the terminal velocity of a sphere can be strongly influenced by the time interval between the successive sphere drop tests. For example, the experimental results of Bisgaard and Hassager (1982) and Bisgaard(1983) show that if spheres are released in a visco-elastic medium every 10th minute, the terminal velocity of the spheres can be up to 30% higher than that of the first. Even after a gap of 1 h, Bisgaard reported a 9% increase in velocity. Cho et al. (1984) speculated that when a sphere settles through a highly visco-elastic fluid, it "opens up" or "ruptures" the polymer network matrix locally and as the sphere moves down the axis of the cylinder, the solvent from the neighboring region fills the "cavity" left by the sphere in the center region immediately. However, the experimental work of Ambeskar and Mashelkar (1990) does not support this hypothesis. In spite of this, Cho and Hartnett (1979), Cho et al. (1984), and Yoo et al. (1994) have used such time-dependent terminal velocity measurements to infer the characteristic time for visco-elastic liquids. Similar time-dependent terminal velocities have also been reported for bubbles moving in visco-elastic media (Barnett et al., 1966; Carreau et al., 1974; De Kee et al., 1986). Tirtaatmadja et al. (1990), however, reported no such effects. The magnitude of the time-interval effect is certainly such as to make it necessary to be careful in conducting falling sphere experiments.

4.2.4 Velocity Overshoot

The simplest and possibly also the most widely used method of performing sphere drop tests is to release a sphere from rest and time its descent by, for example, a stroboscopic technique or a stopwatch. Thus, it can be ascertained if the sphere has attained its terminal velocity. Under most conditions of interest, the distance (or equivalently the time) required for a sphere to reach its terminal velocity in Newtonian fluids is very modest and is of the order of a few sphere diameters (Clift et al., 1978; Bagchi and Chhabra, 1991a; Chhabra et al., 1998). Furthermore, for Newtonian and inelastic non-Newtonian fluids, the build up from rest to the ultimate velocity is monotonic. For visco-elastic fluids, the situation can be much more complex. Walters and Tanner (1992) have presented a photograph of successive positions of a sphere settling in a Boger fluid. In this case, not only the distance traveled by the sphere before reaching the terminal fall condition is of the order of $40R$ but also evident is the noticeable overshoot in the velocity of the sphere before attaining its terminal velocity. There is certainly no monotonic build up to the terminal velocity, and the maximum velocity reached is three times the ultimate terminal velocity. These observations are in qualitative agreement with the theory of King and Waters (1972) who in fact outlined a scheme for evaluating visco-elastic material parameters from such transient experiments. This phenomenon was also demonstrated at the Society of Rheology meeting held in Madison in 1961 (Philippoff, 1961) and by Broadbent and Walters (as cited in King and Waters, 1972). However, none of the other transient analysis (Thomas and Walters, 1966; Lai, 1973, 1974, 1975; Gupta, 1976) predicted either of these effects. Over the past ten years or so, there has been a spurt in transient numerical simulations for a sphere starting from rest, for such studies are useful in benchmarking the effectiveness of time-dependent algorithms and also provide further insights into the development of the eventual steady-state flow field. While Walters and Tanner (1992) presented a series of multiple images of a falling sphere showing a velocity overshoot in a Boger fluid, subsequent more systematic studies show the effect much more clearly, both in corn syrup-based and PIB/PB-based Boger fluids (Jones et al., 1994; Becker et al., 1994). Clearly, additional dimensionless parameters are required to describe the transient behavior of a sphere, in addition to those included in Equation 4.1. A minimum of two such parameters involving time and the inertia of the sphere are required. Indeed, depending upon the relative importance of the polymeric elasticity, particle inertia, and the damping action due to the solvent, it is possible to produce a complex oscillatory behavior as also suggested by King and Waters (1972). The oscillations are over-damped due to high solvent viscosity. However, as the solvent viscosity is reduced, the velocity overshoot and the rate of damping will decrease to a point of under-damped oscillatory behavior (Bodart and Crochet, 1994). Additional complications arise if the confining walls are present (Arigo and McKinley, 1997). All these calculations are based on visco-elastic models with a single characteristic relaxation time and therefore these cannot really capture completely the experimental observations involving rapid accelerations and shorter relaxation modes. Some of these ideas have been explored numerically with limited success (Becker et al., 1994; Rajagopalan et al., 1996; Rasmussen and Hassager, 1993, 1995) whereas analogous developments for integral visco-elastic models have been discussed by Rasmussen (1999) and Peters et al. (2000).

On the other hand, when a fluid also exhibits shear-dependent viscosity, in addition to visco-elasticity, it is possible that the amplitude of oscillations can reach such levels that the accelerating sphere can reverse its direction (Zheng and Phan-Thien, 1992). Harlen et al. (1995) alluded to the possibility of a negative wake in the initial stages of transient behavior in an Oldroyd-B model fluid that, however, disappears as the time progresses and the molecules become highly oriented and elongated in the wake region. Indeed, the overshoot in velocity up to a factor of 7 has been observed in visco-elastic shear-thinning polymer solutions for an aluminum sphere (Arigo and McKinley, 1998).

In addition to the aforementioned studies relating to the initial transient acceleration from rest under a constant force (buoyant weight of the sphere), other transient scenarios have also been investigated. For instance, Ramkissoon and Shifang (1993) analyzed cases of the constant and the exponentially decreasing acceleration in a Maxwell fluid. Mei et al. (1996) studied the combined effects of the walls and the visco-elasticity on the small amplitude oscillatory motion of a sphere. In fact, they proposed a scheme to evaluate the visco-elastic material parameters from such an experiment, and if proven, this technique has potential in the rheometry of biopolymeric systems (Levine and Lubensky, 2001). Similarly, in a recent study, Kumar et al. (2012) analyzed the oscillating settling behavior of steel spheres (3–8 mm in diameter) in worm-like micellar solution. The settling velocity exhibited repeated short oscillatory bursts separated by long periods of near nonoscillatory behavior. These results were rationalized by considering the interplay between the settling-induced flow and visco-elastic stresses which influenced the "structure" of the solution. Qualitatively similar results have been also reported for a sphere falling in Johnson–Segalman model fluids (Lee and Zhang, 2012). Zhang and Muller (2018) have also investigated transient settling of a sphere in worm-like micellar solutions and they argued in terms of a critical strain needed for the onset of instability in the negative wake.

Thus, all in all, extreme caution is required when performing such seemingly simple falling-ball tests and, indeed, the complexity of the sphere motion scales with the complexity of the fluid. It is therefore highly dangerous to extrapolate from one fluid to another fluid or follow the common practice of extrapolating behavior for visco-inelastic to visco-elastic fluids, which may also prove to be hazardous.

4.2.5 Drag-Reducing Fluids

In the fluid mechanics literature, the term "drag reduction" is used for characterizing the reduction of friction in turbulent flow through pipes and noncircular ducts. The main concern here is the drag reduction achieved by adding small doses of a class of high-molecular-weight polymers and other substances, such as soaps, clays, biopolymers, and surfactants, to the turbulently flowing water. Indeed, reductions in frictional losses as high as 80% have been documented in the literature. Owing to their wide-ranging practical applications, drag reduction in circular and noncircular ducts has been studied extensively (Hoyt, 1972; Mashelkar, 1973; White, 1976; Sellin et al., 1982a, b; Kulicke et al., 1989; Watanabe et al., 1998). Several mechanisms including slip at the wall (Agarwal et al., 1994), adsorption of polymeric molecules thereby depleting the polymer solution, nonisotropic viscosity and normal stresses, suppression of turbulence in the wall region, polymer degradation (Brennen and Gadd, 1967), etc. have been postulated to explain this effect (Granville, 1971; Hoyt, 1972; Sellin et al., 1982a). Irrespective of the nature of the mechanism postulated, the visco-elasticity of the dilute polymer solutions is often invoked, directly or indirectly, to explain the phenomenon of drag reduction. In contrast, the behavior of the drag-reducing polymer solutions in external flows has been studied less extensively. Such studies have been motivated by the possibility of reducing hydrodynamic drag on boats, canoes, ship hulls, submarines, etc. (Sellin et al., 1982a, b; Kulicke et al., 1989). To date, only highly idealized shapes such as spheres, cylinders, discs, flat plates, etc. have been used, albeit some results are also available for more realistic shapes of vehicles (White, 1976; Francois et al., 2008; Pipe and Monkewtiz, 2006). The pertinent literature for the flow of drag-reducing solutions past a cylinder has been summarized elsewhere (Chhabra, 2011).

One important feature of the external flows that sets them apart from the internal flows is that the solid flow boundary is of finite extent (in the downstream direction) so that the flow must be treated as a developing boundary layer in which time average steady-state conditions cannot be reached before the solid surface terminates. The other distinct feature of such flows is that they develop in a semi-infinite fluid body in which the growing boundary layer does not interact with those from the neighboring surfaces such as that in conduits.

Drag reduction, as can be seen in Figure 4.7, for freely falling spheres, has been shown to occur in the subcritical Reynolds number region (Crawford and Pruitt, 1963; Ruszczycky, 1965; Lang and Patrick, 1966; Gadd, 1965, 1966; White, 1966, 1967, 1968, 1970; Sanders, 1967, 1970; Hino and Hasegawa, 1968; Carey and Turian, 1970; Puris, 1973; Puris et al., 1981; Watanabe et al., 1998). Results of different investigators seldom agree with each other. One possible reason for this discrepancy can be attributed to the viscosity of the solvent or the solution used in calculating the value of the Reynolds number. The two values are sufficiently close only for dilute polymer solutions. At high concentrations, not only the two viscosity values differ appreciably, but the solutions may also exhibit shear-rate-dependent viscosity and nonzero primary normal stress differences (Ruszczycky, 1965; Sanders, 1967). Although a range of drag-reducing

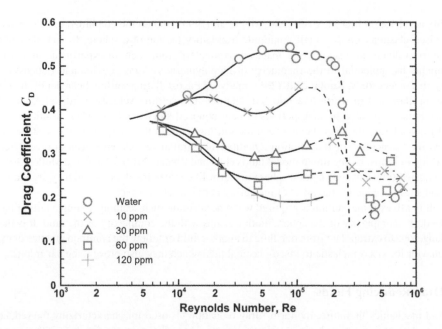

FIGURE 4.7 Typical drag reduction results for spheres falling in PEO (Polyox WSR-301) solutions. (From White, A., Drag of spheres in dilute high polymer solutions, *Nature*, **216**, 994, 1967.)

additives has been used, most work has been carried out with the different grades of polyethylene oxide (Polyox WSR-301, WSR-205, Coagulant) in the concentration range of 10 to 150 ppm. In spite of the lack of quantitative agreement among various workers, the drag reduction for spheres seems to occur when the boundary layer is still in the laminar regime (Re ~ 2×10^3 to 3×10^5).

This is in sharp contrast to the skin friction reduction in turbulent pipe flows. Uhlherr et al. (1974) have reported qualitative changes in the flow patterns around a sphere in the range 50 < Re < 450, but without any drag reduction under these conditions. On the other hand, Usui et al. (1980), Puris et al. (1981), and Kato and Mizuno (1983) have reported significant changes in the velocity and pressure fields close to spheres and cylinders in dilute polymer solutions thereby suggesting a direct link between the detailed kinematics and drag reduction (Chhabra, 2011). Killen and Almo (1969) have documented similar results for a rotating cylinder. Stow and Elliott (1975), on the other hand, did not observe any drag reduction, and Latto et al. (1973) reported slight drag enhancement for freely suspended spheres in upward flowing dilute polymer solutions.

The limited literature on the flow of drag-reducing solutions across cylinders and thin wires is also inconclusive. James and Acosta (1970), James and Gupta (1971), and Sanders (1967, 1970) have reported drag enhancement in the laminar flow conditions (2 < Re < 500). On the other hand, Sarpkaya et al. (1973) did not report any drag reduction. However, some changes in the flow patterns around cylinders and wires in drag-reducing polymer and surfactant solutions have been observed (Kalashnikov and Kudin, 1970; Koniuta et al., 1980; Ogata et al., 2006). Pipe and Monkewtiz (2006) reported the vortex-shedding frequency to be suppressed in dilute Poly Ethylene Oxide (PEO) solutions flowing over a cylinder in the range $50 \leq Re \leq 150$. This is subsequently confirmed by Sahin and Atalik (2019). Analogous studies for flat plates have been carried out by Wu (1969) and Latto and Middleton (1969), among others, who reported drag reduction of 60% and 30%, respectively, with Polyox WSR-301 solutions (100 and 50 ppm) in the range Re ~ 0.5 to 1.5×10^6. Granville (1971) has postulated a possible mechanism for the reduced skin friction on flat plates.

4.2.6 Sphere in Mixed Flows

Over the years, there has been a growing interest in investigating the hydrodynamics of a sphere in composite flows; much of the literature focuses on a simultaneously translating and rotating sphere, or on a sphere settling in sheared visco-elastic fluids. Current interest in such complex flows stems from pragmatic considerations (such as the flow of suspensions, rheometry of suspensions, removal of cuttings in oil wells, separation in microfluidic, etc.) as well as from fundamental considerations (such as to test the validity of the constitutive visco-elastic models and that of the numerical algorithms). Thus, for instance, Castillo et al. (2019) have studied translation and rotation of a sphere in Boger fluids (Oldroyd-B) in the zero Reynolds number limit. Based on their experimental, numerical, and asymptotic regular perturbation analysis results, the rotation causes drag reduction due to the lowering of the pressure contribution to the drag. Similarly, Housiadas (2019) has considered the simultaneous translation and rotation of a sphere in both Oldroyd-B and FENE-P model fluids (Re = 0). His predictions are consistent with that of Castillo et al. (2019). It is useful to recall here that there is no coupling between translation and rotation of a sphere (Re = 0) in Newtonian fluids, whereas this is not so for non-Newtonian fluids due to their shear-dependent properties. Housiadas (2019) noted that the viscous drag on the sphere increases with the increasing Weissenberg number (based on rotation) with a concomitant decrease in pressure component due to the solvent contribution. Similarly, D'Avino et al. (2008) and Snijkers et al. (2009, 2011) have considered rotation of a sphere in multimode Giesekus fluid, both numerically and experimentally. The sphere was seen to slow down with the increasing Weissenberg number under otherwise identical conditions. These findings are in agreement with each other. Based on their extensive experimental results in Newtonian constant viscosity Boger fluids, worm-like micellar solutions, and shear-thinning visco-elastic test fluids, Snijkers et al. (2009) concluded that elasticity slows down the rotation, and the nondimensional velocity correlates rather well with a suitably defined Weissenberg number. These predictions are consistent with predictions made by D'Avino et al. (2008) and Snijkers et al. (2011). Analogous results for a rotating and translating sphere in a power-law fluid have been reported by Godinez et al. (2014).

The sedimentation of a sphere in sheared visco-elastic fluids has also received considerable attention in recent years. For instance, Housiadas and Tanner (2012, 2014) and Tanner et al. (2014) studied the behavior of a sphere falling along the vorticity axis in a sheared visco-elastic fluid and reported the drag to be higher than that for a sphere falling in quiescent medium otherwise under identical conditions, in line with the experimental results of van den Brule and Gheissary (1993).This difference was attributed to the deformation of streamlines and the resulting lift due to the asymmetry of the flow field. Qualitatively similar results have been reported by others, e.g., see Padhy et al. (2013a, b) and Busch and Johansen (2019). Einarsson and Mehlig (2017) considered the steady translation of a sphere in mildly visco-elastic sheared fluid. They reported the shear-induced lift to be of the order ~ We whereas the first correction to the drag is of the order of We^2. Busch and Johansen (2019) studied the settling behavior of a sphere in an orthogonal shear flow of a shear-thinning and mildly elastic fluid. Based on the idea of "effective" shear rate (due to the background flow plus $2V_r/d$ where V_r is the relative velocity between the sphere and the fluid), they found that their non-Newtonian results coincided with the corresponding results for a Newtonian fluid (of same viscosity as the zero-shear viscosity of the visco-elastic fluid). Hu et al. (2015) measured the settling velocity of a single particle in a Couette cell. Their results suggest the settling to be retarded due to the enhanced viscosity, elastic lift, and elastic instability, i.e., the drag showed a positive dependence on both shear rate and fluid relaxation time. However, these results are at variance with the trends reported by Bazilevskii et al. (2010). Furthermore, Padhy et al. (2013a,b) reported shear-thinning to lower the drag whereas elasticity tended to increase it for a sphere in cross-shear flow. Thus, for instance, for a sphere settling in a cross-shearing FENE-P fluid, the drag increased by a factor of two with Weissenberg number up to about We = 15. The predictions are close to the experiments for We ≤ 2, but the agreement is only qualitative for higher values. The confinement effect was accentuated with the increasing Weissenberg number.

Similarly, Childs et al. (2016a,b) have considered the settling of a small particle (so that the shear rate due to the particle velocity is ≈ 0) in a sheared medium. In this a case, the effective viscosity is determined by the background flow of the fluid. Thus, for a shear-thinning fluid, the viscosity would be maximum at the axis of the channel thereby retarding the sedimentation velocity here. Childs et al. (2016a,b) have also presented results on the particle trajectories during the course of their settling. Detailed comparisons between the predictions and experimental results on particle trajectories indicate satisfactory match only in weakly visco-elastic fluids which deteriorates rapidly with the increasing visco-elasticity (Bazilevskii et al., 2010; Zoric et al., 2015; Khatibi et al., 2016). For instance, Bazilevskii et al. (2010) studied the sedimentation of single particles and concentrated suspensions in Newtonian and visco-elastic shear-thinning test fluids. However, no quantitative information about the elasticity (like N_1, G', G'') was obtained and they simply used the power-law model to approximate shear-thinning behavior. They analyzed their data for single (~ spherical) particles by defining an effective shear rate as $\left(\dot{\gamma}_p^2 + \dot{\gamma}_f^2\right)^{1/2}$ where $\dot{\gamma}_p$, particle-induced shear rate ~ V_0/d and $\dot{\gamma}_f$ is the shear rate in the Couette cell. Though the sedimentation velocity increased with the increasing value of $\dot{\gamma}_f$, the experimental increase was much smaller than that predicted based on the use of power-law model. Such underprediction was attributed to the visco-elastic effects. As expected, the shearing did not influence the settling velocity of a particle in Newtonian glycerol solutions.

4.3 Flow Over a Long Circular Cylinder

In contrast to the extensive literature on the flow past a sphere, the corresponding body of information for the flow past a long cylinder is limited, as can be seen in Tables 4.1 and 4.2, and most of it relates to the benchmark case of the flow over a cylinder in between two plane walls with the cylinder-to-slit width ratio of 0.5, that is, $\beta = 0.5$ (Figure 4.8). Additional motivation for these studies also stems from its relevance to active swimming in viso-elastic fluids (Riley and Lauga, 2014, 2015, 2017). Early analytical treatments for slow visco-elastic flow past an unconfined cylinder are described by Ultman and Denn (1971) for a Maxwell fluid and by Mena and Caswell (1974) for a corotational Oldroyd model. These analyses rely on different techniques to match an Oseen-type solution far from the cylinder with that near the cylinder. Both analyses predict a slight reduction in drag due to visco-elasticity but differ in detail. Similarly, Pilate and Crochet (1977) presented a numerical solution for inertial and inertia-less flows of second-order fluids past a cylinder, and they observed a complex interplay between elasticity and inertia. Perhaps the most detailed results for the flow of a Maxwell fluid past an unconfined cylinder are by Hu and Joseph (1990). They reported values of drag coefficient and Nusselt number for moderate values of the Reynolds and Prandtl numbers that are qualitatively consistent with the experimental results of James and Acosta (1970). Overall, their study suggests an increasing downstream shift in streamlines with the increasing values of the elasticity number at a fixed value of the Reynolds number. In subcritical regime, that is, $M < 1$, the drag on the cylinder is unaffected by visco-elasticity whereas for $M > 1$, the drag on the cylinder decreases due to visco-elasticity. These authors also noted a change in the type of governing equations when the fluid velocity exceeds the shear wave velocity. This transition, in turn, manifests in various ways such as delayed die swell (Joseph et al., 1987; Cloitre et al., 1998), anomalous rates of heat and momentum transport (James and Acosta, 1970; James and Gupta, 1971) and anomalous transport properties (Delvaux and Crochet, 1990), stagnant regions near a cylinder (Koniuta et al., 1980), orientation of rods (Cho et al., 1991, 1992; Joseph and Liu, 1993; Chee et al., 1994; Joseph, 1996), visco-elastic instability (Ambari et al., 1984a; Shiang et al., 1997, 2000), etc.

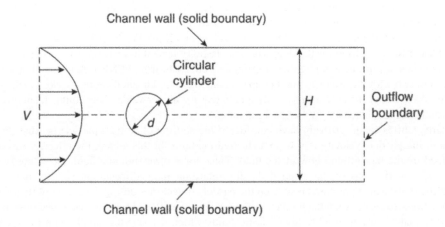

FIGURE 4.8 Schematics of the flow over a cylinder in a plane channel (cylinder diameter to channel height ratio, $\beta = (d/H) = 0.5$).

Some of the recent studies have dealt with the role of inertia (Re = 100 and Re = 300) on the hydrodynamic aspects by performing 3-D simulations of FENE-P model fluids (Richter et al., 2010). With the increasing extensibility parameter (L^2), both the wake size and drag increase, though some of the modes of instability are suppressed by the fluid elasticity. Similarly, Sahin and Atalik (2019) considered a range of visco-elastic models including Oldroyd-B, Giesekus, and FENE-P to explore the nature of hydrodynamic forces, vortex shedding, etc. over the range $80 \leq Re \leq 300$ together with a moderate degree of shear-dependent viscosity. They reported some drag reduction, and the frequency of vortex shedding was seen to increase with the decreasing power-law index whereas the elasticity was seen to increase the drag and suppress vortex shedding. This result is in line with the experimental results of Pipe and Monkewitz (2006) and Francois et al. (2008). In yet another experimental study, Dey et al. (2018) have examined the coupling between the visco-elasticity of worm-like micellar solutions and the flexibility of the cylinder surface. Similar complex interplay between shear-thinning, elasticity, and fluid inertia has been reported by Varshney and Steinberg (2018) by considering the stability of flow between two cylinders. Dollet and Graner (2007) attributed the fore-and-aft asymmetry in the 3-D flow of foam past a cylinder to the elasticity of the working fluid. Similar to the case of a sphere, negative wake can also form in the 2-D flow over a circular cylinder (Dou and Phan-Thien, 2003, 2004). Based on extensive simulations with a range of visco-elastic models (Oldroyd-B, Phan-Thien and Tanner, FENE-CR, FENE-P, Giesekus, etc.), Dou and Phan-Thien concluded that it is the first normal stress difference and its rate of change with the shear rate which are responsible for the formation of a negative wake in this case. Furthermore, shear-thinning behavior seems to delay or suppress this tendency.

In contrast, an enormous amount of research effort has been expended in elucidating the role of elasticity and shear-thinning on the flow field and drag on a cylinder placed in a planar slit with aspect ratio β = 0.5. The corresponding result for a Newtonian fluid (Faxen, 1946) is used for establishing the accuracy and reliability of the numerical solution procedures. Here too, the emerging scenario is as complicated as in the case of a falling sphere in a cylindrical tube. For instance, some studies (Carew and Townsend, 1991; Mitsoulis, 1998b; Sun et al., 1999; Dou and Phan-Thien, 1998, 1999, 2001, 2007; Phan-Thien and Dou, 1999; Wapperom and Webster, 1999; Yurun et al., 1999) predict the drag to decrease initially below its Newtonian value with the increasing value of the Weissenberg or Deborah number, akin to that in the case of a sphere. Within the framework of the Phan-Thien–Tanner (PTT) model, such a decrease is ascribed to shear-thinning whereas drag reduction is attributed to normal stress differences in the context of the UCM model. Qualitatively similar drag reduction has been reported with integral models also (Barakos and Mitsoulis, 1995). As the value of the Deborah number is increased further, large elongational stresses and stress gradients develop that tend to increase the drag (Dou and Phan-Thien, 1999; Sun et al., 1999; Yurun et al., 1999; Kim et al., 2004). At very high values of the Weissenberg number, Renardy (2000a) has advocated the use of the stress boundary layer approach. Suffice it to add here that most algorithms are plagued by a limiting value of the Weissenberg or Deborah number, and it is not at all obvious whether this limitation is physical or numerical in origin. Good summaries of theoretical developments in this field are available in the literature (Sun et al., 1999; Alves et al., 2001, 2021).

The relevant experimental studies have been reported by Baaijens et al. (1994, 1995, 1997) for low-density polyethylene (LDPE) melts and by Hartt and Baird (1996) for a linear LDPE and LDPE melts. While Baaijens et al. presented detailed results on the velocity and stress fields, Hartt and Baird (1996) provided data only on the flow-induced birefringence. Qualitatively, these results are in line with the aforementioned numerical simulations. In a later study, Dou and Phan-Thien (2003) have examined in detail the structure of the flow field for the creeping visco-elastic flow past a cylinder in a plane channel (β = 0.5). Four constitutive equations namely, Oldroyd-B, UCM, PTT, and FENE-CR models were used to elucidate the role of different rheological characteristics. As far as the drag behavior is concerned, drag reduction (with reference to the Newtonian value) occurs at low Deborah numbers whereas drag enhancement occurs at high Deborah numbers. This trend is consistent with the behavior observed for spheres (Section 4.2.1). Although the wall effects tend to increase the drag further, this tendency is suppressed in visco-elastic fluids. Furthermore, Dou and Phan-Thien (2003) also reported the formation of a negative wake when the extensional viscosity is only weakly dependent on strain rate. However, the absence or presence of a negative wake seems to bear no relation to the shifting of streamline patterns. Similarly, the tendency for the velocity overshoot is enhanced by the first normal stress difference, whereas shear-thinning weakens the velocity overshoot. Subsequently, Dou and Phan-Thien (2004) have argued that while the constant shear viscosity FENE-CR model fluid promotes the extent of the velocity overshoot, the formation of a negative wake is delayed by shear-thinning behavior of the fluid. Recent studies suggest that the formation of a negative wake behind a cylinder confined in a planar slit (aspect ratio of 0.5) is facilitated under the uniform flow conditions as opposed to that under the Poiseuille flow conditions (Kim et al., 2005). Indeed, Xiong et al. (2017) have used the Oldroyd-B model to study the flow dynamics for a cylinder up to Reynolds number values of 5000 and We ≤ 8 for the benchmark case of $\beta = 0.5$. They reported drag reduction only in the vortex-shedding regime and drag enhancement was observed when vortex shedding was suppressed. Subsequently, interactions between two side-by-side cylinders in visco-elastic fluids have been explored by Peng et al. (2020) and that of the angle of attack on a hydrofoil in dilute polymer solutions has been numerically studied by Xiong et al. (2018).

In addition to aforementioned literature on the benchmark configuration, many other related aspects including the inertial effects with and without the rotation of cylinder (Townsend, 1980, 1984; Matallah et al., 1998; Hu et al., 2005), nonisothermal rheological effects (Wu et al., 1999, 2003), influence of eccentricity (Dhahir and Walters, 1989), wall effects (Hu, 1995; Huang and Feng, 1995; Oliveira and Miranda, 2005), startup flow (van Heel et al., 1999), unsteady flow (Gilligan and Jones, 1970), and time-dependent simulation of vortex shedding (Oliveira, 2001; Minaeian et al., 2022) have also been explored to some extent.

In spite of the unprecedented efforts directed at obtaining the numerical results for the benchmark flows over a sphere in a tube and over a cylinder in a planar slit, reliable predictions have not been forthcoming, especially at large values of the Deborah number and Weissenberg number. Initially, this failure was ascribed to the inadequacy of numerical schemes and solution procedures. Over the years, it has evolved that the early onset of 3-D steady and time-dependent flow conditions seem to be a norm rather than an exception in the flow of visco-elastic fluids (Oliveira and Miranda, 2005). Further complications arise due to an intricate interplay between the fluid elasticity and fluid inertia. Thus, for instance, the fluid elasticity seems to promote stability in laminar flows for relatively strong inertial flows, while the role of fluid elasticity is completely reversed in the inertia-less flows. Thus, strong nonlinear elastic effects are observed in flows with little or no inertial effects. This has also been demonstrated experimentally for the flow past a cylinder in a slit with a cylinder-to-slit aspect ratio of 0.5 (Shiang et al., 2000; Verhelst and Nieuwstadt, 2004). In their excellent experimental study, Verhelst and Nieuwstadt (2004) demonstrated that for appropriate values of the Deborah number, there was a sort of straining flow in the direction along the cylinder thereby raising some doubts about the utility of 2-D simulations in such cases. Furthermore, their study also revealed strong interaction, even at macroscopic level in terms of the drag between two cylinders aligned in the direction of flow. For the center-to-center distance shorter than 3R, the polymer molecules were not able to relax fully after their encounter with the first cylinder and therefore the drag on the second cylinder was lower than that on the first one. However, this effect disappeared almost completely when the center-to-center separation was increased to 4R. Both the drag and wake size were strongly influenced by the value of the Deborah number.

In an extensive experimental study, James et al. (2016) measured the hydrodynamic drag on an isolated cylinder in the creeping flow of PIB in PB/Kerosene-type Boger fluids. The maximum Reynolds number was of the order of ~0.5 and the blockage ranged from 0.3% to 2.5%. Despite such small blockages, they applied the wall-effect (Faxen, 1946) and end-effect (numerically) corrections to their results in Newtonian fluids. For the symmetric confinement of the cylinder, the first signature of elasticity appears at about De ~ 0.7 and the wake region extended up to ~15 times the cylinder diameter in the downstream direction (Shiang et al., 1997). Duly corrected drag force F_D^{**} results in two Newtonian fluids, namely, PB ($\mu = 35$ Pa.s) and silicone oil ($\mu = 1.1$ Pa.s), which were found to be in excellent agreement with the following expression by Kaplun (1957):

$$F_D^{**} = \frac{F_D}{4\pi\mu V} = \frac{1}{\phi} - \frac{0.87}{\phi^3} \qquad (4.8)$$

where $\phi = -0.0772 - \ln\left(\dfrac{\text{Re}}{8}\right)$

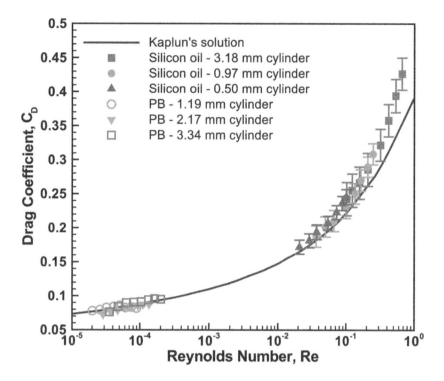

FIGURE 4.9 Comparison of experimental results of fully corrected drag coefficients (with wall and end effects) for cylinders in Newtonian fluids with that of Kaplun's solutions. (Replotted from James et al., *J. Rheo.* **60**, 1137, 2016.)

Figure 4.9 shows a comparison between the predictions of Equation 4.8 and the experimental results of James et al. (2016); the agreement is seen to be excellent up to about $Re \cong 10^{-2}$ and Equation 4.8 underpredicts the experimental results by up to ~15%–20% at $Re \cong 0.5$ or so.

The corresponding results for the Boger fluids are shown in Figure 4.10 in the form of normalized drag force F_D^{**} against the Deborah number ($De = \lambda V/d$ where λ is estimated using the expression $\frac{N_1}{2\dot{\gamma}^2 \mu_P}$). At low Deborah numbers (Figure 4.10), the drag values, consistently below the predictions of Equation 4.8, deviate a little before taking off at $De \sim 0.6$. James et al. (2012) were able to locate this point rather precisely by using a low-speed motor to drive the tank. The results shown in this figure span cylinder diameters ranging from 500 μm to 3.34 mm in the same fluid. This also testifies to the reproducibility of their results. On the other hand, the results for the two Boger fluids overlap well up to about $De \sim 1$, but the results for the two fluids deviate increasingly from each other at $De > \sim 1$, by up to 30%–40% at $De \approx 100$. James et al. (2012) attributed these differences to the different relaxation behavior of the molecules in the two fluids. Included in Figure 4.10 are the predictions of Chilcott and Rallison (1988) for two values of the extensibility parameter (L^2) and viscosity ratio (c). The predictions and experimental results are in qualitative agreement with each other. Their PIV results also revealed significant changes in the upstream and downstream velocity profiles due to visco-elasticity with reference to the corresponding Newtonian velocity profiles.

It is also being recognized that while the flow over a sphere or a cylinder is free from geometric singularities, the singularities of the constitutive equations should also be taken into account. Thus, for instance, the commonly used UCM and Oldroyd-B models exhibit singularities in their extensional viscosity behavior at low rates of extension. This also impinges directly on the flow behavior in the wake regions dominated by the extensional component. Therefore, there is a growing trend to use FENE-CR or extended Pom-Pom type fluid models to study this class of flows (Oliveira and Miranda, 2005; van Os and Phillips, 2005). Some attempts have also been made to carry out the stability analysis to understand the origins of the instabilities encountered in numerical simulations (Hulsen et al., 1997, 2005). Another

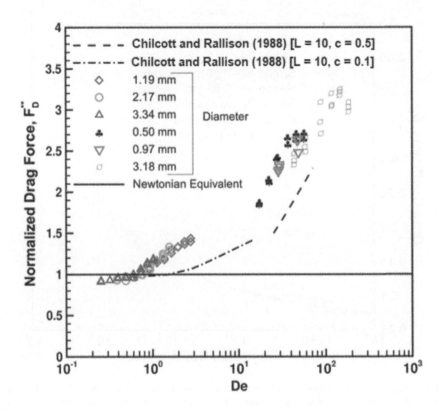

FIGURE 4.10 Normalized drag on cylinders in a Boger fluid. (Modified after James, D.F., Shiau T. and Aldridge, P.M., *J. Rheol.*, **60**, 1137, 2016.)

recent study has examined the streamline patterns past a cylinder with and without the inertial effects at high Weissenberg number (Hu et al., 2005).

4.4 Interaction between Visco-Elasticity, Particle Shape, Multiple Particles, Confining Boundaries, and Imposed Fluid Motion

Until now the discussion has been restricted to the simplest case of the steady and time-independent flow of visco-elastic fluids past a single sphere or a cylinder in relatively simple geometrical configurations. Also, the major thrust of most studies has been to elucidate the role of visco-elasticity on the detailed kinematics of the flow and on global parameters such as drag coefficient, wake sizes, and vortex shedding. A sizeable body of knowledge now also exists on several other interesting phenomena including interaction between two and three or more spheres (Riddle et al., 1977; Brunn, 1977b; van den Brule and Gheissary, 1993; Bot et al., 1998; Daugan et al., 2002a, b, 2004; Ardekani et al., 2008; D'Avino et al., 2013; Freire et al., 2019), orientation of elongated particles (Leal, 1975; Zana, 1975; Leal and Zana, 1975; Tiefenbruck, 1979; Tiefenbruck and Leal, 1980a; Chiba et al., 1986; Joseph and Liu, 1993; Joseph, 1996), agglomeration and clustering of particles in shear flow (Michele et al., 1977; Feng and Joseph, 1996; Lyon et al., 2001; Lee et al., 2003; Won and Kim, 2004), migration of particles (Gauthier et al., 1971a, b; Karnis and Mason, 1966, 1967 ; Bartram et al., 1975; Karis et al., 1984), interaction between multiple particles in visco-elastic fluids (Phillips, 1996; Goel et al., 2002), etc. Some of these works have been reviewed by Leal (1979, 1980), Brunn (1980), McKinley (2002), and recently by D'Avino et al. (2012, 2017, 2019), Eisenberg et al. (2013), etc., and hence only the salient features are summarized here. It is readily conceded that a small departure from the Newtonian fluid behavior, or from spherical shape (Saffman, 1956; D'Avino et al., 2012, 2019; Spanjaards et al., 2019), or from the creeping flow conditions can produce dramatic changes even at macroscopic level including in terms of the orientation of long and rod-like objects, drag, etc. Much of the research effort in this field has been directed toward the understanding of one or more of the following phenomena:

1. The simplest problem involves the prediction of the orientation of a transversely isotropic particle falling through in an unconfined visco-elastic fluid. Both theoretical and experimental results suggest that such particles rotate to attain a stable configuration and tend to settle with their longest axis (almost) parallel to the direction of gravity (Leal, 1975; Brunn, 1977; Leal and Zana, 1975; Tiefenbruck, 1979; Tiefenbruck and Leal, 1980a; Chiba et al., 1986; Joseph and Liu, 1993; Joseph, 1996). However, the presence of a wall alters the situation significantly.

2. Many workers have attempted to predict the lateral position relative to the boundaries in shearing flows, as discussed in Section 2.4. Given an appropriate blend of particle inertia and fluid elasticity, a particle may move away from or toward a wall (Gauthier et al., 1971a, b; Bartram et al., 1975) and some of these observations are qualitatively consistent with the analytical predictions available in the literature (Leal, 1975; Brunn, 1977a, D'Avino et al., 2019). Feng and Joseph (1996) and Feng et al. (1995) have studied the migration of spherical particles in the torsional flow of visco-elastic liquids. They also observed the formation of ring-like structures. Wang et al. (2018) have studied particle migration in Giesekus-model fluids flowing in a rectangular channel. In the limit of creeping flow, depending upon the initial position, a particle may migrate to the axis of the channel or to the closest corner of the channel. However, the corner-attraction region expands with the increasing Weissenberg number. When both inertial and elastic effects are significant, a particle tends to migrate to the face center of shorter sidewalls.

3. Another important phenomenon is the migration of particles across streamlines, thereby inducing formation of structures that clearly impact upon the rheological properties and their measurements for highly loaded systems (Kamal and Mutel, 1985; Metzner, 1985; Moshev, 1989; Schaink et al., 2000). It is worthwhile to recall here that in a Newtonian fluid, particles migrate laterally either due to inertial effects (Segre and Silberberg, 1963; Ho and Leal, 1974, 1976), or due to such deformation that occurs with fluid particles (Brenner, 1966; Chan and Leal, 1977, 1979, 1981), or due to an externally imposed flow (Jefri and Zahed, 1989; Tehrani, 1996). In visco-elastic media, such migration occurs at vanishingly small Reynolds numbers and with rigid spheres in shear and torsional flows (Karnis and Mason, 1966, 1967; Highgate, 1966; Highgate and Whorlow, 1967, 1969, 1970; Ponche and Dupuis, 2005). Similarly, Tehrani (1996) reported that particles migrated to the core region in tube flow of a visco-elastic system. Given the complexity of the problem, some tentative explanations/mechanisms have been postulated (Ho and Leal, 1974, 1976; Chan and Leal, 1977, 1979; Brunn, 1976a, b). Owing to the new emerging applications in the microfluidic-based particle separation, there has been a renewed interest in this field (Wang et al., 2018; D'Avino et al., 2012; Caserta et al., 2011; Liu et al., 2019b; Li et al., 2015, etc.). Broadly speaking, while in Newtonian media a particle moves away from the axis of the tube due to inertial effects, both shear-thinning viscosity and the presence of secondary flow also cause the particle to migrate away from the axis. In contrast, fluid elasticity counters this tendency and thus the equilibrium position of a particle is determined by a balance between these two competing mechanisms. It is interesting that similar "structuring" of particles has also been observed during the flow of suspensions in visco-plastic fluids in a sudden expansion (Jossic et al., 2002; Jossic and Magnin, 2005) and during the fluidization of particles by visco-elastic liquids (see Chapter 7).

4. The simplest nonaxisymmetric geometry of a sphere falling in a visco-elastic fluid next to a plane wall has been considered by Jefri and Daous (1991), Liu et al. (1993), Becker et al. (1996), Singh and Joseph (2000), and Lee et al. (2001). For instance, particles have been observed to experience a force toward the wall, that is, a negative lift force and the so-called anomalous rotation in the opposite direction to the anticipated rolling motion in narrow channels (Liu et al., 1993). Subsequently, Becker et al. (1996) have observed similar effects in much larger containers also. Some of these aspects have been simulated at least qualitatively (Feng et al.,

1996) using 2-D flow approximation. With advancing computational capabilities, Binous and Phillips (1999a, b) presented a Stokesian dynamics algorithm for simulating the behavior of a suspension of FENE dumbbells, and Singh and Joseph (2000) have used a finite element method to analyze the sedimentation of particles in Oldroyd-B fluids. These simulations clearly show that the so-called negative lift force arises only in a 3-D flow.

5. The interaction between two spheres (one behind the other) falling inline in visco-elastic media has also generated some interest (Caswell, 1977; Riddle et al., 1977; Lee et al., 2003; Horsley et al., 2004; Freire et al., 2019; Ardekani et al., 2008). The early study of Riddle et al. (1977) suggested that the two spheres may converge or diverge depending upon the level of the elasticity, shear-thinning, and the initial separation, as confirmed by a recent study (Freire et al., 2019). Subsequent observations in Boger fluids (Bot et al., 1998) and Xanthan solutions (Won and Kim, 2004) show that the trailing sphere settles faster than the leading one. However, as the trailing sphere gets within the wake region (which is elongated due to elasticity) of the leading sphere, the polymer molecules near the nose of the trailing sphere retain an increasing fraction of the orientation induced in the wake region of the first sphere. Once these two competing effects balance, the separation between the two spheres attains a constant value that increases with Deborah number. On the other hand, in concentrated polymer solutions with a negative wake there is no equilibrium separation, and the two spheres aggregate and fall as a doublet that in turn can lead to the formation of long chains of spheres (Liu and Joseph, 1993; Patankar and Hu, 2000, 2001). When the two spheres settle side by side, the two can attract each other, also undergo rotation, and again fall as a doublet (Joseph et al., 1994). Possible physical reasons based on the second-order fluid model behind such migration, rotation, clustering, etc. have been postulated by Joseph and Feng (1995, 1996). Also, similar 2-D calculations for pairs of circular and elliptical particles falling in Oldroyd-B fluids (Feng et al., 1995, 1996; Huang et al., 1997, 1998; Hu, 1998; Peng et al., 2020) and for one and two particles settling in a visco-elastic suspension of FENE dumbbells (Binous and Phillips, 1999a) are now available in the literature. In fact, even a 3-D study of such aggregation in a suspension of FENE dumbbells is available (Binous and Phillips, 1999b). Analogous experimental results involving three spheres in static and shearing fluids have been reported (van den Brule and Gheissary, 1993). Similarly, it has been observed that the trailing sphere from a chain settling in visco-elastic liquids may get detached from the chain. However, this phenomenon is not limited to visco-elastic fluids, as it can occur in Newtonian fluids too under appropriate circumstances. Patankar et al. (2002) have reported 2-D simulations for this case. For chains falling under the effect of gravity, as the weight of the chain scales with the length of the chain, the chain settles faster than a single particle. Due to the wake phenomena coupled with the stress effects, the drag on trailing spheres is reduced. Thus, the long-body effect competes with the wake and normal stress effects, thereby giving rise to the notion of a critical separation distance. For initial separation less than the critical value, particles get attracted, else separation occurs (Patankar et al., 2002; Freire et al., 2019).

6. The interplay between the fluid elasticity and particle shape has been investigated experimentally by Maalouf and Sigli (1984) and analytically by Dairenieh and McHugh (1985), Kim (1986), Ramkissoon (1990), Joseph and Feng (1996), Huang et al. (1998), and Galdi (2000). While preliminary comparisons between the predictions and the observations are encouraging, detailed quantitative comparisons are still awaited. The potential flow of the second-order fluids past a sphere and an ellipse has been studied by Wang and Joseph (2004) thereby elucidating the interplay between viscous, elastic, and inertial effects. Subsequently, Patankar et al. (2002) have shown that the extra stress tensor does not contribute to the normal component of the stress on the surface of a rigid particle translating in an incompressible Oldroyd-B fluid.

7. Additional unexpected and fascinating phenomena involving non-Newtonian liquids and particles are found in the papers of Cheny and Walters (1996), Gheissary and van den Brule (1996), Mollinger et al. (1999), Podgorski and Belmonte (2002, 2004), and Akers and Belmonte (2006). In the first of these, Cheny and Walters (1996) studied the deformation behavior of the free surface of a visco-elastic medium by releasing a sphere from a height into it. They reported the length of the jets so formed to be dramatically reduced in visco-elastic fluids. Gheissary

and van den Brule (1996), on the other hand, reported time dependence of the settling velocity of a sphere and this effect gets accentuated further in case of multiple spheres. Finally, Mollinger et al. (1999) observed an oscillatory behavior in which the settling velocity of a sphere decreases but then it suddenly increases, and thus the cycle repeats itself, as also reported by Kumar et al. (2012) in the context of worm-like micellar solutions. Podgorski and Belmonte (2002, 2004) have studied the shape of the crater created by a sphere settling through the free surface of Newtonian and visco-elastic liquids. While in Newtonian viscous media, deformed free surface behind the sphere is like a funnel, not only is the axisymmetry of the crater lost in visco-elastic liquids, but it also buckles and leads to a sort of pinch-off. Podgorski and Belmonte (2002) explained this phenomenon by postulating the existence of a stress boundary layer. Subsequently, they have developed a so-called finite elasticity membrane model by treating the free surface as a stretched elastic membrane (Podgorski and Belmonte, 2004). Likewise, Akers and Belmonte (2006) have presented similar results with micellar visco-elastic fluids. The penetration depths were shown to scale with the ratio of the initial kinetic energy of the sphere to the elastic modulus of the fluid. Since none of these phenomena is observed in linear fluids, it is yet another warning that extrapolation of the Newtonian thinking to visco-elastic liquids is dangerous!

8. In a detailed experimental study, Jayaraman and Belmonte (2001, 2003) have reported oscillating behavior of spheres in worm-like micellar fluids. Figure 4.11 shows representative results from their study for different size plastic spheres ($\rho_p = 1350$ kg m^{-3}). In the presence of certain organic salts, the surfactant molecules can self-organize to form long tube-like structures that are called "worms" or micelles; these can be as long as 1 μm but are still much smaller than

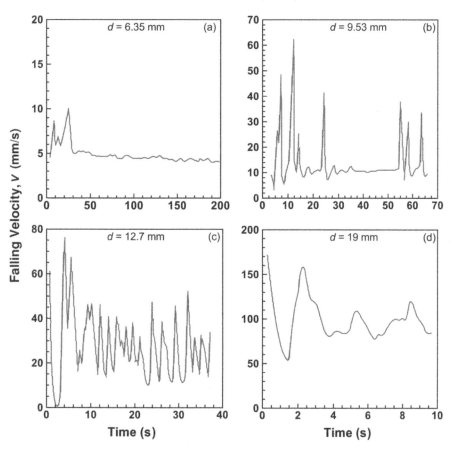

FIGURE 4.11 Unstable time-dependent behavior of a series of plastic spheres in a worm-like micellar solution for different sphere diameters (a) d = 6.35 mm (b) d = 9.53 mm (c) d = 12.7 mm (d) d = 19 mm. (Replotted from Jayaraman, A. and Belmonte, A., *Phys. Rev.*, **E67**, 65301, 2003.)

the macro-dimensions of the balls and the apparatus. The steady and oscillating tests also show some unusual characteristics, like a near plateau in shear stress–shear-rate plots, following the zero-shear viscosity region. Finally, these fluids do show shear-thinning behavior under appropriate conditions. While the elastic modulus G' qualitatively follows the predictions of the Maxwell model (single relaxation time), the loss modulus G'' shows an upward trend with frequency (> ~1 rad s^{-1}). The settling experiments were performed in cylindrical tubes such that $\beta \ll 0.21$. Based on the results shown in Figure 4.11 and the other data obtained with nylon and teflon sphere, the key points can be summarized as follows: small spheres ultimately do reach a constant terminal velocity after some transient oscillation as shown here for $d = 6.35$ mm (Figure 4.11a). For large spheres, it is difficult to say whether they will ever reach the constant falling velocity (Figure 4.11b–d); the sphere sizes of 9.53, 12.7, and 19 mm are shown here. Notwithstanding the lack of perfect periodicity and the irregularity present, one can still define a sort of average frequency as the number of oscillations divided by the time taken to descent the predetermined distance. This value of frequency was seen to increase with the size of the sphere, similar to the case of oscillations seen with bubbles (Belmonte, 2000). Furthermore, while it is difficult to discern a pattern in the amplitude of such oscillations, visual observations showed some common features of a sudden acceleration and a relatively slower deceleration. During the period of a sudden acceleration, a strong negative wake was seen as a recoil in the liquid. For a fixed value of sphere size, there is a transition to oscillations as the sphere density is increased. Based on qualitative scaling considerations, the onset of oscillations seems to correlate with the frequency at which the loss modulus G'' shows an upturn. These oscillations were attributed to the flow-induced structures that are broken down by the passage of a sphere and reformed again, which is accentuated by the complex flow field produced by a falling sphere. Some of these trends have been theoretically captured by Lee and Zhang (2012).

Similar (but more quantitative) observations have been recently made by Chen and Rothstein (2004), Kumar et al. (2012), Mohammadigoushki, and Mullerand (2016), Wang et al. (2020) and Lopez-Aguilar et al. (2022) with worm-like micellar media. Initially, the drag on a sphere reduces mainly due to the shear-thinning behavior. As the Deborah (or Weissenberg) number is progressively increased, strong extensional flow is setup in the rear of the sphere that causes the drag to increase. This steady flow regime continues until a critical value of the Deborah number is reached, beyond which the flow becomes unstable. Indeed, a wide variety of instabilities have been documented by Weidman et al. (2012) for a sphere falling in Hyperbranched Poly Glycerol (HPG) gelling agent liquids, with and without strong wall effects. This is yet another warning that there is much more to this seemingly simple flow than meets the eye!

4.5 Conclusions

The complete picture concerning the drag on a sphere moving in visco-elastic fluids in the creeping flow regime is emerging slowly, and it is clear that with a suitable choice of the Weissenberg number (We) and the drag correction factor (Y), visco-elasticity causes drag reduction for "small" values of Weissenberg numbers followed by a region of drag enhancement at "high" values in the absence of shear-thinning viscosity. It has been possible to simulate the initial drag reduction, albeit there is very little quantitative agreement among the predictions even for the same class of rheological models, namely the so-called continuum models. Over the years, it has been possible to obtain fully convergent values of drag up to about We ~ 15 or so. Also, the internal consistency between results from different algorithms is encouraging and it inspires confidence in such calculations. There have been significant advances on the experimental front also. Not only the currently available drag results encompass the values of Weissenberg numbers as high as 15, but detailed flow visualization studies also show the lengthening of the wake region with the increasing Weissenberg number as predicted by some of the simulations. Despite such a significant progress, there are still many gaps in the falling sphere problem in model visco-elastic fluids, which await the development of appropriate theoretical frameworks for their rationalization.

The limited experimental evidence suggests that in the case of visco-elastic shear-thinning fluids, the drag on a sphere is largely determined by the shear-dependent viscosity and the visco-elasticity appears to exert a little influence, at least in quiescent fluids. This situation changes dramatically in composite flows such as that described in Section 4.2.6. From an engineering applications standpoint, predictive correlations presented in this chapter may be used for estimating the values of drag coefficient in a new application. It is also now clear that the so-called negative wake can only form in visco-elastic shear-thinning liquids. The literature on the transient sphere motion in model visco-elastic liquids is still in its infancy and the limited results available thus far clearly show the hazards and risks involved in extrapolating the results from one fluid to another and the extreme care needed in such experiments. The literature on sphere motion in drag-reducing fluids of polymers, soaps, fine particle suspensions, micelles, etc. is reviewed briefly and since not much has been reported on this topic in recent years, it appears that the matter has been put to rest.

In contrast to the sphere problem, the flow over an unconfined cylinder has received little attention, albeit the benchmark configuration has been investigated in some detail. Qualitatively, the same overall trends can be seen in terms of the detailed kinematics and drag behavior as that observed for a sphere in a tube. The activity in this field has not been very rapid partly due to the fact that experiments are much more difficult to perform than that with a sphere. Ample evidence now exists indicating the onset of time-dependent and 3-D flow characteristics even in such simple flows at zero Reynolds number, whereas the visco-elasticity acts as a stabilizing factor in flows with inertia. It is possible to observe a negative wake in this case as well under appropriate circumstances.

In addition to the aforementioned specific flow configurations, many other related problems involving interactions between multiple particles, particle shape, rheology, confining boundaries, background fluid motion, etc. have also been investigated. Such studies have been motivated by the growing importance of highly filled systems encountered in a range of industrial settings. There are many interesting (bizarre!) observations awaiting the development of suitable frameworks.

Nomenclature

- c Ratio of the solvent viscosity to the zero-shear viscosity (-)
- C_D Drag coefficient (-)
- C_{De} Drag coefficient in visco-elastic fluids (-)
- d Sphere or cylinder diameter (m)
- D Fall tube diameter (m)
- De   Deborah number (-)
- El   Elasticity number (= We/Re) (-)
- F_D Drag force on a cylinder per unit length (N/m)
- F_D^{**} Drag force normalized using the Newtonian value (-)
- H Height of the slit (m)
- L^2 Extensibility parameter (-)
- M Mach number $\left(=\sqrt{We/Re}\right)$ (-)
- n Power-law index for shear viscosity (-)
- N_1 First normal stress difference (Pa)
- p Pressure (Pa)
- p_0 Pressure for Newtonian fluids (Pa)
- p_1, p_2 First- and second-order contributions to pressure due to visco-elasticity, Equation 4.2 (Pa)
- R Sphere or cylinder radius (m)
- Re   Reynolds number for Newtonian fluids (-)
- Re_0 Reynolds number based on zero-shear viscosity, Equation 2.42b (-)

Re_{PL} Reynolds number for power-law fluids, Equation 2.39 (-)
s Power-law index for first normal stress difference (-)
V Free-fall velocity or uniform-fluid velocity (m/s)
V_r Relative velocity between sphere and fluid (m/s)
We Weissenberg number (-)
Y Drag correction factor (-)

Greek Symbols

β Sphere-to-tube diameter ratio or cylinder diameter-to-slit-width ratio (-)
$\dot{\gamma}_p$ Particle-induced shear rate (s^{-1})
$\dot{\gamma}_s$ Shear rate or velocity gradient (s^{-1})
μ Viscosity (Pa.s)
μ_0 Zero-shear viscosity (Pa.s)
μ_s Solvent viscosity (Pa.s)
ρ Fluid density (kg m^{-3})
ρ_p Particle density (kg m^{-3})
$\Delta\rho$ Density difference, (kg m^{-3})
λ Fluid relaxation time (s)
ϕ Factor in Equation 4.8 (-)
ψ Stream function (m^2/s)

5

Fluid Particles in Non-Newtonian Media

5.1 Introduction

Fluid particles—bubbles and drops—are ubiquitous in everyday life, both in nature and in technology (Hughes, 1990; Weaire, 1994; Lohse and Zijm, 2003). The single most important feature of the fluid particles that sets them apart from the rigid particles is their mobile surface and their ability to deform during motion depending upon the relative magnitudes of different forces (surface tension, viscous, inertial, and elastic). Indeed, depending upon the relative magnitudes of the forces present in the continuous phase, fluid particles may exhibit a wide variety of shapes. Furthermore, the shape of a fluid particle may change with time and position during the course of its movement in a piece of equipment. Such shape changes directly impact not only the rates of momentum, heat, and mass transfer but also their breakage and coalescence behavior.

Current interest in the hydrodynamics of fluid particles in a non-Newtonian continuous phase stems from theoretical considerations such as understanding the interplay between the non-Newtonian characteristics and the kinematics of flow as well as from pragmatic considerations such as the fact that reliable quantitative information on the free-rise velocity, heat and mass transfer, breakage, and coalescence behavior is frequently needed for process design calculations. Furthermore, bubble and drop dynamics also play an important role in the production, stability, and rheological behavior of gas–liquid dispersions and liquid–liquid emulsions (Pal, 2006, Schramm, 2014). There are numerous instances where bubbles and drops are encountered in a moving or quiescent continuous phase that exhibits non-Newtonian characteristics. Typical examples include the use of bubble columns (see Hecht et al., 1980; Sada et al., 1983; Chhabra et al., 1996b; Vandu et al., 2004; Gomez-Diaz et al., 2009; Radl et al., 2007; Radl and Khinast, 2007), three-phase fluidized and sparged reactors, and stirred vessels, all of which are used extensively in food, polymer, biochemical, and other processing applications (Fryer et al., 1997). Further examples are found in degassing of molten metals, glass, magmas and other process streams, devolatilization of polymer melts and solutions (Gestring and Mewes, 2002; Divoux et al., 2011) and glasses, the formation of voids due to the presence of moisture/solvent/unreacted monomers in the production of foams, polymeric alloys, and composites, processing of materials in space under low-gravity conditions (Shankar Subramanian and Balasubramanian, 2001), entrapment of bubbles in film coatings (Simpkins and Kuck, 2000), cavitation in turbomachinery (Brennen, 1995; Williams, 2002; Machado and Valente, 2003; Ichihara et al., 2004; Jimenez-Fernandez and Crespo, 2005), use of nanobubbles in ultrasonic chemistry, oxygenation of blood and in other biomedical applications (Allen and Roy, 2000; Khismatullin and Nadim, 2002), production of foams and batters and other aerated foodstuffs (Niranjan, 1999; Massey et al., 2001), movement of magma diapiric and rheology of bubble-laden magmas (Paterson, 1987; Weinberg, 1993; Weinberg and Podladchikov, 1994; Lejeune et al., 1999; Colucci et al., 2017), atomization and sprays of paints (El-Awady, 1978; Hartranft and Settles, 2003; Mulhem et al., 2003), in the production of cosmetics, insecticides and pesticides, coal-water slurries, pulp fiber suspensions, etc. (Lefebvre, 1989; Rozhkov et al., 2003; Ishkintana and Bennington, 2010), migration of gas slugs in drilling muds and magma (Johnson and White, 1993; Carew et al., 1995; von der Lieth and Hort, 2016), gas kicks in horizontal oil wells (Baca et al., 2003), gas-assisted displacement of liquids in channels (Kamisli and Ryan, 1999, 2001; Yamamoto et al., 2004; Delgado et al., 2005), in bubbling of paper pulp water slurries (Lindsay et al., 1995) and sewage sludges (Carne et al., 1982; Gauglitz and Terrones, 2002; Gauglitz et al., 2003), and in the dynamics of falling drops in air (Smolka and Belmonte, 2003). Similarly, the penetration of a single gas bubble in a visco-elastic liquid filled in a tube is relevant to the production of hollow fiber membranes, coating of monoliths, and in gas-assisted injection molding (Huzyak and Koelling, 1997; Gauri and Koelling, 1999a, 1999b). The two other important phenomena of breakage and coalescence are invariably present in such dispersed systems. Clearly, a satisfactory understanding of the underlying physical processes and the formulation of satisfactory schemes for the prediction of the rates of momentum, heat, and mass transfer in such multiphase systems are germane to a rational design and operation of scores of industrially important processes. Similarly, the dynamics of splashing of drops (Newtonian or non-Newtonian) impacting on the different kinds of substrates is also important in numerous applications including ink-jet printing, spray coating, etc. (German and Bertola, 2009; Blackwell et al., 2015). There are situations when bubbles are used to promote mixing and convective transport without inducing large deformations in the processing of shear-sensitive fluids (Hojeij et al., 2020, 2022; Risso, 2018). Yet there are applications where accelerated degassing (removal of bubbles) by stirring is needed (Ma et al., 2020).

There is no question that the dynamics of fluid particles in non-Newtonian media is influenced by a large number of process, including physical and kinematic variables. Consequently, considerable research effort has been directed at elucidating the role of non-Newtonian features, notably shear-thinning, yield stress, and visco-elasticity of the continuous phase on various aspects of bubble and drop phenomena in quiescent and moving media. In recent years, some knowledge has also accrued on the role of the dispersed phase rheology on the hydrodynamics (shape and buoyancy-driven translation) of single droplets in Newtonian and non-Newtonian continuous medium. Clearly, it is not possible to cover here such wide-ranging aspects of the interaction between non-Newtonian characteristics and bubble-drop phenomena. In particular, a substantial body of knowledge is now available on the following aspects of bubble and drop phenomena in rheologically complex continuous media:

1. Formation and growth (or collapse) of bubbles and drops in stagnant and moving liquids, and the stability and disintegration of liquid jets and sheets;
2. Shapes of bubbles and drops in free rise or fall in Newtonian and non-Newtonian continuous phase;
3. Terminal velocity–volume or drag coefficient–Reynolds number relationship for single and ensembles of fluid particles;
4. Coalescence and breakage of fluid particles in different flow fields, and
5. Miscellaneous studies.

Each of these will now be dealt with in some detail in this chapter.

5.2 Formation of Fluid Particles

5.2.1 Bubbles

Perhaps the simplest and possibly also the most widely used method for effecting gas–liquid contacting is the dispersion of gas through submerged nozzles, porous plates, slots, and holes. Hence, much of the effort has been devoted to the formation of single bubbles from submerged orifices in stagnant and flowing liquids. The parameter of central interest is the size of the bubble (volume or diameter) produced under specified conditions, for a given gas–liquid system and the characteristics of the orifice. There are numerous system and physical parameters including physical properties of the two phases, gas flow rate, pressure above the orifice, height of the liquid, etc. that exert varying levels of influence on the size of a bubble (Kumar and Kuloor, 1970). Consequently, a general model encompassing the formation of bubbles under all conditions of interest is yet to emerge, even for Newtonian liquids. Conversely, all available models not only entail varying degrees of approximations but are also limited in their applicability to a range of conditions. There are essentially two models (Davidson and Schuler, 1960a,b; Kumar and Kuloor, 1970) available that have gained wide acceptance in the literature. In recent years, these models (and modifications thereof) have also been extended to describe the formation of bubbles in non-Newtonian media. It is therefore instructive and desirable to first recapitulate the salient features of these two models.

Fluid Particles in Non-Newtonian Media 177

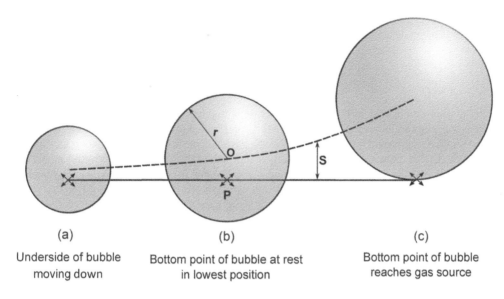

(a) Underside of bubble moving down

(b) Bottom point of bubble at rest in lowest position

(c) Bottom point of bubble reaches gas source

FIGURE 5.1 Schematic representation of bubble formation as postulated by Davidson and Schuler (1960a).

5.2.1.1 Davidson–Schuler Model

In this model, the bubble is assumed to form at a point source where the gas is introduced. With the passage of time as the bubble forms, it gradually moves upward with a velocity determined by the net force acting on the bubble. The detachment of the bubble is assumed to occur when the center of the bubble has moved a distance equal to the sum of the radius of the orifice and that of the bubble, as shown in Figure 5.1. During the formation stage, the bubble is assumed to retain its spherical shape. Depending upon the physical properties of the liquid phase (surface tension, density, viscosity) and the gas flow rate, Davidson and Schuler (1960a,b) identified two regimes of bubble formation. Thus, in low-viscosity systems and at relatively large gas flow rates, the flow in the liquid phase can be assumed to be inviscid and irrotational (Davidson and Schuler, 1960b) and, furthermore, if the surface tension effects are negligible, the only relevant forces acting on the bubble are those due to buoyancy and inertia of the liquid moving with the expanding bubble, that is,

$$V_b(\rho_L - \rho_g)g = \frac{d}{dt}\left[\left(\frac{11}{16}\rho_L + \rho_g\right)V_b\frac{dx}{dt}\right] \tag{5.1}$$

The factor of $(11/16)\rho_L$ in Equation 5.1 accounts for the virtual mass of the liquid moving along with the bubble. For a constant gas flow rate Q, and for the initial conditions, namely, at $t=0$, both x and dx/dt are zero, and the volume of the bubble at the time of detachment is given by the expression

$$V_{bf} = 1.378 Q^{6/5} g^{-3/5} \tag{5.2a}$$

Subsequently, Davidson and Harrison (1963) have replaced the factor of (11/16) by (1/2) for the added mass correction.

At the other extreme is the case of highly viscous liquids, and low gas flow rates, when the inertial force of the liquid being carried by the gas bubble would be negligible and thus the buoyancy force would be balanced by the viscous drag force; the latter can be approximated by the Stokes formula.

In the absence of surface tension effects, these considerations lead to the following expression for bubble volume at the time of detachment:

$$V_{bf} = \left(\frac{4\pi}{3}\right)^{1/4} \left(\frac{15\mu Q}{2\rho_L g}\right)^{3/4} \quad (5.2b)$$

Finally, Davidson and Schuler also considered the case of high gas flow rates wherein the inertial forces are no longer negligible; however, the inclusion of this contribution in the analysis yields a rather cumbersome and implicit relationship for bubble volume, which is available in their original paper (Davidson and Schuler, 1960a). Over the years, it has been widely demonstrated that the predictions of this model are in good agreement with the experimental results under a wide range of conditions, except at extremely low flow rates of gas when the surface tension forces cannot be ignored (Kumar and Kuloor, 1970).

Furthermore, the effects of the circulation in the liquid surrounding the bubble and the momentum of the gas stream have been neglected in deriving Equation 5.2. Also, the above noted analysis assumes that $\rho_g \ll \rho_L$.

5.2.1.2 Kumar–Kuloor Model

In this approach, the bubble is assumed to form in two stages, namely, the growth (or expansion) stage followed by the detachment stage, as shown schematically in Figure 5.2. This model differs from that of Davidson and Schuler in that the bubble stays at the orifice during the growth stage, whereas in the second stage it moves away from (but remains in contact with) the orifice tip via a neck until its detachment. During the second stage, though the bubble moves away from the orifice, it keeps expanding due to the continuous supply of gas. It should be recognized here that a bubble will only lift off from the orifice tip when there is a net upward force acting on it. The forces acting on a bubble during the first stage are: buoyancy (acting in the upward direction) and surface tension and drag forces (acting downward). Thus, the first stage is assumed to end when the net force acting on the bubble is zero. The second stage is characterized mainly by the motion of an expanding bubble and the detachment is assumed to occur when the bubble has traveled a distance equal to its radius at the end of the first stage, that is, the final bubble volume comprises two components:

$$V_{bf} = V_{b1} + Qt_c \quad (5.3)$$

where t_c is the time of detachment, and V_{b1} is the volume of the bubble at the end of the first stage. In order to predict the final volume of the bubble, knowledge of the volume of the bubble at the end of the

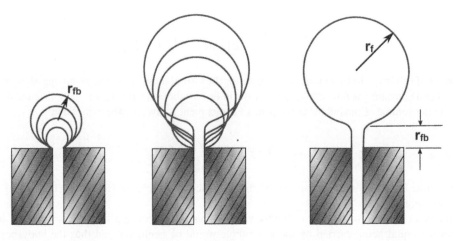

FIGURE 5.2 The model of Kumar–Kuloor (1970) for bubble and drop formation.

first stage is thus required, which is evaluated by writing a macroscopic force balance on the bubble. In the absence of surface tension effects, three forces, namely, buoyancy, drag, and inertial, act on the bubble. Depending upon the viscosity of the liquid phase, one may invoke either the irrotational (inviscid) flow simplification or approximate the drag force by the Stokes expression. While writing the inertial force associated with the expansion of the bubble in terms of the rate of change of momentum of the bubble, Kumar and Kuloor (1970) noted that the upper part of the bubble moves with a velocity equal to its rate of change of diameter whereas the base of the bubble remains stationary. Thus, they evaluated the rate of change of momentum by using the average velocity at the center of the bubble. The condition of the dynamic equilibrium of the bubble yields the following implicit expression for the bubble volume at the end of the first stage (for $\rho_g \ll \rho_L$):

$$V_{b1} = 0.0474 \left(\frac{Q^2 V_{b1}^{-2/3}}{g} \right) + 2.42 \left(\frac{\mu}{\rho_L} \right) V_{b1}^{-1/3} \left(\frac{Q}{g} \right) \tag{5.4}$$

Furthermore, when conditions are such that the flow can be assumed to be inviscid, the second term (accounting for viscous effects) on the right-hand side can be dropped. The equation of motion for the bubble center can now be written as

$$\frac{d}{dt}(MV') = (V_{b1} + Qt_c)(\rho_L - \rho_g)g - 6\pi r \mu V' \tag{5.5}$$

where V', the velocity of the bubble center, is made up of two contributions, namely, the rate of expansion of the bubble (dr/dt) plus the velocity of the base, V_e. Thus

$$V' = \frac{dr}{dt} + V_e = V + V_e \tag{5.6}$$

This allows Equation 5.5 to be rewritten as

$$M\left(\frac{dV}{dt}\right) + V\left(\frac{dM}{dt}\right) = (V_{b1} + Qt_c)(\rho_L - \rho_g)g - 6\pi r \mu V - 6\pi r \mu V_e - \frac{d}{dt}(MV_e) \tag{5.7}$$

Now noting that $M = (\rho_g + (11/16)\rho_L) Qt$ and,

$$V_e = \frac{dr}{dt} = \frac{Q}{4\pi r^2} = \frac{QV^{-2/3}}{4\pi(3/4\pi)^{2/3}}$$

one can finally obtain the following expression that is implicit in V_{b1}

$$(V_{b1} + Qt_c)\left(\rho_g + \frac{11}{16}\rho_L\right)\frac{dV}{dt} + \left(\rho_g + \frac{11}{16}\rho_L\right)QV = (V_{b1} + Qt_c)(\rho_L - \rho_g)g$$

$$- \frac{3\mu Q}{2}(3/4\pi)^{-1/3}(V_{b1} + Qt)^{-1/3}$$

$$- \frac{Q^2(\rho_g + (11/16)\rho_L)}{12\pi(3/4\pi)^{2/3}}(V_{b1} + Qt_c)^{-2/3}$$

$$- 6\pi(3/4\pi)^{1/3}\mu(V_{b1} + Qt_c)^{-1/3}V$$

(5.8)

Kumar and Kuloor (1970) and their coworker have numerically solved Equation 5.8 for various conditions, and subsequently have also incorporated the surface tension effects into their model. Despite the widely different physical backgrounds of these two models, there is a striking similarity in the expression for bubble volume under the inviscid flow conditions and without any surface tension effects, that is, both the aforementioned models reduce to

$$V_b = C\left(\frac{Q^2}{g}\right)^{3/5} \tag{5.9}$$

where C is a constant. Davidson and Schuler (1960a) reported a value of 1.387 that was subsequently modified to 1.138. The two-stage model of Kumar and Kuloor (1970) yields a value of 0.976. All these values compare favorably with the experimental value of 1.722 (van Krevelen and Hoftijzer, 1950). Detailed comparisons between the predictions of these two models and with experiments suggest that the two-stage model of Kumar and Kuloor generally performs better than that of Davidson and Schuler particularly when the surface tension effects are significant. Finally, it is appropriate to point out here that the aforementioned treatment is applicable only when the bubbles are formed under constant flow rate conditions. Bubbles can also be formed under constant pressure conditions. The constant pressure condition occurs when the chamber volume is sufficiently large (~ 0.001 m³) and the pressure in the gas chamber is maintained constant. With the progression of time and the extent of bubble formation, Δp, the pressure drop across the orifice varies, thereby resulting in a variable flow rate. Based on limited experimental results, Costes and Alran (1978) suggested that the bubbles are formed under the constant flow conditions provided the orifice Reynolds number is larger than 1000, and the constant pressure conditions prevail for the orifice Reynolds number smaller than 1000. Excellent reviews on the bubble and drop formation in different regimes both in stagnant and moving Newtonian media are available in the literature (Kumar and Kuloor, 1970; Ponter and Surati, 1997). The effect of liquid cross-flow on bubble formation from an orifice has been studied by Marshall et al. (1993). They used the potential flow theory to account for the liquid inertial forces and reported good correspondence between their predictions and observations.

One would intuitively expect that either of these two models can be extended to bubble formation in purely viscous non-Newtonian media, albeit additional complications may arise in the case of viscoplastic and visco-elastic media (Tsuge and Terasaka, 1989; Terasaka and Tsuge, 1990, 1991, 1997, 2001). For instance, under the constant flow rate conditions, in the case of power-law liquids, one can easily modify the drag force term to obtain the following expression for bubble volume at the end of the expansion stage (neglecting surface tension effects):

$$V_{b1}(\rho_L - \rho_g)g = \frac{Q^2\left(\rho_g + (11/16)\rho_L\right)V_{b1}^{-2/3}}{12\pi(3/4\pi)^{2/3}} + \frac{24YmQ^n V_{b1}^{(2-3n)/3}}{2^{3n+1}\pi^{n-1}(3/4\pi)^{(3n-2)/3}} \tag{5.10}$$

where Y is the drag correction factor, and its value as a function of the power-law index is available in the literature (Hirose and Moo-Young, 1969; Chhabra and Dhingra, 1986; Velez-Cordero et al., 2011). Equation 5.10 can, in turn, be combined with the equation of motion (akin to Equation 5.7 with the drag term suitably modified) to obtain the final equation for the second stage of bubble formation. Preliminary comparisons indicate a good agreement between the predicted and measured values of the bubble volume. A similar extension of these models to power-law liquids has also been attempted by Costes and Alran (1978).

In contrast to the extensive information available on bubble formation in Newtonian fluids, there is a real paucity of analogous results in well-characterized non-Newtonian systems. Owing to their high viscosities, the inviscid flow models are likely to be relevant only at very high gas flow rates. Indeed, the limited experimental results reported by Acharya et al. (1978a) substantiate this assertion; their experimental values of bubble volume are well predicted by Equation 5.9 with $C = \sim 0.976$ to 1.138. Moreover, in this flow regime, even the fluid visco-elasticity was found to exert virtually no influence on the bubble volume. At low flow rates, on the other hand, it has been argued that the fluid visco-elasticity alters the shape of the bubbles at the detachment stage (Rabiger and Vogelpohl, 1986; Ghosh and Ulbrecht, 1989). In this case, the bubbles are elongated but still remain attached to the orifice tip via a rather "drawn neck" thereby resulting in detachment times longer than those predicted by the models of Davidson and Schuler (1960a) and Kumar and Kuloor (1970). Based on photographic evidence, both Rabiger and Vogelpohl (1986) and Ghosh and Ulbrecht (1989) have identified the so-called waiting stage before the final lift off of the bubble from the orifice. Intuitively it appears that this must be a direct consequence of the deformation/relaxation of polymeric molecules; it is, however, not obvious how to quantify this effect.

Costes and Alran (1978) studied the formation of bubbles in one Carboxy Methyl Cellulose (CMC) solution ($m = 3.04$ Pa s^n and $n = 0.68$) under constant pressure and constant flow conditions. The measured values of bubble volumes showed slightly better agreement with the model of Davidson and Schuler (1960a) than that of Kumar and Kuloor (1970). Miyahara et al. (1988) have built upon the model of McCann and Prince (1969) to study the phenomenon of "weeping" on orifices submerged in non-Newtonian liquids. They also found that the volumes of bubbles formed in highly viscous Newtonian and non-Newtonian fluids are generally larger than those produced in low-viscosity systems. Though no comparison has been reported with theoretical predictions, a close look at their results suggests that the bubble volumes are in line with the predictions of Equation 5.9. Subsequently, Tsuge and Terasaka (1989) and Terasaka and Tsuge (1991) have systematically studied the influence of gas chamber volume and rheological characteristics of the liquid phase on the volume of bubbles produced under the constant flow rate and intermediate (neither constant flow nor constant pressure) conditions. Using dimensional analysis, Tsuge and Terasaka (1989) proposed the expression for bubble volume as

$$V_b^* \text{Bo} = \pi N_c \text{Mo}_{\text{PL}}^{0.07} + 2^{3(2-n)} \text{Fr}^{3/8} \text{Mo}_{\text{PL}}^{3/16} \text{Bo}^{(10-3n)/16} \tag{5.11}$$

Equation 5.11 was stated to be applicable in the following ranges of conditions:

$$10^{-3} \leq \text{Mo}_{\text{PL}} \leq 1.7 \times 10^6; \ 0.38 \leq n \leq 1; \ 10^{-3} \leq \text{Fr} \leq 1.8 \times 10^5; \ 0.13 \leq \text{Bo} \leq 1.3; \ 1 \leq N_c \leq 64.$$

In a later study, Terasaka and Tsuge (1991) elucidated the roles of power-law constants, gas flow rate, orifice diameter, and gas chamber size on bubble formation. The bubble volume was seen to increase with the increasing values of m and n, with gas chamber volume, and with the decreasing orifice diameter. Some of these trends are also supported by a recent study (Taghi Esfidani et al., 2017). They also extended their previous model to include power-law fluid behavior (Terasaka and Tsuge, 1990). Subsequently, this approach has been extended to bubble formation in visco-elastic and visco-plastic liquids (Terasaka and Tsuge, 1997, 2001; Alicke et al., 2012). Owing to the distorted shapes observed in visco-elastic fluids, the two-stage model performs less satisfactorily in visco-elastic fluids, and bubble volumes comparable to that in a Newtonian liquid are obtained in yield stress fluids. Similarly, Jiang et al. (2007) studied bubble formation in visco-elastic polyacrylamide aqueous solutions, though no elasticity measurements were reported. However, they reported the bubble size to bear a positive relationship with polymer concentration, orifice size, and gas flow rate whereas the reservoir size had no influence. Scant work is also available on the formation of non-Newtonian drops in the Newtonian medium (Liu et al., 2022). Qualitatively similar results have also been reported by Acharya et al. (1978a) and Li (1999). In a series of papers, Li and coworkers (Li and Qiang, 1998; Li, 1999; Li et al., 2002) have studied the formation of inline bubbles in stagnant power-law fluids. In particular, their studies have focused on the detailed flow field existing around bubbles during their formation stage. They also reported a definitive effect of the injection period on the behavior of bubbles. The evolution of the bubble shape under constant flow rate conditions was numerically calculated, which was supported by detailed experiments. This approach thus does not rely on a priori assumption regarding the shape of the bubble.

An empirical correlation for the mean bubble size of a swarm of bubbles produced from a perforated plate in mildly shear-thinning liquids is also available in the literature (Miyahara and Hayashino, 1995).

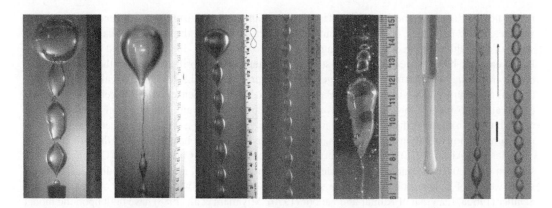

FIGURE 5.3 Formation of bubble chains in 2 and 3% aqueous methocel solutions and in a soap solution. (Based on photographs kindly provided by Professor I.L. Kliakhandler, Michigan Technological University, Houghton, MI.)

Similarly, Briscoe and Chaudhary (1989) studied the formation of nitrogen bubbles in molten LDPE (low-density polyethylene) at elevated temperatures. They observed appreciable distortions in bubble shapes due to visco-elasticity and buoyancy effects. The bubble size stabilized only after approximately 50–100 s. Similarly, the continuous introduction of air into visco-elastic polymer solutions can lead to the formation of stable chains of bubbles ("Sausages") as shown in Figure 5.3 (Kliakhandler, 2002). In the initial stages, a thin "pipe" is attached to the bubble which due to the Rayleigh–Taylor-type instability gives rise to fully developed chains of bubbles, which look like sausages. Similar chaining of bubbles in Boger fluids has also been observed where the initial separation between the two bubbles is smaller than a critical value (Velez-Cordero et al., 2012). They also highlighted the potential difficulties in achieving uniform aeration of visco-elastic fluids.

Bubble formation in moving liquids has received even less attention. Rabiger and Vogelpohl (1986) reported preliminary results on the size of bubbles produced in co-currently and counter-currently flowing liquids and they have also developed an intuitive model to elucidate the effect of the imposed liquid motion. Kawase and Ulbrecht (1981c) and Ghosh and Ulbrecht (1989) have experimentally studied the formation of bubbles and drops in flowing power-law liquids. Little quantitative information is available regarding the role of visco-elasticity in the process of bubble formation (Terasaka and Tsuge, 1997).

5.2.2 Drops

Single or multiple drops, like bubbles, are usually formed by using nozzles, capillaries, microbueretts, spinning disks, or other atomization devices, and by the disintegration of liquid sheets and threads. When a liquid is introduced into another immiscible liquid using a nozzle, the first liquid gets dispersed in the form of drops into the second liquid (continuous phase). The size of the resulting drops is influenced by a large number of physical and operating variables such as the velocity of the drop-forming liquid, viscosity and density of continuous phase, the density of the dispersed phase, interfacial tension, and nozzle diameter. Indeed, the literature abounds with conflicting results with regard to the influence of the above-noted variables on the size of drops even when both phases are Newtonian (Kumar and Kuloor, 1970). Among the various models (Hayworth and Treybal, 1950; Null and Johnson, 1958; Kumar and Kuloor, 1970) available for drop formation in stagnant media, the one by Kumar and Kuloor (1970) has been the most successful in explaining the experimental observations. Since this model has also been extended to non-Newtonian systems, it will be described here briefly. Rao et al. (1966) have extended the aforementioned ideas about bubble formation to the formation of drops in stagnant liquids. This approach stipulates the drop to form in two stages: the first stage is characterized by the expansion or growth of the drop attached to the nozzle tip, and this stage ends when the buoyancy force is balanced

Fluid Particles in Non-Newtonian Media

by that due to the interfacial tension. Thus, these two forces can be equated to obtain the drop volume, at the end of the expansion stage, as

$$V_{d1} = \frac{2\pi R \sigma \phi\left(R/V_d^{1/3}\right)}{(\Delta \rho) g} \tag{5.12}$$

where $\phi(R/V_d^{1/3})$ is the correction factor for the residual drop effect (Harkins and Brown, 1919). In the second stage, the drop rises above the nozzle tip but still remains in contact with the nozzle via a liquid neck, and finally, it breaks away from the tip. Under the constant flow rate conditions, the final drop volume can be expressed as

$$V_d = V_{d1} + Q t_c \tag{5.13}$$

Kumar and Kuloor (1970) proposed two criteria for the evaluation of the detachment time, t_c.

5.2.2.1 Criterion I: Low-Viscosity Systems

Based on the notion that the detaching drop leaves behind a hemispherical residual drop, this model postulates that the drop detachment takes place when the rate of growth of the hemisphere becomes equal to the velocity of the moving drop. This condition leads to the following expression for the rate of growth of the hemisphere as

$$\frac{dr}{dt} = \left(\frac{1}{2}\right) V_c' \tag{5.14}$$

Now, the equation describing the drop motion in the second stage is written as

$$\frac{d}{dt}(MV) = Q(\Delta \rho) g t + Q V_c \rho_d - 6\pi r \mu V \tag{5.15}$$

The virtual mass, M, of the drop, is approximated by $[\rho_d + (11/16)\rho_c] V_{d1}$, where the factor of $(11/16)\rho_c$ accounts for the inertia of the continuous phase. One can thus integrate Equation 5.15 between the limits $t=0$, $V=0$ to $t=t_c$, $V=V_c'/2$, and the value of t_c so obtained can be substituted in Equation 5.13 to calculate the final drop volume. This approach works well in low-viscosity Newtonian systems.

5.2.2.2 Criterion II: High-Viscosity Systems

In high-viscosity systems, it has been observed that the necking occurs before the final lift off of the drop. The length of the neck is assumed to be equal to the diameter of the drop at the end of the first stage. Thus, the time of detachment, t_c, is equal to the time taken for the ascending drop to travel a distance equal to $(6V_{d1}/\pi)^{1/3}$. For a viscous continuous phase, one can neglect the momentum of the continuous phase and the resulting simplified version of Equation 5.15 can be integrated between the limits $x=0$, $t=0$, to $x=(6V_{d1}/\pi)^{1/3}$, $t=t_c$ to obtain the following expression for t_c:

$$t_c = \left[\frac{12\pi r \mu (6V_{d1}/\pi)^{1/3}}{Q(\Delta \rho) g}\right]^{1/2} \tag{5.16}$$

Hence, the final drop volume can again be estimated by inserting this value of t_c in Equation 5.13.

Aside from these two models, based on extensive experimental results, Humphrey (1980) put forward the dimensionless correlation for calculating the detached drop diameter in liquid–liquid (Newtonian) systems as

$$d^* = (d/D_n) = 0.39 \left(\frac{gD_n^2 \Delta \rho}{\sigma} \right)^{-0.49} \left(\frac{\mu_d^2}{D_n \rho_d \sigma} \right)^{-0.09} + 0.39 \qquad (5.17)$$

where the subscripts "d" and "n" refer to the dispersed phase and nozzle conditions, respectively. This correlation is based on experimental data in the ranges $1 < d^* < 5$; $0.01 < \left(gD_n^2 \Delta \rho / \sigma \right) < 1.2$; $5 \times 10^{-6} \leq \left(\mu_d^2 / D_n \rho_d \sigma \right) \leq 6 \times 10^{-3}$ and $(d\rho_d V_n / \mu_d) \leq 100$.

Beyond Reynolds number, Re $(= d\rho_d V_n / \mu_d) > 100$, since Equation 5.17 underpredicts the value of the detached drop diameter, this effect is adequately accounted for by adding the term $0.12 \text{Re}^{0.11}$ on the right-hand side of Equation 5.17. Similarly, the evolution of drop shapes in Newtonian viscous fluids has been studied numerically by Zhang (1999) whereas the effect of an electric field has been investigated by Notz and Basaran (1999).

Kumar and Kuloor (1970) argued that the aforementioned ideas are also applicable to drop formation in power-law type non-Newtonian liquids by modifying the drag force term in Equation 5.15. Indeed, the experimental results reported by Kumar and Saradhy (1972) confirm this assertion as shown in Figure 5.4 for benzene (dispersed phase) –1% Carbopol solution (continuous phase) system. Skelland and Raval (1972) have also successfully used this model to calculate drop sizes in power-law fluids. Kumar and Kuloor (1970) also considered the case of non-Newtonian drops being formed in a Newtonian liquid, and when both the phases are non-Newtonian in behavior. However, no suitable experimental data are available to substantiate or refute these results. Nor is anything known about the role of visco-elastic behavior on drop formation except for the scant work of Shore and Harrison (2005). They found the fluid elasticity to suppress the formation of satellite drops, but greater pulse strength is required to eject

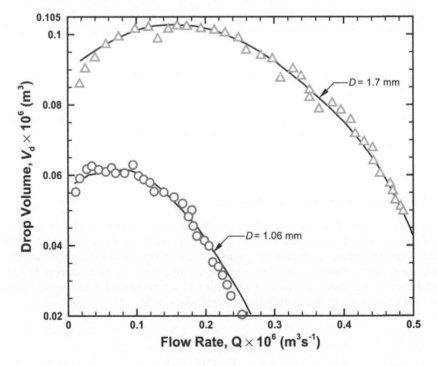

FIGURE 5.4 Comparison between the experimental (○, △) and predicted volumes using Equation 5.15 (shown as lines) for benzene drops in 1% Carbopol solution. (Replotted from Kumar, R. and Saradhy, Y.P., *Ind. Eng. Chem. Fund.*, **11**, 307, 1972.)

droplets from the nozzle. Subsequent studies on the drop characteristics produced using coaxial high-speed air jets suggest that strain hardening (described in terms of a fluid relaxation time) exerts a significant influence on the mean size and size distribution (Christianti and Walker, 2006; Aliseda et al., 2008). Qualitatively similar trends have also been reported in the opposing jet devices used for atomization (Christianti and Walker, 2001, 2002).

Some of these ideas have been extended to multiple drop applications such as drop formation at sieve plate distributors (Saradhy and Kumar, 1976) and under pulsed conditions (Yaparpalvi et al., 1988). Janna and John (1979, 1982) have presented an empirical correlation for the drop size distribution of Bingham plastic fluids formed by using fan-jet pressure nozzles.

5.2.3 Disintegration (or Breakup) of Jets and Sheet

When a liquid jet issues from a nozzle as a continuous column into the air, the cohesive and disruptive forces acting on the free surface of the jet give rise to oscillations and disturbances (instabilities). Under appropriate conditions, such perturbations grow, and the liquid jet disintegrates into drops. This process is called atomization. If the drops so formed exceed a critical size, these further disintegrate into smaller drops, resulting in the so-called secondary atomization. The main parameters of interest in such applications are the continuous length (approximately provides a measure of the growth rate of disturbance) and the drop size (which indicates the wave number of the least stable disturbance). Clearly, both these variables are strongly influenced by a large number of system, process, and material parameters including the size and design of the nozzle, flow rate, and the physical properties of the liquid, notably viscosity and surface tension (Clift et al., 1978; Lefebvre, 1989; Eggers, 1997, 2005; Picot and Kristmanson, 1997). Additional complications arise with non-Newtonian fluids owing to the commonly encountered strain-hardening behavior observed in extensional flows. Dimensional analysis of the drop size for non-Newtonian fluids reveals a large number of dimensionless groups required to establish a universal functional relationship between the drop size and the other system variables for power-law liquids (Teske and Bilanin, 1994; Teske and Thistle, 2000). Owing to the availability of limited experimental results, it is not yet possible to put forward a generalized correlation, albeit a few such expressions are available in the literature for the atomization of Newtonian and non-Newtonian liquids. Thus, for instance, Kaupke and Yates (1966) and Teske and Thistle (2000) examined the drift characteristics of viscosity-modified agricultural sprays whereas Hedden (1961) and Tate and Janssen (1966) put forward tentative correlations for drop size distributions for agricultural sprays and pesticides. Similarly, Stelter et al. (2002) have developed an empirical correlation for the prediction of the Sauter mean diameter of droplets formed by flat-fan and pressure-swirl atomizers for visco-elastic fluids. Kaminski and Persson (1966) have attempted a boundary layer-type analysis to study the distribution of a viscous liquid from a rotating disk. The role of non-Newtonian fluid characteristics on the atomization of a range of formulations has been studied by many investigators (Matta et al., 1983; Mannheimer, 1983; Ellwood et al., 1990; Arcoumanis et al., 1994, 1996; Mansour and Chigier, 1995; Dexter, 1996; Smolinski et al., 1996; Son and Kihm, 1998; Christianti and Walker, 2006; Aliseda et al., 2008; Dravid et al., 2008). Similarly, the stability and breakup of sheets and jets of non-Newtonian fluids being discharged into a gas have been analyzed by many investigators (Goldin et al., 1969, 1972; Chojnacki and Feikema, 1997; Liu et al., 1998b; Mun et al., 1998; Parthasarathy, 1999; Brenn et al., 2000; Cramer et al., 2002; Hartranft and Settles, 2003; Plog et al., 2005). While the breakup of a Newtonian jet is well understood (see Eggers, 1997, for a review), our knowledge about the behavior of non-Newtonian, especially visco-elastic, jets is not only very limited but is also somewhat inconclusive. The most striking feature of the breakup of visco-elastic jets is that these do not disintegrate neatly into single drops, but rather several drops are connected by threads along the length of the jet (Goldin et al., 1969) that ultimately break, but over a much longer distance than that for a Newtonian or an inelastic fluid. The visco-elasticity seems to destabilize the jet relative to a Newtonian jet at small disturbances as also is borne out by subsequent analyses (Goren and Gottlieb, 1982; Bousfield et al., 1986). On the other hand, Kroesser and Middleman (1969) and others (Shirotsuka and Kawase, 1974a, 1975; Kitamura et al., 1982; Kitamura and Takahashi, 1982) reported shorter breakup lengths for power-law liquids. Thus, for instance, Dravid et al. (2008) argued that, at low Reynolds numbers, the satellite drop size exhibits a peak with respect to the power-law index. For a constant wave number, the time of the breakup of jet scales with the jet diameter raised to the power-law index (n). In an attempt to elucidate the role of stretching, Renardy (1994, 1995) used an asymptotic analysis supplemented by numerical simulations and concluded that the finite values of the elongational viscosity indeed exert a strong influence on the breakup of the jet. The molecular dynamic simulations of the rupturing of non-Newtonian filaments due to elongation also allude to similar difficulties in forming drops in visco-elastic systems (Koplik and Banavar, 2003). The current situation can be summarized by noting that the behavior is similar to that of a Newtonian jet in dilute polymer solutions at low velocities. The breakup length increases with the increasing extensional viscosity, and under these conditions the process of secondary drop formation is also suppressed. The effects of polymer rigidity and concentration on spray atomization have been studied by Harrison et al. (1999). Similarly, the extensive literature on near- and far-field breakup and atomization of a liquid (Newtonian) jet by a high-speed annular gas jet has been thoroughly and critically reviewed by Lasheras and Hopfinger (2000). Finally, some attention has also been given to drop formation using capillaries such as that encountered in ink-jet printing, see Wilkes et al. (1999).

5.2.4 Growth or Collapse of Bubbles

The growth (or collapse) of gas bubbles and cavities in a stagnant liquid medium represents an idealization of many industrially important processes. Typical examples include cavitation in turbomachinery, devolatilization of polymer melts (Advani and Arefmanesh, 1993; Favelukis and Albalak, 1996a,b) and in the boiling of polymeric solutions (Levitskiy and Shulman, 1995), degassing of process streams (Ma et al., 2020), and production of polymeric foams. Additional examples are found in geological processes involving the growth of cavities in molten lava due to decompression (Barclay et al., 1995; Durban and Fleck, 1997; Navon et al., 1998), in dilatant soils (Yu and Houlsby, 1991), and in soft biological matter (Dollet et al., 2019). Notwithstanding the fact that most aforementioned examples involve swarms of bubbles, it is readily acknowledged that an adequate understanding of the dynamics of a single bubble serves as a precursor to the modeling of multibubble systems and much of the available literature, therefore, relates to the behavior of a single isolated bubble. It is well known that the underlying physical processes governing the growth or collapse of a bubble represent a complex interplay between mass transfer (or heat transfer) and viscous flow. The coupling between the fluid mechanical and mass transfer aspects arises in three ways: first, via the velocity through the bulk transport term in the species continuity equation; second, due to the volume changes, and this leads to a time-dependent normal velocity near the bubble surface. Finally, the changes in bubble volume are reflected in continually varying velocity via buoyancy effects. The scaling of the field equations and boundary conditions suggests the emergence of a Peclet number, Pe, in addition to the usual fluid mechanical parameters. For large values of Pe, the fluid mechanical analysis is relevant and the species continuity equation is dropped. At the other extreme of small values of Pe, diffusion is the main mode of bubble growth or collapse and under these conditions, it is sufficient to treat it as a problem of mass transfer from a stationary bubble (Favelukis and Albalak, 1996a,b; Venerus and Yala, 1997; Venerus, 2001). For non-Newtonian liquids surrounding a gas cavity, additional difficulties arise due to different shapes of bubbles and the nonlinearity of the rheological models themselves. Indeed, a range of non-Newtonian fluid models has been used to study the effect of rheological parameters on the rate of growth and on the pressure field produced by an expanding or contracting bubble. However, a complete solution involving all aspects is not yet available and each analysis entails varying levels of approximations. The dynamics of the growth of a single spherical bubble in purely viscous model fluids including power law (Yang and Yeh, 1966; Street et al., 1971; Shima and Tsujino, 1976; Burman and Jameson, 1978; Chhabra et al., 1990; Zaitsev and Polyanin, 1992), Bingham plastics (Shima and Tsujino, 1977; De Corato et al., 2019), Powell–Eyring fluid (Shima and Tsujino, 1981), and other generalized Newtonian fluid models (Shima and Tsujino, 1978; Brujan, 1994a,b, 1998, 1999, 2000; Brujan et al., 1996; Brutyan and Krapivsky, 1993; Bloom, 2002) have been analyzed. In particular, the work of De Corato et al. (2019) offers interesting possibilities with regard to the use of an externally oscillating pressure field in the context of removal of gas cavities in food processing and oil-drilling-related applications. Similarly, Iwata et al. (2019) have examined the shape and rising velocity of bubbles in worm-like micellar solution subject to an oscillating pressure field.

They reported up to ten-fold enhancement in rising velocity due to oscillating pressure field. However, only Street et al. (1971) have considered the combined effects of momentum, mass, and heat transfer on the growth of bubbles in power-law fluids. They concluded that the growth of a single bubble is governed predominantly by the viscosity and the molecular diffusivity of the system. Broadly speaking, all else being equal, bubbles in power-law fluids grow more slowly than those in Newtonian media. Furthermore, the smaller the value of the power-law index n, the slower the growth of the bubble. In contrast, much more work has been reported for this kind of flow in visco-elastic media, see Yang and Lawson (1974), Shima and Tsujino (1982), Shima et al. (1986), Venerus et al. (1998), Brown and Williams (1999), etc. Fogler and Goddard (1970) appear to have been the first to develop a theoretical framework elucidating the effect of visco-elasticity on the dynamics of a single bubble with significant inertial effects. They concluded that the visco-elasticity retarded the rate of collapse of bubbles, and under certain conditions even oscillations might occur; the latter finding is qualitatively in line with the other studies (Yoo and Han, 1982). However, no significant deviation from the Newtonian response for spherical shape is predicted for low levels of visco-elasticity (Ting and Ellis, 1974; Ting, 1975; Inge and Bark, 1982). This is so in the case of surface tension-driven oscillations also (Inge and Bark, 1982). Similarly, the analyses of bubble growth in a nonlinear visco-elastic fluid indicate the effects of visco-elasticity to be only minor (Arefmanesh and Advani, 1991; Kim, 1994; Venerus et al., 1998). The rate of growth is bounded by that of the diffusion-induced and diffusion-controlled limiting rates, respectively.

Based on the limited experimental results of Chahine and Fruman (1979) and their theory, Hara and Schowalter (1984) argued that one is likely to observe significant visco-elastic effects for nonspherical bubbles. Even when the departure from the spherical shape is taken into account, these authors found the effect of visco-elasticity on the rate of collapse of a bubble to be rather small. This was surmised by noting that this process is primarily inertia dominated, and strong visco-elastic effects will only manifest themselves when the surrounding liquid is undergoing shearing independently of the collapse or growth of a bubble, as demonstrated by the elegant experiments of Kezios and Schowalter (1986). Tanasawa and Yang (1970) also reported that the effect of viscous damping on the rate of collapse is suppressed in visco-elastic media. Street (1968), on the other hand, has predicted an increase in the rate of growth, at least initially, and also precluded the possibility of oscillations, both of which are at variance with the findings of others.

Significant deviations from the Newtonian response have been predicted when the collapse (or growth) of a bubble in visco-elastic media is accompanied by heat and mass transfer (Zana and Leal, 1974, 1975, 1978; Yoo and Han, 1982), at least during the initial and final stages of the process. Both Yoo and Han (1982) and Zana and Leal (1975, 1978) have considered the dynamics of a bubble in visco-elastic media with mass transfer. Zana and Leal showed that the rate of bubble collapse increased with the decreasing viscosity of the ambient fluid and with the increasing surface tension. The influence of visco-elasticity was found to be somewhat more involved. Depending upon the relaxation and retardation times of the ambient liquid, the rate of collapse may initially show an upward trend followed by a decrease with the increasing levels of visco-elasticity, albeit the magnitude of this effect is also rather small. Yoo and Han (1982), on the other hand, reported that liquid rheology played little role in the case of slow diffusion. Subsequently, Chung (1985) has further considered the effect of diffusion on the inflation of a spherical visco-elastic film.

Many other studies encompassing a range of settings including the combined effects of compressibility and shear-thinning on bubble growth (Brujan, 1998, 1999), the effect of polymer concentration (thus rheology) on the oscillations of a bubble in sound-irradiated liquids (Brujan, 1994b), the dynamics of laser-induced bubbles in polymeric solutions (Brujan et al., 1996), and the behavior of bubbles in boiling polymer solutions (Levitskiy and Shulman, 1995; Shulman and Levitskiy, 1996) are also available in the literature.

Aside from such extensive literature, this flow configuration has also been exploited for measuring extensional viscosities of polymer solutions (Pearson and Middleman, 1978) and of melts (Johnson and Middleman, 1978) and the resulting values (at the rate of deformation $\sim 1s^{-1}$) show good agreement with the numerical simulations of bubble growth in visco-elastic media (Papanastasiou et al., 1984). On the other hand, McComb and Ayyash (1980) have used the pulsating and damping behavior of bubbles in dilute polymer solutions to infer the values of extensional viscosity.

5.3 Shapes of Bubbles and Drops in Free Rise or Fall

The shape of a bubble (or a drop) not only influences its terminal free-rise or -fall velocity, but also plays an important role in determining the rates of heat and mass transfer and their coalescence behavior. Indeed, the shape and rise velocity relationship is entwined even in Newtonian liquids, and it gets accentuated in non-Newtonian media due to the fluid elasticity and yield stress effects. Owing to their mobile interface, bubbles and drops deform when subjected to external flow fields until normal and shear stresses reach an equilibrium at the interface. In contrast to the infinite number of shapes possible for solid particles, bubbles and drops (under steady-state conditions) are limited in the number of possibilities by virtue of the fact that sharp corners or edges are precluded due to the interfacial forces. Therefore, in many cases, bubbles never attained a steady shape and hence the so-called terminal velocity. Often shape and size vary on account of the coalescence and disintegration of bubbles and drops. This also leads to zigzag, spiral, and oscillating trajectories during the buoyancy-induced motion of bubbles and spheres under appropriate kinematic conditions (Zenit and Magnaudet, 2008; Ern et al., 2012; Yan et al., 2017, etc.). These difficulties coupled with that of describing nonspherical shapes have impeded the development of universally applicable shape (pattern) maps and drag correlations for fluid particles, even when the continuous phase is Newtonian, e.g., see the recent works of Loth (2008b), Wegener et al. (2010), Kelbaliyev (2011), Yan et al. (2018), Zhou et al. (2020), Freitas and Pereira (2021), etc. One common measure of shape is the aspect ratio, defined as the ratio of the vertical length (along the direction of bubble rise) to that across the bubble. This has proved to be of some value in correlating experimental results, but it is of limited relevance for complex shapes like drops with long tails, or cusped and/or skirted bubbles. Notwithstanding these difficulties, a sizeable body of knowledge is now available on the shapes (and the other aspects) in the free rise or fall of bubbles and of drops in non-Newtonian continuous media. However, before undertaking a detailed discussion on their behavior in rheologically complex media, it is instructive and desirable to provide a terse description of the analogous problem involving the Newtonian continuous phase, for this serves as a reference in order to draw qualitative inferences regarding the role of non-Newtonian properties of the continuous phase.

5.3.1 Newtonian Continuous Media

Broadly speaking, the observed shapes of bubbles and drops in free motion in the absence of wall effects can be divided mainly into three categories as follows:

1. *Spherical*: At low Reynolds numbers, the interfacial tension and viscous forces essentially govern the shape, and both bubbles and drops deviate very little from the spherical shape. Clift et al. (1978) have arbitrarily defined that the fluid particles having the ratio of major to minor axis between 0.9 and 1.1 (i.e. ±10% distortion) are regarded as being spherical.
2. *Ellipsoidal*: Bubbles and drops that are oblate with a convex interface (viewed from inside) around the entire surface are called ellipsoidal in shape. In practice, the actual shapes do differ considerably from true ellipsoids, and the particle may not exhibit fore-and-aft symmetry. Furthermore, ellipsoidal-shaped bubbles and drops are known to undergo periodic dilation or wobbling motion which further complicates their characterization.
3. *Spherical-cap or ellipsoidal-cap*: Large bubbles and drops usually have flat or indented bases thereby exhibiting no fore-and-aft symmetry. Such fluid particles may resemble segments cut from spheres.

In addition to the aforementioned main types of shapes, numerous other shapes such as "dimpled," "skirts" with or without an open wake, and heart-shaped have been reported in the literature (Clift et al., 1978; Chen et al., 1999; Zhou et al., 2007). However, all these may be regarded as slight variations of the three principal shapes discussed above. Figure 5.5 shows representative shapes of air bubbles in a Newtonian medium (40% aqueous glycerol solution). Besides the physical properties of the continuous phase, the confining walls also exert appreciable influence on the shape of fluid particles, especially

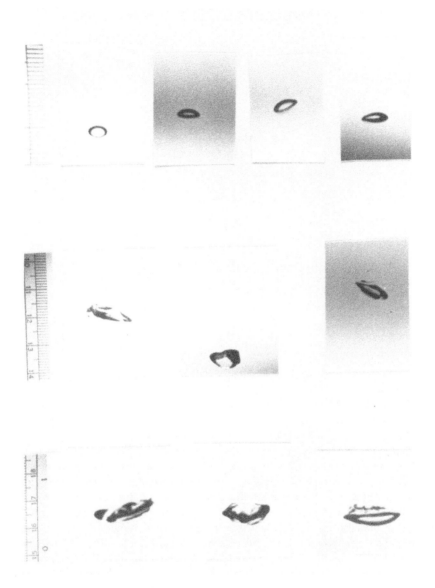

FIGURE 5.5 Photographs of air bubbles rising freely in a 40% aqueous glycerin solution. (From De Kee, D. and Chhabra, R.P., *Rheol. Acta*, **27**, 656, 1988. With permission.)

bubbles (Coutanceau and Hajjam, 1982). Thus, at moderate gas flow rates and in narrow channels, the familiar Taylor bubbles with a hemispherical nose are commonly encountered which are separated from the channel wall with a thin liquid film (Sousa et al., 2006; Abhishek et al., 2015; Majumdar and Das, 2019; Sontti and Atta, 2019, etc.).

In the absence of wall effects, it is generally agreed that the shapes of bubbles and drops rising or falling freely in liquid media are governed by the magnitudes of the following dimensionless parameters:

Reynolds number:

$$\mathrm{Re} = \frac{\rho_c V d}{\mu_c} \tag{5.18a}$$

Eötvös number:

$$\mathrm{Eo} = \frac{\Delta \rho g d^2}{\sigma} \quad (5.18b)$$

Morton number:

$$\mathrm{Mo} = \frac{g \mu_c^4 \Delta \rho}{\rho_c^2 \sigma^3} \quad (5.18c)$$

Viscosity ratio:

$$X_E = \frac{\mu_d}{\mu_c} \quad (5.18d)$$

Density ratio:

$$\gamma = \frac{\rho_d}{\rho_c} \quad (5.18e)$$

where the subscripts c and d refer to the continuous and dispersed phase, respectively. Undoubtedly, several other dimensionless groups including the drag coefficient, Froude number, Weber number, Cauchy number, Bond number, Archimedes number, Galileo number, etc. are also currently in use, but these can be obtained by suitable combinations of the five parameters listed above. For instance, the commonly used drag coefficient can be obtained as $(4/3)(\mathrm{Eo}^{3/2}/\mathrm{Mo}^{1/2}\mathrm{Re}^2)$, etc. On the other hand, however, additional dimensionless groups usually emerge when one or both phases are non-Newtonian. It is also clear that not all groups would always be important. In the case of a bubble rising through a liquid, for instance, $X_E \sim 0$, $\gamma = 0$, and $\Delta \rho = \rho_c$, etc. Based on these dimensionless groups, Grace et al. (1976) and others (Bhaga and Weber, 1981; Grace, 1983; Grace and Wairegi, 1986) have constructed the so-called "shape maps" that have proved to be quite useful in delineating the shapes of bubbles and drops under most conditions of practical interest for their free rise in Newtonian fluids. Some of these ideas have also been corroborated, in parts at least, by subsequent numerical studies (Tripathi et al., 2015a; Sharaf et al., 2017; etc.) While using these maps, it should be borne in mind that all shape maps are based on visual observations, and thus invariably entail a degree of arbitrariness and subjectivity. Furthermore, the transition from one shape to another usually occurs over a range of conditions in contrast to the sharp boundaries shown in all such maps. In broad terms, the spherical shape is observed at extremely small values of Mo, Eo, and Re whereas ellipsoidal particles are encountered at relatively high Reynolds and moderate Eötvös numbers. Finally, the cap shape occurs only at moderately high Eötvös and Reynolds numbers. Similarly, Maxworthy et al. (1996) have attempted to delineate the transition boundaries from one shape to another by obtaining experimental results over wide ranges of conditions, especially a 13 orders of magnitude variation in the value of the Morton number ($\sim 10^{-12} \leq \mathrm{Mo} \leq \sim 10$). They have also outlined the different scaling laws pertinent to each bubble shape regime. Similarly, some numerical work is also available in which the shape of bubbles is calculated as a part of the solution that allows the precise determination of the shape transition criteria, see Chen et al. (1999).

In contrast to the aforementioned qualitative description of bubble shapes, Tadaki and Maeda (1961) developed a simple quantitative shape map using eccentricity ($e = \xi^{-1}$), defined as the maximum width divided by the maximum height, and equivalent sphere volume diameter coordinates, as shown in Figure 5.6 for bubble shapes observed in water. Indeed, this simple approach has proved to be very successful for bubbles rising freely in stagnant Newtonian liquids.

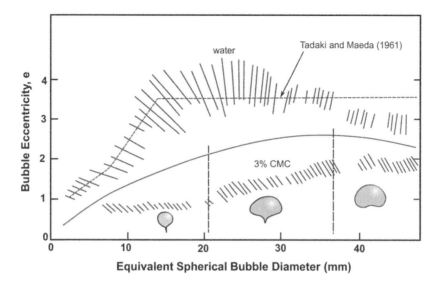

FIGURE 5.6 Eccentricity (*e*) of bubbles rising in water and in a 3% carboxymethyl cellulose solution. The hashed area is indicative of the spread of experimental data.

5.3.2 Non-Newtonian Continuous Media

Of the numerous studies pertaining to the free motion of bubbles and drops in the non-Newtonian continuous phase (see Table 5.1), only a few investigators have reported and discussed the observed shapes of fluid particles. It is now generally agreed that in contrast to the aforementioned limited number of bubble shapes (mainly spherical, ellipsoidal, and spherical/ellipsoidal caps) observed in Newtonian fluids, many more shapes in inelastic and visco-elastic liquids have been documented in the literature (De Kee et al., 1996a,b, 2002; Zhou et al., 2007; Sikorski et al., 2009; Mougin et al., 2012; Ohta et al., 2015, 2019; Ortiz et al., 2016; Norouzi et al., 2019a,b; etc.). The salient findings of these studies can be summarized as follows: at very low Reynolds numbers, surface tension forces tend to maintain the spherical shape. Depending upon the size (volume of a bubble) and the values of the physical properties of the continuous phase, it is possible to observe shape transitions from spherical to prolate-tear, oblate, cusped, oblate, and finally to Davies–Taylor-type spherical caps (Davies and Taylor, 1950). Figure 5.7 shows the shapes of air bubbles in an inelastic polymer solution (1% carboxymethyl cellulose aqueous solution) whereas Figure 5.8 shows the influence of visco-elasticity and surfactant concentration on the shapes of bubbles. Rodrigue and Blanchet (2001, 2002) demonstrated that the formation of a cusp is not a sufficient condition for a jump in velocity to occur (Section 5.4). Tzounakos et al. (2004), on the other hand, have demonstrated that the bubble shapes are not influenced by the addition of surface-active agents to CMC solutions, albeit the drag behavior is modified by surfactants for bubbles smaller than 8 mm in diameter. Similarly, depending upon the physical properties of the dispersed and continuous phases and the size of drops, quantitatively different shapes are observed in rheologically complex liquids. The dispersed phase, in most cases, is Newtonian. For a drop size for which the surface forces are overcome by the viscous forces, the shape changes from spherical to ovate with increasing drop size. Further increase in drop volume results in a tear drop shape with a tailing filament (Mhatre and Kintner, 1959; Warshay et al., 1959; Wilkinson, 1972). Finally, for very large drops, the ratio of vertical to horizontal diameters decreases, and the rear surface begins to fold in. Figure 5.9 schematically shows this sequence of shapes of Newtonian drops whereas good photographs showing the actually observed shapes have been presented among others by Warshay et al. (1959), Fararoui and Kintner (1961), Mohan (1974b), Shirotsuka and Kawase (1975), Rodrigue and Blanchet (2001), and Wanchoo et al. (2003). Figure 5.10 shows the shapes of corn oil (Newtonian) in a 0.025% polyacrylamide in 20/80 glycerin/water solution, which are qualitatively similar to those postulated in Figure 5.9. Acharya et al. (1978b) have also studied the characteristics of the wake behind a drop at high Reynolds numbers.

TABLE 5.1
Summary of Investigations on the Free Motion of Bubbles and Drops in Non-Newtonian Media

Investigator	Details	Fluid Model (Continuous Phase)	Comment
[a]Philippoff (1937)	Air bubble in time-dependent fluids	—	Data on bubble shapes.
[a]Warshay et al. (1959)	Drops in aqueous solutions of CMC and Lytron	—	Wall effects and shapes.
[a]Mhatre and Kintner (1959); [a]Fararoui and Kintner (1961)	Drops in aqueous solutions of CMC and Lytron 890	Power-law model	Shapes and drag data.
[a]Astarita and Apuzzo (1965)	Bubbles in aqueous solutions of CMC, Carbopol, and ET 497	Power-law model	Velocity–bubble size data and shapes of bubbles.
Astarita (1966b)	Gas bubbles	Maxwell fluid model	Qualitative results on the role of visco-elasticity on creeping bubble motion.
[a]Barnett et al. (1966)	Bubbles in aqueous solutions of CMC	Power-law and Ellis models	Shapes of bubbles and mass transfer data.
Nakano and Tien (1968, 1970)	Newtonian fluid spheres	Power-law model	Drag coefficient at low and intermediate Reynolds numbers.
Hirose and Moo-Young (1969)	Single bubbles	Power-law model	Approximate expressions for drag and Sherwood number in creeping flow.
[a]Marrucci et al. (1970)	Drops in non-Newtonian media	Power-law model	Drag coefficient data.
[a]Calderbank et al. (1970)	CO_2 bubbles in aqueous solutions of Polyox	Power-law model	Shapes of bubbles, velocity-size, and mass transfer data.
[a]Leal et al. (1971)	Bubbles in solutions of Separan AP-30	No specific fluid model	Attempted to explain the discontinuity in volume–velocity data.
Wagner and Slattery (1971)	Non-Newtonian droplet in a visco-elastic medium	Third-order Coleman-Noll fluid	Expressions for drag and shape of the droplet.
[a]Wilkinson (1972)	Droplets	—	A tail behind the drops was observed.
[a]Mohan et al. (1972)	Drops in aqueous solutions of PEO, PAA, and CMC	Power-law model	Drag coefficient data.
Mohan (1974a, b); Mohan and Raghuraman (1976b, 1976c); Mohan and Venkateswarlu (1974, 1976)	Drops and bubbles	Power-law and Ellis model fluids	Upper and lower bounds on drag coefficient in creeping flow. Also obtained limited results for finite Reynolds numbers.
[a]Mitsuishi et al. (1972)	Air bubbles and CCl_4 drops in aqueous solutions of CMC and PEO	Sutterby fluid model	Empirical expressions for drag coefficient.

(Continued)

TABLE 5.1 (Continued)
Summary of Investigations on the Free Motion of Bubbles and Drops in Non-Newtonian Media

Investigator	Details	Fluid Model (Continuous Phase)	Comment
* Shirotsuka and Kawase (1973, 1974b)	Fluid spheres	Both phases assumed to be power-law fluids	Drag coefficient and mass transfer data in viscous and visco-elastic fluids
[a]Carreau et al. (1974)	Bubbles in aqueous solutions of CMC and Separan AP-30	—	Velocity–volume data for bubbles in visco-elastic fluids. Some time effects are also reported.
[a]Macedo and Yang (1974)	Air bubbles in aqueous solutions of Separan AP-30	Power-law model	Drag coefficient data.
[a]Zana and Leal (1974, 1975, 1978)	CO_2 gas bubbles in solutions of Separan AP-30	Oldroyd fluid model	Mass transfer and drag data in visco-elastic fluids.
Ajayi (1975)	Fluid spheres	Oldroyd fluid model	Expression for drag coefficient and shape of slightly deformed fluid spheres.
[a]Yamanaka and Mitsuishi (1977)	Air bubbles and CCl_4 drops in aqueous solutions of CMC and PEO	Sutterby fluid model	Drag coefficient data.
[a]Acharya et al. (1977, 1978b)	Bubbles and drops in solutions of CMC, PEO, and PAA	Power-law model	Data on bubble and drops shapes, volume–velocity, and drag coefficient relationship.
Bhavaraju et al. (1978)	Single bubbles and swarms of bubbles	Power-law and Bingham plastic models	Drag coefficient and mass transfer results in creeping flow.
[a]Hassager (1979)	Single bubbles	Rivlin–Ericksen fluid model	Effect of the two normal stress differences on the shape of bubbles. Also observed a negative wake.
[a]Kawase and Hirose (1977)	Drops in non-Newtonian media	Power-law model	Correlation for drag coefficient.
Tiefenbruck and Leal (1980b, 1982)	Solid spheres and gas bubbles	Oldroyd fluid model	Numerical results on the detailed flow field and expression for drag and mass transfer for nonspherical bubbles.
Kawase and Ulbrecht (1981d, 1981g, 1982)	Fluid spheres	Power-law model	Expressions for drag coefficient and Sherwood number in creeping flow. Also attempted to explain the discontinuity in volume–velocity data and the effect of surfactants on the terminal velocity of liquid drops.
[a]Bisgaard and Hassager (1982)	Air bubbles in solutions of PAA	—	Flow visualization and measurement of the axial velocity component.
[a]Coutanceau and Hajjam (1982)	Air bubbles in solutions of PAA, MC, and Polyox	—	Combined effects of walls, elasticity, and shear-thinning on bubble shape and motion.
[a]Gillaspy and Hoffer (1983)	Non-Newtonian drops falling in air	—	Drag results and some hints about internal circulation.
Kawase and Moo-Young (1985)	Bubbles	Carreau and Ellis models	Expression for drag coefficient.

(*Continued*)

TABLE 5.1 (*Continued*)
Summary of Investigations on the Free Motion of Bubbles and Drops in Non-Newtonian Media

Investigator	Details	Fluid Model (Continuous Phase)	Comment
[a]De Kee and Chhabra (1988); De Kee et al. (1986, 1990a)	Bubbles in solutions of CMC and PAA	Power-law and De Kee models	Data on shape, velocity–volume behavior, and coalescence in visco-elastic fluids.
Jarzebski and Malinowski (1986a, 1986b, 1987a, 1987b)	Swarms of bubbles and drops	Power-law and Carreau model fluid	Steady and unsteady flow results on drag and mass transfer
Chhabra and Dhingra (1986); Manjunath and Chhabra (1992); Gummalam and Chhabra (1987); Gummalam et al. (1988); Chhabra (1998)	Bubble swarms and single-fluid spheres	Power-law and Carreau fluid models	Upper and lower bounds on drag and rise velocity of swarms at low and high Reynolds numbers.
[a]Haque et al. (1987, 1988)	Bubbles in solutions of CMC	Power-law model	Velocity–volume data for bubbles.
Quintana et al. (1987, 1992)	Newtonian droplets	Oldroyd fluid model	Effect of elasticity on fall velocity of fluid spheres.
Gurkan (1989, 1990)	Power-law drop in Newtonian media	Newtonian	Numerical solution for outside Reynolds number of 10 to 70.
Ramkissoon (1989a, 1989b)	Fluid spheres	Reiner–Rivlin model	Drag reduction is predicted.
[a]Tsukada et al. (1990)	Air bubbles in CMC solutions	Power-law model	Theoretical predictions and data on shape and rise velocity.
[a]Miyahara and Yamanaka (1993)	Bubbles in CMC solutions	Power-law model	Results on shape, drag, and rocking behavior.
Chan Man Fong and De Kee (1994)	Bubbles	Second-order fluid	Combined effects of thermal gradient, surface tension, and visco-elasticity.
[a]Liu et al. (1995)	Bubbles in PEO solutions	–	Bubble shapes and cusps.
Rodrigue et al. (1996a,b,c, 1997, 1998, 1999a,b); De Kee et al. (1990a, 1996a,b)	Bubbles in solutions of CMC and PAA	Carreau and power-law model fluids	Effect of surfactants on drag and discontinuity in size–velocity behavior.
[a]Chhabra and Bangun (1997)	Drops in CMC solutions	Power-law model	Wall effects on droplets.
[a]Li (1998, 1999) and Li et al. (1997a,b, 2001)	Bubbles in CMC and PAA solutions	–	Bubble formation and coalescence.
[a]Dewsbury et al. (1999)	Bubbles in CMC solutions	Power-law model	Bubble shapes and rise velocities.
[a]Margaritis et al. (1999)	Bubbles in CMC and xanthan solutions	Power-law model	Bubble velocities and drag.
Pillapakkam (1999)	Fluid particles	Olsroyd-B and FENE models	Deformation of bubbles and drops as functions of capillary and Deborah numbers.
Stein and Buggisch (2000)	Bubbles	Bingham plastic	Rise velocity of pulsating bubbles.

(*Continued*)

TABLE 5.1 (Continued)
Summary of Investigations on the Free Motion of Bubbles and Drops in Non-Newtonian Media

Investigator	Details	Fluid Model (Continuous Phase)	Comment
Allen and Roy (2000)	Bubbles	Linear and nonlinear visco-elastic models	Oscillations of bubbles as encountered in medical ultrasound applications.
[a]Belmonte (2000)	Bubbles in worm-like micellar solutions	—	Oscillating behavior of cusped bubbles.
Shosho and Ryan (2001)	Bubbles in CMC, PAA, HEC, PVP solutions	—	Dynamics of bubbles in inclined tubes.
[a]Rodrigue and Blanchet (2002)	Bubbles in PAA solutions	Carreau model	Effect of surfactants on bubble velocities. Also presented a criterion for the critical bubble size.
[a]Wanchoo et al. (2003)	Droplets in CMC, PAA, and PVP solutions	Power-law model	Shapes of drops.
[a]Dziubinski et al. (2001, 2002, 2003)	Bubbles in CMC, PAA, and Carbopol solutions	Power-law model	Drag on bubbles and the use of "bubble" viscometer.
[a]Herrera-Velarde et al. (2003)	Bubbles in PAA solutions	—	PIV measurements of negative wake and jump velocity.
Ohta et al. (2003, 2005, 2006, 2009, 2010a,b, 2011, 2012, 2015, 2019)	Droplets in CMC solutions	Cross–Carreau equation Visco-elastic FENE-CR	3-D numerical and experimental results on fluid spheres.
Boyd and Varley (2004)	Bubbles in xanthan solutions	Power-law model	Acoustic emission measurement of bubble size.
[a]Lin and Lin (2003, 2005)	Bubbles in polyacrylamide solutions	Power-law model	Detailed flow field around a single and two inline bubbles using particle image analyzer.
[a]Frank et al. (2003, 2005), Frank and Li (2005, 2006)	Bubbles and spheres in polyacrylamide solutions and laponite suspensions	Power-law model	Rise velocity and detailed velocity profile data for single bubbles and of periodic bubble chains reported.
Onishi et al. (2003)	Bubbles	Oldroyd-B fluid	Lattice–Boltzmann model for bubbles. Results for a single bubble with a cusp are reported as a limiting case.
[a]Dubash and Frigaard (2004)	Bubbles	Herschel–Bulkley fluid	Shapes and static criterion for bubble motion.
[a]Tzounakos et al. (2004)	Bubbles in CMC solutions with and without surfactants	Power-law model	Effect of surfactants on drag behavior.
Sousa et al. (2004, 2005)	Bubbles in CMC solutions (inelastic and visco-elastic)	—	Detailed PIV measurements in the negative wake of bubbles and for Taylor bubbles.
Albernaz and Cunha (2013)	Bubbles in transient motion	Maxwell fluid	The continuous phase consists of macromolecules or fibers.
[a]Norouzi et al. (2019a,b); Norouzi and Davoodi (2018)	Drops of Newtonian and Boger fluids	Oldroyd-B and Newtonian (air and oil)	Detailed analytical and experimental results on shape and velocity of Boger fluid drops in visco-elastic and Newtonian media.

(Continued)

TABLE 5.1 (Continued)
Summary of Investigations on the Free Motion of Bubbles and Drops in Non-Newtonian Media

Investigator	Details	Fluid Model (Continuous Phase)	Comment
[a]Bohm et al. (2014)	Bubbles in narrow rectangular channels filled with xanthan gum solutions	Xanthan solution's velocity ~0.25 m/s	Shapes and velocities of single bubbles over the Reynolds number range of 10–75 for xanthan solutions.
[a]Sikorski et al. (2009)	Bubbles in Carbopol solutions	Herschel–Bulkley fluid model $n = 0.53, 0.55$	Extensive results on shape, velocity, and stopping criterion.
[a]Wenyuan et al. (2010)	Bubbles in Newtonian and polymer solutions	Glycerine, polyacrylamide, carboxymethyl cellulose solutions (Power-law)	Extensive drag coefficient results as functions of Reynolds, Archimedes, and Eötvös number ($0.56 \leq n \leq 1$).
[a]Li et al. (2012)	Spherical, prolate, oblate, and cap-shaped bubbles in carboxymethyl cellulose solutions	Power-law model ($0.56 \leq n \leq 0.94$)	Drag correlation in the range $0.05 \leq Re_{PL} \leq 300$.
[a]Lopez et al. (2018)	Bubbles in elastic visco-plastic fluids	Carbopol solutions (Herschel–Bulkley model)	Delineates the effects of elasticity and yield stress on bubble dynamics.
Amani et al. (2019)	Taylor bubbles	Carreau viscosity equation	Numerical results on the dynamics of bubbles, film thickness, etc.
Battistella et al. (2020)	Bubbles in power-law fluids	Truncated power-law	Direct numerical simulation of single bubbles. An existing correlation for drag is modified.
Tripathi et al. (2015b)	Axisymmetric bubbles	Bingham plastic fluid	Used the volume of fluid method together with the Bercovier–Engleman linearization to study evolution of shape of single bubbles and their rising behavior. At low Bingham numbers constant rise velocity is reached whereas at high Bingham numbers, velocity and shape oscillate.
Samson et al. (2017)	Bubbles in Carbopol solutions	Herschel–Bulkley model ($0.36 \leq n \leq 0.51$)	Yield stress modifies the pressure distribution, similar to that in an elastomer (Zhu et al., 2011).
Dimakopoulos et al. (2013)	Steady axisymmetric bubbles	Herschel–Bulkley model fluid	Bubble shape and velocity as functions of Bingham, Bond, and Archimedes number are studied. The stopping criterion is independent of fluid index.
Prieto (2015, 2016)	Buoyancy-driven Newtonian drops	Oldroyd-B and FENE models	2-D simulations up to about De ~ 1. As De increases, cusp-like tail and negative wake appear. Subsequently, Brownian-based, micro-macro finite element has been used to study bubble shape and velocity in FENE/Hookean fluids.
[a]Vamerzani et al. (2016)	Visco-elastic drops (xanthan solution modeled as Giesekus model fluid)	Newtonian media (Glycerine–water solutions)	Perturbation analysis (Re = 0) and the drag do not change up to O (De). These results are in line with that of Sostarez and Belmonte (2003).

(Continued)

TABLE 5.1 (Continued)

Summary of Investigations on the Free Motion of Bubbles and Drops in Non-Newtonian Media

Investigator	Details	Fluid Model (Continuous Phase)	Comment
Tsamopoulos et al. (2008)	Bubbles in Bingham plastic fluid	Regularized Bingham model fluid	Shape and rise velocity of bubbles as functions of the Bingham, Bond, and Archimedes numbers. The stopping critical Bingham number is found to be 3/2 times that for a solid sphere (Beris et al. 1985).
[a]Dubash and Frigaard (2007)	Bubbles in Carbopol solutions	Herschel–Bulkley Model fluid ($0.34 \leq n \leq 0.4$)	For small bubbles, rise velocity increases as tube diameter increases. The stopping criterion depends upon the shape of the bubble. Significant wall effects are reported.
[a]Alicke et al. (2012)	Formation of bubbles in weakly elastic visco-plastic liquids	Carbopol solutions	The effect of bubble dispenser on the initial shape of bubbles is studied.
Gollakota and Kishore (2018)	Spheroidal bubbles	Carreau viscosity equation	Rising velocity and deformation of prolate- and oblate-shaped bubbles at finite Reynolds numbers.
[a]Kemiha et al. (2006)	Bubbles and solid spheres in aqueous polyacrylamide solutions	Sixth-order Maxwell model	Detailed PIV data reveal three regions: negative wake which is surrounded by a conical upward flow and an upward flow in the front of the sphere/bubble. Negative wake is ascribed to the visco-elasticity of the liquid.
[a]Lavrenteva et al. (2009)	Newtonian drops (single and pair) in Carbopol solutions	Herschel–Bulkley fluid ($0.32 \leq n \leq 0.35$)	The terminal velocity peaks at about $d/D \sim 0.5$ and the critical drop radius scales $(\tau_0^{HB})^{0.65}$.
[a]Mougin et al. (2012)	History effects on bubble dynamics in Carbopol gels	—	Bubble dynamics (shape and rising trajectories) is strongly influenced by the internal stresses (both of structural effects and mechanical history). Depending upon the concentration, several regimes are possible. Some of these effects can persist even for months.
Mukherjee and Sarkar (2011, 2014)	Visco-elastic drop in stagnant and sheared Newtonian fluid	Newtonian; FENE-MCR	Effects of extensional properties and surface tension on the shape of droplets, which are in line with the experiments (Sostarez and Belmonte, 2003). Also developed a shape regime map in term of De and Ca.
[a]Hassan et al. (2010a,b)	Bubbles in xanthan gum solutions containing polystyrene beads	—	Numerical and experimental results on the zigzag/spiraling motion of bubbles in complex fluids.
You et al. (2008, 2009)	Cusped bubbles and drops in visco-elastic fluids	—	Stability analysis and deformation of fluid particles.
[a]Ortiz et al. (2016)	PIV measurements for drops of model fluid S-1	Aqueous polyacrylamide and PEO solutions	Deformation and breakup dynamics of visco-elastic drops in visco-elastic fluids. While both a tail and negative wake are formed, no jump discontinuity is observed. Conditions are identified if the tail will disintegrate or not.

(Continued)

TABLE 5.1 (*Continued*)
Summary of Investigations on the Free Motion of Bubbles and Drops in Non-Newtonian Media

Investigator	Details	Fluid Model (Continuous Phase)	Comment
Potapov et al. (2006)	Single and pair of droplets	Herschel–Bulkley model fluid	Numerical results indicate that a single drop attains a quasi-steady-state behavior.
[a]Vasil'chenko and Potapov (1996)	Long bubbles in tubes filled with polyamide/Bentonite systems	Elastic/thixotropic behavior	Presented bubble velocity as a function of its length.
[a]Shew and Pinton (2006)	mm size air bubbles	1% polyacrylamide aqueous solution	Reported on rising trajectories at about $De \sim 1$ but with significant inertial effects.
Singh and Denn (2008)	2-D single and multiple fluid spheres including side-by-side and vertical pairs of particles	Bingham plastic fluid	Numerically identified the "unyielded" regions and stopping criterion with shear-free and no-slip condition.
Smagin et al. (2011)	Creeping flow of prolate- and oblate-shaped drops of visco-plastic fluids	Newtonian medium	Shape is gradually stabilized by the increasing Bingham number.
[a]Sousa et al. (2004, 2005, 2006)	PIV and shadowgraphy measurements around Taylor bubbles	0.01–0.8% aqueous polyacrylamide solutions. (Carreau–Yasuda viscosity equation)	Presented correlations for the maximum bubble diameter as function of Re and De.
[a]Sun et al. (2015)	Single bubbles	Xanthan solutions with and without the addition of NaCl/ surfactants ($0.35 \leq n \leq 0.52$)	Presented piecewise correlations for drag ($Re_{PL} \leq 100$).
[a]Sun et al. (2020a)	Interactions between three horizontally aligned bubbles	Shear-thinning fluids	Effects of initial bubble size, separation, and shear-thinning on their dynamics.
[a]Funfschilling and Li (2001, 2006)	Single bubbles	Newtonian, inelastic, and visco-elastic polymer solutions	No effect of injection period in Newtonian and inelastic fluids. Detailed flow field data.
Abhishek et al. (2015)	Taylor bubbles in steady and pulsating liquids	Carreau–Yasuda viscosity equation	Oscillations in the bubble velocity increase with the increasing polymer concentration in the continuous phase.
Majumdar and Das (2019)	Taylor bubbles	Power-law fluid	Based on analytical and numerical results, expression for film thickness is developed.
[a]Tran et al. (2015)	Bubbles and rigid spheres	Volcanic muds modeled as Herschel–Bulkley model fluids ($0.51 \leq n \leq 0.71$; $93 \leq \tau_0^{HB} \leq 127$ Pa)	Terminal velocity and drag results in the creeping flow regime.

(*Continued*)

TABLE 5.1 (Continued)
Summary of Investigations on the Free Motion of Bubbles and Drops in Non-Newtonian Media

Investigator	Details	Fluid Model (Continuous Phase)	Comment
[a]Xu et al. (2021)	CO_2 bubbles mass transfer	Aqueous polyacrylamide solutions (both viscosity and N_1 data fitted to power-law model)	Ellipsoidal bubbles exhibited transient behavior. Bubble size reduced by 20% over 1 m column of liquid.
[a]Xu et al. (2018)	Air bubbles in visco-elastic polymer solution	Aqueous polyacrylamide and Breox solutions	Using the effective shear rate as V/d, presented drag correlation for Newtonian ($11 \le Re \le 273$) and polymer solutions ($12 \le Re_{PL} \le 30$).
Zamankhan et al. (2018)	Long (Taylor) bubbles in tubes and channels	Bingham plastic fluid	2-D numerical study in the axisymmetric regime. The film thickness scales with Bingham number in monotonic/non-monotonic manner depending upon the severity of surface tension effects (capillary number).
[a]Zhou et al. (2007)	Bubbles	Anisotropic micellar solutions	Reports on bubble shapes (including heart-shaped).
[a]Zhang et al. (2010)	Bubbles in aqueous solution of carboxymethyl cellulose, hydroxyl ethyl cellulose, and xanthan gum	Carreau–Yasuda viscosity equation	Extensive experimental results on shape and terminal velocity and stream function obtained numerically.
[a]Sontti and Atta (2019)	Taylor bubbles	Aqueous CMC solutions (power-law model; $0.4 \le n \le 0.87$).	Both numerical and experimental results suggest the bubble length to decrease with the increasing capillary number. Presented a flow pattern map in terms of the gas and liquid velocities.
[a]Premlata et al. (2017a,b)	Bubbles	Carreau–Yasuda viscosity equation	Axisymmetric and 3-D simulations suggest that shear-thinning viscosity suppresses shape distortion and zigzag/spiraling rise.
[a]Zhu et al. (2018)	Interaction between two bubbles	Xanthan gum solutions modeled as power-law fluids	Numerical simulations and experiments suggest three regimes of coalescence: at the nozzle itself, inline, and no coalescence depending upon the initial horizontal separation and the kinematic conditions.
[a]Fan et al. (2014, 2016, 2020)	Bubble dynamics and coalescence behavior	Numerical and aqueous CMC solutions (Power-law; $0.5 \le n \le 1$)	Extensive results on bubble size and shape, coalescence elucidating the role of orifice size, gas flow rate, etc.
[a]Al-Matroushi and Borhan (2009)	Two inline fluid spheres of different sizes in a tube	Aqueous solutions of PEO, CMC, and polyacrylamide to capture weak, no, and strong elastic effects	Coalescence time increases as the ratio of the trailing to leading bubble size increases.
[a]Böhm et al. (2016)	Single bubbles rising in Newtonian and non-Newtonian fluids	Xanthan gum solutions ($n = 0.42$, mildly elastic)	Calculated vorticity, strain, and negative wake features using PIV data over the ranges as $1.2 \le Eo \le 13$; $10 \le Re_{PL} \le 76$; and $1.3 \le We \le 7.2$.
Mukunda krishnan et al. (2009)	Axisymmetric motion of a single bubble	Truncated power-law fluid model	Bubble dynamics in tubes as a function of Reynolds number, Capillary number, Weber and Froude numbers.

(*Continued*)

TABLE 5.1 (Continued)

Summary of Investigations on the Free Motion of Bubbles and Drops in Non-Newtonian Media

Investigator	Details	Fluid Model (Continuous Phase)	Comment
Zhang (2012)	Bubble in a Bingham plastic fluid in a tube	Bingham plastic	Major thrust is on the validation of a new solver.
[a]Islam et al. (2020)	Bubbles in xanthan gum solutions	Power-law fluid model	Numerical and experimental results suggest no zigzag motion is observed, but bubble shapes are different from that in Newtonian fluids.
Liu et al. (2015b)	Coalescence of multiple bubbles	Power-law fluid model	3-D volume of fluid simulations to identify the conditions for coalescence.
Enders et al. (2019)	Bubbles in model visco-elastic fluids to study biopolymer solutions	Giesekus, linear, and exponential Phan-Thien & Tanner models	All three models predict negative wake, though shapes and rise velocities are somewhat model dependent.
[a]Aguirre et al. (2017)	Bubbles in xanthan gum solutions	Giesekus model	Numerical prediction of bubble shape and velocity are consistent with their experimental results.
Descher and Wünsch (2019)	Shape of bubbles	Reiner–Rivlin model fluid	These results suggest that strain hardening leads to the cusp formation in the rear of the bubble.
[a]Xu et al. (2019a)	Bubble shape and velocity in polymer (xanthan and polyacrylamide solutions)	Carreau viscosity equation	No jump in velocity is reported. Developed equations for bubble shape.
[a]Fu et al. (2009, 2011, 2012)	Bubble formation and breakup dynamics of slender bubbles in polymer solutions	Polyacrylamide solutions	Significant deviations in terms of filament diameter and coalescence time are reported for Newtonian and non-Newtonian fluids.
[a]Aminzadeh et al. (2012)	Newtonian and non-Newtonian drops (carboxymethyl cellulose solution, polystyrene-in-toluene solution)	Air and water	Drop rheology is immaterial in water, but zigzag motion and oscillating shapes are observed in air. These results are in line with that of Wegener et al. (2010).
Velez-Cordero et al. (2011, 2012); Velez-Cordero and Zenit (2011)	Single, two, and multiple bubbles in Boger fluids and xanthan gum solutions	—	Reported jump in velocity but no negative wake. The phenomenon of DKT (drifting–kissing–tumbling) is somewhat suppressed.
Pang and Lu (2018)	Single bubble	Carreau viscosity equation	Numerical results on shape and velocity.
Hu and Pang (2022)	Single bubble	Carreau viscosity equation	Effect of the ratio of the two limiting viscosities. A shape regime is presented.
Pilz and Brenn (2007)	Single bubble	Dilute/semi-dilute aqueous solutions of polyethylene oxide and polyacrylamide	Developed a criterion for the critical bubble size for discontinuity.

(*Continued*)

TABLE 5.1 (Continued)
Summary of Investigations on the Free Motion of Bubbles and Drops in Non-Newtonian Media

Investigator	Details	Fluid Model (Continuous Phase)	Comment
Niethammer et al. (2019)	Single bubble	Generic constitutive equation which degenerates into Maxwell, Oldroyd-B, PTT models, etc.	DNS capture the velocity jump, in line with the data of Pilz and Brenn (2007).
Cao et al. (2020)	Single bubble	Herschel–Bulkley fluid model	Numerical results on drag and mass transfer.
Nirmalkar et al. (2021) and Alam et al. (2022)	Fluid sphere	Bingham plastic fluids and Herschel–Bulkley fluids	Extensive numerical results on stability criterion, drag, and mass transfer.
[a]Zhang et al. (2008a,b)	Single bubbles in Carreau viscosity fluids	Aqueous solutions of CMC, Xanthan, hydroxyethyl cellulose, polyacrylamide	Drag correlation reported for steady and unsteady rise of bubbles.
Dhole et al. (2007a,b)	Single bubbles	Power-law fluid	Drag and Sherwood number for spherical bubbles over the range $2 \leq Re_{PL} \leq 500$; $0.5 \leq n \leq 2$.
Holenberg et al. (2013)	Viscous drops	Carbopol solutions	Detailed velocity profile measurement
Fraggedakis et al. (2016)	Single bubble	Exponential Phan-Thien–Tanner model	Numerical results explain the jump in velocity and are consistent with the results of Astarita and Apuzzo (1965) and Pilz and Brenn (2007).
[a]Wang et al. (2019)	Single bubbles	Carbopol solutions (Herschel–Bulkley model)	Stopping criterion and drag correlation.

[a] Denotes experimental work.

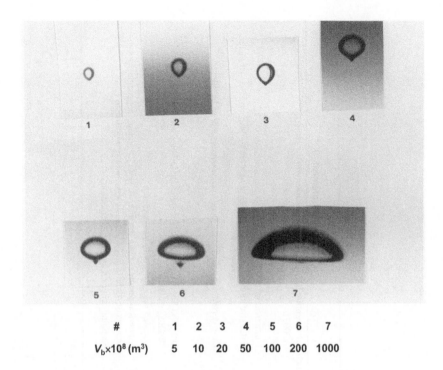

FIGURE 5.7 Photographs of air bubbles in a 1% carboxymethyl cellulose solution. (0.087 < Re < 6). (From De Kee, D. and Chhabra, R.P., *Rheol. Acta*, **27**, 656, 1988. With permission.)

In a few studies, shapes of drops made of non-Newtonian fluids falling in Newtonian or non-Newtonian continuous phase have been investigated. For instance, Norouzi and Davoodi (2018) and Norouzi et al. (2019a,b) reported the small drops of a Boger fluid falling in Newtonian and another immiscible viscoelastic medium to be nearly spherical whereas "dimpled" shapes were recorded for large drops due to the nonuniform normal stresses developed on the interface. These findings are in line with the previous results of Sostarzez and Belmonte (2003), Mukherjee and Sarkar (2011), and Vamerzami et al. (2016). Ortiz et al. (2016) also reported qualitatively similar results when both phases are non-Newtonian including the formation of a long tail and negative wake. Smagin et al. (2011) found the visco-plastic drops to have stable spheroidal shapes translating in a Newtonian medium in the creeping flow regime. In summary, the dispersed phase rheology does not seem to influence the shape to any significant extent.

Similarly, another important distinct feature of bubble shapes in non-Newtonian fluids is the appearance of a "pointed" tail before the transition to hemispherical caps. The early and even subsequent works (Astarita and Apuzzo, 1965; Calderbank, 1967; Calderbank et al., 1970; Budzynski et al., 2003; Iwata et al., 2019) are of qualitative nature without even including rheological measurements on the test liquids used; only a few investigators (see Carreau et al., 1974; De Kee and Chhabra, 1988; Rodrigue and De Kee, 2002; Herrera-Velarde et al., 2003; Li et al., 2012; Ohta et al., 2015) have presented more complete and quantitative information on bubble shapes. It is now generally acknowledged that bubbles rising in stagnant inelastic and visco-elastic fluids remain spherical up to larger volumes than that in Newtonian media (De Kee et al., 1986, 1996a, 1996b, 2002; De Kee and Chhabra, 1988). It is, however, not yet possible to predict quantitatively a priori the shape of a bubble in a new application. Not only are the transition boundaries strongly dependent on the physical and rheological properties, but such visual observations of shapes also have an inherent subjectivity and arbitrariness. Suffice it to say here that the observed shapes are broadly in line with the available approximate theoretical predictions (Ajayi, 1975; Zana and Leal, 1978; Seeling and Yeow, 1992; Ohta et al., 2019). Additional complications such as "tailing" (Barnett et al., 1966), time-dependent effects (Carreau et al., 1974; Frank et al., 2005; Funfschilling and Li, 2006), negative and oscillating wake (Hassager, 1977, 1979, 1988; Bisgaard and

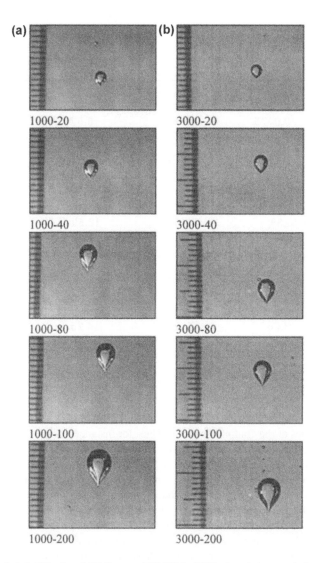

FIGURE 5.8 Shapes of air bubbles in a 0.5% Separan MG-700 in 20/80 glycerin/water solutions. The first and second numbers are the concentration of sodium dodecyl sulfate (SDS) in ppm and the volume of bubble in cubic mm, respectively (a) when the SDS concentration is below the critical micelle concentration (CMC), left column and (b) when the SDS concentration is above the CMC value, right column. (Photographs courtesy: Professor Denis Rodrigue, University of Laval, Quebec.)

Hassager, 1982; Bisgaard, 1983; Belmonte, 2000; Aminzadeh et al., 2012; Islam et al., 2020), oscillatory behavior (Handzy and Belmonte, 2004), cusped bubbles (Liu et al., 1995; Sikorski et al., 2009; Iwata et al., 2019; Ohta et al., 2019), open-ended bubbles (Bird et al., 1987a), and "heart"-shaped bubbles (Zhou et al., 2007) arise for bubbles rising in strongly visco-elastic liquids. It is perhaps fair to say that cusp formation has not been reported in inelastic fluids. Qualitatively, some of these phenomena have been explained by postulating that the fluid displaced by a rising bubble does not recover instantaneously, thereby creating a hole in the rear of the bubble (Barnett et al., 1966). This, in turn, sets up an adverse pressure gradient, thereby resulting in the formation of the so-called negative wake. Others including Philippoff (1937) and Warshay et al. (1959) have attributed these effects to the time-dependent rheological behavior of the continuous medium. Although the increasing levels of visco-elasticity do not seem to influence the shapes per se, the transition from one shape to another appears to be strongly dependent on

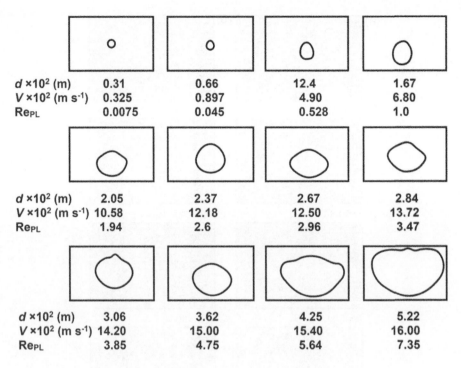

FIGURE 5.9 Schematic representation of shapes of nitrobenzene drops falling in a CMC solution. (Adapted from Fararoui, A. and Kintner, R.C., *Trans. Soc. Rheol.*, **5**, 369, 1961.)

the rheological properties of the continuous phase. Thus, for instance, the deviations from the spherical shape as well as the final transition to spherical cap shape occur at larger and larger volumes of bubbles with increasing levels of visco-elasticity than that in Newtonian media, otherwise under identical conditions. Aside from the foregoing qualitative treatment of bubble shapes, it is perhaps worthwhile to correlate the results in the form of eccentricity (*e*)—bubble volume behavior in the context of the approach of Tadaki and Maeda (1961). Representative results for air bubbles rising in a 3% aqueous carboxymethyl cellulose solution demonstrated by Calderbank (1967) are shown in Figure 5.6. Notwithstanding the inherent differences such as tailing or negative wake, overall, the bubbles appear to be flatter in the 3% CMC polymer solution than in water. Some attempts have also been made to establish quantitative relationships between the bubble eccentricity and the pertinent dimensionless groups (Acharya et al., 1977; Miyahara and Yamanaka, 1993; Wenyuan et al., 2010). Thus, for instance, based on visual observations of bubble shapes in a series of inelastic power-law fluids ($0.64 \leq n \leq 0.9$), Miyahara and Yamanaka (1993) developed the following correlation for aspect ratio (major/minor dimension of a bubble) for air bubbles:

$$b/a = 0.0628 \delta^{0.46} \quad 20 \leq \delta \leq 100 \tag{5.19a}$$

$$b/a = 1.4 \quad \delta \leq 4 \tag{5.19b}$$

$$b/a = 6.17 \delta^{-1.07} \quad 4 \leq \delta \leq 20 \tag{5.19c}$$

where $\delta = Re_{PL} Mo^{0.078}$. The Reynolds number, Re_{PL}, is based on the equal volume sphere diameter, and Mo, the Morton number, is given by $(We^{n+2} Fr^{2-3n}/Re_{PL}^4)$; finally, the Weber number is defined, $We = \rho_c d V^2/\sigma$. The difficulty with this correlation is that it predicts a significant departure from spherical shape even at small values of Reynolds number (hence δ), which is at odds with observations. Similarly,

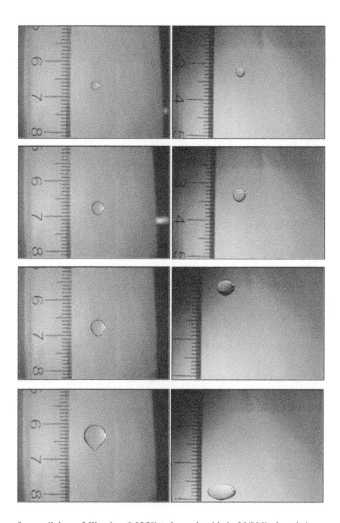

FIGURE 5.10 Shapes of corn oil drops falling in a 0.025% polyacrylamide in 20/80% glycerin/water solvent. The concentration of sodium dodecyl sulfate (SDS) is 10 ppm for the left column and 1000 ppm for the right column. Within each column, the volume of the drops is 25, 70, 120, and 400 mm³. (Photograph courtesy: Professor Denis Rodrigue, University of Laval, Quebec.)

Acharya et al. (1977) argued that at low Reynolds numbers, for prolate-shaped bubbles, the aspect ratio ($e < 1$) should correlate with a mixed dimensionless group, G_1, representing a ratio of elastic and surface tension forces as:

$$e = 0.62 G_1^{-0.17} \tag{5.20a}$$

and

$$G_1 = \frac{A(V/R)^\beta}{\sigma/R} \tag{5.20b}$$

where A and β are the power-law constants for the first normal stress difference, that is, $N_1 = A(\dot{\gamma})^\beta$. The numerical constants appearing in Equation 5.20a are based on data embracing $1 \leq G_1 \leq 8$ and $0.68 \leq e \leq 1$.

On the other hand, for the oblate-shaped bubbles ($e>1$) encountered at high Reynolds numbers, it is fair to anticipate the eccentricity to be influenced by inertial forces also. Thus, Acharya et al. (1977) introduced another composite parameter, G_2, defined as (Wi/ReWe) and correlated their results for oblate-shaped bubbles as

$$e = 1 + 0.00083\, G_2^{-0.87} \tag{5.21}$$

Equation 5.21 applies over a range of variables as $1 \le e \le 1.5$ and $0.001 \le G_2 < 0.015$. Note that while Equation 5.21 correctly reduces to the limits of $e=1$ as $G_2 \to 0$, Equation 5.20a does not approach this limiting behavior. A similar correlation for the shape of drops in power-law liquids has been reported by Wanchoo et al. (2003). On the other hand, based on their experimental results, Wenyuan et al. (2010) developed a relationship between e and Eötvös number (Eo) based on the equal volume sphere diameter.

In some of the theoretical and numerical investigations listed in Table 5.1, the drop or bubble shape has been calculated as a part of the solution for the free rise of a bubble. Though the shapes of fluid particles have defied predictions from first principles even in Newtonian continuous media, the limited results reported by Ajayi (1975), Wagner and Slattery (1971), Seeling and Yeow (1992), Iwata et al. (2019), and Ohta et al. (2019) do not preclude some of the shapes observed experimentally. In most cases, particle shape has been assumed a priori while seeking a solution to the field equations. There is no question that the shapes of particles play a central role in the calculations of the mean residence time and rates of transfer processes in biotechnological and process engineering applications, as conjectured by Barnett et al. (1966), Calderbank et al. (1970), and others (Buchholz et al., 1978; Godbole et al., 1984; Terasaka and Shibata, 2003; Ishkintana and Bennington, 2010; Gomez-Diaz et al., 2009; Radl and Khinast, 2007). It is worthwhile to reiterate here that none of the expressions that purport to predict aspect ratio and eccentricity captures the fine details such as tail and negative wake or the peculiar cigar-like elongated shapes encountered in confined domains (van Wijngaarden and Vossers, 1978; Coutanceau and Thizon, 1981; Coutanceau and Hajjam, 1982). Little is known about the shapes of non-Newtonian droplets falling in a Newtonian fluid, except the preliminary results of Rodrigue and Blanchet (2001, 2002) and Gillaspy and Hoffer (1983) on non-Newtonian droplets falling in air and that of the other continuous media, as referred to earlier in this section.

5.4 Terminal Velocity–Volume Behavior in Free Motion

In Newtonian media, small bubbles are known to adhere to solid-like behavior (no-slip) and the transition from the no-slip to the shear-free condition occurs fairly gradually. In contrast, in non-Newtonian liquids, this transition may manifest as an abrupt jump (or discontinuity) in terminal velocity at a critical bubble size. This is perhaps one of the most striking and fascinating effects associated with bubble motion in non-Newtonian liquids. Thus, when the free-rise velocity of a bubble is measured as a function of its size, while small bubbles conform to the solid-like behavior, a discontinuity could be observed in terminal velocity at a critical bubble size. Indeed, jumps in the free-rise velocity for bubbles in an aqueous solution of a commercial polymer ET-497 as large as six to ten fold (Figures 5.11 and 5.12) have been reported in the literature (Astarita and Apuzzo, 1965; Calderbank et al., 1970; Leal et al., 1971; Acharya et al., 1977; Zana and Leal, 1978; Haque et al., 1988, Rodrigue et al., 1996a, 1998; Rodrigue and De Kee, 1999; Herrera-Velarde et al., 2003). Less dramatic jumps in bubble velocity have also been reported in recent studies (Pilz and Brenn, 2007; Velez-Cordero et al., 2012). In spite of the wide range of polymer solutions used in such studies, the transition seems to occur over a rather narrow range of bubble radii (on an equal volume sphere basis) hovering around 2.5 to 3 mm, with a few exceptions (De Kee et al., 1996a). On the other hand, no such discontinuity in the free-rise velocity of bubbles was reported by Macedo and Yang (1974), De Kee et al. (1986, 1990a), De Kee and Chhabra (1988), Miyahara and Yamanaka (1993), Margaritis et al. (1999), Dewsbury et al. (1999), Ohta et al. (2015), Ortiz et al. (2016), and Xu et al. (2019a). Since the sudden jump in the free-rise velocity occurs with bubble diameters of 5–6 mm, clearly it would not have been possible for Macedo and Yang (1974), Margaritis et al. (1999), and Dewsbury

Fluid Particles in Non-Newtonian Media

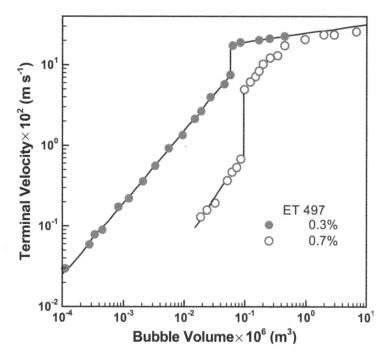

FIGURE 5.11 Terminal velocity–bubble volume data showing an abrupt increase in velocity. (Replotted from Astarita, G. and Apuzzo, G., *AIChE J.*, **11**, 815, 1965.)

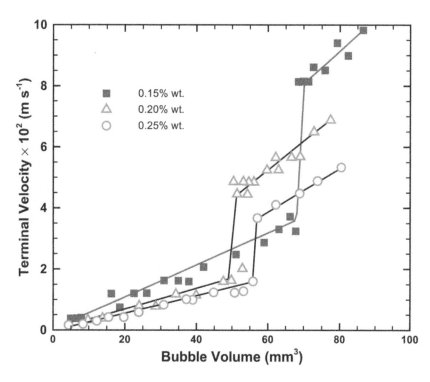

FIGURE 5.12 Bubble terminal velocity–bubble volume relationship showing the jump for air bubbles in different solutions of Separan AP-30 in 50/50 glycerin/water mixtures. (Replotted from Herrera-Velarde, J.R., Zenit, R., Chehata, D., and Mena, B., *J. Non- Newt. Fluid. Mech.*, **111**, 199, 2003.)

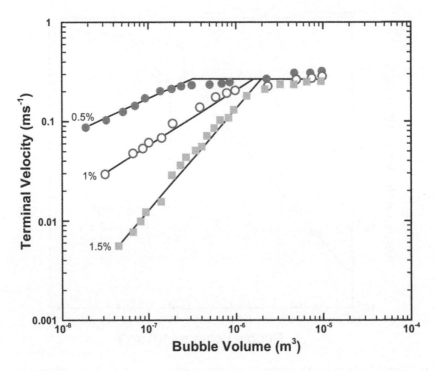

FIGURE 5.13 Terminal velocity–bubble volume data showing gradual change of slope in polyacrylamide solutions. (Based on data of Dajan, A., M. Eng. Sci. Dissertation, University of Windsor, Windsor, ON, Canada, 1985.)

et al. (1999) to see this behavior as the maximum bubble diameter ranged from 400 μm to 3.36 mm in these studies. Similarly, De Kee et al. (1986) and Miyahara and Yamanaka (1993) missed it because the minimum bubble diameter in these studies was about 5 mm. However, it is far from clear why many other investigators (Barnett et al., 1966; Calderbank, 1967; De Kee and Chhabra, 1988; Tsukada et al., 1990) did not see such a discontinuity, as the bubble diameters in these studies were in the range of 2–6 mm (see Figure 5.13). This suggests that, apart from the bubble size, some other factors are clearly at work. In view of the lack of agreement between various studies (Dajan, 1985; De Kee et al., 1986, 1990a; De Kee and Chhabra, 1988; Dewsbury et al., 1999; Margaritis et al., 1999), it is thus not at all clear under what conditions the jump in bubble velocity occurs. However, no such discontinuity has been observed in yield stress fluids so one can surmise that the abrupt increase in velocity must be due to shear-thinning and/or visco-elasticity of the continuous phase. Similarly, Mhatre and Kintner (1959), Warshay et al. (1959), and others (Fararoui and Kintner, 1961; Mohan, 1974a) have reported a steep and gradual change in the terminal velocity of Newtonian drops moving in non-Newtonian media (Figure 5.14). Undoubtedly, this change in bubble size–velocity relationship is real and it is also likely to have a strong bearing on the rate of transfer processes. Thus, a complete understanding of this behavior is desirable and significant research effort has been devoted to this issue (De Kee et al., 1996a; Rodrigue and De Kee, 2002; Caswell et al., 2004).

As noted earlier for bubbles rising in Newtonian liquids, it is well known that small bubbles behave like rigid spheres (i.e., follow the Stokes law) whereas large gas bubbles exhibit a shear-free mobile boundary. These considerations lead to a 50% increase in the rise velocity even for Newtonian liquids. The extension of this idea to bubbles moving in power-law fluids in the creeping flow regime yields the following expression for the ratio (V_b/V_s):

$$\frac{V_b}{V_s} = \left(1.5\frac{Y_s}{Y_b}\right)^{1/n} \quad (5.22)$$

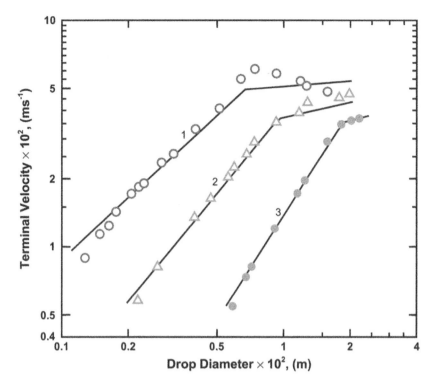

FIGURE 5.14 Variation of terminal velocity with drop diameter for Aniline drops falling in (1) water (2), 0.5% Sodium Alginate solution and (3), 1% CMC solution. (Based on data of Mohan, V., *AIChE J.*, **20**, 180, 1974b).

where Y_s and Y_b are the drag correction factors for solid spheres and bubbles, respectively, and are the functions of the power-law index, n, only. While the values of Y_s are listed in Table 2.4, the values of Y_b are available in Table 5.2. For example, for $n=0.6$, $Y_s=1.381$ and $Y_b=1.332$. Equation 5.22 predicts the ratio $V_b/V_s=2.1$ thereby suggesting that a bubble will rise at twice the velocity of an equivalent sphere. This lends some credibility to the argument that the sudden increase in the velocity, at least in part, is due to the switch-over from the no-slip to shear-free conditions. However, more detailed analyses (Zana and Leal, 1974, 1978) and qualitative dimensional considerations (Leal et al., 1971) suggest that shear-thinning can account only for a small increase in the rise velocity (as also seen in the above example) whereas the introduction of a small degree of fluid visco-elasticity can lead to as large jumps in velocities as documented in the literature. For instance, Liu et al. (1995) attributed the steep increase in the velocity to the onset of cusp formation and they proposed it to occur when the viscous forces are balanced by the surface tension forces, i.e., when the capillary number ($Ca = \mu_0 V/\sigma$) is of order one. However, De Kee et al. (1986, 1990a) and Rodrigue and Blanchet (2002) observed cusp-shaped bubbles but without any dramatic increase in their rise velocity. In an elegant numerical study, Fraggekadis et al. (2016) have argued that the discontinuity in velocity is attributed to major changes in the flow field around the bubble beyond its critical volume. They reported steady-state simulations for a bubble rising in an exponential Phan-Thien–Tanner visco-elastic fluid. They were able to join the two solutions at the critical bubble size by the hysteresis loop. Along such a loop, the rear pole of the bubble transforms into a tip thereby eliminating the rear stagnation point. Such a steep change in curvature also creates a force which enhances the bubble motion. They used their numerical results to rationalize much of the experimental results available in the literature including that of Astarita and Apuzzo (1965) and of Pilz and Brenn (2007).

Based on various considerations, several attempts have been made at postulating schemes for predicting the critical bubble size at which the transition from the no-slip to the shear-free regime occurs. Most of these hinge on the classical work of Bond and Newton (1928) and the available pertinent literature has been reviewed elsewhere (Chhabra, 1988; Chhabra and De Kee 1991; De Kee et al., 1996a; Rodrigue

TABLE 5.2
Analytical Expressions of Drag Correction Factor (Y) for Creeping Bubble Motion

Investigator	Fluid Model	Equation for Y
Hirose and Moo-Young (1969)	Power-law	$2^n 3^{(n-3)/2} \left(\dfrac{13 + 4n - 8n^2}{(2n+1)(n+2)} \right)$
Bhavaraju et al. (1978)	Power-law	$2^n 3^{(n-3)/2} \left[1 - 3.83(n-1) \right]$ for $1 \geq n \geq 0.7$
Kawase and Moo-Young (1985)	Carreau fluid model	$\left(\dfrac{2}{3} \right)(3\Lambda^2)^{(n-1)/2} \left(\dfrac{13 + 4n - 8n^2}{(2n+1)(n+2)} \right)$ for $\Lambda > 10$
	Ellis fluid model	$\left(\dfrac{2}{3} \right)(3\mathrm{El}^{-2})^{(1-\alpha)/2} \left(\dfrac{-11 + 28\alpha - 8\alpha^2}{(4-\alpha)(5-2\alpha)} \right)$ for $\mathrm{El} > 10$
Gummalam and Chhabra (1987)	Power-law model	$3^{-(n+3)/2} \left[\dfrac{2(2n+1)(2-n)}{n^2} \right]^n$ (Lower bound)
Rodrigue et al. (1996c)	Carreau fluid	$\left(\dfrac{2}{3} \right) \left[1 + \dfrac{9}{50}(n-1)\Lambda^2 \right]$
Rodrigue et al. (1999a)	Power-law fluid	$2^n 3^{(n-3)/2} \left\{ \dfrac{1 + 7n - 5n^2}{n(n+2)} \right\}$ for $n > 0$

and De Kee, 2002; Rodrigue and Blanchet 2002; Pilz and Brenn, 2007). The key points are recapitulated here. Acharya et al. (1977) and subsequently others (Haque et al., 1988) reported that the criterion developed by Bond and Newton (1928) worked remarkably well for estimating the critical bubble size, regardless of the rheological characteristics of the ambient medium. This simple criterion is written as

$$R_c = \left(\frac{\sigma}{g \Delta \rho} \right)^{1/2} \tag{5.23}$$

The values of d_{cr} (= $2R_c$) calculated using Equation 5.23 are remarkably close to the experimental values in most cases, with very few exceptions (Rodrigue and De Kee, 2002). It is tempting to conclude that the process is essentially controlled by the surface tension forces and liquid rheology plays little or no role! This is however not tenable as no abrupt change in the bubble velocity has been observed in Newtonian media. Furthermore, detailed PIV (Particle Image Velocimetry) data not only suggest significant changes in the flow field near this transition, but also indicate that the discontinuity appears at smaller and smaller bubble volumes as the polymer concentration is progressively increased (Herrera-Velarde et al., 2003). This finding is clearly at variance with the findings of Rodrigue et al. (1998). Furthermore, Herrera-Velarde et al. (2003) observed that the negative wake appeared only after the bubble size exceeded the critical value corresponding to the velocity jump. In an attempt to explain this phenomenon, Kawase and Ulbrecht (1981d) modified the analysis of Schechter and Farley (1963) to elucidate the role of the continuous phase rheology on the transition from the no-slip to the shear-free regime. In essence, they have evaluated the drag on a bubble in the presence of surface tension gradients present on the bubble surface and have deduced the conditions under which the bubble will act as a rigid sphere, obeying the no-slip boundary condition. Although no suitable experimental results are available to substantiate their predictions, the analysis seems to suggest increasingly big jumps in terminal velocity with a decreasing value of the power-law index. This finding is qualitatively consistent with the experimental results of Haque et al. (1988). Similar types of discontinuities also occur in other flow configurations involving Maxwellian fluids (Astarita and Denn, 1975). It has been conjectured that the discontinuity occurs at a flow velocity equal to the so-called shear-wave velocity $\left(= \sqrt{\mu/\rho\lambda} \right)$, where λ is the Maxwellian relaxation time. Admittedly, this notion has proved to be of value in explaining some of the features of creeping visco-elastic flow past immersed bodies, but it does not seem to be relevant to the discontinuity observed for freely moving bubbles. For instance, Leal et al. (1971) observed the discontinuities to occur at about ~22 and 5 mms^{-1}, respectively, in 0.5% and 1.0% polyacrylamide solutions, whereas the corresponding shear-wave velocities are of the order of 40 and 50 mms^{-1}, albeit these values are strongly dependent on the way the Maxwellian relaxation time λ is evaluated.

Subsequent works (Rodrigue et al., 1996a, 1998; Rodrigue and De Kee, 2002) seem to suggest that the discontinuity in bubble velocities is primarily due to a complex interplay between the surface-active agents (and their distribution) and the visco-elastic effects. Their systematic study clearly shows that the change of slope is smooth and gradual in inelastic solutions, irrespective of whether surface-active agents are present or not. On the other hand, they reported varying levels of the jump in the bubble velocity in visco-elastic liquids, both with and without the addition of a surface-active agent. Both the jump in the velocity and the critical size of the bubble are strongly dependent on the type and the concentration of polymers, visco-elasticity, and the concentration and characteristics of the surfactants, etc. The role of visco-elasticity as postulated here is clearly consistent with the assertions of others also (Leal et al., 1971; Zana and Leal, 1974, 1978; Niethammer et al., 2019), but is at variance with the findings of Liu et al. (1995) who attributed the jump in velocity to the changes in bubble shape and the appearance of a cusp. Additional complications can arise when the molecules of surfactants get detached from the interface due to the intense localized shearing and elongational stresses developed in the continuous phase. Based on the literature data supplemented by their own extensive results, Rodrigue and De Kee (2002) developed the following correlation for the critical bubble size marking the transition from the no-slip to the shear-free regime:

$$\frac{\rho g d_c^2}{\sigma} = 28 \left(\frac{\Delta \sigma}{\sigma} \right)^{0.2} \left(\frac{N_1 d_c}{2\sigma} \right)^{0.4} \tag{5.24}$$

Rodrigue and De Kee (2002) have subsequently also developed another criterion involving a capillary number, Ca ($= \mu_0 V/\sigma$), Deborah number, De ($= a/2m(V/R)^{b-n}$), and Marangoni number, Ma ($= \Delta\sigma/Rm(R/V)^n$), as

$$\alpha_0 = \frac{\text{CaDe}}{\text{Ma}} \tag{5.25}$$

and they asserted $\alpha_0 = 1$ denoted the transition from the no-slip to the shear-free condition. Extensive comparisons between the predictions of Equation 5.25 and experimental results show that while it works reasonably well, it is not completely satisfactory and further refinements are needed. Also, it is not immediately obvious whether the formation of a cusp has any influence on the discontinuity in the velocity (De Kee et al., 2002; Rodrigue and Blanchet, 2001).

In an extensive study, Pilz and Brenn (2007) have argued that the jump in velocity is a signature of strong elongational flow whereas much of the previous literature has attempted to use shear viscosity of the continuous phase to explain this phenomenon. Based on the shear and elongational properties, Pilz and Brenn (2007) put forward the following dimensionless criterion:

$$\text{Eo}_c = 5.37 \left(\frac{\psi - 9.894}{\text{Mo}^{0.33}} \right)^{0.921} \tag{5.26}$$

where the critical value of the Eötvös number (Equation 5.18b), Eo_c, includes the critical bubble diameter corresponding to the velocity jump and the Morton number. Equation 5.18c is based on the zero-shear viscosity of the continuous phase, and the new dimensionless parameter ψ appearing in Equation 5.26 is defined as:

$$\psi = \lambda_E \left(g^3 \rho_c / \sigma_c \right)^{1/4} \tag{5.27}$$

where λ_E is the fluid relaxation time evaluated from extensional data rather than from the steady shear data. Equation 5.27 captured the experimental results for wide-ranging polymer solutions including polyethylene oxide (PEO) and polyacrylamide and their mixtures in water and other solvents.

Thus, in summary, the dependence of the free-rise velocity on bubble size undergoes a transition at a critical bubble size. This switch-over corresponds to the no-slip/shear-free transition. In inelastic liquids, the change in the slope of the velocity-size data is gradual whereas it occurs as a "discontinuity" in visco-elastic systems. The behavior of these systems is further complicated by the presence of surfactants. Depending upon the amount of information available, especially with regard to the concentration of the surfactant and the interfacial tension and its variation over the bubble surface, Equation 5.23, 5.24, 5.25, or 5.26 yields satisfactory predictions of the critical bubble size at the transition point. However, to date, it has not been possible to predict a priori the magnitude of the jump in velocity or whether such a discontinuity will occur or not, in a new application. The analogous effect for Newtonian drops falling in non-Newtonian fluids is rather small and has thus attracted very little attention. Likewise, the reverse case of non-Newtonian drops falling in Newtonian media has also received only very scant attention.

Additional complications arise when the rise velocity of bubbles is also influenced by the injection frequency, i.e., the time elapsed between the release of two successive bubbles. This phenomenon has been reported, among others, by Barnett et al. (1966), Carreau et al. (1974), and others (Frank et al., 2003; Funfschilling and Li, 2006). In an extensive study involving rise velocity and detailed PIV measurements, Frank et al. (2003) systematically varied the injection period from 0.3 to 60s for air bubbles of constant size rising in polyacrylamide solutions and laponite suspensions (Figure 5.15). Unlike the behavior reported for the solid spheres in visco-elastic liquids (Chapter 4), the rise velocity of bubbles decreases with the increasing injection period. It is conceivable that at low injection periods, two inline bubbles interact, thereby modifying their rise behavior. However, in spite of this complexity, Frank et al. (2003) reported that the drag behavior was nearly independent of the injection period. In a subsequent work, Frank et al. (2005) have attempted to predict the rise velocity of a periodic chain of bubbles. No such phenomenon is observed in Newtonian and inelastic fluids (Funfschilling and Li, 2006).

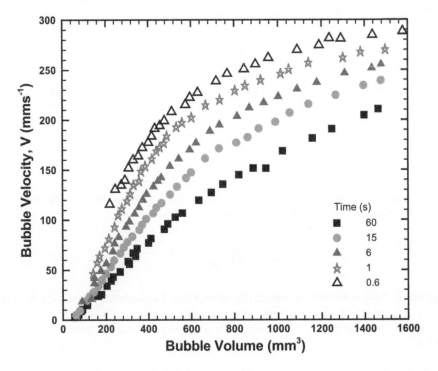

FIGURE 5.15 Effect of injection period on the free-rise velocity of bubbles in a 0.75% aqueous polyacrylamide solution. (Replotted from Frank, X., Li, H.Z., Funfschilling, D., Burdin, F., and Ma, Y, *Can. J. Chem. Eng.*, **81**, 483, 2003.)

5.5 Drag Behavior of Single Particles

5.5.1 Theoretical Developments

The flow configuration considered here is a fluid sphere of radius R moving relative to an incompressible fluid of infinite extent with a steady velocity V, as shown schematically in Figure 2.1. Since the flow is axisymmetric, one can introduce a stream function. The governing equations are identical to those written for a rigid sphere (Equations 2.6 and 2.7). The case of a fluid sphere differs from that of the rigid sphere in so far that similar equations are also required to describe the flow inside the fluid sphere. The boundary conditions for fluid spheres, however, merit special attention. Taking a reference frame attached to the particle with origin at its center, these are

1. No flow across the interface ($r=R$):

$$V_{r_c} = V_{r_d} = 0 \tag{5.28a}$$

2. Continuity of tangential velocity across the interface ($r=R$):

$$V_{\theta_c} = V_{\theta_d} \tag{5.28b}$$

3. Continuity of tangential stress across the interface ($r=R$):

$$\tau_{r\theta_c} = \tau_{r\theta_d} \tag{5.28c}$$

4. Continuity of normal stress across the interface ($r=R$):

$$\left(p - \tau_{rr} + 2\frac{\sigma}{R} \right)_c = \left(p - \tau_{rr} \right)_d \tag{5.28d}$$

5. Far away from the fluid sphere, the uniform velocity in z-direction, that is,

$$r \to \infty \qquad V_r = V\cos\theta$$
$$V_\theta = -V\sin\theta \tag{5.28e}$$

In writing the above-noted boundary conditions, it has been tacitly assumed that there are no surface-active agents present at the interface, so that the interfacial tension (σ) is constant in Equation 5.28d. For the internal flow, the velocity remains finite at the center of the fluid sphere.

The equations of continuity and momentum, written for the continuous and dispersed phases, together with the boundary conditions constitute the theoretical framework for the calculation of the velocity and pressure fields in the inner and outer flow regions for a fluid sphere. The problem description is, however, not complete without specifying a constitutive relation describing the stress–strain rate behavior of the continuous and dispersed phases. For a given fluid behavior, once the velocity and pressure fields are known, one can proceed along the lines similar to those used for rigid spheres (Section 2.2) to obtain drag coefficient, and other derived variables such as stream function and vorticity, for instance.

In the following sections, theoretical advances made in the area of the free motion of Newtonian fluid spheres in unbounded quiescent media are reviewed. In particular, consideration is given to the effect of the rheological characteristics of the continuous phase. However, as usual, we begin with the simplest type of fluid behavior for the continuous phase, namely Newtonian.

5.5.1.1 Newtonian Continuous Phase

Any discussion on the motion of fluid spheres (with clean interface) must inevitably begin with the pioneering work of Hadamard and Rybczynski, see Clift et al. (1978). In the absence of inertial effects in both fluids, the drag coefficient on a Newtonian fluid sphere (with clean interface) translating with a constant velocity V in another immiscible Newtonian fluid is given by

$$C_D = \left(\frac{24}{\text{Re}}\right) Y \tag{5.29}$$

where

$$Y = \frac{2 + 3X_E}{3 + 3X_E} \tag{5.30}$$

$$\text{Re} = \frac{\rho_c V_d d_d}{\mu_c} \quad \text{and} \quad X_E = \frac{\mu_d}{\mu_c}$$

For a gas bubble rising through a liquid, $X_E \ll 1$, and Equation 5.30 yields $Y = (2/3)$, whereas for a solid sphere moving in a liquid $X_E \to \infty$ and Y approaches unity. Thus, Equation 5.29 embraces the complete spectrum of particles ranging from gas bubbles to solid spheres. Equations 5.29 and 5.30 are expected to hold for Re < 1. The creeping flow result given by Equation 5.29 is complemented at a high Reynolds number by the first-order asymptotic solution and boundary layer treatments described by Moore (1959a, 1963, 1965) and Harper and Moore (1968) as

$$C_D = \frac{48}{\text{Re}} \{1 + (3/2) X_E\} \tag{5.31}$$

In between these two limits, no general solution is possible, and therefore most of the progress has been made by numerical solutions of the governing equations together with a large body of experimental results.

For the general case of a clean Newtonian fluid sphere translating through another Newtonian continuous phase, simple dimensional considerations suggest the drag coefficient to be a function of the Reynolds number (Re), viscosity ratio (X_E), and the density ratio ($\gamma = \rho_d/\rho_c$). A cursory glance of the pertinent literature immediately reveals that much of the information relates to the limiting case of gaseous spheres, that is $X_E \to 0$ and $\gamma \to 0$. Thus, for instance, the available numerical simulations of clean spherical gas bubbles reveal that the drag coefficient evolves smoothly between the creeping flow, Equation 5.29, and the high Reynolds number limit given by Equation 5.31. This led Mei et al. (1994) to devise the following empirical expression spanning the complete range of Reynolds numbers:

$$C_D = \frac{16}{\text{Re}} \left[1 + \left\{ \frac{8}{\text{Re}} + \frac{1}{2} \left(1 + 3.315 \text{Re}^{-0.5} \right) \right\}^{-1} \right] \tag{5.32}$$

Furthermore, some of the numerical simulations are based on a priori assumption regarding the shapes of the bubble, while in other cases the shape has been calculated as a part of the numerical simulation.

Thus, for instance, the results of spherical bubbles have been reported among others by Hamielec and Johnson (1962), Rivkind et al. (1973), Abdel-Alim and Hamielec (1975), Rivkind and Ryskin (1976), Oliver and Chung (1987), Hoffmann and van den Bogaard (1995), Takagi and Matsumoto (1996), Juncu (1999), Feng and Michaelides (2001), and Saboni and Alexandrova (2002). Analogous results for ellipsoidal bubbles are reported by Dandy and Leal (1986) and Blanco and Magnaudet (1995). Similarly, toroidal bubbles have been treated by Lundgren and Mansour (1991). On the other hand, Chen et al. (1999) have calculated the shapes of fluid spheres as a part of the numerical solution that allows the deformation of the particle during the course of its rise or fall. Their results on the velocity–shape relationship seem to suggest that a toroidal-shaped bubble moves slower than an elliptical or a mushroom-shaped one whereas a spherical cap moves faster than a skirt-shaped bubble. Similarly, Raymond and Rosant (2000) have also numerically studied the deformation of a 2-D axisymmetric bubble in the Reynolds number range, $Re \leq 200$. In addition to the drag behavior, some efforts have also been directed at predicting the aspect ratio of the bubble. One such analytical expression is explained by Moore (1965) that applies to small Weber numbers. Loth (2008b), Kelbaliyev (2011), Yan et al. (2018) and Freitas and Pereira (2021) have reviewed much of the literature in this field. The result of Moore (1965) is written as:

$$\frac{b}{a} = 1 + (9/64)\text{We} + O(\text{We}^2) \tag{5.33}$$

In contrast to such a voluminous literature on bubbles, analogous results for a Newtonian drop falling in another immiscible Newtonian medium are rather limited (Michaelides, 2006; Kishore et al., 2007a; Loth, 2008b). For fluid spheres, Juncu (1999) has theoretically examined the effects of viscosity ratio ($0.01 \leq X_E \leq 100$) and density ratio ($0.01 \leq \gamma \leq 100$) on drag in the Reynolds number range $Re \leq 500$. The effect of the density ratio was found to be small, which has also been confirmed subsequently by Feng and Michaelides (2001) and Saboni and Alexandrova (2002). While the study of Feng and Michaelides (2001) encompasses the values of the Reynolds number up to 1000, Saboni and Alexandrova (2002), on the other hand, have investigated the effect of X_E much more thoroughly in the range $Re \leq 400$. Broadly all these numerical studies report values of the drag coefficient that are in good agreement with the other studies available in the literature. Saboni and Alexandrova (2002) correlated their numerical results by the expression

$$C_D = \left\{ \left[X_E \left(\frac{24}{\text{Re}} + \frac{4}{\text{Re}^{1/3}} \right) + \frac{14.9}{\text{Re}^{0.78}} \right] + 40 \left(\frac{3X_E + 2}{\text{Re}} \right) + 15 X_E + 10 \right\} \left\{ (1 + X_E)(5 + \text{Re}^2) \right\}^{-1} \tag{5.34}$$

It should be mentioned here that the awkward-looking Equation 5.34 does reduce to Equation 5.29 in the limit of $Re \to 0$.

The aforementioned extensive numerical activity in this field has also been complemented by experimental studies, albeit the bulk of these studies relate to gaseous spheres rising vertically (Duineveld, 1995; Maxworthy et al., 1996; Ybert and DiMeglio, 1998; Takemura and Yabe, 1999; Han et al., 2001; Di Marco et al., 2003; Cieslinski and Mosdorf, 2005; Wegener et al., 2010) and in liquids in inclined tubes (Masliyah et al., 1994; Shosho and Ryan, 2001). The literature is inundated with numerous empirical correlations for bubble-rise velocity or drag coefficient (Mendelson, 1967; Maneri and Mendelson, 1968; Churchill, 1989; Jamialahmadi et al., 1994; Karamanev, 1994; Maneri, 1995; Berdnikov et al., 1997; Nguyen, 1998; Scheid et al., 1999; Wesselingh and Bollen, 1999; Rodrigue, 2001, 2002, 2004; Funada et al., 2005). By way of example, Rodrigue (2004) culled most of the literature data on bubble-rise velocity extending over wide ranges of conditions ($10^{-7} \leq Re \leq 10^4$; $10^{-11} \leq Mo \leq 10^7$; $722 \leq \rho_c \leq 1380$ kg m^{-3}, and $15.9 \leq \sigma \leq 91$ mN m^{-1}) and developed the following correlation for clean bubbles with an accuracy of ±20%:

$$V^* = \frac{F}{12(1 + 0.018F)^{0.75}} \tag{5.35a}$$

where the new coordinates V^* and F are defined as

$$V^* = V\left(\frac{\rho_c^2 d^2}{\sigma \mu_c}\right)^{1/3} \quad (5.35b)$$

and

$$F = g\left(\frac{\rho_c^5 d^8}{\sigma \mu_c^4}\right)^{1/3} \quad (5.35c)$$

where d is the equivalent bubble diameter.

Many other factors influencing the bubble-rise velocity have been explored. The effect of surfactants on bubble behavior has been studied by Fdhila and Duineveld (1996), Zhang and Finch (1999, 2001), and Gollakota and Kishore (2018b). They reported that irrespective of the concentration of Triton X-100, the bubbles attained the same steady-state terminal rise velocity. On the other hand, the effect of microgravity on bubble motion has been studied experimentally by Tomiyama et al. (1998). The velocity behavior of spherical and ellipsoidal bubbles in stagnant distilled water at elevated temperatures has been studied by Okawa et al. (2003). The drag behavior of droplets in stagnant and turbulent gaseous media has been studied and correlated empirically by Warnica et al. (1995). The zigzag and periodic rising behavior of bubbles in stagnant Newtonian liquid have also been explored (Krishna and van Baten, 1999; Ortiz-Villafuerte et al., 2001). The formation and rise of bubble streams in a viscous liquid have been studied by Snabre and Magnifotcham (1998). Greene et al. (1993) studied the free fall of Newtonian drops in a stagnant Newtonian liquid. They reported a good agreement with numerical simulations and previous results for spherical drops. However, at a critical value of the Reynolds number and of the Weber number, droplets exhibited oscillations and their drag behavior began to deviate from that of spherical droplets. Based on their experiments in the range $10^{-3} \le \text{Re} \le 10^4$, they put forward the criterion for the onset of the oscillations as

$$We \times Re^{0.65} = 165 \quad (5.36)$$

Thus, the value of this group $We \times Re^{0.65} < 165$ suggests stable behavior and there is very little deviation from the spherical shape. Mao et al. (1995) have reported scant results on the terminal velocities of spherical droplets as a part of their study on drop breakage in structured packings. On the other hand, Bhavasar et al. (1996) have presented and correlated extensive results of drag on single droplets rising through stagnant Newtonian fluids. Excellent reviews summarizing the current state of the art of bubble and drop motion in Newtonian media are now available (Magnaudet and Eames, 2000; Michaelides, 1997, 2002, 2003, 2006).

5.5.1.2 Shear-Thinning Continuous Phase

A considerable body of literature is now available on the free motion of Newtonian fluid particles in a shear-thinning continuous phase (see Table 5.1). The power-law fluid model (Equation 2.12) has received the greatest amount of attention and the bulk of the literature pertains to the inertialess flows (both Reynolds numbers small). Also, in most cases, the dispersed phase is Newtonian. Finally, the limiting case of gaseous spheres ($X_E = 0$) has been studied much more widely than that of a droplet with finite values of X_E. Owing to the shear-rate-dependent viscosity, the governing equations are hopelessly complex even when the nonlinear inertial terms are neglected in both phases. Thus, an exact solution that can rival the Hadamard–Rybczynski solution, for power-law or any other generalized Newtonian fluid model, is simply not possible. Consequently, numerous investigators (Table 5.1) have obtained results with varying degrees of approximations. Most of the available analyses for the creeping flow regime can be divided into four types depending upon the mathematical technique used and the approximations invoked in seeking such solutions:

1. The velocity and stress variational principles (referred to earlier in Chapter 2) have been used extensively to obtain upper and lower bounds on the drag experienced by a single fluid sphere and ensembles of spherical fluid spheres. This technique has been used for Newtonian fluid spheres moving in various generalized Newtonian fluid models including power-law (Nakano and Tien, 1968; Mohan, 1974a,b; Mohan and Venkateswarlu, 1974, 1976; Mohan and Raghuraman, 1976b,c; Jarzebski and Malinowski, 1986a,b), Carreau viscosity equation (Chhabra and Dhingra, 1986), etc. It is instructive to recall here that strictly speaking this method yields rigorous bounds only in the case of the Newtonian and power-law fluids and for gas bubbles wherein the inner flow does not contribute to the energy dissipation. In view of this, the results obtained for generalized Newtonian fluid models other than the power-law fluid model and for fluid spheres ($X_E \neq 0$) should be treated with reserve. While the exact solution is believed to be enclosed by the upper and lower bounds, in the absence of any definitive information, it is customary to use the arithmetic average of the two bounds. Broadly, the accuracy and reliability of the two bounds deteriorate with the increasing degree of shear-thinning behavior.

2. The second class of solutions is based on the so-called standard perturbation approach. In this method, it is tacitly assumed that the flow field produced by a fluid sphere translating in non-Newtonian fluids can be obtained by adding corrections to the Newtonian kinematics. Thus, the flow variables are written in the form of a series (similar to Equation 2.36) consisting of the Newtonian solution as the first term and the non-Newtonian material parameters as coefficients of subsequent terms. Bhavaraju et al. (1978) and Hirose and Moo-Young (1969) have used this scheme to analyze the motion of single bubbles and swarms of bubbles rising freely through stagnant power-law and Bingham plastic media, respectively. Such solutions are likely to be applicable only for a small degree of departure from the Newtonian fluid behavior. Nevertheless, this approach does afford a closed-form analytical solution in terms of a stream function, which in turn can be used to infer drag.

3. The third category of solutions relies on the linearization of the nonlinear viscous terms appearing in the momentum equations. The Newtonian flow field is used to calculate the shear-rate-dependent viscosity. In broad terms, this approach is also tantamount to the one with the Newtonian response as the base solution to which the effects arising from the non-Newtonian behavior are added. This approach too has been used extensively for analyzing the motion of fluid spheres in various generalized Newtonian fluid models including power-law (Hirose and Moo-Young, 1969; Kawase and Ulbrecht, 1981g; Jarzebski and Malinowski, 1987a,b), Ellis and Carreau viscosity equations (Kawase and Moo-Young, 1985). Intuitively, one would expect this class of solutions also to apply to a small degree of non-Newtonian fluid behavior only.

4. Finally, there has been a spurt in research activity in terms of the complete numerical solutions of the field equations for fluid spheres translating into a non-Newtonian continuous phase. Two aspects have received much attention in this sub-field. First, considerable research effort has been directed at the behavior of bubbles (Potapov et al., 2006; Singh and Denn, 2008; Tsamopolous et al., 2008; Dimakopoulos et al., 2013; Tripathi et al., 2015a; Cao et al., 2020; Nirmalkar et al., 2021). While some of these studies are aimed at elucidating the combined effects of elasticity and yield stress, a great majority of these have focused on developing a stopping criterion and very few have reported drag values for bubbles and drops (Cao et al., 2020; Nirmalkar et al., 2021). The second class of studies have focused on the numerical solution of the complete field equations for the steady translation of a fluid sphere in power law (Kishore et al., 2007b), Carreau–Yasuda (Ohta et al., 2003, 2005), etc.

Virtually all investigations in this field represent highly idealized flow conditions in so far that the fluid particles are assumed to be spherical or axisymmetric in shape, of constant size, and free from surfactants. This is in stark contrast to the fact that in practice a wide variety of shapes of bubbles and drops in non-Newtonian media are encountered (see Section 5.2) and naturally occurring surface-active agents are always present. Such limitations of the theoretical/analytical treatments coupled with the uncontrolled nature of experiments indeed have been the main impediments in performing detailed comparisons between theory and practice in this field. In the ensuing sections, we present some key results available in the literature.

5.5.1.2.1 Low Reynolds Number Regime

In most cases, the results have been expressed in the form of a drag correction factor Y (defined by Equation 5.29). When the continuous phase obeys the power-law behavior, Y is a function of the viscosity ratio (X_E) and the power-law index as

$$Y = f(X_E, n) \tag{5.37}$$

For a power-law fluid,

$$X_E = \frac{\mu_d}{m_c (V/R)^{n-1}} \tag{5.38a}$$

$$Re_{PL} = \frac{\rho_c V^{2-n} d^n}{m} \tag{5.38b}$$

Figure 5.16 shows the typical upper and lower bounds on the drag correction factor, Y, for power-law fluids (Mohan, 1974b). Typical analogous results for fluid spheres moving freely in unbounded Carreau

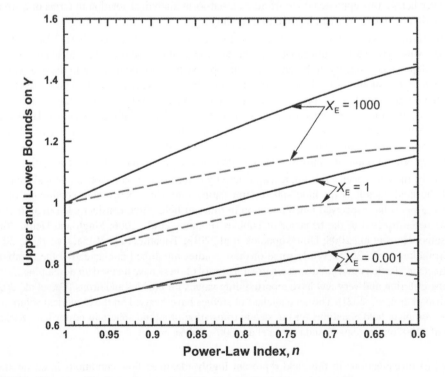

FIGURE 5.16 Drag correction factor for creeping fluid sphere motion in power-law media.

Fluid Particles in Non-Newtonian Media

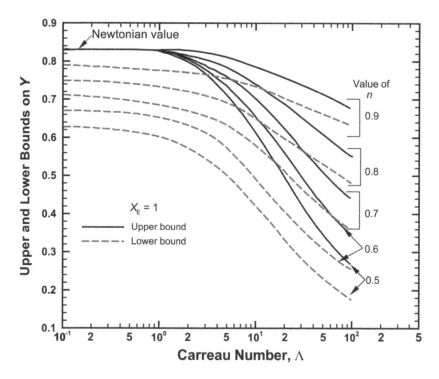

FIGURE 5.17 Typical upper and lower bounds on drag for fluid spheres moving slowly in Carreau model fluids. (Replotted from Chhabra, R.P. and Dhingra, S.C. *Can. J. Chem. Eng.*, **64**, 897, 1986).

model fluids (Chhabra and Dhingra, 1986) are shown in Figure 5.17. In this case, the drag correction factor, Y, is given by

$$Y = f(X_E, n, \Lambda) \tag{5.39}$$

and

$$X_E = \frac{\mu_d}{\mu_0}$$

An inspection of Figures 5.16 and 5.17 reveals that as the extent of shear-thinning increases (i.e., decreasing n, and/or increasing Λ), the two bounds diverge increasingly. As noted earlier, the use of an arithmetic average of the two bounds has often been suggested. Furthermore, it appears that the bounds remain unchanged for $X_E < 0.001$ and $X_E > 1000$ thereby suggesting that these limiting conditions correspond to the cases of gas bubbles and rigid spheres, respectively. In this regard, the results plotted in Figures 5.16 and 5.17 span the complete spectrum of particles, albeit the nature of idealizations involved severely limits the utility of these results for fluid spheres.

The closed-form expressions for the creeping motion of Newtonian fluid spheres obtained by Kawase and Ulbrecht (1981g) and Jarzebski and Malinowski (1987a,b) are really quite involved and cumbersome for intermediate values of the viscosity ratio. However, for the limiting case of gas bubbles ($X_E = 0$), these reduce to particularly simple expressions, as shown in Table 5.2, along with some of the other works, and these are plotted in Figure 5.18 for spherical bubbles rising in stagnant power-law fluids. Clearly, these results apply to the bubble sizes that are beyond the critical bubble diameter at which velocity discontinuity (if any) will occur, that is, to the bubbles with the shear-free surface. Similar comparisons for the Carreau model fluids have been reported elsewhere (Chhabra and Dhingra, 1986). In an

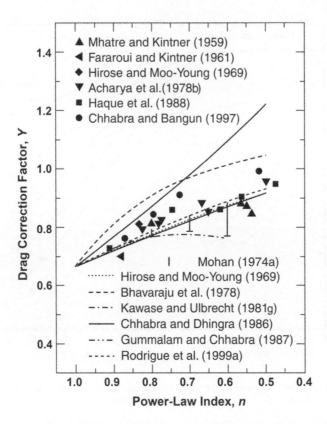

FIGURE 5.18 Comparison between various predictions and experiments for creeping bubble motion in power-law media.

extensive numerical study for a fluid sphere undergoing steady translation in power-law fluids, Kishore et al. (2007b) obtained drag results over wide ranges as: $5 \leq \text{Re}_{PL} \leq 500$; $0.1 \leq X_E \leq 50$; and $0.6 \leq n \leq 1.6$. They consolidated their numerical results by the following expression:

$$C_D = \left(\frac{2}{X_E + 1.85}\right)\left[1.55 X_E \left\{\frac{2^{n+2.6}}{\text{Re}_{PL}^{1.36}} + \frac{2^n}{\text{Re}_{PL}^{0.38}}\right\} + 1.55 \frac{2^{n+2}}{\text{Re}_{PL}^{0.78}}\right] \quad (5.40)$$

Equation 5.40 purports to correlate their numerical results with a mean error of ~13%. In addition to these studies, many investigators have examined the rising velocity behavior of swarms of spherical bubbles (Gummalam and Chhabra, 1987; Gummalam et al., 1988; Zhu, 1995, 2001) and of ensembles of spherical droplets (Chhabra and Dhingra, 1986; Jarzebski and Malinowski, 1986a,b, 1987a,b; Zhu and Deng, 1994; Zhu, 1995; Sun and Zhu, 2004; Kishore et al., 2006, 2008a,b) in stagnant power-law and Carreau model fluids. In the limit of a very small volume fraction (zero) of the dispersed phase, all these studies yield the results for single particles. While it is not possible to obtain closed-form expressions for drag, suffice it to add here that most of these are consistent with the results included in Table 5.2. Aside from these studies, as far as known to us, there is only one analysis available on the creeping motion of gas bubbles in Bingham plastic fluids (Bhavaraju et al., 1978), according to which the drag correction factor Y is given by

$$Y = \frac{2}{3}(1 + 1.61\text{Bi}) \quad (5.41)$$

where Bi, the Bingham number, is defined as $\tau_0^B d / V \mu_0$ and Equation 5.41 applies in the limit of Bi → 0. In a recent numerical study, Cao et al. (2020) numerically studied the buoyancy-driven rise of bubbles in

Herschel–Bulkley fluids up to about $Re_{PL} \sim 8$–9 and $Bi_{HB} \le \sim 7$ and put forward the following expression for Y:

$$Y = \frac{2}{3}\left(1 + 1.35 Bi_{HB}^{0.86}\right) \qquad (5.42)$$

The effective shear rate implicit in their analysis is $\sim V/d$ where d is the equal volume sphere diameter. No additional dependence on the power-law index was reported in the range $0.5 \le n \le 0.9$. Stein and Buggisch (2000) have elucidated the role of pressure oscillations in the liquid phase on the rising behavior of a spherical bubble in Bingham model fluids. This technique can cause the bubbles to move, which will otherwise remain stationary due to the yield stress of the liquid. The theoretical predictions are in qualitative agreement with the preliminary experimental results. In another numerical study, Tripathi et al. (2015b) used the volume of fluid (VOF) method to study the evolution of bubble shapes in a regularized Bingham fluid. At low Bingham numbers/strong surface tension effects, the bubble exhibits a steady-rise behavior with little deviation from a spherical shape. At the other extreme of high Bi/weak surface tension effects, both shape and velocity exhibit oscillating behavior. Tsamopolous et al. (2008), on the other hand, used the exponential regularization and predicted shapes of bubbles as functions of the Archimedes, Bingham, and Bond numbers. The limiting value of the Bingham number Bi ($=\tau_0/\rho_c gR$) was found to be 0.143. This value is independent of the Archimedes number, but it increases with the Bond number, i.e., for instance, $Bi_c = 0.21$ at $Bo = 50$. However, Dimakopoulos et al. (2013) used the Herschel–Bulkley fluid model in the axisymmetric steady flow regime and concluded that the limiting value of the Bi was independent of the power-law index in the Herschel–Bulkley fluid model but it varied with the bubble shape. This finding is at variance with that of Dubash and Frigaard (2004) with respect to the role of the power-law index (n), for their approximate analysis indicates that a spherical bubble of diameter, d, will not move in a Herschel–Bulkley fluid as long as $Bi_{HB} \ge 2^{(1-2n)/2}$. This is, however, a very conservative estimate. For a Bingham plastic fluid, this results in the critical value of the Bingham number of $1/\sqrt{2}$. However, both studies suggest the limiting value to be strongly influenced by the bubble shape.

In another numerical study dealing with the buoyancy-driven motion of spherical droplets in Bingham plastic fluids, Nirmalkar et al. (2021) reported extensive results on the stopping criterion and drag coefficient over wide ranges of conditions as $1 \le Re_B \le 150$; $Bi \le 50$ and $X_E = 0.1$ and $X_E = 10$. For a spherical bubble, they predict the limiting $Bi \sim 0.124$. In contrast, the experimental values ranging from ~ 0.07 to ~ 0.5 (Sikorski et al., 2009; Dubash and Frigaard, 2007) have been reported. Furthermore, this value seems to be influenced by the wall effects as well (Dubash and Frigaard, 2007). A general framework for estimating the critical yield number for arbitrary shape particles has also been postulated in the literature (Frigaard et al., 2017). Wang et al. (2019) reported the critical value of the limiting Bingham number to be ~ 0.153.

Before leaving this section, it is appropriate to mention here that the influence of surfactants on the drag behavior of single bubbles in inelastic and visco-elastic liquids has been examined theoretically (Rodrigue et al., 1997; Kishore et al., 2013) and experimentally (Rodrigue et al., 1996a; Tzounakos et al., 2004). Drag on an oscillating bubble in power-law fluids has recently been studied by Zhang et al. (2023).

5.5.1.2.2 High Reynolds Number Regime

Limited numerical work is available on the motion of Newtonian fluid spheres in shear-thinning media beyond the creeping flow regime. Astarita and Marrucci (1964) used the potential flow theory to evaluate the drag on a spherical bubble rising through quiescent power-law liquids. Under these conditions, the drag coefficient is given by K_n/Re_{PL} where K_n is a function of the power-law index only. For $n = 1$, $K_n = 48$, which is consistent with the well-known result for Newtonian liquids. For other values of n (<1), Table 5.3 shows the variation of K_n with the power-law index. For finite values of the Reynolds number, the governing equations for Newtonian fluid spheres (with clean interface) translating into quiescent power-law fluids have been solved numerically by Nakano and Tien (1970). However, the fact that their predictions in the limiting case of $n = 1$ (i.e., Newtonian continuous phase) do not compare well with the literature values (Feng and Michaelides, 2001) casts some doubts about the accuracy of their drag values for power-law fluids. Notwithstanding this uncertainty, in the range ($5 \le Re_{PL} \le 25$; $0.01 \le X_E \le 2$; $0.6 \le n \le 1$), the main findings of Nakano and Tien (1970) can be summarized as follows:

TABLE 5.3
Values of K_n for Bubbles

n	0	0.1	0.25	0.333	0.5	0.667	0.8	1.0
K_n	30.6	27.0	25.6	26.2	28.8	33.2	38.4	48.0

Source: Astarita, G., and Marrucci, G., Accademia Nazionale dei Lincei, Ser. VIII, 36, 836 (1964).

1. The friction drag coefficient contributes increasingly to the drag as the viscosity of the dispersed phase (i.e., the value of X_E) increases.
2. For a given value of Re_{PL} and X_E, the total drag coefficient is only slightly dependent on the value of the power-law index n.

Similar results for the Ellis model fluids are also available in the literature (Mohan and Raghuraman, 1976c). In another numerical study, Ohta et al. (2003, 2005) have studied the steady translation of a spherical drop in a stagnant shear-thinning liquid approximated by the generalized Cross–Yasuda viscosity equation. Weak inertial effects were also included in this study as the Reynolds number based on the dispersed phase properties was of the order of 10 or so. The drag behavior of the droplet was influenced by shear-dependent viscosity only when the average shear rate was outside the zero-shear viscosity region. These predictions are in qualitative agreement with their preliminary experiments. Two detailed numerical studies concerning the steady axisymmetric translation of spherical fluid spheres (bubbles and drops) are that of Dhole et al. (2007a,b) and Kishore et al. (2007b) in quiescent power-law fluids. However, surface tension forces are assumed to be large so that there is no departure from the spherical shape. The work of Kishore et al. (2007b) has been discussed previously culminating into Equation 5.40 linking the drag coefficient with the other pertinent parameters. In a sense, Dhole et al. (2007a,b) investigated the limiting case of $X_E=0$. Based on their numerical results spanning the conditions ($5 \leq Re_{PL} \leq 500$; $0.5 \leq n \leq 2$), they put forward the following correlation:

$$C_D = \frac{16}{Re_{PL}} \left\{ 1 + 0.27 Re_{PL}^{\left(\frac{1.41n}{(0.335n+3.76)}\right)} \right\} \tag{5.43}$$

The predictions of Equation 5.43 were found to be in moderate agreement with the experimental correlation of Dewsbury et al. (1999). Since the minimum value of $X_E=0.1$ is implicit in Equation 5.40, these two Equations can be used to interpolate the results for the intermediate values of the parameter X_E. Limited overviews of the available analytical, numerical, and experimental results for spherical and spheroidal bubbles have been documented by Sokovnin et al. (2012a,b, 2013) and Gollakota and Kishore (2017).

5.5.1.3 Visco-Elastic Continuous Phase

Some theoretical attempts have been made to assess the importance of visco-elastic effects on the steady translation of (Newtonian) fluid particles in a stagnant visco-elastic continuous phase. Most studies have assumed the particles to be spherical and valid only for the so-called weakly elastic conditions. Thus, for instance, Wagner and Slattery (1971) and Quintana et al. (1987) have analyzed the steady creeping translation of a droplet in visco-elastic media. Wagner and Slattery (1971) assumed both the dispersed and continuous phases to be visco-elastic fluids of grade 3 and have accounted for weak inertial effects. Their predictions about the shapes of droplets are in qualitative agreement with experimental results available in the literature. Quintana et al. (1987), on the other hand, investigated the steady translation of a Newtonian droplet in the four-constant Oldroyd model fluids. This study clearly shows that the terminal velocity of a Newtonian droplet may increase or decrease (hence the drag may reduce or enhance) as compared to the

Hadamard–Rybczynski value depending upon the degree of shear-thinning and the extent of visco-elastic behavior of the continuous phase. In particular, Quintana et al. (1987) demonstrated that for large values of X_E (i.e., approaching the solid sphere limit), the fluid elasticity reduces the drag below its Newtonian value which is consistent with the results for solid spheres reported in Chapter 4. Subsequently, Quintana (1991) and Quintana et al. (1992) asserted that in the limit of low Weissenberg numbers, both the surfactants and visco-elasticity cause a reduction in the settling velocity of a Newtonian fluid sphere, that is, the drag is increased. Pillapakkam (1999) has numerically studied the slow translation of a droplet in Oldroyd-B and FENE dumbbell-type visco-elastic fluids. The major thrust of his study was to examine the steady shapes attained by bubbles and drops in simple shear and Poiseuille flows. When the dispersed phase is Newtonian, these simulations clearly show that there are critical values of the Deborah and capillary numbers below which the bubbles and drops attain a steady (constant) shape while above these critical values, no steady shape is possible. Indeed, this study predicts most of the shapes that have been observed experimentally. However, no drag results have been reported in this work.

In contrast, numerous investigators have considered the analogous problem of gas bubbles. Early analyses are of perturbation type and assume the bubble shape to be spherical, i.e., surface tension effects to be strong. Within the range of their applicability, this approach predicts very little departure from the Newtonian response. For instance, Ajayi (1975) studied the creeping visco-elastic flow around a slightly deformed sphere. Tiefenbruck and Leal (1980b, 1982) have calculated the terminal velocity of a spherical bubble rising through an unconfined four-constant Oldroyd fluid as

$$\frac{V}{V_0} = \left[1 - \left(\frac{We^2}{75}\right)\left(18 - 25\alpha^2 - 2\alpha^2\varepsilon\right)\left(1 - \varepsilon\right)\right] \quad (5.44)$$

where V_0 is the corresponding Hadamard and Rybczynski velocity and α, ε, etc. are material parameters. The fluid elasticity is seen to reduce the rise velocity provided $\alpha > (18/(25 + 2\varepsilon))^{1/2}$. This result is in agreement with the earlier studies of Ajayi (1975), Hassager (1977), and Shirotsuka and Kawase (1974b), but is at variance with that of Hirose and Moo-Young (1969); the latter results, however, appear to be in error. Astarita (1966b) has analyzed the steady rise of spherical bubbles through Maxwell fluids using dimensional arguments. In recent years, Ohta et al. (2009, 2019) have not only presented an account of previous studies but also reported 3-D simulations of the shape and velocities of a buoyancy-driven bubble in a hybrid Carreau and FENE-CR model. In the absence of shear-thinning viscosity, the bubbles became increasingly long and pointed with the rising value of the Deborah number. For a shear-thinning fluid ($n=0.5$), bubbles were more spherical, though depending upon the relative magnitudes of elasticity and shear-thinning, all shapes observed in experiments were found to be plausible. You et al. (2008, 2009), on the other hand, have studied drop motion and deformation by carrying out stability analysis. Similar conclusions have been reached by Prieto (2015) for the 2-D translation of a drop in Oldroyd-B and FENE-type visco-elastic media.

5.5.1.4 Non-Newtonian Drops

The aforementioned discussion is restricted to the situation when the dispersed phase is Newtonian and the continuous phase is non-Newtonian. However, there are situations (such as in ink-jet printing, spraying, polymeric blends, and alloys) when the dispersed phase is non-Newtonian and the continuous phase is Newtonian, such as air or water. In view of the fact that most gases are Newtonian, this discussion is solely limited to the case of droplets. Some literature is available on this subject. For instance, Gurkan (1989, 1990) has studied the translation of power-law liquid droplets falling in a Newtonian continuous medium. In the range $10 \leq Re \leq 50$, $0.6 \leq n \leq 1$, and $0.1 \leq X_E \leq 1000$, while the degree of shear-thinning (i.e., the value of n) only has a minor effect on the drag and mass transfer results, the internal circulation in the fluid sphere is significantly suppressed. Subsequently, this work has been extended to reactive systems (Gurkan, 1990). The observation that the rheology of the dispersed medium has little influence on the drag is also in line with the scant experimental and numerical results (Gillaspy and Hoffer, 1983; Aminzadeh et al., 2012; Vamerzani et al., 2016). Similarly, Tripathi and Chhabra (1992a, 1994) have used the variational principles to obtain approximate upper and lower bounds on the drag of a single droplet and on a swarm of spherical droplets, when both phases exhibit power-law behavior. This study clearly brings out the influence of the rheology of the two phases on the terminal velocity of sedimentation of single and ensembles of droplets. However, the effects arising from the presence of surfactants were not included in this work. Similarly, Ramkissoon (1989a, 1989b) has studied the creeping flow of an incompressible Newtonian fluid over a Reiner–Rivlin fluid sphere and a slightly deformed sphere. Subsequently, this work has been extended to the case of a fluid sphere moving in a spherical container (Ramkissoon and Rahaman, 2001, 2003) to assess the extent of wall effects. In view of the perturbation method used, these analyses are also tantamount to the so-called weak flows. Pillapakkam (1999) showed that a visco-elastic drop moving in a Newtonian fluid deforms only very little as the Deborah number is progressively increased as long as the capillary number is held constant. Preliminary experimental results on the shapes of visco-elastic drops in Newtonian media are not inconsistent with such theoretical/numerical predictions (Rodrigue and Blanchet, 2001; Sostarecz and Belmonte, 2003). Figure 5.19 shows the shapes of visco-elastic drops (0.5% polyacrylamide in 20/80 glycerin/water mixture) falling in corn oil in the presence of sodium dodecyl sulfate. Indeed, these shapes are in line with the observations and predictions of Sostarecz and Belmonte (2003). There has been a renewed interest in

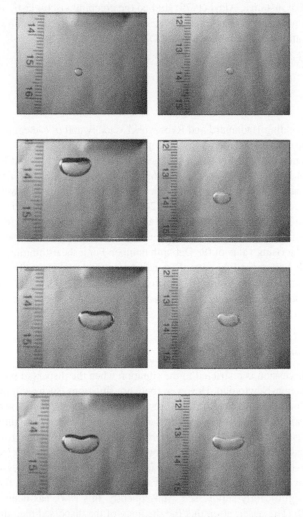

FIGURE 5.19 Shapes of visco-elastic drops falling in still Newtonian corn oil. The dispersed phase is 0.5% Separan MG-700 in 20/80% glycerin/water mixture. The concentration of SDS is 0 and 300 ppm, respectively, for the left and right columns. The volumes of droplets are 30, 100, 700, 900 mm^3 for the left column and 30, 100, 300, 900 mm^3 for the right column. (Photograph courtesy: Professor Denis Rodrigue, University of Laval, Quebec.)

studying the effect of dispersed phase rheology on the dynamics of fluid spheres undergoing steady translation in another medium. Thus, Kishore et al. (2008d) revisited the studies of Tripathi and Chhabra (1992a, 1994) by numerically solving the complete field equations for the case of a fluid sphere (power-law) settling steadily in another immiscible power-law medium up to the (external) Reynolds number, $Re_{PL} \approx 200$. The contribution of the dispersed phase rheology was found to be rather small as far as the terminal velocity (drag coefficient) was concerned. Mukherjee and Sarkar (2011) and others (Norouzi and Davoodi, 2018; Norouzi et al., 2019a,b) have investigated, both numerically and experimentally, the shape of visco-elastic drops in Newtonian or another visco-elastic medium. Norouzi and coworkers concluded that small drops tend to retain their spherical shape whereas large drops tend to develop dimples in the rear due to the nonuniform distribution of normal stress on the interface. The analysis of Mukerjee and Sarkar (2011) suggests the increasing role of extensional flow in the rear of the drop, and the rear of the drop is pulled in and this propensity is countered by the surface tension force. This is in line with the experimental results of Sostarez and Belmonte (2003). When the surface tension is weak, the dimples grow in an unbounded manner leading to doughnut-like shapes. Finally, Mukherjee and Sarkar (2011) used the Deborah number–capillary number coordinates to consolidate their results on the shapes of drops. Some of the trends are in line with the experimental results of Ortiz et al. (2016). Smagin et al. (2011) found the drops made up of visco-plastic liquids falling slowly in the Newtonian continuous phase to attain steady oblate and prolate shapes with the increasing Bingham number.

5.5.2 Experimental Results

A considerable amount of experimental results on the free fall or rise of fluid particles in non-Newtonian media has accrued during the past 50–60 years (see Table 5.1). In most cases, the two-parameter power-law model has been used to describe the shear-rate-dependent viscosity of the continuous phase. More significantly, a thorough examination of the literature data pertaining to liquid drops reveals that the value of viscosity ratio, X_E, lies in the range of 10^{-4} to 10^{-3} and in the creeping flow regime ($Re_{PL} \ll 1$). This is particularly so of the data obtained by Warshay et al. (1959), Mhatre and Kintner (1959), Mohan et al. (1972), Acharya et al. (1978b), Chen (1980), Chhabra and Bangun (1997), and Wanchoo et al. (2003). Hence, these results can be treated effectively as those for gas bubbles, and some of these are included in Figure 5.18 where the agreement is seen to be about as satisfactory as can be expected in view of the idealizations used. In later years, extensive new experimental data have been reported on the freely rising bubbles in stagnant power-law liquids (Chen, 1980; De Kee and Carreau, 1993; Dewsbury et al., 1999; Margaritis et al., 1999; Dziubinski and Orczykowska, 2002; Dziubinski et al., 2002, 2003; Fan et al., 2010; Wenyuan et al., 2010; Li et al., 2012). Likewise, the literature abounds with numerous empirical correlations that purport to predict drag on bubbles and the free-rise velocity of bubbles. Recognizing the fact that the bubbles are necessarily not always spherical, Dewsbury et al. (1999) introduced the so-called horizontal bubble diameter when calculating the projected area of the bubble in the definition of the drag coefficient. Based on their own experimental results embracing the ranges of conditions as $0.16 \leq n \leq 1$ and $Re_{PL} \leq \sim 1000$, they put forward the empirical correlation for drag coefficient for a single bubble as

$$C_D = \frac{16}{Re_{PL}}\left(1 + 0.173 Re_{PL}^{0.66}\right) + \frac{0.413}{1 + 16300 Re_{PL}^{-1.09}} \quad (5.45)$$

They also reported that the drag coefficient seemed to approach a constant value of approximately 0.95 at about $Re_{PL} \approx 60$ and therefore Equation 5.45 is really valid for $Re_{PL} < 60$. Similarly, Rodrigue (2002, 2004) collated much of the literature data on bubbles in power-law fluids and argued that Equation 5.35 developed for bubbles in Newtonian viscous media was applicable to the power-law fluids as long as the characteristic viscosity is calculated at the shear rate of (V/d). Indeed, extensive comparisons between the predictions of Equation 5.35 for power-law fluids and the experimental results of Barnett et al. (1966), Haque et al. (1988), Margaritis et al. (1999), Miyahara and Yamanaka (1993), Rabiger and Vogelpohl (1986), and Rodrigue et al. (1996b, 1999a) showed the average error to be of the order of approximately

23%, albeit the maximum deviations due to increasing departure from spherical shape at high Reynolds numbers (Chen, 1980) could be as high as 100%. This inference is further reinforced by other independent studies (Dziubinski et al., 2003; Karamanev et al., 2005). On the positive side, Equation 5.35 obviates the necessity of the evaluation of the so-called horizontal bubble diameter (required in Equation 5.45), but it does not predict the constant value of C_D beyond $Re_{PL} > 60$ as observed and reported by Dewsbury et al. (1999). The validity of Equation 5.35 has received further support from the recent numerical (Islam et al., 2020) and independent experimental (Zhang et al., 2008a,b) results.

Undoubtedly, over the years, a significant volume of experimental results has been documented in the literature, which includes quasi-steady-state data on bubble shape and velocity in various model non-Newtonian fluids (see Table 5.1). Yet it has not been possible to develop drag correlations and shape maps that can rival the corresponding developments in Newtonian fluids, e.g., see Clift et al. (1978), Michaelides (2006), and Loth (2008b). Some of these studies are summarized here. Zhang et al. (2008b) reported on the bubble dynamics in inelastic and visco-elastic shear-thinning polymer solutions. They used the Carreau viscosity equation (Equation 1.17 without μ_∞) to capture the shear-thinning viscosity and argued that the time parameter λ in Equation 1.17 was sufficient to account for visco-elasticity. Using the equal volume sphere diameter as the bubble size (d) together with the mean shear rate of ($2V/d$), they were able to consolidate their results for inelastic shear-thinning fluids by re-defining the Reynolds number, Re_M, as

$$\text{Re}_M = \frac{\rho_c V d}{\mu_0 \left[1 + (2\lambda V/d)^2\right]^{\frac{n-1}{2}}} \tag{5.46}$$

and the drag coefficient is now given by:

$$C_D = \frac{16}{\text{Re}_M}\left(1 + 0.12\,\text{Re}_M^{0.6}\right) \tag{5.47}$$

An additional correction was introduced for visco-elastic effects. Subsequent comparisons with independent experimental results reveal the limitations of Equations 9.46 and 9.47. Similarly, Wenyuan et al. (2010), Li et al. (2012), and Sun et al. (2015) have developed drag correlations for single bubbles in shear-thinning (inelastic power-law) fluids, but these have not yet been validated extensively and therefore must be treated with reserve. This section is closed by mentioning the different approach adopted by Battistella et al. (2020) for correlating drag for bubbles in power-law liquids. Based on the two limiting regimes wherein the drag coefficient C_D (Re_{PL}) is solely a function of the Reynolds number or of the Eötvös number, C_D(Eo), one can now postulate:

$$C_D = \left[C_D^2(\text{Re}_{PL}) + C_D^2(\text{Eo})\right]^{\frac{1}{2}} \tag{5.48}$$

where the Reynold number-dependent drag coefficient was obtained from the direct numerical simulations for a power-law fluid. The Eötvös number-dependent drag coefficient is given by (Dijkhuizen et al., 2010):

$$C_D(\text{Eo}) = \frac{4\text{Eo}}{9.5 + \text{Eo}} \tag{5.49}$$

Battistella et al. (2020) reported a good match between their numerical predictions and Equation 5.48 in the range $0.5 \leq n \leq 1.5$.

Similarly, in addition to the aforementioned studies dealing with drag, some information is also available on the detailed flow field around a bubble (Hassager, 1979; Bisgaard and Hassager, 1982; Tiefenbruck and Leal, 1982; Belmonte, 2000; Herrera-Velarde et al., 2003). The effect of surfactants on bubble/drop motion in non-Newtonian ambient medium has been considered by Kawase and Ulbrecht (1982), Quintana (1991), Rodrigue et al. (1996a, 1997), Rodrigue and Blanchet (2002), and Tzounakos et al. (2004). The combined effects of shear-thinning and visco-elasticity have been studied experimentally by Acharya et al. (1977, 1978b), Yamanaka et al. (1976a), Yamanaka and Mitsuishi (1977), and Velez-Cordero et al. (2011a,b, 2012).

In recent years, there has been a growing interest in flow visualization and the detailed flow field structure around the bubbles rising in stagnant polymer solutions. Thus, for instance, Sousa et al. (2004, 2006) have used a combination of PIV and shadowgraphy methods to investigate the nature of flow around Taylor bubbles rising in two aqueous CMC solutions (0.8% and 1%). Their work relates to the behavior of a Taylor bubble (100 cc) ascending at the rate of 182 and 160 mm s^{-1}, respectively, in the two solutions. Their results clearly show that the flow in the nose (prolate-spheroid) region is pretty insensitive to the fluid rheology. On the other hand, the flow field in the wake region is significantly different in polymer solutions as compared to that in an equivalent Newtonian fluid. Thus, for instance, the trailing edge is neither flat nor concave (as seen in Newtonian liquids) in polymer solutions. Furthermore, the trailing edge was seen to oscillate, expand, and contract thereby implying the rotational movement of the cusp. Since Sousa et al. (2004) did not measure any visco-elastic characteristics of the CMC solutions, they used the time parameter in the Carreau viscosity equation to estimate the Deborah numbers to be about 0.2 to 0.4. A negative wake is certainly seen in their 1% solution. Subsequently, Sousa et al. (2005) have reported a more detailed study encompassing 0.1%–1% CMC solutions and this time also included the values of the first normal stress difference to evaluate the fluid characteristic time using the approach of Leider and Bird (1974) (Equation 1.45). However, their main findings remain more or less the same as noted above. Figure 5.20 shows representative results from their study for a range of values of the bubble Reynolds number in support of the aforementioned description. Numerical simulations of Taylor bubbles

FIGURE 5.20 Structure of the flow field on the trailing edges for Taylor bubbles rising in CMC solutions. (a) Re = 4, (b) Re = 10, (c) Re = 24, (d) Re = 70, (e) Re = 114, (f) Re = 714. (Photograph courtesy: Professor J.B.L.M. Campos, University of Porto, Porto.)

in generalized Newtonian fluids have been reported among others by Vasil'chenko and Potapov (1996), Abhishek et al. (2015), Zamankhan et al. (2018), Amani et al. (2019), Majumdar and Das (2019), and Sontti and Atta (2019) who have attempted to estimate the film thickness separating the bubble and the tube wall. Similarly, Lin and Lin (2005) have reported the detailed structure of the flow field (in terms of vorticity, shear rate, and viscosity contours) for a 2-D bubble ascending in an aqueous polyacrylamide solution. They reported the angle of the conical negative wake to gradually shrink with the increasing bubble size, and the negative wake finally disappears as the bubble transits to the spherical-cap bubble shape. For a spherical-cap bubble, they also reported the formation of a pair of closed vortices. Evidently, such detailed information would facilitate the development of suitable theoretical frameworks to describe the bubble dynamics in rheologically complex media in the future.

Many excellent experimental studies dealing with the buoyancy-driven motion of bubbles in yield stress fluids have also been reported (Table 5.1). Much of this literature is concerned with the establishment of the stopping criterion and deformation of bubbles in Bingham and Herschel–Bulkley model fluids (Dubash and Frigaard, 2007; Sikorski et al., 2009; Samson et al., 2017; Lopez et al., 2018; etc.) and these results have been discussed earlier along with the corresponding theoretical developments. Alicke et al. (2012) and Mougin et al. (2012) have alluded to the possible visco-elastic effects in such systems on the formation, detachment, and quasi-steady shape and velocity of bubbles. In a significant study, Tran et al. (2015) have collated much of the literature data on the bubble size–velocity relationship and have attempted to develop a drag correlation along the same lines as that for a solid sphere discussed in Chapter 3. Their final correlation takes the form:

$$C_D = \frac{5Y(n)}{Re^*}\left(1 + \frac{3}{16}Re^*\right) \tag{5.50}$$

where $Re^* = \dfrac{Re_{PL}}{1 + 0.823 Bi_{HB}}$ and $Y(n)$ is the drag correction factor for a solid sphere (Chapter 4), though the corresponding results for a bubble are also available (Table 5.2). The factor of $(1+(3/16) Re^*)$ is added empirically here based on the Oseen approximation for a solid sphere in Newtonian fluids to account for weak inertial effects.

In contrast to the reasonable experimental results on bubble motion, little experimental information is available on drops in falling non-Newtonian media (see Ohta et al., 2003), and in most of these studies, the value of X_E seldom exceeded 0.1 and therefore these results relate more to gaseous spheres than to fluid spheres. A few correlations are available in the literature for drag on droplets (Licht and Narsimhamurthy, 1955; Klee and Treybal,1956; Acharya et al., 1977, 1978b; Wanchoo et al., 2003), but none of these has been tested adequately using independent experimental data. Thus, these are too tentative to be included here.

Lavrenteva et al. (2009) have examined the behavior of Newtonian drops in Carbopol solutions (modeled as Herschel–Bulkley fluid) and found that the terminal velocity exhibits its peak value at about 50% blockage (drop diameter/tube diameter). Their previous numerical study (Potapov et al., 2006) suggests that drops attain only a quasi-steady behavior in such fluids.

5.6 Bubble and Drop Ensembles in Free Motion

In contrast to the numerous single-particle studies, the analogous problem involving ensembles of bubbles and drops moving freely in non-Newtonian media (such as that encountered in bubble columns, sparged reactors, and in three-phase fluidized beds, retention of gas bubbles in sewage sludges) has received only scant attention. From a theoretical standpoint, in addition to the field equations, a mathematical description of interparticle interactions is also needed. Thus, theoretical treatments are available only for idealized and preconceived particle arrangements such as cell models (Happel, 1958; Kuwabara, 1959) and periodic cubic arrays (Sangani and Acrivos, 1983; Mo and Sangani, 1994; Chhabra, 1995b) even for Newtonian liquids. For instance, the free surface cell model reported by Happel (1958) has

Fluid Particles in Non-Newtonian Media

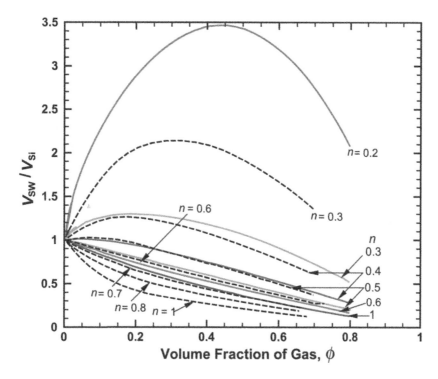

FIGURE 5.21 Dependence of free-rise velocity of bubble swarms on gas fraction in stationary power-law media. - - - - creeping flow region. (From Gummalam, S. and Chhabra, R.P., *Can. J. Chem. Eng.*, **65**, 1004, 1987) —— Potential flow region. (From Chhabra, R.P., *Can. J. Chem. Eng.*, **76**, 137, 1998).

been used extensively for estimating the free-rise velocity of swarms of spherical bubbles at low and high Reynolds numbers (Marrucci, 1965; Manjunath et al., 1994; Chhabra, 1998) and of drop ensembles with and without surfactants in the creeping flow region (Gal-Or and Waslo, 1968), when the continuous phase is Newtonian. Likewise, the zero-vorticity cell model due to Kuwabara (1959) has been used by LeClair and Hamielec (1971) for swarms of spherical bubbles rising in stagnant Newtonian liquids at finite Reynolds numbers. Mo and Sangani (1994) have used the Stokesian dynamic approach to calculate the drag on random arrays of bubbles in the creeping flow regime. Suffice it to add here that in the limit of zero Reynolds number, the cell models and the periodic/random array predictions of bubble drag are not too different, and indeed differ from each other by less than 100%. Some experimental results on bubble swarms are also available (Krishna et al., 1999a). The free surface cell model has also been extended to predict the steady fall or rise of ensembles of fluid spheres in various viscous non-Newtonian media including power-law model fluids (Bhavaraju et al., 1978; Jarzebski and Malinowski, 1986a,b, 1987a; Gummalam and Chhabra, 1987; Manjunath and Chhabra, 1992; Zhu and Deng, 1994), in Carreau viscosity models (Jarzebski and Malinowski, 1987b; Gummalam et al., 1988; Zhu, 1995, 2001), and in Bingham plastic fluids (Bhavaraju et al., 1978). Based on the average of upper and lower bound results (Gummalam and Chhabra, 1987), Figure 5.21 shows the rise velocity of a swarm of spherical bubbles in quiescent power-law liquids as a function of the gas content. These results are in excellent agreement with the analogous numerical results (Zhu, 2001). The maximum in the rise velocity seen in Figure 5.21 is attributable to the competition between the thin film of liquids separating two bubbles at high gas content and the lower effective viscosity due to the enhanced levels of shearing. Indeed, this behavior is in qualitative agreement with the scant experimental results available in the literature (Deckwer, 1992; Job and Blass, 1992, 1994; Radl et al., 2007; Velez-Cordero et al., 2012). Similarly, Chhabra (1998) extended the approach of Marrucci (1965) to obtain the high Reynolds number asymptote for bubble swarms in power-law liquids and these results are also shown in Figure 5.21. The dependence of the swarm velocity on the power-law index and the fractional gas content is qualitatively similar to that in the creeping regime as seen in Figure 5.21, though the power-law rheology is more important in the creeping flow regime than that at high Reynolds numbers. Limited results are also available for bubble swarms for the intermediate values of the Reynolds number up to $Re_{PL} = 200$ in power-law fluids (Manjunath and Chhabra, 1992; Kishore et al., 2008a). A detailed analysis of these results shows that the potential flow limit of drag coefficient is seen to be reached at as low a Reynolds number as 50 in highly shear-thinning liquids. Indeed, the smaller the value of the power-law index n, the lower is the value of the Reynolds number at which the potential flow solution begins to apply. Kishore et al. (2008a) also presented a drag correlation for bubble swarms as a function of the Reynolds number, gas fraction, and power-law index. They were also able to recast their results in order to estimate a priori the swarm rise velocity without necessitating an iterative approach. Subsequently, they (Kishore et al., 2008b) reported similar results for ensembles of mono-dispersed fluid spheres ($0.1 \leq X_E \leq 50$). The effect of contamination on the rise velocity of bubble swarms has been studied by Nalajala et al. (2014) whereas Mao and Chen (2005) have presented a modified cell model.

Needless to say, neither the results for finite values of the viscosity ratio X_E nor those obtained with the other generalized Newtonian fluid models (Ellis or Carreau models) are amenable to such an explicit representation as shown in Figure 5.21 for power-law fluids. Therefore, iterative methods and graphical results must be used to extract the value of swarm velocity in such conditions. This is simply due to the fact that the unknown velocity appears in the viscosity ratio and in the additional dimensionless groups such as the Ellis number or the Carreau number.

Numerous theoretical attempts have also been made at estimating the effective viscosity of dilute dispersions and emulsions containing small spherical gas bubbles or droplets, dispersed in an incompressible and immiscible Newtonian medium. Taylor (1932, 1954) obtained the following theoretical expression that may be regarded as an extension of Einstein's expression for solid suspensions to the viscosity of dilute dispersions of small fluid spheres in another immiscible Newtonian liquid:

$$\mu^* = \mu \left[1 + 2.5\phi \left(\frac{X_E + 0.4}{X_E + 1.0} \right) \right] \tag{5.51}$$

Equation 5.51 clearly includes the Einstein's formula as a special case corresponding to $X_E \to \infty$. On the other hand, for gas-in-liquid dispersion where $X_E \approx 0$, it reduces to

$$\mu^* = \mu [1 + \phi] \tag{5.52}$$

Subsequently, Oldroyd (1953) has estimated the viscous and elastic properties of dilute emulsions. Sangani and Lu (1987) have used the method of singularity distribution for calculating the effective viscosity of an ordered emulsion. The dilatational characteristics of bubbles in Newtonian and non-Newtonian suspending media have been studied by Prud'homme and Bird (1978). Similarly, Ducloue et al. (2015) have studied the rheology of bubble-laden yield stress fluids. They noted that the gas fraction and the stiffness of bubbles (capillary number) exert a significant influence on the bulk rheology of such systems. Excellent reviews of the developments on the rheological behavior of emulsions and aerated liquids are available in the literature (Pal, 2006; Schramm, 2014). Lastly, though a number of investigators (Buchholz et al., 1978; Godbole et al., 1984; Kelkar and Shah, 1985; Haque et al., 1987, 1988; Job and Blass, 1992; Deckwer, 1992) have experimentally examined the hydrodynamic behavior of bubble columns using non-Newtonian continuous phase, sufficient details are, however, not available to enable a direct comparison between theory and practice.

5.7 Coalescence of Bubbles and Drops

Coalescence of bubbles and drops inevitably occurs in applications involving ensembles of bubbles and drops. Depending upon the envisaged application, coalescence may be desirable (such as in promoting separation) or detrimental (such as in chemical reactors where it is desirable to have a large interfacial area) to a process. It is reasonable to expect that coalescence would be significantly influenced by the rheological complexities of the continuous phase and the type and distribution of surface-active agents present in the system. Irrespective of the application, bubble coalescence exerts a strong influence on the value of the volumetric mass transfer coefficients, especially in fermentation and allied applications (Adler et al., 1980; Shuler and Kargi, 2002). The scant information available on these important processes, with special reference to non-Newtonian liquids, is reviewed in the following sections.

5.7.1 Bubble Coalescence

In a process application involving gas–liquid dispersions, bubbles are constantly colliding with each other. Depending upon the bubble size and their velocity and frequency of such collisions, bubbles may separate from each other or coalesce following a collision. It is well known that the phenomenon of coalescence entails three stages: initial contact between bubbles, controlled essentially by the hydrodynamics of the bulk liquid, which results in a film of the thickness of a few microns separating the two bubbles. The second step is the gradual thinning of this film to a few Angstroms. The rate of film thinning or drainage is determined by the hydrodynamics of thin films. The final stage is the rupturing of the film leading to the coalescence of the two bubbles. Evidently, the rate of film drainage and thinning in the second step determines whether coalescence will occur or not. If the time required to drain the film to reach the rupturing thickness level is longer than the period of contact, the two bubbles may separate rather than coalesce. The last step (that is film rupturing) is usually much faster than the other two steps.

Based on these ideas, numerous models (Marrucci, 1969; Dimitrov and Ivanov, 1978; Sagert and Quinn, 1978; Chesters and Hofmann, 1982; Narayanan et al., 1974; Oolman and Blanch, 1986; Anthony et al., 2017) have been proposed in the literature. However, most of these deal with the coalescence of bubbles in Newtonian continuous media. Moreover, the developments are restricted to highly idealized conditions such as two bubbles growing on adjacent orifices or inline bubbles, or involving bubbles in a predetermined geometrical configuration. Despite their limited applicability to real-life applications, such studies have provided useful insights into the physics of the process and have added to our understanding. Broadly speaking, the overall qualitative predictions are in line with experimental observations in this field, at least in low-viscosity systems where generally turbulent conditions prevail. Blass (1990), Trambouze (1993), and Chaudhari and Hofmann (1994) have presented useful reviews of the advances made in this field. Stewart (1995) and Zhang and Fan (2003) have treated the issue of bubble interactions in low-viscosity Newtonian liquids in some detail. In contrast, only a few investigations on the coalescence of bubbles in non-Newtonian media have so far been reported in the literature. Acharya and Ulbrecht (1978) merely reported experimental observations on the coalescence of inline bubbles and drops in visco-elastic media with no accompanying explanations of results. De Kee et al. (1986, 1990a) and De Kee and Chhabra (1988) have also provided extensive results on the coalescence of bubbles in purely shear-thinning and visco-elastic polymer solutions. In this study, the bubbles of three different gases (air, carbon dioxide, and nitrogen) were formed and released from predetermined geometrical configurations of orifices. The volume of a bubble required to achieve coalescence with another bubble (of fixed volume) at a fixed height above the orifice was measured for a range of geometrical parameters and rheological properties of the continuous phase. Whether the coalescence would occur or not was found to be strongly influenced by the initial orifice separation whereas the rheological characteristics and the surfactants influenced the process of coalescence by way of altering the shapes of bubbles, wakes, and their rise velocities. Though no quantitative or predictive framework was developed in these studies, Figure 5.22 shows the process of coalescence between two and three bubbles.

In a series of papers, Li and coworkers (Li et al., 1997a,b, 2001, 2002; Li and Qiang, 1998; Li, 1998, 1999; Funfschilling and Li, 2001) have extensively studied the interactions and coalescence of inline bubbles released through an orifice submerged in inelastic CMC solutions and in highly elastic polyacrylamide solutions. Most of their work relates to the creeping flow regime conditions, and, therefore, they have argued that coalescence, in this case, is inherently different from that seen in low-viscosity liquids where turbulent conditions generally occur (Hinze, 1955). Some of their key findings can be summarized as follows: the rise velocity of a single bubble is influenced by the time interval between two bubbles (or the frequency of bubbles, as seen in Figure 5.15) and this effect is more prominent in

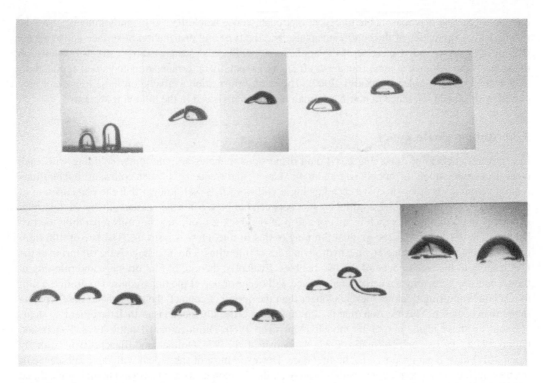

FIGURE 5.22 Qualitative photographs showing coalescence between two and three bubbles in a 1% polyacrylamide solution. Top: $V_{b1}=9.33\,\text{cm}^3$; $V_{b2}=3.53\,\text{cm}^3$; and initial horizontal separation 2.4 mm, Bottom: $V_{b1}=V_{b2}=V_{b3}=7.5\,\text{cm}^3$ with initial horizontal separation 3 mm. (From De Kee, D. and Chhabra, R.P., *Rheol. Acta*, **27**, 656, 1988. With permission.)

visco-elastic systems than that in inelastic liquids. Under suitable conditions, a time interval as large as 90 s is required to prevent interactions between the two successive bubbles. Conversely, coalescence occurred almost always if the bubbles were released at shorter intervals than the critical time interval. Similar time dependence of the rise velocity of bubbles in visco-elastic liquids was documented by Barnett et al. (1966) and Carreau et al. (1974). Based on their PIV and birefringence measurements, Li et al. (2001) and Funfschilling and Li (2006) found the flow fields around a bubble to be strongly influenced by the non-Newtonian properties of the liquid. Thus, for instance, while the flow in the front of the bubble is almost similar in Newtonian and in non-Newtonian liquids, the flow in the main wake is downward, surrounded by a hollow cone of upward flow in visco-elastic liquids. This conical upward flow zone first appears on the two sides of the bubble and progressively extends in the downstream direction (Funfschilling and Li, 2001, 2006; Kemhia et al., 2006). The role of a negative wake is not entirely clear, but it does not seem to prevent interactions and the coalescence of bubbles rising in a chain. Indeed, these intrinsic differences seem to suggest that some new mechanisms need to be postulated to model coalescence in non-Newtonian systems, since neither time dependence nor negative wake phenomena are encountered in Newtonian systems. Similarly, while most of the currently available models purport to explain coalescence in Newtonian systems under turbulent conditions, the analogous non-Newtonian flows occur almost invariably under laminar flow conditions without any turbulence. Li and coworkers have introduced the notion of residual stress. After the passage of a leading bubble, the elasticity (memory?) holds the shear-thinning process for a fixed period so that the local viscosity decreases and causes the trailing bubble to rise faster than the leading bubble. Consequently, the two bubbles interact and coalesce in due course of time. Based on this picture, Li et al. (2001) have attempted a cognitive approach to model bubble coalescence in these systems. The preliminary comparisons between predictions and their observations appear to be encouraging. Subsequently, Lin and Lin (2003, 2005) have reported the detailed flow field for a pair of inline bubbles ascending in a visco-elastic polyacrylamide solution in a 2-D column. They have reported extensive results on the temporal evolution of the shapes of two inline bubbles showing how the bubble shapes change, which ultimately facilitates coalescence. Under appropriate combinations of shear-thinning and visco-elasticity, the trailing bubble accelerates and gets sucked into the wake of the leading bubble. The liquid in between the two bubbles is pushed out radially and circulated back to the trailing bubble, thereby providing an upward push. Naturally, during the course of these events, both bubbles transit from one shape to another.

In recent years, Fan et al. (2016, 2020) and Sun et al. (2017) have studied the interaction between two bubbles released from two adjacent orifices into (inelastic) carboxymethyl cellulose solutions. Based on their visual observations, they postulated four steps leading to coalescence between two bubbles, namely spherical expansion, rapid mergence, overall growth of the merged bubbles and lift off. Thus, the coalescence occurs at the orifice itself. The coalescence efficiency appears to be influenced mainly by orifice spacing and gas flow rate. Naturally, the immobilization of the bubble surface by surfactants will impede coalescence (Zhu et al., 2018). Similarly, Liu et al. (2015b) have carried out 3-D VOF simulations of multiple horizontally aligned bubbles in power-law fluids. They have identified the conditions in terms of bubble size, power-law index, and geometric configuration of orifices for coalescence to occur. On the other hand, Velez-Cordero et al. (2011a,b, 2012) have investigated interactions between two and multiple bubbles rising in constant viscosity Boger fluids. They reported shearing levels to be higher in bubble swarms than that for a single bubble. Sun et al. (2020a) have examined the effect of differences in bubble sizes on the hydrodynamic interactions between three bubbles arranged in a horizontal plane and rising in a shear-thinning fluid. Islam et al. (2015) have studied the free rise of two identical side-by-side bubbles in xanthan solutions. Whether the two bubbles will be attracted or repelled is strongly influenced by their initial separation. For initial separation of ~1.5 to $2\,d$, the bubbles tend to drift apart from each other due to strong repulsion. Some of these trends are also borne out by their numerical simulations.

5.7.2 Drop Coalescence

In this case also, the collision between two drops may lead to coalescence or they may rebound. Similar to the case of bubbles, drop coalescence is always preceded by a period of film drainage in which the film of continuous phase separating the two drops gradually thins further. When the film becomes sufficiently thin, the rupturing takes place. Typical rupturing thickness is of the order of 100–1000 Å (Scheele and Leng, 1971). The rupturing may result in complete or partial coalescence. Rebounding usually occurs when the rate of film drainage is slow.

Owing to its pragmatic significance in liquid–liquid extraction, there has been considerable study of the coalescence between a drop and a plane interface, and the resulting voluminous literature has been reviewed by Hartland and Hartley (1976). Inter-drop coalescence has been investigated less extensively than that for bubbles. Informative and general descriptions of drop coalescence involving Newtonian media are available in a number of references (Kintner, 1963; Tavlarides and Stamatoudis, 1981; Godfrey and Hanson, 1982; Grace, 1983; Mobius and Miller, 1998). As far as we know, there has been no study reported in the literature on drop coalescence in non-Newtonian media, except a very few results reported by Acharya and Ulbrecht (1978). Das et al. (1987) have developed a model for the coalescence of drops in stirred dispersions involving non-Newtonian polymer solutions. Bazhlekov et al. (1999, 2000), on the other hand, have examined the drainage and rupture of thin (Newtonian) films separating two power-law fluid droplets under the action of a constant force. The main finding appears to be that the critical rupture thickness is only weakly dependent on the power-law rheology of the dispersed phase. Similarly, Yildirim and Basaran (2001) have studied the deformation and breakup of stretching bridges (between two disks) of Newtonian and non-Newtonian liquids. The liquid bridge undergoes deformation until it breaks as the two disks are being pulled apart. Notwithstanding the highly idealized conditions, such studies add to our understanding of film thinning in the context of coalescence.

In contrast, Holenberg et al. (2011) and Singh and Denn (2008) have studied interactions between two vertically aligned drops and bubbles in visco-plastic fluids. Depending upon the initial separation (below a critical value). Holenberg et al. (2011) reported both drops to fall slowly with little or no interaction, i.e., the fluid-like regions surrounding each drop do not overlap in this case. As expected, if these two regions overlap, the two drops interact hydrodynamically. This is consistent with the findings of Atapattu et al. (1995), Hariharaputhiran et al. (1998), Gumulya et al. (2007), Al-Matroushi and Borhan (2009),

Lavrenteva et al. (2009), Lavrenteva and Nir (2010), etc. who have experimentally studied coalescence of two inline drops and bubbles (of two different diameters) in inelastic and visco-elastic polymer solutions in a tube. In particular, they studied the effect of rheology, collision velocity, and size ratio on the coalescence time. The coalescence time increases with the increasing ratio of the trailing bubble size to the leading bubble size.

5.8 Breakage of Drops

Drop breakage is important in any process in which surface area must be created, such as in pneumatic atomization and in the production of fine emulsions. The mode by which the breakage occurs depends on the conditions of experimentation. For instance, in stagnant media, drop breakage occurs due to the perturbations prevailing at the surface between a denser phase with the less-dense phase underneath. This perturbation grows in amplitude if the wavelength of the disturbance exceeds a critical value $\left(2\pi\sqrt{(\sigma/\Delta\rho g)}\right)$. As the disturbance grows, the drop surface becomes spiked or indented. Drop splitting occurs if the disturbance grows rapidly enough to allow the disturbance to be carried to the side. Using this picture of breakup, a method has been developed for predicting the maximum stable drop size in Newtonian fluids (Grace, 1983). Another mode of breakage involves the application of shear and elongational flow fields. Rumscheidt and Mason (1961) have carried out extensive experimental studies on the rotation, deformation, and breakage of liquid drops in shear fields and hyperbolic flows. Finally, drop breakage also occurs when the drops are subjected to turbulent conditions in agitated liquid–liquid dispersions and in pipe flows (Guido and Greco, 2004). Excellent review articles on the mechanisms and also summarizing the bulk of the literature on drop breakup in Newtonian media are available (Hinze, 1955; Rallison, 1984; Stone, 1994).

Some of these ideas have also been extended to drop breakage in non-Newtonian media. The scant information available on the drop and bubble deformation and breakage in this field has been reviewed by Zana and Leal (1974). Flumerfelt (1972), Lee and Flumerfelt (1981), and Lee et al. (1981) have studied the deformation of droplets in simple shear fields and the stability of visco-elastic threads, whereas Lagisetty et al. (1986), Koshy et al. (1988a,b, 1989), Muralidhar et al. (1988), and Shimizu et al. (1999) have studied drop breakage in stirred tanks and have obtained expressions for the maximum stable size for viscous Newtonian and non-Newtonian drops in stirred vessels. When the continuous phase is visco-elastic, the maximum stable drop size is observed to be larger than that found in Newtonian media which is attributed to the high extensional viscosity of the continuous phase. Over the past few decades or so, the issue of bubble/drop deformation and breakup has received considerable attention from a theoretical standpoint also, mainly due to the development of robust numerical solution procedures. Consequently, a wealth of information has accrued on this subject, most of which, however, relates to drops. Thus, in a series of articles, Toose et al. (1995, 1996, 1999) have numerically simulated the time-dependent deformation of axisymmetric non-Newtonian (Oldroyd-B) droplets in a Newtonian medium subject to an axisymmetric flow field. At small capillary numbers, deformation is governed by two relaxation times. Subsequently, this work has been extended to compound drops (a Newtonian core encapsulated by a visco-elastic substance). Varanasi et al. (1994) have studied experimentally the deformation of a visco-elastic drop (Boger fluid) in a uniform shear flow. For a fixed value of X_E, there is a critical shear rate above which it was difficult to break the drop. Conversely, for a given shear rate, the larger the value of X_E, the easier it was to break the drop. Li and Renardy (2000) have studied the shear-induced rupturing of drops in visco-plastic fluids. Similarly, Khayat and coworkers have used boundary element methods to study the planar drop deformation in a range of confined geometries including in a convergent–divergent channel or in the screw channel of a mixing extruder (Khayat et al., 1998a,b, 2000; Khayat, 2000a,b). Indeed, this group has investigated the effects of shear and elongational flow fields when both the dispersed/continuous phase are Newtonian or Maxwellian or one Newtonian and the other non-Newtonian. Similarly, Gonzalez–Nunez et al. (1996) have experimentally studied the deformation of nylon drops in polyethylene melts, with and without an interfacial agent, in the presence of an extensional flow. Delaby et al. (1994, 1995) have also studied the deformation of drops dispersed in immiscible polymeric blends. They reported the predictions to be extremely sensitive to the parameters of the rheological model. The influence of uni- and bi-axial extension on the deformation of a Newtonian drop has been analyzed by

Ramaswamy and Leal (1999a,b) and Ha and Leal (2001). The effect of surfactants has been elucidated by Milliken et al. (1993). The deformation and breakup of conducting drops in an electric field have been experimentally investigated by Notz and Basaran (1999) and Ha and Yang (2000). The role of non-Newtonian properties was found to be rather small. Similarly, there is a sizeable body of information now available on the deformation of bubbles in non-Newtonian media (Noh et al., 1993; Favelukis and Albalak, 1996a,b; Favelukis and Nir, 2001). Skelland and Kanel (1990) have experimentally studied the minimum impeller speed required for complete dispersion of non-Newtonian liquids in a stirred tank.

5.9 Motion and Deformation of Bubbles and Drops in Confined Flows

Drops falling freely through another immiscible liquid in the presence of confining walls and with the imposition of external flow fields (e.g. Couette flow, extensional flow, etc.) exhibit a range of complexities including migration away or toward a solid boundary, for instance. The non-Newtonian properties of the ambient fluid seem to further complicate the response of drops in such complex flow geometries. van Wijngaarden and Vossers (1978) and Coutanceau and Hajjam (1982) reported on the hydrodynamic behavior of a single gas bubble rising through a visco-elastic medium in a cylindrical tube. In the absence of inertial and surface tension effects, gas bubbles were found to develop a thin pointed tail, akin to that observed by Wilkinson (1972) in the case of Newtonian droplets falling freely in non-Newtonian liquids. Furthermore, the shear-thinning viscosity together with the wall effects has a tendency to thin down the bubble whereas the visco-elasticity has the reverse influence. Based on their flow visualization experiments, a region of "negative flow" was observed in the vicinity of the rising bubble. Chan and Leal (1979, 1981) and Olbricht and Leal (1982, 1983) have also carried out extensive theoretical and experimental work on the behavior of a single drop in various flow configurations and conditions. Based on theory (Chan and Leal, 1979) and experimental observations (Goldsmith and Mason, 1962; Gauthier et al., 1971a,b), it is now generally agreed that particles in Poiseuille flow of visco-elastic media have a propensity to move toward the centerline of a pipe. Subsequently, Chan and Leal (1981) have studied the behavior of a drop in Newtonian and visco-elastic media undergoing shearing flow between concentric cylinders. The earlier work of Karnis and Mason (1967) suggested that a Newtonian drop, in another immiscible Newtonian medium in Couette flow, attained an equilibrium position close to the centerline. New experimental results encompassing wide ranges of fluid behavior as well as the gap width also suggest the final equilibrium position somewhere in between the centerline and the inner cylinder. The equilibrium position, however, keeps moving toward the inner cylinder with the increasing profile curvature. These observations are in agreement with the theoretical frameworks developed by Chan and Leal (1981) and Olbricht and Leal (1982). The behavior of visco-elastic drops in a Newtonian continuous phase was found to be strongly influenced by the non-Newtonian characteristics of the dispersed phase. The deformable drops suspended in visco-elastic media undergoing Couette flow, on the contrary, display outward migration, attaining an equilibrium position away from the inner cylinder. Olbricht and Leal (1982) have measured the additional pressure drop and some other parameters for the Newtonian droplets suspended in Newtonian and visco-elastic media in a straight tube of comparable diameter. Significant macroscopic differences have been observed in the behavior of Newtonian droplets flowing in Newtonian and visco-elastic media. The visco-elasticity of the suspending medium exerted a strong influence on the shape as well as on the lateral position assumed by the drops that in turn influenced the additional pressure drop and drop velocity. In order to simulate some aspects of the processes encountered in the enhanced oil-recovery applications, Olbricht and Leal (1983) also studied the behavior of single drops suspended in Newtonian and visco-elastic media flowing in tubes with its diameter varying sinusoidally. Irrespective of the suspending medium rheology, the shape of the drop was strongly influenced by the value of the capillary number. In Newtonian systems, the drop deforms to "squeeze through" the throat of the tube and then follows the contour of the tube wall whereas prolate spheroidal-shaped drops are encountered in visco-elastic media. Furthermore, owing to the memory effects, the drop responds slowly to the conduit geometry as compared with the Newtonian suspending medium. Some interesting time effects have also been observed in this flow configuration (Olbricht and Leal, 1983). Mukherjee and Sarkar (2014) have thus studied the migration of a visco-elastic drop in a Newtonian fluid shear flow near a wall. In recent years, considerable interest has also been shown in the dynamics of Taylor bubbles in narrow ducts. The main objective is to be able to predict the liquid film thickness separating the bubble and the duct wall, the shape of the bubble, etc., for example, see the experimental studies of Sousa et al. (2006), Sontti and Atta (2019), and numerical results of Mukundakrishnan et al. (2009), Abhishek et al. (2015), Amani et al. (2019), Majumdar and Das (2019), Zamankahn et al. (2018), etc.

Finally, it is worthwhile to point out here that this chapter has been concerned mainly with the formation and behavior of fluid particles under the influence of the earth's gravitational field. However, a considerable body of literature is also available on the hydrodynamics of bubbles and drops in low-gravity fields, with special reference to a range of potential applications in material processing in low-gravity conditions. This has been treated thoroughly in a monograph (Shankar Subramanian and Balasubramanian, 2001). Other important discussions dealing with the fluid mechanics of bubbles, drops, and compound drops are also available (Johnson and Sadhal, 1985; Sadhal et al., 1997; Crowe et al., 1998; Mobius and Miller, 1998; Sirignano, 1999; Liu, 2000; Frohn and Roth, 2000; Kulkarni and Joshi, 2005).

5.10 Conclusions

This chapter has provided an overview of the activity relating to the formation, shapes, and rising behavior of fluid particles—bubbles and drops—in the non-Newtonian continuous phase. Starting with the formation of bubbles and drops at submerged orifices and nozzles in stagnant Newtonian liquids, the available limited information seems to suggest that some of these ideas can also be extended to non-Newtonian liquids, at least to purely viscous systems that do not entail any visco-elasticity and time-dependent effects. Visco-elasticity seems to add to the complexity due to a different detachment process. The interactions and coalescence of bubbles in viscous Newtonian fluids are strongly influenced by visco-elasticity via the formation of a negative wake. The extensive literature on bubble growth or collapse clearly shows that the process is primarily inertia dominated, thereby rendering the rheological properties of the ambient medium to be of little consequence, though all else being equal, shear-thinning behavior appears to slow down the growth of a bubble. During the free rise/fall in stagnant liquids, entirely different shapes are observed in non-Newtonian systems depending upon the degree of shear-thinning and visco-elasticity of the continuous medium. However, these await suitable theoretical developments. Likewise, the literature on the terminal velocity/volume behavior, and drag coefficient–Reynolds number relationship also abounds with contradictions and conflicting inferences. Even the simplest case of a Newtonian fluid sphere undergoing steady translation in a quiescent power-law fluid has received limited attention. Consequently, there are no results that can rival the classical Hadamard–Rybczynski solution for a fluid sphere sedimenting in a non-Newtonian medium at low Reynolds numbers. Limited results are now available for the clean fluid sphere beyond the creeping flow regime. It is perhaps appropriate to say that the experimental work with bubbles and drops is much more demanding than that for rigid particles. This is so partly due to the additional complications arising from the presence of surface-active agents and due to their ability to deform during their translation. The important areas of multiparticle systems such as ensembles of bubbles and drops encountered in bubble columns and in liquid–liquid extraction and the associated processes of coalescence and breakup have now begun to receive attention. There has been a renewed interest in studying the behavior of bubbles in yield stress fluids, bubble swarms, and Taylor bubbles in generalized Newtonian fluids. Our understanding of the behavior of fluid particles in rheologically complex media is far from being satisfactory and it is nowhere near as coherent and matured as that in Newtonian fluids.

Nomenclature

a, b — Dimensions of a fluid particle in two orthogonal directions (m)

$\text{Bi} = \dfrac{\tau_0^B d}{\mu_0 V}$ — Bingham number (-)

$\mathrm{Bi}_{HB} = \dfrac{\tau_0^H}{m(V/d)^n}$ Modified Bingham number for Herschel–Bulkley fluids (-)

$\mathrm{Bo} = \dfrac{\rho D^2 g}{\sigma}$ Bond number, Equation 5.11 (-)

$\mathrm{Ca} = \dfrac{\mu_0 V}{\sigma}$ Capillary number (-)

C_D Drag coefficient (-)

d Bubble or drop diameter (m)

D Orifice diameter (m)

D_n Nozzle diameter (m)

$\mathrm{De} = \dfrac{N_1}{2\tau}$ Deborah number (-)

Eo Eötvös number, Equation 5.18b (-)

El Ellis number (same as Chapter 2)

e Eccentricity of bubble (-)

F Dimensionless factor, Equation 5.35a (-)

$\mathrm{Fr} = \dfrac{V^2}{gD}$ Froude number, Equation 5.11 (-)

G_1 Dimensionless parameter, Equation 5.20b (-)

G_2 Dimensionless parameter, Equation 5.21 (-)

g Acceleration due to gravity (m/s²)

m Power-law consistency coefficient (Pa s^n)

$\mathrm{Ma} = \dfrac{\Delta\sigma}{Rm}\left(\dfrac{R}{V}\right)^n$ Marangoni number (-)

Mo Morton number, Equation 5.18c (-)

$\mathrm{Mo}_{PL} = \dfrac{g^{3n-2} m^4}{\rho^{2-n}\sigma^{2+n}}$ Morton number for power-law fluids (-)

n Power-law flow behavior index (-)

$N_c = \dfrac{4V_c \rho g}{\pi D^2 p_h}$ Dimensionless chamber volume (-)

N_1 First normal stress difference (Pa)

p Pressure (Pa)

p_h Pressure at orifice (Pa)

Q Volumetric gas flow rate (m³/s)

R Bubble/drop radius (m)

R_c Critical radius of bubble, Equation 5.23 (-)

Re Reynolds number for a Newtonian medium, Equation 5.18a (-)

Re_M Modified Reynolds number, Equation 5.46 (-)

Re* Modified Reynolds number, Equation 5.50 (-)

Re_{PL} Reynolds number for power-law liquids (-)

t Time (s)

t_c Detachment time of bubble or drop (s)

V Free-rise velocity (m/s¹)

V_b Volume of bubble (m³)

$V_b^* = \dfrac{V_b}{D^3}$ Dimensionless bubble volume (-)

V^* Dimensionless rise/fall velocity, Equation 5.35 (-)

V_c Chamber volume (m³)

V_{bf}	Final bubble volume (m³)
V_d	Drop volume (m³)
V_0	Superficial velocity through an orifice plate or Hadamard–Rybczynski velocity (m/s)
We	Weissenberg number, Equation 5.44; Weber number $\rho_c dV^2/\sigma$, Equation 5.36 (-)
x	Distance (m)
X_E	viscosity ratio of dispersed to continuous phase (-)
Y	Drag correction factor (-)

Greek Symbols

α	Fluid model parameter, Equation 5.41; also Ellis model fluid parameter (-)
α_0	Parameter, Equation 5.25 (-)
γ	Density ratio, Equation 5.18e (-)
ε	Fluid parameter, Equation 5.44 (-)
μ	Viscosity (Pa s)
μ^*	Viscosity of mixture (Pa s)
μ_0	Zero-shear viscosity (Pa s)
ρ	Density (kg/m³)
$\Delta\rho$	Density difference (kg/m³)
σ	Surface tension (N/m)
λ	Fluid relaxation time (s)
Λ	Carreau number, Equation 2.45 (-)
τ	Shear stress (Pa)
τ_{0B}	Yield stress of Bingham fluid (Pa)
τ_{0H}	Yield stress of Herschel–Bulkley fluid (Pa)
ϕ	Volume fraction of dispersed phase (-)
ψ	Dimensionless number, Equation 5.27 (-)

Subscripts

b	Bubble
c	Continuous phase or critical value
d	Dispersed phase or drop
E	Pertains to extensional behavior
g	Gas
l	Liquid
r	r-component
s	Solid
θ	θ-component
si	Single bubble
sw	Swarm of bubbles

6

Non-Newtonian Fluid Flow in Porous Media and Packed Beds

6.1 Introduction

In the preceding chapters, the discussion has been restricted to the hydrodynamic behavior of single particles, with occasional reference to the motion of ensembles, in various non-Newtonian fluids. In this chapter, the single- and two-phase flow of non-Newtonian fluids through porous media, a subject of overwhelming practical significance, is considered. Porous media have often been simulated using packed columns of well-characterized particles of regular shape (spheres and cylinders, for instance) as well as that made up of naturally occurring rocks, stones, etc. The flow of fluids in a porous medium is encountered literally everywhere, both in nature and technology. The phenomena of the uptake of water and the flow of sap and gums in trees, the flow of blood in blood vessels and in various organs with porous boundaries, and oxygenation of blood are some examples that immediately come to mind. Further examples are found in the flow of fluids through textile fabrics, pulp fibers, woven and nonwoven mats (see Pierce, 1947; Cheikhrouhou and Sigli, 1988; Picaro and van de Ven, 1995; Epps and Leonas, 1997; Termonia, 1998; Caputo and Pelagagge, 1999; Brasquet and Le Cloirec, 2000), and in coffee and cigarette filters. Likewise, in the technology world, porous media flow applications abound. The classical examples include groundwater hydraulics (Springer et al., 1998; Barr, 2001a, 2001b), flow in fractured rocks (Zimmerman, 2000; Yu, 2008) and their contamination by non-Newtonian oil pollutants (Theodoropoulou et al., 2001; Gioia and Urciuolo, 2004), flow in saturated aquifers (Li and Helm, 1998), aerosol filtration (Brown, 1998; Raynor, 2002), filtration of slurries, sludges, and polymer melts using sand packed beds, screens, and metallic filters (Churaev and Yashchenko, 1966; Lorenzi, 1975; Khamashta and Virto, 1981; Kolodziej, 1986; Kozicki, 1988; Kiljanski and Dziubinski, 1996; Khuzhayorov et al., 2000; Auriault et al., 2002; Fadili et al., 2002; Chase and Dachavijit, 2003, etc.), and in the application of polymer flooding for enhanced oil recovery (Sorbie 1991; Nilsson et al., 2013; Skauge et al., 2018; Afolabi et al., 2019). Of course, the widespread use of packed and fluidized beds to carry out diverse operations in chemical, food, and biochemical process engineering applications may also be mentioned in passing. For instance, drying of wheat and other agricultural products is carried out by passing a hot gas (mostly air) through a bed of grains (Giner and Denisienia, 1996), apples and chicory roots (Verboven et al., 2004), beds of poppy and soybean seeds (Khatchatourian and Savicki, 2004; Sacilik, 2004, etc.), cotton seeds (Tabak et al., 2004), and coffee beans (Agullo and Marenya, 2005). The polymer and ceramic processing industry are inundated with numerous steps wherein a liquid permeates and flows through a porous matrix, for example, during consolidation of warp-knitted reinforced laminates (Rozant et al., 2001; Nielsen and Pitchumani, 2002; Idris et al., 2004), impregnation of compressible fiber mats with a resin (Parnas and Phelan, 1991; Lin et al., 1994; Phelan and Wise, 1996; Choi et al., 1998; Michaud and Manson, 2001; Michaud and Mortensen, 2001; Abrate, 2002), and in the production of fiber composites via the liquid metallurgy route (Reed, 1993; Bhat et al., 1995; Nagelhout et al., 1995). Last but not the least, the flow of polymer solutions on their own and together with a gas is extensively encountered at various stages of the drilling and enhanced oil recovery operations (Burcik, 1968, 1969; Jennings et al., 1971; Dabbous, 1977; Dreher and Gogarty, 1979; Schramm, 2014; Ates and Kelkar, 1998; Baca et al., 2003; Grattoni et al., 2004, etc.). This limited list of porous media flows is concluded by mentioning the use of packed beds as a calming section to eliminate turbulence in a fluid stream before entering a test section (Schmid et al., 1999), or to disentangle polymer melts before processing (Done et al., 1983; Bourgeat and Mikelic, 1996; Goutille and Guillet, 2002), which is also common. The flow of fluids in ligaments, tendons, tumors, and biological tissues (Chen et al., 1998a; Khaled and Vafai, 2003), in paper sheets (Sayegh and Gonzalez, 1995; Reverdy-Bruas et al., 2001), and in gel permeation chromatography columns (Hoagland and Prud'homme, 1989) provides additional applications of porous media flows. In view of such a wide occurrence of porous media flows, it will be no exaggeration to say that the phenomenon is ubiquitous!

It is readily recognized that the understanding and mathematical modeling of transport processes in porous media requires a detailed description of both the porous medium and the process itself. Ever since the pioneering work of Darcy in the nineteenth century, a vast body of knowledge concerning the fluid flow in porous media has accrued, and excellent treatises dealing with various aspects of porous media per se and the phenomenon of fluid flow through it are available in the literature, albeit a large part of this body of knowledge relates to the flow of Newtonian (air and water) fluids (Collins, 1961; Bear, 2018; Scheidegger, 1974; Greenkorn, 1983; Bear and Bachmat, 1990; Adler, 1992; Dullien, 1992; Kaviany, 1995; Nield and Bejan, 1995; Vafai, 2005; Xue et al., 2021). A historical perspective has recently been presented by De Boer (2003). Clearly, a detailed description encompassing all aspects of transfer processes in porous media is beyond the scope of this work. Besides, in keeping with the overall objectives and the general philosophy of the present work, attention is given here mainly to what one might call macroscopic treatment of fluid flow phenomena in porous media. More specifically, our primary concern here is to establish methodologies for the prediction of the resistance to flow in commonly used porous media for the flow of single-phase non-Newtonian and gas/non-Newtonian liquid mixtures, though this flow has also been exploited to extract the detailed pore size distribution by using yield stress fluids. Analogous phenomena of heat and mass transfer in porous media are discussed in detail in Chapter 8.

We begin with the definition and macroscopic characterization of a porous medium, followed by a terse account of the current status of the Newtonian fluid flow in porous media that, in turn, sets the stage for dealing with the flow of non-Newtonian liquids in such systems.

6.2 Porous Medium

6.2.1 Definition of a Porous Medium, Its Classification, and Examples

One can define a porous medium in various ways; the simplest of all being as a solid or a structure with holes in it. More rigorous definitions are also available. For instance, according to Dullien (1992), a structure or a material is called porous if it meets at least one of the following two requirements:

1. It has the so-called "voids" or "pores" that are imbedded in a solid matrix, or
2. It allows various fluids to penetrate through the interstitial spaces. For instance, a fluid should be able to "flow" through a septum made of the material in question. Such a medium is also called permeable porous material.

Notwithstanding the fact that all solids and semisolids have interstitial spaces that are accessible to some ordinary liquids and gases by molecular diffusion mechanism, the distinction between a solid and a porous matrix (or a porous solid) is unambiguous, provided the permeation of a fluid by the viscous flow mechanism, as stipulated in (2) above, is used as the defining criterion for a porous structure. Furthermore, a porous structure that is homogeneous, uniform, and isotropic is known as the ideal porous medium.

With this definition of a porous medium, it is truly remarkable that with the exception of metals, some plastic and glassy substances, and some dense rocks, all materials encountered in everyday life (in technology and in nature) are porous to varying extents. Broadly speaking, porous media can be further classified, somewhat arbitrarily though, as unconsolidated and consolidated, and as being ordered and random media. The unconsolidated porous media is exemplified by glass bead packs, made up of regular packings (such as Raschig rings and Berl saddles), beach sand, catalyst pellets, etc. whereas most of the naturally occurring rocks such as sand-, lime-, and marble-stones are good examples of consolidated media. Also, there are several examples of man-made consolidated media including paper, cloth, bricks, and earthenware. Loosely speaking, these two types of media essentially differ by the "ease of flow,", and the unconsolidated media, being more porous or permeable, permits flow with less resistance. The ordered and random media, on the other hand, differ from each other if there is a discernible sense of order or a correlating factor. Thus, random media are structures without any order; such materials are indeed very scarce as one can always find a correlation factor for a given medium. Bread and cakes, biological tissues, and skin seem to be good examples of random media. The commonly encountered media such as packed columns, wood, coal, textiles, and leather possess a discernible degree of order present in them and hence are classed as ordered media. Finally, there are situations when it is convenient to classify a porous medium as granular or fibrous depending upon whether it is made up of grains or of fibers and rods. Rocks and stones and sand packs are good examples of a granular medium whereas paper, textile, leather, and glass wool are representative of fibrous porous media.

6.2.2 Description of a Porous Medium

Clearly, a porous medium is characterized by specifying the manner in which "voids" are present in the matrix, their location, size or size distribution, shape, interconnectedness, etc. It should be recognized that only the open-ended pores contribute to flow whereas the so-called dead pores (with one end closed) only influence the processes of heat and mass transfer. Since our main concern here is to develop an understanding of the phenomenon of fluid flow in porous media, consideration will only be given to those features of porous media that influence the flow of fluids.

It has long been known that there are two distinct levels of description of a porous medium: microscopic and macroscopic. The microscopic-level description of a porous medium is really equivalent to a statistical description of the pore size distribution, albeit the description itself may be quite arbitrary. The second approach, that is the macroscopic description, involves the use of bulk quantities averaged over scales much larger than the size of pores. Both approaches are complementary to each other, and are used extensively; the choice, however, varies from one application to another depending upon the purpose of modeling. For instance, the use of microscopic description is a necessity while trying to understand the physics of fluid flow at a molecular level whereas the macroscopic description of a porous medium is quite adequate for engineering design calculations involving fluid flow in porous media. Once again, bearing in mind our (biased) goals, the macroscopic approach for describing porous media is more appropriate in the present context. Interested readers, however, are referred to the excellent treatments concerning the techniques used for microscopic-level characterization of porous media (see van Brakel, 1975; Greenkorn, 1983; Adler, 1992; Dullien, 1992; Bear, 2018; von Seckendorff and Hinrichsen, 2021).

Since the macroscopic description of a porous medium is really nothing more than the microscopic description integrated over sizes much larger than the dimensions of a pore, it is important to identify an appropriate scale to reach the point of the so-called macroscopic description. One obvious implication of this approach is that a microscopically nonideal medium may well be treated as an ideal one at the macroscopic level. Numerous workers (Whitaker, 1970, 1998; Slattery, 1972; Liu and Masliyah, 1996a,b, 1998; Travkin and Catton, 1998; Bear, 2018; Xue et al., 2021) have described the averaging procedures used for macroscopic description of porous media and for volume averaging of the field equations whereas Greenkorn and Kessler (1970) have related the nonideal features with the permeability of the porous medium.

Several macroscopic parameters have been proposed to characterize nonideal porous media (Dullien, 1992), but the following four have gained wide acceptance in describing the flow of single-phase and multiphase fluids in a porous medium in terms of gross engineering parameters. Conversely, their values can be evaluated from single-phase fluid (Newtonian) flow experiments.

1. *Porosity* is a measure of the effective pore volume or void volume per unit volume of the matrix. Depending upon the nature of the porous medium, the porosity (also known as voidage) may vary from near zero to almost unity. For instance, certain rocks, sandstones, etc. may have very low values (~ 0.15–0.2) whereas fibrous beds, glass wool, metallic foams, structured packings, for instance, may have high values of voidage (~ 0.9 or even higher). Within the context of fluid flow, it is important to distinguish between the so-called "interconnected" or "effective"

voids (which contribute to fluid flow) and the so-called "nonconnected" voids; the latter obviously are passive pores as far as the flow is concerned. The pores connected only from one side are known as "blind" or "dead-end" pores that make negligible contribution to the flow processes. Evidently, when a single value of the porosity is used to characterize a porous medium, the detailed structure of the matrix is unimportant. Thus, it is quite possible that two porous media having identical values of porosity may have completely different microstructures, for example, pore size distribution, type of networks, etc. Finally, while it is convenient to define a mean (or bulk) voidage for engineering applications, strictly speaking, porosity generally varies in the radial direction, being almost unity near the confining walls. This is particularly so for narrow-packed beds characterized by low values of column diameter to particle diameter ratios (Bu et al., 2014; Yang et al., 2015, 2016; Das et al., 2017; Guo et al., 2017a,b, etc.). A range of experimental methods is available to measure mean porosity and porosity profiles (Bories et al., 1991; Taud et al., 2005; Nguyen et al., 2005). The effects of particle size distribution on porosity have been reviewed by Dias et al. (2004).

2. *Permeability* is simply defined as the conductance of the porous medium via Darcy's law as

$$\frac{Q}{A} = \left(\frac{k}{\mu}\right)\left(\frac{\Delta p}{L}\right) \tag{6.1}$$

Equation 6.1 is a physical (empirical) law, akin to Newton's law of viscosity, relating the volumetric flow rate (Q) through a porous medium having a cross-sectional area normal to the flow (A) under the pressure gradient ($\Delta p/L$); μ is the viscosity of the fluid, and k is the permeability of the porous medium. In other words, it is a measure of the resistance to fluid flow, and generally depends upon the pore size distribution, length, shape, entrances, exits of the pores, etc. Intuitively, it is reasonable to anticipate a connection between the porosity and permeability, since a medium with zero porosity will have zero permeability. However, the porosity cannot be estimated from the permeability alone and vice versa. Additional information about the structure of the porous medium is required to establish the link between these two macroscopic parameters. The permeability of a porous medium is expressed in terms of "darcy." A porous material is said to have the permeability of one darcy if a pressure difference of 1 atm results in a flow of $1\,cm^3/s$ of a fluid having a viscosity of 1 cP through a cube (of porous matrix) having sides 1 cm in length. In SI units, it is expressed as m^2 and 1 darcy $\approx 10^{-12}\,m^2$. Typical values of permeability range from $10^{-11}\,m^2$ for fiber glass to $10^{-14}\,m^2$ for silica powder and limestone. For unconsolidated porous media, the porosity is determined directly from a knowledge of the mass (or volume) of the medium, and its permeability is evaluated by using Equation 6.1 together with experiments. However, some indirect techniques such as those based on the reflected waves of oblique incidence, tomography, etc. are also available (Greenkorn, 1983; Dullien, 1992; Fellah et al., 2003; Acharya et al., 2004; Felix and Munoz, 2005; von Seckendorff and Hinrichsen 2021; Xue et al., 2021). Though Equation 6.1 portrays an empirical relation between flow rate and driving force, as envisioned by Darcy in 1856, it can be obtained via the application of the homogenization theory (Bourgeat and Mikelic, 1996). Finally, the factor (k/μ) is also known as the mobility.

3. *Tortuosity* is a measure of the tortuous zig-zag paths traversed by fluid elements in a porous matrix. It is defined as the ratio of the average length of the flow paths to the actual length of the porous medium in the direction of mean flow. Clearly, tortuosity can be viewed as a macroscopic measure of both the sinuousness of the flow path and the pore size variation along the direction of flow. Once again, though tortuosity bears some correlation with permeability and porosity, one cannot be predicted from the knowledge of the other alone. Clearly, the tortuosity also depends on voidage and approaches unity as the voidage approaches unity. This relationship has been explored in detail among others by Foscolo et al. (1983), Epstein (1984, 1989), Agarwal and O'Neill (1988), Liu and Masliyah (1996b), and Pisani (2011). For instance, Pisani

(2011) suggested the transport coefficient (permeability) to be proportional to voidage and the characteristic size of the void space, and inversely proportional to the tortuosity factor. Since both voidage and tortuosity depend on the structure of the bed, he defined a structure factor as $\left(\frac{\varepsilon}{T}\right)$. Similarly, a particularly simple relation between tortuosity (geometric) and voidage attributed to Bruggeman (1935) is given as:

For a random bed of spheres:

$$T = \left(\frac{1}{\sqrt{\varepsilon}}\right) \qquad (6.2a)$$

For a random bed of cylinders:

$$T = \left(\frac{1}{\varepsilon}\right) \qquad (6.2b)$$

Tjaden et al. (2016), among others, have discussed the limitations of these simple relations. Note, however, that both these expressions meet the limiting requirements of $\varepsilon \to 0$, $T \to \infty$ and $\varepsilon \to 1$, $T \to 1$. Tang and Lu (2014) have also reviewed the dependency of the tortuosity factor on the voidage and power law index for power-law fluids. The tortuosity of a medium is thus strongly influenced by particle shape, size, and their orientation, and the type of packing (loose or tight) in relation to the direction of flow. For instance, for plate-like flat particles, the tortuosity is greater when they are oriented normal to the direction of flow than when the particles are packed parallel to flow. However, unlike porosity and permeability, tortuosity is a model parameter and hence it is not an intrinsic property of the medium. Nor can it be measured directly.

In a recent review, Ghanbarian et al. (2013) have identified four types of tortuosity factors: geometric, hydraulic, electrical, and diffusive. Naturally, the geometric tortuosity factor is expected to be the smallest, followed by the hydraulic tortuosity; evidently the hydraulic tortuosity is relevant to the present discussion. Ghanbarian et al. (2013) have argued that much confusion exists in the literature regarding the meaning of tortuosity because of indiscriminate use of different types of tortuosities.

4. *Specific surface area* of a bed influences both its structure and the resistance it offers to fluid flow. It is defined as the surface area per unit volume of the bed and is directly related to the specific area of the packing via the bed porosity. It is readily seen that, for a given particle shape, the specific surface area is inversely proportional to the particle size. For instance, highly porous fiber glasses have specific surface areas in the range $5-7 \times 10^4 \, m^2/m^3$, while compact lime stones (4%–10% porosity) have specific surface areas in the range ~$2-20 \times 10^5 \, m^2/m^3$. There are occasions when it must be recognized that due to the overlapping of particles (plates oriented normal to flow, for instance) the full geometric surface area is not exposed to the fluid. Some models of fluid flow account for this feature (Comiti and Renaud, 1989, 1991).

Though many other macroscopic and microscopic parameters (mobility, connectivity, formation resistivity, pore size distribution, pore networks, etc.) are currently in use for characterizing a porous material, the aforementioned four macroscopic attributes are adequate for our purpose here. Detailed discussions of experimental techniques for characterizing the flow in a porous medium and the medium per se are available in the literature (Greenkorn, 1983; Bories et al., 1991; Dullien, 1992; Stapf and Han, 2006; Fantazzini et al., 2011). Finally, aside from the aforementioned quantitative features, many qualitative descriptions such as "loosely packed," "tightly packed," "randomly packed," and "dumped particle packed" beds are also used. Evidently while it is difficult to quantify these features, some of these do add to the degree of confusion and complexity in this field.

6.3 Newtonian Liquids

The contemporary literature available on the flow of Newtonian liquids in porous media is indeed voluminous. A cursory inspection of the relevant literature shows that a majority of the research efforts has been directed at elucidating the following macroscopic aspects of the single-phase Newtonian liquid flow in porous media:

1. Flow regimes
2. Pressure loss—throughput relationship
3. Wall effects
4. Effect of particle shape and roughness on pressure drop

6.3.1 Flow Regimes

There is no question that the detailed kinematics and structure of the flow field (or flow pattern) prevailing in a porous medium directly determines the macroscopic effects such as pressure loss, dispersion, rates of heat and mass transfer (Sederman et al., 1997, 1998; Chhabra et al., 2001b). Therefore, many workers have reported on these aspects in porous media flows, for example, see Dybbs and Edwards (1984) and Ren et al. (2005a,b). Indeed, a wide variety of experimental techniques has been used to establish the nature of flow in porous media flows including optical methods and colored plumes (Jolls and Hanratty, 1966; Wegner et al., 1971; Stephenson and Stewart, 1986), electrochemical methods (Latifi et al., 1989; Rode et al., 1994; Seguin et al., 1998; Comiti et al., 2000a; Lesage et al., 2004; Mertens et al., 2006; Yang et al., 2015), hot-wire anemometry to ascertain the level of turbulence (Mickley et al., 1965; van der Merwe and Gauvin, 1971), laser anemometry (Ganoulis et al., 1989; Hall and Hiatt, 1996), particle image velocimetry (PIV) (Saleh et al., 1993; James et al., 2012), photoluminescent volumetric imaging (Montemagno and Gray, 1995), and nuclear magnetic resonance (NMR) imaging (Johns et al., 2000; Gotz et al., 2002; Sheppard et al., 2003; Suekane et al., 2003; Bleyer and Coussot, 2014; Marliere et al., 2015) and Lattice–Boltzmann simulations (Hill and Koch, 2002). While some of these studies have endeavored to provide qualitative insights such as locating the stagnant zones, regions of back flow, or the onset of convection (Jolls and Hanratty, 1966; Karabelas et al., 1973; Shattuck et al., 1995), others have attempted to deduce quantitative information about the velocity distribution in 2-D and 3-D systems and about the onset of turbulence (Kutsovsky et al., 1996; Feinauer et al., 1997; Manz et al., 1999; Baumann et al., 2000). In order to keep the complexity at a tractable level, many of the aforementioned studies have used highly idealized porous medium such as inline or staggered arrays of cylinders (Dybbs and Edwards, 1984) and cubic packing of spheres (Suekane et al., 2003), and also deal exclusively with the flow of Newtonian liquids. Notwithstanding the importance of such detailed information, it is probably adequate for our purpose here to talk in terms of the various flow regimes that relate to the gross nature of dependence of pressure drop on the flow rate or on the mean velocity. Indeed, depending upon the nature of the fluid, the type of porous medium, and the flow rate, different flow patterns have been observed and documented in the literature (Chauveteau and Thirriot, 1967; Ahmed and Sunada, 1969; Dybbs and Edwards, 1984; Coulaud et al., 1986; Hassanizadeh and Gray, 1987; Skjetne and Auriault, 1999a,b; Fourar et al., 2004, 2005; Panfilov and Fourar, 2006; Bagci et al., 2014; Bu et al., 2014). The ultimate objective of all these studies is to establish criteria for predicting the transition from one flow regime to another. This objective is realized in terms of an appropriately defined Reynolds number, which, of course, hardly takes into account the detailed structure of the medium. Based on experimental results for the flow of water past a bundle of rods, the main flow patterns can be summarized as follows (Dybbs and Edwards, 1984):

> 1. *Darcy (creeping flow) regime*: This flow region is dominated by viscous forces and the local velocity profile is determined primarily by the local geometry. In view of the fact that boundary layers begin to develop near the pore walls at about $Re_i \approx 1$, the Darcy regime ceases to exist at

this value of the Reynolds number. At the other extreme, this regime may also breakdown at extremely slow flow rates wherein the pore wall–fluid interactions are more important than the viscous effects.

2. *Inertial flow regime*: As the flow rate is progressively increased such that $Re_i > 1$, boundary layer effects dominate, and an inertial core appears outside the boundary layer. The flow is still steady, but the pressure drop–flow rate relationship veers away from the linear dependence, typical of the Darcy regime. While it is difficult to pinpoint the onset of this steady nonlinear flow regime, it occurs somewhere in the range $1 < Re_i < 10$ and persists up to about $Re_i = 150$. While there might be pockets of local turbulence in some pores, this flow regime is also known as nonlinear laminar flow regime.

3. *Unsteady laminar flow regime*: In the range $150 \leq Re_i \leq {\sim}300$, the flow in a porous medium is characterized by the formation of waves thereby imparting an unsteady character to the flow.

4. *Turbulent flow regime*: At high Reynolds numbers ($Re_i > 300$), one obtains the flow conditions dominated by eddies that closely resemble that of turbulent flow in pipes. In literature, this regime is also known as "highly chaotic" flow pattern.

At this juncture, it is appropriate to make four observations: first, there is always a degree of arbitrariness and subjectivity inherent in the interpretation of flow visualization studies (Barak, 1987). Second, transition from one flow regime to another occurs gradually over a range of conditions rather than abruptly, as is implied by the single value of the Reynolds number. Third, much confusion exists in the literature about the occurrence of turbulence in porous media (Kyle and Perrine, 1971; Schmid et al., 1999; Niven, 2002), and therefore some authors (Scheidegger, 1974) have attributed the deviation from Darcy's law to the distortion of streamlines (thereby emphasizing the role of inertial forces) rather than to the true turbulence (Himbert, 1965; Pech, 1984). Finally, each transition is also strongly influenced by the type of medium and the pore structure, and therefore the critical values of the Reynolds number denoting the transition of flow regimes for one medium may not apply to another dramatically different porous medium. Consequently, considerable confusion exists in the literature about the limit of the Darcy regime. For instance, values of the Reynolds number ranging from 0.1 to 75 have been reported in the literature (Dullien, 1992). A part of this confusion can also be attributed to the variety of definitions of the Reynolds number used in the literature. Thus, for instance, Comiti and Sabiri (1997) have advocated the use of a Reynolds number based on the pore characteristic size and velocity, Re_{pore}, and they asserted that $Re_{pore} = 0.83$ marks the end of Darcy's regime (based on the criterion that the pressure drop derives only 1% contribution from the inertial effects). Similarly, some authors (Wegner et al., 1971) proposed the end of the nonlinear laminar regime to occur somewhere in the range $90 \leq Re_p \leq 120$ whereas the corresponding value of Re_{pore} is believed to be ~180 (Seguin et al., 1998) for this transition. Similarly with the exception of Dybbs and Edwards (1984), the so-called unsteady laminar flow regime has not been reported in the literature and thus one wonders about its physical significance. Finally, for the transition to the fully turbulent flow conditions, the reported values of the Reynolds number range from $Re_p \geq 300$ (Jolls and Hanratty, 1966) to $Re_p \geq 400$ (Latifi et al., 1989) to $Re_{pore} \geq {\sim}900$ (Seguin et al., 1998). With the exception of Dybbs and Edwards (1984), all other studies are based on the beds made up of spherical particles. Some of the recent experimental and numerical studies also confirm these findings, e.g., Panfilov and Fourar (2006), Adler et al. (2013), and Bagci et al. (2014), who analyzed transitions in terms of inertial effects, bending of streamlines, etc. Adler et al. (2013) thus postulated corrections to the linear Darcy law by adding V^2 and V^3 terms to account for nonlinear dependence of Δp arising in different flow regimes. Based on the local Reynolds number $Re_{local} = (\rho V_0 d/\pi \mu)$, they suggested the cubic term to be significant for $Re_{local} < 28$ whereas above this value the quadratic (V^2) term dominates the contribution to the total pressure loss. Similarly, the effect of mixed size particle beds on flow transitions has been studied by Bu et al. (2014).

In summary, in view of the complex interactions between the structure of the porous medium and the resulting flow field, it is really neither justifiable nor possible to offer universally applicable transition criteria in terms of a single value of the Reynolds number. This difficulty is further accentuated by the fact that each porous medium is unique in its characteristics. From an engineering applications' viewpoint, however, suffice it to say here that one can work in terms of the three broad flow regimes, namely, the Darcy regime, non-Darcy (inertial) flow regime, and the turbulent flow regime. Also, it is perhaps fair to say that the transition values of the different numbers mentioned here are not inconsistent with each other.

6.3.2 Pressure Loss–Throughput Relationship

The objective here is to develop reliable and accurate means of estimating the pressure gradient required to maintain a fixed flow rate through a porous medium in an envisaged application, or conversely, to predict the throughput for an available pressure gradient. The results in this field are frequently expressed either in terms of a dimensionless permeability (or in terms of the so-called Kozeny constant) or the usual friction factor—Reynolds number approach. All these approaches will be used here, and it is worthwhile mentioning that all these forms of presentation complement each other rather than being mutually exclusive. Various model approaches (which are not at all mutually exclusive) have been used for this purpose in the literature. Depending upon one's taste and viewpoint, these may be characterized in many ways. Broadly speaking one can discern four distinct strategies: dimensional approach together with empirical considerations (also known as phenomenological models), conduit or capillary flow analogy, submerged object or drag theories, and those based on the averaging of the field equations. Alternatively, Sochi (2010a, b) views these approaches as continuum models, capillary bundles, numerical models, and pore-scale network (PSN) modeling. Naturally, the final objective in each case is to formulate the flow rate–pressure gradient relations for a given porous medium. These approaches are neither mutually exclusive nor free from varying levels of approximation and empiricism. Each of these will now be described briefly. At this point, it is appropriate to introduce the following two dimensionless parameters that would be used extensively in this chapter:

$$\text{Friction factor: } f = \left(\frac{\Delta p}{L}\right)\left(\frac{d}{\rho V_0^2}\right) \tag{6.3a}$$

$$\text{Reynolds number: } \text{Re} = \frac{\rho V_0 d}{\mu} \tag{6.3b}$$

Several other definitions of the Reynolds number are currently in use (some of these will be introduced later), but all of these are interrelated through a function of porosity and sometime of tortuosity in the case of Newtonian fluids. Irrespective of the definition of the Reynolds number and in the absence of significant wall effects, Figure 6.1 shows the generic relationship between the friction factor (pressure drop) and the Reynolds number (flow rate) for a given isotropic porous medium. At very low Reynolds numbers, this relationship is characterized by a slope of −1 and this corresponds to the Darcy regime or the creeping flow conditions. On the other hand, at very high Reynolds numbers, the friction factor is nearly independent of the Reynolds number, but depends on particle roughness, pore structure, etc. (Dullien and Azzam, 1973a,b; Macdonald et al., 1979; Carpinlioglu and Ozahi, 2008; Erdim et al., 2015; von Seckendorff and Hinrichsen, 2021).

In between these two limits, both viscous and inertial effects are important. However, before leaving this section, it is appropriate to say here that much has been written about the so-called nonlinear effects in porous media flows (Wodie and Levy, 1991; Hjelmfelt and Brooker, 1995; Liu and Masliyah, 1999; Skjetne and Auriault, 1999a,b; Skjetne et al., 1999; Fourar et al., 2004, 2005; Montillet, 2004; Panfilov and Fourar, 2006; Adler et al., 2013). Broadly speaking, the pressure loss shows a linear dependence on the velocity in the viscous region and it is assumed to vary as V^2 in the fully turbulent conditions. But given the extent of confusion regarding the occurrence and role of turbulence in such systems, the literature abounds with various ideas and inferences in this regard. For instance, Wodie and Levy (1991), Skjetne and Auriault (1999a) and Koch and Ladd (1997) have shown that the correction to Darcy's law

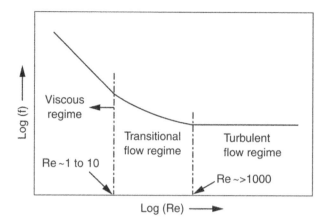

FIGURE 6.1 Schematic representation of friction factor–Reynolds number relationship for a porous medium.

for weak inertial effects yields a V^3 term rather than the usual V^2 term, in line with the study of Adler et al. (2013). Similarly, at high Reynolds numbers, if one invokes the assumptions of the inviscid irrotational flow in the pores and wall boundary layers, this leads to a $V^{3/2}$ term in the pressure loss equations. Therefore, they attribute the presence of a V^2 term to the development of strong localized dissipation zones even in laminar flow conditions. In the next section, we review the progress made in predicting the dependence of pressure loss (friction factor) on the flow rate (Reynolds number), mainly for unconsolidated and isotropic porous media.

6.3.2.1 Dimensionless Empirical Correlations

This category represents the simplest and perhaps the oldest class of descriptions of the flow of Newtonian liquids through random beds of particles. Perhaps the most complete and general dimensional analysis of this flow problem is that of Rumpf and Gupte (1971). Based on extensive experimental work with beds consisting of narrow- and wide-size distribution of spherical particles and embracing a wide range of voidage ($0.35 \leq \varepsilon \leq 0.7$) and Reynolds number ($0.01 < \text{Re} < 100$), Rumpf and Gupte (1971) concluded that the values of friction factor for beds of uniform size particles virtually superimpose (5% discrepancy) over those for wide-size distribution systems, provided a surface average particle diameter ($\bar{d}$) is used as the characteristic linear dimension in the definitions of the friction factor and the Reynolds number. At low Reynolds numbers ($<\sim 1$), their results are well represented by the simple equation

$$f\varepsilon^{5.5} = \frac{5.6A}{\text{Re}} \tag{6.4}$$

where A is a constant and is a function of the particle size distribution with further possible dependence on particle shape and structure of the bed. Rumpf and Gupte (1971) reported the value of A to vary between 1 and 1.05. It is instructive to rewrite Equation 6.4 in the general form

$$f\,\text{Re} = A'f(\varepsilon) \tag{6.5}$$

Indeed, most of the correlations available in the literature are of (or can be reduced to) this form (Erdim et al., 2015). However, the numerical values of A' and the choice of $f(\varepsilon)$ continue to be a matter of disagreement in this area. These uncertainties coupled with the additional possible dependence of the constant A' on particle shape, size distribution and structure of bed, wall effects, etc. continue to be the main impediments in developing a universally applicable form of Equation 6.5. Table 6.1, modified after

TABLE 6.1
Different Forms of $f(\varepsilon)$ in Equation 6.5

$f(\varepsilon)$	Reference
$\dfrac{(1-\varepsilon)^2}{\varepsilon^3}$	Blake (1922), Kozeny (1927), Carman (1937, 1956)
$\dfrac{1-\varepsilon^2}{\varepsilon}$	Zunker (1920)
$\left[\dfrac{(1-\varepsilon)^{1.3}}{(\varepsilon-0.13)}\right]^2$	Terzaghi (1925)
$\{1.115(1-\varepsilon)\varepsilon^{1.5}\}\{(1-\varepsilon)^2+0.018\}$	Rapier (1949)
$69.43-\varepsilon$	Hulbert and Feben (1933)
$\varepsilon^{-3.3}$	Slichter (1898)
$\dfrac{1}{\varepsilon}$	Krüger (1918)
ε^{-6}	Hatch (1934), Mavis and Wilsey (1936)
ε^{-4}	Fehling (1939)
$\varepsilon^{-4.1}$	Rose (1945)
$\varepsilon^{-5.5}$	Rumpf and Gupte (1971)

Source: Reproduced from Dullien, F.A.L., *Porous Media: Fluid Transport and Pore Structure*, 2nd Ed., Academic Press, New York (1992) whereas the original references are given in Rumpf, H. and Gupte, A.R., Einflusse der porositat and Korngrobenverteilung im Widerstandsgesetz der porentstromung, *Chem. Ing. Tech.*, **43**, 367, 1971.

Dullien (1992), illustrates the variety of forms of $f(\varepsilon)$ used in the literature. Obviously, the value of A' depends upon the choice of $f(\varepsilon)$. What is even more frustrating is the fact that even for the same choice of $f(\varepsilon)$, widely different values of A' have been reported in the literature. For instance, the well-known Blake–Kozeny equation (see Bird et al., 2002; Coulson and Richardson, 2002; Chhabra and Gurrappa, 2019) developed for the so-called creeping flow regime (Re $\ll$1) can be expressed as

$$f\left[\frac{\varepsilon^3}{(1-\varepsilon)^2}\right]=\frac{150}{\text{Re}} \tag{6.6}$$

whereas according to the so-called Carman–Kozeny equation,

$$f\left[\frac{\varepsilon^3}{(1-\varepsilon)^2}\right]=\frac{180}{\text{Re}} \tag{6.7}$$

Thus, even the so-called best values of the constant (A') differ by 20%. Indeed, the values ranging from 112 to 368 (even greater) have been reported in the literature (Larkins et al., 1961; Al-Fariss and Pinder, 1987; Fand and Thinakaran, 1990; Nemec and Levec, 2005; Erdim et al., 2015; von Seckendorff and Hinrichsen, 2021). The reasons for such a large discrepancy are not fully known but it (or at least a part of it) has been attributed to the inappropriate choice of the characteristic particle size and to the influence of confining walls and particle shape (Fand and Thinakaran, 1990; Dullien, 1992).

For the flow conditions outside the range of Darcy's law, literature abounds with numerous empirical formulae that purport to provide satisfactory means of estimating pressure drop in unconsolidated porous media. Perhaps the two best-known expressions are due to Ergun (1952) (for Re < ~1000) and the so-called Burke–Plummer equation for Re > 1000 (Bird et al., 2002). The Ergun equation is written as

$$f\left[\frac{\varepsilon^3}{(1-\varepsilon)}\right] = \frac{150(1-\varepsilon)}{\text{Re}} + 1.75 \tag{6.8a}$$

whereas the Burke–Plummer equation is given by

$$f\left[\frac{\varepsilon^3}{(1-\varepsilon)}\right] = 1.75 \tag{6.8b}$$

Both these equations are based on extensive experimental data gleaned in columns packed with uniform packings and with negligible channeling. Though the Ergun equation is purely an empirical development, some attempts have been made to suggest that this form of equation is not inconsistent with the volume averaging of governing equations (Irmay, 1958; du Plessis, 1994; Niven, 2002; Stevenson, 2003).

Other analogous expressions that offer some improvement over these equations and also purport to take into account the roughness of particles are due to Macdonald et al. (1979). For smooth surface particles, their expression is given below:

$$f\left[\frac{\varepsilon^3}{(1-\varepsilon)}\right] = \frac{180(1-\varepsilon)}{\text{Re}} + 1.8 \tag{6.9}$$

Similarly, in an extensive review, Erdim et al. (2015) reported new experimental data and evaluated the relative performance of 38 correlations from the literature. For beds of monosize spheres and in the absence of wall effects, they also proposed the following correlation for friction factor:

$$\left(\frac{\varepsilon^3}{1-\varepsilon}\right)f = \frac{160(1-\varepsilon)}{\text{Re}} + 2.81\left(\frac{1-\varepsilon}{\text{Re}}\right)^{0.096} \tag{6.10}$$

Note the weak dependence of the otherwise constant term (corresponding to the fully inertial flow) in Equation 6.10. Similar modifications have been also proposed by von Seckendorff et al. (2020, 2021) to account for particle shape, roughness, and wall effects in estimating the pressure drop across a packed bed.

The surface roughness of particles contributes to the frictional pressure drop only under highly turbulent conditions, thereby resulting in the values of the so-called inertia parameter (second term on the right-hand side of Equation 6.9) larger than 1.8. For the roughest particles used so far, this value turns out to be 4. Macdonald et al. (1979) asserted that Equation 6.9 can be expected to yield values of the friction factor for a wide variety of unconsolidated media with a maximum uncertainty of 50% whereas the resulting errors are significantly greater in the case of consolidated media. Meyer and Smith (1985) and Dukhan et al. (2006) have evaluated the performance of Equation 6.8 in correlating their results on air flow through consolidated and unconsolidated porous media with varying levels of particle surface roughness.

6.3.2.2 The Conduit or Capillary Models

In this approach, the interstitial void space in a porous medium is envisioned to form tortuous conduits of complicated cross-section but with a constant cross-sectional area on the average. Thus, the flow in a porous medium is equivalent to that in conduits whose length and diameter are chosen such that the resistance to flow is equal to that in the actual porous medium. Undoubtedly, the so-formed conduits or capillaries are interconnected in an irregular manner and form a network of tangled capillaries, but the simplest 1-D models of this class do not take into account this complexity. There are essentially three models and modifications thereof available that fall in this category: Blake, Blake–Kozeny, and Kozeny–Carman models. In the Blake model, the bed is simply replaced by a bundle of straight tubes of complicated cross-section (characterized by a hydraulic radius R_h), and the interstitial velocity (V_i) is related to the superficial velocity (V_0) through the well-known Dupuit equation, that is,

$$V_i = \frac{V_0}{\varepsilon} \qquad (6.11)$$

For a homogeneous and isotropic bed of spherical particles of uniform diameter d, the expression for hydraulic radius is obtained as

$$R_h = \frac{d\varepsilon}{6(1-\varepsilon)} \qquad (6.12)$$

In writing Equation 6.12, the wall effects have been ignored (Mehta and Hawley, 1969). The Blake–Kozeny model, on the other hand, postulates that the effective length of the tangled capillaries is greater than that of the porous medium, thereby introducing a tortuosity factor T defined as (L_e/L). Finally, the Kozeny–Carman model is identical to the Blake–Kozeny model except that it also corrects the average velocity for the tortuous nature of the flow path as

$$V_i = \left(\frac{V_0}{\varepsilon}\right)\left(\frac{L_e}{L}\right) \qquad (6.13)$$

In Equation 6.13, the (hydraulic) tortuosity factor T corrects for the fact that a fluid particle (appearing in the macroscopic flow equations) moving with the superficial velocity V_o traverses the path length L in the same time as an actual fluid particle moving with velocity V_i covers an average effective length of L_e.

The starting point for the development of this class of models is the well-known Hagen–Poiseuille equation for the fully developed, steady, and laminar flow of an incompressible Newtonian fluid in a circular tube, that is,

$$\Delta p = \frac{32VL\mu}{D^2} \qquad (6.14)$$

Equation 6.14 is adapted for the laminar flow through a porous medium by substituting D_h ($=4R_h$) for D, L_e for L, and the interstitial velocity V_i for V that, in turn, yields

$$\Delta p = \frac{72\mu V_0 L}{d^2}\left\{\frac{(1-\varepsilon)^2}{\varepsilon^3}\right\}\left(\frac{L_e}{L}\right)^2 \qquad (6.15)$$

This is the well-known Kozeny–Carman model. It can readily be seen that Equation 6.15 includes both the Blake model ($L_e=L$) and the Blake–Kozeny model (only length is corrected, i.e., (L_e/L) will appear in Equation 6.15 instead of its square) as special cases. Despite the differences inherent in these models, the predictions of the Blake and the Kozeny–Carman models are indistinguishable from each other, whereas those of the Blake–Kozeny differ by about 20%. Carman (1956) suggested a value of $\sqrt{2}$ for (L_e/L) on the premise that the capillaries deviate on average by an angle of 45° from the direction of mean flow. Based on the assumption that the cross section of the capillaries lies somewhere in between that of a circular tube and of a parallel slit, Carman (1956) used a value of 40 instead of 32 in Equation 6.14. With these modifications, Equation 6.15 reduces to

$$\Delta p = \frac{180 \mu V_0 L}{d^2} \left\{ \frac{(1-\varepsilon)^2}{\varepsilon^3} \right\} \tag{6.16}$$

This is the well-known Kozeny–Carman equation, which in dimensionless form is given by Equation 6.7. It must, however, be emphasized here that in spite of the great degree of similarity between the two end results (i.e., Equations 6.16 and 6.8a), the Kozeny–Carman development is based on a presupposed model of the porous medium whereas the Ergun equation is purely an empirical development. However, all of these equations are applicable in the so-called creeping or Darcy regime (Figure 6.1).

This approach has been extended by Comiti and Renaud (1989) to encompass the complete range of Reynolds numbers, as seen in Figure 6.1. By utilizing the notion that the pressure drop varies linearly with fluid velocity in the viscous region and in a quadratic fashion under fully turbulent conditions (ignoring the V_0^3 term), and by further assuming these two contributions to be additive,

$$\frac{\Delta p}{LV_0} = MV_0 + N \tag{6.17}$$

The viscous term, N, was evaluated using Equation 6.15 and is recast as

$$N = 2\gamma \mu T^2 a_{vd}^2 \left(\frac{(1-\varepsilon)^2}{\varepsilon^3} \right) \tag{6.18}$$

In Equation 6.18, γ is a shape factor ((also used by Carman (1956)) and a_{vd} is the specific surface area of the packing exposed to fluid flow. While for beds with point contacts (such as spheres, $a_{vd}=6/d$), a_{vd} will coincide with the geometric surface area (a_{vs}), the two can deviate significantly for beds made up of plate-like particles due to the overlapping of packing. Clearly in this case, the value of a_{vd} will be less than a_{vs}. While $\gamma=1$ for cylindrical pores, Comiti and Renaud (1989) suggested that this value could also be used for other shapes in the first instance. On the other hand, Comiti and Renaud (1989) argued that the value of M essentially accounts for the significant kinetic energy losses encountered at high flow rates. This coupled with the proposition of Himbert (1965) and Pech (1984) that the pores in a porous medium can be viewed as conduits with their roughness $E \approx$ pore diameter. Under these conditions of fully turbulent flow, the friction factor can be approximated by the Nikuradse formula. This, in turn, leads to the expression for M as

$$M = 0.0968 T^3 a_{vd} \rho \frac{(1-\varepsilon)}{\varepsilon^3} \tag{6.19}$$

The two parameters T and a_{vd} can thus be evaluated using experimental pressure drop data encompassing wide ranges of conditions for a Newtonian fluid (air or water) in a bed of known porosity and plotted in accordance with Equation 6.17. These two factors are regarded to be the characteristics of the bed and are therefore strongly influenced by the method of bed preparation and the structure of the resulting bed. The distinct advantage of using a_{vd} rather than the size of the packing is that this approach can be directly used for nonspherical packings and particles with a size distribution. Indeed, this approach has been shown to work well for the packed beds of plates (Comiti and Renaud, 1989), angular-shaped crushed rocks, and irregular-shaped sand particles (Legrand, 2002; Wahyudi et al., 2002). Finally, it is customary to recast all these equations in terms of the usual friction factor–Reynolds number coordinates.

While the values of the numerical constants depend upon the definition of the Reynolds number used, Equations 6.17–6.19 can be written as

$$f_{\text{pore}} = \frac{16}{\text{Re}_{\text{pore}}} + 0.194 \qquad (6.20)$$

where

$$\text{Re}_{\text{pore}} = \frac{4\rho T V_0}{\mu(1-\varepsilon)a_{\text{vd}}} \qquad (6.21a)$$

$$f_{\text{pore}} = \frac{2\Delta p \varepsilon^3}{\rho V_0^2 T^3 (1-\varepsilon) a_{\text{vd}}} \qquad (6.21b)$$

Suffice it to add here that the corresponding dimensionless forms of Equation 6.15 or 6.16 will have a form similar to Equation 6.20, without the constant term on the right-hand side.

Considerable confusion exists regarding the numerical value of the tortuosity factor and the terminology used in the literature (Epstein, 1989; Puncochar and Drahos, 1993; Liu and Masliyah, 1996a,b; Ghanbarian et al., 2013). For instance, based on the premise that a fluid particle follows the surface of a particle, Sheffield and Metzner (1976) proposed a value of $T = (\pi/2)$ for packed beds of spherical particles. This value combined with the numerical constant 32 for circular tubes (in Equation 6.15) yields a value of 178 that is near enough to 180. Epstein (1989), however, pointed out that the term tortuosity factor has often been used as being synonymous with the factor $(L_e/L)^2$. Carman (1956), Scheidegger (1974), and most others, on the other hand, define the tortuosity as (L_e/L). Finally, the aforementioned two values of the tortuosity factor are by no means universally accepted. Indeed based on intuitive arguments, Foscolo et al. (1983) have approximated the tortuosity factor by $(1/\varepsilon)$ which coincides with Equation 6.2b for a random bed made up of cylinders. This assertion has received further support from the work of Agarwal and O'Neill (1988). The latter authors have also compiled the literature values of the tortuosity factor ranging from 1 to 1.65 for beds of spheres that seem to correlate well with the bed voidage. Based on their experimental results for the beds of monosize spheres, the experimental values of tortuosity (via Equation 6.17) Comiti and Renaud (1989) proposed the relationship as

$$T = 1 - P' \ln(\varepsilon) \qquad (6.22)$$

where P' is a constant that depends upon the shape of particles. Based on the literature data, they obtained $P'=0.41$ for spheres and mixtures of spheres, $P'=0.63$ for cubes, $P'=1.6$ for wood chips, etc. For plate-like particles, it shows further dependence on the plate thickness. Equation 6.22 yields the values of T ranging from 1.40 to 3.49. Therefore, there seems to be a merit in using the value of the tortuosity factor T extracted using the approach of Comiti and Renaud (1989) rather than using a constant value for T. In some instances, the tortuosity factor has been postulated to be a function of the Reynolds number, also in addition to bed voidage (Dharamadhikari and Kale, 1985; Mauret and Renaud, 1997; Epstein, 1998; Ciceron et al., 2002b).

Though in general, this simple approach of capillary bundles has enjoyed a good deal of success in the packed bed range ($\varepsilon < \sim 0.6$), it has also come under severe criticism (see Dullien, 1975a,b, 1992; Molerus, 1980; Duda et al., 1983; Puncochar and Drahos, 2000). Often, the lack of agreement between the predictions and experiments is attributed to the inappropriate values of the tortuosity factor (Agarwal and O'Neill, 1988; Dullien, 1992). Aside from this fact, the deficiencies of this simple approach can be classified into two types: the first type relates to what might be called as an *inappropriate* application of the

Hagen–Poiseuille equation that is based on the assumption of the fully developed laminar flow and hence the shear stress being constant at each point on the wall. Second, it is readily recognized that the concept of the hydraulic radius has proved to be successful only under fully turbulent conditions. Clearly, neither of these requirements is met in the case of streamline fluid flow through a packed bed. The second type of criticism pertains to its total inadequacy in modeling the medium itself. For instance, this approach does not take any cognizance of the different kinds of nonuniformities present in the medium (e.g., series and parallel networks) whereby a fluid element is exposed to converging–diverging type of flow, etc. Also, due to their inherently "one-dimensional" character, this approach does not predict anisotropic permeability of heterogeneous porous media. Furthermore, the Kozeny–Carman equation can be rearranged to yield the value of permeability that turns out to be independent of the particle size distribution and the topology of capillaries, etc. This result is at variance with the experimental observations (Dullien, 1992). An excellent critique of capillary models is available in the literature (Scheidegger, 1974).

Numerous investigators have attempted to rectify some of the aforementioned drawbacks of capillary models. For instance, Dullien (1992) has dealt with the influence of the parallel and series-type pore nonuniformities, the distribution of pore entry dimensions, etc. on the permeability of a porous medium. Over the years, various statistical and capillary network models have been developed, for example, see Haring and Greenkorn (1970), Dullien and Dhawan (1975), Dullien (1975b), Chatzis and Dullien (1977), Greenkorn (1981), Kanellopoulos (1985), Mikelic (1997), Thompson and Fogler (1997), Lao et al. (2004), and Zerai et al. (2005). Undoubtedly, these developments have obviated some of the deficiencies inherent in the simple capillary models but, unfortunately, none of these is yet refined to the extent to be used as a basis for design calculations. A detailed discussion of these advances is beyond the scope of this book; attention is, however, drawn to very readable and critical appraisals available in the literature (Dullien, 1975a, 1992; Chatzis and Dullien, 1977; Thompson and Fogler, 1997; Al-Raoush et al., 2003; Balhoff and Thompson, 2006). Likewise, several investigators have solved the complete Navier–Stokes equations for the flow in various periodically varying geometries to elucidate the role of the converging–diverging nature of the flow as encountered in porous media. These studies are reviewed in Section 6.3.2.4.

6.3.2.3 The Submerged Objects Models or Drag Theories

In this approach, the flow through a porous medium or a packed bed is viewed as being equivalent to the flow around an assemblage of submerged objects, and the resulting fluid dynamic drag manifests itself as the frictional pressure drop across the bed. Thus, the central problem here is essentially that of calculating the drag force on a typical particle of the assemblage. Various ideas has been used to achieve this objective; all of which, however, involve the modification of the Stokes drag on a single particle to account for the additional resistance arising from the presence of neighboring particles and hence the required correction is expected to be a function of bed voidage (ε) only, at least in the Stokes regime. Two distinct approaches can be discerned that have been used in the literature to evaluate the drag on a particle in a particle assemblage. In the first case, purely dimensional considerations with varying degrees of empiricism have been used to obtain the modified drag on a particle, without any appeal to the arrangement of particles. The second approach relies on the solution of the governing equations for a preconceived geometrical arrangement of particles. Both approaches would be dealt with briefly by way of referring to the representative studies available in the literature.

The studies of Barnea and coworkers (Barnea and Mizrahi, 1973; Barnea and Mednick, 1975, 1978) and its subsequent modifications by Zimmels (1988) illustrate the successful application of dimensional analysis to the problem of flow in packed beds. In an attempt to collapse the friction factor–Reynolds number data for multiparticle assemblages on to the standard drag curve for a single particle, Zimmels (1988) introduced the following modifications to the traditional definitions of drag coefficient and Reynold number:

$$\mathrm{Re}_1 = \frac{\mathrm{Re}}{(1-\phi)} \exp\left\{\left(\frac{5}{3}\right)\left(\frac{\phi}{(1-\phi)}\right)\right\} \qquad (6.23)$$

$$C_{D_1} = \frac{8F_D(1-\phi)^2}{\pi \rho d^2 V_0^2 \left(1 + \beta \phi^{1/3}\right)} \tag{6.24}$$

Here,

$$\phi = 1 - \varepsilon$$

$$\beta = 1 + \frac{8\phi}{3} \tag{6.25}$$

The modified drag coefficient C_{D_1} is related to the single sphere value C_{D_0} as

$$\frac{C_{D_1}}{C_{D_0}} = \frac{1}{1 + \phi^{1/3} + (8/3)\phi^{4/3}} \tag{6.26}$$

where C_{D_1} is the drag coefficient of multiparticle assemblage of voidage ε ($= 1 - \phi$) and C_{D_0} is the drag coefficient of a single sphere ($\varepsilon = 1$) otherwise under identical conditions. Based on a vast amount of data available in the literature and using the aforementioned definitions of drag coefficient and Reynolds number, Zimmels (1988) presented the expression for friction factor as

$$f = \left(\frac{3}{16}\right)\left\{0.63 + 4.8 \operatorname{Re}_1^{-1/2} + \phi \operatorname{Re}_1^{1/8}/4.8\right\}^2 \left[\frac{\phi(1 + \beta\phi^{1/3})}{(1-\phi)^2}\right] \tag{6.27}$$

This equation was stated to be applicable over wide ranges of condition ($10^{-3} < \operatorname{Re}_1 < 10^4$; $0.3 \leq \varepsilon \leq 0.7$). Unfortunately, the correspondence between the predictions of Equation 6.27 and those of the Ergun equation is disappointingly poor, thereby warranting further refinements in Equation 6.27. Zimmels (1988) has also outlined a scheme for taking into account the nonuniform particle size and the porosity variation in the column.

In contrast to this, the second approach in which the Stokes drag on a single particle is modified due to the presence of neighboring particles has proved to be somewhat more successful. Most of these treatments are, however, limited only to the so-called creeping flow conditions. A wide spectrum of models has appeared in the literature that purport to capture the hydrodynamic influence of neighboring particles on the particle under examination. For instance, in his pioneering work, Brinkman (1947, 1948) calculated the force on a typical particle of the bed by assuming it to be buried in a homogeneous and isotropic porous medium, and obtained the relation for permeability as

$$k_B = \left(\frac{d^2}{72}\right)\left[3 + \frac{4}{(1-\varepsilon)} - 3\left\{\frac{8}{1-\varepsilon} - 3\right\}^{1/2}\right] \tag{6.28}$$

Note that this equation yields the unrealistic result of zero permeability for $\varepsilon = 1/3$. Tam (1969) has subsequently provided a theoretical justification for this result, whereas Lundgren (1972) has improved upon Brinkman's expression by including the effective viscosity of the suspension as

$$k_L = \frac{k_B}{M(\varepsilon)} \tag{6.29}$$

where $M(\varepsilon)$, the effective viscosity function, is related to the porosity as

$$M(\varepsilon) = \left(\frac{4\pi}{3}\right)\left(\frac{\alpha^2 R^2}{(1-\varepsilon)F(\alpha^2 R^2, \alpha R)}\right) \quad (6.30)$$

$$\alpha R = \left(\frac{3}{4}\right)\frac{\left(3-\sqrt{(8/(1-\varepsilon))-3}\right)}{\left((1/(1-\varepsilon))-1.5\right)} \quad (6.31)$$

$F(\alpha^2 R^2, \alpha R)$ is a cumbersome expression involving Bessel functions and Legendre polynomials. Lundgren (1972) has also evaluated the validity of his theory by using experimental data from fluidization and sedimentation tests; the agreement can, at best, be described only as being moderate. Wilkinson (1985) has outlined a scheme to account for the effect of porosity and grain-size distribution on the permeability of a granular porous medium. Renard and LeLoc'h (1996) have presented a simplified renormalization method to estimate the permeability of a porous medium.

In the second approach, the Stokes drag force is corrected for multiparticle effects by using the so-called cell models. Here, the influence of the neighboring particles is simulated by enclosing the particle in question in an artificial cage or cell. Thus, the difficult many-body problem is converted into a conceptually much simpler one-body equivalent. This approach is also not completely devoid of empiricism, especially with regard to the shape of the fluid envelope and the boundary conditions, etc. (Slobodov and Chepura, 1982; Mao, 2002; Zholkovskiy et al., 2007). Consequently, a wide variety of cell models differing in shape and the associated boundary conditions has been proposed in the literature. The available body of knowledge in this field has been reviewed thoroughly by Happel and Brenner (1965) and others (LeClair and Hamielec, 1968a,b; LeClair, 1970; Tal and Sirignano, 1982; Jean and Fan, 1989; Chhabra 1993a,b; Chhabra et al., 2001b; Faltas and Saad, 2011). Among the various cell models available, perhaps the free surface model (Happel, 1958, 1959) and the zero-vorticity cell model (Kuwabara, 1959) have been accorded the greatest amount of attention. Both are of a sphere-in-sphere or a cylinder-in-cylinder configuration and exactly identical in all respects, except with regard to the boundary condition at the cell surface. In the free surface cell model, the cell boundary is assumed to be frictionless whereas Kuwabara (1959) suggested the vorticity to be zero at the cell boundary. In the latter case, since the shear stress does not vanish at the cell boundary, there is an exchange of energy between the cell and the surroundings (i.e., the other cells), thereby violating the noninteracting nature of the cells. In view of this, it has been argued in the literature (Happel and Brenner, 1965; El-Kaissy and Homsy, 1973) that the free surface cell model has a sounder physical basis than the zero-vorticity cell model. Figure 6.2 shows the two-cell model idealizations of flow in a multiparticle assemblage. Each particle is envisioned to be surrounded by a hypothetical spherical envelope of fluid. The particle moves with a velocity equal to the superficial velocity of the liquid in the assemblage. The usual no-slip boundary condition is applied at the particle surface whereas both the radial velocity and the shear stress (or vorticity) vanish at the cell surface. The radius of the cell is chosen so that the voidage of each cell is equal to that of the overall assemblage. In the creeping flow regime, the predictions of both these models are shown in Figure 6.3 in the form of a nondimensional drag correction factor $Y (= C_D/C_{D0} = C_D \text{Re}/24)$ as a function of the voidage for assemblages of spheres.

As expected, the zero-vorticity cell model always predicts higher values of pressure drop than those yielded by the free surface cell model, the difference between the two values being of the order of 10%–20%. Similar results based on a cylindrical cell model are also available in the literature (Tal and Sirignano, 1982). Beside these idealized cell models, Zick and Homsy (1982) and Sangani and Acrivos (1982b) have studied the steady incompressible flow (at zero Reynolds number) past simple, body-centered, and face-centered cubic arrays of uniform size spheres. The predictions of Sangani and Acrivos (1982b) for all three configurations are also included in Figure 6.3 where it is seen that their values seem

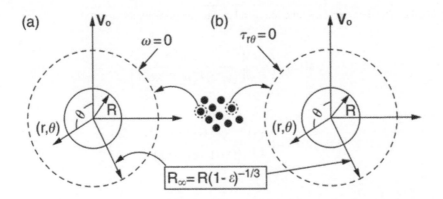

FIGURE 6.2 Cell model idealizations of multiparticle assemblages (a) zero-vorticity model (b) free surface model.

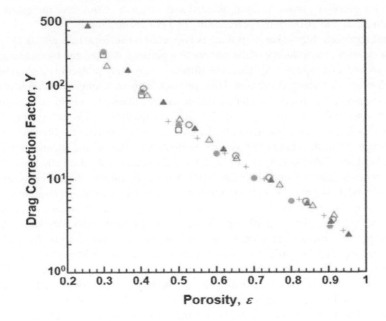

FIGURE 6.3 Various theoretical predictions of drag on particle assemblages in Newtonian media in the creeping flow regime. (○) Zero vorticity cell model (Kuwabara, 1959); (●) Free surface cell model (Happel, 1958); (+) Simple cubic array; (△) Body centered cubic array; (▲) Face centered cubic array (Sangani and Acrivos, 1982); (□) Equation 6.16.

to be a little closer to the predictions of the zero-vorticity model. Finally, by way of validation, the predictions of Equation 6.16 are also included in this figure in the range $0.3 < \varepsilon < 0.5$ that seem to lie below all predictions shown here. The inertial effects during the flow in regular arrays of spheres have been considered among others by Hill et al. (2001a, 2001b) and Gunjal et al. (2005). Similarly, Reddy and Joshi (2008) have numerically estimated the pressure drop across a narrow-packed bed ($D_c/d = 5$) consisting of 151 spheres in eight layers. The effect of slip and nonconcentric particle-in-cell models have been studied by Faltas and Saad (2011) for both boundary conditions.

Though originally developed for the creeping flow conditions, the cell models have also been used in the intermediate Reynolds numbers regime. El-Kaissy and Homsy (1973) presented a perturbation analysis for both cell models, which partially takes into account the inertial effects. However, the maximum value of the Reynolds number up to which this analysis is valid depends upon the value of voidage; the lower the voidage, the higher is the value of the Reynolds number Re; furthermore, this limiting value of the Reynolds number cannot be delineated a priori. Extensive numerical results for drag coefficients of particle assemblages ($0.408 \leq \varepsilon \leq 1$; $0.1 \leq Re \leq 1000$) using the zero-vorticity cell model have been reported by LeClair (1970) and LeClair and Hamielec (1971) whereas the corresponding results for the free surface cell model have been presented by Jaiswal et al. (1991a, 1991b) and Dhole et al. (2004); limited results ($Re \leq 100$) based on a cylindrical cell model are also available (Tal and Sirignano, 1982). Numerical simulations (LeClair and Hamielec, 1971; Jaiswal et al., 1991a, 1991b; Dhole et al., 2004) clearly show that the product ($C_D Re$) remains constant up to about $Re = 1$ for all values of voidage (<0.9), and it is therefore safe to conclude that $Re = 1$ marks the cessation of the so-called creeping flow regime in particle assemblages. Extensive comparisons between the numerical predictions and experimental results as predicted by Equation 6.8a show that the correspondence deteriorates rapidly for $Re > \sim 30$ or so. Furthermore, at low Reynolds numbers, the free surface cell model leads to better predictions whereas the zero-vorticity cell model appears to perform better in the range $Re > \sim 10$, thereby making them complementary in their scope. Nishimura and Ishii (1980) and Fukuchi and Ishii (1982) have combined the boundary layer flow approximation with the cell model idealization to obtain extensive results on drag coefficients as a function of the Reynolds number and bed voidage. Aside from these two well-known cell models, other studies which exploit the particle-in-tube and duct flow analogy with the packed bed flow have also been attempted for the flow of Newtonian (Di Felice, 1996) and power-law fluids (Liu and Masliyah, 1998). These predictions are also in as good an agreement with the experiments as can be expected in this type of work. While several workers have reported 2-D simulations for periodic arrays, very few results are available based on 3-D simulations. By way of example, Larson and Higdon (1989) reported a 3-D analytical and numerical study of a lattice of spheres at zero Reynolds numbers. However, their approach cannot be extended to incorporate inertial effects. Similarly, Nakayama et al. (1995) and Inoue and Nakayama (1998) used a collection of cubes to capture the 3-D nature of the porous media flow. Subsequently, this model was extended to thermal dispersion in porous media (Kuwahara et al., 1996).

Over the years extensive comparisons between drag theories and experiments have revealed that the submerged objects approach describes the flow at high values of voidage whereas the capillary model provides a good description of the flow at low values of ε, with a gray area in between these two limits. It is worthwhile to reiterate here that both these approaches are not mutually exclusive, and indeed some attempts have been made at reconciling them, albeit only limited success has been achieved. (Foscolo et al., 1983; Agarwal, 1988; Agarwal and O'Neill, 1988; Agarwal et al., 1988; Agarwal and Mitchell, 1989; Gmachowski, 1996; Ciceron et al., 2002b).

6.3.2.4 Use of the Field Equations for Flow through a Porous Medium

In Section 6.3.2.2, it was pointed out that the simple capillary tube model fails to take into account the excess pressure loss arising from the convergent–divergent nature of the flow passages in a porous medium. Numerous workers have attempted to estimate the extent of this contribution to the overall pressure loss by solving the Navier–Stokes equations for various constricted tube configurations whereas others have averaged the field equations over a representative region of the matrix. Only a selection of the representative results obtained using these two approaches are presented and discussed in this section to elucidate the strengths and weaknesses of this approach.

6.3.2.5 Flow in Periodically Constricted Tubes (PCTs)

In the simple capillary model, all velocity components but one are neglected whereas it is evident that in convergent/divergent geometry, strictly speaking, the flow is two- (even three-) dimensional. Various conduit cross-sections have been used to simulate this feature of flow and to calculate the resulting excess pressure drop in comparison with that occurring in an equivalent circular tube. Table 6.2 shows a selection of the PCTs used in the literature, the most common being sinusoidal variation (Chaudhary and

TABLE 6.2
Different Types of PCTs Used for Simulating Flow in Porous Media

Geometry	Investigators
	Payatakes et al. (1973)
	Pellerin and Thirriot (1976)
	Deiber and Schowlter (1979,1981) Phan-Thien and Khan (1987) Pilitsis and Beris (1989) Zheng et al. (1990b)
	Duda et al. (1983)
	Magueur et al. (1985)
	Ghoniem (1985)
	Dullien (1992)
	Balhoff and Thompson (2006)

Böhme, 1987; Fedkiw and Newman, 1987; Sisavath et al., 2001). Admittedly, extensive theoretical and experimental results on the detailed flow fields and friction factor are available for the geometries shown in Table 6.2, but unfortunately neither the excess pressure drop attributable to the successive divergent–convergent character of each geometry has been deduced nor is it clear how it can be integrated with the packed bed results explicitly in view of the numerous geometric parameters specific to a particular geometry. Some workers including Batra et al. (1970), Dullien and Azzam (1973a,b), Azzam (1975), and Payatakes and Neira (1977) have asserted that the friction factor obtained by the numerical solution of the complete Navier–Stokes equations (for the Payatakes model) differs, at most, by 30% from that obtained by the Hagen–Poiseuille equation otherwise under identical conditions. This indeed makes one wonder whether the convergent–divergent nature of the flow is worth worrying about (at least in the absence of any memory effects for visco-elastic liquids). This result is, however, not at all surprising. For small to moderate amplitude of undulations, the slow-moving fluid elements in laminar flow regime follow the contour of the tube with a little loss of kinetic energy. On the other hand, this effect should manifest in high Reynolds number flows, even for Newtonian fluids. Lahbabi and Chang (1986) have carried out time-dependent 3-D simulations to predict the transition to inertial flow conditions in a PCT and their results confirm this expectation. At the next level, the so-called numerical modeling entails a detailed description of the porous medium at the pore level with the relevant physics (Sochi 2010a,b). Naturally, this approach warrants enormous computational resources and effort, though it is perhaps the most direct method to study the complex flow phenomena and other transport processes occurring in real rocks, sandstones, etc. The so-called pore-scale network (PSN) approach, on the other hand, is regarded to be a compromise between the rather primitive continuum approach (capillary bundle, submerged object models) and the direct numerical approach described above. In broad terms, the porous medium is idealized as a network of interconnected flow passages (including uniform and variable cross sections) of regular shape, and the flow equations (simplified form of the Navier–Stokes equations) lead to a system of simultaneous equations to determine the flow rate–pressure drop behavior in each duct.

This approach can be customized for 2-D and 3-D networks with a random or regular lattice structure. The general strategy here is to input the physical properties of the fluid and the porous medium description (porosity).The flow rate is known to be proportional to the pressure drop, and the constant of proportionality is known as conductance which includes the fluid/porous medium details. The solution of the simultaneous equations corresponding to each node must satisfy the overall conservation of mass and the pressure at the inlet and outlet of the network. For Newtonian fluids, conductance for a particular capillary is a known function of the fluid viscosity. This approach has been shown to yield satisfactory results, e.g., see Blunt et al. (2002), Sochi (2009), Blunt (2001), etc.

6.3.2.6 Volume Averaging of the Navier–Stokes Equations

The techniques for volume averaging of the Navier–Stokes equations over a representative portion of a rigid porous mass have been extensively dealt with, among others, by Whitaker (1970, 1998), Slattery (1972), and Mikelic (1997). There has been a renewed interest in the use of this approach to describe the flow of Newtonian and non-Newtonian fluids in granular and fibrous porous media (van der Westhuizen and du Plessis, 1994, 1996; Liu et al., 1994; Hayes et al., 1995, 1996; Liu and Masliyah, 1996a, 1996b, 1998, 1999; Wu and Pruess, 1996, 1998; Smit and du Plessis, 1997, 1999, 2000; Travkin and Catton, 1998; Diedericks et al., 1998; Getachew et al., 1998; Woudberg et al., 2006). Some homogenization theories and effective medium approach have also been outlined by Shah and Yortsos (1995), Orgeas et al. (2006), and Airiau and Bottaro (2020), and for averaging in the non-Darcy regime by Tsakiroglou (2002). Here only the representative final results are presented. For steady, incompressible, and creeping flow of a Newtonian medium, Dullien and Azzam (1973a,b) used the volume-averaging technique by Slattery (1972) to obtain the following form of the Navier–Stokes equation:

$$\nabla^* \langle p \rangle = \mu V_0 \left\{ \frac{1}{D^2 \forall} \int \nabla^{*2} V^* d\forall \right\} - \rho V_0^2 \left[\left(\frac{1}{D} \right) \left(\frac{1}{\forall} \right) \left\{ \int_\forall V^* \cdot \nabla^* V^* d\forall + \int_{A_i} P^* n dA \right\} \right] \quad (6.32)$$

where

$$V^* = \frac{V}{V_0}; \quad P^* = \frac{P}{\rho V_0^2}; \quad \nabla^* = D\nabla \quad (6.33)$$

For 1-D flow, Equation 6.32 can be written as

$$-\frac{\Delta p}{L} = \alpha \mu V_0 + \beta \rho V_0^2 \quad (6.34)$$

Equation 6.34 is the well-known Forchheimer equation with α and β being two constants. The two terms on the right-hand side of Equation 6.34 are recognized as the "viscous" and "inertial" contributions. Despite the fact that Equation 6.34 describes experimental data within the limits of experimental errors, some investigators have added another term to Equation 6.34 (see Bear, 1972; Dullien, 1992; Adler et al., 2013). Perhaps the most successful and well-known correlation based on this approach is that of Ahmed and Sunada (1969) who reported values of α and β for wide ranges of physical and kinematic conditions. At low velocities, the second term is negligible, and Equation 6.34 reduces to the Darcy equation whereas at high flow rates, the first term drops out, thereby leading to a constant value of the friction factor. Subsequently, Liu et al. (1994) have reinvestigated the volume-averaging procedure and resolved the differences resulting from the use of the Darcy and Brinkman equations at low flow rates. They put forward the following expression for friction factor (in the absence of wall effects) encompassing both Darcy's and Forchheimer's regimes:

$$f\frac{\varepsilon^{11/3}}{(1-\varepsilon)^2} = \frac{85.2}{\mathrm{Re}} + \frac{0.69 F^3(\varepsilon)\mathrm{Re}^3}{256 + F^2(\varepsilon)\mathrm{Re}^2} \tag{6.35}$$

where

$$F(\varepsilon) = \frac{1 + (1-\sqrt{\varepsilon})^{1/2}}{(1-\varepsilon)\varepsilon^{1/6}} \tag{6.36}$$

They reported a good match between the predictions of Equation 6.35 and experimental data for granular beds ($0.36 \leq \varepsilon \leq 0.6$) and fixed fiber foam porous media ($\varepsilon = 0.93$–0.94) up to about $[F(\varepsilon)\cdot\mathrm{Re}]$ ~6000. Liu et al. (1994) also recognized the role of wall effects to be different in the viscous and inertial flow regimes and thus incorporated these in a modified form of Equation 6.35. The dependence of the friction factor on the porosity in the viscous regime is seen to be quite different in this case. Hayes et al. (1995) have compared the predictions of various similar expressions available in the literature. du Plessis and Masliyah (1988) have presented some results on time-independent laminar flow through a rigid isotropic and consolidated porous medium. Using NMR techniques for a high porosity foam, Givler and Altobelli (1994) have reported the effective viscosity for the Brinkman–Forchheimer model to be about 7.5 times the fluid viscosity.

6.3.3 Wall Effects

In most practical applications, the porous medium or packed bed is of finite size in the radial and axial directions whence the confining walls influence the flow phenomena. The wall effects manifest in two ways: the wall of the tube provides an extra surface that comes in contact with the moving fluid and therefore the frictional losses occur over an area larger than that of the particles itself. The second, rather the more important effect, is due to the fact that the bed voidage in the wall region is significantly higher than that in the center of the tube, therefore resulting in the so-called channeling near the wall. Three different approaches have evolved for taking into account the wall effects; the simplest of all is that the constants appearing in the Ergun equation (or any other such expression) are correlated with the diameter ratio of the particle (d) to that of the column (D_c). This approach has been used for both Newtonian and non-Newtonian flows in packed beds (Reichelt, 1972; Fand et al., 1987, 1993; Fand and Thinakaran, 1990; Srinivas and Chhabra, 1992; Foumeny et al., 1993; Raichura, 1999; Eisfeld and Schnitzlein, 2001; de Klerk, 2003). For instance, Reichelt (1972) asserted that, in the range $1.7 < (D_c/d) < 91$, the first (viscous) constant in Equation 6.8a was insensitive to the value of (D_c/d) whereas the second (inertial) constant showed the dependence on (D_c/d) as

$$\frac{1}{\sqrt{B}} = 1.5\left(\frac{d}{D_c}\right)^2 + 0.88 \tag{6.37}$$

As (d/D_c) decreases, $B \to \sim 1.35$, which is 10% lower than the generally accepted value of 1.5. Subsequently, Chu and Ng (1989) and Guo et al. (2017a,b,c, 2019a,b) have investigated the effect of (D_c/d) on the permeability of beds of spheres under wide range of conditions and have reached qualitatively similar conclusions. Yang et al. (2015, 2016) have reported the wall effects to be negligible for the values of (D_c/d) greater than 10 on the transitions between flow regimes.

In an extensive study, Eisfeld and Schnitzlein (2001) have collated 2300 data points encompassing $0.33 \le \varepsilon \le 0.882$; $1.6 \le (D_c/d) \le 57$ and $0.01 \le \text{Re} \le 1.7 \times 10^4$ for beds composed of variously shaped particles (spheres, cubes, cylinders, granules, etc.) and reaffirmed that the approach of Reichelt (1972) yielded the overall minimum deviation when contrasted with the predictions of the other correlations available in the literature. However, in view of the extensive data, they found it necessary to slightly modify the constants appearing in the Ergun equation to rewrite it as

$$f\left(\frac{\varepsilon^3}{1-\varepsilon}\right) = 155 \frac{A_W^2 (1-\varepsilon)}{\text{Re}} + \frac{A_W}{B_W} \qquad (6.38)$$

where

$$A_W = 1 + \frac{2d}{3D_c (1-\varepsilon)} \qquad (6.39a)$$

$$B_W = \left[1.42(d/D_c)^2 + 0.83\right]^2 \qquad (6.39b)$$

Note that in the limit of $(d/D_c) \to 0$, that is, $D_c \to \infty$, $A_w = 1$ and $B_w = 0.69$, and therefore the resulting values of the two constants are 155 and 1.45, which are only slightly different from the original suggestion of Ergun (1952). This approach has, however, been criticized for its completely empirical character (Tsotsas, 2002). More significantly, Eisfeld and Schnitzlein (2001) suggested the wall effect to be dependent on the flow regime. Thus, the wall effects result in higher pressure drop in the viscous region whereas the severe channeling in the high-voidage region near the wall can lead to the reduction in pressure drop at high Reynolds numbers. Therefore, they suggested that in the low Reynolds number flow regime, wall effects are negligible for $(D_c/d) > 10$, in line with the findings of Yang et al. (2015, 2016). Qualitatively, similar observations have been made by others (Liu and Masliyah, 1998; Di Felice and Gibilaro, 2004).

In the second approach, the contribution of the confining walls to the wetted perimeter is incorporated into the definition of the hydraulic radius. Mehta and Hawley (1969) thus obtained the expression for R_h as

$$R_h = \frac{\varepsilon d}{6(1-\varepsilon)A_w} \qquad (6.40)$$

where A_w is given by Equation 6.39a.

Thus, d/A_w is substituted for the characteristic linear dimension, d, in the Ergun equation or in any other expression of this type. This method has also been extended to non-Newtonian fluids (Park et al., 1975; Hanna et al., 1977; Srinivas and Chhabra, 1992). Interestingly, a similar expression, except for the factor of (2/3) in A_w, was presented much earlier by Carman (1937), Coulson (1949), and later by Dolejs (1978). This method, however, does not take into account the variation of the bed voidage in the radial direction. Furthermore, Cohen and Metzner (1981) argued that one would expect the wall effects to persist only in the wall region whereas Equation 6.40 is applied uniformly across the entire bed. Nor does this approach take into account the fact that the nature of wall effects varies from one flow regime to another. This approach has also been used to consolidate the numerical data for structured wall packing friction factor results for $1 \le D_c/d \le \infty$ for the Reynolds number values up to 5000 (Palle and Aliabadi, 2013).

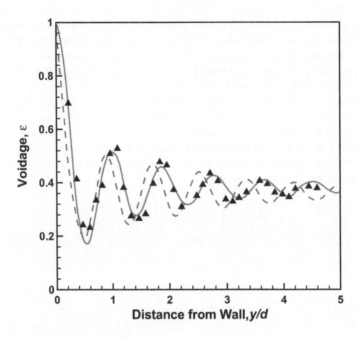

FIGURE 6.4 Typical voidage variation in packed beds: Predictions versus measurements. —— Mueller (1991); - - - - Cohen and Metzner (1981); (▲) Benenati and Brosilow (1962).

The third method takes into account the radial voidage distribution in a bed. The limited literature in this field has been reviewed critically by Cohen and Metzner (1981) and von Seckendorff and Hinrichsen (2021). The voidage is almost unity at the wall, it oscillates about a mean value as one moves away from the wall, and finally, it attains a constant value of the mean bulk voidage (see Figure 6.4). Extensive experimental data elucidating the effects of (d/D_c), particle shape, and particle roughness on the bulk voidage of packed beds as well as on the radial porosity profiles in beds of spherical particles are available in the literature (Roblee et al., 1958; McGeary, 1961; Benenati and Brosilow, 1962; Haughey and Beveridge, 1966; Ridgway and Tarbuck, 1968; Eastwood et al., 1969; Lee, 1970; Marivoet et al., 1974; Pillai, 1977; Chandrasekhara and Vortmeyer, 1979; Crawford and Plumb, 1986; Dixon, 1988; Mueller, 1991, 1992, 1997, 1999, 2005, 2019; Benyahia, 1996; McWhirter et al., 1997; Wang et al., 2001; Lu and Tang, 2015). In fact, Dixon (1988) put forward the following simple relations for the bulk voidage:

For spheres;

$$\varepsilon = 0.4 + 0.05x + 0.412x^2 \qquad x < 0.5$$
$$= 0.528 + 2.464(x - 0.5) \qquad 0.5 \leq x \leq 0.536 \qquad (6.41)$$
$$= 1 - 0.667x^3(2x-1)^{-0.5} \qquad x > 0.536$$

For cylindrical particles,

$$\varepsilon = 0.36 + 0.10x + 0.7x^2 \qquad x < 0.6$$
$$= 0.677 - 9(x - 0.625)^2 \qquad 0.6 \leq x \leq 0.7 \qquad (6.42)$$
$$= 1 - 0.763x^2 \qquad x > 0.7$$

where $x=(d/D_c)$ for spheres, and d is replaced by the equal volume diameter for cylinders. Thus, expressions like Equation 6.41 or 6.42 can be used in conjunction with the capillary model approach to account for the wall effects.

Cohen and Metzner (1981), on the other hand, divided the entire bed into three regions. Based on the measurements for beds of spherical particles available in the literature, they developed the expressions for the local voidage as

$$\frac{1-\varepsilon_y(y)}{1-\varepsilon(y)} = 4.5\left(y-\frac{7}{9}y^2\right) \qquad y \leq 0.25$$

$$\frac{\varepsilon-\varepsilon_y(y)}{1-\varepsilon(y)} = 0.3463e^{-0.4273y}\cos(2.451y-2.2011)\pi \qquad 0.25 \leq y \leq 8 \qquad (6.43)$$

$$\varepsilon_y(y) = \varepsilon \qquad y > 8$$

where $\varepsilon_y(y)$ is the local voidage at distance y (normalized with respect to d) from the walls. Based on extensive comparisons, Cohen and Metzner concluded that if the wall effects are to be avoided, packed beds with (D_c/d) larger than 30 should be used, in line with the finding of Verboven et al. (2004). This suggestion seems to be too stringent as the experimental results point to the limiting value of $D_c/d > 10$ (Yang et al., 2015, 2016). Subsequently, Nield (1983) developed a two-layer model with its predictions close to that of Cohen and Metzner (1981). Using energetic considerations, Tosun and Mousa (1986) have also reached similar conclusions. On the other hand, subsequent measurements of Giese (1998) for spherical particles packed in cylindrical tubes correlate rather well with the following simple form of distribution (Winterberg and Tsotsas, 2000b):

$$\varepsilon(y) = 0.37\left[1+1.36\exp(-5y)\right] \qquad (6.44)$$

Winterberg and Tsotsas (2000a,b) found Equation 6.44 to be satisfactory for modeling fluid flow and heat transfer in packed beds in the range $4 \leq (D_c/d) \leq 40$. The effects of the confining walls both on the axial and the radial porosity profiles for ternary mixtures of spheres have been investigated by Ismail et al. (2002). Similar results on wall effects in annular beds are also available (Sodre and Parise, 1998; Mueller, 1999).

Other detailed models dividing the bed into several concentric cylindrical layers that undoubtedly provide more accurate description of the voidage profiles, especially in the wall region, are also available in the literature (Govindarao and Froment, 1986; Foumeny and Roshani, 1991; Mueller, 1991, 1992, 1997, 1999), and some of the other early works have been reviewed by Ziolkowska and Ziolkowski (1988). We conclude this section by mentioning the methods of Di Felice and Gibilaro (2004) and Cheng (2011) to account for the wall effects. In essence, this is also a two-region model, but they have made use of the detailed velocity measurements in packed beds (Giese, 1998; Johns et al., 2000). They assumed the wall region to extend only up to the distance of $(d/2)$ from the wall, albeit they asserted their predictions to be relatively insensitive to this choice, and these were found to be in agreement with the literature data. Cheng (2011), on the other hand, has decomposed the total pressure drop across a packed bed into two components: due to the flow in an empty tube of radius R_h at a velocity of (V_o/ε) and that due to the particles. The latter contribution is postulated to be proportional to the particle drag which will depend upon the number of particles. This results in the wall correction beyond the proposal of Mehta and Hawley (1969) in terms of the A_w factor (Equation 6.39a). Comiti and Renaud (1989) and Liu and Maslivah (1998) have also put forward modified forms of Equations 6.18, 6.19, and 6.35 to account for wall effects on pressure loss in fixed beds. Similar proposals have also been made by Yang et al. (2015), Das et al. (2017), and Guo et al. (2017b) for beds of uniform and mixed-size spheres.

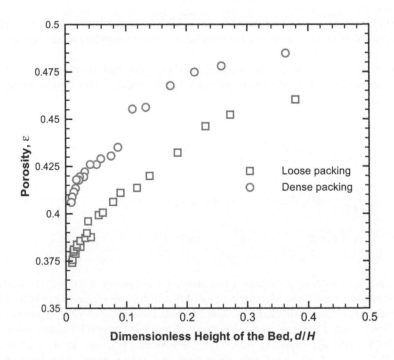

FIGURE 6.5 Thickness effect on the porosity for the ratio of $d/D = 3/74$ for loose and dense packing. (Replotted from Zou, R.P. and Yu, A. B., *Chem. Eng. Sci.*, **50**, 1504–1507, 1995.)

Similarly, the mean voidage of the bed is also influenced by the height of packed bed nondimensionalized using packing diameter. Much of the literature in this field has been reviewed recently by von Seckendorff and Hinrichsen (2021). This effect is readily visualized in terms of the coordination number of a particle in the bed. Notwithstanding the wide scatter, the mean bed porosity increases with the decreasing coordination number. Extensive experimental studies indicate that the bottom plate influences the mean porosity the same way as the column walls do. In an extensive study, Zou and Yu (1995) varied the dimensionless height of the bed from 2.5 to 100 and reported that this effect was negligible for bed heights $>20d$ for the both dense and loose random beds (Figure 6.5).

6.3.4 Effects of Particle Shape, Particle Roughness, and Size Distribution

It has long been known that the voidage (hence pressure drop) of a packed bed is strongly influenced by the particle size distribution, particle shape, and roughness. Particle shape is a much more important variable in determining the value of voidage than roughness, albeit both influence the voidage in the same way. It is readily recognized that the particle shape and orientation are also difficult to characterize. One widely used measure of shape is the sphericity (ψ), which is defined as the ratio of the surface area of a sphere (of the same volume) to that of the particle. This, combined with an equal volume sphere diameter, is used to account for the size and shape of a nonspherical particle. The porosity of a randomly packed bed, in turn, correlates rather well with the sphericity as shown in Figure 6.6 for different types of packings, and such values have been listed by Brown et al. (1950). The effect of particle shape on bulk porosity has been investigated by Eastwood et al. (1969), Milewski (1978), Dixon (1988), Benyahia (1996), Finkers and Hoffmann (1998), Wang et al. (2001), and Benyahia and Ó Neill (2005), among others. Both Crawford and Plumb (1986) and Jordi et al. (1990) reported the particle roughness to result in greater pressure loss than the smooth particles over the complete range of Reynolds numbers, a finding that is at variance with that of Macdonald et al. (1979) and Comiti and Renaud (1989). The role of particle roughness at low Reynolds numbers is less clear than that at high Reynolds numbers. Limited results with the fixed beds of compressible particles indicate nonlinear relation between the pressure drop and

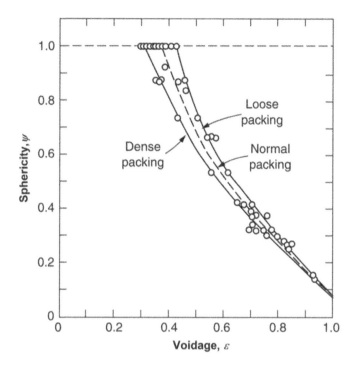

FIGURE 6.6 Dependence of mean porosity on sphericity for beds of nonspherical particles. (Replotted from Brown, G.G. and associates, *Unit Operations*, Wiley, New York, 1950.)

velocity, even at small flow rates (Buchholz and Godelman, 1973; Jönsson and Jönsson, 1992). Likewise, the role of end effects in shallow beds has been studied by Rangel et al. (2001) and in tapered beds by Venkataraman and Mohan Rao (2000).

It is customary to express the effect of particle shape implicitly by modifying Equation 6.15 for a nonspherical particle as

$$V_0 = \frac{1}{K_1}\left(\frac{\varepsilon^3}{a_{vs}^2(1-\varepsilon)^2 \mu}\right)\left(-\frac{\Delta p}{L}\right) \quad (6.45)$$

where K_1 includes the tortuosity factor and the effect of particle shape; a_{vs} is the specific surface area of the particle. Both Coulson (1949) and Wyllie and Gregory (1955) have determined the values of K_1 for a range of shapes, and some of their results are shown in Figure 6.7. Many others (MacDonald et al., 1991; Hamilton, 1997; Li and Park, 1998; Endo et al., 2002) have studied the Newtonian fluid flow in packed beds of polydispersed spheres and have reported their results in terms of a normalized permeability using an average particle size based on the first and second moments of particle size distribution. From engineering design calculations viewpoint, MacDonald et al. (1979) have shown that the widely used Ergun equation yields satisfactory results provided the sphericity factor is included in the definition of the equivalent particle diameter, in line with the suggestion of Li and Ma (2011). However, based on an extensive study with spheres, cylinders, and hollow cylinders (with and without roughness), von Seckendorff et al. (2020) found it necessary to modify the Ergun equation to account for the combined effects of particle shape and roughness and of confining walls, all of which increase the pressure drop across the bed. Other empirical correlations developed specifically for nonspherical particles are also available in the literature (Batishchev, 1986; Foumeny et al., 1996) but these have not been tested as widely as the Ergun equation. In a recent study, Carpinlioglu and Ozahi (2008) have studied the effect of

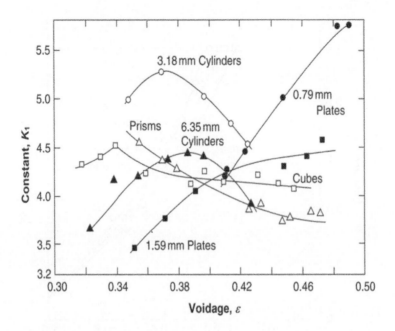

FIGURE 6.7 Variation of K_1 (Equation 6.45) with bed porosity for nonspherical particles. (Replotted from Coulson, J.M. and Richardson, J.F., *Chemical Engineering*, vol. 2, 5th edition, Butterworth-Heinemann, Oxford, 2002.)

particle sphericity (0.55–1) in the turbulent flow regime by using beds of wide-ranging values of porosity ($0.36 \leq \varepsilon \leq 0.56$) at Reynolds number values above 600.

In conclusion, it is important to reiterate here that the foregoing treatment is essentially limited to one-dimensional and steady incompressible fluid flow in unconsolidated (except occasional reference to consolidated medium) and isotropic granular porous media and packed beds. In practice, however, it may not be possible to justify some or all of these simplifications due to the anisotropy of the medium or the compressibility of flow, or the other nonideal features displayed by a porous medium. Additional methods to deal with these situations have been discussed by Greenkorn (1983), Dullien (1992), Wang et al. (1999), Tobis (2000, 2002), Bear (2018), and Xue et al. (2021).

6.3.5 Fibrous Porous Media

As noted earlier, while most porous media are granular in nature, many are composed of very long particles and hence it is appropriate to describe these as fibrous porous media. The permeation and flow of fluids in fibrous porous media is encountered in various settings including aerosol filtration, production, and processing of polymer and metallic composites (Williams et al., 1974; Lin et al., 1994), biotechnological and biomedical applications (Jackson and James, 1982; Chen et al., 1998a), in the processing of textile fibers, paper pulp suspensions, steel wool and cotton batting, gel membranes (Cartier et. al., 1995; Johnson and Deen, 1996), etc. Conversely, fibrous systems have also been studied because fibers can form stable structures of very high porosity, thereby providing high specific surface area that offer relatively low resistance to fluid flow and may also facilitate heat/mass transfer with or without chemical reactions. Finally, it is also customary to include in this category some structures ordinarily not thought of as porous media, such as banks and bundles of heat exchanger tubes and entangled polymer chains in solutions, mats and screens used for cleaning of gases and filtration of slurries. The constituent fibers may be straight or curved, rigid or flexible, man-made or natural, of circular or noncircular cross-section, randomly oriented or arranged in regular arrays, but irrespective of such details, our interest here is limited to those that are sufficiently long so that their aspect ratio is not an issue.

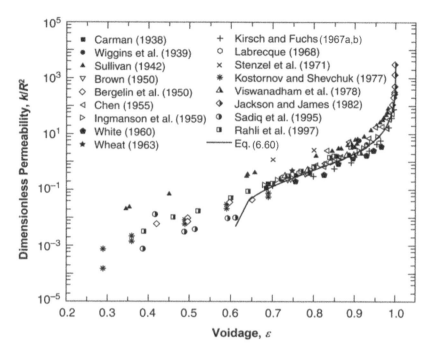

FIGURE 6.8 Dimensionless permeability (k/R^2) plotted against voidage for fibrous media. (Modified after Jackson, G.W. and James, D.F., *Can. J. Chem. Eng.*, **64**, 364, 1986).

Fiber size is thus characterized by a single cross dimension for circular fibers. A significant characteristic that sets the fibrous media apart from the granular media is their relatively low solid fraction, that is, high porosity. Indeed, it is not uncommon for fibrous media to have $\varepsilon > \sim 0.98$. Much of the literature available in this field is concerned with the prediction of the resistance to flow that is expressed using either permeability (Equation 6.1) or the so-called Kozeny constant or simply the usual friction factor–Reynolds number coordinates, and all of these will be used here. Needless to say, one can readily (at least in the viscous flow regime) convert results from one form to another. Excellent reviews of theoretical and experimental developments in this field are available in the literature (Jackson and James, 1986; Levick, 1987; Skartsis et al., 1992a; Mauret and Renaud, 1997). Most advances in this field hinge on the same strategies as that used for granular media presented in Section 6.3.4. Therefore, the thrust in this section is on the results. Most of the low Reynolds number experimental results obtained with a range of fibrous media ranging from glass rods, to hair, to glass wool, to polymer gel, collagen, etc. have been collated by Jackson and James (1986) in terms of a dimensionless permeability, k^* ($= k/R^2$) where R is the radius of the fiber. The scaling arguments suggest the dimensionless permeability to be a function of the porosity, ignoring the detailed structure of the medium. Figure 6.8 shows the bulk of the literature results as synthesized by Jackson and James (1986) supplemented by subsequent limited results (Table 6.3). The results shown in this figure include a wide range of materials that in itself is probably responsible in part for the scatter present. In the literature, the term fiber arrangement is frequently used, which encompasses a number of factors such as the extent of fiber alignment and the homogeneity of the bed (or the porous medium). It is well known that the resistance to flow when the fibers are oriented normal to the direction of flow is twice the value when they are aligned with the direction of flow (Jackson and James, 1986; Skartsis et al., 1992a). Intuitively, it appears that the more homogeneous the medium, the lower is the permeability. Obviously, it is very difficult to make a truly homogeneous fibrous medium, or even to ascertain the degree of homogeneity. Not only can inhomogeneities increase the permeability, but it can also lead to severe channeling.

TABLE 6.3

Summary of Experimental Results on Fibrous Media

Reference	System	R(mm)	Range of ε	Range of k^*
Bergelin et al. (1950)	Transverse flow over tube bundles	4.76 and 9.53	0.42–0.651	0.006–0.046
Brown (1950)	Glass wool	0.046	0.738–0.912	0.25–3.4
Carman (1938)	Stainless-steel wire crimps	0.164	0.681–0.765	0.14–0.38
Chen (1955)	Filter pads	$4.7 \times 10^{-4} - 8.5 \times 10^{-3}$	0.805–0.9936	0.669–128
Davies (1952)	Merino cotton, glass wool, rayon, kapok, down fabric	-	0.7–0.994	Empirical correlations (Table 6.4)
Ingmanson et al. (1959)	Nylon and glass fibers	0.0965 and 0.082	0.68–0.955	0.3–8
Jackson and James (1982)	Hyaluronic acid polymer	4.9×10^{-7}	0.9896–0.99965	75–3000
Johnson and Deen (1996)	Agarose gels	1.9×10^{-6}	0.93–0.98	-
Kirsch and Fuchs (1967a)	Kapron fibers	0.15, 0.225, 0.4	0.7–0.9945	0.14–82
Kostornov and Shevchuck (1977)	Alloy metal fibers	0.025	0.29–0.69	1.43×10^{-4} –0.077
Labrecque (1968)	Nylon fibers	0.0012	0.725–0.84	0.28–0.95
Rahli et al. (1995, 1996, 1997, 1999)	Copper and bronze fibers	150×10^{-3}	0.39–0.89	8.44×10^{-4} –0.877
Sadiq et al. (1995)	Aluminum and nylon rods	0.794 and 3.175	0.388–0.611	5.2×10^{-3}–0.039
Stenzel et al. (1971)	Collagen	1.5×10^{-6}	0.761–0.893	1.25–2.70
Sullivan (1941, 1942)	Goat wool,	0.0195	0.64–0.984	0.317–90
	blond hair,	0.0327	0.45–0.656	0.0685–0.41
	Chinese hair,	0.0367	0.346–0.355	0.0214–0.0233
	glass wool	0.038	0.866	2.99
Viswanadham et al. (1978)	Collagen	$1–1.55 \times 10^{-6}$	0.73–0.9125	0.316–2.08
Wheat (1963)	Glass fibers	3.9×10^{-5} and 7.2×10^{-5}	0.747–0.769	0.41–0.47
White (1960)	Polymer gel	3.5×10^{-7}	0.755–0.965	0.2–3.6
Wiggins et al. (1939)	Glass rods,	0.204	0.685	0.118
	copper wire,	0.0508	0.83	0.846
	glass wool,	9.09×10^{-3}	0.846–0.91	1.37–5.81
	fiber glass,	3.5×10^{-3}	0.885–0.93	1.93–4.42
	yarn	6.7×10^{-3}	0.898–0.904	3.59–3.98

Many investigators, on the other hand, have used the Kozeny constant to report their results (Davies, 1952; Chen, 1955; Ingmanson et al., 1959; Carroll, 1965 as cited by Han (1969)). The Kozeny constant k_k is related to the permeability k via the simple relationship

$$k_k = \frac{\varepsilon^3}{k(1-\varepsilon)^2 a_{vs}} \tag{6.46}$$

where a_{vs} is the specific surface area of the particle or fiber. For a long fiber of radius R, $a_{vs} = (2/R)$. Carman (1937, 1938) interpreted the Kozeny constant to account for the pore shape and tortuosity. For beds of spheres, $k_k \approx 5$ is thought to be a good approximation. For fibrous materials, k_k takes on different values. Table 6.4 provides a summary of representative correlations available in the literature for the prediction of k_k. Based on extensive comparisons between the capillary bundle approach that is germane

TABLE 6.4

Empirical Expressions for Kozeny Constant (k_k)

Reference	Equation for k_k	Observations
Davies (1952)	$\dfrac{A\varepsilon^3}{(1-\varepsilon)^{1/2}}\{1+56(1-\varepsilon)^3\}$	$A=4$ for $\varepsilon<0.98$ $A=4.4$ for $\varepsilon>0.98$
Carroll (1965) (Cited by Han, 1969)	$5+\exp\{14(\varepsilon-0.8)\}$	No details available
Ingmanson et al. (1959)	$\dfrac{3.5\varepsilon^3}{(1-\varepsilon)^{1/2}}\{1+57(1-\varepsilon)^3\}$	Very similar to that of Davies (1952).
Chen (1955)	$\dfrac{0.484\varepsilon^2}{(1-\varepsilon)}\left[\ln\dfrac{0.64}{(1-\varepsilon)^{1/2}}\right]^{-1}$	$\varepsilon>0.7$
Rahli et al. (1997)	$\dfrac{3.6}{\varepsilon}$ (for $l/d \to \infty$)	$0.39\leq\varepsilon\leq0.89$
	$\dfrac{3.6}{\varepsilon}+\dfrac{30}{(l/d)}$	$4.5\leq(l/d)\leq 67$

to the validity of Equation 6.46 and experimental results, it can be said that this approach breaks down for very low porosity systems (close to the maximum packing fraction), high-porosity (> ~0.8) systems, or due to blocked passages, adsorption, and other surface effects. This approach does not account for pore nonuniformities as well. Thus, the physical interpretation of k_k is quite hazy when some of these assumptions are no longer valid. Despite these limitations, the value of k_k lies between 4 and 5 in the porosity range ~$0.4\leq\varepsilon\leq0.7$. Skartsis et al. (1992a,b) also collated most of the literature data and classified the available experimental data into coarse and fine fibers. Quite arbitrarily, the fibers with radii in the range 40–200 µm were labeled as coarse fibers. This included steel wires (Carman, 1938), glass rods (Sullivan and Hertel 1940), copper wires (Wiggins et al., 1939), filter mats (Chen, 1955), fine nylon fibers (Labrecque, 1968), glass wool (Brown, 1950), nylon and glass fibers (Ingamanson et al., 1959), and textile fibers (Anderson and Warburton, 1949; Lord, 1955). On the other hand, fibers with radii in the range 1–9 µm were regarded as fine fibers and these included fine glass wool and fiberglass (Wiggins et al., 1939), filter mats, fine nylon fibers, etc. These results are plotted in Figure 6.9 where the fine fiber data are seen to be scattered more than that for the coarse fibers. Some data on 3-D permeability of fibrous porous media are also available (Weitzenböck et al., 1997).

6.3.6 Theoretical Treatments

The general idea here seems to be to mimic the fibrous porous medium as a matrix of long rods (of circular and noncircular cross-sections) and then seek analytical/numerical solutions of the Navier–Stokes equations in the limit of zero Reynolds numbers (creeping flow), albeit limited results are available that take into account the inertial effects (Edwards et al., 1990; Martin et al., 1998; Asif and Dhiman, 2018; Pravesh et al., 2019a,b). The available body of information can be conveniently divided into three categories depending upon the structure of the matrix (geometrical details) and the direction of flow: (1) flow parallel to cylinders or rods, (2) flow normal to an array of parallel rods, and (3) flow in 3-D arrays. While some results are available for arrays of rectangular (Fardi and Liu, 1992), square (Wang, 1996b), and elliptic cross-sections of cylinders (Raynor, 2002), the ensuing discussion is mainly related to the rods of circular cross-section.

In the first two cases, the rods are arranged in a periodic pattern such as a square or triangular or rectangular configuration and it is thus possible to define a representative unit cell (a polygon with a rod at the center) such that the flow in the cell is equivalent to that in the assembly. Thus the complex multirod problem is reduced to the much simpler one-body problem confined in a complex flow domain. No-slip condition is used on the solid boundaries, and zero velocity gradient or periodicity conditions are used at

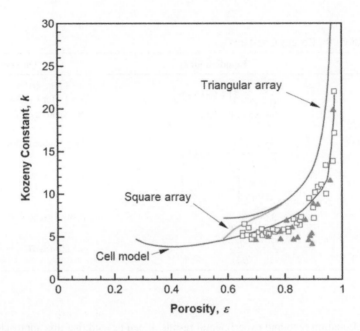

FIGURE 6.9 Typical comparison between predictions and data for fibrous media. (□) Coarse fibers; (▲) Fine fibers. The predictions for the triangular and square arrays are from Drummond and Tahir (1984) and Sangani and Acrivos (1982a). The free surface cell model predictions are from Happel (1959.)

the cell boundary. Clearly, the final result is a function of both the cell shape and the boundary condition used at the cell boundary.

6.3.6.1 Flow Parallel to an Array of Rods

Langmuir (1942) seems to be the first to have tackled this problem, and he solved the problem with a cylinder caged in a cylindrical cell and used the condition of the zero-shear stress at the cell boundary. His final result is

$$k^* = \frac{1}{4(1-\varepsilon)}\left[-\ln(1-\varepsilon) - \frac{3}{2} + 2(1-\varepsilon) - \frac{(1-\varepsilon)^2}{2}\right] \tag{6.47}$$

This result was subsequently rederived independently by Happel (1959). The corresponding result when the zero-shear stress boundary condition is replaced by the zero-vorticity condition as proposed by Kuwabara (1959) coincides with Equation 6.47, see also Kirsch and Fuchs (1967b). Sparrow and Loeffler (1959) presented series solutions for square and triangular arrays, and their results are available in a graphical form. Perhaps the most significant effort in this category is by Drummond and Tahir (1984), and their results are given here:

$$k^* = \frac{1}{4(1-\varepsilon)}\left[-\ln(1-\varepsilon) - K_{DT} + 2(1-\varepsilon) - \frac{(1-\varepsilon)^2}{2}\right] \tag{6.48}$$

where K_{DT} is a constant that varies from one arrangement to another. It takes on values of 1.476 for a square array, 1.498 for an equilateral triangular array, 1.354 for a hexagonal array, and 1.13 for a 2×1 rectangular array. In spite of the inherent differences in the geometry, there is a striking similarity in all these expressions for k^* given by Equations 6.47 and 6.48. Some of these results have been verified by Toll (2001), Mityushev and Adler (2002), and Tamayol and Bahrami (2010) using different solution methods. However, as expected, these results overpredict experimental results due to the simple fact that the fibers in a real medium are seldom oriented parallel to the direction of flow.

6.3.6.2 Transverse Flow over an Array of Rods

Numerous results are available for the 2-D transverse flow of Newtonian liquids past arrays of parallel cylinders oriented normal to the direction of flow. These studies also embrace a wide variety of geometrical arrangements and extend over a wide range of Reynolds numbers. While the creeping flow results are relevant to the applications in filtration, polymer processing, and in biological systems, the flow in tubular heat exchangers invariably tends to be at moderate to high Reynolds numbers. We begin with the creeping flow results for this configuration.

6.3.6.3 Creeping Flow Region

The earliest results for this configuration appear to be those of Happel (1959) and Kuwabara (1959) based on the concentric cylinder cell models. For the zero-shear stress condition at the cell boundary, Happel obtained the expression for the dimensionless permeability as

$$k^* = \frac{1}{8(1-\varepsilon)}\left[-\ln(1-\varepsilon) + \frac{(1-\varepsilon)^2 - 1}{(1-\varepsilon)^2 + 1}\right] \tag{6.49}$$

where the corresponding result for the zero-vorticity boundary condition is given by (Kuwabara, 1959)

$$k^* = \frac{1}{8(1-\varepsilon)}\left[-\ln(1-\varepsilon) - \frac{3}{2} + 2(1-\varepsilon)\right] \tag{6.50}$$

In the same year, Hasimoto (1959) also presented a result for a square array of circular cylinders as

$$k^* = \frac{1}{8(1-\varepsilon)}\left[-\ln(1-\varepsilon) - 1.476 + 2(1-\varepsilon) + \cdots\right] \tag{6.51}$$

Subsequently, this result has been extended by Sangani and Acrivos (1982a) and by Drummond and Tahir (1984) by calculating higher order terms. The extended version of Equation 6.51 by Sangani and Acrivos (1982a) is

$$k^* = \frac{1}{8(1-\varepsilon)}[-\ln(1-\varepsilon) - 1.476 + 2(1-\varepsilon) - 1.774(1-\varepsilon)^2 + 4.076(1-\varepsilon)^3] \tag{6.52}$$

The result of Drummond and Tahir (1984) coincides with Equation 6.52, except for the $(1-\varepsilon)^3$ term.

For a hexagonal array, Sangani and Acrivos (1982a) presented the expression

$$k^* = \frac{1}{8(1-\varepsilon)}\left[-\ln(1-\varepsilon) - 1.49 + 2(1-\varepsilon) - \frac{(1-\varepsilon)^2}{2}\right] \tag{6.53}$$

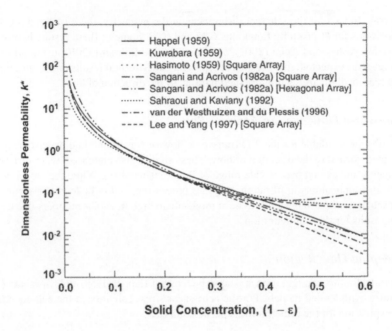

FIGURE 6.10 Comparison between various predictions of permeability of fibrous media.

The results of Sangani and Acrivos (1982a) have been reconfirmed subsequently by numerous numerical studies (Skartsis et al., 1992a; Nagelhout et al., 1995; Koch and Ladd, 1997). The results for different arrangements of cylinders appear to be in moderate agreement, with the notable exception of Equation 6.49, as seen in Figure 6.10.

Based on the assumption that, in random fibrous beds, the total resistance draws (2/3) contribution from the transverse flow, theoretical curves for square arrays (Sangani and Acrivos, 1982a), triangular arrays (Drummond and Tahir, 1984), and the cell model predictions (Happel, 1959) are included in Figure 6.9 where the experimental values are seen to be overpredicted by up to 50% or even more.

Returning to the theoretical studies, Lee and Yang (1997) have calculated the permeability of a staggered array (with equal pitch in both directions), and their numerical results ($0.4345 \leq \varepsilon \leq 0.9372$) can be represented as

$$k^* = \frac{4\varepsilon^3(\varepsilon - 0.2146)}{31(1-\varepsilon)^{1.3}} \tag{6.54}$$

This expression predicts $k^*=0$ at $\varepsilon=0.2146$.

Similarly, while Sahraoui and Kaviany (1992) studied a range of configurations of cylinder arrays, their results ($0.4 \leq \varepsilon \leq 0.8$) for the inline arrangement of rods is given by

$$k^* = \frac{0.0606\pi\varepsilon^{5.1}}{(1-\varepsilon)} \tag{6.55}$$

These two predictions are also included in Figure 6.10. The predictions of Kuwabara (1959), Hasimoto (1959), Sangani and Acrivos (1982a), and Drummond and Tahir (1984) are virtually indistinguishable from each other, especially in the range $\varepsilon \geq \sim 0.6$. A similar trend is also present in the numerical results of Sahraoui and Kaviany (1992) and Lee and Yang (1997). Up to about $\varepsilon \geq 0.6$, most predictions appear to be consistent with each other, but these begin to deviate increasingly from each other for denser systems. From a practical standpoint, however, $\varepsilon < \sim 0.6-0.7$ is of minor importance in the context of fibrous media, albeit some tubular heat exchangers can have voidage in the vicinity of ~0.5–0.55 (Adams and Bell, 1968). Keller (1964), on the other hand, has used lubrication approximation to treat dense systems of evenly spaced and square arrays of cylinders. Hellström et al. (2010) have numerically studied the transverse flow to rows of cylinders in the laminar and turbulent conditions and reported the permeability to be independent of the Reynolds number up to about Re ~10 and this was suggested to be the limit of the Darcy-like viscous flow regime. The effect of the angle of incidence on the permeability of a 3-D array of cylinders has been studied by Chamsri and Bennethum (2015).

Some results on the influence of fiber shape cross-section are also available in the literature. The early results of Epstein and Masliyah (1972) show that for the normal flow with elliptical fibers (with major to minor axes ratio of 5), the permeability for the case of flow parallel to the major axis can be 80% higher than that for circular fibers. On the other hand, permeability for the flow aligned with minor axis was reduced by about 75% below the value of the circular fibers. The shape seems to exert more influence for systems with $\varepsilon > \sim 0.9$. These predictions are qualitatively consistent with the experimental results of Labrecque (1968). Similarly, arrays of fibers of square and rectangular cross-sections have been studied by Fardi and Liu (1992), Wang (1996b), and Raynor (2002), among others, to elucidate the role of fiber shape on permeability in transverse flow. The effect of slip at the surface of a fiber has been studied by Wang (2003). The effect of inhomogeneity for the cross-flow configuration has been assessed by Yu and Soong (1975), Ethier (1991), and Kolodziej et al. (1998). Yu and Soong (1975) divided the bundle into equal-size compartments. In turn, the permeability of each compartment was estimated via the cell models of Kuwabara (1959) and Happel (1959). The overall permeability was estimated by suitably summing the resistances of the individual compartments. On the other hand, Kolodziej et al. (1998) considered the flow normal to a bundle of circular cylinders parallel to each other but placed nonuniformly in space. They attempted an approximate analytical solution. In another study, Spielman and Goren (1968) approached this problem by burying each fiber into an effective porous medium, similar to the model of Brinkman (1947, 1948). Subsequently this approach has also been used by others, for example, see Neale and Masliyah (1975) and Guzy et al. (1983). Suffice it to add here that these predictions are not too different from the ones shown in Figure 6.10. This approach was also extended to a fibrous medium consisting of rods randomly oriented in all three directions. Davis and James (1996) have similarly used an array of thin annular disks arranged in a square or triangular configuration to mimic the flow in fibrous media. Their results show that ring arrays generally have higher permeabilities than equivalent rod arrays, even though the rings create more tortuous flow channels. The effects of the finite length (short) and the curvature of fibers on permeability have been considered analytically by Howells (1998) and experimentally by Rahli et al. (1997).

Finally, we end this section by mentioning the results for randomly oriented fibrous media. The numerical solutions (creeping flow) through arrays of randomly positioned but aligned circular cylinders are due to Sangani and Yao (1988) using a multipole representation of the velocity disturbance caused by each of the cylinders and that of Higdon and Ford (1996), Clague and Phillips (1997), and Koponen et al. (1998). Subsequently, more extensive simulations have been reported by Sangani and Mo (1994) and Alcocer et al. (1999) for staggered arrays. Ghaddar (1995) has used a parallel computational approach (finite element method) to determine the permeability for several values of voidage, but his results are believed to have slightly less statistical accuracy than those of Sangani and Mo (1994). Some of these effects have been investigated experimentally by Kyan et al. (1970). Based on their results for the flow of various liquids (viscosity: 1–22 mPa s) in random beds of nylon and glass fibers, they correlated their results in terms of the Kozeny constant k_k. For long fibers, they presented the following expression:

$$k_k = \frac{\varepsilon^3 \left[62.3 N_e^2 (1-\varepsilon) + 107.4 \right] (1 + f_d \, \text{Re})}{16 N_e^6 (1-\varepsilon)^4} \tag{6.56}$$

where N_e is the so-called effective pore volume, f_d is the normalized friction factor due to the deflection of fibers, and Re is the Reynolds number of flow defined as ($\rho V_0 d/\mu(1-\varepsilon)$).

In many situations in polymer processing, textile, and in hollow membrane reactors, individual fibers are used to form "ropes" and, in turn, several of these ropes form the porous geometry in which the fluid flow occurs. Clearly such systems exhibit double-porosity characteristics. Some results are available for such systems also, for example, see Papathanasiou (1997, 2001) and Spaid and Phelan (1997).

6.3.6.4 Inertial Effects

In contrast to the extensive literature relating to the creeping flow region, little is known about flows with finite inertial effects. For instance, the values of Reynolds number of the order of 10–100 are encountered in pin fin heat exchangers that are used to cool electronic components and in the shells of hollow-fiber filters (containing disordered arrays of aligned fibers). Therefore, cell model and array approaches have also been extended to obtain numerical results at moderate Reynolds numbers. For instance, Ghaddar (1995) computed pressure drop through periodic and random arrays of cylinders in the range $Re \leq 180$. Edwards et al. (1990) and Hellström et al. (2010) examined the effect of Reynolds number on the permeability of periodic arrays of circular cylinders of single and two sizes arranged in a square and hexagonal configuration to ascertain the role of heterogeneity. As expected, they reported the (apparent) permeability to not only exhibit further dependence on the Reynolds number beyond a critical value, but also to be anisotropic. One would, however, intuitively expect an orientation-independent permeability for a sufficiently polydisperse system. Their results are consistent with the scant experimental results available in the literature. Perhaps the most reliable results for periodic and random arrays at finite Reynolds numbers ($Re \leq \sim 180$) are that of Koch and Ladd (1997). In the limit of $Re \ll 1$, the average drag force F_D per unit length of the cylinder in both periodic and random arrays is given by an expression that has two terms—linear in velocity and velocity cubed, respectively, which is in line with the suggestion of Adler et al. (2013). On the other hand, the drag force undergoes a transition from the velocity cubed to velocity squared term somewhere in the range $2 \leq Re \leq 5$ and for $Re > 5$. Koch and Ladd (1997) were able to correlate their numerical results using an Ergun-type expression. Their simulations also reveal the possibilities of time-oscillatory and chaotically varying flow regimes. Likewise, the simple cell models of Kuwabara (1959) and Happel (1959) have also been used extensively to study the transverse flow over bundles of circular rods (LeClair and Hamielec, 1970; Satheesh et al., 1999; Vijaysri et al., 1999; Chhabra et al., 2000; Dhotkar et al., 2000; Shibu et al., 2001). Combined together, these results extend up to $Re = 500$ and $0.4 \leq \varepsilon \leq 0.6$ or so. On the other hand, most of the experimental results in this field have been collated by Prakash et al. (1987), Nishimura et al. (1991), and Ghosh et al. (1994). It is customary to express these results in the form of friction factor–Reynolds number plots. Often, the Ergun-type equation is adequate for this purpose, for example, the one due to Prakash et al. (1987) is written as ($0.4 \leq \varepsilon \leq 0.6$; $Re_2 < \sim 1000$):

$$f_2 = \frac{130}{Re_2} + 0.7 \tag{6.57}$$

where the friction factor and the Reynold number used here are defined as

$$f_2 = \left(\frac{\Delta p}{L}\right)\left(\frac{d}{\rho V_0^2}\right)\left(\frac{\varepsilon^3}{1-\varepsilon}\right) \tag{6.58}$$

$$Re_2 = \frac{\rho V_0 d}{\mu(1-\varepsilon)} \tag{6.59}$$

Similarly, Dybbs and Edwards (1984) quoted the values of the numerical constants as 96 and 1.75 instead of 130 and 0.7, respectively. Needless to add here that these results relate to rigid rod bundles as encountered in heat exchangers. The bulk of the other studies at high Reynolds numbers have been summarized among others by Beale (1999), Beale and Spalding (1998), Ghosh Roychowdhury et al. (2002), Hellström et al. (2010), etc.

In summary, the large scatter seen in Figure 6.8 is primarily due to the bed inhomogeneity, fiber shape, orientation, and slip effects. For arrays of parallel fibers aligned with the direction of flow, the geometrical configuration of the array exerts only a minor influence on the value of the permeability (usually 20%). For the transverse flow, the analytical results can be grouped into two types: arrays and swarm (cell) models. The two results approach each other as the voidage increases. In spite of all these complexities, the results shown in Figure 6.8 can serve as a reasonable basis for estimating the permeability of a fibrous medium. These results can be approximated by the empirical expression

$$k^* = -10.68 + 39.33\varepsilon - 58.75\varepsilon^2 + 32.05\varepsilon^3 \tag{6.60}$$

Equation 6.60 fits the data shown in Figure 6.8 with a regression coefficient of 0.91. But given the number of contributing factors to the variability of the results, the use of Equation 6.60 is suggested only as a first-order approximation. Limited results available at moderate Reynolds number correlate well with a modified form of the Ergun equation.

6.4 Non-Newtonian Fluids

A wealth of information on different aspects of the non-Newtonian fluid flow in porous media is now available. Unfortunately, the growth of the contemporary literature in this rapidly advancing field has been somewhat disjointed, and also the emerging scenario is of highly interdisciplinary character. We begin by providing an exhaustive listing of the pertinent studies on this subject in Table 6.5 to highlight the richness of the literature in this area. An examination of this table shows that various liquid media including polymer melts and solutions, foams, surfactant and micellar solutions, sludges, emulsions and particulate slurries, encompassing wide ranges of fluid characteristics, pseudoplastic, dilatant, viscoplastic, and visco-elastic behavior have been used. Likewise, various porous matrices ranging from simple beds of glass beads (and other types of spheres, pellets, granules, fibers, etc.) to consolidated rocks and cores, granite, sandstones, beds of screens and mats, metallic foams and filters, and 2-D and 3-D arrays of spheres and cylinders, for instance, have been used as model porous media. Each example of a model porous medium is somewhat unique in its geometric morphology, thereby contributing in some measure to the formidable problems of assigning precise geometric description, and of inter-comparisons between results for different media.

Further complications arise from the wide variation in the permeabilities of the nominally similar model porous media. For instance, Christopher and Middleman (1965) used a 25 mm diameter tube packed with glass spheres (710 and 840 μm diameter) and reported the permeability of the order of 450 darcies whereas the glass bead packs (53–300 μm) used by Dauben and Menzie (1967) had an order of 2–18 darcies lower permeabilities. Such large variations even in the macroscopic characteristics of porous media not only make cross-comparisons exceedingly difficult, but also illustrate the complexity of the nature of flow and the role of microstructure in these systems. The characteristics of the consolidated porous media and test liquids used in these studies have been equally diverse.

Table 6.5 clearly shows that much of the research efforts has been expended in elucidating one of the following facets of the non-Newtonian flow phenomena in porous media:

1. To obtain information about the detailed structure of the flow and flow regimes in model porous media including 2-D beds (Gestoso et al., 1999).
2. To derive scale-up relations for predicting pressure loss for a given bed and generalized Newtonian fluid by coupling a specific fluid model with that for a porous medium. It is tacitly assumed that the rheological measurements performed in the well-defined viscometric configurations adequately describe the flow in packed beds and porous media, albeit there is sufficient

TABLE 6.5

Summary of Investigation of Non-Newtonian Flow in Porous Media

Investigator	Test Liquids and Fluid Model	Type of Porous Medium	Remarks
Al-Fariss et al. (1983)[a,d]; Al-Fariss (1989, 1990); Al-Fariss and Pinder (1987)	Crude oils (Herschel–Bulkley) model	Sand beds	Developed semi-theoretical method for friction factor. Also studied temperature effects.
Al Varado and Marsden (1979)[a,d]	Oil/water emulsions	Cores ($0.15 \leq \varepsilon \leq 0.25$)	Correlation for pressure drop.
Aubert and Tirrell (1980)[a]	Polystyrene in tetra hydrofuran	Chromatographic columns packed with 6 µm spheres	Polymer retention in micropores and possible causes.
Baijal and Dey (1982)[a]	Aqueous polyacrylamide solutions	Packs of silica sand	Role of chain length and flexibility on adsorption and permeability.
Barboza et al. (1979)[a]	Solutions of polyacrylamide	Beds of glass and steel spheres, and bundle of cylinders	For dilute solutions, Δp shows a peak due to degradation. Elastic effects correlate with the changes in flow patterns around a bundle of cylinders.
Basu (2001)[a]	CMC solutions ($0.83 \leq n \leq 0.98$)	Ceramic spheres (D_c/d) =3.8	Wall effects in creeping flow region.
Benis (1968)	-	-	Modified Darcy's law using the lubrication flow approximation for visco-elastic media.
Bertin et al. (1998a, 1998b)[a]	Foams	Sandstone cores ($\varepsilon=0.32$)	Transient and cross flow of foams in heterogeneous media.
Brea et al. (1976)[a,d]	Aqueous TiO_2 slurries	Beds of glass, lead and steel spheres	Friction factor/Reynolds number results for fixed and fluidized beds.
Briend et al. (1984)[a,d]	Aqueous solutions of Carbopol and polyacrylamide	Beds of glass, bronze and lead spheres	Friction factor/Reynolds number results for fixed and fluidized beds.
Brunn and Holweg (1988)[a]	Equi-molar mixtures of the cationic surfactant C16TMA-Sal and NaBr	Bed of glass spheres ($d=392$ µm)	Elucidate the importance of shearing for the flow of surfactant solutions.
Burcik (1965, 1968, 1969); Burcik and Ferrer (1968); Burcik and Walrond (1968)	Aqueous solutions of Polyacrylamide	Sandstones	Qualitative results on adsorption and mechanical entrapment of macromolecules, gel formation, etc. and their influence on permeability.
Cakl et al. (1988); Cakl and Machac (1995)[a]	Boger fluids and aqueous solutions of PPA and PEO	Beds of spherical and nonspherical particles	Visco-elastic effects in fixed beds correlate with Deborah number.
Chhabra and Raman (1984)[b]; Chhabra and Srinivas (1991)[a]; Srinivas and Chhabra (1992)[a]; Sharma and Chhabra (1992)[a]	Aqueous solutions of CMC (Carreau and Power-law models)	Beds of Raschig rings, gravel chips, and glass spheres	Approximate upper and lower bounds on drag coefficient using cell model for spherical particles. Experimental data for spherical and nonspherical particles.
Christopher and Middleman (1965)[a,d]	Aqueous solutions of CMC and PIB in toluene (power law)	Beds of spherical glass beads	Modified Blake–Kozeny equation for power-law fluids.
Churaev and Yashchenko (1966)[a]	Aqueous solutions of human sol (visco-elastic media)	Sand packs	Effect of mixed size of particles on Δp.

(Continued)

TABLE 6.5 (*Continued*)

Summary of Investigation of Non-Newtonian Flow in Porous Media

Investigator	Test Liquids and Fluid Model	Type of Porous Medium	Remarks
Ciceron et al. (2002a)	Aqueous CMC solutions (power law)	Bed of glass spheres of mixed sizes	Effect of mixed size particle on Δp.
Cohen and Chang (1984)[a,d]; Cohen and Christ (1986)	Aqueous solutions of HPAM J-333 and water-in-oil emulsions (power law model)	Beds of silica cores and glass beads	New results on the mobility reduction and estimated the effective thickness of absorbed layers.
Cohen and Metzner (1981)	Power-law	-	Developed a three region model to account for the wall effects.
Comiti et al. (2000a, 2000b)[a]; Seguin et al. (1998a, 1998b)	Newtonian fluids and CMC solutions	Beds of spheres and plates, and synthetic foams	Delineation of flow regimes using electrochemical microelectrodes.
Dabbous (1977)[a,d]	Aqueous solutions of a range of commercially available polymers used in flooding	Berea sandstone cores	Qualitative and quantitative results on the mobility and resistance factor.
Dauben and Menzie (1967)[a,e]	Aqueous solutions of PEO (power law)	Glass beads	Significant increase in Δp attributed to in-situ pseudo-dilatant behavior.
Dharmadhikari and Kale (1985)[a,d]	Aqueous solutions of CMC (power law)	Beds of glass beads	Tortuosity factor was found to depend upon the flow rate.
Dhole et al. (2004)[b]	Power-law	Cell model	Numerical results up to Re = 500.
Dolejs and Mikulasek (1997)[a]	Solutions of Natrosol and Methyl cellulose ($0.7 \leq n \leq 0.93$)	Fixed and fluidized beds of spheres	Generalized approach to predict pressure drop.
Dolejs and Siska (2000)[a]; Dolejs et al. (1998)[a]	Herschel–Bulkley and Robertson–Stiff visco-plastic models	-	Modification of the Kozeny–Carman equation.
Dolejs et al. (2002)[a]	Visco-elastic fluids	-	Modification of the Rabinowitsch–Mooney equation for pressure drop prediction.
Dominguez and Willhite (1977)[a]	Polyacrylamide in 2% NaCl solution	Porous teflon core ($\varepsilon = 0.21$)	Permeability reduction.
Done et al. (1983)[a]	Polystyrene and polyethylene terephthalate solutions	Sand beds	Effect of preshearing in porous media on the subsequent flow in capillary tubes. The reduction in Δp across capillary tubes was attributed to changes in rheology.
Unsal et al. (1978); Wang et al. (1979); Hong et al. (1981); Duda et al. (1981, 1983)[a,d]	Aqueous solutions of CMC, HEC, PEO, and xanthan gum	Beds of fine glass beads	Inadequacy of the capillary model and the role of convergent–divergent nature of flow.
Durst et al. (1981, 1987)[a]; Durst and Haas (1982)	Drag reducing polymer solutions	Beds of glass beads	Criterion for the onset of visco-elastic effects and the importance of elongational flow.
Edie and Gooding (1985)[a,c]	Nylon, polyethylene, terephthalate, and polypropylene melts (Power-law)	Sintered metal filters ($\varepsilon = 0.35$-0.44)	Melts exhibited nearly Newtonian behavior ($n > 0.89$) and the results showed good agreement with the predictions of the modified Darcy's law.

(Continued)

TABLE 6.5 (Continued)

Summary of Investigation of Non-Newtonian Flow in Porous Media

Investigator	Test Liquids and Fluid Model	Type of Porous Medium	Remarks
Edwards and Helail (1977)[a]	Polyacrylamide solutions (power law)	Glass spheres	Axial dispersion is little influenced by non-Newtonian behavior.
Elata et al. (1977)[a]	Dilute solutions of Polyox	Beds of spherical particles	Importance of elongational effects and developed a criterion for the onset of visco-elastic effects.
Ershaghi (1972)[a]	Solutions of PAA	Berea sandstones	Mobility of solutions.
Falls et al. (1989)[a]	Foams	Beds of glass spheres	Foams were found to behave like Newtonian fluids in porous media.
Fergui et al. (1998)[a]	Aqueous foams	74 μm sphere packs	Modified Darcy law to interpret transient foam flow.
Flew and Sellin (1993)[a]	Solution of PAA	Arrays of rods, bead packs and PCT	Role of extensional flow.
Gaitonde and Middleman (1967)[a,d]	PIB solutions in toluene	Glass bead packs of uniform and binary sizes	No visco-elastic effects were observed.
Garrouch and Gharbi (1999)[a]	Xanthan gum solutions	Glass beads, sandstone cores ($0.2 \leq \varepsilon \leq 0.39$)	Correlation of Δp.
Gheorghitza (1964)[a,e]	Visco-plastic fluids	-	No qualitative results are given.
Gogarty (1967a, 1967b)[a]	Surfactant stabilized dispersions of water in hydrocarbons and aqueous solutions of PAA (power law)	Berea cores	Expression for effectiveness shear rate is developed. Surface effects and pore blockage are studied.
Greaves and Patel (1985)[a]	Aqueous solutions of polysaccharide biopolymer	Sandstones ($\varepsilon=0.2$)	No plugging of pores was observed.
Gregory and Griskey (1967)[a]; Siskovic et al. (1971); Wampler and Gregory (1972)	Polyethylene and polyethylene terephthalate melts (power law)	Beds of glass and steel spheres	Results for viscous and visco-elastic fluids are in agreement with the power-law form of Darcy's equation.
Gu et al. (1992)[b]	Separan solutions (power law)	Spheres (0.32–2.4 mm)	Numerical results for the free surface cell model and visualization experiments.
Interthal and Haas (1981)[a]; Haas and Kulicke (1984); Kulicke and Haas (1984, 1985)	Dilute solutions of PAA	Glass beads of four different sizes	Role of molecular parameters on the onset of visco-elastic effects. Also, presented an expression for the critical De.
Harrington and Zimm (1968)[a]	Polystyrene in toluene	Fritted disks of pyrex	Severe plugging of disks due to absorption.
Harvey (1968)[a]	Aqueous solution of PAA, PEO, and polysaccharide	Beds of glass beads	An ad-hoc modification of the Ergun equation.
Hassell and Bondi (1965)[a]	Rubber solutions	Beds of glass spheres and York mats	Correlations for friction factor.
Hayes et al. (1996)[c]	CMC solutions	Beds of spheres	Good match between data and volume averaging predictions.
Helmreich et al. (1995)[a]	Xanthan gum solutions (Cross model)	Beds of spheres	Onset of excess pressure drop corresponds to the prevailing shear stress levels at which elastic effects appear in viscometric tests.

(Continued)

TABLE 6.5 (Continued)

Summary of Investigation of Non-Newtonian Flow in Porous Media

Investigator	Test Liquids and Fluid Model	Type of Porous Medium	Remarks
Hirasaki and Pope (1974)[a,d]	Biopolymer and polyacrylamide solutions (Kelzan-M) (power law)	Cores ($\varepsilon \sim 0.2$)	Kelzan-M solutions did not cause any reduction in permeability while adsorption and visco-elastic effects are observed with polyacrylamide solutions.
Hua and Ishii (1981)[a,b]	Power-law	Cell model	Numerical values of drag for particle assemblages at high Reynolds numbers.
Ikoku and Ramey, Jr. (1979, 1980)[a]	Power-law	–	Transient flow of power-law liquids in reservoirs. Weak compressibility effects also included.
Islam and Farouq Ali (1989)[a]; Islam et al. (1989)	Water–oil emulsions and foams	Beds of glass beads	Mechanism for the flow of foam through pores is investigated.
Jaiswal et al. (1991a, 1991b, 1991c, 1992, 1993a, 1993b, 1994)[b]	Power-law and Carreau models	Cell model	Numerical results for shear-thickening and shear-thinning fluids in the range $1 < Re_{PL} < 20$.
James and McLaren (1975)[a]	Dilute aqueous solutions of PEO	Beds of glass beads	Visco-elastic effects in flow through porous media. In spite of significant deviations in Δp from the expected Newtonian value, no discernable changes in flow patterns were observed.
Jennings et al. (1971)[a]	Aqueous solutions of PEO, PIB, and PAA	Berea sandstones	The reduction in mobility does not depend upon adsorption but correlates well with the visco-elastic behavior.
Jones (1979, 1980)[a]; Jones and Maddock (1966, 1969); Jones and Davies (1976); Jones and Ho (1979)	Dilute solutions of PAA, and sodium carboxymethyl cellulose	Beds of glass beads	The increase in Δp attributed to polymer adsorption. Also identified the value of Re above which Darcy's law does not apply.
Kaser and Keller (1980)	Drag-reducing solutions of polyethylene oxide	Glass bead packs	Results show some variance from those of James and McLaren (1975) but are consistent with those of Naudascher and Killen (1977).
Kawase and Ulbrecht (1981a, 1981b)[b]	Power-law	Cell model	Analytical results for drag coefficients of assemblages in low and moderate Re region.
Kemblowski and Mertl (1974)[a,d]; Kemblowski et al. (1974, 1980); Kemblowski and Dziubinski (1978); Kemblowski and Michniewicz (1979)	Aqueous solutions of starch, PVA, kaolin, PEO, and polypropylene melts. (power-law and Carreau models)	Glass bead beds	Extensive Δp results for pseudoplastic, dilatant, and visco-elastic liquids.
Khamashta and Virto (1981)[a]	Plant sludges (power law)	Cylindrical cartridges	Filtration of sludges.
Koshiba et al. (1993, 1999)[a]	Polyacrylamide solutions (power law and N_1)	Cubic packing of spheres ($\varepsilon = 0.33$ and 0.44)	Onset of visco-elastic effects at a critical shear rate.

(Continued)

TABLE 6.5 (Continued)

Summary of Investigation of Non-Newtonian Flow in Porous Media

Investigator	Test Liquids and Fluid Model	Type of Porous Medium	Remarks
Kozicki and Tiu (1973, 1988)[a,d]; Hanna et al. (1977); Kozicki et al. (1967, 1968, 1972, 1984, 1987, 1988); Kozicki (2002)	Aqueous solutions of PEO, CMC, and calcium carbonate slurries (power-law, Ellis, and other models)	Beds of spherical particles	Generalized form of the Rabinowitsch–Mooney equations which involves geometric factors depending upon the shape of the conduit or porous bed parameters. Also elucidated the occurrence of surface effects such as adsorption, gel effects, and slip. Filtration of non-Newtonian slurries has also been studied.
Krüssmann and Brunn (2001, 2002)[a]	Solutions of hydroxyl-propylguar (HPG)	Beds of spheres ($d=392\,\mu m$)	Packed bed tests are used to evaluate viscometric data.
Kumar and Upadhyay (1981)[a,d]	Aqueous solutions of carboxymethyl cellulose and grease/kerosene mixtures (power law)	Beds of spheres and cylinders	Friction factor results at high Re for fixed and fluidized beds.
Lagerstedt (1985)[a]	Dilute solutions of polyethylene oxide and polyacrylamide	Sintered plates of glass beads	Even a 10 ppm solution can yield significant increase in pressure drop.
Larson (1981)[c]	Power law	-	Volume averaging of field equations for power-law liquids.
Laufer et al. (1976)[a]	Dilute solutions of PAA and PEO	Beds of glass spheres	Effects of aging and degradation on Δp.
Lehner (1979)[c]	Power law	-	Averaging of field equations.
Levy (1969)[a]	Aqueous solutions of CMC, PAA, and Carbopol (power law)	Beds of glass spheres	Visco-elastic effects.
Machac and Dolejs (1981, 1982)[a]	Aqueous solutions of PEO, CMC and PAA (Power-law, Ellis, and Carreau fluid models)	Beds of spherical and nonspherical particles	Extensive results on Δp in creeping flow regime.
Machac et al. (1998)[a]	Power-law model	-	Compares the performance of the available models for the prediction of Δp in packed beds.
Maerker (1973, 1975)[a]	500 ppm solutions of polysaccharide	Berea cores	The amount of polymer retained in pores is found to vary with flow rate.
Marshall and Metzner (1967)[a]; Sheffield and Metzner (1976)	Solutions of Carbopol, PIB, ET-497, and microemulsions (power law)	Sintered bronze disks	Role of divergent–convergent character of flow for visco-elastic fluids.
Masuyama et al. (1983-1986)[a,d]	Kaolin slurries (Bingham plastic)	Beds of glass beads and crushed rocks	Empirical correlation for pressure loss.
McAuliffe (1973)[a]	Oil-in-water emulsions	Sandstone cores	Permeability reduction studies.
McKinley et al. (1966)[a]	Dextran solutions	Sandstone cores	Pressure loss data correlated using Darcy's law.
Michele (1977)[a,d]	Aqueous solutions of CMC and PAA (power law)	Beds of glass beads	Modified the Kozeny–Carman equation to correlate Δp results in the range $10^{-2} \leq Re \leq 10^2$.

(Continued)

TABLE 6.5 (Continued)

Summary of Investigation of Non-Newtonian Flow in Porous Media

Investigator	Test Liquids and Fluid Model	Type of Porous Medium	Remarks
Mishra et al. (1975)[a,d]; Mishra and Farid (1983); Singh et al. (1976)	Aqueous solutions of PVA and grease in kerosene	Beds of glass beads	Extensive pressure loss data in fixed and fluidized beds.
Mohan and Raghuraman (1976a, 1976b)[b]	Power-law and Ellis model	Cell model	Upper and lower bounds on pressure drop in creeping flow.
Müller and Brunn (1997, 1999)[a]	Solutions of Schizophyllan biopolymer	Bed of spheres (1.16 mm)	Prediction of Δp for shear-thinning and dilatant fluids.
Müller et al. (1998)[a]	Solutions of polyalphaola-fine (modified Ellis model)	Bed of spheres (9 mm)	Optical studies show severe channeling at high flow rates.
Mungan et al. (1966)[a]; Mungan (1969)	Aqueous solutions of partially hydrolyzed polyacrylamide	Beds of Ottawa sand and silica powder and Berea sandstones	Polymer adsorption under static and dynamic conditions.
Naudascher and Killen (1977)[a]	Aqueous solutions of polyethylene oxide	Beds of glass beads	Onset and saturation of non-Newtonian effects.
Odeh and Yang (1979)[a]	Power law	-	Transient analysis for Power-law flow through porous media.
Park et al. (1975)[a,d]	Aqueous solutions of PAA, polymethyl cellulose, and poly vinylpyrrodine (power law, Ellis, Meter, Herschel–Bulkley, and Spriggs models)	Glass bead packs	Comparisons of experiments and the predictions of the capillary model for various fluid models.
Parker (1977)[a,d]	Polystyrene and PMMA melts and their blends	Glass bead packs	Pressure loss data and surface effects.
Pascal (1983, 1984a, 1984b, 1985, 1986a, 1986b, 1988, 1990a, 1990b)[a,e]; Pascal and Pascal (1985, 1988, 1989a, 1989b)	Power-law, Bingham plastic, Herschel–Bulkley, and Maxwell model fluids	-	Steady and transient flow of non-Newtonian fluids studied using a modified Darcy's law. Stability of the interface between two immiscible phases has also been investigated.
Paterson et al. (1996)[a]	Polymer solutions ($n = 0.23$ and 0.6)	Sintered glass beads ($\varepsilon = 0.30$)	Dispersion coefficients increase due to shear-thinning viscosity.
Payne and Parker (1973)[a]	Aqueous solutions of PEO (power law)	Beds of glass beads	Effect of rheological properties on axial mixing in fixed beds.
Rao and Chhabra (1993)[a]	Aqueous solutions of CMC (power law)	Beds of mixed size spheres	Wall effects and the effect of mixed particle sizes.
Sabiri and Comiti (1995, 1997a, 1997b)[a,d]; Sabiri et al. (1996a, 1996b)	Aqueous solutions of CMC (power law)	Beds of spheres, plates, and cylinder and synthetic foams	Applicability of a generalized capillary model to diverse porous media.

(Continued)

TABLE 6.5 (*Continued*)

Summary of Investigation of Non-Newtonian Flow in Porous Media

Investigator	Test Liquids and Fluid Model	Type of Porous Medium	Remarks
Sadowski and Bird (1965)[a,d]	Aqueous solutions of polyethylene glycol, PVA, and HEC (Ellis model)	Beds of glass beads and lead shots	Constant flow and constant pressure experiments show different behavior due to gel formation.
Sandeep and Zuritz (1996)[a]	Aqueous CMC solutions (power-law model)	Clusters of spheres ($0.89 \leq \varepsilon \leq 0.98$)	Drag on ensembles in tube flow is measured.
Satish and Zhu (1992); Zhu and Satish (1992)	Power law	Cell model	Drag in creeping flow region.
Shvetsov (1979)[a]	Aqueous solutions of PAA	-	Significant differences between bulk and in situ rheology.
Smit and du Plessis (1997, 1999, 2005)[c]	Power law	Granular beds and synthetic foams	Volume averaging for the flow of power-law fluids up to about $Re \approx 50$.
Smith (1970)[a]	Aqueous solutions of partially hydrolyzed polyacrylamide	Alundum and Berea plugs	Critical shear rate for the shear-thickening effect and its role in mobility reduction.
Brown and Sorbie (1989); Sorbie et al. (1987, 1989); Sorbie (1989, 1990)	Solutions of xantham gum (power-law and Carreau model)	Cores	Network models used to account for the depleted layer effects, etc.
Szabo (1975a, 1975b)[a]	Aqueous solutions of partially hydrolyzed polyacrylamide	Sand packs	Quantitative results on polymer retention and adsorption.
Tiu et al. (1997)[a]	Solutions of CMC and polyacrylamide (power law)	Beds of mixed-size spheres and of spheres and cylinders	Combined effects of particle shape and size for visco-elastic liquids.
Abou-Kasseem and Farouq Ali (1986)[a]; Thomas and Farouq Ali (1989)	Oil/water emulsions	Beds of Ottawa sand and glass beads, and Berea sandstones	Qualitative study on in-situ rheology of emulsion in porous media.
Tiu and Moreno (1984)[a,d]; Tiu et al. (1974, 1983)	Aqueous solutions of Methocel and polyacrylamide; Boger fluids (Carreau and power-law models)	Beds of glass beads, cubes, and cylinders	Visco-elastic effects correlate with a modified De and the effect of particle shape on Δp is studied.
van Poolen and Jagron (1969)	Power law	-	Steady and unsteady flow in reservoirs.
Vorwerk and Brunn (1991, 1994)[a]	Fluid A1, Surfactant and HPG solutions (Ellis model)	Bed of steel spheres (794 µm)	Effect of elongation on pressure drop.
Wen and Yim (1971)[a]; Wen and Fan (1973)	Aqueous solutions of polyox (power law)	Beds of glass beads	Axial mixing in fixed and fluidized beds.
White (1967)[a]	Power law	-	Conditions for the applicability of Darcy's law to power-law media are examined.
Wissler (1971)	Power-law visco-elastic fluid model	-	Taking into account the convergent–divergent nature, Darcy's law is adopted for visco-elastic fluids.

(Continued)

TABLE 6.5 (*Continued*)

Summary of Investigation of Non-Newtonian Flow in Porous Media

Investigator	Test Liquids and Fluid Model	Type of Porous Medium	Remarks
Wreath et al. (1990)	Power law	–	Calculation of effective shear rate in porous media.
Yentov and Polishchuk (1979)[a]	Aqueous solution of PAA and microemulsions	–	Onset of visco-elastic effects.
Yu et al. (1968)[a]	Aqueous solution of PEO (power law)	Beds of glass spheres and cubes	Limited amount of Δp data in fixed and fluidized bed regions.
Zhu(1990); Zhu and Chan Man Fong (1988)	Carreau model fluid	–	Approximate solution is obtained for dense cubic packing. Visco-elastic effects are also discussed.
Mertens et al. (2006)[a]	Xanthan gum solutions (visco-elastic)	Glass beads ($d/D_c = 0.085, 0.2$)	NMR imaging of flow field and radial porosity variation.
Odell and Haward (2006)[a]	Dilute solutions of PAA	Beds of spheres in SCC and BCC arrangement	Effect of geometric arrangement on in-situ viscosity.
Broniarz-Press et al. (2007a)	PAA and CMC solutions (Power-law) $0.65 \leq n \leq 1$	Spherical beads ($\varepsilon \approx 0.42$)	Δp-Q relationships in fixed and fluidized bed regions.
Chase and Dachavijit (2008)[a]	Electrorheological fluids	Glass beads	Pressure drop data consistent with the modified Ergun equation and also gave a criterion for the onset of flow.
Sochi and Blunt (2008)[f]	Ellis and Herschel–Bulkley	Model porous medium	Pore network modeling to estimate effective permeability.
Amundarin et al. (2009)[a]	Xanthan and Guar gum mixed solutions	Glass beads ($d = 1.13$mm)	Effective shear rate in a porous medium.
Derksen (2009)	Shear-thinning and thixotropic fluids	Random assemblies of spheres	Calculation of drag in the creeping flow regime.
Morais et al (2009)	Power law	3-D disordered medium	Direct numerical study to develop a universal sealing for permeability–Reynolds number.
Sochi (2009)[f]	Bautista–Manero model	–	Visco-elastic effects investigated via pore network modeling.
Balhoff et al (2012)[f]	Visco-plastic and GNFs	Sphere packs and sandstones	New algorithm for pore network modeling.
Galindo-Rosales et al. (2012)[a]	Xanthan solutions	Corrugated tubes to capture the diverging–converging flow	In microscale channel examines the Δp behavior.
Shende et al. (2021a,b)[d]	Meter, power-law, and Phan-Thien–Tanner models	Capillaries of square wave cross-section	Introduces effective shear rate-based shift factors so that the bulk rheology data coincide with Δp-Q data in porous media.
Singh et al. (2012)	Power law	Beds of cylinders and spheres	Based on ensemble averaging, significant effects of particle interaction for cylinders for $n > 1$ and for sphere for $n > 2$ are predicted. Effective medium approach is used for concentrated systems.

(Continued)

TABLE 6.5 (Continued)

Summary of Investigation of Non-Newtonian Flow in Porous Media

Investigator	Test Liquids and Fluid Model	Type of Porous Medium	Remarks
Chevalier et al (2013, 2014)[e]	Visco-plastic fluid	Glass beads and 2-D model medium	Modified Darcy law and at low voidage, the flow is mainly governed by the microstructure of the medium, in line with the findings of Bleyer and Coussot (2014).
Tosco et al. (2013)	Ellis, Cross, Carreau, and power-law models	2-D porous medium	High Reynolds number results are consistent with the Darcy–Forchheimer formulation.
Lavrov (2014)	Power law	Rough fractures	Radial flow behavior.
Tang and Lu (2014)	Power law	3-D ordered cubic packing of spheres	Proposed modified Ergun equation.
Sobti and Wanchoo (2014)[a]	PAA solutions $0.37 \leq n \leq 0.58$	Glass beads of 4.38 and 5.76 mm size	Empirical correlation for pressure drops in terms of Deborah number for Re < ~1.
Chevalier and Talon (2015)[e]	Bingham fluid	Porous medium	Modification of Darcy's law to predict the flow field and flow rate regimes.
Marliere et al. (2015)[a]	HEC solutions	Glass beads	NMR imaging of the flow to delineate the conditions for jamming and pore blockage.
Rodriguez de Castro and Radilla (2016)[a,e]	Carreau model (xanthan solutions)	Granite and sandstones	Modification of Darcy's law.
Rodriguez de Castro et al. (2016)	Shear-thinning fluids	Microfluidic pore networks	Effect of shear-thinning on residual oil formation.
Fayed et al. (2016)[d]	Carreau fluid model	Capillary bundle (porous medium)	Redefines the effective Reynolds number in terms of pressure gradient.
Roustaei et al (2016)[e]	Visco-plastic fluid	2-D rough fractures	Non-Darcy effects and aims to predict the minimum pressure gradient for the fluid to yield.
Zami–Pierre et al. (2016, 2017, 2018)	Power law	Isotropic porous medium	Simulation to investigate the coupling between rheology and wall interactions, effective slip, etc. Effect of disordered media on flow.
Berg and van Wunnik (2017)	Carreau model fluid	Sandstones	Proposes an expression for effective shear rate.
De et al. (2017a,b,c)	Phan-Thien–Tanner model	3-D random porous medium ($0.3 \leq \varepsilon \leq 0.6$)	Due to the asymmetry in the flow field, pressure drop increases with the increasing Deborah number, similar to the experiments of James et al. (2012).
Malvault et al (2017)[d]	Herschel–Bulkley model	Capillaries of periodically varying cross sections.	Effect of shape and severity of axial formation for the onset of flow is marginal.
Nash and Rees (2017)	Bingham fluid	Model porous medium	1-D capillary model to evaluate the pore-size distribution.
Rodriguez de Castro and Radilla (2017)[a]	Carreau and Herschel–Bulkley fluids	Rough-walled fractures	Effective shear rate and shift factors to match Δp-Q data with viscometric data.
Rodriguez de Castro et al. (2018)	Visco-plastic fluids	–	Use of Δp-Q data for pore-size distribution determination.

(Continued)

TABLE 6.5 (Continued)

Summary of Investigation of Non-Newtonian Flow in Porous Media

Investigator	Test Liquids and Fluid Model	Type of Porous Medium	Remarks
Tian et al. (2018)[a]	PAA solutions ($0.34 \leq n \leq 0.80$)	Cubic assembly of spheres ($\varepsilon=0.48$).	Wall effects, heat transfer and Δp correlation in the range $0.55 \leq Re \leq 1100$.
Bauer et al. (2019)[a]	Aqueous Carbopol solutions (Herschel–Bulkley model)	Sandstone	Creeping flow studied experimentally and using 3-D LBM approach.
Eberhard et al. (2019, 2020)[a]	Xanthan solutions (Carreau model)	8 mm spheres in 50 mm diameter column.	Estimation of effective shear rate. High-resolution velocimetry shows asymmetry in the flow field.
Liu et al. (2019a)[e]	Yield stress fluids	-	Modification of Darcy law to predict effective permeability of the medium.
Rodriguez de Castro et al. (2019)	Yield stress fluids	-	Possible use of Δp-Q data for porosimetry and pore size determination.
Rodriguez de Castro and Agnaou (2019)	Herschel–Bulkley and Carreau fluid models	2-D porous medium pore-size and throat-size distributions	Highlights the deficiencies of the apparent viscosity approach.
Zhang et al. (2019b)[f]	Cross fluid model	Rough fractures	Postulates shift factors (functions of geometry and rheology) for the bulk rheology and porous medium data to match with each other.
Airiau and Bottaro (2020)	Carreau model fluid	3-D porous medium	Multiscale homogenization approach suggests the permeability tensor to become anisotropic which cannot be captured by the use of the effective viscosity method.
Browne et al. (2020)[a]	PEO solutions	Microfluidic porous media	Due to extensional flow component, stable and unstable flow regimes are possible.
Didari et al. (2020)	Bingham plastic model	Sandstone	Pore network modeling combined with ant colony optimization method to delineate conditions for fluid yielding.
Rodriguer de Castro et al. (2020)	Herschel–Bulkley model fluids	2-D porous medium	DNS of interactions between microstructure and rheology.
Rodriguez de Castro and Goyeau (2021)	Visco-elastic effects	Porous medium	Effect of visco-elasticity on pressure drop.
Seybold et al. (2021)[a]	Xanthan solutions (Carreau model)	Disordered porous medium	Local interaction between the nonlinear rheology and heterogeneity of the medium.

Notes: HEC, hydroxy ethyl cellulose; PEO, polyethylene oxide; CMC, carboxymethyl cellulose; PVA, polyvinyl alcohol; PAA, polyacrylamide.
[a] Denotes experimental work.
[b] Cell model approach.
[c] Averaging procedures for field equations.
[d] Capillary model approach.
[e] Modification of Darcy's law.
[f] Pore network modeling.

evidence to suggest this may not be true under certain circumstances (Unsal et al., 1978; Wang et al., 1979; Duda et al., 1981; Wreath et al., 1990; Haward and Odell, 2003). Conversely, some attempts have also been made at developing methodologies for extracting the values of rheological parameters from porous media flow experiments (Krüssmann and Brunn, 2001, 2002). This approach has been moderately successful, at least for the solutions of rigid or semirigid polymers that undergo little flow-induced degradation and for inelastic media. The solutions of hydroxyethyl cellulose (Sadowski and Bird, 1965), carboxymethyl cellulose (Christopher and Middleman, 1965; Duda et al., 1983; Chhabra and Srinivas, 1991; Rao and Chhabra, 1993; Sabiri and Comiti, 1995, 1997a,b, etc.), hydroxypropyl guar (Chakrabarti et al., 1991; Tatham et al., 1995), xanthan (Brown and Sorbie, 1989; Helmreich et al., 1995), scleroglucan, etc. generally conform to this category. Similarly, most particulate suspensions like that of TiO_2 used by Brea et al. (1976) and of kaolin used by Masuyama et al. (1983–1986) also fall in this class of non-Newtonian fluids. This discrepancy between the bulk rheological properties and the porous media experiments is often bridged by developing the so-called shift factors which endeavor to collapse the rheometric data with that deduced from the Δp-Q data obtained in the porous media tests. From a theoretical standpoint, such shift factors are little more than the fudge factors, for the flow in a porous medium is complex and far from being viscometric. Notwithstanding this fundamental issue, the definition of such shift factors hinges on identifying the mean or effective shear rate prevailing in the porous medium. Naturally, this itself depends on the detailed structure of the medium. In the absence of anomalous pore wall–fluid interactions, the effective shear rate frequently takes the form:

$$\dot{\gamma}_{\text{eff}} \sim C_o \frac{V_0}{l_c} \tag{6.61}$$

where the choice of the linear scale l_c is far from obvious and C_o is often a function of the non-Newtonian viscosity model parameters (n for power-law fluids, Λ for the Carreau model, etc.) as well as the properties of the porous medium (ε, k) and of the tortuosity factor (T), for example, see the works of Amundarain et al. (2009), Berg and vanWunnik (2017), Rodriguez de Castro and Radilla (2017), Eberhard et al. (2019), Shende et al. (2021a,b), etc. Similarly, much confusion exists in the literature even regarding the choice of the representative linear scale (l_c); some have assumed it to be proportional to the packing size while others have advocated the use of $\sqrt{k}$. Because of these uncertainties coupled with the fact that each "porous medium" and "non-Newtonian fluid" is somewhat unique, it has not been possible to develop a concrete form of Equation 6.61 spanning a range of fluids and/or porous medium.

These results can be recast in the form of mobility, defined as the ratio of k/μ_{eff}, but this approach also suffers from the aforementioned difficulties surrounding the estimation of $\dot{\gamma}_{\text{eff}}$ needed to evaluate μ_{eff} for a specific non-Newtonian viscosity model and the porous medium (Shahsavari and Mckinley, 2015, 2016) and/or of tortuosity itself (Tang and Lu, 2014). Others have preferred to use permeability to consolidate their numerical and experimental results. In this case too, the permeability of the medium depends upon the rheology of fluid, the microstructure of porous medium and of the Reynolds number beyond the creeping (Darcy) flow regime.

3. Significant research effort has been directed at the flow of visco-plastic (Bingham and Herschel-Bulkley model) fluids in porous media. Much of this effort has been aimed at three distinct but inter-linked aspects: First, some have endeavored to establish flow rate–pressure drop relationship for a given porous medium. In the low Reynolds number regime, this is equivalent to estimating the effective permeability of the medium. This macroscopic approach inherently assumes that the applied or resulting pressure gradient is sufficient to cause fluid yielding everywhere in the porous medium, or takes no cognizance of partial yielding, for example, see Sochi and Blunt (2008), Chevalier and Talon (2015), Bauer et al. (2019), Liu et al. (2019a), etc.

The second class of studies deals with the inverse problem, namely, the prediction of the minimum (threshold) pressure gradient required to initiate fluid yielding everywhere in the porous medium (Chen et al., 2005; Talon and Bauer, 2013; Chevalier and Talon, 2015; Roustaei et al., 2016; Didari et al., 2020, etc.). These features cannot be captured successfully by the elementary capillary bundle models which envisage the flow passages to be of uniform cross section and does not take into account the existence of blind pores, interconnected pores, and/ or the divergent–convergent nature of these flow passages even in 1-D models (Sochi, 2010a,b). Similarly, it has been suggested in the literature to assign one-third of the flow passages to each direction thereby accounting for the anisotropic permeability of the medium. Finally, the standard capillary model predicts a single universal yielding condition at a particular value of ($\Delta p/l$), whereas in practice this transition occurs over a range of ($\Delta p/l$) values. The divergent–convergent character of the flow has been studied for a range of configurations shown in Table 6.2 and many others (Sadowski and Bird, 1965; Sochi, 2010b). Therefore, it is believed that the pore-network modeling approach is likely to lead to more reliable predictions of the so-called minimum threshold pressure. This is equivalent to finding a fluid flow path from the inlet to outlet that minimizes the yielding pressure (Kharabaf and Yortsos, 1997; Sochi, 2010 a,b,c). However, this approach is not devoid of criticism (Sochi, 2010a,b). Another approach to predict the yielding condition uses some elements of percolation theory (Rossen and Gauglitz, 1990; Rossen and Mamun, 1993). These authors studied the flow of foams in porous media. The third subset of studies in this category also tackles the inverse problem of finding the pore-size distribution of a porous medium using the bulk flow measurements (Oukhlef et al., 2010; Nash and Rees, 2017; Rodriguez de Castro et al., 2018, 2019). The porous medium is modeled as a bundle of capillaries whose radii are distributed according to a probability distribution law given by $p(r)$. Since the capillaries are arranged in parallel, the total flow rate Q is related to the individual contributions $q(r)$ as:

$$Q(\Delta p) = \sum q_i = \int q(\Delta p, r) p(r) \, dr \qquad (6.62)$$

Equation 6.62 applies for a fixed value of Δp. The lower limit of the integration is set using the yielding criterion of $r_o = (2\tau_o L/\Delta p)$ where τ_o is the yield stress and the upper limit of the integration is ∞.

The kernel $q(\Delta p, r)$ is simply the Δp-Q relationship for the laminar, steady, incompressible flow in a tube of given radius and length. Of course, for a given value of ($\Delta p/L$), $Q = 0$ below a critical radius of the tube. Therefore, one can solve Equation 6.62 for the unknown $p(r)$, as discussed in the literature (Oukhlef et al., 2010; Rodriguez de Castro et al., 2018, 2019, etc.).

4. The flow behavior of visco-elastic liquids in porous media has been known to differ significantly from that of purely viscous fluids, mainly due to extensional effects (Müller and Saez, 1999; Odell and Haward, 2006; Mertens et al., 2006; James et al., 2012). The literature on this subject, as will be seen, is steadily growing and adding to our understanding.
5. The response of "dilute" drag-reducing polymer solutions in porous media has also received some attention. Such studies have been motivated primarily to gain improved insights into the fundamental aspects of this complex flow at the molecular level.
6. The effects of containing walls and particle shape on pressure drop have not been investigated as extensively for rheologically complex media as done for Newtonian media. Preliminary results on mixing characteristics are also available.
7. Several unusual effects including slip, pore blockage, polymer retention by adsorption and entrapment, mechanical and flow-induced degradation, anisotropy and heterogeneity of porous medium, etc. are also encountered during the flow of macromolecular solutions in porous media and in narrow passages. Owing to their wide-ranging implications in oil-recovery processes, concentrated research efforts have been directed at developing better understanding of these processes. Hereafter, these effects will be referred to collectively as "miscellaneous effect."

6.4.1 Flow Regimes

In contrast to the extensive literature for Newtonian liquids (Section 6.3.1), little is known about the flow regimes for the flow of non-Newtonian liquids in porous media. Bereiziat and Devienne (1999) and Bereiziat et al. (1995) used laser Doppler velocimetry to evaluate wall shear rate and the axial velocity profiles for the flow of inelastic carboxymethyl cellulose solutions in a periodically corrugated channel. Based on their analysis of the wall shear rate–Reynolds number relationship, they concluded that no recirculation zone was observed up to about $Re_g \approx 20$ and this flow regime was simply called the laminar flow. As the flow rate was gradually increased, steady recirculation zones began to form. Therefore, in the range $20 < Re_g \leq \sim 130$, the flow pattern was called laminar flow with steady recirculating zone, and finally, the flow transited to turbulence due to the unsteady recirculating zones appearing at $Re_g \approx 130$, albeit the homogeneous turbulent conditions were realized only at $Re_g \approx 2000$. Undoubtedly, this flow geometry does capture some aspects of flow in a porous medium and this provides useful insights, yet it is not easy to establish a direct link between the flow in a PCT and that in a porous medium. Müller et al. (1998) reported a flow visualization study for the flow of a visco-elastic polymer solution in a bed of spheres using laser Doppler velocimetry, especially in the radial direction. Based on their results, they reported increasing degree of nonuniformity in flow with the increasing flow rate (Deborah number), that is, the formation of preferred passage taken by fluid elements, and these passages themselves fluctuated with time. Using the viscometric measurements for the polymer solutions, the flow was fairly uniform in the radial direction for the values of $(N_1/\tau) \leq O(1)$ and irregularities set in for the conditions such that $(N_1/\tau) > 1$, thereby attributing this phenomenon to the first normal stress differences. Similar observations regarding the formation of preferential flow channels in a porous medium have also been reported by others (Müller et al., 1993; Rodriguez et al., 1993; Saez et al., 1984). There has been only one experimental study dealing with the prediction of transition from one regime to another in terms of the pore Reynolds numbers for power-law liquids. Comiti et al. (2000a) combined the microelectrode data of Seguin et al. (1998a,b) with their extensive pressure drop data and proposed that qualitatively similar flow regimes occur for inelastic power-law liquids during the flow in homogeneous porous media. However, the presently available scant data allow the prediction of the cessation of the creeping (Darcy) regime only that ends at $Re_{pore} = 4.3$. Intuitively, one feels that the delineation and prediction of the flow regimes is going to be extremely difficult for visco-elastic liquids whereas there is little risk involved in the use of Newtonian liquid transition values for inelastic shear-thinning media. In recent years, Mertens et al. (2006) have used NMR imaging technique to study dispersion in glass-bead-packed beds for the flow of visco-elastic xanthan solutions in relatively narrow beds ($D_c/d = 5$ and 12). They reported that the secondary flows appear at much lower Reynolds numbers for their xanthan solution than that in Newtonian fluids. Similar flow features (including formation of aggregates, asymmetric velocity profiles, etc.) with visco-elastic fluids have been reported by others as well (James et al., 2012; Bleyer and Coussot, 2014; Marliere et al., 2015). All these factors allegedly contribute to the increased pressure drop due to visco-elasticity.

6.4.2 Pressure Loss for Generalized Newtonian Fluids

There is no question that the practical problem of estimating the pressure drop required to achieve a given flow rate for a liquid of known rheology, through a bed of known characteristics, has received the greatest amount of attention. In their excellent reviews, Savins (1969), Kemblowski et al. (1987), Chhabra (1993a,b), Chhabra et al. (2001b), and others noted that the four approaches outlined in Section 6.3.1 have also been extended to accomplish this goal: the capillary model, submerged objects model, volume averaging, and the empirical correlations. Recent years have also seen a spurt in the pore network modeling for predicting the flow behavior of a range of non-Newtonian fluids in ordered and disordered (homogenous and heterogeneous) porous media (Sochi, 2010a,b). Broadly speaking, the rheological complexity of the fluid of interest coupled with the complexity of the porous medium and the extent of wall–polymer molecule interactions determine the general applicability of any given method. In the ensuing sections, the aforementioned approaches are described by presenting some of the more successful developments in each area that are restricted primarily to unconsolidated and isotropic porous media.

6.4.2.1 The Capillary Model

6.4.2.1.1 Laminar Flow of Pseudoplastic Fluids

A quick look at Table 6.5 shows the preponderance of the developments based on this simple approach. Evidently, Bird et al. (2002) were the first to allude to the possibility of extending this approach of modeling to the flow of generalized Newtonian fluids in porous media. Early developments in this field make use of the so-called Blake–Kozeny model, whereas over the years it has been argued that the Kozeny–Carman model is to be preferred (Kemblowski and Michniewicz, 1979; Kozicki and Tiu, 1988; Chhabra et al., 2001b). The starting point in both cases is the non-Newtonian equivalent of Equation 6.14, which, in turn, depends upon a specific fluid model for the fluid viscous behavior.

Following Kemblowski et al. (1987), the equivalent of Equation 6.14 for the laminar flow of an incompressible fluid through a conduit of arbitrary cross-section (characterized by hydraulic radius, R_h) can be written as

$$V = \frac{\Delta p R_h^2}{K_o \mu L} \tag{6.63}$$

where K_o is a constant that depends upon the geometry of the conduit, for example, $K_o = 2$ for a circular pipe, and $K_o = 3$ for a planar slit, etc. Equation 6.63 can be rearranged as

$$\left(R_h \frac{\Delta p}{L} \right) = \mu \left(\frac{K_o V}{R_h} \right) \tag{6.64}$$

The left-hand side of Equation 6.64 is identified as the average shear stress at the pore wall and the term within brackets on the right-hand side can be identified as the shear rate at the wall for a Newtonian fluid, and by analogy with the flow in cylindrical pipes, it will be called the nominal shear rate for generalized Newtonian fluids. Thus,

$$\langle \tau_w \rangle = \frac{R_h \Delta p}{L} \tag{6.65a}$$

and

$$\langle \dot{\gamma}_w \rangle_n = \frac{K_o V}{R_h} \tag{6.65b}$$

These equations can now be modified to describe the flow in an isotropic and homogeneous porous medium or packed bed by changing L to L_e and V to V_i, and substituting for R_h from Equation 6.12 to yield for a bed of monosize spheres:

$$\langle \tau_w \rangle = \left(\frac{d}{6} \right) \left(\frac{\varepsilon}{1-\varepsilon} \right) \left(\frac{\Delta p}{TL} \right) \tag{6.66a}$$

$$\langle \dot{\gamma}_w \rangle_n = 6 K_o T \left(\frac{1-\varepsilon}{\varepsilon^2} \right) \left(\frac{V_o}{d} \right) \tag{6.66b}$$

Further progress can only be made by choosing a specific model for the fluid behavior. For a Newtonian fluid, Equations 6.66a and 6.66b combine to yield Equation 6.6 or 6.7, respectively, depending upon the value of K_o. For generalized Newtonian fluids, the shear stress at the wall of a capillary is related to the nominal shear rate at the wall by a relation of the type

$$\langle \tau_w \rangle = m' \langle \dot{\gamma}_w \rangle_n^{n'} \tag{6.67}$$

where m' and n' are the apparent consistency and flow behavior indices, respectively, which are related to the true rheological parameters as outlined by Metzner and Reed (1955). For instance, for a true power-law fluid,

$$n' = n \tag{6.68}$$

and

$$m' = \left(\frac{3n+1}{4n} \right)^n m \tag{6.69}$$

Similar relations for the other generalized Newtonian fluid models are available in the literature, but are much more involved than Equations 6.68 and 6.69 (Metzner, 1956; Skelland, 1967; Govier and Aziz, 1982; Steffe, 1996; Chhabra and Richardson, 2008).

It is worthwhile to add here that the Rabinowitsch–Mooney factor of $((3n+1)/4n)^n$ appearing in Equation 6.69 is strictly applicable for cylindrical tubes. However, at the other extreme, the corresponding factor for a planar slit is $((2n+1)/3n)^n$. In the range $0.1 \le n \le 1$, the values of the Rabinowitsch–Mooney factor for a pipe and a slit differ from each other at most by 2%. Furthermore, the calculations of Miller (1972) suggest that this factor is nearly independent of the conduit geometry, and hence it has been generally used for the flow in porous media also. Thus, for power-law fluids, the Kozeny–Carman equation becomes

$$f\left(\frac{\varepsilon^3}{(1-\varepsilon)} \right) = \frac{180}{\text{Re}^*} \tag{6.70}$$

where

$$\text{Re}^* = \left(\frac{\rho V_0^{2-n} d^n}{m(1-\varepsilon)^n} \right) \left(\frac{4n}{3n+1} \right)^n \left(\frac{15\sqrt{2}}{\varepsilon^2} \right)^{1-n} \tag{6.71}$$

Kemblowski et al. (1987) have recommended the use of $K_o = 2.5$, that is, the mean value of K_o for circular tube and parallel plate geometries. Also, the tortuosity factor T has been taken as $\sqrt{2}$ as suggested by Carman (1956). A typical comparison between the predictions of Equation 6.70 and experimental data for polypropylene melts is shown in Figure 6.11. Subsequently, Kemblowski and Michniewicz (1979) have presented an analogous development for Carreau model fluids. Other significant contributions in this area include the works of Christopher and Middleman (1965), Sadowski and Bird (1965), and Kozicki et al. (1967). Starting from the power-law form of Equation 6.14 and using $T = (25/12)$ and $V_o = \varepsilon V_i$, Christopher and Middleman (1965) obtained the expression

$$f\left(\frac{\varepsilon^3}{1-\varepsilon} \right) = \frac{150}{\text{Re}_{\text{C-M}}} \tag{6.72}$$

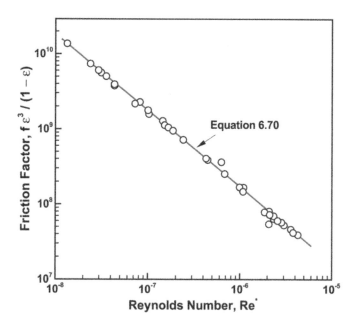

FIGURE 6.11 Typical experimental results demonstrating the applicability of Equation 6.70.

where

$$\text{Re}_{\text{C-M}} = \frac{dG^{2-n}\rho^{n-1}}{H(1-\varepsilon)}$$

$$H = \left(\frac{m}{12}\right)\left(\frac{9n+3}{n}\right)^n (150k'\varepsilon)^{(1-n)/2}$$

$$k' = \frac{d^2\varepsilon^3}{150(1-\varepsilon)^2}$$

Equation 6.72 differs from that of Bird et al. (2002) by a factor of $(25/12)^{n-1}$; this difference is because of the correction of the velocity term or the length term. Christopher and Middleman (1965) also reported a satisfactory agreement between the predictions of Equation 6.72 and their experimental data, as shown in Figure 6.12 for two sets of data.

Sadowski and Bird (1965), on the other hand, modified Darcy's law by introducing an effective viscosity estimated using the Ellis fluid model and noting $V_o = \varepsilon V_i$; $T = 2.5$ to arrive at

$$f\left(\frac{\varepsilon^3}{(1-\varepsilon)^2}\right) = \frac{180}{\text{Re}_{\text{S-B}}}$$

where

$$\text{Re}_{\text{S-B}} = \left(\frac{dV_0\rho}{\mu_o}\right)\left(1 + \left(\frac{4}{\alpha+3}\right)\left(\frac{\langle\tau_w\rangle}{\tau_{1/2}}\right)^{\alpha-1}\right) \qquad (6.73)$$

and $\langle\tau_w\rangle$ is given by Equation 6.66a, with $T = 2.5$.

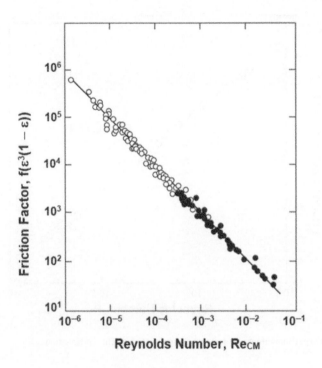

FIGURE 6.12 Typical experimental results demonstrating the applicability of Equation 6.72 (shown as solid line). (○) Christopher and Middleman (1965); (●) Sadowski and Bird (1965).

They found the agreement between their predictions and experiments in general to be satisfactory, but in some cases appreciable deviations were also reported that were tentatively ascribed to the possible wall–polymer molecule interactions and visco-elastic effects. Besides, their data obtained in constant flow and constant pressure experiments also did not superimpose in some cases.

Based on the qualitatively similar forms of the Rabinowitsch–Mooney equation for the laminar flow in circular tubes and in between two parallel plates, Kozicki et al. (1967) proposed the generalized form for 1-D steady flow in a conduit of arbitrary cross-section as

$$\dot{\gamma}_w = f(\langle \tau_w \rangle) = a \langle \tau_w \rangle \left(\frac{d(2V_0 / R_h)}{d \langle \tau_w \rangle} \right) + b \left(\frac{2V_0}{R_h} \right) \tag{6.74}$$

In Equation 6.74, a and b are geometric parameters, for example, for a circular duct, $a=1/4$, $b=3/4$, and for infinite parallel plates $a=1/2$, $b=1$, etc. Furthermore, Kozicki et al. (1967) asserted that this generalization is valid for the laminar flow of any time-independent, incompressible fluid in conduits of arbitrary cross-section characterized by the two geometric parameters, a and b. Since the wall shear stress (τ_w) may not be constant along the contour of the wetted perimeter, the average shear stress at the wall (τ_w) used in Equation 6.74 is defined as

$$\langle \tau_w \rangle = \frac{1}{C} \oint \tau_w \, dc = R_h \left(\frac{dp}{dz} \right) \tag{6.75}$$

where C is the perimeter of the conduit.

For constant values of a and b, that is, fixed geometry, the integration of Equation 6.74 with respect to $\langle \tau_w \rangle$ yields

$$\frac{2V_0}{R_h} = \frac{1}{a}\left(\langle \tau_w \rangle^{-\xi} \int_0^{\langle \tau_w \rangle} \tau^{\xi-1} f(\tau) d\tau \right) \tag{6.76}$$

where $\xi = b/a$.

Since the geometric parameters a and b are postulated to be independent of the fluid model (at least for inelastic fluids), Kozicki et al. (1967) evaluated their values for various cross sections including pipes of rectangular, triangular, and elliptic cross sections, porous media, and open channel, etc. by using the well-known results for Newtonian fluids (Kozicki et al., 1967; Tiu, 1985). Subsequently, for the general case of flow in packed beds, Kozicki and Tiu (1988) have presented the following specific forms of Equation 6.76 depending upon whether one uses the Blake or the Blake–Kozeny or the Kozeny–Carman description for the porous medium:

Blake model

$$\frac{V_i}{R_h} = \frac{(1+\xi)}{K_i} (\langle \tau_w \rangle)^{-\xi} \int_0^{\langle \tau_w \rangle} \tau^{\xi-1} f(\tau) d\tau \tag{6.77a}$$

Blake–Kozeny model

$$\frac{V_i}{R_h} = \left(\frac{1+\xi_o}{K_o}\right)\left(\frac{\langle \tau_w \rangle}{T_{BK}}\right)^{-\xi_o} \int_0^{\langle \tau_w \rangle/T_{BK}} \tau^{\xi_o-1} f(\tau) d\tau \tag{6.77b}$$

Kozeny–Carman model

$$\frac{V_i T_{KC}}{R_h} = \left(\frac{1+\xi_o}{K_o}\right)\left(\frac{\langle \tau_w \rangle}{T_{KC}}\right)^{-\xi_o} \int_0^{\langle \tau_w \rangle/T_{KC}} \tau^{\xi_o-1} f(\tau) d\tau \tag{6.77c}$$

In the Blake model, $K_i = 2(a+b)$ is the well-known Kozeny constant whereas the aspect and shape factors for the other two models are expressed as $\xi_o = (b_o/a_o)$ and $K_o = 2(a_o + b_o)$. The differences between these three models have already been discussed in Section 6.3.2. Undoubtedly, with a suitable choice of a fluid model, that is, $\dot{\gamma} = f(\tau)$, and the geometric parameters, Equation 6.77 includes several known results. For instance, for a power-law fluid characterized by $f(\tau) = (\tau/m)^{1/n}$ and using $K_o = 25/6$ and $T_{BK} = 25/12$, Equation 6.77b yields the well-known result of Christopher and Middleman (1965); likewise, for the Ellis fluid model $f(\tau) = \frac{\tau}{\mu_o}\left\{1+\left(\frac{\tau}{\tau_{1/2}}\right)^{a-1}\right\}$, and noting that $\xi = 3$ and $K_i = 5$, Equation 6.77a reduces to the expression developed by Sadowski and Bird (1965). The cumbersome-looking Equation 6.77 can also be expressed in terms of the usual friction factor and Reynolds number.

Finally, the model of Comiti and Renaud (1989), Equations 6.17–6.21, has been successfully extended to the flow of power-law fluids in isotropic and homogeneous unconsolidated beds of spherical and non-spherical particles by Sabiri and Comiti (1995, 1997a,b). The distinct advantage of this approach is that it covers both viscous and inertial regions, and the values of a_{vd} and T evaluated from Newtonian fluid flow results apply equally well for the flow of inelastic power-law fluids. Furthermore, Sabiri and Comiti (1995) modified the Reynolds number for power-law fluids in such a fashion that a single expression given by Equation 6.20 applies for both Newtonian and power-law liquids. Their so-called pore Reynolds number, Re_{pore}, is defined as

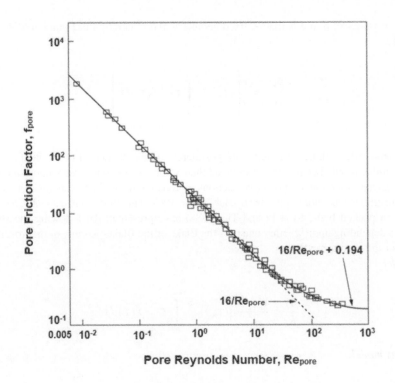

FIGURE 6.13 Predictions of Equation 6.20 for power-law fluids and the range of experimental results. (□) Experimental results relate to the flow of a 0.6% CMC solution, 0.14% and 0.22% Xanthan solutions and a 0.5% Natrosol solution through beds made of spheres, cylinders, polyhedrons, square plates (with thickness to side ratios of 0.1, 0.21, and 0.44). (Modified after Chhabra, R.P. et al., *Chem. Eng. Sci.*, **56**, 1, 2001.)

$$\mathrm{Re}_{\text{pore}} = \frac{\rho \varepsilon^{2n-2}(TV_0)^{2-n}}{2^{n-3}\left[(3n+1)/4n\right]^n m(1-\varepsilon)^n a_{vd}^n} \quad (6.78)$$

Figure 6.13 shows the general applicability of Equation 6.20 (together with Equation 6.78) to the flow of inelastic shear-thinning fluids through packed beds made up of spherical and nonspherical particles without involving any additional measure of shape for nonspherical packings. Subsequently, this approach has also been shown to perform well for beds of mixed size spheres, and reasonably well for synthetic foams viewed as a porous medium (Sabiri et al., 1996b; Ciceron et al., 2002a).

In spite of the fundamental differences inherent in the three capillary models mentioned above, most expressions available in the literature for the creeping flow of power-law fluids are of (or can be reduced to) the form

$$f\left(\frac{\varepsilon^3}{1-\varepsilon}\right) = \frac{A}{\mathrm{Re}^*} \quad (6.79)$$

Though the friction factor is defined unequivocally via Equation 6.3a, the definition of a consistent Reynolds number and the values of A vary from one study to another. In the literature, the predictions of a selection of widely used expressions have been compared in the range $1 \geq n \geq 0.2$ and for $\varepsilon = 0.4$, typical of packed beds (Srinivas and Chhabra, 1992). A summary of the results is shown in Table 6.6. All expressions evaluated were rearranged in the form of Equation 6.79 using the definitions of f and Re^* introduced by Kemblowski and Michniewicz (1979), and the resulting values of A are presented in Table 6.7. An examination of Table 6.6 shows an increasing divergence in predictions with

TABLE 6.6

Comparison between the Predictions of Friction Factor (f/f_N) in Packed Beds ($\varepsilon = 0.4$) Using Different Equations in Creeping Flow

n	Equations 1[a] and 2	Equation 3	Equations 4 and 8	Equation 5	Equation 6	Equation 9
1.0	1.00	1.06	1.2	1.00	0.91	0.91
0.8	1.12	1.21	1.2	1.04	1.21	0.96
0.6	1.25	1.42	1.2	1.08	1.50	1.00
0.4	1.40	1.82	1.2	1.13	1.85	1.06
0.2	1.57	2.93	1.2	1.17	2.60	1.11

Notes:
The above calculations are based on $A_w = 1$. [a] For equation numbers, see Table 6.7.
Equation 7 does not permit the calculation of f without specifying the value of V_0, and hence its predictions are not included in this table.
For equation 9, T is given by Equation 6.22.

the decreasing value of the power-law index, with the exception of the approach of Sabiri and Comiti (1997a), which appears to be fairly robust in accounting for the role of power-law index. Essentially similar observations have also been made by other investigators (Marshall and Metzner, 1967; Metzner, 1977; Kemblowski et al., 1987; Chhabra, 1993a,b; Chhabra et al., 2001b) in predicting the value of frictional pressure loss in packed beds. In addition to the inherent deficiencies of the capillary model and the basic differences in the three models (mentioned in Section 6.3.2.2), several other reasons including experimental uncertainty, the dubious nature of the tortuosity factor (Sheffield and Metzner, 1976; Kemblowski et al., 1987; Agarwal and O' Neill, 1988; Ghanbarian et al., 2013) and its possible dependence on flow rate (Dharmadhikari and Kale, 1985; Ciceron et al., 2002b; Liu et al. 2015a), unaccounted extensional effects (Durst et al., 1981; James, 1984; Gupta and Sridhar, 1985; Jones and Walters, 1989; Tiu et al., 1997; Müller and Saez, 1999; Gonzalez et al., 2005) and visco-elastic effects (Hirasaki and Pope, 1974; Vossoughi and Seyer, 1974; Park et al., 1975; Barboza et al., 1979; Haas and Durst, 1982; Briend et al., 1984; Tiu et al., 1997), wall–polymer molecule interactions (Hanna et al., 1977; Shvetsov, 1979; Cohen and Chang, 1984; Kozicki et al., 1984; Cohen and Christ, 1986; Kozicki and Tiu, 1988), pseudo-dilatant behavior (Dauben and Menzie, 1967; Burcik and Ferrer, 1968; Haward and Odell, 2003), inadequacy of power-law (Duda et al., 1983), wall effects (Cohen and Metzner, 1981; Srinivas and Chhabra, 1992), mechanical degradation of polymer solutions, etc. have been advanced to rationalize the diverse predictions of the scores of methods available in the literature and the rather large discrepancies between theories and experiments. None of these, however, has gained general acceptance. In recent years, anisotropy and heterogeneity of porous medium have also been added to this list (Orgeas et al., 2006).

Aside from these developments, the literature abounds with several other expressions based on the power-law fluid model and a range of the other fluid models including Ellis, Meter, and Carreau viscosity equations (Park et al., 1975; Sochi and Blunt, 2008; Rodriguez de Castro and Radilla, 2017). None of these has, however, been tested extensively and is therefore too tentative to be included here.

6.4.2.1.2 Flow of Visco-Plastic Media

In contrast to the voluminous work available on the flow of fluids without a yield stress, the flow of visco-plastic fluids through packed beds has also received some attention (Chhabra, 1993a,b, 1994; Vradis and Protopapas, 1993; Chhabra et al., 2001b; Sochi., 2010a,b), though as noted earlier, there has been a renewed interest in this domain. Al-Fariss and Pinder (1987) combined the Herschel–Bulkley fluid model with the Blake–Kozeny model to derive an expression of the form of Equation 6.79 with the following definition of Re_{NN}:

TABLE 6.7

Expressions for A in Equation 6.79

Reference	Expression	Value of A for $n=1$	Equation number in Table 6.6
Christopher and Middleman (1965)	$\dfrac{375}{\sqrt{2}}\left(\dfrac{12}{15\sqrt{2}}\right)^n$	150	1
Park et al. (1975)	$\left(\dfrac{375}{\sqrt{2}}\right)^{1-n} 150^n A_w^{1+n}$	$150 A_w^2$	2
Brea et al. (1976)	$160\left[\dfrac{4n}{3n+1}\right]^{n-1}\left(\dfrac{4}{5\sqrt{2}}\right)^{n-1}$	160	3
Kemblowski and Michniewicz (1979)	180	180	4
Al-Fariss and Pinder (1987)	$20\sqrt{3}\left(\dfrac{150}{8}\right)^{n/2}$	150	5
Kawase and Ulbrecht (1981a)	$270\sqrt{2} F_1 \varepsilon^{1-2.65n}\left\{\dfrac{4n}{15\sqrt{2}(1-\varepsilon)(1+3n)}\right\}^n$ $F_1 = 3^{3n-1.5}\left[\dfrac{-22n^2+29n+2}{n(n+2)(2n+1)}\right]$	$18\left(\dfrac{\varepsilon^{-1.65}}{1-\varepsilon}\right)$	6
Dharmadhikari and Kale (1985)	$60\sqrt{2}\left[\dfrac{5n}{2+3n}\right]\left(\dfrac{150}{32}\right)^{n''}\left[\dfrac{4n}{3n+1}\right]^n\left[\dfrac{8}{15\sqrt{2}}\right]^n\left[\dfrac{8V_o}{d}\dfrac{(1-\varepsilon)}{\varepsilon^2}\right]^{0.3(1-n)}$ where $n'' = \left[\dfrac{3n}{2+n}\right]$	150	7
Kozicki and Tiu (1988)	$180 A_w^n$	$180 A_w$	8
Sabiri and Comiti (1997a)	$225\left(\dfrac{2\sqrt{2}}{5}T\right)^{n+1}$	136	9

$$\mathrm{Re}_{NN} = \frac{12\rho V_0^2 \psi}{2md\varepsilon^3 V_0^n + \varepsilon^3 \tau_o \psi} \quad (6.80a)$$

$$\psi = 6\left(\frac{6n}{3n+1}\right)^n \left(\frac{d\varepsilon}{3(1-\varepsilon)}\right)^{n+1} (1-\varepsilon) \quad (6.80b)$$

Using their own experimental data for the flow of waxy crudes through packed beds of glass beads, the best value of A was found to be 150. Qualitatively similar expressions have also been obtained independently by others (Park et al., 1975; Kuang and Kozicki, 1989; Wu et al., 1992). Attention is drawn to the fact that Equation 6.80 includes several known results, for instance, Newtonian results ($\tau_o = 0$, $n = 1$), Bingham plastic behavior ($n = 1$), power-law results ($\tau_o = 0$), etc. Dolejs et al. (1998) and Dolejs and Siska (2000) have recently extended the Kozeny–Carman model to the flow of Robertson–Stiff visco-plastic model fluids in packed beds. Finally, the approach of Kozicki et al. (1967) can also be extended to visco-plastic fluids simply by changing the lower limit of integration in Equation 6.76 and Equation 6.77 to τ_o, the yield stress of the fluid. Based on the experimental data for the flow of Bingham plastic Kaolin slurries through packed beds of glass beads and crushed rocks, Masuyama et al. (1983–1986) also proposed an empirical correlation, which, however, has not been evaluated using independent data.

In an experimental study, Chase and Dachavijit (2008) have studied the flow of electrorheological fluids through a bed of glass beads. Their pressure drop results are in line with the Ergun-type equation and they also put forward a criterion for the initiation of flow. The prediction of the minimum pressure gradient for visco-plastic fluids (Bingham and Herschel–Bulkley model fluids) has received considerable attention over the past 10–15 years. Similarly, many researchers have modified the Darcy law for visco-plastic fluids (Table 6.5). In this case, the effective permeability is determined by a complex interplay between the microstructure of the porous medium and the rheology of the fluid. Thus, it is no longer an intrinsic property of the porous medium.

6.4.2.1.3 Transitional and Turbulent Flow

Owing to their generally high viscosities, the flow conditions for non-Newtonian systems rarely extend beyond the so-called laminar or creeping flow region. There is, however, some work available beyond the creeping flow region. Aside from the approach of Sabiri and Comiti (1997a,b), the capillary bundle approach has also been extended empirically for correlating the friction factor–Reynolds number data in the transitional and turbulent regions. For instance, both Brea et al. (1976) and Kumar and Upadhyay (1981) have used the following expression for estimating the "effective viscosity" in porous media

$$\mu_{\text{eff}} = m' \left\{ \frac{12 V_0 (1-\varepsilon)}{d\varepsilon^2} \right\}^{n'-1} \tag{6.81}$$

to define the particle Reynolds number as

$$Re' = \frac{\rho V_0 d}{\mu_{\text{eff}} (1-\varepsilon)} \tag{6.82}$$

Their results can be expressed as

$$f\left(\frac{\varepsilon^3}{1-\varepsilon}\right) = \frac{\alpha_1}{Re'} + \beta_1 \tag{6.83}$$

where the best value of α_1 and β_1 have been obtained using experimental data; for example, Mishra et al. (1975) and Kumar and Upadhyay (1981) found $\alpha_1 = 150$ and $\beta_1 = 1.75$ (same as in the Newtonian case) whereas Brea et al. (1976) reported $\alpha_1 = 160$ and $\beta_1 = 1.75$. Kemblowski and Mertl (1974), on the other hand, extended their low Reynolds number work to embrace the transitional and turbulent flow regions as

$$f\left(\frac{\varepsilon^3}{1-\varepsilon}\right) = \frac{150}{Re_{\text{C-M}}} + 1.75 \left(\frac{\kappa H^2}{\sqrt{\kappa^2 (H^2-1)^2 + H^2}} \right) \tag{6.84}$$

where κ and H are further correlated as

$$H = \xi \, Re_{\text{C-M}} \tag{6.85}$$

κ and ξ are found to be functions of the flow behavior index, n, as given by Equations 6.86a and b, respectively.

$$\log k = -1.7838 + 5.219n - 6.239n^2 + 1.559n^3 + 2.394n^4 - 1.12n^5 \tag{6.86a}$$

$$\log \xi = -4.9035 + 10.91n - 12.29n^2 + 2.364n^3 + 4.425n^4 - 1.896n^5 \tag{6.86b}$$

Equations 6.84–6.86 are based on experimental data embracing the conditions: $0.5 \le n \le 1.6$ and $0.029 \le \text{Re}_{\text{C-M}} \le 115$. Note that this cumbersome-looking set of equations does reduce to its proper Newtonian limit as the value of n approaches unity. Kemblowski et al. (1987) asserted that this method predicts values of the frictional pressure drop with an uncertainty of about ±30%. Figure 6.14 shows the predictions of Equation 6.84 in a graphical form for a range of values of n, but for a constant value of bed voidage, namely, $\varepsilon = 0.4$. Also, the limited cell model predictions for dilatant ($n > 1$) fluids are consistent with the predictions of Equation 6.84 (Jaiswal et al., 1994).

Similarly, in another study, Tang and Lu (2014) have combined several ideas to modify the form of Ergun equation for the flow of power-law fluids. For instance, they considered the ordered cubic packing of uniform size spheres, similar to that of Wu et al. (2008) and defined a ratio ω of the throat area to the unit cell flow area which is obviously related to the mean voidage of the bed. With a suitably defined Reynolds number (Re′ defined in Equation 6.82 without the factor of 12 appearing in μ_{eff}), they arranged their final expression in the same form as Equation 6.83. However, both constants (like α_1 and β_1 in Equation 6.83) now depend upon the porosity, tortuosity factor, ω, and possibly on the power-law index. For $\varepsilon \approx 0.4$, they reported $\alpha_1 = 198$ and $\beta_1 = 2$ and these values successfully reproduced the experimental results of Yu et al. (1968), Kemblowski and Mertl (1974), Mishra et al. (1975), Brea et al. (1976), etc. For $\varepsilon = 0.5$, it predicts drag reduction for $n < 1$ and drag enhancement for $n > 1$, especially up to about Reynolds number values ≤ 1000; beyond this limit, all the curves merge into a single curve. Similarly, Tosco et al. (2013) have demonstrated that their predictions for the flow of Ellis, Cross, Carreau, and power-law models through 2-D porous medium are consistent with the Darcy–Forchheimer formulation.

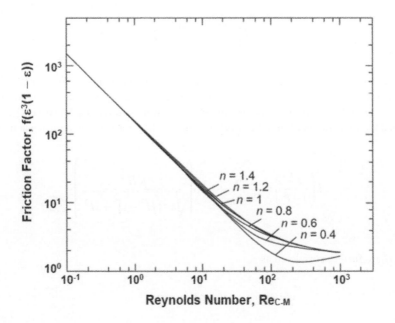

FIGURE 6.14 Predictions of Equation 6.84 for a range of values of power-law index for $\varepsilon = 0.4$.

In another study, Chase and Dachavijit (2005) have empirically modified the Ergun equation for the flow of Bingham plastic fluids as

$$f^+ = \frac{60}{\text{Re}^+}\left[1 - \frac{4}{3}\Omega + \frac{1}{3}\Omega^4\right]^{-1} + 0.6 \tag{6.87}$$

where the modified factor f^+, the Reynolds number Re^+, and Ω are defined as

$$f^+ = \left(\frac{f}{3}\right)\left(\frac{\varepsilon^3}{1-\varepsilon}\right) \tag{6.88a}$$

$$\text{Re}^+ = \frac{\text{Re}}{(1-\varepsilon)} \tag{6.88b}$$

$$\Omega = \frac{3.5 \text{He}}{f^+ \left(\text{Re}^+\right)^2} \tag{6.88c}$$

and finally,

$$\text{He} = \frac{\tau_o^B \rho d^2}{\mu_B^2}\frac{\varepsilon^2}{(1-\varepsilon)^2} \tag{6.89}$$

The applicability of Equation 6.88 was demonstrated for the flow of Carbopol solutions and subsequently for electrorheological fluids (Chase and Dachavijit, 2008). Some anomalous trends have also been documented in the literature on the flow of visco-plastic fluids in microfluidic scale porous media (Talon and Bauer, 2013; Marliere et al., 2015; Waisbord et al., 2019; Seybold et al., 2021).

All aforementioned studies use the simplest of the capillary model involving uniform and single size pores. The effect of pore size distribution on the flow of generalized Newtonian fluids has been treated by Kanellopoulos (1985), Sorbie and Huang (1991), and Di Federico (1998), whereas some network modeling has been attempted by Lopez et al. (2003), Balhoff and Thompson (2004, 2006), and Perrin et al. (2006). Perrin et al. (2006) have used the network modeling to develop a predictive equation for the flow of power-law and Ellis model fluids in packed beds with a single adjustable parameter. On the other hand, Sullivan et al. (2006) have used the Lattice–Boltzmann method to study the flow of power-law fluids in a random porous medium. Additional studies on pore network modeling as well as dealing with heterogeneous and disordered porous media have been summarized by Sochi (2010a, 2010b).

6.4.2.2 Submerged Object Models or Drag Theories

There has been no fundamental study for non-Newtonian fluids in this area that can parallel the pioneering work of Brinkman (1947, 1948) and its subsequent modifications (Tam, 1969; Lundgren, 1972). Nor has there been any concerted empirical effort that can match the developments due to Barnea and Mizrahi (1973), Barnea and Mednick (1975, 1978), and others (Zimmels, 1988), except for the work of Vradis and Protopapas (1993) who used the numerical predictions of Beris et al. (1985) of drag on a single sphere together with a cell model to predict the macroscopic conductivities of Bingham plastic fluids in porous media. Most of the available work in this field is based on the application of the two-cell models, that is, the free surface cell model and the zero-vorticity cell model. For instance, the creeping flow of the power-law, Ellis, and Carreau model fluids through packed beds has been simulated using the free surface cell

TABLE 6.8

Numerical Predictions of Drag Coefficient in terms of Y for Power-Law Fluids in Creeping Flow Region

ε	n	Jaiswal et al. (1992)	Mohan and Raghuraman (1976a)		Kawase and Ulbrecht (1981a)	Manjunath and Chhabra (1991)[a]
			Upper Bound	Lower Bound		
0.4	1.0	85.29 (84.92)	85.12	85.12	85.14	101.82 (101.65)
	0.8	35.13 (36.65)	37.06	37.00	35.81	47.04 (46.01)
	0.6	14.92 (15.82)	17.08	15.92	15.87	20.42 (21.06)
	0.4	5.28	-	-	7.84	8.95
	0.2	1.72	-	-	5.12	-
0.5	1.0	37.82 (37.89)	37.91	37.91	37.92	47.86 (46.77)
	0.8	17.63 (18.48)	18.67	18.50	19.02	22.17 (23.8)
	0.6	7.85 (9.00)	9.17	9.11	9.99	12.00 (12.15)
	0.4	3.34	-	-	5.83	5.92
	0.2	1.19	-	-	4.47	-
0.7	1.0	10.13 (10.15)	10.13	10.13	10.14	12.98 (12.97)
	0.8	5.86 (6.18)	7.22	7.21	7.80	8.54 (8.13)
	0.6	3.22 (3.72)	3.81	3.78	4.72	5.21 (5.07)
	0.4	1.68	-	-	3.58	3.13
	0.2	0.82	-	-	3.53	-
0.9	1.0	3.10 (3.11)	3.11	3.11	3.11	3.81 (3.78)
	0.8	2.36 (2.42)	2.44	2.43	2.74	3.33 (3.00)
	0.6	1.68 (1.84)	1.90	1.88	2.44	2.54 (2.32)
	0.4	1.15	-	-	2.31	1.88
	0.2	0.74	-	-	1.74	-

[a] Based on zero-vorticity cell model. Figures in brackets are from Satish and Zhu (1992).

model to obtain approximate upper and lower bounds on drag coefficient of multiparticle assemblages (Mohan and Raghuraman, 1976a,b; Chhabra and Raman, 1984; Gu et al., 1992; Satish and Zhu, 1992; Zhu and Satish, 1992). These results are often expressed in the form of a drag correction factor Y (defined by Equation 2.37) whereas the corresponding Reynolds number is defined similar to that for a single sphere (Equation 2.39). Kawase and Ulbrecht (1981a) also obtained a closed-form expression for the drag correction factor for both the free surface and the zero-vorticity cell models by linearizing the nonlinear viscous terms for power-law liquids; their results are, however, applicable only for mild shear-thinning behavior. Subsequently, Manjunath and Chhabra (1991), Jaiswal et al. (1991c, 1992, 1993a,b), and Dhole et al. (2004) have solved the complete field equations numerically for power-law fluids in the range ($0.2 \leq n \leq 1$; $0.3 \leq \varepsilon \leq 0.9$; $0.001 \leq Re_{PL} \leq 500$). Table 6.8 presents a summary of the low Reynolds number numerical results and a comparison with the other theoretical predictions available in the literature. Evidently, the finite element and finite difference results (Jaiswal et al., 1992; Dhole et al., 2004; Kishore et al., 2009a,b) are in good agreement with the approximate results available in the literature (Mohan and Raghuraman, 1976a; Kawase and Ulbrecht, 1981a). Suffice it to note here that there seems to be a reasonable agreement between the various results shown in Table 6.8. However, as expected, the two models, namely the free surface and zero-vorticity cell models, yield predictions that can differ up to about 20%. A typical comparison between these simulations and experimental data is shown in Figure 6.15 in the low Reynolds number region. Note the wide scatter present in the available experimental results; the reasons for such large variations have been discussed in detail by Chhabra (1993a,b). The corresponding comparisons at

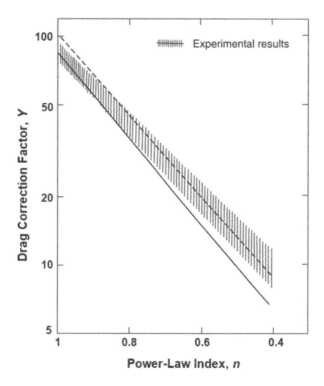

FIGURE 6.15 Typical comparisons between predictions of cell models and experiments (shown as \\\\) for creeping power-law fluid flow in packed beds, $\varepsilon=0.4$. (—) Free surface cell model (Jaiswal et al., 1993a); (- - -) Zero-vorticity cell model (Manjunath and Chhabra, 1991).

high Reynolds numbers are shown in Figure 6.16 with the predictions of Equations 6.20 and 6.84 for three values of bed voidage ($\varepsilon=0.4$, 0.5, and 0.6). The degree of match between the predictions and experiments is seen to improve slightly with the increasing value of bed voidage, but it deteriorates with the increasing Reynolds number. The cell models seem to be more appropriate at high porosities and at low to moderate Reynolds number. The numerical results of Dhole et al. (2004) extend over wide range of conditions ($0.4 \leq n \leq 1.8$; $1 \leq Re_{PL} \leq 500$; $0.6 \geq \varepsilon \geq 0.4$) by the simple expression

$$f\left(\frac{\varepsilon^3}{1-\varepsilon}\right) = \frac{230}{Re^*} + 0.33 \qquad (6.90)$$

Subsequently, Zhu (1990) and Zhu and Chan Man Fong (1988) have reported approximate analytical results for the slow flow of Carreau model fluids through regular arrays of spheres, and their predictions are also in good agreement with those based on the free surface cell model (Chhabra and Raman, 1984).

There has been a renewed interest in this approach to model flow in porous media. Derksen (2009) has calculated the drag on a random assemblage of spheres in the creeping flow regime of shear-thinning thixotropic fluids (modeled using a single structure parameter whose value is dependent on shear rate). Using the Lattice–Boltzmann method, for Newtonian fluids, this approach predicts the drag force to be only a function of the bed voidage or concentration of particles. In the range $0.47 \leq \varepsilon \leq 0.67$, this approach predicts the values of Y which are about 25%–30% lower than those reported in Table 6.8. By introducing suitably normalized drag and shear rate, Derksen (2009) was able to collapse the results for different values of ε on to a master curve. Very recently Fischer et al. (2023) have studied the creeping flow of power-law fluids in three dimensional beds of spheres..

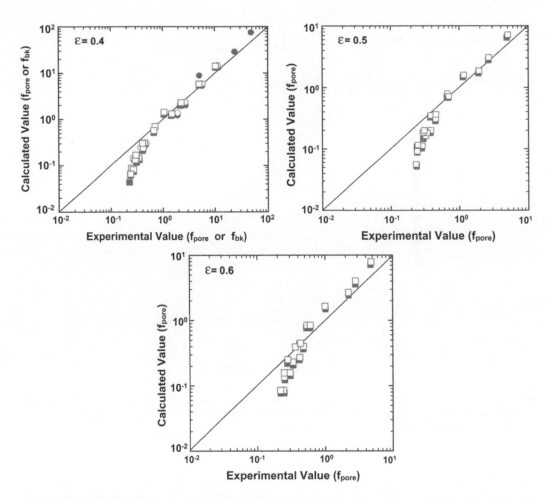

FIGURE 6.16 Comparison between cell model predictions, Equation 6.20 and Equation 6.84 for $\varepsilon = 0.4, 0.5, 0.6$ in terms of f_{pore}. (□) Zero vorticity cell model; (■) Free surface cell model; (●) Predictions of Equation 6.84. (Modified after Dhole, S.D. et al., *Chem. Eng. Res. Des.*,82, 642, 2004.)

Singh et al. (2012), on the other hand, have extended the ensemble averaging of momentum equations to study the drag on a bundle of long cylinders and spheres in power-law fluids in the dilute solid concentration (high ε values). The concentrated system (low ε values) limit was treated via the effective medium theory. In the high-voidage dilute limit systems, they reported an interesting role of power-law index at $n \geq 1$ for cylinders and $n \geq 2$ for spheres. The algebraic growth of the disturbance velocity (due to the effect of neighbors) ceases under these conditions. The resulting predictions for both beds of cylinders and arrays seem to be in agreement with the scant experimental results available in the literature.

Aside from such complete numerical solutions of the field equations, some other approximations together with the cell models have also been used. For instance, by combining the boundary layer flow approximation with the free surface cell model, Hua and Ishii (1981) have obtained drag coefficient results for particle Reynolds numbers up to 1000; their numerical results appear to be in error, as in the limit of a single sphere their values differ from the literature values by as much as a factor of two. Similar results for high Reynolds number flow have also been reported by Kawase and Ulbrecht (1981b).

Finally, this section is concluded by reiterating that these two approaches, namely, the capillary bundle and the submerged objects, complement each other rather than being mutually exclusive. There is a general consensus that the capillary model is more appropriate for packed bed conditions whereas the submerged object model is more suitable at high porosities, when the particles are distended from each other (Mauret and Renaud, 1997; Ciceron et al., 2002b). On the other hand, some attempts have also been made to combine these two ideas to develop a hybrid model. This is exemplified by the approaches put forward by Machac and Dolejs (1981), Dolejs et al. (1995), and Dolejs and Machac (1995). However, this approach not only entails a fair degree of empiricism, but also does not offer significant improvement over any of the methods presented thus far.

6.4.2.3 Volume Averaging of Equations

Since the early works of Lehner (1979) and Larson (1981), there has been a renewed interest in using the volume averaging technique to describe the flow of power-law fluids in granular and fibrous porous media and in synthetic foams (Hayes et al., 1996; Smit and du Plessis, 1997, 1999, 2000; Woudberg et al., 2006). Perhaps the most promising of all such efforts is the approach of Smit and du Plessis (1999), which is a direct modification of the approach developed by du Plessis and Masliyah (1988) for the flow of Newtonian fluids. This analysis depends on the rectangular representation of solid microstructure of the porous matrix. By introducing some simplifications, Smit and du Plessis (1997) were able to develop a predictive model that requires the values of bed porosity, a microstructure length scale (linked to the packing size), and the rheological characteristics of the liquid medium. For a homogeneous and isotropic bed of spheres, this approach yields

$$\varepsilon f = \left(\frac{72}{\text{Re}_{SD}} + 1\right)\left[\left\{1-(1-\varepsilon)^{2/3}\right\}^{n-2}(1-\varepsilon)^{2/3}\right] \quad (6.91)$$

where the Reynolds number Re_{SD} is defined as

$$\text{Re}_{SD} = 12\left(\frac{\rho V_0^{2-n} d^n}{m}\right)\frac{\left\{1-(1-\varepsilon)^{1/3}\right\}^n}{\left\{1-(1-\varepsilon)^{2/3}\right\}^{2-n}}\left(\frac{n}{4n+2}\right)^n \quad (6.92)$$

Smit and du Plessis (1997) reported good agreement between the predictions of Equation 6.91 and the limited results of Chhabra and Srinivas (1991) in the range $0.001 \leq \text{Re}_{SD} \leq 51$ and $0.5 \leq n \leq 1$. Subsequently, this treatment has been extended to other fluid models including visco-elastic liquids and to other types of porous media such as fibrous and synthetic foams (du Plessis et al., 1994; Smit and du Plessis, 1999, 2000; Fourie and du Plessis, 2002; Smit and du Plessis, 2005), and the agreement was reported to be good with the work of Sabiri and Comiti (1997a,b). Other relevant works include the studies of Getachew et al. (1998), Liu and Masliyah (1998), Tsakiroglou (2002), Pearson and Tardy (2002), and Airiau and Bottaro (2020) who have all used this method to model the non-Newtonian fluid flow in porous medium, with and without inertial effects.

6.4.2.4 Other Methods

In addition to the foregoing three distinct schemes, several other ad-hoc methods have also been used. For instance, in a series of papers, Pascal (1983, 1984a,b, 1985, 1986a,b, 1988, 1990a,b) has empirically modified Darcy's law for different types of non-Newtonian fluid models. By way of example, Darcy's law is rewritten for power-law fluids as

$$V_0 = \left(\frac{K}{\mu_{\text{eff}}}\nabla p\right)^{1/n} \quad (6.93)$$

Based on Equation 6.93 and other similar modifications for visco-plastic media, Pascal has extensively investigated the steady and transient flow in porous media and has also attempted to evaluate the rheological model parameters from porous media experiments, etc. This approach has also been extended to study the displacement of one fluid by another as encountered in enhanced oil recovery processes (Pascal and Pascal, 1985, 1988, 1989a,b; Wu and Pruess, 1998). It is worthwhile to reiterate here that though all expressions presented in the two preceding sections can be rearranged in the same form as Equation 6.93, the latter does not involve any description of the porous medium. Similarly, Benis (1968) has utilized the lubrication approximation to modify Darcy's law for the flow of power-law fluids in dense systems whereas Parvazinia and Nassehi (2006) have used the Brinkman model instead of the Darcy equation. Similar modifications of the Darcy law for other fluid models are also available in the literature (e.g., see Chevalier and Talon, 2015; Rodriguez de Castro and Radilla, 2016; Liu et al., 2019a; Shende et al., 2021a,b; etc.). McKinley et al. (1966), White (1967), and Garrouch and Gharbi (1999) have all used dimensional considerations to modify Darcy's law for a generalized Newtonian fluid whereas Hassell and Bondi (1965) presented a completely empirical expression for calculating the frictional pressure gradient through beds of beads, screens, and mats, but all these efforts are severely limited in their scope. Similarly, the unsteady non-Newtonian flow of polymer solutions, foams, and emulsions in oil reservoirs has also been explored by some investigators (van Poollen and Jargon, 1969; Ikoku and Ramey, 1979, 1980; Odeh and Yang, 1979; Cohen and Christ, 1986; Islam et al., 1989; Fletcher et al., 1991; Hejri et al., 1991; Bertin et al., 1998a,b; Fergui et al., 1998). Most of these studies relate to small Reynolds numbers and are directed at the use of polymers and foams in secondary and tertiary oil recovery processes. The terms such as permeability reduction, screen factors, and resistance factors are frequently used to represent the overall fluid flow behavior in petroleum engineering applications.

6.4.3 Visco-Elastic Effects in Porous Media

As remarked in Chapter 1, all materials exhibit a blend of viscous and elastic characteristics to varying extents under appropriate circumstances. It is now well known that the flow of visco-elastic fluids in porous media results in a larger frictional pressure drop than that can be expected from the purely viscous fluid behavior. At low velocities, the frictional pressure gradient for the flow in a porous medium is determined primarily by shear viscosity, and the visco-elastic effects are believed to be negligible (Helmreich et al., 1995; Tiu et al., 1997; James et al., 2012; Sobti and Wanchoo, 2014; De et al., 2017a,b,c, Kawale et al., 2017, etc.). For a given polymer solution, as the flow rate is gradually increased, the visco-elastic effects begin to manifest. Consequently, when the pressure drop (or friction factor or loss coefficient) is plotted against a suitably defined Deborah (or Weissenberg) number, beyond a critical value of the Deborah number, the pressure drop across the bed increases rapidly. This behavior is clearly shown in Figure 6.17a for three different polymer solutions and in Figure 6.17b for a series of hydroxypropyl guar solutions. While numerous workers have obtained qualitatively similar results with various chemically different polymers, there is a little quantitative agreement about the critical value of the Deborah number marking the onset of visco-elastic effects and about the magnitude of the excess pressure loss. One possible reason for this lack of agreement in the reported values of the critical Deborah number is the arbitrariness inherent in the definition of the Deborah number. The oft-used definition of Deborah number is

$$\mathrm{De} = \frac{\theta V_c}{l_c} \qquad (6.94)$$

where θ is a characteristic fluid time, and V_c and l_c are characteristic velocity and linear dimension, respectively, of the system. Considerable confusion exists in the literature regarding the choice of each of these quantities. For instance, some researchers (Sadowski and Bird, 1965; Kemblowski et al., 1974, 1980; Chhabra and Raman, 1984) have evaluated θ from shear viscosity data whereas others (Marshall and Metzner, 1967; Siskovic et al., 1971; Tiu et al., 1983, 1997; Briend et al., 1984) have extracted its value from primary normal stress difference data obtained in steady shear. Other methods such as die

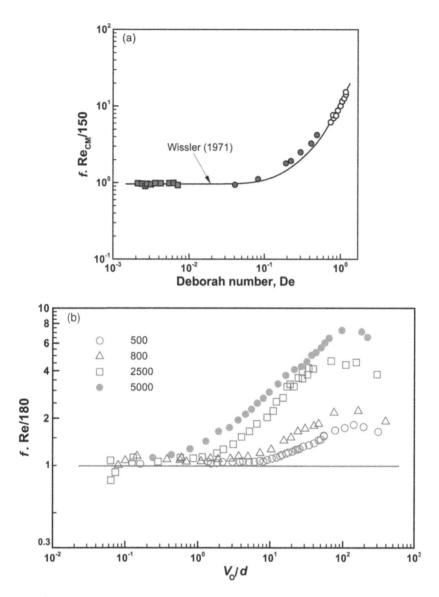

FIGURE 6.17 (a) Typical results showing visco-elastic effects in porous media flow: (□) Carbopol solutions (●) PIB solution (○) ET-597 solution. (Replotted from Kemblowski, Z., Dziubinski, M., and Sek, J., *Advances in Transport Processes*, vol. 5, p. 117, Mashelkar, R.A., Mujumdar, A.S., and Kamal, M.R., Eds., Wiley Eastern Ltd., New Delhi, 1987.) (b) Typical results showing visco-elastic effects during the flow of hydroxypropyl guar (ppm concentration) solutions in a packed bed of glass spheres. (Replotted from Vorwerk, J. and Brunn, P., *J. Non-Newt. Fluid Mech.*, **51**, 79, 1994.)

swell measurements (Kemblowski and Michniewicz, 1979), dilute solution theories (Kemblowski and Dziubinski, 1978), and frequency sweep data (Sobti and Wanchoo, 2014) have also been used to calculate the value of θ. Undoubtedly all these methods yield values of θ, which is a measure of the level of visco-elasticity of a fluid, but different methods are known to yield widely divergent numerical values of θ (Elbirli and Shaw, 1978). Likewise, the characteristic velocity V_c has been taken either as the interstitial velocity or simply the superficial velocity. Though a proper and unambiguous choice for l_c is also far from clear, often the particle diameter has been taken to be the characteristic linear dimension for the flow in packed beds. Some of the recent studies (De et al., 2017a,b; Berg and van Wunnik, 2017, for instance) advocate the use of $\sqrt{k}$ as the characteristic linear dimension to be used in Equation 6.94 or

TABLE 6.9

Definitions and Critical values of Deborah Number Reported in the Literature

Investigator	De	Definition of θ	Range of θ (ms)	Critical De
Sadowski and Bird (1965)	$\dfrac{V_0 \mu_o}{d\tau_{1/2}}$	From Viscosity data via the Ellis model	≤ 77	0.1
Gaitonde and Middleman (1967)	$\dfrac{\theta V_0}{d}$	$\theta = \dfrac{12\mu_o M}{\pi^2 CRT}$ (Bueche theory)	4.7–8.6	1.2
Marshall and Metzner (1967)	$\dfrac{\theta V_0}{\varepsilon d}$	From N_1 data	0.96–34	~ 0.05–0.06
Siskovic et al. (1971)	$\dfrac{\theta V_0}{\varepsilon d}$	From N_1 data	200–4000	0.3
Kemblowski and Dziubinski (1978)	$\dfrac{\theta V_0}{\varepsilon d}$	$\theta = \dfrac{12(\mu_o - \mu_s)M}{\pi^2 CRT}$	33	0.2
Vossoughi and Seyer (1974)	$\dfrac{\theta V_0}{\varepsilon(\Delta z)}$	From N_1 data	-	~0.2
Michele (1977)	$\dfrac{N_1}{2\tau\dot{\gamma}}$	-	-	3
Park et al. (1975)	$\dfrac{\theta V_0}{d}$	From N_1 data	-	> 0.13
Kemblowski and Michniewicz (1979)	$\dfrac{\theta V_0}{\varepsilon d}$ and $\sqrt{2}\dfrac{\theta V_0}{\varepsilon d}$	From die swell data	80–1200	>0.07
Tiu et al. (1983)	$\dfrac{\theta V_0}{\varepsilon d_H}$	From N_1 data	-	No critical value
Manli and Xiaoli (1991)	$\dfrac{2.3\theta V_0}{\varepsilon d}$	From N_1 data	-	~0.05

in calculating the effective shear rate. Table 6.9 gives a concise summary of the different definitions of Deborah number that have been used in the literature.

An examination of Table 6.9 reveals two to three orders of magnitude variation in the reported critical values of the Deborah number, De, denoting the onset of visco-elastic effects. In view of this, it is therefore not at all surprising that the empirical expressions (direct extensions of the expressions presented in the preceding section) that purport to correlate frictional pressure drop are equally diverse in form; a representative sample of these is listed in Table 6.10. It is clearly seen that the majority of the correlations are of the two forms

$$f = \left(\frac{A}{Re}\right) + F(Re, De) \tag{6.95}$$

or

$$f = \left(\frac{A}{Re}\right)\left(1 + B De^2\right) \tag{6.96}$$

While the quadratic dependence on the Deborah number, seen in Equation 6.96, has some theoretical basis (Wissler, 1971; Zhu, 1990), the functional form of Equation 6.95 is completely empirical. Intuitively, it would be desirable for the function F (De, Re) to satisfy the following limiting conditions: for purely

TABLE 6.10

Correlations of f for Visco-Elastic Fluids

Investigator	Equation for f	Observations
Sadowski and Bird (1965)	$f\left(\dfrac{\varepsilon^3}{(1-\varepsilon)^2}\right)\dfrac{\text{Re}_o}{180} = \left\{1+\left(\dfrac{4}{\alpha+3}\right)\left(\dfrac{\tau_w}{\tau_{1/2}}\right)^{\alpha-1} - 4\text{De}\right\}$	Only weak visco-elastic effects.
Wissler (1971)	$\left(\dfrac{\varepsilon^3}{1-\varepsilon}\right) f\,\text{Re}_{CM} = \left(1+10\text{De}^2\right)$	Used data of Metzner and coworkers to evaluate the constants.
Vossoughi and Seyer (1974)	$\left(\dfrac{\varepsilon^3}{1-\varepsilon}\right) f\,\text{Re}_{CM} = \left(1+90\text{De}^2\right)$	-
Michele (1977)	$\psi_M = \left(\dfrac{5}{\text{Re}_M}\right) + \left(\dfrac{0.3}{\text{Re}_M^{1/11}}\right) + \left(0.075\dfrac{\text{De}_M^2}{\text{Re}_M}\right)$	-
Kemblowski and Dziubinski (1978)	$\left(\dfrac{\varepsilon^3}{1-\varepsilon}\right) f\,\text{Re}^* = 150\left(1+8\text{De}^2\right)$	-
Kemblowski and Michniewicz (1979)	$\left(\dfrac{\varepsilon^3}{1-\varepsilon}\right) f\,\text{Re}^* = 180\left(1+\text{De}^2\right)$	-

Note: $\psi_M = \dfrac{1}{6} f\left(\varepsilon^3/(1-\varepsilon)\right)/A_W$; $\text{Re}_M = \rho V_0 d/\left(6(1-\varepsilon)\mu_{\text{eff}} A_W\right)$ where μ_{eff} evaluated at $\dot{\gamma}_{\text{eff}} = \left(6\pi V_0 (2/3)^{0.5}(1-\varepsilon)\right)/d\varepsilon^2$ See Table 6.9 for corresponding definitions of De.

viscous (or negligible elastic) effects, F (Re, De)=0, and for purely elastic liquids, F (Re, De) → 0 as De → 0. Though all expressions listed in Table 6.10 and the others available in the literature attribute the excess pressure drop to the visco-elastic behavior (measured in terms of De), there is also some evidence suggesting that the inclusion of De alone in such correlations is inadequate to account for visco-elastic effects. Furthermore, in recent years, ample evidence has become available suggesting the overwhelming influence of the extensional effects and, therefore, it is an unsound practice to interpret/correlate such results only in terms of the steady shear viscosity (Elata et al., 1977; Durst et al., 1981, 1987; Ghoniem, 1985; Gupta and Sridhar, 1985; Flew and Sellin, 1993; Koshiba et al., 1993, 1999; Vorwerk and Brunn, 1994; Tiu et al., 1997). Numerous experimental studies (Ghoniem, 1985; Ganoulis et al., 1989; Chmielewski et al., 1990b; Chmielewski and Jayaraman, 1992; Odell and Haward, 2006; James et al., 2012) carried out with model porous media such as "in line" and staggered arrays of long cylinders clearly reveal the presence of a strong extensional component in this flow configuration. Detailed PIV measurements in sparse systems ($\varepsilon \geq 0.9$) with Boger fluids indicate the increasingly asymmetric velocity profiles for De $\geq \sim 0.5$ when the visco-elastic effects begin to appear. The experimental pressure drop was found to be several times the purely viscous value. Similarly, Odell and Haward (2006) have reported a significant role of the arrangement of arrays of spheres on the pressure drop, in line with the prediction of De et al. (2017, 2018) for arrays of cylinders. Similarly, in order to highlight the effect of successive contraction/expansion character of the flow geometry, several numerical (Deiber and Schowalter, 1979, 1981) and experimental (Magueur et al., 1985; Phan-Thien and Khan, 1987; Pilitsis and Beris, 1989; James et al., 1990; Zheng et al., 1990b; Huzarewicz et al., 1991; Davidson et al., 1993; Podolsak et al., 1997; Koshiba et al., 1999) investigations have been carried out using the tubes of regular but periodically varying cross sections (Table 6.2), the most common being the sinusoidal variation.

However, considerable controversy surrounds the theoretical predictions and the experimental observations in this area. Besides, the numerical predictions appear to be not only strongly fluid model-dependent, but also weakly sensitive to the numerical scheme used to solve the field equations. In fact, even the suitability of this geometry for simulating visco-elastic effects observed in porous media flows has been questioned (James et al., 1990). Subsequent numerical simulations suggest that even in the presence of significant visco-elastic effects and large amplitude ratio of PCTs, the resulting values of pressure drop are identical to the values as those predicted from the viscosity considerations alone. The general impression is that the steady-state simulations are unlikely to resolve this dilemma.

In spite of the aforementioned inherent difficulties, some attempts have been made to couple the capillary models with specific visco-elastic constitutive equations to gain physical insights into the role of fluid visco-elasticity. For instance, Lopez de Haro et al. (1996) have used the method of volume averaging for the creeping flow of Maxwell fluids. The general framework was specialized for the capillary bundle representation of a homogeneous and isotropic porous medium. Since they transformed the equations in the frequency domain, they were able to deduce an expression for a frequency-dependent permeability that presumably can be linked to the behavior of the fluid under undulating conditions of flow in a porous medium. Similarly, Kozicki (2002) has combined the FENE fluid model with the capillary bundle model. This analysis captures the role of molecular parameters such as molecular weight distribution, polymer concentration, and quality of solvent satisfactorily. Aside from these studies, many others (see Vorwerk and Brunn, 1994; Helmreich et al., 1995) have emphasized that the visco-elastic effects appear only when the ratio of the first normal stress difference to the corresponding shear stress exceeds unity. However, there is a degree of arbitrariness inherent in delineating an appropriate shear rate and elongation rate in packed beds and porous media (Teeuw and Hesselink, 1980; Wreath et al., 1990). Yet some others (Cakl and Machac, 1995; Dolejs et al., 2002) have developed empirical predictive correlations for the estimation of pressure drop. However, almost all such attempts neglect the role of extensional flow. Broadly speaking, visco-elastic effects become increasingly significant with the increasing molecular weight and the flexibility of the polymer and with the decreasing packing size and/or with the increasing degree of polydispersity of packing grains. Additional discussion on this topic can be found in the survey articles of Sochi (2010a, 2010b).

6.4.4 Dilute/Semidilute Drag Reducing Polymer Solutions

The interest in the flow of dilute/semidilute polymer solutions in porous media stems both from theoretical considerations, such as the fact that this flow configuration provides a good "testing ground" for validating the predictions of dilute solution theories, and from practical considerations, such as their applications in enhanced oil recovery, drag reduction, and antimisting processes. It is well known that the efficiency of the oil displacement process improves appreciably if polymer-thickened water is used instead of pure water (Nilsson et al., 2013; Afolabi et al., 2019). The use of dilute polymer solutions that is beneficial at least on two other counts, namely, the reduction in permeability of the rock and the retardation of flow at high flow rates, is brought about by the extensional effects (James, 1984). While the most field applications use the commercially available partially hydrolyzed polyacrylamide (PHPA), the laboratory tests have been carried out with the dilute solutions of polyethylene oxide (PEO) (James and McLaren, 1975; Elata et al., 1977; Naudascher and Killen, 1977; Kaser and Keller, 1980) and of polysaccharide, a biopolymer (Chauveteau, 1982). The main difficulty with PEO solutions appears to be its high susceptibility to mechanical degradation and the scission of PEO molecules in solutions, especially at high flow rates vis-à-vis high deformation rates. This uncertainty has contributed to the degree of confusion (James and McLaren, 1975; Kaser and Keller, 1980).

Broadly speaking, the term "dilute" solution is used to indicate no or little entanglement of polymer molecules in solution. Several quantitative criteria are available to class a solution as being dilute, semidilute, or concentrated, the simplest of all being that a solution is regarded to be dilute and semidilute as long as $[\eta]C < \sim 0.2-0.3$. One can also make another distinction between a concentrated and a dilute/semidilute solution based on the variation of shear viscosity with shear rate. Thus, for instance, most concentrated solutions display pronounced levels of shear-thinning whereas dilute and semidilute solutions ($\sim <100-200$ ppm) are characterized by virtually constant shear viscosities, albeit their behavior in extension may deviate significantly from that of a Newtonian fluid. Based on both these counts, one can say that most studies referred to in the preceding section involved the use of concentrated polymer solutions.

Most investigators in this field have concentrated on the frictional pressure drop associated with the flow of dilute/semidilute polymer solutions in porous media, albeit the stability of flow has also been studied in one instance (Jones and Maddock, 1966). Admittedly, though the first laboratory investigation reporting pressure drops well in excess of those expected from the solution viscosity appeared in the literature more than 60 years ago (Sandiford, 1964, 1977; Pye, 1964), the subject has received systematic attention only in 1970s. Dauben and Menzie (1967) were the first who really worked with dilute/semidilute solutions of PEO in porous media experiments and documented up to 25 times larger values of pressure drop than those that could be attributed to the solution viscosity. Subsequently, qualitatively similar (even more dramatic) results have been reported in the literature both for concentrated (Christopher and Middleman, 1965; Sadowski and Bird, 1965; Marshall and Metzner, 1967) and dilute polymer solutions (Elata et al., 1977; Hanna et al., 1977; Durst et al., 1981, 1987; Interthal and Haas, 1981; Haas and Kulicke, 1984; Kulicke and Haas, 1984, 1985; Flew and Sellin, 1993; Odell and Haward, 2006, etc.)

The emerging overall picture of the frictional pressure drop in a porous medium for dilute polymer solutions can be summarized as follows: for a given polymer solution–porous medium combination, the experimental values of friction factor (or pressure drop) are in line with the expected Newtonian behavior up to a critical Reynolds number (or liquid flow rate). Beyond this value, with the increasing liquid flow rate, the friction factor deviates increasingly from the Newtonian line, goes through a maximum, and finally, shows a weak downward trend, and becomes nearly parallel to the Newtonian curve. Typical experimental results exhibiting all these features are shown in Figure 6.18 for a range of PEO solutions. Qualitatively similar results are available in the literature for PAAm (Poly Acryl Amide) (Durst et al., 1981, 1987; Kulicke and Haas, 1984, 1985) and for polysaccharide solutions (Chauveteau, 1982; Fletcher et al., 1991; Hejri et al., 1991). Numerous plausible mechanisms including gel formation and adsorption (Pye, 1964; Sadowski and Bird, 1965; Burcik and Walrond, 1968; Mungan, 1969; Kozicki et al., 1988), plugging or blockage of pores (Harrington and Zimm, 1968), visco-elastic, especially extensional effects (James and McLaren, 1975; Kaser and Keller, 1980; Haas and Durst, 1982; Jones and Walters, 1989; Vorwerk and Brunn, 1994), slip effects (Zami-Pierre et al., 2017; Browne et al., 2020), etc. have been

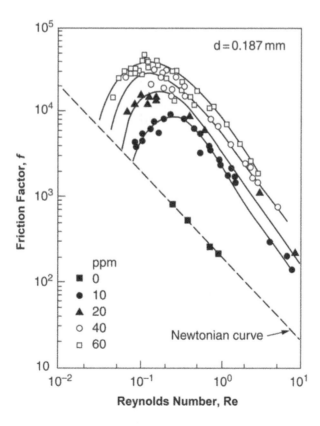

FIGURE 6.18 Typical friction factor–Reynolds number results for a series of drag-reducing polymer solutions. (Replotted from Kaser, F. and Keller, R.J., *J. Eng. Mech. (ASCE)*, **106**, 525, 1980.)

postulated to explain the observed pressure drop behavior. Undoubtedly, the observed pressure drop pattern is solely attributable to the visco-elasticity of the polymer solutions, but how to quantify fluid visco-elasticity for this purpose is far from clear. For instance, the bulk of the attempts at rationalizing the observed pressure drop pattern hinge on the use of a suitably defined Deborah number while some have advocated the use of Trouton ratios. The salient features seen in Figure 6.18 can be qualitatively explained as follows: at sufficiently low flow rates (i.e., small values of Re and De), the relaxation time of the fluid is much shorter than that of the process ($\sim d/V_0$) thereby enabling a fluid element to adjust almost instantaneously to its continually changing surroundings as it traverses the tortuous paths, that is, it is able to relax completely whence no visco-elastic effects are observed. As the liquid velocity is increased, the characteristic time of the flow decreases, and it is no longer possible for a fluid element to respond to its rapidly changing surroundings. This inability of the fluid particles results in the buildup of elastic stresses due to incomplete or no relaxation at all, which, in turn, shows up as an increase in the frictional pressure drop gradient.

Two thorough and systematic studies on the flow of PEO solutions (<80 ppm) in glass bead packs were conducted by Kaser and Keller (1980) and Naudascher and Killen (1977). Based on simple heuristics, these researchers presented the criterion for the onset of visco-elastic effects as

$$\frac{\theta V_0}{d}\sqrt{C} = \text{constant} \tag{6.97}$$

where θ is the fluid relaxation time estimated from the Rouse formula for dilute solutions (Rouse, 1953), and the constant on the right-hand side of Equation 6.97 is a characteristic value for each polymer/solvent combination. Note that the parameter $(\theta V_0/d)\sqrt{C}$, though similar to the Deborah number, De, is not dimensionless. The results shown in Figure 6.18 are plotted in accordance with Equation 6.97 in Figure 6.19 where the agreement between the observed and predicted onset conditions is seen to be fair. The data shown in Figure 6.19 also support the predictions of Naudascher and Killen (1977) that the maximum value of $f \varepsilon^3 \text{Re}/(1-\varepsilon)$ is a linear function of the polymer concentration in the solution. This approach, however, has not been tested with PAM and polysaccharide solutions.

Likewise, much effort has been directed at the flow of PPAM/PAM dilute solutions in porous media. Two research groups have made noteworthy contributions in this field. Durst et al. (1981, 1987), Haas and Durst (1982), Haas and Kulicke (1984), and Kulicke and Haas (1984, 1985) have systematically

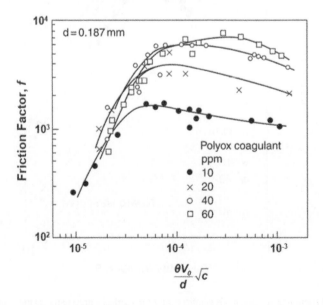

FIGURE 6.19 Results shown in Figure 6.18. (Replotted in accordance with Equation 6.97 [solid lines].)

elucidated the influence of molecular weight and concentration of PAM, degree of hydrolysis, the number and type of ions present in the solution, and of the particle size on the resistance to flow in beds of spherical particles arranged in various geometric configurations. It has been possible to reconcile most of the data using a so-called reduced extensional viscosity and a modified Deborah number based on Bird's FENE model for dilute solutions of flexible macromolecules. While their overall correlation is quite good, it is not completely successful (James, 1984).

The work of the second group (Chauveteau, 1982; Magueur et al., 1985) is also significant because not only have they used the actual sandstone filters but also channels of other geometries to simulate a wide variety of porous media. Besides, instead of using the conventional Ergun coordinates (i.e., f and Re), this group has presented the results in the form of "rheograms" inferred from porous media flow data. Thus, a typical plot for a particular solution may include the true viscosity-shear rate data along with the apparent viscosity data obtained in porous media flow experiments, and, if the two sets of data collapse onto each other, the flow is assumed to be shear dominated and free from all other complications, for example, visco-elastic effects, wall effects, surface effects, slip effects, etc. Conversely, the lack of agreement between the two sets of data implies the presence of visco-elastic, slip effects, degradation, etc. For instance, when the pore size is small, the macromolecules migrate toward the center, and the resistance to flow is found to be lower than that expected from the bulk solution viscosity. Thus, this form of presentation of results facilitates the delineation of shear-thinning, visco-elastic, and wall/surface effects under appropriate circumstances. Qualitatively, similar results have also been documented by others (Wreath et al., 1990; Krussmann and Brunn, 2001, 2002). There have been some theoretical developments that attempt to capture some aspects of the flow behavior of dilute polymer solutions in porous media. Daoudi (1976) estimated the rate of energy dissipation by considering the deformation of randomly coiled macromolecules in a PCT. Elata et al. (1977), on the other hand, invoked the coil-stretch transition hypothesis to explain the abrupt increase in the flow resistance seen with PEO solutions. Similarly, Naudascher and Killen (1977) argued that the strain hardening seen with flexible molecules in porous media flows arises due to the hydrodynamic interactions between the stretched and aligned macromolecules. Preliminary comparisons between the predicted and experimentally observed values (James and McLaren, 1975) of the maximum pressure drop for PEO solutions appear to be encouraging. On the other hand, the stretching and conformation during the flow of polymer solutions in granular and fibrous porous media have been studied to delineate the possible mechanisms for the enhancement of flow resistance (Shaqfeh and Koch, 1992; Evans et al., 1994).

6.4.5 Wall Effects

The influence of containing walls on the frictional pressure drop associated with the non-Newtonian fluid flow in packed beds has not been studied as extensively as for Newtonian media. Owing to the completely geometrical character of the wall effects, it is perhaps reasonable to assume that the non-Newtonian fluid properties (at least for inelastic liquids) do not exert any appreciable influence on the magnitude of wall effects. Indeed, the scant results available in this area (Mehta and Hawley, 1969; Park et al., 1975; Hanna et al., 1977; Kozicki and Tiu, 1988; Srinivas and Chhabra, 1992; Basu, 2001; Tian et al., 2018) lend support to this assertion. Hence, the strategies developed to deal with wall effects in Newtonian systems (see Section 6.3.3) provide a good first-order approximation for the analogous effects in non-Newtonian flows, at least for purely viscous fluids.

Thus, one can replace d by (d/A_w) in the definitions of the Reynolds number and friction factor. As noted elsewhere in this chapter, this approach suffers from two deficiencies: first, while the wall effects are most severe close to the column walls, the factor (d/A_w) is used across the entire bed. Second, this approach also neglects the experimental evidence about the different nature of the wall effects at low and high Reynolds numbers. The approach of Sabiri and Comiti (1997a) circumvents the second difficulty. In their framework, Equation 6.20 is modified to account for wall effects as

$$f_{\text{pore}} = \frac{16\alpha_o}{\text{Re}_{\text{pore}}} + 0.194\beta_o \qquad (6.98)$$

The two constants α_o and β_o, respectively, are given by

$$\alpha_o = 1 + \frac{4}{a_{vd}(1-\varepsilon)D} \tag{6.99}$$

and

$$\beta_o = \left(1 - \frac{d_e}{D}\right)^2 + 0.427\left\{1 - \left(1 - \frac{d_e}{D}\right)^2\right\} \tag{6.100}$$

where $d_e = (6/a_{vd})$ is an effective size of the particle packing. For spherical particles, $a_{vd} = (6/d)$ and hence $\alpha_o = A_w$, the factor introduced by Mehta and Hawley (1969) as given by Equation 6.40.

6.4.6 Effect of Particle Shape and Size Distribution

Not much work has been reported on the effect of particle shape on the pressure loss incurred by non-Newtonian fluids in porous media and packed beds (Yu et al., 1968; Tiu et al., 1974, 1997; Kumar and Upadhyay, 1981; Machac and Dolejs, 1981; Chhabra and Srinivas, 1991; Sharma and Chhabra, 1992). Sharma and Chhabra (1992) have reviewed the pertinent literature, and suffice it to add here that the methods developed for spherical particles also provide an adequate method for correlating/predicting pressure drop data provided the sphere diameter is replaced by $d_e\phi$ where d_e is the volume equivalent diameter and ϕ is the sphericity factor (Li and Ma, 2011). Alternatively, if the values of a_{vd} and T can be evaluated from Newtonian fluid flow data, the approach of Sabiri and Comiti (1995, 1997a,b), Equation 6.20, is probably preferable as it obviates the need of estimating the sphericity which is problematic for irregular shaped packings. Similarly, the scant results available with beds of multisize spheres (Rao and Chhabra, 1993; Ciceron et al., 2002b; Bu et al., 2014) are well represented by Equation 6.20 using the power-law definition of Re_{pore}.

6.4.7 Flow in Fibrous Media

As noted earlier, random-fiber media (mats, screens) are extensively used to filter polymers, monomers, paper pulp suspensions, and sewage sludges (Kaplan et al., 1979). Their chief advantages are high flow rates, low pressure drop, and ease of operation. Similarly, many process streams exhibiting non-Newtonian behavior which are cooled or heated using pin and tubular heat exchangers also flow over a tube bundle (Adams and Bell, 1968; Ghosh et al., 1994). Also, the creeping flow past 2-D periodic arrays has been used extensively to test the validity and efficacy of generalized Newtonian fluid and visco-elastic constitutive equations and of the numerical solution procedures. Additional applications are found in various polymer processing applications such as in resin transfer molding and in assessing the performance of self-compacting concrete in porous matrices (Vasilic et al., 2011, 2016). In spite of such overwhelming pragmatic significance and theoretical importance, in contrast to the extensive literature for the flow of Newtonian fluids in fibrous media, the corresponding body of information for non-Newtonian fluids is quite limited (Table 6.11). An examination of this table shows that the number of analytical/numerical studies far exceeds the experimental results. It is convenient to present the ensuing discussion separately for purely viscous (generalized Newtonian fluid) and visco-elastic fluids.

6.4.7.1 Generalized Newtonian fluids

Most of the available studies relate to the 2-D cross flow of power-law liquids (Tripathi and Chhabra, 1992b; Bruschke and Advani, 1992; Skartsis et al., 1992b; Chen et al., 1998a; Vijaysri et al., 1999; Chhabra et al., 2000; Dhotkar et al., 2000; Shibu et al., 2001; Spelt et al., 2001, 2004, 2005; Woods et al., 2003; Ferreira and Chhabra, 2004). While some investigators have modeled the fibrous medium

TABLE 6.11

Summary of Representative Studies for Transverse Flow Past a Bundle of Rods

Investigator	Fluid Model	Geometry	Remarks
Adams and Bell (1968)[a]	Aqueous CMC solutions (Power-law)	Square and triangular pitch bundles ($\sim 0.4 \leq \varepsilon \leq \sim 0.6$)	Experimental results on pressure drop and heat transfer over wide ranges of Re and Pr.
Alcocer and Singh (2002)	FENE/C-R model	Staggered array (2-D)	Dimensionless permeability shows slight increase with Deborah number.
Barboza et al. (1979)[a]	Aqueous solutions of polyacrylamide	Square arrays	The excess pressure drop relates to changes in the flow patterns at low Re.
Bruschke and Advani (1992)	Power law ($0.5 \leq n \leq 1$)	Square arrays and cell model ($0.15 \leq \varepsilon \leq \sim 0.95$)	Numerical predictions of pressure drop in the creeping flow regime.
Chen et al. (1998a)	Power law	Staggered square array	Limited numerical predictions of pressure drop in the low Re regime.
Chhabra et al. (2000)	Power law ($n > 1$)	Cell model	Numerical results on pressure drop for dilatant fluids.
Chmielewski and Jayaraman (1992); Chmielewski et al. (1990b)	Fluid M-1 and Boger fluids	Square arrays ($\varepsilon = 0.7$)	Effect of polymer characteristics and visco-elasticity on flow patterns and resistance to flow.
Dhotkar et al. (2000)[a]	Power-law	Cell model (Free surface)	Numerical predictions of drag up to about $Re \approx 10$.
Ferreira and Chhabra (2004)	Power-law	Cell models	Closed form analytical solution for creeping flow conditions.
Khomami and Moreno (1997)[a]	Boger fluids and fluid M-1	Inline square array ($\varepsilon = 0.45, 0.86$)	Visualization studies reveal the flow to be 2-D below a critical Weissenberg number, above which it becomes 3-D and time dependent.
Liu et al. (1998a)	Giesekus, FENE, and CR models	Linear arrays confined in a plane channel	Flow and stress fields are qualitatively similar for three fluid models, but the drag is strongly model dependent.
Mangadoddy et al. (2004)	Power-law ($0.5 \leq n \leq 1$)	Cell model	Numerical results on heat transfer under forced convection conditions.
Mendes et al. (2002)[a]	Bingham plastic (grease)	Inline square array	Pressure drop data at low Re.
Nieckele et al. (1998)	Bingham plastics	Staggered arrays	Numerical predictions of pressure drop and yielded/unyielded regions in the creeping flow regime.
Prakash et al. (1987)[a]	Aqueous CMC solutions ($0.56 \leq n \leq 1$)	Staggered and triangular pitch arrangement	Extensive experimental data on Δp which were correlated using a modified capillary model.
Prasad and Chhabra (2001)[a]; Malleswararao and Chhabra (2003)[a]	Aqueous CMC and Methocel solutions	Staggered and inline arrays ($\varepsilon = 0.78, 0.87$)	Pressure drop results over wide ranges of conditions of Re.
Sadiq et al. (1995)	Aqueous Carbopol solutions ($0.39 \leq n \leq 0.54$)	Square arrays ($0.4 \leq \varepsilon \leq 0.6$)	Fair correspondence between the predictions and experiments.
Shibu et al. (2001)	Power law	Cell model	Numerical predictions of drag up to $Re_{PL} = 500$.
Skartsis et al. (1992b)	Power law ($0.33 \leq n \leq \sim 1$)	Square and staggered arrays ($0.43 \leq \varepsilon \leq 0.8$)	Numerical and experimental results on permeability. Visco-elasticity increases the flow resistance.

(Continued)

TABLE 6.11 (*Continued*)

Summary of Representative Studies for Transverse Flow Past a Bundle of Rods

Investigator	Fluid Model	Geometry	Remarks
Souvaliotis and Beris (1992)	UCM	Corrugated tubes and square arrays	2-D /steady simulations do not predict any increase in flow resistance.
Spelt et al. (2001, 2004a,b, 2005)	Power-law and Bingham plastic	Periodic square arrays	Numerical predictions of pressure drop up to Re=100.
Talwar et al. (1994); Talwar and Khomami (1992, 1995)	Carreau model, PTT, UCM, and Oldroyd-B	Square arrays and corrugated tubes	At low Re, 2-D simulations do not predict the increase in Δp.
Tripathi and Chhabra (1992b, 1996)	Power-law and Carreau viscosity	Free surface cell model	Approximate upper and lower bounds on Δp in creeping flow.
Vijaysri et al. (1999)	Power law	Zero-vorticity cell model	Numerical predictions of drag up to Re=10.
Vossoughi and Seyer (1974)[a]	Polyacrylamide solutions	Square arrays	Flow visualization and Δp data for visco-elastic liquids.
Woods et al. (2003)	Power law	Arrays of elliptic cylinders	Numerical predictions of Δp at low Re
Dyakonova et al. (1996)[a]	Polystyrene in tricresyl phosphate	Rows of cylinders with an additional cylinder yielding stagnation point	Increase in Δp is attributed to accumulated strain and elastic instability.
Koshiba et al. (1998)[a]	Aqueous solutions of PAA	Arrays of cylinders ($\varepsilon=0.34$ and $\varepsilon=0.56$)	Extra pressure drop is attributed to the strain-hardening behavior of the solutions.
Orgeas et al. (2006)	Power-law fluid	Model fibrous (periodic) porous medium	Used homogenization method to predict permeability.
Hua and Schieber (1998)	FENE-P and Reptation models	Square array of cylinders ($\varepsilon=0.45$, 0.65)	Reduction in pressure drop which is in contrast to the observations.
Moss and Rothstein (2010)[a]	Micellar solutions	Square array of cylinders	PIV and flow-induced birefringence data, but blockage was ~30%. The Δp initially falls with Deborah number and beyond a critical value, Δp increases with some evidence of leveling off.
Vasilic et al. (2011, 2016)[a]	Self-compacting cement (Bingham fluid)	Simple cubic arrays of cylinders	Permeability prediction with and without slip on the surface of the cylinders.
Yip et al. (2011)[a]	Boger fluids	Square array of cylinders ($\varepsilon=0.9$)	Skewing of velocity profiles beyond a critical Deborah number.
James et al. (2012)[a]	Boger fluids	Tube bundles as model porous medium ($0.9 \leq \varepsilon \leq 0.975$)	PIV measurements show increasing asymmetry in the flow field with increasing De which is used to explain the increase in pressure drop.
Bleyer and Coussot (2014)	Visco-plastic fluids and power-law fluids	2-D porous medium (cylinders)	For low voidage systems, velocity field is invariant to the fluid rheology and is governed mainly by geometry, in line with NMR experiments.
Shahsavari and McKinley (2015, 2016)	Herschel–Bulkley fluid, power-law and Carreau model fluids	Yielding and permeability	Prediction of mobility and permeability. Also, the criterion for yielding is presented.

(*Continued*)

TABLE 6.11 (*Continued*)

Summary of Representative Studies for Transverse Flow Past a Bundle of Rods

Investigator	Fluid Model	Geometry	Remarks
Kolodziej et al. (2015)	Power-law fluids	Hexagonal, triangular, and square arrays of cylinders	Numerical results on flow field and permeability tensor.
De et al. (2017a, b, c)	Phan-Thien–Tanner and FENE type models	Symmetric and asymmetric arrays of cylinder and 3-dimensional random fibrous porous media ($0.3 \leq \varepsilon \leq 0.6$)	Microstructure strongly influences the detailed kinematics even at a fixed value of ε. Beyond a critical value of De, pressure drop increases, in line with the experiments of James et al. (2012) and Galindo-Rosales et al. (2012).
Kawale et al. (2017)[a]	PAA solutions; Carreau viscosity equation	Inline and staggered arrays of circular and square cross-section cylinders	Based on flow visualization experiments, different flow regimes are identified and linked to Weissenberg numbers.
Waisbord et al. (2019)[a]	Carbopol solutions (Herschel–Bulkley fluid)	2-D microfluidic device consisting of 25-50 μm diameter cylinders ($0.46 \leq \varepsilon \leq 0.84$)	Hysteresis effects in Δp-Q relationship.
Xie and Balhoff (2021)	Oldroyd-B and Carreau models	Symmetric and asymmetric arrays of cylinders (as of De et al., 2017a)	2-D LBM simulations predict slight reduction in in-situ viscosity followed by a steep rise with increasing Weissenberg number. Some predictions are at variance with that of De et al. (2017a).

Notes: CMC, carboxymethyl cellulose; CR, Chilcott–Rallison model; UCM, Upper convected Maxwell model; PTT, Phan-Thien–Tanner model.

[a] Denotes experimental work.

as a 2-D periodic array of cylinders such as square arrays (Bruschke and Advani, 1992; Skartsis et al., 1992b; Chen et al., 1998a; Spelt et al., 2001, 2004a,b, 2005), others have used the two commonly used cell models, namely, the free surface and zero-vorticity cell models (Tripathi and Chhabra, 1992b, 1996; Chhabra et al., 2000; Vijaysri et al., 1999; Dhotkar et al., 2000; Shibu et al., 2001; Ferreira and Chhabra, 2004); in a few studies these have been supplemented by flow in corrugated tubes. Similarly, there is a preponderance of results obtained with power-law fluids, albeit limited results for the other GNF models such as Bingham plastics (Nieckele et al., 1998; Mendes et al., 2002; Spelt et al., 2005) and Carreau fluid model (Tripathi and Chhabra, 1996) are also available. Broadly speaking, in the creeping flow regimes, most predictions are fairly close to each other and the agreement with the scant experimental data is also about as good as can be expected in this type of work. Table 6.12 shows a typical comparison between the predictions of Ferreira and Chhabra (2004) and the experimental results of Sadiq et al. (1995), Prasad and Chhabra (2001), Malleswararao and Chhabra (2003), and Singh and Chhabra (2009) for inline and staggered square arrays.

Overall, the experimental values appear to be somewhat closer to the predictions of the zero-vorticity cell model than that of the free surface cell model. On the other hand, while these predictions are based on the assumption of infinitely long cylinders, in most experimental studies the length-to-diameter ratio of the tubes is of the order of 15–20, which might result in slightly larger values of the drag coefficient, C_D, than that expected for $l/d \to \infty$ otherwise under identical conditions. Similarly, experimental results are also influenced by the boundary (blockage) effects to some extent (Zhong et al., 2006).

The effect of fiber cross-section on the permeability of a periodic array has been studied by Woods et al. (2003) for elliptic cylinders. On the other hand, Kaplan et al. (1979) and Åström et al. (1992) have

TABLE 6.12

Comparison between the Predictions and Experimental Values of $(C_D \cdot Re_p)$ in Creeping Region

Source	ε	n	($C_D \cdot Re_p$) Free Surface Model	Zero Vorticity Model	Experimental
Prasad and Chhabra (2001)	0.78	0.54	48.42	60.89	61.39
		0.56	49.28	62.24	62.89
		0.72	58.19	76.31	83.75
		0.81	64.80	86.93	93.00
		0.84	67.27	90.95	103.67
	0.87	0.38	36.54	44.64	38.86
		0.48	36.37	44.86	44.08
		0.52	36.62	45.41	45.36
		0.62	37.82	47.60	46.62
		0.70	39.22	50.05	49.89
		0.72	39.62	50.74	54.29
Sadiq et al. (1995)	0.434	0.33	89.67	99.58	97.64
	0.455	0.33	85.09	95.00	92.80
	0.68	0.33	52.68	62.32	41.80
	0.434	0.39	104.15	117.46	187.7
	0.455	0.39	97.76	110.88	148.1
	0.434	0.53	159.19	186.23	347.00
	0.455	0.53	145.65	171.50	281.6
	0.68	0.53	65.21	81.09	74.37

extended the capillary bundle approach to the creeping flow of power-law liquids. The final expression of Åström et al. (1992) for the superficial velocity is given by

$$V_0 = \frac{\varepsilon^2}{(1-\varepsilon)} \frac{d}{4k_oT} \left\{ \frac{d}{4mT} \left(\frac{\varepsilon}{1-\varepsilon} \right) \frac{\Delta p}{L} \right\}^{1/n} \quad (6.101)$$

where k_o is the shape factor to describe the cross section of the tortuous paths in the fibrous medium. While Åström et al. (1992) did not specify any value for k_o, it varies from 2 (circular tube) to 3 (planar slit). Similarly, no specific value for the tortuosity factor T was indicated in their study.

The numerical study of Nieckele et al. (1998) for the flow of bi-viscosity (Bingham plastic) fluids through a staggered array of cylinders reveals the existence of high viscosity (unyielded?) regions near the throats formed by adjacent cylinders. As expected, there is a critical value of the pressure gradient below which no yielding occurs, and this value is strongly influenced by the tube arrangement and spacing.

There has been a spurt in studying the behavior of visco-plastic fluids in fibrous media, approximated as regular arrays (square or hexagonal). It appears that the transverse flow to square array is capable of capturing macroscopic flow resistance in random porous media. This is believed to be so due to the fact that the bulk of contribution to the overall pressure drop is due to the cross flow. Based on their numerical results, Shahsavari and McKinley (2015, 2016) have developed a universal scaling for predicting the mobility of Herschel–Bulkley fluids in square and hexagonal arrays of cylinders, as described here.

As noted previously, the effective shear rate $\dot{\gamma}_{\text{eff}}$ itself depends upon the choice of characteristic velocity and length scales (Equation 6.61). Shahsavari and McKinley (2015, 2016) expressed it as:

$$\dot{\gamma}_{\text{eff}} = \frac{2V_0}{d} \frac{\left(\dfrac{s}{d}\right)}{\left(\dfrac{s}{d} - 1\right)^2} \tag{6.102}$$

where s is the center-to-center distance between the cylinders.

A nondimensional shear rate $\dot{\gamma}^*$ can be now defined as $\dot{\gamma}^* = \dfrac{\dot{\gamma}_{\text{eff}}}{\left(\dfrac{V_0}{d}\right)}$ and introducing ε, one can express $\dot{\gamma}^*$ as follows:

$$\dot{\gamma}^* = \frac{2\sqrt{a_0(1-\varepsilon)}}{\left(\sqrt{a_0} - \sqrt{1-\varepsilon}\right)^2} \tag{6.103}$$

where $a_0 = \dfrac{\pi}{4}$ for a square array and $\left(\pi\dfrac{\sqrt{3}}{2}\right)$ for a hexagonal array. Using Equation 6.103 together with the Herschel–Bulkley fluid model and comparing it with Darcy's law, followed by dimensional considerations, they obtained the following simple expression for dimensionless pressure gradient:

$$\Delta p^* = 1 + \text{Bi}_{\text{HB}}^* \tag{6.104}$$

where

$$\Delta p^* = \frac{\Delta p}{L}\left(\frac{d^{n+1}}{mV_0^n}\right)\frac{k^*}{(\dot{\gamma}^*)^{n-1}} \tag{6.105a}$$

$$\text{Bi}_{\text{HB}}^* = \left\{\frac{\tau_0^{\text{HB}}}{m\left(\dfrac{V_0}{d}\right)^n}\right\}\frac{1}{(\dot{\gamma}^*)^n} \tag{6.105b}$$

Figure 6.20 shows the validity of Equations 6.104 against their numerical data in the range $0.3 \leq \varepsilon \leq 0.99$ and $0.3 \leq n \leq 0.8$ and $\text{Bi}_{\text{HB}} = \left\{\dfrac{\tau_0^{\text{HB}}}{m\left(\dfrac{V_0}{d}\right)^n}\right\}$ ranging from 0 to 100. Furthermore, their numerical predictions are also consistent with the experimental results of Al-Fariss and Pinder (1987) and Bleyer and Coussot (2014).

Some numerical results showing the influence of fluid inertia are also available for power-law fluids in tube arrays (Shibu et al., 2001; Spelt et al., 2004a,b). The latter analysis predicts the correction due to inertia of the order of Re^2 at small Reynolds numbers. They also identified a critical Reynolds number beyond which no stable steady solution exists. The cell model predictions up to $\text{Re}_{\text{PL}} \approx 500$ for power-law

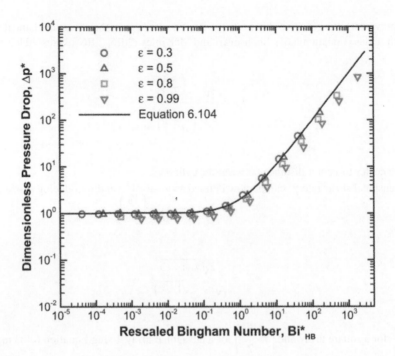

FIGURE 6.20 Relationship between Δp^* and Bi_{HB}^*. (Replotted from Shahsavari, S. and McKinley, G. H., *J. Non-Newt. Fluid Mech.*, **235**, 76, 2016.)

fluids are also available in the literature (Vijaysri et al., 1999; Dhotkar et al., 2000; Shibu et al., 2001). Most of these results and the available scant experimental results have been reviewed by Ghosh et al. (1994), Shibu et al. (2001), Prasad and Chhabra (2001), and subsequently by Malleswararao and Chhabra (2003). The following empirical correlation seems to correlate most of the literature (Adams and Bell, 1968; Prakash et al., 1987; Prasad and Chhabra, 2001) results with reasonable levels of reliability:

$$f\left(\frac{\varepsilon^2}{1-\varepsilon}\right) = \frac{64}{Re_2} + 0.45 \tag{6.106}$$

where

$$Re_2 = \frac{\rho V_0^{2-n} d^n}{m'} \left[\left\{ \frac{8(1-\varepsilon)}{\varepsilon^2} \right\}^{n-1} (1-\varepsilon) \right]^{-1} \tag{6.107}$$

and

$$m' = m\left(\frac{3n+1}{4n}\right)^n$$

Equation 6.106 embraces the following ranges of conditions: $0.5 \leq n \leq 0.94$, $0.2 \leq Re_2 \leq 500$, and $0.45 \leq \varepsilon \leq 0.87$.

6.4.7.2 Visco-Elastic Fluids

A few investigators have numerically studied the creeping flow of visco-elastic fluids past 2-D arrays of circular cylinders. Indeed, various array arrangements including square arrays (Souvaliotis and Beris, 1992; Talwar et al., 1994; Talwar and Khomami 1992, 1995; Khomami and Moreno, 1997; Hua and Schieber, 1998), linear arrays (Liu et al., 1998a), staggered arrays (Alcocer and Singh, 2002) coupled with several visco-elastic models including the upper-convected Maxwell fluid (Talwar et al., 1994), FENE-P and FENE-CR models (Liu et al., 1998a; Alcocer and Singh, 2002), PTT model (De et al., 2017a,b,c), and hybrid models combining Oldroyd-B and Carreau viscosity equation (Xie and Balhoff, 2021) have been used to elucidate the complex interplay between the geometry, fluid rheology, and kinematics. Such numerical studies have also been supplemented by experimental studies (Cressely and Hocquart, 1980; Chmielewski et al., 1990b; Chmielewski and Jayaraman, 1992; Dyakonova et al., 1996; Koshiba et al., 1998; Moss and Rothstein, 2010; Yip et al., 2011; James et al., 2012). In an attempt to simulate the cross-flow of Boger fluids through a square array ($\varepsilon=0.45$, 0.86), Khomami and Moreno (1997) reported the transition from the 2-D steady flow to a 3-D transient flow to occur at a critical value of the Weissenberg number. Similarly, their earlier studies with arrays of cylinders (Talwar and Khomami, 1995) and in a corrugated tube (Talwar et al., 1994) show that the excess pressure drop observed in experiments with visco-elastic fluids is probably not due to the rheological behavior, and it is perhaps linked to the loss of the stability of the flow to a regime exhibiting nonlinear features such as elastic temporal instability (Chmielewski et al., 1993). The calculations of Alcocer and Singh (2002), on the other hand, suggest the strong influence of the geometry of the array in the longitudinal direction. Chmielewski et al. (1990b) and Chmielewski and Jayaraman (1992) reported a comprehensive study for the flow of the fluid M1 in rectangular and triangular arrays with $\varepsilon=0.704$. While the test liquids used by them suffered mechanical degradation (mainly through a loss of elastic characteristics), their results do show much higher pressure drop values (De≥ 0.5) than those anticipated from the viscous effects alone. Their calculations also appear to suggest that the polymer molecules were almost completely stretched at about De~1 in the triangular array. However, shear-thinning viscosity seems to result in a shift in the onset of elastic effects to higher values of Deborah number, De≈ 2. Skartsis et al. (1992b), on the other hand, reported the onset of visco-elastic effects at De≈ 0.01. By analogy, with the flow past a sphere, most simulations suggest little or no deviation from the Newtonian value of the flow resistance at small values of Deborah number, followed by a region of little drag reduction, and finally with further increase in the value of Deborah number, there is a region of substantial increase in the flow resistance. The critical values of the Deborah number corresponding to the transition from one regime to another and the extents of drag reduction and enhancement are, however, strongly dependent on the geometry of the array and the choice of rheological model parameters. Qualitatively, the increase in the flow resistance at high Deborah number is ascribed to the fluctuations in the pressure field downstream of the array (Chmielewski et al., 1990b, 1993; Chmielewski and Jayaraman, 1992), the loss of stability (Talwar and Khomami, 1995; Szady et al., 1995; Liu et al., 1998a), and modifications to the flow field (Kawale et al., 2017). Qualitatively similar results have been reported for the simple geometry consisting of only two cylinders (Jones and Walters, 1989; Georgiou et al., 1991). Heinen et al. (2003) have used magnetic resonance imaging technique to image the flow to gain some insights, and Evans et al. (1994) have analyzed polymer conformation in fibrous media.

In another significant study, James et al. (2012) have not only reported PIV measurements and flow resistance data for the flow of Boger fluids past a square array of cylinders ($\varepsilon=0.9$, 0.95, 0.975), but have also thoroughly reviewed much of the available literature on this subject. They corrected their pressure drop-flow rate data for wall effects using the method of Zhong et al. (2006). The PIV measurements suggest the onset of asymmetry in velocity profiles at De~0.5 which also marks the increase in the pressure drop. Indeed, at De ≈ 4, the pressure drop increased by a factor as high as 10. Under otherwise identical conditions, the pressure drop increased with the decreasing porosity of the array. They attributed such an increase to the flow structures caused by elasticity in the wake region and their heuristic arguments ascribe this to the first normal stress difference rather than to the extensional resistance. James et al. (2012) also highlighted the problems in making cross-comparisons with other studies due to non-overlapping values of the voidage and/or due to different definitions of Deborah number used by various workers. For instance, De et al. (2017a) based the Deborah number on $\sqrt{k}$ as the linear dimension, though the axisymmetry of velocity profiles observed by James et al. (2012) is consistent with the predictions of De et al. (2017a,c) based on the Phan-Thien and Tanner model. Similarly, the role of the first normal stress difference proposed by James et al. (2012) is also in line with the predictions of De et al. (2017c).

6.4.8 Mixing in Packed Beds

Due to dispersion, true plug flow never occurs in a packed bed, except possibly for a visco-plastic liquid. In the limited literature available on this subject, the rate of dispersion has been assumed to be linearly dependent on the concentration gradient while the dispersion coefficients themselves are strongly influenced by the flow conditions (Reynolds number), bed geometry, and the type and degree of non-Newtonian fluid behavior. The constant value of the Peclet number in the creeping flow regime is observed for power-law fluids also (Standish and Bull , 1981; Payne and Parker, 1973). The scant results on axial dispersion available in the literature for weakly shear-thinning liquids ($n > 0.8$) in packed bed conditions ($\varepsilon \sim 0.4 - 0.5$) seem to correlate well with the corresponding Newtonian correlations, at least in the range $7 \leq Re \leq 800$ (Wen and Yim, 1971; Hilal et al., 1991). However, Hilal et al. (1991) reported the particle shape to be a significant factor in influencing the value of Peclet number. On the other hand, Edwards and Helail (1977) reported significantly higher values of the Peclet number for non-Newtonian liquids. The reasons for this discrepancy are not immediately obvious. Paterson et al. (1996) have reported an increase in axial dispersion due to the heterogeneities introduced by multisize grains constituting the porous medium. Chaplain et al. (1992, 1998) have extended the approach of Saffman to predict dispersion behavior of Bingham plastic and power-law fluids in homogeneous and isotropic porous media. Pearson and Tardy (2002) have also developed a general framework for treating dispersion in porous media flows. In the only reported study with non-Newtonian rubber solutions in packed beds, the radial mixing was shown to be impeded with the increasing viscosity (Hassell and Bondi, 1965). A recent review is available in the literature, though much of it relates to Newtonian fluids (Delgado, 2006).

6.5 Miscellaneous Effects

Aside from the voluminous body of knowledge referred to in the preceding sections of this chapter, the contemporary literature on the flow of polymer solutions through unconsolidated and consolidated porous media abounds with several anomalous and hitherto-unexplained effects that are not encountered with the flow of Newtonian fluids (e.g., see Liu and Masliyah (1998) and Muller and Saez (1999)). Conversely, the flow through a porous medium has been used to explore the mechanism of degradation in extensional flows (Farinato and Yen, 1987; Hoagland and Prud'homme, 1989). For instance, the experimental results on pressure drop-flow rate obtained under constant pressure and constant flow rate conditions do not superimpose (Sadowski and Bird, 1965) and the apparent shear stress–shear rate data evaluated from porous media experiments are often at odds with those measured in viscometric flows (see Dauben and Menzie, 1967; Hong et al., 1981; Cohen and Chang, 1984; Wreath et al., 1990; Haward and Odell, 2003). Likewise, several workers (Burcik, 1965, 1968a; Jones and Maddock, 1966, 1969; Barboza et al., 1979; Haward and Odell, 2003) have documented pressure drop values well in excess of those anticipated from the viscous properties, even when the visco-elastic effects are believed to be negligible. All these observations appear to suggest significant differences between the in-situ and bulk rheological characteristics of polymer solutions (Unsal et al., 1978; Wang et al., 1979; Duda et al., 1981; Hong et al., 1981). Several plausible mechanisms including gel effects (Sadowski and Bird, 1965; Burcik, 1969), slip effects (Hanna et al., 1977; Kozicki et al., 1984, 1987; Cohen and Christ, 1986; Kozicki and Tiu, 1988; Waisbord et al., 2019), adsorption (Hirasaki and Pope, 1974; Jones and Ho, 1979; Shvetsov, 1979), wall effects (Pye, 1964; Sandiford, 1964; Omari et al., 1989a,b), pseudo-dilatant behavior (Burcik and Ferrer, 1968; Picaro and van de Ven, 1995; Haward and Odell, 2003), jamming and pore blockage (Marliere et al., 2015), anisotropy and heterogeneity of porous medium (Orgeas et al., 2006; Kolodziej et al., 2015), roughness (Rodriguez de Castro and Radialla, 2017), etc. have been postulated; none, however, has proved to be completely satisfactory.

The presence of a solid boundary is known to alter the rheology of macromolecular solutions; consequently, polymeric liquids display anomalous wall effects (Willhite and Dominguez, 1977; Omari et al., 1989a, 1989b; Schowalter, 1988) including steric hindrance (when the characteristic linear dimension of the flow channel is comparable to, or only slightly larger than, the size of macromolecules), polymer retention, etc. All such phenomena, though poorly understood, are believed to contribute to the observed differences between the *in situ* and bulk rheological properties of polymer solutions in porous media. Excellent reviews are available on these topics (Sandiford, 1977; Willhite and Dominguez, 1977; Dreher and Gogarty, 1979; Cohen, 1988; Sorbie, 1991; Müller and Saez, 1999); only their salient features are recapitulated here.

6.5.1 Polymer Retention in Porous Media

Macromolecules are retained as a polymer solution flows through a porous medium, thereby reducing the solution viscosity and permeability both of which have deleterious effect on the efficiency of oil recovery processes (Wu and Pruess, 1996). Broadly speaking, polymer retention occurs by two mechanisms: adsorption and mechanical entrapment. Often the total amount of polymer retained is estimated from the concentration of the exiting solution, and hence, the individual contributions of adsorption and mechanical entrapment are usually not known, but can be inferred indirectly using additional information such as pore size distribution, size of polymer molecules in solution, etc. Additional complications arise from the fact that adsorption can occur under both static and dynamic conditions as shown by Cohen (1985) by measuring settling velocity of individual spheres in a polyacrylamide solution. For a given solid surface–polymer solution pair, adsorption appears to be influenced by a large number of variables such as the chemical nature of polymer, its molecular weight distribution and concentration, pH, presence/absence of certain ions, porosity, permeability, nature of solid surface, and flow rate. Furthermore, the results obtained under static and dynamic conditions often do not match with (and sometimes even contradict) each other (Willhite and Dominguez, 1977); the adsorption in the latter case is found to be much smaller than that anticipated from the static experiments. Generally, the amount of polymer adsorbed per unit mass of solid increases with polymer concentration with a propensity to level off beyond a critical value of polymer concentration (see Figure 6.21 for typical results). In view of such a large number of influencing variables, the results are strongly system dependent and generalizations should be treated with reserve.

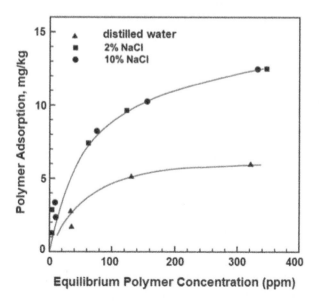

FIGURE 6.21 Static adsorption isotherms for partially hydrolyzed polyacrylamide on silica sand. (Modified from Szabo, 1975b.)

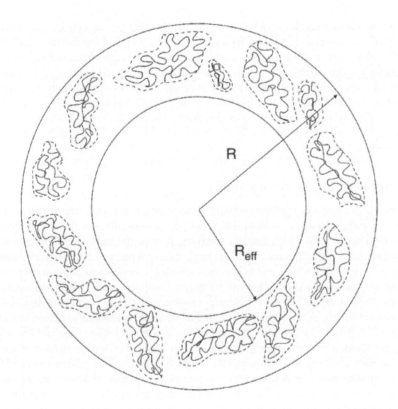

FIGURE 6.22 Schematic representation of effective pore size in presence of polymer adsorption/retention.

Macromolecules are also retained in porous media by the mechanism of mechanical entrapment. Both Gogarty (1967a,b) and Smith (1970) have examined this mode of polymer retention. This mechanism dominates when the polymer molecules in solutions are bigger than the size of pores, for example, plugging of small pores by polymer molecules that are too large to enter the pore; adsorption also promotes a complete or partial blockage of pores. Similar blockage of pores has also been reported by others (Harrington and Zimm, 1968; Aubert and Tirrell, 1980). Thus, the two mechanisms go hand in hand.

From a macroscopic standpoint, regardless of the underlying mechanism, polymer retention has two effects: first, the retained macromolecules occupy a portion, howsoever small, of the void volume thereby reducing its porosity, and hence permeability of the porous medium. Second, the layer of solution in the vicinity of solid walls is depleted in polymer as compared to the bulk solution, thereby altering its rheological properties. Polymer adsorption has been modeled as a kinetic process obeying the Langmuir-type equation whose constants are found to be temperature and system dependent (Willhite and Dominguez, 1977). This approach, however, has not yet been incorporated directly into the models for flow through porous media. On the other hand, based on the assumption of monolayer adsorption, some estimates of the adsorbed layer thickness have been reported in the literature; typically, these are of the order of a few μm. This, in turn, is used to calculate the effective radius (see Figure 6.22) to be used in the usual capillary model for the porous medium (Hirasaki and Pope, 1974; Sorbie et al., 1987, 1989; Brown and Sorbie, 1989; Sorbie, 1989, 1990; Sorbie and Huang, 1991). However, some of the pores plugged at low flow rates may become accessible at high flow rates, presumably due to high level of shear stresses, which can dislodge some of the molecules from the walls of the pores (Rodriguez et al., 1993). For a more detailed description, interested readers are referred to the review paper of Willhite and Dominguez (1977) and the comprehensive book by Sorbie (1991).

6.5.2 Slip Effects

Another phenomenon that has received considerable attention in recent years is the so-called slip effects that arise from the observation that the macromolecular solutions do not seem to satisfy the classical no-slip boundary condition at solid surfaces (Cohen, 1988; Schowalter, 1988; Omari et al., 1989a,b; Archer, 2005; Nilsson et al., 2013; Afolabi et al., 2019). Though, it is not at all clear whether the true slip occurs or not, this notion has proved to be convenient in explaining/interpreting some of the anomalous results reported in the literature. Cohen (1988), Agarwal et al. (1994), and Liu and Masliyah (1998) have presented thorough and thought-provoking reviews of slip effects observed in the flow of polymer solutions in various flow geometries, and how to account for it. In contrast to the polymer retention referred to above and the consequent reduction in the permeability of a porous medium, the slip effects, inferred from the observed abnormally high flow enhancements, result in an increase in the permeability, thereby improving the effectiveness of polymers in enhanced oil recovery. Cohen and Metzner (1985) demonstrated that the slip effects are more pronounced in small pores, which is also conducive for polymer retention by adsorption. It is thus likely that under appropriate conditions these two opposing mechanisms nullify each other, and the permeability may even improve.

Currently, two distinct approaches are available for investigating the slip effects in porous media flows. In the first method, the capillary model equation is modified simply by changing the "no-slip" boundary condition by a nonzero velocity at the wall and the resulting expression for the apparent shear rate can be written as

$$\dot{\gamma}_{app} = \frac{8V_i}{4R_h} = \frac{8V_s}{4R_h} + \frac{4}{\tau_w^3}\int_0^{\tau_w} \tau^2 f(\tau) d\tau \qquad (6.108)$$

The differentiation of Equation 6.108 with respect to $1/R_h$ yields (at constant value of $R_h\tau_w$)

$$\frac{\partial(2V_0/\varepsilon R_h)}{\partial(1/R_h)} = \frac{V_s}{(L_e/L)} \qquad (6.109)$$

Equation 6.109 suggests that one can estimate the slip velocity V_s-provided experimental data are available for a range of values of R_h (i.e., particle diameter). This approach has been successfully exploited by Kozicki et al. (1967, 1968, 1972, 1984, 1987, 1988) and Kozicki and Tiu (1973, 1988).

In the second method, the importance of slip effects is ascertained by comparing the experimental and predicted throughput/pressure drop behavior of a polymer solution. The predictions are based on the choice of a fluid model and a model of the porous medium. For instance, Cohen and Chang (1984) have pursued this line of analysis and asserted that the ratio $(V_{ex}/V_0) < 1$ indicates the presence of slip effects. This assertion was corroborated by using the results obtained with microemulsions flowing in glass bead-packed beds.

As remarked earlier, it is readily recognized that the behavior of macromolecules on and near the solid surface is greatly influenced by steric, repulsive, or attractive surface–polymer interactions. Obviously, attractive forces will lead to polymer adsorption whereas the repulsive and steric effects will cause the molecules to move away from the solid walls, thereby giving rise to depleted layers in the wall region; the latter are thought to be the main mechanism for resulting in slip effects. Such simple ideas coupled with the notions of stress-induced diffusion, thermodynamic equilibrium, etc. have been used to develop various theoretical frameworks to model slip effects in general. While the preliminary results appear to be encouraging, none of these attempts is yet refined to the extent of being completely predictive in character. Detailed critical evaluations of these developments are available in the literature (Cohen, 1988; Agarwal et al., 1994; Nilsson et al., 2013; Afolabi et al., 2019)

6.5.3 Flow-Induced Mechanical Degradation of Flexible Molecules in Solutions

The solutions of flexible macromolecules such as PEO and partially hydrolyzed polyacrylamide (HPAA) exhibit flow-induced mechanical degradation when subjected to extensional flow fields such as in opposed jets (Müller et al., 1988; Narh et al., 1990; Odell et al., 1990) and in porous media flows (Müller et al., 1993; Tatham et al., 1995). Such studies have been motivated by both pragmatic considerations (such as to ascertain the efficacy and viability of polymeric additives in oil recovery, antimisting aerofuel formulations are governed by the life span and stability of polymers) and theoretical considerations (such as the development and validation of theoretical frameworks to model their rheological behavior). Intuitively, the rate and degree of degradation of a flexible molecule in solution is influenced by a large number of variables including the architecture of the molecule, degree of flexibility, molecular weight, polymer–solvent interactions with and without the addition of salts, on the one hand, and the rate of stretching, prevailing stress levels, and the temperature, on the other hand. In simple terms, the stress induced by a shearing motion is not sufficient to uncoil and stretch the molecule leading to its fracture. Owing to the occurrence of significantly higher levels of extensional stresses, mechanical degradation is a phenomenon encountered in extension-dominated flows such as opposed jets, porous medium, spraying and atomization, for instance. Naturally, the scission of a molecule results in the lowering of the molecular weight, which in turn combined with the reduced degree of entanglement (Keller et al., 1987) alters the rheology of the solution. It is thus possible to detect the occurrence of flow-induced degradation through macroscopic measurements such as pressure drop-flow rate data in a porous medium (Müller et al., 1997; Müller and Saez, 1999). The earliest studies recognizing the importance of the flow-induced degradation are by Jennings et al. (1971) and Maerker (1973, 1975). Both reported the results in the form of a screen factor that is simply the ratio of the flow times for a fixed volume of solution to that for the solvent for their flow through a stack of five 100-mesh screens. Maerker (1975) reported that the degradation occurred once the deformation rate exceeded a critical value. While Maerker (1975) reported no effect of polymer concentration for HPAA/salt solutions, subsequent studies suggest a positive effect of concentration up to a critical concentration, followed by a slight drop in degradation beyond the critical concentration (Farinato and Yen, 1987; Müller and Saez, 1999). Similarly, the early studies (James and McLaren, 1975) found a positive correlation between the particle size and the extent of flow-induced degradation that is clearly counterintuitive, as also confirmed by later works for PEO solutions (Kaser and Keller, 1980) and for HPAA in brine solutions (Moreno et al., 1996). Figure 6.23 shows the gradual degradation of 100 ppm and 1000 ppm PEO solutions in a packed bed (of 1 mm spheres) for a range of Reynolds numbers. The ordinate is the loss coefficient, Λ, that is proportional to pressure gradient and the x-axis is a crude measure of "total strain" expressed as a function of the number of passes of the fluid through the packed bed. These results clearly show that in dilute solutions, no degradation occurs at low Reynolds numbers (~low deformation and stresses) whereas the extent of degradation rapidly increases with the increasing concentration and the Reynolds number. Figure 6.24 shows analogous results for a HPAA in a 0.5 M NaCl solution in a bed of 1.1 mm spheres. In both cases, the loss coefficient is seen to approach a limiting value, thereby indicating the possibility of a state of equilibrium, that is, no more scission or fracture is possible. Combined together, the results shown in these figures show the complex interplay between polymer concentration, deformation rate (Reynolds number), and polymer–solvent interactions. When the results of Figures 6.23 and 6.24 are replotted in terms of Λ–Re coordinates, it becomes obvious that Λ is independent of Re up to higher and higher values of Re in degraded solutions. This trend has been reported in numerous other studies also (Haas and Kulicke, 1984; Müller et al., 1988, 1989). Aside from the aforementioned studies, the effect of temperature on the degradation of atactic monodisperse polystyrene in the dilute solution range has been studied by Odell et al. (1990, 1992). Gamboa et al. (1994) have studied the porous media flow behavior of mixtures of PEO and guar solutions. An excellent overview of the pertinent literature on the flow-induced degradation is available (Müller and Saez, 1999). Qualitatively, a different kind of pressure loss–flow rate relationship has also been observed with the flow of surfactant solutions in a porous medium (Brunn and Holweg, 1988; Ruckenstein et al., 1988; Vorwerk and Brunn, 1994). As the flow rate is gradually increased, the pressure drop values begin to deviate from the expected Newtonian behavior. The pressure drop increases rapidly reaching a maximum value, and finally it decreases. This behavior was also ascribed to the formation of shear-induced structures in such surfactant solutions.

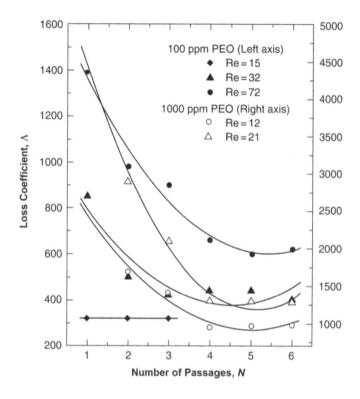

FIGURE 6.23 Flow-induced degradation of PEO solutions in a porous medium. (Replotted from Müller, A.J. and Saez, A.E., *Flexible Polymer Chain Dynamics in Elongational Flow*, Chapter 11, Nguyen, T.Q. and Kausch, H.-H., Eds., Springer, New York, 1999.)

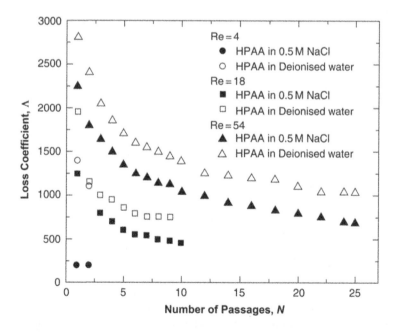

FIGURE 6.24 Flow-induced degradation of partially hydrolyzed, polyacrylamide in NaCl solutions in a porous medium. (Müller, A.J. et al., *App. Mech. Rev.*, **46**, S63, 1993.)

6.6 Two-Phase Gas/Liquid Flow

In modern chemical and process engineering applications, the two-phase flow of a gas and a liquid in a packed bed is encountered quite frequently (Dudukovic et al., 1999, 2002). Depending upon the application, the two phases may flow concurrently in upward or in downward direction or may flow counter currently with gas flowing upward such as in trickle-bed reactors. Consequently, considerable research effort has been expended in developing reliable models and design schemes for such systems. Notwithstanding the significance of the detailed kinematics of the flow, the gross behavior of such systems is generally characterized in terms of flow regimes, liquid holdup, two-phase pressure drop, Peclet number, Nusselt number, etc. The bulk of the literature therefore is devoted to the development of suitable expressions relating these characteristics to the pertinent dimensionless groups. Furthermore, the majority of the available studies relate to the conditions when the liquid phase exhibits the simple Newtonian flow behavior. Excellent reviews summarizing the state of the art are available (Charpentier, 1976; Hofmann, 1986; Larachi et al., 1998, 2003; Dudukovic et al., 1999, 2002; Iliuta and Larachi, 2002a,b; Khan et al., 2002a; Jamialahmadi et al., 2005). Consequently, it is now possible to predict the flow regimes, liquid holdup, two-phase frictional pressure-drop, rates of heat and mass transfer, etc. under most conditions of interest, with reasonable levels of confidence when the liquid phase is Newtonian. On the other hand, the corresponding body of knowledge is indeed very limited when the liquid phase displays non-Newtonian characteristics. Larkins et al. (1961) were the first to report two-phase pressure drop data for the concurrent downflow of air and aqueous methyl cellulose solutions. However, not only no rheological characteristics of the polymer solutions were measured, but they also reported their data for Newtonian and polymer solutions to superimpose onto one curve. It is perhaps likely that the polymer solutions used by them were nearly Newtonian. Sai and Varma (1987) used the well-known Lockhart–Martinelli parameter (Lockhart and Martinelli, 1949) to correlate their pressure loss data for the downward flow of air and carboxymethyl cellulose solutions. Subsequently, Soman et al. (1989) reported the flooding to occur at lower gas velocities with the dilute solutions of PEO and PAA than that for water. Likewise, the liquid holdup was found to increase with liquid viscosity, which is consistent with the subsequent studies (Iliuta and Thyrion, 1997; Khan et al., 2002b). Srinivas and Chhabra (1994) reported extensive data on two-phase frictional pressure drop for the upward flow of air and carboxymethyl cellulose solutions through packed beds of spherical particles. They correlated their data by the simple expression

$$\phi_G = \frac{\Delta P_{TP}/L}{\Delta P_G/L} = 1.27 \chi^{0.906} \quad (6.110)$$

where the Lockhart–Martinelli parameter χ is defined as

$$\chi^2 = \frac{\Delta P_L/L}{\Delta P_G/L} \quad (6.111)$$

In Equation 6.111, the single-phase pressure gradients, $\Delta P_L/L$ and $\Delta P_G/L$, are evaluated using the appropriate expressions at the same mass flow rates as that encountered in the two-phase flow system. Thus, for instance, when the liquid phase exhibits shear-thinning behavior, the value of $(\Delta P_L/L)$ is estimated using the methods outlined in Section 6.5 whereas the same expression may be used to estimate $(\Delta P_G/L)$ with $n = 1$. Equation 6.110 is based on the following ranges of conditions: $0.9 \le \chi \le 104$; $0.54 \le n \le 1$; $3.7 \le Re_G \le 177$; and $10^{-3} \le Re_L^* \le 50$.

Some attempts have also been made to develop phenomenological models (Iliuta and Larachi, 2002a, 2002b) for the two-phase flow of power-law and yield-pseudoplastic liquids in upflow and downflow and in co- and counter-current flow configurations. Several simplifying assumptions are introduced to obtain predictive expressions for flow regimes, liquid holdup, and frictional pressure drop. Similarly, Iliuta et al. (1996) simply used the same correlations for holdup and pressure drop for air/Newtonian and air/

power-law liquids. Unfortunately, the available experimental data are inadequate to substantiate or refute the theoretical developments and to discriminate between the available correlations.

Before concluding this chapter, it is appropriate to list some other important reviews available in this area. Kumar et al. (1981) have presented a succinct account of the developments in the field of single phase non-Newtonian fluid flow in unconsolidated porous media. They have, however, primarily focused on the prediction of pressure drop for concentrated polymer solutions. The voluminous body of knowledge available on the filtration of non-Newtonian solutions and slurries has been summarized by Kozicki (1988). The flow of polymer solutions in porous media as applied to the oil recovery processes has been reviewed by Wu and Pruess (1996), Nilsson et al. (2013), and Afolabi et al. (2019)

6.7 Conclusions

In this chapter, consideration has been given to the complex and important problem of the flow of incompressible fluids in porous media and packed beds. Starting with the definition and methods of macroscopic description of porous media, a terse description of macroscopic fluid flow phenomena, namely wall effects and pressure drop–volumetric throughput relation, pertaining to Newtonian media is presented. This is followed by the analogous treatment for different types of non-Newtonian fluids. The merits and demerits of some of the approaches currently used for modeling porous media flows are revisited. In particular, the capillary and submerged object models have been examined in detail. Over the years, both these approaches have been successfully extended to include time-independent non-Newtonian effects. It is thus possible to predict pressure loss through a bed of known porosity (or permeability) for time-independent non-Newtonian fluids with reasonable levels of accuracy in the absence of anomalous surface effects. Unfortunately, there is no method available to predict a priori whether such effects would occur in an envisaged application. In recent years, some progress has been made in predicting the minimum pressure gradient required for the onset of flow of visco-plastic fluids in porous media. Conversely, this information can also be used to infer some information about the pore size distribution of homogeneous unconsolidated porous media. Our understanding about the flow of visco-elastic fluids in porous media is still incomplete, albeit it is slowly improving. Based on the limited evidence available, the wall effects for non-Newtonian fluid flow are nearly as serious as in the case of Newtonian fluids and one should use columns at least 30–40 particle diameters large to minimize/eliminate the wall effects. The flow of drag-reducing dilute polymer solutions is of immense pragmatic importance in oil recovery processes. Thus, it would be desirable to develop a better understanding of the associated anomalous effects, with the ultimate objective of integrating them into the design methodologies for flow through porous media. The scant literature available on the single-phase flow in fibrous media and on the two-phase gas/liquid flow in granular media has also been reviewed. In recent years, some progress has also been made in the use of arrays of cylinders to simulate the flow in fibrous porous media for generalized Newtonian (visco-plastic) and visco-elastic fluids. Clearly, considerable scope exists for future work in all aspects of porous media flows with non-Newtonian systems.

Nomenclature

a_{vs}	Specific surface area (m^{-1})
a_{vd}	Dynamic specific surface (m^{-1})
a, b	Geometric parameter, Equation 6.74 (-)
a_0	Constant, Equation 6.103 (-)
A	Area for flow; also constants in various expressions for friction factor, for example, Ergun equation (-)
A_w	Wall correction factor, Equation 6.39a (-)
B	Turbulent constant in Ergun equation (-)
B_w	Wall correction factor, Equation 6.39b (-)

Bi_{HB}	Bingham number for Herschel–Bulkley fluid $\left(\tau_0^{HB}/m(V_o/d)^n\right)$ (-)
C	Concentration of polymer in solution (mol/m^3)
C_o	Constant, Equation 6.61 (-)
C_{Do}	Drag coefficient for single particle $\left(8F_D/\rho V_o^2\pi d^2\right)$ (-)
C_{D1}	Drag coefficient for a particle assemblage, Equation 6.24 (-)
d	Sphere or cylinder diameter (m)
d_e	Equivalent diameter (m)
D	Characteristic linear dimension, Equation 6.32 (m)
D_c	Column diameter (m)
De	Deborah number (-)
D_h	Hydraulic diameter (m)
f	Friction factor (-)
f_{pore}	Pore friction factor, Equation 6.21b (-)
f^*	Modified friction factor, Equation 6.88a (-)
G	Mass flow rate (kg/m^2.s)
F_D	Drag force on a sphere (N) or on a cylinder per unit length (N/m)
H	Constant, Equation 6.85 (-)
k'	Constant, Equation 6.72 (m^6/kg^2.s^2)
k^*	Dimensionless permeability (-)
k	Permeability (m^2)
k_B	Brinkman permeability, Equation 6.28 (m^2)
k_L	Modified permeability, Equation 6.29 (m^2)
k_k	Kozeny constant, Equation 6.77 (-)
K_i, K_o	Constants, Equation 6.77 (-)
l	Length of the cylinder (m)
L	Path length, Equation 6.13 (m)
l_c	Characteristic linear dimension (m)
L_e	Average effective length, Equation 6.13 (m)
m	Power-law consistency coefficient (Pa s^n)
M	Defined by Equation 6.19 (Pa/m^3.s^2)
m'	Apparent consistency index (Pa.s^n)
n	Power-law index (-)
n'	Apparent flow behavior index (-)
N	Viscous term, Equation 6.18 (Pa.s/m^2)
N_1	First normal stress difference (Pa)
N_e	Effective pore volume, Equation 6.56 (m^3)
p^*	Pressure (-)
Δp	Pressure drop (Pa)
ΔP_G	Pressure drop, Equation 6.110 (Pa)
ΔP_L	Pressure drop, Equation 6.110 (Pa)
ΔP_{TP}	Pressure drop, Equation 6.111 (Pa)
$p(r)$	Pore size distribution function, Equation 6.62 (-)
q_i	Flow rate through ith pore, Equation 6.62 (m^3 s^{-1})
Q	Volumetric flow rate (m^3/s)
r_o	Minimum radius required to initiate the flow in a tube (m)
R	Radius of the fiber (m)
R_h	Hydraulic radius (m)
Re	Reynolds number (-)
Re'	Modified Reynolds number, Equation 6.82 (-)
Re*	Modified Reynolds number, Equation 6.71 (-)
Re$^+$	Modified Reynolds number, Equation 6.88b (-)

Re_o	Modified Reynolds number based on zero-shear viscosity (-)
Re_1	Modified Reynolds number, Equation 6.23 (-)
Re_2	Modified Reynolds number, Equation 6.59 (-)
Re_{C-M}	Modified Reynolds number, Equation 6.72 (-)
Re_i	Interstitial Reynolds number (-)
Re_M	Modified Reynolds number, Table 6.10 (-)
Re_{NN}	Modified Reynolds number, Equation 6.80a (-)
Re_p	Particle Reynolds number ($\rho V_o^{2-n} d^n/m$) (-)
Re_{PL}	Power-law Reynolds number (-)
Re_{pore}	Pore Reynolds number, Equations 6.21a and 6.78 (-)
Re_{S-B}	Modified Reynolds number, Equation 6.73 (-)
Re_{SD}	Modified Reynolds number, Equation 6.92 (-)
s	Center-to-center gap between two cylinders in arrays (m)
T	Tortuosity factor, absolute temperature (-, K)
T_{BK}, T_{KC}	Tortuosity factor in Blake–Kozeny and Kozeny–Carman equations, respectively (-)
V_c	Characteristic velocity in definition of De (m/s)
V_i	Interstitial velocity (m/s)
V_o	Superficial velocity (m/s)
V_{ex}	Experimental value of V_o (m/s)
y	Radial distance from the tube wall (m)
$Y = C_D/C_{Do}$	Drag correction factor (-)

Greek Symbols

α, β	Constants, Equation 6.34; also α Ellis model parameter (-); β is a constant, Equation 6.25 (-)
α_1, β_1	Constants, Equation 6.83 (-)
$\langle \dot{\gamma}_w \rangle_n$	Nominal shear rate at the wall, Equation 6.65b (s^{-1})
ε	Bed porosity or voidage (-)
$[\eta]$	Specific viscosity (Pa.s.mole/m^3)
θ	Fluid characteristic time (s)
μ	Newtonian viscosity (Pa.s)
μ_B	Bingham viscosity (Pa.s)
μ_{eff}	Effective viscosity of a non-Newtonian fluid, Equation 6.81 (Pa.s)
μ_o	Zero-shear viscosity (Pa.s)
μ_s	Solvent viscosity (Pa.s)
ξ	Constant, Equation 6.86b (-)
k	Constant, Equation 6.86a (-)
Λ	Loss coefficient ($= Re.f$) (-)
ρ	Fluid density (kg/m^3)
τ	Shear stress (Pa)
$\langle \tau_w \rangle$	Average shear stress at wall (Pa)
$\tau_{1/2}$	Ellis model parameter (Pa)
τ_o^B	Bingham yield stress (Pa)
τ_o^{HB}	Herschel–Bulkley model parameter (Pa)
ϕ	Solid volume fraction ($=(1-\varepsilon)$) (-)
ϕ_G	Pressure drop ratio, Equation 6.110 (-)
ψ	Sphericity factor for nonspherical particles (-)
χ	Lockhart–Martinelli parameter, Equation 6.111 (-)

Subscripts

eff   Effective
local   Local

Superscript

*:   Normalized quantity

7

Fluidization and Hindered Settling

7.1 Introduction

When a liquid flows upward through a bed of particles, one can discern three distinct flow regimes depending upon the flow rate of liquid. At sufficiently low flow rate or velocity, it gives rise to a fixed bed; but if the velocity of liquid is sufficiently high, the solid particles will be freely supported in the liquid to give rise to what is known as a fluidized bed. At very high velocities, the solid particles will be transported from the system. The bed in which the conditions cease to exist as a fixed bed is described as the *incipiently fluidized bed* and the value of the liquid velocity corresponding to this point is known as the *minimum fluidization velocity*. When the flow rate of liquid is increased above this value, the bed continues to expand so that the average distance between the particles increases. The behavior of this kind is known as "particulate fluidization." It is now generally agreed that this type of fluidization occurs with most solid–liquid systems (except when the solids are too heavy) and in gas–solid systems over a limited range of conditions, especially with fine particles. The ensuing discussion is, however, largely pertinent to particulate fluidization. Fluidized beds are extensively used as heat/mass exchangers and chemical reactors in chemical and processing industries. Significantly enhanced heat and mass transfer rates can be achieved due to the vigorous mixing between the liquid and solids in a fluidized bed. Existing and potential applications of this mode of contacting involving Newtonian fluids have been dealt with by several authors (Davidson and Harrison, 1971; Davidson et al., 1985; Fan, 1989; Kunii and Levenspiel, 1990; Jamialahmadi and Müller-Steinhagen, 2000; Epstein, 2003) whereas the corresponding limited information for non-Newtonian systems has been discussed by Baker et al. (1981), Joshi (1983), Tonini (1987), Chhabra (1993a, 1993b), Shilton and Niranjan (1993), and Chhabra et al. (2001b).

In recent years, there has been a growing recognition of the fact that there are numerous applications in biotechnology employing three-phase fluidization (Shuler and Kargi, 2002; Yang, 2003; Arastoopour et al., 2021). For instance, many industrially important bioprocesses frequently use agricultural wastes as cheap substrates and nutrients with significant amounts of solids, which can lead to the clogging of packed beds.

Conversely, the sedimentation of solids in bubble column reactors is also an issue. Some of these difficulties can be obviated by fluidizing the solids with the upward flow of liquid, and then bubbling the gas through it, thereby resulting in an effective contacting of the three phases. This mode of operation is called the three-phase fluidization. The specific examples of the practical applications of this technique have been summarized by Shuler and Kargi (2002) and Schugerl (1997). Airlift tower-loop and sparged reactors also share some of the phenomena characterizing the hydrodynamic behavior of a three-phase fluidized bed contactor.

It is readily acknowledged that the minimum fluidization velocity and the extent of bed expansion as a function of the liquid velocity (beyond the point of incipient fluidization) represent the two most important hydrodynamic design parameters for the sizing of the two-phase and three-phase fluidized beds in an envisaged application; consequently, these aspects have received the greatest amount of attention in the literature. Other important aspects required for formulating comprehensive models of fluidized beds include mixing and flow patterns, heat and mass transfer characteristics, detailed particle trajectories, residence time distribution, particle attrition, conversion and selectivity, etc., but these have been studied less extensively, even in the case of Newtonian media. Excellent books and reviews are available on this subject (Davidson and Harrison, 1971; Davidson et al., 1985; Fan, 1989; DiFelice, 1995; Kim and Kang, 1997; Yang, 2003).

Other types of fluid–particle systems that are usually considered along with the fluidized beds are packed beds and settling suspensions or hindered settling in concentrated suspensions of noninteracting particles. While the possible link between a fixed and fluidized bed has already been alluded to in the preceding section, the settling behavior of concentrated suspensions also shows a great deal of similarity with a particulately fluidized bed. In this chapter, this analogy would be extended to the case of inelastic non-Newtonian media. However, the settling behavior of a swarm of non-Brownian particles in viscoelastic liquids is very different from that in inelastic fluids.

This chapter thus presents a critical evaluation of the available literature on the hydrodynamic aspects of the two-phase and three-phase fluidized beds and of the settling of concentrated suspensions of noninteracting particles, especially when the liquid phase exhibits non-Newtonian behavior. In particular, consideration will be given to the prediction of the minimum fluidization velocity, bed expansion characteristics, and the rate of sedimentation of concentrated suspensions under the influence of gravity. However, it appears desirable and instructive to include a brief discussion on the current scene with regard to the Newtonian fluids for each of these parameters which will not only lay the stage for non-Newtonian fluids but also serves as a reference to draw qualitative inferences regarding the influence of nonlinear flow characteristics of the liquid phase.

7.2 Two-Phase Fluidization

7.2.1 Minimum Fluidization Velocity

7.2.1.1 Definition

If a fluid is passed upward through a bed of particles, the pressure drop across the bed, Δp will initially increase as the superficial velocity, V, of the liquid is gradually increased as long as the bed behaves like a fixed bed (see Figure 7.1). When the liquid velocity has reached such a value that the frictional pressure drop (Δp) is equal to the buoyant weight of the particles per unit area, any further increase in the velocity results in the rearrangement of particles such that the resistance to flow remains the same, that is, as the velocity increases, the bed expands, but the pressure drop across the bed remains essentially constant provided there is no severe channeling and the wall effects are negligible. This is the point of incipient fluidization and the corresponding velocity is designated as the minimum fluidization velocity, V_{mf}, and the corresponding bed voidage is denoted by ε_{mf}. Thus, for $V > V_{mf}$, the pressure drop across the bed remains constant. If the velocity is gradually decreased, the pressure drop remains constant up to the point of incipient fluidization, but the pressure drop values in the fixed bed region turn out to be

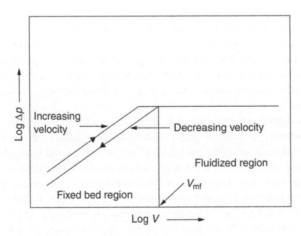

FIGURE 7.1 Ideal pressure drop–velocity curve for flow through a bed of particles.

somewhat lower than that recorded while the velocity was being increased. This difference is attributed to the slight change in the value of porosity on account of "repacking" of the bed. In practice, however, departures from the aforementioned ideal behavior are observed mainly due to interlocking of particles, channeling, wall effects, etc. Besides, the transition from the fixed to the fluidized bed conditions occurs gradually over a range of velocities rather than abruptly as shown in Figure 7.1 (Richardson, 1971). As the minimum fluidization velocity has no absolute significance, the generally accepted standard method for its determination from Δp-V plots involves drawing separate lines through the fixed and fluidized bed regions, and the point of intersection of these two lines yields the value of V_{mf}, as shown in Figure 7.1. Over the years, considerable research effort has been devoted to the development of reliable methods for the estimation of the minimum fluidization velocity for a liquid–solid combination. A brief account of the progress made thus far is presented in the next section.

7.2.1.2 Prediction of V_{mf}

Experimental determination of the minimum fluidization velocity is neither always possible nor desirable. Hence, the need for its prediction often arises while performing process design calculations for fluidizing systems. For a given liquid–solid combination, the value of the minimum fluidization velocity is influenced by a large number of variables including the particle size (and distribution), shape and orientation, density and viscosity of the liquid medium, type and design of distributor, and particle-to-column diameter ratio. Thus, most advances in this field have been made by using dimensional considerations aided by experimental observations.

Most attempts at developing predictive expressions for the estimation of V_{mf} hinge upon the fact that at the point of incipient fluidization, the pressure drop per unit length of the bed is given by its buoyant weight, which in turn is equated to the value obtained by assuming the bed to behave like a fixed bed with a mean voidage of ε_{mf}, that is,

$$\left(\frac{\Delta p}{L}\right) = (1 - \varepsilon_{mf})(\rho_s - \rho)g \tag{7.1}$$

While an incipiently fluidized bed represents a slightly loosened bed, it is customary to treat it as a fixed bed since the particles are still in contact with each other, though this assumption has been questioned by Barnea and Mizrahi (1973) and Barnea and Mednick (1975). Thus, one can use a suitable method (as outlined in Chapter 6) for calculating the pressure drop across a fixed bed. For instance, one can combine the Ergun equation (Equation 6.8a) with Equation 7.1 to yield an incipiently fluidized bed:

$$(1 - \varepsilon_{mf})(\rho_s - \rho)g = \frac{150(1 - \varepsilon_{mf})^2 \mu V_{mf}}{\varepsilon_{mf}^3 d^2} + \frac{1.75 \rho V_{mf}^2 (1 - \varepsilon_{mf})}{d \varepsilon_{mf}^3} \tag{7.2}$$

It is customary to rewrite Equation 7.2 in a dimensionless form using the Galileo number, Ga_{mf}, as

$$Ga_{mf} = 150\left(\frac{1 - \varepsilon_{mf}}{\varepsilon_{mf}^3}\right) Re_{mf} + 1.75 \frac{Re_{mf}^2}{\varepsilon_{mf}^3} \tag{7.3}$$

where the Galileo number is defined as

$$Ga_{mf} = \left[\frac{gd}{V_{mf}^2}\left(\frac{\rho_s - \rho}{\rho}\right)\right]\left(\frac{\rho^2 V_{mf}^2 d^2}{\mu^2}\right) = \frac{3}{4} C_{Dmf} Re_{mf}^2 \tag{7.4}$$

and the Reynolds number as,

$$Re_{mf} = \frac{\rho V_{mf} d}{\mu} \tag{7.5}$$

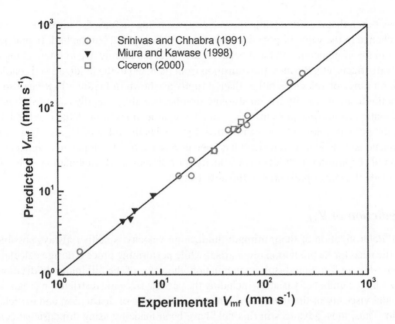

FIGURE 7.2 Typical comparison between experimental and predicted values of V_{mf} using Equation 7.3 (Newtonian fluids).

Thus, for a given liquid–solid system, one can readily calculate the value of Galileo number that in turn facilitates the calculation of Re_{mf} or V_{mf} via Equation 7.3 provided the value of bed voidage is known at the point of incipient fluidization. A typical comparison between the predictions of Equation 7.3 and experimental values is shown in Figure 7.2; the average and maximum deviations are 11.8% and 20%, respectively, for 18 data points. The literature abounds with numerous expressions for the estimation of V_{mf} for Newtonian liquid systems. A selection of widely used correlations has been compiled by Couderc (1985), Jamialahmadi and Müller-Steinhagen (2000), and more recently by Chhabra and Gurappa (2019) who all have also concluded that it is now possible to predict the value of V_{mf} for spherical particles with an accuracy of about 15%–20% in a new application, though somewhat larger errors are encountered for nonspherical particles (Lima-Ballesteros et al., 1982; Flemmer et al., 1993) and for mixed-size particle beds.

7.2.1.3 Non-Newtonian Systems

Table 7.1 gives a succinct summary of the studies pertaining to the flow of non-Newtonian liquids (primarily power-law fluids) in fluidized beds. Evidently, only a few workers have dealt with the measurement and prediction of the minimum fluidization velocity, and a listing of the correlations proposed for power-law fluids is provided in Table 7.2. Suffice it to add here that except the works of Kawase and Ulbrecht (1985b) and Machac et al. (1986, 1988a,b), all other expressions have been obtained by combining the Ergun equation in its modified form for power-law fluids (see Section 6.4) with Equation 7.1. Kawase and Ulbrecht (1985b), on the other hand, modified the Stokes equation to include the effects of non-Newtonian behavior and the bed voidage. They have obtained an approximate analytical solution to the governing equations in conjunction with the free surface cell model (Happel, 1958), whereas Machac et al. (1986) have used purely empirical considerations to arrive at their expression in the form of ($V_{mf}/V_{t\infty}$). While Jaiswal et al. (1992) improved upon the estimates of the drag correction factor Y given by Kawase and Ulbrecht (1985b), Dhole et al. (2004) have extended these calculations to high Reynolds numbers (up to 500) and to shear-thickening fluid behavior ($n > 1$). Aside from these, Miura and Kawase (1998) extended the approach of Kawase and Ulbrecht (1985b) to high Reynolds number flows, but it is limited to $0.74 \leq n \leq 1$. While Sabiri et al. (1996a) extended their capillary model (Sabiri and Comiti, 1995) to predict the minimum fluidization velocity and bed expansion behavior of beds of spherical and nonspherical

TABLE 7.1

Summary of Studies on Fluidization with Non-Newtonian Media

Investigator	Type of Work	Details	Main Results
Yu et al. (1968)	Experimental	$D=100$ $d=2.5, 8.4, 4.8, 9$ $\rho_s=1050-2450$ Aqueous solution of polyox	Ad hoc empirical modification of the Richardson-Zaki correlation for predicting V_{mf} and V-ε behavior. Reasonable agreement between experimental and predictions in the range $0.81 \leq n \leq 1$.
Wen and Fan (1973)	Experimental	$D=50$ $d=0.12-1.43$ $\rho_s=1520-11,300$; Aqueous solutions of CMC	Axial dispersion coefficient almost identical to Newtonian values ($0.86 \leq n \leq 1$).
Mishra et al. (1975)	Experimental	$D=80$ $d=4.3, 6$ $\rho_s=1200, 2500$; PVA in water and grease/kerosene	Weak non-Newtonian effects are observed.
Brea et al. (1976)	Experimental	$D=50$ $d=1.1-3.1$ $\rho_s=2500-8900$; Titanium dioxide slurries	Data on V_{mf} and bed expansion characteristics which are in line with the modified correlation of Richardson and Zaki (1954).
Tonini et al. (1981)	Experimental	$D=50$ $d=1.80$ $\rho_s=2940$; Aqueous solution of CMC	Mainly concerned with the mass transfer aspects in electrochemical reactions. Only two moderately non-Newtonian test fluids were used.
Kumar and Upadhyay (1981)	Experimental	$D=56, 80, 126$ $d=0.52-3.05$ $\rho_s=1300-2500$ One CMC solution ($n=0.850$)	No results on V_{mf}; limited results on V-ε behavior.
Kawase and Ulbrecht (1985b)	Theoretical	–	Cell models are used to derive expression for V_{mf} and V-ε and agreement with the literature data was stated to be moderately good, especially for weakly shear-thinning fluids.
Briend et al. (1984)	Experimental	$D=102$ $d=0.23-1.86$ $\rho_s=2480-11350$; Aqueous solutions of Carbopol and Separan	Preliminary results on V_{mf} and V-ε behavior which are in line with a non-Newtonian form of the Blake-Kozeny equation.
Machac et al. (1986, 1988)	Experimental	$D=20, 40$ $d=1.46-3.98$ $\rho_s=2500-16000$; Aqueous solutions of Natrosol, CMC, and Separan	Correlation of V_{mf} and V-ε behavior for power-law and Carreau model fluids.
Lali et al. (1989)	Experimental	$D=86$ $d=1.65, 3.1$ $\rho_s=2500$; Aqueous solutions of CMC	Bed expansion behavior is in line with Newtonian results.
Srinivas and Chhabra (1991)	Experimental	$D=50, 100$ $d=1.28, 2.6, 3.58$ $\rho_s=2500$; Aqueous solutions of CMC	Extensive results on V_{mf} and bed expansion characteristics. Detailed comparison with existing correlations.
Jaiswal et al. (1992)	Theoretical	–	Cell model predictions; good agreement with the literature data for V-ε behavior and V_{mf}.
Sharma and Chhabra (1992)	Experimental	$D=50, 100$ Nonspherical gravel chips fluidized with aqueous CMC solutions	Extensive data on V_{mf} and V-ε behavior.

(Continued)

TABLE 7.1 (*Continued*)

Summary of Studies on Fluidization with Non-Newtonian Media

Investigator	Type of Work	Details	Main Results
Dolejs et al. (1995), Dolejs and Mikulasek (1997)	Empirical	-	Hybrid model based on a combination of the submerged objects and the capillary bundle approach.
Sabiri et al. (1996a)	Experimental	$D = 90$ Spheres and plates fluidized with CMC solutions	Extensive data on V_{mf} and bed expansion.
Machac et al. (1997, 1999)	Experimental	2-D fluidized beds $1.47 \leq d \leq 4.12$ mm $0.36 \leq n \leq 1$	Effects of shear-thinning and visco-elasticity on fluidization.
Broniarz-Press et al. (1999, 2007a)	Experimental	$D = 90$ $1.8 \leq d \leq 4.4$ $0.56 \leq n \leq 1$	Wall effects and non-Newtonian effects on V_{mf}. Proposed a flow regime map.
Ciceron et al. (2002b)	Experimental	$D = 90$ $1.95 \leq d \leq 6.87$ CMC solutions	Extension of the capillary model to predict bed expansion data.
Machač et al. (2003a,b)	Experimental	$D = 90$ $1.9 \leq d \leq 6.87$	Effect of elasticity on fluidization.
Aghajani et al. (2004)	Experimental	$d = 2$–4 and short cylinders of Aluminum, Brass, Steel, Tantalum Aqueous CMC solutions ($0.5 \leq n \leq 0.95$)	Major thrust on wall heat transfer correlation.
Das et al. (2010)	Experimental	$D = 47, 72$ $3.13 \leq d \leq 5.64$ $900 \leq \rho_s \leq 944$ CMC solutions ($0.6 \leq n \leq 0.9$)	Inverse fluidization behavior.

Note: D and d are in mm; ρ_s in kg m^{-3}; CMC is carboxymethyl cellulose.

particles fluidized by inelastic carboxymethyl cellulose (CMC) solutions. In a subsequent detailed study, the range of applicability of this approach was delineated (Ciceron et al., 2002b). However, this approach yields the functional relationship similar to that obtained using the power-law version of the Ergun equation. Finally, with the exception of the limited data of Yu et al. (1968), Sharma and Chhabra (1992), and Sabiri et al. (1996a), all other studies pertain to the beds of spherical particles.

The relative predictive performance of some of the formulae listed in Table 7.2 has been evaluated (Chhabra, 1993c). Altogether, 70 independent measurements culled from various sources (as indicated in Figure 7.3) and encompassing wide ranges of conditions ($0.34 \leq n \leq 1$; $0.23 \leq d \leq 15.8$ mm; $1,050 \leq \rho_s \leq 11,350$ kg m^{-3}) were used. A comparative summary of the results is shown in Table 7.3. Evidently, none of these methods seems to work particularly well. However, in assessing the results presented in this table, it should be borne in mind that the value of V_{mf} is extremely sensitive to the value of bed voidage at the incipient fluidized conditions. Besides, the errors of the order of 50%–100% are not uncommon in the experimental determination of V_{mf} even with Newtonian liquids. Finally, a part of the discrepancy must be attributable to the unaccounted possible visco-elastic effects and wall effects. In view of this, the predictions of the method I of Machac et al. (1986) and of Jaiswal et al. (1992) appear to entail the minimum average deviations. Figure 7.3 contrasts the experimental and predicted values of V_{mf} for a wide range of conditions.

Preliminary results for nonspherical particles also appear to correlate with similar levels of accuracy with the equations listed in Table 7.2. However, the approaches of Sabiri et al. (1996a) and Ciceron et al. (2002b) are particularly suited to nonspherical particles as it obviates the necessity of estimating any shape factors and/or equivalent particle diameter. Finally, analogous developments for other fluid

TABLE 7.2
Expressions available for Predicting V_{mf} in Power-Law Liquids

Investigator	Expression	Remarks
Yu et al. (1968)	$\mathrm{Re}_{mf} = (\alpha_y \mathrm{Ga}_{mf})^{(2-n)/n}$ $\alpha_Y = \dfrac{\varepsilon_{mf}^{2n+1}}{12.5\left[((9n+3)/n)(1-\varepsilon_{mf})^n\right]}$	For creeping flow only
Mishra et al. (1975)[a]	$7.143\, \mathrm{Ga}_{mf} \beta^{2/2-n} = (\alpha_B \mathrm{Re}_{mf})^{2/(2-n)} + 85.714(\alpha_B \mathrm{Re}_{mf})^{n/(2-n)}$ $\beta = \dfrac{\varepsilon_{mf}^{(n+2)/2}}{(1-\varepsilon_{mf})^n} \left(\dfrac{4n}{3n+1}\right)^n \left(\dfrac{12\sqrt{2}}{5}\right)^{1-n}$	No upper limit on the value of Re_{mf} was stated
Brea et al. (1976)	$\varepsilon_{mf}^3 \mathrm{Ga}_{mf} = \left(\dfrac{160}{\alpha_B}\right) \mathrm{Re}_{mf}^{n/(2-n)} + 1.75\, \mathrm{Re}_{mf}^{2/(2-n)}$ $\alpha_B = \left(\dfrac{4n}{3n+1}\right)^n (1-\varepsilon_{mf})^{-1} \left(\dfrac{\varepsilon_{mf}^2}{12(1-\varepsilon_{mf})}\right)^{n-1}$	No upper limit on the value of Re_{mf}
Kumar and Upadhyay (1981)	Same as Brea et al., except substitute 150 for 160	No upper limit on the value of Re_{mf}
Kawase and Ulbrecht (1985b)	$V_{mf} = \left(\dfrac{\Delta \rho g d^{n+1}}{18 Y m}\right)^{1/n}$ where $Y = f(n, \varepsilon)$ is available in the original publication	Creeping flow only
Machac et al. (1986)	$V_{mf} = 0.019 V_t$ $V_{mf} = 0.015\left(1 + 0.73\left(\dfrac{d}{D}\right)\right) V_t$	Creeping and transitional flow regimes
Jaiswal et al. (1992) and Dhole et al. (2004)	Same as Kawase and Ulbrecht (1985b)	The drag correction factor $Y(n, \varepsilon)$ is available up to $\mathrm{Re}_{PL} = 500$

Notes: $\mathrm{Re}_{mf} = \dfrac{\rho V_{mf}^{(2-n)} d^n}{m}$; $\mathrm{Ga}_{mf} = \left(\dfrac{3}{4}\right) C_{Dmf} \mathrm{Re}_{mf}^{2/(2-n)}$.

[a] It has been subsequently corrected by Kumar and Upadhyay (1981).

models, namely Bingham plastic, Carreau model, and Ellis model fluids, are also available in literature (Mishra et al., 1975; Kawase and Ulbrecht, 1985b; Dolejs et al., 1995).

In summary, the accuracy of the predictions of the minimum fluidization velocity in non-Newtonian liquids is somewhat poorer than that in Newtonian liquids. This is so partly due to the additional complications arising from possible visco-elastic behavior and shear-induced mechanical degradation of macromolecules (Broniarz-Press et al., 1999, 2007a).

7.2.2 Bed Expansion Behavior

As mentioned earlier, once the superficial velocity of the liquid exceeds the minimum fluidization velocity, the mean voidage of the bed gradually increases and the frictional pressure drop across the bed remains constant at a value equal to its buoyant weight. It has been tacitly assumed that this kind of behavior generally occurs and data are seldom reported to confirm this expectation. Figure 7.4 illustrates this kind of behavior for the beds of 3–3.58 mm glass spheres being fluidized by non-Newtonian polymer solutions ($\rho_s > \rho$), as reported by Srinivas and Chhabra (1991) and by Ciceron (2000). This behavior is qualitatively similar to that for Newtonian liquids and has been documented in the literature by others also (Sabiri et al., 1996a; Miura et al., 2001a; Ciceron et al., 2002b).

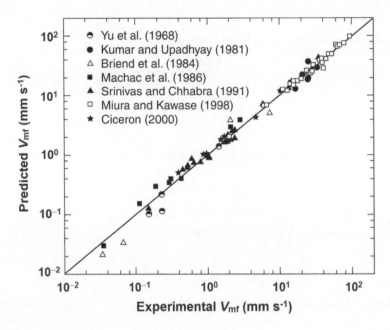

FIGURE 7.3 Typical comparison between the predicted and experimental values of V_{mf} for power-law liquids. (Modified after Chhabra, R.P., *Powder Technol.*, **76**, 225, 1993c.)

TABLE 7.3
Average and Maximum Deviations in Predicting V_{mf} for Spherical Particles in Power-Law Liquids

Investigator or Method	% Error[a]	
	Mean	Maximum
Yu et al. (1968)	44	146
Mishra et al. (1975)	87	96
Brea et al. (1976)	55	174
Kawase and Ulbrecht (1985b)	32	105
Kumar and Upadhyay (1981)	44	146
Machac et al. (1986)		
Method I	28	86
Method II	37	111
Jaiswal et al. (1992)	28	99
Sabiri et al. (1996a)	31	103

[a] %error = 100(experimental − predicted)/predicted.

As anomalous effect has been documented by Machac et al. (1986) who found that, depending upon the value of the power-law index, n, the pressure drop across the bed may drop to a value lower than the buoyant weight of the bed (see Figure 7.5). This type of behavior, in turn, limits the maximum achievable bed voidage (ε_{max}) for a given liquid–solid system. The value of ε_{max} was found to decrease with the decreasing value of the power-law index, n. This type of behavior appears to occur in visco-elastic liquids and is indicative of segregation.

It is customary to depict the bed expansion behavior of fluidized beds by plotting dimensionless velocity ratio ($V/V_{t\infty}$) against bed voidage. Figures 7.6 and 7.7 show typical bed expansion results for beds of 3.57 and 15.8 mm glass spheres being fluidized by non-Newtonian carboxymethyl cellulose solutions of varying levels of pseudoplasticity. Similar results have been reported by several other workers in

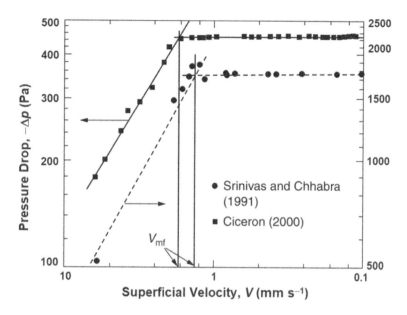

FIGURE 7.4 Experimental pressure drop–velocity curve and determination of V_{mf} for $n = 0.845$. ●— Srinivas and Chhabra (1991); ■ — Ciceron (2000).

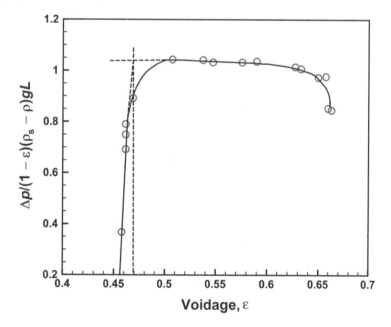

FIGURE 7.5 Anomalous pressure drop–velocity behavior observed in fluidization by visco-elastic liquids, as reported by Machac et al. (1986).

this field (Brea et al., 1976; Lali et al., 1989; Broniarz-Press et al., 1999, 2007a). In the case of fluidized beds, one is usually interested in the extent of bed expansion rather than the value of pressure drop, and therefore often such data have been represented and correlated in terms of the following dimensionless variables for Newtonian systems:

$$f\left(\frac{V}{V_t}, \text{Re}_t, \varepsilon, \frac{d}{D}\right) = 0 \tag{7.6}$$

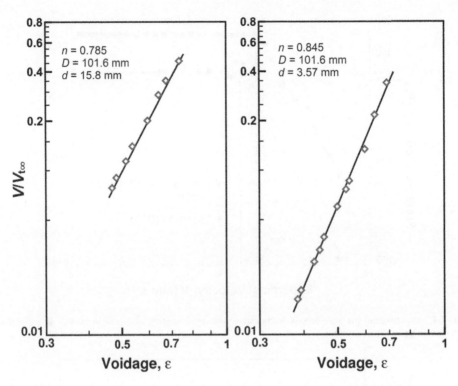

FIGURE 7.6 Typical bed expansion data for $n = 0.785$ and $n = 0.845$. (From Srinivas, B. K., and Chhabra, R.P., *Chem. Eng. Process.*, **29**, 121, 1991.)

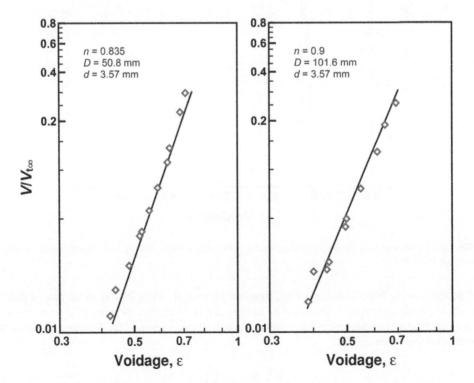

FIGURE 7.7 Typical bed expansion data for $n = 0.835$ and $n = 0.9$. (From Srinivas, B.K. and Chhabra, R.P., *Chem. Eng. Process.*, **29**, 121, 1991.)

Many expressions of varying forms and complexity have been proposed to depict the functional relationship embodied in Equation 7.6, the simplest of all being

$$\frac{V}{V_{t\infty}} = \varepsilon^Z \tag{7.7}$$

The literature abounds with numerous empirical expressions that purport to predict the value of the index, Z, under most conditions of practical interest. A thorough and critical review of the pertinent literature was presented by Khan and Richardson (1989). In particular, two methods for the prediction of the index, Z, in Equation 7.7 have gained wide acceptance and these will be described briefly here. Based on a large body of experimental data, Richardson and Zaki (1954) proposed a set of correlations for calculating the value of Z, which are given below in their modified form:

$$Z = 4.65 + 20\left(\frac{d}{D}\right), \qquad Re_{t\infty} < 0.2 \tag{7.8a}$$

$$= \left(4.40 + 18\left(\frac{d}{D}\right)\right) Re_{t\infty}^{-0.3}, \qquad 0.2 \leq Re_{t\infty} \leq 1 \tag{7.8b}$$

$$= \left(4.40 + 18\left(\frac{d}{D}\right)\right) Re_{t\infty}^{-0.1}, \qquad 1 \leq Re_{t\infty} \leq 200 \tag{7.8c}$$

$$= 2.40, \qquad Re_{t\infty} > 200 \tag{7.8d}$$

The second method is due to Garside and Al-Dibouni (1977) who worked in terms of the ratio (V/V_t) rather than $(V/V_{t\infty})$ as

$$\frac{V}{V_t} = \varepsilon^Z \tag{7.9}$$

where $Z(Re_t)$ is given by

$$Z = \frac{5.09 + 0.2839 Re_t^{0.877}}{1 + 0.104 Re_t^{0.87}}, \qquad 10^{-3} \leq Re_t \leq 10^4 \tag{7.10}$$

One can, however, rewrite Equation 7.10 in terms of $(V/V_{t\infty})$ by introducing a function of (d/D), as proposed by Garside and Al-Dibouni (1977) below

$$\frac{V}{V_{t\infty}} = \left[1 + 2.35\left(\frac{d}{D}\right)\right]^{-1} \varepsilon^Z \tag{7.11}$$

In essence, the new factor in front of the ε^Z term corrects the single sphere terminal velocity for wall effects. Scores of other schemes to predict the value of Z in Newtonian liquids are available in the literature and these have been reviewed by DiFelice (1995) and Jamialahmadi and Müller-Steinhagen (2000). The approach of Richardson and Zaki (1954) has been extended to multisize sphere systems also, see Asif (1998).

7.2.2.1 Inelastic Non-Newtonian Systems

Owing to the qualitatively similar nature of the bed expansion curves observed for Newtonian and inelastic non-Newtonian liquids, it is natural to explore the possibility of extending the aforementioned two formulae to power-law fluids. In this instance, one would expect the index Z to show a possible additional dependence on non-Newtonian model parameters. For power-law type fluids, the relevant definition of Re_t becomes

$$Re'_t = \frac{\rho V_t^{2-n} d^n}{m} \qquad (7.12)$$

And $Z = Z(n, Re'_t, d/D)$.

To delineate any possible dependence of the fluidization index Z on the power-law index, n, in the first instance, the experimental values of Z culled from different sources are compared with those calculated using the methods of Richardson and Zaki (1954) and Garside and Al-Dibouni (1977) in Table 7.4 and the resulting discrepancies, if any, can be unambiguously ascribed to the non-Newtonian behavior of liquids. An examination of Table 7.4 suggests that for large values of (d/D) (see the first three entries), the experimental values of Z are closer to the predictions of Garside and Al-Dibouni (1977) whereas for the remaining cases, that is, small values of (d/D), there is a good correspondence with the predictions of Richardson and Zaki (1954). Attention is drawn to the fact that exceptionally large values of Z have been documented in the literature when the liquid exhibits a visco-elastic behavior (Briend et al., 1984; Srinivas and Chhabra, 1991) for which neither of the aforementioned schemes seem to be applicable. Additional complication also arise with visco-elastic fluids due to the nonuniform expansion of the bed (Briend et al., 1984). Based on the comparisons presented in Table 7.4, it is thus reasonable to conclude that either of the abovementioned two methods may be used to predict the value of Z for viscoinelastic power-law fluids. The limited results available for nonspherical particles also provide further support to this conclusion (Sharma and Chhabra, 1992). Brea et al. (1976), Tonini et al. (1981), and Miura et al.

TABLE 7.4

Values of Z for fluidization with Power-Law liquids

			Values of Z				
n	(d/D)	$Re'_{t\infty}$	Experimental Values	Equation 7.8	Equations 7.9–7.12	Equation 7.13[a]	Data Source
0.882	0.156	3.66	4.81	6.33	4.52	8.04	Srinivas and
0.84	0.156	1.42	6.35	6.95	4.81	8.11	Chhabra
0.886	0.156	8.13	4.09	5.92	4.21	8.03	(1991)
0.845	0.0351	0.33	4.93	5.20	5.01	5.54	
0.835	0.0803	0.55	5.98	5.86	4.98	6.25	
0.900	0.0351	1.03	4.73	5.03	4.22	5.46	
0.941	0.0803	1.68	4.93	5.38	4.80	6.12	
0.382	0.0351	0.58	8.30	5.11	4.96	6.98	
0.382	0.0803	0.58	8.88	5.85	4.98	8.68	
0.603	0.0351	5.85	5.22	4.22	4.38	6.01	
0.603	0.0803	5.85	5.85	4.85	4.42	6.81	
0.603	0.0256	2.06	6.05	4.52	4.83	5.82	
0.603	0.0512	2.06	5.84	4.95	4.73	6.33	
0.68	0.0523	0.33	6.10	5.52	5.02	6.18	Tonini et al.
0.88	0.0523	0.33	5.80	5.52	5.02	5.83	(1981)
0.89	0.0408	22.8	3.58	3.86	3.68	5.59	Lali et al.
0.89	0.0218	5.0	3.98	4.08	4.41	5.21	(1989)
0.82	0.0408	6.14	3.95	4.28	4.34	5.69	
0.82	0.0217	1.58	5.02	4.58	4.89	5.31	
0.696	0.0408	1.48	4.93	4.94	4.81	5.90	
0.84	0.0218	0.61	4.80	4.86	4.95	5.44	
0.85	0.0023	0.0134	4.90	4.80	5.09	5.03	Briend et al.[b]
0.86	0.0023	0.029	6.60	4.80	5.09	5.03	(1984)
0.88	0.0125	16.08	5.10	3.49	3.83	5.20	
0.64	0.0125	19.60	5.80	3.43	3.83	5.46	

[a] Strictly applicable for $Re'_{t\infty} \le \sim 0.2$.
[b] The value of Vt_∞ has been estimated using the numerical results of Gu and Tanner (1985).

(2001a), on the other hand, have presented the following or similar ad hoc modifications of Equation 7.8a for power-law fluids:

$$Z = 4.65 + 20\left(\frac{d}{D}\right) + \frac{(1-n)}{n}, \qquad Re'_{t\infty} \leq 0.2 \tag{7.13}$$

Though in most cases the value of $Re'_{t\infty} \leq 0.2$, it is interesting to note that the predictions of Equation 7.13, also included in Table 7.4, show good match with the experimental values of Z in most cases, even for visco-elastic liquids (Briend et al., 1984; Srinivas and Chhabra, 1991). This agreement is, however, believed to be fortuitous.

Similarly, Machač et al. (1986, 1993) have proposed the explicit empirical correlation for bed expansion in the low Reynolds number regime ($Re'_{t\infty} \leq 0.3$) as

$$\varepsilon = \left(\frac{V}{V_{t\infty}}\right)^{0.218 - 0.404(d/D)} - 0.862(1-n)\left(\frac{V}{V_{t\infty}}\right)^{0.802 - 1.35(d/D)} \tag{7.14a}$$

In the intermediate Reynolds number range ($0.3 \leq Re'_{t\infty} \leq 165$), they employed Equation 7.7 with relationship for Z as

$$Z = \left(4.7 + 8.8\left(\frac{d}{D}\right)\right)\left(Re'_{t\infty}\right)^{-0.1} \tag{7.14b}$$

Equation 7.14b has only slightly different values of the constants as compared with that in Equation 7.8c. Subsequently, Machac et al. (1986) have extended the applicability of Equation 7.14 to include Carreau model fluids.

In the second approach, the drag coefficient values, based on the free surface cell model (Happel, 1958), have been used to calculate the drag force experienced by an assemblage of spherical particles in relative motion with power-law liquids to predict the velocity–voidage or bed expansion behavior of fluidized beds (Kawase and Ulbrecht, 1985b; Jaiswal et al., 1992). Using the numerical values of drag, one can prepare the so-called fluidization charts by introducing the normalized diameter (d^+) and velocity (V^+) as

$$V^+ = \left(\frac{(Re'_{t\infty})^{1/n}}{C_D}\right)^{n/(2+n)} = V\left\{\frac{\left(3\rho/(4g\Delta\rho)\right)^n \rho}{m}\right\}^{1/(2+n)} \tag{7.15}$$

$$d^+ = \left(C_D^{2-n}\left(Re'_{t\infty}\right)^2\right)^{1/(2+n)} = d\left\{\left(\frac{4}{3}\right)\left(\frac{g\Delta\rho}{\rho}\right)^{(2-n)/(2+n)}\left[\left(\frac{\rho}{m}\right)^{2/(2+n)}\right]\right\} \tag{7.16}$$

Figures 7.8–7.10 show the results plotted in the form of V^+ vs. d^+ for three different values of $n = 1, 0.8$, and 0.6. For a given liquid–solid system (i.e., for known values of $\rho, \Delta\rho, d, n, m$), the value of d^+ is known and $Re'_{t\infty}$ can be estimated, thereby fixing a point on $\varepsilon = 1$ curve. One can thus generate $V/V_{t\infty}$ vs. ε curves simply by drawing a line parallel to the y-axis and passing through the point (d^+, $Re'_{t\infty}$) on $\varepsilon = 1$ line. In the absence of wall effects, the results so obtained are compared with the predictions of Equations 7.8 and 7.10 for a range of conditions in Figures 7.11 and 7.12 for Newtonian fluids (Jaiswal et al., 1991a), respectively, whereas Figure 7.13 depicts a typical comparison with experimental data for $n = 0.6$. In both cases, the agreement between theory and experiments is about as good as can be expected in this field. Thus, this approach provides a theoretical vantage point for the prediction of bed expansion behavior.

Finally, it is also possible to use the capillary bundle approach to predict the bed expansion behavior. This approach is exemplified by the work of Mishra et al. (1975). While all the aforementioned three approaches for predicting the bed expansion behavior of a homogeneous fluidized bed have enjoyed varying levels of success in correlating limited data, none has proved to be completely satisfactory

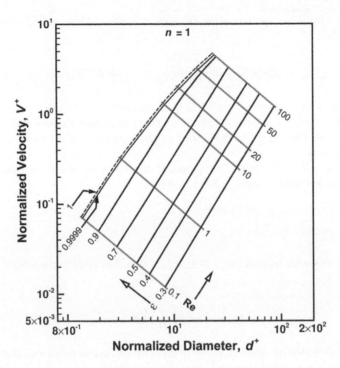

FIGURE 7.8 V^+–d^+ plot for $n = 1$. (From Jaiswal, A.K., Sundararajan, T., and Chhabra, R.P., *Numerical Heat Transfer*, **21A**, 275, 1992.)

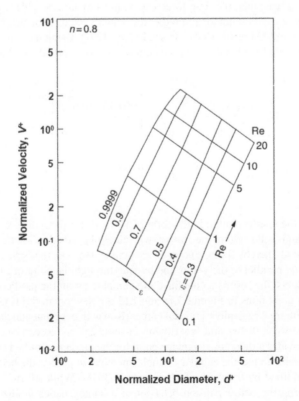

FIGURE 7.9 V^+–d^+ plot for $n = 0.7$. (From Jaiswal, A.K., Sundararajan, T., and Chhabra, R.P., *Numerical Heat Transfer*, **21A**, 275, 1992.)

Fluidization and Hindered Settling

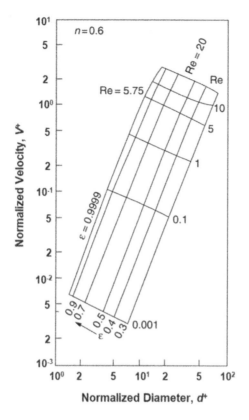

FIGURE 7.10 V^+–d^+ plot for $n=0.6$. (From Jaiswal, A.K., Sundararajan, T., and Chhabra, R.P., *Numerical Heat Transfer,* **21A**, 275, 1992.)

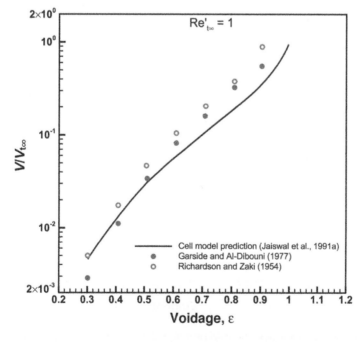

FIGURE 7.11 Typical comparison between the calculated, shown as line and experimental values of $(V/V_{t\infty})$ for $Re'_{t\infty} = 1$ in Newtonian systems.

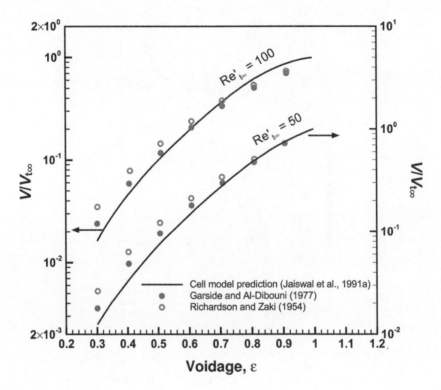

FIGURE 7.12 Typical comparison between the calculated, shown as line and experimental values of $(V/V_{t\infty})$ for $Re'_{t\infty} = 50$ and 100 in Newtonian systems.

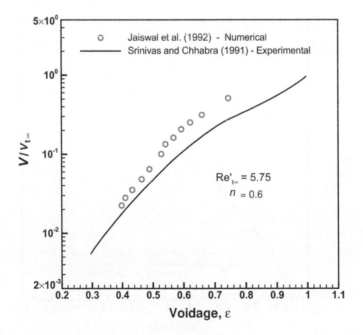

FIGURE 7.13 Typical comparison between the calculated (from Jaiswal et al., 1992) and experimental (Srinivas and Chhabra, 1991) values of $(V/V_{t\infty})$ in power-law liquids.

over the entire range of bed voidage and the Reynolds number. As remarked in Chapter 6, intuitively it appears that the capillary bundle representation should be more relevant in low-to-moderate porosity systems whereas the drag theories or the submerged object approach is clearly more appropriate for the beds of moderate-to-high values of voidage (Dolejs et al., 1995). Ciceron et al. (2002b) have reported a comprehensive study based on a combination of the capillary bundle approach of Comiti and Renaud (1989) and the submerged object model of Mauret and Renaud (1997). For a power-law fluid, one can easily combine Equations 6.20, 6.78, and 7.1 to obtain the dimensionless equation

$$Ar_n = 6(1-0.49\ln\varepsilon)^{n+1}\frac{(1-\varepsilon)^n}{\varepsilon^{2n+1}}\left(\frac{9n+3}{n}\right)^n Re_p^{n/(2-n)} + 0.581\left[\frac{1-0.49\ln\varepsilon}{\varepsilon}\right]^3 Re_p^{2/(2-n)} \qquad (7.17)$$

where the Archimedes number, Ar_n, is defined as

$$Ar_n = d^{(2+n)/(2-n)}\rho^{n/(2-n)}(\Delta\rho)gm^{2/(n-2)} \qquad (7.18)$$

and the particle Reynolds number, Re_p, is given by

$$Re_p = \frac{\rho V^{2-n} d^n}{m} \qquad (7.19)$$

For a given liquid–solid system, d, $\Delta\rho$, ρ, m, and n are all known and therefore Equation 7.17 permits the calculation of bed voidage for a fixed value of V ($>V_{mf}$). It is appropriate to mention here that Sabiri and Comiti (1997a,b) and Ciceron et al. (2002b) pointed out that it is seldom possible to approximate the flow curve of a pseudoplastic liquid with a single set of power-law constants, that is, m and n. It is therefore preferable to fit a series of power-laws to the flow curve and one should thus use appropriate values of (m, n) depending upon the shear rate at the pore wall. For this purpose, the shear rate at the pore wall can be estimated using the expression

$$\dot\gamma = \left(\frac{9n+3}{n}\right)\frac{VT}{d\varepsilon^2}(1-\varepsilon) \qquad (7.20)$$

where T is the tortuosity factor and is given by Equation 6.22, that is, $(1-0.49\ln\varepsilon)$. Figures 7.14 and 7.15 show typical comparisons between the predictions of Equation 7.17 and the experimental results drawn from literature. Based on the extensive comparisons ($Re_p \leq 240$) shown in Figures 7.14 and 7.15 and others not reported herein, it was abundantly clear that the predictions of Equation 7.17 began to diverge increasingly from the experimental results, once the voidage exceeded a limiting value, $\varepsilon_{lim} \approx 0.65$, which itself may be a function of the Reynolds number.

For $\varepsilon \geq \varepsilon_{lim}$ conditions, Ciceron et al. (2002b) developed a submerged object model, similar to that of Mauret and Renaud (1997). This approach yields the relationship

$$g\Delta\rho = \frac{3}{4}C_{Dm}\frac{\tau_H^3}{\varepsilon^3}\frac{\rho V^2}{d} \qquad (7.21)$$

where τ_H is the hydraulic tortuosity that is assumed to be a function of the Reynolds number and porosity (Mauret and Renaud, 1997; Epstein, 1998; Ghanbarian et al., 2013). At low Reynolds numbers, the hydraulic tortuosity is independent of the Reynolds number. The values of τ_H for a range of bed voidages and the Reynolds number are calculated using the approach of Molerus (1980) for Newtonian fluids. In the first instance, it is assumed that the same values can be used for power-law fluids. While it might be a reasonable idea for inelastic liquids, it would be difficult to justify this approximation for visco-elastic liquids. The other unknown in Equation 7.21 is the drag coefficient C_{Dm} for a sphere. Based on the numerical results of Tripathi et al. (1994) and Tripathi and Chhabra (1995), Darby (1996) put forward the correlation for C_{Dm} as

$$C_{Dm} = \left(B_D + \frac{A_D}{\sqrt{Re'_p}}\right)^2 \qquad (7.22)$$

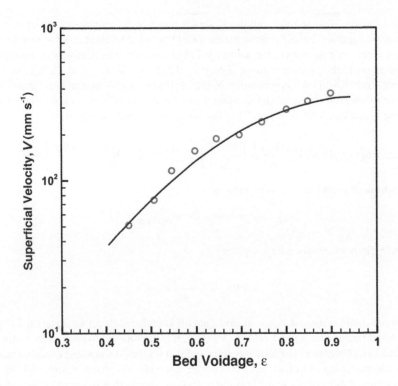

FIGURE 7.14 Comparison between the predictions of Equation 7.17 and experimental data for 3-mm steel spheres fluidized by a TiO$_2$ slurry (Brea et al., 1976).

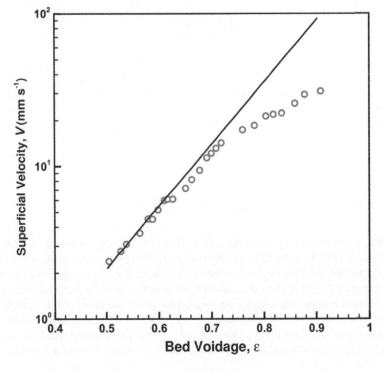

FIGURE 7.15 Comparison between the predictions of Equation 7.17 and experimental data for 2.46-mm glass spheres fluidized by Tylose solution. (Machac, I., Balcar, M., and Lecjaks, Z., *Chem. Eng. Sci.*, **41**, 591, 1986; Machac et al., 1993.)

Fluidization and Hindered Settling

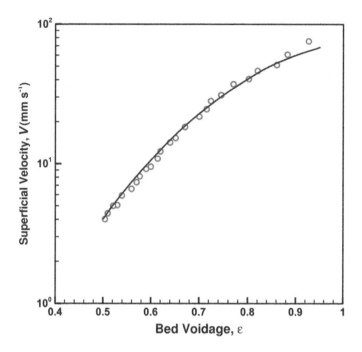

FIGURE 7.16 Comparison between the predictions of Equation 7.21 and experimental results for 4.93-mm glass spheres fluidized by a CMC solution. (Ciceron, D., Comiti, J., Chhabra, R.P., and Renaud, M., *Chem. Engng. Sci.*, **57**, 3225, 2002.)

where

$$A_D = 4.8 \left(\frac{1.33 + 0.37n}{1 + 0.7n^{3.7}} \right)^{0.5} \quad (7.23a)$$

$$B_D = \left[\left(\frac{1.82}{n} \right)^8 + 34 \right]^{-1/8} \quad (7.23b)$$

Equation 7.23 is valid for $Re'_p \leq 100$ and $0.4 \leq n \leq 1.6$. The modified Reynolds number Re'_p is given by Equation 7.19 except that the characteristic velocity is given by $(V\tau_H/\varepsilon)$ instead of V. One can again introduce the shear rate range-dependent power-law constants by ascertaining the value of the mean shear rate using the expression

$$\dot{\gamma}_{\text{mean}} = \frac{2\tau_H V}{d\varepsilon} \quad (7.24)$$

Thus, for known values of the physical and rheological properties, this approach is iterative in nature. Figures 7.16 and 7.17 show typical comparisons between the predictions of Equation 7.21 and experimental results for a range of conditions.

Based on the extensive comparisons for bed expansion data obtained with Newtonian and inelastic power-law liquids, the key findings of Ciceron et al. (2002b) can be summarized as follows: the capillary model yields acceptable predictions in the range of conditions as $\varepsilon_{mf} \leq \varepsilon \leq 0.96$ for $Re'_p > 5$ and $0.4 \leq n \leq 0.96$ whereas the complementary submerged objects model works well for $\varepsilon \geq 0.6$, $n \geq 0.4$ and $Re_p' \leq 100$. The data of Broniarz-Press et al. (1999, 2007a) spanning both packed and fluidized bed regimes are also in line with that of Sabiri and Comiti (1997a,b), and Ciceron et al. (2002b), as also noted by Chhabra et al. (2001a).

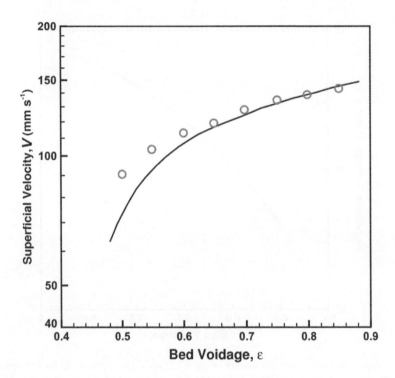

FIGURE 7.17 Comparison between the predictions of Equation 7.21 and experimental data for 2.2-mm lead spheres fluidized by a TiO$_2$ slurry (Brea et al., 1976).

7.2.3 Effect of Visco-elasticity

In some of the studies listed in Table 7.1, it is likely that some of the polymer solutions used as test fluids may have exhibited some degree of visco-elastic characteristics in addition to their pseudo-plastic behavior, for example, the polyacrylamide solutions used by Briend et al. (1984), Srinivas and Chhabra (1991), and Broniarz-Press et al. (2007). While the effect of visco-elasticity on the minimum fluidization velocity is far from clear, inhomogeneities and the formation of preferred flow channels have been reported for fluidization with visco-elastic liquids (Machac et al., 1986, 2003a), similar to that seen in fixed beds (Muller et al., 1998) and hindered settling of concentration suspensions (Section 7.4 in this chapter). Therefore, bed expansion data for these systems does not yield a straight line on log–log coordinates when the bed voidage is plotted against the superficial liquid velocity. In some cases, even the maximum achievable bed voidage is well below the theoretical limit of $\varepsilon = 1$ or when the pressure drop across the bed is graphed against bed voidage under fluidized conditions, the pressure drop may drop below the value $(1 - \varepsilon)(g\Delta\rho)$ after reaching the critical value of bed voidage, as shown in Figure 7.5. This effect has also been observed in two-dimensional fluidized beds. Figure 7.18 clearly shows the formation of structures in the beds of particles being fluidized by highly elastic Boger fluids (Machač et al., 2003a).

Aside from the aforementioned studies on fluidization, Wen and Fan (1973) have reported limited measurements on axial dispersion in beds of spherical particles fluidized by non-Newtonian polymer solutions. They concluded that the Newtonian formulae may be used as a first approximation for calculating the value of axial dispersion coefficient for power-law liquids.

The foregoing treatment is obviously relevant when the density of the particles is larger than that of the fluidizing medium. In many situations, it is not so, and the solids are lighter than the liquid. In such cases, the liquid is introduced from the top and the velocity is gradually increased until the bed is fluidized. This is called inverse fluidization (Fan, 1989). The hydrolysis of milk protein and the bio-oxidation of ferrous iron by *Thiobacillus* is carried out in inverse fluidized bed reactors, for instance. A reasonable

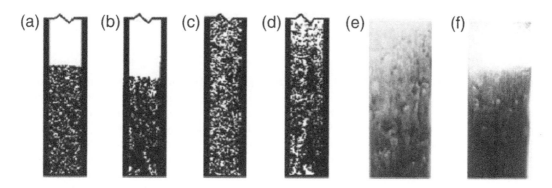

FIGURE 7.18 Formation of structures in sedimentation (e–f) (Allen, E. and Uhlherr, P.H.T., *J. Rheol.*, **33**, 627, 1989) and in fluidization (a–d). (Machac, I., Comiti, J., Brokl, P., and Siska, B., *Chem. Eng. Res. Des.*, **81A**, 1217, 2003. With permission.)

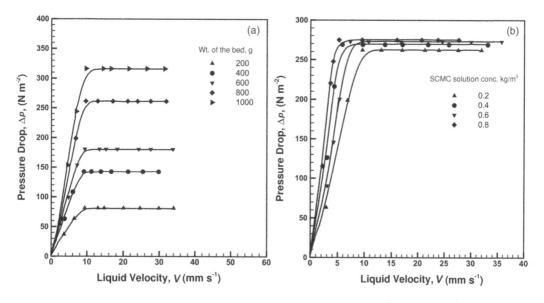

FIGURE 7.19 Variation of pressure drop with superficial velocity of various bd weights for mono-particle system (Solid: LDPE-2; Column diameter: 0.042 m) (a) SCMC solution conc.: 0.2 kg/m³ (b) weight of bed: 800 g. (Replotted from Das et al., *Chem. Eng. Proce.*, **49**, 1169, 2010.)

body of knowledge is available on the hydrodynamics of two-phase and three-phase inverse fluidized beds when the liquid phase is Newtonian, for example, see Legile et al. (1992), Yasser Ibrahim et al. (1996), Ulaganathan and Krishnaiah (1995), etc. In contrast, little is known about the hydrodynamics of inverse fluidized beds with a non-Newtonian liquid phase. Femin Bendict et al. (1998) have reported scant results on the bed expansion and pressure drop characteristics for the fluidization of 6 mm LDPE and PP plastic particles by aqueous CMC solutions ($0.80 \leq n \leq 0.86$). Subsequently, Vijaya Lakshmi et al. (2000) and Das et al. (2010) have studied the effect of particle and column size on the minimum fluidization velocity of plastic particles by similar CMC solutions. The minimum fluidization velocity was shown to decrease with the increasing particle density and the liquid viscosity, and with the decreasing particle size. Figure 7.19 shows representative results on the pressure drop-superficial velocity relationship for inverse fluidization. The macroscopic characteristics, namely minimum fluidizing velocity and bed expansion behavior, are qualitatively similar for both situations, i.e., $\rho > \rho_s$ and $\rho < \rho_s$.

7.3 Three-Phase Fluidized Beds

7.3.1 Introduction

If a gas is passed through a liquid–solid fluidized bed, it is possible to disperse the gas in the form of small bubbles and thereby obtain good contact between the liquid, the solid, and the gas present in such three-phase systems. This mode of contacting is often referred to as three-phase fluidized beds. Three-phase fluidized bed reactors have potential applications in the area of hydrogenation of coal slurries, Fischer–Tropsch synthesis and the heterogeneous catalytic hydrodesulphurization and hydrocracking of oil fractions, and the removal of acidic components from dusty gases. An important application is in a biological fluidized bed reactor in which oxygen transfer to the biomass takes place, first by its dissolution from air that is bubbled through the bed, and then its subsequent transfer from the solution to the biomass particles. Many other existing and potential applications of such systems have been listed by Epstein (1981), Darton (1985), Muroyama and Fan (1985), Fan (1989), and Fan and Yang (2003).

The hydrodynamic behavior of three-phase fluidized systems is much more complex than that of a two-phase liquid–solid fluidized bed, even when the liquid phase is Newtonian, and it is therefore not possible to present a detailed treatment here. Attention is, however, drawn to the excellent books (Fan, 1989; Yang, 2003) and review articles (Epstein, 1981, 2003; Darton, 1985; Muroyama and Fan, 1985) available on this subject.

Notwithstanding the significance of the detailed bubble and wake phenomena and flow patterns, it is customary to characterize these systems at a macroscopic level in terms of the minimum fluidizing velocity, bed expansion characteristic, gas holdup, etc. While a sizeable body of knowledge is available on this aspect when the liquid phase is Newtonian, very little is known about the effect of non-Newtonian liquid behavior on these aspects (Burru and Briens, 1989, 1991; Zaidi et al., 1989,1990; Miura and Kawase, 1997, 1998; Miura et al., 1998, 2001a,b). The available limited information is summarized here.

7.3.2 Minimum Fluidization Velocity

If a gas is introduced into a bed of solids fluidized by a liquid, it is generally observed that the minimum fluidizing velocity of the liquid is reduced by the presence of the gas stream. This is so in part due to the kinetic energy of the gas that facilitates the fluidization of particles (Zhang et al., 1995). Owing to the fluctuating nature of the flow, accurate measurements are rather difficult. Though there exists some confusion about the precise definition of the minimum fluidization velocity in three-phase fluidized systems, frequently the value of the minimum fluidization velocity for a three-phase fluidized system is evaluated in a manner similar to that in the absence of the gas. Figure 7.20 shows the pressure drop across the bed versus the superficial velocity of the liquid phase with and without the introduction of the gas into the bed (of 5 mm glass spheres being fluidized by a non-Newtonian CMC solution; Miura and Kawase, 1997). The value of the minimum fluidization velocity is seen to decrease almost by 50% due to the introduction of the gas. Figures 7.21 and 7.22 show the representative results on the combined effects of the particle size and the gas flow rate on the minimum fluidization velocity in three-phase systems for water and a power-law liquid, respectively. Qualitatively similar trends are present in both these figures. Following the initial drop, the minimum fluidization velocity seems to level off to a constant value that is nearly independent of the gas flow rate. This trend is clearly seen in Figures 7.21 and 7.22 for small particles but is less evident for the larger glass beads. The limited data available thus far seems to suggest that the reduction in the value of the minimum fluidization velocity progressively diminishes with the decreasing value of the power-law index and with the increasing viscosity of the liquid phase (Miura and Kawase, 1997, 1998). Finally, Miura and Kawase (1997, 1998) also concluded that their data on the minimum fluidization velocity is well predicted by the approach of Zhang et al. (1995) modified suitably for power-law liquids.

7.3.3 Bed Expansion Behavior

Depending upon the inertia of solids, the expansion of the bed may increase or decrease when a gas is introduced into a liquid–solid fluidized bed. At relatively low gas flow rates and with particles of

Fluidization and Hindered Settling

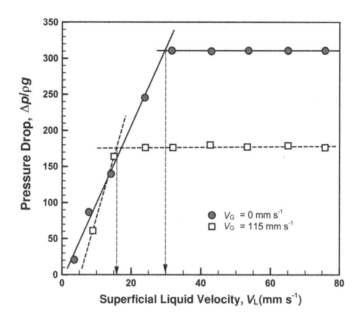

FIGURE 7.20 Experimental determination of the minimum fluidization velocity in three-phase fluidized beds with a polymer solution, $n = 0.79$. (Replotted from Miura, H. and Kawase, Y., *Chem. Eng. Sci.*, **52**, 4095, 1997.)

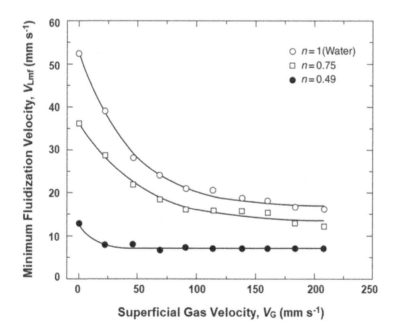

FIGURE 7.21 Effect of power-law rheology and gas flow rate on the minimum (liquid) fluidization velocity for a bed of 5 mm glass spheres. (Modified after Miura, H. and Kawase, Y., *Chem. Eng. Sci.*, **52**, 4095, 1997.)

large inertia (~5–6 mm glass beads being fluidized by water), large gas bubbles are split up by the presence of the solids to produce a dispersion of fine bubbles that present a high interfacial area for mass transfer. Even smaller bubbles are produced if the surface tension of the liquid is reduced. On the other hand, smaller particles are unable to overcome the surface tension forces and do not penetrate into the large gas bubbles. Furthermore, large gas bubbles in fluidized beds have large wakes thereby drawing

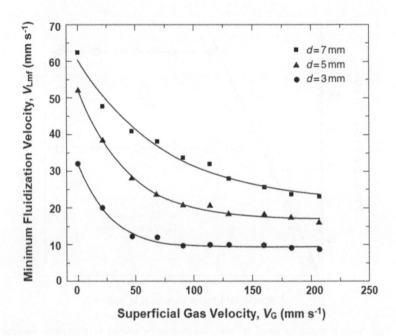

FIGURE 7.22 Effect of gas velocity and particle size on the minimum (liquid) fluidization velocity for beds of glass beads being fluidized by water. (Modified after Miura, H. and Kawase, Y., *Powder Technol.*, **97**, 124, 1998.)

liquid rapidly into their wake. This in turn reduces the flow of liquid in the remainder of the bed, causing the bed to contract. However, while this mechanism of bed contraction is pertinent for the low-viscosity Newtonian fluids, such contraction is not observed with highly viscous non-Newtonian liquids. Figures 7.23 and 7.24 show the effects of particle size and gas flow rate on the bed voidage for water and a non-Newtonian solution. Broadly speaking, for a fixed particle size, the bed expansion behavior is not very sensitive to the rheology of the liquid (Figure 7.23). Likewise, the effect of particle size (Figure 7.24) is also qualitatively similar in water and non-Newtonian solutions.

7.3.4 Gas Holdup

Figures 7.25 and 7.26 show the variation of gas holdup with the increasing flow rate for glass beads of different sizes being fluidized by water and a non-Newtonian polymer solution. Once again, the gas holdup shows qualitatively similar dependence on particle size, gas, and liquid flow rates, thereby suggesting negligible influence of the shear-thinning characteristics of the liquid phase. In view of this, Miura et al. (2001a,b) reported their results on gas holdup for power-law fluids to be in line with the following expression due to Begovich and Watson (1978) initially developed for Newtonian liquids (in SI units)

$$\varepsilon_g = 1.61 V_g^{0.72} d^{0.168} D^{-0.125} \qquad (7.25)$$

Note that this expression does not involve the viscosity of liquid.

Broadly speaking, in a three-phase fluidized system, the minimum fluidizing velocity of the liquid phase progressively decreases as the gas flow rate is gradually increased. This can be explained by noting that the energy input to the system in terms of the kinetic energy of the incoming gas phase facilitates the transition from the fixed bed state to the fluidized bed conditions. Similarly, the liquid holdup also decreases with the increasing gas flow rate. However, the reduction in the value of the liquid fluidization velocity is strongly dependent on the particle size and it generally decreases with the decreasing particle size and with the increasing degree of pseudoplasticity. Similarly, the bed voidage progressively increases with the rising gas flow rate at a constant liquid flow rate. However, the net increase in the bed voidage is also small for small particles.

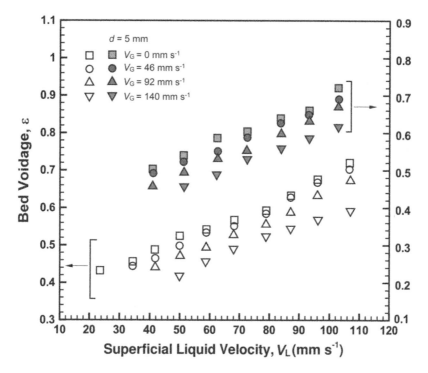

FIGURE 7.23 Effect of gas flow rate on bed expansion behavior of a bed of 5-mm glass beads being fluidized by water (open symbols) and by a polymer solution ($n = 0.84$, filled symbols). (Modified from Miura, H. and Kawase, Y., *Chem. Eng. Sci.*, **52**, 4095, 1997.)

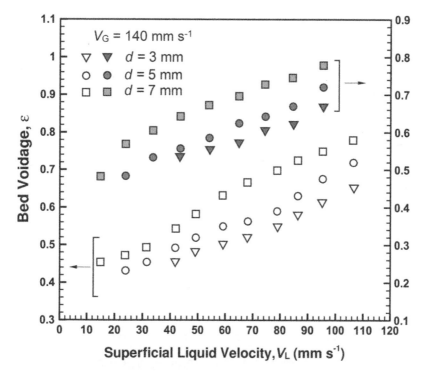

FIGURE 7.24 Effect of particle size on bed expansion behavior of a three-phase fluidized system. Open symbols for water and filled symbols for a polymer solution ($n = 0.72$). (Modified from Miura et al., 2001b.)

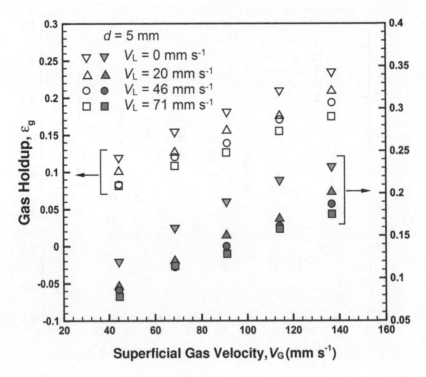

FIGURE 7.25 Effect of gas flow rate and liquid rheology on gas holdup in a fluidized bed of 5-mm glass beads. Open symbols for water and filled symbols for a polymer solution ($n = 0.72$). (Modified from Miura et al., 2001b.)

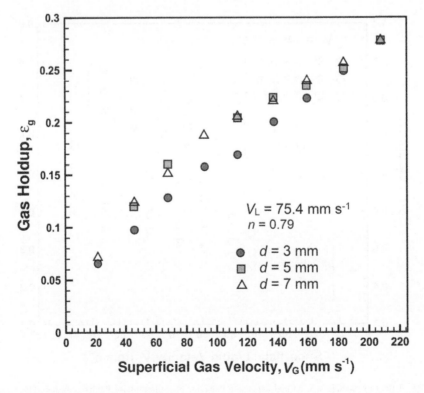

FIGURE 7.26 Effect of particle size on gas holdup in a three-phase fluidized system. (Modified after Miura, H. and Kawase, Y., *Chem. Eng. Sci.*, **52**, 4095, 1997.)

Fluidization and Hindered Settling

In summary, it is perhaps fair to conclude that the available scant data on three-phase fluidized systems with shear-thinning liquids suggests the non-Newtonian effects to be rather weak. It is thus possible to predict the minimum fluidization velocity, bed expansion, and gas holdup characteristics by suitable modifications of the existing frameworks for Newtonian fluids.

7.4 Sedimentation or Hindered Settling

Hindered settling or sedimentation under static and dynamic conditions of nonflocculated homogeneous suspensions of uniform size spherical inert particles represents an idealization of numerous industrially important processes encountered in chemical and processing applications, especially relating to liquid–solid separation using hydrocyclones for slurries (Ortega-Rivas and Svarovsky, 1998; Ortega-Rivas, 2016; Fernando Concha, 2014) and in the pipeline transportation of mineral slurries (DeAngelis et al., 1993; DeAngelis and Mancini, 1997; Rosso, 2000). Undoubtedly, the variable of central interest here is the rate of sedimentation in the initial constant concentration zone. Reliable knowledge of the sedimentation velocities is required for the design of equipment for handling suspensions and slurries such as thickeners, centrifuges, etc. It is readily recognized that the hindered settling velocity of a suspension of spheres is influenced by a large number of variables: liquid viscosity and density, particle shape, particle size and size distribution, density, particle to tube diameter ratio, and the fractional volume concentration of particles even for noninteracting particles. Further complications arise in the case of fine particles due to van der Waals and other surface forces; these systems are not considered here but have been discussed in detail by others, for example, see Russel (1980), Fernando Concha (2014), etc.

Based on heuristic and dimensional considerations, Richardson and Zaki (1954) were probably the first to recognize that, in spite of the obvious differences in the detailed flow fields, a great deal of similarity exists at a macroscopic level between the settling characteristics of a concentrated suspension of noninteracting particles and the expansion behavior of a particulately fluidized bed of uniform size spheres. Moreover, they asserted that their formulae, initially developed for fluidization (Equations 7.7 and 7.8), were equally successful in representing the hindered settling data also. It is thus customary to use similar type of expressions for both fluidization and sedimentation.

Figure 7.27 shows typical hindered settling data for two sizes of glass beads in a Newtonian castor oil. Attention is drawn particularly to the qualitatively similar nature of these plots to those for fluidization

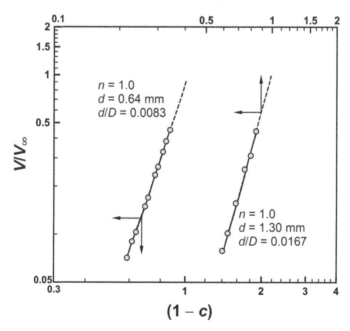

FIGURE 7.27 Hindered settling data in Newtonian media. (From Chhabra et al., 1992.)

(see Figure 7.6). Indeed, the values of Z calculated from Equation 7.8 are in excellent agreement with the corresponding experimental values. Similar conclusions have been reached by numerous investigators during the last 50 years or so (Zeidan et al., 2003). Good reviews on the hindered settling of noncolloidal particles in Newtonian media at low Reynolds numbers are also available (Scott, 1984; Davis and Acrivos, 1985; Zeidan et al., 2003).

7.4.1 Non-Newtonian Studies

Very little theoretical and experimental work is available on the sedimentation of multiparticle particles and of clusters of particles in quiescent non-Newtonian liquids. Caswell (1977) investigated theoretically the settling of two spheres in a generalized visco-elastic model fluid. Depending upon the choice of fluid parameters and the initial separation, the two spheres may converge or diverge; subsequently, this behavior has been experimentally observed by Riddle et al. (1977). Subsequently, similar results for two-, three-, and four-sphere systems in shear-thinning fluids have also been reported by Zhu and Clark (1998), Daugan et al. (2002a,b), Zhu et al. (2003), Huang et al. (2006), Alghalibi et al. (2018, 2020), Yu et al. (2002, 2006), Phillips and Talini (2007), Sulaymon et al. (2010,2013), Mosely et al. (2019), Zhang et al. (2020) and Ferrari et al (2022). All these studies seem to suggest a complex interplay between the rheology and the initial configuration of the particles that may lead to the segregation of particles. Qualitatively similar inhomogeneities and formation of structures in concentrated suspensions in non-Newtonian suspending media have been reported, both under static and dynamic conditions (Allen and Uhlherr, 1989; Darcovich et al., 1996; Bobroff and Phillips, 1998; Siska et al., 1995, 1996; Daugan et al., 2004), as can be seen in Figure 7.18 for the hindered settling of sand particles in visco-elastic polyacrylamide solutions. Note the striking similarity between the heterogeneities observed in the fluidized beds and in the hindered settling of concentrated suspensions of noninteracting particles. Furthermore, the formation of structures may show additional dependence on time, thereby resulting in enhancement or hindrance in settling (Bobroff and Phillips, 1998; Daugan et al., 2004). While the exact reasons for such inhomogeneities are not fully known, either shear-thinning or visco-elasticity, or both are often invoked to explain these phenomena qualitatively (Bobroff and Phillips, 1998). The available numerical and experimental results suggest aggregation of particles in the form of vertical columns during the settling of uniform size particles in visco-elastic fluids even when the suspension is initially homogeneous (Phillips and Talini, 2007; Mora et al., 2005, etc.) Therefore, it is not always possible to achieve steady state. Horizontal migration of particles in these systems is in line with the analytical and experimental studies available in the literature (Michele et al., 1977; Joseph et al., 1994; Lyon et al., 2001; Scirocco et al., 2004). Similarly, Phillips (2010) has argued that such heterogeneities are unlikely to occur in generalized Newtonian fluids. According to his analysis, fluid elasticity promotes aggregation of particles and this tendency is countered by the hydrodynamic dispersive force. For a second-order fluid, he was able to predict vertical stratification under appropriate conditions. Indeed, this analysis not only predicted the time scale for the onset of such instabilities but his predictions regarding the number of vertical columns of particles are also in line with the experimental results (Daugan et al., 2004). In a recent numerical study dealing with the settling in dilute suspensions in power-law fluids, Ferrari et al. (2022) have identified two settling regimes, namely viscosity-dominated regimes in which preferential trajectories were observed in dilute systems. This tendency was suppressed with the increasing particle concentration. Their results also suggest that the interparticle collisions are suppressed in shear-thickening fluids whereas these are accentuated in shear-thinning fluids. Datt and Elfring (2018) have provided the first correction for the viscosity of suspensions of spheres in shear-thinning fluids.

On the other hand, Kawase and Ulbrecht (1981e) employed the Happel's free surface cell model for predicting the hindered settling velocity of suspensions of spheres in power-law fluids. They have obtained a closed form solution that is believed to be applicable for weakly non-Newtonian behavior, that is, the value of power-law index n not too different from unity. Unfortunately, their final expression for $n=1$ shows deviations almost of the order of 100% from the well-established results for Newtonian fluids and hence their predictions for non-Newtonian fluid behavior must be treated with reserve.

There have been only a few experimental investigations relating to the hindered settling of concentrated suspensions in non-Newtonian systems. Balakrishna et al. (1971) reported preliminary results on

Fluidization and Hindered Settling

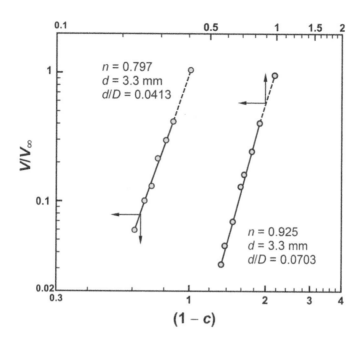

FIGURE 7.28 Hindered settling data in power-law liquids. (From Chhabra et al.,1992.)

the settling of fine sand particles and concluded that there were no non-Newtonian effects present in their results. In spite of this, a cumbersome empirical expression for the velocity ratio ($V/V_{t\infty}$) was presented. Allen and Uhlherr (1989) have reported the rates of sedimentation for glass beads of various sizes settling in aqueous solutions of polyacrylamide. They observed regions of heterogeneities and segregation, akin to that observed in gas–solid fluidization and in liquid–solid fluidization with visco-elastic liquids (Machač et al., 2003a) and in the shearing flow of suspensions (Lyon et al., 2001). Furthermore, when their results are plotted in the form of $V/V_{t\infty}$ vs. ε on log–log coordinates, two distinct power law regions are encountered. Though the exact reasons for this behavior are not immediately obvious, Allen and Uhlherr (1989) have ascribed both these anomalous observations to the visco-elastic behavior of test fluids. Subsequently, these results have been reconfirmed by Bobroff and Phillips (1998) and Daugan et al. (2004). In fact, the temporal variation of particle distribution may even lead to the enhanced rather than hindered settling in such systems (Daugan et al., 2004). Scant results on hindered settling of non-interacting spheres in mildly shear-thinning liquids (Figure 7.28) do seem to conform to Equations 7.7 and 7.9 (Chhabra et al., 1992).

Finally, it is appropriate to report on some of the recent work on the sedimentation of rod-like particles (fibers) in Newtonian suspending media. Typical examples of the sedimentation of rod-like particles are found in paper pulp suspensions, red blood cells, etc. (Kumar and Ramarao, 1991; Chen and Chen, 1997). The interparticle interactions in rod-like systems begin to manifest at much lower volumetric concentrations of fibers than in particulate suspensions. For instance, in a Newtonian liquid at a 3% volume fraction of spheres, the hindered settling velocity is reduced by 20% whereas for elongated bodies, this order of hindrance occurs at much lower concentration of fibers (Turney et al., 1995). Needless to say that with nonspherical particles, it is neither justified to assume the homogeneity of the system nor is it necessary that the interfiber interactions will impede their settling. Indeed, it is possible to obtain enhanced rates of sedimentation in such systems (Herzhaft et al., 1996). However, the Monte-Carlo simulations of Mackaplow and Shaqfeh (1998) predict the degree of hindrance to be linear in fiber concentration provided the system is isotropic and homogeneous. On the other hand, their dynamic simulations suggested spatial and orientational distributions to become anisotropic and heterogeneous. Obviously, the sedimentation of concentrated suspensions of rod-like particles is much more complex than that of spherical particles. It is unlikely that the simple expressions like Equation 7.7 or Equation 7.8 will ever suffice for such systems.

In a recent study, Koblitz et al. (2018) have reported direct numerical simulation (DNS) study of cylindrical particles settling in a quiescent Bingham plastic medium. Their main objective was to establish a criterion for the yielding of the fluid. For large values of yield stress, they reported enhanced settling with the increasing solid concentration in the suspension. On the other hand, the rate of sedimentation decreased with the increasing yield stress for a fixed solid concentration over the range 1%–5%. The enhancement in settling is consistent with the numerical results of Alghalibi et al. (2018, 2020) for spherical particles in Carreau model fluids. Furthermore, Alghalibi et al. (2018, 2020) also suggested that though the shear-thinning behavior ($n=0.6$, $\lambda=10s$, $\rho_s/\rho=1.5$, $\phi=1\%-20\%$) led to enhanced settling velocity, it also suppressed horizontal aggregation. But still, bulk of the particles migrated away from the container walls and settled in the form of a vertical chain in the middle of the column similar to the experimental results of Mora et al. (2005) for a 20% suspension of 95 μm spheres ($\rho_s=3760\,kg/m^3$) in a power-law xanthan solution ($n=0.45$). Detailed flow visualization and concentration profiles obtained via Gamma ray attenuation for sand (4%) and calcium carbonate (9%) suspensions in aqueous xanthan solutions ($n=0.34$) further confirm the formation of vertical arrays of particles.

7.5 Conclusions

In this chapter, the gross flow characteristics of settling suspensions and bed expansion characteristics of a bed of spherical particles fluidized by power- law fluids have been considered. In particular, different methods available to predict the minimum fluidization velocity and velocity–voidage relationships for both fluidization and sedimentation have been examined. Based on the critical analysis of the literature data, it is safe to conclude that satisfactory empirical and theoretical methods are available for estimating the value of the minimum fluidization velocity for power-law liquids whereas velocity–voidage behavior of fluidization involving power-law fluids is well predicted by suitable modifications of the corresponding Newtonian formulae. It is obvious that the capillary model works well up to a limiting value of bed voidage, beyond which a submerged objects approach affords a somewhat better representation. The limiting value of the bed voidage delineating the region of the applicability of these two models is about 0.6–0.65, though it is somewhat dependent on the Reynolds number and the power-law index.

The settling of multiple particles and of concentrated suspensions in non-Newtonian liquids deviates significantly from their Newtonian analogs. For instance, model studies involving two, three, or four spheres show a tendency to segregate even in the creeping flow region whereas this type of effect is caused only by inertial effects in Newtonian fluids. The temporal variation of particles in concentrated systems also leads to the formation of inhomogeneous structures that may even be time-dependent. Thus, extreme caution should be exercised in using the Newtonian formulae to predict the rate of sedimentation when the suspending medium is non-Newtonian.

Very little is known about the role of visco-elasticity as well as about the other facets of liquid–solid fluidization including mixing patterns, detailed velocity profiles, etc. all of which strongly influence the mass/heat transfer processes and chemical reactions in a fluidized bed. The scant literature available on the sedimentation and fluidization of fibrous materials suggest their behavior to be significantly different from that of suspensions of spherical particles even at the gross macroscopic level. Finally, the limited available data for three-phase fluidized beds seem to suggest the non-Newtonian effects to be no more serious than that in the two-phase fluidized beds.

Nomenclature

- A_D: Constant, Equation 7.23a (-)
- Ar_n: Archimedes number, equation 7.18 (-)
- B_D: Constant, Equation 7.23b (-)
- C: Volume fraction of solids (= $1-\varepsilon$) (-)
- C_D: Drag coefficient (-)

Fluidization and Hindered Settling

C_{Dm}: Modified drag coefficient, Equation 7.22 (-)
d: Particle diameter (m)
d^+: Normalized particle diameter, Equation 7.16 (-)
D: Tube diameter (m)
g: Acceleration due to gravity (ms^{-2})
Ga: Galileo number, Equation 7.4 (-)
L: Height of packed bed (m)
m: Power-law consistency index (Pas^n)
n: Flow behavior index (-)
Δp: Pressure drop (Pa)
Re: Reynolds number for Newtonian fluids (-)
Re′: Reynolds number for power-law fluids (-)
Re$_p$: Particle Reynolds number, Equation 7.19 (-)
Re$_t$: Reynolds number based on terminal falling velocity of sphere (-)
T: Tortuosity factor (-)
V: Superficial velocity (ms^{-1})
V^+: Normalized superficial velocity, Equation 7.15 (-)
V_t: Free settling velocity (ms^{-1})
Y: Drag correction factor (= C_DRe′/24) (-)
Z: Fluidization and sedimentation index, Equation 7.7 (-)

Greek Symbols

$\dot{\gamma}_{mean}$: Mean shear rate for a sphere, Equation 7.24 (s^{-1})
$\dot{\gamma}_{wall}$: Pore wall shear rate, Equation 7.20 (s^{-1})
ε: Mean voidage or holdup (-)
μ: Newtonian viscosity (Pa s)
ρ: Fluid density (kg m^{-3})
ρ_s: Solid density (kg m^{-3})
λ: Parameter in Carreau viscosity model, Equation (1.17) (s)
ϕ: Volumetric fractional concentration of solid (-)
$\Delta\rho$: Density difference (kg m^{-3})
τ_H: Hydraulic tortuosity, Equation 7.21 (-)

Subscripts

G: Gas
L: Liquid
mf: Minimum fluidization
∞: Absence of wall effects

8

Heat and Mass Transfer in Particulate Systems: Forced Convection

8.1 Introduction

Most unit operations and processing steps encountered in handling and processing of rheologically complex systems involve nonuniform temperature and concentration fields, thereby resulting in the net transport of heat (or mass) from a region of high temperature (or concentration) to that of a low temperature (or concentration) within the flow domain. Furthermore, there are instances (such as devolatilization and vaporization of solvents, etc.), which entail simultaneous heat and mass transfer. In either event, the conservation equations are coupled, thereby adding further to the complexity of the analysis. Consequently, considerable research effort has hitherto been devoted toward developing a better understanding of heat and mass transfer phenomena in non-Newtonian systems. In general, heat transfer processes occurring in a variety of geometries and under different conditions of practical interest have been investigated much more extensively than analogous mass transfer processes. From a cursory inspection of the available reviews (Metzner, 1956, 1965; Skelland, 1966, 1967; Porter, 1971; Astarita and Mashelkar, 1977; Irvine and Karni, 1987; Chhabra, 1993a,b; Ghosh et al., 1994; Chhabra, 2011), it is abundantly clear that the heat and mass transfer processes occurring in external flows (e.g., between a fluid and solid particles and stationary or moving non-Newtonian media) have received much less attention than that accorded to these phenomena in internal or confined flows, such as in pipes and slits (Polyanin and Vyaz'min, 1995). Even some of the available books barely touch upon this subject (Bird et al., 1987a; Chhabra and Richardson, 2008; Chhabra 2018; Tanner, 2000; Morrison, 2001; Polyanin et al., 2002). Excellent reviews of the free and forced thermal convection in non-Newtonian systems with special reference to boundary layer flows have appeared in the literature (Shenoy and Mashelkar, 1982; Nakayama, 1988; Shenoy, 1988, 2017; Patel et al., 2017, 2018). No such review for mass transfer in non-Newtonian systems however seems to be available.

In this chapter, consideration is given to the phenomena of heat and mass transfer occurring between solid and fluid particles and flowing non-Newtonian media such as that encountered in bubble columns, fixed and fluidized beds and three-phase fluidized bed systems, sparged reactors, boundary layer flows, etc. Further examples are found in emulsion polymerization, the production of polymer alloys via liquid-phase route, devolatilization of films, degassing of polymeric and other multiphase systems, food processing applications, etc. By analogy with the developments for Newtonian fluids, heat and mass transfer to and from single particles of highly idealized shapes, such as a flat plate, a sphere, and a cylinder, are believed to provide the necessary background for understanding the analogous but more complex phenomena in multiparticle systems. From a theoretical standpoint, one must solve the momentum and thermal energy (or species continuity) equations to obtain the velocity and temperature (or concentration) fields, which in turn can be used to infer the values of the Nusselt or Sherwood numbers as functions of the pertinent dimensionless parameters. The coupling between the flow and temperature (or concentration) fields stems from three sources: temperature-dependent thermophysical and rheological properties, temperature (or concentration)-dependent density, and viscous dissipation effect. Hence, for Newtonian fluids, the momentum and thermal energy equations can be decoupled from each other if one assumes the thermophysical properties (density, viscosity) to be temperature- or concentration-independent and when there is negligible viscous dissipation. This applies to the flow of non-Newtonian fluids also.

DOI: 10.1201/9780429260759-9

Additional complications may arise if the thermophysical properties (such as thermal conductivity) are shear rate-dependent, for example, see Lee and Irvine (1997), Chhabra (1999b), Lin et al. (2003), and Broniarz-Press et al. (2007b). The limited available data seem to suggest a weak positive influence of shear rate on thermal conductivity. Similarly, much confusion exists in the literature regarding the effect of temperature on the rheological properties, especially yield stress and elasticity (Poole, 2018). However, in this work, both the thermophysical and rheological properties are assumed to be independent of both temperature and shear rate unless stated otherwise. Even with these simplifying assumptions, general solutions of the governing equations are not at all possible even for Newtonian fluids, let alone for non-Newtonian liquids. But nonetheless, significant advances have been made in this area that is broadly based on either boundary layer approximation, or other approximations such as low Reynolds number and low Peclet number or low Reynolds number and high Peclet number conditions, etc. In recent years, numerical solutions of the complete field equations have also been reported for a few cases. Furthermore, theoretical/numerical predictions have been supplemented by experimental developments, thereby resulting in a reasonable body of knowledge in this field. It is also convenient to treat the different modes of heat transfer with and without buoyancy effects separately. Thus, pure forced convection is treated in this chapter whereas the free- and mixed convections are dealt with in Chapter 9. Combined together, these two chapters provide an overview of the current state of the art in this area by way of presenting representative theoretical treatments and the reliable empirical correlations available in the literature, especially for a sphere, cylinder, and plate.

In keeping up with the practice followed in the preceding chapters, we begin by providing a concise summary of the theoretical and numerical studies available in this field (Table 8.1). The contents of this table are limited primarily to highly idealized shapes such as plates (vertical or horizontal), circular and square cross-section cylinders, and spheres. The corresponding information about the relevant experimental studies is provided in Table 8.2. Combined together, these two tables seem to suggest that a reasonable body of knowledge now exists in this field. The subsequent sections in this chapter will focus on analogous treatments for fluid spheres, packed and fluidized beds, tube bundles, etc.

An examination of Tables 8.1 and 8.2 shows that the bulk of the literature relates to the following aspects of the behavior of non-Newtonian fluids in external flows:

1. Extensive attempts have been made at developing appropriate frameworks for analyzing the momentum, thermal, and species transport in laminar boundary layers for inelastic non-Newtonian fluids. Notwithstanding the general applicability of some analyses, the results have been reported either for a flat plate or a circular cylinder or a sphere. The ultimate goal of these works is to be able to predict skin friction, Nusselt or Sherwood number as functions of the pertinent dimensionless parameters. Limited results are also available for turbulent boundary layers.
2. Over the past 15–20 years, there has been a spurt in numerical activity in studying external flows for finite values of the governing parameters, namely Reynolds and Prandtl (or Schmidt) numbers as opposed to the boundary layer treatments which implicitly assume "large values" of such parameters.
3. While adequate numerical results have been reported for forced convection heat transfer in visco-plastic fluids, analogous results for visco-elastic fluids are rather limited.
4. Limited experimental results are available for the three geometries, namely a flat plate (vertical and horizontal), a cylinder (horizontal or vertical), and a sphere. These results have often been correlated empirically, with virtually no cross-comparisons with other works.

Quite arbitrarily, the available body of information has been organized according to the geometry (plate, cylinder, or sphere).

TABLE 8.1

Summary of Theoretical and Numerical Non-Newtonian Fluids

Investigator	Geometry	Fluid Model	Observations
Acrivos et al. (1960, 1965)	Arbitrary two-dimensional shapes	Power-law	Approximate analysis of laminar momentum and thermal boundary layers for friction and heat transfer from isothermal surfaces.
Agarwal et al. (2002)	Thin needle	Power-law	Extensive numerical results for friction and heat transfer from isothermal and isoflux needles under laminar flow conditions.
Andersson and Toften (1989)	Two-dimensional surface	Power-Law	Points out the inadequacy of the results of Acrivos et al. (1960, 1965) for $n>1$.
Bharti et al. (2008)	Elliptic cylinders	Power-law	Numerical results in the steady flow regime.
Bharti et al. (2007a,b)	Cylinder	Power-law	Numerical predictions of Nusselt number for a single unconfined and confined cylinder.
Bizzell and Slattery (1962)	Two-dimensional and axisymmetric bodies	Power-law	Extended Von Karman–Pohlhausen integral method to study the role of the power-law index on the flow field.
Chauhan et al. (2021)	Sphere-in-a-tube	FENE - P	Effects of blockage and elasticity on heat transfer.
Chen and Kubler (1978)	Thin needles	Power-law	Laminar momentum boundary layer flow using a similarity transformation.
Chen and Radulovic (1973)	Wedge	Power-law	Laminar momentum boundary layer analysis.
Chhabra (1999a)	Horizontal plate	Power-law	Approximate integral analysis for heat transfer for isothermal and isoflux conditions.
Denier and Dabrowski (2004)	Horizontal plate	Power-law	Similarity solutions revisited and a finite width boundary layer is predicted for $n>1$.
Despeyroux et al (2010)	Sphere	Power-law	Effect of nonsymmetric confinement on heat transfer.
Dhole et al. (2006b)	Sphere	Power-law	Extensive numerical results on Nusselt number at finite Reynolds numbers.
El Defrawi and Finlayson (1972)	Horizontal plate	Power-law	Uses the analogy between the developing flow in a planar slit and the laminar boundary layer over a plate. The results for small values of n tend to be less accurate.
Fox et al. (1969)	Moving horizontal plate	Power-law	Laminar momentum and thermal boundary layer on a plate moving in a stagnant liquid. For weak non-Newtonian effects, exact and approximate results almost coincide.
Gupta and Chhabra (2016a)	Spheroids	Bingham plastic fluids	Extensive numerical results on Nusselt number over range of conditions.
Gupta et al. (2017b)	Disk	Bingham plastic fluid	Effect of Bingham number and thickness on heat transfer.
Hassanien (1996)	Flat plate	Power-law	Laminar momentum and thermal boundary layer analysis for a moving plate.
Howell et al. (1997)	Moving plate	Power-law	Employs the Merk-Chao series method to study momentum and heat transfer from a moving surface.
Hsu (1969)	Flat plate	Power-law	Used series expansion and steepest descent method to analyze laminar boundary layer.
Huang and Chen (1984, 1990); Huang and Lin (1992, 1993)	Flat plate	Power-law	Numerical analysis of laminar forced, free, and mixed convection from horizontal and vertical plates.

(Continued)

TABLE 8.1 (*Continued*)
Summary of Theoretical and Numerical Non-Newtonian Fluids.

Investigator	Geometry	Fluid Model	Observations
Kapur and Srivastava (1963)	Wedge and flat plate	Power-law	General treatment for laminar boundary layer flows.
Khan et al. (2006)	Horizontal cylinder	Power-law	Approximate boundary layer analysis which predicts heat transfer enhancement for $n<1$.
Khan et al. (2020)	Rotating cylinder in uniform flow	FENE-P	Effect of visco-elasticity, rotation, and polymer extensibility on Nusselt number.
Kim and Esseniyi (1993)	Rotating axisymmetric surfaces	Power-law	Enhanced heat transfer due to rotation is more significant for shear-thinning fluids than for dilatant fluids.
Kim et al. (1983)	Flat plate, sphere, horizontal cylinder	Power-law	Used Merk-Chao series method for forced convection heat transfer.
Kim and Lee (1989)	Horizontal cylinder	Power-law	Laminar thermal and momentum boundary layers at large Prandtl numbers. Results at large Pr are at variance with those of Shah et al. (1962).
Lal (1968)	Flat plate	Power-law	Similarity solution for momentum boundary layer flow.
Lemieux et al. (1971)	Flat plate	Power-law	Used the variational principle to obtain approximate results. The skin friction values are better than that based on Karman–Pohlhausen method but not as good as those of Acrivos et al. (1960).
Lin and Chern (1979)	Two-dimensional axisymmetric surfaces	Power-law	Used Merk-Chao series method for laminar momentum boundary layer flow. Limited results for spheres and cylinders.
Lin and Fan (1972)	Flat plate	Power-law	Used initial value method to solve the laminar boundary layer flow problems.
Lin and Shih (1980a,b)	Flat plate	Power-law	Forced convection from a horizontal static and moving plate and mixed convection from a vertical plate.
Luikov et al. (1966)	Flat plate	Power-law	General treatment for laminar boundary layers.
Luning and Perry (1984)	Flat plate	Power-law	Issues related to the convergence of numerical iterative methods are discussed, especially for $n>1$.
McDonald and Brandt (1966)	Flat plate	Power-law	Used the experimental velocity profile in a pipe to approximate the turbulent boundary layer flow.
Mishra et al. (1976)	Flat plate	Power-law	Laminar and turbulent mass boundary layer analysis using integral approach.
Mishra et al. (2018)	Cone	Bingham plastic fluid	Effect of orientation and cone angle on heat transfer.
Mishra et al. (2019d)	Disk	Power-law	Effect of the disk thickness on heat transfer.
Mitwally (1979)	Vertical plate	Power-law	Treatment for the far wake region behind a vertical plate.
Mizushina and Usui (1978)	Cylinder	Power-law	Approximate integral treatment for laminar forced convection heat transfer.
Myers (2010)	Plate	Power-law	Nusselt number results for a range of thermal boundary conditions on the plate.
Na (1994); Na and Hansen (1966)	Flat plate and wedge	Power-law and Reiner-Philippoff fluids	Possible similarity solutions.

(*Continued*)

TABLE 8.1 (*Continued*)

Summary of Theoretical and Numerical Non-Newtonian Fluids.

Investigator	Geometry	Fluid Model	Observations
Nachman and Taliaferro (1979)	Permeable flat plate	Power-law	Examines the specific forms of suction/injection velocity profiles for similarity solutions.
Nakayama et al. (1986); Nakayama and Koyama (1988)	Two-dimensional axisymmetric surfaces	Power-law	Developed a general approach to study laminar thermal boundary layer flows under a range of conditions.
Nirmalkar et al. (2013a,b)	Sphere	Herschel-Bulkley fluids	Extensive numerical results on Nusselt number in the steady axisymmetric regime.
Nirmalkar and Chhabra (2014)	Cylinder	Bingham plastic fluid	Numerical prediction of Nusselt number in the steady flow regime.
Oldroyd (1947)	Flat plate	Bingham Plastic	Obtained a similarity solution for momentum boundary layers.
Patnana et al. (2010)	Cylinder	Power-law	Numerical results on Nusselt numbers in the vortex-shedding regime.
Patel and Chhabra (2013, 2014; 2016b)	Elliptic cylinder	Bingham Plastic	Extensive numerical results on Nusselt number.
Piau (2002)	-	Bingham Plastic	General ideas on two-dimensional boundary layer flows.
Pimenta and Alves (2021)	Sphere	PTT model	Nusselt number for a sphere in the creeping flow with conduction inside the sphere. Under appropriate conditions, heat transfer increases in mildly elastic fluids.
Pittman et al. (1994, 1999)	Flat plate	Power-law	Extensive numerical and experimental study on laminar free and forced convection.
Ramteke and Kishore (2018)	Sphere	Power-law	Numerical study of effect of slip on heat transfer.
Rao et al. (1999)	Flat plate	Power-law	Used the Merk-Chao series method to analyze the laminar boundary layer over a plate moving in the direction opposite to that of fluid stream.
Reddy and Kishore (2013)	Sphere on the axis of a tube	Power-law	Numerical prediction of Nusselt number for $1 \leq Re_{PL} \leq 100$; $1 \leq Pr_P \leq 1000$; $0.4 \leq n \leq 1.8$; and $0.2 \leq \lambda \leq 0.5$. Heat transfer increases due to confinement.
Raja et al. (2015)	Plate	Bingham plastic	Numerical results on heat transfer.
Rotem (1966)	Flat plate	Power-law	Similarity solution valid in the range $0.5 \leq n \leq 1$.
Roy (1972)	Flat plate	Power-law	Points out the inadequacy of integral methods for small values of n.
Sahu et al. (2000)	Flat plate	Power-law	Used Merk-Chao method to obtain numerical results on friction and heat transfer from a moving plate.
Sasmal et al. (2013)	Hemisphere	Power-law	Effect of orientation on heat transfer.
Sasmal et al. (2020)	Cylinder	FENE-P	Effects of inertia and elasticity on heat transfer.
Schowalter (1960)	Two-dimensional and three-dimensional surfaces	Power-law	General framework for two-dimensional and three-dimensional boundary layers. For three-dimensional shapes, the possibility of similar solutions is strongly dependent on the fluid model.
Serth and Kiser (1967)	Two-dimensional surface	Power-law	Used the Görtler series method to treat laminar boundary layers.
Shenoy and Nakayama (1986)	Axisymmetric shapes	Power-law	General treatment for arbitrary shapes, and good match with the results of Bizzell and Slattery (1962) and of Lin and Chern (1979).

(*Continued*)

TABLE 8.1 (Continued)

Summary of Theoretical and Numerical Non-Newtonian Fluids.

Investigator	Geometry	Fluid Model	Observations
Skelland (1966)	Flat plate	Power-law	Integral momentum balance approach for turbulent boundary layer flow.
Soares et al. (2005a, 2010)	Circular cylinder	Power-law	Numerical results on Nusselt number in the steady flow regime and effect of temperature-dependent viscosity.
Srinivas and Ramesh (2014)	Spheroids	Power-law ($0.4 \leq n \leq 1.6$)	Numerical results on heat transfer.
Sreenivasulu et al. (2014)	Spheroids	Power-law	Numerical values of Nusselt number.
Tuoc and Keey (1992)	Flat plate	Power-law/ Bingham plastic/ Herschel-Bulkley fluids	Transform the startup flow of a moving plate into a boundary layer flow. Predictions consistent with those of Acrivos et al. (1960).
Van Atta (1967)	Two-dimensional surfaces	Power-law	Amended the analysis of Rotem (1966).
Weidman and van Atta (2001)	Two-dimensional surface	Power-law	Analysis of two-dimensional wake flow behind a slender symmetric object.
Wells, Jr. (1964)	-	Power-law	General ideas on similarity solutions.
Wolf and Szewczyk (1966)	Two-dimensional axisymmetric shapes	Power-law	Skin friction and heat transfer in laminar boundary layer flows.
Yao and Molla (2008)	Plate	Truncated power-law	Approximate boundary layer analysis.
Zhizhin (1987)	-	-	Some ideas on self-similar solutions.

TABLE 8.2

Summary of Experimental Studies on Heat and Mass Transfer from Plates, Spheres, and Cylinders Submerged in Non-Newtonian Liquids

Investigator	Geometry	Fluid Model	Observations
Al Taweel et al. (1978)	Rotating cylinders	Dilute solutions of PEO, Separan and CMC	Rate of mass transfer is reduced due to polymer addition.
Balasubramaniam and Sastry (1994a,b)	Spheres	CMC solutions ($0.60 \leq n \leq 0.86$)	Forced convection in slightly inclined vertical tubes with particles located on- and off-center positions.
Baptista et al. (1997)	Aluminum spheres	CMC solutions ($0.55 \leq n \leq 0.75$)	Good review of the previous literature and presented empirical correlations encompassing free and forced convection conditions.
Bhamidipati and Singh (1995)	Cylinder	CMC solutions	Extensive data and correlations for forced convection.
Chandarana et al. (1990)	Cube	Starch solutions ($0.78 \leq n \leq 1$)	Forced convection results.
Clegg and Whitmore (1966)	Knife edge	China clay suspensions ($15 \leq \tau_0^B \leq 54$ Pa)	Confirms the boundary layer thickness predictions of Oldroyd (1947).
Garg and Tripathi (1981)	Rotating horizontal cylinder	CMC and PVA solutions ($0.8 \leq n \leq 0.95$)	Boiling heat transfer data correlate with Peclet, Prandtl, and dimensionless rotation numbers.

(Continued)

TABLE 8.2 (*Continued*)

Summary of Experimental Studies on Heat and Mass Transfer from Plates, Spheres, and Cylinders Submerged in Non-Newtonian Liquids

Investigator	Geometry	Fluid Model	Observations
Ghosh et al. (1986a,b, 1992)	Vertical plate, crossflow over a cylinder and a sphere	CMC solutions ($n \sim 0.8$ to 0.85)	Extensive data and presented empirical correlations.
Hoyt and Sellin (1989)	Cylinder	Polyacrylamide solutions	Heat transfer is not influenced by dilute polymer solutions.
Hyde and Donatelli (1983)	Sphere	CMC solutions ($0.77 \leq n \leq 1$)	Creeping flow mass transfer results are under-predicted by the available analyses.
Keey et al. (1970)	Spheres suspended in mixing vessels	Wall paper paste solutions ($0.34 \leq n \leq 1$)	Empirical correlation for Sherwood number.
Kumar et al. (1980a, 1980b)	Spheres and cylinders	CMC solutions ($n = 0.77, 0.85$)	Empirical correlations for forced convection.
Kushalkar and Pangarkar (1995)	Spheres suspended in agitated vessels with and without air sparging	CMC solutions ($0.63 \leq n \leq 0.98$)	Empirical correlations for liquid–solid mass transfer.
Lal and Updhyay (1981)	Spheres in agitated vessels	CMC solutions ($0.68 \leq n \leq 0.89$)	Correlated mass transfer data using an effective viscosity.
Luikov et al. (1969a, 1969b)	Plate and cylinder	CMC solutions ($0.89 \leq n \leq 1$)	Extensive mass transfer data on forced convection.
McHale and Richardson (1985)	Horizontal cylinder	Polymer solution ($n = 0.92$)	Preliminary results on forced convection heat transfer.
Mishra et al. (1976)	Plate	CMC solutions ($n \sim 0.88$)	Correlations for Sherwood number in forced convection regime.
Mizushina and Usui (1975)	Horizontal cylinder in crossflow	Dilute PEO solutions	Reduction in heat transfer due to visco-elasticity.
Mizushina et al. (1978)	Horizontal cylinder in crossflow	CMC solutions ($0.72 \leq n \leq 0.96$)	Extensive results on pressure and velocity, separation angle, and Nusselt number.
Ogawa et al. (1984)	Sphere and cylinder	CMC and polyacrylamide solutions ($0.29 \leq n \leq 0.87$)	Empirical correlations for Sherwood number are presented.
Pittman et al. (1994, 1999)	Horizontal plate	CMC and Carbopol solutions ($n \sim 0.5$ to 1)	Experimental and numerical results for forced convection heat transfer.
Rao (2000a, 2003); Rao et al. (1996)	Horizontal cylinder in crossflow	Carbopol solutions	Empirical correlations for forced convection heat transfer.
Sabiri et al. (2012)	Cylinder	CMC solutions	Measurements of shear rate on the surface of cylinder.
Shah et al. (1962)	Cylinders	CMC solutions ($0.58 \leq n \leq 0.79$)	Temperature profiles and Nusselt number data.
Sobolik et al. (1994)	Cylinder	Polyacrylamide solutions	Visco-elasticity can augment the rate of heat transfer in the rear of the obstacle.
Takahashi et al. (1977, 1978)	Cylinder	CMC solutions ($0.78 \leq n \leq 0.91$)	Empirical correlations for forced convection heat transfer.
Zitoun and Sastry (1994)	Cubes	CMC solutions ($0.53 \leq n \leq 0.75$)	Correlations of forced convection data.
Zuritz et al. (1990)	Mushroom shaped aluminum particles	CMC solution ($n = 0.71, 0.95$)	Forced convective heat transfer data and correlation.

8.2 Boundary Layer Flows

Over the past 50 years or so, much effort has been directed at elucidating the influence of non-Newtonian flow characteristics on the hydrodynamics and the rate of convective transport in laminar boundary layers developed over submerged objects. At the outset, it is instructive to point out here that the bulk of the literature relates to the simple power-law fluid, with very little consideration for visco-plastic and viscoelastic fluids. Similarly, much more attention has been accorded to the study of momentum boundary layers than that of thermal and concentration boundary layers. In view of the generally high viscosity levels, the assumption of a high Prandtl (or Schmidt) number has invariably been made in most analytical/numerical studies. Therefore, the boundary layers tend to be thin, thereby justifying the curvature effects to be negligible.

8.2.1 Plates

Figure 8.1 shows the schematics of the flow for a two-dimensional boundary layer over a horizontal plate aligned with the direction of flow. For the standard boundary layer approximations (Schlichting, 1968; Schowalter, 1978), the field equations for the steady, incompressible, and laminar flow of an inelastic fluid (without yield stress) for constant thermophysical properties and in the absence of viscous dissipation and free convection are simplified as

Continuity equation:

$$\frac{\partial V_x}{\partial x} + \frac{\partial V_y}{\partial y} = 0 \tag{8.1}$$

x-momentum equation:

$$V_x \frac{\partial V_x}{\partial x} + V_y \frac{\partial V_y}{\partial y} = -\frac{\partial p}{\partial x} + \frac{1}{\rho} \frac{\partial \tau_{yx}}{\partial y} \tag{8.2}$$

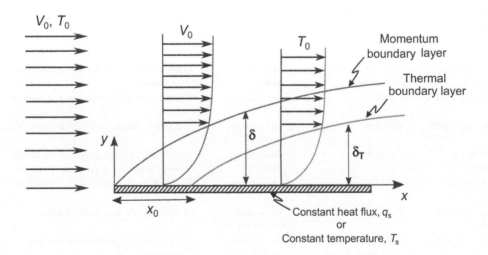

FIGURE 8.1 Schematics of the laminar boundary layer flow over a plate Pr >1.

Energy equation:

$$V_x \frac{\partial T}{\partial x} + V_y \frac{\partial T}{\partial y} = \left(\frac{k}{\rho C_p}\right) \frac{\partial^2 T}{\partial y^2} \tag{8.3}$$

The standard boundary conditions for this flow are that of no-slip at the solid surface, uniform flow, V_0, far away from the plate. Depending upon whether the immersed object is maintained at a uniform temperature or is subject to a constant heat flux, an appropriate boundary condition can be specified for this flow. Thus, the usual boundary conditions are

$$y = 0, V_x = 0, V_y = 0, T = T_s \text{ or } -k\frac{\partial T}{\partial y}\bigg|_{y=0} = q_s \text{ for } x \geq x_o \tag{8.4a}$$

$$x = 0, V_x = V_0, T = T_0 \tag{8.4b}$$

$$y \to \infty, V_x = V_0, T = T_0 \tag{8.4c}$$

where T_0 is the temperature of the fluid in the free stream and V_0 is the velocity in the x-direction. The boundary condition given by Equation 8.4c is often simplified by assuming that at $y=\delta$, $V_x=V_0$ and $y=\delta_T$, $T=T_0$.

For a specific viscosity model, one can write the stress component τ_{yx} in terms of the velocity gradient $(\partial V_x/\partial y)$. Since the bulk of the literature relates to the flow of power-law fluids,

$$\tau_{yx} = m\left(\frac{\partial V_x}{\partial y}\right)^n \tag{8.5}$$

Thus, the solutions of these equations will yield information about the velocity field $V_x(x, y)$ and the temperature field $T(x, y)$ within the boundary layer which in turn can be manipulated to deduce the values of the skin friction and Nusselt number. It is well known that the completely general solutions are hard to come by even for Newtonian fluids. Therefore, a variety of approximations have been introduced for the flow of power-law fluids.

The earliest and perhaps the simplest class of solutions are based on the straightforward extension of the integral momentum and energy balances (due to von Karman and Pohlhausen). Equations 8.2 and 8.3 can be integrated for a finite size control volume to obtain the following integral equations for momentum and energy.

$$\frac{d}{dx}\left[\int_0^{\delta_T} \rho(V_0 - V)V_x \, dy\right] = \tau_{yx}\bigg|_{y=0} \tag{8.6}$$

$$\frac{d}{dx}\left[\int_0^{\delta_T} V_x(T_0 - T) \, dy\right] = \alpha \frac{dT}{dy}\bigg|_{y=0} \tag{8.7}$$

Further progress can only be made by specifying the forms of $V_x(y)$ and $T(y)$ neglecting their dependence on x. The simplest choice is by using the same forms that have been found adequate for Newtonian fluids

(Schlichting, 1968). Besides, ample evidence now exists that the integral parameters like skin friction and Nusselt number are relatively insensitive to the detailed forms of $V_x(y)$ and $T(y)$, at least for the flow over a plate. The commonly used approximations are

$$\frac{V_x}{V_0} = \left(\frac{3}{2}\right)\left(\frac{y}{\delta}\right) - \frac{1}{2}\left(\frac{y}{\delta}\right)^3 \quad (8.8)$$

$$\frac{T - T_s}{T_0 - T_s} = \left(\frac{3}{2}\right)\left(\frac{y}{\delta_T}\right) - \frac{1}{2}\left(\frac{y}{\delta_T}\right)^3 \text{ for } x \geq x_o \quad (8.9)$$

The constant temperature, T_s, condition at the surface of the plate has been assumed here for $x \geq x_o$. For a flat plate, the skin friction coefficient C_f and the Nusselt number, respectively, are given as functions of the pertinent variables as Wu and Thompson (1996), Chhabra and Richardson (2008), and Chhabra (1999a).

$$C_f = \frac{\tau_{yx}\big|_{y=0}}{\frac{1}{2}\rho V_0^2} = C(n)\text{Re}_L^{-1/(n+1)} \quad (8.10)$$

$$C(n) = 2(n+1)\left(\frac{3}{2}\right)^n \left[\frac{280}{39}(n+1)\left(\frac{3}{2}\right)^n\right]^{n/(n+1)} \quad (8.11)$$

For the case when the plate is maintained at a constant temperature T_s, the local Nusselt number Nu_x is given by

$$\text{Nu}_x = \frac{hx}{k} = \phi(n)\text{Pr}_x^{1/3}\text{Re}_x^{(n+2)/3(n+1)} \quad (8.12)$$

The Prandtl number, Pr_x, is defined here as

$$\text{Pr}_x = \left(mC_p/k\right)\left(V_0/x\right)^{n-1} \quad (8.13)$$

and the function $\phi(n)$ is given as follows:

$$\phi(n) = \frac{3}{2}\left[\frac{30(n+1)}{(2n+1)}\left\{\frac{280}{39}\left(\frac{3}{2}\right)^n(n+1)\right\}^{1/(n+1)}\right]^{-1/3} \quad (8.14)$$

These approximate results are compared with the numerical results of Acrivos et al. (1960) for a flat plate in Tables 8.3 and 8.4, respectively.

The values of $C(n)$ due to Acrivos et al. (1960, 1965) are generally seen to be about 10% higher for $n<1$ and 10% lower for $n>1$ than the approximate values given by Equation 8.11. The corresponding comparison for heat transfer results is shown in Table 8.4 for the isothermal plate case. Included in this table are also the results of Nakayama and Koyama (1988). Once again, the approximate and numerical results are seen to be within 10% of each other. Finally, Chhabra (1999a) also presented an approximate analytical result for heat transfer when a constant heat flux is imposed at the surface of the plate.

TABLE 8.3
Values of $C(n)$ in Equation 8.11

n	Acrivos et al. (1960)	Equation 8.11
0.1	2.132	1.892
0.2	2.094	1.794
0.3	1.905	1.703
0.5	1.727	1.554
1.0	1.328	1.292
1.5	1.095	1.128
2.0	0.967	1.014
3.0	0.776	0.872

TABLE 8.4
Values of $\phi(n)$ in Equation 8.14

n	Acrivos et al. (1960)	Nakayama and Koyama (1988)	Equation 8.14
0.2	0.306	0.299	0.273
0.4	0.316	0.317	0.292
0.5	0.325	0.325	0.300
0.6	0.328	0.333	0.307
0.8	0.335	0.346	0.320
1.0	0.339	0.358	0.331
1.2	0.347	0.368	0.341
1.6	0.360	0.385	0.358
2.0	0.374	0.399	0.372
2.5	0.387	0.413	0.386
3.0	0.397	0.425	0.415

In this case, the local Nusselt number is given by the expression

$$\text{Nu}_x = \left(\frac{3}{2}\right) \sqrt{\frac{\frac{5}{2}\sqrt{\frac{8n+7}{6n+3}}}{\sqrt{\frac{2n+2}{2n+1}}}} \phi(n) \text{Pr}_x^{1/3} \text{Re}_x^{(n+2)/3(n+1)} \tag{8.15}$$

Table 8.5 compares the predictions of Equation 8.15 with the numerical results of Huang and Chen (1984) for $\text{Pr}_x = 10$. Once again, the two values are seen to be within 10% of each other. Notwithstanding the fact that the numerical predictions of the detailed velocity and temperature profiles are undoubtedly more accurate than the crude approximations of Equations 8.8 and 8.9, suffice it to add here that for the purpose of process engineering calculations, the closed-form expressions given by Equations 8.10, 8.14, and 8.15, are probably quite adequate. Besides, most of the other results available in the literature for a flat plate are comparable to those listed in Tables 8.3–8.5 (Berkowski, 1966; Wolf and Szewczyk, 1966; Lemieux et al., 1971; El Defrawi and Finlayson, 1972; Roy, 1972; Andersson and Toften, 1989; Kim and Lee, 1989; Pittman et al., 1994). However, the accuracy of the integral balance approach deteriorates rapidly for $n < \sim 0.3$, as is also seen in Tables 8.3–8.5 (Andersson, 1988; Huang and Lin, 1992; Liao, 2005).

The second approach to the solutions of the boundary layer equations for power-law fluids relies on the identification of a suitable similarity variable. This issue has been studied extensively, among others, by Schowalter (1960), Kapur and Srivastava (1963), Wells, Jr. (1964), Lee and Ames (1966), Berkowski (1966), Van Atta (1967), Lal (1968), Chen and Radulovic (1973), and Zhizhin (1987). For instance,

TABLE 8.5
Values of $[Nu_x Re_x^{-1/(n+1)}]$ for a Plate Under Constant Heat Flux Condition

n	(x/L)	Huang and Chen (1984)	Equation 8.15
0.5	0.01	0.602	0.556
	0.1	0.781	0.718
	1	1.01	0.928
1.5	0.01	1.382	1.389
	0.1	1.185	1.192
	1	1.017	1.022

Schowalter (1960) identified the types of potential flows for the existence of a similar solution for a two-dimensional case. On the other hand, for three-dimensional boundary layers, the possibility of a similar solution is strongly influenced by the form of the non-Newtonian constitutive equation. The predictions based on this approach are usually not too different from that of the integral method outlined in the preceding section.

Apart from these two approaches, many other approximations including the Merk-Chao series method (Lin and Chern, 1979; Howell et al., 1997; Rao et al., 1999), the local and pseudo-similarity transformation methods (Haugen, 1968; Lin and Shih, 1980a,b; Huang and Chen, 1984), a variational approach (Lemieux et al., 1971), the Görtler series method (Serth and Kiser, 1967), the initial value method (Lin and Fan, 1972), the series expansion coupled with steepest descent method (Hsu, 1969), and an iterative scheme (Luning and Perry, 1984), etc., have also been used to numerically solve the laminar boundary layer equations.

Mishra et al. (1976) used the velocity profile of Equation 8.8 and the concentration profile given by Equation 8.9 to study convective mass transfer from a flat plate and obtained the following expression for Sherwood number averaged over the length of the plate:

$$Sh_L = C_0(n)\left(Re_L^*\right)^{1/(n+1)}\left(Sc_2^*\right)^{1/3} \tag{8.16}$$

where

$$C_0(n) = \frac{3}{2}(n+1)\left(\frac{3}{2n+1}\right)^{2/3}(4.64)^{-2/(n+1)}\left(\frac{2}{n+1}\right)^{1/(n+1)} \tag{8.17}$$

Both Ghosh et al. (1986a) and Luikov et al. (1969a,b) have recast the heat transfer results of Acrivos et al. (1960) in terms of the equivalent mass transfer case. While the expression of Acrivos et al. (1960) is of the form similar to Equation 8.16, the one due to Luikov et al. (1969a,b) is of the following form:

$$Sh_L = \left(0.474 + 0.436n - 0.12n^2\right)\left(Re_L\right)^{1/(n+1)}\left(Sc_2\right)^{1/3} \tag{8.18}$$

Luikov et al. (1969a,b) reported the agreement of both Equations 8.16 and 8.18 with their limited data to be reasonable.

While most analyses assume the plate to be infinitely wide, it is not possible to satisfy this condition in experiments and the effect of (B/L) needs to be accounted for, at least at low Reynolds numbers. Thus, based on the literature and their own data, Ghosh et al. (1986a) developed the following correlation:

$$Sh_L = 3.23\left(Re_L^* Sc_L^* \frac{L}{B}\right)^{1/3} \qquad 0.2 \leq Re_L^* \leq 100 \tag{8.19a}$$

Heat and Mass Transfer in Particulate Systems

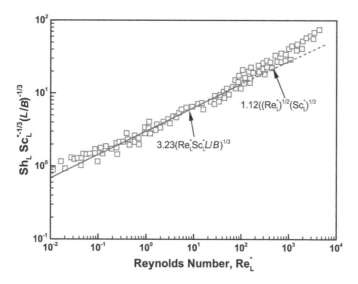

FIGURE 8.2 Correlation of mass transfer to power-law (0.89 ≤ n ≤ 1) fluids from flat plates of different lengths from 10 to 90 mm in accordance with Equation 8.19.

and

$$\mathrm{Sh}_L = 1.12 \left(\mathrm{Re}_L^*\right)^{1/2} \left(Sc_L^*\right)^{1/3} \quad 100 \leq \mathrm{Re}_L^* \leq 5000 \quad (8.19b)$$

Equation 8.19 is based on a wide range of values of the Schmidt number (840 ≤ Sc_L^* ≤ 2×10^6), but unfortunately is limited to only weakly shear-thinning fluids (0.89 ≤ n ≤ 1). Figure 8.2 shows the overall correlation for mass transfer from plates of different lengths, 10 ≤ L ≤ 90 mm.

Analogous treatments for a continuously moving plate in flowing or stagnant power-law fluids have been studied among others by Fox et al. (1969), Lin and Shih (1980b), Gorla (1992), Rao et al. (1999), Sahu et al. (2000), and Zheng and Zhang (2002). Similarly, the role of suction/injection on the laminar boundary layer of power-law fluids has been studied by Nachman and Taliaferro (1979) and Rao et al. (1999). The possibility of a similar solution for Reiner–Philipoff model fluids has been investigated by Na (1994). Turbulent boundary layers over a plate have been studied by McDonald and Brandt (1966), Skelland (1966), and Mishra et al. (1976). In all the three studies, turbulent velocity profiles near the wall in circular tubes have been employed, thereby yielding nearly identical predictions. On the other hand, Pittman et al. (1994) have presented a finite element study of heat transfer based on the solution of full field equations for a plate (with constant heat flux) immersed in power-law fluids over the range of conditions as 30 ≤ $\mathrm{Re}_{(pipe)}$ ≤ 2000 and 50 ≤ Pr ≤ 4000.

8.2.2 Cylinders

The general framework developed by Acrivos et al. (1960, 1965) and Shah et al. (1962) to treat convective transport in two-dimensional boundary layers is also applicable to the specific case of a circular cylinder oriented normal to the direction of flow. The resulting expression for the surface averaged Nusselt number is of the form similar to that for a plate, except for the fact that the diameter is used as the characteristic linear dimension instead of the plate length, that is,

$$\mathrm{Nu}_1 = C_1(n) \mathrm{Re}_{PL}^{1/n+1} \mathrm{Pr}_{PL}^{1/3} \quad (8.20)$$

where the constant $C_1(n)$ is a function of the flow behavior index. Subsequently, these predictions have been substantiated by numerous other studies based on a variety of approximations and solution procedures (Wolf and Szewczyk, 1966; Serth and Kiser, 1967; Lin and Chern, 1979; Kim et al., 1983; Nakayama and Koyama, 1988; Wang and Kleinstreuer, 1988a,b; Kim and Lee, 1989; Meissner et al., 1994; Khan et al., 2006). While cross-comparisons show the internal consistency of the aforementioned analyses, there is a general propensity for the results to diverge for small values of the power-law index and of the Prandtl number. While most non-Newtonian liquids exhibit large values of the Prandtl number, many polymer melts and solutions can have as small a value of the flow behavior index as 0.2–0.3. Under these conditions, these predictions must be treated with reserve. Furthermore, Mizushina and Usui (1978) and Mizushina et al. (1978) have criticized the aforementioned boundary layer analyses due to the fact that these suffer from the so-called zero velocity defect at the separation and stagnation points. They obviated this difficulty by combining the approaches of Karman–Pohlhausen (Schlichting, 1968) for the momentum boundary layer and that of Dienemann (1953) for the thermal boundary layer to obtain approximate results for power-law and Powell–Eyring model fluids.

In parallel, there have been experimental developments in this area. For instance, Shah et al. (1962), Mizushina et al. (1978), and Mizushina and Usui (1978) all reported good agreements between the measurements and the predictions of the local heat/mass transfer coefficients, especially for large values of Prandtl number. Based on these data, the constant in Equation 8.20 for a circular cylinder may be approximated as

$$C_1 = 0.72(n)^{-0.4} \tag{8.21}$$

Mizushina and Usui (1978) also reported a similar correlation for heat transfer to Powell–Eyring model fluids. On the other hand, Luikov et al. (1969a,b) reported that the separation point shifted downstream to $\theta = 155°$ for the crossflow of power-law fluids. While their results generally conform to the functional dependence of Equation 8.20, the value of $C_1(n)$ was reported to range from 0.31 to 0.5 as the value of n dropped from 1 (Newtonian) to 0.88. Similarly, the value of the exponent of the Reynolds number in Equation 8.20, that is, the value of $1/(n+1)$ varied from 0.52 to 0.39, thereby suggesting a trend which is clearly at variance with that predicted by the boundary layer analyses as discussed in Section 8.2. Similar conclusions have also been reached by Takahashi et al. (1977, 1978) who reported the exponent of the Reynolds number closer to (1/3) than to $1/(n+1)$ in the range $0.78 \leq n \leq 1$. Kumar et al. (1980a) have elucidated the influence of confining walls and of the length-to-diameter ratio on mass transfer from circular cylinders in carboxymethyl cellulose solutions. Based on the notion of an effective viscosity (corrected for wall effects), Kumar et al. (1980a) presented an empirical correlation for mass transfer from circular cylinders in crossflow configuration which also correlated with the heat transfer data of Mizushina and Usui (1978) satisfactorily. However, the available literature on convective heat/mass transfer to power-law fluids from circular cylinders oriented normal to the direction of flow has been reviewed, among others by Ghosh et al. (1986b, 1994) and Chhabra (1999b, 2011). Extensive numerical results on skin friction and forced convection heat transfer that are based on the solution of the boundary layer equations from a thin needle oriented parallel to the direction of flow in Newtonian (Eckert and Shadid, 1989), in power-law fluids (Chen and Kubler, 1978; Agarwal et al., 2002), and in Carreau fluids (Shadid and Eckert, 1992) are also available in the literature.

Irrespective of the definitions of the Reynolds and Prandtl numbers employed in the literature, the most predictive correlations for the Nusselt or Sherwood number for convective transport to power-law fluids from a heated circular cylinder are of the following form:

$$\mathrm{Nu}_1 \left(\text{or } \mathrm{Sh}_1 \right) = A \, \mathrm{Re}_{\mathrm{PL}}^B \left(\mathrm{Pr}_{\mathrm{PL}} \text{ or } \mathrm{Sc}_{\mathrm{PL}} \right)^{1/3} \tag{8.22}$$

For instance, for the definitions used by Shah et al. (1962), the best values of the constants are: $A = 2.18$; $B = 1/3$ for $\mathrm{Re}_{\mathrm{PL}} < 10$ and $A = 0.759$; $B = 1/2$ for $\mathrm{Re}_{\mathrm{PL}} > 10$. The corresponding values according to the

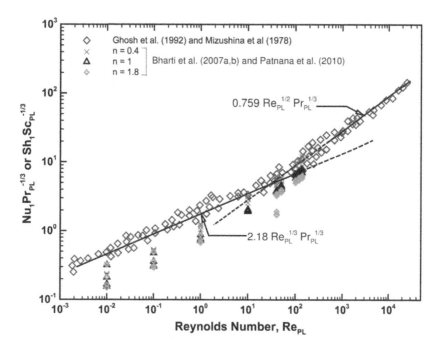

FIGURE 8.3 Overall correlation of heat transfer from a cylinder in crossflow configuration in power-law fluids. (Numerical results from Bharti, R.P., Chhabra, R.P., Eswaran, V., *Int. J. Heat Mass Transfer*, **50**, 977, 2007 and Patnana, V.K., Bharti, R.P., Chhabra, R.P., *Int. J. Heat Mass Transfer*, **53**, 19, 2010.) Experimental results from Ghosh et al. (1992) and Mizushina et al. (1978). (Modified from Ghosh, U.K., Upadhyay, S.N., and Chhabra, R.P., *Adv. Heat Transfer*, **25**, 251, 1994.)

effective viscosity approach of Ghosh et al. (1986b) are: $A=2.26$; $B=1/3$ for $Re_{PL} < 10$ and $A=0.785$; $B=1/2$ for $Re_{PL} > 10$. Figure 8.3 shows the overall correlation using the definitions of Shah et al. (1962), for both mass transfer and heat transfer data where the maximum Reynolds number is seen to be of the order of 25,000. Interestingly, included in this figure are also the data for $n=1$ whence Equation 8.22 successfully brings together the data for Newtonian and power-law fluids. At high Reynolds numbers, while the values of B are in line with the experimental results of Luikov et al. (1969a,b) and Takahashi et al. (1977), the values of A vary from one study to another. In the range of conditions as $40 \leq \rho V^{2-n} d^n / m = Re_{PL} \leq 4000$, $0.784 \leq n \leq 0.914$, Takahashi et al. (1978) reported a value of $A=0.7$. In recent years, there has been some activity on the flow and heat transfer from long squares and rectangles immersed in power-law fluids, in confined and unconfined conditions (Gupta et al., 2003; Paliwal et al., 2003; Nitin and Chhabra, 2005b). In the range of conditions ($5 \leq Re_{PL} \leq 40$; $1 \leq Pe_{PL} \leq 400$) and $0.6 \leq n \leq 1.4$, heat transfer is facilitated by shear-thinning ($n<1$) behavior and, as expected, it is impeded in shear-thickening ($n>1$) fluids.

In a series of papers, Rao et al. (1996) and Rao (2000a, 2003) have studied heat transfer from a circular cylinder (maintained at a uniform heat flux) to polymer solutions, some of which were found to show viscoelastic behavior. The characteristic viscosity evaluated at a mean shear rate of ($2V/d$) was used to calculate the pertinent values of the Reynolds and Prandtl numbers. Here, V is the area-averaged approach velocity, that is, the volumetric flow rate divided by the cross-section area of the duct in which the test cylinder is mounted. With these definitions, he reported his data to correlate well with the available equations for Newtonian fluids. Similarly, by neglecting the internal thermal resistance (Biot number $\ll 1$), many investigators (McHale and Richardson, 1985; Chandarana et al., 1990; Sastry et al., 1990; Zuritz et al., 1990; Awuah et al., 1993; Aström and Bark, 1994; Balasubramaniam and Sastry, 1994a,b; Zitoun and Sastry, 1994; Bhamidipati and Singh, 1995; Baptista et al., 1997; Mankad et al., 1997; Alhamdan and Sastry, 1990, 1998; Ramaswami and Zareifard, 2000) have reported the values of convective heat transfer coefficients from variously shaped objects, as encountered in food processing applications. Unfortunately, neither any of these correlations have been tested extensively nor are the complete details available for these results to be recalculated in the form to allow comparisons with other studies. The problem of making cross-comparisons is further compounded due to the differences in geometries used by different investigators. For instance, the value of the blockage ratio in the studies of Shah et al. (1962), Mizushina et al. (1978), and of Rao (2000a) are 0.25, 0.083, and 0.48, respectively, thereby making cross-comparisons virtually impossible and also unjustified. Similarly, in some of the aforementioned studies, rheological measurements are made at or near room temperature while the heat transfer tests entail the use of liquids above room temperature, thereby casting doubt about the universal applicability of the correlations based on such rheological characteristics. Finally, by blocking the surfaces of a cylinder, it is possible to delineate the contributions of the individual rates of mass transfer from the front-, lateral-, and rear-end surface area to the overall rate of mass transfer. This approach has been successfully employed by Parmaj et al. (1989), Lohia et al. (1995), and by Venkatesh et al. (1994) for cylinders oriented parallel and transverse in streaming Newtonian liquids (water), but it has not been extended to non-Newtonian liquids to delineate the relative contributions of the individual surfaces of the cylinder.

Over the past 20 years or so, significant effort has been expended in seeking complete numerical solutions to the energy equation for the flow of power-law and Bingham plastic fluids past cylinders of different cross-sections. While the corresponding momentum transfer aspects have been discussed in Chapters 2–4, forced convection heat (mass) transfer is discussed here. For instance, Soares et al. (2005a) and Bharti et al. (2007a) have reported extensive results on the distribution of the local Nusselt number along the surface of the cylinder and the average Nusselt number as a function of the Reynolds and Prandtl numbers, and flow behavior index for an isothermal and isoflux cylinder. However, both these studies are restricted to the steady axisymmetric flow regime (Sivakumar et al., 2006).

Figures 8.4 and 8.5 show representative results on the variation of the local Nusselt number along the surface of the cylinder for a range of values of Prandtl number, power-law index, and for two values of the Reynolds number, $Re_{PL}= 5$ and 40, and for the two standard boundary conditions, namely, constant temperature (CWT) and constant heat flux (CHF) prescribed on the surface of the cylinder. The positive relationship between the Nusselt number and Reynolds and Prandtl number is in line with the scaling arguments. However, the peak value of the Nusselt number is seen to be displaced from the front stagnation point in shear-thinning fluids. This is due to the two competing mechanisms, that is, the variation of the viscosity (depends on the value of the power-law index) and the temperature gradient (or temperature for CHF case) along the surface. Other than this complexity, the trends seen in these figures are qualitatively similar to that seen for Newtonian fluids. Overall, shear-thinning behavior facilitates heat transfer and shear-thickening tends to suppress it.

Subsequently, these results have been extended to the laminar vortex-shedding regime (Patnana et al., 2009, 2010). Overall, numerical results for the average Nusselt number for a circular cylinder in power-law fluids are available over the range of conditions as: $0.2 \leq n \leq 1.8$; $0.01 \leq Re_{PL} \leq 140$; $1 \leq Pr_{PL} \leq 100$. By analogy with the trends seen in Newtonian fluids, it is convenient to introduce the Colburn j-factor defined as:

$$j = \frac{Nu}{Re_{PL} Pr_{PL}^{1/3}} \quad (8.23)$$

Figure 8.6 shows the numerical results in terms of the j-factor versus Reynolds number (Re_{PL}) whereas the numerical results are also included in Figure 8.3 where the experimental results for mass transfer are seen to be consistently underpredicted. Possible reasons for this include wall effects and the continually changing size of cylinders due to their dissolution. Figure 8.7 shows a more direct comparison between the numerical predictions and the experimental results on mass transfer (Ghosh et al., 1994). The deviations seem to be within the error band of experimental results.

The role of blockage on forced convection heat transfer in power-law fluids from a symmetrically confined cylinder in between two parallel walls has been investigated, among others by Bharti et al. (2007c,d) and Rao et al. (2011a) whereas the case of asymmetric confinement has been studied by Nirmalkar and

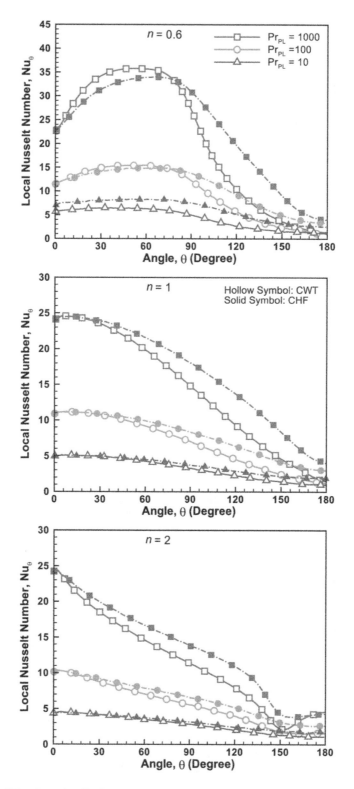

FIGURE 8.4 Local Nusselt number distribution on the surface of the cylinder for $Re_{PL}=5$, over the ranges of n and Pr_{PL}. (Hollow symbols: CWT, Solid symbols: CHF). (Numerical results are from Bharti, R.P., Chhabra, R.P., Eswaran, V., *Int. J. Heat Mass Transfer*, **50**, 977, 2007.)

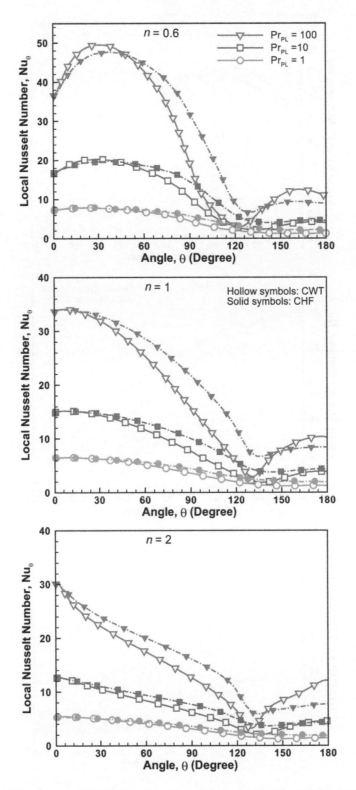

FIGURE 8.5 Local Nusselt number distribution on the surface of the cylinder for $Re_{PL}=40$, over the ranges of n and Pr_{PL}. (Hollow symbols: CWT, Solid symbols: CHF). (Numerical results are from Bharti, R.P., Chhabra, R.P., Eswaran, V., *Int. J. Heat Mass Transfer*, **50**, 977, 2007.)

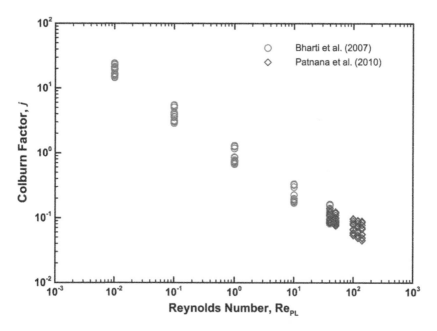

FIGURE 8.6 Dependence of the Colburn factor, j on the Reynolds number, Re_{PL} for an isothermal cylinder in power-law fluids. (Bharti, R.P., Chhabra, R.P., Eswaran, V., *Int. J. Heat Mass Transfer*, **50**, 977, 2007 and Patnana, V.K., Bharti, R.P., Chhabra, R.P., *Int. J. Heat Mass Transfer*, **53**, 19, 2010.)

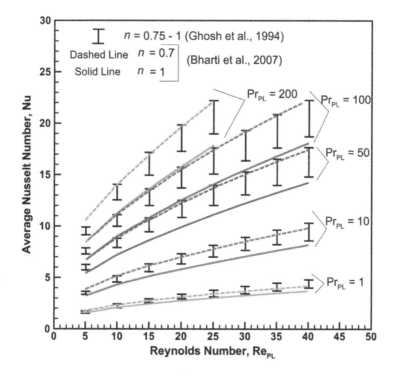

FIGURE 8.7 Comparison between numerical results of heat transfer (from Bharti, R.P., Chhabra, R.P., Eswaran, V., *Int. J. Heat Mass Transfer*, **50**, 977, 2007) and mass transfer from cylinders in power-law fluids. (Ghosh, U.K., Gupta, S.N., Kumar, S., Upadhyay, S.N., *Int. J. Heat Mass Transfer*, **29**, 955, 1986.)

Chhabra (2012). Due to the sharpening of the temperature gradients by blockage, it is possible to further promote heat transfer under these conditions. Interactions between two cylinders confined in between two plates in power-law fluids have been studied by Patil et al. (2008a,b). Analogous results for other two-dimensional shapes on heat transfer in power-law fluids are also available in the literature. For instance, forced convection from a square cylinder has been studied by Dhiman et al. (2007a, 2008a), Sahu et al. (2009a,b,c, 2010b), Rao et al. (2011a), and Sasmal and Chhabra (2012c). Analogous results for a 45°-tilted square cylinder have been reported by Rao et al. (2011b) and for semi-circular cylinders by Chandra and Chhabra (2011a,b,c, 2013), Tiwari and Chhabra (2014b), etc. In each case, the role of the power-law index on the Nusselt number is qualitatively similar to that for a circular cylinder.

In recent years, there has been a spurt in activity on studying heat transfer from a cylinder in viscoplastic (mostly Bingham plastic fluids) systems. As noted in Chapter 3, one of the distinguishing features of such systems is that the flow domain is spanned by fluid-like (yielded) and solid-like (unyielded) regions that need not be contiguous. It thus stands to reason that convection will be limited to the fluid-like subdomains only and heat transfer would occur solely by conduction in unyielded parts. In steady state operations, conduction could thus be the overall limiting factor. However, most of the results available in the literature use the regularized Bingham fluid model wherein the unyielded fluid is assigned a very high viscosity as opposed to the infinite viscosity of a solid. Also, in most cases, the rheological parameters (yield stress and plastic viscosity) have been assumed to be independent of temperature; this assumption is open to criticism (Poole, 2018). Furthermore, the thermo-physical properties other than the viscosity are assumed to have the same values throughout the flow domain irrespective of whether the substance is yielded or not. Notwithstanding these limitations, Figures 8.8 and 8.9 show the variation of the local Nusselt number along the circumference of a circular cylinder for scores of values of Prandtl number (Pr_B) and Bingham number (Bi) for the two standard boundary conditions specified on the surface of the cylinder, namely CWT and CHF. At low Peclet numbers (= $Re_B Pr_B$), conduction dominates, and hence there is very little variation in the values of the local Nusselt number along the surface

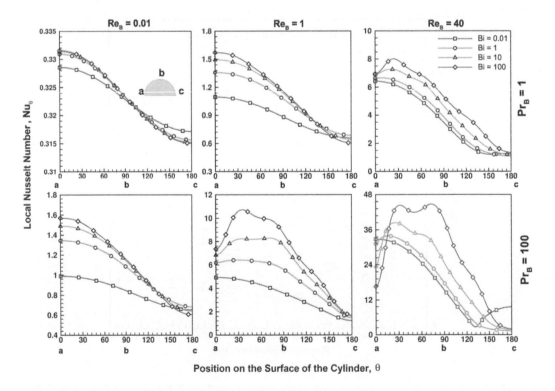

FIGURE 8.8 Local Nusselt number distribution on the surface of the cylinder over the ranges of Re_B, Bi and Pr_B for CWT. (From Patel, S.A., Chhabra, R.P., *Int. J. Heat Mass Transf.*, **73**, 671, 2014.)

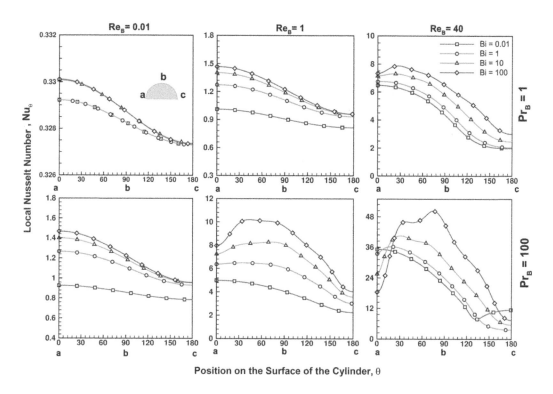

FIGURE 8.9 Local Nusselt number distribution on the surface of the cylinder over the ranges of Re_B, Bi and Pr_B for CHF. (From Patel, S.A., Chhabra, R.P., *Int. J. Heat Mass Transf.*, **73**, 671, 2014.)

of the cylinder and there is very little effect of Bi as well. With the increasing Peclet number, both these features appear prominently. It is useful to recall here that the fluid-like regions grow with the increasing Reynolds number whereas this tendency is countered by the Bingham number. The complex trends seen at $Re_B=40$ are determined by an intricate interplay between three boundary layers: momentum, temperature, and stress. Under appropriate conditions, it is thus possible that the fluid may be completely or partially yielded within the momentum and thermal boundary layers. On the other hand, for a fixed Bi, the Nusselt number bears a positive dependence on both Reynolds and Prandtl numbers. Also, the CHF boundary condition transfers heat more efficiently than the CWT boundary condition, in line with the trends observed in Newtonian fluids. Figure 8.10 shows the typical dependence of the surface average Nusselt number on the Reynolds, Prandtl, and Bingham numbers for the two boundary conditions. Evidently, the average Nusselt number bears a positive relationship with each of these parameters. Patel and Chhabra (2013) consolidated their numerical results in the form of the Colburn *j*-factor (=Nu/$Re_B Pr_B^{1/3}$) as:

$$j = a(1+\text{Bi})^b Re_B^c \tag{8.24}$$

where, $a=0.94$, $b=0.1$, $c=-0.640$ for CWT
$a=1.00$, $b=0.1$, $c=-0.605$ for CHF

This equation is applicable over the range of conditions as $1 \leq Re_B \leq 40$; $1 \leq Pr_B \leq 100$; and $0.01 \leq Bi \leq 100$. In contrast, over the range of conditions of $1 \leq Re_B \leq 40$; $1 \leq Pr_B \leq 100$; and $Bi \leq 10^4$, Nirmalkar and Chhabra (2014) proposed the following expression for heat transfer from an isothermal cylinder:

$$j^* = \frac{2.73}{\left(Re^*\right)^{2/3}} \tag{8.25}$$

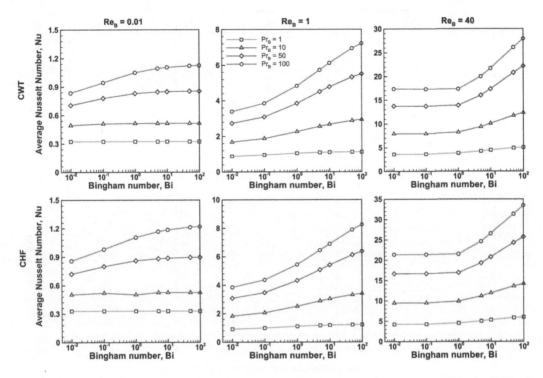

FIGURE 8.10 Dependence of Average Nusselt number of the cylinder over the ranges of Re_B, Bi and Pr_B for CWT and CHF. (From Patel, S.A., Chhabra, R.P., *Int. J. Heat Mass Transf.*, **73**, 671, 2014.)

where $Re^* = \dfrac{Re_B}{(1+Bi)}$ and $Pr^* = Pr_B(1+Bi)$, which are used to define j^*. Both predictions are within 1%–2% of each other. Two more observations are in order here: First, both Equations 8.24 and 8.25 suggest the scaling of Nu ~ $Pe^{1/3}$ for Bingham fluid also, as is observed in Newtonian fluids; second, in the limit of Bi=0, Re^*, Pr^*, j^* all coincide with their Newtonian forms.

Similar results on forced convection in yield stress fluids from some other two-dimensional shapes are also available, e.g., for a square cylinder (Nirmalkar et al., 2013c), a triangular cross-section cylinder (Prhashanna et al., 2011; Bose et al., 2014), elliptical cylinder (Patel and Chhabra, 2014), and a semi-circular cylinder (Tiwari and Chhabra, 2015b). The boundary layer flow created by an axially moving cylinder in a visco-plastic fluid has been studied by Boujlel et al. (2012). The combined effects of yield stress and flow pulsations on heat transfer from a circular cylinder have been studied by Zierenberg et al. (2007) and Gupta et al. (2020b, 2021) and for a semi-circular cylinder by Srivastava and Dhiman (2019). The effect of rotation on heat transfer from a cylinder has also been explored recently (Panda and Chhabra, 2010, 2011; Thakur et al., 2016, 2018, 2019a,b). Heat transfer from two circular cylinders in Bingham fluids has been studied numerically by Gupta et al. (2020a) whereas the corresponding case of two square cylinders has been studied by Shyam and Chhabra (2013a,b, 2014). Similarly, Soares et al. (2010) concluded that the temperature-dependent power-law viscosity further augments heat transfer over and above that for the constant properties case under otherwise identical conditions. Furthermore, the smaller the value of the power-law index, the stronger is this effect.

8.2.3 Spheres

Some attempts have been made to establish the role of power-law rheology on the rates of heat and mass transfer from/to a sphere, most of which are limited to the creeping flow regime and high Prandtl

TABLE 8.6
Expressions for Y_m of Mass Transfer from a Sphere to Power-Law Fluids in Creeping Flow Regime

Investigator	Expression for Y_m
Tomita (1959)	$0.865 n^{2/3}$
Slattery (1962)	$0.866 \left\{ \dfrac{3}{2} - 6 \left(0.293 + \dfrac{0.375}{n} - \dfrac{0.668}{n^2} \right) \right\}^{1/3}$
Wasserman and Slattery (1964)	$0.866 \left\{ 3.686 n^2 - 11.712 n + 9.3025 \right\}^{1/3}$
Acharya et al. (1976); Kawase and Ulbrecht (1981g)	$0.866 \left\{ \dfrac{3}{2} - \dfrac{9n(n-1)}{2n+1} \right\}^{1/3}$

(Schmidt) numbers, that is, the so-called thin boundary layer approximation introduced by Lochiel and Calderbank (1964). The Sherwood number, $\mathrm{Sh}_1 (= k_c d/D)$, is given by

$$\mathrm{Sh}_1 = 0.641 \mathrm{Pe}^{1/3} \left[\int_0^\pi \left(V_\theta^* \sin^3 \theta \right)^{1/2} d\theta \right]^{2/3} \qquad (8.26)$$

where V_θ^* is the dimensionless tangential velocity on the surface of the sphere. As seen in Chapter 2, a variety of stream functions have been used in conjunction with the velocity and stress variational principles to deduce the value of drag on a sphere. Indeed, Kawase and Ulbrecht (1981g) combined some of these stream functions with Equation 8.26 to estimate the rate of mass transfer from a sphere immersed in a streaming power-law fluid. The results are frequently expressed in terms of the ratio of $(\mathrm{Sh}_1/\mathrm{Pe}^{1/3})$ for a power-law to that for a Newtonian liquid, that is,

$$Y_m = \frac{\left(\mathrm{Sh}_1 \mathrm{Pe}^{-1/3} \right)}{\left(\mathrm{Sh}_1 \mathrm{Pe}^{-1/3} \right)_{n=1}} \qquad (8.27)$$

While Table 8.6 lists the analytical expressions for Y_m, Figure 8.11 shows the variation of mass transfer factor, Y_m, with the power-law index suggesting enhancement in mass transfer in the creeping flow regime, except for Tomita (1959). For $n = 1$, the expected limiting behavior of $Y_m = 1$ is reached only by the expressions due to Acharya et al. (1976) and Slattery (1962). Scant experimental results in this regime are due to Moo-Young et al. (1970) and Hyde and Donatelli (1983). The results of Moo-Young et al. (1970) are included in Figure 8.11 where the experimental data are grossly underestimated by all expressions listed in Table 8.6. Indeed, similar observations have been made by Hyde and Donatelli (1983) who attributed such underestimation to the free convection effects which are neglected altogether in the analysis.

The assumption of a high Peclet number was relaxed by Westerberg and Finlayson (1990) who solved the field equations for the creeping flow of Nylon-6 melt over a sphere. They assumed the thermophysical properties to be temperature-independent and also elucidated the role of several other parameters. Though no detailed results are presented in their paper, the Nusselt number was found to be influenced in decreasing order, respectively, by the value of the Peclet number, Brinkman number (viscous dissipation), shear-thinning, temperature-dependent viscosity, elasticity, and temperature-dependent thermal conductivity. The observation that the Nusselt number is little influenced by fluid visco-elasticity is also consistent with the other treatments developed for weakly elastic conditions (Sharma and Bhatnagar, 1975; Kawase et al., 1982). In order to increase the utility of their results, Westerberg and Finlayson (1990) presented the following predictive expressions (Re $\ll 1$; Pe < 1000):

$$\frac{1}{(\mathrm{Nu}_1 - 2)_0} = \frac{2}{\mathrm{Pe}} + \frac{1}{0.89 \mathrm{Pe}^{1/3}} \qquad (8.28a)$$

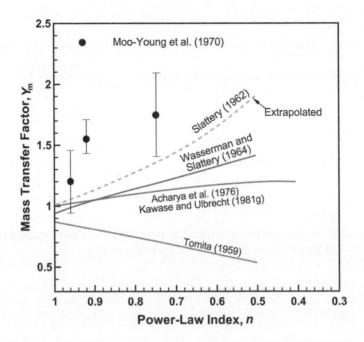

FIGURE 8.11 Theoretical predictions of mass transfer from a single sphere falling in power-law fluids in the creeping flow regime.

TABLE 8.7
Values of δ_1

Pe	δ_1
0.2	0.592
1	0.474
10	0.248
100	0.0865
1000	0.0287

and the effect of temperature-dependent properties is expressed as

$$\frac{(\mathrm{Nu}_1 - 2)_T}{(\mathrm{Nu}_1 - 2)_0} = \left(\frac{\mu_{T0}}{\mu_{Ts}}\right)^{C_0 + D_0 \log Pe} \tag{8.28b}$$

where $C_0=0.0774$ and $D_0=0.0408$ for $T_s > T_0$ (hot sphere) and $C_0=0.0848$ and $D_0=0.0377$ for $T_s < T_0$ (cold sphere) and $\mathrm{Nu}_1 = (1 \pm \delta_1 \mathrm{Br})\mathrm{Nu}_0$ where the numerical values of δ_1 are listed in Table 8.7. These results indicate the diminishing role of viscous dissipation with the increasing Peclet number.

In an interesting study, Morris (1982) has analyzed the flow patterns around a heated sphere undergoing steady translation (Re → 0) in Newtonian and power-law fluids with strongly temperature-dependent viscosity. Significantly, the drag showed the usual linear dependence on velocity in the absence of strong convection, but this dependence was seen to undergo a transition to the fourth power of velocity in the limit of large Peclet numbers, that is, under strong forced convection conditions.

Outside the creeping flow, the boundary layer treatments for heat/mass transfer from/to a sphere are also limited (Acrivos et al., 1960; Bizzell and Slattery, 1962; Lin and Chern, 1979; Nakayama et al., 1986;

Heat and Mass Transfer in Particulate Systems

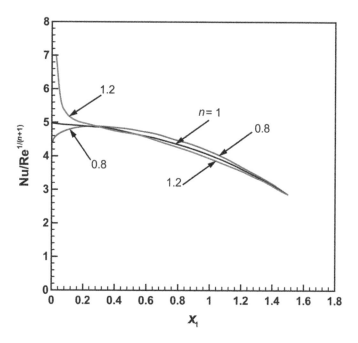

FIGURE 8.12 Effect of power-law index on the variation of Nusselt number on the surface of a sphere in forced convection regime. (Replotted from Nakayama, A., Shenoy, A.V., and Koyama, H., *Warme-und Stoffubertragung*, **20**, 219, 1986.)

Shenoy and Nakayama, 1986; Nakayama and Koyama, 1988). Almost all of these are based on the use of integral methods applied to power-law fluids at high Prandtl or Schmidt numbers. Figure 8.12 shows representative results on the role of the power-law index on the variation of the heat transfer group over the surface of an isothermal sphere. While for shear-thinning fluids ($n \sim 0.8$), the Nusselt number is seen to go through a maximum value, it is seen to decrease monotonically for a shear-thickening fluid ($n=1.2$). Aside from this approach, Kawase and Ulbrecht (1983a,b,d) employed the modified penetration model (Carberry, 1960; Mixon and Carberry, 1960) to develop a general framework for convective mass transfer to power-law fluids. For the specific case of a sphere, this analysis yields the following expression for Sherwood number:

$$\text{Sh}_1 = A_1(n) \text{Re}_P^{(n+2)/3(n+1)} \text{Sc}_P^{1/3} \tag{8.29}$$

The constant $A_1(n)$ is given by

$$A_1(n) = \frac{9(n+1)}{(2n+1)\sqrt{1/3}} \left[\frac{(n+2)}{6(n+1) \left[\frac{280}{39}(n+1)(1.5)^n \right]^{1/(n+1)}} \right]^{1/3} \tag{8.30}$$

For the limiting case of Newtonian fluid behavior, $n=1$, $A_1=0.847$. Interestingly, the analogous expression based on the assumption of a thin boundary layer approximation (Kawase and Ulbrecht, 1981a) is almost identical to Equation 8.29, except for the slightly different value of the constant. Equation 8.29 also predicts a slight enhancement in the value of the Sherwood number due to the shear-thinning fluid behavior, similar to that seen in Figure 8.11 in the creeping flow regime.

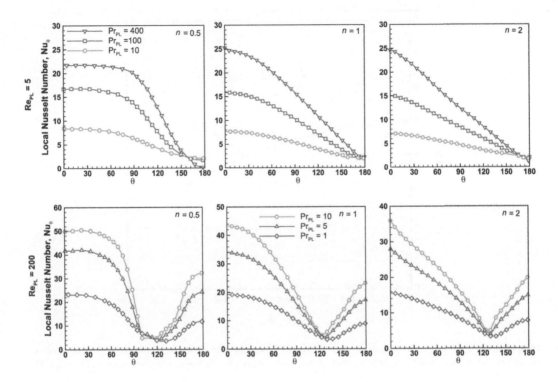

FIGURE 8.13 Local Nusselt number distribution on the surface of the sphere (CWT) over the ranges of Re_{PL}, n, and Pr_{PL}. (From Dhole, S.D., Chhabra, R.P., Eswaran, V., *AICHE J.*, **52**, 3659, 2006.)

Over the past 15–20 years, numerical solutions have been reported for heat transfer from a heated sphere in unconfined (Dhole et al., 2006b) and confined (Song et al., 2010, 2012; Despeyroux et al., 2010; Reddy and Kishore, 2013) power-law fluids. Most of these results are limited to the case of a steady axisymmetric flow regime and that of constant thermo-physical properties and are consistent with each other. Figures 8.13 and 8.14 represent results on the distribution of the local Nusselt number along the surface of the sphere for scores of values of Prandtl number, Reynolds number, and power-law index for the CWT and CHF conditions prescribed on the surface of the sphere. Broadly, the Nusselt number is maximum at the front stagnation point in every case and it continually decreases all the way up to the rear stagnation point at low Reynolds numbers wherein the flow does not detach itself from the sphere. At high Reynolds numbers, the adverse pressure gradient is established (Mishra and Chhabra, 2022) and the Nusselt number decreases from the front stagnation point to the point of separation and beyond this point, heat transfer exhibits some recovery up to the rear stagnation point due to enhanced mixing in this region. In shear-thinning fluids (Figure 8.13, $n=0.5$), the local Nusselt number varies little along the front half of the sphere whereas it decreases rather steeply in a shear-thickening fluid ($n=2$) and the results for a Newtonian fluid are in between these two trends. Qualitatively similar trends are seen in Figure 8.13 at $Re_{PL}=200$. Irrespective of the type of boundary condition, CWT or CHF, on the surface of the sphere, shear-thinning is seen to promote heat transfer with respect to that in a Newtonian fluid under otherwise identical conditions whereas shear-thickening behavior has a detrimental influence. This is seen much more clearly in Figure 8.15 where the dependence of the average Nusselt number on Re_{PL}, Pr_{PL}, and the power-law index (n) is shown. Based on their numerical results spanning the range of conditions as $5 \leq Re_{PL} \leq 200$, $1 \leq Pr_P \leq 400$, and $0.5 \leq n \leq 2$, Dhole et al. (2006b) put forward the following correlations for the average Nusselt number:

$$Nu_l = 0.922 + Re_{PL}^{1/(n+2)} Pr_P^{1/3} + 0.1 Re_{PL}^{2/3} Pr_P^{1/3} \qquad (8.31)$$

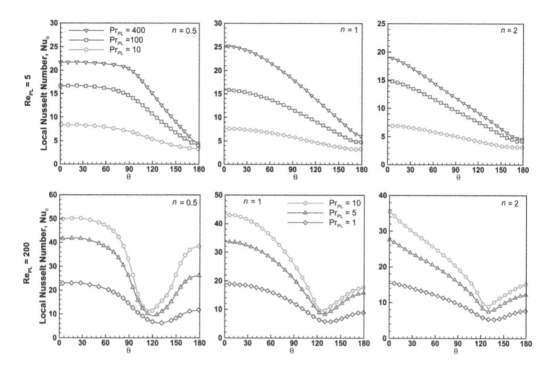

FIGURE 8.14 Local Nusselt number distribution on the surface of the sphere (CHF) over the ranges of Re_{PL}, n, and Pr_{PL}. (From Dhole, S.D., Chhabra, R.P., Eswaran, V., *AICHE J.*, **52**, 3659, 2006.)

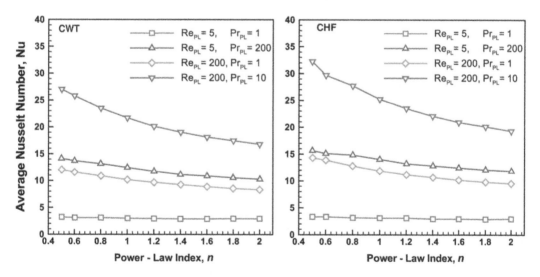

FIGURE 8.15 Average Nusselt number dependence for the sphere over the ranges of Re_{PL}, n, and Pr_{PL}. (From Dhole, S.D., Chhabra, R.P., Eswaran, V., *AICHE J.*, **52**, 3659, 2006.)

for the CWT condition, and

$$Nu_1 = 1 + Re_{PL}^{1/(n+2)} Pr_P^{1/3} + 0.158 Re_{PL}^{2/3} Pr_P^{1/3} \tag{8.32}$$

for the CHF condition.

The resulting average and maximum deviations are of the order of 3% and 10%, respectively. Furthermore, in the limit of Newtonian fluid behavior, Equation 8.31 coincides with the result of Feng and Michaelides (2000) and is in agreement with the widely used heat transfer correlation of Whitaker (1972) as given below:

$$\mathrm{Nu}_1 = 2 + \left(0.4 \mathrm{Re}^{1/2} + 0.06 \mathrm{Re}^{2/3}\right) \mathrm{Pr}^{0.4} \tag{8.33}$$

However, neither Equation 8.31 nor Equation 8.32 approaches the conduction limit as of $\mathrm{Nu}_1 = 2$ $\mathrm{Re}_{PL} \to 0$. This is so simply due to the fact that the results of Dhole et al. (2006b) do not extend to sufficiently low Reynolds numbers for this limit to be reached. Finally, Equation 8.31 also reproduces the maximum value of the local Nusselt number at the front stagnation point by replacing the numerical value of 0.1 by 0.44.

In recent years, heat transfer from a heated sphere submerged in yield stress fluids has also been investigated (Nirmalkar et al., 2013a,b; Gupta and Chhabra, 2014; Thumati et al., 2018) Thus, Nirmalkar et al. (2013a,b) numerically solved the momentum and energy equations for the regularized Bingham and Herschel-Bulkley fluid models and reported extensive results on the local Nusselt number and its surface averaged values for an unconfined sphere over the range of conditions as $1 \leq \mathrm{Re}_B \leq 100$, $1 \leq \mathrm{Pr}_B \leq 100$, $\mathrm{Bi} \leq 10^4$, and $0.2 \leq n \leq 1$. Figure 8.16 shows a typical variation of the local Nusselt number on the surface of an isothermal sphere for Bingham fluids. The limiting case of Newtonian fluid behavior is included here as $\mathrm{Bi}=0$. In this case, this dependence is strongly influenced by the formation of the unyielded (solid-like) regions close to the sphere surface. Due to the yield stress effects, the flow remains attached to the surface of the sphere and thus the Nusselt number does not show any recovery in the rear of the sphere such as that seen in Newtonian fluids due to wake formation (Figure 8.16, Bi=0). Also, the peak value does not occur at the front stagnation point and is displaced to about $\theta \sim 60°$–$70°$ depending upon the rate of change of viscosity and the temperature gradient coupled with the possibility of the formation of unyielded regions. Broadly speaking, the Nusselt number bears a positive dependence on Re_B, Pr_B, and Bi. While this is consistent with the boundary layer considerations due to the progressive thinning of the thermal boundary layer with the increasing Reynolds and Prandtl numbers. The positive dependence of the Nusselt number on the Bingham number is ascribed to the shrinking fluid-like region on the surface of the sphere, thereby sharpening the temperature gradient.

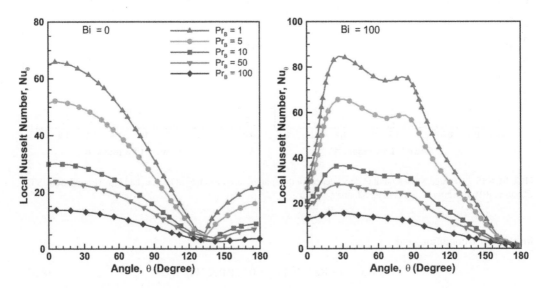

FIGURE 8.16 Local Nusselt number distribution on the surface of the sphere over the ranges of Re_B, Bi, and Pr_B. (From Nirmalkar, N., Chhabra, R.P., Poole, R.J., *Ind. Eng. Chem. Res.*, **52**, 6848, 2013.)

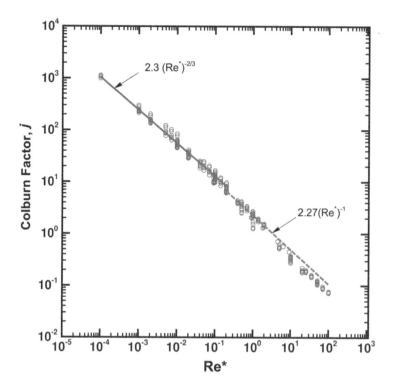

FIGURE 8.17 Dependence of the Colburn factor, j on the modified Reynolds number, Re* for an isothermal sphere in Bingham plastic fluids. (From Nirmalkar, N., Chhabra, R.P., Poole, R.J., *Ind. Eng. Chem. Res.*, **52**, 6848, 2013.)

Figure 8.17 shows their numerical results in terms of the Colburn factor, j, which are well approximated by the following equations:

For $10^{-4} \leq \text{Re}^* \leq 0.2$:

$$j = 2.3\left(\text{Re}^*\right)^{-2/3} \tag{8.34a}$$

For $0.2 < \text{Re}^* \leq 100$:

$$j = \frac{2.27}{\text{Re}^*} \tag{8.34b}$$

While the resulting mean errors are of the order of 18% which rose to ~40% in a few cases, especially at low Bingham numbers.

In the case of the Herschel-Bulkley model, the j-factor shows additional positive dependence on the flow behavior index (n), as seen in Figure 8.18. In this case, Equation 8.34a can be used but the numerical constant of 2.3 varies from 13.35 at $n = 0.2$ to 6.05 at $n = 0.5$, and coincides with 2.3 at $n = 1$. Additionally, at low Bingham numbers, $\text{Bi}_{\text{HB}} = \dfrac{\tau_o^{\text{HB}}}{m\left(V_o/d\right)^n}$, the predictions of the Herschel-Bulkley model coincided with that of the power-law model discussed in the previous section.

Few experimental studies are available in which the rate of mass transfer has been inferred from the rate of dissolution of spheres made of benzoic acid, oxalic acid, etc. that are exposed to the flow of Newtonian and power-law liquids in tubes (Kumar et al., 1980b; Ghosh et al., 1992). Analogous heat transfer results in connection with food process engineering applications have been reported among others by Yamanaka et al. (1976b), Balasubramaniam and Sastry (1994a,b), Baptista et al. (1997), Alhamdan

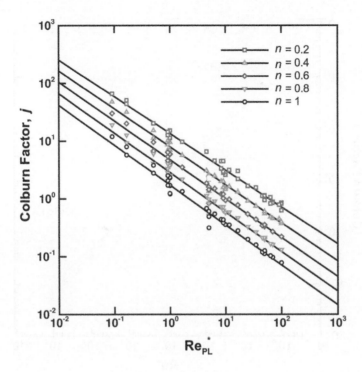

FIGURE 8.18 Colburn factor, j for an isothermal sphere over the ranges of Re, Bi, n, and Pr. (From Nirmalkar, N., Chhabra, R.P., Poole, R.J., *Ind. Eng. Chem. Res.*, **52**, 13490, 2013.)

and Sastry (1998), Aström and Bark (1994), etc. At the outset, it needs to be emphasized here that while most analytical/numerical treatments address the problem of heat/mass transfer from an unconfined sphere, experimental studies always entail, howsoever small, wall effects due to the finite size of the tube in which sphere is exposed to an oncoming fluid steam. Depending upon the proximity of the tube wall, the velocity fields and temperature (and hence velocity and temperature gradients) are significantly altered and therefore, it is not really justifiable to compare these results with theoretical predictions unless the sphere-to-tube diameter ratio is very small, for example, <0.3 (Ghosh et al., 1992). Notwithstanding this inherent difficulty, in most of the experimental studies, empirical correlations have been developed. Ghosh et al. (1992) have collated most of the literature data on heat/mass transfer from spheres and short cylinders (crossflow) to develop the following correlation:

For $Re_P \leq 4$:

$$Y_1 = 1.428\, Re_P^{1/3} \tag{8.35a}$$

For $Re_P > 4$:

$$Y_1 = Re_P^{1/2} \tag{8.35b}$$

where Y_1 is defined as

$$Y_1 = (Nu - 2)\left(\frac{m_s}{m_b}\right)^{1/(3n+1)} Pr_P^{-1/3} \text{ for heat transfer, and}$$

$Y_1 = (Sh_1 - 2)Sc_P^{-1/3}$ for mass transfer

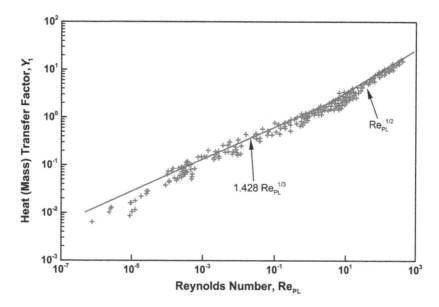

FIGURE 8.19 Generalized correlation of heat and mass transfer from spheres and cylinders to power-law fluids in accordance with Equation 8.35. (Replotted from Ghosh, U.K., Kumar, S. and Upadhyay, S. N., *Polym Plast. Technol. Eng.*, **31**, 271, 1992. Data from Ghosh, U.K., Ph.D. Thesis, Department of Chemical Engineering, Banaras Hindu University, Varanasi, India (1992), and Yamanaka et al., 1976b.)

The characteristic linear dimension for cylinders used in these correlations is the equal volume sphere diameter, which is unlikely to prove satisfactory for cylinders with large length-to-diameter ratios. Figure 8.19 shows the overall correlation including the data for spheres and pellets (Yamanaka et al., 1976b; Kumar et al., 1980b; Ghosh et al., 1992). Due to the lack of details available in the other pertinent studies, it is not possible to include the other literature results in Figure 8.19. The pure conduction limit of Sh (Nu) = 2 is built into Equations 8.35a and 8.35b by virtue of the definition of Y_1. Ghosh et al. (1986a, 1992) asserted that in their studies and that of Yamanaka et al. (1976b), the contribution of free convection was less than 10% of the overall mass transfer.

Similar results for other axisymmetric shapes like a circular disk, cone, hemisphere, etc. have been reported in the literature, most of which are restricted to the steady axisymmetric flow regime. For instance, heat transfer from a heated thin disk in power-law and Bingham plastics has been numerically investigated by Mishra et al. (2019d) and Gupta et al. (2017b), respectively. Limited experimental results for Newtonian fluids have been reported by Kobus and Shumway (2006). Analogous heat transfer results for a hemisphere are also available in the literature (Prakash et al., 2017; Patel et al., 2015; Sasmal et al., 2013). Forced convection heat transfer from cones with various apex angles and orientations has been investigated by Mishra et al. (2018) and over a two-dimensional plane surface by Raja et al. (2015). The numerical predictions of the Nusselt number for spheroidal particles in power-law and yield stress fluids have been documented in the literature by Gupta and Chhabra (2014, 2016a), Reddy and Kishore (2014), and Sreenivasulu et al. (2014). The effect of slip on heat transfer from a sphere in power-law fluids has been explored by Ramteke and Kishore (2018) and that of flow pulsations by Mishra et al. (2020) and by Mishra and Chhabra (2021, 2022, 2023). Flow pulsations can further augment convective heat transfer.

Before leaving this section, it is worthwhile to reiterate here that the foregoing description has primarily focused on the three most widely used geometries, namely plate, cylinder, and sphere and in almost all situations, the simple power-law fluid model has been used to mimic the shear-dependent viscosity, neglecting altogether the other non-Newtonian characteristics notably visco-elasticity. While the role of visco-elasticity in boundary layer flows is briefly dealt with in Section 8.3, the scant literature dealing with the boundary layer flows for the other geometries embedded in a porous medium, and with suction/injection is included here. For instance, some work is available on the momentum and heat transfer to power-law fluids from a wedge (Chen and Radulovic, 1973; Gorla, 1982). Similarly, the flow around and mass transfer from a circular disk rotating in power-law fluids has received some attention (Hansford and Litt, 1968; Lal et al., 1980; Tsay and Chou, 1983; Gorla, 1991b,c; Kim and Esseniyi, 1993). These studies have been motivated mainly by the fact that a rotating disk device affords easy measurement of molecular diffusivity in polymeric solutions (Coppola and Bohm, 1986). Similar results on mass transfer from rotating cylinders to power-law fluids (Kawase and Ulbrecht, 1983d), to drag-reducing polymeric solutions (Al Taweel et al., 1978; Nassar et al., 1989) and to boiling polymer solutions (Yang and Wanat, 1968; Garg and Tripathi, 1981; Shulman and Levitskiy, 1996) are also available. The boundary layer flows and the conjugate heat transfer problems with and without permeable boundaries have been dealt with among others by Chaoyang et al. (1988), Chaoyang and Chuanjing (1989), Chen and Chen (1988), Kleinstreuer and Wang (1988, 1989), Wang and Kleinstreuer (1987, 1990), Nakayama and Shenoy (1992b, 1993a,b), and Chamkha (1997). The role of a transverse magnetic field on the boundary layer flow of a conducting power-law medium has been explored by Pavlov (1979). The behavior of power-law fluids in a two-dimensional axisymmetric wake has been explored by Weidman and van Atta (2001). Finally, the generalized framework developed by Kawase and Ulbrecht (1983b) has been extended (Kawase and Ulbrecht, 1983a) to estimate the rate of mass transfer from spheres suspended in mechanically agitated vessels, by choosing the sphere diameter and the root mean square velocity as the characteristic quantities. The resulting predictions are in line with the scant experimental results available in the literature (Keey et al., 1970; Lal and Upadhyay, 1981; Kushalkar and Pangarkar, 1995).

8.3 Visco-elastic Effects in Boundary Layers

The current interest in the boundary layer flows of visco-elastic fluids stems from three distinct but interrelated objectives: first, due to the occurrence of large strains and strain rates, these flows can be used to test the efficacy and applicability of visco-elastic constitutive equations. Second, these flows are encountered in a range of process engineering applications, particularly in the use of thin wires as measuring probes and therefore, a good fundamental understanding of the underlying phenomena is a prerequisite to analyze the signals from such probes. Finally, a knowledge of skin friction and of the rates of heat and mass transfer from variously shaped objects submerged in visco-elastic media is frequently needed in process design calculations. Indeed, much work has also been stimulated by anomalous transport properties observed experimentally in visco-elastic fluids (Joseph, 1990). In recent years, there has been a recognition of the fact that akin to the momentum, thermal, and concentration boundary layers, stress boundary layers also exist in external flows and a satisfactory resolution of such boundary layers directly impinges on the effectiveness of numerical simulations of external flows for large Weissenberg numbers, as seen in Chapter 4. This section provides an overview of the activity in this field.

The early works (Jain, 1955; Srivastava, 1958; Bhatnagar, 1960; Rajeswari, 1962; Rajeswari and Rathna, 1962) mostly focused on the formulation of the two-dimensional boundary layer equations for the second-order fluid model, with particular reference to the behavior near the stagnation point. Essentially, the same problem was subsequently revisited by Beard and Walters (1964) and Astin et al. (1973) in a slightly different fashion. Most of these studies suggest that the velocity distribution in the boundary layer is similar to that of a Newtonian fluid and the so-called cross-viscosity manifests itself in the form of an increase in pressure at each point of the boundary layer (Bhatnagar, 1960). The use of the standard Prandtl theory of boundary layers for visco-elastic fluids has been questioned by Astarita and Marrucci (1966), and their heuristic analysis clearly shows that the usual assumption of the zero boundary layer thickness at the point of incidence for a plate is incorrect for visco-elastic fluids. In view of the inadequacy of the second-order fluid model to capture the large strain/strain rates in such flows, White and Metzner (1965a,b) developed more complex constitutive equations incorporating both shear-dependent viscosity and second-order visco-elastic effects. They were able to delineate the range of applicability of their analysis and relate it, particularly to the purely viscous (power-law) fluid treatments (Acrivos et al., 1960; Schowalter, 1960). A significant finding at that time was that the purely viscous solutions are valid up to rather large values of the Weissenberg number based on the boundary layer thickness, but up to small values of the Weissenberg number based on the distance along the surface from the forward stagnation point. Therefore, only highly elastic behavior is likely to produce any discernable change in the flow field. While White and Metzner (1965a,b) were able to identify many external flows which would admit the transformation of the partial differential equations (boundary layer equations) into ordinary differential equations, it was not possible in the case of the problem of interest involving constant free stream velocity outside the boundary layer. Denn (1967) made some progress in this direction, but was not fully successful. He employed the usual power-law model for both viscosity and the first normal stress difference in steady shear. Lockett (1969) showed the nonuniqueness of the solution of Denn (1967) in the limit of the second-order fluid behavior and indeed the correct solution for this limiting behavior was presented by Davis (1967). Despite this weakness, the qualitative conclusions reached by Denn (1967), whether the wall shear stress will increase or decrease due to visco-elasticity depends upon the power-law index for the first normal stress difference, remains valid (Serth, 1973). This result is, however, at variance with that of White (1966). Furthermore, the analysis of Davis (1967) suggests that the visco-elastic effects progressively diminish far downstream on the plate and one can thus use purely viscous results away from the leading edge.

Many other investigators have developed and used boundary layer equations for second-grade fluids (Srivastava and Maiti, 1966; Mishra, 1966a,b; Srivastava and Saroa, 1971, 1978; Sarpkaya and Rainey, 1971; Rajagopal et al., 1980, 1983; Garg and Rajagopal, 1990, 1991; Pakdemirli and Suhubi, 1992). The restrictions to retain a conventional boundary layer for second-grade fluids were presented by Rajagopal et al. (1980) and the possibility of multiple inner expansions has been investigated by Pakdemirli (1994a). The boundary layer flows of third-grade fluids have been studied by Pakdemirli (1994b), Pakdemirli et al. (1996), and Yurusoy and Pakdemirli (1997). Three-dimensional boundary layers have been studied among others by Timol and Kalthia (1986) and Verma (1977). Some qualitative results near the stagnation point for the flow of the Walters fluid model over a sphere have been presented by Verma (1977). Rochelle and Peddieson Jr. (1980) have considered the boundary layers in Maxwell-Oldroyd type fluids for Wedges and cones, though no detailed results are presented.

There has been some numerical activity in this field. Thus, Pimenta and Alves (2021) have considered heat transfer from a conducting sphere in the creeping flow of the Phan-Thien Tanner model. Their results spanning Deborah number (0–100), Prandtl number (1–10^5) with the sphere to fluid thermal conductivity ratio (0.1–10), and Brinkman number (up to 100) suggest that a moderately elastic fluid promotes heat transfer up to a critical Deborah number beyond which heat transfer is little influenced by elasticity. This behavior seems to be consistent with the experimental results of James and Gupta (1971). Similarly, Sasmal et al. (2020) have studied heat transfer from a two-dimensional cylinder in Oldroyd-B and FENE-P models at moderate Reynolds numbers of $0.1 \leq Re \leq 20$ and elasticity numbers of $0 \leq El \leq 0.5$. The formation of momentum and thermal boundary layers was seen to be completely suppressed in Oldroyd-B fluids whereas it was accelerated in FENE-P fluids under appropriate conditions with reference to the behavior in Newtonian fluids. For the Oldroyd-B fluids, the cylinder is encapsulated in a quasi-stagnant region and thereby the average Nusselt number becomes independent of the Reynolds number beyond a critical value. On the other hand, for a fixed elasticity number, the average Nusselt number bears a positive relationship with the Reynolds number in FENE-P fluids. However, both the polymer extensibility parameter and the viscosity ratio have a deleterious effect on heat transfer. Subsequently, this work has been extended to heat transfer from a rotating cylinder in FENE-P fluids (Khan et al., 2020). Similarly, Chauhan et al. (2021) have explored numerically the combined effects of blockage and inertia on heat transfer from a sphere undergoing steady translation at the axis of a tube in FENE-P fluids. Irrespective of the values of the Weissenberg number, polymer extensibility, and blockage ratio (sphere diameter/tube diameter), the average Nusselt number shows a positive dependence on the Reynolds number. However, the role of blockage and polymer extensibility is modulated by the values of the Reynolds and Weissenberg numbers. Thus, at high Reynolds numbers, heat transfer increases with the blockage, whereas at low Reynolds numbers, heat transfer flattens out beyond a critical value of the blockage.

Many developments in this field have also occurred through the use of dimensional and heuristic considerations. Such studies have received impetus from the anomalous transport behavior observed in external flows with dilute polymer solutions, notably the experimental works of James and Acosta

(1970), James and Gupta (1971), Ultman and Denn (1970), Ambari et al. (1984a), etc., as reviewed elsewhere (Chhabra, 2011). In particular, these studies, dealing with heat or mass transfer to dilute polymer solutions from a circular cylinder in crossflow orientation, show that the drag coefficient and heat (or mass) transfer coefficient show qualitatively a similar dependence on velocity as that in Newtonian fluids below a critical velocity, and these coefficients become nearly independent of the liquid velocity above the critical value of the velocity. The asymptotic values depend only on the rheological properties of the fluid and the size of the obstacle (cylinder). Although some experimental results obtained with dilute polyacrylamide solutions are at variance with these trends (Hoyt and Sellin, 1989), many investigators have attempted to explain this type of behavior by postulating the so-called "elastic" boundary layer (Mochimaru and Tomita, 1978; Mashelkar and Marrucci, 1980; Ruckenstein and Ramagopal, 1985; Harnoy, 1987; Ruckenstein, 1994). It is generally accepted that the integral momentum balance as applied to purely viscous boundary layers must include a normal stress component and this immediately leads to the result that the condition of zero boundary layer thickness at the leading edge of a plate cannot be satisfied (Metzner and Astarita, 1967; Mashelkar and Marrucci, 1980; Ruckenstein and Ramagopal, 1985). Assuming the two contributions to be simply additive, Ruckenstein (1994) postulated the boundary layer thickness for a visco-elastic fluid to be given by the following expression:

$$\delta^2 = A_0 \frac{x\mu_0}{\rho V_0} + B_0 \frac{\mu_0 \theta_f}{\rho} \tag{8.36}$$

where A_0 and B_0 are two unknown constants, μ_0 is the zero-shear viscosity, and θ_f is the fluid relaxation time. Equation 8.36 does portray the fact that as V_0 grows large, the boundary layer thickness δ (and hence the transfer coefficient) becomes independent of the velocity (Reynolds number). In other words, when the quantity $(B_0/A_0)(\theta_f V_0/x)$ or (B_0/A_0)De is sufficiently large (>>1), δ becomes independent of the velocity. Ruckenstein (1987, 1994) further showed that the mass transfer coefficient is given by the expression

$$k_c \sim \frac{D_{AB}^{2/3}}{(\mu_0/\rho)^{1/6}} \left\{ \alpha_0 \left(1 + (B_0/A_0)\text{De}\right)^{-1/2} + \frac{1}{4}\left(1 + (B_0/A_0)\text{De}\right)^{-3/2} \right\}^{1/3} \sqrt{\frac{V_0}{x}} \tag{8.37}$$

where α_0 is a constant.

In the limiting case of purely viscous case (De=0), Equation 8.37 reduces to

$$k_{cN} \sim \frac{D_{AB}^{2/3}}{(\mu_0/\rho)^{1/6}} \sqrt{\frac{V_0}{x}} \tag{8.38}$$

On the other hand, as De $\to \infty$, $k_c \to 0$, albeit the decay to zero is extremely slow as it follows $k_c \sim k_{cN}\text{De}^{-1/6}$. This scaling result is in line with the more rigorous analysis based on the use of the differential equations. This analysis was further supplemented by the fact that the experimental results of James and Acosta (1970) and Ambari et al. (1984a) relate to moderate Reynolds numbers and high Prandtl/Schmidt numbers. Under these conditions also, the mass transfer coefficient shows similar dependence on the Deborah number as seen above in the analysis of Ruckenstein (1994). Some of these trends are qualitatively in line with the recent predictions for a cylinder in FENE-P and Oldroyd-B fluids (Sasmal et al., 2020).

Apart from the boundary layer approximations, some analytical and numerical results are also available for heat and mass transfer from a cylinder or a sphere to visco-elastic fluids. Sharma and Bhatnagar (1975) employed the velocity field due to Caswell and Schwarz (1962) for the creeping flow of a Rivlin-Ericksen fluid past a sphere to solve the related thermal energy equation. They sought a solution by using the matched asymptotic expansions for both temperature and Nusselt number. Within the range of its validity, the mean Nusselt number may increase above or decrease below the corresponding Newtonian value, depending upon the rheological parameters. However, the effect of visco-elasticity is predicted to be very weak. Similarly, Kawase et al. (1982) combined the stream function due to Leslie and Tanner (1961) with the general short-range diffusion equations (in the limit of large Peclet number) to elucidate the role of visco-elasticity on convective mass transfer from a sphere in the creeping flow region. Though the visco-elasticity seems to augment the rate of mass transfer, the increase is too small to be measured experimentally. Mizushina and Usui (1975) presented limited numerical results for the steady two-dimensional flow of Maxwell fluids across a circular cylinder. In the range $1 \leq Re \leq 20$ and $0.01 \leq We \leq 0.2$ (such that $ReWe \leq 1$), the total drag and Nusselt number were found to decrease with the increasing value of the Weissenberg number. This trend is also substantiated by their own experimental results on drag and mass transfer with polyethylene oxide solutions, albeit the experimental data show considerable scatter, especially at low Reynolds numbers. Similarly, Ogawa et al. (1984) have reported mass transfer data for a sphere and a cylinder to aqueous solutions of polyacrylamide and carboxy-methyl cellulose by using the electrochemical technique. The effect of visco-elasticity was quantified by using the so-called elasticity number El defined as $(\rho V_0^2/G)$ where G is the modulus of shear elasticity of the fluid. In the range of conditions, $1 \leq Re_P \leq 200$ and $El \leq 600$, they presented the following empirical correlation for a sphere:

$$Sh_1 = 1.5 Re_P^{(n+2)/3(n+1)} Sc_P^{1/3} El^{-0.15} \tag{8.39}$$

For a cylinder, they suggested the values of 1.9 and 0.21 instead of 1.5 and 0.15, respectively, in Equation 8.39. The negative index of the elasticity number suggests that the mass transfer is adversely influenced by the fluid elasticity, a trend which is consistent with the findings of Mizushina and Usui (1975). Subsequently, Sobolik et al. (1994) used a segmented electro-diffusion velocity probe to obtain the distribution of Sherwood number over the surface of a cylinder immersed in aqueous solutions of polyacrylamide. These data show that above a critical value of the Weissenberg number, the rate of mass transfer from the rear surface of the cylinder is larger than that from the front surface. This seems to be consistent with the notion of a nonzero thickness of the boundary layer at the point of incidence. Later, Wu et al. (2003) and Lin et al. (2004) have numerically simulated the flow of the White–Metzner model fluid past a circular cylinder. They also reported the shear-thinning to enhance heat transfer, but the elasticity to suppress it.

In external flows of visco-elastic liquids at low Reynolds numbers, but at high Weissenberg numbers, the notion of elastic boundary layer has been found to be useful (Renardy, 1997, 2000a,b). The basic reason for the development of elastic boundary layers is due to the behavior of the convected derivatives in the visco-elastic fluid models at the wall. While these derivatives vanish at a solid surface (due to the no-slip condition which requires the velocity and all its tangential derivatives to be zero), but at high Weissenberg numbers, these terms grow significantly at a short distance (in the transverse direction) away from the solid surface, thereby resulting in the so-called "stress" boundary layers, akin to the momentum, thermal, or concentration boundary layers. This kind of behavior has been observed in the flow past a sphere or a cylinder, for the flow in eccentric cylinders (Hagen and Renardy, 1997; Renardy, 1997) and indeed the inadequate resolution of such boundary layers appears to be the prime reason for the lack of convergence of numerical solutions beyond a critical value of the Weissenberg or Deborah number, as was seen in Chapter 4. Hagen and Renardy (1997) showed that this problem is most acute for the upper convected Maxwell model (UCM) which leads to a stress boundary layer of the order of We^{-1}, in contrast to $We^{-1/3}$ dependence for the Phan-Thien Tanner (PTT) model and $We^{-1/2}$ dependence for the Giesekus fluid model. Therefore, the Giesekus and PTT models are believed to be far less troublesome in this regard. Indeed, Renardy (2000a) and Wapperom and Renardy (2005) have numerically investigated the stress boundary layers for the flow over a circular cylinder to explore the effect of high Weissenberg numbers. This is in stark contrast to the difficulties encountered in numerical solutions at low-to-moderate Weissenberg numbers.

8.4 Bubbles

The hydrodynamic and mass transfer characteristics of single bubbles and bubble swarms directly influence the efficiency and size of process equipment used to carry out a diverse range of industrially important processes. Therefore, a satisfactory understanding of the underlying physical processes is germane to the development of realistic models for such processes. Typical examples include the omnipresent bubble column and three-phase fluidized bed reactors used extensively in biotechnological, food and agro-product processing applications whereas other potential applications include the production of foamed plastics, degassing of polymeric melts, aeration of food batters and cake mixes, etc. Consequently, some research effort has been expended in elucidating the role of the rheology of the continuous phase on mass (heat) transfer to/from stationary and freely rising single bubbles and bubble swarms. The available body of information is reviewed here.

From a theoretical standpoint, in the presence of mass (or heat) transfer, the momentum and continuity equations must be supplemented by the species continuity relations for the diffusing (or dissolving component) or the thermal energy equation. The coupling between the fluid mechanical and the mass transfer processes arises in three ways: first, via the velocity through the bulk transport term in the species continuity (or the thermal energy) equation; second, the changes in bubble volume due to the transfer of a component from/to it. This results in a time-dependent normal velocity adjacent to the bubble surface. Finally, such coupling also stems from temperature- or concentration-dependent thermophysical properties of the continuous phase. Finally, as the bubble size changes, the buoyancy force (and hence its velocity) will continually change with time. By scaling of the pertinent equations for a single bubble, one can readily show that a new nondimensional group, the Peclet number, Pe, emerges in addition to the Reynolds number, Weber number, Weissenberg number, power-law index, etc., as introduced in Chapter 5. A slight rearrangement identifies the Peclet number, Pe, to be the product of the Reynolds number, Re, and the Schmidt number, Sc ($= \mu/\rho\, D_{AB}$) and as such it denotes the ratio of mass transfer by convection to that by molecular diffusion. By similar reasoning, the Peclet number for heat transfer turns out to be the product of the Reynolds and Prandtl numbers. Alternatively, one can view it as a measure of the relative importance of the bulk liquid velocity to that induced by diffusion. Thus, for large values of Pe, the flow field in the ambient liquid is mainly determined by the free stream velocity of the liquid (or the buoyancy-driven rise velocity of a bubble in a quiescent liquid). Thus, the commonly used approximation under such circumstances is to drop the species continuity equation, and the liquid rheology influences the rate of mass transfer by altering the flow field. This approximation has been used extensively in modeling steady and transient mass transfer to/from bubbles and drops moving slowly (Re $\ll 1$) in non-Newtonian fluids, for example, see Hirose and Moo-Young (1969), Moo-Young and Hirose (1972a,b), Bhavaraju et al. (1978), Jarzebski and Malinowski (1986a,b, 1987a,b), and Zhu and Deng (1994). On the other hand, small values of Peclet number (Pe $\ll 1$) indicate that the mass transport by bulk flow is negligible in comparison with that by diffusion. Under these circumstances, the flow field is dominated by the motion induced by the growing/collapsing bubble. This limiting behavior has been successfully analyzed by approximating it as mass transfer from a stationary bubble. The liquid rheology thus enters directly into the analysis. The pertinent scant literature is briefly reviewed in the ensuing sections.

8.4.1 Large Peclet Number (Pe $\gg$ 1)

Both Hirose and Moo-Young (1969) and Bhavaraju et al. (1978) have obtained closed-form expressions for mass transfer from a single bubble rising slowly ($Re_{PL} \ll 1$) through power-law liquids. By analogy with the drag behavior, it is customary to express the mass transfer results in terms of the deviation from the Newtonian result. It is thus instructive to recall that the Sherwood number for a spherical bubble with a clean surface in a Newtonian liquid is given as (Levich, 1962)

$$Sh_1 = 0.65 Pe^{1/2} \qquad (8.40)$$

For power-law liquids, Hirose and Moo-Young (1969) obtained

$$Sh_1 = 0.65 Pe^{1/2} \left(\frac{-4n^2 + 6n + 1}{2n + 1} \right)^{1/2} \tag{8.41}$$

And one can thus introduce an enhancement factor, Y_m, for mass transfer as

$$Y_m = \left(\frac{-4n^2 + 6n + 1}{2n + 1} \right)^{1/2} \tag{8.42}$$

The corresponding expression due to Bhavaraju et al. (1978) is

$$Y_m = \left[1 + 1.62(1 - n) \right]^{1/2} \tag{8.43}$$

Both Equations 8.42 and 8.43 are applicable for small deviations from the Newtonian fluid behavior only and suggest an enhancement in mass transfer due to shear-thinning behavior. Figure 8.20 shows the extent of enhancement in mass transfer attributable to the power-law behavior. Bhavaraju et al. (1978) also reported a similar increase in the value of the mass transfer coefficient for bubbles rising (Re <<1) in Bingham plastic systems in the limit of Bi → 0.

$$Y_m = (1 + 0.25 Bi)^{1/2} \tag{8.44}$$

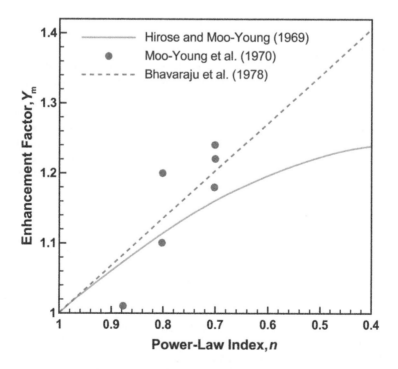

FIGURE 8.20 Predicted and observed dependence of mass transfer factor Y_m for single bubbles in creeping motion in power-law fluids.

Similarly, Moo-Young and Hirose (1972a,b) used the perturbation approach to obtain the approximate stream function for bubble motion in a Maxwell liquid (Re <<1, We <<1), which, in turn, was used to deduce the following expression for Y_m:

$$Y_m = 1 + 0.16 \text{We}^2 \qquad (8.45)$$

Thus, it appears that on all counts, higher mass transfer is predicted in viscoinelastic and in visco-elastic systems, in the limits of diminishing Reynolds number and weak non-Newtonian effects. On the other hand, Tiefenbruck and Leal (1980b, 1982) suggested that the rate of mass transfer may increase or decrease depending upon the values of the visco-elastic material parameters of the continuous phase.

The assumption of the low Reynolds number intrinsic in most of the aforementioned studies has been relaxed subsequently. Thus, Dhole et al. (2007b) predicted mass transfer for clean spherical bubbles in power-law fluids over the range of conditions as $5 \le \text{Re}_{PL} \le 100$, Schmidt number ≤ 1000, and $0.7 \le n \le 2$. Over these conditions, shear-thinning slightly facilitates mass transfer and shear-thickening impedes it. But the effect is very small. Subsequently, Kishore et al. (2007b) studied numerically mass transfer from a clean fluid sphere undergoing steady translation in power-law fluids at intermediate Reynolds numbers (5–200), Schmidt number ($\le 10^3$), power-law index ($0.6 \le n \le 1.6$), and viscosity ratio $0.1 \le X_E \le 50$. Under otherwise identical conditions, shear-thinning fluid behavior fosters convective transport whereas such transport is impeded by shear-thickening behavior with respect to that in Newtonian fluids. This trend also applies to the case of mass transfer from swarms of fluid spheres including bubble swarms studied within the framework of concentric sphere cell models (Kishore et al., 2007c, 2008a,b,c). In the limit of $X_E \to 0$, their results are in line with that of Dhole et al. (2007b) for a gas bubble.

Several other researchers (Barnett et al., 1966; Calderbank, 1967; Calderbank et al., 1970; Moo-Young et al., 1970; Zana and Leal, 1978; Dietrich and Hebrard, 2018; Xu et al., 2018, 2021) have carried out mass transfer studies involving bubbles moving in stagnant power-law and visco-elastic liquids. The resulting values of Y_m from the study of Moo-Young et al. (1970) are shown in Figure 8.20, where the correspondence between predictions and data is seen to be moderately good. The experimental results of Zana and Leal (1978) show even greater enhancements in the value of Sherwood number for carbon dioxide bubbles in visco-elastic polyacrylamide solutions. Qualitatively, this finding is consistent with Equation 8.45, but the resulting values of the Weissenberg number are too large for Equation 8.45 to be used for quantitative predictions. However, more recent studies by Xu et al. (2018, 2021) on mass transfer from CO_2 bubbles in elastic polyacrylamide solutions point to a significant reduction in mass transfer coefficients on account of visco-elasticity of the continuous phase.

When the instantaneous values of the mass transfer coefficient are plotted against time, the data of Barnett et al. (1966) show a rapid initial decrease of the mass transfer coefficient, eventually leveling off after about 15 s. On the other hand, when the same data is plotted against bubble size (equivalent- volume sphere diameter), the mass transfer coefficient exhibits one or two peaks at bubble sizes which approximately correspond to the transition in bubble shapes. Qualitatively, similar results have been reported by Calderbank (1967), Calderbank et al. (1970), and Aiba and Okamoto (1965).

8.4.2 Small Peclet Number (Pe <<1)

As mentioned previously in Chapter 5, small values of the Peclet number denote the situation when the flow field is largely determined by the growth/collapse of gas bubbles due to diffusion, and it is reasonable to model this behavior as that of mass transfer from a stationary gas bubble. Street (1968) and Street et al. (1971) investigated the rate of growth of a spherical cavity in a three-constant Oldroyd fluid. Subsequently, Fogler and Goddard (1970) considered the collapse of a spherical cavity in a generalized Maxwellian fluid. In both cases, the driving force for the growth/collapse was assumed to be the difference between the actual and the equilibrium values of internal pressures. In the absence of mass transfer, a further simplification results because the cavity pressure can be assumed to be constant. Zana and Leal (1974, 1975, 1978), on the other hand, solved the coupled problem of diffusion and collapse-induced flow. In this case, the coupling arises via the varying internal pressure. The ambient visco-elastic liquid was modeled by the eight-constant Oldroyd model. They presented numerical results on the concentration and bubble radius as functions of time for a range of values of the pertinent variables, but for large values of the Schmidt number. Depending on the values of the rheological parameters and of the surface tension, the bubble decay rate may decrease or increase with time or may even display an "overshoot." Increasing the surface tension accelerates the collapse rate, the effect being particularly striking at large times for small bubbles. At sufficiently high values of surface tension, the collapse rate becomes so large that the mass transfer is unable to keep up, thereby resulting in a rapid increase in internal pressure. Because the collapse-induced motion is unidirectional extension, the shear viscosity plays a relatively minor role. The visco-elasticity does not always inhibit the collapse rate, as would be expected from the increased resistance to extensional deformation.

Subsequently, Advani and Arefmanesh (1993) have numerically studied the growth and collapse of gas bubbles encapsulated in a spherical shell of finite size filled with a visco-elastic liquid. They solved the coupled-diffusion, continuity, and momentum equations for the three-constant Oldroyd fluid model. In particular, the influence of the Weber number, Henry's constant, the rheological parameters, and the proximity of the surrounding shell was elucidated on the rate of collapse of bubbles. Striking differences were observed in the rate of change of bubble size in the case of a relatively thin-size shell with a limited amount of solute available. Significant visco-elastic effects were also encountered when an unlimited supply of gas was available, such as when dealing with an infinite expanse of liquid or when the process time is small such as that in foam injection molding. These authors have also outlined possible applications of such model studies.

The only study in which the combined effects of non-Newtonian characteristics and temperature gradient on bubble motion have been explored is that of Chan Man Fong and De Kee (1994). They studied the migration of bubbles in the presence of a thermal gradient for second-order and Carreau model fluids. They found that the surface tension effects are only important for small bubbles. Likewise, Dang et al. (1972) have elucidated the role of non-Newtonian characteristics in reactive systems.

8.5 Drops

Few investigators (Wellek and Huang, 1970; Shirotsuka and Kawase, 1973; Gurkan and Wellek, 1976; Wellek and Gurkan, 1976; Kawase and Ulbrecht, 1981g; Gurkan, 1989, 1990; Saboni, 2008; Saboni et al., 2010) have studied mass transfer from spherical drops translating in stationary power-law media continuous phase. However, no surface tension and Marangoni effects have been considered, and all developments are based on the assumption of the thin concentration boundary layer thickness that is a large Peclet number (Lochiel and Calderbank, 1964). For the creeping motion of spherical droplets, an approximate analysis (Kawase and Ulbrecht, 1981g) based on the linearized equations of motion yields the expression for Sherwood number, expressed as

$$\text{Sh}_1 = 0.921 \text{Pe}^{1/2} \left[\frac{\beta}{\beta + X_E} \left(\frac{1}{2} - \frac{4}{3} \alpha_0 \right) \right]^{1/2} \tag{8.46}$$

with

$$\alpha_0 = \left(\frac{3}{4} \right) \left(\frac{3X_E + 2}{X_E + 1} \right) \frac{n(n-1)}{(2n+1)} \tag{8.47}$$

$$\beta = \left(\frac{3}{4} \left(\frac{3X_E + 2}{X_E + 1} \right)^2 \right)^{(n-1)/2} \tag{8.48}$$

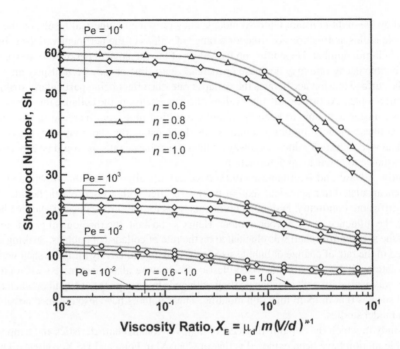

FIGURE 8.21 Effect of power-law index (n), viscosity ratio (X_E) and Peclet number (Pe) on Sherwood number for falling drops in power-law fluids in the creeping flow region. (Replotted from Wellek, R.M. and Huang, C.-C., *Ind. Eng. Chem. Fundam.*, **9**, 480, 1970.)

Equation 8.46 was stated to be applicable for

$$\text{Pe} > 7.08 \frac{(3X_E + \beta)^2 (\beta + X_E)}{(3 - 8\alpha_0)\beta^3} \tag{8.49}$$

In the limiting case of a Newtonian continuous phase ($n=1$), Equation 8.46 reduces to the expected limiting behavior. Likewise, the correct expression for gas bubbles is retrieved by setting $X_E=0$ in Equation 8.46 (Hirose and Moo-Young, 1969). Finally, this theory also predicts an enhancement in the rate of interphase mass transfer for drops in pseudoplastic media as compared with the Newtonian systems under otherwise identical conditions. Wellek and Huang (1970) obtained a numerical solution to the diffusion equation for all values of the Peclet number. For this purpose, they employed the previously suggested velocity profiles (Nakano and Tien, 1968) for the creeping motion of Newtonian fluid spheres in power-law fluids. Theoretical estimates of Sherwood number as a function of the power-law index, viscosity ratio, and Peclet number were reported. The effect of internal circulation was also elucidated. Figure 8.21 shows this functional dependence graphically. It is seen that the Sherwood number increases with the Peclet number under all conditions. As the Peclet number increases, the effect of n becomes more pronounced. For instance, for Pe > ~500, the Sherwood number increases by 25% as n decreases from 1 to 0.6: however, nearly half of this enhancement occurs when the value of n drops from 1 to 0.9. Likewise, the value of the Sherwood number is also influenced strongly by the viscosity ratio only for large values of the Peclet number. However, these results must be used only for qualitative inferences due to the numerical uncertainty inherent in the determination of velocity profiles of Nakano and Tien (1968), as mentioned in Chapter 5. Subsequently, Gurkan and Wellek (1976) demonstrated that the previously estimated values of the Sherwood number are virtually insensitive to the choice of velocity profiles.

Moreover, the influence of the non-Newtonian characteristics of the continuous phase on the dispersed phase mass transfer was also shown to be negligible. Later on, this treatment has been extended to the intermediate Reynolds number regime (Wellek and Gurkan, 1976) and more recently by Kishore et al. (2007b) and Trivedi et al. (2023) in power-law fluids. Qualitatively similar trends are seen as far as the effect of the power-law index on the mean Sherwood number is concerned. The corresponding results for a mass transfer from a spherical drop in Bingham plastic fluids and Herschel-Bulkley fluids at finite Reynolds numbers have been reported by Nirmalkar et al. (2021) and Alam et al. (2022). They reported the mean Sherwood number to increase with the increasing Reynolds, Schmidt, and Bingham numbers, similar to the case of a solid sphere, as discussed in Section 8.2.3. Similar conclusions have been also reached for the case of a circulating power-law drop falling in Newtonian media in the intermediate Reynolds number regime ($10 \leq \text{Re} \leq 50$) (Gurkan, 1989, 1990; Kishore et al., 2008d). Unfortunately, no appropriate experimental data on mass transfer in such systems are available in the literature to refute or substantiate these predictions.

8.6 Ensembles of Bubbles and Drops

Undoubtedly, the single bubble/drop studies provide useful insights into the interphase transport processes, but it is readily acknowledged that one encounters ensembles of bubbles and drops in most applications rather than isolated bubbles or drops. In spite of the overwhelming pragmatic importance of these systems, particularly in biotechnological processes, very little is known about the interphase mass transfer between ensembles of fluid spheres and a non-Newtonian continuous phase. Approximate theoretical results are available for power-law and Carreau model fluids for fluid particles in the creeping flow regime in the absence of any surfactant effects. Bhavaraju et al. (1978) extended their analysis for single bubbles to a swarm of bubbles moving in quiescent power-law media by using the free surface cell model. The approximate flow field, obtained by linearizing the momentum equations, was used together with the thin (concentration) boundary layer assumption to derive the following expression for the liquid-phase Sherwood number in power-law fluids:

$$\text{Sh}_l = 0.65 Y_{m,\text{SW}} \text{Pe}_{\text{SW}}^{1/2} \tag{8.50}$$

where $Y_{m,\text{SW}} = F(\phi, n)$ is available in the original paper of Bhavaraju et al. (1978). The chief finding of this study is that the rate of mass transfer in swarms (at a fixed gas fraction) decreases with the increasing extent of pseudoplastic behavior. This is in stark contrast to the case of a single bubble wherein the rate of mass transfer is enhanced due to the shear-thinning behavior of the continuous phase. Figure 8.22 shows the extent of enhancement in mass transfer for a range of values of gas holdups and flow behavior index. Subsequently, these results have been extended to finite Reynolds numbers (Kishore et al., 2008c). The role of the power-law index seems to be only secondary due to the diminishing significance of viscous effects. This conclusion is in qualitative agreement with the limited experimental results for bubble columns (Buchholz et al., 1978; Deckwer et al., 1982; Godbole et al., 1984; Suh et al., 1991; Deckwer, 1992) and in wetted wall columns with aqueous polymer solutions and fermentation broths, respectively (Aiba and Okamoto, 1965).

In a series of papers, Jarzebski and Malinowski (1986a,b, 1987a,b) have employed variational principles to obtain approximate upper and lower bounds on the terminal velocity of ensembles of Newtonian fluid spheres in power-law and Carreau model fluids. The resulting approximate tangential velocity on the surface of the fluid sphere was used to obtain an expression for the liquid-side Sherwood number via the standard thin boundary layer formalism (Baird and Hamielec, 1962; Lochiel and Calderbank, 1964). Figure 8.23 shows representative results elucidating the relative mass transfer (for a swarm in comparison with a single drop) as a function of the dispersed phase holdup, viscosity ratio, and the flow behavior index. The mass transfer enhancement (with reference to that for a single drop) decreases with the increasing degree of shear-thinning as well as with the decreasing gas fraction in the swarm. On the other hand, it bears a positive relationship with the increasing viscosity of the dispersed phase

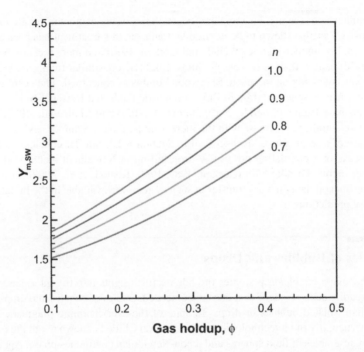

FIGURE 8.22 Effect of power-law index and gas holdup on the mass transfer factor $Y_{m,SW}$ for a bubble swarm in power-law fluids in creeping flow region. (Replotted from Bhavaraju, S.M., Mashelkar, R.A., and Blanch, H.W., *AIChE J.*, **24**, 1067, 1978.)

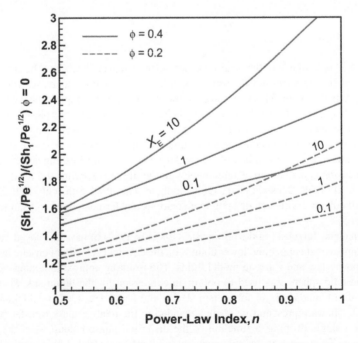

FIGURE 8.23 Effect of power-law index, gas volume fraction (ϕ), and viscosity ratio (X_E) on liquid–liquid mass transfer for drop ensembles falling slowly in power-law continuous phase. (After Jarzebski, A.B. and Malinowski, J.J., *Chem. Eng. Sci.*, **41**, 2569, 1986.)

which is somewhat counter-intuitive. These results have been supplemented by numerical results at finite Reynolds numbers (Kishore et al., 2008c). Admittedly, the aforementioned theoretical studies involving bubbles and drops do provide useful insights, but their utility is severely limited by the main assumptions regarding the absence of surfactant impurities, whereas in practice these would always be present, which lead to the immobilization of the surface of bubble or drop.

8.7 Fixed Beds

Much effort has been devoted to predict the liquid–solid interphase mass transfer in fixed beds with non-Newtonian liquids. As is the case with pressure drop calculations presented in Chapter 6, most of the work is based either on the use of cell models or the capillary bundle representation of packed beds. The early works of Pfeffer (1964) and Pfeffer and Happel (1964) demonstrate the utility of the cell model approach to predict interphase heat/mass transport in concentrated particulate systems in the limit of low Reynolds numbers and high Peclet numbers for Newtonian fluids. This approach has been extended to moderate Reynolds numbers and Peclet numbers by LeClair and Hamielec (1968b), El-Kaissy and Homsy (1973), and Mao and Wang (2003). Combined together, these results encompass the range of conditions as: $1 \leq Re_p \leq 500$; $0.4 \leq \varepsilon \leq 0.9$; and $1 \leq Pe \leq 300$. While this approach has come under some criticism (Sirkar, 1975; Ocone and Astarita, 1991), comparisons with experimental results appear to be reasonably good for Newtonian and power-law liquids. This approach has been extended to power-law fluids in the low Reynolds number and high Peclet number regime by Kawase and Ulbrecht (1981a,b) who obtained approximate closed-form expressions for Sherwood number. The effect of the power-law rheology was found to be rather weak, albeit shear-thinning behavior facilitated mass transfer. Subsequently, qualitatively similar results have been reported by others (Satish and Zhu, 1992; Zhu and Satish, 1992; Shukla and Chhabra, 2004). The restriction of the low Reynolds number was relaxed by Shukla et al. (2004) and Kishore et al. (2009a) who presented extensive numerical results on the local and average Nusselt number as functions of the power-law index, bed voidage, Reynolds number, and Prandtl number. As it will be seen later, these predictions are also consistent with the scant experimental results on the liquid–solid mass transfer results available in the literature. Some results based on the use of boundary layer approximation are also available for the flow of Newtonian (Hayes, 1990) and power-law fluids (Wang et al., 1988) in packed beds.

The second approach to the prediction of interphase heat/mass transport hinges on the capillary bundle model coupled with the boundary layer approximation for a flat plate to develop predictive expressions for Sherwood number. In a series of papers, Kawase and Ulbrecht (1983b, 1985a,b) have pursued this line of analysis in the creeping flow regime. They obtained the following expression for mass transfer in terms of the j-factor:

$$\varepsilon j = 1.85\left[(1-\varepsilon)\varepsilon\right]^{1/3}\left(\frac{3n+1}{4n}\right)^{1/3} Re_p^{-2/3} \qquad (8.51)$$

where the j-factor is defined as

$$j = \frac{Sh_l}{Pe} Sc^{2/3} \qquad (8.52)$$

Equation 8.51 predicts only a very minor effect of the power-law rheology on the j-factor, which is also qualitatively consistent with the cell model predictions (Satish and Zhu, 1992; Shukla and Chhabra, 2004). Comiti et al. (2000b, 2002) extended their version of the capillary model to develop the expression for Sherwood number in the Darcy (creeping) regime as

$$Sh_l = \frac{1.16}{\varepsilon} Pe^{1/3}\left(\frac{3n+1}{4n}\right)^{1/3} \qquad (8.53)$$

Note the striking similarity between Equations 8.51 and 8.53 except for the term $[(1-\varepsilon)\varepsilon]^{1/3}$. Indeed, in the range $0.4 \leq \varepsilon \leq 0.8$, the two predictions differ only by 10%. For the creeping flow of carboxymethyl cellulose solutions ($0.27 \leq n \leq 0.98$) through fixed beds of nonspherical gypsum particles, Peev et al. (2002) put forward the following correlation for mass transfer:

$$\varepsilon j = 0.541 \left(\frac{3n+1}{4n} \right) \text{Re}_2^{-0.73} \tag{8.54}$$

where the modified Reynolds number Re_2 is based on the effective viscosity calculated as $m'(8V_0/\varepsilon d)^{n-1}$. In Equation 8.54, the exponent of 0.73 is close to the value of Kumar and Upadhyay (1981) whereas the other constant, 0.541, is close to that of Hilal et al. (1991), thereby suggesting these results to be consistent with those of Kumar and Upadhyay (1981) and of Hilal et al. (1991). Kawase (1992) used the boundary layer result for a flat plate and made corrections for the tortuosity and interstitial velocity to develop a semi-empirical expression for Sherwood number as

$$\text{Sh}_1 = \alpha(n) \varepsilon^{-1/(n+1)} \text{Re}_p^{(n+2)/3(n+1)} Sc_p^{1/3} \tag{8.55}$$

where

$$\alpha(n) = \frac{9(n+1)}{2(2n+1)} \left\{ \frac{30(n+1)}{(2n+1)} \alpha_0(n) \right\}^{-1/3} \left(\frac{3n+1}{n+1} \right)^{1/(n+1)} (2)^{(1-n)/6(n+1)} \tag{8.56a}$$

and

$$\alpha_0(n) = \left[\frac{280}{39}(n+1) \left(\frac{3}{2} \right)^n \right]^{1/(n+1)} \tag{8.56b}$$

Unlike Equation 8.51 or Equation 8.53 or Equation 8.54, Equation 8.55 does apply to flows at finite Reynolds numbers. Qualitatively, a similar expression for Sherwood number has also been developed by Kawase and Ulbrecht (1983b), except for the constant $\alpha(n)$. While there are no data available on liquid–solid heat transfer, limited experimental results are available on the related mass transfer problem with packed beds made up of spheres (Potucek and Stejskal, 1989; Kumar and Upadhyay, 1981; Hilal et al., 1991) and of nonspherical particles (Wronski and Szembek-Stoeger, 1988; Hwang et al., 1993; Peev et al., 2002). On the other hand, Coppola and Bohm (1985) used the capillary bundle approach to correlate their mass transfer results for beds of stacked screens. Most of these studies have been reviewed recently (Chhabra et al., 2001b; Shukla et al., 2004) and therefore only the main findings are summarized here. Based on the mass transfer data with one polymer solution ($n=0.85$), Kumar and Upadhyay (1981) developed the following empirical correlation:

$$\varepsilon j = \frac{0.765}{\text{Re}_1^{0.82}} + \frac{0.365}{\text{Re}_1^{0.386}} \tag{8.57}$$

Equation 8.57 is based on experimental results covering the range of conditions as $10^{-4} \leq \text{Re}_1 \leq 40$ and $800 \leq \text{Sc}_1 \leq 7.2 \times 10^4$. Figure 8.24 shows the predictions of Equations 8.55 and 8.57. As seen here, suffice it to add that these results are consistent with that of Kawase and Ulbrecht (1983b). Based on the use of the equal volume sphere diameter, Wronski and Szembek-Stoeger (1988) correlated their mass transfer data for cylindrical pellets as

$$\varepsilon j = \left(0.097 \text{Re}_1^{0.30} + 0.75 \text{Re}_1^{0.61} \right)^{-1} \tag{8.58}$$

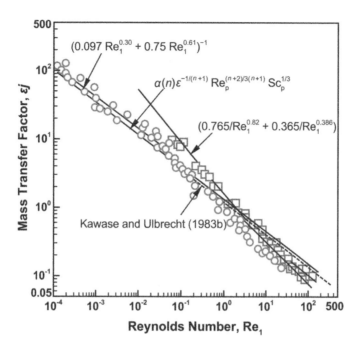

FIGURE 8.24 Liquid–solid mass transfer in fixed beds: predictions versus experiments. (○) Kumar and Upadhayay (1981); (□) Wronski and Szembek-Stoeger (1988).

and the predictions of this equation are also included in Figure 8.24. The agreement between the two sets of independent experimental results is seen to be very good. The subsequent work reported by Hwang et al. (1993) for cylindrical benzoic acid pellets (with length ≈ diameter) fluidized by CMC solutions ($0.63 \leq n \leq 0.92$; $0.01 \leq Re_1 \leq 600$) is also consistent with the predictions of Equations 8.57 and 8.58, except for the low values of the Reynolds number. In spite of the moderate agreement as seen in Figure 8.24, Hwang et al. (1993), however, put forward the correlation which purports to offer an improved fit to data as

$$\log(\varepsilon j) = 0.169 - 0.455 \log Re_1 - 0.0661 (\log Re_1)^2 \tag{8.59}$$

Furthermore, Figure 8.25 shows the utility of the cell model predictions of Shukla et al. (2004) by comparing them with the predictions of Equation 8.58 for three values of voidage, that is, $\varepsilon = 0.4$, 0.5, and 0.6 and for a wide range of power-law index as $0.5 \leq n \leq 1.8$, but for $Pe > 50$, without any discernable trends. Surprisingly while Equation 8.58 is based on data for pseudoplastic liquids, Figure 8.25 seems to suggest that it is applicable also for dilatant fluids, $n > 1$.

In an interesting paper, Coppola and Bohm (1985) have carried out a similar mass transfer study with power-law fluids in a packed bed of screens. Based on two different viewpoints, namely flow around a cylinder or the capillary bundle approach, two separate correlations of the generic form are presented as

$$Sh_1 = A Re_p^B Sc^C \tag{8.60}$$

For the flow around a cylinder approach

$$Sc = \frac{V_0 d}{D_{AB} Re_p^{2/(n+1)}}$$

$A = 0.838$; $B = 0.33$; $C = 0.37$

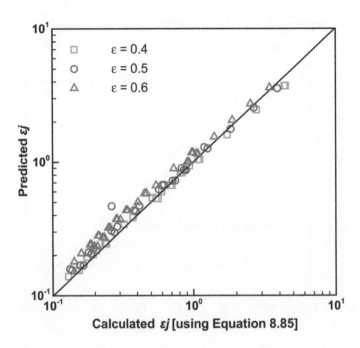

FIGURE 8.25 Comparison between the cell model predictions, y-axis values (Shukla et al., 2004) and Equation 8.58.

For the capillary bundle model

$$\text{Re} = \frac{\rho V_0^{2-n} d}{\mu_{\text{eff}}}; \quad \text{Sc} = \frac{\mu_{\text{eff}} V_0^{n-1}}{\rho D_{\text{AB}}}$$

$$\mu_{\text{eff}} = 2mM^{-(n+1)/2}\left(\frac{3n+1}{n}\right)^n \left\{\frac{d^2\varepsilon^2}{16M(1-\varepsilon)^2}\right\}^{(1-n)/2}$$

$A=0.908$; $B=0.33$; $C=0.34$

where M, the so-called Kozeny constant, is a function of voidage. The best values of the constants A, B, and C were evaluated using experimental data obtained for two values of n only ($n=0.74$ and 0.81). Despite the two entirely different approaches, the resulting values of A, B, and C are nearly the same; however, due to the narrow range of experimental conditions, it is not possible to discriminate between these two approaches.

The effect of the drag-reducing polymer solutions on the rate of mass transfer in fixed bed reactors has been examined by Sedahmed et al. (1987), Zarraa (1998), and Fadali (2003). The rates of mass transfer were measured for the cementation of copper from dilute copper sulfate solutions containing trace quantities of polyethylene oxide in a fixed bed of zinc pellets. Depending upon the polymer concentration and the value of the Reynolds number, the rate of mass transfer decreased by up to 50% below the corresponding value for copper sulfate solutions without any polymer addition. Also, the mass transfer j-factor was seen to show a slightly stronger dependence on the Reynolds number in polymeric solutions. Also, it appears that for a given polymer solution, the rate of mass transfer progressively decreases with the increasing Reynolds number, going through a minimum at a critical Reynolds number and followed by a region of increase in the rate of mass transfer, ultimately reaching the pure water limit (Fadali, 2003). The critical value of the Reynolds number is likely to be strongly dependent on the type of polymer and its concentration. Fadali (2003) worked with polyethylene oxide (WSR-301) solutions in the range of

10–300 ppm and reported the critical Reynolds number to be about ~1400. In view of the fact that such dilute solutions are highly prone to mechanical degradation, extrapolation of results outside the limit of experimental conditions must be treated with reserve.

Aside from these studies, in a series of papers, Rao (2000b, 2001, 2002) has studied the wall-to-power-law fluid heat transfer in packed beds of spherical particles, with rather significant wall effects ($1.6 \leq D_c/d \leq 4.5$). Using aqueous solutions of Carbopol and POLYOX (WSR-301), he was able to cover a wide range of parameters as: $50 \leq \text{Re} \leq 4 \times 10^4$; $6 \leq \text{Pr} \leq 250$; $0.6 \leq n \leq 0.8$ and the bed porosity, $0.3 \leq \varepsilon \leq 0.7$. He found the following correlation to be adequate for his data:

$$\text{Nu} = \frac{hd}{k} = 0.85 n^{-0.5} \text{Re}_B^{1/3} \text{Pr}_B^{0.37} \tag{8.61}$$

The corresponding Reynolds and Prandtl numbers are defined as

$$\text{Re}_B = \frac{\rho V_0 d}{\mu_e (1-\varepsilon)}$$

$$\text{Pr}_B = \frac{C_p \mu_e}{k}$$

where the effective viscosity μ_e used in the definitions of Re_B and Pr_B was evaluated at the effective (nominal) shear rate of $(8V_0/d)$, as if it were an empty tube. Within the range of values of the power-law index ($0.6 \leq n \leq 0.8$), Equation 8.61 predicts enhanced rates of heat transfer in shear-thinning fluids.

8.8 Liquid–Solid Fluidized Beds

Based on the assumption that the sole effect of the presence of neighboring particles is to alter the flow field around each particle, Kawase and Ulbrecht (1985b) modified the correlation for a single particle by using the hindered settling velocity correction due to Richardson and Zaki (1954) to arrive at the following empirical expression for the interphase particle–liquid mass transfer in fluidized beds for power-law liquids:

$$\varepsilon j = 1.50 \varepsilon^{0.2-0.24n} \left[\left(\frac{3n+1}{4n} \right)^n \left\{ \frac{12(1-\varepsilon)}{\varepsilon^2} \right\}^{n-1} \right]^{-0.063} \text{Re}_1^{-0.73} \tag{8.62}$$

A limited amount of experimental results on particle–liquid (Kumar and Upadhyay, 1980, 1981; Burru and Briens, 1989, 1991; Hwang et al., 1993) and on the wall-to-bed (Tonini et al., 1981) mass transfer in fluidized beds is available. Kumar and Upadhyay (1980, 1981) used only one test fluid with $n=0.85$ and found that these results were in line with the predictions of Equation 8.57. Figure 8.26 shows a comparison between the predictions of Equations 8.57, 8.58, and 8.59. The correspondence is seen to be good in the overlapping range of these predictions, but more data is needed to discriminate between them. The only other experimental study on mass transfer in fluidized beds is due to Tonini et al. (1981). They have measured the wall-to-bed mass transfer using an electrochemical technique. The particles were contained in the annular space in between the outer and inner walls. In this manner, mass transfer coefficients were measured for ten different solutions ($1 \geq n \geq 0.68$) and in the bed porosity range of 0.45–0.90. Based on the capillary model, the results were correlated via the following relation:

$$\text{Sh}_1 = 1.45 \text{Re}_1^{0.42} \left(\text{Sc}^* \right)^{0.33} (1-\varepsilon)^{-0.42} \tag{8.63}$$

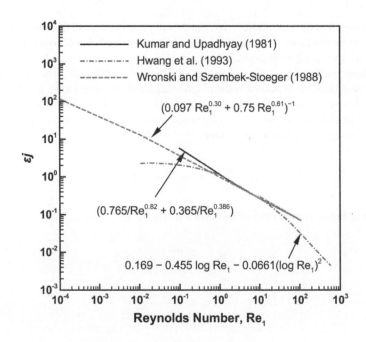

FIGURE 8.26 Correlation of particle–liquid mass transfer in fluidized beds.

The average deviation between their experiments and predictions was stated to be about 10% in the following ranges of conditions: $4.9 \times 10^{-3} \leq Re_1 \leq 190$; $1813 \leq Sc^* \leq 3.7 \times 10^5$; $5.8 \leq Sh_1 \leq 72$.

Similarly, Ehsani et al. (2004) have measured wall-to-bed heat transfer for cylindrical and spherical particles fluidized by non-Newtonian CMC solutions.

8.9 Three-Phase Fluidized Bed Systems

Three-phase fluidized bed (TPFB) systems are widely employed in a range of biotechnological applications. Consequently, several studies (Kato et al., 1981; Kang et al., 1985; Patwari et al., 1986; Schumpe and Deckwer, 1987; Burru and Briens, 1989, 1991; Schumpe et al., 1989; Zaidi et al., 1989, 1990; Miura and Kawase, 1997, 1998) dealing with the hydrodynamics, heat, and mass transfer processes with a non-Newtonian continuous phase in TPFB have been reported in the literature. Prior work with the Newtonian liquid phase has been thoroughly reviewed by Fan (1989) and subsequently by Kim and Kang (1997). It is widely recognized that fermentation broths, xanthan gums, etc. display complex non-Newtonian behavior. Indeed, these as well as the other similar systems of industrial significance have provided much impetus to the recent activity in this field. Consequently, extensive results on gas–liquid mass transfer in TPFB systems with carboxymethyl cellulose and xanthan solutions are now available in the literature (Patwari et al., 1986; Schumpe et al., 1989). Generally speaking, the addition of solid particles to gas–liquid systems leads to higher values of the volumetric mass transfer coefficient and heat transfer coefficients due to enhanced mixing. Usually, this enhancement is attributed to the increased levels of turbulence and mixing brought about by rupturing bubbles. However, the mass transfer coefficient shows complex dependence upon the particle size as it goes through a maximum value at a critical particle diameter. Schumpe et al. (1989) proposed the empirical correlation for predicting the gas–liquid mass transfer coefficient in such three-phase systems, expressed as

$$\frac{k_L a}{\sqrt{D_{AB}}} = 2988 V_G^{0.44} V_L^{0.42} \mu_{\text{eff}}^{-0.34} V_o^{0.75} \qquad (8.64a)$$

where μ_{eff}, the effective viscosity of the liquid, which is evaluated at the effective shear rate $\dot{\gamma}_{\text{eff}}$, estimated using the following expression (albeit considerable confusion exists regarding the estimation of $\dot{\gamma}_{\text{eff}}$ in TPFB systems):

$$\dot{\gamma}_{\text{eff}} = 2800\left(V_G - \frac{\varepsilon_G}{\varepsilon_L}V_L\right) + \frac{12V_L\varepsilon_s}{d\varepsilon_L^2}\left(\frac{3n+1}{4n}\right) \quad (8.64b)$$

Note that Equations 8.64a and 8.64b are not dimensionless and all quantities are in SI units. Equation 8.64a was stated to be applicable over the following ranges of conditions: $0.017 < V_G < 0.118$ m s^{-1}; $0.03 < V_L < 0.16$ m s^{-1}; $1 \leq \mu_{\text{eff}} \leq 119$ mPa s, and $0.08 < V_o < 0.6$ m s^{-1}.

Similarly, Burru and Briens (1989) have also measured particle–liquid mass transfer in TPFB systems. These workers reported a decrease in particle–liquid mass transfer due to the non-Newtonian viscosity. Likewise, limited results on heat transfer from a surface immersed in a TPFB system employing non-Newtonian liquids are also available in the literature (Kato et al., 1981; Kang et al., 1985; Zaidi et al., 1989, 1990). Qualitatively, the heat transfer coefficient increases with the increasing gas velocity and decreases with the increasing viscosity of the liquid phase and with the particle size. Furthermore, the heat transfer coefficient shows a maximum value with respect to the liquid velocity and bed voidage ($\varepsilon \sim 0.5$–0.6).

Zaidi et al. (1990) presented the following empirical correlation for the wall-to-bed heat transfer in TPFB systems:

$$\text{Nu}_1 = 0.042\,\text{Re}_{\text{Liq}}^{0.72}\,\text{Pr}_{\text{liq}}^{0.86}\,\text{Fr}_G^{0.067} \quad (8.65)$$

and it was stated to be applicable in the following ranges of conditions: $0.0081 \leq V_G \leq 0.144$ m s^{-1}; $0.0127 \leq V_L \leq 0.09$ m s^{-1}; $3.7 \leq \mu_{\text{eff}} \leq 300$ mPa s and $d = 3$ and 5 mm. The effective viscosity was evaluated at the shear rate given by Equation 8.64b. It is somewhat surprising that in their previous study, these authors put forward the following dimensional predictive expression:

$$h = 1800 V_G^{0.11} \mu_{\text{eff}}^{-0.14} V_L^{1.03(0.65-\varepsilon)} d^{0.58(\varepsilon-0.68)} \quad (8.66)$$

Equation 8.66 also encompasses the same ranges of variables as Equation 8.65. A discussion on the current state of the art in this field as well as on the aspects meriting further work has recently been presented by Kim and Kang (1997).

8.10 Heat Transfer from Tube Bundles

A wealth of analytical (Sangani and Acrivos, 1982a; Sangani and Yao, 1988; Wang and Sangani, 1997), numerical (Masliyah, 1973; Spalding and Ramachandra, 1982; Martin et al., 1998; Satheesh et al., 1999; Wilson and Bassiouny, 2000; Ghosh Roychowdhury et al., 2002; Mandhani et al., 2002; Comini and Croce, 2003; Mangadoddy et al., 2004), and experimental (Zukauskas, 1987) information is available on heat transfer from tube bundles to Newtonian fluids in crossflow configuration. In contrast, there has been very little activity on heat (or mass) transfer from tube bundles with non-Newtonian liquids (Ghosh et al., 1994; Chhabra, 1999b). Ferreira and Chhabra (2004) paralleled the approach of Kawase and Ulbrecht (1981a) to obtain an approximate closed-form expression for mass transfer from a rod bundle to power-law fluids in the limit of zero Reynolds number and large Peclet number. Subsequently, the range of these results has been extended numerically to finite Reynolds number (1–500) and Peclet (1–5000) numbers using the free surface cell model (Mangadoddy et al., 2004; Soares et al., 2005b). Both these predictions yield varying levels of enhancement in heat transfer depending upon the values of the power-law index ($n < 1$), Reynolds number, and Prandtl number. Broadly, the larger the Reynolds and

Prandtl number, the greater is the enhancement in heat transfer as compared to that in Newtonian fluids. In spite of the moderate values of the Peclet number embraced by these numerical studies, the values of the Reynolds and Prandtl numbers encountered in practical situations (Adams and Bell, 1968; Prakash, 1985) tend to be much larger than those spanned by the current numerical studies. Using the capillary bundle approach, Ghosh (1992) collated much of the literature data for tube banks to develop the following correlation for Nusselt number for the crossflow of power-law fluids past bundles of circular tubes:

$$\text{Nu}_T \left(\Delta \frac{D_E}{L} \text{Pr}_C \right)^{-1/3} \left(\frac{m_w}{m_b} \right)^{0.14} = 4.8 + 0.74 \text{Re}_C^{0.667} \tag{8.67}$$

This correlation is based on the following ranges of data: $0.56 \leq n \leq 1$; $7 \leq \text{Pr}_C \leq 5200$; and $0.3 \leq \text{Re}_C \leq 7000$. The effective viscosity used in the evaluation of the Reynolds and Prandtl numbers is estimated as

$$\mu_\text{eff} = m' \left(\frac{\tau_w}{m'} \right)^{(n-1)/n} \tag{8.68a}$$

where

$$m' = m\Delta^n \tag{8.68b}$$

$$\tau_w = \frac{D_E}{4} \left(-\frac{\Delta P}{l} \right) \tag{8.68c}$$

and in turn, the hydraulic diameter D_E is given by

$$D_E = \frac{d\varepsilon}{(1-\varepsilon)} \tag{8.68d}$$

Within this framework, the relevant Prandtl and Reynolds numbers are defined as

$$\text{Pr}_C = \frac{C_p \mu_\text{eff}}{k} \tag{8.69}$$

$$\text{Re}_C = \frac{\rho V_o d}{(1-\varepsilon)\mu_\text{eff}} \tag{8.70}$$

Limited comparison between the cell model predictions and experimental results are found to be affirmative and encouraging (Ferreira and Chhabra, 2004; Mangadoddy et al., 2004). Forced convection from a square array of cylinders ($0.7 \leq \varepsilon \leq 0.99$) in power-law fluids has been numerically studied by Pravesh et al. (2019b).

8.11 Conclusions

This chapter has been a bit of a mixed bag of ideas and results on convective heat/mass transport in the forced convection regime with non-Newtonian liquids from immersed objects. In particular, consideration has been given to the boundary layer flows over a plate, cylinder, and sphere. These have led to the development of useful expressions for the prediction of Nusselt (or Sherwood) numbers as functions of the relevant dimensionless groups (Reynolds and Prandtl numbers) and of rheological parameters from the immersed surfaces to non-Newtonian liquids. Barring a few exceptions, the simple power-law and Bingham plastic models have been used almost universally to denote shear rate-dependent viscosity. Based on these treatments and further combined with the limited numerical solutions and experimental results, it is possible to predict the rate of heat/mass transfer to power-law fluids in external flows (at least for a plate or a cylinder or a sphere) with reasonable levels of accuracy in the laminar flow regime. The visco-elasticity introduces changes both in the detailed structure of the flow and at the macroscopic level by way of yielding anomalous transport behavior. Our understanding about the role of visco-elasticity is still in its infancy. Similarly, the presence of yield stress also adds to the complexity of analysis whence only limited information is available even for the simplest case of two-dimensional boundary layer flow of a visco-plastic fluid over a flat plate. Likewise, the corresponding problem of a fluid sphere has also received limited attention, most of which is restricted to the zero Reynolds number situations only.

Some information is also available on liquid–solid mass transfer in packed and fluidized beds of spherical and nonspherical particles, mainly for power-law fluids. The visco-elasticity appears to impede the rate of mass transfer in these systems. Finally, very little information is available for three-phase fluidized beds, and for tube bundles. Undoubtedly, the literature is inundated with empirical correlations, some of which have been checked using independent data, but all are restricted to a rather limited range of conditions. Therefore, extreme caution needs to be exercised while using these expressions beyond the range of their applicability.

Nomenclature

B	Width of plate (m)
$Bi = (\tau_0^B d/V_0 \mu_B)$	Bingham number (-)
$Bi_{HB} = \tau_0^{HB}/m(V_0/d)^n$	Bingham number for Herschel-Bulkley fluids (-)
$Br = (\mu_0 V_0^2/k\Delta T)$	Brinkman number (-)
C_f	Skin friction coefficient, Equation 8.10 (-)
$C(n)$	Constant, Equation 8.11 (-)
$C_0(n)$	Constant, Equation 8.16 (-)
C_p	Specific heat (J/kg K)
d	Particle/drop/bubble/cylinder diameter (m)
D	Tube diameter (m)
De	Deborah number (-)
D_{AB}	Molecular diffusivity (m² s⁻¹)
$Fr = (V_0^2/gd)$	Froude number (-)
g	Acceleration due to gravity (m s⁻²)
h	Heat transfer coefficient (W m⁻² K⁻¹)
$j = (k_c/V_0)(Sc^*)^{2/3}$	Mass transfer factor (-)
k	Thermal conductivity of fluid (W m⁻¹ K⁻¹)
k_c	Mass transfer coefficient (m s⁻¹)
$k_L a$	Volumetric mass transfer coefficient (s⁻¹)
m	Power-law consistency index (Pa sⁿ)
$m' = m\Delta^n$	Apparent power-law consistency index (Pa sⁿ)
n	Power-law flow behavior index (-)
$Nu = (hR/k)$	Nusselt number (-)
$Nu_1 = 2Nu = (hd/k)$	Nusselt number (-)
$Nu_2 = Nu_1(\varepsilon_L/\varepsilon_s)$	Modified Nusselt number for TPFB systems (-)
Nu_o	Nusselt number based on temperature independent properties, equation 8.28 (-)
Nu_L	Nusselt number averaged over the length of the plate (-)
$Nu_T = Nu_1(\varepsilon/(1-\varepsilon))$	Modified Nusselt number for tube bundles (-)

Nu_x	Local Nusselt number, Equation 8.12 (-)
$Pe = (dV_0/D_{AB}) = Re \cdot Sc$	Peclet number (-)
$Pr_C = (1/\alpha)(m/\rho)^{(5/(n+4))}$ $\times L^{2(n-1)/(n+4)} \times (g\beta q_s/k)^{(3(n-1))/(n+4)}$	Modified Prandtl number (-)
$Pr_{Liq} = (C_p \mu_{eff}/k)_L$	Prandtl number for TPFB systems (-)
$Pr_p = m(V_0/d)^{n-1} C_p/k$	Prandtl number (-)
$Pr_{pl} = (\rho C_p dV_0/k) Re_p^{-2/(n+1)}$	Modified Prandtl number (-)
$Pr_R = (\rho C_p RV_0/k) Re_R^{-2/(n+1)}$	Modified Prandtl number (-)
Pr_x	Local Prandtl Number, Equation 8.13 (-)
q_s	Constant heat flux specified at the surface (W m^{-2})
R	Particle/bubble/drop/cylinder radius (m)
$Re^* = Re_B/(1+Bi)$	Modified Reynolds number for Bingham plastic fluids, Equation 8.25 and Equations 8.34a, 8.34b (-)
$Re_{HB}^* = Re_{PL}/(1+Bi_{HB})$	Modified Reynolds number for Herschel-Bulkley fluids (-)
$Re_1 = Re_p(4n/(3n+1))^n$	Modified Reynolds number for packed and fluidized beds $\times\{12(1-\varepsilon)/\varepsilon^2\}^{1-n}$(-)
$Re_2 = \{\rho V_0 d/m'(8V_0/\varepsilon d)^{n-1}\}$	Modified Reynolds number, Equation 8.54 (-)
$Re_L = (\rho V_0^{2-n} L^n/m)$	Reynolds number based on the length of the plate (-)
$Re_L^* = (2/3)^{n-1} Re_x (L/x)^{n/(n+1)}$	Modified Reynolds number, Equation 8.16 (-)
$Re_{Liq} = (\rho V_L d/(1-\varepsilon_L)\mu_{eff}$	Modified Reynolds number for TPFB systems (-)
$Re_p = (\rho V_0^{2-n} d^n/m)$	Particle Reynolds number, Equation 8.20 and Equation 8.29
$Re_R = Re_p/2^n$	Modified Reynolds number (-)
$Re_x = (\rho V_0^{2-n} x^n/m)$	Local Reynolds number for a plate (-)
$Sc^* = Sc_p((3n+1)/4n)^n$ $\times\{12(1-\varepsilon)/\varepsilon^2\}^{n-1}$	Modified Schmidt number for fixed and fluidized beds (-)
$Sc_2 = (V_0 L/D_{AB}) Re_L^{n-2}$	Modified Schmidt number, Equation 8.18 (-)
$Sc_L^* = (m/\rho D_{AB}) \times \{(0.105 Re_x)^{1/(n+1)} x\}$ $\times (L/x)^{(1-n)/(1+n)}$	Modified Schmidt number, Equation 8.16 (-)
$Sc_P = (m/\rho D_{AB})(d/V_0)^{1-n}$	Particle Schmidt number (-)
$Sh = (k_c R/D_{AB})$	Mean Sherwood number (-)
$Sh_1 = 2Sh = (k_c d/D_{AB})$	Mean Sherwood number (-)
$Sh_L = (k_c L/D_{AB})$	Sherwood number averaged over the length of the plate (-)
T	Temperature of the fluid (K)
T_b	Bulk temperature (K)
T_s	Heated surface temperature (K)
T_o	Free stream temperature (K)
V_0	Free stream velocity or superficial velocity (m s^{-1})
V_θ^*	Dimensionless angular velocity (-)
V_{sw}	Swarm velocity (m s^{-1})
V_t	Free settling velocity of a solid particle (m s^{-1})
V_x	x-component of velocity (m s^{-1})
V_y	y-component of velocity (m s^{-1})
x	Distance along the surface (m)
x_1	Dimensionless distance along a heated surface (-)
X_E	Dispersed to continuous phase viscosity ratio (-)
y	Distance normal to the surface (m)
Y_m	Mass transfer enhancement factor for a single particle (-)
$Y_{m,sw} = Y_m (V_0/V_{sw})^{1/2}$	Mass transfer enhancement factor for a swarm of bubbles or drops (-)

Greek Symbols

$\alpha = (k/\rho C_p)$	Thermal diffusivity ($m^2\ s^{-1}$)
$\dot{\gamma}_{eff}$	Effective shear rate in TPFB system (s^{-1})
δ	Momentum boundary layer thickness (m)
δ_T	Thermal boundary layer thickness (m)
$\Delta = (3n + 1/(4n))$	Rabinowitsch-Mooney factor (-)
ε	Porosity (voidage) of a fixed or fluidized bed (-)
$\varepsilon_L, \varepsilon_G, \varepsilon_S$	Liquid/gas/solid hold ups in a TPFB system (-)
η	Similarity parameter (-)
$\theta = (T - T_o)/(T_s - T_o)$	Dimensionless temperature (-)
μ_B	Bingham plastic viscosity (Pa.s)
μ_{eff}	Apparent viscosity at $\dot{\gamma} = \dot{\gamma}_{eff}$ (Pa.s)
μ_o	Zero shear viscosity (Pa.s)
μ_{To}	Viscosity evaluated at $T = T_o$ (Pa.s)
μ_{Ts}	Viscosity evaluated at $T = T_s$ (Pa.s)
ξ	Angle between the normal to the surface and the direction of gravity (-)
ρ	Fluid/continuous phase density (kg m^{-3})
τ_o^B	Bingham yield stress (Pa)
τ_{xy}	x–y component of extra stress tensor (Pa)
ϕ	Volume fraction of the dispersed phase (-)
$\phi(n)$	Function of n, Equation 8.14 (-)

Subscripts

B	Evaluated at bulk temperature
G	Gas
L	Liquid
S	Solid
SW	Swarm
0	Either zero-shear condition or refer to temperature independent properties
W	Evaluated at wall temperature
x	Local value
θ	Local value

9

Heat and Mass Transfer in Particulate Systems: Free and Mixed Convection

9.1 Introduction

The density of every fluid decreases with temperature, which gives rise to buoyancy-induced flow resulting in heat transfer from the region of high temperature to that of low temperature. Under these conditions, in the absence of imposed flow, heat transfer occurs by free (or natural) convection. Conversely, as long as a temperature gradient exists in a fluid, free convection is, howsoever feeble, always present even in the forced-convection regime with an externally imposed flow (as discussed in Chapter 8). In confined geometries, such as the reheating of canned foodstuffs, free convection is the sole mechanism of heat transfer. In a given application, the relative importance of the free- and forced-convection contributions to the overall heat transfer is determined by the value of the familiar nondimensional parameter, Richardson number, $Ri = Gr/Re^2$, which is a ratio of the buoyancy forces and inertial forces. Thus, vanishingly small values of $Ri \to 0$ correspond to the pure forced convection (negligible role of free convection). At the other extreme is the pure free convection characterized by large values of $Ri \to \infty$ wherein the forced convection does not contribute to the overall heat transfer. The values of $Ri \sim 1$ denote the situation where the two contributions to the flow are of comparable magnitudes. While the preceding discussion hinges on the variation of the fluid density with temperature, analogous density gradient in solutions also occurs due to the variation of the density with the solute concentration, thereby leading to the buoyancy-induced flow resulting in convective mass transfer without any imposed flow. This chapter is concerned with the convective transport of energy and mass by free- and mixed-convection flows. In this case too, heat transfer aspects have been investigated much more extensively than the corresponding mass transfer characteristics. Also, there exists a voluminous literature on the so-called internal flows whereas the external flows (such as that considered in this book) have received little attention, even in Newtonian fluids let alone in non-Newtonian fluids. Excellent reviews of the free- and mixed-convection heat and mass transfer studies are available in the literature (Martynenko and Khramtsov, 2005; Oosthuizen and Kalendar, 2012, 2013, 2018, Boetcher, 2014; Chhabra, 2018, etc.). Limited reviews (though somewhat dated now) for free and mixed convection in non-Newtonian fluids are that of Shenoy and Mashelkar (1982) and Shenoy (1988, 2017), etc.

In this chapter, attention is given to the free- and mixed-convection heat transfer in external flows, especially for the three model configurations of a vertical plate, a sphere, and a horizontal cylinder immersed in different types of non-Newtonian fluids. We begin with the free convection followed by the mixed-convection regimes for each of these geometries.

9.2 Governing Equations

For an incompressible fluid, the velocity and temperature fields are obtained by solving the momentum and energy equations for a given shape of the object and the type of boundary conditions prescribed on the surface of the solid body. These equations are included in their general dimensionless forms, which is useful in identifying the relevant dimensionless parameters in a given situation. These equations, namely, continuity, Cauchy's momentum equation, and the thermal energy equations are written in their compact forms as:

Continuity equation:

$$\nabla \cdot V = 0 \qquad (9.1)$$

Momentum equation:

$$\frac{DV}{Dt} = -\nabla p + N_1 \nabla \cdot \tau + N_2 \xi \delta_{ij} \qquad (9.2)$$

Energy equation:

$$\frac{D\xi}{Dt} = N_3 \nabla^2 \xi + N_4 \mu \phi_v \qquad (9.3)$$

In writing these equations, characteristic length (l_c), velocity (V_c), and temperature difference (ΔT_{ref}) have been used. The specific choice of these scales may vary from one problem to another depending upon the shape and boundary conditions together with the non-Newtonian viscosity model and the heat transfer regime (free or mixed). Since much of the literature in this field hinges on the use of the all too familiar power-law and Bingham plastic fluid models, the specific form of the nondimensional coefficients N_1, N_2, N_3, and N_4 appearing in Equations 9.2 and 9.3 are presented in Table 9.1.

Clearly, the dimensionless constant N_4 here can be identified to be the familiar Brinkman number (Br) which is a measure of the significance of viscous dissipation contribution. In most studies reported to date, the resulting values of the Brinkman number are much smaller than unity and thus this term is often neglected while seeking solutions to these equations.

As noted earlier, the variation of fluid density with temperature induces buoyancy flow and it is customary to use the well-known Boussinesq approximation for this purpose as follows:

$$\rho = \rho_o \{1 - \beta(T - T_o)\} \qquad (9.4)$$

where ρ_o is the fluid density at the reference temperature T_o and β is the isobaric coefficient of volumetric expansion of fluid given as:

$$\beta = -\frac{1}{\rho}\frac{\partial \rho}{\partial T} \qquad (9.5)$$

TABLE 9.1

Definitions of N_1, N_2, N_3, and N_4 for Power-Law and Bingham Fluids

	Power-law	Bingham plastic
N_1	$\dfrac{1}{\sqrt{Gr_{PL}}}$	$\dfrac{1}{\sqrt{Gr_B}}$
N_2	1 (for free convection) Ri_{PL} (for mixed convection)	1 (for free convection) Ri_B (for mixed convection)
N_3	$\dfrac{1}{Pr_{PL} Gr_{PL}^{1/n+1}}$	$\dfrac{1}{Pr_B \sqrt{Gr_B}}$
N_4	$\dfrac{m(V_c/l_c)^n}{\rho C_p \Delta T_{ref}}$	$\dfrac{\mu_B (V_c/l_c)}{\rho C_p \Delta T_{ref}}$

Within the framework of the Boussinesq approximation, the fluid density is treated as a constant everywhere except for the buoyancy term appearing in the momentum equation (δ_{ij} is the kronecker delta).

For free convection, the velocity scale is deduced by equating the buoyancy forces with the corresponding inertial forces and this leads to the following definition of the characteristic velocity,

$$V_c = \left(l_c g \beta \Delta T_{\text{ref}}\right)^{1/2} \tag{9.6}$$

Other definitions of V_c are possible by equating the buoyancy forces with the viscous forces (relevant in the laminar flow regime) but this scale is easily related to the one given by Equation 9.6 (Shyam et al., 2013b). Hence, in order to maintain consistency with the literature studies, Equation 9.6 has been used here. The choice of ΔT_{ref} depends upon the type of thermal boundary conditions prescribed on the surface of the heated object. The two common situations are of the isothermal surface (maintained at T_s) and of dissipating heat at a constant rate (q_s, W/m²). In these two cases, ΔT_{ref} is defined as $(T_s - T_o)$ and $(q_s l_c / k)$, respectively, where T_o is the initial fluid temperature which is often known. Now turning our attention to the definitions of the Prandtl and Grashof numbers introduced in Table 9.1.

For power-law fluids:

$$\text{Grashof number;} \quad \text{Gr}_{\text{PL}} = \frac{\rho^2 \left(g\beta \Delta T_{\text{ref}}\right)^{2-n} l_c^{2+n}}{m^2} \tag{9.7}$$

$$\text{Prandtl number:} \quad \text{Pr}_{\text{PL}} = \frac{\rho C_p}{k}\left(\frac{m}{\rho}\right)^{\frac{2}{n+1}} l_c^{\frac{1-n}{1+n}} \left(l_c g \beta \Delta T\right)^{\frac{3(n-1)}{2(n+1)}} \tag{9.8}$$

By nondimensionalizing the momentum equation by using l_c and V_c (given by Equation 9.6) leads to the definition of the Grashof number Gr_{PL} given by Equation 9.7. However, other definitions of the Prandtl number have been used in the literature. For instance, one common definition of the Prandtl number is given as:

$$\text{Pr}_o = \frac{m C_p}{k}\left(\frac{g\beta \Delta T_{\text{ref}}}{l_c}\right)^{\frac{(n-1)}{2}} \tag{9.9}$$

But it can readily be shown that this definition can easily be constructed from Gr_{PL} and Pr_{PL} as follows:

$$\text{Pr}_o = \text{Pr}_{\text{PL}}\, \text{Gr}_{\text{PL}}^{\frac{1-n}{2(n+1)}} \tag{9.10}$$

Thus, it is fairly straightforward to convert the results from one format to another. Other definitions will also be used as and when required in order to make comparisons with the literature studies.

Another dimensionless number, the so-called Rayleigh number (Ra), is also used in the context of free convection, and it is defined as the product of the Grashof and Prandtl numbers.

The Richardson number, Ri (N_2 in Equation 9.2), for mixed convection is simply given as:

$$\text{Ri}_{\text{PL}} = \frac{\text{Gr}_{\text{PL}}}{\left(\text{Re}_{\text{PL}}\right)^2} \tag{9.11}$$

where the Reynolds number, $\text{Re}_{\text{PL}} = \rho V_o^{2-n} l_c^n / m$, is based on the forced-convection velocity of the fluid V_o and Gr_{PL} is given by Equation 9.7.

For Bingham plastic fluids, the corresponding definitions are given as:

$$\text{Grashof number,} \quad Gr_B = \left(\frac{\rho^2}{\mu_B^2}\right)(g\beta\Delta T_{\text{ref}})l_c^3 \tag{9.12}$$

$$\text{Prandtl number,} \quad Pr_B = \frac{\mu_B C_p}{k} \tag{9.13}$$

In this case, the Richardson number is given by

$$Ri_B = \frac{Gr_B}{Re_B^2} \tag{9.14}$$

and $Re_B = \dfrac{\rho V_o l_c}{\mu_B}$

It needs to be added here that the power-law behavior index (n) is an independent dimensionless parameter on its own in the case of power-law fluids in addition to its appearance in the definitions of the Prandtl and Grashof numbers. Similarly, for Bingham fluids, the relevant nondimensional parameter is the familiar Bingham number Bi defined as:

$$Bi = \frac{\tau_o^B l_c}{\mu_B V_c} \tag{9.15}$$

The incorporation of the Bingham number into the Reynolds and Prandtl number leads to the following modified definitions of Re_B^* and Pr_B^*:

$$Re_B^* = \frac{Re_B}{1+Bi} \tag{9.16}$$

$$Pr_B^* = Pr_B(1+Bi) \tag{9.17}$$

Equations (9.16) and (9.17) are tantamount to using the viscosity scale of $\left(\dfrac{l_c \tau_o^B}{V_c} + \mu_B\right)$ as opposed to μ_B implicit in the definitions of Re_B and Pr_B. One can similarly define the Grashof and Richardson numbers to incorporate the effect of yield stress in these definitions. This will be done as and when needed in this chapter. Needless to say, the preceding definitions of the dimensionless parameters reduce to their standard forms for Newtonian fluids in the limits of $n=1$ and $\tau_o^B = 0$ for power-law and Bingham fluids, respectively.

Finally, before leaving this section, it is appropriate to make some observations regarding the coupled nature of the velocity and temperature fields. This coupling stems from three sources: buoyancy term in the free- and mixed-convection regimes, temperature-dependent thermophysical properties, and lastly from the viscous dissipation effects. The latter two are relevant to the forced-convection regime also (Chapter 8). Thus, the assumptions of Br $\ll 1$ and constant physical properties lead to one-way coupling, that is, the velocity field can be obtained independently of the temperature field in a sequential manner. Even with the use of these two assumptions, these equations must be solved simultaneously in the free- and mixed-convection regimes. This makes closed-form solutions virtually impossible unless further idealizations are introduced even for the Newtonian fluids. In the next section, we begin with the discussion of the results for the three geometries, namely a vertical plate, a horizontal cylinder, and a sphere. In most cases, the assumptions of constant physical properties and small Brinkman number are applicable. Free convection regime is discussed first followed by the mixed-convection regime. At the outset, Tables 9.2 and 9.3 provide an overview of the research activities in this field.

TABLE 9.2

Summary of Theoretical and numerical studies in non-Newtonian fluids

Investigator	Geometry	Fluid model	Observations
Acrivos (1960)	Horizontal cylinder/ flat plate	Power-law	General treatment for laminar-free convection at high Prandtl numbers for two-dimensional surfaces.
Aherwar et al. (2023)	Vertical plate	Power-law and Bingham plastic fluids	Extensive numerical results on Nusselt numbers for isothermal and isoflux boundary conditions.
Bose et al. (2015)	Cylinder	Bingham plastic fluid	Aiding-buoyancy mixed convection.
Chen and Wollersheim (1973)	Vertical plate	Power-law	Similarity transformation for the free convection from a plate at constant flux conditions. Results almost identical to that of Acrivos (1960) for the constant temperature case.
Erbas and Ece (2001)	Vertical plate	Power-law	Laminar-free convection for a variable surface temperature condition and shear-rate-dependent thermal conductivity. Also, identified the conditions for the existence of a similarity solution.
Fujii et al. (1972, 1973)	Isothermal vertical plate	Sutterby model fluid	Numerical solution for laminar-free convection.
Gorla (1982, 1986, 1991a, b, c, 1992)	Wedge, horizontal cylinder and plate, rotating disk	Power-law and Ellis model fluids	Laminar-free convection from a horizontal cylinder; unsteady state heat transfer from a wedge; heat transfer from a moving plate and a rotating disk.
Gupta and Chhabra (2016a, b, c)	Spheroid	Power-law and Bingham plastic fluids	Aiding-buoyancy mixed convection.
Gupta et al. (2014, 2017a)	Spheroid	Power-law and Bingham plastic fluid	Laminar-free convection.
Huang and Chen (1984, 1990); Huang and Lin (1992, 1993)	Flat plate	Power-law	Numerical analysis of laminar-forced, free, and mixed convection from horizontal and vertical plates.
Kleppe and Marner (1972)	Vertical plate	Bingham plastic fluids	Time-dependent and steady laminar-free convection from an isothermal plate.
Marner and Kleppe (1976)	Vertical plate	Bingham plastic fluids	Mixed-convection heat transfer from an isothermal plate.
Meissner et al. (1994)	Flat plate, sphere, and horizontal cylinder	Power-law	Used Merk–Chao series method to study mixed convection. In the limiting cases of free and forced convection, results are in line with that of Acrivos (1960) and of Acrivos et al. (1960).

(Continued)

TABLE 9.2 (*Continued*)
Summary of Theoretical and numerical studies in non-Newtonian fluids

Investigator	Geometry	Fluid model	Observations
Mishra et al. (2019c)	Cone	Power-law and Bingham plastic fluids	Laminar-free convection.
Nakayama and Shenoy (1991, 1992a)	Arbitrary geometries	Power-law	Extended the approach of Nakayama and Koyama (1988) to turbulent-free convection in drag reducing fluids.
Nalluri et al. (2015)	Hemisphere	Bingham plastic fluid	Laminar-free convection.
Nirmalkar and Chhabra (2013)	Sphere	Power-law	Aiding-buoyancy mixed convection.
Nirmalkar et al. (2014a, b, c)	Sphere and cylinder	Bingham plastic fluid	Heat transfer in free- and mixed-convection regimes.
Patel and Chhabra (2015, 2016a, 2019)	Elliptic cylinder	Bingham plastic fluid	Heat transfer in laminar-free and mixed-convection regime.
Prhashanna and Chhabra (2010, 2011)	Sphere and cylinder	Power-law	Laminar-free convection.
Sairamu et al (2013)	Cylinder in a square duct	Bingham plastic fluid	Effect of confinement on free-convection heat transfer.
Sasmal and Chhabra (2012a, 2014)	Horizontal elliptical cylinder and hemisphere	Power-law	Laminar-free convection.
Shyam et al. (2013a)	Cylinder in a square duct	Power-law	Effect of confinement on free-convection heat transfer.
Shenoy and Mashelkar (1978a)	Vertical plate	Power-law	Turbulent-free convection based on the approach of Eckert and Jackson (1950).
Shulman et al. (1976)	Vertical plate	Power-law	Used the method of coupled asymptotic expansion to analyze the transient-free convection transport.
Soares et al. (2009)	Cylinder	Power-law	Cross-buoyancy-mixed convection in steady regime.
Som and Chen (1984)	Two-dimensional shapes	Power-law	Laminar-free convection for variable surface conditions. Also, attempts to delineate shapes for which similarity solutions are possible.
Srinivas et al. (2009)	Cylinder	Power-law	Aiding-buoyancy mixed convection.
Stewart (1971)	Sphere	Power-law	Improved upon the analysis of Acrivos (1960) for laminar-free convection.
Tien (1967), Tien and Tsuei (1969)	Vertical plate	Power-law and Ellis model fluids	Laminar-free convection from isothermal surfaces.
Wang (1993, 1995)	Vertical and horizontal plate	Power-law	Mixed-convection analysis. In the limit of pure forced convection, the predictions are in line with that of Acrivos et al. (1960), Huang and Chen (1984), and Kim et al. (1983).
Wang and Kleinstreuer (1988a, b)	Sphere and horizontal cylinder	Power-law	Mixed-convection results.

TABLE 9.3

Summary of Experimental Studies on Natural and Mixed Convection from Plates, Spheres, and Cylinders in Non-Newtonian Liquids

Investigator	Geometry	Fluid Model	Observation
Abou-Ziyan et al. (2017)	Horizontal cylinder ($L/d=8$)	Power-law fluids	Laminar-free convection in weakly shear-thinning CMC solutions.
Alhamdan and Sastry (1990)	Mushroom-shaped aluminum particles	CMC solutions ($0.67 \leq n \leq 1$)	Correlations for Nusselt number as functions of Rayleigh and Fourier number for free convection. Results for heating and cooling tests do not coincide.
Amato and Tien (1970, 1972, 1976)	Isothermal spheres and plate	CMC and PEO solutions ($0.59 \leq n \leq 0.95$)	Correlations for free convection. Detailed information on the temperature and velocity profiles.
Awuah et al. (1993)	Cylinders cut from potatoes and carrots in cross-flow	CMC solutions ($0.52 \leq n \leq 1$)	At low velocities, free-convection-dominated heat transfer. Presented empirical correlations for Nusselt number.
Chhabra (1997)	Vertical short cylinders	CMC solutions ($0.62 \leq n \leq 1$)	Experimental results on mass transfer are consistent with the analysis of Acrivos (1960).
Dale and Emery (1972)	Vertical plate	Carbopol solutions ($0.24 \leq n \leq 0.9$)	Numerical solution and experimental correlations for Nusselt number in free convection.
Emery et al. (1971)	Vertical plate	CMC and carbopol solutions	Free-convection heat transfer results consistent with the analysis of Acrivos (1960).
Fujii et al. (1973)	Isothermal vertical plate	CMC and PEO solutions (Sutterby model)	Numerical and experimental results on laminar-free convection.
Gentry and Wollersheim (1974)	Isothermal horizontal cylinder	CMC solutions ($0.64 \leq n \leq 0.93$)	Free-convection results consistent with the analysis of Acrivos (1960).
Kim and Wollersheim (1976)	Horizontal cylinder	Power-law	Experimental results on laminar-free convection.
Lee and Donatelli (1989)	Sphere	CMC solutions ($0.58 \leq n \leq 1$)	Free-convection mass transfer. Results are consistent with the analysis of Acrivos (1960), except in the limit of pure diffusion.
Liew and Adelman (1975)	Isothermal sphere	Natrosol, CMC and carbopol solutions ($0.3 \leq n \leq 1$)	Extensive data on laminar-free convection and developed empirical correlations.
Lyons et al. (1972)	Horizontal cylinder	PEO solutions	Reductions in heat transfer by polymer addition in the laminar-free convection regime.
Ng and Hartnett (1986, 1988); Ng et al. (1986, 1988)	Thin wires in cross-flow	Natrosol, Carbopol and Polyox solutions	Extensive data on laminar-natural convection in pseudoplastic and visco-elastic media. With the use of zero-shear viscosity, one can use the Newtonian formulae.
Reilly et al. (1965)	Vertical plate	Carbopol solutions ($0.72 \leq n \leq 1$)	Extensive data and correlation on laminar-free convection. Data is consistent with the results of Acrivos (1960).
Sharma and Adelman (1969)	Vertical plate	Carbopol solutions ($0.2 \leq n \leq 1$)	Empirical correlations for Nusselt number under isoflux surface conditions for free convection.
Yamanaka et al. (1976b)	Sphere	CMC solutions ($n \sim 1$)	Mixed convection data on Nusselt number correlated as f (Gr, Re, Pr).
Yamanaka and Mitsuishi (1978)	Spheres	Several polymer solutions ($0.3 \leq n \leq 0.93$)	Empirical correlations for mixed convection.

9.3 Vertical Plate

9.3.1 Free Convection

9.3.1.1 Newtonian Fluids

It is useful to begin with a discussion of the early developments in this field which in turn facilitates the presentation and discussion of the new results for non-Newtonian fluids. As noted earlier, free-convection flow arises from the density stratification due to the temperature- or concentration-gradients within the fluid. Even with the feeble flow, the resulting values of the convective heat and mass transfer coefficients are somewhat larger than the corresponding conduction (diffusion) limiting values, but these values are much smaller than those in the forced-convection regime. Consequently, a wealth of information has accrued on the free-convection heat transfer in model configurations (vertical plate, horizontal cylinder, and sphere) for the two standard thermal boundary conditions, namely the isothermal surface or the one dissipating heat at a constant rate (constant flux). Indeed, one of the earliest references can be traced back to the work of Pohlhausen in 1921 who presented an approximate solution of the boundary layer equations for free convection from an isothermal vertical surface for air (Pr=0.733). Subsequently, this study has been extended to other values of Prandtl numbers (e.g., see Ostrach (1972) for $0.1 \leq Pr \leq 1000$), uniform heat flux (e.g., see Sparrow and Gregg, 1956), and this problem continues to attract attention even now (e.g., see Yildiz and Basaran, 2019). Extensive reviews of the pertinent literature are available in Martynenko and Kharamtsov (2005) and Patel et al. (2017). Thus, based on a combination of approximate and numerical results and experimental results, it is now possible to estimate the average Nusselt number for a vertical plate for both boundary conditions under most conditions of practical interest. For instance, Churchill and Chu (1975) collated much of the literature data on the average Nusselt number for an isothermal vertical plate as:

For $Ra_L < 10^9$:

$$\mathrm{Nu} = 0.68 + 0.67 Ra_L^{1/4} \left[1 + \left(\frac{0.492}{\mathrm{Pr}} \right)^{9/16} \right]^{-4/9} \tag{9.18a}$$

For $Ra_L > 10^9$:

$$\sqrt{\mathrm{Nu}} = 0.825 + 0.387 \mathrm{Ra}_L^{1/6} \left[1 + \left(\frac{0.492}{\mathrm{Pr}} \right)^{9/16} \right]^{-8/27} \tag{9.18b}$$

For the constant wall flux condition, Churchill and Ozoe (1973) collated much of the literature data and proposed the following correlation for the local Nusselt number in the laminar regime,

$$\mathrm{Nu}_x = \frac{0.563 \mathrm{Ra}_x^{1/4}}{\left[1 + \left(0.437/\mathrm{Pr} \right)^{9/16} \right]^{4/9}} \tag{9.19}$$

Equation 9.19 is valid over the full range of Prandtl number. While much confusion exists in the literature regarding the laminar-turbulent transition, Prandtl number is believed to influence this transition. Notwithstanding this uncertainty, the limited correlations available for turbulent boundary layers have been reviewed in Martynenko and Khramtsov (2005).

9.3.1.2 Power-Law Fluids

Early analyses in this field are based on the direct extension of the integral methods to solve the boundary layer equations in an approximate manner, which are based on the assumed forms of the velocity and temperature profiles. The scaling arguments suggest the free-convective transport to be governed by the values of the so-called Grashof number (ratio of buoyancy to viscous forces) and the Prandtl number for heat transfer or Schmidt number for mass transfer for a given boundary condition. Intuitively, one would expect the free convection to be weak in highly viscous liquids, which is generally the case for polymer melts and solutions (St. Pierre and Tien, 1963; Yang and Yeh, 1965; Liang and Acrivos, 1970; Ozoe and Churchill, 1972). However, this must be juxtaposed with the fact that owing to their high viscosity, polymer melts are also processed at very low flow rates (creeping flows) and therefore, the resulting Reynolds number is vanishingly small so the Richardson number can be of order 1. Under such circumstances, it is thus not always justified to neglect free convection effects. Once again returning to Tables 9.2 and 9.3, it is obvious that adequate information, analytical and experimental, seems to be available for free-thermal convection in non-Newtonian fluids, at least for the three model geometries, that is, a plate, cylinder, and sphere. The effect of particle shape on free-convection mass transfer in Newtonian fluids at high Rayleigh numbers has been studied by Weber et al. (1984). Most of the information pertains to the power-law fluids, albeit meager information is also available for plates immersed in Ellis model fluid (Tien and Tsuei, 1969), on the shape of convection currents in visco-elastic liquids (Liang and Acrivos, 1970; Garifulin et al., 1982), etc. Figure 9.1 shows the schematics of the flow in this case. Once again, the submerged plate which acts as a source of heat can be maintained either at a constant temperature (different from that of the liquid), or at a temperature which varies along the plate, or at a constant heat flux. Since a thorough review of the free convection in non-Newtonian systems is available in the literature (Shenoy and Mashelkar, 1982; Shenoy, 1988, 2017), emphasis is given here to the key results and the developments which have occurred after the publication of these reviews. Also, the assumptions of the laminar flow, constant properties, and negligible viscous dissipation are inherent in most studies.

Any discussion on free convection in non-Newtonian liquids must inevitably begin with the pioneering study of Acrivos (1960). Utilizing a Mangler-type transformation, the analysis of Acrivos (1960) yields the following expression for the mean Nusselt number when a vertical plate is maintained at a constant temperature and when Pr > 10.

$$\mathrm{Nu} = \frac{hL}{k} = C_3(n) \mathrm{Gr}_{\mathrm{PL}}^{\frac{1}{2(n+1)}} \mathrm{Pr}_{\mathrm{PL}}^{\frac{n}{(3n+1)}} \qquad (9.20)$$

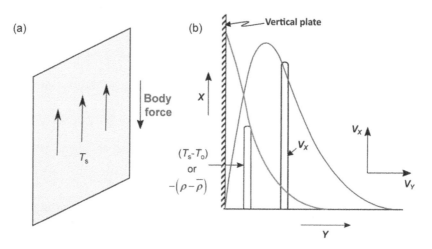

FIGURE 9.1 Schematics of free convection from a vertical plate.

where $C_3(n)$ is a weak function of the power-law index only. Subsequently, almost identical results have been reported by Akagi (1966). The plate length L is used as the characteristic linear dimension ($l_c = L$) in the definitions of the Grashof and Prandtl numbers. Recent numerical results of Aherwar et al. (2023) are also consistent with these predictions at high Rayleigh numbers.

However, the approach of Acrivos (1960) is not applicable when the vertical plate is maintained at the constant heat flux condition instead of the constant temperature. Chen and Wollersheim (1973) transformed the partial differential equations into ordinary differential equations. Their final expression for the mean Nusselt number (for the constant heat flux condition) for a vertical plate is given by

$$\text{Nu} = \frac{1}{\xi(0)} \left(\frac{2}{3n+2} \right)^{n/(3n+2)} \text{Gr}_C^{\frac{1}{(n+4)}} \text{Pr}_C^{\frac{n}{(3n+2)}} \tag{9.21}$$

where the dimensionless temperature difference $\xi(0)$ is evaluated at the surface of the plate and it depends on the value of the power-law index, n. For $n = 0.1, 0.5, 1,$ and 1.5, the value of $\xi(0)$ changes as 1.32, 1.303, 1.147, and 1.03, respectively. When these values of $\xi(0)$ are combined with the factor $(2/(3n+2))^{n/(3n+2)}$, there is a slight increase (~3%) in heat transfer in shear-thickening fluids and a similar deterioration in heat transfer in shear-thinning fluids, a trend which is somewhat counter intuitive. Note that the mean Nusselt number in this case is specified based on the mean heat transfer coefficient, $\bar{h}$, defined as follows:

$$\bar{h} = \frac{q_s}{(T_s - T_o)_{x = L/2}} \tag{9.22}$$

However, if Equation 9.21 is rewritten in terms of the definition of Grashof and Prandtl numbers used by Acrivos (1960), with $\Delta T = \overline{(T_s - T_o)}_L$ and using Equation 9.22, the agreement between the results for the constant temperature condition and the constant heat flux condition is very good. One can thus write both results in the form of Equation 9.20. The resulting values of $C_3(n)$ differ at most by 1%–2% in the range $0.1 \leq n \leq 1.5$ (Chen and Wollersheim, 1973). However, it needs to be emphasized here that the definition of the average heat transfer coefficient is different in these two cases.

Substantially similar results have been reported subsequently by Soundalgekar (1964, 1971, 1972), Akagi (1966), Tien (1967), Kawase and Ulbrecht (1984), Huang and Chen (1990), Pantangi et al. (2003), and Esfahani and Bagherian (2012) for power-law fluids, by Tien and Tsuei (1969) for Ellis model fluids, and by Fujii et al. (1972, 1973) for Sutterby model fluids. Since the results of Huang and Chen (1990) reproduce prior results very well, it is useful to present here their final expressions for the mean Nusselt number (averaged over the length of the plate) for a vertical plate.

For the uniform temperature condition ($0.5 \leq n \leq 1.5$),

$$\text{Nu}_L = \sim 0.66 \, \text{Ra}_L^{1/(3n+1)} \tag{9.23a}$$

where the Rayleigh number, Ra_L, is defined as

$$\text{Ra}_L = \frac{\rho g \beta (\Delta T) L^{2n+1}}{m \alpha^n} \tag{9.23b}$$

For the uniform heat flux condition ($0.5 \leq n \leq 1.5$; $10 \leq \text{Pr}_L^* \leq 2000$),

$$\text{Nu}_L = \sim A(n) \left(\text{Ra}_L^* \right)^{1/(3n+2)} \tag{9.24a}$$

In this case, $(q_s L/k)$ is used in lieu of (ΔT) in Equation 9.23b to modify the definitions of the Rayleigh and Prandtl number as follows:

$$\text{Ra}_L^* = \frac{\left[\rho g \beta (q_s/k)\right] L^{2(n+1)}}{m \alpha^n} \tag{9.24b}$$

and

$$\text{Pr}_L^* = \frac{1}{\alpha}\left(\frac{m}{\rho}\right)^{5/(n+4)} L^{2(n-1)/(n+4)} \left(\frac{g\beta q_s}{k}\right)^{3(n-1)/(n+4)} \tag{9.24c}$$

In Equation 9.24a, the value of the constant $A(n)$ varies both with the power-law index (n) and the Prandtl number, Pr_L^* for $n=0.5$, $10 \leq \text{Pr}_L^* \leq 2000$, $A(n) \approx 0.72$; for $n=1.0$, $A(n) \approx 0.77$ and for $n=1.5$, $A(n) \approx 0.80$. Furthermore, an extensive numerical and experimental study (Pittman et al., 1999) also predicts the dependence of the Nusselt number on the Rayleigh number identical to Equation 9.24a, with the value of $A(n)$ varying from 0.49 to 0.61 as the value of n increases from 0.48 to 1.

Aside from these fairly rigorous treatments based on the differential forms of the governing equations, some attempts have also been made at employing the so-called integral approach. Since the merits and demerits of this approach have been outlined in detail by Shenoy and Mashelkar (1982), the emphasis here is on the results and their range of applicability. For the constant-temperature boundary condition, Shenoy and Ulbrecht (1979) presented the following expression for the mean Nusselt number for a vertical plate submerged in a power-law fluid:

$$\text{Nu} = f_1(n)\left(\frac{3n+1}{2n+1}\right)^{(2n+1)/(3n+1)} \text{Gr}_L^{1/2(n+1)} \text{Pr}_L^{n/(3n+1)} \tag{9.25}$$

where

$$f_1(n) = 2\left[\frac{f_2(A_n)}{2 A_n (10/3)^{1/n}}\right]^{n/3n+1} \tag{9.26}$$

$$A_n = \frac{1}{3n}$$

$$f_2(A_n) = \sum_{j=1}^{6} \alpha_j A_n^j$$

where $\alpha_1 = 1/15$; $\alpha_2 = -5/42$; $\alpha_3 = 3/28$; $\alpha_4 = -1/18$; $\alpha_5 = 1/63$; $\alpha_6 = -3/1540$.

Table 9.4 presents a comparison between the results of Acrivos (1960), Tien (1967), Shenoy and Ulbrecht (1979), and Huang and Chen (1990) in terms of $[\text{Nu} \times \text{Gr}^{-1/2(n+1)} \times \text{Pr}^{-n/(3n+1)}]$; an excellent agreement is seen to exist between all these studies. This also inspires confidence in the use of various approximations inherent in these analyses. For instance, Shenoy and Ulbrecht (1979) employed the integral approach whereas Huang and Chen (1990) invoked the local similarity approximation.

A typical comparison between the predictions and the experimental results for temperature distribution for an isothermal vertical plate submerged in a CMC power-law solution ($n=0.96$) is shown in Figure 9.2 where, once again good correspondence is seen to exist, albeit the polymer solution is nearly Newtonian in this case. The experimental values of the mean Nusselt number of Reilly et al. (1965) and Emery et al. (1971) are within 5%–10% of those predicted by Acrivos (1960) in the range $0.72 \leq n \leq 1.0$.

TABLE 9.4
Comparison Between Different Predictions of $[\mathrm{Nu} \times \mathrm{Gr}^{-1/2(n+1)} \times \mathrm{Pr}^{-n/3n+1}]$ for an Isothermal Vertical Plate

n	Acrivos (1960)	Tien (1967)	Shenoy and Ulbrecht (1979)	Huang and Chen (1990)
1.5	0.71	0.723	0.72	0.701
1.0	0.67	0.684	0.68	0.670
0.5	0.63	0.610	0.60	0.610

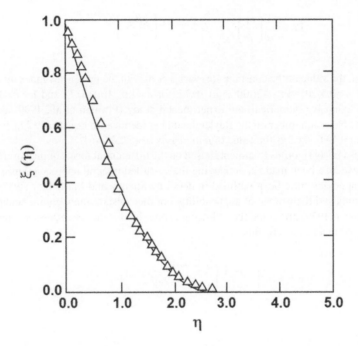

FIGURE 9.2 Comparison between the predicted (Shenoy and Ulbrecht, 1979) and their experimental temperature profiles (at $x = 100$ mm) for an isothermal ($T_s = 47°C$) vertical plate immersed in a carboxymethyl cellulose solution ($n = 0.96$).

For the analogous case of the constant heat flux condition, Shenoy (1977) presented the following expression for the mean Nusselt number (averaged over the length of the plate):

$$\mathrm{Nu} = f_2(n) \left(\frac{3n+2}{2n+2} \right)^{n/(3n+2)} \mathrm{Gr}_C^{1/n+4} \mathrm{Pr}_C^{n/3n+2} \tag{9.27}$$

where

$$f_2(n) = 2 \left[\frac{f_2(A_n)}{2 A_n (20/3)^{1/n}} \right]^{n/3n+2}$$

and A_n and $f_2(n)$ are still given by the expressions introduced in the preceding section.

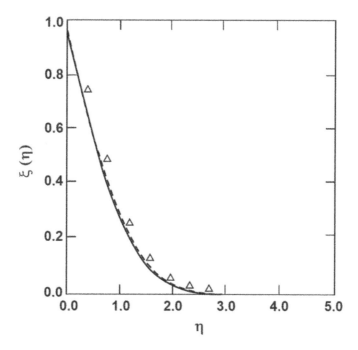

FIGURE 9.3 Comparison between predicted and experimental dimensionless temperature profiles on the surface of a vertical plate ($q_s = 833$ W m^{-2}; $x = 300$ mm; $L = 500$ mm; $n = 0.89$) immersed in a power-law polymer solution. (Replotted from Shenoy, A.V. and Mashelkar, R.A., *Adv. Heat Transfer*, 15, 143, 1982.)

While the results of Chen and Wollersheim (1973), Som and Chen (1984), and Huang and Chen (1990) are fairly close to each other, the predictions of Equation 9.25 deviate increasingly as the degree of shear-thinning increases, that is, as the value of *n* decreases below unity.

For the uniform heat flux condition imposed on a vertical plate, Figure 9.3 shows a typical comparison between the predictions (Chen and Wollersheim, 1973; Shenoy, 1977) of temperature profile $\xi(\eta)$ and the experimental results of Dale and Emery (1972) in a carbopol solution ($n = 0.89$); the correspondence between the two predictions as well as between the predictions and the experiments is seen to be good. Similarly, their experimental values of the local Nusselt number are consistent with the predictions of Tien (1967) and Shenoy (1977).

The general case of the laminar-free convection in power-law fluids from a vertical plate with variable surface temperature and variable surface heat flux has been treated by Som and Chen (1984). Similarly, Erbas and Ece (2001) have examined the role of shear rate-dependent thermal conductivity and identified the limitations of the nature of shear rate-dependency of thermal conductivity, *k*, and power-law viscosity for the existence of a similarity solution. Finally, Shulman et al. (1976) have studied the effect of temperature-dependent power-law consistency coefficient, *m*, on the unsteady laminar-free convection from a vertical plate submerged in power-law fluids. The role of suction and injection has been studied by Sahu and Mathur (1996). Capobianchi and Aziz (2012) have used the extended modified power-law (EMPL) fluid model to study free convection from an isothermal vertical plate for $n > 1$ and $n < 1$. The EMPL viscosity model also includes the zero-shear and infinite-shear limiting viscosities whereas Moulic and Yao (2009) have used the truncated power-law model for the same problem.

Little work has been reported on the free convection from a vertical plate in the turbulent flow regime even for Newtonian liquids (Ruckenstein, 1998; Martynenko and Khramtsov, 2005), let alone for power-law fluids. For instance, using the approach of Eckert and Jackson (1950), Shenoy and Mashelkar (1978a,b) used the integral momentum and energy balances together with the Colburn analogy to obtain an approximate expression for the Nusselt number, which assumes that the boundary layer is turbulent over the entire plate, $0 \leq x \leq L$. Subsequently, Kawase and Ulbrecht (1984) employed a surface renewal model to extend their laminar flow analysis to the turbulent conditions. While their laminar regime analysis is consistent with the other predictions (Acrivos, 1960; Tien, 1967; Shulman et al., 1976), their predictions in the turbulent regime for a vertical plate in power-law fluids are also in fair agreement with the scant experimental results (Reilly et al., 1965; Sharma and Adelman, 1969).

9.3.1.3 Bingham Plastic Fluids

The free convection in yield stress fluids such as Bingham or Herschel-Bulkley or Casson model fluids is fundamentally different from that in fluids without a yield stress (Vikhansky, 2011; Karimfazli et al., 2015, 2016). Due to its yield stress, such a fluid does not yield until the shearing stress induced by buoyancy-driven flow exceeds the fluid yield stress. Thus, there exists a threshold value of the Rayleigh number or Grashof number for a fixed Bingham number for the onset of convection in these fluids. Conversely, there exists a critical Bingham number for a fixed Rayleigh number at which free convection ceases. Notwithstanding this inherently different nature of free convection in such fluids, there has been very little work reported on free convection in Bingham plastic fluids from a vertical plate. Kleppe and Marner (1972) studied transient laminar-free convection from a vertical isothermal plate. They reported that there was an initial period in which heat transfer occurred solely by conduction as the shear stress caused by the density-induced flow was lower than the fluid yield stress. This initial so-called warm up period (t_c^*) scaled with the Prandtl number, Grashof number, and Hedstrom number as:

$$t_c^* = \frac{\pi}{4} \text{Pr}_B \left(\text{He} / \text{Gr}_B^{3/4} \right)^2 \tag{9.28}$$

Note that the Hedstrom number ($= \rho \tau_o^B L^2 / \mu_B$) is akin to the Prandtl number in so far that it contains only the system properties and it is in fact given by the product of the Reynolds number (Re_B) and the Bingham number (Bi). Equation 9.28 suggests the time elapsed before the onset of flow to increase with the increasing Prandtl number and Hedstrom number and with the decreasing value of the Grashof number. Figure 9.4 shows a typical temporal evolution of the average Nusselt number for three values of Prandtl number ($\text{Pr}_B = 1$, 10, and 100) and for a few values of the composite parameter $\text{He}/\text{Gr}_B^{3/4}$. Naturally, $\text{He}/\text{Gr}_B^{3/4} = 0$ line corresponds to the Newtonian fluid behavior and the broken line shows heat transfer by conduction. For small values of t_c^*, the Nusselt number results coincide with the conduction values and beyond the initial warm period before reaching the steady state values. Table 9.5 summarizes the steady state values as a function of Pr_B and $\left(\text{He}/\text{Gr}_B^{3/4}\right)$. No other results seem to be available for free convection from a vertical plate in yield stress fluids. Except for recent study of Aherwar et al (2023). New extensive results on the mean Nusselt number for for a vertical place (isothermal and isoflux) in Bingham plastic fluids have been presented and correlated.

9.3.2 Mixed Convection

In most heat transfer processes, forced- and free-convection mechanisms contribute to the overall rate of heat transfer to varying extents. Depending upon the values of the dimensionless groups like Grashof number and Reynolds number or combinations thereof (like Rayleigh or Richardson numbers), the effects of free convection may or may not be negligible. Furthermore, depending upon the direction of flow and the orientation of the submerged surface, the free convection may be assisting or opposing the flow such as for the upward or downward flow along the surface. In the case of cross-flow over submerged surfaces, the free-convection effects would set up the circulation patterns which would be in the direction normal to the imposed velocity. Figure 9.5 shows schematically the three types of mixed-convection regimes for the case of heat transfer from a heated surface to the cold fluid. For Newtonian fluids, the value of the group ($\text{Ri} = \text{Gr}/\text{Re}^2$) $\ll 1$ indicates little free convection, (Gr/Re^2) $\gg 1$ suggests strong free convection effects and when $\text{Ri} \sim O(1)$, heat transfer occurs in the so-called mixed-convection regime. Since not much is known about the delineation of the flow regimes in non-Newtonian fluids, it is suggested that $\text{Ri} \sim O(1)$ can be used as a crude approximation, at least for inelastic non-Newtonian fluids to denote the mixed-convection regime.

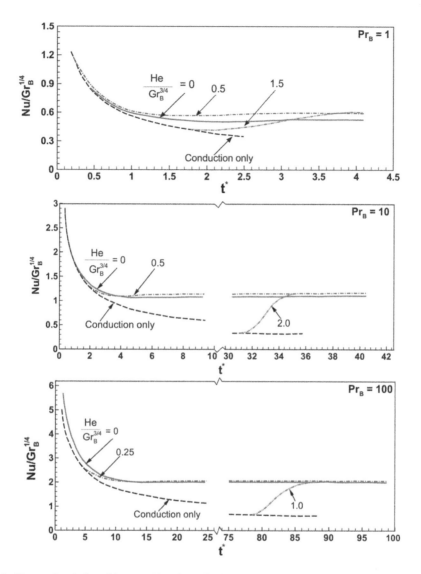

FIGURE 9.4 Temporal evolution of the mean Nusselt number for natural convection from a vertical (isothermal) plate in Bingham plastic fluids. (Replotted from Kleppe, J. and Marner, W.J., *J. Heat Transf.*, **4**, 371, 1972.)

TABLE 9.5

Average Nusselt Number at Steady State for Free Convection in Bingham Plastic Fluids on an Isothermal Vertical Plate

	$Nu/Gr_B^{1/4}$		
$He/Gr_B^{3/4}$	$Pr_B=1$	$Pr_B=10$	$Pr_B=100$
0	0.525	1.105	2.000
0.25	-	-	2.060
0.5	0.593	1.165	-
1.0	-	-	2.000
1.5	0.604	-	-
2.0	-	1.165	-

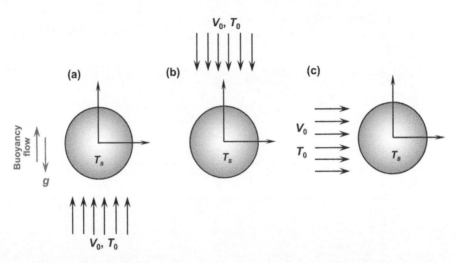

FIGURE 9.5 Schematic representation of mixed convection phenomena from a heated surface to cold fluid (a) aiding buoyancy (b) opposing buoyancy (c) cross buoyancy.

9.3.2.1 Newtonian Fluids

Mixed convection from a vertical plate in Newtonian fluids has been studied extensively, e.g., see the early works of Acrivos (1958, 1966) and Sparrow and Gregg (1959), Eshghy (1964), Merkin (1969), etc. and much of this literature has been reviewed by Martynenko and Khramtsov (2005). Based on the thorough review of the literature, Martynenko and Khramtsov (2005) recommend the following correlation of Churchill (1977) for the buoyancy-assisted mixed convection from a vertical plate, horizontal cylinder, and sphere. The surface average Nusselt number is given by:

$$\left[\frac{(\mathrm{Nu}-\mathrm{Nu}_o)f_1(\mathrm{Pr})}{A_1 \mathrm{Re}^{1/2} \mathrm{Pr}^{1/3}}\right]^3 = 1 + \left[\frac{A_2 Ra^{1/4} f_1(\mathrm{Pr})}{A_1 \mathrm{Re}^{1/2} \mathrm{Pr}^{1/3} f_2(\mathrm{Pr})}\right]^3 \qquad (9.29)$$

where

$$f_1(\mathrm{Pr}) = \left[1 + \left(\frac{A_3}{\mathrm{Pr}}\right)^{9/16}\right]^{4/9} \qquad (9.30a)$$

$$f_2(\mathrm{Pr}) = \left[1 + \left(\frac{A_4}{\mathrm{Pr}}\right)^{2/3}\right]^{1/4} \qquad (9.30b)$$

The values of the various constant like $\mathrm{Nu}_o, A_1, A_2, A_3, A_4$, etc. along with the characteristic linear dimension to be used in the definition of the Nusselt number for each shape are summarized in Table 9.6.

In addition, Martynenko and Khramtsov (2005) also recommend an approximate expression for these three shapes with $A_1 = A_2 = 0.67$ and $A_3 = A_4 = 0.45$ except for a vertical plate for which $A_3 = 0.035$ and $\mathrm{Nu}_o = 0.5, 1$, and π for the vertical plate, cylinder, and sphere, respectively. While Equation 9.29 purports to span the complete range of Prandtl number ($0 \leq \mathrm{Pr} \leq \infty$), it extends over the range of conditions as: $15 \leq \mathrm{Re} \leq 10^4$; $2300 \leq \mathrm{Ra} \leq 10^9$ and the physical properties are evaluated at the mean film temperature in this case.

TABLE 9.6
Values of Constants Appearing in Equation 9.29 (Martynenko and Khramtsov, 2005)

Shape	Boundary Condition on Surface	l_c	A_1	A_2	A_3	A_4	Nu_o
Vertical plate	Isothermal	L	0.677	0.670	0.0468	0.492	0
	Constant heat flux	$L/2$	0.656	0.669	0.0205	0.437	0
Horizontal cylinder	Isothermal	πd	1.08	0.69	0.412	0.559	1.0
	Constant heat flux	πd	-	0.694	0.442	-	-
Sphere	Isothermal	$\pi d/2$	0.69	0.659	-	-	π

9.3.2.2 Power-Law Fluids

Kubair and Pei (1968) were seemingly the first to tackle the problem of mixed convection to power-law fluids from a vertical plate. Not only has this analysis been shown to be erroneous (Shenoy and Mashelkar, 1982) but it also imposes unrealistic restrictions on the functional dependence of the temperature difference on the x-coordinate, similar to that of Na and Hansen (1966). There has been a little activity in this area. Lin and Shih (1980a) employed the local similarity solution procedure to study mixed (assisted) convection from a stationary as well as a moving plate to quiescent or moving power-law fluids. They also considered both types of boundary conditions, that is, uniform heat temperature or constant heat flux condition imposed on the vertical plate. However, both the temperature difference $(T_s - T_0)$ in the first case and the heat flux q_s in the second case were assumed to vary as power-law functions of the distance along the surface of the plate, a condition germane to the existence of a suitable similarity variable. However, only scant results were presented to demonstrate the applicability of this approach. The other studies pertaining to the mixed convection from a vertical plate are due to Pop et al. (1991), Ramamurthy (1995), and Wang (1995). Pop et al. (1991) sought an exact similarity solution for a stationary and moving plate maintained at a surface temperature which was inversely proportional to the distance from the leading edge of the plate. Furthermore, even the thermal conductivity was assumed to be a function of shear rate, as was done by Shvets and Vishnevskiy (1987). However, this analysis is also very restrictive in its scope due to the rather unrealistic requirements imposed on the power-law indices for viscosity and thermal conductivity of the fluid. Wang (1995), on the other hand, developed a nonsimilarity analysis for mixed convection from a stationary vertical surface (at a constant temperature) immersed in power-law fluids. He considered both aiding and opposing flow conditions. He found it more appropriate to use the following composite parameter which is similar to the reciprocal of the Richardson number (Ri) as the criterion for ascertaining the importance of free convection in a given situation,

$$\zeta = \frac{Ra_L^{1/3n+1}}{Re_L^{1/n+1} + Ra_L^{1/3n+1}} \tag{9.31}$$

where the Rayleigh number is defined by Equation 9.23b.

Thus, for pure free convection, $Re_L \to 0$, that is, $\zeta = 1$, and for pure forced convection, $Ra_L \to 0$, that is, $\zeta \to 0$. Under appropriate conditions of ζ, the results of Wang (1995) are in line with that of Acrivos et al. (1960) and Kim et al. (1983) for $\zeta = 1$ and with that of Huang and Chen (1984) for $\zeta = 0$, and that of Huang and Lin (1993) and of Ramamurthy (1995) for the mixed-convection conditions. Mixed convection to power-law liquids from a horizontal plate in cross-flow configuration has been studied by Gorla (1986) and Wang (1993). Gorla (1986) used a parameter similar to the Richardson number (Gr/Re^2) to express the relative importance of the free and forced convection and solved the governing equations numerically for a fixed value of Prandtl number of 10. He reported the skin friction coefficient to be more sensitive to the buoyancy effects than the heat transfer coefficient. Furthermore, for large values of Rayleigh number, an overshoot in velocity beyond the free stream value was also observed. Wang (1993), on the other hand, sought a numerical solution to suitably transformed governing equations. His results for the forced-convection regime are consistent with that of Huang and Chen (1984). This subsection is finally concluded by presenting the practically useful approach of Shenoy (1980) for mixed (buoyancy-assisted) laminar convection to power-law fluids from a vertical plate which is maintained at a constant temperature. For large values of Prandtl number, Shenoy (1980) showed that the approach initially developed by Churchill (1977) for Newtonian fluids and subsequently modified by Ruckenstein (1978) can also be applied to mixed convection in power-law fluids from a vertical plate which may be expressed as

$$(Nu_{xM})^3 = (Nu_{xF})^3 + (Nu_{xN})^3 \tag{9.32}$$

where the local values of the Nusselt number for the forced, Nu_{xF}, and the natural convection, Nu_{xN}, can be estimated using the methods presented in the preceding sections. These, in turn, combined with Equation 9.32 will yield the value of the Nusselt number, Nu_{xM}.

9.3.2.3 Visco-plastic Fluids

As far as we know, there has been very little activity on the boundary layer flows of visco-plastic fluids (Chhabra and Richardson, 2008). In a pioneering study, Oldroyd (1947) predicted the boundary layer thickness for an infinitely long thin knife moving in an ideal Bingham fluid to be given by

$$\frac{\delta}{x} = 3\left(\frac{\mu_B V_o}{2\tau_0^B x}\right)^{1/3} \tag{9.33}$$

Equation 9.33 was deduced under the assumptions of small Reynolds number and large Bingham number conditions. The predictions of Equation 9.33 have been confirmed experimentally by Clegg and Whitmore (1966). Subsequently, Piau (2002) has developed a much more rigorous framework to analyze the low inertia boundary layer flows for Bingham plastic fluids, with Equation 9.33 being as a limiting case under appropriate conditions. The general features of free-thermal convection in a Bingham plastic fluids have been studied by Yang and Yeh (1965), Kleppe and Marner (1972) and Marner and Kleppe (1976) as discussed earlier. Yang and Yeh (1965) studied free-thermal convection in a Bingham plastic fluid contained in between two vertical plates maintained at different temperatures. They found that no "fluid-like" regions were present whenever the dimensionless yield stress exceeded the value of the parameter $(GrPr)^{1/4}$, that is, $Ra_B^{1/4}$. Marner and Kleppe (1976) extended their previous work on free convection to aiding-buoyancy mixed convection from an isothermal vertical plate. Their results span the range of Richardson number from 0 to 1, $Pr_B = 1$ and two values of the composite parameter $He/Gr_B^{3/4} = 0$ and 0.5. Over these range of conditions, they reported very little difference in the velocity profiles near the plate in Newtonian and Bingham plastic fluids and the average Nusselt number for Newtonian and Bingham plastic fluids were within 3% each other.

9.4 Horizontal Cylinders

9.4.1 Free Convection

9.4.1.1 Newtonian Fluids

Much has been written about the free convection heat transfer from a horizontal cylinder in Newtonian fluids (Martynenko and Khramtsov, 2005; Prhashanna and Chhabra, 2011). Early attempts at obtaining temperature and velocity profiles for the laminar-free convection from a horizontal cylinder are based on the approximate solutions of the boundary layer equations with constant physical properties, negligible viscous dissipation, and the two-dimensional flow regime. Due to the assumptions of high Prandtl and Grashof numbers, thin-boundary layer approximation is invoked and thus the curvature effects can be neglected. Similarly, the small Rayleigh number limit can be approached by the matched asymptotic expansion technique. Thus, the early studies (Merk and Prins, 1953; Narain, 1976; Yao et al., 1978; Farouk and Guceri, 1982, etc.) illustrate the applicability of this approach. Evidently, neither of these are relevant to the intermediate flow regime characterized by moderate values of the Grashof and/or Prandtl numbers (Peterka and Richardson, 1969; Kuehn and Goldstein, 1980; Wang et al., 1990). The latter authors solved the complete governing equations for an isothermal horizontal cylinder in the range $1 \leq Ra \leq 10^7$. These results clearly revealed the inadequacies of the boundary layer approach. Subsequently, similar numerical results have been reported among others by Moon et al. (1987) and Saitoh et al. (1993) for an isothermal cylinder. Wilks (1972) and Ahmad and Qureshi (1992) have considered the case of constant heat flux condition on the surface of the cylinder. Extensive numerical activity in this field has also been matched by excellent experimental studies, e.g., see Fand et al. (1977), Kitamura et al. (1999), Fujii et al. (1970), Atayilmaz and Teke (2009), and Fand and Brucker (1983), etc. Extensive reviews of the pertinent studies are available in the literature itself (Morgan, 1975; Martynenko and Khramtsov, 2005). The effect of temperature-dependent viscosity on heat transfer has been studied by Cheng (2006).

Figures 9.6 and 9.7 shows typical temperature profiles in the vicinity of an isothermal cylinder in Newtonian fluids and the corresponding local Nusselt number distribution along the surface of the cylinder culled from different numerical studies in Newtonian fluids which testify to the reliability of the available numerical results on free convection from a horizontal cylinder in Newtonian fluids.

This section is concluded by presenting a widely used correlation for estimating the average Nusselt number for a horizontal cylinder. For instance, Churchill and Chu (1975) presented the following correlation for an isothermal cylinder (spanning both laminar and turbulent flow regimes) for $Ra_D > 10^{-6}$:

$$\mathrm{Nu} = \frac{hd}{k} = 0.60 + 0.387 \left\{ \frac{Ra_D}{\left[1 + \left(\frac{0.559}{\mathrm{Pr}}\right)^{9/16}\right]^{16/9}} \right\}^{1/6} \tag{9.34}$$

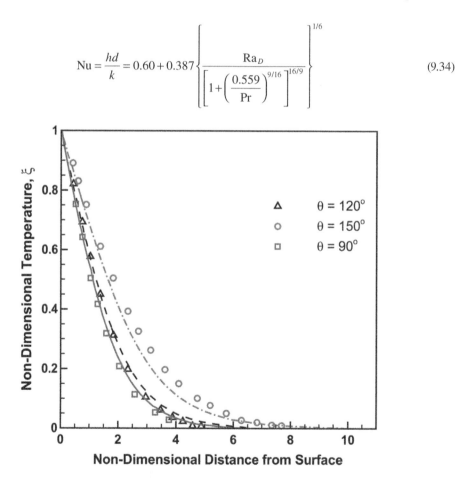

FIGURE 9.6 Temperature profiles at various angles in the vicinity of the cylinder at Ra = 12500, Pr = 0.7. Symbols: Kuehn and Goldstein (1980); lines: Prhashanna and Chhabra (2011). (Replotted from Prhashanna, A., and Chhabra, R.P., *Ind. Eng. Chem. Res.*, **50**, 2424, 2011.)

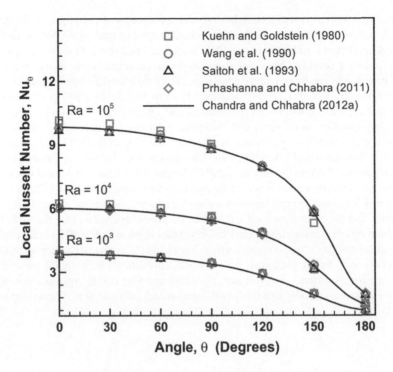

FIGURE 9.7 Distribution of the local Nusselt number, Nu_θ on the surface of the circular cylinder with literature values in air (Pr=0.7). (Replotted from Chandra, A. and Chhabra, R.P., *Int. J. Heat Mass Transf.*, 55, 2934, 2012a.)

The corresponding expression for the constant heat flux condition on the cylinder surface is due to Churchill and Thelen (1975):

$$Nu_l^{1/2} = \left(\frac{hd}{k}\right)^{1/2} = 0.60 + \left\{\frac{0.00333 Ra_D}{\left[1+\left(\frac{0.44}{Pr}\right)^{9/16}\right]^{16/9}}\right\}^{1/6} \quad (9.35)$$

Equation 9.35 is valid in both laminar and turbulent flow regimes over the range of conditions as $0.022 \leq Pr \leq 7640$ and $Ra_D > 1$ and the physical properties are evaluated at the mean fluid temperature.

9.4.1.2 Power-Law Fluids

The high Prandtl number analysis of Acrivos (1960) for two-dimensional surfaces can readily be specialized for the laminar-free convection heat/mass transfer to power-law fluids from a horizontal (isothermal) cylinder. The local Nusselt number (based on the radius of the cylinder), Nu, is given by the expression:

$$Nu = -\frac{\xi'(0)\left[(2n+1)/(3n+1)\right]^{n/(3n+1)} Gr^{1/2(n+1)} Pr^{n/3n+1}(\sin\varsigma)^{1/2n+1}}{\left[\int_0^\xi (\sin x_1)^{1/2n+1} dx_1\right]^{n/(3n+1)}} \quad (9.36)$$

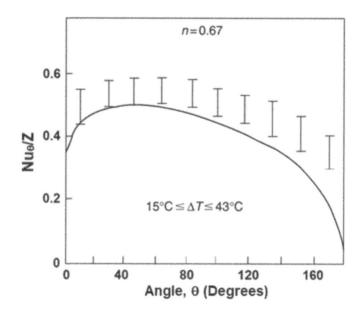

FIGURE 9.8 Typical comparison between the predicted, shown as solid lines, (Acrivos, 1960) and experimental results for the local Nusselt number for free convection from a cylinder to a carbopol solution ($n = 0.67$). (Replotted from Gentry, C.C. and Wollersheim, D.E., *J. Heat Transf.*, **96**, 3, 1974.)

In the stagnation zone, $0 \leq x \leq (\pi/6)$, one can approximate $\sin x_1 \approx x_1$ and hence the integral in the denominator of Equation 9.36 can be evaluated analytically. This yields the following expression for the local Nusselt number in this region:

$$\text{Nu} \approx \xi'(0)\left(\frac{2n+1}{3n+1}\right)^{n/(3n+1)} \text{Gr}^{1/2(n+1)} \text{Pr}^{n/(3n+1)} \varsigma^{(1-n)/(3n+1)} \tag{9.37}$$

On the other hand, these values can be integrated over the surface of the cylinder to obtain the mean Nusselt number which is of the same form as Equation 9.20, except that the constant term takes on different values depending upon the geometry. Subsequently, Ng and Hartnett (1986) have rewritten the result of Acrivos (1960) in terms of the Rayleigh number as,

$$\text{Nu} = C_0(n) \text{Ra}^{1/3n+1} \tag{9.38}$$

where $C_0(n)$ is a weak function of the flow-behavior index. The high Prandtl number limitation inherent in the analysis of Acrivos (1960) gets translated into $\text{Ra}^{1/(3n+1)} \gg 1$. Figure 9.8 shows a typical comparison between the experimental results of Gentry and Wollersheim (1974) and the predictions of Acrivos (1960) for $n = 0.67$. While the experimental results of Gentry and Wollersheim (1974) covering the range $0.64 \leq n \leq 0.93$ are in line with the predictions of Acrivos (1960) as far as the local Nusselt number is concerned, the observed average values of the Nusselt number for carbopol solutions are not too different from the corresponding Newtonian values. This lends support to the assertion of Ng and Hartnett (1986) that $C_0(n)$ is only a weak function of n. Therefore, one can use the single value of $C_0(n)$, corresponding to $n = 1$, and the effect of power-law behavior is completely accounted for by using the modified definitions of the Grashof and Prandtl numbers.

For the case of the constant heat flux condition, Kim and Wollersheim (1976) paralleled the treatment of Gentry and Wollesheim (1974). These predictions were supported by their own experiments for dilatant fluids in the range $1 \leq n \leq 1.46$. They also presented the following expression for the mean Nusselt number based on their data

$$\mathrm{Nu} = 3.54 \left(\mathrm{Gr}_K \, \mathrm{Pr}_K \right)^{0.071} \tag{9.39}$$

where the Grashof number and Prandtl number are defined as

$$\mathrm{Gr}_K = \left(\frac{\rho}{m} \right)^2 R^4 \left(\frac{g\beta q_s}{k} \right)^{2-n} \tag{9.40a}$$

$$\mathrm{Pr}_K = \frac{\rho C_P}{k} \left(\frac{m}{\rho} \right)^{2/n+1} R^{2(n-1)/n+1} \left(\frac{g\beta q_s}{k} \right)^{3(n-1)/2(n+1)} \tag{9.40b}$$

Aside from these studies, Ng and Hartnett (1986, 1988) and Ng et al. (1988) have reported extensive results on the laminar-free convection from horizontal thin wires (similar to thermocouples) in pseudoplastic liquids. In some cases, the diameter of the wire was comparable to the expected boundary layer thickness. Based on their data, they proposed the following modified form of Equation 9.38:

$$\mathrm{Nu}_1 = \left(0.761 + 0.413n \right) \mathrm{Ra}^{1/2(3n+1)} \tag{9.41}$$

Equation 9.41 was stated to be valid in the range $10^{-3} \leq \mathrm{Ra} \leq 1$, which is clearly outside the limits of applicability of the analysis of Acrivos (1960) and of Gentry and Wollersheim (1974). Furthermore, Ng et al. (1988) also noted that the circulation currents setup by buoyancy were so weak that the Nusselt number approached the conduction limit in some of the polymer solutions used by them. Undoubtedly, many other correlations for heat transfer from cylinders and nonspherical shaped objects submerged in power-law fluids are available in the literature (e.g., Alhamdan and Sastry, 1998), but none of these have been substantiated by independent experiments (Barigou et al., 1998). Finally, Gorla (1991a) has further extended the results of Acrivos (1960) to laminar-free convection from a horizontal cylinder to Ellis model fluids.

In an extensive numerical study, Prhashanna and Chhabra (2011) elucidated the role of power-law rheology on the steady laminar-free convection from an isothermal horizontal cylinder. Over the range of $\mathrm{Gr}_{\mathrm{PL}}$ (with l_c=radius of the cylinder) from 10 to 10^5, $\mathrm{Pr}_{\mathrm{PL}}$ (with l_c=radius of the cylinder) from 0.72 to 100, they reported extensive results on the local and surface-averaged Nusselt number (hd/k) over the range of power-law index ($0.3 \leq n \leq 1.8$). Figures 9.9 and 9.10 show representative trends for scores of values of the Prandtl number and Grashof number for the two extreme values of the power-law index, namely $n=0.3$ and $n=1.8$, respectively. Similar to the behavior seen in forced convection (Chapter 8), the maximum value of the local Nusselt number is shifted away from the front stagnation point in shear-thinning fluids (Figure 9.9) whereas the rate of decrease of the local Nusselt number is gradual up to about $\theta \sim 150°$ before dropping rather steeply to its minimum value at the rear stagnation point (Figure 9.10). Other than this, shear-thinning behavior is seen to promote heat transfer whereas shear-thickening is seen to impede it with reference to the corresponding value in Newtonian fluids otherwise under identical conditions. This trend is seen clearly in Figure 9.11 in terms of average Nusselt number. Finally, they consolidated their results by introducing a composite parameter, Z, proposed by Acrivos (1960) and subsequently employed by Amato and Tien (1976) in the context of free convection from an isothermal sphere, defined as

$$Z = \mathrm{Gr}_{\mathrm{PL}}^{\frac{1}{2(n+1)}} \mathrm{Pr}_{\mathrm{PL}}^{\frac{n}{3n+1}} \tag{9.42a}$$

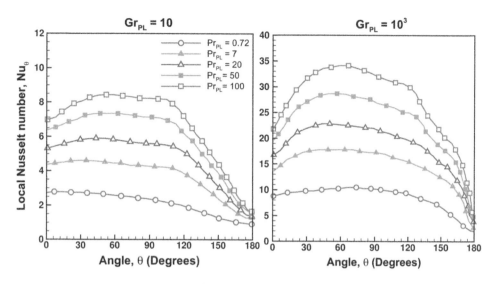

FIGURE 9.9 Distribution of the local Nusselt number over the surface of a cylinder at $n=0.3$. (Replotted from Prhashanna, A., and Chhabra, R.P., *Ind. Eng. Chem. Res.*, **50**, 2424, 2011.)

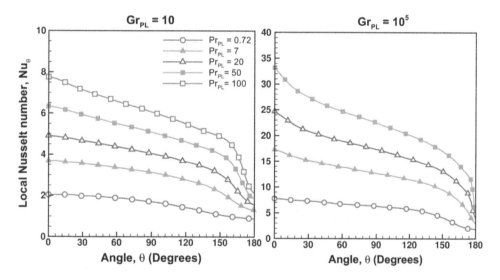

FIGURE 9.10 Distribution of the local Nusselt number over the surface of a cylinder at $n=1.8$. (Replotted from Prhashanna, A., and Chhabra, R.P., *Ind. Eng. Chem. Res.*, **50**, 2424, 2011.)

The resulting correlation of Prhashanna and Chhabra (2011) is given as:

$$\mathrm{Nu} = 1.19 Z^{0.89} \tag{9.42b}$$

Over the range of conditions spanned by the study of Prhashanna and Chhabra (2011), the mean and maximum deviations are of the order of ~7% and 21% without any discernable trends. Furthermore, Table 9.7 compares their results with those of Acrivos (1960) based on the approximate boundary layer analysis. Clearly, the two values are close for $Z \geq \sim 10$, thereby delineating the range of applicability of the predictions of Acrivos (1960).

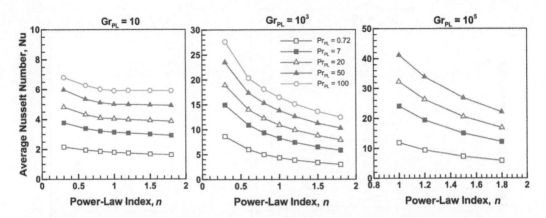

FIGURE 9.11 Dependence of the average Nusselt number on Grashof number, Prandtl number, and power-law index (Replotted from Prhashanna, A., and Chhabra, R.P., *Ind. Eng. Chem. Res.*, **50**, 2424, 2011.)

TABLE 9.7
Comparison between the Numerical Results (Prhashanna and Chhabra, 2011) and Approximate Results (Acrivos, 1960)

		Value of Nu	
n	Z	Prhashanna & Chhabra (2011)	Acrivos (1960)
1.5	1.45	1.67	1.30
	5.56	5.91	5.01
	13.98	13.58	12.58
	29.06	26.80	26.16
1	1.64	1.79	1.38
	5.18	4.30	4.35
	16.38	11.86	13.76
	47.29	41.23	39.72

A close inspection of the definition of the composite factor Z shows it to be similar to the Rayleigh number. In the limit of Newtonian fluid behavior ($n=1$), $Z=Ra^{1/4}$. Conversely, the definition of Z indicates the role of Grashof number and Prandtl number to be modulated by the value of the power-law index. For instance, for $n=0.5$, it yields $Z=Gr_{PL}^{1/3} Pr_{PL}^{1/5}$, that is, a stronger influence of the Grashof number. On the other hand, for $n=1.5$, $Z=Gr_{PL}^{1/5} Pr_{PL}^{0.27}$, that is, a stronger dependence on the Prandtl number.

Similar results on the laminar-free convection in power-law fluids from horizontal isothermal cylinders of various cross-sections are also available in the literature. Thus, for instance, Sasmal and Chhabra (2011, 2012a,b) studied the cases of square and elliptical cross-sections in different orientations and of various aspect ratios. Chandra and Chhabra (2012a) and Tiwari and Chhabra (2013) studied two different orientations of a semicircular cylinder. The case of a triangular bar has been explored by Tiwari and Chhabra (2014a). Sairamu and Chhabra (2013) and Gupta et al. (2019) have explored the effect of confinement on free convection from a horizontal cylinder.

9.4.1.3 Bingham Plastic Fluids

Laminar-free convection in Bingham plastic fluids from an isothermal horizontal cylinder has been studied by Nirmalkar et al. (2014b) over the range of conditions as $30 \leq Pr_B \leq 100$, $0 \leq Bi \leq 10^4$ and $10^2 \leq Ra_B \leq 10^6$. Patel and Chhabra (2016a) paralleled this study for elliptical cylinders and in the limit

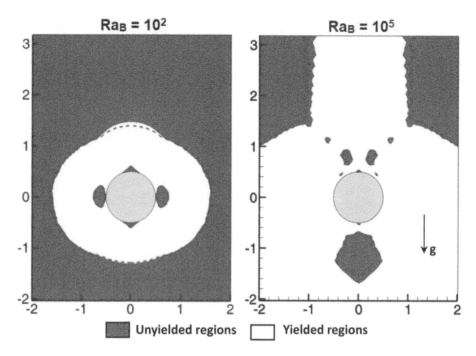

FIGURE 9.12 Yielded/Unyielded regions in the vicinity of a heated cylinder at Bi=10, Pr_B=30 (From Nirmalkar, N., Bose, A. and Chhabra, R.P., *Int. J. Therm. Sci.*, **83**, 33, 2014.)

of $E=1$ (minor axis=major axis), their results coincide with that of Nirmalkar et al. (2014b). Broadly, since the fluid-like regions expand spatially with the increasing Rayleigh number, this tendency is suppressed by the increasing Bingham number. For a given Rayleigh number, therefore there exists a limiting Bingham number beyond which the fluid is, so to say, frozen everywhere and thus heat transfer occurs solely by conduction. Conversely, for a given Bingham number, there exists a minimum value of Rayleigh number required for the onset of convection whereas below this value, conduction is the sole mechanism of heat transfer. This is in line with the propositions of Vikhansky (2011) and Karimfazli et al. (2015) and with the findings of Kleppe and Marner (1972). Intuitively, it appears that the average Nusselt number will continually decrease from its maximum value in a Newtonian fluid (Bi=0) with the increasing Bingham number eventually approaching the conduction limit at sufficiently high values of the Bingham number denoting the fully plastic flow limit.

Similarly, since there are regions of unyielded material present close to the surface of the cylinder (e.g., see Figure 9.12), the local Nusselt number distribution over the surface of cylinder shows deviations from the trends seen in Newtonian and power-law liquids (Figures 9.9 and 9.10). Bearing in mind these factors, Figures 9.13 and 9.14 show typical results for two values of the Rayleigh number but scores of values of the Bingham number. Included in these figures are the conduction results as well. Figure 9.15 consolidates the numerical results of Nirmalkar et al. (2014b) and Patel and Chhabra (2016a) in the form of the average Nusselt number versus the composite parameter $Bi/\sqrt{Gr_B}$ where the results for different values of Pr_B and Ra_B are seen to collapse into one curve which approaches the conduction limit. Furthermore, the average Nusselt number values are well correlated as follows:

$$\mathrm{Nu}_l = \mathrm{Nu}_\infty + 0.638 \mathrm{Ra}^{*0.32} \mathrm{Pr}^{*-0.24} \tag{9.43}$$

where, $\mathrm{Nu}_\infty = \dfrac{2}{\ln(D_\infty/d)}$; $\mathrm{Ra}^* = \dfrac{\mathrm{Ra}_B}{1+\mathrm{Bi}}$; $\mathrm{Pr}^* = \mathrm{Pr}_B(1+\mathrm{Bi})$

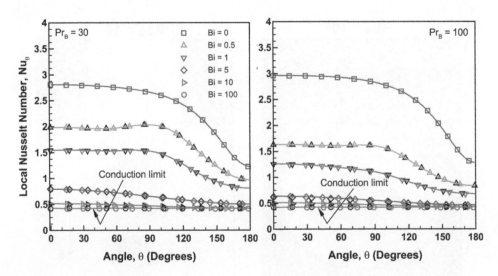

FIGURE 9.13 Distribution of the local Nusselt number (Nu_x) along the surface of the cylinder at $Ra_B = 10^2$ (colored symbols: Papanastasiou model: black color symbols: prediction from the bi-viscous model). (Replotted from Nirmalkar, N., Bose, A. and Chhabra, R.P., *Int. J. Therm. Sci.*, **83**, 33, 2014.)

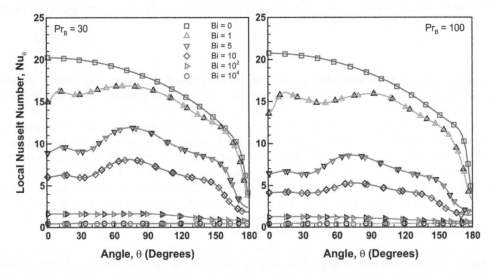

FIGURE 9.14 Distribution of the local Nusselt number (Nu_x) along the surface of the cylinder at $Ra_B = 10^6$ (colored symbols: Papanastasious model: black color symbols: prediction from the bi-viscous model). (Replotted from Nirmalkar, N., Bose, A. and Chhabra, R.P., *Int. J. Therm. Sci.*, **83**, 33, 2014.)

Equation 9.43 correlates 237 data points over the range of Rayleigh number, $10^2 \leq Ra_B \leq 10^6$, Bingham number, $0 \leq Bi \leq 10^7$, and Prandtl number, $10 \leq Pr_B \leq 100$ with the mean and maximum deviations of 9% and 40%, respectively, from the numerical results of Nirmalkar et al. (2014b), Dutta et al. (2018), and Patel and Chhabra (2016a). Similar results for free convection in Bingham fluids from semicircular and elliptical cylinders are also available in the literature (Tiwari and Chhabra, 2016; Patel and Chhabra, 2016a).

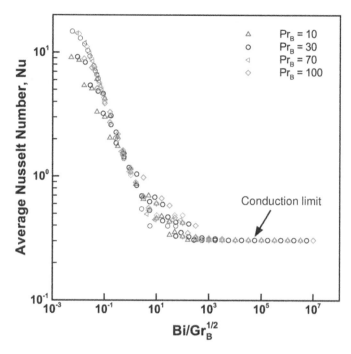

FIGURE 9.15 Average Nusselt number versus the composite parameter $Bi/\sqrt{Gr_B}$. (Based on the results of Nirmalkar, N., Bose, A. and Chhabra, R.P., *Int. J. Therm. Sci.*, **83**, 33, 2014 and Patel S.A. and Chhabra, R.P., *J. Thermophysics & Heat Transfer*, 30, 153, 2016a.)

9.4.2 Mixed Convection

9.4.2.1 Newtonian Fluids

As noted earlier, mixed convection heat transfer occurs for the intermediate values of the Richardson number. Only limited literature is available on mixed-convection heat transfer from a horizontal cylinder even in Newtonian fluids (Martynenko and Khramtsov, 2005; Soares et al., 2009, 2012; Srinivas et al., 2009). Broadly, irrespective of the type of mixed convection (aiding, opposing or cross-buoyancy), the rate of heat transfer is enhanced in this regime with reference to the pure free- ($Ri \to \infty$) and forced ($Ri \to 0$)-convection limits. However, the underlying mechanisms vary from one regime to another, especially in terms of the modifications to the coupled velocity and temperature fields in the close proximity of the cylinder. For instance, the aiding-buoyancy flow tends to stabilize the flow by suppressing the wake formation (Gupta et al., 2016). Thus, the wake contribution to the heat transfer is reduced or completely eliminated under appropriate operating conditions. Some of this is compensated for by the enhanced effective velocity as both velocities are additive in the aiding-buoyancy regime. In contrast, both the cross-buoyancy and opposing-buoyancy flows induce asymmetry in the temperature field and advances the cessation of the steady flow regime (Kieft et al., 1999; Hu and Kochesfahani, 2005; Khyati et al., 2007). One of the early studies in this field is that of Collis and Willams (1959) who presented an approximate criterion for the onset of buoyancy effects as $Re = 1.85 \, Gr^{0.35}$ which can be rearranged approximately as $Gr/Re^{2.86} \geq 5.81$ in the cross-buoyancy mixed-convection regime. In the same regime, Fand and Keswani (1973) reported extensive experimental results with water as the working fluid. Depending upon the value of the Richardson number, they identified four flow regimes as follows: For $Ri < 0.5$, heat transfer occurs primarily by forced convection; next, in the range $0.5 \leq Ri \leq 2$, free convection only contributes about 10% to the overall heat transfer; in the third regime, $2 \leq Ri \leq 40$, the heat transfer coefficient exhibits oscillating behavior, and finally, $Ri > 40$, free convection dominates the overall heat transfer. Subsequent experimental results (Villimpoc et al., 1990) indicate the forced-convection effects persist even up to $Ri = 4.4$ for high Prandtl number fluids. Similarly, Sarma and Sukhatme (1969) experimentally investigated the combined effects of free-convection and free stream turbulence on the local heat transfer from a cylinder to air in the range $50 \leq Re \leq 7000$ and $0.02 \leq Gr/Re^{2.5} \leq 32$. Early experimental studies also include the works of Hatton et al. (1970) and Jackson and Yen (1971). For the aiding-buoyancy mixed-convection regime, Hatton et al. (1970) consolidated their results by using the net velocity (sum of imposed and buoyancy-induced velocity) to define an effective Reynolds number Re_{eff}:

$$Nu = 0.384 + 0.581 Re_{eff}^{0.439} \qquad (9.44a)$$

where Re_{eff} is related to the Reynolds number based on the imposed forced velocity (Re) as:

$$Re_{eff} = Re\left[1 \pm 2.06\chi + 1.06\chi^2\right] \qquad (9.44b)$$

where $\chi = \dfrac{Ra^{0.48}}{Re}$

In Equation 9.44, the positive sign is used for the aiding-buoyancy and the negative sign applies for the opposing-buoyancy. Their experiments extend over the ranges of conditions as $10^{-3} \leq Ra \leq 10$ with air as the working fluid (Pr=0.71). Similarly, the aiding-buoyancy experiments of Jackson and Yen (1971) encompass Reynolds number in the range (100–3000) and Grashof number ($2.5 \times 10^4 \leq Gr \leq 3 \times 10^6$) with air (Pr=0.71) and they consolidated their data as:

$$\frac{Nu}{Nu_{Ri=0}} = (1+Ri)^{0.2133} \qquad (9.45)$$

Early analytical studies include that of Sparrow and Lee (1976) for the aiding-buoyancy flow from a horizontal cylinder by expanding the velocity and temperature profiles. In a series of papers, Badr (1983, 1984, 1985) has analyzed the aiding-, opposing- and cross-buoyancy configurations with air as the working fluid ($1 \leq Re \leq 40$; $0 \leq Ri \leq 5$) by numerically solving coupled Navier-Stokes and energy equations. Some of these results have been subsequently corroborated by Ahmad and Qureshi (1992, 1993) for both constant wall temperature and constant heat flux boundary conditions. Analogous studies in the vortex shedding regime ($Re > \sim 45$) are due to Chang and Sa (1990) and Patnaik et al. (1999) who have focused on the wake characteristics and temporal evolution of the hydrodynamic forces exerted on the cylinder. The effect of wide-ranging Prandtl numbers on cross-buoyancy mixed-convection flow and heat transfer from a horizontal cylinder has been explored by Ajith Kumar et al. (2016) and Hema Sundar Raju et al. (2018). Many workers have studied mixed convection from a horizontal elliptical cylinder (Ahmad and Badr, 2001; D'Alessio et al., 2003; Alnakeeb et al., 2017; Paul et al., 2013, 2016). In the limit of the aspect ratio of unity, these results also correspond to the case of a circular cylinder.

From the foregoing discussion, it is abundantly clear that the literature on mixed convection from a horizontal cylinder is indeed very limited. It is thus not possible to put forward predictive correlations even for the average Nusselt number, as can be also seen in the thorough review of Martynenko and Khramtsov (2005).

9.4.2.2 Power-Law Fluids

Mixed convection from a horizontal circular cylinder to power-law fluids has received very limited attention (Wang and Kleinstreuer, 1988a,b; Meissner et al., 1994; Bhowmick et al., 2014a,b). Wang and Kleinsteruer (1988a) studied the mixed convection from an isothermal horizontal cylinder in power-law fluids. Depending upon whether the cylinder was being heated or being cooled down, they studied both the cases, that is, free convection assisting or opposing the forced convection process. They used the edge velocity distribution based on the experimental measurements in Newtonian fluids for power-law

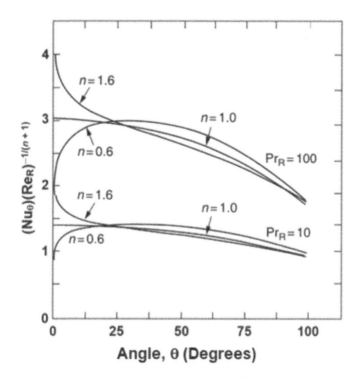

FIGURE 9.16 Effect of Prandtl number and power-law index on heat transfer from a horizontal heated cylinder in a mixed convection regime for $Ri_{PL}=2$. (Replotted from Wang, T.-Y. and Kleinstreuer, C., *Int. J. Heat Fluid Flow*, **9**, 182, 1988.)

fluid also. The relative importance of the free convection was quantified by using the following modified definition of the Richardson number, Ri_{PL}:

$$Ri_{PL} = \frac{Gr_{PL}}{Re_{PL}^{2/(2-n)}} \qquad (9.46)$$

Evidently, in the limit of $n=1$, Equation 9.46 reduces to the usual definition of the Richardson number, that is, (Gr/Re^2). Note that the Reynolds number here is based on the cylinder radius rather than the diameter $(= \rho V_0^{2-n} R^n / m)$.

Their numerical results for $Ri_{PL}=0$ and for $n=0.52$ and $n=1.6$ are in line with that of Kim et al. (1983). Figure 9.16 shows the influence of the power-law index and Prandtl number on the variation of the heat transfer group $(Nu/Re_R^{1/(n+1)})$ when the free convection aids the forced convection for $Ri_{PL}=2$. Subsequently, Meissner et al. (1994) have employed the Merk–Chao series method to study the mixed convection from two-dimensional and axisymmetric bodies (plate, sphere, and cylinder) to power-law fluids, when the submerged surface is maintained at a constant temperature. They quantified the relative importance of the free and forced convection using the ratio $(V_0/(V_0+V_f))$, where V_f given by $(g\beta R\Delta T)^{1/2}$ is the characteristic velocity due to buoyancy effects. This ratio can be rearranged as

$$Ri_m = \frac{1}{1+\sqrt{Ri}} = \frac{V_0}{V_0+V_f} \qquad (9.47)$$

Thus, $Ri_m=0$, that is, $V_0=0$ will imply pure free convection. Their results for $Ri_m=0$ are in excellent agreement with that of Acrivos (1960) over the most part of the cylinder surface, except close to the stagnation point. Likewise, for $Ri_m=1$, (when $V_f=0$), their results are consistent with that of Kim et al.

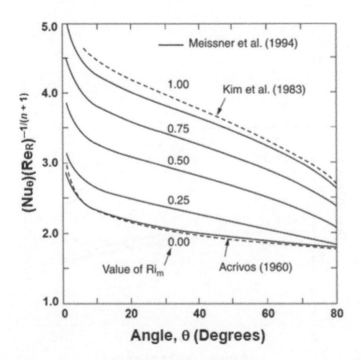

FIGURE 9.17 Effect of Richardson number (Ri_m) on local Nusselt number from a horizontal heated cylinder to a power-law fluid in a dilatant fluid ($n=1.6$, $Pr=100$). (Replotted from Meissner, D. L., Jeng, D.R., and DeWitt, K.J., *Int. J. Heat Mass Transf.*, **37**, 1475, 1994.)

(1983). Figure 9.17 shows representative results for the effect of the modified Richardson number, Ri_m, on the heat transfer group, $NuRe_R^{-1/(n+1)}$, for $n=1.6$ and Prandtl number of 100.

Wang and Kleinstreuer (1988b) have also studied mixed convection to power-law fluids from a slender vertical cylinder. In this case, both flow and heat transfer are strongly influenced by the curvature effects. Besides the earlier-noted boundary layer analyses, there have been a few numerical studies of the aiding-buoyancy (Srinivas et al., 2009) and cross-buoyancy (Soares et al., 2009) mixed convection in power-law fluids from an isothermal horizontal cylinder. Figure 9.18 shows the influence of the Prandtl number (Pr_{PL}), Reynolds number (Re_{PL}), and power-law index (n) on the local Nusselt number variation on the surface of the cylinder at $Ri_{PL}=2$. Heat transfer is seen to increase with the decreasing value of the power-law index and with the increasing values of the Reynolds and Prandtl numbers. The effect of Richardson number on the average heat transfer coefficient was found to be stronger at low Prandtl numbers and/or in highly shear-thinning fluids (e.g., $n=0.2$). Heat transfer was found to be relatively insensitive to the Richardson number in shear-thickening fluids. Figure 9.19 shows the dependence of the Nusselt number normalized with reference to the corresponding value for the forced convection on the power-law index and the other parameters where varying levels of enhancement in heat transfer are evident. Similarly, Soares et al. (2009) have investigated the case of the cross-buoyancy mixed convection in power-law fluids from an isothermal horizontal cylinder over the range of conditions as $n=0.6, 0.8, 1$, and 1.6, $Pr_{PL}=1, 100$, $Ri_{PL}=0, 1, 3$, and $1 \leq Re_{PL} \leq 30$. In this case also, the average Nusselt number exhibits qualitatively similar dependence on n, Ri_{PL}, Pr_{PL}, and Re_{PL} as that reported by Srinivas et al. (2009) for the aiding-buoyancy configuration. Subsequently, Soares et al. (2012) have extended their study for the constant heat flux boundary condition but only for Newtonian fluids. Singh et al. (2015) have numerically studied aiding-buoyancy mixed convection for an isothermal cylinder in nanofluids (approximated as a power-law fluid) in the range of $0.5 \leq n \leq 0.88$ in the steady flow regime ($1 \leq Re_{PL} \leq 40$) for Richardson number values up to $Ri_{PL}=5$. Their results are consistent with that of Srinivas et al. (2009).

Similar limited results are available on cross-buoyancy mixed convection in power-law fluids from a square cross-section cylinder (Dhiman et al., 2007b, 2008b), trapezoidal cylinder (Dhiman et al., 2019),

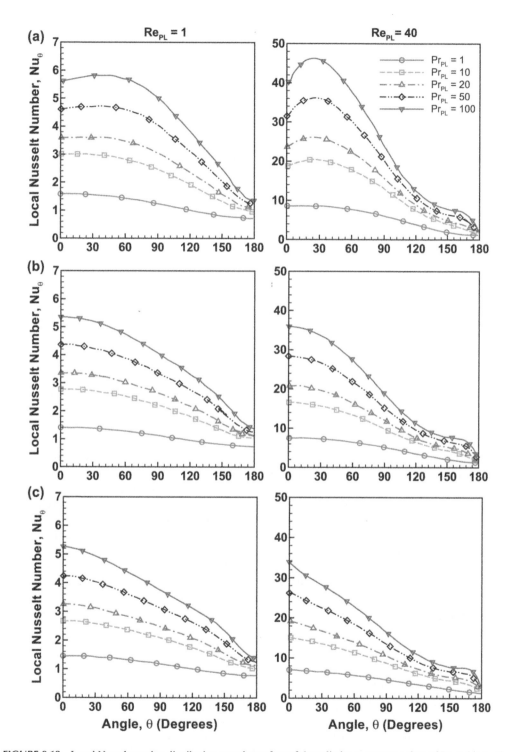

FIGURE 9.18 Local Nusselt number distribution over the surface of the cylinder at extreme values of Reynolds number ($Re_{PL} = 1, 40$) for a range of values of Prandtl number (Pr_{PL}) and for power-law index of (a) $n = 0.6$, (b) $n = 1$, and (c) $n = 1.4$ at $Ri_{PL} = 2$. (Replotted from Srinivas, A.T., Bharti, R.P. and Chhabra, R.P., *Ind. Eng. Chem. Res.*, **48**, 9735, 2009.)

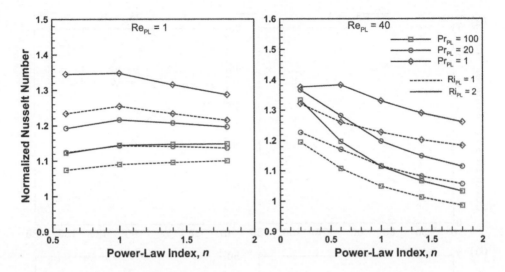

FIGURE 9.19 Dependence of Normalized Nusselt number (with respect to the corresponding values of forced convection) on power-law index, Richardson number and Prandtl number. (Replotted from Srinivas, A.T., Bharti, R.P. and Chhabra, R.P., *Ind. Eng. Chem. Res.*, **48**, 9735, 2009.)

and the aiding-buoyancy regime (Sivakumar et al., 2014). Scant numerical studies dealing with a horizontal semicircular cylinder are also available (Chandra and Chhabra, 2012b; Tiwari and Chhabra, 2015a). Mixed convection from a rotating cylinder in power-law fluids has been studied by Sasmal et al. (2019).

9.4.2.3 Bingham Plastic Fluids

Bose et al. (2015) have numerically investigated the aiding-buoyancy mixed convection from an isothermal horizontal cylinder in Bingham plastic fluids. Their results span the ranges of conditions as $0.1 \leq Re_B \leq 40$; $1 \leq Pr_B \leq 100$; $Bi \leq 10$; and $Ri_B \leq 2$. They found the momentum and thermal boundary layers to progressively become thinner with the increasing values of each of these parameters, that is, Re_B, Pr_B, Bi, and Ri_B. Thus, the average Nusselt number was shown to bear a positive dependence on all these parameters. While the effect of imposing the buoyancy-induced flow on the forced convection flow was maximum in Newtonian fluids (Bi=0) and beyond a critical value of Bi, the yield stress effects completely overshadowed the feeble buoyancy-induced forces thereby contributing very little to the overall yielding of the fluid and heat transfer over the range of conditions spanned in their work.

9.5 Spheres

9.5.1 Free Convection

9.5.1.1 Newtonian Fluids

As noted by Martynenko and Khramtsov (2005) and others (Jafarpur and Yovanovich, 1992; Patel et al., 2017), much research effort has been directed at studying free convection from a sphere in Newtonian fluids. The high Grashof and/or Rayleigh number limit is treated via the use of the standard boundary layer equations. The works of Merk and Prins (1953–1954), Chiang et al. (1964), Stewart (1971), and Potter and Riley (1980), for instance, exemplify the utility of this approach. This has led to the scaling of the Nusselt number with the Grashof and Prandtl numbers which has received support from the relevant experimental results, albeit this approach completely overlooks the contribution of wake. Similarly, the low Grashof number limit is approached by employing the matched asymptotic expansion method. The studies of Fendell (1968), Hossain and Gebhart (1970), and Heiber and Gebhart (1969) illustrate the utility of this approach. Neither of these results are applicable in the intermediate region between these two limits. This gap has been bridged by the numerical studies of Geoola and Cornish (1981, 1982), Singh and Hasan (1983), Riley (1986), Takamatsu et al. (1988), Jia and Gogos (1996a,b), Yang et al. (2007), and Suri et al. (2023b) etc. Owing to the rich variety of the flow phenomena observed in this simple case, this topic continues to receive attention in the literature, e.g., see the recent studies of Kitamura et al. (2015) and Lee and Chung (2017), etc. Based on their experiments in air and water, Kitamura et al. (2015) proposed that the turbulence initiates at the top of the sphere at about $Ra \sim 3 - 3.5 \times 10^8$, and this observation is in line with the previous studies (Martynenko and Khramtsov 2005). Similarly, Lee and Chung (2017) performed mass transfer experiments for a sphere at extremely high Rayleigh numbers ranging from $\sim 1.7 \times 10^8$ to $\sim 3 \times 10^{11}$.

Based on the extensive literature results, Martynenko and Khramtsov (2005) suggest the following two equations for the laminar-free convection from an isothermal sphere:

Churchill (1983) equation:

$$Nu_1 = 2 + \frac{0.589 Ra^{1/4}}{\left[1 + \left(\frac{0.43}{Pr}\right)^{9/16}\right]^{4/9}} \tag{9.48}$$

Jafarpur and Yovanovich (1992) equation:

$$Nu_1 = 2 + \frac{0.6 Ra^{1/4}}{\left[1 + \left(\frac{0.5}{Pr}\right)^{9/16}\right]^{4/9}} \tag{9.49}$$

The slight differences to the numerical constants appearing in these two equations have been prompted by the numerical results reported after the formulation of Equation 9.48. As can be seen in Table 9.7, the two predictions are not only extremely close to each other but also close to the recent numerical results as well.

At low Grashof numbers, Figures 9.20 and 9.21 show comparisons between the experimental temperature profile and average Nusselt number with the recent numerical results of Suri et al. (2023b). The correspondence seen in these figures inspires confidence in the reliability of such numerical predictions. Similar results for free convection from a hemisphere (Liu et al., 2018; Zhang et al., 2019a; Martynenko and Khramtsov, 2005), from spheroids (Ke et al., 2018), and from a cone (Oosthuizen and Donaldson, 1972) have also been reported in the literature. In addition, Arabi and Jafarpur (2016, 2017), Eslami and

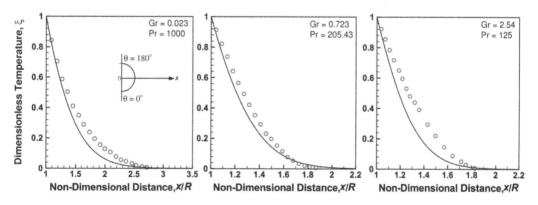

FIGURE 9.20 Comparison between the experimental results of temperature profiles (line) at $\theta = 90°$ (Hossain and Gebhart, 1970) and numerical results of Suri et al. (2023b) in Newtonian fluids.

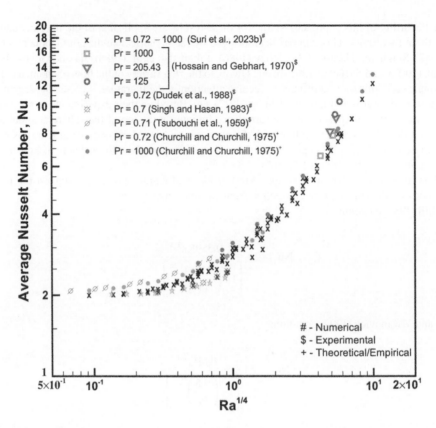

FIGURE 9.21 Comparison of the average Nusselt number results at low Grashof number with literature for Pr = 0.7. (From Suri et al., 2023b.)

Jafarpur (2012), and Kalendar et al. (2017) have attempted to develop a unified framework to consolidate the average Nusselt number for scores of convex shapes in the laminar-free convection regime.

Results on the hydrodynamic drag on a sphere in free-convection regime are indeed very scarce. Limited results have been reported by Dudek et al. (1988), Mograbi and Bar-Ziv (2005a,b), and these have been reviewed in detail by Suri et al. (2023b).

9.5.1.2 Power-Law Fluids

Using a Mangler-type transformation, Acrivos (1960) extended his treatment for the case of an isothermal sphere to obtain the local Nusselt number as

$$\mathrm{Nu} = -\xi'(0) \left\{ \frac{(2n+1)}{(3n+1)} \right\}^{n/3n+1} \frac{(\sin \varsigma)^{n+1/(2n+1)}}{\left\{ \int_0^{x_1} (\sin \varsigma)^{3n+2/2n+1} dx_1 \right\}^{n/3n+1}} \times Z \quad (9.50)$$

which, for the front part of the sphere, that is, $x_1 \to 0$, reduces to

$$\mathrm{Nu} = -\xi'(0) \left\{ \frac{2n+1}{3n+1} \right\}^{n/3n+1} x_1^{(1-n)/(3n+1)} \times Z \quad (9.51)$$

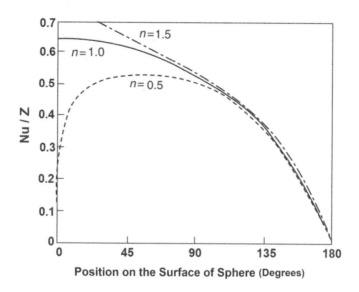

FIGURE 9.22 Effect of power-law index on the local Nusselt number on the surface of a sphere. (Modified after Acrivos, A., *AIChE., J.*, 6, 584, 1960.)

Equation 9.51 has been shown to be quite accurate in the range $0 \leq x_1 \leq \pi/6$. The resulting typical variation of the Nusselt number over the surface of a sphere is shown in Figure 9.22 for a range of values of the flow-behavior index. In another significant study based on a modification of the earlier analysis, Stewart (1971) solved the laminar-boundary layer equations for free convection from a range of geometric surfaces including a sphere. Unlike the treatment of Acrivos (1960), the modification due to Stewart (1971) is uniformly valid over the entire surface of the sphere. Owing to the different choices of the characteristic velocity and the linear dimension, the resulting expressions for the surface-averaged Nusselt numbers from these two analyses are slightly different, presented as follows:

Acrivos equation:

$$\text{Nu} = C_1(n) Z \tag{9.52}$$

Stewart equation:

$$\text{Nu}_1 = C_2(n)(\text{Gr}_1 \text{Pr}_1)^{1/3n+1} \tag{9.53}$$

where $C_1(n)$ and $C_2(n)$ are weak functions of the power-law flow-behavior index. While Equation 9.53 was stated to be applicable under most conditions but is expected to be particularly suitable for large values of the Prandtl number, an assumption which is also inherent in the derivation of Equation 9.52.

Experimental results on the free-thermal convection from isothermal spheres to power-law polymer solutions have been reported by Liew and Adelman (1975) and Amato and Tien (1972, 1976). In an extensive experimental study, Liew and Adelman (1975) studied free-convective heat transfer from electrically heated copper spheres to aqueous solutions of carbopol, carbose, and Natrosol. Based on their data in the range $0.66 < n < 1$, Liew and Adelman (1975) found that the single value of $C_1(n) = 0.561$ in Equation 9.52 described their results with an average error of 5.9%. Interesting enough, Liew and Adelman (1975) also reported that their experimental results correlated equally well ($\pm 6.2\%$) with the empirical correlation:

$$\text{Nu} = 0.611(\text{Gr} \text{Pr}_2)^{0.241} \tag{9.54}$$

where $Pr_2 = ((\rho C_p)/k)(m/\rho)^{1/(2-n)} R^{(2n-2)/(n-2)}$.

Note the similarity between Equations 9.53 and 9.54, except for the slightly different definition of the Prandtl number. Furthermore, the value of the exponent ($1/(3n+1)$) in Equation 9.53 ranges from 0.25 to 0.33 for the experimental conditions (Liew and Adelman, 1975) which is qualitatively similar to the experimental value of 0.241 in Equation 9.54.

In an extensive experimental investigation, Amato and Tien (1972, 1976) also studied natural thermal convection from electrically heated isothermal copper spheres to water and aqueous solutions of carboxymethyl cellulose and polyox ($0.592 \leq n \leq 0.948$). In addition to reporting the values of the surface-averaged Nusselt numbers, Amato and Tien (1976) also reported detailed temperature and velocity profiles in the vicinity of the heated spheres. A typical comparison between the measured and predicted (Acrivos, 1960) temperature profiles is shown in Figure 9.23; the agreement is seen to be almost perfect. A typical comparison between the predicted and measured variations of the local heat transfer rate on the surface of a sphere is shown in Figure 9.24; again, the agreement is seen to be as good as can be expected in this type of work. Amato and Tien (1976) also examined the applicability of Equation 9.52 by plotting the measured mean values of the Nusselt number versus the composite dimensionless group Z (= $Gr^{1/(2n+2)}$ $Pr^{n/(3n+1)}$) as shown in Figure 9.25. While in general the agreement is seen to be satisfactory, there are two distinct regions separated by $Z = \sim 10$. Amato and Tien (1976) presented the following separate correlations to cover these two regions:

For $Z < 10$,

$$Nu = (0.996 \pm 0.12) Z^{0.682 \pm 0.062} \tag{9.55}$$

While in the range $10 \leq Z \leq 40$,

$$Nu = (0.489 \pm 0.005) Z \tag{9.56}$$

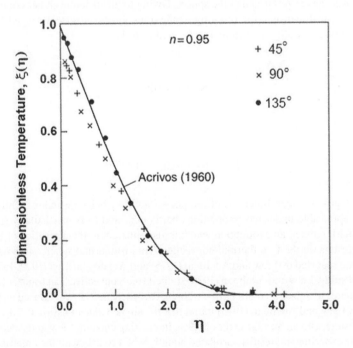

FIGURE 9.23 Representative comparison between the predicted and measured temperature distribution for an isothermal sphere submerged in a polymer solution ($n=0.95$). Modified after Amato, W.S. and Tien, C., *Int. J. Heat Mass Transf.*, **19**, 1257, 1976.)

Equations 9.55 and 9.56 correlate much of their experimental results with a mean deviation of 8%. Furthermore, the numerical constant of 0.489 in Equation 9.56 is remarkably close to the value of 0.49, predicted by Acrivos (1960) for large values of the Prandtl number. Further analysis of the results of Amato and Tien (1976) also suggests that almost two-thirds of the total heat transfer takes place from the bottom half of the sphere and that the heat transfer is maximum at the stagnation point. The detailed temperature and velocity profiles can also be used to derive useful information about the momentum and thermal boundary layer thicknesses. The ratio of the calculated thermal-to-momentum boundary layer

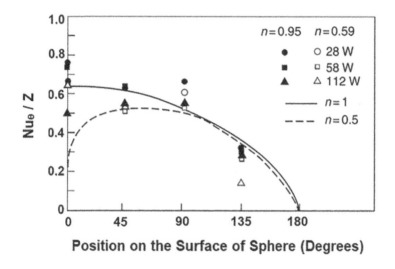

FIGURE 9.24 Representative comparison between the predicted and experimental Nusselt number variation for a sphere (under constant flux condition) for a Newtonian and shear-thinning fluid ($n=0.5$). (Modified after Amato, W.S. and Tien, C., *Int. J. Heat Mass Transf.*, **19**, 1257, 1976.)

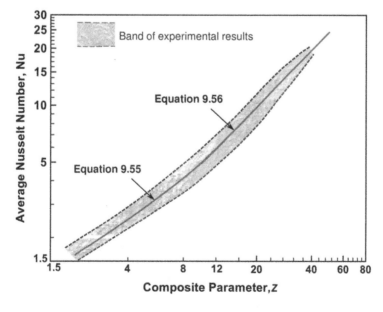

FIGURE 9.25 Dependence of average Nusselt number on the composite parameter, Z. (Modified after Amato, W.S. and Tien, C., *Int. J. Heat Mass Transfer.*, **19**, 1257, 1976.)

thicknesses in polymer solutions is of the same order as that in water, except for that at the stagnation point. In conclusion, perhaps this is the only study available in the literature which affords a complete validation of the analysis of Acrivos (1960).

The analogous mass transfer results for spheres in free-convection regime have been reported by Lee and Donatelli (1989). These investigators have measured the rate of dissolution of benzoic acid spheres suspended in quiescent aqueous solutions of carboxymethyl cellulose encompassing the range $0.58 < n < 1$. By invoking the usual analogy between mass and heat transfer, the mass transfer analogs of Equations 9.52 and 9.53 can be written as

$$\text{Sh} = C_1'(n) \text{Gr}_m^{1/2n+1} \text{Sc}^{n/3n+1} \tag{9.57}$$

and

$$\text{Sh}_1 = C_2'(n) \left(\text{Gr}_{1m} \text{Sc}_1 \right)^{1/3n+1} \tag{9.58}$$

For the range of conditions, namely n, covered by the study of Lee and Donatelli (1989), C_1' and C_2' deviate from their Newtonian ($n=1$) values by less than 10%. Similarly, one can also generalize the correlations of Liew and Adelman (1975) and of Amato and Tien (1976) to interpret free-convective mass transfer data. Lee and Donatelli (1989) have also adapted the empirical correlation due to Churchill (1983) (originally developed for Newtonian systems) for power-law fluids as

$$\text{Sh}_1 = 2 + \frac{C_2' \left(\text{Gr}_{1m} \text{Sc}_1 \right)^{1/3n+1}}{\left[1 + \left(\frac{0.43}{\text{Sc}_1} \right)^{9/16} \right]^{4/9}} \tag{9.59}$$

Figures 9.26 and 9.27 show comparisons between the experimental and predicted mean values of the Sherwood number using Equations 9.55–9.59. In general, the agreement is seen to be satisfactory. Lee

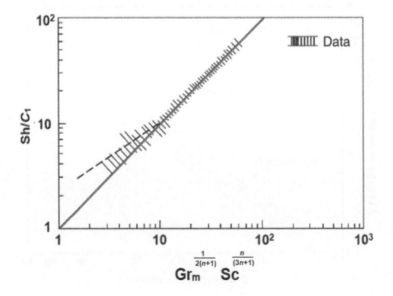

FIGURE 9.26 Comparison between the predictions of Acrivos (1960) and data of Amato and Tien (1976) for free convection from a sphere. (-------- Equation 9.55 and ——— Equation 9.57.)

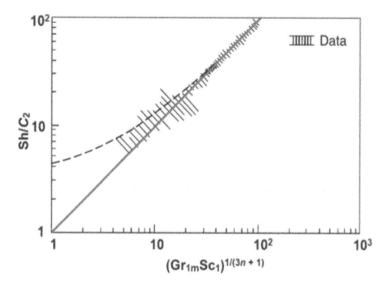

FIGURE 9.27 Comparison between the predictions of Stewart (1971), Churchill (1983), and data of Lee and Donatelli (1989) for free convection from a sphere. (——— Equation 9.58 and --------- Equation 9.59.)

and Donatelli (1989) reported that their results are in line with Equations 9.55 and 9.56 with the mean deviations of 17% and 13%, respectively. However, as can be seen from Figures 9.26 and 9.27, the experiments show appreciable deviations from Equations 9.57 and 9.58 in the low Grashof number regime. One would intuitively expect that as the free convection effects diminish with the decreasing value of the Grashof number, the Sherwood number should approach its asymptotic value of 2, corresponding to pure diffusion limit. Owing to the thin-boundary layer assumption inherent in analyses, this limiting behavior is not predicted by Equation 9.57 or Equation 9.58. On the other hand, Equation 9.59 includes this limit. Owing to the generally high viscosities, the value of Schmidt number for polymeric solutions is generally high, and Equation 9.59 thus simplifies to the more familiar form:

$$Sh_1 \approx 2 + C_2'(n)\left(Gr_{1m}\ Sc_1\right)^{1/3n+1} \tag{9.60}$$

Finally, due to the fact that C_2' varies only little with the power-law index, n, Lee and Donatelli (1989) found that their experiments were in good agreement with Equation 9.59 or Equation 9.60 even when the value of the constant $C_2'=0.589$ for Newtonian fluids is used for power-law fluids. Thus, the modified form of the correlation due to Churchill (1983) includes all known limiting conditions.

In addition to the preceding approximate analyses, there have been two extensive numerical studies dealing with the solutions of the complete field equations for the laminar-free convection heat transfer in power-law fluids from an isothermal sphere in the steady axisymmetric flow regime (Prhashanna and Chhabra, 2010; Suri et al., 2023b). Combined together, these two studies span the ranges of conditions as: $Gr_{PL} \leq 10^7$; $0.72 \leq Pr_{PL} \leq 1000$ and $0.1 \leq n \leq 2$. Figures 9.28 and 9.29 show typical variation of the local Nusselt number over the surface of the sphere for a range of conditions in terms of Gr_{PL}, Pr_{PL}, and power-law index (n). At low Grashof/Prandtl numbers, there is very little variation along the sphere surface due to the dominance of conduction.

The buoyancy-driven flow strengthens with the increasing Grashof number and/or Prandtl number, the local Nusselt number shows the usual variation from its maximum value at the front stagnation point decreasing all the way to the rear stagnation point provided there is no wake formation. As expected, the local Nusselt number exhibits a positive dependence on both Grashof and Prandtl numbers for a given power-law index. On the other hand, for fixed values of Gr_{PL} and Pr_{PL}, Nusselt number is seen to be

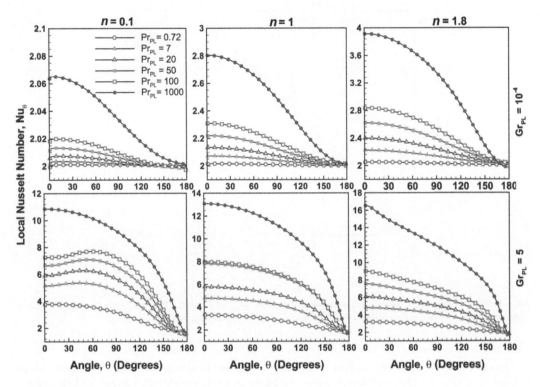

FIGURE 9.28 Distribution of local Nusselt number on the surface of an isothermal sphere in power-law fluids over low Grashof number range. (Based on the results of Suri et al., 2023b).

higher in shear-thinning fluids ($n<1$) and lower in shear-thickening fluids ($n>1$) with reference to that in Newtonian fluids otherwise under identical conditions. Figure 9.30 shows the multivariable dependence of the average Nusselt number on the Grashof and Prandtl numbers and the power-law index. It is possible to consolidate these numerical results by employing the composite parameter Z as follows:

$0.0177 \leq Z \leq 10$

$$\mathrm{Nu}_1 = 2 + 0.900 \left(\frac{3n+1}{4n}\right)^{0.358} Z^{1.070} \tag{9.60a}$$

$10 < Z \leq 730$

$$\mathrm{Nu}_1 = Z \tag{9.60b}$$

Attention is drawn to the fact that these equations are of the form of Equations 9.55 and 9.56, respectively, thereby providing a theoretical justification to the correlations of Amato and Tien (1976). Equation 9.60a correlates 1566 data points encompassing the values of the Grashof number ($10^{-4} \leq \mathrm{Gr}_{PL} \leq 10$), Prandtl number ($7 \leq \mathrm{Pr}_{PL} \leq 1000$), and power-law index ($0.1 \leq n \leq 2$) with a mean error of 4% which rises to a maximum of 35% in a few cases, without any discernable trends. For Z >10, total 140 individual results covering the range of parameters as $10 \leq \mathrm{Gr}_{PL} \leq 10^7$, $0.72 \leq \mathrm{Pr}_{PL} \leq 100$, and $0.4 \leq n \leq 1.8$ best correlated by Equation 9.60b with 0.34% average and 11% maximum deviations from the numerical results. Figure 9.31 shows a comparison between the predictions of Equation 9.60a and 9.60b and the experimental results for heat transfer (Amato and Tien, 1976; Liew and Adelman, 1975) and for mass transfer

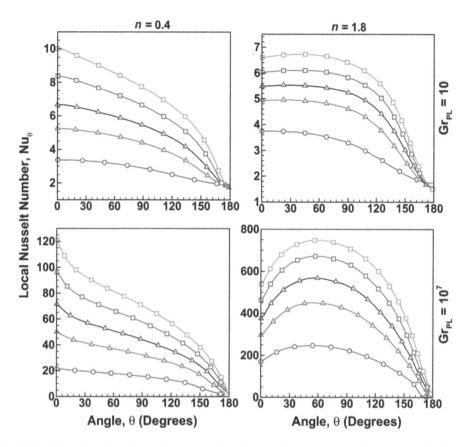

FIGURE 9.29 Distribution of local Nusselt number on the surface of sphere in power-law fluids over low Grashof number range (Modified after Prhashanna, A. and Chhabra, R.P., *Chem. Eng. Sci.*, **65**, 6190, 2010).

(Lee and Donatelli, 1989). The numerical results are seen to approach the expected limiting value of Nu=2 for sufficiently small values of Z. Similar results on free convection in power-law fluids from a cone, hemisphere, disk, and spheroid have also been reported in the literature (Mishra et al., 2019c; Sasmal and Chhabra, 2014; Gupta et al., 2014; Siddiqa et al., 2016)

9.5.1.3 Bingham Plastic Fluids

Only limited results are available on free convection in Bingham plastic fluids from heated spheres and spheroids for the constant wall temperature condition. Nirmalkar et al. (2014c) have studied numerically the laminar-free convection from an isothermal sphere in the steady axisymmetric flow regime. Their results on drag and heat transfer extend over the range of conditions as $10^2 \leq Ra_B \leq 10^6$, $10 \leq Pr_B \leq 100$, and $Bi \leq 10^4$. Subsequently, these results have been corroborated by independent studies of Gupta and Chhabra (2016c) and Gupta et al. (2017a). Figure 9.32 shows typical morphology of the yield surfaces separating the fluid-like and solid-like regions as functions of the Rayleigh number, Prandtl number, and Bingham number. As expected, the entire fluid is, so to say, frozen and heat transfer occurs by conduction under these conditions. Conversely, there exists a limiting Bingham number (Bi_{max}) for each pair of Rayleigh and Prandtl numbers beyond which conduction is the sole mechanism of heat transfer, as can be seen in Figure 9.33. In the range of their study, Nirmalkar et al. (2014c) correlated their results as:

$$Bi_{max} = 0.215 + 0.06 Ra_B^{0.65} Pr_B^{-0.47} \qquad 10^2 < Ra_B \leq 10^4 \tag{9.61a}$$

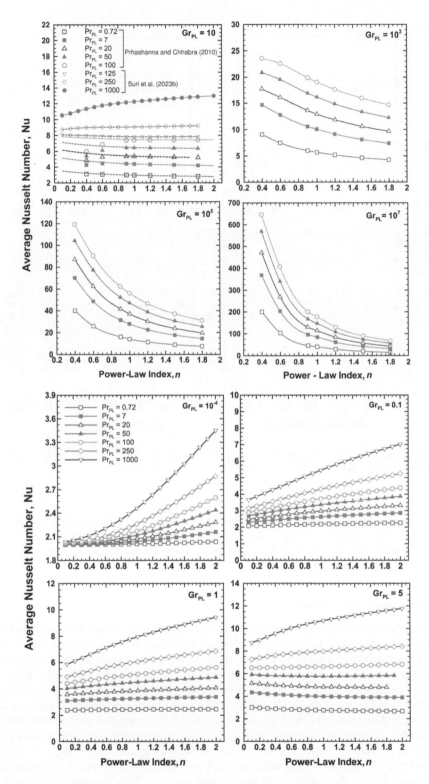

FIGURE 9.30 (a) Dependence of the average Nusselt number on the power-law index based on the results of Prhashanna and Chhabra (2010). (Broken lines for $Gr_{PL} = 10$ are from Suri et al., 2023b.) (b) Dependence of the average Nusselt number on the power-law index in the range of low Grashof number. (From Suri et al., 2023b.)

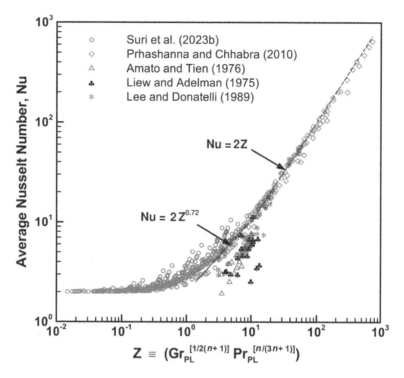

FIGURE 9.31 Comparison of the numerical results of the average Nusselt number with experimental results of Amato and Tien (1976), Liew and Adelman (1975) and Lee and Donatelli (1989).

and

$$Bi_{max} = 1.04 \left(\frac{Ra_B}{Pr_B} \right)^{0.35} \quad 10^4 < Ra_B \leq 10^6 \quad (9.61b)$$

Figure 9.34 shows a representative distribution of the Nusselt number along the surface of the sphere for a range of values of Ra_B, Pr_B, and Bi including the limiting case of Newtonian fluids (Bi=0). At high Bingham numbers, the local Nusselt number remains constant over the entire surface ($Nu_\theta \approx 2$) corresponding to the conduction limit and it shows some up and down behavior for the intermediate conditions, which is attributed to the changes in the unyielded regions close to the sphere (Figure 9.32). Finally, Figures 9.35a and b show the dependence of the average Nusselt number on the Rayleigh number and the composite parameter $\left(Bi\sqrt{Pr_B} \right)$ and $\left(Bi/\sqrt{Gr_B} \right)$. The numerical results shown in Figure 9.35 are best approximated as:

$$Nu_1 = 2 \quad \text{for } Bi \geq Bi_{max} \quad (9.62a)$$

and

$$Nu_1 = 2 + \frac{0.55 Ra^{1/4} \left(1 - \frac{Bi}{Bi_{max}} \right)^{4.66}}{(1+Bi)^{0.18}} \quad \text{for } Bi < Bi_{max} \quad (9.62b)$$

Equation 9.62 consolidates their numerical results with the mean deviation of 4.7% which rises to a maximum of 30%. In the limit of Bi → 0, the familiar scaling of Nu ~ $Ra^{1/4}$ is predicted by these equations. The effect of viscous dissipation and temperature-dependent viscosity on heat transfer from spheres and spheroids in yield stress fluids has been found to be rather small (Gupta et al., 2017a). Laminar-free convection from a cone in Bingham plastics has been studied by Mishra et al. (2019c).

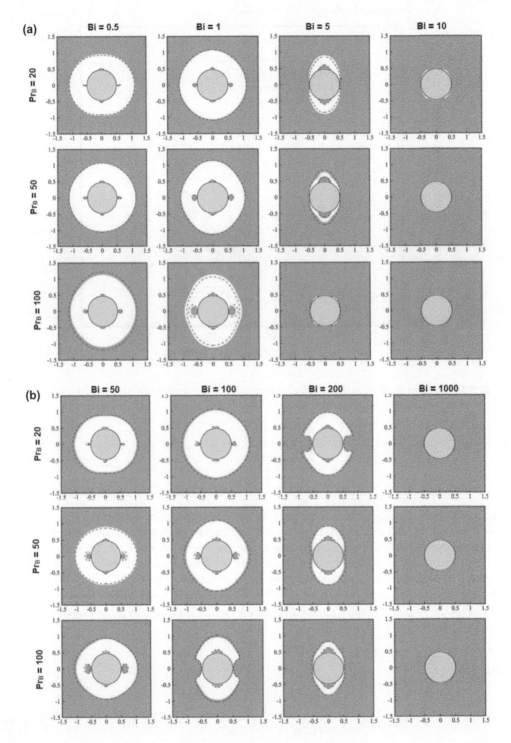

FIGURE 9.32 Morphology of the yielded (unshaded)/unyielded (shaded) region at (a) $Ra_B = 10^2$ and (b) $Ra_B = 10^6$. (Nirmalkar, N., Gupta, A.K. and Chhabra, R.P., *Ind. Eng. Chem. Res.*, **53**, 17818, 2014.)

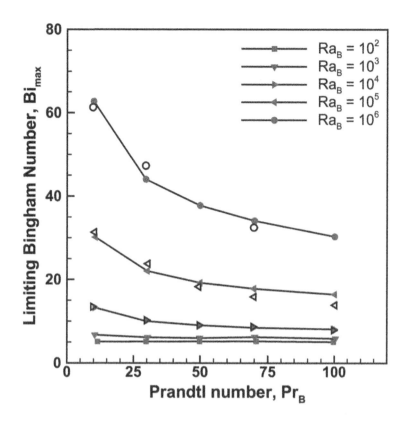

FIGURE 9.33 Dependence of the limiting Bingham number (Bi_{max}) on Rayleigh and Prandtl numbers (filled symbols show the fit of Equations 9.61a and 9.61b). (Nirmalkar, N., Gupta, A.K. and Chhabra, R.P., *Ind. Eng. Chem. Res.*, 53, 17818, 2014.)

9.5.2 Mixed Convection

9.5.2.1 Newtonian Fluids

Early attempts at analyzing mixed convection from a heated sphere hinge on the use of the boundary layer equations. Within the framework of the other simplifications inherent in such analyses, these results do not account for the wake contributions and also neglect the curvature effects. The early studies of Acrivos (1960) and Chen and Mucoglu (1977, 1978) illustrate the utility of this approach. These results are applicable at high Reynolds and Grashof numbers. At the other extreme of low Reynolds and/or Grashof numbers, one can linearize the inertial terms in the momentum and energy equations, and this is amenable to the matched asymptotic expansion approach (Hieber and Gebhart, 1969). Both these approaches have been shown to yield reliable scaling of the skin friction and Nusselt number with the pertinent parameters. Chen and Mucglou (1977, 1978) have considered both aiding and opposing buoyancy as well as the constant wall temperature and constant heat flux condition on the surface of the sphere. One must however resort to numerical solution of the coupled momentum and energy equations for the intermediate values of the Reynolds, Grashof, and Prandtl numbers. Thus, Wong et al. (1986) solved the steady equations for the aiding-buoyancy mixed convection from an isothermal sphere. With air as the working fluid, their results embrace the range of conditions as $5 \leq Re \leq 100$ and Richardson number values up to 10. They reported the surface pressure to decrease significantly near the rear stagnation point whereas it increased slightly near the front stagnation point. Subsequently, Nguyen et al. (1993) introduced two additional features, namely the transient effects and heat transfer inside the sphere, that is, the conjugate heat transfer problem. In a detailed numerical study, Bhattacharyya and Singh (2008) investigated the aiding-buoyancy mixed convection from an isothermal sphere in the range $1 \leq Re \leq 200$

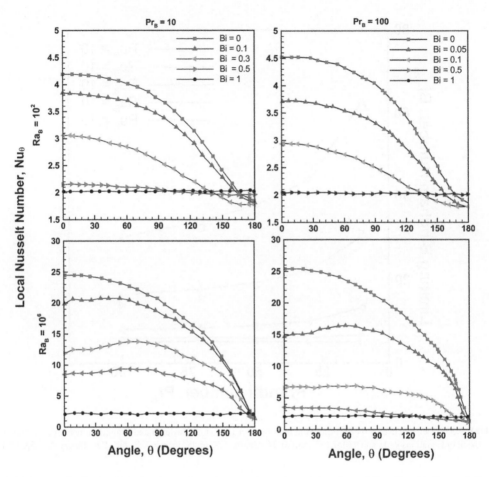

FIGURE 9.34 Distribution of the local Nusselt number over the surface of the sphere. (Nirmalkar, N., Gupta, A.K. and Chhabra, R.P., *Ind. Eng. Chem. Res.*, **53**, 17818, 2014.)

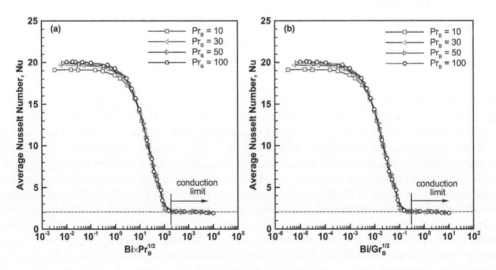

FIGURE 9.35 Functional dependence of the average Nusselt number on the dimensionless groups (a) $Bi \times Pr_B^{1/2}$ and (b) $Bi/Gr_B^{1/2} = 10^2$ at $Ra_B = 10^6$ (Nirmalkar, N., Gupta, A.K. and Chhabra, R.P., *Ind. Eng. Chem. Res.*, **53**, 17818, 2014.)

and Gr≤6×10⁴ with air as the fluid medium (Pr=0.72) in the steady axisymmetric regime. Their results are consistent with that of Nazar et al. (2002) for water (Pr=6.8). Much of this literature has been critically reviewed by Kotouc et al. (2008, 2009a, b) and Nath at el. (2019). These authors have examined the flow regime transitions in the mixed convection from an isothermal sphere. Thus, they reported that the flow remains attached up to about Re~960 in air even at Ri=0.7. The corresponding value is much smaller, Re~261 in water (Pr=7). Furthermore, in the aiding-buoyancy case, the primary bifurcation of flow is always regular whereas at Ri≥0.6, it appears in azimuthal subspaces with high azimuthal wave numbers. At high Richardson numbers (aiding flow), steady axisymmetric flow persists up to Re=1400 for air (Pr=0.721 and Ri=0.7). The aforementioned analytical and numerical activity in this field has also been supported by limited experimental studies. Yamanaka et al. (1976b) reported extensive experimental results on mixed convection from a sphere over the ranges of conditions (based on sphere diameter) as $2\times10^{-4} \leq Re \leq 900$, $7 \leq Pr \leq 2.4\times10^4$, and $5.7\times10^{-3} \leq Gr \leq 10^7$ and they proposed the following correlation for the aiding-buoyancy case:

$$Nu_1 = 2 + \left\{ \left[\frac{(126\,Re + 57\,Re^{3/2})Pr^{1/3}}{1 + 52\,Re^{1/2} + 100\,Re} \right]^{3/2} + 0.44\,Ra^{3/2} \right\}^{2/3} \left(\frac{\mu_b}{\mu_s} \right)^{1/4} \quad (9.63)$$

Equation 9.63 purports to describe the original experimental data with a mean deviation of ~14%, though the deviations as high as 54% can also be seen in their work. Similarly, Tang and Johnson (1990, 1992) and Tang et al. (1991) have reported flow visualization results for the aiding- and opposing-buoyancy mixed convection regimes. They argued that the available sphere surface for the development and growth of the free-convection flow is minimum for the aiding- and maximum for the cross-buoyancy regime. They employed the notion of an effective sphere diameter to correlate their experimental results in the range $220 \leq Re \leq 604$ and $3\times10^5 \leq Gr \leq 9\times10^5$ in air (Pr=0.7). Lastly, Mograbi and Bar-Ziv (2005a, b) and Mograbi et al. (2002) have reported experimental results of drag on a sphere in the aiding and opposing-buoyancy cases for Gr<~ 10⁻⁴ and Re≤~ 0.3. These results were used to develop scaling relationships similar to the Stokes law in the forced convection regime (Ziskind et al., 2001).

9.5.2.2 Power-Law Fluids

Little numerical and experimental work is available on heat and mass transfer from a sphere to power-law fluids in the mixed-convection regime. The studies of Wang and Kleinstreuer (1988a) and Meissner et al. (1994) referred to in Section 9.4.2.2 also presented limited numerical results for mixed convection from an isothermal sphere to power-law fluids. Figures 9.36 and 9.37 show the effects of Prandtl number, power-law index, and Richardson number on the variation of the Nusselt number on the surface of a sphere including the limiting case of pure free convection; in the latter case, the results of Meissner et al. (1994) coincide with the asymptotic analysis of Acrivos (1960) as well as with the numerical results of Nirmalkar and Chhabra (2013). Yamanaka et al. (1976b) reported an experimental study of mixed-convective heat transfer from isothermal spheres in a series of dilute solutions of methyl cellulose and carboxymethyl cellulose in water. However, since these solutions displayed only weak non-Newtonian characteristics, Yamanaka et al. (1976b) treated them as Newtonian fluids, with constant values of shear viscosity and proposed Equation 9.63. Therefore, it is virtually impossible to establish the role of non-Newtonian characteristics on mixed convection from a sphere using their experimental data. Subsequently, Yamanaka and Mitsuishi (1978) have reported experimental results on mixed convection from isothermal spheres to power-law fluids. Using the definitions of Grashof and Prandtl numbers of Acrivos (1960), with the sphere diameter as the characteristic length, Yamanaka and Mitsuishi (1978) put forward the following correlation:

$$Nu_1 = 2 + \left[\left(0.866\sigma^{2/3}Pe^{1/3} - 0.553\sigma - 0.341 \right)^{3/2} + \left(0.44\,Gr_2^{1/2(n+1)}\,Pr_2^{n/3n+1} \right)^{3/2} \right]^{2/3} \left(\frac{m_b}{m_s} \right)^{1/(3n+1)} \quad (9.64)$$

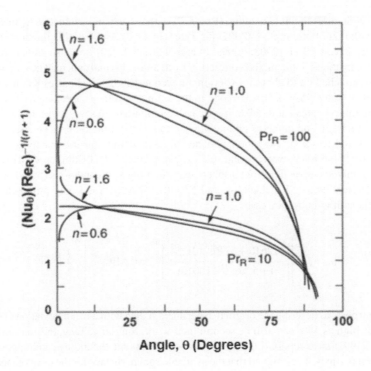

FIGURE 9.36 Effect of Prandtl number and power-law index on heat transfer from a heated sphere in a mixed convection regime for $Ri_{PL}=2$ for upward flow of power-law fluids. (Replotted from Wang, T.-Y. and Kleinstreuer, C., *Int. J. Heat Mass Transf.*, **31**, 91, 1988b.)

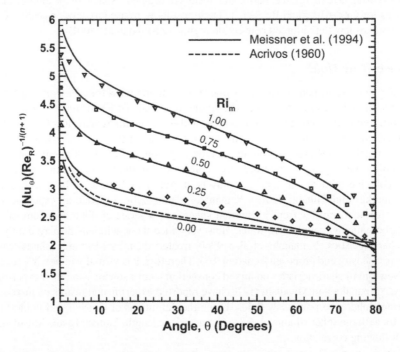

FIGURE 9.37 Effect of Richardson number (Ri_m) on the local Nusselt number for heat transfer from a heated sphere to a power-law fluid flowing upward for $n = 1.6$ and $Pr_{PL} = 100$ in mixed convection regime (symbols represents the results of Nirmalkar and Chhabra, 2013). (Replotted from Meissner, D.L., Jeng, D.R., and DeWitt, K.J., *Int. J. Heat Mass Transf.*, **37**, 1475, 1994.)

where σ is empirically related to the power-law index via the following relationship:

$$\sigma = -2.475n^3 + 6.738n^2 - 7.668n + 4.74 \tag{9.65}$$

Equations 9.64 and 9.65 encompass the following ranges of conditions:

$$1 \leq \text{Pe} \leq 10^3; 1.6 \times 10^{-6} \leq \text{Gr}_2 \leq 0.44; 2.6 \times 10^4 \leq \text{Pr}_2 \leq 6.4 \times 10^5 \text{ and } 0.3 \leq n \leq 0.93.$$

The average and maximum deviations are of the order of 30% and 68%, respectively.

In addition to the preceding approximate studies, Nirmalkar and Chhabra (2013) solved the complete governing equations for the laminar aiding-buoyancy mixed convection from a sphere in the steady axisymmetric regime in power-law fluids over the ranges of conditions as $0.2 \leq n \leq 2$, $\text{Ri}_{PL} \leq 2$, $1 \leq \text{Re}_{PL} \leq 100$, and $1 \leq \text{Pr}_{PL} \leq 100$. They reported both the drag coefficient and the average Nusselt number to increase with the Richardson number thereby suggesting slight enhancement on both counts, namely the aiding-buoyancy flow as well as the shear-thinning viscosity of the fluid. Using the modified definitions of the Reynolds and Prandtl numbers as (Meissner et al., 1994):

$$\text{Re}^* = \text{Re}_{PL}\left(1 + \sqrt{\text{Ri}_{PL}}\right)^{2-n} \tag{9.66a}$$

$$\text{Pr}^* = \text{Pr}_{PL}\left(1 + \sqrt{\text{Ri}_{PL}}\right)^{n-1} \tag{9.66b}$$

and introducing the heat transfer factor, $j = \text{Nu}/\text{Re}^*(\text{Pr}^*)^{1/3}$, Nirmalkar and Chhabra (2013) consolidated their numerical results as follows:

$$j = 0.88\left(\text{Re}^*\right)^{-2/3}\left(\frac{3n+1}{4n}\right)^{0.74} \tag{9.67}$$

Figure 9.38 shows a parity plot between the predictions of Equation 9.67 and the numerical results of Nirmalkar and Chhabra (2013).

9.5.2.3 Bingham Plastic Fluids

In a subsequent numerical study, Nirmalkar et al. (2014a) extended their power-law work to mixed convection from a heated sphere in Bingham plastic fluids in the aiding-buoyancy regime. Their results encompass the ranges of conditions as $0.1 \leq \text{Re}_B \leq 100$, $10 \leq \text{Pr}_B \leq 100$, $0 \leq \text{Ri}_B \leq 2$, and $0 \leq \text{Bi} \leq 10$ for the constant wall temperature condition (isothermal sphere). The influence of the aiding-buoyancy was seen to be maximum in Newtonian fluids (Bi=0) and it diminished with the increasing Bingham number. Conversely, though the average Nusselt number exhibited a positive dependence on the Richardson number, the latter's contribution seldom exceeded ~10% over and above the value of the average Nusselt number in the forced convection regime at the same values of Re_B, Pr_B, and Bi. Subsequent analogous studies for isothermal hemispheres (Nalluri et al., 2015; Patel et al., 2018) and for spheroids (Gupta and Chhabra, 2016) also indicate similar weak influence of the Richardson number in visco-plastic fluids.

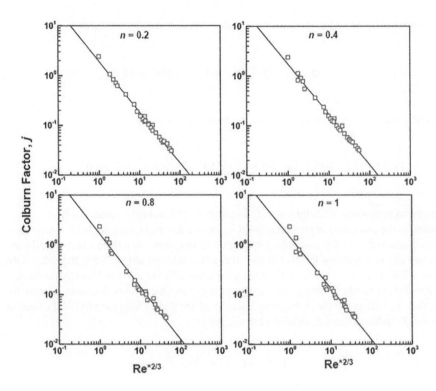

FIGURE 9.38 Variation of j-factor with the modified Reynolds number and power-law index. (Nirmalkar, N. and Chhabra, R.P., *Chem. Eng. Sci.*, **89**, 49 2013.)

9.6 Visco-elastic Effects in Boundary Layers

Little is known about the role of visco-elasticity in the free- and mixed-convection regimes. The bulk of the available literature relates either to a vertical plate or to a horizontal cylinder. Mishra (1966a,b) was seemingly the first to analyze free-convection heat transfer to second-order and to Walters-B model fluids from a vertical plate. A similarity solution was attempted which necessitated the temperature of the plate to vary linearly along the wall. Interestingly, this analysis predicts the momentum and thermal boundary layer thicknesses which are invariant with reference to the distance along the plate; obviously, this is unrealistic. Subsequently, Amato and Tien (1970) revisited this problem for an Oldroyd model fluid. Their analysis appears to be in error, especially the way the normal stress term has been introduced into the momentum balance. Also, their final expression for Nusselt number seems to depend only on material properties. While the latter finding is similar to that seen in the preceding section, the reasons for this behavior are not immediately obvious. It is now well established that a similarity solution is not possible for any reasonable visco-elastic fluid model and therefore, recourse to a complete numerical analysis is required to address this problem of laminar-free convection from a vertical plate (Shenoy and Mashelkar, 1978b). Soundalgekar (1971, 1972) has examined the role of unsteady free convection from a vertical plate with constant suction to visco-elastic fluids without viscous dissipation effects.

The analysis of Shenoy and Mashelkar (1978b) is also applicable to laminar-free convection from curved surfaces. When this treatment is specifically developed for a horizontal cylinder, the visco-elasticity is seen to adversely influence the rate of heat transfer. This prediction is well supported, at least qualitatively by the limited results of Lyons et al. (1972). Scant results are also available on turbulent-free convection in visco-elastic liquids from a vertical plate (Shenoy and Mashelkar, 1978a) and from a horizontal cylinder or a sphere (Nakayama and Shenoy, 1991, 1992a), especially as applied to the external flow of drag reducing dilute polymers solutions.

9.7 Conclusions

This chapter has been a bit of a mixed bag of ideas and results on free- and mixed-convection energy and mass transport in non-Newtonian liquids from immersed objects. In particular, consideration has been given to the boundary layer flows due to buoyancy driven with and without the imposed flow for a plate, cylinder, and sphere. These have led to the development of useful expressions for the prediction of Nusselt (or Sherwood) number as functions of the relevant dimensionless groups (Reynolds, Prandtl, and Grashof or combinations thereof) and of rheological parameters from the immersed surfaces to non-Newtonian liquids, primarily for the power-law and Bingham plastic fluids. Most of these results relate to the steady flow regime and with the assumption of constant physical properties. Based on these treatments and further combined with the limited numerical solutions and experimental results, it is possible to predict the rate of heat/mass transfer to power-law fluids in external flows (at least for a plate or a cylinder or a sphere) with reasonable levels of accuracy in the laminar flow regime in power-law fluids. In contrast, only numerical predictions are available for visco-plastic fluids, at least as of now. The visco-elasticity introduces changes both in the detailed structure of the flow and at macroscopic level by way of yielding anomalous transport behavior. Our understanding about the role of visco-elasticity on free and mixed-convective transport of energy and mass is still in its infancy.

Nomenclature

$A(n), A_1(n)$: Functions of n, Equation 9.24a
$Bi = (\tau_o^B d / V_{o\mu B})$: Bingham number (-)
Bi_{max}: Limiting Bingham number for pure conduction heat transfer
$Br = (\mu_o V_o^2 / k\Delta T)$: Brinkman number (-)
$C_1(n), C_2(n)$: Constant functions of n, Equation 9.52
$C_1'(n), C_2'(n), C_3'(n)$: Equations 9.53, 9.20, 9.57, and 9.58, respectively (-)
C_2': Constant, Equation 9.59 (-)
C_p: Specific heat of fluid (J/kg K)
d: Sphere or cylinder diameter (m)
D_∞: Diameter of computational domain (m)
D_{AB}: Molecular diffusivity (m^{2}s^{-1})
g: Acceleration due to gravity (m s^{-2})
$Gr = (\rho/m)^2 R^{n+2} (\beta g \Delta T)^{2-n}$: Grashof number for heat transfer (-)
$Gr_c = (\rho/m)^2 L^4 (\beta g q_s/k)^{2-n}$: Modified Grashof number for constant wall flux case (-)
$Gr_m = (\rho/m)^2 R^{n+2} (g \Delta \rho/\rho)^{2-n}$: Grashof number for mass transfer (-)
$Gr_1 = (\rho^2 d^3 g \beta \Delta T/m^2)(d^2/\alpha)^{2n-2}$: Grashof number for heat transfer (-)
$Gr_2 = (\rho/m)^2 d^{n+2} (g\beta \Delta T)^{2-n}$: Modified Grashof number, Equation 9.64 (-)
$Gr_{1m} = (\rho/m)^2 d^3 (g\Delta\rho/\rho)(d^2/D_{AB})^{2n-2}$: Grashof number for mass transfer (-)
Gr_B: Bingham plastic Grashof number (-)
Gr_{PL}: Power-law Grashof number (-)
h: Heat transfer coefficient (W m^{-2} K)
$He = \rho \tau_o^B L^2 / \mu_B$: Hedstrom number (-)
j: Heat transfer factor (-)
k: Thermal conductivity of fluid (W m^{-1} K)
k_c: Mass transfer coefficient (ms^{-1})
L: Length of plate (m)
l_c: Characteristic length (m)
m: Power-law consistency index (Pa s^n)

m_b: Power-law consistency index at bulk temperature (Pa s^n)

m_s: Power-law consistency index at surface temperature (Pa s^n)

n: Power-law flow-behavior index (-)

N_1, N_2, N_3, N_4: Coefficients, Equations 9.2 and 9.3 (-)

Nu ($= hR/k$): Nusselt number based on radius (-)

Nu$_1 = 2$Nu $= (hd/k)$: Nusselt number based on diameter (-)

Nu$_L$: Nusselt number averaged over the length of the plate (-)

Nu$_x = hx/k$: Local Nusselt number (-)

Nu$_\infty$: Average Nusselt number in conduction limit (-)

p: Pressure (-)

Pe $= (dV/D_{AB}) = $ Re Sc $=$ Re Pr: Peclet number (-)

Pr $= (1/\alpha)(m/\rho)^{(2/(n+1))} R^{(1-n)/(1+n)}(Rg\beta\Delta T)^{(3(n-1))/(2(n+1))}$: Prandtl number based on radius, Equation 9.50 (-)

Pr$_1 = (mC_p/k)(\alpha/d^2)^{n-1}$: Prandtl number, Equation 9.53 (-)

Pr$_2 = (1/\alpha)(m/\rho)^{(2/(n+1))} d^{(1-n)/(1+n)} (dg\beta\Delta T)^{(3(n-1))/(2(n+1))}$: Modified Prandtl number based on diameter, Equation 9.64 (-)

Pr$_C = (1/\alpha)(m/\rho)^{(5/(n+4))} L^{2(n-1)/(n+4)} (g\beta q_s/k)^{(3(n-1))/(n+4)}$: Modified Prandtl number (-)

Pr$_B$: Bingham plastic Prandtl number (-)

Pr$_o$: Modified Prandtl number, Equations 9.9 and 9.10 (-)

Pr$_{PL}$: Power-law Prandtl number (-)

q_s: Constant heat flux specified at the surface (W m^{-2})

R: Sphere or cylinder radius (m)

Ra $=$ Gr$_1$Pr$_1$: Rayleigh number, Equation 9.38 (-)

Ra$_D = g\beta\Delta T d^3/\alpha\nu$: Rayleigh number, Equation 9.34 (-)

Re: Reynolds number (-)

Re$_{eff}$: Effective Reynolds number, Equation 9.44b (-)

Re$_{PL}$: Power-law Reynolds number, Equation 9.46 (-)

Ri$_{PL}$: Richardson number, Equation 9.46 (-)

Ri$_m$: Modified Richardson number, Equation 9.47 (-)

Ri$_B$: Bingham plastic Richardson number (-)

Ri$_{PL}$: Power-law Richardson number (-)

Sc $= (m/\rho)^{2/(n+1)} R^{(1-n)/(1+n)} (Rg\Delta\rho/\rho)^{3(n-1)/(2(n+1))}/D_{AB}$: Schmidt number based on radius, Equation 9.57 (-)

Sc$_1 = m(D_{AB}/d^2)^{n-1}/(\rho D_{AB})$: Schmidt number, Equation 9.58 (-)

Sh $= (k_c R/D_{AB})$: Mean Sherwood number based on radius (-)

Sh$_1 = 2$Sh $= (k_c d/D_{AB})$: Mean Sherwood number based on diameter (-)

$t^* = (tv/L^2)$Gr$_L^{1/2}$: Time (-)

t^*_c: Length of initial period, Equation 9.28 (-)

t: Time (s)

ΔT: Temperature difference (K)

T: Temperature of fluid (K)

T_b: Bulk temperature (K)

T_s: Heated surface temperature (K)

T_o: Free stream temperature (K)

T_∞: Reference temperature

V: Velocity vector (-)

V_c: Characteristic velocity (m s^{-1})

V_f: Characteristic velocity due to buoyancy effects based on radius, Equation 9.6 (m s^{-1})

V_0: Free stream imposed velocity (m s^{-1})
x: Distance along the surface (m)
x_1: Dimensionless distance along a heated surface (-)
$Z = Gr^{1/(2n+2)}Pr^{n/(3n+1)}$: Dimensionless group, Equation 9.36 (-)

Greek Symbols

$\alpha = (k/\rho C_p)$: Thermal diffusivity (m^{2}s^{-1})
β: Thermal expansion coefficient (K^{-1})
δ: Momentum boundary layer thickness, Equation 9.33 (m)
∇: Del operator
$\Delta\rho$: Density difference in free convection mass transfer (kg m^{-3})
ζ: Composite parameter, Equation 9.31 (-)
η: Similarity parameter (-)
θ: Position on the surface of the geometry, (degrees)
$\xi = (T - T_o)/(T_s - T_o)$: Temperature (-)
μ_B: Bingham plastic viscosity (Pa s)
μ: Nondimensional viscosity (-)
ς: Angle between the normal to the surface and the direction of gravity (-)
ρ: Fluid density (kg m^{-3})
τ_o^B: Bingham yield stress (Pa)
ν: Kinematic viscosity (m^2/s)
ϕ_v: Viscous dissipation function (-)
χ: Composite parameter, Equation 9.44 (-)

Subscripts

b: Evaluated at bulk temperature
c: characteristic scale
B: Refers to Bingham plastic fluids
eff: effective, Equation 9.44
F: forced convection
L: based on the plate length
PL: Refers to power-law fluids
M: mixed convection
max: limiting value
N: natural (free) convection
o: Refers to the free stream conditions
x: Local value
θ: Local value

10
Wall Effects

10.1 Introduction

The steady motion of rigid and fluid particles through quiescent unbounded (or infinite expanse of) fluids has been a subject of theoretical study for many years, dating back to the pioneering studies of Newton (1687), Hadamard and Rybczynski (Clift et al., 1978), and Stokes (1851) to mention a few. Owing to the finite size of vessels and tubes employed in such experimental studies, it is clearly not possible to realize the unbounded flow conditions in experimental or numerical studies. In practice, therefore, the confining walls exert an extra retardation effect on a particle, whether settling freely in a quiescent fluid or being suspended in an upward flowing stream of fluid; this effect has also been studied for centuries (Newton, 1687; Munroe, 1888; Ladenburg, 1907). The effect is caused by the upward flux of the fluid displaced by the particle; the smaller the gap between the particle and the boundary, the more severe is the effect. At sufficiently high Reynolds numbers, interactions between the boundary layers developed on the particle and on the tube wall also intensify the wall effects. Furthermore, sorting of particles and droplets using microfluidic channels and networks also entail significant wall effects. Similarly, under certain conditions, confinement is employed to augment the rates of heat and mass transfer in narrow packed beds as well as in classification of particles using electric and magnetic fields.

In addition to these phenomena, blockage also influences the wake characteristics such as the onset of flow separation, wake length, and separation angle, etc. A knowledge of this so-called wall effect for both rigid and fluid particles is therefore necessary for a rational understanding and interpretation of experimental data in a number of situations of overwhelming pragmatic significance. For a rigid sphere, typical examples include falling ball viscometry, hydrodynamic chromatography, membrane transport, hydraulic and pneumatic transport of coarse particles in pipes, etc. Furthermore, in recent years, the problem of flow around solid particles in a tube has received further impetus from the use of electric fields to achieve enhanced rates of transport phenomena and of separations in multiphase systems. On the other hand, in the case of fluid particles, not only is their velocity influenced by the presence of boundaries but their shapes are also greatly altered due to the extra dissipation at the rigid walls. Conversely, their free surface enables them to negotiate their way through the narrow throats in undulating tubes and in porous media, as encountered in the enhanced oil recovery processes. It is thus much more difficult to quantify the severity of wall effects for bubbles and drops than that for rigid particles. In view of the significant differences in the velocity field in the immediate vicinity of the particle, the rate of interphase heat and mass transfer is also influenced (generally enhanced) due to the confining walls in relation to the unconfined case. Evidently, the magnitude of the wall effect will depend upon the size and the shape of the confining walls, that is, whether the particle is moving axially or nonaxially in circular or noncircular ducts (e.g., square, triangular, elliptic cross-sections, planar slit, or annular region between two concentric cylinders, for instance) or it is sedimenting toward or parallel to a plane wall, etc.

Similarly, sphere-in-sphere geometry has also been used to gain some knowledge about the extent of wall effects in a confined domain (Happel and Brenner, 1965). There is no question that the case of a rigid or a fluid sphere falling at the axis of a cylindrical tube represents the most commonly used and studied configuration, both theoretically and experimentally, and hence this chapter is mainly concerned with the extent of wall effects on a rigid or a fluid sphere falling along the axis of a cylindrical tube and on the rate of heat transfer from a sphere. Additional complications arise for nonspherical particles due to their preferred orientation which may get accentuated in the presence of boundaries. The available body of information for nonspherical particles in cylindrical tubes as well as that for spherical particles in noncircular tubes is rather limited, as will be seen in this chapter.

DOI: 10.1201/9780429260759-11

10.2 Definition

Due to the extra retardation effect exerted by the confining walls, the terminal falling velocity of a rigid particle is lower in a confined geometry than that in an unbounded fluid under otherwise identical conditions. Conversely, the drag experienced by the particle in a confined medium is higher than that without any walls being present otherwise under identical conditions. There are several ways to quantify this effect; perhaps, the simplest of all is by defining a wall factor, f, as the ratio of the two velocities as

$$f = V/V_\infty \tag{10.1}$$

V is the terminal velocity of a sphere (of diameter d) falling on the axis of a tube of diameter D and V_∞ is the terminal velocity of the same sphere in an unbounded fluid medium. Obviously, the wall factor, f, defined here, will take on values between zero and unity. Other definitions of the wall factor involving the ratio of drag forces in the absence and presence of the confining walls, the reciprocal of f, the ratio of viscosities calculated using the Stokes formula with and without walls (Bacon, 1936; Sutterby, 1973a, b; Clift et al., 1978), etc. have also been used in the literature. Obviously, all these definitions are not mutually exclusive, though beyond the creeping flow region, the interrelationships between them is far from being straightforward and in fact is quite involved (Happel and Brenner, 1965; Clift et al., 1978; Kim and Karrila, 1991). The scaling of the field equations and the pertinent boundary conditions for a spherical particle moving axially in an incompressible Newtonian medium in a cylindrical tube reveals the wall factor f to be a function of the sphere Reynolds number (Re), the sphere-to-tube diameter ratio, λ ($= d/D$), and the viscosity ratio, X_E, that is,

$$f = \phi(\text{Re}, \lambda, X_E) \tag{10.2}$$

Additional dimensionless groups will emerge depending upon the choice of a rheological model to approximate the non-Newtonian flow behavior of the fluid medium and/or the shape of the container. For instance, for the case of the simple power-law fluid model, the power-law flow behavior index, n, or a Bingham number for Bingham plastic model fluids, or the Weissenberg or the Deborah number for visco-elastic fluids, etc. will appear in Equation 10.2. Likewise, other dimensionless groups will also emerge for particles falling in off-center locations in cylindrical tubes or in noncircular tubes. In this chapter, it is endeavored to present the current status of the progress made in establishing the functional dependence depicted in Equation 10.2 for Newtonian and non-Newtonian fluids. We begin with the case of rigid particles, (i.e., $X_E \to \infty$), followed by that of bubbles ($X_E \to 0$) and finally that of droplets falling/rising freely at the axis of cylindrical tubes so that no additional complexities are involved due to eccentricity and/or noncircular cross-section of the sedimentation vessels.

10.3 Rigid Spheres

It is instructive to begin with the nature of wall effects on a solid sphere settling in Newtonian fluids, and this, in turn, sets the stage for presenting the analogous treatment for spheres falling in non-Newtonian fluids, in the ensuing section.

10.3.1 Newtonian Fluids

10.3.1.1 Theoretical Treatments

From a theoretical standpoint, the effect of confining walls is to change the boundary conditions for the equations of motion and continuity for the continuous phase. In place of the condition of a uniform flow far away from the sphere, confining walls impose conditions which must be satisfied at definite

Wall Effects

boundaries. Further complications arise from the prevailing velocity profiles (uniform or Poiseuille) in the tube. The available experimental and analytical or numerical results clearly indicate that the wall factor is a function of λ only under both creeping and fully turbulent conditions, whereas it depends on both λ and Re in the intermediate Reynolds number range. It is thus convenient to present the available body of information separately for each flow region.

10.3.1.1.1 Creeping Flow Region

In the creeping flow region (Re <<< 1), accurate analytical solutions have been obtained by using the so-called method of reflections for the system of a sphere falling in a cylindrical tube. Reflection solutions due to Faxen (1923), Happel and Byrne (1954), and Wakiya (1957) give good predictions for the wall factor, which are restricted to small values of the sphere – to – tube diameter ratios λ ($= d/D$) <<0.1. For instance, the celebrated result of Faxen (1923) is given by

$$f = 1 - 2.104\lambda + 2.09\lambda^3 - 0.95\lambda^5 + \cdots \tag{10.3}$$

Subsequently, Bohlin (1960) has extended these results to higher values of λ up to about ≈ 0.6, as follows:

$$f = 1 - 2.10443\lambda + 2.08877\lambda^3 - 0.94813\lambda^5 - 1.372\lambda^6 + 3.87\lambda^8 - 4.19\lambda^{10} \cdots \tag{10.4}$$

The coefficients of the first three terms in Equations 10.3 and 10.4 are virtually identical. Subsequently, Bohlin (1960) also presented an expression for the conditions when the Poiseuille flow is imposed on the sphere motion in a cylinder.

Almost concurrently but independently, Haberman and Sayre (1958) provided the following analytical expression for the wall factor f for a sphere settling in a stationary liquid filled in a cylinder:

$$f = \frac{1 - 2.105\lambda + 2.0865\lambda^3 - 1.7068\lambda^5 + 0.72603\lambda^6}{1 - 0.75857\lambda^5} \tag{10.5}$$

This expression was stated to be applicable in the range $0 \leq \lambda \leq 0.8$. Subsequently, the accuracy of Equation 10.5 has been demonstrated by many numerical studies (Paine and Scherr, 1975; Tullock et al., 1992; Bowen and Sharif, 1994; Higdon and Muldowney, 1995; Wham et al., 1996; Vidyasagar et al., 2017a) up to about $\lambda \approx 0.9$, thereby covering almost the complete range of interest $0 \leq \lambda \leq \sim 1$.

As will be seen in the following section, Equation 10.5 is in excellent agreement with the experimental results available in the literature (Happel and Brenner, 1965; Clift et al., 1978; Chhabra et al., 2003).

10.3.1.1.2 Intermediate Reynolds Number Region

It is not clear beyond what value of the Reynolds number the expressions presented in the preceding section cease to apply. Broadly speaking, the larger the value of λ, the higher is the value of the particle Reynolds number marking the end of the creeping flow region. This issue will be dealt with in more detail later in this section. There are a few numerical studies available in the literature in which the effect of the particle Reynolds number on the wall factor has been investigated. Faxen (1923) applied the Oseen-type linearization to a sphere moving axially in a tube, but the resulting predictions are no more reliable than that for an unbounded fluid (Happel and Bryne, 1954; Haberman and Sayre, 1958). Johansson (1974) numerically solved the Navier–Stokes equations for the Poiseuille flow around an axially fixed sphere in a cylinder. For $\lambda = 0.1$, he calculated the values of drag coefficient for Reynolds number (based on the mean velocity and the tube diameter) up to about 150. The corresponding particle Reynolds number will be one-tenth of this value. Based on limited comparisons with experimental results, Johansson (1974) concluded that the wall effects become negligible for Re > 50 for $\lambda = 0.1$, which is also consistent with the subsequent findings. Subsequently, Oh and Lee (1988) treated the same problem for $\lambda=0.5$ and 0.74 and for the values of the particle Reynolds number (based on sphere diameter) up to about 200. However, it is virtually impossible to use the numerical results from both these studies as sufficient details are not available enabling the recalculation of their results in the form of the wall factor f, as defined here. Xu and Michaelides (1996) studied the flow over ellipsoidal particles placed axially in cylindrical tubes. For a sphere, they compared their predictions with the analysis of Bohlin (1960). Unfortunately, however, neither the value of λ nor the range of the Reynolds number employed by them is known. This indeed severely limits the utility of their results. Wham et al. (1996) have investigated the effect of cylindrical confining walls on freely falling spheres in quiescent liquids as well as on spheres suspended in upward moving liquids. Their results encompass the values of diameter ratio in the range $0.08 \le \lambda \le 0.70$ and the sphere Reynolds number up to 200. After some rearrangement, their final expression can be recast in the following form:

$$f = \frac{1+0.03708(0.5\,\mathrm{Re}_\infty)^{A_0}}{\left[1+.0.3708(0.5f\,\mathrm{Re}_\infty)^B\right]C} \tag{10.6}$$

where

$$A_0 = 1.514 - 0.1016\ln(0.5\,\mathrm{Re}_\infty) \tag{10.7a}$$

$$B = 1.514 - 0.1016\ln(0.5f\,\mathrm{Re}_\infty) \tag{10.7b}$$

$$C = \frac{1-0.75857\lambda^5}{1-K\lambda+2.0865\lambda^3-1.7068\lambda^5+0.72603\lambda^6} \tag{10.7c}$$

and

$$K = 0.6628 + 1.458\exp(-0.028175f\,\mathrm{Re}_\infty) \tag{10.7d}$$

It should be noted that Equations 10.6 and 10.7 are implicit in f and thus an iterative procedure is needed to estimate the value of f for known values of λ and Re_∞ (based on the sphere diameter and V_∞). It is also appropriate to add here that in the limit of vanishingly small Reynolds numbers, these expressions do reduce to Equation 10.5. It can be easily shown from the predictions of Equations 10.6 and 10.7 that the wall factor indeed is independent of the Reynolds number in the viscous region, though the critical value of Re below which the wall factor is independent of the Reynolds number is seen to vary appreciably. Based on the criterion of 5% deviation from the constant limiting value of f, Equations 10.6 and 10.7 were used to establish the critical values of Re_∞ below which the creeping flow approximation might be applicable. Unfortunately, this approach predicts the critical value of Re to decrease with the increasing value of λ which is exactly contrary to the experimental observations, as will be seen in the next section. Subsequently, Henschke et al. (2000) have also reported numerical results for the wall effects on a sphere settling in cylindrical tubes in the range of conditions as $0.01 \le \lambda \le \sim 0.97$ and $10^{-3} \le \mathrm{Re}_\infty \le 10^4$. However, smaller the value of λ, wider was the range of Reynolds number of numerical simulations. The agreement between their predictions and experiments was reported to be good. Feng and Michaelides (2002) have considered the effect of cross-sections of tubes on drag of a sphere in the creeping and finite Reynolds number regimes.

10.3.1.1.3 Turbulent Flow Regions

Under these conditions, the flow in dominated by inertial forces, and Newton (1687) (also see Barr, 1931) presented the following expression for the wall factor for a sphere settling in a vessel of cross-sectional area, A:

$$f = (1-\beta)(1-0.5\beta)^{0.5} \tag{10.8}$$

where $\beta = (\pi\, d^2/4A)$, and clearly for the special case of a sphere settling in a cylindrical vessel, $\beta = \lambda^2$, and hence Equation 10.8 becomes:

$$f = (1-\lambda^2)(1-0.5\lambda^2)^{0.5} \tag{10.9}$$

Although virtually, no analytical details are available for the reasoning behind this expression, it will be seen in a later section that indeed Equation 10.8 provides excellent agreement with the experimental results available in the literature for this flow regime (DiFelice et al., 1995; Chhabra et al., 1996a).

From the foregoing discussion, it is thus abundantly clear that analytical and numerical predictions of the wall factor are available for up to about $\lambda \leq 0.97$ and $Re_\infty < 10^4$, though not for all values of λ, the numerical results extend up to $Re = 10^4$. With the notable exception of Equation 10.8, all works on wall effects at high Reynolds numbers are based on experimental data reported by different investigators, as will be seen in the next section.

10.3.1.2 Experimental Results and Correlations

Detailed discussion on the experimental determination of the wall factor, f, experimental uncertainty, etc. is available elsewhere (Chhabra, 2002b). Table 10.1 provides a succinct summary of the experimental studies available on this subject where it is clearly seen that indeed experimental results are now available for the complete range of λ and up to about $Re \approx 2 \times 10^6$. Figures 10.1 and 10.2, based on the work of Uhlherr and Chhabra (1995), show representative results illustrating the nature of variation of the wall factor with λ and the Reynolds number in the three flow regimes, namely viscous, transition, and fully turbulent (inertial), respectively. Figure 10.1 employs the usual drag coefficient (C_D), Reynolds number (Re) coordinates, and illustrates the universal form of relationship for a range of values of λ. With the increasing value of λ, the viscous region seems to persist up to larger and larger values of the Reynolds number (Re) and this is also accompanied by a rather late transition to the fully turbulent conditions thereby resulting in a substantial intermediate transition region. Suffice it to add here that in the overlapping range of conditions, there is a good internal consistency of data culled from different sources. Figure 10.2, on the other hand, confirms the expectation that the wall factor is a function of λ only both in the viscous and in the inertial regions, as asserted earlier. It is thus convenient to deal with each region separately.

10.3.1.2.1 Creeping Flow Region

At the outset, it is perhaps appropriate to address the issue of the limiting value of the Reynolds number denoting the cessation of the creeping flow. Based on an (arbitrary) 5% criterion, the critical Reynolds number is defined here as the value at which the value of the wall factor is 5% higher than the corresponding value given by Equation 10.5. Based on this criterion applied to the experimental results available in the literature (McNown et al., 1948; Fidleris and Whitmore, 1961; Uhlherr and Chhabra, 1995; Kehlenbeck and DiFelice, 1999; DiFelice and Kehlenbeck, 2000), the resulting limiting values of the sphere Reynolds number are summarized in Table 10.2. It is clearly seen that the so-called creeping flow persists up to a higher value of the Reynolds number, Re_∞, as the value of λ increases.

TABLE 10.1
Summary of Experimental Studies on Wall Effects for a Sphere in Newtonian Fluids

Investigator	Range of λ	Flow Regime	Remarks
Munroe (1888–1889)	0.11–0.83	Inertial region	Equation 10.17
Lunnon (1928)	<0.7	Inertial region	Equation 10.18
Francis (1933)	<0.97	Viscous	Equation 10.10
Bacon (1936)	<0.3155	Viscous	Emphasis on falling ball viscometry
Lee (1947)	<0.168	Viscous	Empirical correlation
McPherson (1947)	<0.1571	Re<317	Graphs
Engez (1948)	<0.9	Re<1000	Empirical correlation
Fayon and Happel (1960)	0.13–0.313	$0.1 \leq Re \leq 40$	Empirical correlation
Fidleris and Whitmore (1961)	<0.6	$0.05 \leq Re \leq 20,000$	Graphs
Sutterby (1973b)	<0.13	Re<4	Emphasis on falling ball viscometry
Achenbach (1974)	0.5–0.92	$3 \times 10^4 \leq Re \leq 2 \times 10^6$	Correlation
Iwaoka and Ishii (1979)	<0.9	Viscous	Empirical correlation
Chhabra and Uhlherr (1980c)	<0.5	Viscous and intermediate	No correlation
Modi and Akutsu (1984)	$0.03 \leq \beta \leq 0.3$	$30 \leq Re \leq 2000$	Effect of blockage on surface pressure and drag
Lali et al. (1989)	0.05–0.78	$0.002 \leq Re_\infty \leq 200$	Empirical correlation
Humphrey and Murata (1992)	0.442, 0.757, 0.882	Viscous	Vertical and inclined tubes
Bougas and Stamatoudis (1993)	≤ 0.7	$13,500 \leq Re \leq 70,000$	Accelerating spheres
Uhlherr and Chhabra (1995)	0.03–0.9	$0.038 \leq Re \leq 47,000$	Graphs
Ataide et al. (1999)	<0.55	Re<311	Empirical correlation
Kehlenbeck and DiFelice (1999)	0.1–0.9	$2 \leq Re_\infty \leq 185$	Equation 10.13 and Equation 10.14
Arsenijevic et al. (2010)	≤ 0.88	$53 \leq Re \leq 15,100$	–

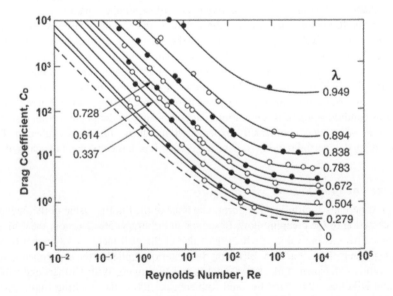

FIGURE 10.1 Effect of λ on drag coefficient — Reynolds number relationship in Newtonian fluids for a sphere. (Modified from Uhlherr, P.H.T. and Chhabra, R.P., *Can. J. Chem. Eng.*, **73**, 918, 1995.)

Wall Effects

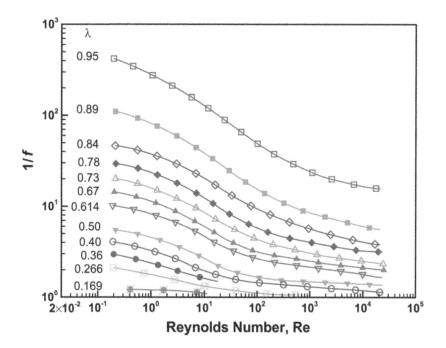

FIGURE 10.2 Variation of wall factor with Reynolds number and diameter ratio. (Based on the results shown in Figure 10.1.)

TABLE 10.2
Limiting Values of Reynolds Number for Creeping Flow Conditions

λ	Re_∞	Re
0.1	0.027	0.021
0.2	0.04	0.023
0.3	0.05	0.020
0.4	0.083	0.023
0.5	0.18	0.035
0.6	0.52	0.069
0.7	2.1	0.18
0.8	8.4	0.42
0.9	25.17	0.56

Hereafter, the creeping flow region is assumed to occur for the values of Re or Re_∞ smaller than those listed in Table 10.2. In this region, perhaps the simplest (and possibly also the most widely used) experimental correlation is that of Francis (1933) given as follows:

$$f = \left[\frac{1-\lambda}{1-0.475\lambda} \right]^4 \qquad (10.10)$$

This expression covers virtually the entire range of the diameter ratio up to $\lambda \leq 0.97$.

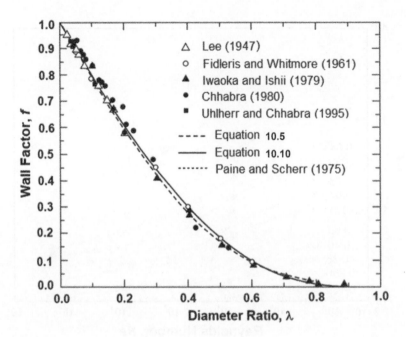

FIGURE 10.3 Representative comparison between various predictions and experimental results on wall factor in the creeping flow region.

By using a fluid dynamic analogy between a single particle in a tube and a multiparticle suspension, DiFelice (1996) was able to obtain satisfactory predictions of the wall factor and his expression is given as

$$f = \left[\frac{1-\lambda}{1-0.33\lambda}\right]^\alpha \tag{10.11}$$

where α is obtained from the following relationship:

$$\frac{3.3-\alpha}{\alpha-0.85} = 0.1 \text{Re}_\infty \tag{10.12}$$

The numerical constants appearing in Equation 10.12 are based on the experimental results of Fidleris and Whitmore (1961) and thus this expression is limited to the range of conditions implicit in the study of Fidleris and Whitmore (1961). In the viscous region, it predicts $\alpha \approx 3.3$.

Figure 10.3 contrasts various predictions of the wall factor together with representative experimental data taken from different sources in the creeping flow region. Overall, good correspondence is seen to exist between various predictions themselves and also between the predictions and experimental results in this region. Suffice it to add here that it is virtually impossible to discriminate between Equations 10.5, 10.10, and 10.11, as all perform equally well over the entire range of λ in the region of low Reynolds number.

10.3.1.2.2 Intermediate Reynolds Number Region

As mentioned previously, only a few numerical studies are available on wall effects beyond the viscous flow region and out of these, only the study of Wham et al. (1996), Equations 10.6–10.7, is convenient for calculating the values of the wall factor, f, as a function of λ and Re (or Re_∞). There is a paucity of empirical correlations also in this flow region. Subsequent to the paper by DiFelice (1996), Kehlenbeck and DiFelice (1999) revisited this problem. Detailed comparisons between the predictions of Equations

Wall Effects

10.11 and 10.12 with the new extensive experimental data of their own and those of Barr (1931), Francis (1933), McNown et al. (1948), Fidleris and Whitmore (1961), Okuda (1975), and Uhlherr and Chhabra (1995) revealed inadequacies of their earlier equation, namely Equation 10.11. Hence, Kehlenbeck and DiFelice (1999) proposed the following new correlation:

$$f = \frac{1 - \lambda^p}{1 + (\lambda/\lambda_o)^p} \tag{10.13}$$

where both λ_o and p are functions of Re_∞ as

$$\frac{\lambda_o - 0.283}{1.2 - \lambda_o} = 0.041 Re_\infty^{0.524} \tag{10.14}$$

For $Re_\infty \leq 35$,

$$p = 1.44 + 0.5466 Re_\infty^{0.434} \tag{10.15a}$$

and for $Re_\infty \geq 35$,

$$p = 2.3 + 37.3 Re_\infty^{0.434} \tag{10.15b}$$

They also proposed a simpler version of Equation 10.13 by assigning a constant value $p=2.2$ and then λ_o is obtained from the following relation:

$$\frac{\lambda_o - 0.27}{1.2 - \lambda_o} = 0.05 Re_\infty^{0.65} \tag{10.16}$$

Based on the detailed comparisons between the predictions of Equations 10.6 and 10.7, Equations 10.11–10.15, and the experimental results in the common range of conditions, the main findings can be summarized as follows: Equations 10.6 and 10.7 based on the numerical predictions yield values of the wall factor which are within 5%–6% of the corresponding experimental values only for $\lambda \leq 0.3$ and $Re_\infty \leq 200$. As the value of λ increases, the correspondence deteriorates rather rapidly. Thus, for instance, Equations 10.6 and 10.7 yield acceptable values of the wall factor only for $Re_\infty < 10$ when $\lambda = 0.4$. Similarly, for $\lambda \geq 0.5$ and for all values of Re_∞ (≤ 200), these equations result in values of the wall factor which deviate from experimental values by as much as 200%. Furthermore, for $\lambda = 0.7$ and $Re_\infty > 100$, these equations predict $f > 1$ which is clearly inadmissible. On the other hand, Equations 10.13–10.15 predict bulk of the literature data with an accuracy of about 10% or so under most conditions (Chhabra et al., 2003).

10.3.1.2.3 Fully Turbulent Region

As mentioned previously and seen in Figures 10.1 and 10.2 under fully turbulent conditions ($Re \rightarrow \infty$), the wall factor again becomes independent of the Reynolds number. In general, the effect progressively diminishes as the Reynolds number is increased. An inspection of Table 10.1 shows that many empirical correlations are available in the literature for the estimation of the wall factor in this regime. Perhaps the simplest and also the most successful of these are the Newton's formula, Equation 10.8 or Equation 10.9, and that of Munroe (1888–1889) which is based on his own experimental results in the range $0.11 \leq \lambda \leq 0.83$. The Munroe's equation is given by

$$f = 1 - \lambda^{1.5} \tag{10.17}$$

The other commonly cited correlation in this flow region is that of Mott (1951) which in fact is based on the experimental results of Lunnon (1928). Mott (1951) correlated Lunnon's limited data by the following two expressions:

For $0.15 \leq \lambda \leq 0.5$;

$$f = (1 + 3.2\lambda^2)^{-0.5} \tag{10.18a}$$

and in the range $0.4 \leq \lambda \leq 0.7$

$$f = (1 + 16\lambda^4)^{-0.5} \tag{10.18b}$$

Finally, it can readily be seen that in the limit of $Re_\infty \to \infty$, the correlations of Kehlenbeck and DiFelice (1999), Equations 10.13–10.16, respectively reduce to

$$f = \frac{1 - \lambda^{2.3}}{1 + (\lambda/1.2)^{2.3}} \tag{10.19}$$

Similarly, in the limit of $Re_\infty \to \infty$, Equations 10.11 and 10.12 yield

$$f = \left(\frac{1 - \lambda}{1 - 0.33\lambda}\right)^{0.85} \tag{10.20}$$

Figure 10.4 shows a comparison between various predictions, namely Equations 10.8, 10.17, 10.18, 10.19, and 10.20. In general, there appears to be an overall good agreement, though Equations 10.8 and 10.17 have the added advantage of their particularly simple forms. Included in this figure are the available experimental data culled from various sources including some results for wall effects in square channels (Awbi and Tan, 1982). Furthermore, the data included in this figure relates to the freely falling spheres (Uhlherr and Chhabra, 1995) as well as to the spheres fixed in flow streams (Achenbach, 1974;

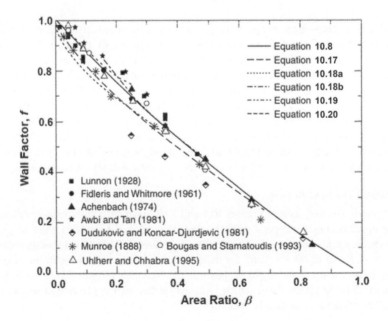

FIGURE 10.4 Comparison between various predictions and experimental results on wall factor of a sphere under fully turbulent conditions.

TABLE 10.3
Values of Reynolds Number for the Onset of Fully Turbulent Conditions

λ	0.1	0.2	0.3	0.4	0.5	0.6	0.7	0.8	0.85
Re	60	110	200	500	2000	6700	10^4	1.25×10^4	1.5×10^4

Dudukovic and Koncar-Djurdjevic, 1981); clearly, there are no discernable trends present in this figure. This is so presumably owing to the generally turbulent conditions in the tube and therefore the velocity profile of the oncoming stream is relatively flat. Evidently, excellent correspondence is seen to exist between most experimental results and the predictions of Equation 10.8, except for the results of Dudukovic and Koncar-Djurdjevic (1981) which lie slightly below the other results. This is so probably due to the fact that the value of the Reynolds number is not sufficiently high for the fully turbulent conditions to exist in this work. Subsequent results of Arsenijevic et al. (2010) for fully turbulent flow conditions are also consistent with Equation 10.9.

Finally, it would be useful to delineate the limiting values of the sphere Reynolds number beyond which the fully turbulent conditions can be assumed to prevail and therefore, Equation 10.8 or Equation 10.17 can be used to calculate the value of the wall factor. As seen in Figures 10.1 and 10.2, for a fixed value of λ, the wall factor is a function of the Reynolds number and only approaches the values predicted by Equation 10.8 or Equation 10.17 asymptotically. In view of the fact that the wall factor can only be estimated with an accuracy of about 10% in this regime, it is appropriate to define the critical Reynolds number as the value at which the value of the wall factor reaches the 95% of the value predicted by Equation 10.8 for a fixed value of λ. Based on this criterion, the critical values of the Reynolds number have been extracted from the literature data (Fidleris and Whitmore, 1961; Uhlherr and Chhabra, 1995); the resulting values are summarized in Table 10.3 (Chhabra et al., 1996a; Uhlherr and Chhabra, 1995). An inspection of this table shows the strong dependence of the critical Reynolds number on λ. Owing to the limited experimental data in this region, it is not possible to obtain the corresponding values for $\lambda > 0.85$ (Chhabra et al., 1996a). However, the value of 60 for $\lambda = 0.1$ is in good agreement with the value of 50 reported by Johansson (1974).

In summary, it is clear from the foregoing description that based on a combination of theoretical, numerical, and experimental results, it is possible to estimate the value of the wall factor for spheres falling axially in cylindrical tubes under most conditions of practical interest, that is, $0 \leq \lambda \leq$ ~1 and $Re_\infty \leq Re_{critical}$. In the low and high Reynolds number regions, the wall factor is a function of λ only whereas it depends on both the diameter ratio (λ) and the Reynolds number (Re or Re_∞) in the intermediate Reynolds number region.

Apart from influencing the terminal fall velocity, the presence of boundaries also alters the detailed structure of the flow field prevailing around a sphere moving or fixed in a cylindrical tube. For instance, Coutanceau (1971) reported that the formation of the wake is delayed due to the confining walls. Modi and Akutsu (1984) also determined the influence of blockage on the surface pressure and wake features. For $\beta \leq 0.05$, the effect of Reynolds number manifested beyond the point of separation. At about $Re \approx 1000$ (based on the average channel velocity), higher the value of β, higher is the surface pressure in the front of the sphere; however, this trend was reversed beyond the separation point. Also, the friction contribution decreased with the increasing Reynolds number but the ratio of the friction to the total drag was relatively insensitive to the value of β in line with the fact that the wall effects diminish with the increasing Reynolds number. Finally, they concluded that the blockage tends to stabilize the flow by suppressing flow separation and vortex shedding. Similarly, Krishnan and Kannan (2011) have examined the wall effect on a sphere translating off-axis of cylindrical tubes in the range of $1 \leq Re \leq 500$ up to $\lambda \approx 0.5$ and up to eccentricity E (distance from axis/ radius of the tube) of 0.6. The drag coefficient was found to increase with both λ an E at a fixed Reynolds number.

Similarly, Cliffe et al. (2000) have reported that the transition from axisymmetric to asymmetrical type of flow is also delayed, thereby suggesting that the walls tend to stabilize the flow. Notwithstanding the fact that the cylindrical tubes represent the most commonly employed geometry, some analytical and numerical results on the extent of wall effects (in the creeping region) on spheres settling in ducts of square and triangular cross-sections, and in planar slits, are also available (Brenner, 1961; Cox and Brenner, 1967; Happel and Bart, 1974; Tullock et al., 1992). The corresponding experimental results have been reported by Miyamura et al. (1981), Chow et al. (1989), Ilic et al. (1992), and Balaramakrishna and Chhabra (1992). The effects of eccentricity (spheres falling off-center) on the falling velocity and drag of a sphere have been studied, among others, by Tozeren (1983), Higdon and Muldowney (1995), and Krishnan and Kannan (2011) as noted above. Under otherwise identical conditions, eccentricity often leads to an increased drag on the sphere. Similarly, the wall effects on a sphere falling in the presence of curved boundaries such as in the annular gap formed by two concentric cylinders have been studied analytically by Hasimoto (1976), Shinohara and Hasimoto (1980), Alam et al. (1980), Fukumoto (1985), and experimentally by Zheng et al. (1992). Preliminary comparisons between the observations and the predictions for some of these configurations appear to be encouraging. Likewise, the case of a sphere settling toward or parallel to a plane wall has been studied by Cooley and O'Neill (1969), Adamczyk et al. (1983), Ambari et al. (1983, 1984b), Lecoq et al. (1993), and Gondret et al. (1999) among others.

10.3.2 Inelastic Non-Newtonian Liquids

10.3.2.1 Theoretical and Numerical Treatments

Only limited theoretical and numerical work has been carried out on the effect of containing walls on sphere motion in purely viscous fluids without a yield stress. Kawase and Ulbrecht (1983c) assessed the significance of the wall effects by studying the creeping motion of a sphere in a spherical envelope of power-law fluid and concluded that shear-thinning behavior suppresses the effect of walls. Gu and Tanner (1985), on the other hand, numerically solved the field equations for the creeping motion of power-law fluids both for a sphere-in-a-cylinder and for a sphere-in-a-sphere configurations. Both studies concluded that the wall effects are less severe in power-law fluids ($n<1$) than that in Newtonian fluids otherwise under identical conditions, which is consistent with the reported experimental findings, as will be seen later in this section. In an extensive numerical, study Missirlis et al. (2001) studied the wall effects on the settling velocity of a sphere falling freely at the axis of a cylindrical tube in the creeping flow regime. Their results encompass wide ranges of conditions as: $1 \geq n \geq 0.1$ and $0.50 \geq \lambda \geq 0.02$. The wall effect is predicted to decrease with the decreasing values of diameter ratio, λ, and the power-law index, n.

Table 10.4 summarizes the numerical values of the drag correction factor $Y(n, \lambda)$, introduced in Chapter 2, including the extrapolated values for $n \to 0$ and $\lambda \to 0$. In the creeping flow regime, the wall factor, f, is related to the drag correction factor via the following relationship:

$$f = \left[\frac{Y(n, \lambda = 0)}{Y(n, \lambda)} \right]^{1/n} \tag{10.21}$$

TABLE 10.4

Values of the Drag Correction Factor $Y(n, \lambda)$

$n\downarrow$	λ						
	0.5	0.25	0.125	0.10	0.05	0.02	0[a]
1.0	5.944	1.979	1.354	1.267	1.119	1.046	1.004
0.9	5.093	1.925	1.402	1.330	1.214	1.163	1.137
0.8	4.366	1.869	1.443	1.387	1.301	1.269	1.257
0.7	3.740	1.809	1.477	1.434	1.376	1.360	1.355
0.6	3.201	1.745	1.500	1.469	1.435	1.428	1.429
0.5	2.740	1.678	1.510	1.493	1.475	1.473	1.476
0.4	2.340	1.607	1.507	1.495	1.489	1.490	1.492

[a] Extrapolated

Source: Missirlis, K. A., Assimacopoulos, D., Mitsoulis, E., and Chhabra, R. P., *J. Non-Newt. Fluid Mech.* **96**, 459 (2001).

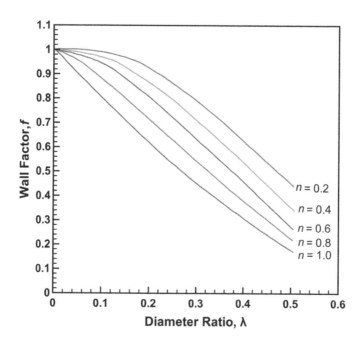

FIGURE 10.5 Dependence of the wall factor on λ and n in creeping flow regime. (Based on the results of Missirlis, K.A., Assimacopoulos, D., Mitsoulis, E. and Chhabra, R.P., *J. Non-Newt. Fluid Mech.*, **96**, 459, 2001.)

Obviously, Equation 10.21 becomes indeterminate in the limit of $n=0$. Figure 10.5 shows the dependence of the theoretical estimates of the wall factor on the power-law index and the diameter ratio λ. The wall factor is seen to approach the limiting value of $f=1$ at finite values of the diameter ratio, λ, as the value of n is gradually reduced below unity. Also, for each value of n, the wall factor shows almost linear dependence on the diameter ratio, λ.

A typical comparison between these predictions and the experimental results is shown in Figure 10.6 where a satisfactory correspondence is seen to exist. However, the match between the theory and experiments deteriorates rather rapidly with the decreasing value of the power-law index, n. Furthermore, while the theory predicts a strong influence of the power-law index n on the wall factor, the available experimental results suggest it to be negligible (Chhabra et al., 1977; Chhabra and Uhlherr, 1980c).

Over the past 15 years or so, some numerical results on wall effects for spheres falling along the axis or in off-axis positions in power-law, Bingham plastic, and Carreau model fluids have been reported in the literature. Table 10.5 provides a brief overview of such studies. Song et al. (2009, 2011), Thumati et al. (2018) and Vidyasagar et al. (2017a,b, 2022) have extensively studied the effect of symmetric confinement (sphere at the axis of tube) up to $\lambda \approx 0.9$. Figures 10.7 and 10.8 show representative results elucidating the influence of n, λ, and Re_{PL} on the drag coefficient of a sphere undergoing steady translation in a tube (Figure 10.8) and for a sphere held stationary in the fully developed Poiseuille flow of power-law fluids (Figure 10.7), respectively. In both cases, the drag is increased with the increasing blockage, though this effect weakens with the decreasing value of the power-law index and/or increasing value of the Reynolds number. As noted previously, it is really not convenient to convert these results in terms of the wall factor f. Suffice it to say that these results are also consistent with the other numerical studies available in the literature.

Figure 10.9 shows the effect of blockage on the wake length for the two cases where the wakes are seen to be somewhat shorter for a confined sphere than that for an unconfined sphere. The effect of blockage on separation angle is rather weak (Song et al., 2009, 2011) (Table 10.6). All these studies are based on the assumption of the steady axisymmetric flow regime.

For the spheres falling in off-axis positions, Krishnan and Kannan (2012) resorted to three-dimensional simulations for a sphere held fixed in the fully developed flow condition. They reported both lift and drag

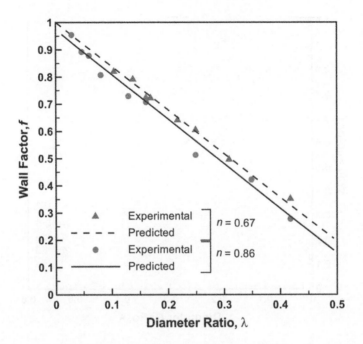

FIGURE 10.6 Typical comparison between the predicted (Missirlis et al., 2001) and experimental (Chhabra, 1980) values of the wall factor for power-law fluids in the creeping flow regime.

coefficient values as functions of λ and E. The lift was always found to be positive which decreased with the increasing Reynolds number. In contrast, for fixed values of λ and Re_{PL}, the drag decreased with the increasing value of E, albeit this effect also diminished with the increasing Reynolds number.

10.3.2.2 Experimental Studies

Among the numerous experimental studies available on the free settling of spheres (see Chapter 2), very few investigators have addressed the question of wall effects on the sphere motion falling in cylindrical tubes. Wall effects in inelastic non-Newtonian liquids have been either ignored (Slattery and Bird, 1961) or the Newtonian expressions (Turian, 1967; Acharya et al., 1976; Uhlherr et al., 1976) have been used for correcting the terminal falling velocity data in power-law fluids without any justification. Over the years, considerable experimental and theoretical evidence has accumulated which suggests that the wall effects are generally less severe in generalized Newtonian fluids than those in Newtonian media (Sato et al., 1966; Turian, 1967; Chhabra et al., 1977; Chhabra and Uhlherr, 1980c; Muratsune and Ishii, 1981; Gu and Tanner, 1985; Missirlis et al., 2001; Song et al., 2009, 2011; Thumati et al., 2018; Vidyasagar et al., 2017a). If power-law model is employed to depict the shear-thinning behavior, the flow behavior index (n) seems to exert virtually no influence on the extent of wall effects, which is in stark contrast to the numerical predictions as discussed earlier. By analogy with the wall effects in Newtonian fluids, it is generally recognized that the wall factor exhibits a functional dependence on the Reynolds number (Re_{PL}) and λ, which is qualitatively similar to that observed for spheres in Newtonian fluids. For power-law fluids in the creeping flow regime, the wall factor is independent of the sphere Reynolds number (suitably modified for power-law fluids) and varies linearly with the diameter ratio as

$$f_0 = 1 - A\lambda \tag{10.22}$$

Equation 10.22 implies that for a given sphere/fluid combination, the fall velocity decreases linearly with the sphere-to-tube diameter ratio. This expectation is borne out by the representative results shown

TABLE 10.5

Summary of Numerical Studies on Wall Effects for a Sphere in Cylindrical Tubes

Reference	Fluid Model	Range of Conditions	Remarks
Missirlis et al. (2001)	Power-law	$0.02 \leq \lambda \leq 0.5$ $0.1 \leq n \leq 1$ ($Re_{PL}=0$)	Wall effect decreases with the decreasing λ and n.
Yu and Wachs (2007)	Bingham plastic fluid	$\lambda = 0.25$ $Re_B = 0$	Drag results.
Machac et al. (2009) Strnadel and Machac (2009), Strnadel et al. (2011)	Carreau fluid model	$0.5 \leq \lambda \leq \sim 1$ $0.3 \leq n \leq 1$ Creeping flow	Reported good agreement between their predictions and experiments in shear-thinning and visco-elastic fluids.
Song et al. (2009, 2010, 2011, 2012)	Power-law	$0 \leq \lambda \leq 0.5$ $0.2 \leq n \leq 1$ $1 \leq Re_{PL} \leq 100$	Both quiescent and Poiseuille flow cases are considered for drag and heat transfer.
Despeyroux et al. (2010)	Power-law	$Re_{PL} = 0$ Several values of λ and E.	Heat transfer results for a sphere.
Krishnan and Kannan (2011, 2012)	Newtonian and power-law fluids	$0.01 \leq \lambda \leq 0.5$ $n = 0.57, 0.76, 0.94$ $\leq Re_{PL} \leq 40$ $0 \leq E \leq 0.6$	Effect of E and λ on drag and lift forces.
Reddy and Kishore (2013)	Power-law fluid	$\leq \lambda \leq 0.5$ $0.4 \leq n \leq 1.8$ $1 \leq Re_{PL} \leq 100$ $1 \leq Pr_{PL} \leq 1000$	Effect of blockage on heat transfer.
Das et al. (2015)	Bingham plastic fluid	$0 \leq \lambda \leq 0.5$ $0 \leq Bi \leq 100$ $1 \leq Pr_B \leq 100$ $1 \leq Re_B \leq 100$	Wall effect on drag and Nusselt number.
Vidyasagar et al. (2017a, 2022), Thumati et al. (2018)	Power-law and Bingham plastic fluids	$0.5 \leq \lambda \leq 0.95$ $0.2 \leq n \leq 2$ $\leq Re_{PL} \leq 100$ $1 \leq Re_B \leq 100$ $5 \leq Pr_B \leq 100$ $0.01 \leq Bi \leq 100$	Wall effect on drag coefficient and Nusselt number for both quiescent and Poiseuille flow of fluids.
Tian (2018)	Power-law (Poiseuille flow)	$0.02 \leq \lambda \leq 0.5$ $1 \leq n \leq 1.8$ $10^{-3} \leq Re_{PL} \leq 100$	No wall effects for $\lambda \leq 0.1$ and at $Re_{PL} = 100$. Also, see the comments of Pantokratoras (2021) and Tian (2022).

in Figure 10.10. Despite some confusion (Turian, 1964, 1967) regarding the value A in Equation 10.22, most of the literature data (Karino et al., 1972; Chhabra, 1980; Lali et al., 1989; Zhang et al., 2015) covering the ranges of conditions as $0 \leq \lambda \leq 0.5$; $0.52 \leq n \leq 0.95$; $Re_{PL} \leq \sim 1$ is well correlated (Chhabra et al., 1977) with a single value of $A = 1.6$. By comparing Equation 10.22 with the corresponding expression for Newtonian fluids (Equation 10.5 for instance), the wall effects are seen to be smaller in power-law fluids than those in Newtonian media, at least in the viscous region. This inference is qualitatively consistent with the findings of Tanner (1964), Caswell (1970), Gu and Tanner (1985), and of Missirlis et al. (2001).

Similarly, the value of the wall factor is believed to reach a constant value (f_∞) at high Reynolds numbers. However, the critical value of the latter denoting the attainment of a constant value in the fully turbulent region is strongly dependent upon the diameter ratio and possibly on the power-law index. For instance, for $\lambda = 0.1$, this transition seems to take place ca. $Re_{PL} \sim 50$ whereas the corresponding value is about 1000 for $\lambda = 0.5$. The fact that neither of these values is significantly different from the corresponding transition values for Newtonian fluids (Table 10.3) seems to suggest that one could use the same

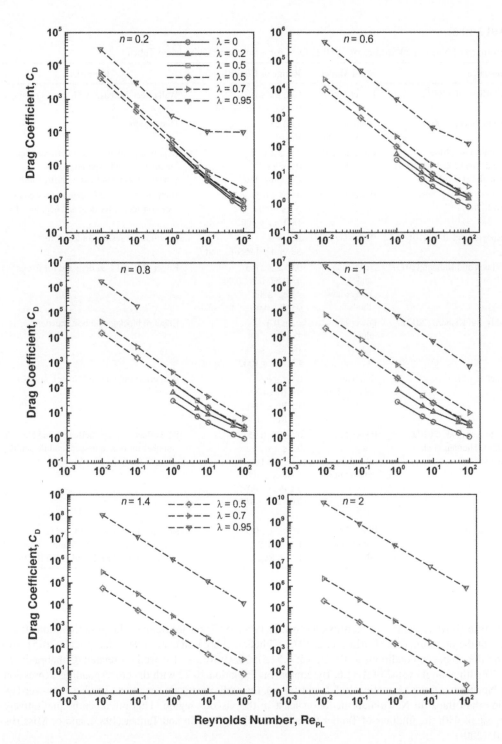

FIGURE 10.7 Dependence of the drag coefficient on diameter ratio, power-law index and Reynolds number in Poiseuille flow past a sphere. (From Song, D, Gupta, R.K. and Chhabra, R.P., *Ind. Eng. Chem. Res.*, **50**, 13105, 2011; Thumati, V.T. M.Tech Thesis, IIT Kanpur, India, 2016 (dotted lines).)

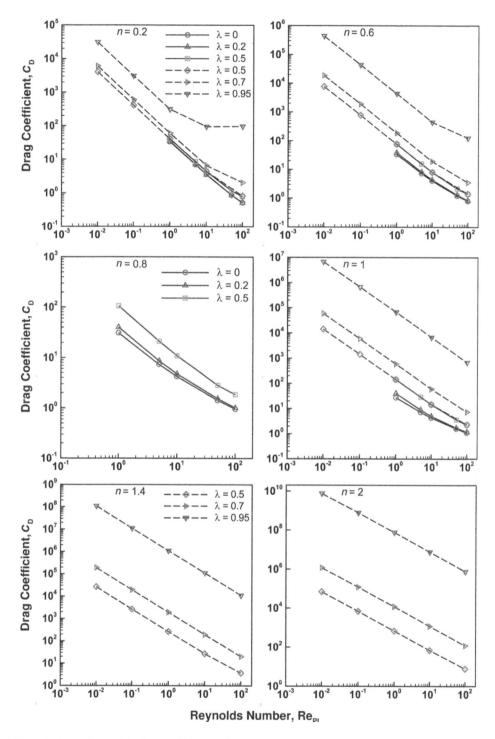

FIGURE 10.8 Dependence of the drag coefficient on diameter ratio, power-law index and Reynolds number for a sphere falling in quiescent power-law fluid. (From Song, D, Gupta, R.K. and Chhabra, R.P., *Ind. Eng. Chem. Res.*, **48**, 5845, 2009. (solid lines); Thumati, V.T. M.Tech Thesis, IIT Kanpur, India, 2016 (dotted lines).)

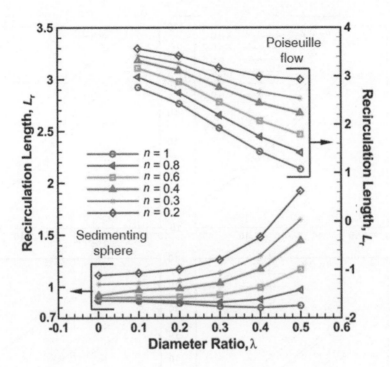

FIGURE 10.9 Effect of diameter ratio and power-law index on the recirculation length of the confined sphere. (From Song, D, Gupta, R.K. and Chhabra, R.P., Ind. Eng. Chem. Res., 48, 5845, 2009, Song, D, Gupta, R.K. and Chhabra, R.P., *Ind. Eng. Chem. Res.*, **50**, 13105, 2011.)

TABLE 10.6
Angle of Separation (in degrees) for Confined Sphere at $Re_{PL} = 100$

$n\downarrow$	$\lambda\rightarrow$					
	0.5	0.4	0.3	0.2	0.1	0.0
Poiseuille Flow[a]						
0.2	105.4	104.6	102.7	101.7	103.2	-
0.3	109.3	107.2	105.3	104.4	105.0	-
0.4	112.5	109.4	107.4	106.3	106.6	-
0.6	118.5	114.7	111.6	110.2	109.8	-
0.8	124.7	119.4	115.7	113.8	113.1	-
1	130.9	124.2	119.4	117.1	116.3	-
Sedimenting Sphere[b]						
0.2	108.4	109.3	109.1	110.1	111.7	115.7
0.3	113.5	114.8	115.3	116.3	117.2	118.5
0.4	117.3	118.3	118.9	119.6	120.1	123.1
0.6	122.7	123.3	123.4	123.6	123.8	124.1
0.8	126.7	126.7	126.1	125.8	125.7	126.0
1	130.1	129.2	127.9	127.1	126.8	126.5

Source: From [a] Song, D., Gupta, R.K. and Chhabra, R.P., Ind. Eng. Chem. Res., **50**, 13105 (2011).
[b] Song, D., Gupta, R.K. and Chhabra, R.P., *Ind. Eng. Chem. Res.,* **48**, 5845 (2009).

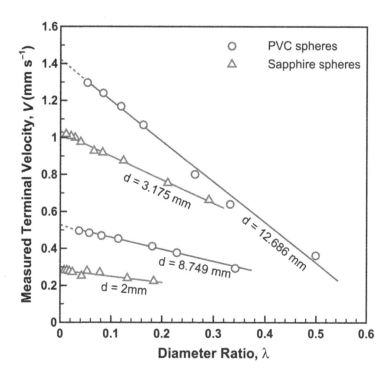

FIGURE 10.10 Typical variation of sphere fall velocity with diameter ratio in a 1% aqueous methocel solution in the low Reynolds number regime. (Based on the data from Chhabra, R.P., Ph. D Thesis, Monash University, Melbourne, Australia, 1980.)

values for power- law fluids also, at least as a rough guide. Based on the meager amount of data ($\lambda \leq 0.5$), Chhabra and Uhlherr (1980c) gave the following simple relation for f_∞

$$f_\infty = 1 - 3\lambda^{3.5} \tag{10.23}$$

In the intermediate transition regime, the wall factor depends upon both the diameter ratio and the Reynolds number (i.e., the terminal fall velocity no longer decreases linearly with diameter ratio as shown in Figure 10.11). There is only limited data available in the literature demonstrating the presence of all the three flow regions (Chhabra et al., 1977; Chhabra and Uhlherr, 1980c), and these are reproduced in Figure 10.12; the corresponding Newtonian results are also included in this figure for a qualitative comparison. The functional dependence of the wall factor on the Reynolds number and the diameter ratio is well approximated by the following relationship (Chhabra and Uhlherr, 1980c):

$$\frac{(1/f)-(1/f_\infty)}{(1/f_0)-(1/f_\infty)} = \left\{1 + 1.3\mathrm{Re}_{\mathrm{PL}}^2\right\}^{-0.33} \tag{10.24}$$

where f_0 and f_∞ are the limiting values of the wall factor in the low and high Reynolds number regions estimated using Equations 10.22 and 10.23, respectively. Equation 10.24 encompasses the following ranges of conditions: $10^{-2} \leq \mathrm{Re}_{\mathrm{PL}} \leq 10^3$; $0 \leq \lambda \leq 0.5$; $0.53 \leq n \leq 0.95$. The resulting overall average deviation is of the order of 8%. Other similar expressions (Turian, 1967; Lali et al., 1989; Ataide et al., 1998, 1999) based on the power-law fluid model are also available in the literature, but none of these have been tested as rigorously as the ones presented here. For instance, based on their experimental data over the ranges

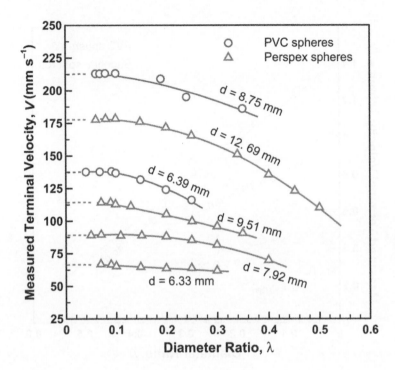

FIGURE 10.11 Typical variation of sphere fall velocity with diameter ratio in a 0.5% aqueous methocel solution in the intermediate Reynolds number region. (Based on the data from Chhabra, R.P., Ph. D Thesis, Monash University, Melbourne, Australia, 1980.)

of conditions as $0.56 \leq n \leq 0.84$ and $1 \leq Re_{PL} \leq 200$, Lali et al. (1989) put forward the following empirical correlation for wall effects:

$$f = (1-\lambda)^{A_1} Re_{PL,\infty}^{B_1} \qquad (10.25)$$

where

$$A_1 = \frac{1.8 + 0.2(n-1)}{1 + 0.2(n-1)} \qquad (10.26a)$$

$$B_1 = \frac{0.1}{1 + 0.2(n-1)} \qquad (10.26b)$$

Admittedly, Equation 10.25 has implicit in it the dependence of the wall factor on the power-law index via Equation 10.26; in the overlapping range of conditions, the predictions of Equations 10.24 and 10.25 differ at most by 7%–8% which is well within the limits of experimental accuracy of the wall factor, f. However, for a few combinations of λ and $Re_{PL,\infty}$, Equation 10.25 yields values of the wall factor which are greater than unity which are clearly physically unrealistic.

Unfortunately, sufficient information is not yet available to infer the values of the Reynolds number to delineate the boundaries of the flow regimes for power-law liquids. In addition to the wall effects in cylindrical tubes, scant results are also available on wall effects for a sphere falling in vessels of square cross-section (Balaramakrishna and Chhabra, 1992; Machac and Lecjaks, 1995). The wall effects appear

Wall Effects

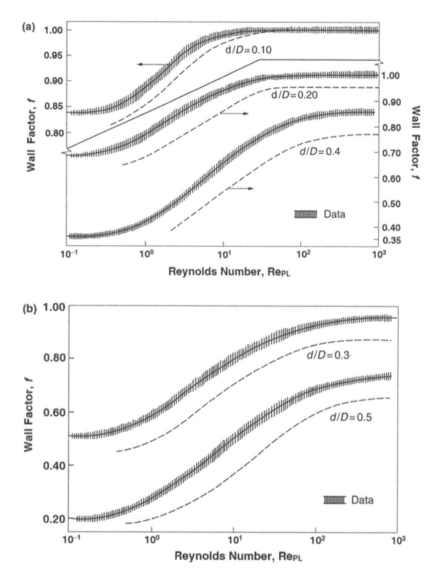

FIGURE 10.12 Wall factor as a function of the Reynolds number and diameter ratio for spheres falling in power-law liquids. The prediction of Equation 10.24 is shown as solid lines; The corresponding Newtonian results are shown as broken lines. (Based on the data from Chhabra, R.P., Ph. D Thesis, Monash University, Melbourne, Australia, 1980.)

to be less severe in square ducts than in a cylindrical tube for power-law liquids. Dewsbury et al. (2002b) have reported completely different type of wall effects on the (light) solid spheres rising through viscous Newtonian and inelastic polymer solutions. Owing to the nonvertical rise, it is much more difficult to quantify the wall effects in these systems.

10.3.3 Visco-plastic Liquids

Unlike in the case of Newtonian and pseudoplastic fluids, intuitively it appears that a sphere moving in a visco-plastic medium will experience the effect of the containing walls only if the boundary intersects with the sheared zone moving with the sphere (Carreau et al., 2021). This seems to suggest that for a given visco-plastic medium-sphere combination, there must be a critical value of the sphere-to-tube diameter

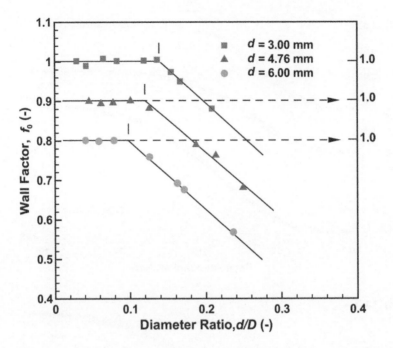

FIGURE 10.13 Wall factor as a function of diameter ratio and yield parameter in visco-plastic media. (Modified from Atapattu, D.D., Chhabra, R.P. and Uhlherr, P.H.T., *Proceedings 9th Australasian Fluid Mech. Conf.*, Auckland, New Zealand, p. 338, 1986.)

ratio (value of λ) below which there would be no wall effects, namely the wall factor would be unity. This expectation is borne out by the numerical study of Blackery and Mitsoulis (1997) and confirmed in Figure 10.13, where the values of wall factor are shown for three steel spheres as reported by Atapattu et al. (1986). Based on their experimental results encompassing the range $0.0091 \leq Y_G \leq 0.053$, Atapattu et al. (1986) proposed the following predictive expression for wall effects in visco-plastic media:

$$f_0 = 1 \qquad \lambda \leq \lambda_{crit}$$
$$f_0 = 1 - 1.7(\lambda - \lambda_{crit}) \qquad \lambda \geq \lambda_{crit} \qquad (10.27)$$

where $\lambda_{crit} = 0.055 + 3.44 Y_G$.

Note that this simple correlation is based solely on the yield stress of the fluid. Next, one should anticipate a relation between the critical sphere-to-tube diameter ratio and the size of the sheared zone (δ) moving along with the sphere; this contention is examined in Figure 10.14 where the predicted values of (d/δ) are contrasted with the experimentally determined values of (λ_{crit}) and the two seem to differ by almost a factor of two. Undoubtedly, one reason for this discrepancy could be the difference between the ideal Bingham fluid and the real fluid behavior. The other contributing factor could be the extent of back flow of the displaced liquid which inevitably occurs with the descent of a sphere in a cylindrical fall tube, whereas no such effect is encountered in theoretical predictions which assume the infinite extent of fluid. Subsequent numerical work (Beaulne and Mitsoulis, 1997; Blackery and Mitsoulis, 1997) suggests the wall effects to be more significant up to about Bi ≈ 10 beyond which the drag results for $0.02 \leq \lambda \leq 0.5$ collapse on to a single drag curve. This is qualitatively consistent with the form of Equation 10.27.

In recent numerical studies, the creeping flow analysis has been extended to high Reynolds numbers and the values of λ, as detailed in Table 10.5. Figure 10.15 shows both these situations, and it thus stands to reason that there will be no wall effects on the drag for the conditions shown in Figure 10.15b. Notwithstanding this aspect, Figures 10.16 and 10.17 show the functional relationship between drag

Wall Effects

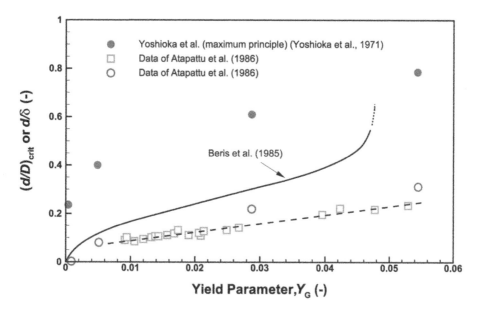

FIGURE 10.14 Relationship between the critical diameter ratio (λ_{crit}) and (d/δ) as a function of the yield parameter.

coefficient, Reynolds number, and λ. Here also, the drag is seen to bear a positive dependence on the severity of blockage. These results are also consistent with the limited results of Yu and Wachs (2007) for a fixes value of λ.

10.3.4 Visco-elastic Liquids

As seen in Chapter 4, the available literature (limited to the creeping region) on the sedimentation of a sphere in visco-elastic liquids relates either to the usual polymer solutions which exhibit both shear-dependent viscosity and visco-elasticity (finite primary normal stress difference) or to the so-called Boger fluids. In spite of the fact that several studies are available on the drag of spheres in shear-thinning visco-elastic polymer solutions, only Cho et al. (1980) and Chhabra et al. (1981b) have presented detailed results on wall effects. Depending upon whether the primary normal stress difference data is available or not, one of the following empirical expressions can be used for estimating the value of the wall factor, f_0, in visco-elastic fluids:

$$f_0 = 1 - 1.3\lambda^{0.94}(\Lambda)^{-0.077} \tag{10.28a}$$

or

$$f_0 = 1 - 0.94\lambda^{0.80}(We)^{-0.073} \tag{10.28b}$$

These equations are valid for the following ranges of conditions: $\lambda \leq 0.5$; $0.9 \leq \Lambda \leq 300$; $0.4 \leq n \leq 0.55$; $0.02 \leq We \leq 11$; and $Re_{PL} < 1$. The dimensionless Carreau number (Λ) and the Weissenbeerg number (We) are measures of visco-elastic effects, as discussed in Chapters 1 and 4. In the presence of shear-thinning behavior, elastic effects further suppress the wall effects, though the effect is very small (Tanner, 1964; Caswell, 1970). This trend is also borne out by the recent experimental results of Zhang et al. (2015).

10.3.4.1 Boger Fluids

The early limited experimental work on spheres in Boger fluids suggested the Newtonian wall correction to be adequate for $\lambda \leq \sim 0.25$ provided the sphere motion occurred in the constant relaxation time range, that is, at low Weissenberg numbers (Chhabra and Uhlherr, 1988b; Chmielewski et al., 1990a; Tirtaatmadja et al., 1990). The wall effects were reported to diminish progressively with the increasing values of the Weissenberg number, We. While the extensive numerical results for $\lambda = 0.5$ have already been presented in Chapter 4, limited numerical and experimental results are also available for the other values of λ on both sides of 0.5. In most cases, the values of the drag correction factor, Y, have been presented as functions of We and λ. Indeed, the Y — We map is strongly influenced by the value of λ and therefore extrapolations from one value of λ to another may not yield even qualitatively correct results, as can be seen in Figure 10.18 for three values of λ. It is clearly seen that the value of the drag correction

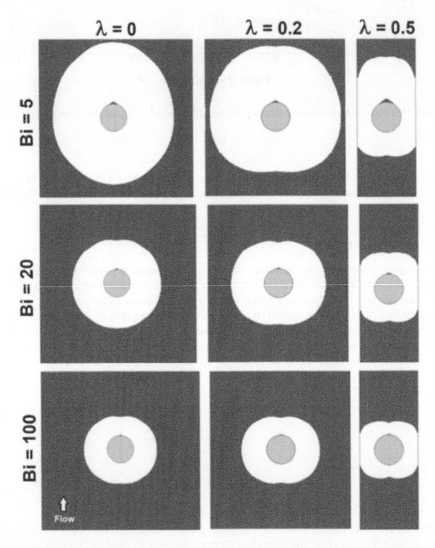

FIGURE 10.15a (a) Effect of confinement on the structure of fluid-like (unshaded) and solid-like (shaded) regions for a flow past a sphere in Bingham plastic fluids at $Re_B = 100$. (From Das. P.K., Gupta, A.K., Nirmalkar, N. and Chhabra, R.P., *Korea Aust. Rheol. J.*, **27**, 75, 2015.) (b) Effect of confinement on the structure of fluid-like (unshaded) and solid-like (shaded) regions for a flow past a sphere in Bingham plastic fluids at $Re_B = 100$. (From Thumati, V.T., Patel, S.A., Gupta, A.K., and Chhabra, R.P., *J. Chem. Eng. Jpn.*, **51**, 899, 2018.)

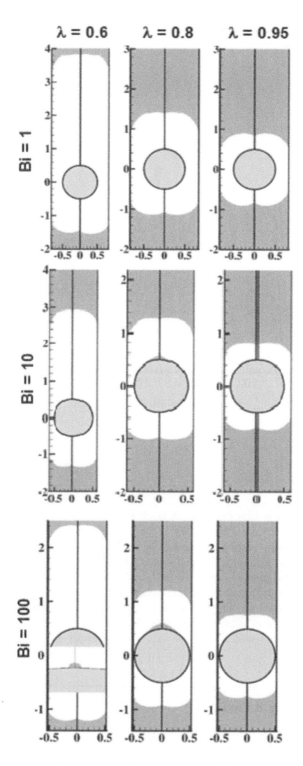

FIGURE 10.15b (*CONTINUED*) (a) Effect of confinement on the structure of fluid-like (unshaded) and solid-like (shaded) regions for a flow past a sphere in Bingham plastic fluids at $Re_B = 100$. (From Das. P.K., Gupta, A.K., Nirmalkar, N. and Chhabra, R.P., *Korea Aust. Rheol. J.*, **27**, 75, 2015.) (b) Effect of confinement on the structure of fluid-like (unshaded) and solid-like (shaded) regions for a flow past a sphere in Bingham plastic fluids at $Re_B = 100$. (From Thumati, V.T., Patel, S.A., Gupta, A.K., and Chhabra, R.P., *J. Chem. Eng. Jpn.*, **51**, 899, 2018.)

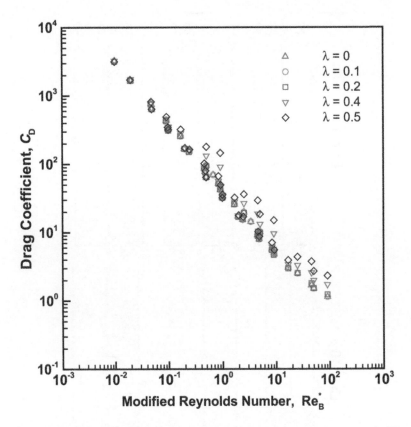

FIGURE 10.16 Dependence of drag coefficient on modified Reynolds number and diameter ratio. (From Das. P.K., Gupta, A.K., Nirmalkar, N. and Chhabra, R.P., *Korea Aust. Rheol. J.*, **27**, 75, 2015.)

factor, Y, may be smaller or greater than unity depending upon the value of λ. This appears reasonable because as the value of λ increases, the stresses in the nip region must rise steeply, thereby leading to significant drag enhancements for large values of λ due to the strong extensional flow component. This assertion is in line with the scant experimental results (Oh and Lee, 1992; Degand and Walters, 1995) and some of the numerical simulations for $\lambda \sim 0.9$ (Mitsoulis, 1998a). It is difficult to quantify the severity of wall effects over the entire range of λ in a simple manner; however, both experiments and predictions suggest the wall effects to be negligible up to about $\lambda \leq \sim 0.15$ and the wall effects are further suppressed even with a moderate degree of shear-thinning behavior. Such experiments, however, require special care to ensure that the sphere settles along the axis of the cylinder and without undergoing rotation (Tanner, 1964).

10.4 Nonspherical Rigid Particles

Additional complications arise in quantifying the extent of wall effects for nonspherical rigid particles owing to the inherent difficulties in an unambiguous description of their size, shape, and orientation in the free settling motion. This problem is further accentuated by the loss of symmetry with respect to the shape of sedimentation vessels. Notwithstanding these intrinsic difficulties, the current status of the available body of information is reviewed in the ensuing sections, adhering to the same pattern as followed in the preceding section for spherical particles. Other pertinent reviews are available in the books like Happel and Brenner (1965), Clift et al. (1978), Kim and Karrila (1991), and Michaelides (2006).

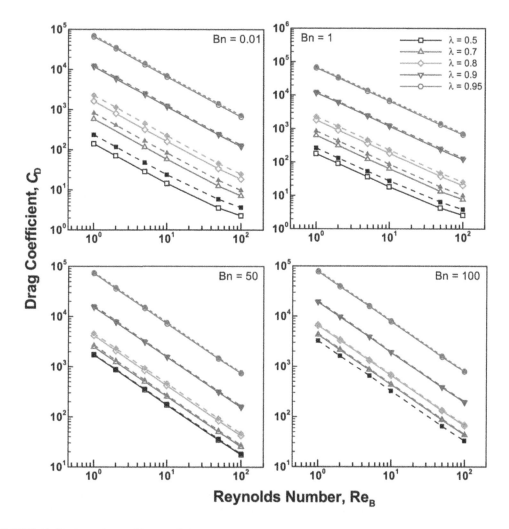

FIGURE 10.17 Dependence of drag coefficient Reynolds number, Bingham number and diameter ratio. (From Thumati, V.T., Patel, S.A., Gupta, A.K., and Chhabra, *R.P., J. Chem. Eng. Jpn,.* 51, 899, 2018.)

10.4.1 Newtonian Liquids

In the absence of inertial effects (Re → 0), attempts have been made to parallel the Faxen-type analysis for an arbitrarily shaped and oriented particle translating axially in long cylindrical tubes (Brenner, 1962; Rallison, 1978; Hirschfeld et al., 1984; Kasper, 1987). This approach yields results that are valid when the ratio of the particle size to that of the sedimentation vessel is small (<<<1). As expected, for a fixed shape and orientation, the terminal settling velocity decreases linearly with the size ratio. Similarly, the wall effects on circular thin disks falling freely (oriented normal to the direction of gravity) at the axis of cylindrical tubes have been studied analytically by Shail and Norton (1969), Pulley et al. (1996), Wang (1996a), Zimmerman (2002), and Nitin and Chhabra (2005a). The theoretical and numerical results have been supplemented by experimental studies by Schmiedel (1928), Squires and Squires (1937), Chhabra (1995c), and Rami et al. (2000). In the limit of creeping flow and for $\lambda < \sim 0.08$, the analytical results of Shail and Norton (1969) can be rearranged as (Trahan et al., 1989):

$$f = 1 - 1.7861\lambda + 1.278\lambda^3 - 0.3582\sqrt{\lambda} \tag{10.29}$$

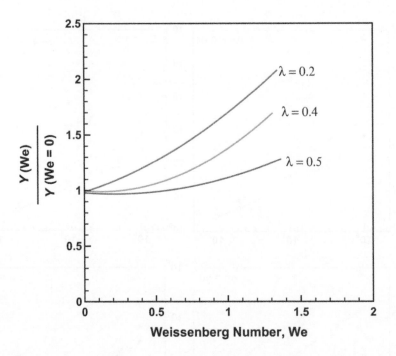

FIGURE 10.18 Wall effects on a falling sphere in visco-elastic liquids in low Reynolds number regime. (Based on the results of Jones, W.M., Price, A.H. and Walters, K., *J. Non-Newt. Fluid Mech.*, **53**, 175, 1994.)

The predictions of Equation 10.29 are in line with the experimental results of Squires and Squires (1937) and the calculations of Wang (1996a). The additional effects arising from the ends of the tube have been studied by Trahan (2006) and Trahan et al. (1989). The sedimentation of a disk toward a plane boundary has been studied by Trahan et al. (1987) and Davis (1990). Most of these and the other related studies have been reviewed by Zimmerman (2002). Subsequently, Nitin and Chhabra (2005a) have numerically studied wall effects on a circular disk moving broadside at the axis of a cylindrical tube in the range $0.02 \leq \lambda \leq 0.5$ and $1 \leq Re \leq 100$. Qualitatively speaking, the wall effects diminish with the increasing Reynolds number and the decreasing diameter ratio. Their results can be recast in terms of the wall factor f as

$$f = \left(1 + a\lambda + b\lambda^2\right)^{1/2} \quad (10.30)$$

where the values of the two constants a and b are given in Table 10.7 as functions of the Reynolds number.

Also, the wall effects on circular cylinders of finite aspect ratio settling with their axis aligned with the direction of motion have been evaluated theoretically by Allan and Brown (1986), Trahan et al. (1989)

TABLE 10.7
Values of a and b in Equation 10.30

Re	-a	-b
1	1.293	0.441
2	0.961	1.032
5	0.467	1.807
10	0.192	2.106
20	0.0546	2.147
50	0.0112	2.152
100	0.008	2.200

and experimentally by Huner and Hussey (1977), Curtis and Marr (1978), Ui et al. (1984), Unnikrishnan and Chhabra (1990, 1991), and Chhabra (1995c). Analogous results for cross-flow configuration have been provided by Bohlin (1960), Stalnaker and Hussey (1979), Chakraborty et al. (2004), and Semin et al. (2009). Recently, similar numerical results for ellipsoidal-, prolate-, and oblate-shaped particles have also been documented in the literature (Xu and Michaelides, 1996; Shahcheraghi and Dwyer, 1998; Reddy and Kishore, 2013). On the other hand, limited experimental results on wall effects for cones, prisms, cubes, needles, and cylinders sedimenting in cylindrical tubes are also available (Heiss and Coull, 1952; Sharma and Chhabra, 1991; Venumadhav and Chhabra, 1995; Chhabra, 1995c; Ataide et al., 1998) and these have been correlated empirically (Chhabra, 1995c). Broadly speaking, the low Reynolds number predictions for discs and cylinders are in line with experimental results. But clearly, all such attempts are shape-specific and extrapolations beyond the range of conditions are not recommended. Furthermore, though the presently available predictive formulae have been tested only in a limited way, a summary is presented in Table 10.8. Some results for wall effects on the straight chains of spheres and conglomerates of spheres falling at the axis of cylindrical tubes are also available (Kasper et al., 1985; Chhabra et al., 1995). Limited results on the terminal falling velocities of porous particles (flocs) in cylindrical tubes have been reported by Hsu and Hsieh (2004).

10.4.2 Inelastic Non-Newtonian Liquids

Even less is known about the wall effects on nonspherical particles sedimenting in stagnant time-independent fluids. As far as we know, only Tanner (1993) has reported that the wall effects for power-law fluids flowing normal to a cylinder are less severe than those in a Newtonian fluid in the creeping flow region. Later, Nitin and Chhabra (2006) have reported the extent of wall effects on a disk translating in power-law fluids in a cylindrical tube. The shear-thinning behavior was found to suppress the wall effects with reference to that in Newtonian fluids. This study embraces the range of conditions as: $0.02 \leq \lambda \leq 0.5$, $Re_{PL} < 100$, and $0.4 \leq n \leq 1.0$. Hsu et al. (2005b, 2006) have examined the combined effects of blockage and length-to-diameter ratio on the sedimentation of a cylinder in Carreau model fluids at $Re_\infty = 100$. Likewise, only scant experimental results for prisms, needles, cylinders, cones, and disks falling in cylindrical vessels filled with power-law fluids are available in the literature (Unnikrishnan and Chhabra, 1990; Sharma and Chhabra, 1991; Venumadhav and Chhabra, 1994; Chhabra, 1996b; Ataide et al., 1998; Rami et al., 2000; Siman et al., 2002). Based on data from all these sources, for a fixed shape and orientation, the wall factor, f_0, in power-law fluids can be related linearly with λ as given by Equation 10.22 where $\lambda = d_s/D$; A is a constant that appears to be independent of the power-law index, n, but it varies with the particle shape. The values of A for the particle shapes studied thus far are summarized in Table 10.8. Also included are the corresponding values of A for Newtonian fluids (Chhabra, 1995c). For each case, the wall effects are seen to be less severe in power-law fluids than that in Newtonian media within the range of conditions (creeping flow) studied so far. The value of $A = 1.7$ for discs in Newtonian media is

TABLE 10.8

Values of A in Equation 10.22 for Power-Law and Newtonian Fluids

| | Value of A | | |
Shape	Newtonian	Power-Law	Range of λ
Discs and plates	1.7	1.52	≤ 0.32
Cylinders			
Short ($L/d < 10$)	1.33	1.2	≤ 0.19
Long ($L/d > 10$)	10.58	3.0	
Rectangular prisms	1.42	1.28	≤ 0.60
Cones	2.11	1.63	≤ 0.37
Hollow cylinders (Ataide et al., 1998)	-	1.13–1.25	$\leq \sim 0.5$

Source: From Chhabra, R.P., Transport Processes in Bubbles, Drops and Particles, De Kee, D. and Chhabra, R.P., ed., p. 316, Taylor & Francis, New York (2002b).

remarkably close to the theoretical value of 1.6 (Shail and Norton, 1969; Wang, 1996a). The other scant results on wall effects for cubes and hollow cylinders in power-law fluids suggest the values of A in the range 1.13–1.25, which are comparable to those listed in Table 10.8. The limited results for linear chains of spheres and conglomerates of spheres settling in power-law fluids also show linear dependence of the settling velocity on the reciprocal of tube diameter (Chhabra et al., 1995).

The effect of visco-elasticity on wall effects for a cylinder confined in a planar slit has been studied by Huang and Feng (1995). They used the Oldroyd-B fluid model to delineate the roles of weak inertia (Re < 10) and of elasticity (We < ~3) on the drag of the cylinder for a range of the blockage ratios ($\lambda \leq$ ~0.6). Both visco-elasticity and shear-thinning were shown to suppress the wall effects, a finding which is in qualitative agreement with the experiments of Dhahir and Walters (1989) for $\lambda = 0.6$. Limited results on wall effects for a square cylinder confined between two plates with power-law fluids have been reported by Bouaziz et al. (2010).

10.5 Effect of Blockage on Heat Transfer from a Sphere

As noted earlier, the confinement can also amend the temperature field prevailing in the vicinity of a sphere undergoing steady translation in a fluid-filled tube and hence the temperature gradient normal to the surface of the sphere. This, in turn, manifests in terms of the improved heat transfer in Newtonian and power-law fluids whereas the blockage effect in the visco-plastic fluids is governed by the fact that whether the fluid-like region extends up to the tube wall or not.

Only limited results are available on the effect of symmetric and asymmetric confinement on heat transfer from a sphere. For instance, Despeyroux et al. (2010) studied heat (mass) transfer from a sphere translating slowly (creeping flow) at the axis and off-axis positions in power-law fluids in a long tube. They found the heat transfer to be promoted or impeded depending upon the value of λ, E, Peclet number, power-law index, and the type of boundary condition prescribed on the tube wall (Neumann or Dirichlet type). In diffusion (conduction) regime, Pe ≤ ~0.1, the Nusselt number was independent of the fluid rheology and some enhancement in mass transfer was seen in shear-thinning fluids at high Peclet numbers. In fact, the scaling of Nu ~ $Pe^{1/3}$ was observed for Pe > ~100 or so irrespective of the values of n and λ. The trend of Nusselt number with the eccentricity was rather complex. There have been a few studies at finite Reynolds numbers for a confined sphere in Newtonian (Krishnan and Kannan, 2010) and in power-law fluids (Song et al, 2010, 2012; Reddy and Kishore, 2013; Vidyasagar et al, 2017b; Thumati et al, 2018, etc.) as summarized in Table 10.5. Figures 10.19 and 10.20 show the combined effects of power-law index and blockage at $Re_{PL} = 100$ and for two values of the Prandtl number (Pr_{PL}) for an isothermal sphere and adiabatic tube wall. Here the surface mean value of the Nusselt number has been normalized with respect to the corresponding value for Newtonian fluids ($n = 1$) otherwise under identical conditions (Figure 10.19). Based on these results (Song et al., 2010), the effect of blockage on heat transfer from a sphere can be summarized as follows: the blockage does promote heat transfer in shear-thinning fluids over and above that in Newtonian fluids. However, the role of shear-thinning viscosity causing the augmentation in heat transfer is reduced here with reference to that for an unconfined sphere. In other words, the maximum enhancement in heat transfer on account of shear-thinning viscosity is realized only for an unconfined sphere ($\lambda = 0$). However, the degree of augmentation increases with the rising degree of advection at high Reynolds and/or Prandtl numbers. Two competing processes are at play here. On the one hand, the boundary layers tend to be thinner in power-law fluids for $n < 1$ than that in Newtonian fluids and it is thus likely that there is no interaction between the boundary layers formed on the surface of the sphere and the tube wall. On the other hand, the thin boundary layers sharpen the temperature gradients which lead to high values of the heat transfer coefficient in shear-thinning fluids. Considering all these factors, Song et al. (2010) reported 60%–70% increase in Nusselt number over the range of conditions spanned in their study. Qualitatively, similar trends have been reported for the case (Song et al., 2012) when the isothermal sphere is held fixed in the Poiseuille flow of power-law fluids. Based on their numerical results over the ranges of conditions as $5 \leq Re_{PL} \leq 100$, $1 \leq Pr_{PL} \leq 100$, and $0.2 \leq n \leq 1$, Song et al. (2012) presented the following predictive expression for the mean Nusselt number for the Poiseuille flow case:

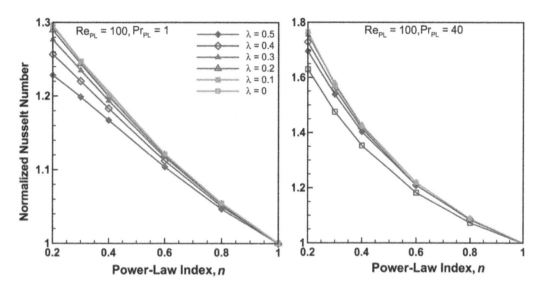

FIGURE 10.19 Effect of power-law index and diameter ratio on normalized Nusselt number. (From Song, D, Gupta, R.K. and Chhabra, R.P., *Ind. Eng. Chem. Res.*, **49**, 3849, 2010.)

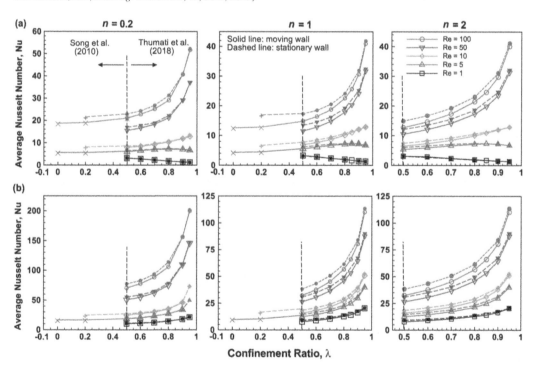

FIGURE 10.20 Effect of power-law index and diameter ratio on Average Nusselt number. (From Thumati, V.T., Patel, S.A., Gupta, A.K., and Chhabra, R.P., *J. Chem. Eng. Jpn,.* **51**, 899, 2018.)

$$Nu = b_1 \, \text{Re}_{\text{PL}}^{b_2} \, \text{Pr}_{\text{PL}}^{b_3} \left(\frac{3n+1}{4n} \right)^{b_4} \tag{10.31}$$

where the fitted constants b_1, b_2, b_3, and b_4 show further dependence on the blockage ratio (λ) as shown in Table 10.9. Equation 10.31 consolidates the numerical results of Song et al. (2012) with a mean deviation

TABLE 10.9
Values of b_1, b_2, b_3 and b_4 in Equation 10.31 (From Song et al., 2012)

λ	b_1	b_2	b_3	b_4
0	1.434	0.367	0.303	0.567
0.05	1.546	0.416	0.333	0.365
0.1	1.544	0.416	0.333	0.367
0.2	1.545	0.414	0.334	0.374
0.3	1.593	0.407	0.336	0.370
0.4	1.653	0.399	0.341	0.353
0.5	1.696	0.396	0.353	0.326

of ~5% which reaches the maximum value of ~20%. The dependence of the Nusselt number is seen to be somewhere in between $Re_{PL}^{1/3}$ to $Re_{PL}^{1/2}$, though the dependence on the Prandtl number conforms to the familiar $Pr_{PL}^{1/3}$ scaling. For $n < 1$, the Rabinowitsch-Mooney factor $((3n+1)/4n) > 1$ and the positive values of b_4 reinforce the positive influence of the power-law index on heat transfer. As discussed earlier, the effect diminishes with the increasing value of λ.

The analogous studies in Bingham plastic fluids are due to Chen et al. (1998b), Das et al. (2015) and Thumati et al. (2018). While Das et al. (2015) studied heat transfer from an isothermal sphere translating at the axis of a tube in quiescent fluid with adiabatic tube wall, Thumati et al. (2018) considered both situations, that is, falling sphere and sphere held fixed in the Poiseuille flow of Bingham fluids in an adiabatic tube. Combined together, these two studies span wide-ranging Reynolds numbers ($1 \leq Re_B \leq 100$), Prandtl numbers ($1 \leq Pr_B \leq 100$), Bingham numbers ($0.01 \leq Bi \leq 100$), and confinement ($0 \leq \lambda \leq 0.95$). Figure 10.21 shows representative results elucidating the multivariable dependence of the mean Nusselt number on Re_B, Pr_B, λ, and Bi. The key trends can be summarized in the following section. The fluid-like region grows with the increasing fluid inertia (i.e., Reynolds number Re_B), but this tendency is countered by the increasing yield stress (i.e., Bingham number, Bi). The wall effects on drag appear only when the yielded fluid-like regions intersect with the tube wall. However, heat transfer is influenced by the tube wall even under these conditions due to the conduction thermal resistance of the unyielded material present in this region. Thus, the imposition of boundaries sharpens the temperature gradients on the surface of the sphere and hence the Nusselt number bears a positive dependence on the Bingham number. However, some of this gain is offset by the fact that wake formation is suppressed in yield stress fluids. Qualitatively, similar trends are observed in the case of a sphere fixed in the Poiseuille flow and when the constant temperature condition on the surface of the sphere is replaced by the constant heat flux condition. Limited results are also available on the combined effects of blockage and yield stress on heat transfer from an array of three spheres (Dasgupta et al., 2019). Wall effects on heat transfer to nano-fluids from a heated sphere have been studied by Gupta et al. (2018). It needs to be emphasized here that the aforementioned heat transfer studies pertain to the base case of constant properties and negligible dissipation. More importantly, heat transfer calculation for yield stress fluids is based on the implicit assumption of continuous thermal properties (thermal conductivity, heat capacity, thermal diffusivity) in the yielded and unyielded regions. The two regions are differentiated only by vastly different values of the viscosity.

10.6 Drops and Bubbles

Much less is known about the wall effects on the terminal falling velocity of a bubble or a droplet even when the continuous phase is Newtonian, let alone for non-Newtonian media. It is generally believed that the sole effect of the confining walls is to retard the particle as long as the drop-to-tube diameter ratio $\lambda \leq \sim 0.6$ (Clift et al., 1978). Conversely, under these conditions, the confining walls cause little deformation beyond that which may be present in an unbounded medium. In some sense, therefore, the treatment for rigid particles presented in the preceding section forms a good starting point for the discussion on wall effects for fluid spheres for $\lambda \leq \sim 0.6$, whereas owing to the increasing degree of deformation

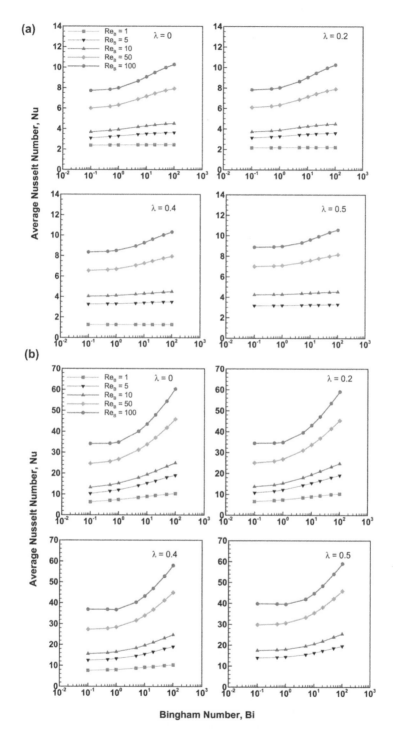

FIGURE 10.21 Effect of Bingham number, Reynolds number and diameter ratio on Average Nusselt number (a) $Pr_B = 1$ (b) $Pr_B = 100$. (From Das. P.K., Gupta, A.K., Nirmalkar, N. and Chhabra, R.P., *Korea Aust. Rheol. J*, **27**, 75, 2015.)

due to walls, the case of $\lambda > 0.6$ must obviously be treated separately. However, since the results on wall effects for fluid particles falling in non-Newtonian continuous phase are indeed scant and the corresponding literature in Newtonian fluids has been extensively reviewed by Clift et al. (1978), only a short discussion on wall effects for $\lambda \leq \sim 0.6$ is included here. Furthermore, it is convenient to deal with the low Reynolds number and high Reynolds number regimes separately. As usual, we begin with the behavior in Newtonian liquids.

10.6.1 Newtonian Continuous Phase

10.6.1.1 Low Reynolds Number Regime

It is useful to recall here that fluid particles at low Reynolds numbers in unconfined media tend to retain spherical shape, and furthermore, owing to the presence of surfactants or large values of the viscosity ratio, X_E, the interface is generally immobilized (or stagnant). Indeed, only minor errors occur in using the wall correction factors for rigid spheres up to about $\lambda \approx 0.3$. For a fluid sphere with clean interface translating at the axis of a long cylindrical tube, Haberman and Sayre (1958) derived the following approximate expression for wall correction factor:

$$f = \frac{1 - a_1\lambda - a_2\lambda^3 + a_3\lambda^5 - a_4\lambda^6}{1 + a_5\lambda^5} \tag{10.32}$$

where the constants a_1 to a_5 are given by

$$a_1 = 0.702\left(\frac{2 + 3X_E}{1 + X_E}\right) \tag{10.33a}$$

$$a_2 = 2.087\left(\frac{X_E}{1 + X_E}\right) \tag{10.33b}$$

$$a_3 = 0.569\left(\frac{2 - 3X_E}{1 + X_E}\right) \tag{10.33c}$$

$$a_4 = 0.726\left(\frac{1 - X_E}{1 + X_E}\right) \tag{10.33d}$$

$$a_5 = 2.276\left(\frac{1 - X_E}{2 + 3X_E}\right) \tag{10.33e}$$

Needless to say that Equation 10.32 does reduce to Equation 10.5 in the limit of $X_E \to \infty$. Likewise, subsequent numerical studies for spherical bubbles (Hartholt et al., 1994; Higdon and Muldowney, 1995; Wham et al., 1997) confirm the validity of Equation 10.32 in the limit of $X_E \to 0$. Haberman and Sayre (1958) also reported that the scant results for drops (of aqueous glycerin solutions or silicone oil) falling in castor oil were consistent with the predictions of Equation 10.32 for two values of $X_E \approx 0.005$ and $X_E \sim 0.07$ and $\lambda \leq \sim 0.5$ and beyond this value of λ, they reported appreciable deformation of fluid spheres due to the confining walls. The effect of slip condition on the particle surface undergoing steady translation in a confined medium has also been examined by Ramkissoon and Rahaman (2003).

10.6.1.2 High Reynolds Number Regime

Wham et al. (1997) extended the analysis of Haberman and Sayre (1958) to finite values of Reynolds number (≤ 100) but limited to $0.1 \leq \lambda \leq 0.2$ and $3 \leq X_E \leq 10$. While the low values of λ correspond to

Wall Effects

spherical shape, the range of values of the viscosity ratio does capture the drop-like behavior. Based on their numerical results, Wham et al. (1997) presented a modified form of Equation 10.32 which purports to encompass the full range of particle behavior. Furthermore, Wham et al. (1997) also observed no wake formation at all for $X_E \leq \sim 2.75$. Sherwood (2001), on the other hand, has invoked the irrotational flow approximation to elucidate the role of weak wall effects (small values of λ on the drag of a bubble rising through a liquid in a cylindrical tube and modified the drag expression (Moore, 1959a,1963)) for an unconfined bubble. Similarly, Bozzi et al. (1997) have numerically studied the extent of deformation due to wall and inertial effects for $\lambda = 0.5$, $X_E = 1$, and $Re \leq \sim 540$.

While experimental studies on the drag on drops and bubbles reported in the literature (see Chapter 5) have been carried out using vessels or containers of finite size, very few workers have actually set out to glean data on wall effects per se by using containers of different diameters. Admittedly, the presence of confining walls influences shape, wake structures, and breakup behavior (Hetsroni et al., 1970; Borhan and Pallinti, 1998, 1999), but most of the experimental literature on wall effects is limited to establishing the influence of λ on the terminal velocity, that is, to delineate the relationship implied by Equation 10.2. The studies of Strom and Kintner (1958), Harmathy (1960), Salami et al. (1965), Eaton and Hoffer (1970), Mao et al. (1995), Bhavasar et al. (1996), and Chhabra and Bangun (1997) are representative of those in which wall effects on falling drops have been studied. The corresponding representative studies for bubbles are due to Uno and Kintner (1956), Harmathy (1960), Maneri and Mendelson (1968), Tsuge and Hibino (1975), and Krishna et al. (1999b). Much of the available data on bubbles and drops relates to the conditions of $Re > 1$ and $E_0 < 40$. Clift et al. (1978) found that the experimental data for wall effects on fluid particles were in line with that for rigid spheres, at least up to about $\lambda \leq \sim 0.6$. Furthermore, Clift et al. (1978) proposed that the value of the wall factor, f, was greater than 0.98 (i.e., 2% reduction in velocity) under the following conditions:

$$\lambda \leq 0.06, \quad Re \leq 0.1 \tag{10.34a}$$

$$\lambda \leq 0.08 + 0.02 \log Re, \quad 0.1 \leq Re \leq 100 \tag{10.34b}$$

$$\lambda \leq 0.12, \quad Re > 100 \tag{10.34c}$$

On the other hand, for $Re > 100$, the wall factor becomes independent of the Reynolds number and Clift et al. correlated the literature data as:

$$f = \left(1 - \lambda^2\right)^{1.5} \tag{10.35}$$

The equation due to Strom and Kintner (1958) is similar to Equation 10.35 except for the value of the exponent being 1.43. There has not been much activity in this field since the aforementioned earlier studies, except the scant data of Mao et al. (1995) which is in line with Equation 10.35.

Beyond these conditions, the wall effects on spherical gas bubbles have been discussed by Clift et al. (1978). Okhotskii (2001) has used the wave analogy to develop a semi-empirical expression for wall effects on bubbles. Based on extensive experiments with air–water and air–Tellus oil ($\mu = 75$ mPa s) systems, Krishna et al. (1999b) reported extensive results on wall effects on the free rise velocity of bubbles up to about $\lambda \sim 0.7$ or so, and these data are also consistent with Equation 10.34.

10.6.2 Non-Newtonian Continuous Phase

Little is known about the influence of non-Newtonian rheology of the continuous phase on the confined free rise/fall of fluid particles. Mukundakrishnan et al. (2007, 2009) have numerically studied hydrodynamics of freely rising bubbles in a truncated power-law fluid in the range $0.5 \leq \lambda \leq 0.9$ for two values of the Reynolds number (based on infinite shear viscosity and the tube diameter) of 0.06 and 40. At low

Reynolds numbers, the drag coefficient increased with the rising values of λ and this effect was stronger in power-law fluids than that in Newtonian fluids. However, it diminished with the increasing Reynolds number. Chhabra and Bangun (1997) reported preliminary results on the terminal velocity of droplets of distilled water, ethylene glycol, chloroform, 1,2-dichloroethane, chlorobenzene, and furfural falling in aqueous solutions of carboxymethyl cellulose and polyacrylamide ($0.52 \leq n \leq 0.87$) in the creeping flow regime (Reynolds number, $Re_{PL} < 0.07$) and $\lambda \leq \sim 0.45$. For a fixed diameter of drop, the terminal falling velocity showed linear dependence on λ. Since the maximum value of the viscosity ratio, X_E, was only 0.0027, these results are tantamount to that for bubbles. In these range of conditions, the wall factor was well approximated by the expression, $f = 1.0 - 1.6\lambda$, that is, Equation 10.22. Later, Dziubinski et al. (2001) have studied wall effects on single bubbles rising through power-law liquids ($0.43 \leq n \leq 0.83$; $\lambda \leq \sim 0.84$; $3.3 \leq Re_{PL} \leq 1580$; $0.6 \leq E_o \leq 10.8$). Qualitatively, the wall effects were reported to be less severe in power-law fluids than that for Newtonian liquids, which is consistent with the trends seen for solid spheres.

Coutanceau and Hajjam (1982) investigated the wall effects on a bubble rising through visco-elastic and viscoinelastic liquids. For values of $\lambda \sim 0.7-0.8$ (defined using equal volume sphere diameter) and in creeping flow, they reported significant departure from spherical shape. Indeed, bubbles of inverted bottle-like shapes were reported. Similarly, scant data on slug velocities of bubbles in non-Newtonian liquids in vertical and inclined tubes have been reported by Johnson and White (1993), Carew et al. (1995), and Baca et al. (2003), but no general predictive expressions are yet available for establishing the role of confining walls on moving bubbles under these conditions. Harris (1993, 1996) has investigated the behavior of bubbles in Newtonian media near solid structures.

10.7 Conclusions

In this chapter, the available body of knowledge on the severity of wall effects on the rigid and fluid particles falling in Newtonian and non-Newtonian media has been presented. Even for the simplest case of a solid sphere sedimenting at the axis of a cylindrical tube filled with a power-law liquid, the results are nowhere near as extensive or conclusive as that for the Newtonian liquids. Broadly speaking the shear-thinning behavior suppresses wall effects for solid spherical and nonspherical particles and for gas bubbles. Virtually, no experimental results are available for fluid spheres with finite values of the viscosity ratio. Visco-elasticity further lowers the extent of wall effects, at least for falling spheres. However, owing to the complex interplay between particle shape and visco-elasticity, extrapolations outside the range of conditions must be treated with reserve. Finally, it is now possible to estimate the effects of λ on settling velocity of a rigid sphere falling in power-law liquids in cylindrical tubes up to $\lambda \leq 0.9$ and $Re_{PL} \sim 1000$. Scant results on heat/mass transfer to power-law and Bingham plastic fluids from a heated sphere confined symmetrically and asymmetrically in a tube have been reviewed. Broadly, blockage tends to promote heat transfer with reference to that from an unconfined sphere in both power-law and Bingham plastic fluids.

Nomenclature

A: Constant, Equation 10.22 (-)
A_0: Constant, Equation 10.6 (-)
A_1: Constant, Equation 10.25 (-)
a: Constant, Equation 10.30 (-)
$a_1 - a_5$: Constant, Equation 10.32 (-)
B: Constant, Equation 10.6 (-)
$b_1 - b_4$: Fitted constants, Equation 10.31 (-)
Bi: Bingham number $\left(\dfrac{\tau_0^B d}{\mu_B V} \right)$ (-)

B_1: Constant, Equation 10.25 (-)
b: Constant, Equation 10.30 (-)
C: Constant, Equation 10.6 (-)
C_D: Drag coefficient (-)
C_p: Fluid heat capacity (J/kg K)
D: Tube/Vessel diameter (m)
d: Sphere diameter (m)
d_S: Equal volume sphere diameter (m)
E: Eccentricity (distance from axis/tube radius) (-)
Eo: Eotvos number ($=g\Delta\rho d^2/\sigma$) (-)
f: Wall factor (-)
f_0: Wall factor in the creeping flow region (-)
f_∞: Wall factor in the fully turbulent region (-)
g: Acceleration due to gravity (m s^{-2}) (-)
h: Heat transfer coefficient (W/m^{2}K)
K: Constant, Equation 10.6 (-)
k: Thermal conductivity of fluid (W/mk)
m: Power-law consistency coefficient (Pa s^n)
n: Power-law index (-)
Nu: Nusselt number (hd/k)
N_1: First normal stress difference (Pa)
p: Constant, Equation 10.13 (-)
Pr_B: Prandtl number ($=\mu_B C_p/k$) (-)
Pr_{PL}: Prandtl number ($mc_p(V_\infty/d)^{n-1}/k$) (-)
Re: Reynolds number ($=\rho V d/\mu$) (-)
Re_B: Reynolds number ($\rho V_\infty d/\mu_B$) (-)
Re_{PL}: Reynolds number ($\rho V^{2-n} d^n/m$) (-)
V: Terminal velocity in a confined medium (m s^{-1})
We: Weissenberg number ($=\lambda_f V/d$) (-)
X_E: Viscosity of dispersed phase divided by viscosity of continuous phase (-)
$Y(n, \lambda)$: Drag correction factor ($= C_D Re/24$) (-)
Y_G: yield parameter ($= \tau_0/gd(\rho_p-\rho)$) (-)

Greek Symbols

α: Constant, Equation 10.11 (-)
β: Ratio of projected area of particle and cross-sectional area of vessel (-)
$\dot{\gamma}$: Shear rate (s^{-1})
δ: Size of yielded region (m)
μ: Viscosity (Pa s)
ρ: Continuous phase density (kg m^{-3})
ρ_p: Particle density (kg m^{-3})
Λ: Carreau number ($\lambda_C V/d$) (-)
λ: Particle to tube diameter ratio (-)
λ_C: Carreau model parameter (s)
λ_{crit}: Critical value of λ, Equation 10.27
λ_f: Fluid characteristic time $\left(= N_1/(2\tau\dot{\gamma})\right)$ (s)
λ_o: Constant, Equation 10.13 (-)
σ: Surface tension (N m^{-1})
τ: Shear stress (Pa)
τ_o: Yield stress (Pa)

Subscripts

∞: Unconfined medium
B: Bingham model

11
Falling Object Rheometry

11.1 Introduction

In a quest to develop simple and reliable methods for the measurement of viscosity and other non-Newtonian fluid parameters, considerable effort has been made in examining and establishing the suitability of the flows produced by falling objects (sphere, needle, cylinder), rolling ball and rotating sphere, vibrating sphere, bubble viscometer, etc. Over the years, it has been demonstrated convincingly that while most of these devices yield reliable values of shear viscosities for Newtonian fluids, their applicability to non-Newtonian fluids is severely limited by the fact that all such flows are nonviscometric. Therefore, neither the shear stress or the shear rate is uniform, nor are these known a priori. However, most non-Newtonian fluids (except for visco-plastic and time-dependent fluids) approach the Newtonian fluid behavior in the limit of vanishingly small Reynolds and Deborah (or Weissenberg) numbers. It is thus possible to evaluate some characteristics of visco-inelastic and visco-elastic fluids from suitable data obtained from falling object tests under appropriate conditions, especially in the limit of zero-shear rate. This chapter presents an overview of the developments in this field. In particular, consideration is given to the falling ball method, the falling cylinder method, and the rolling ball method. Other less common methods such as rotating and vibrating sphere devices, bubble viscometer, etc. are also mentioned briefly. In order to maintain the consistency of the style of presentation, a short section on the use of each device to measure the viscosity of Newtonian fluids is included, which serves as the background for the subsequent discussion for non-Newtonian fluids.

11.2 Falling Ball Method

11.2.1 Newtonian Fluids

The incompressible and steady flow generated by a sphere translating in an infinite expanse of a fluid presents a viscosity-measuring device which is simple to fabricate and operate. The underlying principle of the falling sphere viscometry is the well-known Stokes velocity formula, written as

$$\mu_s = \frac{gd^2(\rho_s - \rho)}{18V} \tag{11.1}$$

It is useful to recall here that the Stokes formula for drag is based on the assumptions of the creeping flow (small values of Reynolds number), constant falling velocity region, and unbounded domain (i.e., no wall and end effects). In practice, however, usually, the terminal settling velocity is measured in a cylindrical fall tube and thus the measured velocity should be corrected for wall and end effects before calculating the value of viscosity using Equation 11.1. The creeping flow assumption is verified by calculating the value of the particle Reynolds number which should not exceed 0.1. Beyond this value, a correction for inertial effects is also needed. While the end effects are known to be negligible (Maude, 1961; Tanner, 1963; Flude and Daborn, 1982) as long as the terminal velocity is measured three-to-four sphere diameters away from the top and bottom ends of the fall tube, the corrections due to wall and inertial effects have been dealt with in Chapter 10 and by others such as Sutterby (1973a,b), Dimova et al. (1999), and Ben-Richou et al. (2003).

The issue of the time and/or distance required for the terminal velocity to be reached has been addressed by Ferreira and Chhabra (1998), Michaelides (2006), and Malvandi et al. (2014), among others. With appropriate corrections, this technique has been used extensively for absolute viscosity measurements for a range of materials including silicate melts (Kahle et al., 2003), biodiesel (Camas-Anzueto et al., 2017), whole blood and plasma (Doffin et al., 1984), and of suspensions of small spheres and rods by treating them as a pseudo-homogeneous fluids in comparison with the size of the falling ball (Milliken et al., 1989a, 1989b; Harlen et al., 1999; Kaiser et al., 2004), though some unusual phenomena can also be encountered in these systems (Kaiser et al., 2004). Gupta (2014) and Kono (2018) have dealt with the calibration and application of the falling ball viscometer in high temperature and pressure settings.

Naturally when dealing with highly viscous substances like epoxy resins, greases, pastes, etc., the motive force due to gravity is very small in comparison with the drag force and therefore the falling velocity of the ball will be extremely small. This difficulty has been obviated by developing the so-called magnetic sphere viscometer, in which the translational velocity of the sphere can be regulated via the strength of the imposed magnetic field (Adam et al., 1984; Sobczak, 1986; Gahleitner and Sobczak, 1987, 1988; Tran-Son-Tay et al., 1988, 1990; Hermann and Sobczak, 1989; Hilfiker et al., 1989). Linliu et al. (1994) and Nyrkova et al. (1997) have also described the design and operation of a centrifuge ball viscometer and tested its reliability for scores of Newtonian liquids and polymer melts. On the other hand, Ringhofer and Sobczak (1997) have attempted to relate the values of viscosity from a magneto-viscometer to that of the melt indexer.

The aforementioned corrections due to inertial, wall, and end effects are necessary only when the absolute values of viscosity are needed Kono, 2018; Suri et al., 2023a). When only a relative value is desired (or the falling ball device has been calibrated) for quality control purposes, Equation 11.1 can be simplified as

$$\mu_s = k(\Delta\rho)t \qquad (11.2)$$

where k is the system constant and t is the time of fall of the ball between two marks in the viscometer tube. Obviously, under these conditions, it is expected that the resulting values of the Reynolds number do not vary too much in the range of operation, else the constant k will also depend upon the value of the Reynolds number. The commercially available falling ball viscometers have been described in detail by van Wazer et al. (1963), Heywood (1985) and Gupta (2014).

11.2.2 Non-Newtonian Fluids

As mentioned earlier, most time-independent (excluding visco-plastic) and visco-elastic substances exhibit Newtonian flow behavior at very low shear rates. The limiting value of the apparent viscosity in the low shear rate region is termed the zero-shear viscosity, denoted by μ_0. The range of shear rates over which the transition from the zero-shear viscosity to the so-called power-law region occurs is quite narrow for polymers of narrow molecular weight distribution and vice versa, and it is also a function of the polymer concentration, temperature and the type of solvent, etc. It is, however, not yet possible to predict a priori the range of shear rate (or shear stress) for the onset of shear-thinning behavior in a new application.

The determination of the zero-shear viscosity is of both fundamental and pragmatic importance (Shaw, 2021). Most rheological equations of state include this limiting behavior as a model parameter and thus its determination by an independent method is desirable rather than treating it as an adjustable parameter (Shaw, 2021). From an engineering standpoint, a knowledge of the zero-shear viscosity is also useful in the analysis of flow configurations which entail low shear rates and involve stagnation points such as the flow around an isolated sphere or a cylinder and flow past an assemblage of particles, etc. Unfortunately, it is generally not possible to achieve sufficiently low shear rates in the conventional rotational and capillary viscometers to measure the zero-shear viscosity directly, especially for dilute polymer solutions. Admittedly, although some modern instruments are capable of producing extremely low shear rates, the corresponding shear stress produced is too small to be measured with the desirable level of accuracy. In this context, the historic success of the falling ball method for measuring the viscosity of Newtonian fluids has motivated several workers to explore the possibility of extending this method to non-Newtonian systems. Likewise, one of the main characteristics of visco-plastic media is the existence of a yield stress, and the fluid deformation occurs only when the applied shear stress (due to the buoyant weight) exceeds this value. Intuitively, it appears that the falling ball method can be employed to measure the yield stress for a given system by continually changing the size/density of a sphere until the cessation or commencement of sphere motion is achieved. Thus, the falling ball method has been employed to extract a range of quantitative rheological information (including zero-shear and shear-dependent viscosities, fluid characteristics time for visco-elastic fluids, yield stress, etc.) as well as for qualitative quality control purposes in a spectrum of industrial settings.

11.2.2.1 Zero-Shear Viscosity

While the value of μ_s obtained from Equation 11.1 is the true viscosity of a Newtonian fluid, the significance and meaning of this quantity (μ_s) for a non-Newtonian fluid is far from clear, the viscosity (due to its shear-dependence) of the liquid is not uniform around the sphere surface. Strictly speaking, the measurements carried out in the so-called second-order region only would yield the true value of the zero-shear viscosity (Bird et al., 1987a). For a given liquid, two or more spheres of different size and/or density can be used to establish whether the measurements relate to the constant viscosity region or not. Admittedly, it is possible, in principle at least, to achieve arbitrarily small values of shear rate by varying the size and density of the sphere for a given liquid. However, owing to the practical considerations, such as too small a size and impracticable values of sphere density, it is not always possible to perform experiments in the constant viscosity region. In view of these difficulties, it is not at all surprising to find an abundance of extrapolation procedures for the evaluation of zero-shear viscosity using falling sphere data. In essence, all such methods involve extrapolating the value of viscosity, calculated using Equation 11.1, to zero-shear rate or zero-shear stress conditions. Some of the procedures are based on sound theoretical considerations while others are simply empirical in nature. Extensive studies contrasting the performance and reliability of the different methods proposed in the literature are available in those literature (Subbaraman et al., 1971; Chhabra and Uhlherr, 1979; Gottlieb, 1979). In an extensive study, Shaw (2021) has contrasted the performance of scores of non-Newtonian viscosity models in extracting reliable values of the zero-shear viscosity. Here, only a selection of the widely used extrapolation procedures is included, and the resulting values of zero-shear viscosity, μ_0, are compared with the directly measured values for six polymer solutions. The corresponding falling sphere data obtained in these polymer solutions are free from the end, wall and inertial effects.

11.2.2.1.1 Theoretical Extrapolation Methods

As discussed in Chapters 2 and 4, numerous theoretical studies of the slow (in the limit of zero Reynolds number) non-Newtonian flow around a sphere are available in the literature. Some of these (Caswell, 1962, 1970; Caswell and Schwarz, 1962; Giesekus, 1963) provide a sort of theoretical framework for analyzing the falling sphere data with a view to evaluate the zero-shear viscosity. However, the theory only identifies as to what variables should be used for extrapolation to zero-shear rate or stress conditions and does not provide any theoretical basis for the extrapolation per se.

11.2.2.1.1.1 Caswell Method (1962) Based on a perturbation analysis for the creeping motion of a third-order Rivlin–Ericksen fluid past a sphere, Caswell (1962) presented the following expression:

$$\frac{\mu_s}{W} = \mu_0 + \frac{4\lambda_1}{W}\left(\frac{V}{d}\right)^2 \tag{11.3}$$

In this equation, W (dimensionless) is the combined correction factor for the wall and end effects and λ_1 is a combination of Rivlin–Ericksen fluid model parameters. Evidently, this method involves plotting μ_s against $(V/d)^2$ and the zero-shear viscosity μ_0 is obtained as the intercept on the y-axis. However, experimental data often deviate from the linear dependence as suggested by Equation 11.3. Figure 11.1 shows data for three polymer solutions plotted in accordance with Equation 11.3; similar curvatures in such plots have been observed by others (Subbaraman et al., 1971; Gottlieb, 1979). Obviously, the

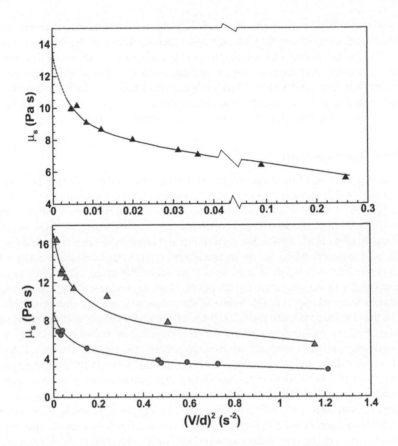

FIGURE 11.1 Determination of zero-shear viscosity using the method due to Caswell (1962).

extrapolation to $(V/d)^2 = 0$ is not very meaningful in such instances. However, the data in the region $(V/d)^2 < 0.1 \text{ s}^{-2}$ conform to the linear variation and the resulting values of μ_0 are summarized in Table 11.1.

11.2.2.1.1.2 Giesekus Method (1963) In a fundamental study, Giesekus theoretically investigated the simultaneous translation and rotation of a sphere in a four-constant Oldroyd model fluid in the absence of inertial effects. For the case of pure translation motion, his expression, in its corrected form (Bird et al., 1987a), can be rearranged as

$$\mu_s = \mu_0 - \left\{\mu_0 \lambda_1^2 \phi(\lambda_1, \lambda_2)\right\}\left(\frac{V}{d}\right)^2 + \cdots \tag{11.4}$$

where λ_1, λ_2, etc. are the Oldroyd model parameters and this analysis is applicable only for $(2\lambda_1 V/d) \ll 1$. From the extrapolation viewpoint, this method involves plotting μ_s against $(V/d)^2$ (if the higher order terms are neglected) whence it is equivalent to the Caswell's method in this regard.

11.2.2.1.1.3 Modified Caswell-Schwarz Method (1962) Based on the analysis of the creeping flow of a Rivlin–Ericksen fluid over a sphere, Caswell and Schwarz (1962) obtained the following expression for viscosity μ_s:

$$\mu_s = \mu_0 + (3/16)\rho \, dV + C\left(\frac{V}{d}\right)^2 + \cdots \tag{11.5}$$

Falling Object Rheometry

TABLE 11.1

Comparison of True Zero-Shear Viscosities with the Values Obtained by Extrapolation Procedures (From Chhabra and Uhlherr, 1979)

Values of μ_o (Pa s)	Test Fluid Number					
	1	2	3	4	5	6
Actual value (Chhabra, 1980)[a]	3.30	8.20	10.50	19.90	28.80	59.00
Caswell method (1962)	3.20	9.10	14.00	17.30	33.00	64.00
Williams method (1965b)	3.46	11.00	12.75	25.00	44.00	90.00
Sutterby method (1966)	3.39	9.40	10.26	21.50	34.00	71.50
Caswell method (1970)	3.18	8.40	10.40	21.40	31.10	70.30
Cygan and Caswell Method (1971)	3.39	13.20	13.70	62.50	66.70	105.00
$(1/\mu_s)$ vs (V/d) Method	3.23	9.40	12.50	23.80	37.00	83.30
Symbols used in Figures 11.1–11.7	□	○	▲	△	■	●

[a] Measured using Weissenberg Rheogoniometer; (1) 1.46% Methocel, (2) 0.85% Separan, (3) 0.75% Separan/0.75% CMC, (4) 1.25% Separan, (5) 1.63% Separan, (6) 2% Separan (All in water).

where C (dimensional) is a model parameter.

Turian (1967) simply rearranged Equation 11.5 and used $\left(\mu_s - \frac{3}{16}\rho V d\right)$ versus $(V/d)^2$ plots for extrapolating the falling sphere data to the zero-shear rate condition. Though some minor numerical errors in Equation 11.5 have been pointed out, the functional form is correct, however (Verma and Sacheti, 1975).

In most cases, the term $\frac{3}{16}\rho V d$ is negligible in comparison with μ_s and thus this method is identical to method (1) above. For instance, the largest value of $\frac{3}{16}\rho V d$ is only about 0.02% of μ_s for the six polymer solutions examined herein and by Chhabra and Uhlherr (1979).

11.2.2.1.1.4 Caswell's Second Method (1970) In a subsequent paper, Caswell (1970) elucidated the extent of retardation exerted by cylindrical boundaries on the creeping sphere motion in Rivlin–Ericksen model fluids. His final result can be rearranged as

$$\frac{1}{\mu_s} = \frac{1}{\mu_0} - C_1\left(\frac{\mu_s V}{d}\right)^2 + O\left(\frac{\mu_s V}{d}\right)^4 \qquad (11.6)$$

where C_1 is a constant involving the fluid model parameters. The data shown in Figure 11.2 confirms the validity of Equation 11.6 provided the higher order terms are neglected. Similar linear dependence has been reported by other investigators also (Subbaraman et al., 1971; Broadbent and Mena, 1974). The resulting values of the zero-shear viscosity (corresponding to $(\mu_s V/d)^2 = 0$) are also listed in Table 11.1.

11.2.2.1.2 Empirical Extrapolation Methods

Most of the procedures in this category entail the extrapolation of either the average Newtonian shear stress ($\sim d\Delta\rho g$) or shear rate ($\sim V/d$) to zero. A variety of combinations and/or coordinates have been employed for extrapolating the falling sphere data to evaluate zero shear viscosity.

11.2.2.1.2.1 Williams Method (1965b) This method involves plotting log μ_s^N versus τ_N, the maximum shear stress for a Newtonian fluid, i.e.,

$$\tau_N = \frac{d(\rho_s - \rho)g}{6} \text{ and } \mu_s^N = \mu_s f_B f_w f_I \qquad (11.7)$$

where f_B, f_w, and f_I are the correction factors for the end, wall, and inertial effects, respectively, all of which are, however, redundant in the context of the present results for the six polymer solutions and

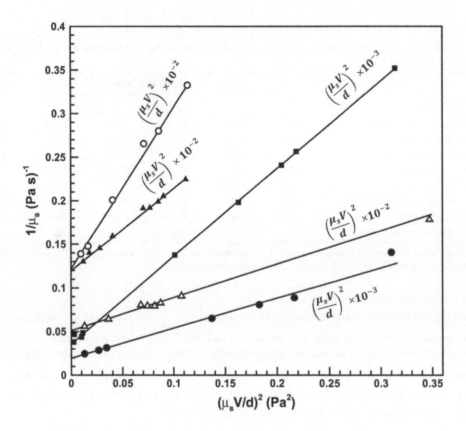

FIGURE 11.2 Evaluation of zero-shear viscosity using the second method due to Caswell (1970).

therefore $\mu_s^N = \mu_s$. Figure 11.3 shows representative plots of μ_s versus τ_N on semi-logarithmic coordinates. Similar pronounced curvatures in such plots have also been reported by Subbaraman et al. (1971). However, the results for $\tau_N < 35$ Pa conform to Equation 11.7 and the resulting values of the zero-shear viscosity are included in Table 11.1.

11.2.2.1.2.2 Sutterby Method (1966) There is no reason why shear stress should be used as the independent variable rather than the shear rate. Sutterby (1966) suggested plotting μ_s versus $(3V/d)$ and the intercept on ordinate yields the value of the zero-shear viscosity. This method has also been used by numerous other workers (Torrest, 1982, 1983; Gahleitner and Sobczak, 1987, 1988). Though there is no question that the average shear rate around a sphere may be expressed as $(\alpha V/d)$, considerable uncertainty exists regarding the exact value of α. Strictly speaking, α is a function of the rheological parameters itself, for example, the power-law flow behavior index (Hirota and Takada, 1959; Uhlherr et al., 1976; Cho et al., 1984; Butcher and Irvine, Jr., 1990; Sato et al., 1966; Suri et al. 2023a). Based on the limited experiments in carboxymethyl cellulose and hydroxyethyl cellulose solutions, Sato et al. (1966) proposed $\alpha = 0.6$ whereas some data (Gahleitner and Sobczak, 1987) obtained for polypropylenes seems to suggest $\alpha = 1$. The latter value is also in line with the theoretical estimates available in the literature (El Kayloubi et al., 1987). However, such an uncertainty in α will only shift the x-axis without affecting the intercept on the ordinate. In this method, then, the zero-shear viscosity is evaluated from $\log \mu_s$ versus (V/d) plots. Figure 11.4 shows the representative data for five polymer solutions plotted in this manner, appreciable curvatures are seen to be present. The linear variation of $\log \mu_s$ with (V/d) is obtained only in the range $(V/d) < 0.2\,\text{s}^{-1}$ and the resulting values of the zero-shear viscosity are included in Table 11.1.

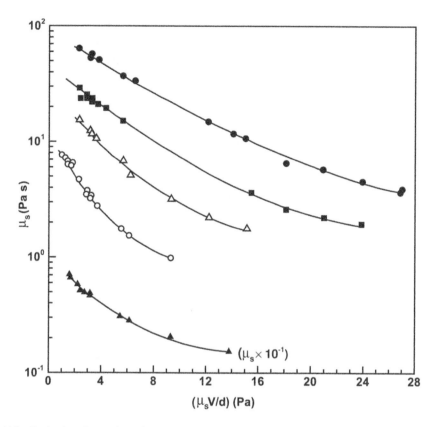

FIGURE 11.3 Evaluation of zero-shear viscosity by the Williams method (1965b).

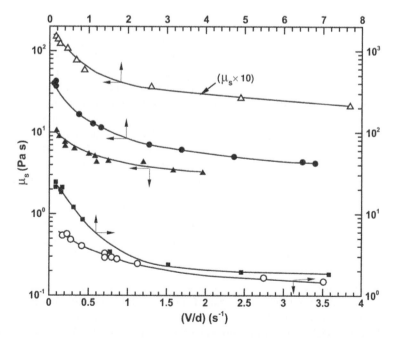

FIGURE 11.4 Evaluation of zero-shear viscosity using the extrapolation method due to Sutterby (1966).

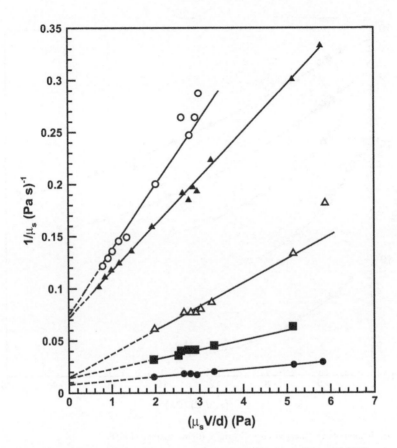

FIGURE 11.5 Determination of zero-shear viscosity using the method of Cygan and Caswell (1971).

11.2.2.1.2.3 Cygan and Caswell Method (1971) These investigators carried out a systematic experimental study with a view to verify the validity of Equation 11.6 by gleaning falling sphere data for polyisobutylene (PIB) in toluene solutions. Their data, however, did not conform to Equation 11.6. Instead, data resulted in straight lines when the fluidity ($1/\mu_s$) is plotted against ($\mu_s V/d$) rather than ($\mu_s V/d)^2$, as predicted by the theory (Caswell, 1970). Figure 11.5 supports this assertion and the corresponding values of the zero-shear viscosity are listed in Table 11.1. In this case also, one can conceivably replace the apparent shear stress ($\mu_s V/d$) by the apparent shear rate (V/d). This is demonstrated in Figure 11.6 and the values of the zero-shear viscosity so obtained are summarized in Table 11.1. Another method (Turian, 1967) uses extrapolating $\log \mu_s$ vs. ($\mu_s V/d$) plots, but this method is equivalent to that of Williams (1965b).

An inspection of Table 11.1 clearly reveals that most extrapolation procedures generally overestimate the value of the zero-shear viscosity by varying amounts; the predictions of the Caswell method (1970) being the closest to the actual values. However, it is difficult to delineate a priori the most appropriate range of conditions over which falling sphere data must be gleaned to obtain reliable values of the zero-shear viscosity, as this will vary from one polymer solution to another. Notwithstanding these limitations, this configuration does offer a simple device to estimate the value of the zero-shear viscosity for analytical as well as comparative purposes.

The falling ball method has been used to study the pressure dependence of the zero-shear viscosity of polypropylene melts (Foltz et al., 1978; Linliu et al., 1994; Mattischek and Sobczak, 1997), evaluation of the leveling characteristics of latex paints (Quach and Hansen, 1974) and for measuring the viscosity of blood (Doffin et al., 1984) and other biological systems (Munzel and Schaub, 1961). In a series of papers, Sobczak and coworkers (Sobczak, 1986; Hermann and Sobczak, 1989) and others (Gauthier-Manuel et al., 1984; Hilfiker

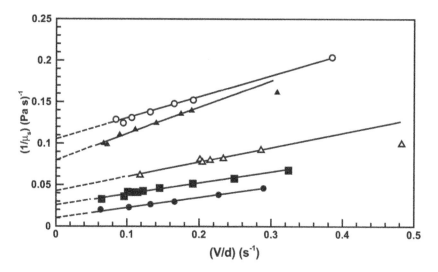

FIGURE 11.6 Evaluation of zero-shear viscosity by extrapolating $(1/\mu_s)$ versus (V/d) data.

et al., 1989) have presented detailed descriptions of the various detection methods used for timing the sphere descent as well as new designs based on the sphere motion in magnetic fields to facilitate the measurement of unusually high viscosities and/or for opaque liquids. On the other hand, Maclean-Fletcher and Pollard (1980) have used the falling ball method to monitor the kinetics of the gelation of *Acanthamoeba*.

11.2.2.2 Shear-Dependent Viscosity

Few investigators (Hirota and Takada, 1959; Uhlherr et al., 1976; Cho et al., 1984; Butcher and Irvine, 1990) have attempted to extend the range of utility of the falling sphere configuration to determine shear stress–shear rate data for purely viscous fluids. Owing to the nonviscometric nature of the flow around a sphere, both shear stress and shear rate vary from point to point on the surface of the sphere and therefore all such attempts rely on the use of their surface-averaged values. The first attempt at obtaining the shear stress–shear rate data for pseudoplastic polymer solutions using the falling ball method is due to Hirota and Takada (1959) who presented the following definitions of the shear stress and shear rate, respectively.

$$\tau_m = \frac{1}{6}d(\rho_s - \rho)g \tag{11.8}$$

$$\dot{\gamma}_{av} = \left(\frac{3V}{d}\right)\left[1 + 2.5(\delta - 2) - 0.75(\delta - 2)^2\right] \tag{11.9}$$

where $\delta = \dfrac{d\log V}{d\log R}$

Note that Equation 11.8 gives the maximum value of the shear stress prevailing on the surface of a sphere falling in Newtonian media in the creeping regime. Without giving sufficient details for their results to be recalculated, Hirota and Takada (1959) concluded that the predictions of Equations 11.8 and 11.9 agreed very well with the corresponding capillary viscometer data for six polymer solutions. Subsequently, both Uhlherr et al. (1976) and Cho et al. (1984) have outlined the schemes for the evaluation of shear stress and shear rate data for power-law model fluids using falling sphere data. The method of Uhlherr et al. (1976) assumes that the surface average shear stress for a sphere is the same as that in a Newtonian fluid, that is,

$$\tau_{av} = \frac{1}{9} g d (\rho_s - \rho) \tag{11.10}$$

and the corresponding average shear rate is obtained, using the approximate flow field used by Slattery (1962) to evaluate the drag of a sphere, as

$$\dot{\gamma}_{av} = \frac{2V}{d}(1+4A) \tag{11.11}$$

where the values of A, a function of the flow behavior index n, are available in their original paper (Slattery, 1962). Satisfactory agreement was reported between the viscometric and falling ball results for two weakly pseudoplastic ($n=0.825$ and $n=0.865$) polymer solutions. However, not only is this method restricted to a rather narrow range of the power-law index ($1 \geq n \geq 0.763$) due to the limitation of approximate analysis of Slattery (1962) but also found to be unsuitable for visco-elastic fluids. Cho et al. (1984), on the other hand, proposed corrections to both the average shear stress and shear rate relevant for a sphere in Newtonian media. Based on the assumed velocity and stress profiles (Wasserman and Slattery, 1964; Cho and Hartnett, 1983a,b), the following semi-empirical expressions for the upper- and lower bounds for shear stress and shear rate were derived as

$$\tau_{av} = F_1(n) \frac{g d (\rho_s - \rho)}{9} \tag{11.12}$$

where $F_1(n)_{UB} = 0.2827 + 0.8744n + 0.4562n^2 - 0.7486n^3$ \hfill (11.13a)

and $F_1(n)_{LB} = 0.6388 + 0.6418n - 0.4344n^2 + 0.1560n^3$ \hfill (11.13b)

$$\dot{\gamma}_{av} = F_2(n) \frac{2V}{d} \tag{11.14}$$

where $F_2(n)_{UB} = -1.731 + 41.28n - 116n^2 + 123.9n^3 - 46.72n^4$ \hfill (11.15a)

and $F_2(n)_{LB} = -2.482 + 54.35n - 160.1n^2 + 178.2n^3 - 69.04n^4$ \hfill (11.15b)

$$\text{and} \quad \mu = \frac{\tau_{av}}{\dot{\gamma}_{av}} \tag{11.16}$$

While in actual calculations, one should use the mean values of $F_1(n)$ and $F_2(n)$. These expressions are retained here in terms of the upper- and lower bounds to give an idea about the variability of the predictions of this method. Equations 11.12–11.16 are applicable in the complete range of power-law index, that is, $1 \geq n \geq 0.1$. However, their use necessitates a prior knowledge of the value of the power-law index, n. Cho et al. (1984) suggested that it can be evaluated from the slope of a log–log plot as $d \log\{R(\rho_s - \rho)\} / d \log(V/R)$ for a true power-law fluid. This approach has recently been improved significantly (Suri et al., 2023a).

It is well known that the power-law model does not predict the transition from the zero-shear viscosity region to power-law region. In a subsequent numerical study (Butcher and Irvine, 1990), this difficulty

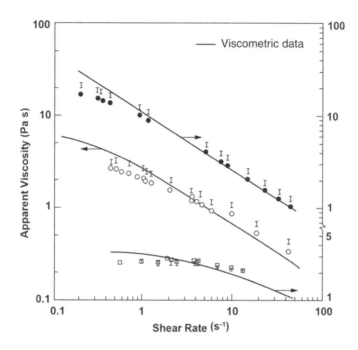

FIGURE 11.7 Typical comparison between the predicted and measured apparent viscosity in the power-law region. ○, □, ●- Predictions of Butcher and Irvine, Jr. (1990); I - Predictions of Equations 11.12–11.16.

was obviated by employing a three parameter truncated form of the power-law model. Again, the average shear stress and shear rate were obtained in the form of corrections to the corresponding Newtonian values. The correction factors, which depend only upon the value of n, are available both in tabular and graphical forms in the original paper (Butcher and Irvine, 1990). Figure 11.7 shows a typical comparison between the predictions (Cho et al., 1984; Butcher and Irvine, 1990) and viscometric measurements for three polymer solutions. An examination of this figure shows that the method of Cho et al. (1984) always overpredicts the values of viscosity, and the deviations between the falling sphere results and viscometric measurements increase, as the shear rate decreases. This is not at all surprising as the predictions assume a constant value of the power-law index, n, to be applicable over the whole range, whereas strictly speaking it varies somewhat with shear rate. Despite these limitations, the predictions rarely seem to differ from the viscometric data by more than 50% and hence this method does offer an attractive alternative for measuring shear-dependent viscosities, at least for qualitative comparison purposes. This approach also seems to work moderately well for weakly visco-elastic polymer solutions (Kanchanalakshana and Ghajar, 1986), however this coincidence is fortuitous. It needs to be emphasized here that all such efforts (Uhlherr et al., 1976; Cho et al., 1984; Butcher and Irvine, 1990, etc.) implicitly assume that the viscous drag constitutes 2/3rd of the total drag or estimate its value using the upper- and lower bound calculations; the latter is based on the overall energy dissipation and it is thus not justified to use only one component of the energy functional in the context of variational principles. This is true only for Newtonian fluids and this ratio itself is indeed a function of the power-law index (Dhole et al., 2006a, Suri et al., 2023a).

11.2.2.3 Yield Stress

As noted in Chapter 3, by virtue of its yield stress, a visco-plastic substance in an unsheared state has the capacity to support the weight of an embedded particle for a sufficiently long time. This idea has been used to evaluate the yield stress of visco-plastic media. This method hinges on the fact that for a given medium, it is possible to find the heaviest ball which will not move under its own weight. Obviously, the criterion to delineate whether a ball is falling or not implicitly assumes a timescale of observation (Barnes and Walters, 1985). Notwithstanding this arbitrariness and subjectivity, since the sphere is not moving, the only relevant forces are due to its weight and due to the yield stress of the medium. Many investigators (Maclean-Fletcher and Pollard, 1980; Hartnett and Hu, 1989; Schurz, 1990; Wunsch, 1990) have used this configuration to evaluate the value of fluid yield stress. While Wunsch (1990) made use of the numerical results of Beris et al. (1985) to estimate the value of τ_o, all other methods are based on heuristic considerations.

However, much confusion exists regarding the calculation of the force due to the yield stress, F_o, acting on a sphere. For instance, Maclean-Fletcher and Pollard (1980) and many others (such as Johnson, 1970) have approximated it as

$$F_o = C_o \tau_o \left(\frac{\pi d^2}{4} \right) = \frac{\pi d^3}{6} (\rho_s - \rho) g \tag{11.17}$$

where C_o is a constant, the values ranging from 1.75 to 3.14 (Maclean-Fletcher and Pollard, 1980; Chhabra and Uhlherr, 1988a) have been used in the literature. This uncertainty coupled with the practical difficulties of finding a ball which will just remain suspended has led to the adaptations of the falling ball method. For instance, Wunsch (1992, 1994) developed a method based on an oscillating sphere and outlined a scheme which allows the evaluation of the yield stress. Unlike Equation 11.17, the method of Wunsch (1994) is obviously based on the choice of a specific fluid model. However, Wunsch (1994) reported good agreement between the predicted and measured values for a few Carbopol solutions. Similarly, Singh et al. (1991) have outlined a method for evaluating the fluid parameters of a Herschel–Bulkley fluid from the behavior of a sphere moving in visco-plastic media subject to vibrations.

Uhlherr and coworkers (Uhlherr, 1986; Guo and Uhlherr, 1996) have developed the so-called pendulum methods which can be used to measure the yield stress simply from the knowledge of the equilibrium position of the pendulum (see Figure 11.8). This method does not require a priori knowledge of the fluid model. While initially Uhlherr (1986) introduced the sphere–pendulum method, subsequently he argued that the deformation is more uniform around a cylinder with hemispherical ends than that around a sphere (Guo and Uhlherr, 1996).

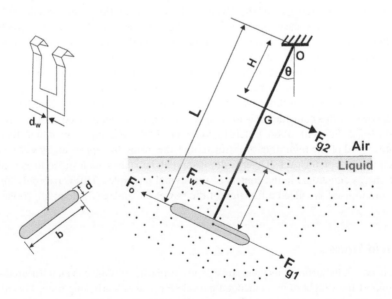

FIGURE 11.8 Schematics of a cylinder pendulum. (Redrawn from Guo, J. and Uhlherr, P.H.T., *Proceedings of the XII International Congress on Rheology*, p. 73, Quebec City, PQ, Canada, 1996.)

By writing a moment balance about the hinge (o) at equilibrium, it can easily be shown that for a given system, the equilibrium angle (θ) depends upon the depth of immersion, l (shown as X in Figure 11.8), as

$$\sin\theta = -A_1 l^2 + B_1 l + C_1 \tag{11.18}$$

From a knowledge of the intercept C_1, one can readily extract the value of the yield stress τ_o as

$$\tau_o = \frac{\left[m_c + m_p \dfrac{H}{L} - \left(\dfrac{\pi}{4}d^2 b + \dfrac{\pi d^3}{6}\right)\rho\right] g \sin\theta_{\min}}{\pi db + \left(\dfrac{\pi d}{2}\right)^2} \tag{11.19}$$

where m_c and m_p are the masses of the cylinder and the plate-hook assembly, respectively; $\theta_{\min}$ is the equilibrium angle from the vertical direction. Guo and Uhlherr (1996) reported a good match between the values of the yield stress τ_o obtained using the pendulum and the well-established vane technique for scores of polymer solutions and silica suspensions. Likewise, a plate pendulum was successfully used for measuring the yield stress of fire-fighting foams (Gardiner et al., 1998). De kee (2021) has recently reviewed the pertinent literature on the measurement of the yield stress using different methods including an inter-laboratory study (Nguyen et al., 2006).

In summary, this method does yield values of the yield stress (static) comparable to the vane technique, provided due care is taken to eliminate the wall effects (large container), the end effects (by using several values of l), and the slip effects. Notwithstanding the theoretical pitfalls, this method offers a simple and convenient tool for comparative- and quality control purposes when yield stress is an important characteristic of the fluid.

11.2.2.4 Characteristic Time for Visco-elastic Fluids

Some anomalous effects associated with the motion of spheres in concentrated polymer solutions have been observed which are attributed to the visco-elasticity of polymer solutions. For instance, Walters and Tanner (1992) presented photographic evidence of the initial oscillatory motion of a sphere released from rest into a visco-elastic fluid before attaining its terminal falling velocity. Likewise, Cho and Hartnett (1979) and Cho et al. (1984) found the terminal velocity of spheres to depend upon the time elapsed between two successive experiments. In some cases, up to 30 min were required before the original value of terminal velocity could be reproduced. The time-dependent terminal velocity measurements were explained qualitatively by hypothesizing that the elastic energy produced by the sphere motion must be gradually relaxed by the polymer solution for it to return to its undeformed virgin state. Though these authors also suggested that this information can be used to evaluate the fluid characteristic time, the available limited experimental results, however, do not seem to support this contention (Tanner, 1964; Ambeskar and Mashelkar, 1990). Similarly, Yoo et al. (1994) have attempted to evaluate the fluid characteristic and diffusion times of polymer solution using the falling sphere and needle data. Adam et al. (1984) have also outlined a method of evaluating the longest relaxation time of polymer solutions using a magnetic sphere rheometer.

11.3 Rolling Ball Method

11.3.1 Newtonian Fluids

Similar to the falling ball method, the velocity of a sphere rolling along the wall of a cylindrical tube filled with a liquid and inclined at a fixed angle to the horizontal gives a measure of the viscosity (Flowers, 1914). While the flow field produced is quite complex (especially when the ball diameter is almost equal to the tube diameter, as is common for this device), Hubbard and Brown (1943a) used dimensional arguments to show that for a given arrangement, the rolling velocity and the viscosity of a Newtonian fluid are linked as

$$\mu = \frac{k'(\rho_s - \rho)}{V_R} \quad (11.20)$$

where V_R is the rolling (without sliding) velocity of the ball. This expression has been used successfully to measure the pressure and temperature-dependent viscosity of scores of Newtonian fluids (Hubbard and Brown, 1943b; Harrison and Gosser, 1965; Geils and Keezer, 1977; Medani and Hasan, 1977; Hasan, 1983). While Equation 11.20 is really applicable when the streamline (laminar) flow conditions exist, Hasan (1983) has provided the necessary correction for inertial effects. Obviously, the instrument constant k^r includes the system specifics such as tube and ball diameter, acceleration due to gravity, the angle of inclination, etc.

Lewis (1953) improved upon the treatment of Hubbard and Brown (1943a) by considering it as one-dimensional pressure-driven flow in the constriction formed between the tube wall and the surface of the ball. He presented the following expression for viscosity:

$$\mu = \frac{D^2}{12\pi J_1} \frac{g \sin \beta}{V_R} (\rho_s - \rho) \left(\frac{D-d}{D}\right)^{2.5} \quad (11.21)$$

where J_1 is a constant which equals 0.398. In contrast to Equation 11.20, Equation 11.21 yields absolute values of shear viscosity and no calibration is needed (Bagchi and Chhabra, 1991b). Commercially available rolling ball viscometers have been discussed by van Wazer et al. (1963), Sherman (1970) and by Gupta (2014).

11.3.2 Non-Newtonian Fluids (Shear-Dependent Viscosity)

In view of the exceedingly complex nature of the flow field created by a rolling ball, little effort has been devoted to the extension of this method to measure non-Newtonian viscosity. Bird and Turian (1964) extended the approach of Lewis (1953) for power-law fluids. Subsequently, Sestak and Ambros (1973) refined the analysis of Bird and Turian (1964) and also presented limited data to demonstrate the applicability of their analysis for a few power-law fluids. Their analysis is, however, restricted to $[(D-d)/d] \ll 1$, that is, for a closely fitting ball. Without any justification, Stastna and De Kee (1987) argued that only 5/7th of the buoyant weight of the sphere balances the resistance force. According to Stastna and De Kee (1987), the power-law index n of a fluid can be evaluated using the rolling velocity as a function of the angle of inclination as

$$n = \frac{\log(\sin \beta_1 / \sin \beta_2)}{\log(V_{R_1} / V_{R_2})} \quad (11.22)$$

and the power-law consistency, m, is given by the following expression due to Sestak and Ambros (1973):

$$m = \frac{D^n g d (\rho_s - \rho) \left(\frac{D-d}{D}\right)^{\frac{4n+1}{2}} \sin \beta}{3\left[\{(2n+1)/n\} \pi V_R\right]^n J_n} \quad (11.23)$$

The values of J_n for a range of values of n are available in the paper of Sestak and Ambros (1973). In the limit of $d \approx D$ and $n = 1$, Equation 11.23 does reduce approximately to Equation 11.21. Bagchi and Chhabra (1991b) reported moderate agreement between the values of m and n obtained using this method and those from a rotational viscometer for a few polymer solutions. Finally, much confusion exists in the literature regarding the calculation of the mean shear rate in the rolling ball device. For instance, Sestak and Ambros (1973) derived the following expression for the average shear rate at the surface of the ball.

$$\dot{\gamma}_b = \frac{2DV_R}{(D-d)^2} \frac{(n+1)\,(2n+1)^2}{n(2+n)\,(3n+2)} \tag{11.24}$$

whereas the corresponding expression due to Stastna and De kee (1987) is slightly different as

$$\dot{\gamma}_b = \frac{2\pi d^2 V_R}{D(D-d)^2} \frac{(n+1)\,(2n+1)^2}{n(n+2)\,(3n+2)} \tag{11.25}$$

In the limit of $n=1$, Equation 11.24 yields for a Newtonian fluid, $\dot{\gamma}_{bn} = (12/5) DV_R/(D-d)^2$ whereas Equation 11.24 leads to a different result as it does not impose any restriction on the value of $[(D-d)/d]$. Bryan and Silman (1990), on the other hand, simply postulated the shear rate to be proportional to $\sim (V_R/d)$ and reported good agreement between the rolling ball viscometer and Brookfield viscometer data for a few fermentation broths. Similarly, following the early work of Scott Blair and Oosthuizen (1960), Briscoe et al. (1992b, 1994) have postulated the mean shear rate to be given by

$$\dot{\gamma}_b = \frac{4DV_R}{(D^2 - d^2)} \tag{11.26}$$

and they also reported good agreement between the rolling ball data and the viscometric data for a few bentonite suspensions and polyox polymer solutions. In essence, the approaches of Bryan and Silman (1990) and Briscoe et al. (1992b, 1994) hinge on identifying appropriate shift factors for shear stress–shear rate data from the rolling ball device to superimpose on to the corresponding data from a viscometer. Undoubtedly, this simple approach is probably quite adequate for comparative purposes, but is difficult to justify on theoretical grounds for the shear rate itself is expected to be a function of the power-law index, n.

11.3.3 Yield Stress

By analogy with the falling ball and the pendulum techniques presented in the preceding sections, some attempts have been made for using the rolling ball method to evaluate the yield stress of visco-plastic substances. Gruber et al. (1973) and Schurz (1990) outlined the design and operation of a rolling ball device in which a sphere rolled down a curved groove (a circular arc, see Figure 11.9). The underlying principle is that if the material under study is indeed visco-plastic, the rolling ball should come to an equilibrium position when the buoyant weight equals the force due to the yield stress acting on the ball, akin to the sphere–pendulum method.

Therefore, the yield stress can be shown to be proportional to sin θ, θ being the angle of equilibrium from the vertical direction. Some data was presented supporting the validity of this method. Similarly, Briscoe et al. (1992b, 1994) have also outlined an empirical scheme to extract the values of the Bingham plastic model parameters by extrapolating the rolling ball data.

11.4 Rotating Sphere Viscometer

Some attempts have also been made to use a rotating sphere to evaluate the zero-shear viscosity and characteristic relaxation time for visco-elastic fluids (Walters and Waters, 1963; Bourne, 1965; Walters and Savins, 1965; Hermes, 1966; Mashelkar et al., 1972; Kelkar et al., 1973; Acharya and Maaskant, 1978). In the absence of secondary flows and inertial effects, all one needs to measure is the torque as a function of the speed of rotation to evaluate viscous and elastic characteristics, albeit these results are not only model dependent but are also restricted to the vanishingly small values of the Reynolds and

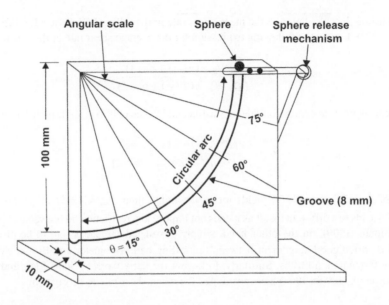

FIGURE 11.9 Rolling ball device of Schurz (1990).

Deborah (Weissenberg) numbers. For a power-law fluid, Bourne (1965) solved the governing equations for a sphere of diameter d rotating in a power-law fluid filled in a spherical vessel of diameter (D). In the absence of wall effects, that is, $D \gg d$, the shear rate at the surface of the sphere is given by

$$\dot{\gamma} = \frac{3\Omega \sin\theta}{n} \tag{11.27}$$

which is seen to vary from 0 at the poles ($\theta=0$, π) to a maximum at the equator ($\theta=\pi/2$). The value of power-law index n is obtained as the slope of torque vs. speed plots (for the same sphere) on log–log coordinates. Acharya and Maaskant (1978) and Cairncross and Hansford (1978) have presented some experimental results on the feasibility of this method for visco-elastic fluids. Subsequently, Schatzmann et al. (2003) have developed a ball measuring system to measure the flow curves for debris and muds which yielded shear stress–stress rate data comparable to those obtained from a rotational viscometer.

Another variant of the ball devices is the so-called oscillating sphere configuration which has also received some attention in the literature (Sellers et al., 1987; Tran-Son-Tay et al., 1988, 1990). There is no doubt that in principle all these devices can yield reliable values of viscosity for Newtonian fluids, especially at high pressures and temperatures, but the interpretation of results obtained for non-Newtonian fluids is far from straightforward and must therefore be treated with caution, at least beyond the zero-shear limit. Also, the simple design and operation of the falling ball and the rolling ball configurations makes them much more attractive than the rotating and oscillating ball methods.

11.5 Falling Cylinder Viscometer

11.5.1 Newtonian Fluids

Considerable attention has been devoted to the development of suitable analytical frameworks to evaluate the viscosity of Newtonian fluids from a falling cylinder configuration. While detailed descriptions of various designs of this kind of viscometer are available in the literature (Sherman, 1970; Irving and Barlow, 1971; Mclachlan, 1976; Claesson et al., 1983; Park and Irvine, 1984, 1988; Uchida et al., 1984; Cristescu et al., 2002), Figure 11.10 shows the basics of this device. The earliest attempts to develop an

Falling Object Rheometry

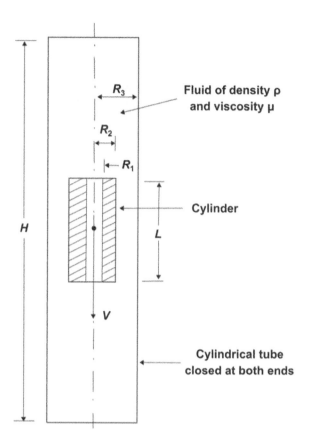

FIGURE 11.10 Schematics of a falling cylinder viscometer.

analytical expression relating the viscosity to the other pertinent variables are due to Smith (1957) and Lohrenz et al. (1960). For a circular cylinder of radius R_2 falling vertically in a fluid at the axis of a tube of radius R_3, the viscosity is given by

$$\mu = \left(\frac{g\Delta\rho}{V}\right) \frac{R_2^2 \left\{\left(1+k_2^2\right)\ln k_2 - \left(k_2^2 - 1\right)\right\}}{2\left(1+k_2^2\right)} \tag{11.28}$$

where $k_2 = (R_3/R_2)$.

The assumptions underlying the development of Equation 11.28 include no end effects, no eccentricity, long cylinder (large length-to-diameter ratio) and small Reynolds numbers. Obviously, for given values of R_2 and R_3, the viscosity is given by

$$\mu = \frac{k_s(\rho_s - \rho)}{V} \tag{11.29}$$

where k_s is a system constant.

In practice, however, end effects are always present and as is the degree of eccentricity, howsoever small. Therefore, the use of Equation 11.29 requires calibration of the falling cylinder viscometer. Also, the constant k_s not only represents the system geometry but also indirectly accounts for the end effects, and for the finite size of the cylinder. It is tacitly assumed that the value of k_s is constant in the narrow range of conditions of interest.

In order to minimize the possibility of eccentricity, some novel designs of cylinders (or slugs) including one with hemispherical ends (Jackson and Bedborough, 1978; Chan and Jackson, 1985; Chu and Hilfiker, 1989), a cylinder with stabilizers on both ends (Lohrenz and Kurata, 1962), conical bottom similar to rotameter floats (Dandridge and Jackson, 1981), cylinders with central holes or hollow cylinders (Irving and Barlow, 1971; Mclachlan, 1976), with cone at one end and a propeller at the other end (Claesson et al., 1983), rounded corners (Lescarboura and Swift, 1968), etc. have been developed and used to examine the temperature and pressure (up to ~350 MPa) dependence of the Newtonian viscosities up to about 10^6 Pa s. While Equation 11.29 does apply to all such variants of the falling cylinder (also called sinker or slug) with an appropriate value of k_s determined via calibration, rigorous analytical treatments are clearly not possible. However, for the case of a hollow cylinder of inner radius R_1, Irving and Barlow (1971) presented the following expression for k_s:

$$k_s = \frac{gR_2^2\left(1-k_1^2\right)\left\{\left(k_2^4-1+k_1^4\right)\ln k_2 - \left(k_2^2-1\right)^2\right\}}{2\left(k_2^4-1+k_1^4\right)} \qquad (11.30)$$

Clearly, in the limit of $k_1\left(=R_1/R_2\right) \to 0$, Equation 11.30 reduces to the limiting behavior for a solid cylinder as given by Equation 11.28.

The applicability of the aforementioned expressions for Newtonian fluids has been amply demonstrated among others by McDuffie and Barr (1969), Irving and Barlow (1971), Claesson et al. (1983), Chan and Jackson (1985), Dandridge and Jackson (1981), and Mclachlan (1976). The extent of end effect has been studied by Cristescu et al. (2002).

While Equations 11.28 and 11.30 are valid over the entire range of k_2, Park and Irvine, Jr. (1984, 1988) have advocated the use of the so-called falling needle viscometer, i.e., $k_2 \to 0$. In this case, the system constant k_s (for $k_2 < 0.033$) is given by (Park and Irvine, Jr., 1984):

$$k_s = \left(\frac{gR_2^2}{2}\right)\left\{-(1+\ln k_2)\right\} \qquad (11.31)$$

The experimental verification of Equation 11.31 has been provided by Park and Irvine, Jr. (1984) who measured the viscosity of aqueous glycerol solutions using such a falling needle viscometer. Subsequently, a theoretical justification has been provided by Davis and Brenner (2001) which also eliminates the necessity of experimental determination of the system constant k_s for the falling needle viscometers.

Some effort has also been directed at ascertaining the role of eccentricity (Heyda, 1959; Chen et al., 1968; Lescarboura and Swift, 1968; Liu et al., 2004), end effects (near the top and bottom of the falling cylinder) (Chen and Swift, 1972; Park and Irvine, 1988; Wehbeh et al., 1993) and the effects arising from the bottom end of the outer cylinder (Wehbeh et al., 1993; Cristescu et al., 2002). For a needle with hemispherical ends, Park and Irvine (1988) presented the following end correction factor (ECF) for the settling velocity to be used in Equation 11.28:

$$\text{ECF} = \frac{V_m}{V} = 1 + \frac{2}{3L^+}\left[1 + \frac{3}{2f_w L^+}\left\{\frac{k_2^2(1-\ln k_2)-(1+\ln k_2)}{1+k_2^2}\right\}\right]^{-1} \qquad (11.32)$$

where the wall correction factor, f_w, arises from the hemispherical ends and is the same as that given by Faxen (1923) and discussed in Chapter 10. It is a function of k_2 (<<0.1) as

$$f_w = 1 - 2.104k_2 + 2.09k_2^3 + \cdots \qquad (11.33)$$

The dimensionless length

$$L^+ = \left(\frac{L}{2R_2}-1\right)$$

Subsequently, Kim et al. (1994) reported a flow visualization study and found the flow disturbance due to the end effects to persist only up to about 6% of the needle length. This finding is also consistent with the theoretical analysis of Chen and Swift (1972) and experiments of Wehbeh et al. (1993). Thus, by an appropriate choice of (L/R_2), one can minimize the end effects. Chen et al. (1968) have also carried out a theoretical analysis to elucidate the role of eccentricity on the fall velocity and hence on the measured viscosity. They presented their results in the form of an eccentricity ratio, ER, defined as

$$\mathrm{ER} = \frac{\varepsilon}{R_3(1-k_2)} \tag{11.34}$$

where ε is the distance between the axes of the falling cylinder and the tube. While they have tabulated extensive results as functions of eccentricity ratio ($0.05 \leq \mathrm{ER} \leq 0.95$) and $0.80 \leq k_2 \leq 0.99$, the influence of eccentricity is negligible up to about ER ~ 0.20. In fact, the effect becomes increasingly significant with the increasing value of k_2. Conversely, the role of eccentricity is expected to be minimal in a falling needle viscometer. These predictions are supported by the limited experimental results of Lescarboura and Swift (1968). The role of the end effects due to the finite length (closed end) of the tube has been assessed by Wehbeh et al. (1993) and Cristescu et al. (2002).

11.5.2 Non-Newtonian Fluids

11.5.2.1 Shear-Dependent Viscosity

Some analytical attempts have also been made to explore the use of a falling cylinder viscometer to obtain shear stress–shear rate curves for power-law fluids (Ashare et al., 1965; Eichstadt and Swift, 1966; Park and Irvine, 1988; Park et al., 1990; Phan-Thien et al., 1993; Zheng et al., 1994; Yamamoto and Shibata, 1999), for Bingham plastic fluids (Eichstadt and Swift, 1966) and for visco-elastic fluids (Phan-Thien et al., 1993; Zheng et al., 1994; Tigoiu, 2004). The early studies of Ashare et al. (1965) and Eichstadt and Swift (1966) are straightforward extensions of the analysis of Lohrenz et al. (1960). For large values of k_2, the narrow annular gap between the falling cylinder and the tube wall is treated as a planar slit and then the Newtonian corrections for the curvature and the end effects are used for power-law and Bingham plastic fluids. In spite of these developments, the interpretation of falling cylinder data for non-Newtonian fluids is far from straightforward for large values of k_2. On the other hand, the falling needle configuration has been found to be promising. Park and Irvine (1988) and Park et al. (1990) have outlined a theoretical procedure that allows the determination of the shear stress–shear rate curves for power-law fluids. Subsequently, some weaknesses of this analysis have been pointed out by Phan-Thien et al. (1993) and Zheng et al. (1994). The latter authors have also studied this configuration numerically in detail and their results are presented here. The average shear rate at the surface of the needle is given by

$$\dot{\gamma}_{\mathrm{av}} = \alpha \frac{V}{R_2} \tag{11.35}$$

where α is a function of both n and k_2 as follows:

$$\log \alpha = \alpha_o - \alpha_1 \log n \tag{11.36}$$

$$\alpha_o = 0.4579 + 3.2305(k_2 - 0.1)^2 \tag{11.37a}$$

$$\alpha_1 = 0.07044(k_2)^{-0.585} \tag{11.37b}$$

Equation 11.37 correlates their numerical results with an average error of less than 2% over wide ranges of n (0.1–0.9) and k_2 (0.05–0.3).

The corresponding shear stress is estimated from a knowledge of the drag force on a needle falling at the same velocity in a power-law fluid and in an equivalent Newtonian fluid (of viscosity μ_N). The drag force F_{DN} in a Newtonian fluid is given by the formula

$$F_{DN} = \frac{2\pi\mu_N VL\left(1+k_2^2\right)}{\left(1-k_2^2\right)+\left(1+k_2^2\right)\ln k_2} \tag{11.38}$$

The drag on the needle in a power-law fluid F_D is given as:

$$F_D = \left(\frac{F_{DN}}{\mu_N}\right)m\left(\alpha\frac{V}{R_2}\right)^{n-1} \tag{11.39}$$

Note that at the terminal fall condition, the drag force F_D is simply equal to the buoyant weight of the needle and is thus known for a needle-liquid combination. Thus, the slope of log $(F_D/(F_D/\mu_N))$ versus log (V/R_2) plot will yield the value of n. This, in turn, will allow the evaluation of α and hence the shear rate via Equations 11.35–11.37. A comparison of Equations 11.39 with the power-law model suggests the corresponding shear stress to be given by $\left(F_D\mu_N/F_{DN}\right)\dot{\gamma}_{av}$ and thereby yielding the desired shear stress–shear rate curve for a fluid. This analysis, however, assumes the needle to be infinitely long and Zheng et al. (1994) recommend the value of $(L/R_2)>40$ for the end effects to be negligible.

11.5.2.2 Yield Stress

The falling cylinder configuration (also known as a penetrometer) has also been exploited to evaluate the static yield stress of visco-plastic materials. Park et al. (1988) outlined an extrapolation procedure to evaluate the value of the yield stress from a series of falling velocity data for needles of the same size, but with different values of $\Delta\rho$. Obviously, this method is equivalent to the extrapolation of viscometric shear stress–shear rate data to zero-shear rate. On the other hand, Uhlherr et al. (2002) have proposed a novel design of a hollow penetrometer whose mass can be changed by addition of tungsten powder or mercury to locate the motion/no motion point, that is, when the penetrometer does not move. Under such static equilibrium conditions, the buoyant weight must equal the vertical component of the force due to the yield stress of the medium acting on the penetrometer (shown schematically in Figure 11.11). The static yield stress is calculated from the expression

$$\tau_o = \left[\frac{m_1 - \rho\pi d^2\left(\frac{h}{4}+\frac{d}{12}\right)}{\pi d\left(h+\frac{\pi d}{8}\right)}\right]g \tag{11.40}$$

Uhlherr et al. (2002) reported good agreement between the penetrometer and the vane values of the yield stress for a few Carbopol solutions and titania suspensions.

Before concluding this chapter, it is appropriate to mention here that many other devices based on the principles presented herein are available including the so-called air bubble viscometer (Barr, 1926, 1931), diamond anvil cell viscometer (Piermarini et al., 1978; King et al., 1992; Herbst et al., 1993; Cook et al., 1993) and a vibrating sphere viscometer (Ray and Biswas, 1971; Allen and Baird, 1973). However, while their utility for Newtonian fluids is well established, none of these have yet been extended to the viscometry of non-Newtonian fluids.

11.6 Concluding Summary

In this chapter, consideration has been given to the use of falling, rolling, oscillating and rotating sphere viscometry and of falling cylinder, and falling needle viscometry for Newtonian and non-Newtonian media. While their utility for the measurement of viscosity of Newtonian liquids at high temperatures

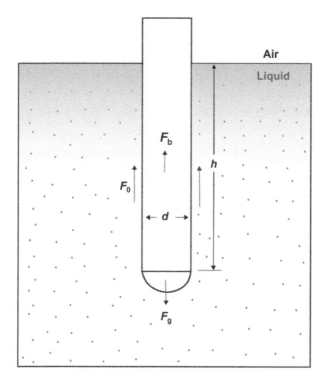

FIGURE 11.11 Schematics of a penetrometer. (From Uhlherr, P.H.T., Guo, J., Fang, T.-N., and Tiu C., *Korea-Australia Rheol. J.*, 14, 17, 2002.)

and pressures is established beyond doubt, their application to the measurement of non-Newtonian fluid characteristics is severely limited by the fact that the flow is nonviscometric in each case. It is, however, possible to extract some information under appropriate experimental conditions. For instance, falling ball method can be used to obtain reasonable estimates of the zero-shear viscosity. On the other hand, this method yields less reliable values of shear-dependent viscosity. The falling needle method offers an attractive alternative for shear-thinning fluids simply because the shear rate is more uniform in a falling needle device than that in a falling ball rheometer. The rolling ball method can certainly be used for quality control purposes. The pendulum and penetrometer methods offer simple and reliable means of evaluating static yield stress of visco-plastic materials. Finally, contrary to many attempts and suggestions, it is not at all easy to obtain visco-elastic characteristics from such devices.

Nomenclature

A:	Function of n, Equation 11.11 (-)
b:	Length of cylinder in pendulum in Equation 11.19, Figure 11.8 (m)
C:	Constant, Equation 11.5 (kg.s/m)
C_o:	Constant, Equation 11.7 (-)
C_I:	Constant, Equation 11.6 (kg/ms³)
d:	Cylinder or sphere diameter (m)
D:	Tube diameter (m)
ECF:	End correction factor, Equation 11.32 (-)
ER:	Eccentricity ratio, Equation 11.34 (-)
f_B:	Correction due to bottom end (-)
f_I:	Correction due to inertia (-)

f_w:	Wall correction factor (-)
F_D:	Drag force on a needle (N)
F_{DN}:	Drag force on a needle in a Newtonian medium (N)
F_o:	Force on a sphere due to yield stress (N)
$F_1(n)$, $F_2(n)$:	Functions of n, Equations 11.12 and 11.14 (-)
g:	Acceleration due to gravity (m/s^2)
H:	Center of mass, Figure 11.8 (m)
h:	Depth of immersion of penetrometer (m)
k:	Constant, Equation 11.2 (m^2/s^2)
k':	Constant, Equation 11.20 (m^3/s^2)
k_1:	Ratio of radii (= R_1/R_2) (-)
k_2:	Ratio of radii (= R_2/R_3) (-)
k_s:	System constant, Equation 11.29 (m^3/s^2)
l:	Immersed length, Equation 11.18 (m)
L:	Length of needle or cylinder (m)
m:	Power-law consistency index (Pa.s^n)
m_c:	Mass of cylinder, Figure 11.8 (kg)
m_p:	Mass of plate-hook assembly, Equation 11.19 (kg)
m_I:	Mass of penetrometer, Equation 11.40 (kg)
n:	Power-law index (-)
R_1:	Inner radius of hollow cylinder (m)
R_2:	Outer radius of hollow cylinder (m)
R_3:	Radius of cylindrical tube (m)
t:	Fall time (s)
V:	Falling velocity free from wall, end and eccentricity effects (m/s)
V_m:	Measured fall velocity (m/s)
V_R:	Rolling velocity of ball (m/s)

Greek Symbols

α:	Constant, Equation 11.36 (-)
α_o, α_1:	Constants, Equation 11.37 (-)
β_1, β_2:	Angle of inclination from horizontal (-)
$\dot{\gamma}_{av}$:	Surface average shear rate (s^{-1})
μ_0:	True zero-shear viscosity (Pa s)
μ_N:	Viscosity of a Newtonian fluid (Pa s)
μ_s:	Viscosity defined by Equation 11.1 (Pa s)
Ω:	Rotational velocity (rad/s)
ρ:	Fluid density (kg/m^3)
ρ_s:	Density of falling ball or cylinder (kg/m^3)
$\Delta\rho$:	Density difference ($\rho_s - \rho$) (kg/m^3)
θ:	Inclination of pendulum from vertical or spherical coordinate (-)
τ_{av}:	Surface average shear stress (Pa)
τ_m:	Maximum shear stress (Pa)
τ_o:	Yield stress (Pa)

Subscripts

N:	Newtonian
LB:	Lower bound
UB:	Upper bound

References

Abdel-Alim, A.H. and Hamielec, A.E., A theoretical and experimental investigation of the effect of internal circulation on the drag of spherical droplets falling at terminal velocity in liquid media, *Ind. Eng. Chem. Fund.*, **14**, 308 (1975).

Abedijabari, A. and Khomami, B., Sedimentation of a sphere in a viscoelastic fluid: A multiscale simulation approach, *J. Fluid Mech.*, **694**, 78 (2012).

Abhishek, S., King. A.J.C. and Narayanaswamy, R., Dynamics of a Taylor bubble in steady and pulsatile co–current flow of Newtonian and shear- thinning liquids in a vertical tube, *Int. J. Multiph. Flow*, **74**, 148 (2015).

Abou-Kasseem, J.H. and Farouq Ali, S.M., Flow of non-Newtonian fluids in porous media, paper presented at the *SPE Eastern Regional Meeting*, Columbus, OH (Nov. 12–14, 1986).

Abou-Ziyan, H., Kalendar, A., Shedid, M. and Abdel-Hameed H., Experimental investigation of free convection from short horizontal cylinder to Newtonian and power-law liquids of large Prandtl numbers, *Exp. Therm. Fluid Sci.*, **86**, 102 (2017).

Abrate, S., Resin flow in fiber preforms, *App. Mech. Rev.*, **55**, 579 (2002).

Acharya, A., Particle transport in viscous and viscoelastic fracturing fluids, *SPE Prod. Eng.*, **1**, 104 (1986).

Acharya, A., Viscoelasticity of cross-linked fracturing fluids and proppant transport, *SPE Prod. Eng.*, **3**, 483 (1988).

Acharya, A. and Maaskant, P., The measurement of the material parameters of viscoelastic fluids using a rotating sphere and a rheogoniometer, *Rheol. Acta*, **17**, 377 (1978).

Acharya, A., Mashelkar, R.A. and Ulbrecht, J., Flow of inelastic and viscoelastic fluids past a sphere, *Rheol. Acta*, **15**, 454 (1976).

Acharya, A., Mashelkar, R.A. and Ulbrecht, J., Mechanics of bubble motion and deformation in non-Newtonian media, *Chem. Eng. Sci.*, **32**, 863 (1977).

Acharya, A., Mashelkar, R.A. and Ulbrecht, J., Bubble formation in non-Newtonian liquids, *Ind. Eng. Chem. Fund.*, **17**, 230 (1978a).

Acharya, A., Mashelkar, R.A. and Ulbrecht, J., Motion of liquid drops in rheologically complex fluids, *Can. J. Chem. Eng.*, **56**, 19 (1978b).

Acharya, A. and Ulbrecht, J., Note on the influence of viscoelasticity on the coalescence rate of bubbles and drops, *AIChE J.*, **24**, 348 (1978).

Acharya, R.C., van der Zee, S.E.A.T.M. and Leijnse, A., Porosity-permeability properties generated with a new two-parameter three-dimensional hydraulic pore-network model for consolidated and unconsolidated porous media, *Adv. Water Resour.*, **27**, 707 (2004).

Achenbach, E., The effects of surface roughness and tunnel blockage on the flow past sphere. *J. Fluid Mech.*, **65**, 113 (1974).

Acrivos, A., Combined laminar free-and forced-convection heat transfer in external flows, *AIChE J.*, **4**, 285 (1958).

Acrivos, A., A theoretical analysis of laminar natural convection heat transfer to non-Newtonian fluids, *AIChE J.*, **6**, 584 (1960).

Acrivos, A., On the combined effect of forced and free convection heat transfer in laminar boundary layer flows, *Chem. Eng. Sci.*, **21**, 343 (1966).

Acrivos, A., Shah, M.J. and Petersen, E.E., Momentum and heat transfer in laminar boundary layer flows of non-Newtonian fluids past external surfaces, *AIChE J.*, **6**, 312 (1960).

Acrivos, A., Shah, M.J. and Petersen, E.E., On the solution of the two-dimensional boundary-layer flow equations for a non-Newtonian power law fluid, *Chem. Eng. Sci.*, **20**, 101 (1965).

Adachi, K., *Ph. D Dissertation*, Kyoto University, Kyoto, Japan (1973).

Adachi, K. and Yoshioka, N., On creeping flow of a viscoplastic fluid past a circular cylinder, *Chem. Eng. Sci.*, **28**, 215 (1973).

Adachi, K., Yoshioka, N. and Sakai, K., An investigation of non-Newtonian flow past a sphere, *J. Non-Newt. Fluid Mech.*, **3**, 107 (1977/78).

Adachi, K., Yoshioka, N. and Yamamoto, K., On non-Newtonian flow past a sphere, *Chem. Eng. Sci.*, **28**, 2033 (1973).

Adam, M., Delsanti, M., Pieransky, P. and Meyer, R., Magnetorheometre a bille, *Revue Phys. Appl.*, **19**, 253 (1984).

Adamczyk, Z., Adamczyk, M. and van de Ven, T.G.M., Resistance coefficient of a solid sphere approaching plane and curved boundaries, *J. Colloid Interface Sci.*, **96**, 204 (1983).

Adams, D. and Bell, K.J., Fluid friction and heat transfer for flow of sodium carboxymethyl cellulose solutions across banks of tubes, *Chem. Eng. Prog. Symp. Ser.*, **64**, No. 82, 133 (1968).

Adler, I., Buchholz, H., Voigt, J., Wittler, R. and Schugrel, K., Bubble coalescence behaviour in biological media, *Eur. J. Appl. Microbiol. Biotechnol.*, **9**, 249 (1980).

Adler, P.M., *Porous Media, Geometry and Transport*, Butterworth-Heinemann, Boston, MA (1992).

Adler, P.M., Malevich, A.E. and Mityushev, V.V., Nonlinear correction to Darcy's law for channels with wavy walls, *Acta Mech.*, **224**(8), 1823 (2013).

Advani, S.G. and Arefmanesh, A., Bubble growth and collapse in viscoelastic liquids, *Adv. Transp. Process*, **9**, 445 (1993).

Afolabi, R.O., Oluyemi, G.F., Officer, S. and Ugwu, J.O., Hydrophobically associating polymers for Enhanced Oil Recovery – Part A: A review on the effects of some key reservoir conditions. *J. Pet. Sci. Eng.*, **180**, 681 (2019).

Agarwal, M., Chhabra, R.P. and Eswaran, V., Laminar momentum and thermal boundary layers of power law fluids over a slender cylinder, *Chem. Eng. Sci.*, **57**, 1331 (2002).

Agarwal, N. and Chhabra, R.P., Settling velocity of cubes in Newtonian and power law liquids, *Powder Technol.*, **178**, 17 (2007).

Agarwal, P.K., Transport phenomena in multi-particle systems II. Particle-Fluid Heat and Mass Transfer, *Chem. Eng. Sci.*, **43**, 2501 (1988).

Agarwal, P.K. and Mitchell, W.J., Generalised Reynolds number, drag curve and interphase transport phenomena in viscous flow, *Chem. Eng. Sci.*, **44**, 405 (1989).

Agarwal, P.K., Mitchell, W.J. and La Nauze, R.D., Transport phenomena in multi-particle systems – III. Active particle mass transfer in fluidized beds of inert particles, *Chem. Eng. Sci.*, **43**, 1511 (1988).

Agarwal, P.K. and O'Neill, B.K., Transport phenomena in multi-particle systems – I. Pressure drop and friction factor: Unifying the hydraulic-radius and submerged object approaches, *Chem. Eng. Sci.*, **43**, 2487 (1988).

Agarwal, U.S., Dutta, A. and Mashelkar, R.A., Migration of macromolecules under flow: The physical origins and engineering implications, *Chem. Eng. Sci.*, **49**, 1693 (1994).

Aghajani, M., Muller-Steinhagen, H. and Jamialahmadi, M., Heat transfer of liquid/solid fluidized beds for Newtonian and non-Newtonian fluids, *Iran. J. Chem Chem. Eng.*, **23**, 119 (2004).

Aguirre, V.A., Castillo, B.A. and Narvaez, C.P., Numerical simulation of a bubble rising in an environment consisting of xanthan gum, *AIP Conf. Proc.*, **1872**(1), 020022 (2017).

Agullo, J.O. and Marenya, M.O., Airflow resistance of parchment Arabica coffee, *Biosys. Eng.*, **91**, 149 (2005).

Agwu, D.E., Akpabio, J.U., Alabi, S.B. and Dosunmu, A., Settling velocity of drill cuttings in drilling fluids: A review of experimental, numerical simulations and artificial intelligence studies, *Powder Technol.*, **339**, 728 (2018).

Aherwar, K., Raja, A.H., Patel, S.A. and Chhabra, R.P., Free convection from a vertical plate to generalized Newtonian fluids. *J. Non-Newt. Fluid Mech.*, **317**, 105046 (2023).

Ahmad, E.A. and Badr, H.M., Mixed convection from an elliptic tube placed in a fluctuating free stream, *Int. J. Eng. Sci.*, **39**, 669, (2001).

Ahmad, R.A. and Qureshi, Z.H., Laminar mixed convection from a uniform heat flux horizontal cylinder in a cross flow, *J. Thermophys. Heat Transf.*, **6**, 277 (1992).

Ahmad, R.A. and Qureshi, Z.H., Buoyancy effects on forced convection from a horizontal cylinder in a crossflow, *J. Thermophys. Heat Transf.*, **7**, 574 (1993).

Ahmadi, A. and Karimfazli, I., A quantitative evolution of viscosity regularization in predicting transient flows of viscoplastic fluids, *J. Non-Newt. Fluid Mech.*, **287**, 104429 (2021).

Ahmed, H.A.M., Computational study of the flow around a spherical particle in Newtonian and shear-thinning fluids, *PhD Thesis*, University of Surrey, Guildford, U.K. (2002).

Ahmed, N. and Sunada, D.K., Non-linear flow in porous media, *J. Hyd. Div.*, **95**, 1847 (1969).
Ahonguio, F., Jossic, L. and Magnin, A., Influence of surface properties on the flow of a yield stress fluid around spheres, *J. Non-Newt. Fluid Mech.*, **206**, 57 (2014).
Ahonguio, F., Jossic, L. and Magnin, A., Influence of slip on the flow a yield-stress fluid around a flat plate, *AIChE J.*, **62**, 1356 (2016a).
Ahonguio, F., Jossic, L., Magnin, A. and Dufour, F., Flow of an elasto-viscoplastic fluid around a flat plate: Experimental and numerical data, *J. Non-Newt. Fluid Mech.*, **238**, 131 (2016b).
Aiba, S. and Okamoto, R., Power requirements of agitation and capacity coefficient of mass transfer in bubble aeration, *J. Ferment. Technol.*, **43**, 609 (1965).
Airiau, C. and Bottaro, A., Flow of shear-thinning fluids through porous media, *Adv. Water Resour.*, **143**, 103658 (2020).
Ajayi, O.O., Slow motion of a bubble in a viscoelastic fluid, *J. Eng. Math.*, **9**, 273 (1975).
Ajith Kumar, S., Mathur, M., Sameen, A. and Anil Lal, S., Effects of Prandtl number on the laminar cross flow past a heated cylinder, *Phys. Fluids*, **28**, 113603 (2016).
Akagi, S., Free convection heat transfer of non-Newtonian fluid, *Trans. Jpn. Soc. Mech. Eng.*, **32**, 919 (1966).
Akers, B. and Belmonte, A., Impact dynamics of a solid sphere falling into a viscoelastic micellar fluid, *J. Non-Newt. Fluid Mech.*, **135**, 97 (2006).
Al Taweel, A.M., Sedahmed, H.G., Abdel-Khalik, A. and Farag, H.A., Mass transfer between solid surfaces and drag reducing fluids, *Chem. Eng. J.*, **15**, 81 (1978).
Al Varado, D.A. and Marsden, S.S., Flow of oil-in-water emulsions through tubes and porous media, *Soc. Pet. Eng. J.*, **17**, 369 (1979).
Alam, M.J., Nirmalkar, N. and Gupta, A.K., Stability criteria and convective mass transfer from the falling spherical drops, Part 2: Herschel-Bulkley fluids, *Can. J. Chem. Eng.*, **100**, 1640 (2022).
Alam, M.S., Ishii, K. and Hasimoto, H., Slow motion of a small sphere outside of a circular cylinder, *J. Phys. Soc. Jpn.*, **49**, 405 (1980).
Albernaz, D.L. and Cunha, F.R., Unsteady motion of a spherical bubble in a complex fluid: Mathematical modelling and simulations, *Appl. Math. Model.*, **37**, 8972 (2013).
Alcocer, F.J., Kumar, V. and Singh, P., Permeability of periodic porous media, *Phys. Rev. E*, **59**, 711 (1999).
Alcocer, F.J. and Singh, P., Permeability of periodic arrays of cylinders for viscoelastic flows, *Phys. Fluids*, **14**, 2578 (2002).
Al-Fariss, T., Pascal, H. and Pinder, K.L., Flow through porous media of a shear-thinning liquid with a yield value, Paper presented at the *55th Annual Meeting of the Soc. Rheol.*, Knoxville, TN (1983).
Al-Fariss, T. and Pinder, K.L., Flow through porous media of a shear thinning liquid with yield stress, *Can. J. Chem. Eng.*, **65**, 391 (1987).
Al-Fariss, T.F., A new correlation for non-Newtonian flow through porous media, *Comp. Chem. Eng.*, **13**, 475 (1989).
Al-Fariss, T.F., Flow of Polymer solutions through porous media, *Ind. Eng. Chem. Res.*, **29**, 2150 (1990).
Alghalibi, D., Fornari, W., Rosti, M.E. and Brandt, L., Sedimentation of finite-size particles in quiescent wall-bounded shear-thinning and shear-thickening fluids, *Int. J. Multiph. Flow*, **129**, 103291 (2020).
Alghalibi, D., Lashgari, I., Brandt, L. and Hormozi, S., Interface-resolved simulations of particle suspensions in Newtonian, shear-thinning and shear-thickening fluids, *J. Fluid Mech.*, **852**, 329 (2018).
Alhamdan, A. and Sastry, S.K., Natural convection heat transfer between non-Newtonian fluids and an irregular-shaped particle, *J. Food Process Eng.*, **13**, 113 (1990).
Alhamdan, A. and Sastry, S.K., Bulk average heat transfer coefficient of multiple particles flowing in a holding tube, *Trans. Inst. Chem. Eng.*, **76C**, 95 (1998).
Alicke, A.A., da Senhora, F.V., Cid, M. and de Souza Mendes, P.R., Motion of gas bubbles in viscoplastic fluids, *Proc. of the ENCIT 2012*, Nov 18–22, Rio de Janeiro, Brazil (2012).
Aliseda, A., Hopfinger, E.J., Lasheras, J.C., Kremer, D.M., Berchielli, A. and Connolly, E.K., Atomization of viscous and non-Newtonian liquids by a coaxial, high speed gas jet. Experiments and droplet size modelling, *Int. J. Multiph. Flow*, **34**, 161 (2008).
Allahham, A., Stewart, P., Marriott, J. and Mainwaring, D., Factors affecting shear- thickening behaviour of a concentrated injectable suspension of Levodopa, *J. Pharma. Sci.*, **94**, 2393 (2005).
Allan, Y.M. and Brown, J.M.B., Creeping motion of a cylinder through incompressible fluid bounded by a coaxial closed cylinder, *Comput. Fluids*, **14**, 283 (1986).
Allen, D.J. and Baird, M.H.I., Vibrating sphere viscometry, *Can. J. Chem. Eng.*, **51**, 776 (1973).

Allen, E. and Uhlherr, P.H.T., A kinetic based equation with easily evaluated parameters, *4th Proc. Nat. Cong. Rheol.*, Adelaide, p. 63 (1986).

Allen, E. and Uhlherr, P.H.T., Non-homogeneous sedimentation in viscoelastic fluids, *J. Rheol.*, **33**, 627 (1989).

Allen, J.S. and Roy, R.A., Dynamics of gas bubbles in viscoelastic fluids. I. Linear viscoelasticity, *J. Acoust. Soc. Am.*, **107**, 3167 (2000). Also see *ibid*, **108**, 1640 (2000).

Al-Matroushi, E. and Borhan, A., Coalescence of drops and bubbles rising through a non-Newtonian fluid in a tube, *Ann. N. Y. Acad. Sci.*, **1161**(1), 225 (2009).

Alnakeeb, M.A., El-Maghlany, W.M., Teamah, M.A. and Sorour, M.M., Experimental study of mixed convection from horizontal isothermal elliptic cylinders at different aspect ratios, *Proc. ISER 81st Int. Conf.*, Mecca, Saudi Arabia, 23–24 Oct. 2017, pp. 9–14 (2007).

Al-Raoush, R., Thompson, K. and Willson, C.S., Comparison of network generation techniques for unconsolidated porous media, *Soil Sci. Soc. Am. J.*, **67**, 1687 (2003).

Alvarez, E., Cancela, M.A. and Maceiras, R., Effect of temperature on rheological properties different jams, *Int. J. Food Prop.*, **9**, 135 (2018).

Alves, M.A., Oliveira, P.J. and Pinho, F.T., Numerical methods for viscoelastic fluid flows, *Annu. Rev. Fluid Mech.*, **53**, 509 (2021).

Alves, M.A., Pinho, F.T. and Oliveira, P.J., The flow of viscoelastic fluids past a cylinder: Finite-volume high-resolution methods, *J. Non-Newt. Fluid Mech.*, **97**, 207 (2001).

Amani, E., Ahmadpour, A. and Tohidi, M., A numerical study of the rise of a Taylor bubble through a sudden/gradual expansion in Newtonian and shear-thinning liquids, *Int. J. Mech. Sci.*, **152**, 236 (2019).

Amaratunga, M., Rabenjafimanantsoa, H.A. and Time, R.W., Influence of low frequency oscillatory motion on particle settling in Newtonian and shear-thinning non-Newtonian fluids, *J. Pet. Sci. Eng.*, **196**, 107786 (2020).

Amato, W.S. and Tien, C., Natural convection heat transfer from a vertical plate to an Oldroyd fluid, *Chem. Eng. Prog. Sym. Ser.*, **66** (102), 92 (1970).

Amato, W.S. and Tien, C., Free convection heat transfer from isothermal spheres in water, *Int. J. Heat Mass Transf.*, **15**, 327 (1972).

Amato, W.S. and Tien, C., Free convection heat transfer from isothermal spheres in polymer solutions, *Int. J. Heat Mass Transf.*, **19**, 1257 (1976).

Ambari, A., Deslouis, C. and Tribollet, B., Coil-stretch transition of macromolecules in laminar flow around a small cylinder, *Chem. Eng. Commun.*, **29**, 63 (1984a).

Ambari, A., Gautheir-Manuel, B. and Guyon, E., Effect of a plane wall on a sphere moving parallel to it, *J. Phys. Lett. (Paris)*, **44**, 143 (1983).

Ambari, A., Gautheir-Manuel, B. and Guyon, E., Wall effects on a sphere translating at constant velocity, *J. Fluid Mech.*, **149**, 235 (1984b).

Ambeskar, V.D. and Mashelkar, R.A., On the role of stress-induced migration on time dependent terminal velocities of falling spheres, *Rheol. Acta*, **29**, 182 (1990).

Aminzadeh, M., Maleki, A., Firoozabadi, B. and Afshin, H., On the motion of Newtonian and non-Newtonian drops, *Scientia Iranica*, **19**, 1265 (2012).

Amundarain, J.L., Castro, L.J., Rojas, M.R., Siquier, S., Ramírez, N., Müller, A.J. and Sáez, A.E., Solutions of xanthan gum/guar gum mixtures: Shear rheology, porous media flow, and solids transport in annular flow. *Rheol. Acta*, **48**(5), 491 (2009).

Anderson, S.L. and Warburton, F.L., The porous plug and fiber diameter measurement: Effect of fibre orientation and use of plugs of randomized fibres, *Textile Inst. J.*, **41**, T749 (1949).

Andersson, H.I., The Nakayama-Koyama approach to laminar forced convection heat transfer to power law fluids, *Int. J. Heat Fluid Flow*, **9**, 343 (1988), Also see *ibid* **10**, 86 (1989).

Andersson, H.I. and Toften, T.H., Numerical solution of the laminar boundary layer equations for power-law fluids, *J. Non-Newt. Fluid Mech.*, **32**, 175 (1989).

Andrade, E.N.da C. and Fox, J.W., The mechanism of dilatancy, *Proc. Phys. Soc.*, **62B**, 483 (1949).

Andres, U.Ts., Equilibrium and motion of spheres in a viscoplastic liquid, *Sov. Phys. – Doklady (U.S.A.)* **5**, 723 (1961).

Ansley, R.W. and Smith, T.N., Motion of spherical particles in a Bingham plastic, *AIChE J.*, **13**, 1193 (1967).

Anthony, C.R., Kamat, P.M., Thete, S.S., Munro, J.P., Lister, J.R., Harris, M.T. and Basaran, O.A., Scaling laws and dynamics of bubble coalescence, *Phys. Rev. Fluids*, **2**, 083601 (2017).

Arabi, P. and Jafarpur, K., Effect of different flow regimes on free convection heat transfer from isothermal convex bodies over all range of Rayleigh and Prandtl numbers, *Heat Mass Transf.* **52**, 1665 (2016).

Arabi, P. and Jafarpur, K., Criteria for predicting transitions in free convection heat transfer from isothermal convex bodies in fluids with any Prandtl number: A new analytical model, *Heat Transf. Eng.*, **38**, 578 (2017).

Arastoopour, H., Gidaspow, D. and Lyczkowski, R.W., *Transport Phenomena in Multiphase Systems*, Springer, New York (2021).

Archer, L.A., Wall slip: Measurement and modeling issues, *Polymer Processing Instabilities: Control and Understanding*, edited by S. G. Hatzikiriakos and K. B. Migler, Marcel Dekker, New York (2005).

Arcoumanis, C., Khezzar, L., Whitelaw, D.S. and Warren, B.C.H., Breakup of Newtonian and non-Newtonian fluids in air jets, *Exp. Fluids*, **17**, 405 (1994).

Arcoumanis, C., Whitelaw, D.S. and Whitelaw, J.H., Breakup of droplets of Newtonian and non-Newtonian fluids, *At. Sprays*, **6**, 245 (1996).

Ardekani, A.M., Rangel, R.H. and Joseph, D.D., Two spheres in a free stream of a second –order fluid, *Phys. Fluids*, **20**, 063101 (2008).

Ardakani, H.A., Mitsoulis, E. and Hatzikiriakos, S.G., Thixotropic flow of toothpaste through extrusion dies, *J. Non- Newt. Fluid Mech.*, **166**, 1262 (2011).

Arefmanesh, A. and Advani, S.G., Diffusion-induced growth of a gas bubble in a viscoelastic fluid, *Rheol. Acta*, **30**, 274 (1991).

Arigo, M.T. and McKinley, G.H., The steady and transient motion of a sphere through a viscoelastic fluid, *Cavitation and Multiphase Flow*, FED Vol. 194, 139 (1994).

Arigo, M.T. and McKinley, G.H., The effects of viscoelasticity on the transient motion of a sphere in a shear-thinning fluid, *J. Rheol.*, **41**, 103 (1997).

Arigo, M.T. and McKinley, G.H., An experimental investigation of negative wakes behind spheres settling in a shear-thinning viscoelastic fluid, *Rheol. Acta*, **37**, 307 (1998).

Arigo, M.T., Rajagopalan, D.R., Shapley, N.T. and McKinley, G.H., Sedimentation of a sphere through an elastic fluid: Part I Steady motion, *J. Non-Newt. Fluid Mech.*, **60**, 225 (1995).

Arnipally, S.K. and Kuru, E., Settling velocity of particles in viscoelastic fluids: A comparison of the shear-viscosity and elasticity effects, *SPE J.*, **23**, 1689 (2018).

Arsenijevic, Z.Lj., Grbavcic, Z.B., Garic-Grulovic, R.V. and Boskovic-Vragolovic, N.M., Wall effects on the velocities of a single sphere settling in a stagnant and counter- current fluid and rising in a co-current fluid, *Powder Technol.*, **203**, 237 (2010).

Arzate, A., Ascanio, G., Carreau, P.J. and Tanguy, P.A., Extensional viscosity of coating colors and its relation with jet coating performance, *Appl. Rheol.*, **14**, 240 (2004).

Aschenbrenner, B.C., A new method of expressing particle sphericity, *J. Sed. Petrol.*, **26**, 15 (1956).

Ashare, E., Bird, R.B. and Lescarboura, J.A., Falling cylinder viscometer for non-Newtonian fluids, *AIChE J.*, **11**, 910 (1965).

Asif, M., Generalized Richardson-Zaki correlation for liquid fluidization of binary solids, *Chem. Eng. Technol.*, **21**, 77 (1998).

Asif, M. and Dhiman, A., Analysis of laminar flow across a triangular periodic array of heated cylinders, *J. Braz. Soc. Mech. Sci. Eng.*, **40**, 350 (2018).

Astarita, G., Letter to the Editor, *Can. J. Chem. Eng.*, **44**, 59 (1966a).

Astarita, G., Spherical gas bubble motion through Maxwell liquids, *Ind. Eng. Chem. Fund.*, **5**, 548 (1966b).

Astarita, G., Dimensional analysis of flow of viscoelastic fluids, *Chem. Eng. Sci.*, **29**, 1273 (1974).

Astarita, G., Variational principles and entropy production in creeping flow of non-Newtonian fluids, *J. Non-Newt. Fluid Mech.*, **2**, 343 (1977).

Astarita, G., On the relationship between a dimensional and a thermodynamic peculiarity of the power-law constitutive equation, *J. Non-Newt. Fluid Mech.*, **13**, 223 (1983).

Astarita, G., Letter to the editor: The engineering reality of the yield stress, *J. Rheol.*, **34**, 275 (1990).

Astarita, G., Dimensional analysis, scaling, and orders of magnitude, *Chem. Eng. Sci.*, **52**, 4681 (1997).

Astarita, G. and Appuzzo, G., Motion of gas bubbles in non-Newtonian liquids, *AIChE J.*, **11**, 815 (1965).

Astarita, G. and Denn, M.M., The effect of the non-Newtonian properties of polymer solutions on the flow fields, *Theoretical Rheology*, edited by J.F. Hutton, J.R.A. Pearson and K. Walters, p. 333, Applied Science Publishers, London (1975).

Astarita, G. and Marrucci, G., Moto di una bolla di gas in un liquido non-Newtoniano a legge di potenza, *Accademia Nazionale dei Lincei, Ser. VIII*, **36**, 836 (1964).

Astarita, G. and Marrucci, G., Boundary layer flows of viscoelastic materials, *Lincei-Rend. Sc. fis. mat. e nat.*, **XLI**, 355 (November 1966).

Astarita, G. and Marrucci, G., *Principles of non-Newtonian Fluid Mechanics*, McGraw-Hill, New York (1974).

Astarita, G. and Mashelkar, R.A., Heat and mass transfer in non-Newtonian fluids, *The Chem. Eng.*, 100 (1977).

Astin, J., Jones, R.S. and Lockyer, P., Boundary layers in non-Newtonian fluids, *J. Mecanique*, **12**, 527 (1973).

Aström, A. and Bark, G., Heat transfer between fluid and particles in aseptic processing, *J. Food Eng.*, **21**, 97 (1994).

Åström, B.T., Pipes, R.B. and Advani, S.G., On flow through aligned fiber beds and its application to composite processing, *J. Comp. Mater.*, **26**, 1351 (1992).

Ataide, C.H., Barrozo, M.A.S. and Pereira, F.A.R., Study of the motion of particles in suspensions of hydroxymethyl cellulose: Wall effect and drag coefficient, *World Congress on Particle Technology-3*, Brighton (UK), pp. 2776–2787 (1998).

Ataide, C.H., Pereira, F.A.R. and Barrozo, M.A.S., Wall effects on the terminal velocity of spherical particles in Newtonian and non-Newtonian fluids, *Braz. J. Chem. Eng.*, **16**, 387 (1999).

Atapattu, D.D., *Ph.D Dissertation*, Monash Uni., Melbourne, Australia (1989).

Atapattu, D.D., Chhabra, R.P., Tiu, C. and Uhlherr, P.H.T., The effect of cylindrical boundaries for spheres falling in fluids having a yield stress, *Proc. 9th Australasian Fluid Mech. Conf.*, Auckland, p. 338 (1986).

Atapattu, D.D., Chhabra, R.P. and Uhlherr, P.H.T., Particle drag and equilibrium in viscoplastic fluids, *Proc. Int. Conf. Hyd. Transport.*, Bhubaneswar, India, p. 342 (1988).

Atapattu, D.D., Chhabra, R.P. and Uhlherr, P.H.T., Wall effect for spheres falling at small Reynolds number in a viscoplastic medium, *J. Non-Newt. Fluid Mech.*, **38**, 31 (1990).

Atapattu, D.D., Chhabra, R.P. and Uhlherr, P.H.T., Creeping sphere motion in Herschel-Bulkley fluids: Flow field and drag, *J. Non-Newt. Fluid Mech.*, **59**, 245 (1995).

Atapattu, D.D. and Uhlherr, P.H.T., Creeping motion of spheres in viscoplastic fluids, *Proc. Xth Int. Cong. Rheol.*, **1**, 350 (1988). Sydney.

Atayilmaz, S.O. and Teke, I., Experimental and numerical study of the natural convection from a heated horizontal cylinder, *Int. Comm. Heat Mass Transf.*, **36**, 731 (2009).

Ates, H. and Kelkar, M., Two-phase pressure-drop predictions across gravel pack, *SPE Prod. Facil.*, **18**, 104 (1998).

Atkinson, H., From ketchup to car parts, *Chem. Britain*, **37**, 32 (Feb. 2001).

Aubert, J.H. and Tirrell, M., Flows of dilute polymer solutions through packed porous chromatographic columns, *Rheol. Acta*, **19**, 452 (1980).

Auriault, J.-L., Royer, P. and Geindreau, C., Filtration law for power law fluids in anisotropic porous media, *Int. J. Eng. Sci.*, **40**, 1151 (2002).

Awbi, H. and Tan, S.H., Effect of wind tunnel walls on the drag of a sphere. *J. Fluids Eng. (ASME)*, **103**, 461 (1981).

Awuah, G.B., Ramaswamy, H.S. and Simpson, B.K., Surface heat transfer coefficients associated with heating of food particles in CMC solutions, *J. Food Process Eng.*, **16**, 39 (1993).

Azzam, M.I.S., *Ph. D Dissertation*, University of Waterloo, Waterloo, ON (1975).

Baaijens, F.P.T., Application of low-order discontinuous Galerkin methods to the analysis of viscoelastic flows, *J. Non-Newt. Fluid Mech.*, **52**, 37 (1994).

Baaijens, F.P.T., Mixed finite element methods for viscoelastic flow analysis: A review, *J. Non-Newt. Fluid Mech.*, **79**, 361 (1998).

Baaijens, F.P.T., Baaijens, H.P.W., Peters, G.W.M. and Meijer, H.E.H., An experimental and numerical investigation of a viscoelastic flow around a cylinder, *J. Rheol.*, **38**, 351 (1994).

Baaijens, F.P.T., Selen, S.H.A., Baaijens, H.P.W., Peters, G.W.M. and Meijer, H.E.H., Viscoelastic flow past a confined cylinder of a low-density polyethylene melt, *J. Non-Newt. Fluid Mech.*, **68**, 173 (1997).

Baaijens, H.P.W., Peters, G.W.M., Baaijens, F.P.T. and Meijer, H.E.H., Viscoelastic flow past a confined cylinder of a polyisobutylene solution, *J. Rheol.*, **39**, 1243 (1995).

Baca, H., Nikitopoulos, D.E., Smith, J.R. and Bourgoyne, A.T., Counter-current and co-current gas kicks in horizontal wells: Non-Newtonian rheology effects, *J. Energy Resour. Technol. (ASME)*, **125**, 51 (2003).

Bacon, L.R., Measurement of absolute viscosity by the falling sphere method. *J. Franklin Inst.*, **221**, 251 (1936).

Badr, H.M., A theoretical study of laminar mixed convection from a horizontal cylinder in a cross stream, *Int. J. Heat Mass Transf.*, **26**, 639 (1983).

Badr, H.M., Laminar combined convection from a horizontal cylinder—parallel and contra flow regimes, *Int. J. Heat Mass Transf.*, **27**, 15 (1984).

Badr, H.M., On the effect of flow direction on mixed convection from a horizontal cylinder, *Int. J. Num. Methods Fluids*, **5**, 1 (1985).

Bagchi, A. and Chhabra, R.P., Accelerating motion of spherical particles in power-law type non-Newtonian liquids, *Powder Technol.*, **68**, 85 (1991a).

Bagchi, A. and Chhabra, R.P., Rolling ball viscometry for Newtonian and power law liquids, *Chem. Eng. Process.*, **30**, 11 (1991b).

Bağcı, Ö., Dukhan, N. and Özdemir, M., Flow regimes in packed beds of spheres from pre-Darcy to turbulent. *Trans. Porous Media*, **104** (3), 501 (2014).

Bagley, E.B. and Dintzis, F.R., Shear-thickening and flow induced structures in foods and biopolymer systems, *Advances in the Flow and Rheology of Non-Newtonian Fluids: PART A*, edited by D. Siginer, D. De Kee, and R.P. Chhabra, p. 63, Elsevier, Amsterdam (1999).

Baijal, S.K. and Dey, N.C., Role of molecular parameters during flow of polymer solutions in unconsolidated porous media, *J. Appl. Polym. Sci.*, **27**, 121 (1982).

Bailoor, S., Seo, J.-H. and Mittal, R., Vortex shedding from a circular cylinder in shear-thinning Carreau fluids, *Phys. Fluids*, **30**, 011703 (2019).

Bair, S., The high pressure, high-shear stress rheology of a polybutene, *J. Non-Newt. Fluid Mech.*, **97**, 53 (2001).

Baird, M.H.I. and Hamielec, A.E., Forced convection transfer around spheres at intermediate Reynolds numbers, *Can. J. Chem. Eng.*, **40**, 119 (1962).

Baker, C.G.J., Margaritis, A. and Bergougnou, M.A., Fluidization principles and applications to biotechnology, *Adv. Biotechnol.*, **1**, 635 (1981).

Balakrishna, M., Murthy, M.S. and Kuloor, N.R., Sedimentation of fine particles in non-Newtonian fluids, *Indian Chem. Eng.*, **13**, T12 (1971).

Balan, C., Pure material instability and the concept of yield stress, *Appl. Rheol.*, **9**, 58 (1999).

Balaramakrishna, P.V. and Chhabra, R.P., Sedimentation of a sphere along the axis of a long square duct filled with non-Newtonian liquids, *Can. J. Chem. Eng.*, **70**, 803 (1992).

Balasubramaniam, V.M. and Sastry, S.K., Convective heat transfer at particle-liquid interface in continuous tube flow at elevated fluid temperatures, *J. Food Sci.*, **59**, 675 (1994a).

Balasubramaniam, V.M. and Sastry, S.K., Liquid-to-particle convective heat transfer in non-Newtonian carrier medium during continuous tube flow, *J. Food Eng.*, **23**, 169 (1994b).

Balhoff, M., Sanchez-Rivera, D., Kwok, A., Mehmani, Y. and Prodanović, M., Numerical algorithms for network modeling of yield stress and other non-Newtonian fluids in porous media. *Trans. Porous Media*, **93** (3), 363 (2012).

Balhoff, M.T. and Thompson, K.E., Modeling the steady flow of yield-stress fluids in packed beds, *AIChE J.*, **50**, 3034 (2004).

Balhoff, M.T. and Thompson, K.E. A macroscopic model for shear-thinning flow in packed beds based on network modeling, *Chem. Eng. Sci.*, **61**, 698 (2006).

Ballistella, A., van Schijndel, S.J.G., Baltussen, M.W., Roghair, I. and van Sint Annaland, M., On the terminal velocity of single bubbles rising in non-Newtonian power-law liquids, *J. Non-Newt. Fluid Mech.*, **278**, 104249 (2020).

Balmforth, N.J., Frigaard, I.A. and Ovarlez, G., Yielding to stress: Recent developments in viscoplastic fluid mechanics, *Annu. Rev. Fluid Mech.*, **46**, 121 (2014).

Banfill, P.F.G., *Rheology of Fresh Cement and Concrete*, Spon E & FN, London (1991).

Baptista, P.N., Oliveira, F.A.R., Oliveria, J.C. and Sastry, S.K., Dimensionless analysis of fluid-to-particle heat transfer coefficients, *J. Food Eng.*, **31**, 199 (1997).

Barabati, A.C., Desroches, J., Robisson, A. and McKinley, G.H., Complex fluids and hydraulic fracturing, *Annu. Rev. Chem. Biomol. Eng.*, **7**, 415 (2016).

Barak, A.Z., Comments on high velocity flow in porous media by Hassanizadeh and Gray, *Trans. Porous Media*, **2**, 533 (1987).

Barakos, G. and Mitsoulis, E., Numerical simulations of viscoelastic flows around spheres, *Recent Developments in Structured Continua III*, Montreal, May 26–28 (1993).

Barakos, G. and Mitsoulis, E., Numerical simulation of viscoelastic flow around a cylinder using an integral constitutive equation, *J. Rheol.*, **39**, 1279 (1995).

Barboza, M., Rangel, C. and Mena, B., Viscoelastic effects in flow through porous media, *J. Rheol.*, **23**, 281 (1979).

Barclay, J., Riley, D.S. and Sparks, R.S.J., Analytical models for bubble growth during decompression of high viscosity magmas, *Bull. Volcanol.*, **57**, 422 (1995).

Barentin, C. and Liu, A.J., Shear-thickening in dilute solutions of wormlike micelles, *Europhys. Lett.*, **55**, 432 (2001).

Barigou, M., Fairhurst, P.G., Fryer, P.J. and Pain, J.P., Concentric flow regime of solid-liquid food suspensions: Theory and experiment, *Chem. Eng. Sci.*, **58**, 1671 (2003).

Barigou, M., Mankad, S. and Fryer, P.J., Heat transfer in two-phase solid-liquid food flows: A review, *Trans. Inst. Chem. Eng.*, **76C**, 3 (1998).

Barnea, E. and Mednick, R.L., Correlation for minimum fluidization velocity, *Trans. Inst. Chem. Eng.*, **53**, 278 (1975).

Barnea, E. and Mednick, R.L., A generalized approach to the fluid dynamics of particulate systems, *Chem. Eng. J.*, **15**, 215 (1978).

Barnea, E. and Mizrahi, J., A generalized approach to the fluid dynamics of particulate systems Part I: General correlation for fluidization and sedimentation in solid multi particle systems, *Chem. Eng. J.*, **5**, 171 (1973).

Barnes, H.A., Review of shear-thickening (dilatancy) in suspensions of non-aggregating solid particles dispersed in Newtonian liquids, *J. Rheol.*, **33**, 329 (1989).

Barnes, H.A., The yield stress myth revisited, *Proc. XIth Int. Cong. Rheol.*, Brussels, Belgium, p. 576 (1992).

Barnes, H.A., Rheology for the chemical engineer, The Chem. Engineer (I. Chem. E., U. K.), June 24, 17 (1993).

Barnes, H.A., A review of the slip (wall depletion) of polymer solutions, emulsions and particle suspensions in viscometers; its cause, character and cure, *J. Non-Newt. Fluid Mech.*, **56**, 221 (1995).

Barnes, H.A., Thixotropy – a review, *J. Non-Newt. Fluid Mech.*, **70**, 1 (1997).

Barnes, H.A., The yield stress - a review or "παντα ρει" - everything flows?, *J. Non-Newt. Fluid Mech.*, **81**, 133 (1999). Barnes, H.A., *A Handbook of Elementary Rheology*, University of Wales Institute of Non-Newtonian Fluid Mechanics, Aberystwyth, Great Britain (2000).

Barnes, H.A., A principled but pragmatic view of the yield stress, *Rheol. Bull. Br. Soc. Rheol.*, **44**, 10 (2001).

Barnes, H.A., Hutton, J.F. and Walters, K., *An Introduction to Rheology*, Elsevier, Amsterdam (1989).

Barnes, H.A. and Nguyen, Q.D., Rotating vane rheometry – a review, *J. Non-Newt. Fluid Mech.*, **98**, 1 (2001).

Barnes, H.A., Schimanski, H. and Bell, D., 30 years of progress in viscometers and rheometers, *Appl. Rheol.*, **9**, 69 (1999).

Barnes, H.A. and Walters, K., The yield stress myth? *Rheol. Acta*, **24**, 323 (1985).

Barnett, S.M., Humphery, A.E. and Litt, M., Bubble motion and mass transfer in non-Newtonian fluids, *AIChE J.*, **12**, 253 (1966).

Barr, D.W., Coefficient of permeability determined by measurable parameters, *Ground Water*, **39**, 356 (2001a).

Barr, D.W., Turbulent flow through porous media, *Ground Water*, **39**, 646 (2001b).

Barr, G., The air bubble viscometer, *Phil. Mag. Ser. 7*, **1**, 395 (1926).

Barr, G., *A Monograph of Viscometry*, Oxford University Press, Oxford (1931).

Barree, R.D. and Conway, M.W., Experimental and numerical modeling of convective proppant transport, *J. Pet. Technol.*, **47**, 216 (1995).

Bartram, E., Goldsmith, H.L. and Mason, S.G., Particle motions in non-Newtonian media III: Further observations in elastic-viscous fluids, *Rheol. Acta*, **14**, 776 (1975).

Basset, A.B., On the motion of a sphere in a viscous liquid, *Phil. Trans. Roy. Soc. London*, **179**, 43 (1888).

Basu, S., Wall effect in laminar flow of non-Newtonian fluid through a packed bed, *Chem. Eng. J.*, **81**, 323 (2001).

Batishchev, Ya.F., Experimental investigation of steady state flow in tubes filled with fixed non-spherical packing, *Fluid Mech. – Sov. Res.*, **15** (3), 118 (1986).

Batra, V.K., Fulford, G.D. and Dullien, F.A.L., Laminar flow through periodically convergent – divergent tubes and channels, *Can. J. Chem. Eng.*, **48**, 622 (1970).

Battistella, A., van Schijndel, S.J.G., Baltussen, M.W., Roghair, I. and van Sint Annaland, M., On the terminal velocity of single bubbles rising in non-Newtonian power-law liquids, *J. Non-Newt. Fluid Mech.*, **278**, 104249 (2020).

Baudez, J., Markis, F., Eshtiaghi, N. and Slatter, P., The rheological behavior of anaerobic digested sludge, *Water Res.*, **45**, 5675 (2011).

Bauer, D., Talon, L., Peysson, Y., Ly, H.B., Batot, G., Chevalier, T. and Fleury, M., Experimental and numerical determination of Darcy's law for yield stress fluids in porous media, *Phys. Rev. Fluids*, **4**, 063301 (2019).

Baumann, T., Petsch, R. and Niessner, R., Direct 3-D measurement of the flow velocity in porous media using magnetic resonance tomography, *Environ. Sci. Technol.*, **34**, 4242 (2000).

Bazhlekov, I.B., van de Vosse, F.N. and Chesters, A.K., Numerical simulation of coalescence of viscoelastic drops in Newtonian fluids, Paper presented at the *PPS 15th Annual Mtg.*, Hertogenbosch, The Netherlands (May 31-June 4, 1999).

Bazhlekov, I.B., van de Vosse, F.N. and Chesters, A.K., Drainage and rupture of a Newtonian film between two power-law liquid drops interacting under a constant force, *J. Non-Newt. Fluid Mech.*, **93**, 181 (2000).

Bazilevskii, A.V., Koroteev, D.A., Rozhkov, A.N. and Skobeleva, A.A., Sedimentation of particles in shear flows of viscoelastic fluids, *Fluid Dyn.*, **45** (4), 626 (2010).

Beale, S.B., Potential flow in tube banks, *Trans. Can. Soc. Mech. Eng.*, **23**, 353 (1999).

Beale, S.B. and Spalding, D.B., Numerical study of fluid flow and heat transfer in tube banks with stream-wise periodic boundary conditions, *Trans. Can. Soc. Mech. Eng.*, **22**, 397 (1998).

Bear, J., *Dynamics of Fluids in Porous Media*, Elsevier, New York (1972).

Bear, J., *Modeling Phenomena of Flow and Transport in Porous Media*, Springer International Publishing, Cham (2018).

Bear, J. and Bachmat, Y., *Introduction to Modeling of Transport Phenomena in Porous Media*, Kluwer Academic, Dordrecht (1990).

Beard, D.W. and Walters, K., Elastico-viscous boundary layer flows I. Two-dimensional flow near a stagnation point, *Proc. Camb. Phil. Soc.*, **60**, 667 (1964).

Beaulne, M. and Mitsoulis, E., Creeping motion of a sphere in tubes filled with Herschel-Bulkley fluids, *J. Non-Newt. Fluid Mech.*, **72**, 55 (1997).

Becker, L.E., McKinley, G.H., Rasmussen, H.K. and Hassager, O., The unsteady motion of a sphere in a viscoelastic fluid, *J. Rheol.*, **38**, 377 (1994).

Becker, L.E., McKinley, G.H. and Stone, H.A., Sedimentation of a sphere near a plane wall: Weak non-Newtonian and inertial effects, *J. Non-Newt. Fluid Mech.*, **63**, 201 (1996).

Begovich, J.M. and Watson, J.S., Hydrodynamic characteristics of three-phase fluidized beds, *Fluidization*, edited by J.F. Davidson and D.L. Kearns, p. 190, Cambridge University Press, Cambridge, UK (1978).

Belmonte, A., Self-oscillations of a cusped bubble rising through a micellar solution, *Rheol. Acta*, **39**, 554 (2000).

Benenati, R.F. and Brosilow, C.B., Void fraction distribution in beds of spheres, *AIChE J.*, **8**, 359 (1962).

Benilov, E.S., Cummins, C.P. and Lee, W.T., Why do bubbles in Guinness sink? *Am. J. Phys.*, **81**, 88 (2013).

Benis, A.M., Theory of non-Newtonian flow through porous media in narrow three dimensional channels, *Int. J. Non-Linear Mech.*, **3**, 31 (1968).

Benn, D.I. and Ballantyne, C.K., Pebble shape (and size!) – Discussion, *J. Sed. Petrol.*, **62**, 1147 (1992).

Ben-Richou, A., Ambari, A. and Naciri, J.K., Correction factor of the Stokes force undergone by a sphere in the axis of a cylinder in uniform and Poiseuille flows, *Eur. Phys. J. Appl. Phys.*, **24**, 153 (2003).

Benyahia, F., On the global and local structural properties of packed beds of nonequilateral cylindrical particles, *Particul. Sci. Technol.*, **14**, 221(1996).

Benyahia, F. and O'Neill, K.E., Enhanced voidage correlations for packed beds of various particle shapes and sizes. *Particul. Sci. Technol.*, **23**,169 (2005).

Bercovier, M. and Engleman, M., A finite element method for incompressible non – Newtonian flows, *J. Comp. Phys.*, **36**, 313 (1980).

Berdnikov, V.I., Gudim, Y.A. and Karteleva, M.I., A generalized formula for calculating the speed at which solid particles, bubbles and droplets move in liquid and gaseous media, *Steel Transl.*, **27**, 22 (1997).

Bereiziat, D. and Devienne, R., Experimental characterization of Newtonian and non-Newtonian fluid flows in corrugated channels, *Int. J. Eng. Sci.*, **37**, 1461 (1999).

Bereiziat, D., Devienne, R. and Lebouche, M., Local flow structure for non-Newtonian fluids in a periodically corrugated wall channel, *J. Enhanc. Heat Transf.*, **2**, 71 (1995).

Berg, S. and van Wunnik, J., Shear rate determination from pore-scale flow fields. *Trans. Porous Media*, **117** (2), 229 (2017).

Bergelin, O.P., Brown, G.A., Hull, H.L. and Sullivan, F.W., Heat transfer and fluid friction during viscous flow across banks of tubes – III: A study of tube spacing and tube size, *Trans. ASME*, **72**, 881 (August, 1950).

Beris, A.N., Tsamopoulos, J., Armstrong, R.C. and Brown, R.A., Creeping motion of a sphere through a Bingham plastic, *J. Fluid Mech.*, **158**, 219 (1985).

Berkowski, B.M., A class of self-similar boundary layer problems for rheological power law fluids, *Int. Chem. Eng.*, **6**, 187 (1966).

Berney, B.M. and Deasy, P.B., Evaluation of Carbopol 934 as a suspending agent for sulphadimidine suspensions, *Int. J. Pharm.*, **3**, 73 (1979).

Bernstein, B., Kearsley, E.A. and Zapas, L., A study of stress relaxation with finite strain, *Trans. Soc. Rheol.*, **7**, 391 (1963).

Bertin, H.J., Apaydin, O.G., Castanier, L.M. and Kovscek, A.R., Foam flow in heterogeneous porous media: Effect of crossflow, Paper presented at *1998 SPE/DOE Improved Oil Recovery Symposium*, p. 255 (1998a).

Bertin, H.J., Quintard, M.Y. and Castanier, L.M., Modelling transient foam flow in porous media using a bubble population correlation, Paper presented at *1998 SPE Annual Technical Conference and Exhibition*, New Orleans, LA (1998b). (Paper # SPE 49020).

Bertsch, P., Savorani, L. and Fischer, P., Rheology of Swiss cheese fundue, *ACS Omega*, **4**, 1103 (2019).

Best, A.C., Empirical formulae for the terminal velocity of water drops falling through the atmosphere, *Quart. J. Roy. Meteor. Soc.*, **76**, 302 (1950).

Bethmont, S., Aloia, L.D., Stefani, C. and Leroy, R., Defining the stability criterion of a sphere suspended in a cement paste: A way to study the segregation risk in self-compacting concrete (SCC), *Proc. 3rd Int. RILEM Sym. Self-Compacting Concrete*, Reykjavik, pp. 94–105 (2003).

Bhaga, D. and Weber, M.E., Bubbles in viscous liquids: Shapes, wakes and velocities, *J. Fluid Mech.*, **105**, 61 (1981).

Bhamidipati, S. and Singh, R.K., Determination of fluid-particle convective heat transfer coefficient, *Trans. ASAE*, **38**, 857 (1995).

Bharti, R.P., Chhabra, R.P. and Eswaran, V., Steady flow of power-law fluids across a circular cylinder, *Can. J. Chem. Eng.*, **84**, 406 (2006).

Bharti, R.P., Chhabra, R.P. and Eswaran, V., A numerical study of the steady forced convection heat transfer from an unconfined circular cylinder, *Heat Mass Transf.*, **43**, 639 (2007a).

Bharti, R.P., Chhabra, R.P. and Eswaran, V., Steady forced convection heat transfer from a heated circular cylinder to power-law fluids, *Int. J. Heat Mass Transf.*, **50**, 977 (2007b).

Bharti, R.P., Chhabra, R.P. and Eswaran, V., Two-dimensional steady Poiseuille flow of power –law fluids across a circular cylinder in a plane confined channel: Wall effect and drag coefficient, *Ind. Eng. Chem. Res.*, **46**, 3820 (2007c).

Bharti, R.P., Chhabra, R.P. and Eswaran, V., Effect of blockage on heat transfer from a cylinder to power law liquids, *Chem. Eng. Sci.*, **62**, 4729 (2007d).

Bharti, R.P., Sivakumar, P. and Chhabra, R.P., Forced convection heat transfer from an elliptical cylinder to power law fluids, *Int. J. Heat Mass Transf.*, **51**, 1838 (2008).

Bhat, M.S., Poirier, D.R. and Heinrich, J.C., A permeability length scale for cross-flow through model structures, *Met. Trans.*, **26B**, 1091 (1995).

Bhatnagar, P.L., On two-dimensional boundary layer in non-Newtonian fluids with constant coefficients of viscosity and cross-viscosity, *Proc. Ind. Acad. Sci.*, **53A**, 95 (1960).

Bhattacharyya, S. and Singh, A., Mixed convection from an isolated spherical particle, *Int. J. Heat Mass Transf.*, **51**, 1034 (2008).

Bhavaraju, S.M., Mashelkar, R.A. and Blanch, H.W., Bubble motion and mass transfer in non-Newtonian fluids, *AIChE J.*, **24**, 1067 (1978). Corrections *ibid*, **26**, 528 (1980).

Bhavasar, P.M., Jafarabad, K.R., Pandit, A.B., Sawant, S.B. and Joshi, J.B., Drop volumes and terminal velocities in aqueous two-phase systems, *Can. J. Chem. Eng.*, **74**, 852 (1996).

Bhowmick, S., Molla, M.M. and Yao, L.-S., Non-Newtonian mixed convection flow along an isothermal horizontal circular cylinder, *Num. Heat Transf.*, **66A**, 509 (2014a).

Bhowmick, S., Molla, M.M., Mia, M. and Saha, S.C., Non-Newtonian mixed convection flow from a horizontal circular cylinder with uniform surface heat flux, *Procedia Eng.*, **90**, 510 (2014b).

Bingham, E.C., An ancient problem in rheology, *J. Rheol.*, **3**, 341 (1932).

Binnington, R.J. and Boger, D.V., Constant viscosity elastic liquids, *J. Rheol.*, **29**, 887 (1985).

Binnington, R.J. and Boger, D.V., Remarks on nonshear-thinning elastic fluids, *Polym. Eng. Sci.*, **26**, 133 (1986).
Binous, H. and Phillips, R.J., Dynamic simulation of one and two particles sedimenting in viscoelastic suspensions of FENE dumbbells, *J. Non-Newt. Fluid Mech.*, **83**, 93 (1999a).
Binous, H. and Phillips, R.J., The effect of sphere-wall interactions on particle motion in a viscoelastic suspension of FENE dumbbells, *J. Non-Newt. Fluid Mech.*, **85**, 63 (1999b).
Bird, R.B., New variational principle for incompressible non-Newtonian fluids, *Phys. Fluids*, **3**, 539 (1960).
Bird, R.B., Experimental tests of generalized Newtonian fluid models containing a zero-shear viscosity and a characteristic time, *Can. J. Chem. Eng.*, **43**, 161 (1965).
Bird, R.B., Useful non-Newtonian models, *Ann. Rev. Fluid Mech.*, **8**, 13 (1976).
Bird, R.B., Armstrong, R.C., Curtiss, C.F. and Hassager, O., *Dynamics of Polymeric Liquids, Vol. II: Kinetic Theory*, 2nd ed., Wiley, New York (1987b).
Bird, R.B., Armstrong, R.C. and Hassager, O., *Dynamics of Polymeric Liquids Vol. 1: Fluid Dynamics*, 2nd ed., Wiley, New York (1987a).
Bird, R.B., Dai, G.C. and Yarusso, B.J., The rheology and flow of viscoplastic materials, *Rev. Chem. Eng.*, **1**, 1 (1983).
Bird, R.B., Stewart, W.E. and Lightfoot, E.N., *Transport Phenomena*, 2nd ed., Wiley, New York (2002).
Bird, R.B. and Turian, R.M., Non-Newtonian flow in a rolling ball viscometer, *Ind. Eng. Chem. Fund.*, **3**, 87 (1964).
Bird, R.B. and Wiest, J.M., Constitutive equations for polymeric liquids, *Annu. Rev. Fluid Mech.*, **27**, 169 (1995).
Bisgaard, C., Velocity field around spheres and bubbles investigated by laser doppler anemometry, *J. Non-Newt. Fluid Mech.*, **12**, 283 (1983).
Bisgaard, C. and Hassager, O., An experimental investigation of velocity fields around spheres and bubbles moving in non-Newtonian liquids, *Rheol. Acta*, **21**, 537 (1982).
Bizzell, G.D. and Slattery, J.C., Non-Newtonian boundary layer flow, *Chem. Eng. Sci.*, **17**, 777 (1962).
Bjerkholt, J.T., Cumby, T.R. and Scotford, I.M., The properties of combined horizontal flows of air and farm livestock slurries in a tubular loop aerator, *Biosys. Eng.*, **91**, 77 (2005a).
Bjerkholt, J.T., Cumby, T.R. and Scotford, I.M., Pipeline design procedures for cattle and pig slurries using a large-scale pipeline apparatus, *Biosys. Eng.*, **91**, 201 (2005b).
Bjerkholt, J.T., Cumby, T.R. and Scotford, I.M., The effects of air injection on the pipeline transport of cattle and pig-slurries, *Biosys. Eng.*, **91**, 361 (2005c).
Blackery, J. and Mitsoulis, E., Creeping motion of a sphere in tubes filled with a Bingham plastic material, *J. Non-Newt. Fluid Mech.*, **70**, 59 (1997).
Blackwell, B.C., Deetjen, M.E., Gaudio, J.E. and Ewoldt, R.H. Sticking and splashing in yield stress fluid drop impacts on coated surfaces, *Phys. Fluids*, **27**, 043101 (2015).
Blake, F.C., The resistance of packing to fluid flow, *Trans. Amer. Inst. Chem. Eng.*, **14**, 3 (1922).
Blanco, A. and Magnaudet, J., The structure of the axisymmetric high Reynolds number flow around an ellipsoidal bubble of fixed shape, *Phys. Fluids*, **7**, 1265 (1995).
Blass, E., Formation and coalescence of bubbles and droplets, *Int. Chem. Eng.*, **30**, 206 (1990).
Bleyer, J. and Coussot, P., Breakage of non-Newtonian character in flow through a porous medium: Evidence from numerical simulation, *Phys. Rev. E*, **89**(6), 063018 (2014).
Bloom, F., Bubble stability in a class of non-Newtonian fluids with shear dependent viscosities, *Int. J. Non-Linear Mech.*, **37**, 527 (2002).
Blunt, M.J., Flow in porous media—pore-network models and multiphase flow. *Curr. Opin. Colloid Interface Sci.*, **6**,197 (2001).
Blunt, M.J., Jackson, M.D., Piri, M. and Valvatne, P.H., Detailed physics, predictive capabilities and macroscopic consequences for pore-network models of multiphase flow. *Adv.Water Resourc.*, **25**, 1069 (2002).
Boardman, G. and Whitmore, R.L., Yield stress exerted on a body immersed in a Bingham fluid, *Nature*, **187**, 50 (1960).
Boardman, G. and Whitmore, R.L., The static measurement of yield stress, *Lab. Prac.*, **10**, 782 (1961).
Bobinski, T., Goujon-Durand, S. and Wesfreid, J.E., Instabilities in the wake of a circular disk, *Phys. Rev., E.* **89**, 053021 (2014).
Bobroff, S. and Phillips, R.J., NMR imaging investigation of sedimentation of concentrated suspensions in non-Newtonian fluids, *J. Rheol.*, **42**, 1419 (1998).

Bodart, C. and Crochet, M.J., The time-dependent flow of a viscoelastic fluid around a sphere, *J. Non-Newt. Fluid Mech.*, **54**, 303 (1994).

Boersma, W.H., Laven, J. and Stein, H.N., Shear-thickening (dilatancy) in concentrated suspensions, *AIChE J.*, **36**, 321 (1990).

Boetcher, S.K.S., *Natural Convection from Circular Cylinders*, in Springer Briefs in Thermal Engineering and Applied Science, Springer, Cham (2014).

Boger, D.V., A highly elastic constant viscosity fluid, *J. Non-Newt. Fluid Mech.*, **3**, 87 (1977a).

Boger, D.V., Demonstration of upper and lower Newtonian fluid behaviour in a pseudoplastic fluid, *Nature*, **265**, 126 (1977b).

Boger, D.V., Dilute Polymer solutions and their use to model polymer processing flows, *Inter-relations between Processing, Structure and Properties of Polymeric Materials*, edited by J.C. Seferis and P.S. Theocaris, p. 307, Elsevier, Amsterdam (1984).

Boger, D.V., Viscoelastic fluid mechanics-Interaction between prediction and experiments, *Exp. Thermal Fluid Sci.*, **12**, 234 (1995).

Boger, D.V., Tiu, C. and Uhlherr, P.H.T., *Introduction to the Flow Properties of Polymers*, Royal Australian Chemical Institute and British Soc. Rheol. (Australian Branch) (1980).

Boger, D.V. and Walters, K., *Rheological Phenomena in Focus*, Elsevier, Amsterdam (1992).

Bohlin, T., On the drag on a rigid sphere moving in a viscous fluid inside a cylindrical tube, *Trans. Roy. Inst. Tech. Stockholm*, No. 155 (1960).

Böhm, L., Brehmer, M. and Kraume, M., Comparison of the single bubble ascent in a Newtonian and a non-Newtonian liquid: A phenomenological PIV study, *Chem. Eng. Technol.*, **88**, 93 (2016).

Böhm, L., Kurita, T., Kimura, K. and Kraume, M., Rising behaviour of single bubbles in narrow rectangular channels in Newtonian and non-Newtonian liquids, *Int. J. Multiph. Flow*, **65**, 11 (2014).

Bonadonna, C., Connor, C.B., Houghton, B.F., Connor, L., Byrne, M., Laing, A. and Hincks, T.K., Probabilistic modeling of tephra dispersal: Hazard assessment of a multiphase rhyolitic eruption at Tarawera, New Zealand, *J. Geophys. Res.*, **110**, B03203 (2005).

Bond, W.N. and Newton, D.A., Bubbles, drops and particles, *Phil. Mag.*, **5**, 794 (1928).

Bonn, D. and Denn, M.M., Yield stress fluids slowly yield to analysis, *Science*, **324**, 1401 (2009).

Bonn, D., Denn, M.M., Berthier, L. and Divoux, T., Yield stress materials in soft condensed matter, *Rev. Mod. Phys.*, **89**, 035005 (2017).

Borah, A. and Chhabra, R.P., Drag on freely falling cones in Newtonian and power law fluids, *Can. J. Chem. Eng.*, **83**, 559 (2005).

Borhan, A. and Pallinti, J., Pressure-driven motion of drops and bubbles through cylindrical capillaries: Effect of buoyancy, *Ind. Eng. Chem. Res.*, **37**, 3748 (1998).

Borhan, A. and Pallinti, J., Breakup of drops and bubbles translating through cylindrical capillaries, *Phys. Fluids.*, **11**, 2846 (1999).

Bories, S.A., Charrier-Mojtabi, M.C., Houi, D. and Raynaud, P.G., Non-invasive measurement techniques in porous media, *Convective Heat and Mass Transfer in Porous Media*, edited by S. Kakac, et al., p. 883, Kluwer Academic, Netherlands (1991).

Borwankar, R. and Shoemaker, C.F., *Rheology of Foods*, Elsevier, Amsterdam (1992).

Bose, A., Nirmalkar, N. and Chhabra, R.P., Forced convection from a heated equilateral triangular cylinder in Bingham plastic fluids, *Numer. Heat Transf. Part A*, **66**, 107 (2014).

Bose, A., Nirmalkar, N. and Chhabra, R.P., Effect of aiding-buoyancy on mixed convection from a heated cylinder in Bingham plastic fluids, *J. Non-Newt. Fluid Mech.*, **220**, 3 (2015).

Bot, E.T.G., Hulsen, M.A. and van den Brule, B.H.A.A., The motion of two spheres falling along their line of centers in a Boger fluid, *J. Non-Newt. Fluid Mech.*, **79**, 191 (1998).

Bouaziz, M., Kessentini, S. and Turki, S., Numerical prediction of flow and heat transfer of power-law fluids in a plane channel with a built-in heated square cylinder, *Int. J. Heat Mass Transf.*, **53**, 5420 (2010).

Bougas, A. and Stamatoudis, M., Wall factor acceleration and terminal velocity of falling spheres at high Reynolds numbers, *Chem. Eng. Technol.*, **16** 314 (1993).

Boujlel, J., Maillard, M., Lindner, A., Ovarlez, G., Chateau, X. and Coussot, P., Boundary layer in pastes - Displacement of a long object through a yield stress fluid, *J. Rheol.*, **56**, 1083 (2012).

Bourgeat, A. and Mikelic, A., Homogenization of a polymer through a porous medium, *Nonlinear Anal. Theory Meth. Appl.*, **26**, 1221 (1996).

Bourne, J.R., Further relationships for rotating sphere viscometer, *Brit. J. Appl. Phys.*, **16**, 1411 (1965).

Bousfield, D.W., Keunings, R., Marrucci, G. and Denn, M.M., Nonlinear analysis of the surface tension driven breakup of viscoelastic filaments, *J. Non-Newt. Fluid Mech.*, **21**, 79 (1986).
Boussinesq, J. *Applications a l etude des potentials*, Blanchard, Paris (1885).
Bowen, B.D. and Masliyah, J.H., Drag force on isolated particles in Stokes flow, *Can. J. Chem. Eng.*, **51**, 8 (1973).
Bowen, W.R. and Sharif, A.O., Transport through microfiltration membranes – Particle hydrodynamics and flux reduction, *J. Colloid Interfacial Sci.*, **168**, 414 (1994).
Boyd, J.W.R. and Varley, J., Acoustic emission measurement of low velocity plunging jets to monitor bubble size, *Chem. Eng. J.*, **97**, 11 (2004).
Bozzi, L.A., Feng, J.Q., Scott, T.C. and Pearlstein, A.J., Steady axisymmetric motion of deformable drops falling or rising through a homo-viscous fluid in a tube at intermediate Reynolds number, *J. Fluid Mech.*, **336**, 1 (1997).
Brasquet, C. and Le Cloirec, P., Pressure drop through textile fabrics – experimental data modeling using classical models and neural networks, *Chem. Eng. Sci.*, **55**, 2767 (2000).
Brea, F.M., Edwards, M.F. and Wilkinson, W.L., The flow of non-Newtonian slurries through fixed and fluidized beds, *Chem. Eng. Sci.*, **31**, 329 (1976).
Breakey, D.E.S., Vaezi, G.F., Masliyah, J.H. and Sanders, R.S., Side-view-only determination of drag coefficient and settling velocity for non-spherical particles, *Powder Technol.*, **339**, 182 (2018).
Brenn, G., Liu, Z. and Durst, F., Linear analysis of the temporal instability of axisymmetrical non-Newtonian liquid jets, *Int. J. Multiph. Flow*, **26**, 1621 (2000).
Brennen, C.E., *Cavitation and Bubble Dynamics*, Cambridge University Press, New York (1995).
Brennen, C. and Gadd, G.E., Aging and degradation in dilute polymer systems, *Nature*, **215**, 1368 (1967).
Brenner, H., The slow motion of a sphere through a viscous fluid towards a plane surface, *Chem. Eng. Sci.*, **16**, 242 (1961).
Brenner, H., Dynamics of a particle in a viscous fluid, *Chem. Eng. Sci.*, **17**, 435 (1962).
Brenner, H., Hydrodynamic resistance of particles at small Reynolds numbers, *Adv. Chem. Eng.*, **6**, 287 (1966).
Briedis, D., Moutrie, M.F. and Balmer, R.T., A study of the shear viscosity of human whole saliva, *Rheol. Acta*, **19**, 365 (1980).
Briend, P., Chavarie, C., Tassart, M. and Hlavacek, B., Comportement des lits fluidizes en milieu viscoelastique, *Can. J. Chem. Eng.*, **62**, 26 (1984).
Briens, C.L., Correlation for the direct calculation of the terminal velocity of spherical particles in Newtonian and Pseudoplastic (power-law) fluids, *Powder Technol.*, **67**, 87 (1991).
Brinkman, H.C., A calculation of the viscous force exerted by a flowing fluid on a dense swarm of particles, *App. Sci. Res.*, **A1**, 27 (1947).
Brinkman, H.C., On the permeability of media consisting of closely packed porous particles, *App. Sci. Res.*, **A1**, 81 (1948).
Briscoe, B.J. and Chaudhary, B.I., Bubble dynamics in polymer melts, *Proc. MRS Symposium on Interfaces between Polymers, Melts and Ceramics*, San Diego, CA (1989).
Briscoe, B.J., Glaese, M., Luckham, P.F. and Ren, S., The falling of spheres through Bingham fluids, *Colloids Surf.*, **65**, 69 (1992a).
Briscoe, B.J., Luckham, P.F. and Ren, S.R., An assessment of a rolling ball viscometer for studying non-Newtonian fluids, *Colloids Surf.*, **62**, 153 (1992b).
Briscoe, B.J., Luckham, P.F. and Ren, S.R., The settling of spheres in clay suspensions, *Powder Technol.*, **76**, 165 (1993).
Briscoe, B.J., Luckham, P.F. and Ren, S.R., The properties of drilling muds at high pressures and high temperatures, *Phil. Trans. R. Soc. Lond. A*, **348**, 179 (1994).
Briscoe, B.J., Luckham, P. and Zhu, S., Pressure influences upon shear-thickening of poly (acrylamide) solutions, *Rheol. Acta*, **38**, 224 (1999).
Broadbent, J.M. and Mena, B., Slow flow of an elastico-viscous fluid past cylinders and spheres, *Chem. Eng. J.*, **8**, 11 (1974).
Broer, L.J.R., On the hydrodynamics of viscoelastic fluids, *App. Sci. Res.*, **6**, 226 (1957).
Broniarz-Press, L., Agacinski, P. and Rozanski, J., Effect of fluidization process on rheological properties of polymer solutions, *Rec. Prog. Genie des Procedes*, **13**, 205 (1999).
Broniarz-Press, L., Agacinski, P. and Rozanski, J., Shear-thinning fluids flow in fixed and fluidized beds, *Int. J. Multiph. Flow*, **33**, 675 (2007a).

Broniarz-Press, L., Pralat, K. and Pye, K.W., Experimental analysis of thermal conductivity of carboxymethyl Sodium salt aqueous solutions in coaxial cylinder system, *Proc. European Cong. Chem. Eng.* (ECCE-6), Copenhagen, 16–20 September (2007b).
Brookes, G.F. and Whitmore, R.L., The static drag on bodies in Bingham Plastics, *Rheol. Acta*, **7**, 188 (1968).
Brookes, G.F. and Whitmore, R.L., Drag forces in Bingham plastics, *Rheol. Acta*, **8**, 472 (1969).
Brower, L.E. and Ferraris, C.F., Comparison of concrete rheometers, *Concrete Int.*, **25**, 41 (August 2003).
Brown, D.J., Vickers, G.T., Collier, A.P. and Reynolds, G.K., Measurement of the size, shape and orientation of convex bodies, *Chem. Eng. Sci.*, **60**, 289 (2005).
Brown, E., Forman, N.A., Orellana, C.S., Zhang, H., Maynor, B.W., Betts, D.E., De Simone, J.M. and Jaegar, H.M., Generality of shear-thickening in dense suspensions, *Nature Mater.*, **9**, 220 (2010).
Brown, G.G. and associates, *Unit Operations*, John Wiley & Sons, New York (1950).
Brown Jr., J.C., Determination of the exposed specific surface of pulp fibers from air permeability measurements, *TAPPI*, **33**, 130 (1950).
Brown, P.P. and Lawler, D.F., Sphere drag and settling velocity revisited, *ASCE J. Environ. Eng.*, **129**, 222 (2003).
Brown, R.A. and McKinley, G.H., Report on the VIIIth international workshop on numerical methods in viscoelastic flows, *J. Non-Newt. Fluid Mech.*, **52**, 407 (1994).
Brown, R.A., Szady, M.J., Northey, P.J. and Armstrong, R.C., On the numerical stability of mixed finite-element methods for viscoelastic flows governed by differential constitutive equations, *Theoret. Comput. Fluid Dyn.*, **5**, 77 (1993).
Brown, R.C., Airflow through filters – beyond single fibre theory, *Advances in Aerosol Filtration*, edited by K.R. Spurny, Lewis Publishers, New York, Chapter 8 (1998).
Brown, S.W.J. and Williams, P.R., Bubble collapse and liquid jet formation in non-Newtonian liquids, *AIChE J.*, **45**, 2653 (1999).
Brown, W.D. and Sorbie, K.S., Dispersion and polydispersity effects in the transport of Xanthan in porous media, *Macromolecules*, **22**, 2835 (1989).
Browne, C.A., Shih, A. and Datta, S.S., Bistability in the unstable flow of polymer solutions through pore constriction arrays. *J. Fluid Mech.*, **890**, A2–1 (2020).
Bruggeman, D.A.G., Dielectric constant and conductivity of mixtures of isotropic materials. *Ann. Phys. (Leipzig)*, **24**, 636 (1935).
Brujan, E.-A., The behaviour of bubbles in Bueche model fluids, *Polym. Eng. Sci.*, **34**, 1550 (1994a).
Brujan, E.-A., The effect of polymer concentration on the non-linear oscillation of a bubble in a sound-irradiated liquid, *J. Sound Vib.*, **173**, 329 (1994b).
Brujan, E.-A., A first-order model for bubble dynamics in a compressible viscoelastic liquid, *J. Non-Newt. Fluid Mech.*, **84**, 83 (1999).
Brujan, E.-A., Bubble dynamics in a compressible shear-thinning liquid, *Fluid Dyn. Res.*, **23**, 291 (1998).
Brujan, E.-A., Collapse of cavitation bubbles in blood, *Europhys. Letters*, **50**, 175 (2000).
Brujan, E.-A., Ohl, C.-D., Lauterborn, W. and Philipp, A., Dynamics of laser-induced cavitation bubbles in polymer solutions, *Acta Acustica*, **82**, 423 (1996).
Brummer, R., *Rheology Essentials of Cosmetic and Food Emulsions*, Springer, Berlin (2006).
Brunn, P., The slow motion of a sphere in a second order fluid, *Rheol. Acta*, **15**, 163 (1976b). Errata *ibid.*, 16, 324 (1977).
Brunn, P., The slow motion of a rigid particle in a second order fluid, *J. Fluid Mech.*, **82**, 529 (1977a).
Brunn, P., Interaction of spheres in a viscoelastic fluid, *Rheol. Acta*, **16**, 461 (1977b).
Brunn, P., The motion of a slightly deformed sphere in a viscoelastic fluid, *Rheol. Acta*, **18**, 229 (1979).
Brunn, P., The motion of rigid particles in viscoelastic fluids, *J. Non-Newt. Fluid Mech.*, **7**, 271 (1980).
Brunn, P.O., The behaviour of a sphere in non-homogeneous flows of a viscoelastic fluid, *Rheol. Acta*, **15**, 589 (1976a).
Brunn, P.O. and Holweg, J., The effect of shear in porous medium flow of surfactant solutions, *J. Non-Newt. Fluid Mech.*, **30**, 317 (1988).
Bruschke, M.V. and Advani, S.G., Flow of generalized Newtonian fluids across a periodic array of cylinders, *J. Rheol.*, **37**, 479 (1992).
Brutyan, M.A. and Krapivsky, P.L., Collapse of spherical bubbles in fluids with nonlinear viscosity, *Quart. Appl. Math.*, **L1**, 745 (1993).

Bryan, W.L. and Silman, R.W., Rolling-sphere viscometer for *in situ* monitoring of shake-flask fermentations, *Enzyme Microb. Technol.*, **12**, 818 (1990).

Bu, S., Yang, J., Dong, Q. and Wang Q., Experimental study of transition flow in packed beds of spheres with different particle sizes based on electrochemical microelectrodes measurement, *App. Therm. Eng.*, **73**, 1525 (2014).

Buchholz, H., Buchholz, R., Lucke, J. and Schugerl, K., Bubble swarm behaviour and gas absorption in non-Newtonian fluids in sparged vessels, *Chem. Eng. Sci.*, **33**, 1061 (1978).

Buchholz, K. and Godelmann, B., Pressure drop across compressible beds, *Enzyme Eng.*, **4**, 89 (1973).

Budzynski, P., Dziubinski, M. and Orczykowska, M., Shape of gas bubbles rising in non-Newtonian liquids, *Inzynieria Chemiczna I Procesowa*, **24**, 533 (2003).

Bui, C.M. and Ho, T.X., Numerical study of an unsteady flow of thixotropic liquids past a cylinder, *AIP Adv.*, **9**(11), 115002 (2019).

Bui, C.M. and Ho, T.X., Flow of a thixotropic Bingham fluids over a cylinder in a stationary and non-stationary regimes, *J. Appl. Fluid Mech.*, **13**, 1527 (2020).

Bulina, I.G., Myasnikov, V.P. and Savin, V.G., Experimental investigation of flow around blunted bodies by a flat flow of a plastic medium, *J. Appl. Mech. Tech. Phys.*, **203**, (May 1967) (English translation).

Bullivant, S.A. and Jones, T.E.R., Elastico-viscous properties of deflocculated china clay suspensions – concentration effects, *Rheol. Acta*, **20**, 64 (1981).

Burcik, E.J., A note on the flow behaviour of polyacrylamide solutions in porous media, *Prod. Monthly*, **29** (6), 14 (1965).

Burcik, E.J., The use of polymers in the recovery of petroleum, *Earth Min. Sci.*, **37** (7), 57 (1968).

Burcik, E.J., The mechanism of microgel formation in partially hydrolyzed polyacrylamide, *J. Pet. Tech.*, **21**, 373 (1969).

Burcik, E.J. and Ferrer, J., The mechanism of pseudo dilatant flow, *Prod. Monthly*, **32** (3), 7 (1968).

Burcik, E.J. and Walrond, K.W., Microgel in polyacrylamide solutions and its role in mobility control, *Prod. Monthly*, **32** (9), 12 (1968).

Burman, J.E. and Jameson, G.J., Growth of spherical gas bubbles by solute diffusion in non-Newtonian (Power law) liquids, *Int. J. Heat Mass Transf.*, **21**, 127 (1978).

Burru, I.G. and Briens, C.L., Particle-liquid mass transfer in three phase fluidized beds at high velocities with non-Newtonian liquids, *Fluidization VI*, edited by J.R. Grace, L.W. Shemilt and M.A. Bergougnou, Engineering Foundation, American Institute of Chemical Engineers, New York, p. 499 (1989).

Burru, I.G. and Briens, C.L., Particle-liquid mass transfer in three-phase fluidized beds Part I- Newtonian liquids, *Powder Technol.*, **68**, 243 (1991). Also see *ibid* 255.

Busch, A. and Johansen, S.T., An Eulerian-Lagrangian CFD study of a particle settling in an orthogonal shear flow of a shear-thinning, mildly viscoelastic fluid, *J. Non-Newt. Fluid Mech.*, **263**, 77 (2019).

Bush, M.B., The stagnation flow behind a sphere, *J. Non-Newt. Fluid Mech.*, **49**, 103 (1993).

Bush, M.B., On the stagnation flow behind a sphere in a shear-thinning viscoelastic liquid, *J. Non-Newt. Fluid Mech.*, **55**, 229 (1994).

Bush, M.B. and Phan-Thien, N., Drag force on a sphere in creeping motion through a Carreau model fluid, *J. Non-Newt. Fluid Mech.*, **16**, 303 (1984).

Butcher, T.A. and Irvine Jr., T.F., Use of the falling ball viscometer to obtain flow curves for inelastic non-Newtonian fluids, *J. Non-Newt. Fluid Mech.*, **36**, 51 (1990).

Cairncross, E.K. and Hansford, G.S., An experimental investigation of the flow about a sphere rotating in a Rivlin-Ericksen fluid, *J. Non-Newt. Fluid Mech.*, **3**, 203 (1977/1978).

Cakl, J. and Machac, I., Pressure drop in the flow of viscoelastic fluids through fixed beds of particles, *Collect. Czech. Chem. Commun.*, **60**, 1124 (1995).

Cakl, J., Machac, I. and Lecjaks, Z., Flow of viscoelastic liquids through fixed beds, *Prog. Trends in Rheol. II*, 266 (1988).

Calderbank, P.H., *Gas Absorption from Bubbles*, p. 209, Institution of Chemical Engineers, London (October 1967).

Calderbank, P.H., Johnson, D.S.L. and Loudon, J., Mechanics and mass transfer of single bubbles in free rise through some Newtonian and non-Newtonian liquids, *Chem. Eng. Sci.*, **25**, 235 (1970).

Camacho, M.M., Martinez-Navarrete, N. and Chiralt, A., Rheological characterization of experimental dairy creams formulated with locust bean gum (LBG) and λ- carrageenan combinations, *Int. Dairy J.*, **15**, 243 (2005).

Camas-Anzueto, J.L., Gomez-Perez, J., Meza-Gordillo, R., Anzueto-Sanchez, G., Perez-Patricio, M., Lopez-Estrada, F.R., Abud-Archilla, M. and Rios-Rojas, C., Measurement of the viscosity of biodiesel by using an optical viscometer, *Flow Meas. Instrum.*, **54**, 82 (2017).

Cao, B., Fan, J., Sun, X. and Li, S., Numerical simulation of mass transfer characteristics of a bubble rising in yield stress fluids, *ACS Omega*, **5**, 13878 (2020).

Capobianchi, M. and Aziz, A., Laminar natural convection from an isothermal vertical surface to pseudoplastic and dilatant fluids, *J. Heat Transf.*, **134**, 122502-1 (2012).

Caputo, A.C. and Pelagagge, P.M., Flow modeling in fabric filters, *J. Porous Media*, **2**, 191 (1999).

Carberry, J.J., A boundary layer model of fluid particle mass transfer in fixed beds, *AIChE J.*, **6**, 460 (1960).

Carew, E.O.A. and Townsend, P., Non-Newtonian fluid flow past a sphere in a long cylindrical tube, *Rheol. Acta*, **27**, 125 (1988).

Carew, E.O.A. and Townsend, P., Slow visco-elastic flow past a cylinder in a rectangular channel, *Rheol. Acta*, **30**, 58 (1991).

Carew, P.S., Thomas, N.H. and Johnson, A.B., A physically based correlation for the effects of power law rheology and inclination on slug bubble rise velocity, *Int. J. Multiph. Flow*, **21**, 1091 (1995).

Carey, W.W. and Turian, R.M., Settling of spheres in drag reducing polymer solutions, *Ind. Eng. Chem. Fund.*, **9**, 185 (1970).

Carman, P.C., Fluid flow through granular beds, *Trans. Inst. Chem. Eng.*, **15**, 150 (1937).

Carman, P.C., The determination of the specific surface of powders I", *Soc. Chem. Ind. (Trans. Commun.)*, **57**, 225 (1938).

Carman, P.C., *Flow of Gases Through Porous Media*, Butterworths, London (1956).

Carne, R., Clough, G.F. and Newcombe, S., Oxygen injection into a rising main: Its effect on pumping, *Water Pollut. Control.*, **81**, 399 (1982).

Carpinlioglu, M.O. and Ozahi, E., A simplified correlation for fixed bed pressure drop, *Powder Technol.*, **187**, 94 (2008).

Carreau, P.J., Rheological equations from molecular network theories, *Trans. Soc. Rheol.*, **16**, 99 (1972).

Carreau, P.J., De Kee, D. and Chhabra, R.P., *Rheology of Polymeric Systems*, 2nd ed., Hanser, Munich (2021).

Carreau, P.J., Devic, M. and Kapellas, M., Dynamique des bulles en milieu viscoelastique, *Rheol. Acta*, **13**, 477 (1974).

Cartier, S., Horbett, T.A. and Ratner, B.D., Glucose-sensitive membrane coated porous filters for control of hydraulic permeability and insulin delivery from a pressurized reservoir, *J. Membrane Sci.*, **106**, 17 (1995).

Caserta, S., D'Avino, G., Greco, F., Guido, S. and Maffettone, P.L., Migration of a sphere in a viscoelastic fluid under planar shear flow: Experiments and numerical predictions, *Soft Matter*, **7**, 1100 (2011).

Castillo, A., Murch, W.L., Einarsson, J., Mena, B., Shaqfeh, E.S.G. and Zenit, R., Drag coefficient for a sedimenting and rotating sphere in a viscoelastic fluid, *Phys. Rev. Fluids*, **4**, 063302 (2019).

Caswell, B., Fluid mechanics of non-Newtonian fluids, *Ph. D Thesis*, Stanford University, CA (1962).

Caswell, B., The effect of finite boundaries on the motion of particles in non-Newtonian fluids, *Chem. Eng. Sci.*, **25**, 1167 (1970).

Caswell, B., The stability of particle motion near a wall in Newtonian and non-Newtonian fluids, *Chem. Eng. Sci.*, **27**, 373 (1972).

Caswell, B., Sedimentation of particles in non-Newtonian fluids, *The Mechanics of Viscoelastic Fluids*, R.S. Rivlin (ed.), ASME, AMD, 22, 19 (1977).

Caswell, B., Report on the IXth International workshop on Numerical Methods in Non-Newtonian Flows, *J. Non-Newt. Fluid Mech.*, **62**, 99 (1996).

Caswell, B., Manero, O. and Mena, B., Recent developments on the slow viscoelastic flow past spheres and bubbles, *Rheology Reviews*, p. 197, published by the British Soc. of Rheology (2004).

Caswell, B. and Schwarz, W.H., The creeping motion of a non-Newtonian fluid past a sphere, *J. Fluid Mech.*, **13**, 417 (1962).

Ceylan, K., Hardem, S. and Abbasov, T., A theoretical model for estimation of drag force in the flow of non-Newtonian fluids around spherical solid particles, *Powder Technol.*, **102**, 286 (1999).

Chafe, N.P. and de Bruyn, J.R., Drag and relaxation in a bentonite clay suspension, *J. Non-Newt. Fluid Mech.*, **130**, 129 (2005).

Chahine, G.L. and Fruman, D.H., Dilute polymer solutions effects on bubble growth and collapse, *Phys. Fluids*, **22**, 1406 (1979).

Chakrabarti, A. and Chaudhury, M.K., Direct measurement of the surface tension of a soft elastic hydrogel: Exploration of elasto-capillary instability in adhesion, *Langmuir*, **29**, 6926 (2013a).

Chakrabarti, A. and Chaudhury, M.K., Surface-folding-induced attraction and motion of particles in a soft elastic gel: Cooperative effects of surface tension, elasticity, and gravity, *Langmuir*, **29**, 15543 (2013b).

Chakrabarti, S., Seidl, B., Vorwerk, J. and Brunn, P.O., The rheology of hydroxypropyl guar (HPG) solutions and its influence on the flow through a porous medium and turbulent flow (Part I), *Rheol. Acta.*, **34**, 114 (1991). Also see *ibid*, 124 (1991).

Chakraborty, J., Verma, N. and Chhabra, R.P., Wall effects in flow past a circular cylinder in a plane channel: A numerical study, *Chem. Eng. Process.*, **43**, 1529 (2004).

Chamkha, A.J., Similarity solution for thermal boundary layer on a stretched surface of a non-Newtonian fluid, *Int. Comm. Heat Mass Transf.*, **24**, 643 (1997).

Chamsri, K. and Bennethum, L.S., Permeability of fluid flow through a periodic array of cylinders, *App. Math. Modell.*, **39**, 244 (2015).

Chan Man Fong, C.F. and De Kee, D., The effect of a thermal gradient on the motion of a bubble in a viscoelastic fluid, *J. Non-Newt. Fluid Mech.*, **53**, 165 (1994).

Chan, P.C.-H. and Leal, L.G., A note on the motion of a spherical particle in a general quadratic flow of a second order fluid, *J. Fluid Mech.*, **82**, 549 (1977).

Chan, P.C.-H. and Leal, L.G., The motion of a deformable drop in a second order fluid, *J. Fluid Mech.*, **92**, 131 (1979).

Chan, P.C.-H. and Leal, L.G., An experimental study of drop migration in shear flow between concentric cylinders, *Int. J. Multiph. Flow*, **7**, 83 (1981).

Chan, R.K.Y. and Jackson, D.A., An automated falling-cylinder high pressure laser-doppler viscometer, *J. Phys. E: Sci. Instrum.*, **18**, 510 (1985).

Chandarana, D.I., Gavin III, A. and Wheaton, F.W., Particle/fluid interface heat transfer under UHT conditions at low particle/fluid relative velocities, *J. Food Process Eng.*, **13**, 191 (1990).

Chandra, A. and Chhabra, R.P., Flow over and forced convection heat transfer in Newtonian fluids from a semi-circular cylinder, *Int. J. Heat Mass Transf.*, **54**, 225 (2011a).

Chandra, A. and Chhabra, R.P., Momentum and heat transfer characteristics of a semi-circular cylinder immersed in power-law fluids in the steady flow regime, *Int. J. Heat Mass Transf.*, **54**, 2734 (2011b).

Chandra, A. and Chhabra, R.P., Influence of power-law index on transitional Reynolds numbers for flow over a semi-circular cylinder, *App. Math. Modell.*, **35**, 5766 (2011c).

Chandra, A. and Chhabra, R.P., Laminar free convection from a horizontal semicircular cylinder to power-law fluids, *Int. J. Heat Mass Transf.*, **55**, 2934 (2012a).

Chandra, A. and Chhabra, R.P., Mixed convection from a heated semi-circular cylinder to power-law fluids in the steady flow regime, *Int. J. Heat Mass Transf.*, **55**, 214 (2012b).

Chandra, A. and Chhabra, R.P., Momentum and heat transfer from a semi-circular cylinder to power-law fluids in the vortex shedding regime, *Numer. Heat Transf., Part A*, **63**, 489 (2013).

Chandrasekhara, B.C. and Vortmeyer, D., Flow model for velocity distribution in fixed porous beds under isothermal conditions, *Wärme-ünd Stöffubertragung*, **12**, 105 (1979).

Chang, C. and Smith, P.A., Rheological characterization of nuclear waste slurries, *Particulate Sci. Technol.*, **14**, 165 (1996).

Chang, J.Y., Oh, Y.-K., Choi, H., Kim, Y.B. and Kim, C.-K., Rheological evaluation of thermosensitive and mucoadhesive vaginal gels in physiological conditions, *Int. J. Pharm.*, **241**, 155 (2002).

Chang, K.-S. and Sa, J.-Y., The effect of buoyancy on vortex shedding in the near wake of a circular cylinder, *J. Fluid Mech.*, **220**, 253 (1990).

Chaoyang, W. and Chuanjing, T., Boundary layer flow and heat transfer of non-Newtonian fluids in porous media, *Int. J. Heat Fluid Flow*, **10**, 160 (1989).

Chaoyang, W., Chuanjing, T. and Ling, Y., Boundary layer flow and heat transfer of power-law fluids in packed beds, *Heat Transf.* (2nd U.K. Nat. Conf.), **2**, 1431 (1988).

Chaplain, V., Allain, C. and Hulin, J.P., Tracer dispersion in power law fluids through porous media: Evidence of cross-over from a logarithmic to a power law behaviour, *Eur. Phys. J.*, **B6**, 225 (1998).

Chaplain, V., Mills, P., Guiffant, G. and Cerasi, P., Model for the flow of a yield fluid through a porous medium, *J. Phys. II France*, **2**, 2145 (1992).

Charpentier, J.C., Recent progress in two-phase gas-liquid mass transfer in packed bed, *Chem. Eng. J.*, **11**, 161 (1976).

Chase, G.D., *Ph.D Dissertation*, Temple University, Philadelphia, PA (1955).
Chase, G.D. and Dachavijit, P., Incompressible cake filtration of a yield stress fluid, *Sep. Sci. Technol.*, **38**, 745 (2003).
Chase, G.G. and Dachavijit, P., A correlation for yield stress fluid flow through packed beds, *Rheol. Acta*, **44**, 495 (2005).
Chase, G.G. and Dachavijit, P., An experimental study of electrorheological fluid flow through a packed bed of glass beads. *Trans. Porous Media*, **72** (1), 25 (2008).
Chatzis, I. and Dullien, F.A.L., Modelling pore structure by 2-D and 3-D networks with application to sandstones, *J. Can. Pet. Tech.*, **16** (1), 97 (1977).
Chaudhari, R.V. and Hofmann, H., Coalescence of gas bubbles in liquids, *Rev. Chem. Eng.*, **10**, 131 (1994).
Chaudhary, R.C. and Böhme, G., Zur stromung Viskoser flussigkeiten durch gewellte rohre, *Rheol. Acta*, **26**, 272 (1987).
Chauhan, A., Sasmal, C. and Chhabra, R.P., Effects of blockage and fluid inertia on drag and heat transfer of a solid sphere translating in FENE-P viscoelastic fluids in a tube. *J. Non-Newt. Fluid Mech.*, **294**, 104593 (2021).
Chauveteau, G., Rodlike polymer solution flow through fine pores: Influence of pore size on rheological behaviour, *J. Rheol.*, **26**, 111 (1982).
Chauveteau, G. and Thirriot, E.C., Regimes d'ecoulement en millieu poreux et limite de la loi de Darcy, *L. Houille Balanche*, **2**, 141 (1967).
Chauviere, C. and Owens, R.G., How accurate is your solution? Error indicators for viscoelastic flow calculations, *J. Non-Newt. Fluid Mech.*, **95**, 1 (2000).
Cheddadi, I., Saramito, P. and Graner, F., Steady Couette flows of elasto-visco-plastic fluids are nonunique, *J. Rheol.* **56**, 213 (2012).
Chee, D.-W., Cho, Y.I., Cho, K. and Lu, P., DNA orientation during electrophoresis in a viscoelastic solution, *Sep. Technol.*, **4**, 55 (1994).
Cheikhrouhou, M. and Sigli, D., Influence of the structure of fabric filters on the velocity and stress fields of filtration flows in their vicinity, *Text. Res. J.*, **58**, 371 (1988).
Chen, C.-T., Malkus, D.S. and Vanderby Jr., R., A fiber matrix model for interstitial fluid flow and permeability in ligaments and tendons, *Biorheol.*, **35**, 103 (1998a).
Chen, C.Y., Filtration of aerosols by Fibrous media, *Chem. Rev.*, **55**, 595 (1955).
Chen, H.-T. and Chen, C.-K., Natural convection of a non-Newtonian fluid about a horizontal cylinder and a sphere in a porous medium, *Int. Comm. Heat Mass Transf.*, **15**, 605 (1988). Also see *Chem. Eng. Commun.*, **69**, 29 (1988).
Chen, I.-H., An experimental study of the behaviour of gas bubbles and liquid drops in non-Newtonian fluids, *MS Dissertation*, State University of New York at Buffalo, Buffalo, NY (1980).
Chen, J.L.S. and Kubler, E.A., Non-Newtonian flow along needles, *Phys. Fluids*, **21**, 749 (1978).
Chen, J.L.S. and Radulovic, P.T., Heat transfer in non-Newtonian flow past a wedge with nonisothermal surfaces, *J. Heat Transf. (ASME)*, **95**, 498 (1973).
Chen, K.-F. and Chen, S.-M., Fluidization properties of high – consistency fiber suspensions, *Expt. Thermal Fluid Sci.*, **14**, 149 (1997).
Chen, L., Garimella, S.V., Reizes, J.A. and Leonardi, E., The development of a bubble rising in a viscous liquid, *J. Fluid Mech.*, **387**, 61 (1999).
Chen, L., Reizes, J.A., Leonardi, E. and Li, Y., Ball falling through viscoplastic fluid in a cylinder, *Proc. 13th Australasian Fluid Mech. Conf.*, Melbourne, p. 297 (1998b).
Chen, M., Rossen, W. and Yortsos, Y.C., The flow and displacement in porous media of fluids with yield stress. *Chem. Eng. Sci.*, **60** (15), 4183 (2005).
Chen, M.C.S., Lescarboura, J.A. and Swift, G.W., The effect of eccentricity on the terminal velocity of the cylinder in a falling cylinder viscometer, *AIChE J.*, **14**, 123 (1968).
Chen, M.C.S. and Swift, G.W., Analysis of entrance and exit effects in a falling cylinder viscometer, *AIChE J.*, **18**, 146 (1972).
Chen, S. and Rothstein, J.P., Flow of a wormlike micelle solution past a falling sphere, *J. Non-Newt. Fluid Mech.*, **116**, 205 (2004).
Chen, S., Sun, Q., Jin, F. and Liu, J., Simulations of Bingham plastic flows with the multiple-relaxation-time lattice Boltzmann model. *Sci. China: Phys. Mech. Astron.*, **57**, 532 (2014).

Chen, S.-G., Zhang, C.-H., Feng, Y.-T., Sun, Q.-C. and Jin, F., Three dimensional simulations of Bingham plastic flows with multiple relaxation time lattice Boltzmann model, *Eng. Appl. Comput. Fluid Mech.*, **10**, 346 (2016).

Chen, T.S. and Mucoglu, A., Analysis of mixed forced and free convection about a sphere, *Int. J. Heat Mass Transf.*, **20**, 867 (1977).

Chen, T.S. and Mucoglu, A., Mixed convection about a sphere with uniform surface heat flux, *J. Heat Transf.*, **100**, 542 (1978).

Chen, T.V.W. and Wollersheim, D.E., Free convection at a vertical plate with uniform flux condition in non-Newtonian power law fluids, *J. Heat Transf. (ASME)*, **95**, 123 (1973).

Chen, Y.R. and Hashimoto, A.G., Rheological properties of aerated poultry waste slurries, *Trans. ASAE*, **19**, 128 (1976).

Cheng, C.-Y., The effect of temperature dependent viscosity on the natural convection heat transfer from a horizontal isothermal cylinder of elliptic cross section, *Int. Comm. Heat Mass Transf.*, **33**, 1021 (2006).

Cheng, D.C.-H., Measurement techniques for thixotropic properties, *Brit. Ceramic Soc. Convention*, York, England (1979).

Cheng, D.C.-H., Yield stress: A time dependent property and how to measure it, *Rheol. Acta*, **25**, 542 (1986).

Cheng, D.C.-H., Thixotropy, *Int. J. Cosmetic Sci.*, **9**, 151 (1987).

Cheng, D.C.-H., Characterization of thixotropy revisited, *Rheol. Acta*, **42**, 372 (2003).

Cheng, N.-S., Comparison of formulas for drag coefficient and settling velocity of spherical particles, *Powder Technol.*, **189**, 395 (2009).

Cheng, N.S., Wall effect on pressure drop in packed beds, *Powder Technol.*, **210**, 261 (2011).

Cheng, X., Mc Coy, J.H., Israelachvili, J.N. and Cohen, I., Imaging the microscopic structure of shear-thinning and thickening colloidal suspensions, *Science*, **333**, 1276 (2011).

Cheny, J.M. and Walters, K., Extravagant viscoelastic effects in the Worthington jet experiment, *J. Non-Newt. Fluid Mech.*, **67**, 125 (1996).

Chester, W. and Breach, D.R., On the flow past a sphere at low Reynolds number, *J. Fluid Mech.*, **37**, 75 (1969).

Chesters, A.K. and Hoffman, G., Bubble coalescence in pure liquids, *App. Sci. Res.*, **38**, 353 (1982).

Chevalier, T., Chevalier, C., Clain, X., Dupla, J.C., Canou, J., Rodts, S. and Coussot, P., Darcy's law for yield stress fluid flowing through a porous medium. *J. Non-Newt. Fluid Mech.*, **195**, 57 (2013).

Chevalier, T., Rodts, S., Chateau, X., Chevalier, C. and Coussot, P., Breaking of non-Newtonian character in flows through a porous medium, *Phys. Rev. E*, **89**, 023002 (2014).

Chevalier, T. and Talon, L., Generalization of Darcy's law for Bingham fluids in porous media: From flow-field statistics to the flow-rate regimes. *Phys. Rev. E*, **91** (2), 023011 (2015).

Chhabra, R.P., *PhD thesis*, Monash University, Melbourne, Australia (1980).

Chhabra, R.P., Some remarks on drag coefficients of a slowly moving sphere in non-Newtonian fluids, *J. Non-Newt. Fluid Mech.*, **13**, 225 (1983).

Chhabra, R.P., Steady Non-Newtonian flow about a rigid sphere, *Encyclopedia of Fluid Mechanics*, edited by N.P. Cheremisinoff, Vol. 1, Gulf, Houston, Chapter 19 (1986).

Chhabra, R.P., Hydrodynamics of bubbles and drops in rheologically complex liquids, *Encycl. Fluid Mech.*, **7**, 253 (1988).

Chhabra, R.P., Non-Newtonian flow past a sphere: Evaluation of three viscosity models, *Int. J. Eng. Fluid Mech.*, **3**, 17 (1990a).

Chhabra, R.P., Motion of spheres in power-law (viscoinelastic) fluids at intermediate Reynolds numbers: A unified approach, *Chem. Eng. Process.*, **28**, 89 (1990b).

Chhabra, R.P., Settling behavior of spherical particles in power-law non-Newtonian fluids, *Freight Pipelines*, edited by H. Liu, and G.F. Round, p. 259, Hemisphere, New York (1990c).

Chhabra, R.P., Settling of cylinders in power-law liquids, *Can. J. Chem. Eng.*, **70**, 385 (1992).

Chhabra, R.P., Fluid flow, heat and mass transfer in non-Newtonian fluids: Multiphase systems, *Adv. Heat Transf.*, **23**, 187 (1993a).

Chhabra R.P., Transport processes in particulate systems with non-Newtonian fluids, *Adv. Transp. Process.*, **9**, 501 (1993b).

Chhabra, R.P., Estimation of the minimum fluidization velocity for the beds of spherical particles fluidized by power law fluids, *Powder Technol.*, **76**, 225 (1993c).

Chhabra, R.P., Macroscopic conductivities for flow of Bingham plastics in porous media, *J. Hyd. Eng.*, **120**, 994 (1994).

Chhabra, R.P., Calculating velocities of particles, *Chem. Eng.*, **102**, 133 (1995a).
Chhabra, R.P., Further remarks on the drag of a swarm of bubbles, *Int. J. Eng. Sci.*, **33**, 1849 (1995b).
Chhabra, R.P., Wall effects on free settling velocity of non-spherical particles in viscous media in cylindrical tubes, *Powder Technol.*, **85**, 83 (1995c).
Chhabra, R.P., Hydrodynamics of Non-spherical particles in non-Newtonian fluids, *Handbook of Applied Polymer Processing Technology*, edited by N.P. Cheremisinoff and P.N. Cheremisinoff, Marcel Dekker, New York, Chapter 1 (1996a).
Chhabra, R.P., Wall effects on terminal velocity of non-spherical particles in non-Newtonian polymer solutions, *Powder Technol.*, **88**, 39 (1996b).
Chhabra, R.P., Free convective mass transfer from vertical short cylinders to non-Newtonian fluids, *Proc. 3 ISHMT-ASME Heat & Mass Transfer Conf.*, p.507, Narosa Publishing Co., New Delhi (1997).
Chhabra, R.P., Rising velocity of a swarm of spherical bubbles in non-Newtonian power-law fluids at high Reynolds numbers, *Can. J. Chem. Eng.*, **76**, 137 (1998).
Chhabra, R.P., Laminar boundary layer heat transfer to power law fluids: An approximate analytical solution, *J. Chem. Eng. Jpn.*, **38**, 812 (1999a).
Chhabra, R.P., Heat and mass transfer in rheologically complex systems, *Advances in the Flow and Rheology of Non-Newtonian Fluids*, edited by D. Siginer, D. De Kee and R.P. Chhabra, Elsevier, Amsterdam, Chapter 39 (1999b).
Chhabra, R.P., Sedimentation of particles in non-Newtonian media, *HYDROTRANSPORT-15*, 153 (2002a).
Chhabra, R.P., Wall effects on spheres falling axially in cylindrical tubes, *Transport Processes in Bubbles, Drops and Particles*, D. De Kee and R.P. Chhabra, p. 316, Taylor and Francis, New York (2002b).
Chhabra, R.P., Fluid mechanics and heat transfer with non-Newtonian liquids in mechanically agitated vessels, *Adv. Heat Transf.*, **37**, 77 (2003).
Chhabra, R.P., Non-Newtonian fluids: An introduction, *Rheology of Complex Fluids*, eds. A.P. Deshpande, J. Murali Krishnan and P.B. Sunil Kumar, Springer, Munich, Chapter 1 (2010).
Chhabra, R.P., Fluid flow and heat transfer from circular and non-circular cylinders submerged in non-Newtonian liquids, *Adv. Heat Transf.*, **43**, 289 (2011).
Chhabra, R.P., Rheology: From simple fluids to complex suspensions, *Lignocellulosic Fibers and Wood Handbook*, edited by N. Belgacem and A. Pizzi, p. 407, Scrivener, New York (2016).
Chhabra, R.P., *The CRC Handbook of Thermal Engineering*, 2nd ed., CRC Press, Boca Raton, FL (2018).
Chhabra, R.P., Agarwal, L. and Sinha, N.K., Drag on non-spherical particles: An evaluation of available methods, *Powder Technol.*, **101**, 288 (1999).
Chhabra, R.P., Agarwal, S. and Chaudhary, K., A note on wall effect on the terminal falling velocity of a sphere in quiescent Newtonian media in cylindrical tubes, *Powder Technol.*, **129**, 53 (2003).
Chhabra, R.P. and Bangun, J., Wall effects on terminal velocity of small drops in Newtonian and non-Newtonian fluids, *Can. J. Chem. Eng.*, **75**, 817 (1997).
Chhabra, R.P., Comiti, J. and Machac, I., Flow of non-Newtonian fluids in fixed and fluidized beds: A review, *Chem. Eng. Sci.*, **56**, 1 (2001b).
Chhabra, R.P. and De Kee, D., Fluid particles in rheologically complex media, *Transport Processes in Bubbles, Drops and Particles*, 1st ed., Hemisphere, New York, Chapter 2 (1991).
Chhabra, R.P. and Dhingra, S.C., Creeping motion of a Carreau fluid past a Newtonian fluid sphere, *Can. J. Chem. Eng.*, **64**, 897 (1986). Addendum: *ibid*, **66**, 176 (1988).
Chhabra, R.P., Dhotkar, B.N., Eswaran, V., Satheesh, V.K. and Vijaysri, M., Steady flow of Newtonian and dilatant fluids over an array of long circular cylinders, *J. Chem. Eng. Jpn.*, **33**, 832 (2000).
Chhabra, R.P., Ghosh, U.K., Kawase, Y. and Upadhyay, S.N., Non-Newtonian effects in bubble column reactors, *Multiphase Reactor and Polymerization System Hydrodynamics*, edited by N.P. Cheremisinoff, pp. 539–570, Gulf, Houston, TX (1996b).
Chhabra, R.P. and Gurappa, B., *Coulson Richardson's Chemical Engineering: Volume 2A: Particulate Systems and Particle Technology*, Butterworth-Heinemann, Oxford, UK (2019).
Chhabra, R.P., Machac, I. and Uhlherr, P.H.T., Some further observations on the creeping motion of spheres through Ellis model fluids, *Rheol. Acta*, **23**, 457 (1984).
Chhabra, R.P., McKay, A. and Wong, P., Drag on discs and square plates in pseudoplastic polymer solutions, *Chem. Eng. Sci.*, **51**, 5353 (1996c).
Chhabra, R.P. and Peri, S.S., A simple method for the estimation of free fall velocity of spherical particles in power-law liquids, *Powder Technol.*, **67**, 287 (1991). Errata *ibid*, **77**, 219 (1993).

Chhabra, R.P. and Raman, J.R., Slow non-Newtonian flow past an assemblage of rigid spheres, *Chem. Eng. Commun.*, **27**, 23 (1984).

Chhabra, R.P., Rami, K. and Uhlherr, P.H.T., Drag on cylinders in shear-thinning viscoelastic liquids, *Chem. Eng. Sci.*, **56**, 2221 (2001a).

Chhabra, R.P. and Richardson, J.F., *Non-Newtonian Flow and Applied Rheology*, 2nd ed., Butterworth-Heinemann, Oxford, UK (2008).

Chhabra, R.P., Sharma, R.B. and Rai, S., Collapse of gas bubbles in non-Newtonian power law liquids, *Int. J. Eng. Fluid Mech.*, **3**, 27 (1990).

Chhabra, R.P., Singh, T. and Nandrajog, S., Drag on chains and agglomerates of spheres in viscous Newtonian and power-law fluids, *Can. J. Chem. Eng.*, **73**, 566 (1995).

Chhabra, R.P., Soares, A.A. and Ferreira, J.M., A numerical study of the accelerating motion of a dense rigid sphere in non-Newtonian power-law fluids, *Can. J. Chem. Eng.*, **76**, 1051 (1998).

Chhabra, R.P., Soares, A.A. and Ferreira, J.M., Steady non-Newtonian flow past a circular cylinder: A numerical study, *Acta Mech.*, **172**, 1 (2004).

Chhabra, R.P. and Srinivas, B.K., Non-Newtonian (purely viscous) fluid flow through packed beds: Effect of particle shape, *Powder Technol.*, **67**, 15 (1991).

Chhabra, R.P., Tiu, C. and Uhlherr, P.H.T., Wall effect for sphere motion in inelastic non-Newtonian fluids, *Proc. 6th Australasian Hydraulics and Fluid Mech. Conf.*, Adelaide, p. 435 (1977).

Chhabra, R.P., Tiu, C. and Uhlherr, P.H.T., Shear-thinning effects in creeping flow about a sphere, *Rheology*, edited by G. Astarita, G. Marrucci and L. Nicolais, Vol. 2, p. 9, Plenum, New York (1980b).

Chhabra, R.P., Tiu, C. and Uhlherr, P.H.T., Creeping motion of spheres through Ellis model fluids, *Rheol. Acta*, **20**, 346 (1981a).

Chhabra, R.P., Tiu, C. and Uhlherr, P.H.T., A study of wall effects on the motion of a sphere in viscoelastic fluids, *Can. J. Chem. Eng.*, **59**, 771 (1981b).

Chhabra, R.P. and Uhlherr, P.H.T., Estimation of zero shear viscosity of polymer solutions from falling sphere data, *Rheol. Acta*, **18**, 593 (1979).

Chhabra, R.P. and Uhlherr, P.H.T., Creeping motion of spheres through shear-thinning elastic fluids described by the Carreau viscosity equation, *Rheol. Acta*, **19**, 187 (1980a).

Chhabra, R.P. and Uhlherr, P.H.T., Sphere motion through non-Newtonian fluids at high Reynolds number, *Can. J. Chem. Eng.*, **58**, 124 (1980b).

Chhabra, R.P. and Uhlherr, P.H.T., Wall effect for high Reynolds number motion of spheres in shear-thinning fluids, *Chem. Eng. Commun.*, **5**, 115 (1980c).

Chhabra, R.P. and Uhlherr, P.H.T., Shortcomings of the power-law in describing creeping flow about a sphere, *Proc. 2nd Nat. Conf. Rheol.*, Sydney, p. 89 (1981).

Chhabra, R.P. and Uhlherr, P.H.T., Static equilibrium and motion of spheres in viscoplastic liquids, *Encyclopedia of Fluid Mechanics*, edited by N.P. Cheremisinoff, Gulf, Houston, Chapter 21 (1988a).

Chhabra, R.P. and Uhlherr, P.H.T., The influence of fluid elasticity on wall effects for creeping sphere motion in cylindrical tubes, *Can. J. Chem. Eng.*, **66**, 154 (1988b).

Chhabra, R.P., Uhlherr, P.H.T. and Boger, D.V., The influence of fluid elasticity on the drag coefficient for creeping flow around a sphere, *J. Non-Newt. Fluid Mech.*, **6**, 187 (1980a).

Chhabra, R.P., Uhlherr, P.H.T. and Richardson, J.F., Some further observations on the hindered settling velocity of spheres in the inertial flow regime, *Chem. Eng. Sci.*, **51**, 4531 (1996a).

Chhabra, R.P., Unnikrishnan, A. and Unnikrishnan Nair, V.R., Hindered settling in non-Newtonian power law liquids, *Can. J. Chem. Eng.*, **71**, 716 (1992).

Chiang, T., Ossin, A. and Tien, C.L., Laminar free convection from a sphere, *J. Heat Transf.*, **86**, 537 (1964).

Chiba, K., Song, K.-W. and Horikawa, A., Motion of a slender body in quiescent polymer solutions, *Rheol. Acta*, **25**, 380 (1986).

Chien, N. and Wan, Z., *Mechanics of Sediment Movement*, Science Press, Beijing, PRC (in Chinese) (1983).

Chien, S.-F., Settling velocity of irregularly shaped particles, *SPE Drill. Complet.*, **9**, 281 (1994).

Chilcott, M.D. and Rallison, J.M., Creeping flow of dilute polymer solutions past cylinders and spheres, *J. Non-Newt. Fluid Mech.*, **29**, 381 (1988).

Childs, L.H., Hogg, A.J. and Pritchard, D., Dynamic settling of particles in shear flows of shear-thinning fluids, *J. Non-Newt. Fluid Mech.*, **235**, 83 (2016a); Also see *ibid* **238**, 158 (2016b).

Chmiel, H. and Walitza, E., *On the Rheology of Blood and Synovial Fluids*, Wiley, London (1980).

Chmielewski, C. and Jayaraman, K., The effect of polymer extensibility on cross flow of polymer solutions through cylinder arrays, *J. Rheol.*, **36**, 1105 (1992).

Chmielewski, C., Nichols, K.L. and Jayaraman, K., A comparison of the drag coefficients of spheres translating in corn syrup based and polybutene based Boger fluids, *J. Non-Newt. Fluid Mech.*, **35**, 37 (1990a).

Chmielewski, C., Petty, C.A. and Jayaraman, K., Cross flow of elastic liquids through arrays of cylinders, *J. Non-Newt. Fluid Mech.*, **35**, 309 (1990b).

Chmielewski, C., Petty, C.A. and Jayaraman, K., Elastic instability in cross-flow of polymer solutions through periodic arrays of cylinders, *J. Non-Newt. Fluid Mech.*, **48**, 285 (1993).

Cho, K., Cho, Y.I. and Park, N.A., Hydrodynamics of a vertically falling thin cylinder in non-Newtonian fluids, *J. Non-Newt. Fluid Mech.*, **45**, 105 (1992).

Cho, Y.I. and Hartnett, J.P., The falling ball viscometer – A new instrument for viscoelastic fluids, *Lett. Heat and Mass Transf.*, **6**, 335 (1979).

Cho, Y.I. and Hartnett, J.P., Drag coefficients of a slowly moving sphere in non-Newtonian fluids, *J. Non-Newt. Fluid Mech.*, **12**, 243 (1983a).

Cho, Y.I. and Hartnett, J.P., Reply to Chhabra's remarks, *J. Non-Newt. Fluid Mech.*, **13**, 229 (1983b).

Cho, Y.I., Hartnett, J.P. and Kwack, E.Y., A study of wall effects for viscoelastic fluids in the falling ball viscometer, *Chem. Eng. Commun.*, **6**, 141 (1980).

Cho, Y.I., Hartnett, J.P. and Lee, W.Y., Non-Newtonian viscosity measurements in the intermediate shear rate range with the falling ball viscometer, *J. Non-Newt. Fluid Mech.*, **15**, 61 (1984).

Cho, Y.I., Kim, I. and Tsou, F.-K., Fluid dynamics of large aspect-ratio cylinders in a viscoelastic fluid, *Proc. 2nd Int. Symposium on Multiphase Flow and Heat Transfer*, Chen, X.-J., Veziroglu, T.N. and Tien, C.L., eds., Hemisphere, New York, Vol.1, p. 25 (1991).

Choi, M.A., Lee, M.H., Chang, J. and Lee, S.J., Permeability modeling of fibrous media in composite processing, *J. Non-Newt. Fluid Mech.*, **79**, 585 (1998).

Chojnacki, K.T. and Fiekema, D.A., Instability of liquid sheets formed by impinging pseudoplastic liquid Jets, *J. Flow Visual. Image Process.*, **4**, 317 (1997).

Choplin, L, Carreau, P.J. and Ait-Kadi, A., Highly elastic-constant viscosity fluids, *Polym. Eng. Sci.*, **23**, 459 (1983).

Chow, L.C., Leland, J.E., Beam, J.E. and Mahefkey, E.T., The drag coefficient of a sphere in a sphere channel. *J. Fluids Eng.* (ASME), **111**, 229 (1989).

Christanti, Y. and Walker, L.M., Surface tension driven jet breakup of strain hardening polymer solutions, *J. Non-Newt. Fluid Mech.*, **100**, 9 (2001).

Christanti, Y. and Walker, L.M., Effect of fluid relaxation time of dilute polymer solutions on jet breakup due to forced disturbance, *J. Rheol.*, **46**, 733 (2002).

Christanti, Y. and Walker, L.M., Qualifying air atomization of viscoelastic fluids through fluid relaxation times, *At. Sprays*, **16**, 777 (2006).

Christopher, R.H. and Middleman, S., Power law flow through a packed tube, *Ind. Eng. Chem. Fund.*, **4**, 422 (1965).

Chrust, M., Bouchet, G. and Dusek, J., Numerical simulation of the dynamics of freely falling discs, *Phys. Fluids*, **25**, 044102 (2013).

Chu, B. and Hilfiker, R., Magnetic needle rheometer, *Rev. Sci. Instrum.*, **60**, 3047 (1989).

Chu, C.F. and Ng, K.M., Flow in packed tubes with a small tube to particle diameter ratio, *AIChE J.*, **35**, 148 (1989).

Chung, B.J. and Vaidya, A., On the slow motion of a sphere in fluids with non- constant viscosities, *Int. J. Eng. Sci.*, **48**, 78 (2010).

Chung, T.-S., The effect of diffusion on the inflation of a spherical viscoelastic film, *Chem. Eng. Sci.*, **40**, 1608 (1985).

Churaev, N.V. and Yashchenko, A., Experimental study of filtration of viscoplastic liquids through porous bodies, *Kolloid Zhurnal (USSR)*, **28**, 302 (1966).

Churchill, S.W., A comprehensive correlating equation for laminar, assisting, forced and free convection, *AIChE J.*, **23**, 16 (1977).

Churchill, S.W., Comprehensive, theoretically based, correlating equations for free convection from isothermal spheres, *Chem. Eng. Commun.*, **24**, 339 (1983).

Churchill, S.W., A theoretical structure and correlating equation for the motion of single bubbles, *Chem. Eng. Process.*, **26**, 269 (1989).

Churchill, S.W. and Chu, H.H.S., Correlating equations for laminar and turbulent free convection from a vertical plate, *Int. J. Heat Mass Transf.*, **18**,1323 (1975).
Churchill, S.W. and Ozoe. H., Correlations for laminar forced convection with uniform heating in flow over a plate and in developing and fully developed flow in a tube, *ASME J. Heat Transf.*, **95**, 78 (1973).
Churchill, S.W. and Thelen, H.-J., Eine Allegemeine Korrelationsgleichung fü den Wärme- und Stoffübergang bei freier Konvection, *Chem. Ing. Tech.*, **47**, 453 (1975).
Ciceron, D. Ph D Thesis, GEPEA, Universite de Nantes, St Nazaire, France (2000).
Ciceron, D., Comiti, J. and Chhabra, R.P., Pressure drops for purely viscous non-Newtonian fluid flow through beds packed with mixed size spheres, *Chem. Eng. Commun.*, **189**, 1403 (2002a).
Ciceron, D., Comiti, J., Chhabra, R.P. and Renaud, M., Non-Newtonian fluidisation of spherical particles, *Chem. Eng. Sci.*, **57**, 3225 (2002b).
Cieslinski, J.T. and Mosdorf, R., Gas bubble dynamics- experiment and fractal analysis, *Int. J. Heat Mass Transf.*, **48**, 1808 (2005).
Claesson, S., Ali, S. and McAtee Jr., J.L., New types of viscometer plummets for measuring viscosity of low viscous liquids under pressure, *Ind. Eng. Chem., Proc. Des. Dev.*, **22**, 633 (1983).
Clague, D.S. and Phillips, R.J., A numerical calculation of the hydraulic permeability of three-dimensional disordered porous media, *Phys. Fluids*, **9**, 1562 (1997).
Clark, N.N., A new scheme for particle shape characterization based on fractal harmonics and fractal dimensions, *Powder Technol.*, **51**, 243 (1987).
Clark, P.E. and Guler, N., Prop transport in vertical fractures: Settling velocity correlation, *SPE/DOE Symposium on Low Permeability*, Denver, CO (1983). (Paper # SPE/DOE 11636).
Clark, P.E., Halvaci, M., Ghaeli, H. and Parks, C.F., Proppant transport by xanthan and xanthan-hydroxypropyl guar solutions: Alternatives to crosslinked fluids, *SPE/DOE 1985 Low Permeability Gas Reservoirs*, Denver, CO, May 19-22 (1985). (Paper # SPE/DOE 13907).
Clark, P.E. and Quadir, J.A., Prop transport in hydraulic fractures: A critical review of particles settling velocity equations, Paper # SPE 9866 presented at the *1981 SPE/DOE Low Permeability Symposium*, Denver, CO (1981).
Clegg, D.B. and Power, G., The instantaneous slow flow of a viscoelastic fluid between two concentric spheres, *Appl. Sci. Res.*, **13A**, 423 (1964).
Clegg, D.B. and Whitmore, R.L., Boundary layers in Bingham plastics, *Rheol. Acta*, **5**, 130 (1966).
Cliffe, K.A., Spence, A. and Tavener, S.J., O (2) – symmetry breaking bifurcation, with application to the flow past a sphere in a pipe, *Int. J. Num. Methods in Fluids*, **32**, 175 (2000).
Clift, R., Grace, J.R. and Weber, M.E., *Bubbles, Drops and Particles*, Academic, New York (1978).
Cloitre, M., Hall, T., Mata, C. and Joseph, D.D., Delayed-die swell and sedimentation of elongated particles in worm-like micellar solutions, *J. Non-Newt. Fluid Mech.*, **79**, 157 (1998).
Coelho, P.M. and Pinho, F.T., Vortex shedding in cylinder flow of shear-thinning fluids. I. Identification and demarcation of flow regimes, *J. Non-Newt. Fluid Mech.*, **110**, 143 (2003). Also see *ibid*, 177.
Cohen, Y., The effect of polymer adsorption on particle settling in polymer solutions. *AIChE J.*, **31**(4), 695 (1985).
Cohen, Y., Apparent slip flow of polymer solutions, *Encyclopedia of Fluid Mechanics*, 7, Gulf, Chap. 14 (1988).
Cohen, Y. and Chang, C.-N., the flow of microemulsions through packed beds and capillary tubes, *Chem. Eng. Commun.*, **28**, 73 (1984).
Cohen, Y. and Christ, F.R., Polymer retention and adsorption in the flow of polymer solutions through porous media, *SPE Reservoir Eng.*, **1**, 113 (1986).
Cohen, Y. and Metzner, A.B., Wall effects in laminar flow of fluids through packed beds, *AIChE J.*, **27**, 705 (1981).
Cohen, Y. and Metzner, A.B., Apparent slip flow of polymer solutions, *J. Rheol.*, **29**, 67 (1985).
Coleman, B.D. and Noll, W., An approximation theorem for functionals, with applications in continuum mechanics, *Arch. Rat. Mech. Anal.*, **6**, 355 (1960).
Coleman, B.D. and Noll, W., Foundations of linear viscoelasticity, *Rev. Mod. Phys.*, **33**, 239 (1961).
Collins, R.E., *Flow of Fluids Through Porous Materials*, Van Nostrand, New York (1961).
Collis, D.C. and Williams, M.J., Two-dimensional convection from heated wires at low Reynolds numbers, *J. Fluid Mech.*, **6**, 357 (1959).
Colucci, S., Papale, P. and Montagna, C.P., Non-Newtonian flow of bubbly magma in volcanic conduits, *J. Geophys. Res. Solid Earth*, **122**, 1789 (2017).

Comini, G. and Croce, G., Numerical simulation of convective heat and mass transfer in banks of tubes, *Int. J. Numer. Meth. Eng.*, **57**, 1755 (2003).

Comiti, J., Mauret, E. and Renaud, M., Mass transfer in fixed beds: Proposition of a generalized correlation based on an energetic criterion, *Chem. Eng. Sci.*, **55**, 5545 (2000b).

Comiti, J., Montillet, A., Seguin, D. and Hilal, M., Modelling of power law liquid-solid mass transfer in packed beds in Darcy regime, *Chem. Eng. J.*, **89**, 29 (2002).

Comiti, J. and Renaud, M., A new model for determining mean structure parameters of fixed beds from pressure drop measurements – Application to beds packed with parallelepipedal particles, *Chem. Eng. Sci.*, **44**, 1539 (1989).

Comiti, J. and Renaud, M., Liquid-solid mass transfer in packed beds of parallelepipedal particles: Energetic correlation, *Chem. Eng. Sci.*, **46**, 143 (1991).

Comiti, J. and Sabiri, N.E., Limit of Darcy's law validity in packed beds, *Proc. First European Conf. Chem. Eng. (ECCEI)*, Florence, Italy, Volume 3, pp. 1863–1866 (1997).

Comiti, J., Sabiri, N.E. and Montillet, A., Experimental characterization of flow regimes in various porous media-III. Limit of Darcys or Creeping Flow regime for Newtonian and purely viscous non-Newtonian fluids, *Chem. Eng. Sci.*, **55**, 3057 (2000a).

Cook, R.L., Herbst, C.A. and King Jr., H.E., High-pressure viscosity of glass-forming liquids measured by the centrifugal force diamond anvil cell viscometer, *J. Phys. Chem.*, **97**, 2355 (1993).

Cooley, M.D.A. and O'Neill, M.E., On the slow motion generated in a viscous fluid by approach of a sphere to a plane wall or stationary sphere, *Mathematika*, **16**, 37 (1969).

Coppola, L. and Bohm, U., Mass transfer to packed beds of screens from non-Newtonian fluids, *Chem. Eng. Sci.*, **40**, 1594 (1985).

Coppola, L. and Bohm, U., Diffusivity of polymeric solutions, *Int. Comm. Heat Mass Transf.*, **13**, 77 (1986).

Costes, J. and Alran, C., Models for the formation of gas bubbles at a single submerged orifice in a non-Newtonian fluid, *Int. J. Multiph. Flow*, **4**, 535 (1978).

Couderc, J.-P., Incipient Fluidization and particulate systems, *Fluidization*, J.F. Davidson, R. Clift and D. Harrison, 2nd ed., Academic Press, Cambridge, MA, Chapter 1 (1985).

Coulaud, O., Morel, P. and Caltagirone, J.-P., Nonlinear effects for flow in porous media, *C.R. Acad. Sci. Paris, Ser. II*, **302**, 263 (1986).

Coulson, J.M., The flow of fluids through granular beds: Effects of particle shape and voids, *Trans. Inst. Chem. Eng.*, **27**, 237 (1949).

Coulson, J.M. and Richardson, J.F., *Chemical Engineering*, Vol. II, 5th Ed., Butterworth-Heinemann, Oxford (2002).

Coussot, P., *Mudflow Rheology and Dynamics*, Balkema, Amsterdam (1997).

Coussot, P., *Rheometry of Pastes, Suspensions and Granular Materials*, Wiley, New York (2005).

Coussot, P., Rheophysics of pastes: A review of microscopic modeling approaches, *Soft Matter*, **3**, 528 (2007).

Coussot, P., Yield stress fluid flows: A review of experimental data, *J. Non-Newt. Fluid Mech.*, **211**, 31 (2014).

Coussot, P., Bingham's heritage, *Rheol. Acta*, **56**, 163 (2017).

Coussot, P. and Boyer, S., Determination of yield stress fluid behavior from inclined plane test, *Rheol. Acta*, **34**, 534 (1995).

Coussot, P., Malkin, A.Y. and Ovarlez, G., Introduction: Yield stress - or 100 years of rheology, *Rheol. Acta*, **56**, 161 (2017).

Coutanceau, M., Sur le calcul du champ hydrodynamique autour d'une sphere qui se deplace dans l'axe d'un e'coulement de Poiseuille, *C.R. Acad. Sci.*, **273A**, 1097 (1971).

Coutanceau, M. and Hajjam, M., Viscoelastic effect on the behaviour of an air bubble rising axially in a tube, *App. Sci. Res.*, **38**, 199 (1982).

Coutanceau, M. and Thizon, P., Wall effect on the bubble behaviour in highly viscous liquids, *J. Fluid Mech.*, **107**, 339 (1981).

Couto, S.M., dos Santos, A.E.O., Vieira, S.M.J. and da Silva, D.J.P., Determination of drag force and terminal velocity of coffee fruits using finite element technique, *Rev. Bras. Eng. Agricola. Ambient.*, **8**, 274 (2004).

Cox, R.G. and Brenner, H., Effect of finite boundaries on the Stokes resistance of an arbitrary particle, *J. Fluid Mech.*, **28**, 391 (1967).

Cox, S.J., Dollet, B. and Graner, F., Foam flow around an obstacle: Simulations of obstacle-wall interaction, *Rheol. Acta*, **45**, 403 (2006).

Cramer, C., Berüter, B., Fischer, P. and Windhab, E.J., Liquid jet stability in a laminar flow field, *Chem. Eng. Technol.*, **25**, 499 (2002).

Crawford, C.W. and Plumb, O.A., The influence of surface roughness on resistance to flow through packed beds, *J. Fluids Eng. (ASME)*, **108**, 343 (1986).

Crawford, H.R. and Pruitt, G.T., Drag reduction of dilute polymer solutions, *Sym. Non-Newtonian Fluid Mech., 56th Annual meeting A. I. Ch. E.*, Houston, Texas (1963).

Craze, D.J., On the near wake behind a circular disk, *Proc. 6th Australasian Hydraulics and Fluid Mech. Conf.*, Adelaide, 5–9th December, pp. 282–286 (1977).

Cressely, R. and Hocquart, R., Birefringence D' ecoulmente localize induite a l' arriere d' obstacles. *Opt. Acta*, **27**, 699 (1980).

Cristescu, N., *Rock and Soil Rheology*, Springer-Verlag, Berlin (1988).

Cristescu, N.D., Conrad, B.P. and Trans-Son-Tay, R., A closed form solution for falling cylinder viscometers, *Int. J. Eng. Sci.*, **40**, 605 (2002).

Crochet, M.J., The flow of Maxwell fluids around a sphere, *Finite Elem. Fluids*, **4**, 573 (1982).

Crochet, M.J., Numerical simulation of highly viscoelastic flows, *Proc. Xth Int. Cong. Rheol.*, **1**, 19 (1988).

Crochet, M.J., Numerical simulation of viscoelastic flow: A Review, *Rubber Chem. Technol.*, **62**, 426 (1989).

Crochet, M.J., Davies, A.R. and Walters, K., *Numerical Simulation of Non-Newtonian Flow*, Elsevier, Amsterdam, Chapter 9 (1984).

Crochet, M.J., Delavaux, V. and Marchal, J.M., On the convergence of the streamline-upwind mixed finite element, *J. Non-Newt. Fluid Mech.*, **34**, 261 (1990).

Crochet, M.J. and Walters, K., Numerical methods in non-Newtonian fluid mechanics, *Ann. Rev. Fluid Mech.*, **15**, 241 (1983).

Crochet, M.J. and Walters, K., Computational rheology: A new science, *Endeavour*, **17**, 64 (1993).

Cross, M.M., Rheology of non-Newtonian fluids: A new flow equation for pseudoplastic systems, *J. Colloid Sci.*, **20**, 417 (1965).

Cross, R., Elastic and viscous properties of silly putty, *Am. J. Phys.*, **80**, 870 (2012).

Crowe, C., Sommerfeld, M. and Tsuji, Y., *Multiphase Flows with Droplets and Bubbles*, CRC Press, Boca Raton, FL (1998).

Curtis, A.R. and Marr, G.R., The viscous drag on cylinders falling symmetrically between parallel walls, *J. Phys. D: Appl. Phys.*, **11**, 1173 (1978).

Cygan, D.A. and Caswell, B., Precision falling sphere viscometry, *Trans. Soc. Rheol.*, **15**, 663 (1971).

D' Alessio, S.J.D., Private communication (2003).

D'Alessio, S.J.D. and Finlay, J.A., Power law flow past a cylinder at large distance, *Ind. Eng. Chem. Res.*, **43**, 8407 (2005).

D'Alessio, S.J.D. and Pascal, J.P., Steady flow of a power-.law fluid past a cylinder, *Acta Mech.*, **117**, 87 (1996).

D'Álsessio, S.J.D., Saunders, M.G. and Harmsworth, D.L., Forced and mixed convective heat transfer from accelerated flow past an elliptic cylinder, *Int. J. Heat Mass Transf.*, **46**, 2927 (2003).

D'Avino, G., Greco, F. and Maffettone, P.L., Particle migration due to viscoelasticity of the suspending liquid and its relevance in microfluidic devices, *Annu. Rev. Fluid Mech.*, **49**, 341 (2017).

D'Avino, G., Hulsen, M.A., Greco, F. and Maffettone, P.L., Numerical simulations on the dynamics of a spheroid in a viscoelastic liquid in a wide-slit microchannel, *J. Non-Newt. Fluid Mech.*, **263**, 33 (2019).

D'Avino, G., Hulsen, M.A. and Maffettone, P.L., Dynamics of pairs and triplets of particles in a viscoelastic fluid flowing in a cylindrical channel, *Comp. Fluids*, **86**, 45 (2013).

D'Avino, G., Hulsen, M.A., Snijkers, F., Vermant, J., Greco, F. and Maffettone, P.L., Rotation of a sphere in a viscoelastic liquid subjected to a shear flow. Part 1: Simulation results, *J. Rheol.*, **52**, 1331 (2008).

D'Avino, G., Snijkers, F., Pasquino, R., Hulsen, M.A., Greco, F., Maffettone, P.L. and Vermant, J., Migration of a sphere suspended in viscoelastic liquids in Couette flow: Experiments and simulations, *Rheol. Acta*, **51**, 215 (2012).

Dabbous, M.K., Displacement of polymers in water-flooded porous media and its effects on a surfactant micellar flood, *Soc. Pet. Eng. J.*, **15**, 358 (1977).

Dail, R.V. and Steffe, J.F., Dilatancy in starch solutions under low acid aseptic processing conditions, *J. Food Sci.*, **55**, 1764 (1990).

Dairenieh, I.S. and McHugh, A.J., Viscoelastic fluid flow past a submerged spheroidal body, *J. Non-Newt. Fluid Mech.*, **19**, 81 (1985).

Dajan, A., *M.Eng. Sci. Dissertation*, University of Windsor, Windsor, ON, Canada (1985).

Dale, J.D. and Emery, A.F., The free convection of heat from a vertical plate to several non-Newtonian pseudoplastic fluids, *J. Heat Transf. (ASME)*, **94**, 64 (1972).
Dallon, D.S., *PhD Dissertation*, Univ. of Utah, Salt Lake City, UT (1967).
Dandridge, A. and Jackson, D.A., Measurements of viscosity under pressure: A new method, *J. Phys. D: Appl. Phys.*, **14**, 829 (1981).
Dandy, D.S. and Leal, L.G., Boundary layer separation from a smooth slip surface, *Phys. Fluids*, **29**, 1360 (1986).
Daneshy, A.A., Numerical simulation of sand transport in hydraulic fracturing, *J. Pet. Technol.*, **30**, 132 (1978).
Dang, A., Ooi, L., Fales, J. and Stroeve, P., Yield stress measurements of magneto-rheological fluids in tubes, *Ind. Eng. Chem. Res.*, **39**, 2269 (2000).
Dang, V.V., Gill, W.N. and Ruckenstein, E., Unsteady mass transfer between bubbles and non-Newtonian liquids (Power law model) with chemical reactions, *Can. J. Chem. Eng.*, **50**, 300 (1972).
Daoudi, S., Regimes transitioires simples de fortes deformations d'une macromolecule, *J. De Physique*, **36**, 1285 (1975); Also see *J. De Physique Letters*, **37**, L41 (1976).
Darby, R., Determining settling rates of particles, *Chem. Eng.*, **103**(12), 109 (1996).
Darby, R., Pressure drop for non-Newtonian slurries: A wider path, *Chem. Eng.*, **107** (5), 64 (2000).
Darby, R. and Chhabra, R.P., *Chemical Engineering Fluid Mechanics*, 3rd ed., CRC Press, Boca Raton, FL (2017).
Darcovich, K., Gierer, C. and Capes, C.E., The application of dynamic clustering data to the sedimentation rates of concentrated suspensions, *Adv. Powder Technol.*, **7**, 1 (1996).
Darton, R.C., The physical behaviour of three-phase fluidized beds, *Fluidization*, edited by J.F. Davidson, R. Clift and D. Harrison, 2nd ed., Cambridge University Press, Cambridge, UK, Chapter 15 (1985).
Das, B., Ganguly, U.P. and Das, S.K., Inverse fluidization using non-Newtonian liquids, *Chem. Eng. Process.*, **49**, 1169 (2010).
Das, P.K., Gupta, A.K., Nirmalkar, N. and Chhabra, R.P., Effect of confinement on forced convection from a heated sphere in Bingham plastic fluids, *Korea-Aust. J. Rheol.*, **27**(2), 75 (2015).
Das, P.K., Kumar, R. and Ramkrishna, D., Coalescence of drops in stirred dispersions: A white noise model for coalescence, *Chem. Eng. Sci.*, **42**, 213 (1987).
Das, S., Deen, N.G. and Kuipers, J.A.M., A DNS study of flow and heat transfer through slender fixed-bed reactors randomly packed with spherical particles. *Chem. Eng. Sci.*, **160**, 1 (2017).
Dasgupta, N., Borah, R., Nirmalkar, N., Mishra, P., Gupta, A.K. and Chhabra, R.P., Combined effects of blockage and yield stress on drag and heat transfer from an in-line array of three spheres, *J. Disp. Sci. Technol.*, **40**(6), 855 (2019).
Datt, C. and Elfring, G.J., Dynamics and rheology of particles in shear-thinning fluids, *J. Non-Newt. Fluid Mech.*, **262**, 107 (2018).
Dauben, D.L. and Menzie, D.E., Flow of polymer solutions through porous media, *J. Pet. Tech.*, **19**, 1065(1967).
Daugan, S., Talini, L., Herzhaft, B. and Allain, C., Aggregation of particles settling in shear-thinning fluids. Part 1. Two particle aggregation, *Eur. Phys. J.*, **E7**, 73 (2002a).
Daugan, S., Talini, L., Herzhaft, B. and Allain, C., Aggregation of particles settling in shear-thinning fluids. Part 2. Three-particle aggregation, *Eur. Phys. J.*, **E9**, 55 (2002b).
Daugan, S., Talini, L., Herzhaft, B., Peysson, Y. and Allain, C., Sedimentation of suspensions in shear-thinning fluids, *Oil Gas Sci. Technol.-Rev. IFP*, **59**, 71 (2004).
Daunais, C.-A., Barbeau, L and Blais, B., An extensive study of shear-thinning flow around a spherical particle for power-law and Carreau fluids, *J. Non-Newt. Fluid Mech.*, **311**, 104951, (2023).
Davaadorj, B.-E., Kim, Y. and Lee, J., Settling velocity of irregularity shaped particles in Newtonian Fluids, *Geosys. Eng.*, **16**, 225 (2013).
Davenport, T., *Rheology of Lubricants*, Elsevier, London, UK (1973).
Davidson, D.L., Graessley, W.W. and Schowalter, W.R., Velocity and stress fields of polymeric liquids flowing in a periodically constricted channel Part 1: Experimental methods and straight channel validations, *J. Non-Newt. Fluid Mech.*, **49**, 317 (1993). Also see *ibid* 345.
Davidson, J.F. and Harrison, D., *Fluidized Particles*, Cambridge University Press, Cambridge, UK (1963).
Davidson, J.F. and Harrison, D. (eds.), *Fluidization*, Academic Press, New York (1971).
Davidson, J.F. and Schuler, B.O.G., Bubble formation at an orifice in a viscous fluid, *Trans. Inst. Chem. Eng.*, **38**, 145 (1960a).

Davidson, J.F. and Schuler, B.O.G., Bubble formation at an orifice in an inviscid liquid, *Trans. Inst. Chem. Eng.*, **38**, 335 (1960b).

Davidson, J.F., Clift, R. and Harrison, D., *Fluidization*, 2nd ed., Academic Press, London (1985).

Davies, C.N., The separation of airborne dust and particles, *Proc. Inst. Mech. Eng. (U.K.)*, **B1**, 185 (1952).

Davies, R.M. and Taylor, G.I., The mechanics of large bubbles rising through extended liquids and through liquids in tubes, *Proc. Roy. Soc.*, **200A**, 375 (1950).

Davis, A.M.J., Stokes drag on a disk sedimenting toward a plane or with other disks: Additional effects of a side wall or free surface, *Phys. Fluids*, **A2**, 301 (1990).

Davis, A.M.J. and Brenner, H., The falling-needle viscometer, *Phys. Fluids*, **13**, 3086 (2001).

Davis, A.M.J. and James, D.F., Slow flow through a model fibrous porous medium, *Int. J. Multiph. Flow*, **22**, 969 (1996).

Davis, R.H. and Acrivos, A., Sedimentation of non-colloidal particles at low Reynolds numbers, *Ann. Rev. Fluid Mech.*, **17**, 91 (1985).

Davis, R.T., Boundary layer theory for viscoelastic liquids, *Proc. 10th Midwest. Mechanics Conf.*, 4, 1145 (1967).

de Angelis, E., Fasano, A., Primicerio, M., Rosso, F., Carniani, E. and Ercolani, D., Modeling sedimentation in CWS, *Proc. 12th Int. Conf. Slurry Handling and Pipeline Transport*, p. 399, edited by C.A. Shook, MEP Publication (1993).

de Angelis, E. and Mancini, A., A model for the evolution of sedimentation beds in the dynamic of a pipelined non-Newtonian fluid, *Math. Comput. Model.*, **25**, 65 (1997).

de Boer, R., Reflections on the development of the theory of porous media, *App. Mech. Rev.* **56** (6), R27 (2003).

de Bruyn, J.R., Transient and steady-state drag in foam, *Rheol. Acta*, **44**, 150 (2004).

de Bruyn, J.R., Age dependence of the drag force in an aqueous foam, *Rheol. Acta*, **45**, 803 (2006).

de Corato, M., Saint- Michel, B., Makrigiorgos, G., Dimakopoulos, Y., Tsamopoulos, J. and Garbin, V., Oscillations of small bubbles and medium yielding in elastoviscoplastic fluids, *Phys. Rev. Fluids*, **4**, 073301 (2019).

de Gennes, P.G., *Scaling Concepts in Polymer Physics*, Cornell University Press, Ithaca, NY (1984).

De Kee D., Yield stress measurement techniques: A review, *Phys. Fluids*, **33**, 111301 (2021).

De Kee, D. and Carreau, P.J., Friction factor and bubble dynamics in polymer solutions, *Can. J. Chem. Eng.*, **71**, 183 (1993).

De Kee, D., Carreau, P.J. and Mordarski, J., Bubble velocity and coalescence in viscoelastic liquids, *Chem. Eng. Sci.*, **41**, 2273 (1986).

De Kee, D., Chan Man Fong, C.F. and Yao, J., Bubble shape in non-Newtonian fluids, *J. Appl. Mech. (ASME)*, **69**, 703 (2002).

De Kee, D. and Chhabra, R.P., A photographic study of shapes of bubbles and coalescence in non-Newtonian polymer solutions, *Rheol. Acta*, **27**, 656 (1988).

De Kee, D., Chhabra, R.P. and Dajan, A., Motion and coalescence of gas bubbles in non-Newtonian polymer solutions, *J. Non-Newt. Fluid Mech.*, **37**, 1 (1990a).

De Kee, D., Chhabra, R.P., Powley, M.B. and Roy, S., Flow of viscoplastic fluids on an inclined plane: Evaluation of yield stress, *Chem. Eng. Commun.*, **96**, 229 (1990b).

De Kee, D., Rodrigue, D. and Chan Man Fong, C.F., The motion of bubbles in non-Newtonian fluids, *Rheology and Fluid Mechanics of Nonlinear Materials*, AMD-Vol. 217, 37 (1996b).

De Kee, D., Rodrigue, D. and Chhabra, R.P., Hydrodynamics of free-rise of bubbles in non-Newtonian polymer solutions, *Handbook of Applied Polymer Processing Technology*, edited by N.P. Cheremisinoff and P.N. Cheremisinoff, pp. 87–123, Marcel Dekker, New York (1996a).

De Kee, D., Turcotte, G., Fildey, K. and Harrison, B., New method for determination of yield stress, *J. Texture Studies*, **10**, 281 (1980).

de Klerk, A., Voidage variation in packed beds at small column to particle diameter ratio, *AIChE J.*, **49**, 2022 (2003).

de Kruijf, A. Roodhart, L.P. and Davies, D.R., The relation between chemistry and flow mechanics of borate cross-linked fracturing fluids, *SPE Int. Sym. on Oilfield Chemistry*, New Orleans, LA, March 2–5 (1993). (Paper # SPE 25206).

de Larrard, F., Ferraris, C.F. and Sedran, T., Fresh concrete: A Herschel-Bulkley material, *Mater. Struct.*, **31**, 494 (1998).

de Souza Mendes, P.R. and Thompson, R.L., A critical overview of elasto-viscoplastic thixotropic modelling, *J. Non-Newt. Fluid Mech.*, **187**, 8 (2012).

De, S., Koesen, S.P., Maitri, R.V., Golombok, M., Padding, J.T. and van Santvoort, J.F.M., Flow of viscoelastic surfactants through porous media, *AIChE J.*, **64**, 773 (2018).

De, S., Kuipers, J.A.M., Peters, E.A.J.F. and Padding, J.T., Viscoelastic flow simulations in model porous media, *Phys. Rev Fluids*, **2**, 053303 (2017a).

De, S., Kuipers, J.A.M., Peters, E.A.J.F. and Padding, J.T., Viscoelastic flow past mono-and bidisperse random arrays of cylinders: Flow resistance, topology and normal stress distribution. *Soft Matter*, **13**(48), 9138 (2017b).

De, S., Kuipers, J.A.M., Peters, E.A.J.F. and Padding, J.T., Viscoelastic flow simulations in random porous media. *J. Non-Newton. Fluid Mech.*, **248**, 50 (2017c).

Dean, E.J., Glowinski, R. and Guidoboni, G., On the numerical simulation of Bingham viscoplastic flow: Old and new results, *J. Non-Newt. Fluid Mech.*, **142**, 36 (2007).

Debbaut, B. and Crochet, M.J., Extensional effects in complex flows, *J. Non-Newt. Fluid Mech.*, **30**, 169 (1988).

Deckwer, W.-D., *Bubble Column Reactors*, Wiley, New York (1992).

Deckwer, W.-D., Nguyen-Tien, K., Schumpe, A. and Serpemen, Y., Oxygen mass transfer into aerated CMC solutions in a bubble column, *Biotech. Bioeng.*, **24**, 461 (1982).

Dedegil, M.Y., Drag coefficient and settling velocity of particles in non-Newtonian suspensions, *J. Fluids Eng. (ASME)*, **109**, 319 (1987).

Degand, E. and Walters, K., On the motion of a sphere falling through an elastic liquid contained in a tightly-fitting cylindrical container, *J. Non-Newt. Fluid Mech.*, **57**, 103 (1995).

Deglo De Besses, B., Magnin, A. and Jay, P., Viscoplastic flow around a cylinder in an infinite medium, *J. Non-Newt. Fluid Mech.*, **115**, 27 (2003).

Deglo De Besses, B., Magnin, A. and Jay, P., Sphere drag in a viscoplastic fluid, *AIChE J.*, **50**, 2627 (2004).

Deiber, J.A. and Schowalter, W.R., Flow through tubes with sinusoidal axial variations in diameter, *AIChE J.*, **25**, 638 (1979).

Deiber, J.A. and Schowalter, W.R., Modelling the flow of viscoelastic fluids through porous media, *AIChE J.*, **27**, 912 (1981).

Delaby, I., Ernst, B., Germain, Y. and Muller, R., Drop deformation in polymer blends during uniaxial elongational flow: Influence of viscosity ratio for large capillary numbers, *J. Rheol.*, **38**, 1705 (1994).

Delaby, I., Ernst, B. and Muller, R., Drop deformation during elongational flow in blends of viscoelastic fluids. Small deformation theory and comparison with experimental results, *Rheol. Acta*, **34**, 525 (1995).

Delgado, J.M.P.Q., A critical review of dispersion in packed beds, *Heat Mass Transf.*, **42**, 279 (2006).

Delgado, M.A., Franco, J.M., Partal, P. and Gallegos, C., Experimental study of grease flow in pipelines: Wall slip and air entrainment effects, *Chem. Eng. Process.*, **44**, 805 (2005).

Delvaux, V. and Crochet, M.J., Numerical prediction of anomalous transport properties in viscoelastic flow, *J. Non-Newt. Fluid Mech.*, **37**, 297 (1990).

Denier, J.P. and Dabrowski, P.P., On the boundary-layer equations for power-law fluids, *Proc. R. Soc.*, **460A**, 3143 (2004).

Denn, M.M., Boundary layer flows for a class of elastic liquids, *Chem. Eng. Sci.*, **22**, 395 (1967).

Denn, M.M. and Bonn, D., Issues in the flow of yield-stress liquids, *Rheol. Acta*, **50**, 307 (2011).

Denn, M.M. and Porteous, K.C., Elastic effects in flow of viscoelastic fluids, *Chem. Eng. J.*, **2**, 280 (1971).

Derakhshandeh, B., Rheology of low to medium consistency pulp fibre suspensions, *PhD Thesis*, University of British Columbia, Van Couver, BC (2011).

Derakhshandeh, B., Kerekes, R.J., Hatzikiriakos, S.G. and Bennington, C.P.J., Rheology of pulp fibre suspensions: A critical review, *Chem. Eng. Sci.*, **66**, 3460 (2011).

Derakhshandeh, B., Vlassopoulos, D. and Hatzikiriakos, S.G., Thixotropy, yielding and ultrasonic Doppler velocimetry in pulp fiber suspensions, *Rheol. Acta*, **51**, 201 (2012).

Derksen, J.J., Drag on random assemblies of spheres in shear-thinning and thixotropic liquids, *Phys. Fluids*, **21**, 083302 (2009).

Descher, S. and Wünsch, O., On the influence of strain hardening on a rising bubble, *Proc. Appl. Math. Mech.*, **19**, e201900347 (2019).

Deshpande, R., Kanti, V., Desai, A. and Mittal, S., Intermittency of laminar separation bubble on a sphere during drag crisis, *J. Fluid Mech.*, **812**, 815 (2017).

Despeyroux, A., Ambari, A., Ben Richou, A. and Champmartin, S., Consequences of the confinement on the mass or heat transfer on a spherical particle in non-Newtonian fluids, *FEDSM - ICNMM2010–30506*, Aug 1–5, Montreal, Canada (2010).

Dewsbury, K.H., Karamanev, D.G. and Margaritis, A., Hydrodynamic characteristics of free rise of light solid particles and gas bubbles in non-Newtonian liquids, *Chem. Eng. Sci.*, **54**, 4825 (1999).

Dewsbury, K.H., Karamanev, D.G. and Margaritis, A., Dynamic behaviour of freely rising buoyant solid spheres in non-Newtonian liquids, *AIChE J.*, **46**, 46 (2000).

Dewsbury, K.H., Karamanev, D.G. and Margaritis, A., Rising solid sphere hydrodynamics at high Reynolds numbers in non-Newtonian fluids, *Chem. Eng. J.*, **87**, 129 (2002a).

Dewsbury, K.H., Tzounakos, A., Karamanev, D.G. and Margaritis, A., Wall effect for the free rise of solid spheres in moderately viscous liquids, *Can. J. Chem. Eng.*, **80**, 974 (2002b).

Dexter, R.W., Measurement of extensional viscosity of polymer solutions and its effect on atomization from a spray, *At. Sprays*, **6**, 167 (1996).

Dey, A.A., Modarres-Sadeghi, Y. and Rothstein, J.P., Viscoelastic fluid- structure interactions between a flexible cylinder and wormlike micelle solution, *Phys. Rev. Fluids*, **3**, 063301 (2018).

Dhahir, S.A. and Walters, K., On non-Newtonian flow past a cylinder in a confined flow, *J. Rheol.*, **33**, 781 (1989).

Dharmadhikari, R.V. and Kale, D.D., Flow of non-Newtonian fluids through porous media, *Chem. Eng. Sci.*, **40**, 527 (1985).

Dhiman, A.K., Anjaiah, N., Chhabra, R.P. and Eswaran, V., Mixed convection from a heated square cylinder to Newtonian and power-law fluids, *J. Fluids Eng. (Trans ASME)*, **129**, 506 (2007b).

Dhiman, A.K., Chhabra, R.P. and Eswaran, V., Flow and heat transfer across a confined square cylinder in the steady flow regime: Effect of Peclet number, *Int. J. Heat Mass Transf.*, **48**, 4598 (2005).

Dhiman, A.K., Chhabra, R.P. and Eswaran, V., Steady flow of power-law fluids across a square cylinder, *Chem. Eng. Res. Des.*, **84**, 300 (2006).

Dhiman, A.K., Chhabra, R.P. and Eswaran, V., Heat transfer to power law fluids from a heated square cylinder, *Num. Heat Transf.*, **52**, 185 (2007a).

Dhiman, A.K., Chhabra, R.P. and Eswaran, V., Steady flow across a confined square cylinder: Effects of power-law index and of blockage, *J. Non-Newt. Fluid Mech.*, **148**, 141 (2008a).

Dhiman, A.K., Chhabra, R.P. and Eswaran, V., Steady mixed convection from a confined square cylinder, *Int. Comm. Heat Mass Transf.*, **35**, 47–55 (2008b).

Dhiman, A., Ghosh, R. and Baranyi, L., Hydrodynamic and thermal study of a trapezoidal cylinder placed in shear-thinning and shear-thickening non-Newtonian liquid flows, *Int. J. Mech. Sci.*, **157–158**, 304 (2019).

Dhole, S.D., Chhabra, R.P. and Eswaran, V., Power law fluid flow through beds of spheres at intermediate Reynolds numbers: Pressure drop in fixed and distended beds, *Chem. Eng. Res. Des.*, **82**, 642 (2004).

Dhole, S.D., Chhabra, R.P. and Eswaran, V., Flow of power-law fluids past a sphere at intermediate Reynolds numbers, *Ind. Eng. Chem. Res.*, **45**, 4773 (2006a).

Dhole, S.D., Chhabra, R.P. and Eswaran, V., Forced convection heat transfer from a sphere to non-Newtonian power law fluids, *AIChE J.*, **52**, 3658 (2006b).

Dhole, S.D., Chhabra, R.P. and Eswaran, V., Drag of a spherical bubble rising in power law fluids at intermediate Reynolds numbers, *Ind. Eng. Chem. Res.*, **46**, 939 (2007a).

Dhole, S.D., Chhabra, R.P. and Eswaran, V., Mass transfer from a spherical bubble rising in power-law fluids at intermediate Reynolds numbers, *Int. Comm. Heat Mass Transf.*, **34**, 971 (2007b).

Dhotkar, B.N., Chhabra, R.P. and Eswaran, V., Flow of non-Newtonian polymeric solutions through fibrous media, *J. Appl. Polym. Sci.*, **76**, 1171 (2000).

Dias, R.P., Teixeira, J.A., Mota, M.G. and Yelshin, A.I., Particulate binary mixtures: dependence of packing porosity on particle size ratio, *Ind. Eng.Chem. Res.*, **43**, 7912 (2004).

Di Federico, V., Non-Newtonian flow in a variable aperture fracture, *Trans. Porous Media*, **30**, 75 (1998).

Di Felice, R., Hydrodynamics of liquid fluidization, *Chem. Eng. Sci.*, **50**, 1213 (1995).

Di Felice, R., A relationship for the wall effect on the settling velocity of a sphere at any flow regime, *Int. J. Multiph. Flow*, **22**, 527 (1996). Also see *ibid* 515.

Di Felice, R. and Gibilaro, L.G., Wall effects for the pressure drop in fixed beds, *Chem. Eng. Sci.*, **59**, 3037 (2004).

Di Felice, R., Gibilaro, L. and Foscolo, P.U., On the hindered settling velocity of spheres in the inertial flow regime, *Chem. Eng. Sci.*, **50**, 3005 (1995).

Di Felice, R. and Kehlenbeck, R., Sedimentation velocity of solids in finite size vessels, *Chem. Eng. Technol.*, **23**, 1123 (2000).

Di Marco, P., Grassi, W. and Memoli, G., Experimental study on rising velocity of nitrogen bubbles in FC-72, *Int. J. Thermal Sci.*, **42**, 435 (2003).

Di Vaira, N.J., Performance evaluation of an implicitly regularized model for viscoplastic particle suspensions, *B. Eng. Thesis*, University of Queensland, St. Lucia, Australia (2016).

Didari, H., Aghdasinia, H., Salami Hosseini, M., Ebrahimi, F. and Sahimi, M., Identifying the optimal path and computing the threshold pressure for flow of bingham fluids through heterogeneous porous media, *Trans. Porous Media*, **135**(3), 779 (2020).

Diedericks, G.P.J., du Plessis, J.P., Montillet, A., Comiti, J. and Legrand, J., Flow through a highly porous anisotropic multifilament knit, *Chem. Eng. Commun.*, **167**, 21 (1998).

Dienemann, W., Berechnung des Wärmeüberganges an laminar umströmten Körpern mit konstanter und ortsveränderlicher Wandtemperatur, *Z. Angew. Math. Mech.*, **33** (3), 89 (1953).

Dietrich, N. and Hebrard, G., Visualization of gas-liquid mass transfer around a rising bubble in a quiescent liquid using an oxygen sensitive dye, *Heat & Mass Transf.*, **54**, 2163 (2018).

Dietrich, W.E., Settling velocity of natural particles, *Water Resour. Res.*, **18**, 1615 (1982).

Dijkhuizen, W., Roghair, I., Annaland, M.V.S. and Kuipers, J.A.M., DNS of gas bubbles behaviour using an improved 3D front tracking model—Drag force on isolated bubbles and comparison with experiments, *Chem. Eng. Sci.*, **65**, 1415 (2010).

Dimakopoulos, Y., Pavlidis, M. and Tsamopoulos, J., Steady bubble rise in Herschel-Bulkley fluids and comparison of predictions via the augmented Lagrangian method with those via the Papanastasiou model, *J. Non-Newt. Fluid Mech.*, **200**, 34 (2013).

Dimitrov, D.S. and Ivanov, I.B., Hydrodynamics of thin liquid films: On the rate of thinning of microscopic films with deformable interfaces, *J. Coll. Interfac. Sci.*, **64**, 97 (1978).

Dimova, R., Dietrich, C., Hadjiisky, A., Danov, K. and Pouligny, B., Falling ball viscometry of giant vesicle membranes: Finite-size effects, *Eur. Phys. J.*, **B12**, 589 (1999).

Dinkgreve, M., Denn, M.M. and Bonn, D., "Everything flows?": Elastic effects on startup flows of yield-stress fluids, *Rheol. Acta*, **56**, 189 (2017).

Divoux, T., Vidal, V., Ripepe, M. and Geminard, J.C., Influence of non-Newtonian rheology on magma degassing, *Geophys. Res. Lett.*, **38**, L12301 (2011).

Dixon, A.G., Correlations for wall and particle shape effects on fixed bed bulk voidage, *Can. J. Chem. Eng.*, **66**, 705 (1988).

Doffin, J., Perrault, R. and Garnaud, G., Blood viscosity measurements in both extensional and shear flow by a falling ball viscometer, *Biorheology Suppl.*, **1**, 89 (1984).

Doi, M., *Introduction to Polymer Physics*, Clarendon, Oxford (1997).

Doi, M. and Edwards, S.F., *The Theory of Polymer Dynamics*, Clarendon, Oxford (1986).

Dolecek, P., Bendova, H., Siska, B. and Machac, I., Fall of spherical particles through a Carreau fluid, *Chem. Pap.*, **58**(6), 397 (2004).

Dolecek, P., Machacova, L., Machac, I. and Lecjaks, Z., Vypocet padove rychlosti kulovych castic v Ellisove kapaline, *Proc. Nat. CHISA Cong.*, Strabske Pleasco (1983).

Dolejs, V., Pressure drop in viscous flow of Newtonian liquid through a fixed random bed of spherical particles, *Int. Chem. Eng.*, **18**, 718 (1978).

Dolejs, V., Cakl, J., Siska, B. and Dolecek, P., Creeping flow of viscoelastic fluid through fixed beds of particles, *Chem. Eng. Process.*, **41**, 173 (2002).

Dolejs, V. and Machac, I., Pressure drop during the flow of a Newtonian fluid through a fixed bed of particles, *Chem. Eng. Process.*, **34**, 1 (1995).

Dolejs, V. and Mikulasek, P., Creeping flow of generalized Newtonian fluid through a fixed and fluidized bed of spherical particles, *Chem. Eng. Process.*, **36**, 111 (1997).

Dolejs, V., Mikulasek, P. and Dolecek, P., Approximate solution of momentum transfer in system generalized Newtonian fluid-Fluidized bed of spherical particles using modified Rabinowitsch-Mooney equation, *Collect. Czech. Chem. Commun.*, **60**, 1281 (1995).

Dolejs, V. and Siska, B., Flow of viscoplastic fluids through fixed beds of particles: Comparison of three approaches, *Chem. Eng. Process.*, **39**, 417 (2000).

Dolejs, V., Siska, B. and Dolecek, P., Modification of Kozeny-Carman concept for calculating pressure drop in flow of viscoplastic fluids through fixed beds, *Chem. Eng. Sci.*, **53**, 4155 (1998).

Dollet, B., Durth, M. and Graner, F., Flow of foam past an elliptical obstacle, *Phys. Rev. E*, **73**, 061404 (2006).
Dollet, B., Elias, F., Quilliet, C., Huillier, A., Aubouy, M. and Graner, F., Two-dimensional flows of foam: Drag exerted on circular obstacles and dissipation, *Colloids Surf. A.*, **263**, 101 (2005a).
Dollet, B., Elias, F., Quilliet, C., Raufaste, C., Aubouy, M. and Graner, F., Two-dimensional flow of foam around an obstacle: Force measurements, *Phys. Rev. E.*, **71**, 031403 (2005b).
Dollet, B. and Graner, F., Two-dimensional flow of foam around a circular obstacle: Local measurement of elasticity, plasticity and flow, *J. Fluid Mech.*, **585**, 181 (2007).
Dollet, B., Marmottant, P. and Garbin, V., Bubble dynamics in soft and biological matter, *Annu. Rev. Fluid Mech.*, **51**, 331 (2019).
Dolz-Planas, M., Roldan-Garcia, C., Herraez-Dominguez, J.V. and Belda-Maximino, R., Thixotropy of different concentrations of microcrystalline cellulose: Sodium Carboxymethyl Cellulose gels, *J. Pharma. Sci.*, **80**, 75 (1991).
Dominguez, J.G. and Willhite, G.P., Retention and flow characteristics of polymer solutions in porous media, *Soc. Pet. Eng. J.*, **15** (4), 111 (1977).
Donald, A.M., Physics of foodstuffs, *Rep. Prog. Phys.*, **57**, 1081 (1994).
Done, D.S., Baird, D.G. and Everage, A.E., The influence of porous media on the flow of polymer melts in capillaries, *Chem. Eng. Commun.*, **21**, 293 (1983).
Doraiswamy, D., The origins of rheology: A short historical excursion, *Rheol. Bull.*, **71,** 1 (2002).
Dou, H.-S. and Phan-Thien, N., Parallelisation of an unstructured finite volume implementation with PVM: Viscoelastic flow around a cylinder, *J. Non-Newt. Fluid Mech.*, **77**, 21 (1998).
Dou, H.-S. and Phan-Thien, N., The flow of an Oldroyd-B fluid past a cylinder in a channel: Adaptive viscosity vorticity (DAVSS-ω) formulation, *J. Non-Newt. Fluid Mech.*, **87**, 47 (1999).
Dou, H.-S. and Phan-Thien, N., Numerical difficulties at high elasticity for viscoelastic flow past a confined cylinder, *Int. J. Comput. Eng. Sci.*, **2**, 249 (2001).
Dou, H.-S. and Phan-Thien, N., Negative wake in the uniform flow past a cylinder, *Rheol. Acta*, **42**, 383(2003).
Dou, H.-S. and Phan-Thien, N., Criteria of negative wake generation behind a cylinder, *Rheol. Acta*, **43**, 203 (2004).
Dou, H.-S. and Phan-Thien, N., Viscoelastic flow past a confined cylinder: Instability and velocity inflection, *Chem. Eng. Sci.*, **62**, 3909 (2007).
Dratler, D.I., Schowalter, W.R. and Hoffman, R.L., Dynamic simulation of shear-thickening in concentrated colloidal suspensions, *J. Fluid Mech.*, **353**, 1 (1997).
Dravid, V., Loke, P.B., Corvalan, C.M. and Sojka, P.E., Drop formation in non-Newtonian jets at low Reynolds numbers, *J. Fluids Eng.*, **130**, 0815041 (2008).
Dreher, K.D. and Gogarty, W.B., An overview of mobility control in micellar/polymer enhanced oil recovery processes, *J. Rheol.*, **23**, 209 (1979).
Drummond, J.E. and Tahir, M.I., Laminar viscous flow through regular arrays of parallel solid cylinders, *Int. J. Multiph. Flow*, **10**, 515 (1984).
du Plessis, J.P., Analytical quantification of coefficients in the Ergun equation for fluid friction in a packed bed, *Trans. Porous Media*, **16**, 189 (1994).
du Plessis, M.P. and Ansley, R.W., Settling parameters in solids pipelining, *J. Pipeline Div. (ASCE)*, **93**, 1 (1967).
du Plessis, J.P. and Masliyah, J.H., Mathematical modeling of flow through consolidated isotropic porous media, *Tran. Porous Media*, **3**, 145 (1988).
du Plessis, J.P., Montillet, A., Comiti, J. and Legrand, J., Pressure drop prediction for flow through high porosity metallic foams, *Chem. Eng. Sci.*, **49**, 3545 (1994).
Duan, Z., He, B. and Duan, Y., Sphere drag and heat transfer, *Sci. Rep.*, **5**(1), 1 (2015).
Dubash, N. and Frigaard, I., Conditions for static bubbles in viscoplastic fluids, *Phys. Fluids*, **16**, 4319 (2004).
Dubash, N. and Frigaard, I.A., Propagation and stopping of air bubbles in Carbopol solutions, *J. Non-Newt. Fluid Mech.*, **142**, 123 (2007).
Duckworth, R.A., Pullum, L., Addie, G.R. and Lockyear, C.F., The pipeline transport of coarse materials in a non-Newtonian carrier fluid, *Proc. Hydrotransport 10*, Innsbruck, Austria (1986).
Ducloue, L., Pitois, O., Goyon, J., Chateau, X. and Ovarlez, G., Rheological behaviour of bubbles suspensions of yield-stress fluids, *J. Non-Newt. Fluid Mech.*, **215**, 31 (2015).
Duda, J.L., Hong, S.-A. and Klaus, E.E., Flow of polymer solutions in porous media: Inadequacy of the capillary model, *Ind. Eng. Chem. Fund.*, **22**, 299 (1983).

Duda, J.L., Klaus, E.E. and Fan, S.K., Influence of polymer-molecule/Wall interactions on mobility control, *Soc. Pet. Eng. J.*, **19**, 613 (1981).

Dudek, D.R., Fletcher, T.H. and Longwell, J.P., Natural convection induced drag forces at low Grashof numbers: Comparison of theory with experiment, *Int. J. Heat Mass Transf.*, **31**, 863 (1988).

Dudukovic, A.P. and Koncar-Djurdjevic, S.K. The effect of tube walls on drag coefficients of coaxially placed objects. *AIChE J.*, **27**, 837 (1981).

Dudukovic, M.P., Larachi, F. and Mills, P.L., Multiphase reactors-revisited, *Chem. Eng. Sci.*, **54**, 1975 (1999).

Dudukovic, M.P., Larachi, F. and Mills, P.L., Catalytic reactors: A perspective on current knowledge and future trends, *Catal. Rev.*, **44**, 123 (2002).

Duineveld, P.C., The rise velocity and shape of bubbles in pure water at high Reynolds numbers, *J. Fluid Mech.*, **292**, 325 (1995).

Dukhan, N., Picon- Feliciano, R. and Alvarez- Hernandez, A.R., Air flow through compressed and uncompressed aluminum foam: Measurements and correlations, *J. Fluids Eng.*, **128**, 1004 (2006).

Dullaert, K. and Mewis, J., Thixotropy: Build-up and breakdown curves during flow, *J. Rheol.*, **49**, 1213 (2005).

Dullaert, K. and Mewis, J., A structural kinetics model for thixotropy, *J. Non-Newt. Fluid Mech.*, **39**, 21 (2006).

Dullien, F.A.L., Single phase flow through porous media and pore structure, *Chem. Eng. J.*, **10**, 1 (1975a).

Dullien, F.A.L., New network permeability model of porous media, *AIChE J.*, **21**, 299 (1975b).

Dullien, F.A.L., *Porous Media: Fluid Transport and Pore Structure*, 2nd ed., Academic Press, New York (1992).

Dullien, F.A.L. and Azzam, M.I.S., Flow rate-pressure gradient measurements in periodically non-uniform capillary tubes, *AIChE J.*, **19**, 222 (1973a).

Dullien, F.A.L. and Azzam, M.I.S., Effect of geometric parameters on the friction factor in periodically constricted tubes, *AIChE J.*, **19**, 1035 (1973b).

Dullien, F.A.L. and Dhawan, G.K., Bivariate pore size distributions of some sand stones, *J. Interface Colloid Sci.*, **52**, 129 (1975).

Dunand, A., Guillot, D. and Soucemarianadin, A., Viscous properties of glass bead suspensions and falling sphere experiments in hydroxypropyl guar solutions, *Proc. IX Int. Cong. Rheol.*, Acapulco, Mexico, p. 623 (1984).

Dunand, A. and Soucemarianadin, A., Concentration effects on the settling velocities of proppant slurries, Paper presented at *60th Annual Tech. Conf. & Exhibition*, Soc. Petr. Engrs., Las Vegas, NV, September 22–25 (1985). (Paper # SPE 14259).

Dunn, P.F. and Picologlou, B.F., Variation in human semen viscoelastic properties with respect to time post ejaculation and frequency of ejaculation, *Int. J. Fertil.*, **22**, 217 (1977a).

Dunn, P.F. and Picologlou, B.F., Investigation of the rheological properties of human semen, *Biorheology*, **14**, 277 (1977b).

Dupret, F., Marchal, J.M. and Crochet, M.J., On the consequence of discretization errors in the numerical calculation of viscoelastic flows, *J. Non-Newt. Fluid Mech.*, **18**, 173 (1985).

Durban, D. and Fleck, N.A., Spherical cavity expansion in a Drucker-Prager fluid, *J. Appl. Mech. (ASME)*, **64**, 743 (1997).

Durst, F. and Haas, R., Dehnstromungen mit verdunnten polymerlosungen: Ein theoretisches model und seine experimentelle verification, *Rheol. Acta*, **20**, 179 (1981); Also see *ibid*, **21**, 150 (1982).

Durst, F., Haas, R. and Interthal, W., The nature of flows through porous media, *J. Non-Newt. Fluid Mech.*, **22**, 169 (1987).

Durst, F., Haas, R. and Kaczmar, B.U., Flows of dilute hydrolyzed polyacrylamide solutions in porous media under various solvent conditions, *J. Appl. Polym. Sci.*, **26**, 3125 (1981).

Dutta, A., Gupta, A.K., Mishra, G. and Chhabra, R.P., Effect of fluid yield stress and of angle of tilt on natural convection from a square bar in a square annulus, *Comput. Fluids*, **160**, 138 (2018).

Dwyer, H.A. and Dandy, D.S., Some influences of particle shape on drag and heat transfer, *Phys. Fluids*, **A2**, 2110 (1990).

Dyakonova, N.E., Odell, J.A., Brestkin, Y.V., Lyulin, A.V. and Saez, A.E., Macromolecular strain in periodic models of porous media flows. *J. Non-Newt. Fluid Mech.*, **67**, 285 (1996).

Dybbs, A. and Edwards, R.V., A new look at porous media fluid mechanics – Darcy to turbulent, *Fundamentals of Transport Phenomena in Porous Media Flow*, p. 199, Martinus Nishoff, Dordrecht (1984).

Dziubinski, M. and Orczykowska, M., The drag coefficient of bubbles rising in non-Newtonian liquids, *Inzynieria Chemiczna I Procesowa*, **23**, 67(2002).

Dziubinski, M., Orczykowska, M. and Budzynski, P., Average value of shear stress and shear rate at the surface of gas bubble rising in non-Newtonian liquids, *Inzynieria Chemiczna I Procesowa*, **23**, 341 (2002).

Dziubinski, M., Orczykowska, M. and Budzynski, P., Comments on bubble rising velocity in non-Newtonian liquids, *Chem. Eng. Sci.*, **58**, 2441 (2003).

Dziubinski, M., Orczykowska, M., Kiljanski, T. and Budzynski, P., Wall effects for bubbles flow in Newtonian and non-Newtonian liquids, *Inzynieria Chemiczna I Proceswa*, **22**, 403 (2001).

Eastwood, J., Matzen, E.J.P., Young, M.J. and Epstein, N., Random loose porosity of packed beds, *Brit. Chem. Eng.*, **14**, 1542 (1969).

Eaton, L.R. and Hoffer, T.E., Experiments on droplets in free fall I: Terminal velocity and wall effects, *J. App. Meterol.*, **9**, 269 (1970).

Eberhard, U., Seybold, H.J., Floriancic, M., Bertsch, P., Jiménez-Martínez, J., Andrade Jr., J.S. and Holzner, M., Determination of the effective viscosity of non-Newtonian fluids flowing through porous media, *Front. Phys.*, **7**, 71 (2019).

Eberhard, U., Seybold, H.J., Secchi, E., Jiménez-Martínez, J., Rühs, P.A., Ofner, A. and Holzner, M., Mapping the local viscosity of non-Newtonian fluids flowing through disordered porous structures, *Sci. Rep.*, **10**(1), 1 (2020).

Eckert, E.R.G. and Jackson, T.W., Analysis of turbulent free-convection boundary layer on flat plate, NACA Technical Note 2207 (1950).

Eckert, E.R.G. and Shadid, J.N., Viscous heating of a cylinder with finite length by a high viscosity fluid in steady longitudinal flow. Newtonian fluids, *Int. J. Heat Mass Transf.*, **32**(2), 321 (1989).

Edgeworth, R., Dalton, B.J. and Parnell, T., The pitch drop experiment, *Eur. J. Phys.*, **5**, 198 (1984).

Edie, D.D. and Gooding, C.H., Prediction of pressure drop for the flow of polymer melts through sintered metal filters, *Ind. Eng. Chem. Proc. Des. Dev.*, **24**, 8 (1985).

Edwards, D.A., Shapiro, M., Bar-Yoseph, P. and Shapira, M., The influence of Reynolds number upon the apparent permeability of spatially periodic arrays of cylinders, *Phys. Fluids*, **A2**, 45 (1990).

Edwards, M.F. and Helail, T.R., Axial dispersion in porous media, *Proc. 2nd Eur. Conf. Mixing*, E2–9, 30th Mar – 1st April, Cambridge, U.K. (1977).

Eesa, M. and Barigou, M., Horizontal laminar flow of coarse nearly neutrally buoyant particles in non-Newtonian conveying fluids: CFD and PEPT experiments compared, *Int. J. Multiph. Flow*, **34**, 997 (2008).

Eesa, M. and Barigou, M., CFD investigation of the pipe transport of coarse solids in laminar power-law fluids, *Chem. Eng. Sci.*, **64**, 322 (2009).

Eggers, J., Nonlinear dynamics and breakup of free-surface flows, *Rev. Mod. Phys.*, **69**, 865 (1997).

Eggers, J., Drop formation – an overview, *ZAMM*, **85**, 400 (2005).

Ehsani, A., Behbahani, B. and Jami, A.M., Bed voidage and heat transfer in non-Newtonian liquid-solid fluidized bed, *Iran. J. Chem. Chem. Eng.*, **23**, 73 (2004).

Eichstadt, F.J. and Swift, G.W., Theoretical analysis of the falling cylinder viscometer for power law and Bingham plastic fluids, *AIChE J.*, **12**, 1179 (1966).

Einarsson, J. and Mehlig, B., Spherical particle sedimenting in weakly viscoelastic shear flow, *Phys. Rev. Fluids*, **2**, 063301 (2017).

Eisenberg, D.A., Klink, I.M. and Phillips, R.J., Axisymmetric sedimentation of spherical particles in a viscoelastic fluid: Sphere-wall and sphere-sphere interactions, *J. Rheol.*, **57**, 857 (2013).

Eisfeld, B. and Schnitzlein, K., The influence of confining walls on the pressure drop in packed beds, *Chem. Eng. Sci.*, **56**, 4321 (2001).

El Defrawi, M. and Finlayson, B.A., On the use of the integral method for flow of power-law fluids, *AIChE J.*, **18**, 251 (1972).

El Fadili, Y., Drag coefficient model for single particle settling in non-Newtonian pseudoplastic fluids, MS dissertation, University of Oklahoma, Norman, OK (2005).

El Kayloubi, A., Kaddioui, N. and Sigli, D., Structure of the rate of deformation field of Newtonian and viscoelastic fluids in a falling ball experiment, *J. Non-Newt. Fluid Mech.*, **22**, 335 (1987).

Elata, C., Burger, J., Michlin, J. and Taksermani, U., Dilute polymer solutions in elongational flow, *Phys. Fluids*, **20**(10), Pt. II, S49 (1977).

El-Awady, M.N., An atomization theory for swirl nozzles, *Trans. ASAE*, **21**, 70 (1978).

Elbirli, B. and Shaw, M.T., Time constants from shear viscosity data, *Trans. Soc. Rheol.*, **22**, 561 (1978).

El-Kaissy, M.M. and Homsy, G.M., A theoretical study of pressure drop and transport in packed beds at intermediate Reynolds numbers, *Ind. Eng. Chem. Fund.*, **13**, 82 (1973).

Ellwood, K.R.J., Georgiou, G.C., Papanastasiou, T.C. and Wilkes, J.O., Laminar jets of Bingham plastic liquids, *J. Rheol.*, **34**, 787 (1990).

Emady, H., Caggioni, M. and Spicer, P., Colloidal microstructure effects on particle sedimentation in yield stress fluids, *J. Rheol.*, **57**, 1761 (2013).

Emery, A.F., Chi, H.S. and Dale, J.D., Free convection through vertical plane layers of non-Newtonian power law fluids, *J. Heat Transf. (ASME)*, **93**, 164 (1971).

Enders, F., Merker, D., Kolano, M., Bohm, L. and Kraume, M., Numerical characterization of the bubble rise behavior in viscoelastic liquids, *Chem. Eng. Technol.*, **42**, 1295 (2019).

Endo, Y., Chen, D.-R. and Pui, D.Y.H., Theoretical consideration of permeation resistance of fluid through a particle packed layer, *Powder Technol.*, **124**, 119 (2002).

Engez, S.M., An extension of the study of boundary influence on the wall velocity of spheres. MS Thesis, Univ. of Iowa, Ames (1948).

Epps, H.H. and Leonas, K.K., The relationship between porosity and air permeability of woven textile fabrics, *J. Test. Eval.*, **25**, 108 (Jan. 1997).

Epstein, N., Three-phase fluidization: Some knowledge gaps, *Can. J. Chem. Eng.*, **59**, 649 (1981).

Epstein, N., Letter to the Editor, *Chem. Eng. Sci.*, **39**, 1533 (1984). Also see author's reply *ibid*, 1819.

Epstein, N., On tortuosity and the tortuosity factor in flow and diffusion through porous media, *Chem. Eng. Sci.*, **44**, 777 (1989).

Epstein, N., Letter to the Editor, *Chem. Eng. Sci.*, **53**, 1469 (1998).

Epstein, N., Liquid-Solid Fluidization, *Handbook of Fluidization and Fluid-Particle Systems*, edited by W.-C., Yang, Marcel Dekker, New York, Chapter 26 (2003).

Epstein, N. and Masliyah, J.H., Creeping flow through clusters of spheroids and elliptic cylinders, *Chem. Eng. J.*, **3**, 169 (1972).

Erbas, S. and Ece, M.C., An analysis of free convection to power law fluids from a vertical plate of variable surface temperature, *Trans. Can. Soc. Mech. Eng.*, **25**, 1 (2001).

Erdim, E., Akgiray, Ö. and Demir, İ., A revisit of pressure drop-flow rate correlations for packed beds of spheres. *Powder Technol.*, **283**, 488 (2015).

Ergun, S., Fluid flow through packed columns, *Chem. Eng. Prog.*, **48** (2), 89 (1952).

Ern, P., Risso, F., Fabre, D. and Magnaudet, J., Wake-induced oscillatory paths of bodies freely rising or falling in fluids, *Annu. Rev. Fluid Mech.*, **44**, 97 (2012).

Ershaghi, I., *Ph.D. Dissertation*, University of Southern California, Los Angeles (1972).

Esfahani, J.A. and Bagherian, B., Similarity solution for unsteady free convection from a vertical plate at constant temperature to power- law fluids, *J. Heat Transf.*, **134**, 1025011 (2012).

Eshghy, S., Forced-flow effects on free-convection flow and heat transfer, *J. Heat Transf.*, **86**, 290 (1964).

Eshtiaghi, N., Markis, F., Yap, S.D., Baudez, J.-C. and Slatter, P., Rheological characterization of municipal sludge: A review, *Wat. Res.*, **47**, 5493 (2013).

Eslami, M. and Jafarpur, K., Laminar free convection heat transfer from isothermal convex bodies of arbitrary shape: A new dynamic model, *Heat Mass Transf.*, **48**, 301 (2012).

Ethier, C.R., Flow through mixed fibrous material, *AIChE J.*, **37**, 1227 (1991).

Evans, A.R., Shaqfeh, E.S.G. and Frattini, P.L., Observations of polymer conformation during flow through a fixed fibre bed, *J. Fluid Mech.*, **281**, 319 (1994).

Evans, I.D., On the nature of the yield stress, *J. Rheol.*, **36**, 1313 (1992).

Ewoldt, R.H. and McKinley, G.H., Mapping thixo-elasto-visco-plastic behavior, *Rheol. Acta*, **56**, 195 (2017).

Fabris, D., Muller, S.J. and Liepmann, D., Wake measurements for flow around a sphere in a viscoelastic fluid, *Phys. Fluids*, **11**, 3599 (1999).

Fadali, O.A., Effect of drag-reducing polymer on the rate of cementation of Copper ion on Zinc pellets, *Chem. Eng. Technol.*, **26**, 491 (2003).

Fadili, A., Tardy, P.M.J. and Pearson, J.R.A., A 3-D filtration law for power-law fluids in heterogeneous porous media, *J. Non-Newt. Fluid Mech.*, **106**, 121 (2002).

Fahs, H., Ovarlez, G. and Chateau, X., Pair-particle trajectories in a shear flow of a Bingham fluid, *J. Non. Newt. Fluid Mech.*, **261**, 171 (2018).

Fairhurst, P.G., Barigou, M., Fryer, P.J. and Pain, J.-P., Particle passage time distributions in vertical pipe flow of solid-liquid food mixtures, *Trans. Inst. Chem. Eng.*, **77C**, 293 (1999).

Fairhurst, P.G. and Pain, J.-P., Passage time distributions for high solid fraction solid-liquid food mixtures in horizontal flow: Unimodal size particle distributions, *J. Food Eng.*, **39**, 345 (1999).

Falls, A.H., Musters, J.J. and Ratulowski, J., The apparent viscosity of foam in homogeneous bead packs, *SPE Reservoir Eng.*, **4**, 155 (1989).

Faltas, M.S. and Saad, E.I., Stokes flow past an assemblage of slip eccentric spherical particle-in-cell models. *Math. Methods Appl. Sci.*, **34**(13), 1594 (2011).

Fan, D., Du, M., Sun, Y. and Chen, H., Coalescence characteristics of side-by-side growing bubbles in carboxymethyl cellulose solutions, *Chem. Eng. Technol.*, **43**, 230 (2020).

Fan, L.-S., *Gas-Liquid-Solid Fluidization Engineering*, Butterworths, Stoneham, MA (1989).

Fan, L.-S. and Yang, G., Gas-liquid-solid three-phase fluidization, *Handbook of Fluidization and Fluid-Particle Systems*, edited by W.-C. Yang, Marcel Dekker, New York, Chapter 27 (2003).

Fan, W., Ma, Y., Jiang, S., Yang, K. and Li, H.-Z., An experimental investigation for bubble Rising in non-Newtonian fluids and empirical correlation of drag coefficient, *ASME J. Fluids Eng.*, **132**, 021305 (2010).

Fan, W., Qi, T., Sun, Y., Zhu, P. and Chen, H., Coalescence deformation of bubble pairs generated from twin nozzles in CMC solutions, *Chem. Eng. Technol.*, **39**, 1895 (2016).

Fan, D., Sun, Y. and Chen, H., Bubble volume and aspect ratio generated in non-Newtonian fluids, *Chem. Eng. Technol.*, **37**, 1566 (2014).

Fand, R.M. and Brucker, J., A correlation for heat transfer by natural convection from horizontal cylinders that accounts for viscous dissipation, *Int. J. Heat Mass Transf.*, **26**, 709 (1983).

Fand, R.M. and Keswani, K.K., Combined natural and forced convection heat transfer from horizontal cylinders to water, *Int. J. Heat Mass Transf.*, **16**, 1175 (1973).

Fand, R.M., Kim, B.Y.K., Lam, A.C.C. and Phan, R.T., Resistance to the flow of fluids through simple and complex porous media whose matrices are composed of randomly packed spheres, *J. Fluids Eng.*, **109**, 268 (1987).

Fand, R.M., Morris, E.W. and Lum, M., Natural convection heat transfer from horizontal cylinders to air, water and silicone oils for Rayleigh numbers between 300 and 2×10^7, *Int. J. Heat Mass Transf.*, **20**, 1173 (1977).

Fand, R.M., Sundaram, M. and Murali, V., Incompressible fluid flow through pipes packed with spheres at low dimension ratios, *J. Fluids Eng.*, **115**, 169 (1993).

Fand, R.M. and Thinakaran, R., The influence of the wall on flow through pipes packed with spheres, *J. Fluids Eng.*, **112**, 84 (1990).

Fantazzini, P., Bortolotti, V., Kärger, J. and Galvosas, P., Magnetic resonance in porous media: *Proceedings of the 10th International Bologna Conference on Magnetic Resonance in Porous Media (MRPM10), including the 10th Colloquium on Mobile Magnetic Resonance (CMMR10)*. Magnetic Resonance in Porous Media, 1330 (2011).

Fararoui, A. and Kintner, R.C., Flow and shape of drops in non-Newtonian fluids, *Trans. Soc. Rheol.*, **5**, 369 (1961).

Faridi, H.A., *Dough Rheology and Baked Product Texture*, Aspen, Gaithersburg, MD (1989).

Fardi, B. and Liu, B.Y.H., Flow field and pressure drop of filters with rectangular fibers, *Aerosol Sci. Technol.*, **17**, 36 (1992). Also see *ibid* 45.

Farinato, R.S. and Yen, W.S., Polymer degradation in porous media flow, *J. Appl. Polym. Sci.*, **33**, 2353 (1987).

Faroughi, S.A., Frenandes, C., Noberga, J.M. and McKinley, G.H., A closure model for the drag coefficient of a sphere translating in a viscoelastic fluid, *J. Non-Newt. Fluid Mech.*, **277**, 104218 (2020).

Farouk, B. and Guceri, S.I., Natural and mixed convection heat transfer around a horizontal cylinder within confining walls, *Num. Heat Transf.*, **5**, 329 (1982).

Favelukis, M. and Albalak, R.J., Bubble growth in viscous Newtonian and non-Newtonian liquids, *Chem. Eng. J.*, **63**, 149 (1996a).

Favelukis, M. and Albalak, R.J., Fundamentals of bubble growth, *Polymer Devolatiliztion*, edited by R.J. Albalak, Marcel Dekker, New York, Chapter 5 (1996b).

Favelukis, M. and Nir, A., Deformation of a slender bubble in a non-Newtonian liquid in an extensional flow, *Chem. Eng. Sci.*, **56**, 4643 (2001).

Faxen, H., Die bewegung einer starren kugel langs der achse eines mit zaher flussigkeit gefullten rohres, *Ask. Mat. Arstron. Fys.*, **17**, 1 (1923).

Faxen, O.H., Forces exerted on a rigid cylinder in a viscous fluid between two parallel fixed planes, *Proc. Roy. Swedish Acad. Eng. Sci.*, **187**, 1 (1946).

Fayed, H.E., Sheikh, N.A. and Iliev, O., On laminar flow of non-Newtonian fluids in porous media, *Trans. Porous Media*, **111** (1), 253 (2016).

Fayon, A.M. and Happel, J., Effect of a cylindrical boundary on a fixed rigid sphere in a moving viscous fluid, *AIChE J.*, **6**, 55 (1960).
Fdhila, R.B. and Duineveld, P.C., The effect of surfactant on the rise of a spherical bubble at high Reynolds and Peclet numbers, *Phys. Fluids*, **8**, 310 (1996).
Fedkiw, P.S. and Newman, J., Friction factors for creeping flow in sinusoidal periodically constricted tubes, *Chem. Eng. Sci.*, **42**, 2962 (1987).
Fehling, R., Der Strömungswiderstand ruhender Schüttungen. *Feuerungstechn*, **27**, 33 (1939).
Feinauer, A., Altobelli, S.A. and Fukushima, E., NMR measurement of flow profiles in a coarse bed of packed spheres, *Magn. Resonance Imaging*, **15**, 479 (1997).
Feldman, G.A. and Brenner, H., Experiments on the pressure drop created by a sphere settling in a viscous liquid. Part 2. Reynolds number from 0.2 to 21,000, *J. Fluid Mech.*, **32**, 705 (1968).
Felix, L.C.M. and Munoz, L.A.B., Representing a relation between porosity and permeability based on inductive rules, *J. Pet. Sci. Eng.*, **47**, 23 (2005).
Fellah, Z.E.A., Berger, S., Lauriks, W., Depollier, C., Aristegui, C. and Chapelon, J.-Y., Measuring the porosity and the tortuosity of porous materials via reflected waves of oblique incidence, *J. Accoust. Soc. Am.*, **113**, 2424 (2003).
Feltham, D.L., Sea ice rheology, *Annu. Rev. Fluid Mech.*, **40**, 91 (2008).
Femin Bendict, R.J., Kumaresan, G. and Velan, M., Bed expansion and pressure drop studies in a liquid-solid inverse fluidized bed reactor, *Bioprocess Eng.*, **19**, 137 (1998).
Fendell, F.E., Laminar natural convection about an isothermally heated sphere at small Grashof number, *J. Fluid Mech.*, **34**, 163 (1968).
Feng, J., Huang, P.Y. and Joseph, D.D., Dynamic simulation of sedimentation of solid particles in an Oldroyd-B fluid, *J. Non-Newt. Fluid Mech.* **63**, 63 (1996).
Feng, J. and Joseph, D.D., The motion of solid particles suspended in viscoelastic liquids under torsional shear, *J. Non-Newt. Fluid Mech.*, **324**, 199 (1996).
Feng, J., Joseph, D.D., Glowinski, R. and Pan, T.W., A three-dimensional computation of the force and torque on an ellipsoid settling slowly through a viscoelastic fluid, *J. Fluid Mech.*, **283**, 1 (1995).
Feng, Z.G. and Michaelides, E.E., A numerical study on the transient heat transfer from a sphere at high Reynolds and Peclet numbers, *Int. J. Heat Mass Transf.*, **43** (2), 219–229 (2000).
Feng, Z.-G. and Michaelides, E.E., Drag coefficients of viscous spheres at intermediate and high Reynolds numbers, *J. Fluids Eng. (ASME)*, **123**, 841 (2001).
Feng, Z.-G. and Michaelides, E.E., Hydrodynamic force on spheres in cylindrical and triangular enclosures, *Int. J. Multiph. Flow*, **28**, 479 (2002).
Fergui, O., Bertin, H. and Quintard, M., Transient aqueous foam flow in porous media: Experiments and modeling, *J. Pet. Sci. Eng.*, **20**, 9 (1998).
Fernando Concha, A., *Solid-Liquid Separation in the Mining Industry*, Springer, Berlin (2014).
Ferrari, M.A., Lugarini, A. and Franco, A.T., On the settling of spherical particles in power-law fluid at moderate Reynolds number, *Powder Technol.*, **405**, 117510 (2022).
Ferraris, C.F., Measurement of the rheological properties of high-performance concrete: State of the art report, *J. Res. Natl. Inst. Stand. Technol.*, **104**, 461 (1999).
Ferreira, J.M. and Chhabra, R.P., Accelerating motion of a vertically falling sphere in incompressible Newtonian media: An analytical solution, *Powder Technol.*, **97**, 6 (1998).
Ferreira, J.M. and Chhabra, R.P., Analytical study of drag and mass transfer in creeping power law flow across tube banks, *Ind. Eng. Chem. Res.*, **43**, 3439 (2004).
Ferreira, J.M., Duarte Naia, M. and Chhabra, R.P., An analytical study of the transient motion of a dense rigid sphere in an incompressible Newtonian fluid, *Chem. Eng. Commun.*, **168**, 45 (1998).
Ferroir, T., Huynh, H.T., Chateau, X. and Coussot, P., Motion of a solid object through a pasty (thixotropic) fluid, *Phys. Fluids*, **16**, 594 (2004).
Ferry, J.D., *Viscoelastic Properties of Polymers*, 3rd ed., Wiley, New York (1980).
Fidleris, V. and Whitmore, R.L., Experimental, determination of the wall effects for spheres falling axially in cylindrical vessels, *Brit. J. Appl. Phys.*, **12**, 490 (1961).
Field, S.B., Klaus, M., Moore, M.G. and Nori, F., Chaotic dynamics of falling disks, *Nature*, **388**, 252 (17 July 1997).
Finkers, H.J. and Hoffman, A.C., Structural ratio for predicting the voidage of binary particle mixtures, *AIChE J.*, **44**, 495 (1998).

Fischer, N., Becker T. and Fattahi, E., Correction of Stokes drag for non-Newtonian flow through pack of spheres, *Comput. Fluids*, 105895 (2023).

Flemmer, R.L.C., Pickett, J. and Clark, N.N., An experimental study on the effect of particle shape on fluidization behaviour, *Powder Technol.*, **77**, 123 (1993).

Fletcher, A.J.P., Flew, S.R.G., Lamb, S.P., Lund, T., Bjornestad, E., Stavland, A. and Gjovikli, N.B., Measurements of polysaccharide polymer properties in porous media, SPE 21018, Presented at *SPE Int. Sym. On Oilfield Chemistry*, Anaheim, CA (1991).

Flew, S. and Sellin, R.H.J., Non-Newtonian flow in porous media-a laboratory study of polyacrylamide solutions, *J. Non-Newt. Fluid Mech.*, **47**, 169 (1993).

Flowers, A.E., Viscosity measurement and a new viscometer, *Proc. Amer. Soc. Test. Mater.*, **14**, 565 (1914).

Flude, M.J.C. and Daborn, J.E., Viscosity measurement by means of falling spheres compared with capillary viscometry, *J. Phys. E: Sci. Instrum.*, **15**, 1313 (1982).

Flumerfelt, R.W., Drop breakup in simple shear fields of viscoelastic fluids, *Ind. Eng. Chem. Fundam.*, **11**, 312 (1972).

Fogler, H.S. and Goddard, J.D., Collapse of spherical cavities in viscoelastic fluids, *Phys. Fluids*, **13**, 1135 (1970).

Foltz, R.G., Wang, K.K. and Stevenson, J.F., An experiment to measure the pressure dependence of the zero-shear rate viscosity, *J. Non-Newt. Fluid Mech.*, **3**, 347 (1978).

Fonseca, C., Frey, S., Naccache, M.F. and de Souza Mendes, P.R., Flow of elasto-viscoplastic thixotropic liquids past a confined cylinder, *J. Non-Newt. Fluid Mech.*, **193**, 80 (2013).

Foo, J.J., Liu, K.K. and Chan, V., Viscous drag of deformed vesicles in optical trap: Experiments and simulations, *AIChE J.*, **50**, 249 (2004).

Ford, J.T., Oyeneyin, M.B., Williamson, R.S. and Peel, L.C., The formulation of milling fluids for efficient hole cleaning: An experimental investigation, paper presented at the *European Petroleum Conference*, (SPE 28819), London, October 25–27 (1994).

Fornberg, B., Steady viscous flow part a sphere at high Reynolds numbers, *J. Fluid Mech.*, **190**, 471 (1988).

Fortin, A., Bertrand, F., Fortin, M., Chamberland, E., Boulanger-Nadeau, P.E., Maliki, A.E. and Najeh, N., An adaptive remeshing strategy for shear-thinning fluid flow simulations, *Comp. Chem. Eng.*, **28**, 2363 (2004).

Foscolo, P.U., Gibilaro, L.G. and Waldram, S.P., A unified model for particulate expansion of fluidized beds and flow in fixed porous media, *Chem. Eng. Sci.*, **38**, 1251 (1983).

Foster, R.D. and Slattery, J.C., Creeping flow past a sphere of a Reiner-Rivlin fluid, *Appl. Sci. Res.*, **12A**, 213 (1962).

Foumeny, E.A., Benyahia, F., Castro, J.A.A., Moallemi, H.A. and Roshani, S., Correlations of pressure drop in packed beds taking into account the effect of confining walls, *Int. J. Heat Mass Transf.*, **36**, 536 (1993).

Foumeny, E.A., Kulkarni, A., Roshani, S. and Vatani, A., Elucidation of pressure drop in packed-bed systems, *Appl. Therm. Eng.*, **16**, 195 (1996).

Foumeny, E.A. and Roshani, S., Mean voidage of packed beds of cylindrical particles, *Chem. Eng. Sci.*, **46**, 2363 (1991).

Fourar, M., Lenormand, R., Karimi-Fard, M. and Horne, R., Inertia effects in high-rate flow through heterogeneous porous media, *Trans. Porous Media*, **60**, 353 (2005).

Fourar, M., Radilla, G., Lenormand, R. and Moyne, C., On the non-linear behaviour of a laminar single-phase flow through two- and three- dimensional porous media, *Adv. Water Resour.*, **27**, 669 (2004).

Fourie, J.G. and du Plessis, J.P., Pressure drop modeling in cellular metallic foams, *Chem. Eng. Sci.*, **57**, 2781 (2002).

Fox, V.G., Erickson, L.E. and Fan, L.T., The laminar boundary layer on a moving continuous sheet immersed in a non-Newtonian fluid, *AIChE J.*, **15**, 327 (1969).

Fraggedakis, D., Dimakopoulos, Y. and Tsamopoulos, J., Yielding the yield-stress analysis: A study focussed on the effects of elasticity on the settling of a single spherical particle in simple yield-stress fluids, *Soft Matter*, **12**, 5378 (2016).

Francis, A.W., Wall effect in falling ball method for viscosity, *Physics*, **4**, 403 (1933).

Francois, N., Lasne, D., Amarouchene, Y., Lounis, B. and Kellay, H., Drag enhancement with polymers, *Phys. Rev. Lett.*, **100**, 018302 (2008).

Frank, X. and Li, H.-Z., Complex flow around a bubble rising in a non-Newtonian fluid, *Phys. Rev.*, **71E**, 036309 (2005).

Frank, X. and Li, H.-Z., Negative wake behind a sphere rising in viscoelastic fluids: A lattice Boltzmann investigation, *Phys. Rev.*, **74E**, 056307 (2006).

Frank, X., Li, H.Z. and Funfschilling, D., An analytical approach to the rise velocity of periodic bubble trains in non-Newtonian fluids, *Eur. Phys. J.*, **E16**, 29 (2005).

Frank, X., Li, H.Z., Funfschilling, D., Burdin, F. and Ma, Y., Bubble motion in non-Newtonian fluids and suspensions, *Can. J. Chem. Eng.*, **81**, 483 (2003).

Freire, D., Sarasua, L.G., Vernet, A., Varela, S., Usera, G., Cabeza, C. and Marti, A.C., Separation regimes of two spheres falling in shear-thinning viscoelastic fluids, *Phys. Rev. Fluids*, **4**, 023302 (2019).

Freitas, L.G. and Pereira, L.A.A., Comparison among drag coefficient models of single bubbles under high and low Morton number regimes, *Chem. Eng. Sci.*, **236**, 116473 (2021).

Freund, J.B., Kim, J. and Ewoldt, R.H., Field sensitivity of flow predictions to rheological parameters, *J. Non-Newt. Fluid Mech.*, **257**, 71 (2018).

Freundlich, H. and Juliusburger, F., Thixotropy, influenced by the orientation of anisometric particles in sols and suspensions, *Trans. Faraday Soc.*, **31**, 920 (1935).

Frigaard, I., Iglesias, J.A., Mercier, G., Poschl, C. and Scherzer, O., Critical yield numbers of rigid particles settling in Bingham fluids and Cheeger sets, *SIAM J. App. Math.*, **77**, 638 (2017).

Frigaard, I. and Nouar, C., On the usage of viscosity regularisation methods for viscoplastic fluid flow computation, *J. Non-Newt. Fluid Mech.*, **127**, 1 (2005).

Frohn, A. and Roth, N., *Dynamics of Droplets*, Springer, Munich (2000).

Fryer, P.J., Pyle, D.L. and Reilly, C.D., *Chemical Engineering for the Food Industry*, Blackie Academic and Professional, London, UK (1997).

Fu, T., Ma, Y., Funfschilling, D. and Li, H.Z., Bubble formation and break up in microfluidic flow focusing device, *Chem. Eng. Sci.*, **64**, 2392 (2009).

Fu, T., Ma, Y., Funfschilling, D. and Li, H.Z., Gas-liquid flow stability and bubble formation in non-Newtonian fluids in microfluidic flow-focusing devices, *Microfluid. Nanofluid.*, **10**, 1135 (2011).

Fu, T., Ma, Y., Funfschilling, D., Zhu, C. and Li, H.Z., Breakup dynamics of slender bubbles in non-Newtonian fluids in microfluidic flow-focusing devices, *AIChE J.*, **58**, 3560 (2012).

Fujii, T., Miyatake, O., Fujii, M. and Tanaka, H., A numerical analysis of natural convection heat transfer to non-Newtonian Sutterby fluids, *Trans. Jap. Soc. Mech. Eng.*, **38**, 2883 (1972).

Fujii, T., Miyatake, O., Fujii, M., Tanaka, H. and Murakami, K., Natural convective heat transfer from a vertical isothermal surface to a non-Newtonian Sutterby fluid, *Int. J. Heat Mass Transf.*, **16**, 2177 (1973).

Fujii, T., Takeuchi, M., Fujii, M., Suzaki, K. and Uehara, H., Experiments on natural convection heat transfer from the outer surface of a vertical cylinder to liquids, *Int. J. Heat Mass Transf.*, **13**, 753 (1970).

Fukuchi, T. and Ishii, T., An analysis of transport phenomena for multi-solid particle systems at higher Reynolds numbers by a 5th order polynomial Karman-Pohlhausen method, *Int. J. Eng. Sci.*, **20**, 121 (1981).

Fukumoto, Y., Slow motion of a small sphere in a viscous fluid between two concentric circular cylinders, *J. Phys. Soc. Jpn.*, **54** 1322 (1985).

Funada, T., Joseph, D.D., Maehara, T. and Yamashita, S., Ellipsoidal model of the rise of a Taylor bubble in a round tube, *Int. J. Multiph. Flow*, **31**, 473 (2005).

Funatsu, K., Kajiwara, T. and Shiraishi, Y., Stress distribution in creeping flow around a falling sphere and its drag force – Measurement by flow birefringence technique, *Kogaku Kagaku Ronbunshu*, **12**, 582 (1986).

Funfschilling, D. and Li, H.-Z., Effects of the injection period on the rise velocity and shape of a bubble in a non- Newtonian fluid, *Chem. Eng. Res. Des.*, **84**, 875 (2006).

Funfschilling, D. and Li, H.Z., Flow of non-Newtonian fluids around bubbles: PIV measurements and birefringence visualization, *Chem. Eng. Sci.*, **56**, 1137 (2001).

Gabelnick, H.L. and Litt, M., *Rheology of Biological Systems*, Charles C. Thomas, Springfield, CT (1973).

Gadd, G.E., Turbulence damping and drag reduction produced by certain additives in water, *Nature*, **206**, 463 (1965).

Gadd, G.E., Reduction of turbulent friction in liquids by dissolved additives, *Nature*, **212**(#5065), 874 (1966).

Gahleitner, M. and Sobczak, R., Viscosity measurements with a magneto-viscometer in the zero shear and transition region of polypropylenes, *Rheol. Acta*, **26**, 371 (1987).

Gahleitner, M. and Sobczak, R., A new apparatus for measuring high viscosities, *J. Phys. E: Sci. Instrum.*, **21**, 1074 (1988).

Gaitonde, N.Y. and Middleman, S., Flow of viscoelastic fluids through porous media, *Ind. Eng. Chem. Fundam.*, **6**, 145 (1967).

Galdi, G.P., Slow steady fall of rigid bodies in a second-order fluid, *J. Non-Newt. Fluid Mech.*, **93**, 169 (2000).
Galindo-Rosales, F.J., Campo-Deano, L., Pinho, F.T., Van Bokhorst, E., Hamersma, P.J., Oliveira, M.S. and Alves, M.A., Microfluidic systems for the analysis of viscoelastic fluid flow phenomena in porous media. *Microfluid. Nanofluidics*, **12**(1), 485 (2012).
Galindo-Rosales, F.J., Rubio-Hernandez, F.J. and Sevilla, A., An apparent viscosity function for shear-thickening fluids, *J. Non-Newt. Fluid Mech.*, **166**, 321 (2011a).
Galindo-Rosales, F.J., Rubio-Hernandez, F.J., Sevilla, A. and Ewoldt, R.H., How Dr. Malcom M. Cross may have tackled the development of "An apparent viscosity function for shear-thickening fluids", *J. Non-Newt. Fluid Mech.*, **166**, 1421 (2011b).
Galindo-Rosales, F.J., Rubio-Hernandez, F.J. and Velazquez-Navarro, J.F., Shear-thickening behavior of Aerosil® R816 nanoparticles suspensions in polar organic liquids, *Rheol. Acta*, **48**, 699 (2009).
Gal-Or, B. and Waslo, S., Hydrodynamics of an ensemble of drops and bubbles in the presence or absence of surfactants, *Chem. Eng. Sci.*, **23**, 1431 (1968).
Gamboa, A.C., Saez, A.E. and Müller, A.J., Flow of solutions of hydroxypropyl guar-poly (ethylene oxide) mixtures through a porous medium, *Polym. Bull.*, **33**, 317 (1994).
Ganguli, S. and Lele, S.K., Drag of a heated sphere at low Reynolds numbers in the absence of buoyancy, *J. Fluid Mech.*, **869**, 264 (2019).
Ganoulis, J., Brunn, P.O., Durst, F., Holweg, J. and Wunderlich, A., Laser measurements and computations of viscous flows through cylinders, *J. Hyd. Eng. Proc. ASCE*, **115**, 1223 (1989).
Ganser, G.H., A rational approach to drag prediction of spherical and non-spherical particles, *Powder Technol.*, **77**, 143 (1993).
Garcia-Morales, M., Partal, P., Navarro, F.J., Martinez-Boza, F., Mackley, M.R. and Gallegos, C., The rheology of recycled EVA/LDPE modified bitumen, *Rheol. Acta*, **43**, 482 (2004).
Gardiner, B.S., Dlugogorski, B.Z., Jameson, G.J. and Chhabra, R.P., Yield stress measurements of aqueous foams in the dry limit, *J. Rheol.*, **42**, 1437 (1998).
Garduno, I.E., Tamaddon-Jahromi, H.R. and Webster, M.F., The falling sphere problem and capturing enhanced drag with Boger fluids, *J. Non-Newt. Fluid Mech.*, **231**, 26 (2016).
Garg, N.S. and Tripathi, G., Boiling heat transfer from rotating horizontal cylinder to non-Newtonian fluid, *Ind. J. Technol.*, **19**, 131 (1981).
Garg, V.K. and Rajagopal, K.R., Stagnation point flow of a non-Newtonian fluid, *Mech. Res. Commun.*, **17**, 45 (1990).
Garg, V.K. and Rajagopal, K.R., Flow of non-Newtonian fluid past a wedge, *Acta Mech.*, **88**, 113 (1991).
Garifulin, F.A., Zapparov, F.I., Mingaleyev, N.Z., Norden, P.A. and Tazukov, F.Kh., Convection in a horizontal layer of a viscoelastic fluid, *Heat Transf. Sov. Res.*, **14**, 121 (1982).
Garlaschelli, L., Ramaccini, F. and Della Sala, S., A miracle diagnosis, *Chem. Britain*, **30**(2), 123 (1994).
Garraud, A., Velez, C., Shah, Y., Kozissnik, B., Yarmola, E.G., Allen, K.D., Dobson, J. and Arnold, D.P., Investigation of the capture of magnetic particles from high-viscosity fluids using permanent magnets, *IEEE Trans. Biomed. Eng.*, **63**, 372 (2016).
Garrouch, A.A. and Gharbi, R.B., An empirical investigation of polymer flow in porous media, *Ind. Eng. Chem. Res.*, **38**, 3564 (1999).
Garside, J. and Al-Dibouni, M.R., Velocity-voidage relationship for fluidization and sedimentation in liquid-solid systems, *Ind. Eng. Chem. Proc. Des. Dev.*, **16**, 206 (1977).
Gauglitz, P.A. and Terrones, G., Estimated maximum gas retention from uniformly dispersed bubbles in K basin sludge stored in large- diameter containers, *Report #PNNL -13893*, Pacific Northwest National Laboratory, Oak Ridge, TN (2002).
Gauglitz, P.A., Terrones, G., Muller, S.J., Denn, M.M. and Rossen, W.R., Mechanics of bubbles in sludges and slurries, Final Report, Project No # 60451, U. S. Department of Energy, Washington, DC. (2003).
Gauri, V. and Koelling, K.W., Gas-assisted displacement of viscoelastic fluids: Flow dynamics at the bubble front, *J. Non-Newt. Fluid Mech.*, **83**, 183 (1999a).
Gauri, V. and Koelling, K.W., The motion of long bubbles through viscoelastic fluids in capillary tubes, *Rheol. Acta*, **38**, 458 (1999b).
Gauthier, F., Goldsmith, H.L. and Mason, S.G., Particle motions in non-Newtonian media II: Poiseuille Flow, *Trans. Soc. Rheol.*, **15**, 297 (1971a).
Gauthier, F., Goldsmith, H.L. and Mason, S.G., Particle motions in non-Newtonian media – I. Couette Flow, *Rheol. Acta*, **10**, 344 (1971b).

Gauthier-Manuel, B., Meyer, R. and Pieranski, P., The sphere rheometer: I Quasi-static measurements, *J. Phys. E: Sci. Instrum.*, **17**, 1177 (1984); Also see *ibid*, **17**, 1183 (1984).

Gavignet, A.A. and Sobey, I.J., Model aids cuttings transport prediction, *J. Pet. Tech.*, **41**, 916 (1989).

Geils, R.H. and Keezer, R.C., Small-volume, inclined, falling-ball viscometer, *Rev. Sci. Instrum.*, **48**, 783 (1977).

Gentry, C.C. and Wollersheim, D.E., Local free convection to non-Newtonian fluids from a horizontal, isothermal cylinder, *J. Heat Transf.*, **96**, 3 (1974).

Geoola, F. and Cornish, A.R.H., Numerical solution of steady-state free convective heat transfer from a solid sphere, *Int. J. Heat Mass Transf.*, **24**, 1369 (1981).

Geoola, F. and Cornish, A.R.H., Numerical solution of free convective heat transfer from a sphere, *Int. J. Heat Mass Transf.*, **25**, 1677 (1982).

Georgiou, G.C. and Crochet, M.J., The simultaneous use of 4 × 4 and 2 × 2 bilinear stress elements for viscoelastic flows, *Comput. Mech.*, **11**, 341 (1993).

Georgiou, G.C., Momani, S., Crochet, M.J. and Walters, K., Newtonian and non-Newtonian flow in a channel obstructed by an antisymmetry array of cylinders, *J. Non-Newt. Fluid Mech.*, **40**, 231 (1991).

German, G. and Bertola, V., Impact of shear-thinning and yield stress drops on solid substrates, *J. Phys. Cond. Matter*, **21**, 375111 (2009).

Gervang, B., Davies, A.R. and Phillips, T.N., On the simulation of viscoelastic flow past a sphere using spectral methods, *J. Non-Newt. Fluid Mech.*, **44**, 281 (1992).

Gestoso, P., Müller, A.J. and Saez, A.E., Two-dimensional flow of polymer solutions through porous media, *J. Porous Media*, **2**, 251 (1999).

Gestring, I. and Mewes, D., Degassing of molten polymers, *Chem. Eng. Sci.*, **57**, 3415 (2002).

Getachew, D., Minkowycz, W.J. and Poulikakos, D., Macroscopic equations of non-Newtonian fluid flow and heat transfer in a porous matrix, *J. Porous Media*, **1**, 273 (1998).

Ghaddar, C.K., On the permeability of unidirectional fibrous media: A parallel computational approach, *Phys. Fluids*, **7**, 2563 (1995).

Ghanbarian, B., Hunt, A.G., Ewing, R.P. and Sahimi, M., Tortuosity in porous media: A critical review, *Soil Sci. Soc. America J.*, **77**, 1461 (2013).

Gheissary, G. and van den Brule, B.H.A.A., Unexpected phenomena observed in particle settling in non-Newtonian media, *J. Non-Newt. Fluid Mech.*, **67**, 1 (1996).

Gheorghitza, St.I., On the non-steady motion of visco-plastic liquids in porous media, *J. Fluid Mech.*, **20**, 273 (1964).

Ghoniem, S.A.-A., Extensional flow of polymer solutions through porous media, *Rheol. Acta*, **24**, 588 (1985).

Ghosh, U.K., Ph D Thesis, Department of Chemical Engineering, Banaras Hindu University, Varanasi, India (1992).

Ghosh, A.K. and Ulbrecht, J., Bubble formation from a sparger in polymer solutions – I. Stagnant liquids, *Chem. Eng. Sci.*, **44**, 957 (1989); Also see *ibid*, 969.

Ghosh, U.K., Dey, K.N., Gupta, S.N., Kumar, S. and Upadhyay, S.N., Mass transfer from flat plates to power law fluids in laminar flow, *Chem. Eng. Commun.*, **43**, 335 (1986a).

Ghosh, U.K., Gupta, S.N., Kumar, S. and Upadhyay, S.N., Mass transfer in cross flow of non-Newtonian fluid around a circular cylinder, *Int. J. Heat Mass Transf.*, **29**, 955 (1986b).

Ghosh, U.K., Kumar, S. and Upadhyay, S.N., Mass transfer from spherical and non-spherical particles to non-Newtonian fluids, *Polym.- Plast. Technol. Eng.*, **31**, 271 (1992).

Ghosh, U.K., Upadhyay, S.N. and Chhabra, R.P., Heat and mass transfer from immersed bodies to non-Newtonian fluids, *Adv. Heat Transf.*, **25**, 251 (1994).

Ghosh Roychowdhury, D., Das, S.K. and Sundararajan, T., Numerical simulation of laminar flow and heat transfer over banks of staggered cylinders, *Int. J. Numer. Meth. Fluids*, **39**, 23 (2002).

Giese, M., Stromung in porosen medien unter berucksichtigung effektiver viskositaten, PhD Dissertation, Technical University, Munchen, Germany (1998).

Giesekus, H., Die simultane translations und rotationsbewegung einer kugel in einer elastovisksen flussigkeit, *Rheol. Acta*, **3**, 59 (1963).

Giesekus, H., Die bewegung von teilchen in stromungen nicht-Newtonscher flussigkeiten, *ZAMM*, **58**, 26 (1978).

Gilchrist, I.C.R. and Chandler, H.D., Thixotropy in flocculated slurries, *Hydrotransport*, **13**, 115 (1996).

Gillaspy, P.H. and Hoffer, T.E., Experimental measurements of the effect of viscosity on drag for liquid drops, *AIChE J.*, **29**, 229 (1983).

Gilligan, S.A. and Jones, R.S., Unsteady flow of an elastico-viscous fluid past a circular cylinder, *ZAMP*, **21**, 786 (1970).

Giner, S.A. and Denisienia, E., Pressure drop through wheat as affected by air velocity, moisture content and fines, *J. Agric. Eng. Res.*, **63**, 73 (1996).

Gioia, F. and Urciuolo, M., The containment of oil spills in unconsolidated granular porous media using xanthan/Cr (III) and xanthan/Al (III) gels, *J. Hazardous Mater.*, **116B**, 83 (2004).

Givler, R.C. and Altobelli, S.A., A determination of the effective viscosity for the Brinkman-Forchheimer flow model, *J. Fluid Mech.*, **258**, 355 (1994).

Gkormpatsis, S.D., Gryparis, E.A., Housiadas, K.D. and Beris, A.N., Steady sphere translation in a viscoelastic fluid with slip on the surface of the sphere, *J. Non-Newt. Fluid Mech.*, **275**, 104217 (2020).

Glowinski, R., Finite element methods for incompressible viscous flow, *Handbook Numer. Anal.*, **9**, 3 (2003).

Glowinski, R. and Wachs, A., On the numerical simulation of viscoplastic fluid flow, *Handbook of Numerical Analysis*, edited by P.G. Ciarlet, Gulf Professional Publishing, Houston, Vol. 16, p. 483 (2011).

Gmachowski, L., Flow drag in heterogeneous systems over wide intervals of porosity and Reynolds number, *J. Chem. Eng. Jpn.*, **29**, 897 (1996).

Godbole, S.P., Schumpe, A., Shah, Y.T. and Carr, N.L., Hydrodynamics and mass transfer in non-Newtonian solutions in a bubble column, *AIChE J.*, **30**, 213 (1984).

Goddard, J.D. and Bashir, Y.M., On Reynolds dilatancy, *Recent Developments in Structured Continua*, edited by De Kee, D. and Kaloni, P.N., Longman, London, Chapter 2 (1990).

Godfrey, J.C. and Hanson, C., Liquid-liquid systems, *Handbook of Multiphase Systems*, edited by G. Hetsroni, McGraw-Hill, New York, Chapter 4 (1982).

Godinez, F.A., de la Calleja, E., Lauga, E. and Zenit, R., Sedimentation of a rotating sphere in a power-law fluid, *J. Non-Newt. Fluid Mech.*, **213**, 27 (2014).

Goel, N., Shah, S.N. and Grady, B.P., Correlating viscoelastic measurements of fracturing fluid to particles suspension and solids transport, *J. Pet. Sci. Eng.*, **35**, 59 (2002).

Gogarty, W.B., Mobility control with polymer solutions, *Soc. Pet. Eng. J.*, **5**, 161 (June 1967a).

Gogarty, W.B., Rheological properties of pseudoplastic fluids in porous media, *Soc. Pet. Eng. J.*, **5**, 149 (June 1967b).

Göğüs, M., Ipekci, O.N. and Kökpinar, M.A., Effect of particle shape on fall velocity of angular particles, *J. Hyd. Eng.*, **127**, 860 (2001).

Goharzadeh, A., Khezzar, L. and Molki, A., PIV measurements of laminar flow around a hemisphere, *Adv. Fluid Mech. IX*, **74**, 29 (2012).

Goldin, M., Pfeffer, R. and Shinnar, R., Breakup-of a capillary jet of a non-Newtonian fluid having a yield stress, *Chem. Eng. J.*, **4**, 8 (1972).

Goldin, M., Yerushalmi, J., Pfeffer, R. and Shinnar, R., Breakup of a laminar capillary jet of a viscoelastic fluid, *J. Fluid Mech.*, **38**, 689 (1969).

Goldsmith, H.L. and Mason, S.G., The flow of suspensions through tubes – I. Single spheres, rods and discs, *J. Colloid Sci.*, **17**, 448 (1962).

Goldstein, S., The steady flow of viscous fluid past a fixed spherical obstacle at small Reynolds number, *Proc. Roy. Soc.*, **123A**, 216 (1929).

Gollakota, A.R.K. and Kishore, N., CFD study on rise and deformation characteristics of buoyancy-driven spheroid bubbles in stagnant Carreau model non-Newtonian fluids, *Theor. Comput. Fluid Dyn.*, **32**, 35 (2017).

Gollakota, A.R.K. and Kishore, N., Flow behavior and drag coefficients of spherical bubbles in surfactant-laden Carreau model fluids, *Prog. Comput. Fluid Dyn.*, **18**, 257 (2018).

Gomez-Diaz, D., Navaza, J.M., Quintans-Riveiro, L.C. and Sanjuji, B., Gas absorption in bubble column using a non-Newtonian liquid phase, *Chem. Eng. J.*, **146**, 16 (2009).

Gondret, P., Hallouin, E., Lance, M. and Petit, L., Experiments on the motion of a solid sphere toward a wall: From viscous dissipation to elastohydrodynamic bouncing, *Phys. Fluids*, **11**, 2803 (1999).

Gonzalez, J.M., Muller, A.J., Torres, M.F. and Saez, A.E., The role of shear and elongation in the flow of solutions of semi-flexible polymers through porous media, *Rheol. Acta*, **44**, 396 (2005).

Gonzalez-Nunez, R., Chan Man Fong, C.F., Favis, B.D. and De Kee, D., Deformation of drops in extensional viscoelastic flow, *J. Appl. Polym. Sci.*, **62**, 1627 (1996).

Goossens, W.R.A., Review of the empirical correlations for the drag coefficient of rigid spheres, *Powder Technol.*, **352**, 350 (2019).

Goossens, W.R.A., A new explicit equation for the terminal velocity of a settling sphere, *Powder Technol.*, **362**, 54 (2020).

Goren, S.L. and Gottlieb, M., Surface-tension-driven breakup of viscoelastic liquid threads, *J. Fluid Mech.*, **120**, 245 (1982).

Gorla, R.S.R., Unsteady heat transfer in laminar non-Newtonian boundary layer over a wedge, *AIChE J.*, **28**, 56 (1982).

Gorla, R.S.R., Combined forced and free convection in boundary layer flow of non-Newtonian fluid on a horizontal plate, *Chem. Eng. Commun.*, **49**, 13 (1986).

Gorla, R.S.R., Free convection to Ellis fluids from a horizontal isothermal cylinder, *Polym.-Plast. Technol. Eng.*, **30**, 37 (1991a).

Gorla, R.S.R., Heat transfer to a non-isothermal rotating disk in a non-Newtonian fluid, *Polym.-Plast. Technol. Eng.*, **30**, 75 (1991b). Also see *ibid* 89 (1991c).

Gorla, R.S.R., Heat transfer from a continuous surface to a non-Newtonian fluid, *Polym.-Plast. Technol. Eng.*, **31**, 241 (1992).

Gottlieb, M., Zero shear rate viscosity measurements for polymer solutions by falling ball viscometry, *J. Non-Newt. Fluid Mech.*, **6**, 97 (1979).

Götz, J., Kick, K., Heinen, C. and König, T. Visualization of flow processes in packed beds with NMR imaging: Determination of the local porosity, velocity vector and local dispersion coefficients, *Chem. Eng. Process.*, **41**, 611 (2002).

Goutille, Y. and Guillet, J., Disentanglement of polymer melts flowing through porous medium before entering a capillary die, *J. Rheol.*, **46**, 1307 (2002).

Govier, G.W. and Aziz, K., *The Flow of Complex Mixtures in Pipes*, R.E. Krieger, Malabar, FL (1982).

Govindarao, V.M.H. and Froment, G.F., Voidage profiles in packed beds of spheres, *Chem. Eng. Sci.*, **41**, 533 (1986). Also see *ibid* **43**, 2544 (1988).

Grace, J.R., Hydrodynamics of liquid drops in immiscible liquids, *Handbook of Fluids in Motion*, edited by N.P. Cheremisinoff, and R. Gupta, Ann Arbor Sci., Michigan (1983), Chap. 38.

Grace, J.R. and Wairegi, T., Properties and characteristics of drops and bubbles, *Encycl. Fluid Mech.*, **3**, 43 (1986).

Grace, J.R., Wairegi, T. and Nguyen, T.H., Shapes and velocities of single drops and bubbles moving freely through immiscible liquids, *Trans. Inst. Chem. Eng.*, **54**, 167 (1976).

Graessley, W.W., The entanglement concept in polymer rheology, *Adv. Polym. Sci.*, **16**, 3 (1974).

Graessley, W.W., *Polymeric Liquids and Networks: Structure and Properties*, Taylor and Francis, New York (2004).

Graham, D.I. and Jones, T.E.R., Settling and transport of spherical particles in power-law fluids at finite Reynolds number, *J. Non-Newt. Fluid Mech.*, **54**, 465 (1994).

Granville, P.S., Maximum drag reduction at high Reynolds number for a flat plate immersed in polymer solutions, *Naval Ship Research and Development Center*, Tech. Note 205, Washington, DC (1971).

Grattoni, C.A., Luckham, P.F., Jing, X.D., Norman, L. and Zimmerman, R.W., Polymers as relative permeability modifiers: Adsorption and the dynamic formation of thick polyacrylamide layers, *J. Pet. Sci. Eng.*, **45**, 233 (2004).

Greaves, M. and Patel, K., Flow of Polymer solutions in porous media, *Chem. Eng. Res. Des.*, **63**, 199 (1985).

Green, A.E. and Rivlin, R.S., The mechanics of nonlinear materials with memory: Part 1, *Arch. Rat. Mech. Anal.*, **1**, 1 (1957).

Green, A.E. and Rivlin, R.S., The mechanics of nonlinear materials with memory: Part 3, *Arch. Rat. Mech. Anal.*, **4**, 387 (1960).

Green, A.E., Rivlin, R.S. and Spencer, A.J.M., The mechanics of nonlinear materials with memory: Part 2, *Arch. Rat. Mech. Anal*, **3**, 82 (1959).

Greene, G.A., Irvine Jr., T.F., Gyves, T. and Smith T., Drag relationships for liquid droplets settling in a continuous liquid, *AIChE J.*, **39**, 37 (1993).

Greenkorn, R.A., Steady flow through porous media, *AIChE J.*, **27**, 529 (1981).

Greenkorn, R.A., *Flow Phenomena in Porous Media*, Marcel-Dekker, New York (1983).

Greenkorn, R.A. and Kessler, D.P., Dispersion in heterogeneous non-uniform, anisotropic porous media, *Flow Through Porous Media*, American Chemical Society, Washington, DC, Chapter 8 (1970).

Gregory, D.R. and Griskey, R.G., Flow of molten polymers through porous media, *AIChE J.*, **13**, 122 (1967).
Griffiths, R.W., The dynamics of lava flows, *Ann. Rev. Fluid Mech.*, **32**, 477 (2000).
Grimm, R.J., Squeezing flow of polymeric liquids, *AIChE J.*, **24**, 427 (1978).
Griskey, R.G., Nechrebecki, D.G., Notheis, P.J. and Balmer, R.T., Rheological and pipeline flow behaviour of corn starch suspensions, *J. Rheol.*, **29**, 349 (1985).
Groisman, A. and Steinberg, V., Elastic turbulence in a polymer solution flow, *Nature*, **405** (May 4), 53 (2000).
Grove, A.S., Shair, F.H. and Petersen, E.E., An experimental investigation of the steady separated flow past a circular cylinder, *J. Fluid Mech.*, **19**, 60 (1964).
Gruber, V.E., Sezen, M.C. and Schurz, J., Ein Fallkugel viskosimeter mit geneigtem rohr fur sehr kleine probenmengen, *Die. Ang. Mak. Chem.*, **28**, 57 (1973).
Gu, D. and Tanner, R.I., The drag on a sphere in a power law fluid, *J. Non-Newt. Fluid Mech.*, **17**, 1 (1985).
Gu, D.Z., Li, Z.X. and Wu, D.C., Non-Newtonian flow through assemblage of uniform spheres, *J. Chengdu Uni. Sci. Tech.*, **4**, 1 (1992).
Gueslin, B., Talini, L., Herzhaft, B., Peysson, Y. and Allain, C., Flow induced by a sphere settling in an aging yield-stress fluid, *Phys. Fluids*, **18**, 103101 (2006).
Guido, S. and Greco, F., Dynamics of a liquid drop in a flowing immiscible liquid, *Rheology Reviews*, p. 99, published by the British Soc. Rheology (2004).
Gummalam, S. and Chhabra, R.P., Rising velocity of a swarm of spherical bubbles in a power law non-Newtonian liquid, *Can. J. Chem. Eng.*, **65**, 1004 (1987). Also see *Chem. Eng. Sci.*, **43**, 399 (1988).
Gummalam, S., Narayan, K.A. and Chhabra, R.P., Rise velocity of a swarm of spherical bubbles through a non-Newtonian fluid: Effect of zero shear viscosity, *Int. J. Multiph. Flow*, **14**, 361 (1988).
Gumulya, M., Horsley, R.R. and Pareek, V., Numerical simulation of the settling behavior of particles in thixotropic fluids, *Phys. Fluids.*, **26**, 023102 (2014).
Gumulya, M.M., Horsley, R.R. and Wilson, K.C., The settling of consecutive spheres in viscoplastic fluids. *Int. J. Mineral Proc.*, **82**, 106 (2007).
Gunjal, P.R., Ranade, V.V. and Chaudhari, R.V., Computational study of a single-phase flow in packed beds of spheres, *AIChE J.*, **51**, 365 (2005).
Guo, J. and Uhlherr, P.H.T., Static yield stress using a pendulum with cylindrical bob, *Proc. XII Int. Cong. Rheology*, p. 731, Quebec City, PQ, Canada (1996).
Guo, Z., Sun, Z., Zhang, N. and Ding, M., Influence of confining walls on pressure drop and particle-to-fluid heat transfer in packed beds with small D/d ratios under high Reynolds number, *Chem. Eng. Sci.*, **209**, 115200 (2019a).
Guo, Z., Sun, Z., Zhang, N., Ding, M. and Cao, X., Experimental characterization of pressure drop in slender packed bed ($1 < D/d < 3$), *Chem. Eng. Sci.*, **173**, 578 (2017a).
Guo, Z., Sun, Z., Zhang, N., Ding, M. and Cao, X., Radial porosity peak at the centerline of packed beds with small tube to particle diameter ratios, *Powder Technol.*, **319**, 445 (2017b).
Guo, Z., Sun, Z., Zhang, N., Ding, M. and Liu, J., Pressure drop in slender packed beds with novel packing arrangement, *Powder Technol.*, **321**, 286 (2017c).
Guo, Z., Zhang, N., Cao, X. and Ding, M., Mean porosity variations in packed bed of monosized spheres with small tube-to-particle diameter ratios, *Powder Technol.*, **354**, 842 (2019b).
Gupta, R.K., Unsteady motion of a spheroid in an elastico-viscous liquid, *ZAMP*, **27**, 273 (1976).
Gupta, R.K., *Polymer and Composites Rheology*, Marcel-Dekker, New York (2000).
Gupta, S.V., *Viscometry for Liquids: Calibration of Viscometers*, Springer, Berlin (2014).
Gupta, A.K. and Chhabra, R.P., Spheroids in viscoplastic fluids: Drag and heat transfer, *Ind. Eng. Chem. Res.*, **53**, 18943 (2014).
Gupta, A.K. and Chhabra, R.P., Combined effects of fluid shear-thinning and yield stress on heat transfer from an isothermal spheroid, *Int. J. Heat Mass Transf.*, **93**, 803 (2016a).
Gupta, A.K. and Chhabra, R.P., Effect of buoyancy-assisted flow on convection from an isothermal spheroid in power-law fluids, *Korea-Austr. Rheol. J.*, **28**, 87 (2016b).
Gupta, A.K. and Chhabra, R.P., Mixed convection from a spheroid in Bingham plastic fluids: Effect of buoyancy-assisted flow, *Numer. Heat Transf. Part A*, **69**, 898 (2016c).
Gupta, A.K., Gupta, S. and Chhabra, R.P., Natural convection in Bingham plastic fluids from an isothermal spheroid: Effects of fluid yield stress, viscous dissipation and temperature-dependent viscosity, *Korea-Austr. Rheol. J.*, **29**, 163 (2017a).

Gupta, A.K., Mishra, G., Nirmalkar, N. and Chhabra, R.P., Effect of confinement on heat transfer in aqueous nano-fluids from a heated sphere, *Powder Technol.*, **325**, 576 (2018).

Gupta, A.K., Mishra, P. and Chhabra, R.P., Momentum and heat transfer characteristics of a thin circular disk in Bingham plastic fluids, *Numer. Heat Transf. Part A*, **72**, 844 (2017b).

Gupta, A.K., Sasmal, C, Sairamu, M. and Chhabra, R.P., Laminar and steady free convection in power-law fluids from a heated spheroidal particle: A numerical study, *Int. J. Heat Mass Transf.*, **75**, 592 (2014).

Gupta, A.K., Sharma, A., Chhabra, R.P. and Eswaran, V., Two-dimensional steady flow of a power-law fluid past a square cylinder in a plane channel: Momentum and heat transfer characteristics, *Ind. Eng. Chem. Res.*, **42**, 5674 (2003).

Gupta, R.K., Ryan, M.E. and Sridhar, T., On the formulation of highly elastic constant viscosity liquids, *J. Rheol.*, **30**, 1181 (1986).

Gupta, R.K. and Sridhar, T., Viscoelastic effects in non-Newtonian flows through porous media, *Rheol. Acta*, **24**, 148 (1985).

Gupta, R.K. and Sridhar, T., Elongational Rheometers, *Rheological Measurements*, A.A. Collyer and D.W. Clegg, 2nd ed., pp. 516–549, Chapman and Hall, London (1998).

Gupta, S., Gupta, A.K. and Chhabra, R.P., Effect of confinement on natural convection in power-law fluids from a circular cylinder, *Comput. Thermal Sci.*, **11**, 445 (2019).

Gupta, S., Patel, S.A. and Chhabra, R.P., Flow and thermal characteristics of two interacting cylinders in yield stress fluids, *Comput. Thermal Sci.*, **12**, 133 (2020a).

Gupta, S., Patel, S.A. and Chhabra, R.P., Pulsatile flow of power-law fluids over a heated cylinder: Flow and heat transfer characteristics. *Int. J. Therm. Sci.*, **152**, 106330 (2020b).

Gupta, S., Patel, S.A. and Chhabra, R.P., Effect of sinusoidally varying flow of yield stress fluid on heat transfer from a cylinder. *J. Heat Transf.*, **143** (6), 061802 (2021).

Gupta, S.K., Ray, S. and Chatterjee, D., Influence of aiding buoyancy on the suppression of flow separation for power-law fluids around a circular object, *Heat Transf. Eng.*, **37**, 1267 (2016).

Gürgen, S., *Shear Thickening Fluid: Theory and Applications*, Springer, New York (2023).

Gürgen, S., Kushan, M.C. and Li, W.H., The effect of carbide particle additives on rheology of shear-thickening fluids, *Korea-Aust. Rheol. J.*, **28**, 121 (2016a).

Gürgen, S., Li, W.H. and Kushan, M.C., The rheology of shear-thickening fluids with various ceramic particle additives, *Mater. Des.*, **104**, 312 (2016b).

Gurkan, T., Motion of a circulating power law drop translating through Newtonian fluids at intermediate Reynolds numbers, *Chem. Eng. Commun.*, **80**, 53 (1989). Also see *ibid*, **89**, 73 (1990).

Gurkan, T. and Wellek, R.M., Mass transfer in dispersed and continuous phases for creeping flow of fluid spheres through power law fluids, *Ind. Eng. Chem. Fund.*, **15**, 45 (1976).

Gutierrez-Lemini, D., *Engineering Viscoelasticity*, Springer, New York (2014).

Guzy, C.J., Bonano, E.J. and Davis, E.J., The analysis of flow and colloidal particle retention in fibrous porous media, *J. Colloid Interface Sci.*, **95**, 523 (1983).

Ha, J.-W. and Leal, L.G., An experimental study of drop deformation and breakup in extensional flow at high capillary number, *Phys. Fluids*, **13**, 1568 (2001).

Ha, J.-W. and Yang, S.-M., Deformation and breakup of Newtonian and non-Newtonian conducting drops in an electric field, *J. Fluid Mech.*, **405**, 131 (2000).

Haas, R. and Durst, F., Viscoelastic flow of dilute polymer solutions in regularly packed beds, *Rheol. Acta*, **21**, 566 (1982).

Haas, R. and Kulicke, W.-M., Flow behaviour of dilute polyacrylamide solutions through porous media. 2. Indirect determination of extremely high molecular weights and some aspects of viscosity decrease over long-time intervals, *Ind. Eng. Chem. Fund.*, **23**, 316 (1984).

Haberman, W.L. and Sayre, R.M., Motion of rigid and fluid spheres in stationary and moving liquids inside cylindrical tubes, *David Taylor Model Basin Report*, No. 1143, Dept. of Navy, Washington, DC (1958).

Haddow, J.B. and Luming, H., An extension of one of the extremum principles for a Bingham solid, *Appl. Sci. Res.*, **15A**, 81 (1965).

Haddow, J.B. and Luming, H., The application of the extremum principles for a Bingham solid, *Appl. Sci. Res.*, **16A**, 469 (1966).

Hagen, T. and Renardy, M., Boundary layer analysis of the Phan-Thien-Tanner and Giesekus model in high Weissenberg number flow, *J. Non-Newt. Fluid Mech.*, **73**, 181 (1997).

Haider, A. and Levenspiel, O., Drag coefficient and terminal velocity of spherical and non-spherical particles, *Powder Technol.*, **58**, 63 (1989).

Hailemariam, H. and Mulugeta, G., Temperature-dependent rheology of bouncing putties as rock analogs, *Tectonophysics*, **294**, 131 (1998).

Hall, M.J. and Hiatt, J.P., Measurements of pore scale flows within and exiting ceramic foams, *Exp. Fluids*, **20**, 433 (1996).

Hamielec, A.E. and Johnson, A.I., Viscous flow around fluid spheres at intermediate Reynolds numbers, *Can. J. Chem. Eng.*, **40**, 41 (1962).

Hamilton, R.T., Darcy constant for multi-sized spheres with no arbitrary constants, *AIChE J.*, **43**, 835 (1997).

Han, E., Wyart, M., Peters, I.R. and Jaeger, H.M., Shear fronts in shear-thickening suspensions, *Phys. Rev. Fluids*, **3**, 073301 (2018).

Han, S.-I., Stapf, S. and Blumich, B., NMR imaging of falling water drops, *Phys. Rev. Lett.*, **87**(1), 144501 (2001).

Han, S.T., Compressibility and permeability of fiber mats, *Pulp & Paper Mag. Canada*, T134 (May 2, 1969).

Handzy, N.Z. and Belmonte, A., Oscillatory rise of bubbles in wormlike micellar fluids with different microstructures, *Phys. Rev. Lett.*, **92**(12), 124501 (2004).

Hanehara, S. and Yamada, K., Rheology and early age properties of cement systems, *Cement Concrete Res.*, **21**, 175 (2008).

Hanks, R.W. and Sen, S., The influence of yield stress and fluid rheology on particle drag coefficients, *Proc. 9th Int. Tech. Conf. Slurry Transportation*, p. 71 (1983).

Hanna, M.R., Kozicki, W. and Tiu, C., Flow of drag reducing fluids through packed beds, *Chem. Eng. J.*, **13**, 93 (1977).

Hannah, R.R. and Harrington, L.J., Measurement of dynamic proppant fall rates in fracturing gels using a concentric cylinder tester, *J. Pet. Technol.*, **33**, 909 (1981).

Hansford, G.S. and Litt, M., Mass transport from a rotating disk into power law liquids, *Chem. Eng. Sci.*, **23**, 849 (1968).

Happel, J., Viscous flow in multiparticle systems: Slow motion of fluids relative to beds of spherical particles, *AIChE J.*, **4**, 197 (1958).

Happel, J., Viscous flow relative to arrays of cylinders, *AIChE J.*, **5**, 174 (1959).

Happel, J. and Bart, E., The settling of a sphere along the axis of a long square duct at low Reynolds number, *Appl. Sci. Res.*, **29**, 241 (1974).

Happel, J. and Brenner, H., *Low Reynolds Number Hydrodynamics*, Prentice Hall, Englewood Cliffs, NJ (1965).

Happel, J. and Byrne, B.J., Motion of a sphere and fluid in a cylindrical tube. *Ind. Eng. Chem.*, **46**, 1181 (1954).

Haque, M.W., Nigam, K.D.P., Srivastava, V.K. and Joshi, J.B., Studies on gas holdup and bubble parameters in bubble columns, *Ind. Eng. Chem. Res.*, **26**, 82 (1987).

Haque, M.W., Nigam, K.D.P., Viswanathan, K. and Joshi, J.B., Studies on bubble rise velocity in bubble columns employing non-Newtonian solutions, *Chem. Eng. Commun.*, **73**, 31 (1988).

Hara, S.K. and Schowalter, W.R., Dynamics of non-spherical bubbles surrounded by a viscoelastic fluid, *J. Non-Newt. Fluid Mech.*, **14**, 249 (1984).

Hariharaputhiran, M., Shankar Subramanian, R., Campbell, G.A. and Chhabra, R.P., The settling of spheres in a viscoplastic medium, *J. Non-Newt. Fluid Mech.*, **79**, 87 (1998).

Haring, R.E. and Greenkorn, R.A., A statistical model of a porous medium with non-uniform pores, *AIChE J.*, **16**, 477 (1970).

Harkins, W.D. and Brown, F.E., The determination of surface tension (free surface energy) and the weight of falling drops: The surface tension of water and benzene by the capillary height method, *J. Amer. Chem. Soc.*, **41**, 499 (1919).

Harlen, O.G., High Deborah number flow of a dilute polymer solution past a sphere falling along the axis of a cylindrical tube, *J. Non-Newt. Fluid Mech.*, **37**, 137 (1990).

Harlen, O.G., The wake behind a sphere sedimenting in a viscoelastic fluid, *Proc. XIIIth Int. Cong. Rheol.*, **2**, 288, Cambridge, U.K. (2000).

Harlen, O.G., The negative wake behind a sphere sedimenting through a viscoelastic fluid, *J. Non-Newt. Fluid Mech.*, **108**, 411 (2002).

Harlen, O.G., Rallison, J.M. and Chilcott, M.D., High Deborah number flow of dilute polymer solutions, *J. Non-Newt. Fluid Mech.*, **34**, 319 (1990).

Harlen, O.G., Rallison, J.M. and Szabo, P., A split Lagrangian-Eulerian method for simulating transient viscoelastic flows, *J. Non-Newt. Fluid Mech.*, **60**, 81 (1995).

Harlen, O.G., Sundararajakumar, R.R. and Koch, D.L., Numerical simulation of a sphere settling through a suspension of neutrally buoyant fibres, *J. Fluid Mech.*, **388**, 355 (1999).

Harmathy, T.Z., Velocity of large drops and bubbles in media of infinite or restricted extent, *AIChE J.*, **6**, 281 (1960).

Harnoy, A., An investigation into the flow of elastico-viscous fluids past a circular cylinder, *Rheol. Acta*, **26**, 493 (1987).

Harper, J.F. and Moore, D.W., The motion of a spherical liquid drop at high Reynolds number, *J. Fluid Mech.*, **32**, 367 (1968).

Harrington, L.J., Hannah, R.R. and Williams, D., Dynamic experiments on proppant settling in crosslinked fracturing fluids, Paper # SPE 8342 presented at the *54th Annual Fall Technical Conference and Exhibition of the Soc. Pet. Engrs. of AIME*, Las Vegas (1979).

Harrington, R.E. and Zimm, B.H., Anomalous plugging of sintered glass filters by high molecular weight polymers, *J. Polym. Sci.*, (Part A-2), **6**, 294 (1968).

Harris, P.C., Walters, H.G. and Bryant, J., Prediction of proppant transport from rheological data, *SPE Prod. Oper.*, **24**, 550 (2009).

Harris, P.J., A numerical method for predicting the motion of a bubble close to a moving rigid structure, *Comm. Num. Methods Eng.*, **9**, 81 (1993).

Harris, P.J., A numerical method for modeling the motion of a spherical bubble, *Int. J. Num. Methods Fluids*, **22**, 1125 (1996).

Harrison, D.E. and Gosser, R.B., Rolling ball viscometer for use at temperatures to 400 C under pressures to 5 kilobar, *Rev. Sci. Instrum.*, **36**, 1840 (1965).

Harrison, G.M., Lawson, N.J. and Boger, D.V., The measurement of the flow around a sphere settling in a rectangular box using 3-dimensional particle image velocimetry, *Chem. Eng. Commun.*, **188**, 143 (2001).

Harrison, G.M., Mun, R., Cooper, G. and Boger, D.V., A note on the effect of polymer rigidity and concentration on spray atomization, *J. Non-Newt. Fluid Mech.*, **85**, 93 (1999).

Hart, S.A., Moore, J.A. and Hale, W.F., Pumping manure slurries, management of farm animal wastes, *Proc. Nat. Symp. Animal Waste Management*, ASAE Publication. No. SP-0366, p. 34 (1966).

Hartholt, G.P., Hoffmann, A.C., Janssen, L.P.B.M. and Moes, J.H., Finite element calculations of flow past a spherical bubble rising on the axis of a cylindrical tube, *ZAMP*, **45**, 733(1994).

Hartland, S. and Hartley, R.W., *Axisymmetric Fluid-Liquid Interfaces*, Elsevier, Amsterdam (1976).

Hartman, M., Trnka, O. and Svoboda, K., Free settling of non-spherical particles, *Ind. Eng. Chem. Res.*, **33**, 1979 (1994).

Hartnett, J.P. and Hu, R.Y.Z., The yield stress: An engineering reality, *J. Rheol.*, **33**, 671 (1989).

Hartranft, T.J. and Settles, G.S., Sheet atomization of non-Newtonian liquids, *At. Sprays*, **13**, 191 (2003).

Hartt, W.H. and Baird, D.G., The confined flow of polyethylene melts past a cylinder in a planar channel, *J. Non-Newt. Fluid Mech.*, **65**, 247 (1996).

Harvey, A.H., *Ph. D Dissertation*, University of Oklahoma, Stillwater, OK (1968).

Hasan, M.A., Calibration of rolling ball viscometer in the intermediate Reynolds number range, *Can. J. Chem. Eng.*, **61**, 607 (1983).

Hasimoto, H., On the periodic fundamental solutions of the Stokes equations and their application to viscous flow past a cubic array of spheres, *J. Fluid Mech.*, **5**, 317 (1959).

Hasimoto, H., Slow motion of a small sphere in a cylindrical domain, *J. Phys. Soc. Jpn.*, **41**, 2143 (1976).

Hashimoto, A.G. and Chen, Y.R., Rheology of livestock waste slurries, *Trans. ASAE*, **19**, 930 (1976).

Hassager, O., Bubble motion in structurally complex fluids, *Chem. Eng. with Per Stoft*, Teknisk for lagas, Kbenhavn, 105 (1977).

Hassager, O., Negative wake behind bubbles in non-Newtonian liquids, *Nature*, **279**, 402 (1979).

Hassager, O., Working group on numerical techniques, *J. Non-Newt. Fluid Mech.*, **29**, 2 (1988).

Hassager, O. and Bisgaard, C., A Lagrangian finite element method for the simulation of flow of non-Newtonian fluids, *J. Non-Newt. Fluid Mech.*, **13**, 153 (1983). Also see *Rheol. Acta*, **21**, 534 (1982).

Hassan, N.M.S., Khan, M.M.K. and Rasul, M.G., A modelling and experimental study of the bubble trajectory in a non-Newtonian crystal suspension, *Fluid Dyn. Res.*, **42**, 065502 (2010b).

Hassan, N.M.S., Khan, M.M.K., Rasul, M.G. and Rackemann, D.W., Bubble rise velocity and trajectory in xanthan gum crystal suspension, *Appl. Rheol.*, **20**, 065102 (2010a).

Hassanien, I.A., Flow and heat transfer on a continuous flat surface moving in a parallel free stream of power-law fluid, *Appl. Math. Model.*, **20**, 779 (1996).
Hassanizadeh, S.M. and Gray, W.G., High velocity flow in porous media, *Trans. Porous Media*, **2**, 52 (1987).
Hassell, H.L. and Bondi, A., Mixing of viscous non-Newtonian fluids in packed beds, *AIChE J.*, **11**, 217 (1965).
Hatch, H.H., Tests for hydraulic-fill dams. *Trans. ASCE*, **99**, 206 (1934).
Hatton, A.P., James, D.D. and Swire, H.W., Combined forced and natural convection with low-speed air flow over horizontal cylinders, *J. Fluid Mech.*, **42**, 17 (1970).
Hatzikiriakos, S. and Vlassopoulos, D., Brownian dynamics simulations of shear-thickening in dilute polymer solutions, *Rheol. Acta*, **35**, 274(1996).
Haugen, R., Laminar flow along a vertical wall, *J. App. Mech. (ASME)*, **35**, 631 (1968).
Haughey, D.P. and Beveridge, G.S.G., Local voidage variations in a randomly packed bed of equal size spheres, *Chem. Eng. Sci.*, **21**, 905(1966).
Haward, S.J. and Odell, J.A., Viscosity enhancement in non-Newtonian flow of dilute polymer solutions through crystallographic porous media, *Rheol. Acta*, **42**, 516 (2003).
Haward, S.J. and Odell, J.A., Molecular orientation in non-Newtonian flow of dilute polymer solutions around spheres, *Rheol. Acta*, **43**, 350 (2004).
Haward, S.J., Odell, J.A., Berry, M. and Hall, T., Extensional rheology of human saliva, *Rheol. Acta*, **50**, 869 (2011).
Hayes, R.E., Forced convection heat transfer at the boundary layer of a packed bed, *Trans. Porous Media*, **5**, 231 (1990).
Hayes, R.E., Afacan, A. and Boulanger, B., An equation of motion for an incompressible Newtonian flow in a packed bed, *Trans. Porous Media*, **18**, 185 (1995).
Hayes, R.E., Afacan, A., Boulanger, B. and Shenoy, A.V., Modelling the flow of power law fluids in a packed bed using a volume averaged equation of motion, *Trans. Porous Media*, **23**, 175 (1996).
Hayworth, C.B. and Treybal, R.E., Drop formation in two liquid phase systems, *Ind. Eng. Chem.*, **42**, 1174 (1950).
He, Y, Li, L., Taniguchi, T., Tuinier, R. and Fan, T.-H., Flow induced by an oscillating sphere in probing complex viscosity of polymer solutions, *Phys. Rev. Fluids*, **5**, 013302 (2020).
He, Y.B., Laskowski, J.S. and Klein, B., Particle movement in non-Newtonian slurries: The effect of yield stress on dense medium separation, *Chem. Eng. Sci.*, **56**, 2991 (2001).
Hecht, V., Voigt, J. and Schugerl, K., Absorption of oxygen in counter current multistage bubble columns – III, *Chem. Eng. Sci.*, **35**, 1325 (1980).
Hedden, O.K., Spray drop sizes and size distribution in pesticides sprays, *Trans. ASAE*, **4**, 159 (1961).
Heinen, C., Buggisch, H. and Guthausen, G., Flow of Newtonian/non-Newtonian fluids in a bundle of tubes and in a packing of beads by MRI, *Mag. Resonance Imag.*, **21**, 377 (2003).
Heiss, J.F. and Coull, J., The effect of orientation and shape on the settling velocity of non-isometric particles in a viscous medium, *Chem. Eng. Prog.*, **48**(3), 133 (1952).
Hejri, S., Willhite, G.P. and Green, D.W., Development of correlations to predict biopolymer mobility in porous media, *SPE Reservoir Eng.*, **6**(2), 91 (1991).
Heldman, D.R. and Lund, D.B., *Handbook of Food Engineering*, Marcel Dekker, New York, NY (1992).
Hellström, J.G.I., Jonsson, P.J.P. and Lundström, T.S., Laminar and turbulent flow through an array of cylinders, *J. Porous Media*, **13**, 1073 (2010).
Helmreich, A., Vorwerk, J., Steger, R., Muller, M. and Brunn, P.O., Non-viscous effects in the flow of xanthan gum solutions through a packed bed of spheres, *Chem. Eng. J.*, **59**, 111 (1995).
Hema Sundar Raju, B., Nath, D. and Pati, S., Effect of Prandtl Number on thermo-fluidic transport characteristics for mixed convection past a sphere, *Int. Comm. Heat Mass Transf.*, **98**, 191 (2018).
Henschke, M., Waheed, A. and Pfennig, A., Wandeinfluss auf die sedimentations – geschwindigkeit von kuglen, *Chem. Ing. Tech.*, **72**, 1376 (2000).
Herbst, C.A., Cook, R.L. and King Jr., H.E., High pressure viscosity of glycerol measured by centrifugal-force viscometry, *Nature*, **361**, 518 (Feb 11, 1993).
Hermann, W. and Sobczak, R., A new inductive detection for the magneto-viscometer, *J. Appl. Polym. Sci.*, **37**, 2675 (1989).
Hermes, R.A., Measurement of the limiting viscosity with a rotating sphere viscometer, *J. Appl. Polym. Sci.*, **10**, 1793 (1966).

Herrera-Velarde, J.R., Zenit, R., Chehata, D. and Mena, B., The flow of non-Newtonian fluids around bubbles and its connection to the jump discontinuity, *J. Non-Newt. Fluid Mech.*, **111**, 199 (2003).

Herzhaft, B., Guazzelli, E., Mackaplow, M.B. and Shaqfeh, E.S.G., Experimental investigation of the sedimentation of a dilute fiber suspension, *Phys. Rev. Lett.*, **77**, 290 (1996).

Hetsroni, G., Haber, S. and Wacholder, E., The flow fields in and around a droplet moving axially within a tube, *J. Fluid Mech.*, **41**, 689 (1970).

Heyda, J.F., A Green's function solution for the case of laminar incompressible flow between non-concentric circular cylinders, *J. Franklin Inst.*, **267**, 25 (1959).

Heywood, N.I., Selecting a viscometer, *The Chemical Engineer*, No. 415, 16 (1985).

Hieber, C.A. and Gebhart, B., Mixed convection from a sphere at small Reynolds and Grashof numbers, *J. Fluid Mech.*, **38**, 137 (1969).

Higdon, J.J.L. and Ford, G.D., Permeability of three-dimensional models of fibrous porous media, *J. Fluid Mech.*, **308**, 341 (1996).

Higdon, J.J.L. and Muldowney, G.P., Resistance functions for spherical particles, droplets and bubbles in cylindrical tubes, *J. Fluid Mech.*, **298**, 193 (1995).

Higgins, B., *Coatings Fundamentals: Suspension Rheology*, TAPPI Press, Atlanta, GA (1997).

Highgate, D.J., Particle migration in cone-plate viscometry of suspensions, *Nature*, **211**, 1390 (1966).

Highgate, D.J. and Whorlow, R.H., The viscous resistance to motion of a sphere falling through a sheared non-Newtonian liquid, *Brit. J. App. Phys.*, **18**, 1019 (1967).

Highgate, D.J. and Whorlow, R.H., End effects and particle migration effects in concentric cylinder rheometry, *Rheol. Acta*, **8**, 142 (1969).

Highgate, D.J. and Whorlow, R.H., Rheological properties of suspensions of spheres in non-Newtonian media, *Rheol. Acta*, **9**, 569 (1970).

Hilal, M., Brunjail, D. and Comiti, J., Electro diffusion characterization of non-Newtonian flow through packed beds, *J. Appl. Electrochem.*, **21**, 1087 (1991).

Hilfiker, R., Chu, B. and Shook, J., Magnetic sphere rheometer, *Rev. Sci. Instrum.*, **60**, 760 (1989).

Hill, C.T., *Ph. D Dissertation*, University of Wisconsin, Madison, WI (1969).

Hill, K.B. and Shook, C.A., Pipeline transport of coarse particles by water and by fluids with yield stress, *Part. Sci. Technol.*, **16**, 163 (1998).

Hill, R. and Power, G., Extremum principles for slow viscous flow and the approximate calculation of drag, *Quart. J. Mech. Appl. Math.*, **9**, 313 (1956).

Hill, R.J. and Koch, D.L., The transition from steady to weakly turbulent flow in a close-packed ordered array of spheres. *J. Fluid Mech.*, **465**, 59 (2002).

Hill, R.J., Koch, D.L. and Ladd, A.J.C., The first effects of fluid inertia on flows in ordered and random arrays of spheres, *J. Fluid Mech.*, **448**, 213 (2001a).

Hill, R.J., Koch, D.L. and Ladd, A.J.C., Moderate Reynolds number flows in ordered and random arrays of spheres, *J. Fluid Mech.*, **448**, 243 (2001b).

Himbert, F., Emploi due facteur de tortuosite dans l'etude des ecoulements de fluide a travers un lit granule, *Genie Chim. (Paris)*, **93**, 110 (1965).

Hino, M. and Hasegawa, S., Some hydrodynamical experiments on dilute polymer solutions, *Tech. Rep. # 6*, Dept. Civil Eng., Tokyo Inst. Tech., Tokyo, p. 123 (1968).

Hinze, J.P., Fundamentals of the hydrodynamic mechanisms of splitting in dispersion processes, *AIChE J.*, **1**, 289 (1955).

Hirasaki, G.J. and Pope, G.A., Analysis of factors influencing mobility and adsorption in the flow of polymer solution through porous media, *Soc. Pet. Eng. J.*, **12**, 337 (1974).

Hirose, T. and Moo-Young, M., Bubble drag and mass transfer in non-Newtonian fluids: Creeping flow with power-law fluids, *Can. J. Chem. Eng.*, **47**, 265 (1969).

Hirota, S. and Takada, M., Analysis of non-Newtonian flow by falling sphere method, *Bull. Chem. Soc. Jpn.*, **32**, 1191 (1959).

Hirschfeld, B.R., Brenner, H. and Falade, A., First- and second-order wall effects upon the slow viscous asymmetric motion of an arbitrarily – shaped, - positioned and –oriented particle within a circular cylinder, *Physicochem. Hydrodynamics*, **5**, 99 (1984).

Hjelmfelt Jr., A.T. and Brooker, D.B., Nonlinear flow through an anisotropic porous media, *Trans. ASAE*, **38**, 863 (1995).

Ho, B.P. and Leal, L.G., Inertial migration of rigid spheres in two-dimensional unidirectional flow, *J. Fluid Mech.*, **65**, 365 (1974).

Ho, B.P. and Leal, L.G., Migration of rigid spheres in a two-dimensional unidirectional shear flow of a second order fluid, *J. Fluid Mech.*, **76**, 783 (1976).

Hoagland, D.A. and Prud'homme, R.K., Hydrodynamic chromatography as a probe of polymer dynamics during flow through porous media, *Macromolecules*, **22**, 775 (1989).

Hoerner, S.F., *Fluid Dynamic Drag*, Hoerner Fluid Dynamics, Bricktown, NJ (1965).

Hoffman, A.C. and van den Bogaard, H.A., A numerical investigation of bubbles rising at intermediate Reynolds and large Weber numbers, *Ind. Eng. Chem. Res.*, **34**, 366 (1995).

Hofmann, H., Hydrodynamics and hydrodynamic models of fixed bed reactors, *Multiphase Chemical Reactors – Theory, Design and Scale Up*, edited by A. Gianetto and P.L. Silveston, Hemisphere, New York, Chapter 8 (1986).

Hojeij, A., Jossic, L., Magnin, A., Bleses, D. and Hattou, S., Gas injection in a yield stress fluid, *AIChE J.*, **66**(4), e16878 (2020).

Hojeij, A., Jossic, L., Sechet, P., Magnin, A. and Hattou, S., Agitation of yield stress fluids by gas injection, *AIChE J.*, **68** (4), e17562 (2022).

Holdsworth, S.D., *Aseptic Processing and Packaging of Food Products*, Elsevier Applied Science, London, UK (1992).

Holdsworth, S.D., Rheological models used for the prediction of the flow properties of food products: A literature review, *Trans. I. Chem. E.*, **71C**, 139 (1993).

Holenberg, Y., Lavrenteva, O.M., Liberzon, A., Shavit, U. and Nir, A., PTC and PIV study of the motion of viscous drops in yield stress material, *J. Non-Newt. Fluid Mech.*, **193**, 129 (2013).

Holenberg, Y., Lavrenteva, O.M. and Nir, A., Interaction of viscous drops in a yield stress material, *Rheol. Acta*, **50**, 375 (2011).

Holenberg, Y., Lavrentova, O.M., Shavit, U. and Nir, A., Particle tracking velocimetry and particle image velocitmetry study of the slow motion of rough and smooth solid spheres in a yield-stress fluid, *Phys. Rev. E.*, **86**, 066301 (2012).

Holzwarth, G., Bonin, K. and Hill, D.B., Forces required of kinesin during processive transport through cytoplasm, *Biophys.*, **82**, 1784 (2002).

Homann, F., Einfluss grosser ziihigkeit bei stromung um zylinder, *E'orsch. Ingwes.*, **7**, 1 (1936).

Hong, S.-A., Duda, J.L. and Klaus, E.E., The influence of bulk rheology on the flow of polymer solutions in porous media, *Polymer Preprints*, **22**(2), 20 (1981).

Hopke, S.W. and Slattery, J.C. Upper and lower bounds on the drag coefficient of a sphere in an Ellis model fluid, *AIChE J.*, **16**, 224 (1970a).

Hopke, S.W. and Slattery, J.C., Note on the drag coefficient for a sphere, *AIChE J.*, **16**, 317 (1970b).

Horowitz, M. and Williamson, C.H.K., The effect of Reynolds number on the dynamics and wakes of freely rising and falling spheres, *J. Fluid Mech.*, **651**, 251 (2010).

Horsley, M.R., Horsley, R.R., Wilson, K.C. and Jones, R.L., Non-Newtonian effects on falling velocities of pairs of vertically aligned spheres, *J. Non-Newt. Fluid Mech.*, **124**, 147 (2004).

Hossain, M.A. and Gebhart, B., Natural convection about a sphere at low Grashof number, *Int. Heat Transfer Conference, Digital Library*, Begell House, New York, 1970.

Hou, J., Liu, Z., Zhang, S., Yue, X. and Yang, J., The role of viscoelasticity of alkali/surfactant/ polymer solutions in enhanced oil recovery, *J. Pet. Sci. Eng.*, **47**, 219 (2005).

Housiadas, K.D., Steady sedimentation of a spherical particle under constant rotation, *Phys. Rev. Fluids*, **4**, 103301 (2019).

Housiadas, K.D., The singularity of the UCM/Oldroyd-B models at a finite Weissenberg number for the steady sphere translation with Navier slip on the sphere, *J. Non-Newt. Fluid Mech.*, **298**, 104679 (2021).

Housiadas, K.D. and Tanner, R.I., The drag of a freely sedimenting sphere in a sheared weakly viscoelastic fluid, *J. Non-Newt. Fluid Mech.*, **183–184**, 52 (2012).

Housiadas, K.D. and Tanner, R.I., Rheological effects in the 3D creeping flow past a sedimenting sphere subject to orthogonal shear, *Phys. Fluids*, **26**, 013102 (2014).

Housiadas, K.D. and Tanner, R.I., A high-order perturbation solution for the steady sedimentation of a sphere in a viscoelastic fluid, *J. Non-Newt. Fluid Mech.*, **233**, 166 (2016).

Houska, M., Ph D thesis, Czech Technical University, Prague (1981).

Houwink, R. and de Decker, J.K., *Elasticity, Plasticity and Structure of Matter*, 3rd Ed., Cambridge University Press, Cambridge (1971).

Howell, T.G., Jeng, D.R. and DeWitt, K.J., Momentum and heat transfer on a continuous moving surface in a power law fluid, *Int. J. Heat Mass Transf.*, **40**, 1853 (1997).

Howells, I.D., Drag on fixed beds of fibres in slow flow, *J. Fluid Mech.*, **355**, 163 (1998).

Hoyt, J.W., The effect of additives on fluid friction, *J. Basic Eng., Trans. ASME*, **94**, 258 (1972).

Hoyt, J.W. and Sellin, R.H.J., Cylinder cross-flow heat transfer in drag reducing fluid, *Exp. Heat Transf.*, **2**, 113 (1989).

Hsu, C.C., A simple solution for boundary layer flow of power law fluids past a semi-infinite flat plate, *AIChE J.*, **15**, 367 (1969).

Hsu, J.-P., Chen, W.-J., Tseng, S. and Chen, C.-J., Sedimentation of a cylindrical particle along the axis of a cylindrical tube filled with Carreau fluid, *Powder Technol.*, **166**, 1 (2006).

Hsu, J.P. and Hsieh, Y.H., Boundary effect on the drag force on a non-homogeneous floc, *J. Colloid Interface Sci.*, **264**, 517 (2004). Also see *ibid*, **275**, 309 (2004).

Hsu, J.P., Shie, C.F. and Teng, S., Sedimentation of a cylindrical particle in a Carreau fluid, *J. Coll. Interface Sci.*, **286**, 392 (2005b).

Hsu, J.P., Hsieh, Y.H. and Tseng, S., Drag force on a rigid spheroidal particle in a cylinder filled with Carreau fluid, *J. Colloid Interface Sci.*, **284**, 729 (2005a).

Hu, B. and Pang, M., Numerical study on the influence of liquid viscosity ratio on the hydrodynamics of a single bubble in shear-thinning liquid, *Appl. Math. Modelling*, **103**, 122 (2022).

Hu, H. and Kochesfahani, M., The wake behind a heated cylinder in forced and mixed convection regimes, *2005 ASME Summer Heat Transfer Conference*, San Francisco, CA, (2005).

Hu, H.H., Motion of a circular cylinder in a viscous liquid between parallel plates, *Theo. Comp. Fluid Dynamics*, **7**, 441 (1995).

Hu, H.H., Numerical simulation of particle motion in viscoelastic fluids, *Proc. IUTAM Symposium on Lubricated Transport of Viscous Materials*, Ramkissoon, H., ed., p. 177, Kluwer, Dordrecht (1998).

Hu, H.H. and Joseph, D.D., Numerical simulation of viscoelastic flow past a cylinder, *J. Non-Newt. Fluid Mech.*, **37**, 347 (1990).

Hu, H.H. and Joseph, D.D., Lift on a sphere near a plane wall in a second-order fluid, *J. Non-Newt. Fluid Mech.*, **88**, 173 (1999).

Hu, X., Ding, Z. and Lee, L.J., Simulation of 2-D transient viscoelastic flow using the CONNFFESSIT approach, *J. Non-Newt. Fluid Mech.*, **127**, 107 (2005).

Hu, Y.T., Chung, H., Maxey, J.E., What is more important for proppant transport, viscosity or elasticity? *Proc. SPE Hydraulic Fracturing Technology Conference*, SPE, The Woodlands, TX (2015).

Hua, C.C. and Schieber, J.D., Viscoelastic flow through fibrous media using the CONNFFESSIT approach. *J. Rheol.*, **42** (3), 477 (1998).

Hua, T.N. and Ishii, T., Momentum transfer for multi-solid-particle power-law fluids systems at high Reynolds number, *J. Non-Newt. Fluid Mech.*, **9**, 301 (1981).

Huang, M.-J. and Chen, C.K., Numerical analysis for forced convection over a flat plate in power law fluids, *Int. Comm. Heat Mass Transf.*, **11**, 361 (1984).

Huang, M.-J. and Chen, C.-K., Local similarity solutions of free convective heat transfer from a vertical plate to non-Newtonian power law fluids, *Int. J. Heat Mass Transf.*, **33**, 119 (1990).

Huang, M.-J. and Lin, B.L., Forced convective flow over a flat plate in non-Newtonian power law fluids, *Warme-und Stoffubertragung.*, **27**, 399 (1992).

Huang, M.J. and Lin, B.L., Mixed convection from a vertical plate to power-law fluids, *J. Thermophys. Heat Transf.*, **7**, 171 (1993).

Huang, P.Y. and Feng, J., Wall effects on the flow of viscoelastic fluids around a circular cylinder, *J. Non-Newt. Fluid Mech.*, **60**, 179 (1995).

Huang, P.Y., Feng, J., Hu, H.H. and Joseph, D.D., Direct simulation of the motion of solid particles in Couette and Poiseuille flows of viscoelastic fluids, *J. Fluid Mech.*, **343**, 73 (1997).

Huang, P.Y., Hu, H.H. and Joseph, D.D., Direct simulation of the sedimentation of elliptic particles in Oldroyd-B fluids, *J. Fluid Mech.*, **362**, 297 (1998).

Huang, W., Li, H., Xu, Y. and Lian, G., Hydrodynamic force on two hard spheres tangentially translating in a power-law fluid, *Chem. Eng. Sci.*, **61**, 1480 (2006).

Hubbard, R.M. and Brown, G.G., The rolling ball viscometer, *Ind. Eng. Chem. Anal. Edition*, **15**, 212 (1943a).

Hubbard, R.M. and Brown, G.G., Viscosity of n-Pentane, *Ind. Eng. Chem.*, **35**, 1276 (1943b).
Hughes, S., Popping the champagne bubble, *New Scientist*, **41**, (22/29 December 1990).
Huilgol, R.R., On the concept of the Deborah number, *Trans. Soc. Rheol.*, **19**, 297 (1975).
Huilgol, R.R., *Fluid Mechanics of Viscoplasticity*, Springer, Berlin (2015).
Huilgol, R.R. and Kefayati, G.H.R., Natural convection problem in a Bingham fluid using operator-splitting method, *J. Non-Newt. Fluid Mech.*, **220**, 22 (2015).
Huilgol, R.R., Panizza, M., Phan-Thien, N. and Zheng, R., Application of the boudnary integral method to the sedimentation of irregular shaped particles, *J. Non-Newt. Fluid Mech.*, **57**, 49 (1995).
Hulbert, R. and Feben, D., Hydraulics of rapid filter sand, *J. Amer. Water Works Assoc.*, **25**, 19 (1933).
Hulsen, M.A., Fattal, R. and Kupferman, R., Flow of viscoelastic fluids past a cylinder at high Weissenberg number: Stabilized simulations using matrix logarithms, *J. Non-Newt. Fluid Mech.*, **127**, 27 (2005).
Hulsen, M.A., van Heel, A.P.G. and van den Brule, B.H.A.A., Simulation of viscoelastic flows using Brownian configuration fields, *J. Non-Newt. Fluid Mech.*, **70**, 79 (1997).
Humphrey, J.A.C., Note on drop formation at low velocity in quiescent liquids, *Chem. Eng. Sci.*, **35**, 1452 (1980).
Humphrey, J.A.C. and Murata, H., On the motion of solid spheres falling through viscous fluids in vertical and inclined tubes. *J. Fluids Eng.*, **114**, 2 (1992).
Huner, B. and Hussey, R.G., Cylinder drag at low Reynolds numbers, *Phys. Fluids*, **20**, 1211 (1977).
Huzarewicz, S., Gupta, R.K. and Chhabra, R.P., Elastic effects in flow of fluids through sinuous tubes, *J. Rheol.*, **35**, 221 (1991).
Huzyak, P.C. and Koelling, K.W., The penetration of a long bubble through a viscoelastic fluid in a tube, *J. Non-Newt. Fluid Mech.*, **71**, 73 (1997).
Hwang, S.J., Liu, C.B. and Lu, W.J., Solid-liquid mass transfer in a non-Newtonian liquid fluidized bed, *Chem. Eng. J.*, **52**, 131 (1993).
Hyde, M.A. and Donatelli, A.A., Mass transfer from a solid sphere to power law fluids in creeping flow, *Ind. Eng. Chem. Fund.*, **22**, 500 (1983).
Iaanniruberto, G. and Marrucci, G., Falling spheres in polymeric solutions: Limiting results of the two-fluid theory of migration, *J. Non-Newt. Fluid Mech.*, **54**, 231 (1994).
Ichihara, M., Ohkunitani, H., Ida, Y. and Kameda, M., Dynamics of bubble oscillation and wave propagation in viscoelastic liquids, *J. Volcan. Geothermal Res.*, **129**, 37 (2004).
Idris, Z., Oregas, L., Geindreau, C., Bloch, J.-F. and Auriault, J.-L., Microstructural effects on the flow law of power-law fluids through fibrous media, *Model. Simul. Mater. Sci. Eng.*, **12**, 995 (2004).
Ikoku, C.U. and Ramey Jr., H.J., Well bore storage and skin effects during the transient flow of non-Newtonian power law fluids in porous media, *Soc. Pet. Eng. J.*, **17**, 164 (June 1979). Also see ibid **18**, 25 (1980).
Ilic, V., Tullock, D., Phan-Thien, N. and Graham, A.L., Translation and rotation of spheres settling in square and circular conduits: Experiments and numerical predictions, *Int. J. Multiph. Flow*, **18**, 1061 (1992).
Iliuta, I. and Larachi, F., Hydrodynamics of power-law fluids in trickle-flow reactors: Mechanistic model, experimental verification and simulations, *Chem. Eng. Sci.*, **57**, 1931 (2002a).
Iliuta, I. and Larachi, F., Hydrodynamic models for rheologically complex fluids in co-and counter current gas-liquid packed bed bioreactors, *Ind. Eng. Chem. Res.*, **41**, 2096 (2002b).
Iliuta, I. and Thyrion, F.C., Flow regimes, liquid holdups and two-phase pressure drop for two-phase cocurrent downflow and upflow through packed beds: Air/Newtonian and non-Newtonian liquid systems, *Chem. Eng. Sci.*, **52**, 4045 (1997).
Iliuta, I., Thyrion, F.C. and Muntean, O., Hydrodynamic characteristics of two-phase flow through fixed beds: Air/Newtonian and non-Newtonian liquids, *Chem. Eng. Sci.*, **51**, 4987 (1996).
Illenberger, W.K., Pebble shape (and size!) – Reply, *J. Sed. Petrol.*, **62**, 1151 (1992).
Inge, C. and Bark, F.H., Surface tension driven oscillations of a bubble in a viscoelastic liquid, *Appl. Sci. Res.*, **38**, 231 (1982).
Ingmanson, W.L., Andrews, B.D. and Johnson, R.C., Internal pressure distributions in compressible mats under fluid stress, *TAPPI*, **42**, 840 (1959).
Inoue, M. and Nakayama, A., Numerical modeling of non-Newtonian fluid flow in a porous medium using a three-dimensional periodic array, *J. Fluids Eng.*, **120**, 131 (1998).
Interthal, W. and Haas, R., Effects of dilute polymer solutions on porous media flows Part I: Basic concepts and experimental results, *Proc. Euromech.*, **143**, Delft, p. 157 and p. 163 (Sep 2–4, 1981).

Ireland, P.M. and Jameson, G.J., Drag force on a spherical particle moving through a foam: The role of wettability, *Int. J. Min. Process.*, **102**, 78 (2012).

Irmay, S., On the theoretical derivation of Darcy and Forchheimer equations, *Trans. Am. Geo. Union*, **39**, 702 (1958).

Irvine Jr., T.F. and Karni, J., Non-Newtonian fluid flow and heat transfer, *Handbook of Single-Phase Convective Heat Transfer*, edited by S. Kakac, R.K. Shah and W. Aung, Wiley, New York, Chapter 20 (1987).

Irving, J.B. and Barlow, A.J., An automated high-pressure viscometer, *J. Phys. E: Sci. Instrum.*, **4**, 232 (1971).

Ishkintana, L.K. and Bennington, C.P.J., Gas holdup in pulp fibre suspensions: Gas voidage profiles in a batch-operated sparged tower, *Chem. Eng. Sci.*, **65**, 2569 (2010).

Islam, M.R. and Farouq Ali, S.M., Numerical simulation of emulsion flow through porous media, Preprint # 89-40-63, *Pet. Soc. Can. Inst. of Mining* (1989).

Islam, M.R., Selby, R.J. and Farouq Ali, S.M., Mechanics of foam flow in porous media and applications, *J. Can. Pet. Tech.*, **28**, 88 (1989).

Islam, M.T., Ganesan, P. and Cheng, J., A pair of bubbles' rising dynamics in a Xanthan gum solution: A CFD study, *RSC Adv.*, **5**, 7819 (2015).

Islam, M.T., Ganesan, P.B., Cheng, J. and Uddin, M.S., Single bubble rising behaviors in Newtonian and non-Newtonian fluids with validation of empirical correlations: A computational fluid dynamics study, *Eng. Rep.*, **2**, e12100 (2020).

Ismail, J.H., Fairweather, M. and Javed, K., Structural properties of beds packed with ternary mixtures of spherical particles Part 1 – Global properties, *Chem. Eng. Res. Des.*, **80A**, 637 (2002). Also see *ibid* 645.

Ito, S. and Kajiuchi, T., Drag force on a sphere moving in plastic fluid, *J. Chem. Eng. Jpn.*, **2**, 19 (1969).

Iwaoka, M. and Ishii, T., Experimental wall correction factors of single solid spheres in circular cylinders, *J. Chem. Eng. Jpn.*, **12**, 239 (1979).

Iwata, S., Takahashi, T., Onuma, T., Nagumo, R. and Mori, H., Local Flow around a tiny bubble under a pressure-oscillation field in a viscoelastic worm-like micellar solution, *J. Non-Newt. Fluid Mech.*, **263**, 24 (2019).

Jackson, D.A. and Bedborough, D.S., An application of intensity fluctuation spectroscopy to the falling slug method of viscosity determination, *J. Phys. D App. Phys.*, **11**, L135 (1978).

Jackson, G.W. and James, D.F., The hydrodynamic resistance of hyaluronic acid and its contribution to tissue permeability, *Biorheology*, **19**, 317 (1982).

Jackson, G.W. and James, D.F., The permeability of fibrous porous media, *Can. J. Chem. Eng.*, **64**, 364 (1986).

Jackson, K.P., Walters, K. and Williams, R.W., A. rheometrical study of Boger fluids, *J. Non-Newt. Fluid Mech.*, **14**, 173 (1984).

Jackson, T.W. and Yen, H.H., Combining forced and free convective equations to represent combined heat-transfer coefficients for a horizontal cylinder, *J. Heat Transf.*, **93**, 247 (1971).

Jafarpur, K. and Yovanovich, M.M., Laminar free convective heat transfer from isothermal spheres: A new analytical method, *Int. J. Heat Mass Transf.*, **35**, 2195 (1992).

Jain, M.K., Boundary layer effects in non-Newtonian fluids, *ZAMM*, **35**, 12 (1955).

Jaiswal, A.K., Sundararajan, T. and Chhabra, R.P., Flow characteristics of settling suspensions and fluidized beds of spherical particles, *Chem. Eng. Commun.*, **106**, 139 (1991a).

Jaiswal, A.K., Sundararajan, T. and Chhabra, R.P., Hydrodynamics of Newtonian fluid flow through assemblages of rigid spherical particles in intermediate Reynolds number regime, *Int. J. Eng. Sci.*, **29**, 693 (1991b).

Jaiswal, A.K., Sundararajan, T. and Chhabra, R.P., Flow of power law liquids through particle assemblages at intermediate Reynolds numbers, *Can. J. Chem. Eng.*, **69**, 1235 (1991c).

Jaiswal, A.K., Sundararajan, T. and Chhabra, R.P., Simulation of non-Newtonian fluid flow through fixed and fluidized beds of spherical particles, *Numer. Heat Transf.*, **21A**, 275 (1992).

Jaiswal, A.K., Sundararajan, T. and Chhabra, R.P., Hydrodynamics of creeping flow of power law liquids through particle assemblages, *Int. J. Eng. Sci.*, **31**, 293 (1993a).

Jaiswal, A.K., Sundararajan, T. and Chhabra, R.P., Slow non-Newtonian flow through packed beds: Effect of zero shear viscosity, *Can. J. Chem Eng.*, **71**, 646 (1993b).

Jaiswal, A.K., Sundararajan, T. and Chhabra, R.P., Pressure drop for the slow flow of dilatant fluids through a fixed bed of spherical particles, *Can. J. Chem. Eng.*, **72**, 352 (1994).

James, D.F., Non-Newtonian effects in porous media flow, *Proc. IX Int. Cong. Rheol.*, Mexico, **I**, p. 279 (1984).

James, D.F., Boger fluids, *Annu. Rev. Fluid Mech.*, **41**, 129 (2009).

James, D.F. and Acosta, A.J., The laminar flow of dilute polymer solutions around circular cylinders, *J. Fluid Mech.*, **42**, 269 (1970).

James, D.F. and Gupta, O.P., Drag on circular cylinders in dilute polymer solutions, *Chem. Eng. Prog. Sym. Ser.*, **67**(111), 62 (1971).

James, D.F. and McLaren, D.R., The laminar flow of dilute polymer solutions through porous media, *J. Fluid Mech.*, **70**, 733 (1975).

James, D.F., Phan-Thien, N., Khan, M.M.K., Beris, A.N. and Pilitsis, S., Flow of test fluid M1 in corrugated tubes, *J. Non-Newt. Fluid Mech.*, **35**, 405 (1990). Also see, *ibid* **39**, 375 (1991).

James, D.F., Shiau, T. and Aldridge, P.M., Flow of a Boger fluid around an isolated cylinder, *J. Rheol.*, **60**, 1137 (2016).

James, D.F. and Walters, K., A critical appraisal of available methods for the measurement of extensional properties of mobile systems, *Techniques in Rheological Measurements*, edited by A.A. Collyer, p. 33, Elsevier, Amsterdam (1994).

James, D.F., Yip, R. and Currie, I.G., Slow flow of Boger fluids through model fibrous porous media, *J. Rheol.*, **56**, 1249 (2012).

Jamialahmadi, M., Branch, C. and Müller-Steinhagen, H., Terminal bubble rise velocity in liquids, *Trans. Inst. Chem. Eng.*, **72A**, 119 (1994).

Jamialahmadi, M. and Müller-Steinhagen, H., Hydrodynamics and heat transfer of liquid fluidized bed systems, *Chem. Eng. Commun.*, **179**, 35 (2000)

Jamialahmadi, M., Muller-Steinhagen, H. and Izadpanah, M.R., Pressure drop, gas holdup and heat transfer during single- and two-phase flow through porous media, *Int. J. Heat Fluid Flow*, **26**, 156 (2005).

Janna, W.S. and John, J.E.A., Drop size distribution of Bingham liquid (Paint) sprays produced by fan-Jet pressure nozzles, *J. Eng. Ind., Trans. ASME*, **101**, 449 (1979). Also see *ibid* **103**, 402 (1981).

Jarzebski, A.B. and Malinowski, J.J., Drag and mass transfer in multiple drop slow motion in a power law fluid, *Chem. Eng. Sci.*, **41**, 2569 (1986a).

Jarzebski, A.B. and Malinowski, J.J., Transient heat and mass transfer from drops or bubbles in slow non-Newtonian flows, *Chem. Eng. Sci.*, **41**, 2575 (1986b).

Jarzebski, A.B. and Malinowski, J.J., Drag and mass transfer in a creeping flow of a Carreau fluid over drops or bubbles, *Can. J. Chem. Eng.*, **65**, 680 (1987a).

Jarzebski, A.B. and Malinowski, J.J., Drag and mass transfer in slow non-Newtonian flows over an ensemble of Newtonian spherical drops or bubbles, *Chem. Eng. Commun.*, **49**, 235 (1987b).

Jayaraman A. and Belmonte, A., Presentation at the Annual Meeting of the Society of Rheology, Hilton Head, Feb. (2001).

Jayaraman, A. and Belmonte, A., Oscillations of a solid sphere falling through a worm-like micellar fluid, *Phys. Rev.*, **E67**, 65301 (2003).

Jean, B., *Rheological Properties of Lubricants*, Gulf, Houston (1989).

Jean, R.-H. and Fan, L.-S., A fluid mechanic-based model for sedimentation and fluidization at low Reynolds numbers, *Chem. Eng. Sci.*, **44**, 353 (1989).

Jeffrey, G.B., The motion of ellipsoidal particles immersed in a viscous fluid, *Proc. Roy. Soc. Lond.*, **102A**, 161 (1922).

Jefri, M.A. and Daous, M.A., The normal force exerted by a second order fluid on a small sphere touching a plane, *Chem. Eng. Commun.*, **100**, 65 (1991).

Jefri, M.A., Nichols, K.L. and Jayaraman, K., Sedimentation of two contacting spheres in dilute polymer solutions, *Recent Developments in Structured Continua*, De Kee, D. and Kaloni, P.N., eds, p. 21 (1985).

Jefri, M.A. and Zahed, A.H., Elastic and viscous effects on particle migration in plane-Poiseuille flow, *J. Rheol.*, **33**, 691 (1989).

Jennings, R.R., Rogers, S.M. and West, T.J., Factors influencing mobility control by polymer solutions, *J. Pet. Tech.*, **23**, 391 (Mar. 1971).

Jenny, M., Bouchet, G. and Dusek, J., Nonvertical ascension or fall of a free sphere in a Newtonian fluid, *Phys. Fluids*, **15**, L9 (2003).

Jenny, M., Dusek, J. and Bouchet, G., Instabilities and transition of a sphere falling or ascending freely in a Newtonian fluid, *J. Fluid Mech.*, **508**, 201 (2004).

Jenson, V.G., Viscous flow around a sphere at low Reynolds number (<40), *Proc. Roy. Soc.*, **249A**, 346(1959).

Jia, H. and Gogos, G., Laminar natural convection heat transfer from isothermal spheres, *Int. J. Heat Mass Transf.*, **39**, 1603 (1996a).

Jia, H. and Gogos, G., Transient laminar natural convection heat transfer from isothermal spheres, *Num. Heat Transf.*, **29A**, 83 (1996b).

Jiang, S., Fan, W., Zhu, C., Ma, Y. and Li, H.-Z., Bubble formation in non-Newtonian fluids using laser image measurement system, *Chin. J. Chem. Eng.*, **15**, 611 (2007).

Jiang, W., Sun, Y., Xu, Y., Peng, C., Gong, X. and Zhang, Z. Shear-thickening behavior of polymethyl methacrylate particle suspensions in glycerin- water mixtures, *Rheol. Acta*, **49**, 1157 (2010).

Jie, P. and Ke-Qin, Z., Drag force of interacting coaxial spheres in viscoplastic fluids, *J. Non-Newt. Fluid Mech.*, **135**, 83 (2006).

Jimenez-Fernandez, J. and Crespo, A., Bubble oscillation and inertial cavitation in viscoelastic fluids, *Ultrasonics*, **43**, 643 (2005).

Jin, H., Phan-Thien, N. and Tanner, R.I., A finite element analysis of the flow past a sphere in a cylindrical tube: PTT fluid model, *Comp. Mech.*, **8**, 409 (1991).

Jin, L. and Chenevert, M.E., A study of particle settling in non-Newtonian fluids- Part I: A new method for the study of particle settling in drilling and fracturing fluids, *J. Energy Resour. Technol. (ASME)*, **116**, 10 (1994).

Jin, L. and Penny, G.S., Dimensionless methods for the study of particle settling in non-Newtonian fluids, *J. Pet. Tech.*, **47**, 223 (1995).

Job, C. and Blass, E., Methods for characterization of the hydrodynamic behaviour of fermenter broths in liquid-liquid extractors, *Solvent Extraction 1990*, edited by T. Sekine, Elsevier, p. 1893 (1992).

Job, C. and Blass, E., Problems concerning liquid-liquid extraction of extracellular products directly from fermenter broths, *Chem. Eng. J.*, **56**, B1 (1994).

Johansson, H., A numerical solution of the flow around a sphere in a circular cylinder, *Chem. Eng. Commun.*, **1**, 271 (1974).

Johns, M.L., Sederman, A.J., Bramley, A.S., Gladden, L.F. and Alexander, P., Local transitions in flow phenomena through packed beds identified by MRI, *AIChE J.*, **46**, 2151 (2000).

Johnson, A.B. and White, D.B., Experimental determination of gas migration velocities with non-Newtonian fluids, *Int. J. Multiph. Flow*, **19**, 921 (1993).

Johnson, A.M., *Physical Processes in Geology*, Freeman, Cooper & Co., San Francisco, CA (1970).

Johnson, E.D. and Middleman, S., Elongational flow of polymer melts, *Polym. Eng. Sci.*, **18**, 963 (1978).

Johnson, E.M. and Deen, W.M., Hydraulic permeability of agarose gels, *AIChE J.*, **42**, 1220 (1996).

Johnson Jr., M.W., Some variational theorems for non-Newtonian fluids, *Phys. Fluids*, **3**, 871 (1960).

Johnson Jr., M.W., On variational principles for non-Newtonian fluids, *Trans. Soc. Rheol.*, **5**, 9 (1961).

Johnson, R.E. and Sadhal, S.S., Fluid mechanics of compound multiphase drops and bubbles, *Annu. Rev. Fluid Mech.*, **17**, 289 (1985).

Johnson, T.A. and Patel, V.C., Flow past a sphere up to a Reynolds number of 300, *J. Fluid Mech.*, **378**, 19 (1999).

Jolls, K.R. and Hanratty, T.J., Transition to turbulence for flow through a dumped bed of spheres, *Chem. Eng. Sci.*, **21**, 1185 (1966).

Jones, W.M., The flow of dilute aqueous solutions of macromolecules in various geometries: VI. Properties of the solutions, *J. Phys. D Appl. Phys.*, **12**, 369 (1979).

Jones, W.M., Polymer additives in reservoir flooding for oil recovery: Shear thinning or shear thickening? *J. Phys. D.*, **13** (5), L87 (1980).

Jones, W.M. and Davies, O.H., The flow of dilute aqueous solutions of macromolecules in various geometries. III. Bent pipes and porous materials. *J. Phys. D.*, **9**(5), 753 (1976).

Jones, W.M. and Ho, S.P., The flow of dilute aqueous solutions of macromolecules in various geometries. VII. Mechanisms of resistance in porous media, *J. Phys. D.*, **12**, 383 (1979).

Jones, W.M. and Maddock, J.L., Onset of instabilities and reduction of drag in the flow of relaxing liquids through tubes and porous beds. *Nature*, **212** (5060), 388 (1966).

Jones, W.M. and Maddock, J.L., Relaxation effects in the flow of dilute polymer solutions through tubes and granular beds. *J. Phys. D.*, **2**(6), 797 (1969).

Jones, D.M. and Walters, K., The behaviour of polymer solutions in extension-dominated flows with application to enhanced oil recovery, *Rheol. Acta*, **28**, 482 (1989).

Jones, D.M., Walters, K. and Williams, P.R., On the extensional viscosity of mobile polymer solutions, *Rheol. Acta*, **26**, 20 (1987).

Jones, W.M., Price, A.H. and Walters, K., The motion of a sphere falling under gravity in a constant-viscosity elastic liquid, *J. Non-Newt. Fluid Mech.*, **53**, 175 (1994).

Jönsson, K.A.-S. and Jönsson, B.T.L., Fluid flow in compressible porous media: I: Steady-state conditions, *AIChE J.*, **38**, 1340 (1992). Also see *ibid* 1349.

Jordi, R.G., Young, B.D. and van Vliet, B.M., The effect of surface roughness on pressure drop in a packed bed, *Chem. Eng. Commun.*, **89**, 137 (1990).

Joseph, D.D., Flow induced microstructure in Newtonian and viscoelastic fluids, *Proc. 5th World Cong. Chem. Eng.*, San Diego, CA, p. 3 (1996).

Joseph, D.D., *Fluid Dynamics of Viscoelastic Liquids*, Springer-Verlag, New York (1990).

Joseph, D.D., Historical perspectives in the elasticity of liquids, *J. Non-Newt. Fluid Mech.*, **19**, 237 (1986).

Joseph, D.D. and Feng, J., The negative wake in a second order fluid, *J. Non-Newt. Fluid Mech.*, **57**, 313 (1995). Corrections *ibid* **63**, 263 (1996a).

Joseph, D.D. and Feng, J., A note on the forces that move particles in a second-order fluid, *J. Non-Newt. Fluid Mech.*, **64**, 299 (1996b).

Joseph, D.D. and Liu, Y.J., Orientation of long bodies falling in a viscoelastic liquid, *J. Rheol.*, **37**, 961 (1993).

Joseph, D.D., Liu, Y.J., Poletto, M. and Feng, J., Aggregation and dispersion of spheres falling in viscoelastic liquids, *J. Non-Newt. Fluid Mech.*, **54**, 45 (1994).

Joseph, D.D., Matta, J.E. and Chen, K., Delayed die-swell, *J. Non-Newt. Fluid Mech.*, **24**, 31 (1987).

Joshi, J.B., Solid-Liquid Fluidized beds: Some design aspects, *Chem. Eng. Res. Des.*, **61**, 143 (1983).

Jossic, L., Briguet, A. and Magnin, A., Segregation under flow of objects suspended in a yield stress fluid and NMR imaging visualization, *Chem. Eng. Sci.*, **57**, 409 (2002).

Jossic, L. and Magnin, A., Drag and stability of objects in a yield stress fluid, *AIChE J.*, **47**, 2666 (2001).

Jossic, L. and Magnin, A., Structuring under flow of suspensions in a gel, *AIChE J.*, **50**, 2691 (2004).

Jossic, L. and Magnin, A., Structuring of gelled suspensions flowing through a sudden three-dimensional expansion, *J. Non-Newt. Fluid Mech.*, **127**, 201 (2005).

Jossic, L. and Magnin, A., Drag on an isolated cylinder and interactions between two cylinders in yield-stress fluids, *J. Non-Newt. Fluid Mech.*, **164**, 9 (2009).

Juncu, G., A numerical study of steady viscous flow past a fluid sphere, *Int. J. Heat Fluid Flow*, **20**, 414 (1999).

Kadijk, S.E. and Van Den Brule, B.H.A.A., On the pressure dependency of the viscosity of molten polymers, *Polym. Eng. Sci.*, **34**, 1535 (1994).

Kahle, A., Winkler, B. and Hennion, B., Is Faxen's correction function applicable to viscosity measurements of silicate melts with the falling sphere method? *J. Non-Newt. Fluid Mech.*, **112**, 203 (2003).

Kaiser, A.E., Graham, A.L. and Mondy, L.A., Non-Newtonian wall effects in concentrated suspensions, *J. Non-Newt. Fluid Mech.*, **116**, 479 (2004).

Kalashnikov, V.N. and Kudin, A.M., Karman vortices in the flows of drag reducing polymer solutions, *Nature*, **225**, 445 (1970).

Kalendar, A., Karar, S., Kalendar, A. and Oosthuizen, P., Correlations for natural convective heat transfer from isothermal surface of octagonal and hexagonal shaped of different aspect ratios, *Heat Transf. Asian Res.*, **46**, 362 (2017).

Kalman, H. and Matana, E., Terminal velocity and drag coefficient for spheres, *Powder Technol.*, **396**, 181 (2022).

Kaloni, P.N. and Stastna, J., Flow of particles in viscoelastic fluids, *Transport Processes in Bubbles, Drops and Particles*, edited by R.P. Chhabra and D. De Kee, Hemisphere, New York, Chapter 7 (1992).

Kalra, T.R. and Uhlherr, P.H.T., Geometry of bluff body wakes, *Can. J. Chem. Eng.*, **51**, 655 (1973).

Kamal, M.R. and Mutel, A., Rheological properties of suspensions in Newtonian and non-Newtonian fluids, *J. Polym. Eng.*, **5**, 293 (1985).

Kaminski, T.L. and Persson, S., Distribution of a viscous liquid by a rotating disk, *Trans. ASAE*, **9**, 875 (1966).

Kamisli, F. and Ryan, M.E., Perturbation method in gas-assisted power-law fluid displacement in a circular tube and a rectangular channel, *Chem. Eng. J.*, **75**, 167 (1999).

Kamisli, F. and Ryan, M.E., Gas-assisted non-Newtonian fluid displacement in circular tubes and non-circular channels, *Chem. Eng. Sci.*, **56**, 4913 (2001).

Kanaris, N., Kassinos, S.C. and Alexandrou, A.N., On the transition to turbulence of a viscoplastic fluid past a confined cylinder: A numerical study, *Int. J. Heat Fluid Flow*, **55**, 65 (2015).

Kanchanalakshana, D. and Ghajar, A.J., An improved falling sphere viscometer for intermediate concentrations of viscoelastic fluids, *Int. Comm. Heat Mass Transf.*, **13**, 219 (1986).

Kanellopoulos, N.N., Capillary models for porous media: Newtonian and non-Newtonian flow, *J. Colloid Interface Sci.*, **108**, 11 (1985).

Kang, Y., Suh, I.-S. and Kim, S.D., Heat transfer characteristics of three phase fluidized beds, *Chem. Eng. Commun.*, **34**, 1 (1985).

Kaplun, S., Low Reynolds number flow past a circular cylinder, *J. Math. Mech*, **6**, 595 (1957).

Kaplan, S.J., Morland, C.D. and Hsu, S.C., Predict non-Newtonian fluid pressure drop across random-fiber filters, *Chem. Eng.*, **86**, 93 (Aug 27, 1979).

Kapur, J.N. and Srivastava, R.C., Similar solutions of the boundary layer equations for power law fluids, *ZAMP*, **14**, 381 (1963).

Karabelas, A.J., Wegner, T.H. and Hanratty, T.J., Flow pattern in close packed cubic array of spheres near the critical Reynolds number, *Chem. Eng. Sci.*, **28**, 673 (1973).

Karamanev, D.G., Rise of gas bubbles in quiescent liquids, *AIChE J.*, **40**, 1418 (1994).

Karamanev, D.G., The study of free rise of buoyant spheres in gas reveals the universal behaviour of free rising rigid spheres in fluid in general, *Int. J. Multiph. Flow*, **27**, 1479 (2001).

Karamanev, D.G., Chavarie, C. and Mayer, R.C., Dynamics of the free rise of a light solid sphere in liquid, *AIChE J.*, **42**, 1789 (1996).

Karamanev, D.G., Dewsbury, K. and Margaritis, A., Comments on the free rise of gas bubbles in non-Newtonian liquids, *Chem. Eng. Sci.*, **60**, 4655 (2005).

Karamanev, D.G. and Nikolov, L.N., Free rising spheres do not obey Newton's law for free settling, *AIChE J.*, **38**, 1843 (1992).

Karimfazli, I., Frigaard, I.A. and Wachs, A., A novel heat transfer switch using the yield stress, *J. Fluid Mech.*, **783**, 526 (2015).

Karimfazli, I., Frigaard, I.A. and Wachs, A., Thermal plumes in viscoplastic fluids: Flow onset and development, *J. Fluid Mech.*, **787**, 474 (2016).

Karino, I., Kanno, H., Ozeki, E. and Egawa, S., *Proc. Soc. Chem. Engrs. (Japan)* (1972).

Karis, T.E., Prieve, D.C. and Rosen, S.L., Anomalous lateral migration of a rigid sphere in torsional flow of a viscoelastic fluid, *J. Rheol.*, **28**, 381 (1984).

Karnis, A., Goldsmith, H.L. and Mason, S.G., The flow of suspensions through tubes. V. Inertial effects, *Can. J. Chem. Eng.*, **44**, 181 (1967).

Karnis, A. and Mason, S.G., Particle motions in sheared suspensions – XIX, Viscoelastic media, *Trans. Soc. Rheol.*, **10**, 571 (1966).

Karnis, A. and Mason, S.G., Particle motions in sheared suspensions – XXIII Migration of fluid drops, *J. Colloid Interface Sci.*, **24**, 164 (1967).

Kaser, F. and Keller, R.J., Flow of dilute polymers through porous media, *J. Eng. Mech., (ASCE)*, **106**, 525 (1980).

Kasper, G., Dynamics and measurement of smokes I: Size characterization of non-spherical particles, *Aerosol Sci. Technol.*, **1**, 187 (1982).

Kasper, G., Wall correction to the Stokes resistance of arbitrarily shaped particles, *J. Aerosol Sci.*, **18**, 457 (1987).

Kasper, G., Niida, T. and Yang, M., Measurements of viscous drag on cylinders and chains of spheres with aspect ratios between 2 and 50, *J. Aerosol Sci.*, **16**, 535 (1985).

Kato, H. and Mizuno, Y., An experimental investigation of viscoelastic flow past a circular cylinder, *Bull. JSME*, **26**, 529 (1983).

Kato, H., Tachibana, M. and Oikawa, K., On the drag of a sphere in polymer solutions, *Bull. JSME*, **15**, 1556 (1972).

Kato, Y., Uchida, K., Kago, T. and Morooka, S., Liquid holdup and heat transfer coefficient between bed and wall in liquid-solid and gas-liquid-solid fluidized beds, *Powder Technol.*, **28**, 173 (1981).

Kaupke, C.R. and Yates, W.E., Physical properties and drift characteristics of viscosity modified agricultural sprays, *Trans. ASAE*, **9**, 797 (1966).

Kaur, A., Sobti, A., Toor, A.P. and Wanchoo, R.K., Motion of spheres and cylinders in viscoelastic fluids: Asymptotic behaviour, *Powder Technol.*, **345**, 82 (2019).

Kaviany, M., *Principles of Heat Transfer in Porous Media*, 2nd ed., Springer-Verlag, New York (1995).

Kawale, D., Marques, E., Zitha, P.L., Kreutzer, M.T., Rossen, W.R. and Boukany, P.E., Elastic instabilities during the flow of hydrolyzed polyacrylamide solution in porous media: Effect of pore-shape and salt, *Soft Matter*, **13**(4), 765 (2017).

Kawase, Y., Particle-fluid heat/mass transfer: Newtonian and non-Newtonian fluids, *Warme-und Stoffubertragung*, **27**, 73 (1992).

Kawase, Y. and Hirose, T., Motion of drops in non-Newtonian fluid system at low Reynolds number, *J. Chem. Eng. Jpn.*, **10**, 68 (1977).

Kawase, Y., Mashelkar, R.A. and Ulbrecht, J., Particle liquid mass transfer in viscoelastic media, *Int. J. Multiph. Flow*, **8**, 433 (1982).

Kawase, Y. and Moo-Young, M., Approximate solutions for drag coefficients of bubbles moving in shear-thinning elastic fluids, *Rheol. Acta*, **24**, 202 (1985).

Kawase, Y. and Moo-Young, M., Approximate solutions for power-law fluid flow past a particle at low Reynolds numbers, *J. Non-Newt. Fluid Mech.*, **21**, 167 (1986).

Kawase, Y. and Ulbrecht, J.J., Drag and mass transfer in non-Newtonian flows through multi-particle systems at low Reynolds numbers, *Chem. Eng. Sci.*, **36**, 1193 (1981a).

Kawase, Y. and Ulbrecht, J.J., Motion of and mass transfer from an assemblage of solid spheres moving in a non-Newtonian fluid at high Reynolds numbers, *Chem. Eng. Commun.*, **8**, 233 (1981b).

Kawase, Y. and Ulbrecht, J.J., Formation of drops and bubbles in flowing liquids, *Ind. Eng. Chem. Proc. Des. Dev.*, **20**, 636 (1981c).

Kawase, Y. and Ulbrecht, J.J., On the abrupt change of velocity of bubbles rising in non-Newtonian liquids, *J. Non-Newt. Fluid Mech.*, **8**, 203 (1981d).

Kawase, Y. and Ulbrecht, J.J., Sedimentation of particles in non-Newtonian Fluids, *Chem. Eng. Commun.*, **13**, 55 (1981e).

Kawase, Y. and Ulbrecht, J.J., A power-law fluid flow past a porous sphere, *Rheol. Acta*, **20**, 128 (1981f).

Kawase, Y. and Ulbrecht, J.J., Newtonian fluid sphere with rigid or mobile interface in a shear-thinning liquid: Drag and mass transfer, *Chem. Eng. Commun.*, **8**, 213 (1981g).

Kawase, Y. and Ulbrecht, J.J., The effect of surfactant on terminal velocity of and mass transfer from a fluid sphere in a non-Newtonian fluid, *Can. J. Chem. Eng.*, **60**, 87 (1982).

Kawase, Y. and Ulbrecht, J.J., Mass transfer from spheres submerged in Newtonian and non-Newtonian fluids, *Electrochimica Acta*, **28**, 643 (1983a).

Kawase, Y. and Ulbrecht, J.J., Non-Newtonian fluid-particle mass transfer in granular beds, *AIChE J.*, **29**, 689 (1983b).

Kawase, Y. and Ulbrecht, J.J., The influence of walls on the motion of a sphere in non-Newtonian fluids, *Rheol. Acta*, **22**, 27 (1983c).

Kawase, Y. and Ulbrecht, J.J., Mass transfer from cylinders rotating in Newtonian fluids and dilute polymer solutions, *J. App. Electrochem.*, **13**, 289 (1983d).

Kawase, Y. and Ulbrecht, J.J., Approximate solution to the natural convection heat transfer from a vertical plate, *Int. Comm. Heat Mass Transf.*, **11**, 143 (1984).

Kawase, Y. and Ulbrecht, J.J., A new approach for heat and mass transfer in granular beds based on the capillary model, *Ind. Eng. Chem. Fund.*, **24**, 115 (1985a).

Kawase, Y. and Ulbrecht, J.J., Mass and Momentum transfer with non-Newtonian fluids in fluidized beds, *Chem. Eng. Commun.*, **32**, 263 (1985b).

Ke, C., Shu, S., Zhang, H. and Yuan, H., Drag coefficient and averaged Nusselt number of a scalene prolate ellipsoid, *App. Math. Modell.*, **64**, 556 (2018).

Keentok, M., The measurement of the yield stress of liquids, *Rheol. Acta*, **21**, 325 (1982).

Keentok, M., Georogescu, A.G., Sherwood, A.A. and Tanner, R.I., The measurement of the second normal stress difference for some polymer solutions, *J. Non-Newt. Fluid Mech.*, **6**, 303 (1980).

Keey, R.B., Mandeno, P. and Tuoc, T.K., Dissolution of freely suspended solid particles in an agitated non-Newtonian fluid: Rheology and mass transfer, *CHEMECA-70*, 53 (1970).

Kehlenbeck, R. and Di Felice, R., Empirical relationships for the terminal settling velocity of spheres in cylindrical columns, *Chem. Eng. Technol.*, **21**, 303 (1999).

Kelbaliyev, G.I., Drag coefficients of variously shaped solid particles, drops and bubbles, *Theor. Found. Chem. Eng.*, **45**, 248 (2011).

Kelessidis, V.C., An explicit equation for the terminal velocity of solid spheres falling in pseudoplastic liquids, *Chem. Eng. Sci.*, **59**, 4435 (2004).

Kelessidis, V.C. and Mpandelis, G., Measurements and prediction of terminal velocity of solid spheres falling through stagnant pseudoplastic liquids, *Powder Technol.*, **147**, 117 (2004).

Kelkar, B.G. and Shah, Y.T., Gas hold-up and back mixing in bubble columns with polymer solutions, *AIChE J.*, **31**, 700 (1985).

Kelkar, J.V., Mashelkar, R.A. and Ulbrecht, J., A rotating sphere viscometer, *J. Appl. Polym. Sci.*, **17**, 3069 (1973).

Keller, A., Muller, A.J. and Odell, J.A., Entanglements in semi-dilute solutions as revealed by elongational flow studies, *Prog. Colloid Polym. Sci.*, **75**, 179 (1987).

Keller, D.S. and Keller Jr., D.V., An investigation of the shear-thickening and antithixotropic behaviour of concentrated coal-water dispersions, *J. Rheol.*, **34**, 1267 (1990).

Keller, J.B., Viscous flow through a grating or lattice of cylinders, *J. Fluid Mech.*, **18**, 94 (1964).

Kemblowski, Z. and Dziubinski, M., Resistance to flow of molten polymers through granular beds, *Rheol. Acta*, **17**, 176 (1978).

Kemblowski, Z., Dziubinski, M. and Sek, J., Flow of non-Newtonian fluids through granular media, *Advances in Transport Processes*, edited by R.A. Mashelkar, A.S. Mujumdar and M.R. Kamal, Wiley Eastern, New Delhi, vol. 5, p. 117 (1987).

Kemblowski, Z. and Mertl, J., Pressure drop during the flow of Stokesian fluids through granular beds, *Chem. Eng. Sci.*, **29**, 213 (1974).

Kemblowski, Z., Mertl, J. and Dziubinski, M., The resistance to flow of elasticoviscous fluids through granular beds, *Chem. Eng. Sci.*, **29**, 1343 (1974).

Kemblowski, Z. and Michniewicz, M., A new look at the laminar flow of power law fluids through granular beds, *Rheol. Acta*, **18**, 730 (1979).

Kemblowski, Z., Michniewicz, M. and Torzecki, J., Flow of molten polymers used in the synthetic fibre industry through granular beds, *Rheology*, **1**, 159 (1980).

Kemiha, M., Frank, X., Poncin, S. and Li, H.-Z., Origin of the negative wake behind a bubble rising in non-Newtonian fluids, *Chem. Eng. Sci.*, **61**, 4041 (2006).

Kenchington, J.M., Prediction of critical conditions for pipeline flow of settling particles in a heavy medium, *Proc. Hydrotransport 4*, Paper D3 (1976).

Keunings, R., A survey of computational rheology, *Proc. XIIIth Int. Cong. Rheol.*, **1**, 7 Cambridge, U.K. (2000).

Kezios, P.S. and Schowalter, W.R., Rapid growth and collapse of single bubbles in polymer solutions undergoing shear, *Phys. Fluids*, **29**, 3172 (1986).

Khaled, A.-R.A. and Vafai, K., The role of porous media in modeling flow and heat transfer in biological tissues, *Int. J. Heat Mass Transf.*, **46**, 4989 (2003).

Khamashta, M. and Virto, L., Flow of water treatment plant sludges through porous media, *Proc. Euromech*, 143, Delft, p. 173 (September 2–4, 1981).

Khan, A., Khan, A.A. and Varma, Y.B.G., Prediction of two-phase frictional pressure drop for the concurrent gas/Newtonian liquids upflow through packed beds, *Chem. Eng. Technol.*, **25**, 51 (2002a).

Khan, A., Khan, A.A. and Varma, Y.B.G., Influence of liquid viscosity and bed porosity on two-phase pressure drop in concurrent gas-liquid upflow through packed beds, *Can. J. Chem. Eng.*, **80**, 313 (2002b).

Khan, A.R. and Richardson, J.F., The resistance to motion of a solid sphere in a fluid, *Chem. Eng. Comm.*, **62**, 135 (1987).

Khan, A.R. and Richardson, J.F., Fluid particle interactions and flow characteristics of fluidized beds and settling suspensions of spherical particles, *Chem. Eng. Comm.*, **78**, 111 (1989).

Khan, M.B., Sasmal, C. and Chhabra, R.P., Flow and heat transfer characteristics of a rotating cylinder in a FENE-P type viscoelastic fluid, *J. Non-Newt. Fluid Mech.*, **282**, 104333 (2020).

Khan, W.A., Culham, J.R. and Yovanovich, M.M., Fluid flow and heat transfer in power-law fluids across circular cylinders: Analytical study, *J. Heat Transf.*, **128**, 870 (2006).

Kharabaf, H. and Yortsos, Y.C., Invasion percolation with memory, *Phys. Rev. E*, **55**, 7177 (1997).

Khatchatourian, O.A. and Savicki, D.L., Mathematical modeling of airflow in an aerated soyabean store under non-uniform conditions, *Biosyst. Eng.*, **88**, 201 (2004).

Khatibi, M., Time, R.W. and Rabenjafimanantsoa, H.A., Particles settling through viscoelastic non-Newtonian flows in a horizontal rectangular channel analyzed with PIV and PIV techniques, *J. Non-Newt. Fluid Mech.*, **235**, 143 (2016).

Khayat, R.E., A boundary-only approach to the deformation of a shear-thinning drop in extensional Newtonian flow, *Int. J. Numer. Meth. Fluids*, **33**, 559 (2000a).

Khayat, R.E., Three-dimensional boundary element analysis of drop deformation in confined flow for Newtonian and viscoelastic systems, *Int. J. Numer. Meth. Fluids.*, **34**, 241 (2000b).

Khayat, R.E., Huneault, M.A., Utracki, L.A. and Duquette, R., A boundary element analysis of planar drop deformation in the screw channel of a mixing extruder, *Eng. Anal. Bound. Elem.*, **21**, 155 (1998a).

Khayat, R.E., Luciani, A. and Utracki, L.A., Boundary-element analysis of planar drop deformation in confined flow, *Eng. Anal. Bound. Elem.*, **19**, 279 (1997). Also see *ibid* 22, 291 (1998b).

Khayat, R.E., Luciani, A., Utracki, L.A., Godbille, F. and Picot, J., Influence of shear and elongation on drop deformation in convergent-divergent flows, *Int. J. Multiph. Flow*, **26**, 17 (2000).

Khismatullin, D.B. and Nadim, A., Radial oscillations of encapsulated microbubbles in viscoelastic liquids, *Phys. Fluids*, **14**, 3534 (2002).

Khomami, B. and Moreno, L.D., Stability of viscoelastic flow around periodic arrays of cylinders, *Rheol. Acta*, **36**, 367 (1997).

Khomikovskii, P.M. and Shilov, D.J., The suspending ability of clayey washing liquids used in drilling, *J. Appl. Chem. (U.S.S.R.)*, **19**, 684 (1946) (in Russian).

Khuzhayorov, B., Auriault, J.-L. and Royer, P., Derivation of macroscopic filtration law for transient linear viscoelastic fluid flow in porous media, *Int. J. Eng. Sci.*, **38**. 487 (2000).

Khyati, B.V., Hui, H., Wang, Z.J., Numerical investigation of effect of buoyancy on the wake instability of a heated cylinder in contra flow, *AIAA 2007*, 0801, 1–19.

Kieft, R.N., Rindt, C.C.M. and Steenhoven, A.A.V., The wake behaviour behind a heated horizontal cylinder, *Exp. Therm. Fluid Sci.*, **19**, 183 (1999).

Kiljanski, T., Letter to the editor, *Chem. Eng. Sci.*, **59**, 731 (2004).

Kiljanski, T. and Dziubinski, M., Resistance to flow of molten polymers through filtration screens, *Chem. Eng. Sci.*, **51**, 4533 (1996).

Killen, J.M. and Almo, J., An experimental study of the effects of dilute solutions of polymer additives on boundary layer characteristics, *Viscous Drag Reduction*, edited by C.S. Wells, p. 447, Plenum, New York (1969).

Kim, B.K. and Lee, H.S., Boundary layer analysis of power law fluids, *Kor. J. Chem. Eng.*, **6**, 227 (1989).

Kim, C., Collapse of spherical bubbles in Maxwell fluids, *J. Non-Newt. Fluid Mech.*, **55**, 37 (1994).

Kim, C.B. and Wollersheim, D.E., Free convection heat transfer to non-Newtonian, dilatant fluids from a horizontal cylinder, *J. Heat Transf.*, **98**, 144 (1976).

Kim, D. and Choi, H., Laminar flow past a hemisphere, *Phys. Fluids*, **15**, 2457 (2003).

Kim, H.W. and Esseniyi, A.J., Forced convection of power law fluids flow over a rotating non-isothermal body, *J. Thermophys. Heat Transf.*, **7**, 581 (1993).

Kim, H.W., Jeng, D.R. and DeWitt, K.J., Momentum and heat transfer in power law fluid flow over two-dimensional or axisymmetrical bodies, *Int. J. Heat Mass Transf.*, **26**, 245 (1983).

Kim, I., Elghobashi, S. and Sirignano, W.A., On the equation for spherical-particle motion: Effect of Reynolds and acceleration numbers, *J. Fluid Mech.*, **367**, 221 (1998).

Kim, I., Irvine, Jr., T.F. and Park, N.A., Experimental study of the velocity field around a falling needle viscometer, *Rev. Sci. Instrum.*, **65**, 224 (1994).

Kim, J. and Park, J.D., The non-homogeneous flow of a thixotropic fluid around a sphere, *App. Math. Model.*, **82**, 848 (2020).

Kim, J.M., Kim, C., Ahn, K.H. and Lee, S.J., An efficient iterative solver and high-resolution computations of the Oldroyd-B fluid past a confined cylinder, *J. Non-Newt. Fluid Mech.*, **123**, 161 (2004).

Kim, J.M., Kim, C., Chung, C., Ahn, K.H. and Lee, S.J., Negative wake generation of FENE-CR fluids in uniform and Poiseuille flows past a cylinder, *Rheol. Acta*, **44**, 600 (2005).

Kim, S., The motion of ellipsoids in a second order fluid, *J. Non-Newt. Fluid Mech.*, **21**, 255 (1986).

Kim, S. and Karrila, S., *Microhydrodynamics: Principles and Selected Applications*, Butterworth-Heinemann, Boston, MA (1991).

Kim, S.D. and Kang, Y., Heat and mass transfer in three-phase fluidized-bed reactors – An overview, *Chem. Eng. Sci.*, **52**, 3639 (1997).

King Jr., H.E., Herbolzheimer, E. and Cook, R.L., The diamond-anvil cell as a high-pressure viscometer, *J. App. Phys.*, **71**, 2071 (1992).

King, M.J. and Waters, N.D., The unsteady motion of a sphere in an elastico-viscous liquid, *J. Phys. D. Appl. Phys.*, **5**, 141 (1972).

Kintner, R.C., Drop phenomena affecting liquid extraction, *Adv. Chem. Eng.*, **4**, 51 (1963).

Kirkby, L.L. and Rockefeller, H.A., Proppant settling velocities in non-flowing slurries, *SPE/DOE Low Permeability Gas Reservoirs Symposium*, Denver, CO (1985). (Paper # SPE/DOE 13906).

Kirsch, A.A. and Fuchs, N.A., Studies on fibrous aerosol filters-II. Pressure drops in systems of parallel cylinders, *Ann. Occup. Hyg.*, **10**, 23 (1967a).

Kirsch, A.A. and Fuchs, N.A., The fluid flow in a system of parallel cylinders perpendicular to the flow direction at small Reynolds numbers, *J. Phys. Soc. Jpn.*, **22**, 1251 (1967b).

Kishore, N., Chhabra, R.P. and Eswaran, V., Sedimentation in emulsions of mono-size droplets at moderate Reynolds numbers, *Chem. Eng. Res. Des.*, **84**, 1180 (2006).

Kishore, N., Chhabra R.P. and Eswaran, V., Drag on a single fluid sphere translating in power law liquids at moderate Reynolds numbers, *Chem. Eng. Sci.*, **62**, 2422 (2007a).

Kishore, N., Chhabra, R.P. and Eswaran, V., Mass transfer from a single fluid sphere to power-law liquids at moderate Reynolds numbers, *Chem. Eng. Sci.*, **62**, 6040 (2007b).

Kishore, N., Chhabra, R.P. and Eswaran, V., Mass transfer from ensembles of Newtonian fluid spheres at moderate Reynolds and Peclet numbers, *Chem. Eng. Res. Des.*, **85**, 1203 (2007c).

Kishore, N., Chhabra, R.P. and Eswaran, V., Bubble swarms in power-law liquids at moderate Reynolds numbers: Drag and mass transfer, *Chem. Eng. Res. Des.*, **86**, 39 (2008a).

Kishore, N., Chhabra, R.P. and Eswaran, V., Drag on ensembles of fluid spheres translating in a power law liquid at moderate Reynolds numbers, *Chem. Eng. J.*, **139**, 224 (2008b).

Kishore, N., Chhabra, R.P. and Eswaran, V., Mass transfer from ensembles of fluid spheres to a power-law liquid at moderate Reynolds and Peclet numbers, *Chem. Eng. Sci.*, **63**, 2484 (2008c).

Kishore, N., Chhabra, R.P. and Eswaran, V., Effect of dispersed phase rheology on the drag of single and of ensembles of fluid spheres at moderate Reynolds numbers, *Chem. Eng. J.*, **141**, 387 (2008d).

Kishore, N., Dhole, S.D., Chhabra, R.P. and Eswaran, V., Momentum and heat transfer phenomena for power-law liquids in assemblages of solid spheres of moderate to large void fractions, *Numer. Heat Transf. Part A*, **56**, 970 (2009a).

Kishore, N., Nalajala, V.S. and Chhabra, R.P., Effects of contamination and shear-thinning fluid viscosity on drag behaviour of spherical bubbles, *Ind. Eng. Chem. Res.*, **50**, 6049 (2013).

Kishore, N., Patnana, V.K. and Chhabra, R.P., Flow of power-law liquids past a solid sphere with and without radial mass flux at moderate Reynolds numbers, *J. Chem. Eng. Jpn.*, **42**, 545 (2009b).

Kitamura, K., Kami-Iwa, F. and Misumi, T., Heat transfer and fluid flow of natural convection around large horizontal cylinders, *Int. J. Heat Mass Transf.*, **42**, 4093 (1999).

Kitamura, K., Mitsuishi, A., Suzuki, T. and Misumi, T., Fluid flow and heat transfer of high-Rayleigh-number natural convection around heated spheres, *Int. J. Heat Mass Transf.*, **86**, 149 (2015).

Kitamura, Y., Mishima, H. and Takahashi, T., Stability of jets in liquid-liquid systems, *Can. J. Chem. Eng.*, **60**, 723 (1982).

Kitamura, Y. and Takahashi, T., Breakup of jets in power-law non-Newtonian-Newtonian liquid systems, *Can. J. Chem. Eng.*, **60**, 732 (1982).

Klee, A.J. and Treybal, R.E., Rate of rise or fall of liquid drops, *AIChE J.*, **2**, 444 (1956).

Kleinstreuer, C. and Wang, T.-Y., Heat transfer between rotating spheres and flowing power-law fluids with suction and injection, *Int. J. Heat Fluid Flow*, **9**, 328 (1988).

Kleinstreuer, C. and Wang, T.-Y., Mixed convection heat and surface mass transfer between power law fluids and rotating permeable bodies, *Chem. Eng. Sci.*, **44**, 2987 (1989).

Kleppe, J. and Marner, W.J., Transient free convection in a Bingham plastic in a vertical flat plate, *J. Heat Transf.*, **94**, 371 (1972).

Kliakhandler, I.L., Continuous chain of bubbles in concentrated polymeric solutions, *Phys. Fluids*, **14**, 3375 (2002).

Knoll, M., Gerhardter, H., Prieler, R., Muhlbock, M., Tomazic, P. and Hochenauer, C., Particle classification and drag coefficients of irregularly-shaped combustion residues with various size and shape, *Powder Technol.*, **345**, 405 (2019).

Koblitz, A.R., Lovett, S. and Nikiforakis, N., Direct numerical simulation of particle sedimentation in a Bingham fluid, *Phys. Rev. Fluids*, **3**, 093302 (2018).

Kobus, C.J. and Shumway, G., An experimental investigation into impinging forced convection heat transfer from stationary isothermal circular disks, *Int. J. Heat Mass Transf.*, **49**, 411 (2006).

Koch, D.L. and Ladd, A.J.C., Moderate Reynolds number flows through periodic and random arrays of aligned cylinders, *J. Fluid Mech.*, **349**, 31 (1997)

Koizumi, A., *PhD Dissertation*, University of Delaware, Newark, Delaware (1974).

Kolodziej, J.A., Filtration resistance of a system of parallel cylinders at a transverse creeping flow, *Mech. Teort. Stosowana*, **24**, 537 (1986).

Kolodziej, J.A., Dziecielak, R. and Konczak, Z., Permeability tensor of heterogeneous porous medium of fiber type, *Trans. Porous Media*, **32**, 1 (1998).

Kolodziej, J.A., Meirzwiczak, M. and Cialkowski, M., Power law fluid flow through a bundle of regular fibres, *App. Math. Modell.*, **39**, 6425 (2015).

Komar, P.D. and Reimers, C.E., Grain shape effects on settling rates, *J. Geology*, **86**, 193 (1978).

Koniuta, A., Adler, P.M. and Piau, J.-M., Flow of dilute polymer solutions around circular cylinders, *J. Non-Newt. Fluid Mech.*, **7**, 101 (1980).

Kono, Y., Viscosity measurement, *Magmas Under Pressure*, edited by Y. Kono and C. Sanloup, Elsevier, Amsterdam, Chapter 10 (2018).

Koplik, J. and Banavar, J.R., Extensional rupture of model non-Newtonian fluid filaments, *Phys. Rev.*, **E67**, 011502 (2003).

Koponen, A., Kandhai, D., Hellen, E., Alava, M., Hoekstra, A., Kataja, M., Niskansen, N., Sloot, P. and Timonen, J., Permeability of three-dimensional random fiber webs, *Phys. Rev. Lett.*, **80**, 716 (1998).

Korhonen, M., Lehtonen, J., Hellen, L., Hirvonen, J. and Yliruusi, J., Rheological properties of three component creams containing sorbitan monoesters as surfactants, *Int. J. Pharm.*, **247**, 103 (2002).

Korhonen, M., Nisakanen, H., Kiesvaara, J. and Yliruusi, J., Determination of optimal combination of surfactants in creams using rheology measurements, *Int. J. Pharm.*, **197**, 143 (2000).

Koshiba, T., Mori, N. and Nakamura, K., Measurement of pressure loss in the flow of polymer solutions through packed beds of particles, *J. Soc. Rheol. Jpn.*, **21**, 163 (1993).

Koshiba, T., Mori, N. and Nakamura, K., Measurement of pressure loss and flow visualization in the flow of viscoelastic fluids through cylinder arrays, *Nihon Reoroji Gakkaishi (J. Soc. Rheol. Jpn.)*, **26**, 27 (1998).

Koshiba, T., Mori, N., Sugiyama, S. and Nakamura, K., Elongational effects in the flow of viscoelastic fluid through a wavy channel, *Rheol. Acta*, **38**, 375 (1999).

Koshy, A., Das, T.R. and Kumar, R., Effect of surfactants on drop breakage in turbulent liquid dispersions, *Chem. Eng. Sci.*, **43**, 649 (1988a).

Koshy, A., Das, T.R., Kumar, R. and Gandhi, K.S., Breakage of viscoelastic drops in turbulent stirred dispersions, *Chem. Eng. Sci.*, **43**, 2625 (1988b).

Koshy, A., Kumar, R. and Gandhi, K.S., Effect of drag reducing agents on drop breakage in stirred dispersions, *Chem. Eng. Sci.*, **44**, 2113 (1989).

Kostornov, A.G. and Shevchuk, M.S., Hydraulic characterstics and structure of porous metal fiber materials III. Laws of liquid permeability of materials, *Poroshkovaya Metallurgiya*, **9**(177), 50 (1977).

Koteswara Rao, P., Sahu, A.K. and Chhabra, R.P., Flow of Newtonian and power-law fluids past an elliptic cylinder: A numerical study, *Ind. Eng. Chem. Res.*, **49**, 6649 (2010).

Kotouč, M., Bouchet, G. and Dušek, J., Loss of axisymmetry in the mixed convection, assisting flow past a heated sphere, *Int. J. Heat Mass Transf.*, **51**, 2686 (2008).

Kotouč, M., Bouchet, G. and Dušek, J., Transition to turbulence in the wake of a fixed sphere in mixed convection, *J. Fluid Mech.*, **625**, 205 (2009a).

Kotouč, M., Bouchet, G. and Dusek, J., Drag and flow reversal in mixed convection past a heated sphere, *Phys. Fluids*, **21**, 054104 (2009b).

Kozeny, J., Uber kapillare Leitung des Wassers im Boden-Aufstieg, Versickerung und Anwendung auf die Bewasserung, Sitzungsberichte der Akademie der Wissenschaften Wien. *Mathematisch Naturwissenschaftliche Abteilung*, **136**, 271 (1927).

Kozicki, W., Filtration in non-Newtonian media, *Encyclopedia of Fluid Mechanics*, **5**, Gulf, Chap. 23 (1988).

Kozicki, W., Flow of a FENE fluid in packed beds or porous media, *Can. J. Chem. Eng.*, **80**, 818 (2002).

Kozicki, W., Hanna, M.R. and Tiu, C., Polymer adsorption in packed bed flow, *J. Rheol.*, **32**, 593 (1988).

Kozicki, W., Hsu, C.J. and Pasari, S.N., Evaluation of polymer adsorption-gel formation and slip in polymer solution flows, *Chem. Eng. Commun.*, **59**, 137 (1987).

Kozicki, W., Hsu, C.J. and Tiu, C., Non-Newtonian flow through packed beds and porous media, *Chem. Eng. Sci.*, **22**, 487 (1967).

Kozicki, W., Rao, A.R.K. and Tiu, C., Filtration of polymer solutions, *Chem. Eng. Sci.*, **27**, 615 (1972).

Kozicki, W., Son, J.E. and Hanna, M.R., Characterisation of adsorptional phenomena in polymer solution flows, *Chem. Eng. J.*, **29**, 171 (1984).

Kozicki, W. and Tiu, C., Flow of complex fluids in open or closed conduits, packed beds and porous media, *Can. J. Chem. Eng.*, **51**, 359 (1973).

Kozicki, W. and Tiu, C., A unified model for non-Newtonian flow in packed beds and porous media, *Rheol. Acta*, **27**, 31 (1988).

Kozicki, W., Tiu, C. and Rao, A.R.K., Filtration of non-Newtonian fluids, *Can. J. Chem. Eng.*, **46**, 313 (1968).

Koziol, K. and Glowacki, P., Determination of the free settling parameters of spherical particles in power-law fluids, *Chem. Eng. Process.*, **24**, 183 (1988).

Krishna, R., Urseanu, M.I., van Baten, J.M. and Ellenberger, J., Rise velocity of a swarm of large gas bubbles in liquids, *Chem. Eng. Sci.*, **54**, 171 (1999a).

Krishna, R., Urseanu, M.I., van Baten, J.M. and Ellenberger, J., Wall effects on the rise of single bubbles in liquids, *Int. Comm. Heat Mass Transf.*, **26**, 781 (1999b).

Krishna, R. and van Baten, J.M., Simulating the motion of gas bubbles in liquids, *Nature*, **398**, 208 (March 1999).

Krishnan, S. and Kannan, A., Effect of blockage ratio on drag and heat transfer from a centrally located sphere in pipe flow, *Eng. Appl. Comput. Fluid Mech.*, **4**, 396 (2010).

Krishnan, S. and Kannan, A., Effects of a particle diameter and position on hydrodynamics around a confined sphere, *Ind. Eng. Chem. Res.*, **50**, 13137 (2011).

Krishnan, S. and Kannan, A., Effects of a particle blockage and eccentricity in location on the non-Newtonian fluid hydrodynamics around a sphere, *Ind. Eng. Chem. Res.*, **51**, 14867 (2012).

Kroesser, F.W. and Middleman, S., Viscoelastic jet stability, *AIChE J.*, **15**, 383 (1969).

Kroger, M., Simple models for complex non-equilibrium fluids, *Phys. Rep.*, **390**, 453 (2004).

Krüger, E., Die Grundwasserbewegung. *Internationale Mitteilungen für Bodenkunde*, **8**, 105 (1918).

Krüssmann, R. and Brunn, P.O., Elastische effekte bei der strömung viskoelastischer fluide durch Haufwerke, *Chem. Ing. Tech.*, **73**, 1002 (2001). Also see, *Chem. Eng. Technol.*, 25, 259 (2002).

Kuang, P.Q. and Kozicki, W., Two phase flow of Bingham fluids through porous media, *Mech. Res. Commun*, **16**, 333 (1989).

Kubair, V.G. and Pei, D.C.T., Combined laminar free and forced convection heat transfer to non-Newtonian fluids, *Int. J. Heat Mass Transf.*, **11**, 855 (1968).

Kuehn, T.H. and Goldstein, R.J., Numerical solutions to the Navier–Stokes equations for laminar natural convection, *Int. J. Heat Mass Transf.*, **23**, 971 (1980).

Kulicke, W.-M. and Classen, C., *Viscosimetry of Polymers and Polyelectrolytes*, Springer, Munich (2004).

Kulicke, W.-M. and Haas, R., Flow behaviour of dilute polyacrylamide solutions through porous media. 1. Influence of chain length, concentration, and thermodynamic quality of the solvent, *Ind. Eng. Chem. Fund.*, **23**, 308 (1984).

Kulicke, W.-M. and Haas, R., Characterization of dilute polyacrylamide and polystyrene solutions by means of porous media flow, *The Influence of Polymer Additives on Velocity and Temperature Fields*, edited by B. Gampert, Springer, Berlin, p. 119 (1985).

Kulicke, W.-M., Kotter, M. and Grager, H., Drag reduction phenomenon with special emphasis on homogeneous polymer solutions, *Adv. Polym. Sci.*, **89**, 1 (1989).

Kulicke, W.-M. and Wallabaum, U., Determination of first and second normal stress differences in polymer solutions in steady shear flow and limitations caused by flow irregularities, *Chem. Eng. Sci.*, **40**, 961 (1985).

Kulkarni, A.A. and Joshi, J.B., Bubble formation and bubble rise velocity in gas-liquid systems: A review, *Ind. Eng. Chem. Res.*, **44**, 5873 (2005).

Kulkarni, S., Jamison, D.E. and Gollapalli, J.S., Modelling sag in the wellbore for oil-based drilling fluids, *Paper presented at the IADC/SPE Drilling Conference and Exhibition*, Fort Worth, Tx, 1–3 March (2016a). Paper # IADC/SPE-178832-MS.

Kulkarni, S., Jamison, D.E., Teke, K.D. and Savari, S., Managing suspension characteristics of lost circulation materials in a drilling fluid, *SPE Drill. Complet.*, **30**, 310 (2016b).

Kulkarni, S., Savari, S., Teke, K., Murphy, R.J. and Jamison, D.E., Hindrance effect on Barite sag in non-aqueous drilling fluids, *Paper presented at the American Association of Drilling Engineers*, Houston, TX, April10–11, 2012. Paper# AADE-12-FTCE-23.

Kumar, N., Majumdar, S., Sood, A., Govindarajan, R., Ramaswamy, S. and Sood, A.K., Oscillatory settling in worm-like micelle solutions: Bursts and a long-time scale, *Soft Matter*, **8**, 4310 (2012).

Kumar, P. and Ramarao, B.V., Enhancement of the sedimentation rates of fibrous suspensions, *Chem. Eng. Commun.*, **108**, 381 (1991).

Kumar, R. and Kuloor, N.R., The formation of bubbles and drops, *Adv. Chem. Eng.*, **8**, 255 (1970).

Kumar, R. and Saradhy, Y.P., Drop formation in non-Newtonian fluids, *Ind. Eng. Chem. Fund.*, **11**, 307 (1972).

Kumar, S., Kishore, J., Lal, P. and Upadhyay, S.N., Non-Newtonian flow through packed beds and porous media, *J. Sci. Ind. Res.*, **40**, 238 (1981).

Kumar, S., Mall, B.K. and Upadhyay, S.N., On the mass transfer in non-Newtonian fluids II, Transfer from cylinders to power law fluids, *Lett. Heat Mass Transf.*, **7**, 55 (1980a).

Kumar, S., Tripathi, P.K. and Upadhyay, S.N., On the mass transfer in non-Newtonian fluids I. Transfer from spheres to power law fluids, *Lett. Heat Mass Transf.*, **7**, 43 (1980b).

Kumar, S. and Upadhyay, S.N., Mass transfer to power law fluids in fluidized beds of large particles, *Lett. Heat Mass Transf.*, **7**, 199 (1980).

Kumar, S. and Upadhyay, S.N., Mass and momentum transfer to Newtonian and non-Newtonian fluids in fixed and fluidized beds, *Ind. Eng. Chem. Fund.*, **20**, 186 (1981).

Kunii, D. and Levenspiel, O., *Fluidization Engineering*, 2nd ed., Butterworth, Stoneham, MA (1990).

Kushalkar, K.B. and Pangarkar, V.G., Particle-liquid mass transfer in three-phase mechanically agitated contactors: Power law fluids, *Ind. Eng. Chem. Res..*, **34**, 2485 (1995).

Kutsovsky, Y.E., Scriven, L.E., Davis, H.T. and Hammer B.E., NMR imaging of velocity profiles and velocity distributions in bead packs, *Phys. Fluids*, **8**, 863 (1996).

Kuwabara, S., The forces experienced by randomly distributed parallel circular cylinders or spheres in a viscous flow at small Reynolds numbers, *J. Phys. Soc. Jpn.*, **14**, 527 (1959).

Kuwahara, F., Nakayama, A. and Koyama, H., A numerical study of thermal dispersion in porous media, *J. Heat Transf.*, **118**, 756 (1996).

Kyan, C.P., Wasan, D.T. and Kintner, R.C., Flow of single-phase fluids through fibrous beds, *Ind. Eng. Chem. Fund.*, **9**, 956 (1970).

Kyle, C.R. and Perrine, R.L., An experimental model for visual studies on turbulent flow in porous materials, *Can. J. Chem. Eng.*, **40**, 19 (1971).

Laba, D., *Rheological Properties of Cosmetics and Toiletries*, Marcel Dekker, New York (1993).

Labrecque, R.P., The effects of fiber cross-sectional shape on the resistance to the flow of fluids through fiber mats, *TAPPI*, **51**, 8 (1968).

Ladenburg, R., Uber den einflub von wanden auf die bewegung einer kugel in einer reibenden flussigkeit, *Ann. Phys., (Leipzig)*, **23**, 447 (1907).

Lagerstedt, T., An experimental study of polymer induced drag for flows through porous media, *The Influence of Polymer Additives on Velocity and Temperature Fields*, (IUTAM Symposium, Essen, 1984), edited by B. Gampert, Springer Verlag, Berlin, p. 29 (1985).

Lagisetty, J.S., Das, P.K., Kumar, R. and Gandhi, K.S., Breakage of viscous and non-Newtonian drops in stirred dispersions, *Chem. Eng. Sci.*, **41**, 65 (1986).

Lahbabi, A. and Chang, H.-C., Flow in periodically constricted tubes: Transition to inertial and nonsteady flows, *Chem. Eng. Sci.*, **41**, 2487 (1986).

Lai, R.Y.S., Translatory accelerating motion of a circular disk in a viscous fluid, *Appl. Sci. Res.*, **27A**, 440 (1973).

Lai, R.Y.S., Drag on a sphere accelerating rectilinearly in a Maxwell fluid, *Int. J. Eng. Sci.*, **12**, 645 (1974).

Lai, R.Y.S., Accelerating motion of a sphere in a Maxwell fluid, *Appl. Sci. Res.*, **30**, 208 (1975).

Lakes, R.S., *Viscoelastic Solids*, CRC Press, Boca Raton, FL (1999).

Lal, K., Similarly solutions for the power law fluids flow behind a flat plate, *Ind. J. Phys.*, **42**, 590 (1968).

Lal, P., Mishra, P. and Upadhyay, S.N., Diffusion and mass transfer from a rotating disk in aqueous polymeric solutions, *Lett. Heat Mass Transf.*, **7**, 65 (1980).

Lal, P. and Upadhyay, S.N., Mass Transfer from suspended solid spheres in agitated power law fluids, *Chem. Eng. Sci.*, **36**, 1865 (1981).

Lali, A.M., Khare, A.S., Joshi, J.B. and Nigam, K.D.P., Behaviour of solid particles in viscous non-Newtonian solutions: Settling velocity, wall effects and bed expansion in solid-liquid fluidized beds, *Powder Technol.*, **57**, 39 (1989).

Lang, T.G. and Patrick, H.V.L., Drag of blunt bodies in polymer solutions, Paper (# 66WA/FE33) presented at the winter annual meeting of ASME (1966).

Langins, J., Weber, M.E. and Pliskin, I., The pressure drop created by cones settling on the axis of a pipe, *Chem. Eng. Sci.*, **26**, 693 (1971).

Langmuir, I., Report of smokes and filters, Part IV of a report for the office of scientific research and development (OSRD), No. 865, Ser.No.353, Filtration of Aerosols and the Development of Filter Materials, by Rodebusch, W.H. et al. (September 1942).

Lao, H.W, Neeman, H.J. and Papavassiliou, D.V., A pore network model for the calculation of non-Darcy coefficients in fluid flow through porous media, *Chem. Eng. Commun.*, **191**, 1285 (2004).

Lapasin, R. and Pricl, S., *Rheology of Industrial Polysacchrides: Theory and Applications*, Blackie Academic and Professional, London, UK (1995).

Larachi, F., Belfares, L., Iliuta, I. and Grandjean, B.P.A., Heat and mass transfer in cocurrent gas-liquid packed beds. Analysis, recommendations, and new correlations, *Ind. Eng. Chem. Res.*, **42**, 222 (2003).

Larachi, F., Bensetiti, Z., Grandjean, B.P.A. and Wild, G., Two-phase frictional pressure drop in flooded-bed reactors: A state-of-the-art correlation, *Chem. Eng. Technol.*, **21**, 887 (1998).

Lareo, C. and Fryer, P.J., Vertical flow of solid-liquid food mixtures, *J. Food Eng.*, **36**, 417 (1998).

Lareo, C., Fryer, P.J. and Barigou, M., The fluid mechanics of two-phase solid-liquid food flows: A review, *Trans. Inst. Chem. Eng.*, **75C**, 73 (1997).

Larkins, R.P., White, R.R. and Jeffrey, D.W., Two-phase cocurrent flow in packed beds, *AIChE J.*, **7**, 231 (1961).

Larson, R.E. and Higdon, J.J.L., A periodic grain consolidation model of porous media, *Phys. Fluids*, **A1**, 38 (1989).

Larson, R.G., Derivation of generalized Darcy equation for creeping flow in porous media, *Ind. Eng. Chem. Fund.*, **20**, 132 (1981).

Larson, R.G., *Constitutive Equations for Polymer Melts and Solutions*, Butterworth, London (1988).

Larson, R.G., *The Structure and Rheology of Complex Fluids*, Oxford University Press, New York (1998).

Larson, R.G., Turbulence without inertia, *Nature*, **405**, 27 (2000).

Larson, R.G., Constitutive equations for thixotropic fluids, *J. Rheol.*, **59**, 595 (2015).

Larson, R.G. and Wei, Y., A review of thixotropy and its rheological modelling, *J. Rheol.*, **63**, 477 (2019).

Lasheras, J.C. and Hopfinger, E.J., Liquid jet instability and atomization in a coaxial gas stream, *Annu. Rev. Fluid Mech.*, **32**, 275 (2000).

Lashgari, I., Pralits, J.O., Giannetti, F. and Brandt, L., First instability of the flow of shear-thinning and shear-thickening fluids past a circular cylinder, *J. Fluid Mech.*, **701**, 201 (2012).

Lasso, I.A. and Weidman, P.D., Stokes drag on hollow cylinders and conglomerates, *Phys. Fluids*, **29**, 3921 (1986).

Latifi, M., Midoux, N., Storck, A. and Gence, J.M., The use of microelectrodes in the study of flow regimes in a packed bed reactor with single phase liquid flow, *Chem. Eng. Sci.*, **44**, 2501 (1989).

Latto, B. and Middleton, J.A., Effect of dilute polymer solutions on external boundary layers, *Proc. Sym. Turbulence Measurements in Liquids*, Patterson, G.K. and Zakin, J.L. eds., Missouri-Rolla, p. 116 (1969).

Latto, B., Round, G.F. and Anzenavs, R., Drag coefficients and pressure drops for hydrodynamically suspended spheres in a vertical tube with and without polymer addition, *Can. J. Chem. Eng.*, **51**, 536 (1973).

Lau, E.M., Zhang, J.-D., Jia, Y-X., Huang, W.-X. and Xu, C.-X., Vortical structures in the wake of falling plates, *J. Vis.*, **22**, 15 (2019).

Laufer, G., Gutfinger, C. and Abuaf, N., Flow of dilute polymer solutions through a packed bed, *Ind. Eng. Chem. Fund.*, **15**, 74 (1976).

Lauga, E., Life at high Deborah number, *Euro. Phys. Lett.*, **86**, 64001 (2009).

Lavrenteva, O.M., Holenberg, Y. and Nir. A., Motion of viscous drops in tubes filled with yield stress fluid, *Chem. Eng. Sci.*, **64**, 4772 (2009).

Lavrenteva, O.M. and Nir. A., Viscoplastic flows with free boundaries and interfaces, *Rev. Chem. Eng.*, **26**, 149 (2010).

Lavrov, A., Radial flow of non-Newtonian power-law fluid in a rough-walled fracture: Effect of fluid rheology. *Trans. Porous Media*, **105**(3), 559 (2014).

Lawson, N.J., Tatum, J.A., Finnis, M.V. and Harrison, G.M., Stereoscopic particle image velocimetry: Applications to a non-Newtonian flow field generated by a sedimenting sphere, *Proc. 12th Int. Sym. on applications of Laser Techniques to Fluid Mechanics*, Lisbon, July 12–15 (2004).

Laxton, P.B. and Berg, J.C., Gel trapping of dense colloids, *J. Colloid Interface Sci.*, **285**, 152 (2005).

Le Roux, J.P., Settling velocity of spheres: A new approach, *Sediment. Geol.*, **81**, 11 (1992).

Le Roux, J.P., Settling velocity of ellipsoidal grains as related to shape entropy, *Sediment. Geol.*, **101**, 15 (1996).

Le Roux, J.P., Application of the Hofmann shape entropy to determine the settling velocity of irregular, semi-ellipsoidal grains, *Sediment. Geol.*, **149**, 237 (2002).
Le Roux, J.P., A. hydrodynamic classification of grain shapes, *J. Sed. Res.*, **74**, 135 (2004).
Leal, L.G., The slow motion of slender rod-like particles in a second-order fluid, *J. Fluid Mech.*, **69**, 305 (1975).
Leal, L.G., The motion of small particles in non-Newtonian fluids, *J. Non-Newt. Fluid Mech.*, **5**, 33 (1979).
Leal, L.G., Particle motions in a viscous fluid, *Ann. Rev. Fluid Mech.*, **12**, 435 (1980).
Leal, L.G., Skoog, J. and Acrivos, A., On the motion of gas bubbles in a viscoelastic liquid, *Can. J. Chem. Eng.*, **49**, 569 (1971).
Leal, L.G. and Zana, E., A preliminary experimental investigation of the motion of slender particles in Separan AP-30, *J. Fluid Mech.*, **69**, 334 (1975). Also see *ibid.*, **83**, 273 (1977).
LeClair, B.P., *PhD Thesis*, McMaster University, Hamilton, ON (1970).
LeClair, B.P. and Hamielec, A.E., Viscous flow through particle assemblages at intermediate Reynolds numbers, *Ind. Eng. Chem. Fund.*, **7**, 542 (1968a).
LeClair, B.P. and Hamielec, A.E., Viscous flow through particle assemblages at intermediate Reynolds numbers: Heat or mass transport, *Institution of Chemical Engineers Sym. Ser.*, No. 30, 197 (1968b).
LeClair, B.P. and Hamielec, A.E., Viscous flow through particle assemblages at intermediate Reynolds numbers: Steady state solutions for flow through assemblages of cylinders, *Ind. Eng. Chem. Fund.*, **9**, 608 (1970).
LeClair, B.P. and Hamielec, A.E., Viscous flow through particle assemblages at intermediate Reynolds numbers – A cell model for transport in bubble swarms, *Can. J. Chem. Eng.*, **49**, 713 (1971).
LeClair, B.P., Hamielec, A.E. and Pruppacher, H.R., A numerical study of the drag on a sphere at low and intermediate Reynolds number, *J. Atm. Sci.*, **27**, 308 (1970).
Lecoq, N., Feuillebois, F., Anthore, N., Anthore, R., Bostel, F. and Petipas, C., Precise measurement of particle-wall hydrodynamic interactions at low Reynolds number using laser interferometry, *Phys. Fluids.*, **5**, 3 (1993).
Lee, D.I., Packing of spheres and its effect on the viscosity of suspensions, *J. Paint Technol.*, **42**, 579 (1970).
Lee, D.-Y. and Chung, B.-J., Visualization of natural convection heat transfer on a sphere, *Heat Mass Transf.*, **53**, 3613 (2017).
Lee, D.-Y. and Irvine, T.F., Shear rate dependent thermal conductivity measurements of non-Newtonian fluids, *Exp. Thermal Fluid Sci.*, **15**, 16 (1997).
Lee, H.M., A modification of Stokes law to account for boundary influence. MS Thesis, Univ. of Iowa, Ames (1947).
Lee, S.J. and Yang, J.H., Modeling of Darcy-Forchheimer drag for fluid flow across a bank of circular cylinders, *Int. J. Heat Mass Transf.*, **40**, 3149 (1997).
Lee, S.W., Ryu, S.H. and Kim, C., Studies on the axisymmetric sphere – sphere interaction problem in Newtonian and non-Newtonian fluids, *J. Non-Newt. Fluid Mech.*, **110**, 1 (2003).
Lee, S.W., Sohn, S.-M., Ryu, S.H., Kim, C. and Song, K.-W., Experimental studies on the axisymmetric sphere-wall interaction in Newtonian and non-Newtonian fluids, *Korea-Aust. Rheol. J.*, **13**, 141 (2001).
Lee, S.Y. and Ames, W.F., Similarity solutions for non-Newtonian fluids, *AIChE J.*, **12**, 700 (1966).
Lee, T.-L. and Donatelli, A.A., Mass transfer by natural convection from a solid sphere to Power law fluids, *Ind. Eng. Chem. Res.*, **28**, 105 (1989).
Lee, W.K. and Flumerfelt, R.W., Instability of stationary and uniformly moving cylindrical fluid bodies I. Newtonian systems, *Int. J. Multiph. Flow*, **7**, 363 (1981).
Lee, W.K., Yu, K.L. and Flumerfelt, R.W., Instability of stationary and uniformly moving cylindrical fluid bodies II. Viscoelastic threads and experimental observations, *Int. J. Multiph. Flow*, **7**, 385 (1981).
Lee, W.T., Kaar, S. and O'Brien, S.B.G., Sinking bubbles in stout beers, *Am. J. Phys.*, **86**, 250 (2018).
Lee, Y.-J. and Zhang, C.S., Self- sustaining oscillations of a falling sphere through Johnson- Segalman fluids, *arXIV: 1209.1328V2* [math-ph] (2012).
Lefebvre, A.H., *Atomization and Sprays*, Taylor and Francis, New York (1989).
Legile, P.G., Menard, G., Laurent, C., Thomas, D. and Bernis, A., Contribution to the study of an inverse three-phase fluidized bed operating counter-currently, *Intl. Chem. Eng. J.*, **32**, 41 (1992).
Legrand, J., Revisited analysis of pressure drop in flow through crushed rocks, *J. Hyd. Eng. (ASCE)*, **128**, 1027 (2002).
Lehner, F.K., A derivation of the field equations for slow viscous flow through a porous medium, *Ind. Eng. Chem. Fund.*, **18**, 41(1979).

Leider, P.J. and Bird, R.B., Squeezing flow between parallel disks I: Theoretical analysis, *Ind. Eng. Chem. Fund.*, **13**, 336 (1974). Also see *ibid* 342 for experimental results.

Leigh, D.C., Non-Newtonian fluids and the second law of thermodynamics, *Phys. Fluids*, **5**, 501 (1962).

Lejeune, A.M., Bottinga, Y., Trull, T.W. and Richet, P., Rheology of bubble-bearing magmas, *Ear. Planet. Sci. Lett.*, **166**, 71 (1999).

Lemieux, P.F., Dubey, R.N. and Unny, T.E., Variational method for a pseudoplastic fluid in a laminar boundary layer over a flat plate, *J. App. Mech. (ASME)*, **38**, 345 (1971).

Leonov, A.I., Extremum principles and exact two side bounds of potential: Functional and dissipation for slow motions of nonlinear viscoplastic media, *J. Non-Newt. Fluid Mech.*, **28**, 1 (1988).

Leonov, E.G. and Isayev, V.I., *Hydroair Mechanics in Drilling Operations*, Nedra, Moscow (1989). (in Russian).

Lesage, F., Midoux, N. and Latifi, M.A., New local measurements of hydrodynamics in porous media, *Expt. Fluids*, **37**, 257 (2004).

Lescarboura, J.A. and Swift, G.W., The effect of eccentricity on the terminal velocity of the cylinder in a falling cylinder viscometer: Experimental verification for Newtonian fluids, *AIChE J.*, **14**, 651 (1968).

Leslie, F.M. and Tanner, R.I., The slow flow of a viscoelastic liquid past a sphere, *Quart. J. Mech. Appl. Maths.*, **14**, 36 (1961).

Levich, V.G., *Physicochemical Hydrodynamics*, Prentice Hall, Englewood Cliffs, NJ (1962).

Levick, J.R., Review article: Flow through interstitium and other fibrous matrices, *Quart. J. Exp. Phys.*, **72**, 409 (1987).

Levine, A.J. and Lubensky, T.C., Response function of a sphere in a viscoelastic two-fluid medium, *Phys. Rev. E*, **63**, 1510 (2001).

Levitskiy, S.P. and Shulman, Z.P., *Bubbles in Polymeric Liquids*, Technomic, Lancaster, PA (1995).

Levy, G.L., *PhD Dissertation*, University of Denver, Boulder, CO (1969).

Lewis, H.W., Calibration of the rolling ball viscometer, *Anal. Chem.*, **25**, 507 (1953).

Li, G., McKinley, G.H. and Ardekani, A.M., Dynamics of particle migration in channel flow of viscoelastic fluids, *J. Fluid Mech.*, **785**, 486 (2015).

Li, H.Z., Chaotic behaviour of bubble coalescence in non-Newtonian fluids, *Chem. Eng. Technol.*, **21**, 983 (1998).

Li, H.Z., Bubbles in non-Newtonian fluids: Formation, interactions and coalescence, *Chem. Eng. Sci.*, **54**, 2247 (1999).

Li, H.Z., Frank, X., Funfshilling, D. and Mouline, Y., Towards the understanding of bubble interactions and coalescence in non-Newtonian fluids: A cognitive approach, *Chem. Eng. Sci.*, **56**, 6419 (2001).

Li, H.Z., Mouline, Y., Choplin, L. and Midoux, N., Chaotic bubble coalescence in non-Newtonian fluids, *Int. J. Multiph. Flow*, **23**, 713 (1997a).

Li, H.Z., Mouline, Y., Choplin, L. and Midoux, N., Interactions et coalescence des bulles dans les fluides rheologiquement complexes, *C.R. Acad. Sci. Paris*, **324**, 491 (1997b).

Li, H.Z., Mouline, Y. and Midoux, N., Modelling the bubble formation dynamics in non-Newtonian fluids, *Chem. Eng. Sci.*, **57**, 339 (2002).

Li, H.Z. and Qiang, S., Formation des bulles dans les fluids newtoniens et non-newtoniens, *C.R. Acad. Sci. Paris*, **326**, 301 (1998).

Li, J. and Helm, D., Viscous drag, driving forces, and their reduction to Darcy's law, *Wat. Resour. Res.*, **34**, 1675 (1998).

Li, J. and Renardy, Y.Y., Shear-induced rupturing of a viscous drop in a Bingham liquid, *J. Non-Newt. Fluid Mech.*, **95**, 235 (2000).

Li, L., Larson, R.G. and Sridhar, T., Brownian dynamics simulations of dilute polystyrene solutions, *J. Rheol.*, **44**, 291 (2000).

Li, L. and Ma, W. Experimental study on the effective particle diameter of a packed bed with non-spherical particles. *Trans. Porous Media*, **89**(1), 35 (2011).

Li, S., Ma, Y., Jiang, S., Fu, T. and Li, H.-Z., The drag coefficient and the shape for a single bubble rising in non-Newtonian fluids, *J. Fluids Eng.*, **134**, 084501 (2012).

Li, Y. and Kuru, E., Numerical modeling of cuttings transport with foam in horizontal wells, *J. Can. Pet. Tech.*, **42**(10), 54 (2003).

Li, Y. and Park, C.-W., Permeability of packed beds filled with polydisperse spherical particles, *Ind. Eng. Chem. Res.*, **37**, 2005 (1998).

Liang, S.F. and Acrivos, A., Experiments on buoyancy driven convection in non-Newtonian fluid, *Rheol. Acta*, **9**, 447 (1970).
Liao, S.-J., An analytic approximation of the drag coefficient for the viscous flow past a sphere, *Int. J. Non-Linear Mech.*, **37**, 1 (2002).
Liao, S.-J., A challenging nonlinear problem for numerical techniques, *J. Comp. Appl. Maths.*, **181**, 467 (2005).
Licht, W. and Narsimhamurthy, G.S.R., Rate of fall of single liquid droplets, *AIChE J.*, **1**, 366 (1955).
Liddell, P.V. and Boger, D.V., Yield stress measurements with the vane, *J. Non-Newt. Fluid Mech.*, **63**, 235 (1996).
Liew, K.S. and Adelman, M., Laminar natural heat transfer from an isothermal sphere to non-Newtonian fluids, *Can. J. Chem. Eng.*, **53**, 494 (1975).
Lima-Ballesteros, R., Riba, J.P. and Couderc, J.P., Expansion de couches de particules non spheriques fluidisees par un liquide, *Entropie*, **106**, 37 (1982).
Lin, C.L. and Miller, J.D., 3D characterization and analysis of particle shape using X-ray microtomography (XMT), *Powder Technol.*, **154**, 61 (2005).
Lin, F.N. and Chern, S.Y., Laminar boundary layer flow of non-Newtonian fluid, *Int. J. Heat Mass Transf.*, **22**, 1323 (1979).
Lin, H.L., Ranganathan, S. and Advani, S.G., Consolidation of continuous-fiber systems, *Flow and Rheology in Polymer Composites Manufacturing*, edited by S.G. Advani, Elsevier, Amsterdam (1994) (Chapter 9).
Lin, H.-T. and Shih, Y.-P., Combined free and forced convection from a vertical plate to power law fluids, *Chem. Eng. Commun.*, **7**, 327 (1980a).
Lin, H.-T. and Shih, Y.-P., Laminar boundary layer heat transfer to power law fluids, *Chem. Eng. Commun.*, **4**, 557 (1980b).
Lin, S.H. and Fan, L.T., Examples of the use of initial value method to solve non-linear Boundary value problems, *AIChE J.*, **18**, 654 (1972).
Lin, S.X.Q., Chen, X.D., Chen, Z.D. and Bandopadhyay, P., Shear rate dependent thermal conductivity measurement of two fruit juice concentrates, *J. Food Eng.*, **57**, 217 (2003).
Lin, T.-J. and Lin, G.-M., The mechanisms of bubble coalescence in a non-Newtonian fluid, *Can. J. Chem. Eng.*, **81**, 476 (2003).
Lin, T.-J. and Lin, G.-M., An experimental study on flow structures of a single bubble rising in a shear-thinning viscoelastic fluid with a new measurement technique, *Int. J. Multiph. Flow*, **31**, 239 (2005).
Lin, Y.-M., Wu, G.H. and Ju, S.H., Non-isothermal flow of a polymeric liquid passing an asymmetrically confined cylinder, *Int. J. Heat Mass Transf.*, **47**, 1989 (2004).
Lindsay, J.D., Ghiaasiaan, S.M. and Abdel-Khalik, S.I., Macroscopic flow structures in a bubbling paper pulp-water slurry, *Ind. Eng. Chem. Res.*, **34**, 3342 (1995).
Linliu, K., Yeh, F., Shook, J.W., Tuminello, W.H. and Chu, B., Development of a centrifuge ball viscometer for polymer melts, *Rev. Sci. Instrum.*, **65**, 3823 (1994).
Lipscomb, G.G. and Denn, M.M., Flow of Bingham fluids in complex geometries, *J. Non-Newt. Fluid Mech.*, **14**, 337 (1984).
Litt, M.J., Rheology-past, present and future, *Chem. Eng. Prog.*, **85**(12), 54 (1989).
Liu, A.W., Bornside, D.E., Armstrong, R.C. and Brown, R.A., Viscoelastic flow of polymer solutions around a periodic, linear array of cylinders: Comparisons of predictions for microstructure and flow fields, *J. Non-Newt. Fluid Mech.*, **77**, 153 (1998a).
Liu, B., Lin, J., Ku, X. and Yu, Z., Migration of spherical particles in a confined shear flow of Giesekus fluid, *Rheol. Acta*, **58**, 639 (2019).
Liu, B.T., Muller, S.J. and Denn, M.M., Convergence of a regularization method for creeping flow of a Bingham material about a rigid sphere, *J. Non-Newt. Fluid Mech.*, **102**, 179 (2002).
Liu, B.T., Muller, S.J. and Denn, M.M., Interactions of two rigid spheres translating collinearly in creeping flow in a Bingham material, *J. Non-Newt. Fluid Mech.*, **113**, 49 (2003).
Liu, C., De Luca, A., Rosso, A. and Talon, L., Darcy's law for yield stress fluids, *Phys. Rev. Lett.*, **122**, 245502 (2019).
Liu, H., *Science and Engineering of Droplets: Fundamentals and Applications*, Noyes, Norwich, NY (2000).
Liu, H., Bau, H.H. and Hu, H., On the translation of a cylinder in a long tube, *Phys. Fluids*, **16**, 998 (2004).
Liu, H.L., Um, M.K. and Hwang, W.R., A scaling rule for the flow mobility of a power-law fluid through unidirectional fibrous porous media. *J. Non-Newt. Fluid Mech.*, **224**, 40 (2015a).

Liu, J., Zhao, C.-J., Liu, H. and Lu, W.-Q., Numerical study of laminar natural convection heat transfer from a hemisphere with adiabatic plane and isothermal hemispherical surface, *Int. J. Therm. Sci.*, **131**, 132 (2018).

Liu, J., Zhu, C., Wang, X., Fu, T., Ma, Y. and Li, H.Z., Three-dimensional numerical simulation of coalescence and interactions of multiple horizontal bubbles rising in shear-thinning fluids, *AIChE J.*, **61**, 3528 (2015b).

Liu, S., Afacan, A. and Masliyah, J.H., Steady incompressible laminar flow in porous media, *Chem. Eng. Sci.*, **49**, 3565 (1994).

Liu, S. and Masliyah, J.H., Principles of single-phase flow through porous media, *Adv. Chem. Ser.*, **251**, 227 (1996a).

Liu, S. and Masliyah, J.H., Single phase flow in porous media, *Chem. Eng. Commun.*, **148–150**, 653 (1996b).

Liu, S. and Masliyah, J.H., On non-Newtonian fluid flow in ducts and porous media, *Chem. Eng. Sci.*, **53**, 1175 (1998).

Liu, S. and Masliyah, J.H., Non-linear flows in porous media, *J. Non-Newt. Fluid Mech.*, **86**, 229 (1999).

Liu, W.-W., Peng, J., Zhu, K.-Q. and Zhang, J.-J., A microstructural criterion for yielding: Linking thixotropy and the yield stress, *Rheol. Acta*, **55**, 957 (2016).

Liu, X., Liu, Z., Jiang, S., Zhu, C., Ma, Y. and Fu, T., Formation of droplets of shear-thinning non-Newtonian fluids in a step-emulsification microdevice, *AIChE J.*, **68**, e17395 (2022).

Liu, Y.J. and Joseph, D.D., Sedimentation of particles in polymer solutions, *J. Fluid Mech.*, **255**, 565 (1993).

Liu, Y.J., Liao, T.Y. and Joseph, D.D., A two-dimensional cusp at the trailing edge of an air bubble rising in a viscoelastic liquid, *J. Fluid Mech.*, **304**, 321 (1995).

Liu, Y.J., Nelson, J., Feng, J. and Joseph, D.D., Anomalous rolling of spheres down an inclined plane, *J. Non-Newt. Fluid Mech.*, **50**, 305 (1993).

Liu, Z., Brenn, G. and Durst, F., Linear analysis of the instability of two-dimensional non-Newtonian liquids Sheets, *J. Non-Newt. Fluid Mech.* **78**, 133 (1998b).

Lochiel, A.C. and Calderbank, P.H., Mass transfer in the continuous phase around axisymmetric bodies of revolution, *Chem. Eng. Sci.*, **19**, 471 (1964).

Lockett, F.J., Boundary layer flow of a viscoelastic fluid, National Phys. Lab. Report, Ma 73 (1969).

Lockhart, R.W. and Martinelli, R.C., Proposed correlation of data for isothermal two-phase, two component flow in pipes, *Chem. Eng. Prog.*, **45**, 39 (1949).

Lockyer, M.A., Davies, J.M. and Jones, T.E.R., The importance of rheology in the determination of the carrying capacity of oil-drilling fluids, *Rheology*, edited by G. Astarita, G. Marrucci and L. Nicolais, Vol. 2, Plenum, New York, p. 127 (1980).

Loewenberg, M., Stokes resistance, added mass, and Basset force for arbitrary oriented, finite-length cylinders, *Phys. Fluids*, **5**, 765 (1993).

Lohia, P.K., Narayan, K.A. and Chhabra, R.P., Mass transfer from highly soluble cylinders in cross-flow: Some experimental results, *Can. J. Chem. Eng.*, **73**, 263 (1995).

Lohrenz, J. and Kurata, F., Design and evaluation of a new body for falling cylinder viscometers, *AIChE J.*, **8**, 190 (1962).

Lohrenz, J., Swift, G.W. and Kurata, F., An experimentally verified theoretical study of the falling cylinder viscometer, *AIChE J.*, **6**, 547 (1960).

Lohse, D. and Zijm, W.H.M., Bubble puzzles, *Phys. Today*, **56**(2), 36 (2003).

Lomba, R.F.T., de Sa, C.H.M. and Brandao, E.M., A new approach to evaluate temperature effects on rheological behavior of formate-based fluids, *J. Energy Res. Technol.*, **124**, 141 (2002).

Lopez, W.F., Naccache, M.F. and de Souza Mendes, P.R., Rising bubbles in yield stress materials, *J. Rheol.*, **62**, 209 (2018).

Lopez, X., Valvatne, P.H. and Blunt, M.J., Predictive network modeling of single phase non-Newtonian flow in porous media, *J. Colloid Interface Sci.*, **264**, 256 (2003).

Lopez-Aguilar, J.E., Resendiz-Tolentino, O., Tamaddon-Jahromi, H.R., Ellero, M. and Manero, O., Flow past a sphere: Numerical predictions of thixo-viscoelastoplastic wormlike micellar solutions, *J. Non-Newt. Fluid Mech.*, **309**, 104902 (2022).

Lopez De Haro, M., Del Rio, J.A.P. and Whitaker, S., Flow of Maxwell fluids in porous media, *Trans. Porous Media*, **25**, 167 (1996).

Lord, E., Air flow through plugs of textile fibres, Part I – General flow relations, *Textile Inst. J.*, **46**, T191 (1955).

Lorenzi, A., Laminar, turbulent and transition flow in porous sintered media, *Meccanica*, **10**, 75 (1975).

Losenno, C. and Easson, W.J., PIV measurements of free-falling irregular particles, *Proc. 11th Int. Sym. on Applications of Laser Technology to Fluid Mechanics*, 8–11 July, Lisbon (2002).

Loth, E., Drag of non-spherical particles of regular and irregular shape, *Powder Technol.*, **182**, 342 (2008a).

Loth, E., Quasi-steady shape and drag of deformable bubbles and drops, *Int. J. Multiph. Flow*, **34**, 523 (2008b).

Lovato, S., Toxopeus, S., Settles, J. and Keetels, G., Application of maritime CFD code to a benchmark problem for non-Newtonian fluids: The flow around a sphere, *Int. Shipbuild. Prog.*, **69**, 93 (2022).

Lu, Y.B. and Tang, G.H., Radial voidage variation in packed beds of uniformly sized spheres: Theory and experiments, *J. Porous Media*, **18**(7), 689 (2015).

Luikov, A.V., Shulman, Z.P. and Berkovsky, B.M., Heat and mass transfer in a boundary layer of non-Newtonian fluids, *Proc. 3rd Int. Conf. Heat Transf.*, Chicago, IL, p. 377 (1966).

Luikov, A.V., Shulman, Z.P. and Puris, B.I., External convective mass transfer in non-Newtonian fluid, *Int. J. Heat Mass Transf.*, **12**, 377 (1969b).

Luikov, A.V., Shulman, Z.P., Puris, B.I. and Zhadanovich, N.V., Experimental study of rheodynamics and mass transfer in a forced non-Newtonian fluid flow around bodies, *Progress in Heat Mass Transfer*, edited by T.F. Irvine Jr., W.E. Ibele, J.P. Hartnett and R.J. Goldstein, Vol. 2, p. 267, Pergamon, Oxford (1969a).

Lundgren, T.S., Slow flow through stationary random beds and suspensions of spheres, *J. Fluid Mech.*, **51**, 273 (1972).

Lundgren, T.S. and Mansour, N.N., Vortex ring bubbles, *J. Fluid Mech.*, **224**, 177 (1991).

Luning, C.D. and Perry, W.L., An iterative method for solution of a boundary value problem problem in non-Newtonian fluid flow, *J. Non-Newt. Fluid Mech.*, **15**, 145 (1984).

Lunnon, R.G., Fluid resistance to moving spheres. *Proc. Roy. Soc.* **A118**, 680 (1928).

Lunsmann, W.J., Brown, R.A. and Armstrong, R.C., *6th Workshop on Numerical Methods in Non-Newtonian Flows*, Denmark, June (1989).

Lunsmann, W.J., Genieser, L., Armstrong, R.C. and Brown, R.A., Finite element analysis of steady viscoelastic flow around a sphere in a tube: Calculations with constant viscosity models, *J. Non-Newt. Fluid Mech.*, **49**, 63 (1993).

Luo, X.L., Operator splitting algorithm for viscoelastic flow and numerical analysis for the flow around a sphere in a tube, *J. Non-Newt. Fluid Mech.*, **63**, 121 (1996).

Luo, X.L., An incremental difference formulation for viscoelastic flows and high-resolution FEM solutions at high Weissenberg numbers, *J. Non-Newt. Fluid Mech.*, **79**, 57 (1998).

Luo, X.L. and Tanner, R.I., A streamline element scheme for solving viscoelastic flow problems Part I: Differential constitutive equations, *J. Non-Newt. Fluid Mech.*, **21**, 179 (1986).

Lyon, M.K., Mead, D.W., Elliott, R.E. and Leal, L.G., Structure formation in moderately concentrated suspensions in shear flow, *J. Rheol.*, **45**, 881 (2001).

Lyons, D.W., White, J.W. and Hatcher, J.D., Laminar natural convection heat transfer in dilute aqueous polymer solutions, *Ind. Eng. Chem. Fund.*, **11**, 586 (1972).

Ma, Q., Shang, C., Zhang, G. and Fang, H., Numerical investigation of degassing behavior of highly viscous non-Newtonian fluids under stirring, *Chem. Eng. Technol.*, **43**, 157 (2020).

Maalouf, A. and Sigli, D., Effects of body shape and viscoelasticity on the slow flow around an obstacle, *Rheol. Acta*, **23**, 497 (1984).

Macdonald, I.F., El-Sayed, M.S., Mow, K. and Dullien, F.A.L., Flow through packed beds – The Ergun equation revisited, *Ind. Eng. Chem. Fund.*, **18**, 199 (1979).

MacDonald, M.J., Chu, C.-F., Guilloit, P.P. and Ng, K.M., A generalized Blake-Kozeny equation for multisized spherical particles, *AIChE J.*, **37**, 1583 (1991).

Macedo, I.C. and Yang, W.J., The drag of air bubbles rising in non-Newtonian liquids, *Jap. J. Appl. Phys.*, **13**, 529 (1974).

Machac, I., Balcar, M. and Lecjaks, Z., Creeping motion of non-Newtonian liquids through fluidized beds of spherical particles, *Chem. Eng. Sci.*, **41**, 591 (1986).

Machac, I., Balcar, M. and Lecjaks, Z., Creeping flow of non-Newtonian liquids through fluidized beds of spherical particles, *Prog. & Trends in Rheol. II*, Proc. Second Conf. European Rheologists, Steinkopff, 268 (1988).

Machac, I., Cakl, J., Comiti, J. and Sabiri, N.E., Flow of non-Newtonian fluids through fixed beds of particles: Comparison of two models, *Chem. Eng. Process.*, **37**, 169 (1998).

Machac, I., Cakl, J. and Lecjaks, Z., Fall of spheres through non-Newtonian liquids in transition region, *Proc. CHISA Cong.*, Prague, The Czech Republic (1987).

Machac, I., Comiti, J., Brokl, P. and Siska, B., Fluidization of spherical particle beds with Boger fluids, *Chem. Eng. Res. Des.*, **81A**, 1217 (2003).

Machac, I. and Dolejs, V., Flow of generalized Newtonian liquids through fixed beds of non-spherical particles, *Chem. Eng. Sci.*, **36**, 1679 (1981).

Machac, I. and Dolejs, V., Flow of viscoelastic liquids through fixed beds of particles, *Chem. Eng. Commun.*, **18**, 29 (1982).

Machac, I. and Lecjaks, Z., Wall effect for a sphere falling through a non-Newtonian fluid in a rectangular duct, *Chem. Eng. Sci.*, **50**, 143 (1995).

Machač, I., Mikulášek, P. and Ulbrichová, I., Non-Newtonian fluidization of spherical-particle beds. *Chem. Eng. Sci.*, **48**, 2109 (1993).

Machac, I., Siska, B. and Lecjaks, Z., Inhomogeneities in fluidization of spherical particle beds with non-Newtonian polymer solutions, *Chem. Papers*, **53**, 390 (1999).

Machac, I., Siska, B., Lecjaks, Z. and Bena, J., Fluidization of spherical particle beds with non-Newtonian fluids in columns of rectangular cross-sections, *Chem. Eng. Sci.*, **52**, 3409 (1997).

Machac, I., Siska, B. and Machacova, L., Terminal falling velocity of spherical particles moving through a Carreau model fluid, *Chem. Eng. Process.*, **39**, 365 (2000).

Machac, I., Siska, B. and Teichman, R., Fall of non-spherical particles in a Carreau model liquid, *Chem. Eng. Process.*, **41**, 577 (2002).

Machač, I., Strnadel, J. and Surý, A., Wall effects on a single spherical particle moving through a Carreau model fluid. *AIP Conf. Proc.*, **1152**(1), 110 (2009).

Machac, I., Ulbrichova, I., Elson, T.P. and Cheesman, D.J., Fall of spherical particles through non-Newtonian suspensions, *Chem. Eng. Sci.*, **50**, 3323 (1995).

Machado, J.C.V. and Aragao, A.F.L. Gel strength as related to carrying capacity of drilling fluids, SPE 21106, *Paper presented at SPE Latin American Petroleum Eng. Conf.*, Oct 14–19, Rio de Janeiro, Brazil (1990).

Machado, J.C.V. and Valente, J.S., Ultrasonic scattering cross-sections of shell-encapsulated gas bubbles immersed in a viscoelastic liquid: First and second harmonics, *Ultrasonics*, **41**, 605 (2003).

Maciejewski, W., Lord, E., Gillies, R. and Shook, C.A., Pipeline transport of large ablating particles in a non-Newtonian carrier, *Powder Technol.*, **94**, 223 (1997).

Maciel, G.F., des Santos, H.K. and Ferreira, F.deO., Rheological analysis of water-clay composition in order to investigate mudflows developing in canals, *J. Braz. Soc. Mech. Sci. Eng.*, **31**, 64 (2009).

Mackaplow, M.B. and Shaqfeh, E.S.G., A numerical study of the sedimentation of fibre suspensions, *J. Fluid Mech.*, **376**, 149 (1998).

Mackay, M.E. and Boger, D.V., An explanation of the rheological properties of Boger fluids, *J. Non-Newt. Fluid Mech.*, **22**, 235 (1987).

Maclean-Fletcher, S.D. and Pollard, T.D., Viscometric analysis of the gelation of *Acanthamoeba* extracts and purification of two gelation factors, *J. Cell Biol.*, **85**, 414 (1980).

Macosko, C.W., *Rheology: Principles, Measurements and Applications*, Wiley-VCH, New York (1994).

Madraki, Y., Hormozi, S., Ovarlez, G., Guazzelli, E. and Pouliquen, O., Enhancing shear-thickening, *Phys. Rev. Fluids*, **2**, 033301 (2017).

Maerker, J.M., Dependence of polymer retention on flow rate, *J. Pet. Tech.*, **25**, 1307 (1973).

Maerker, J.M., Shear degradation of partially hydrolyzed polyacrylamide solutions, *Soc. Pet. Eng. J..*, **13**, 311 (August 1975).

Magarvey, R.H. and Bishop, R.L., Transition ranges for three-dimensional wakes, *Can. J. Phys.*, **39**, 1418 (1961).

Magda, J.J. and Larson, R.G., A transition occurring in ideal elastic liquids during shear flow, *J. Non-Newt. Fluid Mech.*, **30**, 1 (1988).

Magnaudet, J. and Eames, I., The motion of high-Reynolds-number bubbles in inhomogeneous flows, *Ann. Rev. Fluid Mech.*, **32**, 659 (2000).

Magueur, A., Moan, M. and Chauveteau, G., Effect of successive contractions and expansions on the apparent viscosity of dilute polymer solutions, *Chem. Eng. Commun.* **36**, 351 (1985).

Mahaut, F., Chateau, X., Coussot, P. and Ovarlez, G., Yield stress and elastic modulus of suspensions of non-colloidal particles in yield stress fluids, *J. Rheol.*, **52**, 287 (2008).

Mahmood, R., Kousar, N., Usman, K. and Mehmood, A., Finite element simulations for stationary Bingham fluid flow past a circular cylinder, *J. Braz. Soc. Mech. Sci. Eng.*, **40**(9), 459 (2018).

Majumdar, A. and Das, P.K., Rise of Taylor bubbles through power-law fluids-Analytical modelling and numerical simulations, *Chem. Eng. Sci.*, **205**, 83 (2019).

Majumder, A.K., Settling velocities of particulate systems –a critical review of some useful models, *Miner. Metall. Process.*, **24**, 237 (2007).

Maklad, O. and Poole, R.J., A review of the second normal stress difference; its importance in various flows, measurement techniques, results for various complex fluids and theoretical predictions, *J. Non-Newt. Fluid Mech.*, **292**, 104522 (2021)

Maleki-Jirsaraei, N., Hassani, S. and Azizi, S., Settling of spherical objects through thixotropic fluids: A statistical approach, *Modern Appl. Sci.*, **12**(5), 72 (2018).

Malhotra, S. and Sharma, M.M., Experimental measurement of settling velocity of spherical particles in unconfined and confined surfactant-based shear thinning viscoelastic fluids, *J. Vis. Expts.*, **83**, e50749 (2014).

Malkin, A.Y., Non-linearity in rheology- an essay on classification, *Rheol. Acta*, **34**, 27 (1995).

Malkin, A.Y. and Isayev, A.I., *Rheology: Concepts, Methods and Applications*, ChemTec, Toronto, Canada (2006).

Malkin, A.Y., Kulichikhin, V. and Ilyin, S., A modern look on yield stress fluids, *Rheol. Acta*, **56**, 177 (2017).

Malkin, A.Y., Masalova, I., Slatter, P. and Wilson, K., Effect of droplet size on the rheological properties of highly concentrated emulsions, *Rheol. Acta*, **43**, 584 (2004).

Malleswararao, T.V. and Chhabra, R.P., A note on pressure drop for the cross flow of power law liquids and air/power law liquid mixtures past a bundle of circular rods, *Chem. Eng. Sci.*, **58**, 1365 (2003).

Malvandi, A., Moshizi, S.A. and Ganji, D.D., An analytical study on unsteady motion of vertically falling spherical particles in quiescent power-law shear-thinning fluids, *J. Mol. Liq.*, **193**, 166 (2014).

Malvault, G., Ahmadi, A. and Omari, A., Numerical simulation of yield stress fluid flow in capillary bundles: Influence of the form and the axial variation in the cross section. *Trans. Porous Media*, **120**(2), 255 (2017).

Mandhani, V.K., Chhabra, R.P. and Eswaran, V., Forced convection heat transfer in tube banks in cross flow, *Chem. Eng. Sci.*, **57**, 379 (2002).

Maneri, C., New look at wave analogy for prediction of bubble terminal velocities, *AIChE J.*, **41**, 481 (1995).

Maneri, C.C. and Mendelson, H.D., The rise velocity of bubbles in tubes and rectangular channels as predicted by wave theory, *AIChE J.*, **14**, 295 (1968).

Manero, O. and Mena, B., On the slow flow of viscoelastic liquids past a circular cylinder, *J. Non-Newt Fluid Mech.*, **9**, 379 (1981).

Manero, O., Mena, B. and de Vargas, L., A note on the translation of a thin rod inside a cylinder, *Rheol. Acta*, **26**, 266 (1987). Corrections *ibid*, **26**, 577 (1987).

Manero, O., Mena, B. and Leal, L.G., The influence of wall effects on the motion of a sphere in viscoelastic media, *Theoretical and Applied Developments in Rheology*, edited by C. Rangel-Nafaile, C.A. Cruz Ramos and E.J.V. Carter, Mexican Society of Rheology, Mexico, p. 29 (1986).

Mangadoddy, N., Prakash, R., Chhabra, R.P. and Eswaran, V., Forced convection in cross flow of power law fluids over a tube bank, *Chem. Eng. Sci.*, **59**, 2213 (2004).

Manjunath, M. and Chhabra, R.P., Flow of non-Newtonian power law liquids through packed and fluidized beds, *Chem. Eng. Commun.*, **106**, 33 (1991).

Manjunath, M. and Chhabra, R.P., Free rise velocity of a swarm of spherical gas bubbles through a quiescent power law liquid, *Int. J. Eng. Sci.*, **30**, 871 (1992).

Manjunath, M., Tripathi, A., Chhabra, R.P. and Sundararajan, T., Numerical simulation of the drag on a swarm of bubbles, *Int. J. Eng. Sci.*, **32**, 927 (1994).

Mankad, S., Nixon, K.M. and Fryer, P.J., Measurements of particle-liquid heat transfer in systems of varied solids fraction, *J. Food Eng.*, **31**, 9 (1997).

Manli, T. and Xiaoli, G., The viscoelastic effect of non-Newtonian fluid through the porous media and its characterization, *Proc. China-Japan Int. Conf. Rheol.*, Beijing, China (1991) p. 259.

Mannheimer, R.J., Rheological and mist ignition properties of dilute polymer solutions, *Chem. Eng. Commun.*, **19**, 221 (1983).

Mansour, A. and Chigier, N., Air-blast atomization of non-Newtonian liquids, *J. Non-Newt. Fluid Mech.*, **58**, 161 (1995).

Manz, B., Gladden, L.F. and Warren, P.B., Flow and dispersion in porous media: Lattice-Boltzmann and NMR studies, *AIChE J.*, **45**, 1845 (1999).

Mao, Z., Numerical simulation of viscous flow through spherical particle assemblage with the modified cell model, *Chinese J. Chem. Eng.*, **10**, 149 (2002).

Mao, Z.-Q., Godfrey, J.C. and Slater, M.J., Single liquid drop velocities and breakage mechanism in sections of structured packings, *Chem. Eng. Technol.*, **18**, 33 (1995).

Mao, Z.-S. and Chen, J., An attempt to improve the cell model for motion and external mass transfer of a drop-in swarms at intermediate Reynolds numbers, *Int. Solvent Extraction Conference* (ISEC 2005), Beijing, China Paper # A417 (2005).

Mao, Z.-S. and Wang, Y., Numerical simulation of mass transfer of a spherical particle assemblage with the cell model, *Powder Technol.*, **134**, 145 (2003).

Marchal, J.M. and Crochet, M.J., The flow of an Oldroyd-B fluid around a sphere. Unpublished manuscript (1988).

Marchal, J.M., Crochet, M.J. and Keunings, R., Adaptive refinement for calculating viscoelastic flows, *Proc. 5th Symposium on Finite Elements and Flow Problems*, p.473, Univ. of Texas, Austin (1984).

Margaritis, A., te Bokkel, D.W. and Karamanev, D.G., Bubble rise velocities and drag coefficients in non-Newtonian polysacchride solutions, *Biotech. Bioeng.*, **64**, 257 (1999).

Maris, H.J., Note on the history effect in fluid mechanics, *Am. J. Phys.*, **87**, 643 (2019).

Marivoet, J., Teodovair, P. and Wajc, S.J., Porosity, velocity, and temperature profiles in cylindrical packed beds, *Chem. Eng. Sci.*, **29**, 1836 (1974).

Markovitz, H., The emergence of rheology, *Physics Today*, **21**(4), 23 (1968).

Markovitz, H., Boltzmann and the beginning of linear viscoelasticity, *Trans. Soc. Rheol.*, **21**, 381 (1977).

Markovitz, H., Rheology: In the beginning, *J. Rheol.*, **29**, 777 (1985).

Marlière, C., Faure, P., Coussot, P., Vlassopoulos, D., Larsen, A. and Loppinet, B., Jamming of cellulose ether solutions in porous medium. *AIChE J.*, **61**(11), 3923 (2015).

Marner, W.J. and Kleppe, J., Combined free and forced laminar convection in a Bingham plastic on a vertical flat plate, *J. Fluids Eng.*, **98**, 119 (1976).

Marrucci, G., Rising velocity of a swarm of spherical bubbles, *Ind. Eng. Chem. Fund.*, **4**, 224 (1965).

Marrucci, G., A theory of coalescence, *Chem. Eng. Sci.*, **24**, 975 (1969).

Marrucci, G., Appuzzo, G. and Astarita, G., Motion of drops in non-Newtonian systems, *AIChE J.*, **16**, 538 (1970).

Marrucci, G. and Ianniruberto, G., Molecular theories of polymer viscosity, *Complex Flows in Industrial Processes*, edited by A. Fasano, Chapter 1, Birkhaüser, Boston (2000).

Marshall, R.J. and Metzner, A.B., Flow of viscoelastic fluids through porous media, *Ind. Eng. Chem. Fund.*, **6**, 393 (1967).

Marshall, S.H., Chudacek, M.W. and Bagster, D.F., A model for bubble formation from an orifice with liquid cross-flow, *Chem. Eng. Sci.*, **48**, 2049 (1993).

Martin, A.R., Saltiel, C. and Shyy, W., Frictional losses and convective heat transfer in sparse, periodic cylinder arrays in cross-flow, *Int. J. Heat Mass Transf.*, **41**, 2383 (1998).

Martynenko, O.G. and Khramtsov, P.P., *Free Convective Heat Transfer*, Springer, Berlin (2005).

Masalova, I., Malkin, A.Y., Ferg, E., Kharatiyan, E., Taylor, M. and Haldenwang, R., Evolution of rheological properties of highly concentrated emulsions with aging- emulsion- to- suspension transition, *J. Rheol.*, **50**, 435 (2006).

Mashelkar, R.A., Drag reduction in external rotational flows, *AIChE J.*, **19**, 382 (1973).

Mashelkar, R.A., Kale, D.D., Kelkar, J.V. and Ulbrecht, J.J., Determination of material parameters of viscoelastic fluids by rotational non-viscometric flows, *Chem. Eng. Sci.*, **27**, 973 (1972).

Mashelkar, R.A. and Marrucci, G., Anomalous transport phenomena in rapid external flows of viscoelastic fluids, *Rheol. Acta*, **19**, 426 (1980).

Masliyah, J.H., Steady wakes behind oblate spheroids: Flow visualization, *Phys. Fluids*, **15**, 1144 (1972).

Masliyah, J.H., Viscous flow across banks of circular and elliptical cylinders: Momentum and heat transfer, *Can. J. Chem. Eng.*, **51**, 550 (1973).

Masliyah, J.H., Jauhari, R. and Gray, M., Drag coefficients for air bubbles rising along an inclined surface, *Chem. Eng. Sci.*, **49**, 1905 (1994).

Massey, A.H., Khare, A.S. and Niranjan, K., Air inclusion into a model cake batter using a pressure whisk: Development of gas hold-up and bubble size distribution, *J. Food Sci.*, **66**, 1152 (2001).

Masuyama, T., Ishihara, T., Quan, G. and Noda, K., Pressure loss of Bingham plastic fluid flow through packed beds, *Soc. Chem. Eng. Jpn.*, **1144**, 463 (1983–86).

Matallah, H., Townsend, P. and Webster, M.F., Recovery and stress-splitting schemes for viscoelastic flows, *J. Non-Newt. Fluid Mech.*, **75**, 139 (1998).

Mathieu, J.-B., Beaudoin, G. and Martel, S., Method of propulsion of a ferromagnetic core in the cardiovascular system through magnetic gradients generated by an MRI system, *IEEE Trans. Biomed. Eng.*, **53**, 292 (2006).

Matijasic, G. and Glasnovic, A., Measurement and evaluation of drag coefficient for settling of spherical particles in pseudoplastic fluids, *Chem. Biochem. Eng. Quart.*, **15**, 21 (2001).

Matta, J.E., Tytus, R.P. and Harris, J.L., Aerodynamic atomization of polymeric solutions, *Chem. Eng. Commun.*, **19**, 191 (1983).

Mattischek, J.-P. and Sobczak, R., High-pressure cell for measuring the zero-shear viscosity of polymer melts, *Rev. Sci. Instrum.*, **68**, 2101 (1997).

Maude, A.D., End effects in a falling-sphere viscometer, *Brit. J. App. Phys.*, **12**, 293 (1961).

Maul, C., Kim. S., Ilic, V., Tullock, D. and Phan-Thien, N., Sedimentation of hexagonal flakes in a half-space: Numerical predictions and experiments in Stokes flow, *J. Imaging Sci. Technol.*, **38**, 241 (1994).

Mauret, E. and Renaud, M., Transport phenomena in multi-particle systems – I. Limits of applicability of capillary model in high voidage beds – application to fixed beds of fibers and fluidized beds of spheres, *Chem. Eng. Sci.*, **52**, 1807 (1997).

Mavis, F.T. and Wilsey, E.F., A study of the permeability of sand. *Univ. Iowa Studies in Engineering, Bull.*, **7** (1936).

Maxworthy, T., Gnann, C., Kurten, M. and Durst, F., Experiments on the rise of air bubbles in clean viscous liquids, *J. Fluid Mech.*, **321**, 421 (1996).

McAuliffe, C.D., Oil-in-water emulsions and their flow properties in porous media, *J. Pet. Tech.*, **25**, 727 (1973).

McCann, D.J. and Prince, R.G.H., Bubble formation and weeping at a submerged orifice, *Chem. Eng. Sci.*, **24**, 801 (1969).

Mc Clements, D.J., *Food Emulsions: Principles, Practice and Techniques*, 2nd ed., CRC, Boca Raton, FL (2004).

Mc Comb, W.D. and Ayyash, S., The production, pulsation and damping of small air bubbles in dilute polymer solutions, *J. Phys. D App. Phys.*, **13**, 773 (1980).

McDonald, A.T. and Brandt, H., Skin friction of power law fluids in turbulent flow over a flat plate, *AIChE J.*, **12**, 637 (1966).

McDuffie, G.E. and Barr, T., Pressure viscometer for viscosities between 1 and 10^4 P, *Rev. Sci. Instrum.*, **40**, 653 (1969).

McGeary, R.K., Mechanical packing of spherical particles, *J. Amer. Ceramic Soc.*, **44**, 513 (1961).

McHale, S. and Richardson, J.F., Heat transfer from immersed surfaces to aqueous liquids, *Chem. Eng. Sci.*, **40**, 2154(1985).

McKinley, G.H., Steady and transient motion of spherical particles in viscoelastic liquids, *Transport Processes in Bubbles, Drops and Particles*, edited by R.P. Chhabra and D. De kee, 2nd ed., p. 338, Taylor and Francis, New York (2002).

McKinley, G.H., Armstrong, R.C. and Brown, R.A., The wake instability in viscoelastic flow past confined circular cylinders, *Phil. Trans. R. Soc. Lond. A.*, **344**, 265 (1993).

McKinley, G.H. and Sridhar, T., Filament stretching rheometry of complex fluids, *Ann. Rev. Fluid Mech.*, **34**, 375 (2002).

McKinley, R.M., Jahns, H.O., Harris, W.W. and Greenkorn, R.A., Non-Newtonian flow in porous media, *AIChE J.*, **12**, 17 (1966).

Mclachlan, R.J., A new high pressure viscometer for viscosity range 10 to 10^6 Pa.s, *J. Phys. E: Sci. Instrum.*, **9**, 391 (1976).

McMechan, D.E. and Shah, S.N., Static proppant-settling characteristics of non-Newtonian fracturing fluids in a large-scale test model, *SPE Prod. Eng.*, **6**, 305 (1991).

McNown, J.S., Lee, H.M., McPherson, M.B. and Engez, S.M., Influence of boundary proximity on the drag of spheres, *Proc. 7th Int. Cong. App. Mech.*, Volume 2, pp. 17–29 (1948).

McPherson, M.B., Boundary influences on the fall velocity of spheres at Reynolds numbers beyond the Stokes range. MS Thesis, Univ. of Iowa, Ames (1947).

McWhirter, J.D., Crawford, M.E. and Klein, D.E., Wall region porosity distributions for packed beds of uniform spheres with modified and unmodified walls, *Trans. Porous Media*, **27**, 99 (1997).

Medani, M.S. and Hasan, M.A., Viscosity of organic liquids at elevated temperatures and the corresponding vapour pressures, *Can. J. Chem. Eng.*, **55**, 203 (1977).
Mehta, D. and Hawley, M.C., Wall effect in packed columns, *Ind. Eng. Chem. Proc. Des. Dev.*, **8**, 280 (1969).
Mei, R., Klausner, J.F. and Lawrence, C.J., A note on the history force on a spherical bubble at finite Reynolds numbers, *Phys. Fluids*, **6**, 418 (1994).
Mei, R.W., Xiong, J. and Tran Son Tay, R., Motion of a sphere oscillating at low Reynolds numbers in a viscoelastic-fluid-filled cylindrical tube, *J. Non-Newt. Fluid Mech.*, **66**, 169 (1996).
Meissner, D.L., Jeng, D.R. and DeWitt, K.J., Mixed convection to power-law fluids from two-dimensional or axisymmetric bodies, *Int. J. Heat Mass Transf.*, **37**, 1475 (1994).
Mena, B. and Caswell, B., Slow flow of an elastico-viscous fluid past an immersed body, *Chem. Eng. J.*, **8**, 125 (1974).
Mena, B., Manero, O. and Leal, L.G., The influence of rheological properties on the slow flow past spheres, *J. Non-Newt. Fluid Mech.*, **26**, 247 (1987).
Mendelson, H.D., The prediction of bubble terminal velocities from wave theory, *AIChE J.*, **13**, 250 (1967).
Mendes, P.R.S., Naccache, M.F., Braga, C.V.M., Nieckele, A.O. and Ribeiro, F.S., Flows of Bingham Materials through ideal porous media: An experimental and theoretical study, *J. Braz. Soc. Mech. Sci.*, **24**, 1 (2002).
Mendoza-Fuentes, A.J., Montiel, R., Zenit, R. and Manero, O., On the flow of associative polymers past a sphere: Evaluation of negative wake criteria, *Phys. Fluids*, **21**, 033104 (2009).
Merk, H.J. and Prins, J.A., Thermal convection in laminar boundary layers. I. *App. Sci. Res.*, **A4**, 11 (1953).
Merk, H.J. and Prins, J.A., Thermal convection in laminar boundary layers II. *App. Sci. Res.*, **A4**, 195 (1954).
Merkak, O., Jossic, L. and Magnin, A., Spheres and interactions between spheres moving at very low velocities in a yield stress fluid, *J. Non-Newt. Fluid Mech.*, **133**, 99 (2006).
Merkak, O., Jossic, L. and Magnin, A., Dynamics of particles suspended in a yield-stress fluid flowing in a pipe, *AIChE J.*, **54**, 1129 (2008).
Merkak, O., Jossic, L. and Magnin, A., Migration and sedimentation of spherical particles in a yield-stress fluid flowing in a horizontal cylindrical pipe, *AIChE J.*, **55**, 2515 (2009).
Merkin, J.H., The effect of buoyancy forces on the boundary-layer flow over a semi-infinite vertical flat plate in a uniform free stream, *J. Fluid Mech.*, **35**, 439 (1969).
Mertens, D., Heinen, C., Hardy, E.H. and Buggisch, H.W., Newtonian and non-Newtonian low Re number flow through bead packings, *Chem. Eng. Technol.*, **29**, 854 (2006).
Metzger, T.G., *The Rheology Handbook*, Vincentz Network, Hannover (2011).
Metzner, A.B., Non-Newtonian technology: Fluid mechanics, mixing and heat transfer, *Adv. Chem. Eng.*, **1**, 79 (1956).
Metzner, A.B., Heat transfer in non-Newtonian fluids, *Adv. Heat Transf.*, **2**, 357 (1965).
Metzner, A.B., Flow of polymeric solutions and emulsions through porous media-current status, *Improved Oil Recovery by Surfactant and Polymer Flooding*, edited by D.O. Shah and R.S. Schechter, Academic Press, New York, p. 439 (1977).
Metzner, A.B., Rheology of suspensions in polymeric liquids, *J. Rheol.*, **29**, 739 (1985).
Metzner, A.B. and Astarita, G., External flows of viscoelastic materials: Fluid property restrictions on the use of velocity-sensitive probes, *AIChE J.*, **13**, 550 (1967).
Metzner, A.B. and Reed, J.C., Flow of non-Newtonian fluids-correlation of laminar, transition and turbulent flow regime, *AIChE J.*, **1**, 434 (1955).
Metzner, A.B., White, J.L. and Denn, M.M., Constitutive equations for viscoelastic fluids for short deformation periods, and for rapidly changing flows: Significance of the Deborah number, *AIChEs J.*, **12**, 863 (1966).
Metzner, A.B. and Whitlock, M., Flow behaviour of concentrated (dilatant) suspensions, *Trans. Soc. Rheol.*, **2**, 239 (1958).
Mewis, J., Thixotropy – A general view, *J. Non-Newt. Fluid Mech.*, **6**, 1 (1979).
Mewis, J. and Wagner, N.J., Thixotropy, *Adv. Colloid Interface Sci.*, **147–148**, 214 (2009).
Mewis, J. and Wagner, N.J., *Colloidal Suspension Rheology*, Cambridge University Press, New York (2012).
Meyer, B.A. and Smith, D.W., Flow through porous media: Comparison of consolidated and unconsolidated materials, *Ind. Eng. Chem. Fund.*, **24**, 360 (1985).
Mhatre, M.V. and Kintner, R.C., Fall of liquid drops through pseudoplastic liquids, *Ind. Eng. Chem.*, **51**, 865 (1959).
Michael, P., Steady motion of a disk in a viscous fluid, *Phys. Fluids*, **9**, 466 (1966).

Michaelides, E.E., Review – The transient equation of motion for particles, bubbles, and droplets, *J. Fluids Eng.*, **119**, 233 (1997).

Michaelides, E.E., Analytical expressions for the motion of particles, *Transport Processes in Bubbles, Drops and Particles*, edited by D. De Kee and R.P. Chhabra, 2nd ed., Chapter 2, Taylor and Francis, New York (2002).

Michaelides, E.E., Hydrodynamic force and heat/mass transfer from particles, bubbles and drops – The Freeman Scholar Lecture, *J. Fluids Eng.* **125**, 209 (2003).

Michaelides, E.E., *Particles, Bubbles, and Drops: Their Motion, Heat and Mass Transfer*, World Scientific, Singapore (2006).

Michaelides, E. E. and Feng, Z.-G., Review- Drag coefficients of non-spherical and irregular shaped particles, *J. Fluids Eng.*, **145**, 0608011 (2023).

Michaelides, E.E. and Roig, A., A reinterpretation of the Odar and Hamilton data on the unsteady equation of motion of particles, *AIChE J.*, **57**, 2997 (2011).

Michaud, V. and Månson, J.-A.E., Impregnation of compressible fiber mats with a thermoplastic resin. Part I: Theory, *J. Comp. Mat.*, **35**, 1150 (2001). Also see *ibid* 1174.

Michaud, V. and Mortensen, A., Infiltration processing of fibre reinforced composites: Governing phenomena, *Compos. Part A*, **32**, 981 (2001).

Michele, H., Zur durchflubcharakteristik von schuttungen bei der durchstromung mit verdunnten losungen aus longkettigln hochpolymeren, *Rheol. Acta*, **16**, 413 (1977).

Michele, J., Patzold, R. and Donis, R., Alignment and aggregation effects in suspensions of spheres in non-Newtonian media, *Rheol. Acta*, **16**, 317 (1977).

Mickley, H.S., Smith, K.A. and Korchak, E.I., Fluid flow in packed beds, *Chem. Eng. Sci.*, **20**, 237 (1965).

Mikelić, A., Non-Newtonian flow, edited by U. Hornung, *Homogenization and Porous Media*, pp. 77–94. Springer, New York (1997).

Milewski, J.V., The combined packing of rods and spheres in reinforcing plastics. *Ind. Eng. Chem. Prod. Res. Dev.*, **17**, 363 (1978).

Miller, C., Predicting non-Newtonian flow behavior in ducts of unusual cross-sections, *Ind. Eng. Chem. Fund.*, **11**, 524 (1972).

Miller, S.C. and Drabik, B.R., Rheological properties of poloxamer vehicles, *Int. J. Pharm.*, **18**, 269 (1984).

Milliken, W.J., Gottlieb, M., Graham, A.L., Mondy, L.A. and Powell, R.L., The viscosity-volume fraction relation for suspensions of rod-like particles by falling ball rheometry, *J. Fluid Mech.*, **202**, 217 (1989a).

Milliken, W.J., Mondy, L.A., Gottlieb, M., Graham, A.L. and Powell, R.L., The effect of the diameter of falling balls on the apparent viscosity of suspensions of spheres and rods, *Physico-Chem. Hydrodynam.*, **11**, 341 (1989b).

Milliken, W.J., Stone, H.A. and Leal, L.G., The effect of surfactant on the transient motion of Newtonian drops, *Phys. Fluids*, **A5**, 69 (1993).

Minaeian, A., Nili-Ahmadabadi, M., Norouzi, M. and Kim, K.C., Effects of viscoelasticity on the onset of vortex shedding and forces applied on a cylinder in unsteady flow regime, *Phys. Fluids*, **34**, 013106 (2022).

Mirzaagha, S., Pasquino, R., Iuliano, E., D'Avino, G., Zonfrilli, F., Guida, V. and Grizzutti, N., The rising motion of spheres in structured fluids with yield stress, *Phys. Fluids*, **29**, 093101 (2017).

Mishra, G. and Chhabra, R.P., Influence of flow pulsations and yield stress on heat transfer from a sphere. *Appl. Math. Model.*, **90**, 1069–1098 (2021).

Mishra, G. and Chhabra, R.P., Effect of power-law index and shape on the onset of flow separation, *J. Chem. Eng. Jpn.*, **55**, 132 (2022).

Mishra, G. and Chhabra, R.P., Oscillating flow of power-law fluids over a sphere: Drag and Nusselt number behavior, *Chem. Eng. Tech.*, **45**, 2254 (2022).

Mishra, G., Patel, S.A. and Chhabra, R.P., Pulsatile flow of power-law fluids over a sphere: Momentum and heat transfer characteristics, *Powder Technol.*, **360**, 789–817 (2020).

Mishra, I.M., Singh, B. and Mishra, P., Non-Newtonian boundary layer flow and mass transfer – Laminar flow, *Indian J. Technol.*, **14**, 322 (1976). Also see *ibid*, 375.

Mishra, P. and Farid, M.M., Non-Newtonian flow through porous media, *Ind. Chem. Eng.*, **25**, 36 (1983).

Mishra, P., Gupta, S. and Chhabra, R.P., Effect of orientation on drag characteristics of a cone settling in power-law fluids, *J. Chem. Eng. Jpn.*, **52**, 19 (2019a).

Mishra, P., Nirmalkar, N. and Chhabra, R.P., Free convection from a heated vertical cone in generalized Newtonian fluids, *J. Thermophys. Heat Transf.*, **33**, 932 (2019c).

Mishra, P., Patel, S.A., Trivedi, M. and Chhabra, R.P., Effect of power-law fluid behavior on Nusselt number of a circular disk in the forced convection regime, *J. Heat Transf.*, **141**, 041701 (2019d).

Mishra, P., Singh, D. and Mishra, I.M., Momentum transfer to Newtonian and non-Newtonian fluids flowing through packed and fluidized beds, *Chem. Eng. Sci.*, **30**, 397 (1975). Corrections in *Chem. Eng. Sci.*, **37**, 801 (1982). Also see, D. Singh et al., *Ind. J. Tech.*, **14**, 591 (1976).

Mishra, P., Tiwari, A.K. and Chhabra, R.P., Effect of orientation on forced convection heat transfer from a heated cone in Bingham plastic fluids, *Int. Comm. Heat Mass Transf.*, **93**, 34 (2018).

Mishra, P., Tiwari, A.K. and Chhabra, R.P., Effect of orientation on drag of a cone settling in Bingham plastic fluids, *Particuology*, **43**, 157 (2019b).

Mishra, S.P., Free convection flow of an elastico-viscous liquid past a hot vertical plate, *Indian Chem. Eng.*, **8**, 28 (1966a).

Mishra, S.P., Free convection flow of a second order fluid past a hot vertical plate, *Proc. Ind. Acad. Sci.*, **64A**, 291 (1966b).

Missirlis, K.A., Assimacopoulos, D., Mitsoulis, E. and Chhabra, R.P., Wall effects for motion of spheres in power-law fluids, *J. Non-Newt. Fluid Mech.*, **96**, 459 (2001).

Mitishita, R.S., Oliveira, G.M., Santos, T.G.M. and Negrao, C.O.R., Pressure transmission in yield stress fluids - An experimental analysis, *J. Non-Newt. Fluid Mech.*, **261**, 50 (2018).

Mitsoulis, E., Effect of rheological properties on the drag coefficient for creeping motion around a sphere falling in a tightly-fitting tube, *J. Non-Newt. Fluid Mech.*, **74**, 263 (1998a).

Mitsoulis, E., Numerical simulation of confined flow of polyethylene melts around a cylinder in a planar channel, *J. Non-Newt. Fluid Mech.*, **76**, 327 (1998b).

Mitsoulis, E., On creeping drag flow of a viscoplastic fluid past a circular cylinder: Wall effects, *Chem. Eng. Sci.*, **59**, 789 (2004).

Mitsoulis, E., Atsbha, H. and Kiriakidis, D.G., Flow of viscoelastic liquids around spheres, *Proc. Int. Sym. on Advances in Structured and Heterogeneous Continua*, August 22–26, Moscow, Russia (1993).

Mitsoulis, E. and Tsamopoulos, J., Numerical simulations of complex yield-stress fluid flows, *Rheol. Acta*, **56**, 231 (2017).

Mitsuishi, N., Yamanaka, A. and Miyahara, F., A study on the drag coefficient of the spheres falling in the non-Newtonian fluids, *Tech. Rep. Kyushu Univ.*, **44**, 192 (1971).

Mitsuishi, N., Yamanaka, A. and Sueyasu, Y., Drag force on a moving bubble and droplet in viscoelastic fluids, *Proc. PACHEC*, 316 (1972).

Mittal, R., A Fourier-Chebyshev spectral collocation method for simulating flow past spheres and spheroids, *Int. J. Num. Meth. Fluids*, **30**, 921 (1999).

Mittal, R., Seshadri, V. and Udaykumar, H.S., Flutter, tumble and vortex induced autorotation, *Theor. Comput. Fluid Dyn.*, **17**, 165 (2004).

Mitwally, E.M., The laminar far wake flow of a non-Newtonian power-law fluid, *J. Fluids Eng. (ASME)*, **101**, 331 (1979).

Mityushev, V. and Adler, P.M., Longitudinal permeability of spatially periodic rectangular arrays of circular cylinders I. A single cylinder in the unit cell, *ZAMM*, **82**, 335 (2002).

Miura, H. and Kawase, Y., Hydrodynamics and mass transfer in three-phase fluidized beds with non-Newtonian fluids, *Chem. Eng. Sci.*, **52**, 4095 (1997).

Miura, H. and Kawase, Y., Minimum liquid fluidization velocity in two-and three-phase fluidized beds with non-Newtonian fluids, *Powder Technol.*, **97**, 124 (1998).

Miura, H., Takahashi, T., Ichikawa, J. and Kawase, Y., Bed expansion in liquid-solid two-phase fluidized beds with Newtonian and non-Newtonian fluids over wide range of Reynolds number, *Powder Technol.*, **117**, 239 (2001a).

Miura, H., Takahashi, T. and Kawase, Y., Effect of pseudoplastic behaviour of liquid in co-current three-phase fluidized beds with non-Newtonian fluids, *Powder Technol.*, **97**, 124 (1998).

Miura, H., Takahashi, T. and Kawase, Y., Effect of pseudoplastic behaviour of liquid in co-current three-phase fluidized beds on bed expansion, *Chem. Eng. Sci.*, **56**, 6047 (2001b).

Mixon, F.O. and Carberry, J.J., Diffusion within a developing boundary layer – A mathematical solution for arbitrary velocity distribution, *Chem. Eng. Sci.*, **13**, 30 (1960).

Miyahara, T. and Hayashino, T., Size of bubbles generated from perforated plates in non-Newtonian liquids, *J. Chem. Eng. Jpn.*, **28**, 596 (1995).

Miyahara, T. and Yamanaka, S., Mechanics of motion and deformation of a single bubble rising through quiescent highly viscous Newtonian and non-Newtonian media, *J. Chem. Eng. Jpn.*, **26**, 297 (1993).

Miyahara, T., Wang, W.-H. and Takahashi, T., Bubble formation at a submerged orifice in non-Newtonian and highly viscous Newtonian liquids, *J. Chem. Eng. Jpn.*, **21**, 620 (1988).

Miyamura, A., Iwasaki, S. and Ishii, T., Experimental wall correction factors of single solid spheres in triangular and square cylinders, and parallel plates, *Int. J. Multiph. Flow*, **7**, 41 (1981).

Mizushina, T. and Usui, H., Transport phenomena of viscoelastic fluid in cross flow around a circular cylinder, *J. Chem. Eng. Jpn.*, **8**, 393 (1975).

Mizushina, T. and Usui, H., Approximate solution of the boundary layer equations for the flow of a non-Newtonian fluid around a circular cylinder, *Heat Transf. Jap. Res.*, **7**(2), 83 (1978).

Mizushina, T., Usui, H., Veno, K. and Kato, T., Experiments of pseudoplastic fluid cross flow around a circular cylinder, *Heat Transf. Jap. Res.*, **7**(2), 92 (1978).

Mo, G.B. and Sangani, A.S., A method for computing Stokes flow interactions among spherical objects and its application to suspensions of drops and porous particles, *Phys. Fluids*, **6A**, 1637 (1994).

Mobius, D. and Miller, R., *Drops and Bubbles in Interfacial Research*, Elsevier, Amsterdam (1998).

Mochimaru, Y. and Tomita, Y., Steady wake flow in dilute polymer solutions, *Bull JSME*, **21**, 1364 (1978).

Modi, V.J. and Akutsu, T., Wall confinement effects for spheres in the Reynolds number range of 30–2000, *J. Fluids Eng.*, **106**, 66 (1984).

Mograbi, E. and Bar-Ziv, E., On the mixed convection hydrodynamics force on a sphere, *Aerosol Sci.*, **36**, 1177 (2005a).

Mograbi, E. and Bar-Ziv, E., Dynamics of a spherical particle in mixed convection flow field, *J.Aerosol Sci.*, **36**, 387 (2005b).

Mograbi, E., Ziskind, G., Katoshevski, D. and Bar-Ziv, E., Experimental study of the forces associated with mixed convection from a heated sphere at small Reynolds and Grashof numbers. Part II: Assisting and opposing flows, *Int. J. Heat Mass Transf.*, **45**, 2423 (2002).

Mohammadigoushki, H. and Muller, S.J., Sedimentation of a sphere in wormlike micellar fluids. *J. Rheol.*, **60**, 587 (2016).

Mohan, V., *PhD Dissertation*, Indian Institute of Technology, Madras, India (1974a).

Mohan, V., Creeping flow of a power law fluid over a Newtonian fluid sphere, *AIChE J.*, **20**, 180 (1974b).

Mohan, V., Nagarajan, R. and Venkateswarlu, D., Fall of drops in non-Newtonian media, *Can. J. Chem. Eng.*, **50**, 37 (1972).

Mohan, V. and Raghuraman, J., A theoretical study of pressure drop for non-Newtonian creeping flow past an assemblage of spheres, *AIChE J.*, **22**, 259 (1976a).

Mohan, V. and Raghuraman, J., Bounds on the drag for creeping flow of an Ellis fluid past an assemblage of spheres, *Int. J. Multiph. Flow*, **2**, 581 (1976b).

Mohan, V. and Raghuraman, J., Viscous flow of an Ellis fluid past a Newtonian fluid sphere, *Can. J. Chem. Eng.*, **54**, 228 (1976c).

Mohan, V. and Venkateswarlu, D., Lower bound on the drag offered to a Newtonian fluid sphere placed in a flowing Ellis Fluid, *J. Chem. Eng. Jpn.*, **7**, 243 (1974).

Mohan, V. and Venkateswarlu, D., Creeping flow of a power law fluid past a fluid sphere, *Int. J. Multiph. Flow*, **2**, 563 (1976). Also see *ibid*, 571.

Molerus, O., A coherent representation of pressure drop in fixed beds and of bed expansion for particulate fluidized beds, *Chem. Eng. Sci.*, **35**, 1331 (1980).

Mollinger, A.M., Cornelissen, E.C. and van den Brule, B.H.A.A., An unexpected phenomenon observed in particle settling: Oscillating falling spheres, *J. Non-Newt. Fluid Mech.*, **86**, 389 (1999).

Montemagno, C.D. and Gray, W.G., Photoluminescent volumetric imaging: A technique for the exploration of multiphase flow and transport in porous media, *Geophys. Res. Lett.*, **22**, 425 (1995).

Montillet, A., Flow through a finite packed bed of spheres: A note on the limit of applicability of the Forchheimer-type equation, *J. Fluids Eng.*, **126**, 139 (2004).

Moon, S.H., Johnson, A.T. and Shih, T.M., Numerical analysis of mixed convection from horizontal cylinders, *J. Agric. Eng. Res.*, **38**, 289 (1987).

Moore, D.W., The rise of a gas bubble in a viscous liquid, *J. Fluid Mech.*, **6**, 113 (1959a).

Moore, D.W., The boundary layer on spherical gas bubbles, *J. Fluid Mech.*, **16**, 161 (1963).

Moore, D.W., The velocity of rise of distorted gas bubbles in a liquid of small viscosity, *J. Fluid Mech.*, **23**, 749 (1965).

Moore, F., The rheology of ceramic slips and bodies, *Trans. Brit. Ceramic Soc.*, **58**, 470 (1959b).

Moo-Young, M. and Hirose, T., Bubble mass transfer in creeping flow of viscoelastic fluids, *Can. J. Chem. Eng.*, **50**, 128 (1972a).

Moo-Young, M. and Hirose, T., On mass transfer from bubbles in non-Newtonian fluids at low Reynolds numbers. An appraisal of the thin-boundary-layer approximation, *Ind. Eng. Chem. Fundam.*, **11**, 281 (1972b).

Moo-Young, M., Hirose, T. and Ali, S., Rheological effects on liquid phase mass transfer in two phase dispersions: Results for creeping flow, *Proc. 5th Int. Cong. Rheol.*, Kyoto, 233 (1970).

Mora, S., Talini, L. and Allain, C., Structuring sedimentation in a shear-thinning fluid, *Phys. Rev. Lett.*, **95**, 088301 (2005).

Morais, A.F., Seybold, H., Herrmann, H.J. and Andrade Jr., J.S., Non-Newtonian fluid flow through three-dimensional disordered porous media. *Phys. Rev. Lett.*, **103**(19), 194502 (2009).

Moreno, R.A., Muller, A.J. and Saez, A.E., Flow-induced degradation of hydrolyzed polyacrylamide in porous media, *Polym. Bull.*, **37**, 663 (1996).

Morgan, V.T., The overall convective heat transfer from smooth circular cylinders, *Adv. Heat Transf.*, **11**, 199 (1975).

Morris, J.F., Toward a fluid mechanics of suspensions, *Phys. Rev. Fluids*, **5**, 110519 (2020).

Morris, S., The effects of a strongly temperature dependent viscosity on slow flow past a hot sphere, *J. Fluid Mech.*, **124**, 1 (1982).

Morrison, F.A., *Understanding Rheology*, Oxford University Press, New York (2001).

Morusic-Paloka, E., On the Stokes paradox for power law flow around a cylinder, *ZAMM.*, **81**, 31 (2001).

Mosely, K., Fairweather, M. and Harbottle, D., Settling dynamics of two identical vertically aligned spheres in a thixotropic fluid, *J. Non-Newt. Fluid Mech.*, **271**, 104146 (2019).

Moshev, V.V., *Rheological Behavior of Concentrated Non-Newtonian Suspensions*, Nauka, Moscow (1989). (in Russian).

Moss, G.R. and Rothstein, J.P., Flow of wormlike micelle solutions through a periodic array of cylinders, *J. Non-Newt. Fluid Mech.*, **165**(1–2), 1 (2010).

Mossaz, S., Jay, P. and Magnin, A., Criteria for the appearance of recirculating and non-stationary regimes behind a cylinder in a viscoplastic fluid, *J. Non-Newt. Fluid Mech.*, **165**, 1525 (2010).

Mossaz, S., Jay, P. and Magnin, A., Non-recirculating and recirculating inertial flows of a viscoplastic fluid around a cylinder, *J. Non-Newt. Fluid Mech.*, **177–178**, 64 (2012a).

Mossaz, S., Jay, P. and Magnin, A., Experimental study of stationary inertial flows of a yield stress fluid around a cylinder, *J. Non-Newt. Fluid Mech.*, **189**, 40 (2012b).

Mott, R.A., *Some Aspects of Fluid Flow*, Arnold, London, p. 242 (1951).

Mougin, N., Magnin, A. and Piau, J.-M., The significant influence of internal stresses on the dynamics of bubbles in a yield stress fluid, *J. Non-Newt. Fluid Mech.*, **171–172**, 42 (2012).

Moulic, S.G. and Yao, L.-S., Non-Newtonian natural convection along a vertical flat plate with uniform surface temperature, *J. Heat Transf.*, **131**, 062501 (2009).

Mrokowska, M.M. and Krzton-Maziopa, A., Viscoelastic and shear-thinning effects of aqueous exopolymer solutions on disk and sphere settling, *Nature Sci. Rep.*, **9**, 7897 (2019).

Mueller, G.E., Prediction of radial porosity distributions in randomly packed fixed beds of uniformly sized spheres, *Chem. Eng. Sci.*, **46**, 706 (1991).

Mueller, G.E., Radial void fraction distributions in randomly packed fixed beds of uniformly sized spheres in cylindrical containers, *Powder Technol.*, **72**, 269 (1992).

Mueller, G.E., Numerical simulation of packed beds with monosized spheres in cylindrical containers, *Powder Technol*, **92**, 179 (1997).

Mueller, G.E., Radial void fraction correlation for annular packed beds, *AIChE J.*, **45**, 2458 (1999).

Mueller, G.E., Numerically packing spheres in cylinders, *Powder Technol.*, **159**, 105 (2005).

Mueller, G.E., A modified packed bed radial porosity correlation, *Powder Technol.*, **342**, 607 (2019).

Mujumdar, A., Beris, A.N. and Metzner, A.B., Transient phenomena in thixotropic systems, *J. Non-Newt. Fluid Mech.*, **102**, 157 (2002).

Mukherjee, S. and Sarkar, K., Viscoelastic drop falling through a viscous medium, *Phys. Fluids*, **23**, 013101 (2011).

Mukherjee, S. and Sarkar, K., Lateral migration of a viscoelastic drop in a Newtonian fluid in a shear flow near a wall, *Phys. Fluids*, **26**, 103102 (2014).

Mukundakrishnan, K., Eckmann, D.M. and Ayyaswamy, P.S., Bubble motion through a generalized power-law fluid flowing in a vertical tube, *Ann. N. Y. Acad. Sci.*, **1161**(1), 256 (2009).

Mukundakrishnan, K., Quan, S., Eckmann, D.M. and Ayyaswamy, P.S., Numerical study of wall effects on buoyant gas-bubble rise in liquid-filled finite cylinder, *Phys. Rev. E*, **76**, 036308 (2007).

Mulhem, B., Fritsching, U., Schulte, G. and Bauckhage, K., Effect of solid particle characteristics on suspension atomization, *At. Sprays*, **13**, 321 (2003).

Müller, A.J., Medina, L.I., Perez-Martin, O., Rodriguez, S., Romero, C., Sargenti, M.L. and Saez, A.Z., Flowing polymers through porous media: An experimental study of flow distribution, polymer degradation, and molecular weight effects, *App. Mech. Rev.*, **46**, S63 (1993).

Müller, A.J., Odell, J.A. and Keller, A., Elongational flow and rheology of monodisperse polymers in solution, *J. Non-Newt. Fluid Mech.*, **30**, 99 (1988).

Müller, A.J., Odell, J.A. and Keller, A., Polymer degradation in extensional flow, *Polym. Commun.*, **30**, 298 (1989).

Müller, A.J., Patruyo, L.G., Montano, W., Roversi-M, D., Moreno, R., Ramirez, N.E. and Saez, A.E., Mechanical degradation of polymers in flows through porous media: Effect of flow path length and particle size, *App. Mech. Rev.*, **50**, S149 (1997).

Müller, A.J. and Saez, A.E., The rheology of polymer solutions in porous media, *Flexible Polymer Chain Dynamics in Elongational Flow*, edited by T.Q. Nguyen and H.-H. Kausch, Springer, New York, Chapter 11 (1999).

Müller, M., Vorwerk, J. and Brunn, P.O., Optical studies of local flow behaviour of a non-Newtonian fluid inside a porous medium, *Rheol. Acta*, **37**, 189 (1998).

Müller, M.V. and Brunn, P.O., Auslegung der strömung in Schüttungen bei nicht-newtonschen flüssigkeiten, *Chem. Ing. Tech.*, **69**, 1636 (1997).

Müller, M.V. and Brunn, P.O., Design strategies for flow of non-Newtonian fluids through packed beds, *Chem. Eng. Technol.*, **22**, 110 (1999).

Mun, R.P., Byars, J.A. and Boger, D.V., The effects of polymer concentration and molecular weight on the breakup of laminar capillary jets, *J. Non-Newt. Fluid Mech.*, **74**, 285 (1998).

Mungan, N., Rheology and adsorption of aqueous polymer solutions, *J. Can. Pet. Tech.*, **8** (2), 45 (1969).

Mungan, N., Smith, F.W. and Thompson, J.L., Some aspects of polymer floods, *J. Pet. Tech.*, **18**, 1143 (1966).

Munroe, H.S., The English versus the continental system of jigging – Is close sizing advantageous? *Trans. AIMME*, **17**, 637 (1888/89).

Munshi, B., Chhabra, R.P. and Ghoshdastidar, P.S., A numerical study of steady incompressible Newtonian fluid flow over a disk at moderate Reynolds numbers, *Can. J. Chem. Eng.*, **77**, 113 (1999).

Munzel, K. and Schaub, K., Die bestimmung der viskositats – kon zeutrationskonstauten bzw. der steigungskonstauten k von schleimlosungem, eine einfache methode zur beurteilung der qualitat von schleimstoffen, *Pharma. Acta Helvetiae*, **36**, 647 (1961).

Murali Krishnan, J. and Rajagopal, K.R., Review of the uses and modelling of bitumen from ancient to modern times, *App. Mech. Rev.*, **56**(2), 149 (2003).

Muralidhar, R., Ramkrishna, D., Das, P.K. and Kumar, R., Coalescence of rigid droplets in stirred dispersions II. Band limit force fluctuations, *Chem. Eng. Sci.*, **43**, 1559 (1988).

Muratsune, K. and Ishii, T., Wall effect on velocity of a single solid sphere falling in CMC aqueous solutions, Proceedings of 15th Autumn Meeting of the Society of Chemical Engineers of Japan, Kanazawa, F301, p. 161 (1981).

Muroyama, K. and Fan, L.S., Fundamentals of gas-liquid-solid fluidization, *AIChE J.*, **31**, 1 (1985).

Mutlu, I., Townsend, P. and Webster, M.F., Adaptive solutions for viscoelastic flows, *Commun. Numer. Meth. Eng.*, **12**, 643 (1996).

Myers, T.G., An approximate solution method for boundary layer flow of a power-law fluid over a flat plate, *Int. J. Heat Mass Transf.*, **53**, 2337 (2010).

N' Gouamba, E., Goyon, J. and Coussot, P., Elastoplastic behaviour of yield stress fluids, *Phys. Rev. Fluids*, **4**, 123301 (2019).

Na, T.Y., Boundary layer flow of Reiner-Philippoff fluids, *Int. J. Non-Linear Mech.*, **29**, 871 (1994).

Na, T.Y. and Hansen, A.G., Possible similarity solutions of the laminar natural convection flow of non-Newtonian fluids, *Int. J. Heat Mass Transf.*, **9**, 261 (1966).

Nachman, A. and Taliaferro, S., Mass transfer into boundary layers for power law fluids, *Proc. Roy. Soc. Lond. A.*, **365**, 313 (1979).

Nagelhout, D., Bhat, M.S., Heinrich, J.C. and Poirier, D.R., Permeability for flow normal to a sparse array of fibers, *Mat. Sci. Eng.*, **A191**, 203 (1995).

Nakamura, I., Steady wake behind a sphere, *Phys. Fluids*, **19**, 5 (1976).
Nakano, Y. and Tien, C., Creeping flow of power-law fluid over a Newtonian fluid sphere, *AIChE J.*, **14**, 145 (1968).
Nakano, Y. and Tien, C., Viscous incompressible non-Newtonian flow around a fluid sphere at intermediate Reynolds numbers, *AIChE J.*, **16**, 569 (1970).
Nakayama, A., Integral methods for forced convection heat transfer in power law non-Newtonian fluids, *Encycl. Fluid Mech.*, **7**, 305 (1988).
Nakayama, A. and Koyama, H., An analysis for friction and heat transfer characteristics of power law non-Newtonian fluid flows past bodies of arbitrary geometrical configuration, *Warme-und Stoffubertragung*, **22**, 29 (1988).
Nakayama, A., Kuwahara, F., Kawamura, Y. and Koyama, H., Three-dimensional numerical simulation of flow through a microscopic porous structure, *Proc. ASME/JSME Thermal Eng. Conf.*, **3**, 313 (1995).
Nakayama, A. and Shenoy, A.V., Turbulent free convection heat transfer to power law fluids from arbitrary geometric configurations, *Int. J. Heat Fluid Flow.*, **12**, 336 (1991).
Nakayama, A. and Shenoy, A.V., Turbulent free convection heat transfer to drag-reducing fluids from arbitrary geometric configurations, *J. Heat Transf.*, **114**, 127 (1992a).
Nakayama, A. and Shenoy, A.V., A unified similarity transformation for Darcy and non-Darcy forced-, free-and mixed-convection heat transfer in non-Newtonian inelastic fluid-saturated porous media, *Chem. Eng. J.*, **50**, 33 (1992b).
Nakayama, A. and Shenoy, A.V., Non-Darcy forced-convection heat transfer in a channel embedded in a non-Newtonian inelastic-fluid-saturated porous medium, *Can. J. Chem. Eng.*, **71**, 168 (1993a).
Nakayama, A. and Shenoy, A.V., Combined forced and free convection heat transfer in power-law fluid-saturated porous media, *App. Sci. Res.*, **50**, 83 (1993b).
Nakayama, A., Shenoy, A.V. and Koyama, H., An analysis for forced convection heat transfer from external surfaces to non-Newtonian fluids, *Warme-und Stoffubertragung*, **20**, 219 (1986).
Nalajala, V.S., Kishore, N. and Chhabra, R.P., Effect of contamination on the rise velocity of bubble swarms at moderate Reynolds numbers, *Chem. Eng. Res. Des.*, **92**, 1016 (2014).
Nalluri, S.V., Patel, S.A. and Chhabra, R.P., Mixed convection from a hemisphere in Bingham plastic fluids, *Int. J. Heat Mass Transf.*, **84**, 304 (2015).
Narain, J.P., Free and forced convective heat transfer from slender cylinders, *Lett. Heat Mass Transf.*, **3**, 21 (1976).
Narayanan, S., Goossens, L.H.J. and Kossen, N.W.F., Coalescence of two bubbles rising in line at low Reynolds number, *Chem. Eng. Sci.*, **29**, 2071 (1974).
Narh, K.A., Odell, J.A., Müller, A.J. and Keller, A., Polymer solution degradation: The combined effects of flow and temperature, *Polym. Commun.*, **31**, 2 (1990).
Nash, S. and Rees, D.A.S., The effect of microstructure on models for the flow of a Bingham fluid in porous media: One-dimensional flows. *Trans. Porous Media*, **116**(3), 1073 (2017).
Nassar, M.M., Fadalli, O.A. and Sedahmed, G.H., Effect of drag reducing polymer on the rate of cementation of copper from dilute solutions on rotating zinc cylinder, *Z. Met.* **80**, 60 (1989).
Nath, D., Pati, S. and Raju, B.H.S., Analysis of mixed convection past a heated sphere, *Proc. I Mech E, Part E: J. Process Mech. Eng.*, **233**, 601 (2019).
Naudascher, E. and Killen, J.M., Onset and saturation limit of polymer effects in porous media flows, *Phys. Fluids*, **20**(10), S280 (1977).
Navez, V. and Walters, K., A note on settling in shear-thinning polymer solutions, *J. Non-Newt. Fluid Mech.*, **67**, 325 (1996).
Navon, O., Chekhmir, A. and Lyakhovsky, V., Bubble growth in highly viscous melts: Theory, experiments, and auto explosivity of dome lavas, *Earth Plan. Sci. Lett.*, **160**, 763 (1998).
Nazar, R., Amin, N. and Pop, I., On the mixed convection boundary-layer flow about a solid sphere with constant surface temperature, *Arab. J. Sci. Eng.*, **27**, 117 (2002).
Neale, G. and Masliyah, J.H., Flow perpendicular to mats of randomly arranged cylindrical fibers (Importance of cell models), *AIChE J.*, **21**, 805 (1975).
Nemec, D. and Levec, J., Flow through packed bed reactors: 1. Single-phase flow, *Chem. Eng. Sci.*, **60**, 6947 (2005).
Newton, I., *Principia*. Book II, Prop. XXXIX, Theor. XXXI (1687).

Ng, M.L. and Hartnett, J.P., Natural convection in power-law fluids, *Int. Comm. Heat Mass Transf.*, **13**, 115 (1986).
Ng, M.L. and Hartnett, J.P., Free convection heat transfer from horizontal wires to pseudoplastic fluids, *Int. J. Heat Mass Transf.*, **31**, 441 (1988).
Ng, M.L., Hartnett, J.P. and Hu, R.Y.Z., Natural convection from horizontal wires – The conduction limit, *Int. Comm. Heat Mass Transf.*, **15**, 293 (1988).
Ng, M.L., Hartnett, J.P. and Kwack, E.Y., Natural convection from horizontal wires to viscoelastic fluids, *J. Heat Transf.*, **108**, 790 (1986).
Nguyen, A.V., Prediction of bubble terminal velocities in contaminated water, *AIChE J.*, **44**, 226 (1998).
Nguyen, H. and Boger, D.V., A model viscoelastic fluid, *Polym. Eng. Sci.*, **18**, 1037 (1978).
Nguyen, H.D., Paik, S. and Chung, J.N., Unsteady mixed convection heat transfer from a solid sphere: The conjugate problem, *Int. J. Heat Mass Transf.*, **36**, 4443 (1993).
Nguyen, N.L., van Buren, V., Reimert, R. and von Garnier, A., Determination of porosity and flow distribution in packed beds by magnetic resonance imaging, *Mag. Res. Imaging*, **23**, 395 (2005).
Nguyen, Q.D., Akroyd, T., De Kee, D. and Zhu, L., Yield stress measurements in suspensions: An inter-laboratory study, *Korea-Australia Rheol. J.*, **18**, 15 (2006).
Nguyen, Q.D. and Boger, D.V., Yield stress measurement for concentrated suspensions, *J. Rheol.*, **27**, 321 (1983).
Nguyen, Q.D. and Boger, D.V., Direct yield stress measurement with the vane method, *J. Rheol.*, **29**, 335 (1985).
Nguyen, Q.D. and Boger, D.V., Measuring the flow properties of yield stress fluids, *Annu. Rev. Fluid Mech.*, **24**, 47 (1992).
Nguyen, Q.D. and Uhlherr, P.H.T., Thixotropic behaviour of concentrated red mud suspensions, *Proc. 3rd Nat. Conf. Rheol.*, Uhlherr, P.H.T., ed., Brit. Soc. Rheol. (Victorian Branch), Melbourne, p. 63 (1983).
Nguyen-Chung, T., Plichta, C. and Mennig, G., Flow disturbance in polymer melt behind an obstacle, *Rheol. Acta*, **37**, 299 (1998).
Nieckele, A.O., Naccache, M.F. and de Souza Mendes, P.R., Cross flow of viscoplastic materials through tube bundles, *J. Non-Newt. Fluid Mech.*, **75**, 43 (1998).
Nield, D.A., Alternative model for wall effect in laminar flow of a fluid through a packed column, *AIChE J.*, **29**, 688 (1983).
Nield, D.A. and Bejan, A., *Convection in Porous Media*, 2nd ed., Springer-Verlag, New York (1995).
Nielsen, D.R. and Pitchumani, R., Control of flow in resin transfer molding with real-time preform permeability estimation, *Polym. Compos.*, **23**, 1087 (2002).
Niethammer, M., Brenn, G., Marschall, H. and Bothe, D., An extended volume of fluid method and its application to single bubbles rising in a viscoelastic fluid, *J. Comp. Phys.*, **387**, 326 (2019).
Nilsson, M.A., Kulkarni, R., Gerberich, L., Hammond, R., Singh, R., Baumhoff, E. and Rothstein, J.P., Effect of fluid rheology on enhanced oil recovery in a microfluidic sandstone device. *J. Non-Newt. Fluid Mech.*, **202**, 112 (2013).
Niranjan, K., An introduction to bubble mechanics in foods, *Bubbles in Foods*, edited by G.M. Campbell, C. Webb, S.S. Pandiella and K. Niranjan, Eagan Press, St. Paul, MN, Chapter 1 (1999).
Nirmalkar, N. and Chhabra, R.P., Forced convection in power-law fluids from an asymmetrically confined heated cylinder, *Int. J. Heat Mass Transf.*, **55**, 235 (2012).
Nirmalkar, N. and Chhabra, R.P., Momentum and heat transfer from a heated circular cylinder in Bingham plastic fluids, *Int. J. Heat Mass Transf.*, **70**, 564 (2014).
Nirmalkar, N. and Chhabra, R.P., Mixed convection from a heated sphere in power-law fluids, *Chem. Eng. Sci.*, **89**, 49 (2013). Also see *ibid* **140**, 359 (2016).
Nirmalkar, N., Alam, M.J. and Gupta, A.K., Stability criteria and convective mass transfer from the falling spherical drops, Part1: Bingham plastic fluids, *Can. J. Chem. Eng.* (2021).
Nirmalkar, N., Bose, A. and Chhabra, R.P., Mixed convection from a heated sphere in Bingham plastic fluids, *Numer. Heat Transf. Part A*, **66**, 1048 (2014a).
Nirmalkar, N., Bose, A. and Chhabra, R.P., Free convection from a heated circular cylinder in Bingham Plastic fluids, *Int. J. Thermal Sci.*, **83**, 33 (2014b).
Nirmalkar, N., Chhabra, R.P. and Poole, R.J., On creeping flow of a Bingham plastic fluid past a square cylinder, *J. Non-Newt. Fluid Mech.*, **171–172**, 17 (2012).

Nirmalkar, N., Chhabra, R.P. and Poole, R.J., Numerical predictions of momentum and heat transfer characteristics from a heated sphere in yield stress fluids, *Ind. Eng. Chem. Res.*, **52**, 6848 (2013a).

Nirmalkar, N., Chhabra, R.P. and Poole, R.J., Effect of shear-thinning behavior on heat transfer from a heated sphere in yield-stress fluids, *Ind. Eng. Chem. Res.*, **52**, 13490 (2013b).

Nirmalkar, N., Chhabra, R.P. and Poole, R.J., Laminar forced convection heat transfer from a heated square cylinder in a Bingham plastic fluid, *Int. J. Heat Mass Transf.*, **56**, 625 (2013c).

Nirmalkar, N., Gupta, A.K. and Chhabra, R.P., Natural convection from a heated sphere in Bingham Plastic fluids, *Ind. Eng. Chem. Res.*, **53**, 17818 (2014c).

Nishimura, T., Itoh, H., Ohya, K. and Miyashita, H., Experimetnal validation of numerical analysis of flow across tube banks for laminar flow, *J. Chem. Eng. Jpn.*, **24**, 666 (1991).

Nishimura, Y. and Ishii, T., An analysis of transport phenomena for multi-solid particle systems at higher Reynolds numbers by a standard Karman-Pohlhausen method – I, Momentum Transfer, *Chem. Eng. Sci.*, **35**, 1195 (1980).

Nitin, S. and Chhabra, R.P., Wall effects in two-dimensional axisymmetric flow over a circular disk oriented normal to flow in a cylindrical tube, *Can. J. Chem. Eng.*, **83**, 450 (2005a).

Nitin, S. and Chhabra, R.P., Non-isothermal flow of a power law fluid past a rectangular obstacle (of aspect ratio 1x2) in a channel: Drag and heat transfer, *Int. J. Eng. Sci.*, **43**, 707 (2005b).

Nitin, S. and Chhabra, R.P., Sedimentation of a circular disk in power law fluids, *J. Colloid Interface Sci.*, **295**, 520 (2006).

Niven, R.K., Physical insights into the Ergun and Wen and Yu equations for fluid flow in packed and fluidized beds, *Chem. Eng. Sci.*, **57**, 527 (2002). Also see *ibid* **58**, 5379 (2003) and *ibid* **60**, 299 (2005).

Noel, A.C., Guo, H.-Y., Mandica, M. and Hu, D.L., Frogs use a viscoelastic tongue and non-Newtonian saliva to catch prey, *J. R. Soc. Interface*, **14**, 20160764 (2016).

Noh, D.S., Kang, I.S. and Leal, L.G., Numerical simulations for the deformation of a bubble rising in dilute polymeric fluids, *Phys. Fluids*, **5**, 1315 (1993).

Norouzi, M., Abdolnezhad, H. and Mandani, S., An experimental investigation on inertia motion and deformation of Boger drops falling through Newtonian media, *Meccanica*, **54**, 473 (2019b).

Norouzi, M. and Davoodi, M., Analytical study on motion and shape of creeping Boger drops falling through viscoelastic media, *J. Braz. Soc. Mech. Sci. Eng.*, **40**, 124 (2018).

Norouzi, M., Emamian, A. and Davoodi, M., An analytical and experimental study on dynamics of a circulating Boger drop translating through Newtonian fluids at inertia regime, *J. Non-Newt. Fluid Mech.*, **267**, 1 (2019a).

Notz, P.K. and Basaran, O.A., Dynamics of drop formation in an electric field, *J. Colloid Interface Sci.*, **213**, 218 (1999).

Novotny, E.J., Proppant transport, Paper # SPE 6813. Presented at the *Annu. Tech. Conf. and Exhibition*, Denver, Colorado (1977).

Null, H.R. and Johnson, H.F., Drop formation in liquid-liquid systems from single nozzles, *AIChE J.*, **4**, 273 (1958).

Nyrkova, I.A., Semenov, A.N., Khokhlov, A.R., Linliu, K. and Chu, B., Motion of a probe ball in the fluid under centrifugal acceleration, *J. Phys. II (France)*, **7**, 1709 (1997).

O'Donovan, E.J. and Tanner, R.I., Numerical study of the Bingham squeeze film problem, *J. Non-Newt. Fluid Mech.*, **15**, 75 (1984).

O'Neill, M.E., Small particles in viscous media, *Sci. Prog. (Oxford)*, **67**, 149 (1981).

Ockendon, J.R. and Evans, G.A., The drag on a sphere in low Reynolds number flow, *Aero. Sci.*, **3**, 237 (1972).

Ocone, R. and Astarita, G., Two conceptual problems in heat or mass transfer to arrays of spheres, *Dev. Chem. Eng.*, **1**, 1 (1991).

Odeh, A.S. and Yang, H.T., Flow of non-Newtonian power law fluids through porous media, *Soc. Pet. Eng. J.*, **17**, 155 (June 1979).

Odell, J.A. and Haward, S.J., Viscosity enhancement in non-Newtonian flow of dilute aqueous polymer solutions through crystallographic and random porous media, *Rheol. Acta*, **45**, 853 (2006).

Odell, J.A., Muller, A.J., Keller, A. and Müller, A.J., Thermomechanical degradation of macromolecules, *Colloid Polym. Sci.*, **270**, 307 (1992).

Odell, J.A., Muller, A.J., Narh, K.A. and Keller, A., Degradation of polymer solutions in extensional flows, *Macromolecules*, **23**, 3092 (1990).

Ogata, K., Kuroda, C. and Watanabe, K., Effect of surfactant solutions on the drag and the flow pattern of a circular cylinder, *AIChE J.*, **52**, 49 (2006).

Ogawa, K., Kuroda, C. and Inoue, I., Forced convective mass transfer in viscoelastic fluid around a sphere and a cylinder, *J. Chem. Eng. Jpn.*, **17**, 654 (1984).

Ogunjimi, A., Mannan, S.H., Whalley, D.C. and Williams, D.J., The assembly process for anisotropic conductivity joints- Some new experimental and theoretical results, *J. Electron. Manuf.*, **5**, 263 (1995).

Oh, J.H. and Lee, S.J., A study on the Newtonian fluid flow past a sphere in a tube, *Korean J. Chem. Eng.*, **5**, 190 (1988).

Oh, J.H. and Lee, S.J., A rheological study on the viscoelastic flow past spheres in a cylinder, *J. Mater. Process. Manuf. Sci.*, **1**, 3 (1992).

Ohta, M., Furukawa, T., Yoshida, Y. and Sussman, M., A three-dimensional numerical study on the dynamics and deformation of a bubble rising in a hybrid Carreau and FENE–CR modeled polymeric liquid, *J. Non-Newt. Fluid Mech.*, **265**, 66 (2019).

Ohta, M., Iwasaki, E., Obata, E. and Yoshida, Y., A numerical study of the motion of a spherical drop rising in shear-thinning fluid systems, *J. Non-Newt. Fluid Mech.*, **116**, 95 (2003).

Ohta, M., Iwasaki, E., Obata, E. and Yoshida, Y., Dynamic processes in a deformed drop rising through shear-thinning fluids, *J. Non-Newt. Fluid Mech.*, **132**, 105 (2005).

Ohta, M., Kikuchi, D., Yoshida, Y. and Sussman, M., Robust numerical analysis of the dynamic bubble formation process in a viscous liquid, *Int. J. Multiph. Flow*, **37**, 1059 (2011).

Ohta, M., Kimura, S., Furukawa, T., Yoshida, Y. and Sussman, M., Numerical simulation of a bubble rising through a shear-thickening fluid, *J. Chem. Eng. Jpn.*, **45**, 713 (2012).

Ohta, M., Kobayashi, N., Shigekane, Y., Yoshida, Y. and Iwata, S., The dynamic motion of single bubbles with unique shapes rising freely in hydrophobically modified alkali-soluble emulsion polymer solutions, *J. Rheol.*, **59**, 303 (2015).

Ohta, M., Onodera, K., Yoshida, Y. and Sussman, M., Three-dimensional numerical simulations of a rising bubble in viscoelastic FENE-CR model fluids, *J. Chem. Eng. Jpn.*, **42**, 705 (2009).

Ohta, M., Yamaguchi, S., Yoshida, Y. and Sussman, M., The sensitivity of drop motion due to the density and viscosity ratio, *Phys. Fluids*, **22**, 072102 (2010b).

Ohta, M., Yoshida, Y. and Sussman, M., Three-dimensional computations of the motion of a Newtonian drop rising through immiscible quiescent shear-thinning liquids, *J. Chem. Eng. Jpn.*, **39**, 394 (2006).

Ohta, M., Yoshida, Y. and Sussman, M., A computational study of dynamic processes of a bubble rising in Carreau model fluids, *Fluid Dyn. Res.*, **42**, 025501 (2010a).

Okawa, T., Tanaka, T., Kataoka, I. and Mori, M., Temperature effect on single bubble rise characteristics in stagnant distilled water, *Int. J. Heat Mass Transf.*, **46**, 903 (2003).

Okhotskii, V.B., Rise of single bubbles in a confined medium, *Theor. Found. Chem. Eng.*, **35**, 510 (2001).

Okuda, K., Pipe wall effects on suspension velocities of single freely-suspended spheres and on terminal velocities of single spheres in a pipe. *Bull. JSME*, **18**(124), 1142 (1975).

Olbricht, W.L. and Leal, L.G., The creeping motion of liquid drops through a circular tube of comparable diameter: The effect of density differences between the fluids, *J. Fluid Mech.*, **115**, 187 (1982).

Olbricht, W.L. and Leal, L.G., The creeping motion of immiscible drops through a converging/diverging tube, *J. Fluid Mech.*, **134**, 329 (1983).

Oldroyd, J.G., Two-dimensional plastic flow of a Bingham Solid, *Proc. Camb. Phil. Soc.*, **43**, 383 (1947).

Oldroyd, J.G., On the formulation of rheological equations of state, *Proc. Roy. Soc.*, **A200**, 523 (1950); also see *ibid*, **A245**, 278 (1958).

Oldroyd, J.G., The elastic and viscous properties of emulsions and suspensions, *Proc. Roy. Soc.*, **A218**, 122 (1953).

Oliveira, P.J., Method for time-dependent simulations of viscoelastic flows: Vortex shedding behind cylinder, *J. Non-Newt. Fluid Mech.*, **101**, 113 (2001).

Oliveira, P.J. and Miranda, A.I.P., A numerical study of steady and unsteady viscoelastic flow past bounded cylinders, *J. Non-Newt. Fluid Mech.*, **127**, 51 (2005).

Oliver, D.L.R. and Chung, J.N.C., Flow about a fluid sphere at low to moderate Reynolds numbers, *J. Fluid Mech.*, **177**, 1 (1987).

Oliver, D.R., Influence of particle rotation on radial migration in the Poiseuille flow of suspensions, *Nature*, **194**, 1269 (1962).

Omari, A., Moan, M. and Chauveteau, G., Hydrodynamic behaviour of semi-rigid polymer at a solid-liquid interface, *J. Rheol.*, **33**, 1 (1989a).

Omari, A., Moan, M. and Chauveteau, G., Wall effects in the flow of flexible polymer solutions through small pores, *Rheol. Acta*, **28**, 520 (1989b).

Onishi, J., Chen, Y. and Ohashi, H., A lattice Boltzmann model for multicomponent flows of non-Newtonian fluids, *Jap. J. Multiph. Flow*, **17**, 413 (2003).

Oolman, T. and Blanch, H.W., Bubble coalescence in stagnant liquids, *Chem. Eng. Commun.*, **43**, 237 (1986).

Oosthuizen, P.H. and Donaldson, E., Free convective heat transfer from vertical cones, *J. Heat Transf.*, **94**, 330 (1972).

Oosthuizen, P.H. and Kalendar, A., *Natural Convection Heat Transfer from Narrow Plates*, Springer, Berlin (2012).

Oosthuizen, P.H. and Kalendar, A., *Natural Convection Heat Transfer from Short Inclined Cylinders*, Springer, Berlin (2013).

Oosthuizen, P.H. and Kalendar, A., *Natural Convection Heat Transfer from Horizontal and near Horizontal Surfaces*, Springer, Berlin (2018).

Oppong, F.K. and de Bruyn, J.R., Diffusion of microscopic tracer particles in a yield stress fluid, *J. Non-Newt. Fluid Mech.*, **142**, 104 (2007).

Oppong, F.K., Rubatat, L., Frisken, B.J., Bailey, A.E. and de Bruyn, J.R., Microrheology and structure of a yield-stress polymer gel, *Phys. Rev. E*, **73**, 041405 (2006).

Orgeas, L., Idris, Z., Geindreau, C., Block, J.-F. and Auriault, J.L., Modelling the flow of power-law fluids through anisotropic porous media at low pore Reynolds number, *Chem. Eng. Sci.*, **61**, 4490 (2006).

Ortega-Rivas, E., *Unit Operations of Particulate Systems*, Taylor & Francis, Boca Raton, FL (2016).

Ortega-Rivas, E. and Svarovsky, L., Generalized Stokes number for modeling settling of non-Newtonian slurries in dynamic separators, *Adv. Powder Technol.*, **9**, 1 (1998).

Ortiz-Villafuerte, J., Hassan, Y.A. and Schmidl, W.D., Rocking motion, trajectory and shape of bubbles rising in small diameter tubes, *Exp. Thermal Fluid Sci.*, **25**, 43 (2001).

Oseen, C.W., *Neuere Methoden und ergebnisse in der hydrodynamik*, Akademische Verlagsgesellschaft, Leipzig (1927).

Ostrach, S., Natural convection in enclosures, *Adv. Heat Transf.*, **8**, 161 (1972).

Ouattara, Z., Jay, P., Bleses, D. and Magnin, A., Drag of a cylinder moving near a wall in a yield stress fluid, *AIChE J.*, **64**, 4118 (2018).

Ouattara, Z., Jay, P., and Magnin, A., Flow of Newtonian fluid and a yield stress fluid around a plate inclined at 45° in interaction with a wall, *AIChE J.*, **65**, e16562 (2019).

Oukhlef, A., Ambari, A., Champmartin, S.P. and Despeyroux, A., Yield stress fluid method to measure the pore size distribution of a porous medium. In *International Conference on Nanochannels, Microchannels, and Minichannels*, vol. 54501, 1535–1540 (2010).

Owens, R.G., A posteriori error estimates for spectral element solutions to viscoelastic flow problems, *Comput. Methods Appl. Mech. Eng.*, **164**, 375 (1998).

Owens, R.G. and Phillips, T.N., Decoupled spectral element methods for steady viscoelastic flow past a sphere, *Proc. ICOSAHOM 95, Houston J. Math*, 287 (1996a).

Owens, R.G. and Phillips, T.N., Steady viscoelastic flow past a sphere using spectral elements, *Int. J. Numer. Methods Eng.*, **39**, 1517 (1996b).

Owens, R.G. and Phillips, T.N., *Computational Rheology*, Imperial College Press, London (2002).

Ozoe, H. and Churchill, S.W., Hydrodynamic stability and natural convection in Ostwald-deWaele and Ellis fluid: The development of a numerical solution, *AIChE J.*, **18**, 1196 (1972).

Ozogul, H., Jay, P. and Magnin, A., Slipping of a viscoplastic fluid flowing on a circular cylinder, *J. Fluids Eng.*, **137**, 0712011 (2015).

Ozogul, H., Jay, P. and Magnin, A., Nonrecirculating flow of a yield-stress fluid around a circular cylinder in a Poiseuille flow, *AIChE J.*, **62**, 4554 (2016).

Oztekin, A., Alakus, B. and McKinley, G.H., Stability of planar stagnation flow of a highly viscoelastic fluid, *J. Non-Newt. Fluid Mech.*, **72**, 1 (1997).

Padhy, S., Rodriguez, M., Shaqfeh, E.S.G., Iaccarino, G., Morris, J.F. and Tonmukayakul, N., The effect of shear-thinning and walls on the sedimentation of a sphere in an elastic fluid under orthogonal shear, *J. Non-Newt. Fluid Mech.*, **201**, 120 (2013b).

Padhy, S., Shaqfeh, E.S.G., Iaccarino, G., Morris, J.F. and Tonmukayakul, N., Simulations of a sphere sedimenting in a viscoelastic fluid with cross shear, *J. Non-Newt. Fluid Mech.*, **193**, 48 (2013a).

Paine, P.L. and Scherr, P., Drag coefficients for the movement of rigid spheres through liquid-filled cylindrical pores, *Biophys. J.*, **15**, 1087 (1975).

Pakdel, P. and McKinley, G.H., Digital particle imaging velocimetry of viscoelastic fluids, *AIChE J.*, **43**, 289 (1997).

Pakdemirli, M., Similarity analysis of boundary layer equations of a class of non-Newtonian fluids, *Int. J. Non-Linear Mech.*, **29**, 187 (1994a).

Pakdemirli, M., Conventional and multiple deck boundary layer approach to second and third grade fluids, *Int. J. Eng. Sci.*, **32**, 141 (1994b).

Pakdemirli, M. and Suhubi, E.S., Similarity solutions of boundary layer equations for second order fluids, *Int. J. Eng. Sci.*, **30**, 611 (1992).

Pakdemirli, M., Yurusoy, M. and Kucukbursa, A., Symmetry groups of boundary layer equations of a class of non-Newtonian fluids, *Int. J. Non-Linear Mech.*, **31**, 267 (1996).

Pal, R., *Rheology of Particulate Dispersions and Emulsions*, Taylor & Francis, Boca Raton, FL (2006).

Paliwal, B., Sharma, A., Chhabra, R.P. and Eswaran, V., Power law fluid flow past a square cylinder: Momentum and heat transfer characteristics, *Chem. Eng. Sci.*, **58**, 5315 (2003).

Palle, S. and Aliabadi, S., Direct simulation of structured wall bounded packed beds using hybrid FE/FV methods, *Computers & Fluids*, **88**, 730 (2013).

Panda, S.K. and Chhabra, R.P., Laminar flow of power-law fluids past a rotating cylinder, *J. Non-Newt. Fluid Mech.*, **165**, 1442 (2010).

Panda, S.K. and Chhabra, R.P., Laminar forced convection heat transfer from a rotating cylinder to power-law fluids, *Numer. Heat Transf. Part A*, **59**, 297 (2011).

Panfilov, M. and Fourar, M., Physical splitting of nonlinear effects in high-velocity stable flow through porous media. *Adv. Water Resour.*, **29**(1), 30 (2006).

Pang, M. and Lu, M., Numerical study on dynamics of single bubble rising in shear-thinning power-law fluid in different gravity environment, *Vacuum*, **153**, 101 (2018).

Pantangi, U.S., Ramamurthy, G. and Vanamala, U.M., Heat transfer analysis of free convection of a non-Newtonian power law fluid over a vertical plate with constant heat flux, *Inst. Eng. (India)-Chem. Eng. Div.*, **84**, 25 (2003).

Pantokratoras, A., Steady flow of a non-Newtonian Carreau fluid across an unconfined circular cylinder, *Meccanica*, **54**, 1007 (2016).

Pantokratoras, A., Comments on the paper "Wall effects for a spherical particle in confined shear-thickening fluids", *J. Non-Newt. Fluid Mech.*, **289**, 104488 (2021).

Papanastasiou, A.C., Scriven, L.E. and Macosko, C.W., Bubble growth and collapse in viscoelastic liquids analyzed, *J. Non-Newt. Fluid Mech.*, **16**, 53 (1984).

Papanastasiou, T.C., Flow of materials with yield, *J. Rheol.*, **31**, 385 (1987).

Papathanasiou, T.D., On the effective permeability of square arrays of permeable fiber tows, *Int. J. Multiph. Flow*, **23**, 81 (1997).

Papathanasiou, T.D., Flow across structured fiber bundles: A dimensionless correlation, *Int. J. Multiph. Flow*, **27**, 1451 (2001).

Park, B.J., Park, B.O., Ryu, B.H., Choi, Y.M., Kwon, K.S. and Choi, H.J., Rheological properties of Ag suspended fluid for inkjet printing, *J. Appl. Phys.*, **108**, 102803 (2010).

Park, H.C., Hawley, M.C. and Blanks, R.F., The flow of non-Newtonian solutions through packed beds, *Polym. Eng. Sci.*, **15**, 761 (1975).

Park, K.H., *PhD Dissertation*, Monash University, Melbourne, Australia (1986).

Park, N.A., Cho, Y.I. and Irvine Jr., T.F., Steady shear viscosity measurements of viscoelastic fluids with the falling needle viscometer, *J. Non-Newt. Fluid Mech.*, **34**, 351 (1990).

Park, N.A. and Irvine Jr., T.F., The falling needle viscometer: A new technique for viscosity measurements, *Wärme-und Stoffubertragung*, **18**, 201 (1984).

Park, N.A. and Irvine Jr., T.F., Measurement of rheological properties with the falling needle viscometer, *Rev. Sci. Instrum.* **59**, 2051 (1988).

Park, N.A., Irvine Jr., T.F. and Gui, F., Yield stress measurements with the falling needle viscometer, *Proc. Xth Int. Cong. Rheol. Sydney*, **2**, 160 (1988).

Parker, J.A., *PhD Dissertation*, New Jersey Institute of Technology, Hoboken, NJ (1977).

Parmaj, N.B., Chhabra, R.P. and Narayan, K.A., Mass transfer from a highly soluble single cylinder in parallel flow, *Chem. Eng. J.*, **40**, 139 (1989).

Parnas, R.S. and Phelan Jr., F.R., The effect of heterogeneous porous media on mold filling in resin transfer molding, *SAMPE Quart.*, **22**, 53 (January 1991).

Parthasarathy, M. and Klingenberg, D.J., Electro-rheology: Mechanisms and models, *Mat. Sci. Eng.*, **R17**, 57 (1996).

Parthasarathy, R.N., Linear spatial stability analysis of slurry sheets subjected to gas flow, *At. Sprays*, **9**, 519 (1999).

Parvazinia, M. and Nassehi, V., Study of shear thinning fluid flow through highly permeable porous media. *Int. Comm. Heat Mass Transf.*, **33**(4), 401 (2006).

Pascal, H., Rheological behaviour effect of non-Newtonian fluids on steady and unsteady flow through a porous medium, *Int. J. Num. Analyt. Methods Geomech.*, **7**, 289 (1983).

Pascal, H., Dynamics of moving interface in porous media for power law fluids with yield stress, *Int. J. Eng. Sci.*, **22**, 577 (1984a).

Pascal, H., Rheological effects of non-Newtonian fluids on gravitational segregation mechanism in a porous medium, *Int. J. Eng. Sci.*, **22**, 857 (1984b).

Pascal, H., Some problems related to the quantitative evaluation of physical properties of the porous medium from flow tests with non-Newtonian fluids, *Int. J. Eng. Sci.*, **23**, 307 (1985).

Pascal, H., Rheological effects of non-Newtonian behaviour of displacing fluids on stability of a moving interface in radial oil displacement mechanism in porous media, *Int. J. Eng. Sci.*, **24**, 1465 (1986a).

Pascal, H., Stability of a moving interface in porous medium for non-Newtonian displacing fluids and its applications in oil displacement mechanism, *Acta Mechanica*, **58**, 81 (1986b).

Pascal, H., On the existence of self-similar solutions of the equations governing unsteady flow through a porous medium, *Int. J. Heat Fluid Flow*, **9**, 381 (1988).

Pascal, H., Non-isothermal flow of non-Newtonian fluids through a porous medium, *Int. J. Heat Mass Transf.*, **33**, 1937 (1990a).

Pascal, H., Some self-similar two-phase flows of non-Newtonian fluids through a porous medium, *Stud. Appl. Maths.*, **82**, 305 (1990b).

Pascal, H. and Pascal, F., Flow of non-Newtonian fluid through porous media, *Int. J. Eng. Sci.*, **23**, 571 (1985).

Pascal, H. and Pascal, F., Dynamics of non-Newtonian fluid interfaces in a porous medium: Incompressible fluids, *Int. J. Num. Methods Fluids*, **8**, 1389 (1988).

Pascal, H. and Pascal, F., On viscoelastic effects in non-Newtonian steady flows through porous media, *Trans. Porous Media*, **4**, 17 (1989a).

Pascal, H. and Pascal, J.P., Nonlinear effects of non-Newtonian fluids on natural convection in a porous medium, *Physica D*, **40**, 393 (1989b).

Pashias, N., Boger, D.V., Summers, J. and Glenister, D.J., A fifty cent rheometer for yield stress measurements, *J. Rheol.*, **40**, 1179 (1996).

Patankar, N.A. and Hu, H.H., A numerical investigation of the detachment of the trailing particle from a chain sedimenting in Newtonian and viscoelastic fluids, *J. Fluids Eng. (ASME)*, **122**, 517 (2000).

Patankar, N.A. and Hu, H.H., Rheology of a suspension of particles in viscoelastic fluids, *J. Non-Newt. Fluid Mech.*, **96**, 427 (2001).

Patankar, N.A., Huang, P.Y., Joseph, D.D. and Hu, H.H., Normal stresses on the surface of a rigid body in an Oldroyd-B fluid, *J. Fluids Eng.*, **124**, 279 (2002).

Patel, O.P., Nirmalkar, N., Patel, S.A. and Chhabra, R.P., Mixed convection from a heated hemisphere falling coaxially in a long cylindrical tube filled with Bingham Plastic Fluids, *J. Energy, Heat Mass Transf.*, **40**, 13 (2018).

Patel, O.P., Patel, S.A., Raja, A.H. and Chhabra, R.P., Forced convection heat transfer from a hemisphere in Bingham plastic fluids: Effects of orientation and thermal boundary condition, *J. Energy, Heat Mass Transf.*, **37**, 27 (2015).

Patel, S.A. and Chhabra, R.P., Steady flow of Bingham plastic fluids past an elliptical cylinder, *J. Non-Newt. Fluid Mech.*, **202**, 32 (2013).

Patel, S.A. and Chhabra, R.P., Heat transfer in Bingham plastic fluids from a heated elliptical cylinder, *Int. J. Heat Mass Transf.*, **73**, 671 (2014).

Patel, S.A. and Chhabra, R.P., Effect of aiding buoyancy on heat transfer from an isothermal elliptical cylinder in Newtonian and Bingham plastic fluids, *Int. J. Heat Mass Transf.*, **89**, 539 (2015).

Patel, S.A. and Chhabra, R.P., Laminar free convection in Bingham plastic fluids from an isothermal elliptic cylinder, *J. Thermophys. Heat Transf.*, **30**(1), 153 (2016a).

Patel, S.A. and Chhabra, R.P., Effect of angle of incidence on laminar forced convection from an elliptical cylinder in Bingham plastic fluids, *Numer. Heat Transf. Part A*, **70**, 917 (2016b).

Patel, S.A. and Chhabra, R.P., Buoyancy-assisted flow of yield stress fluids past a cylinder: Effect of shape and channel confinement, *Appl. Math. Model.*, **75**, 892 (2019).

Patel, S.A. and Chhabra, R.P., Non-Newtonian fluids, *Kirk-Othmer Encyclopedia of Chemical Technology*, p. 49, Wiley, New York (2020).

Patel, S.A., Chhabra, R.P., Irvine Jr., T.F. and Capobianchi, M., Convection heat transfer in non-Newtonian fluids. *CRC Handbook of Thermal Engineering*, 2nd ed., p. 315 (2017).

Paterson, A., d'Onofrio, A., Allain, C., Hulin, J.P., Rosen, M. and Gauthier, C., Tracer dispersion in a polymer solution flowing through a double porosity porous medium, *J. Phys. II France*, **6**, 1639 (1996).

Paterson, M.S., Problems in the extrapolation of laboratory rheological data, *Tectonophysics*, **133**, 33 (1987).

Patil, R.C., Bharti, R.P. and Chhabra, R.P., Forced convection heat transfer in power law liquids from a pair of cylinders in tandem arrangement, *Ind. Eng. Chem. Res.*, **47**, 9141 (2008b).

Patil, R.C., Bharti, R.P. and Chhabra, R.P., Steady flow of power law fluids over a pair of cylinders in tandem arrangement, *Ind. Eng. Chem. Re.*, **47**, 1660 (2008a).

Patnaik, B.S.V., Narayana, P.A.A. and Seetharamu, K.N., Numerical simulation of vortex shedding past a circular cylinder under the influence of buoyancy, *Int. J. Heat Mass Transf.*, **42**, 3495 (1999).

Patnana, V.K., Bharti, R.P. and Chhabra, R.P., Two-dimensional unsteady flow of power-law fluid over a cylinder, *Chem. Eng. Sci.*, **64**, 2978 (2009).

Patnana, V.K., Bharti, R.P. and Chhabra R.P., Two-dimensional unsteady forced convection heat transfer in power-law fluids from a cylinder, *Int. J. Heat Mass Transf.*, **53**, 4152 (2010).

Patton, T., *Paint Flow and Pigment*, Wiley, New York (1979).

Patwari, A.N., Nguyen-Tien, K., Schumpe, A. and Deckwer, W.-D., Three phase fluidized beds with viscous liquids: Hydrodynamics and mass transfer, *Chem. Eng. Commun.*, **40**, 49 (1986).

Paul, I., Arul Prakash, K. and Vengadesan, S., Forced convective heat transfer from unconfined isothermal and isoflux elliptic cylinders, *Num. Heat Transf.*, **64A**, 648 (2013).

Paul, I., Arul Prakash, K., Vengadesan, S. and Pulletikurthi, V., Analysis and characterization of momentum thermal wakes of elliptic cylinders, *J. Fluid Mech.*, **807**, 303 (2016).

Pavlov, K.B., Theory of a laminar boundary layer of a conducting power-law non-Newtonian fluid in a transverse magnetic field, *Magnitnaya Gidrodinamika*, **16**, 361 (1979).

Payatakes, A.C. and Neira, M.A., Model of the constricted unit cell type for isotropic granular porous media, *AIChE J.*, **23**, 922 (1977).

Payatakes, A.C., Tien, C. and Turian, R.M., A new model for granular porous media: Part 1. Model formulation, *AIChE J.*, **19**, 58 (1973).

Payne, L.W. and Parker, H.W., Axial dispersion of non-Newtonian fluids in porous media, *AIChE J.*, **19**, 202 (1973).

Pazwash, H. and Robertson, J.M., Fluid dynamic considerations of bottom materials, *J. Hyd. Div. (ASCE)*, **97**, 1317 (1971).

Pazwash, H. and Robertson, J.M., Forces on bodies in Bingham fluids, *J. Hydraulic Res.*, **13**, 35 (1975).

Pearson, G. and Middleman, S., Elongational flow behaviour of viscoelastic liquids: Modelling bubble dynamics with viscoelastic constitutive relations, *Rheol. Acta*, **17**, 500 (1978).

Pearson, J.R.A. and Tardy, P.M.J., Models for flow of non-Newtonian and complex fluids through porous media, *J. Non-Newt. Fluid Mech.*, **102**, 447 (2002).

Pech, D., Etude de la permeabilite des lits compressibles constitues de copeaux de bois partiellement destructures. These de 3eme cycle, INP Grenoble, France (1984).

Peden, J.M. and Luo, Y., Settling velocity of variously shaped particles in drilling and fracturing fluids, *SPE Drilling Eng.*, **2**, 337 (1987).

Peev, G. and Mateeva, N., A study on the solid sphere in pseudoplastic flow, *Chem. Ind. (Bulgaria)*, **54**, 361 (1982).

Peev, G., Nikolova, A. and Todorova, D., Mass transfer from solid particles to power law non-Newtonian fluid in granular bed at low Reynolds numbers, *Chem. Eng. J.*, **88**, 119 (2002).

Pellerin, F. and Thirriot, F., Numerical study of non-Newtonian flows in a periodical succession of convergents – divergents, *Proc. VIIth Int. Cong. Rheol.*, Gothenberg, Sweden, p. 348 (1976).

Peng, S., Xiong, Y.-L., Xu, X.-Y. and Yu, P., Numerical study of unsteady viscoelastic flow past two side-by-side circular cylinders, *Phys. Fluids*, **32**, 083106 (2020).

Pereira, R.G., Additional effects on internal flow of non-Newtonian fluids in the presence of a particle, *J. Non-Newt. Fluid Mech.*, **95**, 85 (2000).

Perrin, C.L., Tardy, P.M.J., Sorbie, K.S. and Crawshaw, J.C., Experimental and modeling study of Newtonian and non-Newtonian fluid flow in pore network micromodels, *J. Colloid Interface Sci.*, **295**, 542 (2006).

Petera, J., A new finite element scheme using the Lagrangian framework for simulation of viscoelastic fluid flows, *J. Non-Newt. Fluid Mech.*, **103**, 1 (2002).

Peterka, J.A. and Richardson, P.D., Effects of sound on separated flows, *J. Fluid Mech.*, **37**, 265 (1969).

Peters, E.A.J.F., Hulsen, M.A. and van den Brule, B.H.A.A., Instationary Eulerian viscoelastic flow simulations using time separable Rivlin-Sawyers constitutive equations, *J. Non-Newt. Fluid Mech.*, **89**, 209 (2000)

Petford, N., Rheology of granitic magmas during ascent and emplacement, *Annu. Rev. Earth Planet. Sci.*, **31**, 399 (2003).

Petrou, M.F., Wan, B., Gadala-Maria, E., Kolli, V.G. and Harries, K.A., Influence of mortar rheology on aggregate settlement, *ACI Mater.*, **97**, 479 (2000).

Pfeffer, R., Heat and mass transport in multiparticle systems, *Ind. Eng. Chem. Fund.*, **3**, 380 (1964).

Pfeffer, R. and Happel, J., An analytical study of heat and mass transfer in multiparticle systems at low Reynolds numbers, *AIChE J.*, **10**, 605 (1964).

Phair, J.W., Lundberg, M. and Kaiser, A., Levelling and thixotropic characteristics of concentrated Zirconia inks for screen printing, *Rheol. Acta*, **48**, 121 (2009).

Phan-Thien, N. and Dou, H.-S., Viscoelastic flow past a cylinder: Drag coefficient, *Comput. Methods Appl. Mech. Eng.*, **180**, 243 (1999).

Phan-Thien, N., Dudek, J., Boger, D.V. and Tirtaatmadja, V., Squeeze film flow of ideal elastic liquids, *J. Non-Newt. Fluid Mech.*, **18**, 227 (1985).

Phan-Thien, N., Jin, H. and Zheng, R., On the flow past a needle in a cylindrical tube, *J. Non-Newt. Fluid Mech.*, **47**, 137 (1993).

Phan-Thien, N. and Khan, M.M.K., Flow of an Oldroyd-type fluid through a sinusoidally corrugated tube, *J. Non-Newt. Fluid Mech.*, **24**, 203 (1987).

Phan-Thien, N., Zheng, R. and Tanner, R.I., Flow along the centerline behind a sphere in a uniform stream, *J. Non-Newt. Fluid Mech.*, **41**, 151 (1991).

Phelan Jr., F.R. and Wise, G., Analysis of transverse flow in aligned fibrous porous media, *Composites*, **27A**, 25 (1996).

Philippoff, W., The viscosity characteristics of rubber solutions, *Rubber Chem. Technol.*, **10**, 76 (1937).

Philippoff, W., An experiment performed at the SOR Meeting in Madison (1961); See Bird, R.B., Armstrong, R.C. and Hassager, O., *Dynamics of Polymeric Liquids*, Vol. I, Wiley, New York, p. 117 (1977).

Phillips, R.J., Dynamic simulation of hydrodynamically interacting spheres in a quiescent second-order fluid, *J. Fluid Mech.*, **315**, 345 (1996).

Phillips, R.J., Structural instability in the sedimentation of particulate suspensions through viscoelastic fluids, *J. Non-Newt. Fluid Mech.*, **165**, 479 (2010).

Phillips, R.J. and Talini, L., Chaining of weakly interacting particles suspended in viscoelastic fluids, *J. Non-Newt. Fluid Mech.*, **147**, 175 (2007).

Piau, J.-M., Viscoplastic boundary layer, *J. Non-Newt. Fluid Mech.*, **102**, 193 (2002).

Picaro, T. and van de Ven, T.G.M., The flow of dilute Polyethylene oxide solutions through packed beds of pulp fibers, *J. Pulp Paper Sci.*, **21**, J13 (1995).

Picot, J.J. and Kristmanson, D.D., *Forestry Pesticide Aerial Spraying: Spray Droplet Generation, Dispersion and Deposition*, Kluwer, Dordrecht (1997).

Pierce, F.T., Geometrical principles applicable to the design of functional fabrics, *Text. Res. J.*, **17**, 123 (1947).

Piermarini, G.J., Forman, R.A. and Block, S., Viscosity measurements in the diamond anvil pressure cell, *Rev. Sci. Instrum.*, **49**, 1061 (1978).

Pilate, G. and Crochet, M.J., Plane flow of a second order fluid past submerged boundaries, *J. Non-Newt. Fluid Mech.*, **2**, 323 (1977).

Pilitsis, S. and Beris, A.N., Calculations of steady state viscoelasic flow in an undulating tube, *J. Non-Newt. Fluid Mech.*, **31**, 231 (1989).

Pillai, K.K., Voidage variation at the wall of a packed bed of spheres, *Chem. Eng. Sci.*, **32**, 59 (1977).

Pillapakkam, S.B., Dynamics of drops and bubbles in Newtonian and viscoelastic flows, *Proc. ASME Fluids Engineering Div.*, FED-250, 169 (1999).

Pilz, C. and Brenn, G., On the critical bubble volume at the rise velocity jump discontinuity in viscoelastic liquids, *J. Non-Newt. Fluid Mech.*, **145**, 124 (2007).

Pimenta, F. and Alves, M.A., Conjugate heat transfer in the unbounded flow of a viscoelastic fluid past a sphere. *Int. J. Heat Fluid Flow*, **89**, 108784 (2021).

Pinelli, D. and Magelli, F., Solids settling velocity and distribution in slurry reactors with dilute pseudoplastic suspensions, *Ind. Eng. Chem. Res.*, **40**, 4456 (2001).

Pinto, F. and Meo, M., Design and manufacturing of a novel shear-thickening fluid composite (STFC) with enhanced out of plane properties and damage suppression, *Appl. Commun. Mater.*, **24**, 643 (2017).

Pipe, C.J. and Monkewitz, P.A., Vortex shedding in flows of dilute polymer solutions, *J. Non-Newt. Fluid Mech.*, **139**, 54 (2006).

Pisani, L., Simple expression for the tortuosity of Porous Media. *Trans. Porous Media*, **88**(2), 193 (2011).

Pittman, J.F.T., Richardson, J.F. and Sherrard, C.P., An experimental study of heat transfer by laminar natural convection between an electrically heated vertical plate and both Newtonian and non-Newtonian fluid, *Int. J. Heat Mass Transf.*, **42**, 657 (1999).

Pittman, J.F.T., Richardson, J.F., Sharif, A.O. and Sherrard, C.P., Heat transfer from a planar surface to a fluid in laminar flow: An experimental and computational study, *Int. J. Heat Mass Transf.*, **37**(Suppl.1), 333 (1994).

Pliskin, I. and Brenner, H., Experiments on the pressure drop created by a sphere settling in a viscous liquid, *J. Fluid Mech.*, **17**, 89 (1963).

Plog, J.P., Kulicke, W.–M. and Clasen, C., Influence of the molar mass distribution on the elongational behaviour of polymer solutions in capillary breakup, *Appl. Rheol.*, **15**, 28 (2005).

Podczeck, F., A shape factor to assess the shape of particles using image analysis, *Powder Technol.*, **93**, 47 (1997).

Podczeck, F. and Newton, J.M., A shape factor to characterize the quality of spheroids, *J. Pharma. Pharmacol.*, **46**, 82 (1994).

Podczeck, F. and Newton, J.M., The evaluation of three-dimensional shape factor for the quantitative assessment of the sphericity and surface roughness of pellets, *J. Pharm.*, **124**, 253 (1995).

Podgorski, T. and Belmonte, A., Surface folds during the penetration of a viscoelastic fluid by a sphere, *J. Fluid Mech.*, **460**, 337 (2002).

Podgorski, T. and Belmonte, A., Surface folding of viscoelastic fluids: Finite elasticity membrane model, *Eur. J. Appl. Math.*, **15**, 385 (2004).

Podolsak, A.K., Tiu, C. and Fang, T.N., Flow of non-Newtonian fluids through tubes with abrupt expansions and contractions (square wave tubes), *J. Non-Newt. Fluid Mech.*, **71**, 25 (1997).

Pohlhausen, E., Der Wärmeaustausch zwischen festen Körpern und Flüssigkeiten mit kleiner Reibung und kleiner Wärmeleitung, *Zeitschrift für Angewandte Mathematik und Mechanik*, **1**(2), 115 (1921).

Polyanin, A.D., Kutepov, A.M., Vyazmin, A.V. and Kazenin, D.A., *Hydrodynamics, Mass and Heat Transfer in Chemical Engineering*, Taylor and Francis, London, UK (2002).

Polyanin, A.D. and Vyaz'min, A.V., Mass and heat transfer to particles in a flow, *Theor. Found. Chem.Eng.*, **29**, 128 (1995). Also see *ibid* 229.

Ponche, A. and Dupuis, D., On instabilities and migration phenomena in cone and plate geometry, *J. Non-Newt. Fluid Mech.*, **127**, 123 (2005).

Ponter, A.B. and Surati, A.I., Bubble emissions from submerged orifices – A critical review, *Chem. Eng. Technol.*, **20**, 85 (1997).

Poole, R.J., The Deborah and Weissenberg numbers, *BSR Rheol. Bull.*, **53**(2), 32 (2012).

Poole, R.J., Thermo-dependent properties of model parameters in yield-stress fluids: A survey, *Rheol. Bull.*, **59**(2), 36 (2018).

Pop, I., Rashidi, M. and Gorla, R.S.R., Mixed convection to power law type non-Newtonian fluids from a vertical wall, *Polym.-Plast. Technol. Eng.*, **30**, 47 (1991).

Pordesimo, L.O., Zuritz, C.A. and Sharma, M.G., Flow behaviour of coarse solid-liquid food mixtures, *J. Food Eng.*, **21**, 495 (1994).

Porter, J.E., Heat transfer at low Reynolds number (Highly viscous liquids in laminar flow), *Trans. Inst. Chem. Eng.*, **49**, 1 (1971).

Potapov, A., Spivak, R., Lavrenteva, O.M. and Nir, A., Motion and deformation of drops in Bingham fluid, *Ind. Eng. Chem. Res.*, **45**, 6985 (2006).

Potter, J.M. and Riley, N., Free convection from a heated sphere at large Grashof number, *J. Fluid Mech.*, **100**, 769 (1980).

Potucek, F. and Stejskal, J., Oxygen absorption in polymeric solutions in a bead column, *Chem. Eng. Sci.*, **44**, 194 (1989).

Pradipasena, P. and Rha, C., Pseudoplastic and rheopectic properties of a globular protein (β-lactoglobulin) solution, *J. Texture Stud.*, **8**, 311 (1977).

Prakash, O., Ph.D Thesis, Department of Mechanical Engineering, Banaras Hindu University, Varanasi, India (1985).

Prakash, O., Gupta, S.N. and Mishra, P., Newtonian and inelastic non-Newtonian flow across tube banks, *Ind. Eng. Chem. Res.*, **26**, 1365 (1987).

Prakash, O., Patel, S.A., Gupta, A.K. and Chhabra, R.P., Coarse particles in homogeneous non-Newtonian slurries: Combined effects of shear thinning viscosity and fluid yield stress on drag and heat transfer from hemispherical particles, *Trans. Indian Inst. Met.*, **70**, 341 (2017).

Prakash, S., *M. Tech. Dissertation*, Banaras Hindu University, Varanasi, India (1976).

Prakash, S., Experimental evaluation of terminal velocity in non-Newtonian fluids in the turbulent region, *Indian Chem. Eng.*, **25**, 1 (1983).

Prakash, S., Momentum transfer to Newtonian and non-Newtonian fluids flow past a sphere, *Indian Chem. Eng.*, **28**, 1 (1986).

Prakash, S., Mukherjee, K. and Mehrotra, S.P., Cold model study of metal droplet descent through fluids including a treatment of the non-Newtonian behavior as exhibited by molten slag system, *Ironmak. Steelmak.*, **32**, 494 (2002).

Prandtl, L., *Uber Flussigkeitshewegung bei schr kleiner Reibung*, Verhandlungen des III Internationlen Mathematiker Kongresses, Heidelberg (1904).

Prasad, D.V.N. and Chhabra, R.P., An experimental investigation of the cross flow of power law liquids past a bundle of cylinders and in a bed of stacked screens, *Can. J. Chem. Eng.*, **79**, 28 (2001).

Prashant and Derksen, J.J., Direct simulations of spherical particle motion in Bingham liquids, *Comp. Chem. Eng.*, **35**, 1200 (2011).

Pravesh, R., Dhiman, A. and Bharti, R.P., Aiding buoyancy mixed convection flow and thermal features across a periodic array of heated cylinders, *Int. J. Heat Mass Transf.*, **130**, 1141 (2019a).

Pravesh, R., Dhiman, A. and Bharti, R.P., Non-Newtonian power-law fluid's thermal characteristics across periodic array of circular cylinders, *J. Braz. Soc. Mech. Sci. Eng.*, **41**, 88 (2019b).

Premlata, A.R., Tripathi, M.K., Karri, B. and Sahu, K.C., Dynamics of an air bubble rising in a non-Newtonian liquid in the axisymmetric regime, *J. Non-Newt. Fluid Mech.*, **239**, 53 (2017a).

Premlata, A.R., Tripathi, M.K., Karri, B. and Sahu, K.C., Numerical and experimental investigations of an air bubble rising in a Carreau-Yasuda shear-thinning liquid, *Phys. Fluids*, **29**, 033103 (2017b).

Prentice, J.H., *Dairy Rheology*, Wiley, London (1992).

Prhashanna, A. and Chhabra, R.P., Free convection in power-law fluids from a heated sphere, *Chem. Eng. Sci.*, **65**, 6190 (2010).

Prhashanna, A. and Chhabra, R.P. Laminar natural convection from a horizontal cylinder in power-law fluids, *Ind. Eng. Chem. Res.*, **50**, 2424 (2011).

Prhashanna, A., Sahu, A.K. and Chhabra, R.P., Flow of power-law fluids past an equilateral triangular cylinder: Momentum and heat transfer characteristics, *Int. J. Thermal Sci.*, **50**, 2027 (2011).

Prieto, J.L., Stochastic particle level set simulations of buoyancy- driven droplets in non-Newtonian fluids, *J. Non-Newt. Fluid Mech.*, **226**, 16 (2015).

Prieto, J.L., An RBF -reconstructed polymer stress tensor for stochastic, particle-based simulations of non-Newtonian multiphase flows, *J. Non-Newt. Fluid Mech.*, **227**, 90 (2016).

Prilutski, G., Gupta, R.K., Sridhar, T. and Ryan, M.E., Model viscoelastic liquids, *J. Non-Newt. Fluid Mech.*, **12**, 233 (1983).

Prokunin, A.N., Bukman, Yu.A. and Gupalo, Yu.P., On the motion of spherical particles along the wall in the shear flow of Newtonian and non-Newtonian fluid, *Proc. XI Int. Cong. Rheol.*, pp. 210–212, Brussels, Belgium (1992).

Proudman, I. and Pearson, J.R.A., Expansions at small Reynolds numbers for the flow past a sphere and a circular cylinder, *J. Fluid Mech.*, **2**, 237 (1957).

Prud'homme, R.K. and Bird, R.B., The dilatational properties of suspensions of gas bubbles in incompressible Newtonian and non-Newtonian fluids, *J. Non- Newt. Fluid Mech.*, **3**, 261 (1977/78).

Pulley, J.W., Hussey, R.G. and Davis, A.M.J., Low non-zero Reynolds number drag of a thin disk settling axisymmetrically within a cylindrical outer boundary, *Phys. Fluids*, **8**, 2275 (1996).

Pullum, L., Boger, D.V. and Sofra, F., Hydraulic mineral waste transport and storage, *Annu. Rev. Fluid Mech.*, **58**, 157 (2018).

Puncochar, M. and Drahos, J., The tortuosity concept in fixed and fluidized bed, *Chem. Eng. Sci.*, **48**, 2173 (1993).

Puncochar, M. and Drahos, J., Limits of applicability of capillary model for pressure drop correlation, *Chem. Eng. Sci.*, **55**, 3951 (2000).

Puris, B.I., The hydrodynamic influence of polymer additives in the external flow around a body, *Fluid Mech.-Sov. Res.*, **2**, 98 (1973).

Puris, B.I., Polesskii, E.P., Galuts, V.V. and Aerov, V.E., Effect of polymer additives on the large-scale structure of the velocity field in the boundary region of a sphere, *J. Eng. Phys.*, **40**, 385 (1981).

Putz, A. and Frigaard, I.A., Creeping flow around particles in a Bingham fluid, *J. Non- Newt. Fluid Mech.*, **165**, 263 (2010).

Putz, A.M.V., Burghelea, T.I., Frigaard, I.A. and Martinez, D.M., Settling of an isolated spherical particle in a yield stress shear thinning fluid, *Phys. Fluids*, **20**, 033102 (2008).

Pye, D.J., Improved secondary recovery by control of water mobility, *J. Pet. Tech.*, **16**(8), 911 (1964). Also see *ibid* 1430 (1964).

Quach, A. and Hansen, C.M., Evaluation of leveling characteristics of some latex paints, *J. Paint Tech.*, **46**, 40 (1974).

Quintana, G.C., The effect of surfactants on flow and mass transport to drops and bubbles, in *Transport Processes in Bubbles, Drops and Particles*, edited by R.P. Chhabra, and D. De Kee, Chapter 4, Hemisphere, New York (1991).

Quintana, G.C., Cheh, H.Y. and Maldarelli, C.M., The translation of a Newtonian droplet in a 4-constant Oldroyd fluid, *J. Non- Newt. Fluid Mech.*, **22**, 253 (1987).

Quintana, G.C., Cheh, H.Y. and Maldarelli, C.M., The effect of viscoelasticity on the translation of a surfactant covered Newtonian drop, *J. Non-Newt. Fluid Mech.*, **45**, 81 (1992).

Rabiger, N. and Vogelpohl, A., Bubble formation and its movement in Newtonian and non-Newtonian liquids, *Encycl. Fluid Mech.*, **3**, 58 (1986).

Radl, S. and Khinast, J.G., Prediction of mass transfer coefficients in non-Newtonian fermentation media using first principle methods, *Biotech. Bioeng.*, **97**, 1329 (2007).

Radl, S., Tryggvason, G. and Khinast, J.G., Flow and mass transfer of fully resolved bubbles in non-Newtonian fluids, *AIChE J.*, **53**, 1861 (2007).

Rae, D., Yield stress exerted on a body immersed in a Bingham fluid, *Nature*, **194**, 272 (1962).

Rahli, O., Tadrist, L. and Blanc, R., Experimental analysis of the porosity of randomly packed rigid fibers, *C.R. Acad. Sci. Paris, Ser. II*, **327**, 725 (1999).

Rahli, O., Tadrist, L. and Miscevic, M., Experimental analysis of fibrous porous media permeability, *AIChE J.*, **42**, 3547 (1996).

Rahli, O., Tadrist, L., Miscevic, M. and Santini, R., Etude experimentale des ecoulements darceens a travers un lit de fibres rigides empilees aleatoirement: Influence de la porosite, *J. Phys. II France*, **5**, 1739 (1995).

Rahli, O., Tadrist, L., Miscevic, M. and Santini, R., Fluid flow through randomly packed monodisperse fibers: The Kozeny-Carman parameter analysis, *J. Fluids Eng.*, **119**, 188 (1997).

Raichura, R.C., Pressure drop and heat transfer in packed beds with small tube-to-particle diameter ratio, *Exp. Heat Transf.*, **12**, 309 (1999).

Raja, A.H., Patel, S.A. and Chhabra, R.P., Laminar forced convection from a two-dimensional transverse plate in Bingham plastic fluids, *Int. J. Heat Mass Transf.*, **83**, 690 (2015).

Rajagopal, K.R., A note on the drag for fluids of grade three, *Int. J. Non-Linear Mech.*, **14**, 361 (1979).

Rajagopal, K.R., On the cavalier attitude towards referencing, *Int. J. Eng. Sci.*, **51**, 338 (2012).

Rajagopal, K.R., Gupta, A.S. and Na, T.Y., A note on the Falkner-Skan flows of a non-Newtonian fluid, *Int. J. Non-Linear Mech.*, **18**, 313 (1983).

Rajagopal, K.R., Gupta, A.S. and Wineman, A.S., On a boundary layer theory for non-Newtonian fluids, *Lett. Appl. Eng. Sci.*, **18**, 875 (1980).

Rajagopalan, D.R., Arigo, M.T. and McKinley, G.H., Sedimentation of a sphere through an elastic fluid: Part II Transient Motion, *J. Non-Newt. Fluid Mech.*, **65**, 17 (1996).

Rajeswari, G.K., Laminar boundry layer on rotating sphere and spheroids in non-Newtonian fluids, *ZAMP*, **13**, 442 (1962).

Rajeswari, G.K. and Rathna, S.L., Flow of a particular class of non-Newtonian viscoelastic and viscoinelastic fluids near a stagnation point, *ZAMP*, **13**, 43 (1962).

Rajitha, P., Chhabra, R.P., Sabiri, N.E. and Comiti, J., Drag on non-spherical particles in power law non-Newtonian media, *Int. J. Min. Process.*, **78**, 110 (2006).

Rajvanshi, S.C. Slow flow of Oldroyd fluid past a sphere, *Arch. Mech. Stos.*, **4**, 585 (1969).

Rallison, J.M., Note on the Faxen relations for a particle in Stokes flow, *J. Fluid Mech.*, **88**, 529 (1978).

Rallison, J.M., The deformation of small viscous drops and bubbles in shear flow, *Ann. Rev. Fluid Mech.*, **16**, 45 (1984).

Ramamurthy, G., On the existence of similar solutions to combined free and forced convection heat transfer to power law fluids past a vertical plate, *Bull. Cal. Math. Soc.*, **87**, 53 (1995).

Ramaswamy, H.S. and Basak, S., Time-dependent stress decay rheology of stirred yogurt, *Int. Dairy J.*, **1**, 17 (1991).

Ramaswamy, H.S. and Zareifard, M.R., Evaluation of factors influencing tube-flow fluid-to-particle heat transfer coefficients using a calorimetric technique, *J. Food Eng.*, **45**, 127 (2000).

Ramaswamy, S. and Leal, L.G., The deformation of a Newtonian drop in the uniaxial extensional flow of a viscoelastic liquid, *J. Non-Newt. Fluid Mech.*, **85**, 127 (1999a). Also see *ibid* 88, 149 (1999b).

Rameshwaran, P., Townsend, P. and Webster, M.F., Simulation of particle settling in rotating and non-rotating flows of non-Newtonian fluids, *Int. J. Numer. Meth. Fluids*, **26**, 851 (1998).

Ramge, P., Proski, T. and Kuhne, H.C., Segregation of coarse aggregates in self-compacting concrete, *Design, Production and Placement of Self-Compacting Concrete*, edited by K.H. Khayat and D. Feys, RILEM Book Series I, Springer, New York, pp. 113–125 (2010).

Rami, K., Chhabra, R.P. and Richardson, J.F., Drag on disks and square plates in Newtonian and Power-law fluids, *Proc. 10th Int. Conf. Transportation and Sedimentation of Solid Particles*, Wroclaw, Poland, p. 269 (2000).

Ramkissoon, H., Slow flow of a non-Newtonian liquid past a fluid sphere, *Acta Mech.*, **78**, 73 (1989a).

Ramkissoon, H., Stokes flow past a Reiner-Rivlin liquid sphere, *ZAMM*, **69**, 259 (1989b).

Ramkissoon, H., Viscoelastic flow past a spheroid, *ZAMP*, **41**, 137 (1990).

Ramkissoon, H. and Rahaman, K., Non-Newtonian fluid sphere in a spherical container, *Acta Mech.*, **149**, 239 (2001).

Ramkissoon, H. and Rahaman, K., Wall effects with slip, *ZAMM*, **83**, 773 (2003).

Ramkissoon, H. and Shifang, H., Unsteady motion of a sphere in an elastico-viscous fluid, *Int. J. Eng. Sci.*, **31**, 19 (1993).

Ramteke, R.R. and Kishore, N., Computational fluid dynamics study on forced convective heat transfer phenomena of spheres in power-law fluid liquids with velocity slip at the interface, *Heat Transf. Eng.*, **39**, 162 (2018).

Ramya, K.A., Srinivasan, R. and Deshpande, A.P., Time-dependent response of thixotropic systems: Insights from small amplitude oscillating shear, *Phys. Fluids*, **32**, 013109 (2020).

Rangel, N., Santos, A. and Pinho, C., Pressure drop in packed shallow beds of cylindrical cork stoppers, *Chem. Eng. Res. Des.*, **79A**, 547 (2001).

Rankin, P.J., Ginder, J.M. and Klingenberg, D.J., Electro – and magneto – rheology, *Curr. Opin. Colloid Interface Sci.*, **3**, 373 (1998).

Rankin, P.J., Horvath, A.T. and Klingenberg, D.J., Magneto-rheology in viscoplastic media, *Rheol. Acta*, **38**, 471 (1999).

Rao, B.K., Heat transfer to non-Newtonian flows over a cylinder in cross flow, *Int. J. Heat Fluid Flow*, **21**, 693 (2000a).

Rao, B.K., Internal heat transfer to viscoelastic flows through porous media, *Exp. Heat Transf.*, **13**, 329 (2000b).

Rao, B.K., Heat transfer to power-law fluid flows through porous media, *J. Porous Media*, **4**, 339 (2001).

Rao, B.K., Internal heat transfer to power-law fluid flows through porous media, *Exp. Heat Transf.*, **15**, 73 (2002).

Rao, B.K., Heat transfer to two-phase air-viscoelastic fluid flows over a hot cylinder, *Exp. Heat Transf.*, **16**, 227 (2003).

Rao, B.K., Phillips, B.J. and Andrews, J., Heat transfer to viscoelastic polymer solutions flowing over a smooth cylinder, *Appl. Mech. Eng.*, **1**, 355 (1996).

Rao, E.V.L.N., Kumar, R. and Kuloor, N.R., Drop formation studies in liquid-liquid systems, *Chem. Eng. Sci.*, **21**, 867 (1966).

Rao, J.H., Jeng, D.R. and DeWitt, K.J., Momentum and heat transfer in a power law fluid with arbitrary injection/suction at a moving wall, *Int. J. Heat Mass Transf.*, **42**, 2837 (1999).

Rao, M.A., *Rheology of Fluid and Semi-Solid Foods*, Aspen, Gaithersburg, MD (1999).

Rao, M.A., Rizvi, S.S.H. and A.K. Datta, *Engineering Properties of Foods*, 3rd ed., Marcel Dekker, New York (2005).

Rao, M.K., Sahu, A.K. and Chhabra, R.P., Effect of confinement on power-law fluid flow past a circular cylinder, *Polymer Eng. Sci.*, **51**, 2044 (2011a).

Rao, P.K., Sahu, A.K. and Chhabra, R.P., Momentum and heat transfer from a square cylinder in power-law fluids, *Int. J. Heat Mass Transf.*, **54**, 390 (2011c).

Rao, P.K., Sasmal, C., Sahu A.K., Chhabra, R.P. and Eswaran V., Effect of power-law fluid behaviour on momentum and heat transfer characteristics of an inclined square cylinder in steady flow regime, *Int. J. Heat Mass Transf.*, **54**, 2854 (2011b).

Rao, P.T. and Chhabra, R.P., Viscous non-Newtonian flow in packed beds: Effect of column walls and particle size distribution, *Powder Technol.*, **77**, 171 (1993).

Rapier, A.C., Communications on Fluid Flow through Beds of Granular Material, *Proc. Inst. Mech. Engrs., London*, **160**, 507 (1949).

Rasmussen, H.K., Time-dependent finite-element method for the simulation of three-dimensional viscoelastic flow with integral models, *J. Non-Newt. Fluid Mech.*, **84**, 217 (1999).

Rasmussen, H.K. and Hassager, O., Simulation of transient viscoelastic flow, *J. Non-Newt. Fluid Mech.*, **46**, 289 (1993).

Rasmussen, H.K. and Hassager, O., Simulation of transient viscoelastic flow with second order time integration, *J. Non-Newt. Fluid Mech.*, **56**, 65 (1995).

Rasmussen, H.K. and Hassager, O., On the sedimentation velocity of spheres in a polymeric liquid, *Chem. Eng. Sci.*, **51**, 1431 (1996).

Rathna, S.L., Slow motion of a non-Newtonian liquid past a sphere, *Quart. J. Mech. Appl. Math.*, **15**, 427 (1962).

Ray, H.S. and Biswas, A.K., A vertically oscillating sphere viscometer, *Can. J. Chem. Eng.*, **49**, 297 (1971).

Raymond, F. and Rosant, J.-M., A numerical and experimental study of the terminal velocity and shape of bubbles in viscous liquids, *Chem. Eng. Sci.*, **55**, 943 (2000).

Raynor, P.C., Flow field and drag for elliptical filter fibres, *Aerosol Sci. Tech.*, **36**, 1118 (2002).

Reddy, C.R. and Kishore, N., Effects of wall confinement and power-law fluid viscosity on Nusselt number of confined spheres. *Chem. Eng. Technol.*, **9**(36), 1568 (2013).

Reddy, C.R. and Kishore, N., Momentum and heat transfer phenomena of confined spheroid particles in power-law liquids. *Ind. Eng. Chem. Res.*, **53**(2), 989 (2014).

Reddy, J.N. and Gartling, D.K., *The Finite Element Method in Heat Transfer and Fluid Dynamics*, 2nd ed., CRC Press, Boca Raton, FL (2001).

Reddy, R.K. and Joshi, J.B., CFD modelling of presence drop and drag coefficient in fixed and expanded beds, *Chem. Eng. Res. Des.*, **86**, 444 (2008).

Reddy, R.K., Joshi, J.B., Nandakumar, K. and Minev, P.D., Direct numerical simulations of a freely falling sphere using fictitious domain method: Breaking of axisymmetric wake, *Chem. Eng. Sci.*, **65**, 2159 (2010).

Ree, T. and Eyring, H., The relaxation theory of transport phenomena, *Rheology*, edited by F.R. Eirich, Vol. 2, Academic, New York, Chapter 3 (1965).

Reed, J.S., Liquid permeability of packed particles: Why perpetuate the Carman-Kozeny model? *J. Am. Ceram. Soc.*, **76**, 547 (1993).

Reichelt, W., Zur berechnung des druckverlustes einphasig durchstromter kugel-und zylinderschuttungen, *Chem. Ing. Tech.*, **44**, 1068 (1972).

Reilly, I.G., Tien, C. and Adelman, M., Experimental study of natural convective heat transfer from a vertical plate in a non-Newtonian fluid, *Can. J. Chem. Eng.*, **43**, 157 (1965).

Reiner, M., The Deborah number, *Phys. Today*, **17**, 62 (1964).

Ren, S.-R., *PhD Thesis*, Department of Chemical Engineering & Chemical Technology, Imperial College, London, U.K. (1991).

Ren, X.-H., Stapf, S. and Blümich, B., Magnetic resonance visualization of flow and pore structure in packed beds with low aspect ratio, *Chem. Eng. Technol.*, **28**, 219 (2005a).

Ren, X.-H., Stapf, S. and Blumich, B., NMR velocimetry of flow in model-fixed bed reactors of low aspect ratio, *AIChE J.*, **51**, 392 (2005b).

Renard, P. and LeLoc'h, G., A new upscaling technique for the permeability of porous media: The simplified renormalization, *C.R. Acad. Sci. Paris, Ser. IIa*, **323**, 859 (1996).

Renardy, M., Some comments on the surface-tension-driven breakup (or the lack of it) of viscoelastic jets, *J. Non-Newt. Fluid Mech.*, **51**, 97 (1994).

Renardy, M., A numerical study of the asymptotic evolution and breakup of Newtonian and viscoelastic jets, *J. Non-Newt. Fluid Mech.*, **59**, 267 (1995).

Renardy, M., High Weissenberg number boundary layers for the upper convected Maxwell fluid, *J. Non-Newt. Fluid Mech.*, **68**, 125 (1997).

Renardy, M., Asymptotic structure of the stress field in flow past a cylinder at high Weissenberg number, *J. Non-Newt. Fluid Mech.*, **90**, 13 (2000a).

Renardy, M., Wall boundary layers for Maxwell liquids, *Arch. Rat. Mech. Anal.*, **152**, 93 (2000b).

Renaud, M., Mauret, E. and Chhabra, R.P., Power law fluid flow over a sphere: Average shear rate and drag coefficient, *Can. J. Chem. Eng.*, **82**, 1066 (2004).

Repetti, R.V. and Leonard, E.F., Segre-Silberberg annulus formation: A possible explanation, *Nature*, **203**, 1346 (1964).

Reverdy-Bruas, N.R., Serra-Tosio, J.-M., Chave, Y. and Bloch, J.-F., Investigation of the transverse permeability of saturated paper sheets, *Drying Technol.*, **19**, 2421 (2001).

Reynolds, P.A. and Jones, T.E.R., An experimental study of the settling velocities of single particles in non-Newtonian fluids, *Int. J. Min. Process.*, **25**, 47 (1989).

Ribeiro, G.S., Vargas, A.S. and Frota, M.N., Pressure drop induced by a sphere settling in non-Newtonian fluids, *Int. J. Multiph. Flow*, **20**, 355 (1994).

Riberio, V.M., Coelho, P.M., Pinho, F.T. and Alves, M.A., Three-dimensional effects in laminar flow past a confined cylinder, *Chem. Eng. Sci.*, **84**, 155 (2012).

Richardson, J.F., Incipient fluidization and particulate systems, *Fluidization*, edited by J.F. Davidson and D. Harrison, p. 25, Academic Press, New York, Chapter 2 (1971).

Richardson, J.F. and Zaki, W.N., Sedimentation and fluidization Part I, *Trans. Inst. Chem. Eng.*, **32**, 35 (1954).

Richter, D., Iaccarino, G. and Shaqfeh, E.S.G., Simulation of three-dimensional viscoelastic flow past a circular cylinder at moderate Reynolds numbers, *J. Fluid Mech.*, **651**, 415 (2010).

Richter, D., Iaccarino, G. and Shaqfeh, E.S.G., Effects of viscoelasticity in the high Reynolds number cylinder wake, *J. Fluid Mech.*, **693**, 293 (2012).

Riddle, M.J., Narvaez, C. and Bird, R.B., Interactions between two spheres falling along their line of centers in a viscoelastic fluid, *J. Non-Newt. Fluid Mech.*, **2**, 23 (1977).

Ridgway, K. and Tarbuck, K.J., Voidage fluctuations in randomly packed beds of spheres adjacent to a containing wall, *Chem. Eng. Sci.*, **23**, 1147 (1968).

Riley, E.E. and Lauga, E., Enhanced active swimming in viscoelastic fluids, *Euro. Phys. Lett.*, **108**, 34003 (2014).

Riley, E.E. and Lauga, E., Small – amplitude swimmers can self-propel faster in viscoelastic fluids, *J. Theor. Biol.*, **382**, 345 (2015).

Riley, E.E. and Lauga, E., Empirical resistive- force theory for slender biological filaments in shear- thinning fluids, *Phys. Rev. E*, **95**, 062416 (2017).

Riley N., The heat transfer from a sphere in free convective flow, *Comp. Fluids*, **14**, 225 (1986).

Ringhofer, M. and Sobczak, R., A comparison between the magneto viscometer and the melt indexer, *Appl. Rheol.*, **7**, 168 (1997).

Risso, F., Agitation, mixing and transfers induced by bubbles, *Annu. Rev. Fluid Mech.*, **50**, 25 (2018).

Rivero, D., Gouveia, L.M., Müller, A.J. and Saez, A.E., Shear-thickening behaviour of high molecular weight Poly (ethylene oxide) solutions, *Rheol. Acta*, **51**, 13 (2012).

Rivkind, V.Y. and Ryskin, G.M., Flow structure in motion of a spherical drop in a fluid medium at intermediate Reynolds number, *Fluid Dyn.*, **1**, 5 (1976).

Rivkind, V.Y., Ryskin, G.M. and Fishbein, G.A., The motion of a spherical drop in the flow of a viscous fluid, *Fluid Mech. - Sov. Res.*, **1**, 142 (1973).

Roblee, L.H.S., Baird, R.M. and Tierney, J.W., Radial porosity variations in packed beds, *AIChE J.*, **4**, 460 (1958).

Rochelle, S.G. and Peddieson Jr., J., Viscoelastic boundary-layer flows past wedges and cones. *Int. J. Eng. Sci.*, **18**(5), 713–726 (1980).

Rode, S., Midoux, N., Latifi, M.A., Storck, A. and Saatdjian, E., Hydrodynamics of liquid flow in packed beds: An experimental study using electrochemical shear rate sensors, *Chem. Eng. Sci.*, **49**, 889 (1994).

Rodrigue, D., Generalized correlation for bubble motion, *AIChE J.*, **47**, 39 (2001).

Rodrigue, D., A simple correlation for gas bubbles rising in power law fluids, *Can. J. Chem. Eng.*, **80**, 289 (2002).

Rodrigue, D., A general correlation for the rise velocity of single gas bubbles, *Can. J. Chem. Eng.*, **82**, 382 (2004).

Rodrigue, D. and Blanchet, J.-F., The motion of drops: Their shapes in the case of one viscoelastic phase, *Proc. 4th Int. Conf. Multiphase Flows (ICMF-2001)*, New Orleans, LA (2001), paper # 933.

Rodrigue, D. and Blanchet, J.-F., Surface re-mobilization of gas bubbles in polymer solutions containing surfactants, *J. Colloid Interface Sci.*, **256**, 249 (2002).

Rodrigue, D. and De Kee, D., Bubble velocity jump discontinuity in polyacrylamide solutions: A photographic study, *Rheol. Acta*, **38**, 177 (1999).

Rodrigue, D. and De Kee, D., Recent developments in the bubble velocity jump discontinuity, *Transport Processes in Bubbles, Drops and Particles*, D. De Kee and R.P. Chhabra, 2nd ed., Taylor & Francis, New York, Chapter 4 (2002).

Rodrigue, D., De Kee, D. and Chan Man Fong, C.F., An experimental study of the effect of surfactants on the free rise velocity of gas bubbles, *J. Non-Newt. Fluid Mech.*, **66**, 213 (1996a).

Rodrigue, D., De Kee, D. and Chan Man Fong, C.F., The motion of bubbles in non-Newtonian fluids, *Rheology and Fluid Mechanics of Nonlinear Materials*, AMD-217, 37 (1996b).

Rodrigue, D., De Kee, D. and Chan Man Fong, C.F., The slow motion of a spherical particle in a Carreau fluid, *Chem. Eng. Commun.*, **154**, 203 (1996c).

Rodrigue, D., De Kee, D. and Chan Man Fong, C.F., Bubble drag in contaminated non-Newtonian solutions, *Can. J. Chem. Eng.*, **75**, 794 (1997).

Rodrigue, D., De Kee, D. and Chan Man Fong, C.F., Bubble velocities: Further developments on the jump discontinuity, *J. Non-Newt. Fluid Mech.*, **79**, 45 (1998).

Rodrigue, D., De Kee, D. and Chan Man Fong, C.F., A note on the drag coefficient of a single gas bubble in a power-law fluid, *Can. J. Chem. Eng.*, **77**, 766 (1999a).

Rodrigue, D., De Kee, D. and Chan Man Fong, C.F., The slow motion of a single gas bubble in a non-Newtonian fluid containing surfactants, *J. Non-Newt. Fluid Mech.*, **86**, 211 (1999b).

Rodrigue, D., De Kee, D. and Chhabra, R.P., Drag on non-spherical particles in non-Newtonian fluids, *Can. J. Chem. Eng.*, **72**, 588 (1994).

Rodriguez de Castro, A, and Agnaou, M., Numerical investigation of the apparent viscosity dependence on darcy velocity during the flow of shear-thinning fluids in porous media. *Trans. Porous Media*, **129**(1), 93 (2019).

Rodriguez de Castro, A., Agnaou, M., Ahmadi-Sénichault, A. and Omari, A., Application of non-toxic yield stress fluids porosimetry method and pore-network modelling to characterize the pore size distribution of packs of spherical beads. *Trans. Porous Media*, **130**(3), 799 (2019).

Rodriguez de Castro, A., Agnaou, M., Ahmadi-Sénichault, A. and Omari, A., Numerical investigation of Herschel-Bulkley fluid flows in 2D porous media: Yielding behavior and tortuosity, *Comput. Chem. Eng.*, **140**, 106922 (2020).

Rodriguez de Castro, A., Ahmadi-Sénichault, A. and Omari, A., Using xanthan gum solutions to characterize porous media with the yield stress fluid porosimetry method: Robustness of the method and effects of polymer concentration. *Trans. Porous Media*, **122**(2), 357 (2018).

Rodriguez de Castro, A. and Radilla, G., Flow of yield stress and Carreau fluids through rough-walled rock fractures: Prediction and experiments. *Water Resour. Res.*, **53**(7), 6197 (2017).

Rodriguez de Castro, A. and Radilla, G., Non-Darcian flow experiments of shear-thinning fluids through rough-walled rock fractures, *Water Resour. Res.*, **52**, 9020 (2016).

Rodriguez de Castro, A., Oostrom, M. and Shokri, N., Effects of shear-thinning fluids on residual oil formation in microfluidic pore networks. *J. Colloid Inter. Sci.*, **472**, 34 (2016).

Rodriguez de Castro, A.and Goyeau, B., A pore network modelling approach to investigate the interplay between local and Darcy viscosities during the flow of shear-thinning fluids in porous media. *J. Colloid Interface Sci.*, **598**, 446 (2021).

Rodriguez, S., Romero, C., Sargenti, M.L., Müller, A.J., Saez, A.E. and Odell, J.A., Flow of polymer solutions through porous media, *J. Non-Newt. Fluid Mech.*, **49**, 63 (1993).

Roodhart, L.P., Proppant settling in non-Newtonian fracturing fluids, Paper presented at the *SPE/DOE 1985 Symposium on Low Permeability Gas Reservoirs*, Denver, CO, May 19–22 (1985). (Paper # SPE/DOE 13905).

Roquet, N. and Saramito, P., An adaptive finite element method for Bingham fluid flows around a cylinder, *Comput. Methods Appl. Mech. Eng*, **192**, 3317 (2003).

Rose, H.E., An investigation into the laws of flow of fluids through beds of granular materials, *Proc. Inst. Mech. Engrs.*, **153**, 141 (1945).

Rosentrater, K.A. and Flores, R.A., Physical and rheological properties of slaughterhouse swine blood and blood components, *Trans ASAE*, **40**, 683 (1997).

Rossen, W.R. and Gauglitz, P.A., Percolation theory of creation and mobilization of foams in porous media, *AIChE J.*, **36**(8), 1176 (1990).

Rossen, W.R. and Mamun, C.K., Minimal path for transport in networks, *Phys. Rev. B*, **47**(18), 11815 (1993).

Rossi, E., de Beristain, I.G., Vazquez-Quesada, A., López-Aguilar, J.E. and Ellero, M., SPH simulations of thixo-viscoplastic fluid flow past a cylinder, *J. Non-Newt. Fluid Mech.*, **308**, 104891 (2022).

Rosso, F., Sedimentation in coal-water slurry pipelining, *Complex Flows in Industrial Processes*, edited by A. Fasano, pp. 25–61, Birkhauser, Boston, MA (2000).

Rotem, Z., A note on boundary layer solution for pseudoplastic fluids, *Chem. Eng. Sci.*, **21**, 618 (1966).

Rouse, P.E., Jr., A theory of the linear viscoelastic properties of dilute solutions of coiling polymers, *J. Chem. Phys.*, **21**, 1272 (1953).

Roussel, N. and Coussot, P., Fifty-cent rheometer for yield stress measurements: From slump to spreading flows, *J. Rheol.*, **49**, 705 (2005).

Roustaei, A., Chevalier, T., Talon, L. and Frigaard, I.A., Non-Darcy effects in fracture flows of a yield stress fluid, *J. Fluid Mech.*, **805**, 222 (2016).

Roy, S., On Blasius flow in non-Newtonian fluid, *AIChE J.*, **18**, 666 (1972).

Rozant, O., Michaud, V., Bourban, P.-E. and Manson, J.-A.E., A model for the consolidation of warp-knitted reinforced laminates, *Polym. Comp.*, **22**, 432 (2001).

Rozhkov, A., Prunet-Foch, B. and Vignes-Adler, M., Impact of drops of polymer solutions on small targets, *Phys. Fluids*, **15**, 2006 (2003).

Ruckenstein, E., Interpolating equations between two limiting cases for the heat transfer coefficient, *AIChE J.*, **24**, 940 (1978).

Ruckenstein, E., Analysis of transport phenomena using scaling and physical models, *Adv. Chem. Eng.*, **13**, 11 (1987).

Ruckenstein, E., Thermal and diffusion boundary layers in viscoelastic flows, *Ind. Eng. Chem. Res.*, **33**, 2331 (1994).

Ruckenstein, E., On the laminar and turbulent free convection heat transfer from a vertical plate over the entire range of Prandtl numbers, *Int. Comm. Heat Mass Transf.*, **25**, 1009 (1998).

Ruckenstein, E., Brunn, P.O. and Holweg, J., Flow-induced creation and destruction of supermicelles in surfactant solutions, *Langmuir*, **4**, 350 (1988).

Ruckenstein, E. and Ramagopal, A., Anomalous heat transfer and drag in laminar flow of viscoelastic fluids, *J. Non-Newt. Fluid Mech.*, **17**, 145 (1985).

Rumpf, H. and Gupte, A.R., Einflusse der porosität und Korngrobenverteilung im Widerstandsgesetz der porentstromung, *Chem. Ing. Tech.*, **43**, 367 (1971).

Rumscheidt, F.D. and Mason, S.G., Deformation and burst of fluid drops in shear and hyperbolic flows, *J. Colloid Interface Sci.*, **16**, 238 (1961).

Russel, W.B., Review of the role of colloidal forces in the rheology of suspensions, *J. Rheol.*, **24**, 287 (1980).

Ruszczycky, M.A., Sphere drop tests in high polymer solutions, *Nature*, **206**, 614 (1965).

Saak, A.W., Jennings, H.M. and Shah, S.P., New methodology for designing self-compacting concrete, *Mater. J.*, **98**, 429 (2001).

Saar, M.O., Manga, M., Cashman, K.V. and Fremouw, S., Numerical models of the onset of yield strength in crystal-melt suspensions, *Ear. Planet. Sci. Lett.*, **187**, 367 (2001).

Sabiri, N.E., Chhabra, R.P., Comiti, J. and Montillet, A., Measurement of shear rate on the surface of a cylinder submerged in laminar flow of power-law fluids, *Exp. Thermal Fluid Sci.*, **39**, 167 (2012).

Sabiri, N.E. and Comiti, J., Pressure drop in non-Newtonian purely viscous flow through porous media, *Chem. Eng. Sci.*, **50**, 1193 (1995).

Sabiri, N.E. and Comiti, J., Experimental validation of a model allowing pressure gradient determination for non-Newtonian purely viscous fluid flow through packed beds, *Chem. Eng. Sci.*, **52**, 3589 (1997a).

Sabiri, N.E. and Comiti, J., Ecoulement de fluides newtoniens et non-newtoniens a travers des lits fixes stratifies, *Can. J. Chem. Eng.*, **75**, 1030 (1997b).

Sabiri, N.E., Comiti, J. and Brahimi, M., Fluidization of various shaped particles by non-Newtonian purely viscous fluids, *6th World Congress of Chemical Engineering*, San Diego, CA, p. 225 (1996a).

Sabiri, N.E., Montillet, A. and Comiti, J., Pressure drops of non-Newtonian purely viscous fluid flow through synthetic foams, *Chem. Eng. Commun.*, **156**, 59 (1996b).

Saboni, A., Mass Transfer from spherical bubbles, drops and particles, *J. Uni. Chem. Technol. Metall.*, **43**(4), 377 (2008).

Saboni, A. and Alexandrova, S., Numerical study of the drag on a sphere, *AIChE J.*, **48**, 2992 (2002).

Saboni A., Alexandrova, S. and Mory, M., Flow around a contaminated fluid sphere, *Int. J. Multiph. Flow*, **36**, 503 (2010).

Sacilik, K., Resistance of bulk poppy seeds to air flow, *Biosyst. Eng.*, **89**, 435 (2004).

Sada, E., Kumazawa, H. and Lee, C.H., Chemical absorption in a bubble column loading concentrated slurry, *Chem. Eng. Sci.*, **38**, 2047 (1983).

Sadhal, S.S., Ayyaswamy, P.S. and Chung, J.N., *Transport Phenomena with Drops and Bubbles*, Springer-Verlag, New York (1997).

Sadiq, T.A.K., Advani, S.G. and Parnas, R.S., Experimental investigation of transverse flow through aligned cylinders, *Int. J. Multiph. Flow*, **21**, 755 (1995).

Sadowski, T.J. and Bird, R.B., Non-Newtonian flow through porous media: Theoretical, *Trans. Soc. Rheol.*, **9**, 243 (1965).

Saez, A.E., Müller, A.J. and Odell, J.A., Flow of monodisperse polystyrene solutions through porous media, *Coll. Polym. Sci.*, **272**, 1224 (1984).

Saffman, P.G., On the motion of small spheroidal particles in a viscous fluid, *J. Fluid Mech.*, **1**, 540 (1956).

Sagert, N.H. and Quinn, M.J., The coalescence of gas bubbles in dilute aqueous solutions, *Chem. Eng. Sci.*, **33**, 1087 (1978).

Saha, A.K., Three-dimensional numerical simulations of the transition of flow past a cube, *Phys. Fluids*, **16**, 1630 (2004).

Saha, A.K., Far- wake characteristics of two- dimensional flow past a normal flat plate, *Phys. Fluids*, **19**, 128110 (2007).

Saha, G., Purohit, N.K. and Mitra, A.K., Spherical particle terminal velocity and drag in Bingham liquids, *Int. J. Miner. Process.*, **36**, 273 (1992).

Sahin, C. and Atalik, K., Comparison of inelastic and elastic non-Newtonian effects on the flow around a circular cylinder in periodic vortex shedding, *J. Non-Newt. Fluid Mech.*, **263**, 1 (2019).

Sahraoui, M. and Kaviany, M., Slip and no-slip boundary conditions at interface of porous, plane media, *Int. J. Heat Mass Transf.*, **35**, 927 (1992).

Sahu, A.K., Chhabra, R.P. and Eswaran, V., Effects of Reynolds and Prandtl numbers on heat transfer from a square cylinder in the unsteady flow regime, *Int. J. Heat Mass Transf.*, **52**, 839 (2009a).

Sahu, A.K., Chhabra, R.P. and Eswaran, V., Two-dimensional unsteady laminar flow of a power law fluid across a square cylinder, *J. Non-Newt. Fluid Mech.*, **160**, 157 (2009b).

Sahu, A.K., Chhabra, R.P. and Eswaran, V., Forced convection heat transfer from a heated square cylinder to power-law fluids in the unsteady flow regime, *Numer. Heat Transf. Part A*, **56**, 109 (2009c).

Sahu, A.K., Chhabra, R.P. and Eswaran, V., Two-dimensional laminar flow of a power-law fluid across a confined square cylinder, *J. Non-New. Fluid Mech.*, **165**, 752 (2010a).

Sahu, A.K., Chhabra, R.P. and Eswaran, V., Effect of blockage on forced convection heat transfer from a heated square to power-law fluids, *Numer. Heat Transf. Part A*, **58**, 641 (2010b).

Sahu, A.K. and Mathur, M.N., Free convection in boundary layer flows of power law fluids past a vertical flat plate with suction/injection, *Indian J. Pure Appl. Math.*, **27**, 931 (1996).

Sahu, A.K., Mathur, M.N., Chaturani, P. and Bharatiya, S.S., Momentum and heat transfer from a continuous moving surface to a power-law fluid, *Acta Mech.*, **142**, 119 (2000).
Sai, P.S.T. and Varma, Y.B.G., Pressure drop in gas-liquid down flow through packed beds, *AIChE J.*, **33**, 2027 (1987).
Sairamu, M. and Chhabra, R.P., Natural convection in power-law fluids from a tilted square in an enclosure, *Int. J. Heat Mass Transf.*, **56**, 319 (2013).
Sairamu, M., Nirmalkar, N. and Chhabra, R.P., Natural convection from a circular cylinder in confined Bingham plastic fluids, *Int. J. Heat Mass Transf.*, **60**, 567 (2013).
Saitoh, T., Sajiki, T. and Maruhara, K., Benchmark solutions to natural convection heat transfer problem around a horizontal circular cylinder, *Int. J. Heat Mass Transf.*, **36**, 1251 (1993).
Salami, E., Vignes, A. and LeGoff, P., Hydrodynamique des dispersions II. Effect de paroi. Mouvement d'une goutte ou d'unebulle dans un fluide immobile contenu dans un tube vertical de petit diametre, *Genie Chimique*, **94**, 67 (Sep.1965).
Saleh, S., Thovert, J.F. and Adler, P.M., Flow along porous media by particle image velocimetry, *AIChE J.*, **39**, 1765 (1993).
Salt, D.L., Ryan, N.W. and Christiansen, E.B., The rheology of carboxymethyl cellulose dispersions in water, *J. Colloid Sci.*, **6**, 146 (1951).
Samson, G., Phelipot-Mardele, A., Lanos, C. and Pierre, A., Quasi-static bubble in a yield stress fluid: Elasto-plastic model, *Rheol. Acta*, **56**, 431 (2017).
Sandeep, K.P. and Zuritz, C.A., Residence time of multiple particles in non-Newtonian holding tube flow: Effect of process parameters and development of dimensionless correlations, *J. Food Eng.*, **25**, 31 (1995).
Sandeep, K.P. and Zuritz, C.A., Drag on multiple sphere assemblies suspended in non-Newtonian tube flow, *J. Food Process Eng.*, **19**, 171 (1996).
Sandeep, K.P., Zuritz, C.A. and Puri, V.M., Mathematical modeling and experimental studies on RTD and heat transfer during aseptic processing of non-Newtonian suspensions, *Proc. 7th Int. Conf. Eng. & Food*, p. E-52, edited by R. Jowitt, Sheffield Academic Press, Sheffield (1997).
Sanders, J.V., Drag coefficient of spheres in polyethylene oxide solutions, *Int. Shipbuild. Prog.*, **14**, 140 (1967).
Sanders, J.V., The effect of drag reducing polymers on separated flow, *Proc. Drag Reduction Workshop*, U.S. Office of Naval Res., Cambridge, MA, p. 59 (1970).
Sandiford, B.B., Laboratory and field studies of water floods using polymer solutions to increase oil recoveries, *J. Pet. Tech.*, **16**, 917 (Aug. 1964).
Sandiford, B.B., Flow of polymers through porous media in relation to oil displacement, *Improved Oil Recovery by Surfactant and Polymer Flooding*, edited by D.O. Shah and R.S. Schechter, Academic Press, Cambridge, MA, p. 487 (1977).
Sangani, A.S. and Acrivos, A., Slow flow past periodic arrays of cylinders with application to heat transfer, *Int. J. Multiph. Flow*, **8**, 193 (1982a).
Sangani, A.S. and Acrivos, A. Slow flow through a periodic array of spheres, *Int. J. Multiph. Flow*, **8**, 343 (1982b).
Sangani, A.S. and Acrivos, A., Creeping flow through cubic arrays of spherical bubbles, *Int. J. Multiph. Flow*, **9**, 181 (1983).
Sangani, A.S. and Lu, W., Effective viscosity of an ordered suspension of small drops, *ZAMP*, **38**, 557 (1987).
Sangani, A.S. and Mo, G., Inclusion of lubrication forces in dynamic simulations, *Phys. Fluids*, **6**, 1653 (1994).
Sangani, A.S. and Yao, C., Transport processes in random arrays of cylinders. I. Thermal Conduction, *Phys. Fluids*, **31**, 2426 (1988). Also see *ibid* 2435.
Saradhy, Y.P. and Kumar, R., Drop formation at a sieve plate distributor, *Ind. Eng. Chem. Proc. Des. Dev.*, **15**, 75 (1976).
Saramito, P., Efficient simulation of nonlinear viscoelastic fluid flows, *J. Non-Newt. Fluid Mech.*, **60**, 199 (1995).
Saramito, P., A new constitutive equation for elastoviscoplastic fluid flow, *J. Non-Newt. Fluid Mech.*, **145**, 1 (2007).
Saramito, P., A damped Newton algorithm for computing viscoplastic flows, *J. Non-Newt. Fluid Mech.*, **238**, 6 (2016).
Saramito, P. and Wachs, A., Progress in numerical simulation of yield stress fluid flows, *Rheol. Acta*, **56**, 211 (2017).

Sarma, G.K. and Sukhatme, S.P., Combined free and forced convection heat transfer from a heated tube to a transverse air stream, *J. Heat Transf.*, **91**, 457 (1969).
Sarpkaya, T. and Rainey, P.G., Stagnation point flow of a second-order viscoelastic fluid, *Acta Mech.*, **11**, 236 (1971).
Sarpkaya, T., Rainey, P.G. and Kell, R.E., Flow of dilute polymer solutions about circular cylinders, *J. Fluid Mech.*, **57**, 177 (1973).
Sasmal, C. and Chhabra, R.P., Laminar natural convection from a heated square cylinder immersed in power-law liquids, *J. Non-Newt. Fluid Mech.*, **166**, 811 (2011).
Sasmal, C. and Chhabra, R.P., Effect of aspect ratio on natural convection in power-law fluids from a heated horizontal elliptic cylinder, *Int. J. Heat Mass Transf.*, **55**, 4886 (2012a).
Sasmal, C. and Chhabra, R.P., Effect of orientation on laminar natural convection from a heated square cylinder in power-law fluids, *Int. J. Thermal Sci.*, **57**, 112 (2012b).
Sasmal, C. and Chhabra, R.P., Momentum and heat transfer characteristics of a long parallelepiped submerged in power-law fluids in the laminar vortex shedding regime, *Int. J. Heat Mass Transf.*, **55**, 2285 (2012c).
Sasmal, C. and Chhabra, R.P., Laminar free convection in power-law fluids from a heated hemisphere, *J. Thermophys. Heat Transf.*, **28**, 750 (2014).
Sasmal, C., Gupta, A.K. and Chhabra, R.P., Natural convection heat transfer in a power-law fluid from a heated rotating cylinder in a square duct, *Int. J. Heat Mass Transf.*, **129**, 975 (2019).
Sasmal, C., Khan, M.B. and Chhabra, R.P., Combined influence of fluid viscoelasticity and inertia on forced convection heat transfer from a circular cylinder. *J. Heat Transf.*, **142** (4), 041801 (2020).
Sasmal, C., Shyam, R. and Chhabra, R.P., Laminar flow of power-law fluids past a hemisphere: Momentum and forced convection heat transfer characteristics, *Int. J. Heat Mass Transf.*, **63**, 51 (2013).
Sastry, S.K., Lima, M., Brim, J., Brunn, T. and Heskitt, B.E., Liquid-to particle heat transfer during continuous tube flow: Influence of flow rate and particle-to-tube diameter ratio, *J. Food Eng.*, **13**, 239 (1990).
Sastry, S.K. and Zuritz, C.A., A review of particle behaviour in tube flow: Applications to aseptic processing, *J. Food Process Eng.*, **10**, 27 (1987).
Satheesh, V.K., Chhabra, R.P. and Eswaran, V., Steady incompressible flow over a bundle of cylinders at moderate Reynolds numbers, *Can. J. Chem. Eng.*, **77**, 978 (1999).
Satish, M.G. and Zhu, J., Flow resistance and mass transfer in slow non-Newtonian flow through multi-particle systems, *J. Appl. Mech. (ASME)*, **59**, 431 (1992).
Sato, T., Taniyama, I. and Shimokawa, S., Flow of non-Newtonian fluid-Drag Coefficient of a sphere, *Kagaku-Kogaku*, **4**, 215 (1966).
Satrape, J.V. and Crochet, M.J., Numerical simulation of the motion of a sphere in a Boger fluid, *J. Non-Newt. Fluid Mech.*, **55**, 91 (1994).
Savins, J.G., Non-Newtonian flow through porous media, *Ind. Eng. Chem.*, **61**(10), 18 (1969).
Savreux, F., Jay, P. and Magnin, A., Flow normal to a flat plate of a viscoplastic fluid with inertia effects, *AIChE J.*, **51**, 750 (2005)
Sayegh, N.N. and Gonzalez, T.O., Compressibility of fibre mats during drainage, *J. Pulp Paper Sci.*, **21**, J255 (1995).
Schaink, H.M., Slot, J.M., Jongschaap, R.J.J. and Mellema, J., The rheology of systems containing rigid spheres suspended in both viscous and viscoelastic media, studied by Stokesian dynamics simulation, *J. Rheol.*, **44**, 473 (2000).
Schatzmann, M., Bezzola, G.R., Minor, H.E. and Fischer, P., The ball measuring system-a new method to determine debris-flow rheology, *Debris-Flow Hazards Mitigation: Mechanics, Prediction and Assessment*, edited by D. Rickenmann and C.-L. Chen, Millpress, Rotterdam, The Netherlands, pp. 387–398 (2003).
Schechter, R.S., On a variational principle for the Reiner-Rivlin fluid, *Chem. Eng. Sci.*, **17**, 803 (1962).
Schechter, R.S. and Farley, R.W., Interfacial tension gradients and droplet behaviour, *Can. J. Chem. Eng.*, **41**, 103 (1963).
Scheele, G.F. and Leng, D.E., An experimental study of factors which promote coalescence of two colliding drops suspended in water, *Chem. Eng. Sci.*, **26**, 1987 (1971).
Scheid, C.M., Puget, F.P., Halasz, M.R.T. and Massarani, G., Fluid dynamics of bubbles in liquids, *Braz. J. Chem. Eng.*, **16**, 351 (1999).
Scheidegger, A.E., *The Physics of Flow through Porous Media*, University of Toronto Press, Toronto (1974).
Schiller, V.L. and Naumann, A., Uber die grundlegenden berechnungen bei der schwerkraft aufbereitung, *ZVDI.*, **77**, 318 (1933).

Schlichting, H., *Boundary Layer Theory*, 6th ed., Mc Graw Hill, New York (1968).
Schmid, R., Stuff, R., Klein, U.K.A., Jamjoon, F.A. and Al-Suwaiyan, A., Low free-stream turbulence in test sections through packed beds and fibrous mats, *Exp. Fluids*, **26**, 451 (1999).
Schmiedel, J., Experimentelle untersuchungen uber die fallbewegung von kuglen and scheiben in reibender flussigkeiten, *Phys. Z.*, **29**, 593 (1928).
Schowalter, W.R., The application of boundary-layer theory to power law pseudoplastic fluids: Similarity solutions, *AIChE J.*, **6**, 24 (1960).
Schowalter, W.R., *Mechanics of Non-Newtonian Fluids*, Pergamon, Oxford, UK (1978).
Schowalter, W.R., The behaviour of complex fluids at solid boundaries, *J. Non-Newt. Fluid Mech.*, **29**, 25 (1988).
Schramm, L., *Emulsions, Foams, Suspensions, and Aerosols: Microscience and Applications*, 2nd ed., Wiley-VCH, New York (2014).
Schugerl, K., Three-phase biofluidisation – application of three-phase fluidization in the biotechnology - a review, *Chem. Eng. Sci.*, **52**, 3661 (1997).
Schumpe, A. and Deckwer, W.-D., Viscous media in tower bioreactors: Hydrodynamic characteristics and mass transfer properties, *Bioprocess Eng.*, **2**, 79 (1987).
Schumpe, A., Deckwer, W.-D. and Nigam, K.D.P., Gas liquid mass transfer in three phase fluidized beds with viscous pseudoplastic liquids, *Can. J. Chem. Eng.*, **67**, 873 (1989).
Schurz, J., The yield stress- an empirical reality, *Rheol. Acta*, **29**, 170 (1990).
Scirocco, R., Vermant, J. and Mewis, J., Effect of the viscoelasticity of the suspending fluid on structure formation in suspensions, *J. Non-Newt. Fluid Mech.*, **117**, 183 (2004).
Scott, K.J., Hindered settling of a suspension of spheres, CSIR Report # CENG 497, Council for Scientific and Industrial Research, Pretoria, South Africa (1984).
Scott Blair, G.W., Classification of rheological properties, *Nature*, **3772**, 197 (14 Feb 1942).
Scott Blair, G.W., The history of rheology, *Biorheology*, **19**, 231 (1982).
Scott Blair, G.W. and Oosthuizen, J.C., Rolling-sphere viscometer for structured fluids, *Brit. J. App. Phys.*, **11**, 332 (1960).
Sedahmed, G.H., Mansour, I.A.S., Fadali, D.A., Nassar, M.M. and El-Shayeb, M.M., The effect of drag reducing polymers on the rate of mass transfer in fixed bed reactors, *J. App. Electrochem.*, **17**, 583 (1987).
Sederman, A.J., Johns, M.L., Bramley, A.S., Alexander, P. and Gladden, L.F., Magnetic resonance imaging of liquid flow and pore structure within packed beds, *Chem. Eng. Sci.*, **52**, 2239 (1997). Also see *ibid* **53**, 2117 (1998).
See, H. and Brian, P., Measuring the yield stress of a particulate suspension under high electric fields, *Powder Technol.*, **160**, 40 (2005).
Seeling, C.H. and Yeow, Y.L., Finite element computation of bubble shape in Newtonian and power law fluids, *Proc. 6th Nat. Conf. Rheology*, p.99, Clayton, Vic., Australia (1992).
Segre, G. and Silberberg, A., Radial particle displacements in Poiseuille flow of suspensions, *Nature*, **189**, 209 (1961).
Segre, G. and Silberberg, A., Non-Newtonian behaviour of dilute suspensions of macroscopic spheres in a capillary viscometer, *J. Colloid Sci.*, **18**, 312 (1963).
Seguin, D., Montillet, A. and Comiti, J., Experimental characterization of flow regimes in various porous media-I: Limit of laminar flow regime, *Chem. Eng. Sci.*, **53**, 3751 (1998). Also see *ibid* 3897 (1998).
Sellers, H.S., Schwarz, W.H., Sato, M. and Pollard, T., Boundary effects on the drag of an oscillating sphere: Applications to the magnetic sphere rheometer, *J. Non-Newt. Fluid Mech.*, **26**, 43 (1987).
Sellin, R.H.J., Hoyt, J.W. and Scrivener, O., The effect of drag reducing additives on fluid flows and their industrial applications Part I: Basic aspects, *J. Hyd. Res.*, **20**, 29 (1982a).
Sellin, R.H.J., Hoyt, J.W., Pollert, J. and Scrivener, O., The effect of drag reducing additives on fluid flows and their industrial applications Part II: Present applications and future proposals, *J. Hyd. Res.*, **20**, 235 (1982b).
Semin, B., Hulin, J.P. and Auradou, H., Influence of flow confinement on the drag force on a static cylinder, *Phys. Fluids*, **21**, 103604 (2009).
Sen, S., *M.S. Dissertation*, Brigham Young University, Provo, Utah (1984).
Serth, R.W., A note on the Blasius problem for viscoelastic fluids, *AIChE J.*, **19**, 1275 (1973).
Serth, R.W. and Kiser, K.M., A solution of the two-dimensional boundary layer equations for an Ostwald-de Waele fluid, *Chem. Eng. Sci.*, **22**, 945 (1967).

Sestak, J. and Ambros, F., On the use of the rolling ball viscometer for the measurement of rheological parameters of power law fluids, *Rheol. Acta*, **12**, 70 (1973).
Seybold, H.J., Eberhard, U., Secchi, E., Cisne Jr., R.L., Jiménez-Martínez, J., Andrade, R.F., Aranjo, A.D., Holzner, M. and Andrade Jr., J.S. Localization in flow of non-Newtonian fluids through disordered porous media. *Front. Phys.*, **9**, 635051 (2021).
Seyed-Ahmadi, A. and Wachs, A., Dynamics and wakes of freely settling and rising cubes, *Phys. Rev. Fluids*, **4**, 074304 (2019).
Sgreva, S., Davaille, A., Kumagai, I. and Kurita, K., Interaction between a falling sphere and the structure of a non-Newtonian yield-stress fluid, *J. Non-Newt. Fluid Mech.*, **284**, 104355 (2020).
Shadid, J.N. and Eckert, E.R.G., Viscous heating of a cylinder with finite length by a high viscosity fluid in steady longitudinal flow—II. Non-Newtonian Carreau model fluids. *Int. J. Heat Mass Transf.*, **35** (10), 2739–2749 (1992).
Shah, C.B. and Yortsos, Y.C., Aspects of flow of power-law fluids in porous media, *AIChE J.*, **41**, 1099 (1995).
Shah, M.J., Petersen, E.E. and Acrivos, A., Heat transfer from a cylinder to a power law non-Newtonian fluid, *AIChE J.*, **8**, 542 (1962).
Shah, S.N., Proppant settling correlations for non-Newtonian fluids under static and dynamic conditions, *Soc. Pet. Eng. J.*, **29**, 164 (1982).
Shah, S.N., Proppant settling correlations for non-Newtonian fluids, *SPE Prod. Eng.*, **1**, 446 (1986).
Shah, S.N., Fadili, El.Y. and Chhabra, R.P., New model for single spherical particle settling velocity in power law (visco-inelastic) fluids, *Int. J. Multiph. Flow*, **33**, 51 (2007).
Shahcheraghi, N. and Dwyer, H.A., Fluid flow and heat transfer over a three-dimensional spherical object in a pipe, *J. Heat Transf.*, **120**, 985 (1998).
Shahsavari, S. and Mckinley, G.H., Mobility of power-law and Carreau fluids through fibrous media, *Phys. Rev. E.*, **92**, 063012 (2015)
Shahsavari, S. and Mckinley, G.H., Mobility and pore-scale fluid dynamics of rate-dependent yield-stress fluids flowing through fibrous porous media, *J. Non-Newt. Fluid Mech.*, **235**, 76 (2016).
Shail, R. and Norton, D.J., On the slow broadside motion of a thin disk along the axis of a fluid-filled circular duct, *Proc. Camb. Phil. Soc.*, **65**, 793 (1969).
Shankar Subramanian, R. and Balasubramanian, R., *The Motion of Bubbles and Drops in Reduced Gravity*, Cambridge Univ. Press, New York (2001).
Shaqfeh, E.S.G. and Koch, D.L., Polymer stretch in dilute fixed beds of fibres or spheres, *J. Fluid Mech.*, **244**, 17 (1992).
Sharaf, D.M., Premlata, A.R., Tripathi, M.K., Karri, B. and Sahu, K.C., Shapes and paths of an air bubble rising in quiescent liquids, *Phys. Fluids*, **29**, 122104 (2017).
Sharma, H.G., Creeping motion of a non-Newtonian fluid past a sphere, *Indian J. Pure Appl. Math.*, **10**, 1565 (1979).
Sharma, K.K. and Adelman, M., Experimental study of natural convection heat transfer from a vertical plate in a non-Newtonian fluid, *Can. J. Chem. Eng.*, **47**, 553 (1969).
Sharma, M.K. and Chhabra, R.P., A experimental study of free fall of cones in Newtonian and non-Newtonian media: Drag coefficient and wall effects, *Chem. Eng. Process.*, **30**, 61 (1991).
Sharma, M.K. and Chhabra, R.P., A experimental study of non-Newtonian fluid flow through fixed and fluidized beds of non-spherical particles, *Can. J. Chem. Eng.*, **70**, 586 (1992).
Sharma, O.P. and Bhatnagar, R.K., Low Reynolds number heat transfer from a sphere in a laminar flow of non-Newtonian fluids, *ZAMM*, **55**, 235 (1975).
Shattuck, M.D., Behringer, R.P., Johnson, G.A. and Geordiadis, J.G., Onset and stability of convection in porous media: Visualization by magnetic resonance imaging, *Phys. Rev. Lett*, **75**, 1934 (1995).
Shaw, M.T., On finding the zero-shear-rate viscosity of polymer melts, *Polym. Eng. Sci.*, **61**, 1166 (2021).
She, K., Trim, L. and Pope, D., Fall velocities of natural sediment particles: A simple mathematical presentation of the fall velocity law, *J. Hyd. Res.*, **43**, 189 (2005).
Sheffield, R.E. and Metzner, A.B., Flow of non-linear fluids through porous media, *AIChE J.*, **22**, 736 (1976).
Shende, T., Niasar, V. and Babaei, M., Effective viscosity and Reynolds number of non-Newtonian fluids using Meter model, *Rheol. Acta*, **60**, 11 (2021a)
Shende, T., Niasar, V. and Babaei, M., Upscaling non-Newtonian rheological fluid properties from pore-scale to Darcy's scale. *Chem. Eng. Sci.*, **239**, 116638 (2021b).

Shenoy, A.R. and Kleinstreuer, C., Flow over a thin circular disk at low to moderate Reynolds numbers, *J. Fluid Mech.*, **605**, 253 (2008).
Shenoy, A.V., PhD Thesis, University of Salford, Salford, U.K. (1977).
Shenoy, A.V., A correlating equation for combined laminar forced and free convection heat transfer to power law fluids, *AIChE J.*, **26**, 505 (1980).
Shenoy, A.V., Natural convection heat transfer to viscoelastic fluids, *Encycl. Fluid Mech.*, **7**, 287 (1988).
Shenoy, A.V., *Heat Transfer to Non-Newtonian Fluids: Fundamentals and Analytical Expressions*. Wiley-VCH, New York, (2017).
Shenoy, A.V. and Mashelkar, R.A., Turbulent free convection heat transfer from a flat vertical plate to a power law fluid, *AIChE J.*, **24**, 344 (1978a).
Shenoy, A.V. and Mashelkar, R.A., Laminar natural convection heat transfer to a viscoelastic fluid, *Chem. Eng. Sci.*, **33**, 769 (1978b).
Shenoy, A.V. and Mashelkar, R.A., Thermal convection in non-Newtonian fluids, *Adv. Heat Transf.*, **15**, 143 (1982).
Shenoy, A.V. and Nakayama, A., Forced convection heat transfer from axisymmetric bodies to non-Newtonian fluids, *Can. J. Chem. Eng.*, **64**, 680 (1986).
Shenoy, A.V. and Ulbrecht, J., Temperature profiles for laminar natural convection flow of dilute polymer solutions past an isothermal vertical flat plate, *Chem. Eng. Commun.*, **3**, 303 (1979).
Sheppard, S., Mantle, M.D., Sederman, A.J., Johns, M.L. and Gladden, L.F., Magnetic resonance imaging study of complex fluid flow in porous media: Flow patterns and qualitative saturation profiling of amphiphilic fracturing fluid displacement in sandstone cores, *Magnet. Reson. Imaging*, **21**, 365 (2003).
Sherman, P., *Industrial Rheology*, Academic Press, London, UK (1970).
Sherwood, J.D., Steady rise of a small spherical gas bubble along the axis of a cylindrical pipe at high Reynolds number, *Eur. J. Mech.*, **20B**, 399 (2001).
Shew, W.L. and Pinton, J.-F., Viscoelastic effects on the dynamics of a rising bubble, *J. Stat. Mech.: Theory and Expt.*, **2006**, P01009 (2006).
Shi, Y-D., Pan, L.-F., Yang, F.-K. and Wang, S.-Q., A preliminary study on the rheological properties of human ejaculate and changes during liquefaction, *Asian J. Androl.*, **6**, 299 (2004).
Shiang, A.H., Lin, J.C., Oztekin, A. and Rockwell, D., Viscoelastic flow around a confined circular cylinder: Measurements using high-image-density particle image velocimetry, *J. Non-Newt. Fluid Mech.*, **73**, 29 (1997).
Shiang, A.H., Oztekin, A., Lin, J.-C. and Rockwell, D., Hydroelastic instabilities in viscoelastic flow past a cylinder in a channel, *Expt. Fluids*, **28**, 128 (2000).
Shibu, S., Chhabra, R.P. and Eswaran, V., Power law fluid flow over a bundle of cylinders at intermediate Reynolds numbers, *Chem. Eng. Sci.*, **56**, 5545 (2001).
Shilton, N.C. and Niranjan, K., Fluidization and its applications to food processing, *Food Struct.*, **12**, 199 (1993).
Shima, A. and Tsujino, T., The behaviour of bubbles in polymer solutions, *Chem. Eng. Sci.*, **31**, 863 (1976).
Shima, A. and Tsujino, T., The behaviour of bubbles in Bingham fluids, *Proc. 7th Int. Ass. Hyd. Res. Cong.*, page A-52 (1977).
Shima, A. and Tsujino, T., The behaviour of gas bubbles in the Casson fluid, *J. App. Mech., Trans. ASME*, **45**, 37 (1978).
Shima, A. and Tsujino, T., The effect of polymer concentration on the bubble behaviour and impulse pressure, *Chem. Eng. Sci.*, **36**, 931 (1981).
Shima, A. and Tsujino, T., On the dynamics of bubbles in polymer aqueous solutions, *Appl. Sci. Res.*, **38**, 255 (1982).
Shima, A., Tsujino, T. and Nanjo, H., Nonlinear oscillations of gas bubbles in viscoelastic fluids, *Ultrasonics*, **24**, 142 (1986).
Shimizu, K., Minekawa, K., Hirose, T. and Kawase, Y., Drop breakage in stirred tanks with Newtonian and non-Newtonian fluid systems, *Chem. Eng. J.*, **72**, 117 (1999).
Shinohara, M. and Hasimoto, H., The force on a small sphere sedimenting in a viscous fluid outside of a circular cylinder, *J. Phys. Soc. Jpn.*, **49**, 1162 (1980).
Shirotsuka, T. and Kawase, Y., Motion and mass transfer of fluid spheres in non-Newtonian systems, *J. Chem. Eng. Jpn.*, **6**, 432 (1973).

Shirotsuka, T. and Kawase, Y., Jetting velocity of a liquid jet in non-Newtonian fluid systems, *J. Chem. Eng. Jpn.*, **7**, 142 (1974a).

Shirotsuka, T. and Kawase, Y., Motion and mass transfer of fluid spheres in viscoelastic fluid systems, *Chem. Eng. (Jpn.)*, **38**, 797 (1974b).

Shirotsuka, T. and Kawase, Y., A Newtonian jet in a non-Newtonian fluid, *Kagaku Kogaku Ronbunshu*, **1**, 219 (1975). Also see *ibid*, 652.

Shmakov, Y.I. and Shamakova, L.M., *Zh. Prikl. Mekl. Tech. Fiz.*, **39**, 81 (1977).

Shokrollahzadeh, A., Terminal settling velocity of a sphere in a non-Newtonian fluid, M.S. Dissertation, University of Alberta, Edmonton (2015).

Shore, H.J. and Harrison, G.M., The effect of added polymers on the formation of drops ejected from a nozzle, *Phys. Fluids*, **17**, 033104 (2005).

Shosho, C.E. and Ryan, M.E., An experimental study of the motion of long bubbles in inclined tubes, *Chem. Eng. Sci.*, **56**, 2191 (2001).

Shukla, R. and Chhabra, R.P., Effect of non-Newtonian characteristics on convective liquid-solid heat transfer in packed and fluidized beds of spherical particles, *Can. J. Chem. Eng.*, **82**, 1071 (2004).

Shukla, R., Dhole, S., Chhabra, R.P. and Eswaran, V., Convective heat transfer for power law fluids in packed and fluidized beds of spheres, *Chem. Eng. Sci*, **59**, 645 (2004).

Shuler, M.L. and Kargi, F., *Bioprocess Engineering*, 2nd ed., Prentice Hal, New Delhi (2002).

Shulman, Z.P., Baikov, V.I. and Zaltsgendler, E.A., An approach to prediction of free convection in non-Newtonian fluids, *Int. J. Heat Mass Transf.*, **19**, 1003 (1976).

Shulman, Z.P. and Levitskiy, S.P., Growth of vapour bubbles in boiling polymer solutions-I. Rheological and diffusional effects, *Int. J. Multiph. Flow*, **39**, 631 (1996).

Shvets, Yu.I. and Vishnevskiy, V.K., Effect of dissipation on convective heat transfer in flow of non-Newtonian fluids, *Heat Transf. Sov. Res.*, **19**, 38 (1987).

Shvetsov, I.A., Investigation of rheological properties of polymer solutions flowing through porous media, *Fluid Mech- Sov. Res.*, **8**(3), 48 (1979).

Shyam, R. and Chhabra, R.P., Effect of Prandtl number on heat transfer from tandem square cylinders immersed in power-law fluids in the low Reynolds number regime, *Int. J. Heat Mass Transf.*, **57**, 742 (2013a).

Shyam, R. and Chhabra, R.P., Natural convection in power-law fluids from two square cylinders in tandem arrangement at moderate Grashof numbers, *Heat Mass Transf.*, **49**, 843 (2013b).

Shyam, R. and Chhabra, R.P., Low Reynolds number flow of power-law fluids over two square cylinders in tandem, *Korean J. Chem. Eng.*, **31**, 1954 (2014).

Shyam, R., Sairamu, M., Nirmalkar, N. and Chhabra, R.P., Free convection from a heated circular cylinder in confined power-law fluids, *Int. J. Thermal Sci.*, **74**, 156 (2013a).

Shyam, R., Sasmal, C. and Chhabra, R.P., Natural convection heat transfer from two vertically aligned circular cylinders in power-law fluids, *Int. J. Heat Mass Transf.*, **64**, 1127 (2013b).

Siddiqa, S., Hossain, M.A. and Gorla, R.S.R., Temperature-dependent density effect on natural convection flow over a horizontal circular disk, *J. Thermophys. Heat Transf.*, **30**, 890 (2016).

Sigli, D., Sur le calcul de l'ecoulement d'un fluide de Reiner-Rivlin autour d'un ellipsoide de revolution, *C.R. Acad. Sci. Paris*, **272**, 1006 (1971).

Sigli, D. and Coutanceau, M., Effect of finite boundaries on the slow isothermal flow of a viscoelastic fluid around a spherical obstacle, *J. Non-Newt. Fluid Mech.*, **2**, 1 (1977).

Sigli, D. and Kaddioui, N., Combined influence of elasticity and inertia on the flow around a rigid sphere, *Prog. Trends Rheol. II*, **142**, p. 122 (1988).

Sigli, D. and Maalouf, A., Comparison of the responses of viscoelastic fluids to a new rheometrical test, *J. Non-Newt. Fluid Mech.*, **9**, 191 (1981).

Sikorski, D., Tabuteau, H. and de Bruyn, J.R., Motion and shape of bubbles rising through a yield stress fluid, *J. Non-Newt. Fluid Mech.*, **159**, 10 (2009).

Siman, R.R., Scheid, C.M. and Massarani, G., Estudo do efeito de parede sobre a velocidade terminal de particulas solidas isometricas em fluido, Nao-Newtoniano III, *Revista Universidade Rural, Serie Ciencias Exatas e da Terra*, **21**, 211 (2002).

Simpkins, P.G. and Kuck, V.J., Air entrapment in coatings by way of a tip-streaming meniscus, *Nature*, **403**, 641 (2000).

Singh, A.K., Harinadha, G., Kishore, N., Barua, P., Jain, T. and Joshi, P., Mixed convective heat transfer phenomena of circular cylinders to non-Newtonian nanofluids flowing upward, *Procedia Eng.*, **127**, 118 (2015).

Singh, D., Prasad, B. and Mishra, P., Flow of Bingham fluid through fixed and fluidized beds, *Ind. J. Tech.*, **14**, 591 (1976).

Singh, J.P. and Denn, M.M., Interacting two-dimensional bubbles and droplets in a yield stress fluid, *Phys. Fluids*, **20**, 040901 (2008).

Singh, J.P., Padhy, S., Shaqfeh, E.S.G. and Koch, D.L., Flow of power-law fluids in fixed beds of cylinders or spheres, *J. Fluid Mech.*, **713**, 491 (2012).

Singh, P. and Joseph, D.D., Sedimentation of a sphere near a vertical wall in an Oldroyd-B fluid. *J. Non-Newt. Fluid Mech.*, **94**, 179 (2000).

Singh, S.N. and Hasan, M.M., Free convection about a sphere at small Grashof number, *Int. J. Heat Mass Transf.*, **26**, 781 (1983).

Singh, S.P., Srivastava, A.K. and Steffe, J., Vibration induced settling of a sphere in a Herschel-Bulkley fluid, *J. Food Eng.*, **13**, 181 (1991).

Singh, U.K. and Chhabra, R.P., Flow of Newtonian and power-law fluids in tube bundles, *Can. J. Chem. Eng.*, **87**, 646 (2009).

Sirignano, W.A., *Fluid Dynamics and Transport of Droplets and Sprays*, Cambridge University Press, New York (1999).

Sirkar, K.K., Transport in packed beds at intermediate Reynolds numbers, *Ind. End. Chem. Fundam.*, **14**, 73 (1975).

Sisavath, S., Jing, X. and Zimmerman, R.W., Creeping flow through a pipe of varying radius, *Phys. Fluids*, **13**, 2762 (2001).

Siska, B., Bendova, H. and Machac, I., Terminal velocity of non-spherical particles falling through a Carreau model fluid, *Chem. Eng. Process*, **44**, 1312 (2005).

Siska, B., Machac, I., Dolecek, P. and Cakl, J., Batch sedimentation of spherical particles in non-Newtonian liquids, *Proc. 8th Int. Conf. on Transport and Sedimentation of Solid Particles*, paper # F2, Jan 24–26, Prague (1995).

Siska, B., Machac, I., Dolecek, P. and Cakl, J., Sedimentation of spherical particles in non-Newtonian fluids in a column of rectangular cross-section, *Paper presented in CHISA-1996*, Prague (1996).

Siskovic, N., Gregory, D.R. and Griskey, R.G., Viscoelastic behaviour of molten polymers in porous media, *AIChE J.*, **17**, 281 (1971).

Sivakumar, P., Bharti, R.P. and Chhabra, R.P., Effect of power-law index on critical parameters for power-law flow across an unconfined cylinder, *Chem. Eng. Sci.*, **61**, 6035 (2006).

Sivakumar, P., Bharti, R.P. and Chhabra, R.P., Steady flow of power law fluids across an unconfined elliptical cylinder, *Chem. Eng. Sci.*, **62**, 1682 (2007).

Sivakumar, S., Nirmalkar, N. and Chhabra, R.P., Effect of orientation on mixed convection from a heated square bar in Newtonian and power-law fluids, *Numer. Heat Transf.*, **65A**, 435 (2014).

Skadsem, H.J., Leulseged, A. and Cayeux, E., Measurement of drilling fluid rheology and modelling of thixotropic behavior, *Appl. Rheol.*, **29**, 1 (2019).

Skartsis, L., Kardos, J.L. and Khomami, B., Resin flow through fiber beds during composite manufacturing processes. Part 1: Review of Newtonian flow through fiber beds, *Polym. Eng. Sci.*, **32**, 221 (1992a). Also see *ibid* 231.

Skartsis, L., Khomami, B. and Kardos, J.L., Polymer flow through fibrous media, *J. Rheol.*, **36**, 589 (1992b).

Skauge, A., Zamani, N., Gausdal Jacobsen, J., Shaker Shiran, B., Al-Shakry, B. and Skauge, T., Polymer flow in porous media: Relevance to enhanced oil recovery. *Colloids Interfaces*, **2**(3), 27 (2018).

Skelland, A.H.P., Momentum, heat and mass transfer in turbulent non-Newtonian boundary layers, *AIChE J.*, **12**, 69 (1966).

Skelland, A.H.P., *Non-Newtonian Flow and Heat Transfer*, Wiley, New York (1967).

Skelland, A.H.P. and Kanel, J.S., Minimum impellor speeds for complete dispersion of non-Newtonian liquid – liquid systems in baffled vessels, *Ind. Eng. Chem. Res.*, **29**, 1300 (1990).

Skelland, A.H.P. and Raval, V.K., Drop size in power law non-Newtonian systems, *Can. J. Chem. Eng.*, **50**, 41 (1972).

Skjetne, E. and Auriault, J.L., New insights on steady non-linear flow in porous media, *Eur. J. Mech. B/Fluids*, **18**, 131 (1999a).

Skjetne, E. and Auriault, J.L., High-velocity laminar and turbulent flow in porous media, *Trans. Porous Media*, **36**, 131 (1999b).
Skjetne, E., Hansen, A. and Gudmundsson, J.S., High-velocity flow in a rough fracture, *J. Fluid Mech.*, **383**, 1 (1999).
Slattery, J.C., *PhD Dissertation*, Univ. of Wisconsin, Madison, WI (1959).
Slattery, J.C., Flow of a simple non-Newtonian fluid past a sphere, *App. Sci. Res.*, **10A**, 286 (1961).
Slattery, J.C., Approximations to the drag force on a sphere moving slowly through either an Ostwald-de Waele or a Sisko fluid, *AIChE J.*, **8**, 663 (1962).
Slattery, J.C., Dimensional considerations in viscoelastic flows, *AIChE J.*, **14**, 516 (1968).
Slattery, J.C., *Momentum, Energy and Mass Transfer in Continua*, McGraw-Hill, New York (1972).
Slattery, J.C. and Bird, R.B., Non-Newtonian flow past a sphere, *Chem. Eng. Sci.*, **16**, 231 (1961).
Slichter, C. S., Theoretical investigation of the motion of ground water, *U.S. Geol. Survey 19th Ann. Rept., Part 2*, 295 (1898).
Slobodov, E.B. and Chepura, I.V., A cellular model of biphasal media, *Theor. Found. Chem. Eng.*, **16**, 235 (1982).
Smagin, I., Pathak, M., Lavrenteva, O.M. and Nir, A., Motion and shape of an axisymmetric visco plastic drop slowly falling through a viscous fluid, *Rheol. Acta*, **50**, 361 (2011).
Smit, G.J.F. and du Plessis, J.P., Pressure drop prediction of power law fluid through granular media, *J. Non-Newt. Fluid Mech.*, **72**, 319 (1997).
Smit, G.J.F. and du Plessis, J.P., Modelling of non-Newtonian purely viscous flow through isotropic high porosity synthetic foams, *Chem. Eng. Sci.*, **54**, 645 (1999).
Smit, G.J.F. and du Plessis, J.P., Modelling of non-Newtonian flow through isotropic porous media, *Math. Eng. Ind.*, **8**, 19 (2000).
Smit, G.J.F. and du Plessis, J.P., On the modelling of non-Newtonian purely viscous flow through high porosity synthetic foams, *Chem. Eng. Sci.*, **60**, 2815 (2005).
Smith, D.E. and Chu, S., The response of flexible polymer coils to a sudden high strain rate flow, *Science*, **281**, 1335 (1998).
Smith, F.W., The behaviour of partially hydrolyzed polyacrylamide solutions in porous media, *J. Pet. Tech.*, **22**, 148 (Feb., 1970).
Smith, G.S., The plunger rheometer-Law of flow for a Newtonian liquid, *J. Inst. Petroleum*, **43**, 227 (1957).
Smith, M.D., Armstrong, R.C., Brown, R.A. and Suresh Kumar, R., Finite element analysis of two- dimensional viscoelastic flows to three dimensional perturbations, *J. Non-Newt. Fluid Mech.*, **93**, 203 (2000).
Smolinski, J.M., Gulari, E. and Manke, C.W., Atomization of dilute polyisobutylene/ mineral oil solutions, *AIChE J.*, **42**, 1201 (1996).
Smolka, L.B. and Belmonte, A., Drop pinch-off and filament dynamics of wormlike micellar fluids, *J. Non-Newt. Fluid Mech.*, **115**, 1 (2003).
Snabre, P. and Magnifotcham, F., Formation and rise of a bubble stream in a viscous liquid, *Eur. Phys. J.*, **B4**, 369 (1998).
Snijkers, F., D'Avino, G., Maffettone, P.L., Greco, F., Hulsen, M.A. and Vermant, J., Rotation of a sphere in a viscoelastic liquid subjected to a shear flow. Part II: Experimental results, *J. Rheol.*, **53**, 459 (2009).
Snijkers, F., D'Avino, G., Maffettone, P.L., Greco, F. and Vermant, J., Effect of viscoelasticity on the rotation of a sphere in shear low, *J. Non-Newt. Fluid Mech.*, **166**, 363 (2011).
Soares, A.A., Anacleto, J., Caramelo, L., Ferreira, J.M. and Chhabra, R.P., Mixed convection from a circular cylinder to power law fluids, *Ind. Eng. Chem. Res.*, **48**, 8219 (2009).
Soares, A.A., Couto, N.D., Duarte Naia, M., Goncalves, N.J. and Rouboa, A., Numerical investigation of effects of buoyancy around a heated circular cylinder in parallel and contra flow, *J. Mech. Sci. Technol.*, **26**, 1501 (2012).
Soares, A.A., Ferreira, J.M., Caramelo, L., Anacleto, J. and Chhabra, R.P., Effect of temperature-dependent viscosity on forced convection heat transfer from a cylinder in cross-flow of power-law fluids, *Int. J. Heat Mass Transf.*, **53**, 4728 (2010).
Soares, A.A., Ferreira, J.M. and Chhabra, R.P., Flow and forced convection heat transfer in cross flow of non-Newtonian fluids over a circular cylinder, *Ind. Eng. Chem. Res.*, **44**, 5815 (2005a).
Soares, A.A., Ferreira, J.M. and Chhabra, R.P., Steady two-dimensional non-Newtonian flow past an array of long circular cylinders up to Reynolds number 500: A numerical study, *Can. J. Chem. Eng.*, **83**, 437 (2005b).

Sobczak, R., Viscosity measurement by spheres falling in a magnetic field, *Rheol. Acta.*, **25**, 175 (1986).

Sobolik, V., Martemyanov, S. and Cognet, G., Study of mass transfer in viscoelastic liquids by segmented electrodiffusion velocity probes, *J. Appl. Electrochem.*, **24**, 632 (1994).

Sobti, A. and Wanchoo, R.K., Creeping flow of viscoelastic fluid through a packed bed, *Ind. Eng. Chem. Res.*, **53**, 14508 (2014).

Sochi, T., Pore- Scale modeling of viscoelastic flow in porous media using a Bautisa – Manero Fluid, *Int. J. Heat Fluid Flow*, **30**, 1202 (2009).

Sochi, T., Non-Newtonian flow in porous media. *Polymer*, **51**(22), 5007 (2010a).

Sochi, T., Flow of non-Newtonian fluids in porous media. *J. Polym. Sci. B: Polym. Phys.*, **48**(23), 2437 (2010b).

Sochi, T., Modelling the flow of yield-stress fluids in porous media. *Trans. Porous Media*, **85**(2), 489 (2010c).

Sochi, T. and Blunt, M.J., Pore-scale network modeling of Ellis and Herschel–Bulkley fluids. *J. Pet. Sci. Eng.*, **60**(2), 105 (2008).

Sodre, J.R. and Parise, J.A.R., Fluid flow pressure drop through an annular bed of spheres with wall effects, *Exp. Therm. Fluid Sci.*, **17**, 265 (1998).

Sokovnin, O.M., Zagoskina, N.V. and Zagoskin, S.N., Hydrodynamics of motion of spherical particles, drops and bubbles in non-Newtonian fluid: Analytical methods of investigation, *Theor. Found. Chem. Eng.*, **46**, 199 (2012a).

Sokovnin, O.M., Zagoskina, N.V. and Zagoskin, S.N., Hydrodynamics of motion of spherical particles, drops and bubbles in non-Newtonian fluid: Numerical methods of investigation, *Theor. Found. Chem. Eng.*, **46**, 464 (2012b).

Sokovnin, O.M., Zagoskina, N.V. and Zagoskin, S.N., Hydrodynamics of motion of spherical particles, drops and bubbles in non-Newtonian fluid: Experimental studies, *Theor. Found. Chem. Eng.*, **47**, 356 (2013).

Solomon, M.J. and Muller, S.J., Flow past a sphere in polystyrene-based Boger fluids: The effect on the drag coefficient of finite extensibility, solvent quality and polymer molecular weight, *J. Non-Newt. Fluid Mech.*, **62**, 81 (1996).

Som, A and Chen, J.L.S., Free convection of non-Newtonian fluids over non-isothermal two-dimensional bodies, *Int. J. Heat Mass Transf.*, **27**, 791 (1984).

Soman, A.M., Pangarkar, V.G., Joshi, J.B. and Kale, D.D., Hydrodynamics in packed columns: Non-Newtonian liquids, *Trans. Indian Inst. Chem. Eng.*, **31**, 61 (1989).

Son, S.Y. and Kihm, K.D., Effect of coal particle size on coal-water slurry (CWS) atomization, *At. Sprays*, **8**, 503 (1998).

Song, T. and Chiew, Y.-M, Settling characteristics of sediments in a moving Bingham fluid, *J. Hyd. Eng.*, **123**, 812 (1997).

Song, D., Gupta, R.K. and Chhabra, R.P., Wall effects on a sphere falling in power-law fluids in cylindrical tubes, *Ind. Eng. Chem. Res.*, **48**, 5845 (2009).

Song, D., Gupta, R.K. and Chhabra, R.P., Effect of blockage on heat transfer from a sphere in power-law fluids, *Ind. Eng. Chem. Res.*, **49**, 3849 (2010).

Song, D., Gupta, R.K. and Chhabra, R.P., Drag on a sphere in Poiseuille flow of power-law fluids, *Ind. Eng. Chem. Res.*, **50**, 13105 (2011).

Song, D., Gupta, R.K. and Chhabra, R.P., Heat transfer to a sphere in tube flow of power-law liquids, *Int. J. Heat Mass Transf.*, **55**, 2110 (2012).

Song, X., Xu, Z., Li, G., Pang, Z. and Zhu, Z., A new method for predicting drag coefficient and settling velocity of spherical and non-spherical particle in Newtonian fluids, *Powder Technol.*, **321**, 242 (2017).

Sontti, S.G. and Atta, A., CFD study on Taylor bubble characteristics in Carreau-Yasuda shear-thinning liquids, *Can. J. Chem. Eng.*, **97**, 616 (2019).

Sorbie, K.S., Network modeling of Xanthan rheology in porous media in the presence of depleted layer effects, Paper presented at *64th Ann. Tech. Conf. and Exhibition*, SPE, San Antonia, TX, Paper # SPE 19651 (October 8–11, 1989).

Sorbie, K.S., Depleted layer effects in polymer flow through porous media, *J. Colloid Interface Sci.*, **139**, 299 (1990); Also see *ibid*, 315.

Sorbie, K.S., *Polymer Improved Oil Recovery*, Blackie and Sons, Glasgow (1991).

Sorbie, K.S., Clifford, P.J. and Jones, E.R.W., The rheology of pseudoplastic fluids in porous media using network modeling, *J. Colloid. Interface Sci.*, **130**, 508 (1989).

Sorbie, K.S. and Huang, Y., Rheological and transport effects in the flow of low concentration Xanthan solutions through porous media, *J. Colloid Interface Sci.*, **145**, 74 (1991).

Sorbie, K.S., Parker, A. and Clifford, P.J., Experimental and theoretical study of polymer flow in porous media, *SPE Res. Eng.*, **2**, 281 (1987).

Sostarecz, M.C. and Belmonte, A., Motion and shape of a viscoelastic drop falling through a viscous fluid, *J. Fluid Mech.*, **497**, 235 (2003).

Soundalgekar, V.M., Laminar free convection flow of a power law fluid from a vertical plate with variable wall temperature, *Ind. J. Pure Appl. Phys.*, **2**, 360 (1964).

Soundalgekar, V.M., Unsteady free convection flow of an elastico-viscous fluid past an infinite plate with constant suction, *Chem. Eng. Sci.*, **26**, 2043 (1971).

Soundalgekar, V.M., Viscous dissipation effects on unsteady free convective flow past an infinite vertical porous plate with constant suction, *Int. J. Heat Mass Transf.*, **15**, 1253 (1972).

Sousa, R.G., Nogueira, S., Pinto, A.M.F.R., Riethmuller, M.L. and Campos, J.B.L.M., Flow in the negative wake of a Taylor bubble rising in viscoelastic carboxymethylcellulose solutions: Particle image velocimetry measurements, *J. Fluid Mech.*, **511**, 217 (2004).

Sousa, R.G., Riethmuller, M.L., Pinto, A.M.F.R. and Campos, J.B.L.M., Flow around individual Taylor bubbles rising in stagnant CMC solutions: PIV measurements, *Chem. Eng. Sci.*, **60**, 1859 (2005).

Sousa, R.G., Riethmuller, M.L., Pinto, A.M.F.R. and Campos, J.B.L.M., Flow around individual Taylor bubbles rising in stagnant polyacrylamide (PAA) solutions, *J. Non-Newt. Fluid Mech.*, **135**, 16 (2006).

Souvaliotis, A. and Beris, A.N., Applications of domain decomposition spectral collocation methods in viscoelastic flows through model porous media, *J. Rheol.*, **36**, 1417 (1992).

Spaid, M.A.A. and Phelan Jr., F.R., Lattice Boltzmann methods for modeling microscale flow in fibrous porous media, *Phys. Fluids*, **9**, 2468 (1997).

Spalding, D. and Ramachandra, V., Turbulent flow and heat transfer in parallel-rod arrays: A numerical treatment, *Physicochem. Hydrodyn.*, **3**, 23 (1982).

Spanjaards, M.M.A., Jaensson, N.O., Hulsen, M.A. and Anderson, P.D., A numerical study of particle migration and sedimentation in viscoelastic Couette flow, *Fluids*, **4**, 25 (2019).

Sparrow, E.M. and Gregg, J.L., Laminar-free-convection heat transfer from the outer surface of a vertical circular cylinder, *Trans. ASME*, **78**(8), 1823 (1956).

Sparrow, E.M. and Lee, L., Analysis of mixed convection about a horizontal cylinder, *Int. J. Heat Mass Transf.*, **19**, 229 (1976).

Sparrow, E.M. and Loeffler Jr., A.L., Longitudinal laminar flow between cylinders arranged in regular array, *AIChE J.*, **5**, 325 (1959).

Spelt, P.D.M., Selerland, T., Lawrence, C.J. and Lee, P.D., Drag coefficient for arrays of cylinders in flows of power law fluids, *Proc. 14th Australasian Fluid Mech. Conf.*, Adelaide, p. 881 (2001).

Spelt, P.D.M., Selerland, T., Lawrence, C.J. and Lee, P.D., Flow of inelastic non-Newtonian fluids through arrays of aligned cylinders Part 1. Creeping flow, *J. Eng. Math.*, **51**, 57 (2004a). Also see *ibid* **51**, 81 (2004b).

Spelt, P.D.M., Yeow, A.Y., Lawrence, C.J. and Selerland, T., Creeping flows of Bingham fluids through arrays of aligned cylinders, *J. Non-Newt. Fluid Mech.*, **129**, 6 (2005).

Spielman, L. and Goren, S.L., Model for predicting pressure drop and filtration efficiency in fibrous media, *Environ. Sci. Technol*, **2**, 279 (1968).

Spratt, K.S., Lee, K.M. and Wilson, P.S., Champagne acoustics, *Phys. Today*, **71**(8), 66 (2018).

Springer, D.S., Loaiciga, H.A., Cullen, S.J. and Everett, L.G., Air permeability of porous materials under controlled laboratory conditions, *Ground Water*, **36**, 558 (1998).

Squires, L. and Squires Jr., W., The sedimentation of thin discs, *Trans. AIChE*, **33**, 1 (1937).

Sreenivasulu, B., Srinivas, B. and Ramesh, K.V., Forced convection heat transfer from a spheroid to a power-law fluid, *Int. J. Heat Mass Transf.*, **70**, 71 (2014).

Sridhar, T., Gupta, R.K., Boger, D.V. and Binnington, R., Steady spinning of the Oldroyd Fluid B: II-Experimental results, *J. Non-Newt. Fluid Mech.*, **21**, 115 (1986).

Srinivas, A.T., Bharti, R.P. and Chhabra, R.P., Mixed convection heat transfer from a cylinder in power-law fluids: Effect of aiding buoyancy, *Ind. Eng. Chem. Res.*, **48**, 9735 (2009).

Srinivas, B. and Ramesh, K.V., Numerical analysis of heat transfer from a spheroidal shaped body to a power-law fluid at finite Reynolds number, *CFD Lett.*, **6**(1), 1–14 (2014).

Srinivas, B.K. and Chhabra, R.P., An experimental study of non-Newtonian fluid flow in fluidized beds: Minimum fluidization velocity and bed expansion characteristics, *Chem. Eng. Process.*, **29**, 121 (1991).

Srinivas, B.K. and Chhabra, R.P., Effect of particle to bed diameter ratio on pressure drop for power law fluid flow in packed beds, *Int. J. Eng. Fluid Mech.*, **5**, 309 (1992).

Srinivas, K.V. and Chhabra, R.P., Pressure drop in two phase cocurrent upward flow in packed beds: Air/non-Newtonian liquid systems, *Can. J. Chem. Eng..*, **72**, 1085 (1994).

Srivastava, A.C., The flow of non-Newtonian liquids near a stagnation point, *ZAMP*, **9**, 80 (1958).

Srivastava, A.C. and Dhiman, A.K., Pulsatile flow and heat transfer of shear-thinning power-law fluids over a confined semi-circular cylinder, *Eur. Phys. J. Plus*, **134**, 144 (2019).

Srivastava, A.C. and Maiti, M.K., Flow of a second order fluid past symmetrical cylinders, *Phys. Fluids*, **9**, 462 (1966).

Srivastava, A.C. and Saroa, M.S., Phenomenon of separation in second-order fluids, *Int. J. Non-Linear Mech.*, **6**, 607 (1971).

Srivastava, A.C. and Saroa, M.S., Heat transfer in a second-order fluid for flow around a circular cylinder, *Int. J. Non-Linear Mech.*, **13**, 243 (1978).

St. Pierre, C. and Tien, C., Experimental investigation of natural convection heat transfer in confined space for non-Newtonian fluid, *Can. J. Chem. Eng.*, **41**, 122 (1963).

Staley, L.M., Tung, M.A. and Kennedy, G.F., Flow properties of dairy waste slurries, *Can. Agri. Eng.*, **15**, 124 (1973).

Stalnaker, J.F. and Hussey, R.G., Wall effects on cylinder drag at low Reynolds number, *Phys. Fluids*, **22**, 603 (1979).

Standish, N. and Bull, G.D., Axial dispersion in mixed and layered beds of spheres, *Chem. Eng. Sci.*, **36**, 774 (1981).

Stapf, S. and Han, S. I. (Eds.)., *NMR imaging in Chemical Engineering*, Wiley, New York (2006).

Stastna, J. and De Kee, D., The rolling ball viscometer and non-Newtonian fluids, *Int. J. Math. Educ. Sci. Technol.*, **18**, 527 (1987).

Steffe, J., *Rheological Methods in Food Process Engineering*, 2nd ed., Freeman Press, East Lansing, MI (1996).

Steg, I. and Katz, D., Rheopexy in some polar fluids and in their concentrated solutions in slightly polar solvents, *J. Appl. Polym. Sci.*, **9**, 3177 (1965).

Stein, S. and Buggisch, H., Rise of pulsating bubbles in fluids with a yield stress, *Z. Angew. Math. Mech.*, **80**, 827 (2000).

Stelter, M., Brenn, G. and Durst, F., The influence of viscoelastic fluid properties on spray formation from flat-fan and pressure-swirl atomizers, *At. Sprays*, **12**, 299 (2002).

Stenzel, K.H., Rubin, A.L., Yamayoshi, W., Miyata, T., Suzucki, T., Sohde, T. and Nishizawa, M., Optimization of collagen dialysis membranes, *Trans. Amer. Soc. Artif. Int. Organs*, **17**, 293 (1971).

Stephenson, J.L. and Stewart, W.E., Optical measurements of porosity and fluid motion in packed beds, *Chem. Eng. Sci.*, **41**, 2161 (1986).

Stevenson, P., Letter to the Editor, *Chem. Eng. Sci.*, **58**, 5379 (2003).

Stewart, C.S., Bubble interactions in low-viscosity liquids, *Int. J. Multiph. Flow*, **21**, 1037 (1995).

Stewart, W.E., Asymptotic calculation of free convection in laminar three-dimensional systems, *Int. J. Heat Mass Transf.*, **14**, 1013 (1971).

Stokes, G.G., On the effect of the internal friction of fluids on the motion of pendulums, *Trans. Camb. Phil. Soc.*, **9**, 8 (1851).

Stokes, J.R., Boehm, M.W. and Baier, S.K., Oral processing, texture and mouthfeel: From rheology to tribology and beyond, *Curr. Opin. Colloid Interface Sci.*, **18**, 349 (2013).

Stone, H.A., Dynamics of drop deformation and breakup in viscous fluids, *Ann. Rev. Fluid Mech.*, **26**, 65 (1994).

Stow, F.S. and Elliott, J.H., Drag on a tethered ball in solutions of drag reducing polymers, *Polymer Lett.*, **8**, 611 (1975).

Street, J.R., The rheology of phase growth in elastic liquids, *Trans. Soc. Rheol.*, **12**, 103 (1968).

Street, J.R., Fricke, A.L. and Reiss, L.P., Dynamics of phase growth in viscous non-Newtonian liquids, *Ind. Eng. Chem. Fund.*, **10**, 54 (1971).

Strnadel, J. and Machač, I., Fall of spherical particles in viscoelastic fluids. *AIP Conf. Proc.*, **1152**, 57–68 (2009).

Strnadel, J., Simon, M. and Machač, I., Wall effects on terminal falling velocity of spherical particles moving in a Carreau model fluid. *Chem. Papers*, **65**(2), 177–184 (2011).

Strom, J.R. and Kintner, R.C., Wall effect for the fall of single drops, *AIChE J.*, **4**, 153 (1958).

Struble, L.J. and Ji, X., Rheology, *Handbook of Analytical Techniques in Concrete Science and Technology*, edited by V.S. Ramachandran and J.J. Beaudon, William Andrew Inc., New York (2000).

Su, X., Xu, Z., Wang, Z., Jin, H., Wu, S. and Lu, Y., Data-driven closure model for the drag coefficient of the creeping flow past a translating sphere in a shear-thinning viscoelastic fluid, *Powder Technol.*, **400**, 117266 (2022).

Subbaraman, V., Mashelkar, R.A. and Ulbrecht, J., Extrapolation procedures for zero shear viscosity with a falling sphere viscometer, *Rheol. Acta*, **10**, 429 (1971).

Subramanayam, N.V. and Chhabra, R.P., The influence of carrier rheology on drag coefficient of nonspherical particles and its significance in hydraulic transport of solids, *Int. J. Bulk Solids Handl.*, **10**, 417 (1990).

Subramaniam, G. and Zuritz, C.A., A study of drag forces on solid spherical particles in power-law bounded flow: Applications to aseptic processing, *J. Food Process Eng.*, **12**, 137 (1990).

Subramaniam, G., Zuritz, C.A. and Ultman, J.S., A drag correlation for single spheres in pseudoplastic tube flow, *Trans. ASAE*, **34**, 2073 (1991).

Suekane, T., Yokouchi, Y. and Hirai, S., Inertial flow structures in a simple-packed bed of spheres, *AIChE J.*, **49**, 10 (2003).

Sugeng, F. and Tanner, R.I., The drag on spheres in viscoelastic fluids with significant wall effects, *J. Non-Newt. Fluid Mech.*, **20**, 281 (1986).

Suh, I.-S., Schumpe, A., Deckwer, W.-D.and Kulicke, W.-M., Gas liquid mass transfer in the bubble column with viscoelastic liquid, *Can. J. Chem. Eng.*, **69**, 506 (1991).

Sulaymon, A.H., Wilson, C.A.M.E. and Alwared, A.I., Experimental determination of the virtual mass coefficient for two spheres accelerating in a power –law fluid, *J. Fluids Eng.*, **132**, 121204-1 (2010).

Sulaymon, A.H., Wilson, C.A.M.E. and Alwared, A.I., An experimental investigation of the settling behaviour of two spheres in a power-law fluid, *J. Non-Newt. Fluid Mech.*, **192**, 29 (2013).

Sullivan, R.R., Further study of the flow of air through porous media, *J. Appl. Phys.*, **12**, 503 (1941).

Sullivan, R.R., Specific surface measurements on compact bundles of parallel fibers, *J. Appl. Phys.*, **13**, 725 (1942).

Sullivan, R.R. and Hertel, K.L., The flow of air through porous media, *J. Appl. Phys.* **11**, 761 (1940).

Sullivan, S.P., Gladden, L.F. and Johns, M.L., Simulation of power-law fluid flow through porous media using lattice Boltzmann techniques. *J. Non-Newt. Fluid Mech.*, **133**(2–3), 91 (2006).

Sumer, B.M. and Fredsoe, J., *Hydrodynamics around Cylindrical Structures*, World Scientific, Singapore (2006).

Sun, A. and Gunasekaran, S. Yield stress in foods: Measurements and applications, *Int. J. Food Prop.*, **12**, 70 (2009).

Sun, B., Guo, Y., Wang, Z., Yang, X., Gong, P., Wang, J. and Wang, N., Experimental study on the drag coefficient of single bubbles rising in static non-Newtonian fluids in wellbore, *J. Nat. Gas Sci. Eng.*, **26**, 867 (2015).

Sun, D. and Zhu, J., Approximate solutions of non-Newtonian flows over a swarm of bubbles, *Int. J. Multiph. Flow*, **30**, 1271 (2004).

Sun, J., Smith, M.D., Armstrong, R.C. and Brown, R.A., Finite element method for viscoelastic flows based on the discrete adaptive viscoelastic stress splitting and the discontinuous Galerkin method: DAVSS-G/DG, *J. Non-Newt. Fluid Mech.*, **86**, 281 (1999).

Sun, J. and Tanner, R.I., Computation of steady flow past a sphere in a tube using a PTT integral model, *J. Non-Newt. Fluid Mech.*, **54**, 379 (1994).

Sun, W., Zhu, C., Fu, T., Ma, Y. and Li, H.Z., Interaction and drag coefficient of three horizontal bubbles with different sizes rising in the shear-thinning fluids, *Int. J. Multiph. Flow*, **125**, 103214 (2020a).

Sun, W., Zhu, C., Fu, T., Yang, H., Ma, Y. and Li, H.Z., The minimum in-line coalescence height of bubbles in non-Newtonian fluid, *Int. J. Multiph. Flow*, **92**, 161 (2017).

Sun, X., Zhang, K., Chen, Y., Li, W. and Qu, J., Study on the settling velocity of drilling cuttings in the power-law fluid, *Powder Technol.*, **362**, 278 (2020b).

Suri, P. and Patel, S.A., Effect of shapes of particle on flow and heat transfer in confined flow, *Sadhana (Springer)*, In Press (2023).

Suri, P., Patel, S.A., Chhabra, R. P., Falling ball method for determining zero shear and shear-dependent viscosity of polymeric systems: Solutions, melts, and composites, *J. Vinyl Additive Technol.*, in press (2023a).

Suri, P., Verma, A., Patel, S.A., and Chhabra, R. P., Free convection from a single and a row of spheres in power-law fluids at very small Grashof numbers, submitted for publication (2023b)

Sutterby, J.L, Laminar converging flow of dilute polymer solutions in conical sections Part I: Viscosity data, new viscosity model, tube flow solution, *AIChE J.*, **12**, 63 (1966).

Sutterby, J.L., Falling sphere viscometer, *J. Phys. E. Sci. Inst.*, **6**, 1001 (1973a).

Sutterby, J.L., Falling sphere viscometry, I. Wall and inertial corrections to Stokes law in long tubes, *Trans. Soc. Rheol.*, **17**, 559 (1973b).

Syrakos, A., Georgiou, G.C. and Alexandrou, A.N., Thixotropic flow past a cylinder, *J. Non-Newt. Fluid Mech.*, **220**, 44 (2015).

Szabo, M.T., Laboratory investigations of factors influencing polymer flood performance, *Soc. Pet. Eng. J.*, **13**, 338 (1975a).

Szabo, M.T., Some aspects of polymer retention in porous media using a C14-tagged hydrolyzed polyacrylamide, *Soc. Pet. Eng. J.*, **13**, 323 (1975b).

Szady, M.J., Salamon, T.R., Liu, A.W., Bornside, D.E., Armstrong, R.C. and Brown, R.A., A new mixed finite element method for viscoelastic flows governed by differential constitutive equations, *J. Non-Newt. Fluid Mech.*, **59**, 215 (1995).

Taamneh, Y. Non- Newtonian fluid flow around ellipsoidal particles, *Int. J. Fluid Mech. Res.*, **38**(2), 111 (2011).

Tabak, S., Askarov, B., Rashidov, U., Tabak, I., Manor, G. and Shmulevich, I., Airflow through granular beds packed with cottonseeds, *Biosys. Eng.*, **88**, 163 (2004).

Tabata, M. and Itakura, K., A precise computation of drag coefficients of a sphere, *Int. J. Comp. Fluid Dyn.*, **9**, 303 (1996).

Tabuteau, H., Coussot, P. and Bruyn, J.R., Drag force on a sphere in steady motion through a yield-stress fluid, *J. Rheol.*, **51**, 125 (2007).

Tadaki, T. and Maeda, S., On the shape and velocity of single air bubbles rising in various liquids, *Chem. Eng. (Tokyo)*, **25**, 254 (1961).

Taghi Esfidani, M., Roza Oshaghi, M. Afshin, H. and Firoozabadi, B., Modeling and experimental investigation of bubble formation in shear-thinning liquids, *J. Fluids Eng.*, **139**, 0713021 (2017).

Takagi, S. and Matsumoto, Y., Force acting on a rising bubble in a quiescent fluid. *Proc. ASME Summer Meeting on Numerical Methods for Multiphase Flow*, San Diego, CA, p. 575 (1996).

Takahashi, K., Maeda, M. and Ikai, S., Experimental study on heat transfer from a cylinder submerged in a non-Newtonian fluid. *Die dem Nahenungsamsatz von k Pohlhausen Genugen, Lilenthal, Bericht 510*, pp. 335–339 (1977).

Takahashi, M., Maeda, M. and Ikai, S., Experimental study on heat transfer from a cylinder submerged in a non-Newtonian fluid, *Proc. 6th Intl. Heat Transf. Conf.*, Vol. 5, p.335, Hemisphere, Washington, DC (1978).

Takamatsu, H., Fujii, M. and Fujii, T., A numerical analysis of free convection around an isothermal sphere (Effects of space and Prandtl number), *JSME Int J., Ser. II*, **31**, 66 (1988).

Takemura, F. and Yabe, A., Rising speed and dissolution rate of a carbon di oxide bubble in slightly contaminated water, *J. Fluid Mech.*, **378**, 319 (1999).

Tal (Thau), R. and Sirignano, W.A., Cylindrical cell model for the hydrodynamics of particle assemblages at intermediate Reynolds numbers, *AIChE J.*, **28**, 233 (1982).

Talmon, A.M., Determination of non-segregating tailings conditions. In *19th World dredging congress; JGS Pennekamp*, WGM van Kesteren (2010).

Talmon, A. and Mastbergen, D., Solids transport by drilling fluids: Concentrated bentonite-sand-slurries. In *Proceedings of the 12th International Conference on Transportation and Sedimentation of Solid Particles*, Institute of Hydrodynamics, Academy of Sciences of the Czech Republic, Prague, p. 641 (2004).

Talmon, A.M. and Huisman, M., Fall velocity of particles in shear flow of drilling fluids, *Tunn. Undergr. Space Technol.*, **20**, 193 (2005).

Talmon, A.M., van Kesteren, W.G., Sittoni, L. and Hedblom, E.P., Shear cell tests for quantification of tailings segregation. *Can. J. Chem. Eng.*, **92**(2), 362 (2013).

Talon, L. and Bauer, D., On the determination of a generalized Darcy equation for yield-stress fluid in porous media using a Lattice-Boltzmann TRT scheme. *Eur. Phys. J. E*, **36**(12), 1 (2013).

Talwar, K.K., Ganpule, H.K. and Khomami, B., A note on selection of spaces in computation of viscoelastic flows using the hp-finite element method, *J. Non-Newt. Fluid Mech.*, **52**, 293 (1994).

Talwar, K.K. and Khomami, B., Application of higher order finite element methods to viscoelastic flow in porous media, *J. Rheol.*, **36**, 1377 (1992).

Talwar, K.K. and Khomami, B., Flow of viscoelastic fluids past periodic square arrays of cylinders: Inertial and shear thinning viscosity and elasticity effects, *J. Non-Newt. Fluid Mech.*, **57**, 177 (1995).

Tam, C.K.W., The drag on a cloud of spherical particles in low Reynolds number flow, *J. Fluid Mech.*, **38**, 537 (1969).

Tam, K.C., Moussa, T. and Tiu, C., Ideal elastic fluids of different viscosity and elasticity levels, *Rheol. Acta*, **28**, 112 (1989).

Tamayol, A. and Bahrami, M., Parallel flow through ordered fibers: An analytical approach, *J. Fluids Eng.*, **132**, 1145021 (2010).

Tanasawa, I. and Yang, W.-J., Dynamic behaviour of a gas bubble in viscoelastic liquids, *J. App. Phys.*, **41**, 4526 (1970).

Taneda, S., Experimental investigation of the wake behind a sphere at low Reynolds numbers, *J. Phys. Soc. Jpn.*, **11**, 1104 (1956a).

Taneda, S., Experimental investigation of the wake behind cylinders and plates at low Reynolds numbers, *J. Phys. Soc. Jpn.*, **11**, 302 (1956b).

Tang, G.H. and Lu, Y.B., A resistance model for Newtonian and power-law non-Newtonian fluid transport in porous media. *Trans. Porous Media*, **104**(2), 435 (2014).

Tang, L. and Johnson, A.T., Flow visualization of mixed convection about a sphere, *Int. Comm. Heat Mass Transf.*, **17**, 67 (1990).

Tang, L. and Johnson, A.T., Mixed convection about fruits, *J. Agric. Eng. Res.*, **51**, 15 (1992).

Tang, L., Johnson, A.T. and Mc Cuen, R.H., Empirical study of mixed convection about a sphere, *J. Agric. Eng. Res.*, **50**, 197 (1991).

Tang, P., Chan, H.-K. and Raper, J., Prediction of aerodynamic diameter of particles with rough surfaces, *Powder Technol.*, **147**, 64 (2004).

Tanner, R.I., End effects in falling ball viscometry, *J. Fluid Mech.*, **17**, 161 (1963).

Tanner, R.I., Observations on the use of Oldroyd-type equations of state for viscoelastic liquids, *Chem. Eng. Sci.*, **19**, 349 (1964).

Tanner, R.I., Non-Newtonian fluid parameter estimation using conical flows, *Ind. Eng. Chem. Fund.*, **5**, 55 (1966).

Tanner, R.I., Engineering rheology, *Chem. Eng. Australia*, **7**, 18 (1982).

Tanner, R.I., Upper and lower bounds for inelastic flows using BEM and FEM, *J. Non-Newt. Fluid Mech.*, **38**, 101 (1990).

Tanner, R.I., Stokes paradox for power-law flow around a cylinder, *J. Non-Newt. Fluid Mech.*, **50**, 217 (1993).

Tanner, R.I., *Engineering Rheology*, 2nd ed., Oxford University Press, New York (2000).

Tanner, R.I., To yield or not to yield-that is the question, *J. Non-Newt. Fluid Mech.*, **310**, 104941 (2022).

Tanner, R.I., Housiadas, K.D. and Qi, F., Mechanism of drag increase on spheres in viscoelastic cross-shear flows, *J. Non-Newt. Fluid Mech.*, **203**, 51 (2014).

Tanner, R.I. and Walters, K., *Rheology: An Historical Perspective*, Elsevier, Amsterdam (1998).

Tarlet, D., Younes, E., Roux, S., Levy, A. and Burghelea, T., Stopping of a solid object in an elasto- visco-plastic material, *J. Non-Newt. Fluid Mech.*, **263**, 120 (2019).

Tate, R.W. and Janssen, L.F., Droplet size data for agricultural spray nozzles, *Trans. ASAE*, **9**, 303 (1966).

Tatersall, G.H., *The Rheology of Fresh Concrete*, Pitman, London (1983).

Tatham, J.P., Carrington, S., Odell, J.A., Gamboa, A.C., Müller, A.J. and Saez, A.E., Extensional behaviour of hydroxypropyl guar solutions: Optical rheometry in opposed jets and flow through porous media, *J. Rheol.*, **39**, 961 (1995).

Tatum, J.A., Finnis, M.V., Lawson, N.J. and Harrison, G.M., 3-D particle image velocimetry of the flow field around a sphere sedimenting near a wall, Part 2: Effects of distance from the wall, *J. Non-Newt. Fluid Mech.*, **127**, 94 (2005).

Taud, H., Martinez-Angeles, R., Parrot, J.F. and Hernandez-Escobedo, L., Porosity estimation method by X-ray computed tomography, *J. Pet. Sci. Eng.*, **47**, 209 (2005).

Tavlarides, L.L. and Stamatoudis, M., The analysis of interphase reactions and mass transfer, *Adv. Chem. Eng.*, **11**, 199 (1981).

Taylor, G.I., The viscosity of a fluid containing small drops of another fluid, *Proc. Roy. Soc.*, **A138**, 40 (1932).

Taylor, G.I., The two coefficients of viscosity for an incompressible fluid containing air bubbles, *Proc. Roy. Soc.*, **A226**, 34 (1954).

Taylor, M.A., Quantitative measures for shape and size of particles, *Powder Technol.*, **124**, 94 (2002).

Taylor, S.E., Rheology and structure of corn starch suspensions in water-poly (propylene glycol) mixtures, *J. Disp. Sci. Technol.*, **34**, 887 (2013).

Teeuw, D. and Hesselink, F., Power law flow and hydrodynamic behaviour of biopolymer solutions in porous media, Paper # SPE 8982, *Presented at the 1980 SPE Int. Sym. Oilfields & Geothermal Chem.*, Stanford, CA (1980).

Tehrani, M.A., An experimental study of particle migration in pipe flow of viscoelastic fluids, *J. Rheol.*, **40**, 1057 (1996).

Ten Cate, A., Nieuwstad, C.H., Derksen, J.J. and Van den Akker, H.E.A., Particle image velocimetry experiments and lattice-Boltzmann simulations on a single sphere settling under gravity, *Phys. Fluids*, **14**, 4012 (2002).

Terasaka, K. and Shibata, H., Oxygen transfer in viscous non-Newtonian liquids having yield stress in bubble columns, *Chem. Eng. Sci.*, **58**, 5331 (2003).

Terasaka, K. and Tsuge, H., Bubble formation at a single orifice in highly viscous liquids, *J. Chem. Eng. Jpn.*, **23**, 160 (1990).

Terasaka, K. and Tsuge, H., Bubble formation at a single orifice in non-Newtonian liquids, *Chem. Eng. Sci.*, **46**, 85 (1991).

Terasaka, K. and Tsuge, H., Bubble formation at orifice in viscoelastic liquids, *AIChE J.*, **43**, 2903 (1997).

Terasaka, K. and Tsuge, H., Bubble formation at a nozzle submerged in viscous liquids having yield stress, *Chem. Eng. Sci.*, **56**, 3237 (2001).

Termonia, Y., Permeability of sheets of nonwoven fibrous media, *Chem. Eng. Sci.*, **53**, 1203 (1998).

Terzaghi, K., Principles of soil mechanics. IV. Settlement and consolidation of clay. *Engineering News-Record*, **95**, 874 (1925).

Teske, M.N. and Bilanin, A.J., Drop-size scaling analysis of non-Newtonian fluids, *At. Sprays*, **4**, 473 (1994).

Teske, M.N. and Thistle, H.W., Droplet size scaling of agricultural spray material by dimensional analysis, *At. Sprays*, **10**, 147 (2000).

Thakur, P., Mittal, S., Tiwari, N. and Chhabra, R.P., The motion of a rotating cylinder in a stream of Bingham Plastic fluids, *J. Non-Newt. Fluid Mech.*, **235**, 29 (2016).

Thakur, P., Tiwari, N. and Chhabra R.P., Forced convection in a Bingham Plastic fluid from a heated rotating cylinder, *J. Chem. Eng. Jpn.*, **52**, 730 (2019a).

Thakur, P., Tiwari, N. and Chhabra, R.P., Momentum and heat transfer from an asymmetrically confined rotating cylinder in a power-law fluid, *Int. J. Thermal Sci.*, **137**, 410 (2019b).

Thakur, P., Tiwari, N. and Chhabra, R.P., Flow of a power-law fluid across a rotating cylinder in confinement, *J. Non-Newt. Fluid Mech.*, **251**, 145 (2018).

Theodoropoulou, M., Karoutsos, V. and Tsakiroglou, C., Investigation of the contamination of fractured formations by non-Newtonian oil pollutants, *Environ. Forensics*, **2**, 321 (2001).

Thomas, A.D., A rational design philosophy for long distance slurry pipelines, *Chem. Eng. Australia*, **ChE 2**, 22 (1977).

Thomas, R.H. and Walters, K., The unsteady motion of a sphere in an elastico-viscous liquid, *Rheol. Acta*, **5**, 23 (1966).

Thomas, S. and Farouq Ali, S.M., Flow of emulsions in porous media, and potential for enhanced oil recovery, *J. Pet. Sci. Eng.*, **3**, 121 (1989).

Thompson, K.E. and Fogler, H.S., Modeling flow in disordered packed beds from pore-scale fluid mechanics, *AIChE J.*, **43**, 1377 (1997).

Thompson, R.L., Sica, L.U.R. and de Souza Mendes, P.R., The yield stress tensor, *J. Non-Newt. Fluid Mech.*, **261**, 211 (2018).

Thompson, R.L. and Soares, E.J., Viscoplastic dimensionless numbers, *J. Non-Newt. Fluid Mech.*, **238**, 57 (2016).

Thompson, T.L. and Clark, N.N., A holistic approach to particle drag prediction, *Powder Technol.*, **67**, 57 (1991).

Thumati, V.T. M.Tech Thesis, IIT Kanpur, India (2016).

Thumati, V.T., Patel, S.A., Gupta, A.K. and Chhabra, R.P., Effect of confinement and fluid yield stress on heat transfer from an isothermal sphere *J. Chem. Eng. Jpn.*, **51**, 899 (2018).

Tian, S., Wall effects for a spherical particle in confined shear-thickening fluids, *J. Non-Newt. Fluid Mech.*, **257**, 13 (2018). Also see *ibid*, **299**, 104712 (2022).

Tian, X.-W., Xu, S.-M., Sun, Z.-H., Wang, P., Xu, L. and Zhang, Z., Experimental study on flow and heat transfer of power law fluid in structural packed porous media of particles, *Exp. Therm. Fluid Sci.*, **90**, 37 (2018).

Tiefenbruck, G., *PhD Dissertation*, Caltech, Pasadena, CA (1979).

Tiefenbruck, G. and Leal, L.G., A note on rods falling near a vertical wall in a viscoelastic liquid, *J. Non-Newt. Fluid Mech.*, **6**, 201 (1980a).

Tiefenbruck, G. and Leal, L.G., A numerical study of the motion of a viscoelastic fluid past rigid spheres and spherical bubbles, *J. Non-Newt. Fluid Mech.*, **10**, 115 (1982).

Tiefenbruck, G.F. and Leal, L.G., A note on the slow motion of a bubble in a viscoelastic liquid, *J. Non-Newt Fluid Mech.*, **7**, 257 (1980b).

Tien, C., Laminar natural convection heat transfer from vertical plate to power-law fluid, *Appl. Sci. Res.*, **17**, 233 (1967).

Tien, C. and Tseui, H.-S., Laminar natural convection heat transfer in Ellis fluids, *Appl. Sci. Res.*, **20**, 131 (1969).

Tigoiu, V.M., Viscoelastic fluid flows in a falling cylinder viscometer and the evaluation of shear viscosity, *Proc. XXI Int. Conf. Theor. Appl. Mech*, Warsaw (2004).

Timol, M.G. and Kalthia, N.L., Similarity solutions of three-dimensional boundary layer equations of non-Newtonian fluids, *Int. J. Non-Linear Mech.*, **21**, 475 (1986).

Ting, R.Y., Viscoelastic effects of polymers on single bubble dynamics, *AIChE J.*, **21**, 810 (1975).

Ting, R.Y. and Ellis, A.T., Bubble growth in dilute polymer solutions, *Phys. Fluids*, **17**, 1461 (1974). Also see *ibid* **20**, 1427 (1977).

Tirtaatmadja, V. and Sridhar, T., A filament stretching device for measurement of extensional viscosity, *J. Rheol.*, **37**, 1081 (1993).

Tirtaatmadja, V., Uhlherr, P.H.T. and Sridhar, T., Creeping motion of spheres in Fluid M-1, *J. Non-Newt. Fluid Mech.*, **35**, 327 (1990).

Tiu, C., Modelling flow with geometric parameters, *Developments in Plastic Technology-2*, A. Whelan and J.L. Craft, Elsevier, Amsterdam, Chapter 7 (1985).

Tiu, C., Low, G.S. and Moreno, L., Flow of viscoelastic fluids through packed beds, *Proc. IIIrd Nat. Conf. Rheol.*, Melbourne, May, p. 95 (1983).

Tiu, C. and Moreno, L., Flow of polymer solutions through packed beds, *Proc. IX Int. Cong. Rheol.*, Mexico, p. 429 (1984).

Tiu, C., Quinn, B.W. and Uhlherr, P.H.T., Flow of non-Newtonian fluids through beds of various packing geometries, *Proc. 5th Australasian Conf. Hyd. Fluid. Mech.*, Christchurch, p. 411 (1974).

Tiu, C., Zhou, J.Z.Q., Nicolae, G., Tunan, N.F. and Chhabra, R.P., Flow of viscoelastic polymer solutions in mixed beds of particles, *Can. J. Chem. Eng.*, **75**, 843 (1997).

Tiwari, A.K. and Chhabra, R.P., Laminar natural convection in power-law liquids from a heated semi-circular cylinder with its flat side oriented downward, *Int. J. Heat Mass Transf.*, **58**, 553 (2013).

Tiwari, A.K. and Chhabra, R.P., Effect of orientation on steady laminar free convection heat transfer in power-law fluids from a heated triangular cylinder, *Numer. Heat Transf. Part A*, **65**, 780 (2014a).

Tiwari, A.K. and Chhabra, R.P., Momentum and heat transfer characteristics for the flow of power-law fluids over a semi-circular cylinder, *Numer. Heat Transf. Part A*, **66**, 1365 (2014b).

Tiwari, A.K. and Chhabra R.P., Mixed convection in power-law fluids from a heated semi-circular cylinder: Effect of aiding buoyancy, *Numer. Heat Transf. Part A*, **67**, 330 (2015a).

Tiwari, A.K. and Chhabra, R.P., Momentum and heat transfer from a semi-circular cylinder in Bingham Plastic fluids, *Appl. Math. Model.*, **39**, 7045 (2015b).

Tiwari, A.K. and Chhabra, R.P., Laminar free convection in Bingham plastic fluids from an isothermal semi-circular cylinder, *J. Thermophys. Heat Transf.*, **30**, 369 (2016).

Tiwari, S.S., Pal, E., Bale, S., Minocha, N., Patwardhan, A.W., Nandakumar, K. and Joshi, J.B., Flow past a single stationary sphere. 1. Experimental and numerical techniques, *Powder Technol.*, **365**, 115 (2020a).

Tiwari, S.S., Pal, E., Bale, S., Minocha, N., Patwardhan, A.W., Nandakumar, K. and Joshi, J.B., Flow past a single stationary sphere. 2. Regime mapping and effect of external disturbances, *Powder Technol.*, **365**, 215 (2020b).

Tjaden, B., Cooper, S.J., Brett, D.J., Kramer, D. and Shearing, P.R., On the origin and application of the Bruggeman correlation for analyzing transport phenomena in electrochemical systems. *Curr. Opin. Chem. Eng.*, **12**, 44 (2016).

Tobis, J., Influence of bed geometry on its frictional resistance under turbulent flow conditions, *Chem. Eng. Sci.*, **55**, 5359 (2000).

Tobis, J., Modelling the pressure drop in the packing of complex geometry, *Ind. Eng. Chem. Res.*, **41**, 2552 (2002).

Tokpavi, D.L., Jay, P., Magnin, A. and Jossic, L., Experimental study of the very slow flow of a yield stress fluid around a circular cylinder, *J. Non-Newt. Fluid Mech.*, **164**, 35 (2009).

Tokpavi, D.L. and Magnin A., Particle settling in yield stress fluids: Limiting time, distance and applications, *J. Non- Newt. Fluid Mech.*, **238**, 189 (2016).

Tokpavi, D.L., Magnin, A. and Jay, P., Very slow flow of Bingham viscoplastic fluid around a circular cylinder, *J. Non-Newt. Fluid Mech.*, **154**, 65 (2008).

Toll, S., A solution technique for longitudinal Stokes flow around multiple aligned cylinders, *J. Fluid Mech.*, **439**, 199 (2001).

Tomboulides, A.G., Orszag, S.A. and Karniadakis, G.E., Direct and large eddy simulation of axisymmetric wake, *31st Aerospace Sciences Meeting*, AIAA Paper-93-0546 (1993).

Tomita, Y., On the fundamental formula of non-Newtonian flow, *Bull. JSME*, **2**, 469 (1959).

Tomiyama, A., Kataoka, I., Zun, I. and Sakaguchi, T., Drag coefficients of single bubbles under normal and micro-gravity conditions, *JSME Int.*, **41B**, 472 (1998).

Tonini, R.D., Fluidization with non-Newtonian fluids, *Encyclopedia of Fluid Mechanics*, Vol. 6, Gulf, Houston, Edited by N. P. Cheremisinoff, p. 495 (1987).

Tonini, R.D., Bohm, U. and Brea, F.M., Fluidisation with highly viscous and non-Newtonian fluids: Mass transfer from the fluidized bed to the inner wall of an annulus, *Chem. Eng. J.*, **22**, 51 (1981).

Toose, E.M., Geurts, B.J. and Kuerten, J.G.M., A boundary integral method for two-dimensional (non)-Newtonian drops in slow viscous flow, *J. Non-Newt. Fluid Mech.*, **60**, 129 (1995).

Toose, E.M., Geurts, B.J. and Kuerten, J.G.M., A 2-D boundary element method for simulating the deformation of axisymmetric compound non-Newtonian drops, *Int. J. Numer. Meth. Fluids*, **30**, 653 (1999).

Toose, E.M., van Den Ende, D., Guerts, B.J., Kuerten, J.G.M. and Zandbergen, P.J., Axisymmetric non-Newtonian drops treated with a boundary integral method, *J. Eng. Math.*, **30**, 131 (1996).

Torobin, L.B. and Gauvin, W.H., Fundamental aspects of solids-gas flow, *Can. J. Chem. Eng.*, **37**, 167 (1959).

Torobin, L.B. and Gauvin, W.H., Fundamental aspects of solids-gas flow Part IV; The effects of particles rotation, roughness and shape, *Can. J. Chem. Eng.*, **38**, 142 (1960).

Torrest, R.S., Rheological properties of aqueous solutions of the polymer Natrosol 250 HHR, *J. Rheol.*, **26**, 143 (1982).

Torrest, R.S., Particle settling in viscous non-Newtonian hydroxyethyl cellulose polymer solutions, *AIChE J.*, **29**, 506 (1983).

Tosco, T., Marchisio, D.L., Lince, F. and Sethi, R., Extension of the Darcy–Forchheimer law for shear-thinning fluids and validation via pore-scale flow simulations. *Trans. Porous Media*, **96**(1), 1 (2013).

Tosun, I. and Mousa, H., Flow through packed bed: Wall effect on drag force, *Chem. Eng. Sci.*, **41**, 2962 (1986).

Toussaint, F., Roy, C. and Jezequel, P.-H., Reducing shear-thickening of cement-based suspensions, *Rheol. Acta*, **48**, 883 (2009).

Townsend, P., A numerical simulation of Newtonian and viscoelastic flow past stationary and rotating cylinders, *J. Non-Newt. Fluid Mech.*, **6**, 201 (1980).

Townsend, P., On the numerical simulations of two-dimensional time-dependent flows of Oldroyd fluids, *J. Non-Newt. Fluid Mech.*, **14**, 268 (1984).

Tozeren, H., Drag on eccentrically positioned spheres translating and rotating in tubes, *J. Fluid Mech.*, **129**, 77 (1983).

Trahan, J.F. Stokes drag on a thin circular disk moving edgewise midway between parallel plane boundaries, *J. Fluids Eng.*, **128**, 887 (2006).

Trahan, J.F., Folse, R.F. and Hussey, R.G., Combined side wall and bottom wall effects on the Stokes velocity of a disk moving broadside, *Phys. Fluids*, **1**, 1625 (1989).

Trahan, J.F., Wehbeh, E.G. and Hussey, R.G., The limits of lubrication theory for a disk approaching parallel plane wall, *Phys. Fluids*, **30**, 939 (1987).

Trambouze, P., Computational fluid dynamics applied to chemical reaction engineering, *Rev. Inst. Fr. Pet.*, **48**, 595 (1993).

Tran, A., Rudolph, M.L. and Manga, M., Bubble mobility in mud and magnetic volcanoes, *J. Volcano. Geotherm. Res.*, **294**, 11 (2015).

Tran, Q.-K., Trinh, D.T., Horsley, R.R. and Reizes, J.A., Drag coefficient and settling velocities of spheres in yield-pseudoplastic slurries, *Developments in Non-Newtonian Flows (ASME)*, D.A. Siginer, W.E. Van Arsdale, M.C. Atan and A.N. Alexandrou, American Society of Mechanical Engineers, New York, AMD-Vol. 175, p. 131 (1993).

Tran-Cong, S., Gay, M. and Michaelides, E.E., Drag coefficients of irregularly shaped particles, *Powder Technol.*, **139**, 21 (2004).

Tran-Son-Tay, R., Beaty, B.B., Acker, D.N. and Hochmuth, R.M., Magnetically driven, acoustically tracked translating-ball rheometer for small, opaque samples, *Rev. Sci. Instrum.*, **59**, 1399 (1988).

Tran-Son-Tay, R., Coffey, B.E. and Hochmuth, R.M., The motion of a ball oscillating in a bounded fluid: Inertial and wall effects, *J. Rheol.*, **34**, 169 (1990).

Trappe, V. and Sandkuhler, P., Colloidal gels-low density disordered solid-like states, *Curr. Opin. Colloid Interface Sci.*, **8**, 494 (2004).

Travkin, V.S. and Catton, I., Porous media transport descriptions – non-local, linear and non-linear against effective thermal/fluid properties, *Adv. Colloid Interface Sci.*, **76/77**, 389 (1998).

Traynis, V.V., *Parameters and Flow Regimes for Hydraulic Transport of Coal by Pipelines*, Terraspace, Rockville, MD (1977).

Tripathi, A. and Chhabra, R.P., Slow flow of a power law liquid drop in another immiscible power law liquid, *Arch. Appl. Mech.*, **62**, 495 (1992a).

Tripathi, A. and Chhabra, R.P., Slow power law fluid flow relative to an array of cylinders, *Ind. Eng. Chem. Res.*, **31**, 2754 (1992b).

Tripathi, A. and Chhabra, R.P., Hydrodynamics of creeping motion of an ensemble of power law liquid drops in an immiscible power law medium, *Int. J. Eng. Sci.*, **32**, 791 (1994).

Tripathi, A. and Chhabra, R.P., Drag on spheroidal particles in dilatant fluids, *AIChE J.*, **41**, 728 (1995).

Tripathi, A. and Chhabra, R.P., Transverse laminar flow of non-Newtonian fluids over a bank of cylinders, *Chem. Eng. Commun.*, **147**, 197 (1996).

Tripathi, A., Chhabra, R.P. and Sundararajan, T., Power-law fluid flow over spheroidal particles, *Ind. Eng. Chem. Res.*, **33**, 403 (1994).

Tripathi, M.K., Sahu, K.C. and Govindarajan, R., Dynamics of an initially spherical bubble rising in quiescent liquid, *Nature Commun.*, **6**, 6268 (2015a).

Tripathi, M.K., Sahu, K.C., Karapetsas, G. and Matar, O.K., Bubble rise dynamics in a viscoplastic material, *J. Non-Newt. Fluid Mech.*, **222**, 217 (2015b).

Trivedi, M., Sharma, A., Agarwal, K., Chhabra R. P., and Nirmalkar, N. A novel method to enhance extraction efficiency and interfacial mass transfer from spherical drops, *Int. J. Heat Mass Transf.*, **203**, 123776 (2023).

Trouton, F.T., The coefficient of viscous traction and its relation to that of viscosity, *Proc. Roy. Soc.*, **A77**, 426 (1906).

Tsakalakis, K.G. and Stamboltzis, G.A., Prediction of the settling velocity of irregularly shaped particles, *Miner. Eng.*, **14**, 349 (2001).

Tsakiroglou, C.D., A methodology for the derivation of non-Darcian models for the flow of generalized Newtonian fluids in porous media, *J. Non-Newt. Fluid Mech.*, **105**, 79 (2002).

Tsamopoulos, J., Dimakopoulos, Y., Chatzidai, N., Karapetsas, G. and Pavlidis, M., Steady bubble rise and deformation in Newtonian and viscoplastic fluids and conditions for bubble entrapment, *J. Fluid Mech.*, **601**, 123 (2008).

Tsay, S.-Y. and Chou, C.-H., Laminar convection to rotating disks in non-Newtonian power law fluids, *Int. Comm. Heat Mass Transf.*, **10**, 377 (1983).

Tsotsas, E., Letter to the Editor, *Chem. Eng. Sci.*, **57**, 1827 (2002).

Tsubouchi, T., Sato, S. and Nagakura, K., Heat transfer from fine wires and particles by natural convection, *Trans. Jpn. Soc. of Mech. Eng. C*, **25**, 798 (1959).

Tsuge, H. and Hibino, S., Effect of column wall on the ascending velocity of gas bubbles in various liquids, *Int. Chem. Eng.*, **15**, 186 (1975).

Tsuge, H. and Terasaka, K., Volume of bubbles formed from an orifice submerged in highly viscous Newtonian and non-Newtonian liquids, *J. Chem. Eng. Jpn.*, **22**, 418 (1989).

Tsukada, T., Mikami, H., Hozawa, M. and Imaishi, N., Theoretical and experimental studies of the deformation of bubbles moving in quiescent Newtonian and non-Newtonian liquids, *J. Chem. Eng. Jpn.*, **23**, 192 (1990).

Tucker, G.S. and Withers, P.M., Determination of residence time distribution of non-settling food particles in viscous food carrier fluids using Hall effect sensors, *J. Food Process Eng.*, **17**, 401 (1994).

Tullock, D.L., Phan-Thien, N. and Graham, A.L., Boundary element simulations of spheres settling in circular, square and triangular conduits. *Rheol. Acta*, **31**, 139 (1992).

Tung, M.A., Watson, E.L. and Richards, J.F., Rheology of Egg Albumen, *Trans. ASAE*, **14**, 17 (1971).

Tuoc, T.K. and R.B. Keey, A time-space transformation for non-Newtonian laminar boundary layers, *Trans. Inst. Chem. Eng.*, **70A**, 604 (1992).

Turian, R.M., *PhD Dissertation*, Univ. of Wisconsin, Madison, WI (1964).

Turian, R.M., An experimental investigation of the flow of aqueous non-Newtonian high polymer solutions past a sphere, *AIChE J.*, **13**, 999 (1967).

Turney, M.A., Cheung, M.K., Powell, R.L. and McCarthy, M.J., Hindered settling of rod-like particles measured with magnetic resonance imaging, *AIChE J.*, **41**, 251 (1995).

Tyabin, N.V., Discussion – Some questions of the theory of viscoplastic flow of disperse systems, *Colloid J. USSR*, **15**, 325 (1953).

Tyabin, N.V., *Trans. Kirov. Inst. Chem. Tech. Kagan*, **14**, 38 (1949); *Proc. Acad. Sci. USSR*, **38**, 1 (1953) (in Russian).

Tzounakos, A., Karamanev, D.G., Margaritis, A. and Bergougnou, M.A., Effect of the surfactant concentration on the rise of gas bubbles in power law non-Newtonian liquids, *Ind. Eng. Chem. Res.*, **43**, 5790 (2004).

Uchida, Y., Ali, S., McAtee Jr., J.L. and Claesson, S., A new type of falling-body viscometer for measuring the effect of pressure on the viscosity of deoxyribonucleic acid solutions, *Acta Polym.*, **35**, 472 (1984).

Uhlherr, P.H.T., A novel method for measuring yield stress in static fluids, *Proc. IVth Nat. Conf. Rheol.*, Adelaide, p. 231 (1986).

Uhlherr, P.H.T., Boger, D.V. and Anderson, T.J., Drag and flow patterns for the flow of a drag reducing polymer solution about spheres, *Proc. 5th Australasian Conf. Fluid Mech. and Hyd.*, Christchurch, New Zealand (1974).

Uhlherr, P.H.T. and Chhabra, R.P., Wall effect for the fall of spheres in cylindrical tubes at high Reynolds numbers, *Can. J. Chem. Eng.*, **73**, 918 (1995).

Uhlherr, P.H.T., Guo, J., Fang, T.-N. and Tiu, C., Static measuremnt of yield stress using a cylindrical penetrometer, *Korea-Aust. Rheol. J.*, **14**, 17 (2002).

Uhlherr, P.H.T., Le, T.N. and Tiu, C., Characterisation of inelastic power-law fluids using falling sphere data, *Can. J. Chem. Eng.*, **54**, 497 (1976).

Uhlherr, P.H.T., Park, K.H., Tiu, C. and Andrews, J.R.G., Yield stress from fluid behaviour on an inclined plane, *Proc. IXth Int. Cong. Rheol.*, Acapulco, Mexico, p.183 (1984).

Ui, T.J., Hussey, R.G. and Roger, R.P., Stokes drag on a cylinder in axial motion, *Phys. Fluids*, **27**, 787 (1984).

Ulaganathan, N., and Krishnaiah, K., Hydrodynamic characteristics of two-phase inverse fluidized bed, *Bioprocess Eng.*, **15**, 159 (1996).

Ultman, J.S. and Denn, M.M., Anomalous heat transfer and a wave phenomenon in dilute polymer solutions, *Trans. Soc. Rheol.*, **14**, 307 (1970).

Ultman, J.S. and Denn, M.M., Slow viscoelastic flow past submerged objects, *Chem. Eng. J.*, **2**, 81 (1971).

Unnikrishnan, A. and Chhabra, R.P., Slow parallel motion of cylinders in non-Newtonian media: Wall effects and drag coefficient, *Chem. Eng. Process.*, **28**, 121 (1990).

Unnikrishnan, A. and Chhabra, R.P., A experimental study of motion of cylinders in Newtonian fluids: Wall effects and drag coefficient, *Can. J. Chem. Eng.*, **69**, 729 (1991).

Uno, S. and Kintner, R.C., Effect of wall proximity on the rate of rise of single air bubbles in a quiescent liquid, *AIChE J.*, **2**, 420 (1956).

Unsal, E., Duda, J.L. and Klaus, E.E., Comparison of solution properties of mobility control polymers, *Chem. Oil Recov. ACS Sym. Ser.*, **91**, 141 (1978).

Usui, H., Li, L. and Suzuki, H., Rheology and pipeline transportation of dense fly-ash slurries, *Korean-Aust. Rheol. J.*, **13**, 47 (2001).

Usui, H., Shibata, T. and Sano, Y., Karman vortex behind a circular cylinder in dilute polymer solutions, *J. Chem. Eng. Jpn.*, **13**, 77 (1980).

Vafai, K., *Handbook of Porous Media*, 2nd ed., CRC Press, Boca Raton, FL (2005).

Valentik, L. and Whitmore, R.L., The terminal velocity of spheres in Bingham plastics, *Brit. J. Appl. Phys.*, **16**, 1197 (1965).

Valverde, J.M., Perez, A.T., Castellanes, A. and Viturro, R.E., Rheological testing of xerographic liquid inks: A need for printing technology, *Appl. Rheol.*, **14**, 190 (2004).

Vamerzani, B.Z., Norouzi, M. and Firoozabadi, B., Theoretical and experimental study on the motion and shape of viscoelastic falling drops through Newtonian media, *Rheol. Acta*, **55**, 935 (2016).

Van Atta, C.W., Laminar wake solutions for dilatant power law fluids, *Chem. Eng. Sci.*, **22**, 478 (1967).

Van Brakel, J., Pore space models for transport phenomena in porous media: Review and evaluation with special emphasis on capillary liquid transport, *Powder Technol.*, **11**, 205 (1975).

van den Brule, B.H.A.A. and Gheissary, G., Effects of fluids elasticity on the static and dynamic settling of a spherical particle, *J. Non-Newt. Fluid Mech.*, **49**, 123 (1993).

van der Merwe, D.F. and Gauvin, W.H., Velocity and turbulence measurements of air flow through a packed bed, *AIChE J.*, **17**, 519 (1971).

van der Westhuizen, J. and du Plessis, J.P., Quantification of unidirectional fiber bed permeability, *J. Compos. Mater.*, **28**, 619 (1994).

van der Westhuizen, J. and du Plessis, J.P., An attempt to quantify fibre bed permeability utilizing the phase average Navier-Stokes equation, *Compos. Part A*, **27A**, 263 (1996).

Van Dyke, M.D., Extension of Goldstein's series for the Oseen drag of a sphere, *J. Fluid Mech.*, **44**, 365 (1970).

Van Hecke, M., Running on corn flour, *Nature*, **487**, 174 (2012).

van Heel, A.P.G., Hulsen, M.A. and van den Brule, B.H.A.A., Simulation of the Doi-Edwards model in complex flow, *J. Rheol.*, **43**, 1239 (1999).

van Krevelen, D.W. and Hoftijzer, P.J., Studies on gas bubble formation, *Chem. Eng. Prog.*, **46**, 29 (1950).

Van Os, R.G.M. and Phillips, T.N., Efficient and stable spectral element methods for predicting the flow of an XPP fluid past a cylinder, *J. Non-Newt. Fluid Mech.*, **129**, 143 (2005).

van Poollen, H.K. and Jargon, J.R., Steady state and unsteady state flow of non-Newtonian fluids through porous media, *Soc. Pet. Eng. J.*, **7**, 80 (March 1969).

van Wazer, J.R., Lyons, J.W., Kim, K.Y. and Colwell, R.E., *Viscosity and Flow Measurement*, Interscience, New York (1963).

van Wijngaarden, L. and Vossers, G., Mechanics and physics of gas bubbles in liquids: A report on Euromech 98, *J. Fluid Mech.*, **87**, 695 (1978).

Vandu, C.O., Koop, K. and Krishna, R., Large bubble sizes and rise velocities in a bubble column slurry reactor, *Chem. Eng. Technol.*, **27**, 1195 (2004).

Varanasi, P.P., Ryan, M.E. and Stroeve, P., Experimental study on the breakup of model viscoelastic drops in uniform shear flow, *Ind. Eng. Chem. Res.*, **33**, 1858 (1994).

Varchanis, S., Makrigiorgos, G., Moschopoulos, P., Dimakopoulos, Y. and Tsamopoulos, J., Modelling the rheology of thixotropic elasto-visco-plastic materials, *J. Rheol.*, **63**, 609 (2019).

Varshney, A. and Steinberg, V., Drag enhancement and drag reduction in viscoelastic flow, *Phys. Rev. Fluids*, **3**, 103302 (2018).

Vasil'chenko, S.V. and Potapov, A.G., Gas bubble dynamics in a viscoelastic – plastic medium, *Heat Transf. Res.*, **27**, 4 (1996).

Vasilic, K., Meng, B., Kühne, H.C. and Roussel, N., Flow of fresh concrete through steel bars: A porous medium analogy, *Cement Concrete Res.*, **41**, 496 (2011).

Vasilic, K., Schmidt, W., Kühne, H.C., Haamkens, F., Mechtcherine, V. and Roussel, N., Flow of fresh concrete through reinforced elements: Experimental validation of the porous analogy numerical method, *Cement Concrete Res.*, **88**, 1 (2016).

Vasukiran, M., Kishore, N. and Yadav, S., Critical Reynolds numbers of shear-thinning fluids past unbounded spheres, *Powder Technol.*, **339**, 747 (2019).

Veldhuis, C.H.J., Biesheuvel, A. and Lohse, D., Freely rising light solid spheres, *Int. J. Multiph. Flow*, **35**, 312 (2009).

Velez-Cordero, J.R., Samano, D., Yue, P., Feng, J.J. and Zenit, R., Hydrodynamic interactions between a pair of bubbles ascending in shear-thinning inelastic fluids, *J. Non-Newt. Fluid Mech.*, **166**, 118 (2011).

Velez-Cordero, J.R., Samano, D. and Zenit, R., Study of the properties of bubbly flows in Boger-type fluids, *J. Non-Newt. Fluid Mech.*, **175–176**, 1 (2012).

Velez-Cordero, J.R. and Zenit, R., Bubble cluster formation in shear-thinning inelastic bubbly columns, *J. Non-Newt. Fluid Mech.*, **166**, 32 (2011).

Venerus, D.C., Diffusion-induced bubble growth in viscous liquids of finite and infinite extent, *Polym. Eng. Sci.*, **41**, 1390 (2001).

Venerus, D.C. and Yala, N., Transport analysis of diffusion – induced bubble growth and collapse in viscous liquids, *AIChE J.*, **43**, 2948 (1997).

Venerus, D.C., Yala, N. and Bernstein, B., Analysis of diffusion-induced bubble growth in viscoelastic liquids, *J. Non-Newt. Fluid Mech.*, **75**, 55 (1998).

Venkataraman, P. and Mohan Rao, P.R., Validation of Forchheimer's law for flow through porous media with converging boundaries, *J. Hyd. Eng. (ASCE)*, **126**, 63 (2000).

Venkatesh, M., Narayan, K.A. and Chhabra, R.P., An experimental study of mass transfer from a sparingly soluble cylinder in cross-flow configuration, *Chem. Eng. Commun.*, **130**, 181 (1994).

Venumadhav, G. and Chhabra, R.P., Settling velocities of single non-spherical particles in non-Newtonian fluids, *Powder Technol.*, **78**, 77 (1994).

Venumadhav, G. and Chhabra, R.P., Drag on non-spherical particles in viscous fluids, *Int. J. Mineral Process.*, **43**, 15 (1995).

Verboven P., Hoang, M.L., Baelmans, M. and Nicolai, B.M., Airflow through beds of apples and chicory roots, *Biosyst. Eng.*, **88**, 117 (2004).

Verhelst, J.M. and Nieuwstadt, F.T.M., Viscoelastic flow past circular cylinders mounted in a channel: Experimental measurements of velocity and drag, *J. Non-Newt. Fluid Mech.*, **116**, 301 (2004).

Verma, A., Suri, P. and Patel, S.A., Forced convection from chains of spheres. *Int. J. Fluid Mech. Res.*, **49**, 43 (2022).

Verma, P.D. and Rajvanshi, S.C., The instantaneous slow flow of a second-order fluid between two concentric spheres, *Arch. Mech.*, **23**, 613 (1971).

Verma, P.D. and Sacheti, N.C., Low Reynolds number flow of a second order fluid past a porous sphere, *J. Appl. Phys.*, **46**, 2065 (1975).

Verma, R.L., Elastico-viscous boundary layer flow on the surface of a sphere, *Rheol. Acta.*, **16**, 510 (1977).

Verrelli, D.I., Convenient formulae for the drag on a prolate ellipsoid moving along its axis of symmetry perpendicular to a plane surface, *Int. J. Multiph. Flow*, **65**, 138 (2014).

Vidyasagar, T., Patel, S.A. and Chhabra, R.P., Wall effect on the sedimentation of a sphere in power law fluids, HYDROTRANSPORT-20, p. **429**, Melbourne, Australia (2017a).

Vidyasagar, T., Patel, S.A. and Chhabra, R.P., Unpublished results (2022).

Vidyasagar, T., Patel, S.A., Gupta, A.K. and Chhabra, R.P., Effects of severe blockage on heat transfer from a sphere in power-law fluids, *Int. Comm. Heat Mass Transf.*, **84**, 27 (2017b).

Vijaya Lakshmi, A.C., Balamurugan, M., Sivakumar, M., Newton Samuel, T. and Velan, M., Minimum fluidization velocity and friction factor in a liquid-solid inverse fluidized bed reactor, *Bioprocess. Eng.*, **22**, 461 (2000).

Vijaysri, M., Chhabra, R.P. and Eswaran, V., Power-law fluid flow across an array of infinite circular cylinders: A numerical study, *J. Non-Newt. Fluid Mech.*, **87**, 263 (1999).

Vikhansky, A., On the stopping of the thermal convection in viscoplastic liquid, *Rheol. Acta*, **50**, 423 (2011).

Villimpoc, V., Cole, R. and Sukanek, P.C., Heat transfer in Newtonian liquids around a circular cylinder, *Int. J. Heat Mass Transf.*, **33**, 447 (1990).

Viswanadham, R., Agarwal, D.C. and Kramer, E.J., Water transport through reconstructed collagen hollow-fiber membranes, *J. Appl. Polym. Sci.*, **22**, 1655 (1978).

Volarovich, M.P. and Gutkin, A.M., Theory of flow of a viscoplastic medium, *Colloid J. USSR*, **15**, 153 (1953).

von der Lieth, J. and Hort, M., Slug ascent and associated stresses during strombolian activity with non-Newtonian rheology, *J. Geophys. Res. Solid Earth*, **121**, 4923 (2016).

von Seckendorff, J., Achterhold, K., Pfeiffer, F., Fischer, R. and Hinrichsen, O., Experimental and numerical analysis of void structure in random packed beds of spheres. *Powder Technol.*, **380**, 613 (2021).

von Seckendorff, J. and Hinrichsen, O., Review on the structure of random packed-beds. *Can. J. Chem. Eng.*, **99**, S703 (2021).

von Seckendorff, J., Szesni, N., Fischer, R. and Hinrichsen, O., Experimental characterization of random packed spheres, cylinders and rings, and their influence on pressure drop. *Chem. Eng. Sci.*, **222**, 115644 (2020).

Vorwerk, J. and Brunn, P.O., Porous medium flow of the fluid A1: Effects of shear and elongation, *J. Non-Newt. Fluid Mech.*, **41**, 119 (1991).

Vorwerk, J. and Brunn, P.O., Shearing effects for the flow of surfactant and polymer solutions through a packed bed of spheres, *J. Non-Newt. Fluid Mech.*, **51**, 79 (1994).

Vossoughi, S. and Seyer, F.A., Pressure drop for flow of polymer solution in a model porous medium, *Can. J. Chem. Eng.*, **52**, 666 (1974).

Vradis, G.C. and Protopapas, A.L., Macroscopic conductivities for flow of Bingham plastics in porous media, *J. Hyd. Eng.*, **119**, 95 (1993).

Vryzas, Z., Wubulikasimu, Y., Gerogiorgis, D.I. and Kelessidis, V.C., Understanding the temperature effect on the rheology of water-Bentonite suspensions, *Ann. Trans. Nordic Rheol. Soc.*, **24**, 199 (2016).

Wachs, A. and Frigaard, I.A., Particle settling in yield stress fluids: Limiting time, distance and applications, *J. Non-Newt. Fluid Mech.*, **238**, 189 (2016).

Wagner, M.G. and Slattery, J.C., Slow flow of a non-Newtonian fluid past a droplet, *AIChE J.*, **17**, 1198 (1971).

Wahyudi, I., Montillet, A. and Khalifa, A.O.A., Darcy and post Darcy flows within different sands, *J. Hyd. Res.*, **40**, 519 (2002).

Waisbord, N., Stoop, N., Walkama, D.M., Dunkel, J. and Guasto, J.S., Anomalous percolation flow transition of yield stress fluids in porous media, *Phys. Rev. Fluids*, **4**, 063303 (2019).

Waitukaitis, S.R. and Jaeger, H.M., Impact-activated solidification of dense suspensions via dynamic jamming fronts, *Nature*, **487**, 205 (2012).

Wakiya, S. Viscous flow past a spheroid, *J. Phys. Soc. Jpn.*, **12**, 1130 (1957).

Wallick, G.C., Savins, J.G. and Arterburn, D.R., Tomita solution for the motion of a sphere in a power-law fluid, *Phys. Fluids*, **5**, 367 (1962).

Walters, K., *Rheometry*, Chapman and Hall, London (1975).

Walters, K., Developments in non-Newtonian fluid mechanics – A personal view, *J. Non-Newt. Fluid Mech.*, **5**, 113 (1979).

Walters, K. and Barnes, H.A., Anomalous extensional flow effects in the use of commercial viscometers, *Rheology, Vol. 1 Principles*, edited by G. Astarita, G. Marrucci and L. Nicolais, Plenum, New York, p. 45 (1980).

Walters, K. and Savins, J.G., A rotating sphere elasto-viscometer, *Trans. Soc. Rheol.*, **9**, 407 (1965).

Walters, K. and Tanner, R.I., The motion of a sphere through an elastic liquid, *Transport Processes in Bubbles, Drops, and Particles*, edited by R.P. Chhabra and D. De Kee, Hemisphere, New York, Chapter 3 (1992).

Walters, K. and Waters, N.D., On the use of a rotating sphere in the measurement of elastico-viscous parameters, *Brit. J. App. Phys.*, **14**, 667 (1963).

Wampler, F.C. and Gregory, D.R., Flow of molten poly (ethylene terephthalate) through packed beds of glass beads, *AIChE J.*, **18**, 443 (1972).

Wan, Z., Settling velocity of particles in a Bingham fluid, *Institute of Hydrodynamic & Hydraulic Eng.*, Tech. Uni. Denmark, Report # **56**, 3 (1982).

Wan, Z., Bed material movement in hyper concentrated flow, *J. Hyd. Eng. (ASCE)*, **111**, 987 (1985).

Wanchoo, R.K., Sharma, S.K. and Gupta, R., Shape of a Newtonian liquid drop moving through an immiscible quiescent non-Newtonian liquid, *Chem. Eng. Process.*, **42**, 387 (2003).

Wang, C., Chunjiing, T. and Ling, Y., Boundary layer flow and heat transfer of power law fluids in packed beds, *Heat Transf. (IInd U.K. Nat. Conf.)*, **2**, 1431 (1988).

Wang, C.Y., Settling of discs inside a vertical fluid-filled tube, *Appl. Sci. Res.*, **56**, 43 (1996a).

Wang, C.Y., Stokes flow through an array of rectangular fibers, *Int. J. Multiph. Flow*, **22**, 185 (1996b).

Wang, C.Y., Stokes slip flow through square and triangular arrays of circular cylinders, *Fluid Dyn. Res.*, **32**, 233 (2003).

Wang, F.H.L., Duda, J.L. and Klaus, E.E., Influences of polymer solution properties on flow in porous media, paper presented at the *54th Annual Fall Tech. Conf. & Exhibition*, SPE of AIME, Las Vegas, Paper No. SPE 8418, (September 23–26, 1979).

Wang, J. and Joseph, D., Potential flow of a second order fluid over a sphere or an ellipse, *J. Fluid Mech.*, **511**, 201 (2004).

Wang, J., Qi, H. and You, C., Experimental study of sedimentation characteristic of spheroidal particles, *Particuology*, **7**, 264 (2009).

Wang, L., Shu, C., Teo, C.J. and Yang, L.M., Numerical study on the freely falling plate: Effects of density ratio and thickness-to-length ratio, *Phys. Fluids*, **28**, 103603 (2016).

Wang, L., Tiu, C. and Liu, T., Effects of nonionic surfactant and associative thickeners on the rheology of polyacrylamide in aqueous glycerol solutions, *Colloid Polym. Sci.*, **274**, 138 (1996).

Wang, P., Kahawita, R. and Nguyen, T. H.; Numerical computation of the natural convection flow about a horizontal cylinder using splines, *Num. Heat Transfer A*, **17**, 191 (1990).

Wang, P., Yu, Z. and Lin, J., Numerical simulations of particle migration in rectangular channel flow of Giesekus viscoelastic fluids, *J. Non-Newt. Fluid Mech.*, **262**, 142 (2018).

Wang, T.-Y., Mixed convection heat transfer from a horizontal plate to non-Newtonian fluids, *Int. Comm. Heat Mass Transf.*, **20**, 431 (1993).

Wang, T.-Y., Mixed convection heat transfer from a vertical plate to non-Newtonian fluids, *Int. J. Heat Fluid Flow*, **16**, 56 (1995).

Wang, T.-Y. and Kleinstreuer, C., Free convection heat transfer between a permeable vertical wall and a power law fluid, *Num. Heat Transf.*, **12**, 367 (1987).

Wang, T.-Y. and Kleinstreuer, C., Local skin friction and heat transfer in combined free-forced convection from a cylinder or sphere to a power law fluid, *Int. J. Heat Fluid Flow*, **9**, 182 (1988a).

Wang, T.-Y. and Kleinstreuer, C., Combined free-forced convection heat transfer between vertical slender cylinders and power law fluids, *Int. J. Heat Mass Transf.*, **31**, 91 (1988b).

Wang, T.-Y. and Kleinstreuer, C., Mixed thermal convection of power law fluids past bodies with uniform fluid injection or suction, *J. Heat Transf. (ASME)*, **112**, 151 (1990).

Wang, W. and Sangani, A.S., Nusselt number for flow perpendicular to arrays of cylinders in the limit of small Reynolds and large Peclet numbers, *Phys. Fluids*, **9**, 1529 (1997).

Wang, X., Thauvin, F. and Mohanty, K.K., Non-Darcy flow through anisotropic porous media, *Chem. Eng. Sci.*, **54**, 1859 (1999).

Wang, Z., Afacan, A., Nandakumar, K. and Chung, K.T., Porosity distribution in random packed columns by gamma ray tomography, *Chem. Eng. Process.*, **40**, 209 (2001).

Wang, Z., Lou, W., Sun, B., Pan, S., Zhao, X. and Liu, H., A model for predicting bubble velocity in yield stress fluids at low Reynolds number, *Chem. Eng. Sci.*, **201**, 325 (2019).

Wang, Z., Wang, S., Xu, L., Dou, Y. and Su, X., Extremely slow settling behavior of particles in dilute wormlike micellar fluid with broad spectrum of relaxation times. *J. Dispers. Sci. Technol.*, **41**, 639 (2020).

Wapperom, P. and Renardy, M., Numerical prediction of the boundary layers in the flow around a cylinder using a fixed velocity field, *J. Non-Newt. Fluid Mech.*, **125**, 35 (2005).

Wapperom, P. and Webster, M.F., Simulation for viscoelastic flow by a finite volume/element method, *Comput. Methods Appl. Mech. Eng*, **180**, 281 (1999).

Warichet, V. and Legat, V., Adaptive high-order prediction of the drag correction factor for the upper-convected Maxwell fluid, *J. Non-Newt. Fluid Mech.*, **73**, 95 (1997).

Warnica, D., Renksizbulut, M. and Strong, A.B., Drag coefficients of spherical liquid droplets, *Expt. Fluids*, **18**, 258 (1995).

Warshay, M., Bogusz, E., Johnson, M. and Kintner, R.C., Ultimate velocity of drops in stationary liquid media, *Can. J. Chem. Eng.*, **37**, 29 (1959).

Warson, H., *Applied Thixotropy and Allied Phenomena*, published by author. ISBN: 0-9544724-0-3 (2003).

Wasserman, M.L. and Slattery, J.C., Upper and lower bounds on the drag coefficient of a sphere in a power model fluid, *AIChE J.*, **10**, 383 (1964).

Watamura, K., Sugiyama, K., Yotsumoto, Y., Suzuki, M. and Wakabayashi, H., Bubble cascade may form not only in Stout beers, *Phys. Rev. E.*, **103**, 063103 (2021).

Watamura, T., Iwatsubo, F., Sugiyama, K., Yamamoto, Y., Yotsumoto, Y. and Shiono, F., Bubble cascade in Guinness beer is caused by gravity current instability, *Sci. Rep.*, **9**, 5718 (2019).

Watanabe, K., Kui, H. and Motosu, I., Drag of a sphere in dilute polymer solutions in high Reynolds number range, *Rheol. Acta*, **37**, 328 (1998).

Weaire, D., Froths, foams and heady geometry, *New Sci.*, **34** (21 May 1994).

Weber, M.E., Astrauskas, P. and Petsalis, S., Natural convection mass transfer to non-spherical objects at high Rayleigh numbers, *Can. J. Chem. Eng.*, **62**, 68 (1984).

Wegener, M., Kraume, M. and Paschedag, A.R., Terminal and transient drop rise velocity of single toluene droplets in water, *AIChE J.*, **56**, 2 (2010).

Wegner, T.H., Karabelas, A.J. and Hanratty, T.J., Visual studies of flow in a regular array of spheres, *Chem. Eng. Sci.*, **26**, 59 (1971).

Wehbeh, E.G., Ui, T.J. and Hussey, R.G., End effects for the falling cylinder viscometer, *Phys. Fluids*, **5**, 25 (1993).

Weidman, P.D., Roberts, B. and Eisen, S., On the instability of spheres settling through a vertical pipe filled with HPG, *J. Appl. Fluid Mech.*, **5**(4) 113 (2012).

Weidman, P.D. and van Atta, C.W., The laminar axisymmetric wake for power-law fluids, *Acta Mech.*, **146**, 239 (2001).

Weinberg, R.F., The upward transport of inclusions in Newtonian and power-law salt diapirics, *Tectonophysics*, **228**, 141 (1993).

Weinberg, R.F. and Podladchikov, Y., Diapiric ascent of magmas through power law crust and mantle, *J. Geophys. Res.* **99**, 9543 (1994).

Weisenborn, A.J. and Ten Bosch, B.I.M., Analytical approach to the Oseen drag on a sphere at infinite Reynolds number, *Siam J. Appl. Math.*, **53**, 601 (1993).

Weissenberg, K., A continuum theory of rheological phenomena, *Nature*, **159**, 310 (1947).

Weitzenböck, J.R., Shenoi, R.A. and Wilson, P.A., Measurement of three-dimensional permeability, *Compos. Part A*, **29A**, 159 (1997).

Wellek, R.M. and Gurkan, T., Mass transfer to drops moving through power law fluids in the intermediate Reynolds number region, *AIChE J.*, **22**, 484 (1976).

Wellek, R.M. and Huang, C.-H., Mass transfer from spherical gas bubbles and liquid droplets moving through power law fluids in the laminar flow regime, *Ind. Eng. Chem. Fund.*, **9**, 480 (1970).

Wells Jr., C.S., Similar solutions of the boundary layer equations for purely viscous non-Newtonian fluids, *NASA Tech. Note* No. D-2262 (1964).

Wen, C.Y. and Fan, L.S., Axial dispersion of non-Newtonian liquids in fluidized beds, *Chem. Eng. Sci.*, **28**, 1768 (1973).

Wen, C.Y. and Yim, J., Axial dispersion of a non-Newtonian liquid in a packed bed, *AIChE J.*, **17**, 1503 (1971).

Wenyuan, F., Youguang, M., Shaokun, J., Ke, Y. and Li, H.Z., An experimental investigation for bubble rising in non-Newtonian fluids and empirical correlation of drag coefficient, *J. Fluids Eng.*, **134**, 084501 (2010).

Wesselingh, J.A. and Bollen, A.M., Single particles, bubbles and drops: Their velocities and mass transfer coefficients, *Chem. Eng. Res. Des.*, **77A**, 89 (1999).

Westerberg, K.W. and Finlayson, B.A., Heat transfer to spheres from a polymer melt, *Num. Heat Transf.*, **17A**, 329 (1990).

Wham, R.M., Basaran, O.A. and Byers, C.H., Wall effects on flow past solid spheres at finite Reynolds numbers, *Ind. Eng. Chem. Res.*, **35**, 864 (1996).

Wham, R.M., Basaran, O.A. and Byers, C.H., Wall effects on flow past fluid spheres at finite Reynolds number: Wake structure and drag correlations, *Chem. Eng. Sci.*, **52**, 3345 (1997).

Wheat, J.A., The air flow resistance of glass fiber filter paper, *Can. J. Chem. Eng.*, **41**, 67 (1963).

Whitaker, S., Advances in theory of fluid motion in porous media, *Flow Through Porous Media*, American Chemical Society, Washington, D.C., Chapter 2 (1970).

Whitaker, S., Forced convection heat transfer correlations for flow in pipes, past flat plates, single cylinders, single spheres, and for flow in packed beds and tube bundles. *AIChE J.*, **18**, 361 (1972).

Whitaker, S., *The Method of Volume Averaging (Theory and Applications of Transport in Porous Media)*, Springer, New York (1998).

White, A., Effect of polymer additives on boundary layer separation and drag of submerged bodies, *Nature*, **211**, 1390 (1966).

White, A., Drag of spheres in dilute high polymer solutions, *Nature*, **216**, 994 (1967).

White, A., Some observations on the flow characteristics for certain dilute macromolecular solutions, *Proc. Sym. Viscous Drag Reduction*, U.S. Office of Naval Res., p. 297 (1968).

White, A, *Drag Reduction by Additives; Review and Bibliography*, BHRA Fluids Engineering, Cranfield, England (1976).

White, D.A., Drag coefficients for spheres in high Reynolds number flow of dilute solutions of high polymers, *Nature*, **212**, 277 (1966); Also see *ibid*, **226**, 72 (1970).

White, D.A., Non-Newtonian flow in porous media, *Chem. Eng. Sci.*, **22**, 669 (1967).

White, J.L., Application of integral momentum methods to viscoelastic fluids: Flow about submerged objects, *AIChE J.*, **12**, 1019 (1966).

White, J.L., *Principles of Polymer Engineering Rheology*, Wiley, New York (1990).

White, J.L. and Metzner, A.B., Constitutive equations for viscoelastic fluids with application to rapid external flows, *AIChE J.*, **11**, 324 (1965a).

White, J.L. and Metzner, A.B., Thermodynamic and heat transport considerations for viscoelastic fluids, *Chem. Eng. Sci.*, **20**, 1055 (1965b).

White, M.L., The permeability of an acrylamide polymer gel, *J. Phys. Chem.*, **64**, 1563 (1960).

Whitmore, R.L., Drag forces in Bingham plastics, *Proc. 5th Int. Cong. Rheol.*, Tokyo, p.472 (1969).

Whitmore, R.L. and Boardman, G., Reply to Rae, *Nature*, **194**, 272 (1962).

Whitney, M.J. and Rodin, G.J., Force-velocity relationships for rigid bodies translating through unbounded shear-thinning power-law fluids, *Int. J. Non-Linear Mech.*, **36**, 947 (2001).

Whorlow, R.W., *Rheological Techniques*, 2nd ed., Ellis Horwood, London (1992).

Wiggins, E.J., Campbell, W.B. and Maass, O., Determination of the specific surface of fibrous materials, *Can. J. Res.*, **17B**, 318 (1939).

Wilkes, E.D., Phillips, S.D. and Basaran, O.A., Computational and experimental analysis of dynamics of drop formation, *Phys. Fluids*, **11**, 3577 (1999).

Wilkinson, D., Modified drag theory of permeability, *Phys. Fluids*, **28**, 1015 (1985).

Wilkinson, W.L., Tailing of drops falling through viscoelastic liquids, *Nature (Phys. Sci.)*, **240**, 44 (1972).

Wilks, G., External natural convection about two-dimensional bodies with constant heat flux, *Int. J. Heat Mass Transf.*, **15**, 351 (1972).

Willets, W.R., Letter to the Editor, *Physics Today*, **20**, 11 (1967).

Willhite, G.P. and Dominguez, J.S., Mechanisms of polymer retention in porous media, *Improved Oil Recovery by Surfactant and Polymer Flooding*, edited by D.O. Shah and R.S. Schechter, p. 511, Academic Press, Cambridge (1977).

Williams, E.M., A method of indicating pebble shape with one parameter, *J. Sed. Petrol.*, **35**, 993 (1965a).

Williams, J.G., Morris, C.E.M. and Ennis, B.C., Liquid flow through aligned fibers, *Polym. Eng. Sci.*, **14**, 413 (1974).

Williams, M.C., Normal stress and viscosity measurements for polymer solutions in steady cone and plate shear, *AIChE J.*, **11**, 467 (1965b).

Williams, P.R., Cavitation and bubble dynamics, *Transport Processes in Bubbles, Drops and Particles*, edited by D. De Kee and R.P. Chhabra, 2nd ed., Taylor and Francis, New York, Chapter 5 (2002).

Wilson, A.S. and Bassiouny, M.K., Modeling of heat transfer for flow across tube banks, *Chem. Eng. Process.*, **39**, 1 (2000).

Wilson, K.C., Addie, G.R., Sellgren, A. and Clift, R., *Slurry Transport using Centrifugal Pumps*, 3rd ed., Springer, New York (2006).

Wilson, K.C., Horsley, R.R., Kealy, T., Reizes, J.A. and Horsley, M., Direct prediction of fall velocities in non-Newtonian materials, *Int. J. Mineral Process.*, **71**, 17 (2003).

Winterberg, M. and Tsotsas, E., Impact of tube-to-particle diameter ratio on pressure drop in packed beds, *AIChE J.*, **46**, 1084 (2000b).

Winterberg, M. and Tsotsas, E., Modelling of heat transport in beds packed with spherical particles for various bed geometries and/or thermal boundary conditions, *Int. J. Therm. Sci.*, **39**, 556 (2000a).

Wissler, E.H., Viscoelastic effects in the flow of non-Newtonian fluids through a porous medium, *Ind. Eng. Chem. Fund.*, **10**, 411 (1971).

Witten, T.A. and Pincus, P.A., *Structured Fluids: Polymers, Colloids, Surfactants*, Oxford University Press, New York (2004).

Wodie, J.-C. and Levy, T., Non-linear rectification of Darcy's law, *C.R. Acad. Sci. Paris Ser. II*, **312**, 157 (1991).

Wolf, C.J. and Szewczyk, A.A., Laminar heat transfer to power-model non-Newtonian fluids from arbitrary cylinders, *Proc. 3rd Int. Heat Transf. Conf.*, Chicago, IL (1966) p. 388.

Won, D. and Kim, C., Alignment and aggregation of spherical particles in viscoelastic fluid under shear flow, *J. Non-Newt. Fluid Mech.*, **117**, 141 (2004).

Wong, K.L., Lee, S.C. and Chen, C.K., Finite element solution of laminar combined convection from a sphere, *J. Heat Transf.*, **108**, 860 (1986).

Woods, J.K., Spelt, P.D.M., Lee, P.D., Selerland, T. and Lawrence, C.J., Creeping flow of power law fluids through periodic arrays of elliptical cylinders, *J. Non-Newt. Fluid Mech.*, **111**, 211 (2003).

Woudberg, S., du Plessis, J.P. and Smit, G.J.F., Non-Newtonian purely viscous flow through isotropic granular porous media, *Chem. Eng. Sci.*, **61**, 4299 (2006).

Wreath, D., Pope, G.A. and Seperhrnoori, K., Dependence of polymer apparent viscosity on the permeable media and flow conditions, *In Situ*, **14**, 263 (1990).

Wronski, S. and Szembek-Stoeger, M., Mass transfer to non-Newtonian fluids in packed beds at low Reynolds numbers, *Inzynieria Chemiczna Processowa*, **4**, 627 (1988).

Wu, G.H., Ju, S.J. and Wu, C.C., Numerical prediction of non-isothermal flow of Nylon-6 past a cylinder between plates, *J. Polymer Eng.*, **19**, 287 (1999).

Wu, G.H., Wu, B.Y., Ju, S.H. and Wu, C.C., Non-isothermal flow of a polymeric fluid past a submerged circular cylinder, *Int. J. Heat Mass Transf.*, **46**, 4733 (2003).

Wu, J., Drag reduction in external flows of additive solutions, *Viscous Drag Reduction*, C.S. Wells, p. 331, Plenum, New York (1969).

Wu, J. and Thompson, M.C., Non-Newtonian shear-thinning flows past a flat plate, *J. Non-Newt. Fluid Mech.*, **66**, 127 (1996).

Wu, J., Yu, B. and Yun, M., A resistance model for flow through porous media. *Trans. Porous Media*, **71**(3), 331 (2008).

Wu, S. and Mohammadigoushki, H., Flow of a model shear-thickening micellar fluid past a falling sphere, *Phys. Rev. Fluids*, **4**, 073303 (2019).

Wu, Y.-S. and Pruess, K., Flow of non-Newtonian fluids in porous media, *Adv. Porous Media*, **3**, 87 (1996).

Wu, Y.-S. and Pruess, K., A numerical method for simulating non-Newtonian fluid flow and displacement in porous media, *Adv. Water Resour.*, **21**, 351 (1998).

Wu, Y.-S., Pruess, K. and Witherspoon, P.A., Flow and displacement of Bingham non-Newtonian fluids in porous media, *SPE Reserv. Eng.*, **7**, 369 (1992).

Wunsch, O., Experimentalle bestimmung Binghamscher stoffparameter, *Rheol. Acta*, **29**, 163 (1990).

Wunsch, O., Kugelschwingung in einem fluid mit fliebgrenze, *ZAMM*, **72**, 349 (1992).

Wunsch, O., Oscillating sedimentation of spheres in viscoplastic fluids, *Rheol. Acta*, **33**, 292 (1994).

Wyllie, M.R.J. and Gregory, K.R., Fluid flow through unconsolidated porous aggregates – effect of porosity and particle shape on Kozeny-Carman constants, *Ind. Eng. Chem.*, **47**, 1379 (1955).

Xie, C. and Balhoff, M.T., Lattice Boltzmann modeling of the apparent viscosity of thinning–elastic fluids in porous media. *Trans. Porous Media*, **137** (1), 63 (2021).

Xiong, Y.L., Bruneau, C.H. and Yang, D., Numerical study on viscoelastic fluid flow past a rigid body, *App. Math. Modell.*, **42**, 188 (2017).

Xiong, Y.-L., Peng, S., Yang, D., Duan, J. and Wang, L., Influence of polymer additives on flow past a hydrofoil: A numerical study, *Phys. Fluids*, **30**, 013104 (2018).

Xu, D., Liu, C.-Y. and Craig, S.L., Divergent shear-thinning and shear-thickening behavior of supramolecular polymer networks in semi-dilute entangled polymer solutions, *Macromolecules*, **44**, 2343 (2011).

Xu, F., Cockx, A., Hebrard, G. and Dietrich, N., Mass transfer and diffusion of a single bubble rising in polymer solutions, *Ind. Eng. Chem. Res.*, **57**, 15181 (2018).

Xu, F., Midoux, N., Li, H.Z., Hebrard, G. and Dietrich, N., Characterization of bubble shapes in non-Newtonian fluids by parametric equations, *Chem. Eng. Technol.*, **42**, 2321 (2019a).

Xu, M. and Wu, D., The analysis of settling characteristic of a spherical particle in a Bingham fluid, *J. Hyd. Eng. (Peking)*, **29**, 11 (1983).

Xu, Q. and Michaelides, E.E., A numerical study of the flow over ellipsoidal objects inside a cylindrical tube, *Int. J. Num. Methods in Fluids*, **22**, 1075 (1996).

Xu, X., Kordorwu, V., Li, Z., Liu, F., Wei, W. and Liu, Z., Experimental study on the dynamics and mass transfer of CO_2 bubbles rising in viscoelastic fluids, *Int. J. Multiph. Flow*, **135**, 103539 (2021).

Xu, Z., Song, X., Li, G., Pang, Z. and Zhu, Z., Settling behaviour of non-spherical particles in power-law fluids: Experimental study and model development, *Particuology*, **46**, 30 (2019b).

Xue, L., Guo, X. and Chen, H., *Fluid Flow in Porous Media: Fundamentals and Applications*, World Scientific, Singapore (2021).

Yaginuma, T. and Ito, H., Drag and wakes of freely falling 60° cones at intermediate Reynolds numbers, *Phys. Fluids*, **20**, 117102 (2008).

Yahia, A. and Khayat, K.H., Analytical models for estimating yield stress of high-performance pseudoplastic grout, *Cement Concrete Res.*, **31**, 731 (2001).

Yamamoto, H. and Shibata, J., Analysis of flow property using falling slender cylindrical needle in power law fluid, *Kagaku Kogaku Ronbunshu*, **25**, 803 (1999).

Yamamoto, T., Suga, T., Nakamura, K. and Mori, N., The gas penetration through viscoelastic fluids with shear-thinning viscosity in a tube, *J. Fluids Eng.*, **126**, 148 (2004).

Yamanaka, A. and Mitsuishi, N., Drag coefficient of a moving bubble and droplet in viscoelastic fluids, *J. Chem. Eng. Jpn.*, **10**, 370 (1977).

Yamanaka, A. and Mitsuishi, N., An experimental study on combined forced and natural heat transfer from spheres to power-law fluids, *Heat Transf.-Jap. Res.*, **6**, 85 (1978).

Yamanaka, A., Yuki, T. and Mitsuishi, N., Combined forced and natural convective heat transfer from spheres at small Reynolds number, *J. Chem. Eng. Jpn.*, **9**, 445 (1976b).

Yamanaka, A., Yuki, T. and Mitsuishi, N., Drag coefficient of a moving sphere in viscoelastic fluids, *Kagaku-Kogaku Ronbunshu*, **2**, 222 (1976a).

Yan, X., Jia, Y., Wang, L. and Cao, Y., Drag coefficient fluctuation prediction of a single bubble rising in water, *Chem. Eng. J.*, **316**, 553 (2017).

Yan, X., Zheng, K., Jia, Y., Miao, Z., Wang, L., Cao, Y. and Liu, J., Drag coefficient prediction of a single bubble rising in liquids, *Ind. Eng. Chem. Res.*, **57**, 5385 (2018).

Yang, B. and Khomami, B., Simulations of sedimentation of a sphere in a viscoelastic fluid using molecular based constitutive models, *J. Non-Newt. Fluid Mech.*, **82**, 429 (1999).

Yang, C. and Mao, Z.-S., Numerical solution of viscous flow of a non-Newtonian fluid past an irregular solid obstacle by the mirror fluid method, *Proc. 3rd Int. Conf. CFD in the Minerals Process Industrial*, p. 391, CSIRO, Melbourne, Australia (2003).

Yang, C. and Mao, Z.-S., Mirror fluid method for numerical simulation of sedimentation of a solid particle in a Newtonian fluid, *Phys. Rev. E*, **71**, 036704 (2005).

Yang, J., Bu, S., Dong, Q., Wu, J. and Wang, Q., Experimental study of flow transitions in random packed beds with low tube to particle diameter ratios. *Exp. Therm. Fluid Sci.*, **66**, 117 (2015).

Yang, J., Wu, J., Zhou, L. and Wang, Q., Computational study of fluid flow and heat transfer in composite packed beds of spheres with low tube to particle diameter ratio, *Nuclear Eng. Des.*, **300**, 85 (2016).

Yang, S., Raghavan, V. and Gogos, G., Numerical study of transient laminar natural convection over an isothermal sphere, *Int. J. Heat Fluid Flow*, **28**, 821 (2007).

Yang, W.-C. (ed.), *Handbook of Fluidization and Fluid-Particle Systems*, Marcel Dekker, New York (2003).

Yang, W.-J. and Lawson, M.L., Bubble pulsation and cavitation in viscoelastic liquids, *J. App. Phys.*, **45**, 754 (1974).

Yang, W.-J. and Wanat, R.L., Nucleate pool boiling of slurries on horizontal plate and cylinder, *Chem. Eng. Prog. Sym. Ser.*, **64**(82), 126 (1968).

Yang, W.-J. and Yeh, H.-C., Free convective flow of Bingham plastic between two vertical plates, *J. Heat Transf.*, **87**, 319 (1965).

Yang, W.-J. and Yeh, H.-C., Theoretical study of bubble dynamics in purely viscous fluids, *AIChE J.*, **12**, 927 (1966).

Yao, L.-S., Catton, I. and McDonough, J.M., Free-forced convection from a heated longitudinal horizontal cylinder, *Num. Heat Transf.*, **1B**, 255 (1978).

Yao, L.-S. and Molla, M.M., Forced convection of non-Newtonian fluids on a heated flat plate, *Int. J. Heat Mass Transf.*, **51**, 5154 (2008).

Yaparpalvi, R., Das, T.R. and Mukherjee, A.K., Drop formation in non-Newtonian liquids under pulsed conditions, *J. Chem. Tech. Biotech.*, **41**, 183 (1988).

Yaremko, Z.M., Fedyushinskaya, L.B. and Gurzel, N.V., Viscous drag of liquid medium in the sedimentation of particle aggregates, *Theo. Found. Chem. Eng.*, **31**, 181 (1997).

Yaron, I. and Ish-Shalom, M., Shear-induced thinning and thickening of raw cement slurries: Effect of simple ions and polyelectrolytes, *Trans. Soc. Rheol.*, **19**, 381 (1975).

Yasser Ibrahim, A.A., Briens, C.L., Margaritis, A. and Bergougnou, M.A., Hydrodynamic characteristics of a three-phase inverse fluidized bed column, *AIChE J.*, **42**, 1889 (1996).

Yasuda, K., *PhD Thesis*, Massachusetts Institute of Technology, Cambridge, MA (1979).

Ybert, C. and Di Meglio, J.-M., Ascending air bubbles in protein solutions, *Eur. Phys. J.*, **B4**, 313 (1998).

Ye, X., Tong, P. and Fetters, L.J., Transport of probe particles in semi-dilute polymer solutions, *Macromolecules*, **31**, 5785 (1998).

Yentov, V.M. and Polishchuk, A.M., Flow of anomalous fluids in porous media, *Fluid Mech. – Sov. Res.*, **8**(3), 35 (1979).

Yildirim, O.E. and Basaran, O.A., Deformation and breakup of stretching bridges of Newtonian and shear-thinning liquids: Comparison of one-and two-dimensional models, *Chem. Eng. Sci.*, **56**, 211 (2001).

Yildiz, S. and Başaran. B., Investigation of natural convection heat transfer along a uniformly heated vertical plate, *Arab. J. Sci. Eng.*, **44**(2), 1685 (2019).

Yip, R., James, D.F. and Currie, I.G., PIV measurements of slow flow of a viscoelastic fluid within a porous medium, *Exp. Fluids*, **51**(3), 801 (2011).

Yoo, H.J. and Han, C.D., Oscillatory behaviour of a gas bubble growing (or collapsing) in viscoelastic liquids, *AIChE J.*, **28**, 1002 (1982).

Yoo, S.-S., Jeon, C.-Y. and Cho, Y.I., Determination of the characteristic and diffusion times of polyacrylamide solutions using falling balls and needles, *Int. J. Heat Mass Transf.*, **37**(Suppl. I), 113 (1994).

Yoshimura, A.S., Prud'homme, R.K., Princen, H.M. and Kiss, A.D., A comparison of techniques for measuring yield stresses, *J. Rheol.*, **31**, 699 (1987).

Yoshioka, N. and Adachi, K., On variational principles for a non-Newtonian fluid, *J. Chem. Eng. Jpn.*, **4**, 217 (1971).

Yoshioka, N. and Adachi, K., Some deductions from the extremum principles for non-Newtonian fluids, *J. Chem. Eng. Jpn.*, **6**, 134 (1973).

Yoshioka, N. and Adachi, K., Problems of non-Newtonian fluid flow, *Kagaku-Kogaku*, **38**, 727 (1974) (in Japanese).

Yoshioka, N., Adachi, K. and Ishimura, H., On creeping flow of a viscoplastic fluid past a sphere, *Kagaku Kogaku*, **10**, 1144 (1971).

Yoshioka, N., Adachi, K., Nakamura, A. and Ishimura, H., An experimental investigation of viscoplastic flow past a circular cylinder at high Reynolds numbers, *Rheol. Acta*, **14**, 993 (1975).

Yoshioka, N. and Nakamura, R., On the creeping flow of generalized Newtonian fluid around a sphere, *Kagaku Kogaku*, **4**, 130 (1966).

Yotsumoto, Y. and Shiona, T., Bubble cascade in Guinness beer is caused by gravity current instability, *Sci. Rep.*, **9**(1), 1 (2019).

You, R., Borhan, A. and Haj-Hajiri, H., A finite volume formulation for simulating drop motion in a viscoelastic two-phase system, *J. Non-Newt. Fluid Mech.*, **153**, 109 (2008).

You, R., Borhan, A. and Haj-Hajiri, H., Stability analysis of cusped bubbles in viscoelastic flows, *J. Fluid Mech.*, **621**, 131 (2009).

Younes, E., Himl, M., Stary, Z., Bertola, V. and Burghelea, T., On the elusive nature of Carbopol gels "model" weakly thixotropic, or time-dependent viscoplastic materials? *J. Non-Newt. Fluid Mech.*, **281**, 104315 (2020).

Young, B.D., Bryson, A.W. and van Vliet, B.M., An evaluation of the technique of polygonal harmonics for the characterization of particle shape, *Powder Technol.*, **63**, 157 (1990).

Yow, H.N., Pitt, M.J. and Salman, A.D., Drag correlation for particles of regular shape, *Adv. Powder Technol.*, **16**, 363 (2005).

Yu, B., Analysis of flow in fractal porous media, *App. Mech. Rev.*, **61**, 050801-1 (2008).

Yu, C.P. and Soong, T.T., A random cell model for pressure drop prediction in fibrous filters, *J. Appl. Mech.*, **42**, 301 (1975).

Yu, H.S. and Houlsby, G.T., Finite cavity expansion in dilatant soils: Loading analysis, *Geotechnique*, **41**, 173 (1991).

Yu, S.-C. and Liu, H.-S., Letter to the editor, *Chem. Eng. Commun.*, **190**, 1439 (2003).

Yu, Y.H., Wen, C.Y. and Bailie, R.C., Power law fluids flow through multiparticle systems, *Can. J. Chem. Eng.*, **46**, 149 (1968).

Yu, Z., Pan, D., Lin, J. and Shao, X., Orientation of a cone settling in a vertical duct, *Fluid Dyn. Res.*, **46**, 015502 (2014).

Yu, Z., Phan-Thien, N., Fan, Y. and Tanner, R.I., Viscoelastic mobility problem of a system of particles, *J. Non-Newt. Fluid Mech.*, **104**, 87 (2002).

Yu, Z. and Wachs, A., A fictitious domain method for dynamic simulation of particle sedimentation in Bingham fluids, *J. Non-Newt. Fluid Mech.*, **145**, 78 (2007).

Yu, Z., Wachs, A. and Peysson, Y., Numerical simulation of particle sedimentation in shear-thinning fluids with a fictitious domain method, *J. Non-Newt. Fluid Mech.*, **136**, 126 (2006).

Yurun, F., Limiting behaviour of the solutions of a falling sphere in a tube filled with viscoelastic fluids, *J. Non-Newt. Fluid Mech.*, **110**, 77 (2003a).

Yurun, F., Solution behaviour of the falling sphere problem in viscoelastic flows, *Acta Mech. Sinica*, **19**, 394 (2003b).

Yurun, F. and Crochet, M.J., High-order finite-element methods for steady viscoelastic flows, *J. Non-Newt. Fluid Mech.*, **57**, 283 (1995).

Yurun, F., Tanner, R.I. and Phan-Thien, N., Galerkin/least-square finite-element methods for steady viscoelastic flows, *J. Non-Newt. Fluid Mech.*, **84**, 233 (1999).

Yurusoy, M. and Pakdemirli, M., Symmetry reductions of unsteady three-dimensional boundary layers of some non-Newtonian fluids, *Int. J. Eng. Sci.*, **35**, 731 (1997).

Zaidi, A., Benchekchou, B., Karioun, M. and Akharaz, A., Heat transfer in three-phase fluidized beds with non-Newtonian pseudoplastic solutions, *Chem. Eng. Commun.*, **93**, 135 (1990).

Zaidi, A., Deckwer, W.-D., Mrani, A. and Benchekchou, B., Hydrodynamics and heat transfer in three phase fluidized beds with highly viscous pseudoplastic solutions, *Chem. Eng. Sci.*, **45**, 2235 (1989).

Zaitsev, V.F. and Polyanin, A.D., Dynamics of spherical bubbles in non-Newtonian liquids, *Theo. Found. Chem. Eng.*, **26**, 185 (1992).

Zakhem, R., Weidman, P.D. and de Groh, H.C., On the drag of model dendrite fragments at low Reynolds number, *Met. Trans.*, **23A**, 2169 (1992).

Zamankhan, P., Takayama, S. and Grotberg, J.B., Steady displacement of long gas bubbles in channels and tubes filled by a Bingham fluid, *Phys. Rev. Fluids*, **3**, 013302 (2018).

Zami-Pierre, F., de Loubens, R., Quintard, M. and Davit, Y., Transition in the flow of power-law fluids through isotropic porous media. *Phys. Rev. Lett.*, **117**(7), 074502 (2016).

Zami-Pierre, F., de Loubens, R., Quintard, M. and Davit, Y., Polymer flow through porous media: Numerical prediction of the contribution of slip to the apparent viscosity. *Trans. Porous Media*, **119**(3), 521 (2017).

Zami-Pierre, F., de Loubens, R., Quintard, M. and Davit, Y., Effect of disorder in the pore-scale structure on the flow of shear-thinning fluids through porous media. *J. Non-Newt. Fluid Mech.*, **261**, 99 (2018).

Zana, E., *PhD Dissertation*, Caltech, Pasadena, CA (1975).

Zana, E. and Leal, L.G., The dynamics of bubbles and drops in a viscoelastic fluid, *Proc. Int. Coll. Drops and Bubbles*, (eds., Collins, Plesset and Saffren), **Vol. II**. (1974).

Zana, E. and Leal, L.G., Dissolution of a stationary gas bubble in a quiescent viscoelastic liquid, *Ind. Eng. Chem. Fund.*, **14**, 175 (1975).

Zana, E. and Leal, L.G., The dynamics and dissolution of gas bubbles in a viscoelastic fluid, *Int. J. Multiph. Flow*, **4**, 237 (1978).

Zana, E., Tiefenbruck, G. and Leal, L.G., A note on the creeping motion of a viscoelastic fluid past a sphere, *Rheol. Acta*, **14**, 891 (1975).

Zapryanov, Z. and Tabakova, S., *Dynamics of Bubbles, Drops and Rigid Particles*, Kluwer, Dordrecht, The Netherlands (1999).

Zarraa, M.A., Effect of drag reducing polymers on the rate of liquid-solid mass transfer in fixed beds of spheres under forced convection conditions, *Chem. Eng. Tech.*, **21**, 301 (1998).

Zarei, M. and Aalaie, J., Application of shear-thickening fluids in material development, *J. Mat. Res. Technol.*, **9**(5), 10411 (2020).

Zdravkovich, M.M., *Flow around Circular Cylinders*. Volume 1: *Fundamentals*, Oxford University Press, New York (1997).

Zdravkovich, M.M., *Flow around Circular Cylinders*. Volume 2: *Applications*, Oxford University Press, New York (2003).

Zeidan, A., Rohani, S., Bassi, A. and Whiting, P., Review and comparison of solids settling velocity models, *Rev. Chem. Eng.*, **19**, 473 (2003).

Zenit, R. and Feng, J.J., Hydrodynamics interactions among bubbles, drops and particles in non-Newtonian liquids, *Annu. Rev. Fluid Mech.*, **50**, 505 (2018).

Zenit, R. and Magnaudet, J., Path instability of rising spheroidal air bubbles: A shape-controlled process, *Phys. Fluids*, **20**, 061702 (2008).

Zenit, R. and Rodriguez-Rodriguez, J., The fluid mechanics of bubbly drinks, *Phys. Today*, **71**(11), 44 (2018).

Zeppenfeld, R., *PhD Thesis*, Technical University München, Munich (1988).

Zerai, B., Saylor, B.Z., Kadambi, J.R., Oliver, M.J., Mazaheri, A.R., Ahmadi, G., Bromhal, G.S. and Smith, D.H., Flow characterization through a network cell using particle image velocimetry, *Trans. Porous Media.*, **60**, 159 (2005).

Zhang, G., Li, M., Li, J., Yu, M., Qi, M. and Bai, Z., Wall effects on spheres settling through non-Newtonian fluid media in cylindrical tubes, *J. Disp. Sci. Technol.*, **36**, 1199 (2015).

Zhang, J., An augmented Lagrangian approach to simulating yield stress fluid flows around a spherical gas bubble, *Int. J. Num. Methods Fluids*, **69**, 731 (2012).

Zhang, J. and Fan, L.-S., On the rise velocity of an interactive bubble in liquids, *Chem. Eng. J.*, **92**, 169 (2003).

Zhang, J., Liu, J. and Lu, W.-Q., Study on laminar natural convection heat transfer from a hemisphere with uniform heat flux surface, *J. Thermal Sci.*, **28**, 232 (2019a).

Zhang, J.-P., Epstein, N., Grace, J.R. and Zhu, J., Minimum liquid fluidization velocity of gas-liquid fluidized beds, *Chem. Eng. Res. Des.*, **73**, 347 (1995).

Zhang, L., Yang, C. and Mao, Z.-S., Unsteady motion of a single bubble in highly viscous liquid and empirical correlation of drag coefficient, *Chem. Eng. Sci.*, **63**, 2099 (2008a).

Zhang, L., Yang, C. and Mao, Z.-S., An empirical correlation of drag coefficient for a single bubble rising in non-Newtonian liquids, *Ind. Eng. Chem. Res.*, **47**, 9767 (2008b).

Zhang, L., Yang, C. and Mao, Z.-S., Numerical simulation of a bubble rising in shear-thinning fluids, *J. Non-Newt. Fluid Mech.*, **165**, 555 (2010).

Zhang, M., Prodanović, M., Mirabolghasemi, M. and Zhao, J., 3D microscale flow simulation of shear-thinning fluids in a rough fracture, *Trans. Porous Media*, **128**(1), 243 (2019b).

Zhang, X., Dynamics of drop formation in viscous flows, *Chem. Eng. Sci.*, **54**, 1759 (1999).

Zhang, X., Liu, H., Zhang, Y. and Wang, L., Direct numerical simulation of the sedimentation of a particle pair in a shear-thinning fluid, *Phys. Rev. Fluids*, **5**, 014304 (2020).

Zhang, X., Sugiyama, K. and Watamura, T., Drag on an oscillatory spherical bubble in shear-thinning fluid, *J. Fluid Mech.*, **959**, A3-1 (2023).

Zhang, X.Z., Li, W.H. and Gong, X.L., The rheology of shear-thickening fluid (STF) and the dynamic performance of an STG-filled damper, *Smart Mater. Struct.*, **17**, 035027 (2008c).

Zhang, Y. and Finch, J.A., Single bubble terminal velocity – Experiment and modeling, *Advances in Floatation Technology*, edited by B.D. Parekh and J.D. Miller, p. 83, Published by Society for Mining, Metallurgy and Exploration, Littleton, CO (1999).

Zhang, Y. and Finch, J.A., A note on single bubble motion in surfactant solutions, *J. Fluid Mech.*, **429**, 63 (2001).

Zhang, Y. and Muller, S.J., Unsteady sedimentation of a sphere in wormlike micellar solutions, *Phys. Rev. Fluids*, **3**, 043301 (2018).

Zheng, G.-H., Powell, R.L. and Stroeve, P., Settling velocity of a sphere falling between two concentric cylinders filled with a viscous fluid, *Ind. Eng. Chem. Res.*, **31**, 1366 (1992).

Zheng, L.C. and Zhang, X.X., Skin friction and heat transfer in power-law fluid laminar boundary layer along a moving surface, *Int. J. Heat Mass Transf.*, **45**, 2667 (2002).

Zheng, R. and Phan-Thien, N., A boundary element simulation of the unsteady motion of a sphere in a cylindrical tube containing a viscoelastic fluid, *Rheol. Acta*, **31**, 323 (1992).

Zheng, R., Phan-Thien, N. and Ilic, V., Falling needle rheometry for general viscoelastic fluids, *J. Fluids Eng.*, **116**, 619 (1994).

Zheng, R., Phan-Thien, N. and Tanner, R.I., On the flow past a sphere in a cylindrical tube: Limiting Weissenberg number, *J. Non-Newt. Fluid Mech.*, **36**, 27 (1990a).

Zheng, R., Phan-Thien, N. and Tanner, R.I., The flow past a sphere in cylindrical tube: Effects of inertia, shear-thinning and elasticity, *Rheol. Acta*, **30**, 499 (1991).

Zheng, R., Phan-Thien, N., Tanner, R.I. and Bush, M.B., Numerical analysis of viscoelastic flow through a sinusoidally corrugated tube using a boundary element method, *J. Rheol.*, **34**, 79 (1990b). Errata *ibid* 785 (1990).

Zhizhin, G.V., Laminar boundary layer in a non-Newtonian fluid. Qualitative discussion, *J. App. Mech. Tech. Phys.*, **28**, 383 (1987).

Zholkovskiy, E.K., Shilov, V.N., Masliyah, J.H. and Bondarenko, M.P., Hydrodynamic cell model: General formulation and comparative analysis of different approaches. *Can. J. Chem. Eng.*, **85**(5), 701 (2008).

Zhong, H., Chen, S. and Lee, C., Experimental study of freely falling thin disks: Transition from planar zigzag to spiral, *Phys. Fluids*, **23**, 011702 (2011).

Zhong, H.-J. and Lee, C.-B., The wake of falling disks at low Reynolds numbers, *Acta Mech. Sin.*, **28**, 367 (2012).

Zhong, W.H., Currie, I.G. and James, D.F., Creeping flow through a model fibrous porous medium, *Exp. Fluids*, **40**(1), 119 (2006).

Zhou, C., Yue, P., Feng, J.J., Liu, C. and Shen, J., Heart-shaped bubbles rising in anisotropic liquids, *Phys. Fluids*, **19**, 041703 (2007).

Zhou, Y., Zhao, C. and Bo, H., Analyses and modified models for bubble shape and drag coefficient covering a wide range of working conditions, *Int. J. Multiph. Flow*, **127**, 103265 (2020).

Zhu, C., Lam, K., Chu, H.H., Tang, X.D. and Liu, G.L., Drag forces of interacting spheres in power law fluids, *Mech. Res. Commun.*, **30**, 651 (2003).

Zhu, C., Yang, H., Fu, T., Gao, X. and Ma, Y., Computational fluid dynamics simulation of generation and coalescence of bubbles in non-Newtonian fluids, *Chem. Eng. Technol.*, **41**, 541 (2018).

Zhu, J., On the flow resistance of viscoelastic fluids through packed beds, *Rheol. Acta*, **29**, 409 (1990).

Zhu, J., Drag and mass transfer for flow of a Carreau fluid past a swarm of Newtonian droplets, *Int. J. Multiph. Flow*, **21**, 935 (1995).

Zhu, J., A note on slow non-Newtonian flows over an ensemble of spherical bubbles, *Chem. Eng. Sci.*, **56**, 2237 (2001).

Zhu, J. and Chan Man Fong, C.F., Flow of Carreau fluids through packed beds, *Progress Trends Rheol.*, **II**, 141 (1988).

Zhu, J. and Deng, Q., Non-Newtonian flow past a swarm of Newtonian droplets, *Chem. Eng. Sci.*, **49**, 147 (1994).

Zhu, J., Li, T., Cai, S. and Suo, Z., Snap-through expansion of a gas bubbles in an elastomer, *J. Adhesion*, **87**, 466 (2011).

Zhu, J. and Satish, M.G., Non-Newtonian effects on the drag of creeping flow through packed beds, *Int. J. Multiph. Flow*, **18**, 765 (1992).

Zhu, L, Papadopoulos, K. and De Kee, D., A slotted plate device for measuring static yield stress, *J. Rheol.*, **45**, 1105 (2001).

Zhu, L, Papadopoulos, K. and De Kee, D., Yield stress measurements of Silicon Nitride suspensions, *Can. J. Chem. Eng.*, **80**, 1175 (2002).

Zhu, Q. and Clark, P.E., Wall effects in multiparticle settling, *Rheology and Fluid Mechanics of non-Newtonian Materials* (ASME) FED, Vol. 246, 7 (1998). Also see *ibid* pp. 69–74.

Zick, A.A. and Homsy, G.M., Stokes flow through periodic arrays of spheres, *J. Fluid Mech.*, **115**, 13 (1982).

Ziegenhagen, A.J., Bird, R.B. and Johnson, M.W., Non-Newtonian flow around a sphere, *Trans. Soc. Rheol.*, **5**, 47 (1961).

Ziegenhagen, A.J., The very slow flow of a Powell-Eyring fluid around a sphere, *Appl. Sci. Res.*, **14A**, 43 (1964).

Zierenberg, J.R., Fujioka, H., Hirschl, R.B., Bartlett, R.H. and Grotberg, J.B., Pulsatile blood flow and oxygen transport past a circular cylinder, *J. Biomech. Eng.*, **129**, 202 (2007).

Zimmels, Y., A generalized approach to flow through fixed beds, fluidization and hindered sedimentation, *Chem. Eng. Commun.*, **67**, 19 (1988).

Zimmerman, R.W., Fluid flow in rock fractures: From the Navier-Stokes equations to the cubic law, *Dynamics of Fluids in Fractured Rock*, p. 213, Geophysical Monograph No. 122, Am. Geophys. Union (2000).

Zimmerman, W.B., The drag on sedimenting discs in broadside motion in tubes, *Int. J. Eng. Sci.*, **40**, 7 (2002).

Ziolkowska, I. and Ziolkowski, D., Fluid flow inside packed beds, *Chem. Eng. Process.*, **23**, 137 (1988).

Zisis, Th. and Mitsoulis, E., Viscoplastic flow around a cylinder kept between parallel plates, *J. Non-Newt. Fluid Mech.*, **105**, 1 (2002).

Ziskind, G., Zhao, B., Katoshevski, D. and Bar-Ziv, E., Experimental study of the forces associated with mixed convection from a heated sphere at small Reynolds and Grashof numbers. I. Cross flow, *Int. J. Heat Mass Transf.*, **44**, 4381 (2001).

Zitoun, K.B. and Sastry, S.K., Determination of convective heat transfer coefficient between fluid and cubic particles in continuous tube flow using non-invasive experimental technique, *J. Food Process Eng.*, **17**, 209 (1994). Also see *ibid* 229 (1994).

Zlokarnik, M., Scale-up processes using material systems with variable properties, *Chem. Biochem. Eng. Quart.*, **15**, 43 (2001).

Zoric, J., Busch, A., Meese, E.A., Khatibi, M., Time, R.W., Johansen, S.T. and Rabenjafimanantsoa, H.A., On pragmatism in industrial modelling -Part II: Workflows and associated data and meta data, *Proc. 11th Int. Conf. CFD in the minerals and process industries*, Melbourne (Australia) (2015).

Zou, R.P. and Yu, A.B., The packing of spheres in a cylindrical column: The thickness effect, *Chem. Eng. Sci.*, **50**, 1504 (1995).

Zukauskas, A., Convective heat transfer in cross flow, *Handbook of Single-Phase Convective Heat Transfer*, edited by S. Kakac, R.K. Shah and W. Aung, pp. 3–38, Wiley-Interscience, New York (1987).

Zunker, F., Das allgemeine Grundwasserfließgesetz. *J. Gasbel. u. Wasserversorg*, **63**, 332 (1920).

Zuritz, C.A., McCoy, S.C. and Sastry, S.K., Convective heat transfer coefficients for irregular particles immersed in non-Newtonian fluid during tube flow, *J. Food Process. Eng.*, **11**, 159 (1990).

Author Index

Aalaie, J. 22
Abbasov, T. 58, 60, 73
Abdel-Alim, A. H. 215
Abdel-Hameed, H. 423
Abdel-Khalik, A. 368, 394
Abdel-Khalik, S. I. 175
Abdolnezhad, H. 191, 196, 202, 225
Abedijabari, A. 145, 148, 150, 151
Abhishek, S. 189, 198, 228, 236
Abou-Kasseem, J. H. 282
Abou-Ziyan, H. 423
Abrate, S. 239
Abuaf, N. 280
Abud-Archilla, M. 510
Acharya, A. 56, 57, 59, 60, 66, 67, 72, 73, 78, 83, 143, 154, 156, 180, 181, 191, 193, 204, 205, 206, 210, 225, 227, 228, 231, 233, 385, 484, 523, 524
Acharya, R. C. 242
Achenbach, E. 476, 480
Achterhold, K. 249
Acker, D. N. 510, 524
Acosta, A. J. 162, 164, 395, 396
Acrivos, A. 192, 206, 209, 211, 228, 255, 256, 270, 271, 272, 358, 365, 366, 368, 369, 372, 373, 374, 375, 376, 377, 378, 386, 394, 411, 421, 422, 423, 425, 426, 427, 428, 430, 432, 433, 436, 437, 438, 439, 440, 445, 450, 451, 452, 453, 454, 461, 463
Adachi, K. 56, 57, 59, 60, 66, 67, 68, 71, 79, 107, 108, 113, 118, 131, 132, 133, 135
Adam, M. 510
Adamczyk, M. 482
Adamczyk, Z. 482
Adams, D. 312
Addie, G. R. 2, 105
Adelman, M. 423, 427, 430, 451, 452, 454, 456, 459
Adler, I. 231
Adler, P. M. 162, 164, 244, 245, 246, 247, 260, 271, 274
Advani, S. G. 186, 187, 266, 268, 312, 313, 315, 316, 401
Aerov, V. E. 161, 162
Afacan, A. 259, 260, 262, 264, 266, 278, 303, 406
Afolabi, R. O. 239, 308, 323, 327
Afshin, H. 181, 200, 203, 223
Agacinski, P. 283, 336, 337, 339, 349, 350
Agarwal, D. C. 268
Agarwal, K. 303
Agarwal, L. 88
Agarwal, M. 365, 376
Agarwal, N. 89
Agarwal, P. K. 242, 252, 257, 295
Agarwal, S. 473, 479
Agarwal, U. S. 161, 323
Aghajani, M. 336
Aghdasinia, H. 285
Agnaou, M. 285, 287

Aguirre, V. A. 200
Agullo, J. O. 239
Agwu, D. E. 72
Aherwar, K. 421, 426, 430
Ahmad, E. A. 444
Ahmad, R. A. 435, 444
Ahmadi, A. 120, 284
Ahmadi, G. 253
Ahmadi-Sénichault, A. 284, 285, 287
Ahmadpour, A. 196, 228, 236
Ahmed, H. A. M. 61
Ahmed, N. 244, 260
Ahn, K. H. 165, 166
Ahonguio, F. 110, 111, 138, 139
Aiba, S. 400, 403
Airiau, C. 259, 285, 303
Ait-Kadi, A. 10, 156
Ajayi, O. O. 193, 202, 206, 223
Ajith Kumar, S. 444
Akagi, S. 426
Akers, B. 170, 171
Akgiray, Ö. 246, 247, 248, 249
Akharaz, A. 352, 410, 411
Akpabio, J. U. 72
Akroyd, T. 41, 42, 128, 521
Akutsu, T. 476, 481
Al Taweel, A. M. 368, 394
Al Varado, D. A. 276
Alabi, S. B. 72
Alakus, B. 151
Alam, M. J. 201, 217, 221, 403
Alam, M. S. 482
Alava, M. 273
Albalak, R. J. 186, 235
Albernaz, D. L. 195
Alcocer, F. J. 273, 313, 319
Al-Dibouni, M. R. 341, 342
Aldridge, P. M. 166, 167, 168
Alexander, P. 244, 263
Alexandrou, A. N. 132, 137
Alexandrova, S. 215, 401
Al-Fariss, T. F. 248, 276, 295, 296, 317
Alghalibi, D. 358, 360
Alhamdan, A. 377, 392, 423, 438
Ali, S. 385, 400, 524, 526
Aliabadi, S. 261
Alicke, A. A. 181, 197, 228
Aliseda, A. 185
Allahham, A. 3
Allain, C. 110, 168, 281, 320, 358, 359, 360
Allan, Y. M. 87, 498
Allen, D. J. 528
Allen, E. 58, 351, 358, 359
Allen, J. S. 175, 195

Allen, K. D. 4
Al-Matroushi, E. 199, 233
Almo, J. 162
Alnakeeb, M. A. 444
Aloia, L. D. 107
Alran, C. 180
Al-Raoush, R. 253
Al Shakry, B. 239
Al-Suwaiyan, A. 239, 245
Al Taweel, A. M. 368, 394
Altobelli, S. A. 244, 260
Al Varado, D. A. 276
Alvarez- Hernandez, A. R. 249
Alvarez, E. 15, 19
Alves, M. A. 51, 148, 165, 367, 395
Alwared, A. I. 358
Amani, E. 196, 228, 236
Amaratunga, M. 155
Amarouchene, Y. 161
Amato, W. S. 423, 438, 451, 452, 453, 454, 456, 459, 466
Ambari, A. 164, 287, 365, 388, 396, 482, 485, 500, 509, 510
Ambeskar, V. D. 130, 154, 159, 521
Ambros, F. 522, 523
Ames, W. F. 373
Amin, N. 463
Aminzadeh, M. 200, 203, 223
Amundarain, J. L. 286
Anacleto, J. 368, 394, 422, 443, 446
Anderson, P. D. 168
Anderson, S. L. 269
Anderson, T. J. 162
Andersson, H. I. 365, 373
Andrade Jr, J. S. 283, 285, 286, 299
Andrade, E. N. dac. 22
Andrade, R. F. 285, 299
Andres, U. Ts. 107, 108, 111, 123
Andrews, B. D. 268, 269
Andrews, J. 369, 374, 377
Andrews, J. R. G. 41
Anil Lal, S. 444
Anjaiah, N. 446
Annaland, M. V. S. 226
Ansley, R. W. 107, 108, 111, 113, 115, 123, 124, 126, 129, 131
Anthony, C. R. 231
Anthore, N. 482
Anthore, R. 482
Anzenavs, R. 162
Anzueto-Sanchez, G. 510
Apaydin, O. G. 276, 304
Appuzzo, G. 192
Arabi, P. 449
Aragao, A. F. L. 45
Aranjo, A. D. 285, 299
Arastoopour, H. 331
Archer, L. A. 323
Arcoumanis, C. 185
Ardakani, H. A. 3
Ardekani, A. M. 145, 155, 168, 169, 170
Arefmanesh, A. 186, 187, 401

Arigo, M. T. 144, 145, 147, 150, 151, 152, 153, 155, 158, 160
Aristegui, C. 242
Armstrong, R. C. 3, 10, 14, 16, 29, 39, 55, 65, 107, 109, 111, 113, 118, 120, 121, 141, 143, 144, 150, 151, 152, 197, 203, 299, 313, 319, 363, 511
Arnipally, S. K. 45
Arnold, D. P. 4
Arsenijevic, Z. Lj. 476, 481
Arterburn, D. R. 57, 59, 62, 83
Arul Prakash, K. 444
Arzate, A. 3
Ascanio, G. 3
Aschenbrenner, B. C. 88
Ashare, E. 527
Asif, M. 269, 341
Askarov, B. 239
Assimacopoulos, D. 482, 483, 484, 485
Astarita, G. 18, 39, 40, 41, 56, 62, 63, 109, 192, 201, 202, 206, 209, 210, 221, 223, 363, 394, 396, 405
Astin, J. 394
Astrauskas, P. 425
Aström, A. 377, 392
Aström, B. T. 315, 316
Ataide, C. H. 68, 476, 489, 499
Atalik, K. 145, 162, 165
Atapattu, D. D. 109, 111, 113, 114, 115, 116, 117, 118, 122, 123, 125, 126, 130, 233, 492
Atayilmaz, S. O. 435
Ates, H. 239
Atkinson, H. 27
Atsbha, H. 142
Atta, A. 199, 228, 236, 189
Aubert, J. H. 276, 322
Aubouy, M. 4, 139
Auradou, H. 499
Auriault, J.-L. 239, 244, 246, 259, 295, 314, 320
Awbi, H. 480
Awuah, G. B. 377, 423
Ayyash, S. 187
Ayyaswamy, P. S. 200, 236, **505**
Aziz, A. 429
Aziz, K. 27
Azizi, S. 112
Azzam, M. I. S. 246, 258, 259

Baaijens, F. P. T. 144, 148, 150, 152, 165
Baaijens, H. P. W. 144, 152, 165
Babaei, M. 283, 285, 286, 304
Baca, H. 144, 152, 165, 506
Bachmat, Y. 240
Bacon, L. R. 472, 476
Badr, H. M. 444
Baelmans, M. 239, 263
Bagchi, A. 82, 160, 522
Bağcı, Ö. 244
Bagherian, B. 426
Bagley, E. B. 22
Bagster, D. F. 180
Bahrami, M. 271
Bai, Z. 485, 493

Baier, S. K. 2, 3, 35
Baijal, S. K. 276
Baikov, V. I. 422, 429, 430
Bailey, A. E. 112
Bailie, R. C. 283, 298, 312, 335, 336, 337, 338
Bailoor, S. 96
Bair, S. 15
Baird, D. G. 165, 240, 277
Baird, M. H. I. 405, 528
Baird, R. M. 262
Baker, C. G. J. 331
Balakrishna, M. 358
Balamurugan, M. 351
Balan, C. 18
Balaramakrishna, P. V. 482, 490
Balasubramaniam, V. M. 368, 377, 391
Balasubramanian, R. 175, 236
Balcar, M. 334, 335, 336, 337, 338, 339, 343, 350
Bale, S. 50, 51, 54
Balhoff, M. T. 253, 283, 299, 315, 319
Ballantyne, C. K. 88
Balmer, R. T. 3, 22
Balmforth, N. J. 18, 19
Baltussen, M. W. 196, 226
Banavar, J. R. 186
Bandopadhyay, P. 364
Banfill, P. F. G. 3
Bangun, J. 194, 225, **505, 506**
Baptista, P. N. 368, 377, 391
Barabati, A. C. 2
Barak, A. Z. 245
Barakos, G. 143, 165
Baranyi, L. 446
Barbeau, L. 72, 76
Barboza, M. 276, 295, 313, 320
Barclay, J. 186
Barentin, C. 23
Barigou, M. 82, 83, 438
Bark, F. H. 187
Bark, G. 377, 392
Barlow, A. J. 524, 526
Barnea, E. 253, 299, 333
Barnes, H. A. 2, 16, 18, 22, 27, 32, 39, 41, 42, 129, 148
Barnett, S. M. 160, 192, 202, 203, 206, 208, 212, 225, 232, 400
Barr, D. W. 239
Barr, G. 475, 479, 528
Barr, T. 526
Barree, R. D. 83
Barrozo, M. A. S. 68, 476, 489, 499
Bart, E. 482
Bartlett, R. H. 384
Bartram, E. 168, 169
Barua, P. 446
Bar-Yoseph, P. 269, 274
Bar-Ziv, E. 450, 463
Basak, S. 3, 25
Başaran, B. 422
Basaran, O. A. 186, 231, 235, 473, 474, 478, 504, 505
Bashir, Y. M. 22
Basset, A. B. 54

Bassi, A. 358
Bassiouny, M. K. 411
Basu, S. 276, 311
Batishchev, Ya. F. 265
Batot, G. 285, 286
Batra, V. K. 258
Battistella, A. 196, 226
Bau, H. H. 87, 526
Bauckhage, K. 175
Baudez, J. 3
Bauer, D. 285, 286, 287, 299
Baumann, T. 244
Baumhoff, E. 239, 308, 323, 327
Bazhlekov, I. B. 233
Bazilevskii, A. V. 163
Beale, S. B. 275
Beam, J. E. 482
Bear, J. 240, 241, 260, 266
Beard, D. W. 394
Beaty, B. B. 510, 524
Beaudoin, G. 4
Beaulne, M. 107, 109, 111, 113, 115, 117, 118, 120, 121, 492
Becker, L. E. 144, 146, 160, 169
Becker, T. 301, 303
Bedborough, D. S. 526
Begovich, J. M. 354
Behbahani, B. 410
Behringer, R. P. 244
Bejan, A. 240
Belda-Maximino, R. 26
Belfares, L. 326
Bell, D. 41
Bell, K. J. 312
Belmonte, A. 152, 155, 170, 171, 172, 175, 195, 197, 202, 203, 224, 225, 226
Bena, J. 336
Benchekchou, B. 352, 410, 411
Bendova, H. 65, 78
Benenati, R. F. 262
Benilov, E. S. 1
Benis, A. M. 276, 304
Benn, D. I. 88
Bennethum, L. S. 273
Bennington, C. P. J. 3, 18, 175, 206
Ben-Richou, A. 365, 388, 485, 499, 500, 509
Bensetiti, Z. 326
Benyahia, F. 260, 262, 264
Berchielli, A. 185
Bercovier, M. 120
Berdnikov, V. I. 215
Bereiziat, D. 288
Berg, J. C. 105
Berg, S. 284, 286, 305
Bergelin, O. P. 268
Berger, S. 242
Bergougnou, M. A. 33, 191, 195, 221, 227, 351
Beris, A. N. 26, 107, 109, 111, 113, 118, 120, 121, 146, 197, 258, 299, 307, 314, 319, 520
Berkovsky, B. M. 366, 373
Berney, B. M. 2, 3
Bernis, A. 351

Bernstein, B. 39, 187
Berry, M. 3
Berthier, L. 18, 19
Bertin, H. 278, 304
Bertin, H. J. 276, 304
Bertola, V. 127, 176
Bertrand, F. 56
Bertsch, P. 3, 285, 286
Berüter, B. 185
Best, A. C. 50
Bethmont, S. 107
Betts, D. E. 22
Beveridge, G. S. G. 262
Bezzola, G. R. 524
Bhaga, D. 190
Bhamidipati, S. 107
Bharatiya, S. S. 367, 375
Bharti, R. P. 93, 94, 96, 97, 269, 289, 365, 367, 377, 378, 379, 381, 382, 412, 422, 443, 446, 447, 448
Bhat, M. S. 239, 272
Bhatnagar, P. L. 394
Bhatnagar, R. K. 385, 396
Bhattacharyya, S. 461
Bhavaraju, S. M. 122, 193, 210, 217, 220, 229, 398, 399, 403, 404
Bhavasar, P. M. 216, 505
Bhowmick, S. 444
Biesheuvel, A. 85
Bilanin, A. J. 185
Bingham, E. C. 1
Binnington, R. J. 156, 157
Binous, H. 170
Bird, R. B. 3, 10, 14, 16, 18, 19, 29, 39, 40, 41, 46, 55, 56, 57, 67, 70, 154, 155, 168, 170, 182, 203, 230, 248, 249, 282, 286, 287, 289, 290, 291, 292, 293, 304, 306, 307, 309, 320, 358, 363, 484, 511, 522, 527
Bisgaard, C. 142, 152, 153, 154, 159, 193, 202, 203
Bishop, R. L. 54
Biswas, A. K. 528
Bizzell, G. D. 365, 367, 386
Bjerkholt, J. T. 3
Bjornestad, E. 304, 309
Blackery, J. 107, 109, 113, 117, 120, 121
Blackwell, B. C. 176
Blais, B. 72, 76
Blake, F. C. 248
Blanc, R. 268
Blanch, H. W. 122, 193, 210, 217, 220, 229, 231, 398, 399, 403, 404
Blanchet, J.-F. 191, 195, 206, 209, 210, 211, 224, 227
Blanco, A. 215
Blanks, R. F. 261, 281, 295, 296, 306, 311
Blass, E. 229, 230, 231
Bleses, D. 137, 176
Bleyer, J. 244, 284, 288, 314, 317
Bloch, J.-F. 240
Block, J.-F. 239, 259, 295, 314, 320
Block, S. 528
Bloom, F. 186

Blümich, B. 215, 244
Blunt, M. J. 259, 283, 286, 295, 299
Bo, H. 188
Boardman, G. 107, 108, 111, 113
Bobinski, T. 91
Bobroff, S. 358, 359
Bodart, C. 143, 160
Boehm, M. W. 2, 3, 35
Boersma, W. H. 22
Boetcher, S. K. S. 417
Boger, D. V. 1, 3, 10, 11, 12, 29, 41, 42, 57, 66, 129, 149, 152, 156, 157, 162, 185, 186
Bogusz, E. 191, 192, 203, 208, 225
Bohlin, T. 473, 474, 499
Bohm, L. 196, 199, 200
Bohm, U. 335, 342, 394, 406, 407, 409
Bohme, G. 258
Bollen, A. M. 215
Bonadonna, C. 45
Bonano, E. J. 273
Bond, W. N. 209, 210
Bondarenko, M. P. 255
Bondi, A. 278, 304, 320
Bonin, K. 141
Bonn, D. 18, 19, 107, 112
Borah, A. 91, 101
Borah, R. 502
Borhan, A. 197, 199, 233, 505
Bories, S. A. 242
Bornside, D. E. 141, 313, 319
Bortolotti, V. 243
Borwankar, R. 3
Bose, A. 132, 384, 421, 422, 440, 441, 442, 443, 448
Boskovic - Vragolovic, N. M. 476, 481
Bostel, F. 482
Bot, E. T. G. 168, 170
Bothe, D. 201, 211
Bottaro, A. 259, 285, 303
Bottinga, Y. 175
Bouaziz, M. 500
Bouchet, G. 85, 91, 101, **463**
Bougas, A. 476
Boujlel, J. 384
Boukany, P. E. 304, 315, 319
Boulanger, B. 259, 260, 278, 303
Boulanger-Nadeau, P. E. 56
Bourban, P.-E. 239
Bourgeat, A. 240, 242
Bourgoyne, A. T. 144, 152, 165, 506
Bourne, J. R. 523, 524
Bousfield, D. W. 185
Boussinesq, J. 54
Bowen, B. D. 87
Bowen, W. R. 473
Boyd, J. W. R. 195
Boyer, S. 42
Bozzi, L. A. 505
Braga, C. V. M. 313, 315
Brahimi, M. 281, 334
Bramley, A. S. 244, 263
Branch, C. 215

Brandao, E. M. 15
Brandt, H. 366, 375
Brandt, L. 97, 358, 360
Brasquet, C. 239
Brea, F. M. 276, 286, 296, 297, 298, 335, 337, 338, 339, 342, 348, 350, 409
Breach, D. R. 49
Breakey, D. E. S. 88
Brehmer, M. 199
Brenn, G. 185, 201, 206, 209, 210, 211
Brennen, C. E. 161, 175
Brenner, H. 85, 86, 87, 169, 255, 471, 472, 473, 482, 496, 497, 526
Brestkin, Y. V. 314, 319
Brett, D. J. 243
Brian, P. 18
Briedis, D. 3
Briend, P. 276, 295, 304, 335, 342, 343, 350
Briens, C. L. 78, 351, 352, 409, 410, 411
Briguet, A. 169
Brim, J. 377
Brinkman, H. C. 254, 273, 299
Briscoe, B. J. 19, 21, 22, 68, 72, 83, 91, 109, 128, 182, 523
Broadbent, J. M. 148, 153, 513
Broer, L. J. R. 40
Brokl, P. 326, 350, 351, 359
Bromhal, G. S. 253
Broniarz-Press, L. 283, 336, 337, 339, 349, 350, 364
Brooker, D. B. 246
Brookes, G. F. 113, 128, 138
Brosilow, C. B. 262
Brower, L. E. 3, 41
Brown, D. J. 88, 101
Brown, E. 22
Brown, F. E. 183
Brown, G. A. 268
Brown, G. G. 264, 521, 522
Brown, J. C. (Jr.), 269
Brown, J. M. B. 87, 498
Brown, P. P. 51
Brown, R. A. 107, 109, 111, 113, 118, 120, 121, 141, 143, 144, 150, 151, 152, 165, 197, 299, 313, 319
Brown, R. C. 239
Brown, S. W. J. 187
Brown, W. D. 282, 286, 322
Browne, C. A. 285, 309
Brucker, J. 435
Bruggeman, D. A. G. 243
Brujan, E.-A. 186, 187
Brummer, R. 3
Bruneau, C. H. 141, 145, 166
Brunjail, D. 320, 406
Brunn, P. O. 142, 151, 168, 169, 244, 276, 278, 280, 281, 282, 286, 304, 305, 307, 308, 309, 311, 324, 350
Brunn, T. 377
Bruschke, M. V. 312, 313, 315
Brutyan, M. A. 186
Bryan, W. L. 523
Bryant, J. 45
Bryson, A. W. 88

Bu, S. 242, 244, 245, 261, 263, 312
Bui, C. M. 112
Buchholz, H. 206, 230, 231, 403
Buchholz, K. 265
Buchholz, R. 206, 230, 403
Budzynski, P. 195, 202, 225, 226, 506
Buggisch, H. W. 244, 283, 287, 288, 319
Bukman, Yu. A. 83, 101
Bulina, I. G. 108, 113
Bullivant, S. A. 22
Burcik, E. J. 239, 276, 295, 309, 320
Burdin, F. 195, 212
Burger, J. 278, 307, 308, 309, 311
Burghelea, T. 107
Burghelea, T. I. 110, 112, 117
Burman, J. E. 186
Burru, I. G. 352, 409, 410, 411
Busch, A. 83, 163
Bush, M. B. 56, 57, 65, 66, 68, 143, 152, 153, 155
Butcher, T. A. 514, 517, 518, 519
Byars, J. A. 185
Byers, C. H. 473, 474, 478, 504, 505
Byrne, B. J. 473
Byrne, M. 45

Cabeza, C. 155, 168, 170
Caggioni, M. 107, 110, 111, 112
Cai, S. 196
Cairncross, E. K. 524
Cakl, J. 67, 75, 276, 277, 280, 308, 358
Calderbank, P. H. 192, 202, 204, 206, 208, 385, 400, 401, 403
Caltagirone, J.-P. 244
Camacho, M. M. 3
Camas-Anzueto, J. L. 510
Campbell, G. A. 109, 126, 130, 131, 233
Campbell, W. B. 268, 269
Campo-Deano, L. 283, 315
Campos, J. B. L. M. 189, 195, 198, 227, 236
Cancela, M. A. 15, 19
Canou, J. 284
Cao, B. 201, 217, 220
Cao, X. 242, 261, 263
Cao, Y. 188, 215
Capes, C. E. 358
Capobianchi, M. 363, 424, 429, 448
Caputo, A. C. 239
Caramelo, L. 368, 394, 422, 443, 446
Carberry, J. J. 387
Carew, E. O. A. 56, 143, 165
Carew, P. S. 175, **506**
Carey, W. W. 161
Carman, P. C. 248, 250, 251, 252, 261, 268, 269, 290
Carne, R. 175
Carniani, E. 357
Carpinlioglu, M. O. 246, 265
Carr, N. L. 206, 230, 403
Carreau, P. J. 3, 10, 14, 16, 37, 39, 42, 147, 156, 160, 193, 202, 206, 208, 209, 212, 225, 231, 232, 491
Carrington, S. 286, 324
Cartier, S. 266

Caserta, S. 169
Cashman, K. V. 3
Castanier, L. M. 276, 304
Castellanes, A. 18, 19
Castillo, A. 145, 158, 163
Castillo, B. A. 200
Castro, J. A. A. 260
Castro, L. J. 286
Caswell, B. 142, 147, 150, 164, 170, 208, 358, 396, 485, 493, 503, 511, 513, 514, 516
Catton, I. 241, 259, 434
Cayeux, E. 3
Cerasi, P. 320
Ceylan, K. 58, 60, 73
Chafe, N. P. 107, 110, 111, 117, 123, 126, 131
Chahine, G. L. 187
Chakrabarti, A. 4
Chakrabarti, S. 286
Chakraborty, J. 499
Chamberland, E. 56
Chamkha, A. J. 394
Champmartin, S. 287, 365, 388, 485, 500
Chamsri, K. 273
Chan Man Fong, C. F. 56, 58, 65, 191, 194, 202, 206, 208, 209, 210, 211, 215, 221, 225, 234, 227, 283, 301, 401, 498
Chan, H.-K. 88, 101
Chan, P. C.-H. 169, 235
Chan, R. K. Y. 526
Chan, V. 141
Chandarana, D. I. 368, 377
Chandler, H. D. 26
Chandra, A. 382, 436, 440, 448
Chandrasekhara, B. C. 262
Chang, C. 3
Chang, C.-N. 277, 295, 320, 323
Chang, H.-C. 259
Chang, J. 239
Chang, J. Y. 3
Chang, K.-S. 444
Chaoyang, W. 394
Chapelon, J.-Y. 242
Chaplain, V. 320
Charpentier, J. C. 326
Charrier-Mojtabi, M. C. 242
Chase, G. D. 67
Chase, G. G. 239, 283, 297, 299
Chateau, X. 27, 110, 131, 230, 284, 384
Chatterjee, D. 443
Chaturani, P. 367, 375
Chatzidai, N. 197
Chatzis, I. 253
Chaudhari, R. V. 231, 256
Chaudhary, B. I. 182
Chaudhary, K. 473, 479
Chaudhary, R. C. 258
Chaudhury, M. K. 4
Chauhan, A. 365, 395
Chauveteau, G. 244, 258, 307, 308, 309, 311, 320, 321, 323
Chauviere, C. 149, 150
Chavarie, C. 84, 276, 295, 304, 335, 342, 343, 350

Chave, Y. 240
Cheddadi, I. 122
Chee, D.-W. 164
Cheesman, D. J. 72, 75, 109, 128
Cheh, H. Y. 194, 222, 223
Chehata, D. 195, 202, 206, 207, 210, 226
Cheikhrouhou, M. 239
Chekhmir, A. 186
Chen, C.-J. 91, 497
Chen, C.-K. 225, 226, 365, 373, 374, 421, 422, 426, 427, 428, 461
Chen, C.-T. 240, 266, 312, 313, 315
Chen, C. Y. 268, 269
Chen, D.-R. 265
Chen, H. 199, 225, 240, 241, 242, 266
Chen, H.-T. 225, 226
Chen, I.-H. 225, 226
Chen, J. 230
Chen, J. L. S. 365, 373, 376, 394, 422, 429
Chen, K. 164
Chen, K.-F. 359
Chen, L. 188, 190, 215, 502
Chen, M. 287
Chen, M. C. S. 526, 527
Chen, S. 91, 120, 132, 135, 137, 155, 172
Chen, S. -G. 107, 110, 111, 120, 121, 123, 132, 135, 137
Chen, S.-M. 359
Chen, T. S. 461
Chen, T. Y. W. 421, 426, 429
Chen, W.-J. 91, 499
Chen, X. D. 364
Chen, Y. 72, 78, 195
Chen, Y. R. 3
Chen, Z. D. 364
Chenevert, M. E. 68
Cheng, C.-Y. 435
Cheng, D. C.-H. 18, 27
Cheng, J. 200, 203, 226, 233
Cheng, N.-S. 51, 263
Cheng, X. 22
Cheny, J. M. 170
Chepura, I. V. 255
Chern, S. Y. 366, 367, 374, 376, 386
Chester, W. 49
Chesters A. K. 231
Cheung, M. K. 359
Chevalier, C. 284
Chevalier, T. 284, 285, 286, 287, 304
Chhabra, R. P. 2, 3, 11, 14, 15, 16, 17, 20, 29, 37, 39, 41, 42, 55, 56, 57, 58, 59, 61, 62, 63, 64, 65, 66, 67, 68, 70, 71, 72, 74, 75, 76, 78, 79, 80, 82, 83, 84, 87, 88, 89, 90, 91, 92, 93, 94, 95, 96, 97, 101, 109, 113, 115, 117, 118, 122, 123, 125, 126, 130, 131, 132, 136, 138, 139, 147, 153, 154, 160, 175, 186, 189, 191, 194, 201, 202, 206, 208, 209, 210, 215, 217, 219, 222, 223, 225, 229, 230, 231, 232, 233, 244, 252, 255, 257, 260, 261, 274, 276, 277, 279, 288, 289, 294, 295, 300, 301, 302, 303, 304, 307, 311, 312, 313, 314, 315, 318, 326, 331, 334, 335, 336, 337, 338, 339, 340, 342, 343, 344, 345, 346, 347, 350, 363, 364, 365, 366, 367, 368,

369, 376, 377, 378, 379, 381, 382, 383, 384, 388,
389, 390, 393, 395, 396, 400, 403, 405, 406,
407, 408, 411, 412, 419, 421, 422, 424, 426, 430,
434, 435, 436, 438, 439, 440, 441, 442, 443,
446, 447, 448, 449, 450, 455, 456, 457, 458, 459,
460, 461, 462, 465, 473, 476, 478, 481, 482,
483, 484, 485, 486, 487, 488, 489, 490, 491,
494, 495, 496, 498, 499, 500, 501, 502, 503,
505, 506, 510, 511, 513, 519, 520, 521, 522
Chi, H. S. 423, 427
Chiang, T. 448
Chiba, K. 91, 168, 169
Chien, N. 105
Chien, S.-F. 72, 88, 89
Chiew, Y.-M. 109, 111
Chigier, N. 185
Chilcott, M. D. 143, 149, 151, 158, 167
Childs, L. H. 163
Chiralt, A. 3
Chmiel, H. 3
Chmielewski, C. 154, 157, 307, 313, 319, 494
Cho, K. 164
Cho, Y. I. 57, 59, 66, 67, 68, 91, 130, 153, 154, 159, 164,
493, 514, 517, 518, 519, 521, 527
Choi, H. 3, 36, 37, 87
Choi, H. J. 3
Choi, M. A. 239
Choi, Y. M. 3, 36, 37
Chojnacki, K. T. 185
Choplin, L. 10, 156, 194, 231
Chou, C.-H. 394
Chow, L. C. 482
Christ, F. R. 277, 295, 304, 320
Christanti, Y. 185
Christiansen, E. B. 15
Christopher, R. H. 275, 276, 286, 290, 291, 292, 293,
296, 309
Chrust, M. 91, 101
Chu, B. 510, 516, 526
Chu, C.-F. 261, 265
Chu, H. H. 170, 358
Chu, H. H. S. 424, 435
Chu, S. 151
Chuanjing, T. 394, 405
Chudacek, M. W. 180
Chung, B. J. 66, 449
Chung, C. 166
Chung, H. 163
Chung, J. N. 236, **461**
Chung, J. N. C. 215
Chung, K. T. 262, 264
Chung, T.-S. 187
Churaev, N. V. 239, 276
Churchill, S. W. 215, 424, 425, 432, 434, 435, 436, 449,
454, 455
Cialkowski, M. 315, 320
Ciceron, D. 252, 257, 277, 294, 295, 303, 312, 334, 336
Cid, M. 181, 197, 228
Cieslinski, J. T. 215
Cisne Jr, R. L. 285, 299
Claesson, S. 524, 526

Clague, D. S. 273
Clain, X. 284
Clark, N. N. 88, 334
Clark, P. E. 68, 78, 83, 91, 101, 358
Clasen, C. 41
Clegg, D. B. 142, 368, 434
Cliffe, K. A. 49, 481
Clifford, P. J. 282, 322
Clift, R. 2, 50, 51, 54, 72, 87, 88, 95, 160, 185, 188, 214,
226, 331, 332, 471, 472, 473, 496, 502, 503, 505
Cloitre, M. 164
Clough, G. F. 175
Cockx, A. 199, 400
Coelho, P. M. 51, 93, 97, 101
Coffey, B. E. 510, 524
Cognet, G. 369, 397
Cohen, I. 22
Cohen, Y. 261, 262, 263, 277, 295, 304, 320, 321, 323
Cole, R. 443
Coleman, B. D. 39
Collier, A. P. 88, 101
Collins, R. E. 240
Collis, D. C. 443
Colucci, S. 175
Colwell, R. E. 41, 510
Comini, G. 411
Comiti, J. 88, 89, 90, 91, 101, 243, 244, 245, 251, 252, 255,
257, 259, 263, 264, 277, 280, 281, 286, 288,
293, 294, 295, 296, 297, 303, 312, 320, 331, 334,
336, 337, 338, 347, 350, 351, 359, 369, 406
Connolly, E. K. 185
Connor, C. B. 45
Connor, L. 45
Conrad, B. P. 524, 526, 527
Conway, M. W. 83
Cook, R. L. 528
Cooley, M. D. A. 482
Cooper, G. 186
Cooper, S. J. 243
Coppola, L. 394, 406, 407
Cornelissen, E. C. 170, 171
Cornish, A. R. H. 449
Corvalan, C. M. 185
Costes, J. 180
Couderc, J.-P. 334
Coulaud, O. 244
Coull, J. 87, 99, 499
Coulson, J. M. 248, 261, 265, 266
Coussot, P. 3, 18, 19, 27, 41, 42, 107, 110, 131, 244, 284,
288, 299, 314, 317, 320, 384
Coutanceau, M. 62, 142, 153, 154, 189, 193, 206, 235, 481,
506
Couto, N. D. 446
Couto, S. M. 45
Cox, R. G. 482
Cox, S. J. 105
Craig, S. L. 23
Cramer, C. 185
Crawford, C. W. 262, 264
Crawford, H. R. 161
Crawford, M. E. 262

Crawshaw, J. C. 299
Craze, D. J. 101
Crespo, A. 175
Cressely, R. 319
Cristescu, N. 3
Cristescu, N. D. 524, 526, 527
Croce, G. 411
Crochet, M. J. 56, 61, 62, 78, 142, 143, 144, 148, 149, 150, 151, 158, 160, 164, 319
Cross, M. M. 15, 16
Cross, R. 11
Crowe, C. 236
Culham, J. R. 366, 376
Cullen, S. J. 239
Cumby, T. R. 3
Cummins, C. P. 1
Cunha, F. R. 195
Currie, I. G. 167, 244, 284, 287, 288, 304, 307, 314, 315, 319, 320
Curtis, A. R. **499**
Curtiss, C. F. 3, 10, 39
Cygan, D. A. 511, 514

D' Alessio, S. J. D. 91, 92, 93, 444
D'Avino, G. 82, 110, 141, 145, 163, 168, 169
d'Onofrio, A. 281, 320
da Senhora, F. V. 181, 197, 228
da Silva, D. J. P. 45
Dabbous, M. K. 239, 277
Daborn, J. E. 509
Dabrowski, P. P. 365
Dachavijit, P. 239, 283, 297, 299
Dai, G. C. 18, 19
Dail, R. V. 22
Dairenieh, I. S. 142, 151, 170
Dajan, A. 194, 206, 208, 209, 231
Dale, J. D. 423, 429
Dallon, D. S. 59, 66, 67, 68, 70, 75, 76
Dalton, B. J. 11
Dandridge, A. 526
Dandy, D. S. 87, 215
Daneshy, A. A. 77
Dang, A. 18
Dang, V. V. 401
Danov, K. 509
Daoudi, S. 311
Daous, M. A. 169
Darby, R. 2, 45, 72, 78, 82, 347
Darcovich, K. 358
Darton, R. C. 352
Das, B. 336, 351
Das, P. K. 189, 198, 228, 233, 234, 236, 242, 485, 494, 495, 501, 502
Das, S. 242, 263
Das, S. K. 275, 336, 351
Das, T. R. 185, 234
Dasgupta, N. 502
Datt, C. 358
Datta, A. K. 3
Datta, S. S. 285, 309
Dauben, D. L. 275, 277, 295, 309, 320

Daugan, S. 168, 358, 359
Daunais, C.-A. 72, 76
Davaadorj, B. –E. 90
Davaille, A. 110, 112, 130, 131
Davenport, T. 3
Davidson, D. L. 307
Davidson, J. F. 176, 177, 178, 180, 181, 332
Davies, A. R. 56, 61, 62, 78, 143, 148, 149
Davies, C. N. 268, 269
Davies, D. R. 83
Davies, J. M. 57
Davies, O. H. 279
Davies, R. M. 191
Davis, A. M. J. 87, 273, 498, 526
Davis, E. J. 273
Davis, H. T. 244
Davis, R. H. 358
Davis, R. T. 395
Davit, Y. 284, 309
Davoodi, M. 196, 202, 225
De Angelis, E. 357
De Boer, R. 240
de Beristain, I. G. 112
de Bruyn, J. R. 4, 107, 110, 111, 112, 117, 123, 126, 131, 191, 196, 203, 221, 228
de Corato, M. 186
de Decker, J. K. 18
de Gennes, P. G. 39
de Groh, H. C. 9
de Klerk, A. 260
De Kruijf, A. 83
de la Calleja, E. 163
de Larrard, F. 21
de Loubens, R. 284, 309
De Luca, A. 285, 286, 304
de Sa, C. H. M. 15
De Simone, J. M. 22
de Souza Mendes, P. R. 21, 27, 132, 181, 197, 228, 313, 315, 316
de Vargas, L. 91, 154
De, S. 284, 304, 305, 307, 315, 319, 320
Dean, E. J. 120
Deasy, P. B. 2, 3
Debbaut, B. 143, 144, 151
Deckwer, W.-D. 229, 230, 352, 403, 410, 411
Dedegil, M. Y. 109, 128
Deen, N. G. 242, 263
Deen, W. M. 266, 268
Deetjen, M. E. 176
Degand, E. 147, 496
Deglo de Besses, B. 113, 118, 128, 132, 133, 135
Deiber, J. A. 307
De Kee, D. 3, 14, 16, 18, 37, 39, 41, 42, 56, 58, 65, 91, 128, 147, 160, 189, 191, 194, 202, 206, 208, 209, 210, 211, 221, 224, 225, 227, 231, 232, 234, **480, 491, 521, 522, 523**
De Kruijf, A. 83
Del Rio P. 196, 308
Delaby, I. 234
Delgado, J. M. P. Q. 320
Delgado, M. A. 175

Della Sala, S. 27
Delsanti, M. 510
Delvaux, V. 148, 151, 164
Demir, İ. 246, 247, 248, 249
Deng, Q. 220, 229, 398
Denier, J. P. 365
Denisienia, E. 239
Denn, M. M. 3, 5, 18, 19, 40, 107, 110, 112, 113, 120, 142, 148, 153, 154, 164, 175, 185, 198, 210, 217, 233, 395, 396
Depollier, C. 242
Derakhshandeh, B. 3, 18
Derksen, J. J. 51, 118, 283, 301
des Santos, H. K. 3
Desai, A. 50, 54
Descher, S. 200
Deshpande, A. P. 27
Deshpande, R. 50, 54
Deslouis, C. 164, 396
Despeyroux, A. 287, 365, 388, 485, 500
Desroches, J. 2
Devic, M. 160, 193, 202, 212, 232
Devienne, R. 288
DeWitt, K. J. 365, 366, 367, 374, 375, 376, 421, 422, 433, 444, 445, 446, 463, 464, 465
Dewsbury, K. H. 85, 194, 206, 208, 222, 225, 226
Dexter, R. W. 185
Dey, A. A. 165
Dey, K. N. 369, 374, 393
Dey, N. C. 276
Dhahir, S. A. 166, 500
Dharamadhikari, R. V. 277, 295, 296
Dhawan, G. K. 253
Dhiman, A. K. 87, 93, 269, 289, 382, 384, 412, 446
Dhingra, S. C. 65, 180, 194, 217, 219, 220
Dhole, S. D. 56, 62, 72, 79, 80, 84, 90, 201, 222, 257, 277, 300, 301, 302, 334, 337, 365, 388, 389, 390, 400, 405, 519
Dhotkar, B. N. 274, 312, 313, 315
Di Federico, V. 299
Di Felice, R. 257, 261, 263, 332, 341, 475, 476, 478, 479, 480
Di Marco, P. 215
Di Meglio, J. M. 215
Di Vaira, N. J. 107, 110, 111, 120, 121, 127, 135, 137
Dias, R. P. 242
Didari, H. 285
Diedericks, G. P. J. 259
Dienemann, Von W. 376
Dietrich, C. 509
Dietrich, N. 199, 400
Dietrich, W. E. 90
Dijkhuizen, W. 226
Dimakopoulos, Y. 27, 110, 117, 122, 186, 196, 197, 201
Dimitrov, D. S. 231
Dimova, R. 509
Ding, M. 242, 261, 263
Ding, Z. 166, 168
Dinkgreve, M. 18
Dintzis, F. R. 22
Divoux, T. 18, 19, 175

Dixon, A. G. 262, 264
Dlugogorski, B. Z. 521
Dobson, J. 4
Doffin, J. 510
Doi, M. 39, 149
Dolecek, P. 65, 67, 68, 70, 277, 296, 303, 308, 336, 337, 347, 358
Dolejs, V. 261, 277, 280, 296, 303, 308, 312, 336, 337, 347
Dollet, B. 4, 86, 105, 138, 139, 165, 186
Dolz-Planas, M. 26
Dominguez, J. G. 277, 321, 322
Donald, A. M. 1
Donaldson, E. 449
Donatelli, A. A. 369, 385, 423, 454, 455, 457, 459
Done, D. S. 240, 277
Dong, Q. 242, 244, 245, 261, 263, 312
Donis, R. 168, 358
Doraiswamy, D. 2, 7
dos Santos, A. E. O. 45
Dosunmu, A. 72
Dou, H.-S. 141, 145, 153, 155, 165, 166
Dou, Y. 172
Drabik, B. R. 2, 3
Drahos, J. 151, 252
Dratler, D. I. 23
Dravid, V. 185
Dreher, K. D. 239, 321
Drummond, J. E. 270, 271, 272
du Plessis, J. P. 249, 259, 260, 282, 303, 363
du Plessis, M. P. 108, 123, 124, 129
Du, M. 199
Duan, J. 166
Duan, Y. 50
Duan, Z. 50
Duarte Naia, M. 55, 446
Dubash, N. 195, 197, 221, 228
Dubey, R. N. 366
Duckworth, R. A. 105
Ducloue, L. 230
Duda, J. L. 252, 258, 277, 286, 295, 320
Dudek, D. R. 450
Dudek, J. 258, 307
Dudukovic, A. P. 481
Dudukovic, M. P. 326
Dufour, F. 138
Duineveld, P. C. 215, 216
Dukhan, N. 244, 249
Dullaert, K. 27
Dullien, F. A. L. 240, 241, 242, 243, 245, 246, 248, 249, 252, 253, 258, 259, 260, 264, 265, 266
Dunand, A. 67, 70
Dunkel, J. 299, 302, 315
Dunn, P. F. 3
Dupla, J. C. 284
Dupret, F. 148
Dupuis, D. 169
Duquette, R. 234
Durban, D. 186
Durst, F. 185, 190, 215, 244, 277, 295, 307, 309, 310
Durth, M. 138
Dusek, J. 85, 91, 101

Dutta, A. 161, 323, 442
Dwyer, H. A. 87, 499
Dyakonova, N. E. 314, 319
Dybbs, A. 244, 245, 274
Dziecielak, R. 273
Dziubinski, M. 195, 202, 225, 226, 239, 279, 304, 305, 306, 307, 506

Eames, I. 216
Easson, W. J. 90
Eastwood, J. 262, 264
Eaton, L. R. 505
Eberhard, U. 285, 286, 299
Ebrahimi, F. 285
Ece, M. C. 421
Eckert, E. R. G. 376, 422, 429
Eckmann, D. M. 200, 236, 505
Edgeworth, R. 11
Edie, D. D. 277
Edwards, D. A. 269, 274
Edwards, M. F. 276, 278, 286, 296, 297, 298, 320, 335, 337, 338, 339, 342, 348, 350
Edwards, R. V. 244, 245, 274
Edwards, S. F. 39
Eesa, M. 83
Egawa, S. 485
Eggers, J. 185
Ehsani, A. 410
Eichstadt, F. J. **525**
Einarsson, J. 145, 158, 163
Eisen, S. 155, 172
Eisenberg, D. A. 168
Eisfeld, B. 260, 261
El Defrawi, M. 365, 373
El Fadili, Y. 72, 78
El Kayloubi, A. 143, 151, 514
Elata, C. 278, 307, 308, 309, 311
El-Awady, M. N. 175
Elbirli, B. 305
Elfring, G. J. 358
Elghobashi, S. 165
Elias, F. 4, 139
El-Kaissy, M. M. 255, 256, 405
Ellenberger, J. 229, **503**
Ellero, M. 112, 172
Elliott, J. H. 162
Elliott, R. E. 168, 358, 359
Ellis, A. T. 187
Ellwood, K. R. J. 185
El-Maghlany, W. M. 444
El-Sayed, M. S. 246, 249, 264, 265
El-Shayeb, M. M. 408
Elson, T. P. 72, 75, 109, 128
Emady, H. 107, 110, 111, 112
Emamian, A. 191, 196, 202, 225
Emery, A. F. 423, 429
Enders, F. 200
Endo, Y. 265
Engez, S. M. **475, 476, 479**
Engleman, M. 120
Ennis, B. C. 266

Epps, H. H. 239
Epstein, N. 242, 252, 262, 264, 273, 331, 347, 352
Erbas, S. 421
Ercolani, D. 357
Erdim, E. 246, 247, 248, 249
Ergun, S. 249, 261
Erickson, L. E. 365, 375
Ern, P. 188
Ernst, B. 234
Ershaghi, I. 278
Esfahani, J. A. 426
Eshghy, S. 432
Eshtiaghi, N. 3
Eslami, M. 449
Esseniyi, A. J. 366, 394
Eswaran, V. 56, 62, 72, 79, 80, 84, 87, 90, 93, 94, 201, 215, 217, 220, 222, 230, 257, 274, 277, 300, 301, 302, 312, 313, 314, 315, 318, 334, 337, 365, 376, 377, 378, 379, 381, 382, 388, 389, 390, 400, 403, 405, 406, 407, 408, 411, 412, 446, 519
Ethier, C. R. 273
Evans, A. R. 311
Evans, G. A. 49
Evans, I. D. 18
Everage, A. E. 240, 277
Everett, L. G. 239
Ewing, R. P. 243, 252, 295, 347
Ewoldt, R. H. 22, 23, 24, 27, 110, 112, 176
Eyring, H. 15

Fabre, D. 188
Fabris, D. 152, 155, 158
Fadali, D. A. 408
Fadalli, O. A. 394, 408
Fadili, A. 239
Fahs, H. 131
Fairhurst, P. G. 83
Fairweather, M. 263, 358
Falade, A. 497
Fales, J. 18
Falls, A. H. 278
Faltas, M. S. 255, 256
Fan, D. 199
Fan, J. 201, 217, 220
Fan, L.-S. 231, 255, 282, 331, 332, 335, 350, 352, 410
Fan, L. T. 365, 366, 374, 375
Fan, S. K. 277, 286, 320
Fan, T.- H. 141, 146
Fan, W. 181, 199, 225
Fand, R. M. 248, 260, 435, 443
Fang, H. 176, 186
Fang, T.-N. 41, 307, 529
Fantazzini, P. 243
Farag, H. A. 368, 394
Fararoui, A. 191, 192, 204, 208
Fardi, B. 269, 273
Farid, M. M. 281
Faridi, H. A. 3
Farinato, R. S. 320, 324
Farley, R. W. 210
Faroughi, S. A. 146, 158

Farouk, B. 435
Farouq Ali, S. M. 279, 282, 304
Fasano, A. 357
Fattahi, E. 301, 303
Fattal, R. 167
Faure, P. 244, 284, 288, 299, 320
Favelukis, M. 186, 235
Favis, B. D. 234
Faxen, H. 473, 526
Faxen, O. H. 165, 166
Fayed, H. E. 284
Fayon, A. M. 476
Fdhila, R. B. 216
Feben, D. 248
Fedkiw, P. S. 258
Fedyushinskaya, L. B. 90
Fehling, R. 248
Feikema, D. A. 185
Feinauer, A. 244
Feldman, G. A. 86
Felix, L. C. M. 242
Fellah, Z. E. A. 242
Feltham, D. L. 3
Femin Bendict, R. J. 351
Fendell, F. E. 448
Feng, J. J. 82, 141, 144, 149, 151, 159, 166, 168, 169, 170, 180, 188, 199, 200, 203, 227, 233, 358, 499
Feng, J. Q. 505
Feng, Y.-T. 107, 110, 111, 121, 123, 132, 135
Feng, Z.-G. 87, 215, 222, 390, 472
Ferg, E. 27
Fergui, O. 278, 304
Fernando Concha, A. 357
Ferrari, M. A. 358
Ferraris, C. F. 3, 21, 41
Ferreira, F. de O. 3
Ferreira, J. M. 55, 82, 91, 93, 160, 312, 313, 315, 368, 378, 384, 394, 411, 412, 422, 443, 446, 510
Ferrer, J. 276, 295, 320
Ferroir, T. 110, 131
Ferry, J. D. 42
Fetters, L. J. 4
Feuillebois, F. 484
Fidleris, V. 475, 476, 478, 479, 481
Field, S. B. 87, 101
Figueroa-Espinoza, B. 191, 198, 202, 206, 225
Fildey, K. 41
Finch, J. A. 216
Finkers, H. J. 264
Finlay, J. A. 93
Finlayson, B. A. 365, 373, 385
Finnis, M. V. 152
Firoozabadi, B. 181, 197, 200, 203, 223
Fischer, N. 301, 303
Fischer, P. 3, 185, 524
Fischer, R. 249, 265
Fishbein, G. A. 215
Fleck, N. A. 186
Flemmer, R. L. C. 334
Fletcher, A. J. P. 304, 309
Fletcher, T. H. 450

Fleury, M. 285, 286
Flew, S. 278, 307, 309
Flew, S. R. G. 304, 309
Flores, R. A. 3
Floriancic, M. 285, 286
Flowers, A. E. 521
Flude, M. J. C. 509
Flumerfelt, R. W. 234
Fogler, H. S. 253, 400
Folse, R. F. 497, 498
Foltz, R. G. 516
Fonseca, C. 132
Foo, J. J. 141
Ford, G. D. 273
Ford, J. T. 72, 75
Forman, N. A. 22
Forman, R. A. 528
Fornari, W. 358, 360
Fornberg, B. 49, 50, 54
Fortin, A. 56
Fortin, M. 56
Foscolo, P. U. 242, 252, 257, 475
Foster, R. D. 57, 385
Foumeny, E. A. 260, 263, 265
Fourar, M. 244, 245, 246
Fourie, J. G. 303
Fox, J. W. 22
Fox, V. G. 365, 375
Fraggedakis, D. 110, 117, 122, 201
Francis, A. W. 476, 477, 479
Franco, A. T. 358
Franco, J. M. 175
Francois, N. 161
Frank, X. 194, 195, 202, 212, 231, 232
Frattini, P. L. 311
Fredsoe, J. 51
Freire, D. 155, 168, 170
Freitas, L. G. 188, 215
Fremouw, S. 3
Frenandes, C. 146, 158
Freund, J. B. 110, 112
Freundlich, H. 26
Frey, S. 132
Fricke, A. L. 186, 187, 400
Frigaard, I. A. 18, 19, 105, 110, 112, 117, 120, 121, 195, 197, 221, 228, 284, 287, 430, 441
Frisken, B. J. 112
Fritsching, U. 175
Frohn, A. 236
Froment, G. F. 263
Frota, M. N. 86
Fruman, D. H. 187
Fryer, P. J. 82, 83, 175, 377, 438
Fu, T. 181, 196, 199, 200, 202, 225, 226, 231, 233
Fuchs, N. A. 268, 270
Fujii, M. 421, 423, 426, 435, 449
Fujii, T. 421, 423, 426, 435, 449
Fujioka, H. 384
Fukuchi, T. 257
Fukumoto, Y. 482
Fukushima, E. 244

Fulford, G. D. 258
Funada, T. 215, 251
Funatsu, K. 62
Funfschilling, D. 194, 195, 198, 200, 202, 212, 231, 232
Furukawa, T. 191, 195, 202, 203, 206, 223

Gabelnick, H. L. 3
Gadala-Maria, E. 107
Gadd, G. E. 161
Gahleitner, M. 510, 514
Gaitonde, N. Y. 278, 306
Galdi, G. P. 170
Galindo-Rosales, F. J. 22, 23, 24, 283, 315
Gallegos, C. 3, 175
Gal-Or, B. 229
Galuts, V. V. 161, 162
Galvosas, P. 243
Gamboa, A. C. 286, 324
Gandhi, K. S. 234
Ganesan, P. B. 200, 203, 226, 233
Ganguli, S. 51
Ganguly, U. P. 336, 351
Ganji, D. D. **510**
Ganoulis, J. 244, 307
Ganpule, H. K. 314, 319
Ganser, G. H. 88
Gao, X. 199, 233
Garbin, V. 4, 186
Garcia-Morales, M. 3
Gardiner, B. S. 521
Garduno, I. E. 145
Garg, N. S. 368, 394
Garg, V. K. 395
Garic-Grulovic, R. V. 476, 481
Garifulin, F. A. 425
Garimella, S. V. 188, 190, 215
Garlaschelli, L. 27
Garnaud, G. 510
Garraud, A. 4
Garrouch, A. A. 278, 304
Garside, J. 341, 342
Gartling, D. K. 141, 149
Gaudio, J. E. 176
Gauglitz, P. A. 3, 175, 287
Gauri, V. 176
Gausdal Jacobsen, J. 239
Gauthier, C. 281, 320
Gauthier, F. 168, 169, 235
Gauthier-Manuel, B. 482, 516
Gauvin, W. H. 51, 85, 244
Gavignet, A. A. 45
Gavin III, A. 368, 377
Gay, M. 90
Gebhart, B. 448, 449, 461
Geils, R. H. 522
Geindreau, C. 239, 259, 295, 314, 320
Geminard, J. C. 175
Gence, J. N. 244
Genieser, L. 143, 144, 150, 151, 152
Gentry, C. C. 423, 437, 438
Geoola, F. 449

Geordiadis, J. G. 244
Georgescu, A. G. 157
Georgiou, G. C. 132, 143, 150, 185, 319
Gerberich, L. 239, 308, 323, 327
Gerhardter, H. 88
Germain, Y. 234
German, G. 176
Gerogiorgis, D. I. 19
Gervang, B. 143, 149
Gestoso, P. 275
Gestring, I. 175
Getachew, D. 259, 303
Geurts, B. J. 234
Ghaddar, C. K. 273, 274
Ghaeli, H. 68, 78, 83
Ghajar, A. J. 153, 154, 517
Ghanbarian, B. 243, 252, 295, 347
Gharbi, R. B. 278, 304
Gheissary, G. 130, 155, 163, 168, 170
Gheorghitza, St. I. 278
Ghiaasiaan, S. M. 175
Ghoniem, S. A.-A. 258, 307
Ghosh Roychowdhury, D. 275
Ghosh, A. K. 182
Ghosh, R. 446
Ghosh, U. K. 175, 274, 312, 318, 363, 369, 374, 376, 377,
   378, 381, 391, 392, 393, 411
Ghoshdastidar, P. S. 87
Giannetti, F. 97
Gibilaro, L. G. 242, 252, 257, 261, 263, 475
Gidaspow, D. 331
Gierer, C. 358
Giese, M. 263
Giesekus, H. ,142, 147, 151, 511, 512
Gilchrist, I. C. R. 26
Gill, W. N. 401
Gillaspy, P. H. 193, 206, 223
Gillies, R. 2
Gilligan, S. A. 142, 166
Ginder, J. M. 18, 111
Giner, S. A. 239
Gioia, F. 239
Givler, R. C. 260
Gjovikli, N. B. 304, 309
Gkormpatsis, S. D. 146
Gladden, L. F. 244, 263, 299
Glaese, M. 109
Glasnovic, A. 68, 72, 73
Glenister, D. J. 42
Glowacki, P. 66, 68, 78
Glowinski, R. 119, 120, 169, 170
Gmachowski, L. 257
Gnann, C. 190, 215
Godbille, F. 234
Godbole, S. P. 206, 230, 403
Goddard, J. D. 22, 400
Godelmann, B. 265
Godfrey, J. C. 216, 233, 505
Godinez, F. A. 163
Goel, N. 168
Gogarty, W. B. 239, 278, 321, 322

Gogos, G. 449
Göğüs, M. 88
Goharzadeh, A. 89
Goldin, M. 185
Goldsmith, H. L. 168, 169, 235
Goldstein, R. J. 435
Goldstein, S. 49
Gollakota, A. R. K. 197, 216, 222
Gollapalli, J. S. 45
Golombok, M. 307, 315
Gomez-Diaz, D. 175, 206
Gomez-Perez, J. 510
Goncalves, N. J. 446
Gondret, P. 482
Gong, P. 198, 226
Gong, X. 22
Gong, X. L. 23
Gonzalez, J. M. 295
Gonzalez, T. O. 240
Gonzalez-Nunez, R. 234
Gooding, C. H. 277
Goossens, L. H. J. 231
Goossens, W. R. A. 50, 51
Goren, S. L. 185, 273
Gorla, R. S. R. 375, 394, 421, 433, 438, 457
Gosser, R. B. 522
Gottlieb, M. 66, 185, 510, 511
Götz, J. 244
Goujon-Durand, S. 91
Goutille, Y. 240
Gouveia, L. M. 22
Govier, G. W. 27
Govindarajan, R. 161, 171, 172, 190, 217
Govindarao, V. M. H. 263
Goyeau, B. 285
Goyon, J. 27, 230
Grace, J. R. 50, 51, 54, 72, 87, 88, 95, 160, 185, 188, 190, 214, 226, 233, 234, 352, 471, 472, 473, 496, 503, 504, 505
Grady, B. P. 168
Graessley, W. W. 14, 42, 307
Grager, H. 161
Graham, A. L. 473, 482, 510
Graham, D. I. 56, 58, 61, 72
Grandjean, B. P. A. 326
Graner, F. 4, 105, 122, 138, 139, 165
Granville, P. S. 161, 162
Grassi, W. 215
Grattoni, C. A. 239
Gray, M. 215
Gray, W. G. 244
Grbavcic, Z. B. 476, 481
Greaves, M. 278
Greco, F. 82, 141, 145, 163, 168, 169, 234
Green, A. E. 39
Green, D. W. 304, 309
Greene, G. A. 216
Greenkorn, R. A. 240, 241, 242, 243, 253, 266, 280, 304
Gregg, J. 424, 432
Gregory, A. R. 165
Gregory, D. R. 278, 304, 306

Gregory, K. R. 265
Griffiths, R. W. 7
Grimm, R. J. 39
Griskey, R. G. 22, 278, 304, 306
Grizzutti, N. 110
Groisman, A. 41
Grotberg, J. B. 384
Grove, A. S. 98
Gruber, V. E. 523
Gryparis, E. A. 146
Gu, D. 56, 61, 342, **482**, **484**, **485**
Gu, D. Z. 278, 300
Guasto, J. S. 299, 302, 315
Guazzelli, E. 27, 359
Guceri, S. I. 435
Gudim, Y. A. 215
Gudmundsson, J. S. 246
Guerts, B. J. 234
Gueslin, B. 110
Gui, F. 41, 528
Guida, V. 110
Guido, S. 169, 234
Guidoboni, G. 120
Guiffant, G. 320
Guillet, J. 240
Guilloit, P. P. 265
Guillot, D. 67, 70
Gulari, E. 185
Guler, N. 78, 91, 101
Gummalam, S. 194, 210, 220, 229
Gumulya, M. 110, 112, 233
Gunasekaran, S. 42
Gunjal, P. R. 256
Guo, H.-Y. 2, 3
Guo, J. 41, 520, 529
Guo, X. 240, 241, 242, 266
Guo, Y. 198, 226
Guo, Z. 242, 261, 264
Gupalo, Yu P. 83, 101
Gupta, A. K. 93, 94, 138, 139, 201, 217, 221, 365, 377, 390, 393, 403, 421, 422, 440, 442, 448, 457, 459, 460, 461, 462, 465, 483, 484, 485, 494, 495, 496, 499, 500, 501, 502, 503
Gupta, A. S. 395
Gupta, O. P. 162, 164, 395, 396
Gupta, R. 191, 195, 206, 225, 228
Gupta, R. K. 3, 10, 29, 33, 39, 41, 42, 56, 61, 62, 72, 79, 80, 83, 142, 156, 157, 160, 295, 307, 388, 483, 484, 485, 486, 487, 488, 499, 500, 502
Gupta, S. 91, 97, 101, 384, 393, 440, 459
Gupta, S. K. 443
Gupta, S. N. 274, 313, 318, 369, 374, 381, 393
Gupta, S. V. 510, 522
Gupte, A. R. 247, 248
Gurappa, B. 248, 334
Gürgen, S. 23
Gurkan, T. 194, 223, 401, 402, 403
Gurzel, N. V. 90
Gutfinger, C. 280
Guthausen, G. 319
Gutierrez-Lemini, D. 39

Gutkin, A. M. 108, 113
Guyon, E. 482
Guzy, C. J. 273
Gyves, T. 216

Ha, J.-W. 235
Haamkens, F. 312, 314
Haas, R. 277, 278, 295, 307, 309, 310, 324
Haber, S. 505
Haberman, W. L. 115, 150, 473, 504
Haddow, J. B. 56
Hadjiisky, A. 509
Hagen, T. 397
Haider, A. 50, 88
Hailemariam, H. 3
Haj-Hajiri, H. 197, 223
Hajjam, M. 189, 193, 206, 235, 506
Halasz, M. R. T. 215
Haldenwang, R. 27
Hale, W. F. 3
Hall, M. J. 244
Hall, T. 3, 164
Hallouin, E. 482
Halvaci, M. 68, 78, 83
Hamersma, P. J. 283, 315
Hamielec, A. E. 49, 215, 229, 255, 257, 274, 403, 405
Hamilton, R. T. 265
Hammer, B. E. 244
Hammond, R. 239, 308, 323, 327
Han, C. D. 187
Han, E. 23
Han, S.-I. 215, 243
Han, S. T. 268, 269
Handzy, N. Z. 203
Hanehara, S. 3
Hanks, R. W. 109, 124, 126, 129
Hanna, M. R. 261, 280, 295, 309, 311, 320
Hannah, R. R. 83
Hanratty, T. J. 244, 245
Hansen, A. 246
Hansen, A. G. 366, 431
Hansen, C. M. 516
Hansford, G. S. 394, 524
Hanson, C. 233
Happel, J. 87, 228, 255, 256, 270, 271, 272, 273, 274, 334, 343, 405, 471, 472, 473, 476, 482, 496
Haque, M. W. 194, 206, 210, 225, 230
Hara, S. K. 187
Harbottle, D. 358
Hardem, S. 58, 60, 73
Hardy, E. H. 244, 283, 287, 288
Hariharaputhiran, M. 109, 126, 130, 131, 233
Harinadha, G. 446
Haring, R. E. 253
Harkins, W. D. 183
Harlen, O. G. 143, 144, 150, 151, 158, 160, 510
Harmathy, T. Z. 505
Harmsworth, D. L. 444
Harnoy, A. 396
Harper, J. F. 214
Harries, K. A. 107

Harrington, L. J. 77, 83
Harrington, R. E. 278, 309, 322
Harris, J. L. 185
Harris, M. T. 231
Harris, P. C. 45
Harris, P. J. 506
Harris, W. W. 280, 304
Harrison, B. 41
Harrison, D. 177, 331, 332
Harrison, D. E. 522
Harrison, G. M. 152, 184, 186
Hart, S. A. 3
Hartholt, G. P. 504
Hartland, S. 233
Hartley, R. W. 233
Hartman, M. 88
Hartnett, J. P. 18, 57, 59, 66, 67, 68, 91, 130, 153, 154, 159, 164, 423, 437, 438, 493, 514, 517, 518, 519, 521, 527
Hartranft, T. J. 175
Hartt, W. H. 165
Harvey, A. H. 278
Hasan, M. A. 522
Hasan, M. M. 449
Hasegawa, S. 161
Hashimoto, A. G. 3
Hasimoto, H. 271, 272, 482
Hassager, O. 3, 10, 14, 16, 29, 39, 55, 142, 144, 146, 150, 151, 152, 159, 160, 193, 202, 203, 223, 226, 363, 511
Hassan, N. M. S. 197
Hassan, Y. A. 216
Hassani, S. 112
Hassanien, I. A. 365
Hassanizadeh, S. M. 244
Hassell, H. L. 278, 304, 320
Hatch, H. H. 248
Hatcher, J. D. 423, 466
Hatton, A. P. 444
Hattou, S. 176
Hatzikiriakos, S. G. 3, 18, 22
Haugen, R. 374
Haughey, D. P. 262
Haward, S. J. 3, 4, 139, 283, 286, 287, 307, 309, 320
Hawley, M. C. 250, 261, 263, 281, 295, 296, 306, 311, 312
Hayashino, T. 181
Hayes, R. E. 259, 260, 278, 303, 406
Hayworth, C. B. 182
He, B. 50
He, Y. 141, 146
He, Y. B. 109
Hebrard, G. 199, 400
Hecht, V. 175
Hedblom, E. P. 105
Hedden, O. K. 185
Heinen, C. 244, 283, 287, 288, 319
Heinrich, J. C. 239, 272
Heiss, J. F. 87, 99, 499
Hejri, S. 304, 309
Helail, T. R. 278, 320
Heldman, D. R. 3

Hellen, E. 273
Hellen, L. 3
Hellström, J. G. I. 273, 274, 275
Helm, D. 239
Helmreich, A. 278, 286, 304, 308
Hema Sundar Raju, B. 444, 463
Hennion, B. 510
Henschke, M. 474
Herbolzheimer, E. 530
Herbst, C. A. 528
Hermann, W. 510, 516
Hermes, R. A. 523
Hernandez-Escobedo, L. 242
Herraez-Dominguez, J. V. 26
Herrera-Velarde, J. R. 195, 202, 206, 207, 210, 226
Herrmann, H. J. 283
Hertel, K. L. 269
Herzhaft, B. 110, 168, 358, 359
Heskitt, B. F. 377
Hesselink, F. 308
Hetsroni, G. 505
Heyda, J. F. 526
Heywood, N. I. 510
Hiatt, J. P. 244
Hibino, S. 505
Hieber, C. A. 461
Higdon, J. J. L. 257, 273, 473, 482, 504
Higgins, B. G. 3
Highgate, D. J. 169
Hilal, M. 320, 406
Hilfiker, R. 510, 516, 526
Hill, C. T. 148
Hill, D. B. 141
Hill, K. B. 105
Hill, R. 56
Hill, R. J. 244, 256
Himbert, F. 245, 251
Himl, M. 128
Hincks, T. K. 45
Hino, M. 161
Hinrichsen, O. 241, 242, 246, 248, 249, 262, 264, 265
Hinze, J. P. 231
Hirai, S. 244
Hirasaki, G. J. 279, 295, 320, 322
Hirose, T. 56, 60, 180, 192, 193, 210, 217, 223, 234, 385, 398, 399, 400, 402
Hirota, S. 514, 517
Hirschfeld, B. R. 497
Hirschl, R. B. 384
Hirvonen, J. 3
Hjelmfelt, Jr. A. T. 246
Hlavacek, B. 276, 295, 304, 335, 342, 343, 350
Ho, B. P. 169
Ho, S. P. 279, 320
Ho, T. X. 112
Hoagland, D. A. 240, 320
Hoang, M. L. 239, 263
Hochenauer, C. 88
Hochmuth, R. M. 510, 524
Hocquart, R. 319
Hoekstra, A. 273

Hoerner, S. F. 87
Hoffer, T. E. 193, 206, 223, 505
Hoffman, R. L. 23
Hoffmann, A. C. 215, 264, **504**
Hoffman, G. 231
Hofmann, H. 231, 326
Hoftijzer, P. J. 180
Hogg, A. J. 163
Hojeij, A. 176
Holdsworth, S. D. 2, 3, 14
Holenberg, Y. 122, 201, 233
Holweg, J. 244, 276, 307, 324
Holzner, M. 285, 286
Holzwarth, G. 141
Homann, F. 98
Homsy, G. M. 255, 256, 405
Hong, S.-A. 252, 258, 277, 286, 295, 320
Hopfinger, E. J. 186
Hopke, S. W. 56, 57, 60, 62, 63, 64, 66, 68
Horbett, T. A. 266
Horikawa, A. 91, 168, 169
Hormozi, S. 27, 358, 360
Horne, R. 244, 246
Horowitz, M. 85
Horsley, M. 73, 78, 128, 129, 170
Horsley, R. R. 73, 78, 109, 110, 112, 128, 129, 170, 233
Hort, M. 175
Horvath, A. T. 18, 111
Hossain, M. A. 448, 449, **457**
Hou, J. 3
Houghton, B. F. 45
Houi, D. 242
Houlsby, G. T. 186
Housiadas, K. D. 145, 149, 158, 163
Houska, M. 28
Houwink, R. 18
Howell, T. G. 365, 374
Howells, I. D. 273
Hoyt, J. W. 161, 369, 396
Hozawa, M. 194, 208
Hsieh, Y. H. 91, 499
Hsu, C. C. 365, 374
Hsu, C. J. 280, 290, 292, 293, 296, 320, 323
Hsu, J. P. 91, 498, 499
Hsu, S. C. 312
Hu, B. 200
Hu, D. L. 2, 3
Hu, H. 87, 443, 526
Hu, H. H. 144, 164, 166, 170
Hu, R. Y. Z. 18, 423, 438, 520
Hu, X. 166, 168
Hu, Y. T. 163
Hua, C. C. 314, 319
Hua, T. N. 71, 279, 302
Huang, C.-H. 401, 402
Huang, M.-J. 365, 373, 374, 421, 422, 426, 427, 428
Huang, P. Y. 144, 166, 169, 170, 499, 500
Huang, W. 358
Huang, W.-X. 91
Huang, Y. 299, 322
Hubbard, R. M. 521, 522

Hughes, S. 175
Hui, H. 443
Huilgol, R. R. 40, 90, 120
Huillier, A. 4, 139
Huisman, M. 105
Hulbert, R. 248
Hulin, J. P. 281, 320, 499
Hull, H. L. 268
Hulsen, M. A. 145, 144, 160, 163, 166, 167, 168, 169, 170
Humphery, A. E. 160, 192, 202, 203, 206, 208, 212, 225, 232, 400
Humphrey, J. A. C. 184, **476**
Huneault, M. A. 234
Huner, B. 87, 499
Hunt, A. G. 243, 252, 295, 347
Hussey, R. G. 87, 497, 498, 499, 526, 527
Hutton, J. F. 16, 42
Huynh, H. T. 110, 131
Huzarewicz, S. 307
Huzyak, P. C. 176
Hwang, S. J. 406
Hwang, W. R. 295
Hyde, M. A. 369, 385

Iaccarino, G. 145, 163, 165
Ianniruberto, G. 38, 144
Ichihara, M. 175
Ichikawa, J. 72, 75, 337, 343, 352, 354
Ida, Y. 175
Idris, Z. 239, 259, 295, 314, 320
Iglesias, J. A. 221
Ikai, S. 369, 376, 377
Ikoku, C. U. 279, 304
Ilic, V. 90, 482, 527, 528
Iliev, O. 284
Iliuta, I. 326
Illenberger, W. K. 88
Ilyin, S. 21
Imaishi, N. 194, 208
Inge, C. 187
Ingmanson, W. L. 268, 269
Inoue, I. 257, 369, 397
Inoue, M. 257
Interthal, W. 277, 278, 307, 309
Ipekci, O. N. 88
Ireland, P. M. 110
Irmay, S. 249
Irvine, Jr. T. F. 41, 363, 364, 424, 448, 514, 517, 518, 519, 524, 526, 527, 528
Irving, J. B. 524, 526
Isayev, V. I. 29, 42, 66
Ishihara, T. 280, 286, 297
Ishii, K. 482
Ishii, T. 71, 257, 279, 302, 476, 482, 484
Ishimura, H. 107, 131
Ishkintana, L. K. 175, 206
Ish-Shalom, M. 3
Islam, M. T. 200, 203, 226, 233
Islam, M. R. 279, 304
Ismail, J. H. 263
Israelachvili, J. N. 22

Itakura, K. 49
Ito, H. 91, 101
Ito, S. 108, 127, 129
Itoh, H. 274
Iuliano, E. 110
Ivanov, I. B. 231
Iwaoka, M. 476
Iwasaki, E. 195, 217, 222, 228
Iwasaki, S. 482
Iwata, S. 186, 191, 195, 202, 203, 206
Iwatsubo, F. 1
Izadpanah, M. R. 326

Jackson, D. A. 526
Jackson, G. W. 266, 267, 268
Jackson, K. P. 157
Jackson, M. D. 259
Jackson, T. W. 422, 429, 444
Jaeger, H. M. 22, 23
Jaensson, N. O. 168
Jafarabad, K. R. 216, 505
Jafarpur, K. 448, 449
Jahns, H. O. 280, 304
Jain, M. K. 394
Jain, T. 446
Jaiswal, A. K. 257, 279, 298, 300, 301, 334, 335, 336, 337, 338, 343, 344, 345, 346
James, D. D. 444
James, D. F. 10, 11, 33, 41, 62, 162, 164, 166, 167, 168, 244, 266, 267, 268, 273, 279, 284, 287, 288, 295, 304, 307, 308, 309, 311, 314, 315, 319, 320, 324, 395, 396
Jameson, G. J. 110, 186, 521
Jami, A. M. 410
Jamialahmadi, M. 215, 326, 331, 334, 336, 341
Jamison, D. E. 45
Jamjoon, F. A. 239, 245
Janna, W. S. 185
Janssen, L. F. 185
Janssen, L. P. B. M. 504
Jargon, J. R. 304
Jarzebski, A. B. 194, 217, 219, 220, 229, 398, 403, 404
Jauhari, R. 215
Javed, K. 263
Jay, P. 113, 117, 118, 121, 128, 132, 133, 135, 136, 137, 138
Jayaraman A. 155, 171
Jayaraman, K. 91, 154, 157, 307, 313, 319, 494
Jean, B. 3
Jean, R.-H. 255
Jeffrey, D. W. 326
Jeffrey, G. B. 82
Jefri, M. A. 91, 169
Jeng, D. R. 365, 366, 367, 374, 375, 376, 421, 422, 433, 444, 445, 446, 463, 464, 465
Jennings, H. M. 107
Jennings, R. R. 239, 279, 324
Jenny, M. 85
Jenson, V. G. 49
Jezequel, P.-H. 3
Ji, X. 16, 25
Jia, H. 449

Jia, Y. 188, 215
Jia, Y-X.,91
Jiang, S. 181, 196, 202, 225, 226, 231
Jiang, W. 22
Jie, P. 120
Jimenez-Fernandez, J. 175
Jiménez-Martínez, J. 285, 286, 299
Jin, F. 107, 110, 111, 120, 121, 123, 132, 135, 137
Jin, H. 143, 146, 149, 527
Jin, L. 68, 78
Jing, X. 258
Jing, X. D. 239
Job, C. 229, 230
Johansen, S. T. 83, 163
Johansson, H. 473, 481
John, J. E. A. 185
Johns, M. L. 244, 263, 299
Johnson, A. B. 175, **506**
Johnson, A. I. 215
Johnson, A. M. 107, **520**
Johnson, A. T. 435, 463
Johnson, D. S. L. 192, 202, 206, 400
Johnson, E. D. 187
Johnson, E. M. 266, 268
Johnson, G. A. 244
Johnson, H. F. 182
Johnson, M. 191, 192, 203, 208, 225
Johnson, M. W. (Jr.), 56, 57
Johnson, R. C. 268, 269
Johnson, R. E. 236
Johnson, T. A. 49, 51, 54
Jolls, K. R. 244, 245
Jones, D. M. 33, 145, 146, 151, 155, 157, 160, 295, 309, 319
Jones, E. R. W. 282, 322
Jones, R. L. 170
Jones, R. S. 142, 166, 394
Jones, T. E. R. 59, 22, 56, 57, 58, 61, 66, 68, 72, 75, 91
Jones, W. M. 279, 308, 320, 498
Jongschaap, R. J. J. 169
Jonsson, B. T. L. 265
Jonsson, K. A.-S. 265
Jonsson, P. J. P. 273, 274, 275
Jordi, R. G. 264
Joseph, D. D. 2, 144, 145, 148, 152, 155, 164, 168, 169, 170, 194, 203, 209, 211, 215, 358, 394
Joshi, J. B. 50, 51, 54, 66, 68, 72, 73, 75, 194, 206, 210, 216, 225, 230, 236, 256, 326, 331, 335, 339, 342, **476, 485, 489, 490, 505**
Joshi, P. 446
Jossic, L. 107, 110, 111, 117, 121, 131, 132, 134, 136, 138, 139, 169, 176
Ju, S. H. 166, 396, 397
Ju, S. J. 166
Juliusburger, F. 26
Juncu, Gh. 215

Kaar, S. 1
Kaczmar, B. U. 277, 295, 307, 309
Kadambi, J. R. 253
Kaddioui, N. 151
Kadijk, S. E. 15

Kago, T. 410, 411
Kahawita, R. 435
Kahle, A. 510
Kaiser, A. 3
Kaiser, A. E. 510
Kajiuchi, T. 108, 127, 129
Kajiwara, T. 62
Kalashnikov, V. N. 162
Kale, D. D. 277, 295, 296, 326, 523
Kalendar, A. 417, 423, 450
Kalman, H. 50
Kaloni, P. N. 151
Kalra, T. R. 52
Kalthia, N. L. 395
Kamal, M. R. 169
Kamat, P. M. 231
Kameda, M. 175
Kami-Iwa, F. 435
Kaminski, T. L. 185
Kamisli, F. 175
Kanaris, N. 132, 137
Kanchanalakshana, D. 153, 154, 519
Kandhai, D. 273
Kanel, J. S. 235
Kanellopoulos, N. K. 253, 299
Kang, I. S. 235
Kang, Y. 332, 410, 411
Kannan, A. **481, 482, 483, 485, 500**
Kanno, H. 485
Kanti, V. 50, 54
Kapellas, M. 160, 193, 202, 212, 232
Kaplan, S. J. 312
Kaplaun, S. 166, 167
Kapur, J. N. 366, 373
Karabelas, A. J. 244, 245
Karamanev, D. G. 84, 85, 191, 194, 195, 206, 208, 215, 221, 222, 225, 226, 227
Karapetsas, G. 196, 197, 221
Karar, S. 450
Kardos, J. L. 267, 269, 272, 312, 313, 315, 319
Kärger, J. 243
Kargi, F. 231, 331
Karimfazli, I. 120, 430, 441
Karimi-Fard, M. 244, 246
Karino, I. 485
Karioun, M. 352, 410, 411
Karis, T. E. 168
Karni, J. 363
Karniadakis, G. E. 54
Karnis, A. 82, 168, 169, 235
Karoutsos, V. 239
Karri, B. 190, 199
Karrila, S. 51, 87, 472, 496
Karteleva, M. I. 215
Kaser, F. 279, 308, 309, 310, 324
Kasper, G. 88, 90, 497, 499
Kassinos, S. C. 132, 137
Kataja, M. 273
Kataoka, I. 216
Kato, H. 59, 66, 67, 72, 153, 154, 162
Kato, T. 369, 376, 377, 378

Kato, Y. 410, 411
Katoshevski, D. 463
Katz, D. 26, 27
Kaupke, C. R. 185
Kaur, A. 149, 153, 155
Kaviany, M. 240, 272
Kawale, D. 304, 315, 319
Kawamura, Y. 257
Kawase, Y. 56, 57, 59, 60, 61, 65, 72, 75, 91, 175, 182, 185,
    186, 191, 193, 210, 217, 219, 223, 227, 234, 279,
    296, 300, 302, 334, 335, 337, 338, 343, 352,
    354, 355, 356, 358, 385, 387, 394, 397, 401, 405,
    406, 409, 410, 411, 426, 429, 482
Kazenin, D. A. 363
Ke, C. 449
Ke, Y. 196, 225, 226
Kealy, T. 73, 78, 128, 129
Kearsley, E. A. 39
Keentok, M. 41, 129, 157
Keetels, G. 121
Keey, R. B. 368, 369, 394
Keezer, R. C. 522
Kefayati, G. H. R. 120
Kehlenbeck, R. 475, 476, 478, 479, 480
Kelbaliyev, G. I. 88, 188, 215
Kelessidis, V. C. 19, 72, 73, 75
Kelkar, B. G. 230
Kelkar, J. V. 523
Kelkar, M. 239
Kell, R. E. 162
Kellay, H. 161
Keller, A. 324
Keller, D. S. 27
Keller, Jr. D. V. 27
Keller, J. B. 273
Keller, R. J. 279, 308, 309, 310, 324
Kemblowski, Z. 279, 288, 289, 290, 294, 296, 297, 298,
    304, 305, 306, 307
Kemiha, M. 232
Kenchington, J. M. 109
Kennedy, G. F. 3
Ke-Qin, Z. 120
Kerekes, R. J. 3, 18
Kessentini, S. 500
Kessler, D. P. 241
Keswani, K. K. 443
Keunings, R. 143, 185
Kezios, P. S. 187
Kharabaf, H. 287
Khaled, A.-R. A. 240
Khalifa, A. O. A. 251
Khamashta, M. 279
Khan, A. 326
Khan, A. A. 326
Khan, A. R. 50, 51, 341
Khan, M. B. 366, 367, 395, 396
Khan, M. M. K. 197, 258, 307
Khan, W. A. 366, 376
Kharabaf, H. 287
Khramtsov, P. P. 417, 424, 429, 432, 433, 434, 435, 443,
    444, 448, 449

Kharatiyan, E. 27
Khare, A. S. 66, 68, 72, 73, 75, 175, 335, 339, 342, 476,
    485, 489, 490
Khatchatourian, O. A. 239
Khatibi, M. 163
Khayat, K. H. 3
Khayat, R. E. 234
Khyati, B. V. 443
Khezzar, L. 89, 185
Khinast, J. G. 175, 206, 227
Khismatullin, D. B. 175
Khokhlov, A. R. 510
Khomami, B. 144, 145, 148, 150, 151, 267, 269, 272, 312,
    313, 314, 315, 319
Khomikovskii, P. M. 108
Khuzhayorov, B. 239
Khyati, B. V. 443
Kick, K. 244
Kieft, R. N. 443
Kiesvaara, J. 3
Kihm, K. D. 185
Kikuchi, D. 195
Kiljanski, T. 107, 195, 239, 506
Killen, J. M. 162, 279, 281, 308, 310, 311
Kim, B. K. 366, 373, 376
Kim, B. Y. K. 260
Kim, C. 165, 166, 168, 169, 170, 187
Kim, C. B. 423, 437
Kim, C.-K., 3
Kim, D. 87
Kim, H. W. 366, 376, 394, 422, 433, 445
Kim, I. 91, 164, 165, **527**
Kim, J. 110, 112
Kim, J. M. 165, 166
Kim, K. C. 166
Kim, K. Y. 41, 510, 522
Kim, S. 51, 87, 90, 170, 472, 496
Kim, S. D. 332, 410, 411
Kim, Y. 90
Kim, Y. B. 3
Kimura, K. 196
Kimura, S. 195
King, Jr. H. E. 528
King, M. J. 142, 160
King, A. J. C. 189, 198, 228, 236
Kintner, R. C. 191, 192, 203, 204, 208, 225, 233, 273, 505
Kiriakidis, D. G. 142
Kirkby, L. L. 91, 101
Kirsch, A. A. 268, 270
Kiser, K. M. 367, 374, 376
Kishore, J. 327
Kishore, N. 80, 197, 215, 216, 220, 221, 222, 225, 230, 300,
    367, 388, 393, 403, 405, 446, 485, 499, 500
Kiss, A. D. 41
Kitamura, K. 435, 449
Kitamura, Y. 186
Klaus, E. E. 252, 258, 277, 286, 295, 320
Klaus, M. 87, 101
Klausner, J. F. 214
Klee, A. J. 228
Klein, B. 109

Author Index

Klein, D. E. 262
Klein, U. K. A. 239, 245
Kleinstreuer, C. 87, 91, 376, 394, 422, 444, 445, 446, **463**, **464**
Kleppe, J. 421, 430, 431, 434, 441
Kliakhandler, I. L. 182
Klingenberg, D. J. 18, 111
Klink, I. M. 168
Knoll, M. 88
Kobayashi, N. 191, 195, 202, 206
Koblitz, A. R. 360
Kobus, C. J. 393
Koch, D. L. 244, 246, 272, 274, 283, 302, 311, 510
Kochesfahani, M. 443
Koelling, K. W. 176
Koesen, S. P. 307, 315
Koizumi, A. 56
Kökpinar, M. A. 88
Kolano, M. 200
Kolli, V. G. 107
Kolodziej, J. A. 239, 273, 315, 320
Komar, P. D. 90
Koncar-Djurdjevic, S. K. 476, 481
Konczak, Z. 273
König, T. 244
Koniuta, A. 162, 164
Kono, Y. **510**
Koop, K. 175
Koplik, J. 186
Koponen, A. 273
Korchak, E. I. 244
Kordorwu, V. 199, 400
Korhonen, M. 3
Koroteev, D. A. 163
Koshiba, T. 279, 307, 314, 319
Koshy, A. 234
Kossen, N. W. F. 231
Kostornov, A. G. 268
Koteswara Rao, P. 93
Kotouc, M. 463
Kotter, M. 161
Kousar, N. 132
Kovscek, A. R. 276, 304
Koyama, H. 257, 367, 372, 373, 376, 386, 387, 422
Kozeny, J. 248
Kozicki, W. 239, 261, 280, 289, 290, 292, 293, 295, 296, 308, 309, 311, 320, 323, 327
Koziol, K. 66, 68, 78
Kozissnik, B. 4
Kramer, D. 243
Kramer, E. J. 268
Krapivsky, P. L. 186
Kraume, M. 188, 196, 199, 200, 215
Kremer, D. M. 185
Kreutzer, M. T. 304, 315, 319
Krishna, R. 175, 216, 229, 505
Krishnaiah, K. 351
Krishnan, S. 481, 482, 483, 485, 500
Kristmanson, D. D. 185
Kroesser, F. W. 185
Kroger, M. 39

Kruger, E. 248
Krüssmann, R. 280, 286, 311
Krzton-Maziopa, A. 153
Ku, X. 169
Kuang, P. Q. 296
Kubair, V. G. 433
Kubler, E. A. 365, 376
Kuck, V. J. 175
Kucukbursa, A. 395
Kudin, A. M. 162
Kuehn, T. H. 435
Kuerten, J. G. M. 234
Kühne, H. C. 107, 312, 314
Kui, H. 146, 161
Kuipers, J. A. M. 226, 242, 263, 284, 304, 305, 307, 315, 319, 320
Kulichikhin, V. 21
Kulicke, W.-M. 31, 41, 161, 185, 278, 309, 310, 324, 403
Kulkarni, A. 265
Kulkarni, A. A. 236
Kulkarni, R. 239, 308, 323, 327
Kulkarni, S. 45
Kuloor, N. R. 176, 178, 179, 180, 181, 182, 183, 184, 358
Kumagai, I. 110, 112, 130, 131
Kumar, N. 161, 171, 172
Kumar, P. 359
Kumar, R. 176, 178, 179, 180, 181, 182, 183, 184, 185, 233, 234
Kumar, S. 280, 297, 312, 327, 335, 337, 338, 369, 374, 376, 377, 381, 391, 392, 393, 406, 407, 409
Kumar, V. 273
Kumaresan, G. 351
Kumazawa, H. 175
Kunii, D. 331
Kupferman, R. 167
Kurata, F. 525, 526, 527
Kurita, K. 110, 112, 130, 131
Kurita, T. 196
Kuroda, C. 162, 369, 397
Kurten, M. 190, 215
Kuru, E. 45
Kushalkar, K. B. 369, 394
Kushan, M. C. 23
Kutepov, A. M. 363
Kutsovsky, Y. E. 244
Kuwabara, S. 228, 229, 255, 256, 270, 271, 272, 273, 274
Kuwahara, F. 257
Kwack, E. Y. 153, 154, 423, 493
Kwok, A. 283
Kwon, K. S. 3, 36, 37
Kyan, C. P. 273
Kyle, C. R. 245

La Nauze, R. D. 257
Laba, D. 3
Labrecque, R. P. 268
Ladd, A. J. C. 244, 246, 272, 274
Ladenburg, R. 471
Lagerstedt, T. 280
Lagisetty, J. S. 234
Lahbabi, A. 259

Lai, R. Y. S. 142, 160
Laing, A. 45
Lakes, R. S. 29
Lal, K. 366, 373
Lal, P. 327, 369, 394
Lali, A. M. 66, 68, 72, 73, 75, 335, 339, 342, 476, 485, 489, 490
Lam, A. C. C. 260
Lam, K. 358
Lamb, S. P. 304, 309
Lance, M. 482
Lang, T. G. 161
Langins, J. 86
Langmuir, I. 270
Lanos, C. 196, 228
Lao, H. W. 253
Lapasin, R. 3, 29
Larachi, F. 326
Lareo, C. 82, 83
Larkins, R. P. 326
Larsen, A. 244, 284, 288, 299, 320
Larson, R. E. 257
Larson, R. G. 3, 14, 27, 28, 29, 39, 41, 151, 157, 257, 280, 303
Lasheras, J. C. 186
Lashgari, I. 97, 358, 360
Laskowski, J. S. 109
Lasne, D. 161
Lasso, I. A. 87
Latifi, M. A. 244
Latto, B. 162
Lau, E. M. 91
Laufer, G. 280
Lauga, E. 40, 163, 164
Laurent, C. 351
Lauriks, W. 242
Lauterborn, W. 187
Laven, J. 22
Lavrenteva, O. M. 122, 197, 198, 201, 202, 225, 228, 234
Lavrov, A. 284
Lawler, D. F. 51
Lawrence, C. J. 214, 312, 314, 315, 317
Lawson, M. L. 187
Lawson, N. J. 152
Laxton, P. B. 105
Le Clair, B. P. 49, 229, 255, 257, 274, 405
Le Cloirec, P. 239
Le Roux, J. P. 51, 88
Le, T. N. 59, 66, 67, 75, 484, 514, 517, 518, 519
Leal, L. G. 142, 147, 148, 154, 157, 159, 168, 169, 187, 192, 193, 202, 206, 209, 211, 215, 223, 226, 234, 235, 358, 359, 400
Lebouche, M. 288
Lecjaks, Z. 67, 68, 70, 75, 276, 334, 335, 336, 337, 338, 339, 343, 350
Lecoq, N. 482
Lee, C.- B. 91
Lee, C. 91
Lee, C. H. 175
Lee, D. I. 262
Lee, D.-Y. 364, 449

Lee, H. M. 475, 476, 479
Lee, H. S. 366, 373, 376
Lee, J. 90
Lee, J. S. 191, 198, 202, 206, 225
Lee, K. M. 4
Lee, L. 444
Lee, L. J. 166, 168
Lee, M. H. 239
Lee, P. D. 312, 314, 315, 317
Lee, S.-C. 461
Lee, S. J. 81, 153, 154, 165, 166, 239, 272, 474, 496
Lee, S. W. 168, 169, 170
Lee, S. Y. 373
Lee, T.-L. 423, 454, 455, 457, 459
Lee, W. T. 1
Lee, W. K. 234
Lee, W. Y. 66, 130, 159, 514, 517, 518, 519, 521
Lee, Y.-J. 145, 153, 161, 172
Lefebvre, A. H. 175, 185
Legat, V. 144, 150
Legile, P. 351
LeGoff, P. **505**
Legrand, J. 251, 259, 303
Lehner, F. K. 280, 303
Lehtonen, J. 3
Leider, P. J. 39, 40
Leigh, D. C. 57
Leijnse, A. 242
Lejeune, A. M. 175
Leland, J. E. 482
Lele, S. K. 51
LeLoc'h, G. 255
Lemieux, P. F. 366
Leng, D. E. 233
Lenormand, R. 244, 246
Leonard, E. F. 82
Leonardi, E. 188, 190, 215, 502
Leonas, K. K. 239
Leonov, A. I. 56, 58
Leonov, E. G. 66
Leroy, R. 107
Lesage, F. 244
Lescarboura, J. A. 526, 527
Leslie, F. M. 142, 147, 153, 154, 397
Leulseged, A. 3
Levec, J. 248
Levenspiel, O. 50, 87, 88, 331
Levich, V. G. 398
Levick, J. R. 267
Levine, A. J. 141, 161
Levitskiy, S. P. 186
Levy, A. 107
Levy, G. L. 280
Levy, T. 246
Lewis, H. W. 522
Li, G. 50, 51, 169
Li, H. 358
Li, H. Z. 181, 194, 195, 196, 198, 200, 202, 212, 225, 226, 231, 232, 233
Li, J. 234, 239, 485, 493
Li, L. 3, 35, 141, 146, 151, 265, 312

Li, M. 485, 493
Li, S. 181, 196, 201, 202, 217, 220, 225, 226, 231
Li, T. 196
Li, W. 72, 78
Li, W. H. 23
Li, Y. 45, 265, **502**
Li, Z. 199, 400
Li, Z. X. 278, 300
Lian, G. 358
Liang, S. F. 425
Liao, S.-J. 49, 51, 373
Liao, T. Y. 194, 203, 209, 211
Liberzon, A. 201
Licht, W. 228
Liddell, P. V. 41
Liepmann, D. 152, 155, 158
Liew, K. S. 423, 451, 452, 454, 456, 459
Lightfoot, E. N. 46, 248, 249, 289, 291
Lima, M. 377
Lima-Ballesteros, R. 334
Lin, B. L. 365, 373, 421, 500
Lin, C. L. 88
Lin, F. N. 366, 367, 374, 376, 386
Lin, G.-M. 195, 228, 232
Lin, H. L. 266
Lin, H.-T. 366, 374, 375, 433
Lin, J. 82, 141, 169
Lin, J. C. 164, 166
Lin, S. H. 366, 374
Lin, S. X. Q. 364
Lin, T.-J. 195, 228, 232
Lin, Y.-M. 397
Lince, F. 284, 298
Lindner, A. 384
Lindsay, J. D. 169, 175, 285, 286, 304
Ling, Y. 394, 405
Linliu, K. 510
Lipscomb, G. G. 113
Lister, J. R. 231
Litt, M. 3, 160, 192, 202, 203, 206, 208, 212, 225, 232, 394, 400
Litt, M. J. 2, 3
Liu, A. J. 23
Liu, A. W. 141, 313, 319
Liu, B. 169
Liu, B. T. 110, 113, 120
Liu, B. Y. H. 269, 273
Liu, C. 188, 199, 203, 285, 286, 304
Liu, C. B. 406
Liu, C.-Y. 23
Liu, F. 199, 400
Liu, G. L. 358
Liu, H. 87, 201, 221, 236, 358, 449, 526
Liu, H. L. 295
Liu, H.-S. 51
Liu, J. 120, 132, 135, 137, 188, 200, 215, 233, 261
Liu, J. G. 107, 120, 137
Liu, K. K. 141
Liu, S. 241, 242, 246, 252, 256, 257, 259, 260, 261, 263, 266, 303, 320, 323
Liu, T. 22
Liu, W.-W. 121
Liu, X. 181
Liu, Y. J. 164, 168, 169, 170, 194, 203, 209, 211, 358
Liu, Z. 3, 181, 185, 199, 400
Loaiciga, H. A. 239
Lochiel, A. C. 385, 401, 403
Lockett, F. J. 395
Lockhart, R. W. 326
Lockyear, C. F. 105
Lockyear, M. A. 57
Lockyer, P. 394
Loeffler, Jr. A. L. 270
Loewenberg, M. 55
Lohia, P. K. 378
Lohrenz, J. 525, 526, 527
Lohse, D. 1, 85
Loke, P. B. 185
Lomba, R. F. T. 15
Longwell, J. P. 450
Lopez De Haro, M. 196, 308
Lopez, W. F. 228
Lopez, X. 299
Lopez-Aguilar, J. E. 112, 172
Lopez-Estrada, F. R. 510
Loppinet, B. 244, 284, 288, 299, 320
Lord, E. 2, 269
Lorenzi, A. 239
Losenno, C. 90
Loth, E. 87, 88, 188, 215, 226
Lou, W. 201, 221
Loudon, J. 192, 202, 206, 400
Lounis, B. 161
Lovato, S. 121
Lovett, S. 360
Low, G. S. 282, 304, 306
Lu, M. 200
Lu, P. 164
Lu, W. 230
Lu, W. J. 406
Lu, W.-Q. 449
Lu, Y. 146
Lu, Y. B. 243, 262, 284, 286, 298
Lubensky, T. C. 141, 161
Luciani, A. 234
Lucke, J. 206, 230, 403
Luckham, P. F. 19, 21, 22, 239
Lugarini, A. 358
Luikov, A. V. 366, 369, 374, 376, 377
Lum, M. 435
Luming, H. 56
Lund, D. B. 3
Lund, T. 304, 309
Lundberg, M. 3
Lundgren, T. S. 215, 254, 255, 299
Lundström, T. S. 273, 274, 275
Luning, C. D. 366, 374
Lunnon, R. G. 476, 480
Lunsmann, W. J. 143, 144, 150, 151, 152
Luo, X. L. 59, 143, 144, 148
Luo, Y. 66, 68, 78, 91
Ly, H. B. 285, 28

Lyakhovsky, V. 186
Lyczkowski, R. W. 331
Lyon, M. K. 168, 358, 359
Lyons, D. W. 423, 466
Lyons, J. W. 41, 510
Lyulin, A. V. 314, 319

Ma, Q. 176, 186
Ma, W. 265, 312
Ma, Y. 181, 195, 196, 199, 200, 202, 212, 225, 226, 231, 233
Maalouf, A. 91, 151, 153, 157, 170
Maaskant, P. 523, 524
Maass, O. 268, 269
Macdonald, I. F. 246, 249, 264, 265
MacDonald, M. J. 265
Macedo, I. C. 193, 206
Maceiras, R. 15, 19
Machac, I. 57, 63, 65, 67, 68, 70, 72, 75, 76, 78, 101, 109, 128, 244, 255, 276, 280, 288, 295, 303, 308, 312, 331, 334, 335, 336, 337, 338, 339, 343, 348, 350, 351, 358, 359, 406, 485
Machacova, L. 67, 68, 70, 78
Machado, J. C. V. 45
Machado, J. C. 175
Maciejewski, W. 2
Maciel, G. F. 3
Mackaplow, M. B. 359
Mackay, M. E. 149
Mackley, M. R. 3
MacLean-Fletcher, S. D. 517, 520
Macosko, C. W. 36, 41, 120, 187
Maddock, J. L. 279, 308, 320
Madraki, Y. 27
Maeda, M. 369, 376, 377
Maeda, S. 190, 204
Maehara, T. 215
Maerker, J. M. 280, 324
Maffettone, P. L. 82, 141, 145, 163, 168, 169
Magarvey, R. H. 54
Magda, J. J. 157
Magelli, F. 72, 75
Magnaudet, J. 188, 215, 216
Magnifotcham, F. 216
Magnin, A. 105, 107, 110, 111, 113, 117, 118, 121, 128, 131, 132, 133, 134, 135, 136, 137, 138, 139, 169, 176, 191, 197, 228
Magueur, A. 258, 307
Mahaut, F. 27
Mahefkey, E. T. 482
Mahmood, R. 132
Maillard, M. 384
Mainwaring, D. 3
Maiti, M. K. 142, 395
Maitri, R. V. 307, 315
Majumdar, A. 189, 198, 228, 236
Majumdar, S. 161, 171, 172
Majumder, A. K. 51
Maklad, O. 31
Makrigiorgos, G. 27, 186
Maldarelli, C. M. 194, 223

Maleki, A. 200, 203, 223
Maleki-Jirsaraei, N. 110, 112
Malevich, A. E. 245, 246, 247, 260, 274
Malhotra, S. 146
Maliki, A. E. 56
Malinowski, J. J. 194, 217, 219, 220, 229, 398, 403, 404
Malkin, A. Y. 19, 21, 27, 29, 39, 42
Malkus, D. S. 240, 266, 312, 313, 315
Mall, B. K. 369, 376
Malleswararao, T. V. 313
Malvandi, A. **510**
Malvault, G. 284
Mamun, C. K. 287
Mancini, A. 357
Mandani, S. 191, 196, 202, 225
Mandeno, P. 369, 394
Mandhani, V. K. 411
Mandica, M. 2, 3
Maneri, C. C. 215, 505
Manero, O. 91, 141, 147, 148, 151, 147, 149, 153, 154, 155, 157, 159, 172, 208
Manga, M. 3, 199, 228
Mangadoddy, N. 313, 411, 412
Manjunath, M. 194, 229, 230, 300, 301
Mankad, S. 377, 438
Manke, C. W. 185
Manli, T. 306
Mannan, S. H. 141
Mannheimer, R. J. 185
Manor, G. 239
Manson, J.-A. E. 239
Mansour, A. 185
Mansour, I. A. S. 408
Mansour, N. N. 215
Mantle, M. D. 244
Manz, B. 244
Mao, Z. Q. 216, 505
Mao, Z.-S. 51, 97, 199, 201, 216, 226, 230, 255, 405
Marchal, J. M. 143, 148, 158
Marchisio, D. L. 284, 298
Marenya, M. O. 239
Margaritis, A. 85, 191, 194, 195, 206, 208, 221, 222, 225, 226, 227, 331, 351
Maris, H. J. 54
Marivoet, J. 262
Markis, F. 3
Markovitz, H. 2
Marliere, C. 244, 284, 288, 299, 320
Marmottant, P. 4, 186
Marner, W. J. 421, 430, 431, 434, 441
Marques, E. 3, 4, 304, 315, 319
Marr, G. R. 499
Marriott, J. 3
Marrucci, G. 38, 39, 144, 185, 192, 221, 229, 231, 394, 396
Marschall, H. 201, 211
Marsden, S. S. 276
Marshall, R. J. 280, 295, 304, 306, 309
Marshall, S. H. 180
Martel, S. 4
Martemyanov, S. 369, 397
Marti, A. C. 155, 168, 170

Martin, A. R. 269
Martinelli, R. C. 326
Martinez, D. M. 110, 112, 117
Martinez-Angeles, R. 242
Martinez-Boza, F. 3
Martinez-Navarrete, N. 3
Martynenko, O. G. 417, 424, 429, 432, 433, 434, 435, 443, 444, 448, 449
Maruhara, K. 435
Masalova, I. 27
Mashelkar, R. A. 56, 57, 59, 60, 66, 67, 72, 73, 122, 130, 143, 154, 156, 159, 161, 180, 181, 191, 193, 204, 205, 206, 210, 217, 220, 225, 227, 228, 229, 323, 363, 385, 396, 397, 398, 399, 403, 404, 417, 422, 425, 427, 429, 433, 466, 484, 510, 521, 523
Masliyah, J. H. 51, 87, 88, 215, 241, 242, 246, 252, 255, 257, 259, 260, 263, 266, 273, 303, 320, 323, 363, 411
Mason, S. G. 82, 168, 169, 234, 235
Massarani, G. 215, 499
Massey, A. H. 175
Mastbergen, D. 105
Masuyama, T. 280, 286, 297
Mata, C. 164
Matallah, H. 149, 151, 166
Matana, E. 50
Matar, O. K. 196, 221
Mateeva, N. 406
Mathieu, J.-B. 4
Mathur, M. 444
Mathur, M. N. 367, 375, 429
Matijasic, G. 68, 72, 73
Matsumoto, Y. 215
Matta, J. E. 164, 185
Mattischek, J.-P. 516
Matzen, E. J. P. 262, 264
Maude, A. D. 509
Maul, C. 90
Mauret, E. 72, 73, 74, 75, 252, 267, 277, 303, 347, 405
Mavis, F. T. 248
Maxey, J. E. 163
Maxworthy, T. 190, 215
Mayer, R. C. 84
Maynor, B. W. 22
Mazaheri, A. R. 253
McAtee, Jr. J. L. 524, 526
McAuliffe, C. D. 280
McCann, D. J. 181
McCarthy, M. J. 359
McClements, D. J. 3
McComb, W. D. 187
McCoy, J. H. 22
McCoy, S. C. 39, 377
McCuen, R. H. 463
McDonald, A. T. 366, 375
McDonough, J. M. 434
McDuffie, G. E. 526
McGeary, R. K. 262
McHale, S. 369
McHugh, A. J. 142, 151, 170
McKay, A. 91, 101

McKinley, G. H. 2, 27, 41, 141, 144, 145, 146, 147, 149, 150, 151, 152, 153, 155, 158, 159, 160, 168, 169, 286, 314, 316, 317
McKinley, R. M. 280, 304
Mclachlan, R. J. 524, 526
McLaren, D. R. 279, 308, 309, 311, 324
McMechan, D. E. 83
McNown, J. S. 475, 479
McPherson, M. B. 475, 476, 479
McWhirter, J. D. 262
Mead, D. W. 168, 358, 359
Mechtcherine, V. 312, 314
Medani, M. S. 522
Medina, L. I. 288, 324
Mednick, R. L. 253, 299, 333
Meese, E. A. 163
Mehlig, B. 163
Mehmani, Y. 283
Mehmood, A. 132
Mehrotra, S. P. 78
Mehta, D. 250, 261, 263, 311, 312
Mei, R. 214
Mei, R. W. 161
Meijer, H. E. H. 144, 152, 165
Meirzwiczak, M. 315, 320
Meissner, D. L. 376, 421, 444, 445, 446, 463, 464, 465
Mellema, J. 169
Memoli, G. 215
Mena, B. 91, 141, 142, 145, 147, 148, 151, 153, 154, 157, 158, 159, 163, 164, 191, 195, 198, 202, 206, 207, 208, 210, 225, 226, 513
Menard, G. 351
Mendelson, H. D. 215, 505
Mendes, P. R. S. 313, 315
Mendoza-Fuentes, A. J. 149, 153, 155
Meng, B. 276, 295, 312, 314, 320
Mennig, G. 141
Menzie, D. E. 275, 277, 295, 309, 320
Meo, M. 22
Mercier, G. 221
Merk, H. J. 434, 448
Merkak, O. 111, 131
Merker, D. 200
Merkin, J. H. 432
Mertens, D. 244, 283, 287, 288
Mertl, J. 279, 297, 298, 304
Metzger, T. G. 42
Metzner, A. B. 22, 26, 40, 169, 252, 261, 262, 263, 277, 280, 290, 295, 304, 306, 309, 323, 363, 394, 395, 396
Mewes, D. 175
Mewis, J. 22, 26, 27, 28, 39, 42, 82, 358
Meyer, B. A. 249
Meyer, R. 510, 521
Meza-Gordillo, R. 510
Mhatre, M. V. 191, 192, 208, 225
Mia, M. 444
Miao, Z. 188, 215
Michael, P. 87
Michaelides, E. E. 49, 54, 87, 90, 215, 216, 222, 226, 390, 474, 496, 499, 510

Michaud, V. 239
Michele, H. 280, 306, 307
Michele, J. 168, 358
Michlin, J. 278, 307, 308, 309, 311
Michniewicz, M. 279, 288, 289, 290, 294, 296, 297, 305, 306, 307
Mickley, H. S. 244
Middleman, S. 185, 187, 275, 276, 278, 286, 290, 291, 292, 293, 296, 306, 309
Middleton, J. A. 162
Midoux, N. 181, 194, 231, 244
Mikami, H. 194, 208
Mikelić, A. 240, 242, 253, 259
Mikulasek, P. 277, 303, 336, 337, 343, 347, 348
Milewski, J. V. 264
Miller, C. 290
Miller, J. D. 88
Miller, R. 233, 236
Miller, S. C. 2, 3
Milliken, W. J. 235, 510
Mills, P. 320
Mills, P. L. 326
Minaeian, A. 166
Minekawa, K. 234
Minev, P. D. 51
Mingaleyev, N. Z. 425
Minkowycz, W. J. 259, 303
Minocha, N. 50, 51, 54
Minor, H. E. 524
Mirabolghasemi, M. 285
Miranda, A. I. P. 166, 167
Mirzaagha, S. 110
Miscevic, M. 268, 269, 273
Mishima, H. 186
Mishra, G. 80, 81, 87, 96, 98, 393, 442, 502
Mishra, I. M. 281, 297, 298, 335, 337, 338, 343, 366, 369, 374, 375
Mishra, P. 91, 97, 101, 138, 139, 274, 281, 297, 298, 313, 318, 335, 337, 338, 343, 365, 366, 369, 374, 375, 393, 394, 422, 457, 459, 502
Mishra, S. P. 395, 466
Missirlis, K. A. 482, 483, 484, 485
Misumi, T. 435, 449
Mitchell, W. J. 257
Mitishita, R. S. 107
Mitra, A. K. 109, 128
Mitsoulis, E. 3, 107, 109, 111, 113, 115, 117, 118, 120, 121, 126, 133, 134, 135, 142, 143, 165, 482, 483, 484, 485, 492, 496
Mitsuishi, A. 449
Mitsuishi, N. 57, 65, 67, 70, 153, 154, 192, 193, 227, 391, 393, 423, 463
Mittal, R. 49, 96, 101
Mittal, S. 50, 54, 384
Mitwally, E. M. 366
Mityushev, V. 271
Mityushev, V. V. 245, 246, 247, 260, 271, 274
Miura, H. 72, 75, 334, 337, 343, 352, 354, 355, 356
Mixon, F. O. 387
Miyahara, F. 57, 65, 67, 70
Miyahara, T. 181, 194, 204, 206, 208, 225

Miyamura, A. 482
Miyashita, H. 274
Miyata, T. 268
Miyatake, O. 421, 423, 426
Mizrahi, J. 253, 299, 333
Mizuno, Y. 162
Mizushina, T. 366, 369, 376, 377, 378, 397
Mo, G. 273
Mo, G. B. 228
Moallemi, H. A. 260
Moan, M. 258, 307, 320, 321, 323
Mobius, D. 233, 236
Mochimaru, Y. 396
Modarres-Sadeghi, Y. 165
Modi, V. J. 476, 481
Moes, J. H. **504**
Mograbi, E. 450, 463
Mohammadigoushki, H. 84, 172
Mohan, V. 57, 59, 60, 191, 192, 208, 217, 218, 222, 225, 281, 300
Mohan Rao, P. R. 265
Mohanty, K. K. 266
Molerus, O. 252, 347
Molki, A. 89
Molla, M. M. 368, 444
Mollinger, A. M. 170, 171
Momani, S. 319
Mondy, L. A. 510
Monkewitz, P. A. 165
Montagna, C. P. 175
Montano, W. 324
Montemagno, C. D. 244
Montiel, R. 149, 153, 155
Montillet, A. 244, 245, 246, 251, 259, 277, 281, 288, 294, 303, 334, 336, 337, 338, 369, 405
Moon, S. H. 435
Moore, D. W. 214, 215, 505
Moore, F. 28
Moore, J. A. 3
Moore, M. G. 87, 101
Moo-Young, M. 56, 57, 59, 60, 61, 65, 180, 192, 193, 210, 217, 223, 385, 398, 399, 400, 402
Mora, S. 358, 360
Morais, A. F. 283
Mordarski, J. 160, 202, 206, 208, 209, 231
Morel, P. 244
Moreno, L. 282, 304, 306
Moreno, L. D. 319
Moreno, R. 324
Moreno, R. A. 324
Morgan, V. T. 435
Mori, H. 186, 202, 203, 206
Mori, M. 216
Mori, N. 175, 279, 307, 314, 319
Morland, C. D. 312
Morooka, S. 410, 411
Moore, J. A. 3
Moore, M. G. 87, 101
Morris, C. E. M. 266
Morris, E. W. 435
Morris, J. F. 23, 163

Morris, S. 386
Morrison, F. A. 14, 29, 39, 42, 363
Mortensen, A. 239
Morusic-Paloka, E. 91
Mory, M. 215, 401
Moschopoulos, P. 27
Mosdorf, R. 215
Mosely, K. 358
Moshev, V. V. 169
Moshizi, S. A. **510**
Moss, G. R. 314, 319
Mossaz, S. 132, 136
Mota, M. G. 242
Motosu, I. 146, 161
Mott, R. A. **480**
Mougin, N. 191, 197, 228
Moulic, S. G. 429
Mouline, Y. 181, 194, 231
Mousa, H. 263
Moussa, T. 10
Moutrie, M. F. 3
Mow, K. 246, 249, 264, 265
Moyne, C. 244, 246
Mpandelis, G. 72, 75
Mrani, A. 352, 410, 411
Mrokowska, M. M. 153
Mucoglu, A. 461
Mueller, G. E. 262, 263
Muhlbock, M. 88
Mujumdar, A. 26
Mukherjee, A. K. 185
Mukherjee, K. 78
Mukherjee, S. 197, 202, 225, 235
Mukundakrishnan, K. 200, 236, 505
Muldowney, G. P. 473, 482, 504
Mulhem, B. 175
Müller, A. J. 22, 275, 288, 286, 295, 322, 324
Müller, M. 278, 281, 286, 304, 308, 324, 350
Müller, M. V. 281
Muller, R. 234
Muller, S. J. 3, 110, 113, 120, 149, 152, 155, 158, 159, 161, 172, 175
Muller-Steinhagen, H. 215, 326, 331, 334, 336, 341
Mulugeta, G. 3
Mun, R. 186
Mun, R. P. 185
Mungan, N. 281, 309
Munoz, L. A. B. 242
Munro, J. P. 231
Munroe, H. S. 471, 476, 479
Munshi, B. 87
Muntean, O. 326
Munzel, K. 516
Murakami, K. 421, 423, 426
Murali Krishnan, J. 3, 29
Murali, V. 260
Muralidhar, R. 234
Murata, H. 476
Muratsune, K. 484
Murch, W. L. 145, 158, 163
Muroyama, K. 352

Murphy, R. J. 45
Murthy, M. S. 358
Musters, J. J. 131
Mutel, A. 169
Mutlu, I. 144, 151
Myasnikov, V. P. 108, 113
Myers, T. G. 366

N'Gouamba, E. 27
Na, T. Y. 366, 375, 395, 433
Naccache, M. F. 132, 228, 313, 315, 316
Nachman, A. 367, 375
Naciri, J. K. 509
Nadim, A. 175
Nagarajan, R. 192, 225
Nagelhout, D. 239, 272
Nagumo, R. 186, 202, 203, 206
Najeh, N. 56
Nakamura, A. 131, 257, 363, 367, 372, 373, 376, 386, 387, 394
Nakamura, I. 51, 54
Nakamura, K. 175, 279, 307, 314, 319
Nakamura, R. 57
Nakano, Y. 59, 60, 192, 217, 221, 402
Nakayama, A. 257, 363, 367, 372, 373, 376, 386, 387, 394, 422, 466
Nalajala, V. S. 221, 230
Nalluri, S. V. 422, 465
Nandakumar, K. 50, 51, 54, 262, 264
Nandrajog, S. 90, 91, 499, 500
Nanjo, H. 187
Narain, J. P. 434
Narayan, K. A. 194, 220, 229, 378
Narayana, P. A. A. 444
Narayanan, S. 231
Narayanaswamy, R. 189, 198, 228, 236
Narh, K. A. 324
Narsimhamurthy, G. S. R. 228
Narvaez, C. 154, 155, 168, 170, 358
Narvaez, C. P. 200
Nash, S. 284, 287
Nassar, M. M. 394, 408
Nassehi, V. 304
Nath, D. 444
Naudascher, E. 279, 281, 308, 310, 311
Naumann, A. 76
Navarro, F. J. 3
Navaza, J. M. 175, 206
Navez, V. 153, 155
Navon, O. 186
Nazar, R. 463
Neale, G. 273
Nechrebecki, D. G. 22
Neeman, H. J. 253
Negrao, C. O. R. 107
Neira, M. A. 258
Nelson, J. 169
Nemec, D. 248
Newcombe, S. 175
Newman, J. 258
Newton Samuel, T. 351

Newton, D. A. 209, 210
Newton, I. 471, 475
Newton, J. M. 88
Ng, K. M. 261, 265
Ng, M. L. 423, 437, 438
Nguyen, A. V. 215
Nguyen, H. 156
Nguyen, H. D. 461
Nguyen, N. L. 242
Nguyen, Q. D. 25, 26, 41, 42, 128, 129, 157, 521
Nguyen, T. H. 190, 435
Nguyen-Chung, T. 141
Nguyen-Tien, K. 403, 410
Niasar, V. 283, 285, 286, 304
Nichols, K. L. 91, 494
Nicolae, G. 288, 295, 304, 307, 312
Nicolai, B. M. 239, 263
Nieckele, A. O. 313, 315, 316
Nield, D. A. 240
Nielsen, D. R. 239
Niessner, R. 244
Niethammer, M. 201, 211
Nieuwstad, C. H. 51
Nieuwstadt, F. T. M. 166
Nigam, K. D. P. 66, 68, 72, 73, 75, 194, 206, 210, 225, 230, 335, 339, 342, 410, 476, 485, 489, 490
Niida, T. 90, 499
Nikiforakis, N. 360
Nikitopoulos, D. E. 144, 152, 165, 506
Nikolov, L. N. 84
Nikolova, A. 406
Nili-Ahmadabadi, M. 166
Nilsson, M. A. 239, 308, 323, 327
Nir, A. 122, 198, 201, 202, 217, 225, 228, 233, 235
Niranjan, K. 175, 331
Nirmalkar, N. 110, 121, 123, 127, 132, 135, 136, 137, 201, 217, 221, 242, 367, 381, 383, 384, 390, 391, 392, 403, 421, 422, 440, 441, 442, 443, 448, 457, 459, 460, 461, 462, 463, 464, 465, 466, 485, 494, 495, 501, 502, 503
Nishimura, T. 274
Nishimura, Y. 257
Nishizawa, M. 268
Niskanen, H. 3
Niskansen, N. 273
Nitin, S. 87, 93, 101, 377, 497, 498, 499
Niven, R. K. 245, 249
Nixon, K. M. 377
Noberga, J. M. 146, 158
Noda, K. 280, 286, 297
Noel, A. C. 2, 3
Nogueira, S. 195, 198, 227
Noh, D. S. 235
Noll, W. 39
Norden, P. A. 425
Nori, F. 87, 101
Norman, L. 239
Norouzi, M. 166, 196, 197, 202, 223, 225
Northey, P. J. 141, 150
Norton, D. J. 87, 497, 500
Notheis, P. J. 22

Notz, P. K. 184, 235
Nouar, C. 120, 121
Novotny, E. J. 77, 83
Null, H. R. 182
Nyrkova, I. A. 510

O'Brien, S. B. G. 1
O'Donovan, E. J. 18, 119
O'Neill, B. K. 242, 252, 257, 295
O'Neill, K. E. 264
O'Neill, M. E. 49, 482
Obata, E. 195, 217, 222, 228
Ockendon, J. R. 49
Ocone, R. 405
Odeh, A. S. 281, 304
Odell, J. A. 3, 4, 139, 283, 286, 287, 288, 307, 309, 314, 319, 320, 322, 324
Officer, S. 239, 308, 323, 327
Ofner, A. 285
Ogata, S. 162
Ogawa, K. 369, 397
Ogunjimi, A. 141
Oh, J. H. 153, 154, 474, 496
Oh, Y.-K. 3
Ohashi, H. 195
Ohkunitani, H. 175
Ohl, C. D. 187
Ohta, M. 191, 195, 202, 203, 206, 217, 222, 223, 228
Ohya, K. 274
Oikawa, K. 59, 66, 67, 72, 153, 154
Okamoto, R. 400, 403
Okawa, T. 216
Okhotskii, V. B. **505**
Okuda, K. 479
Olbricht, W. L. 235
Oldroyd, J. G. 39, 230, 367, 368, 434
Oliveira, F. A. R. 368, 377, 391
Oliveira, G. M. 107
Oliveira, J. C. 368, 377, 391
Oliveira, M. S. 283, 315
Oliveira, P. J. 148, 165, 166, 167
Oliver, D. L. R. 215
Oliver, D. R. 82
Oliver, M. J. 253
Oluyemi, G. F. 239, 308, 323, 327
Omari, A. 284, 285, 287, 320, 321, 323
Onishi, J. 195
Onodera, K. 195, 223
Onuma, T. 186, 202, 203, 206
Ooi, L. 18
Oolman, T. 231
Oosthuizen, J. C. 523
Oosthuizen, P. H. 417, 449, 450
Oostrom, M. 284
Oppong, F. K. 107, 112
Orczykowska, M. 195, 202, 225, 226, 506
Orellana, C. S. 22
Orgeas, L. 239, 259, 295, 314, 320
Oritz, S. L. 191, 198, 202, 206, 225
Orszag, S. A. 54
Ortega-Rivas, E. 357

Ortiz-Villafuerte, J. 216
Osano, Y. 162
Oseen, C. W. 49
Ossin, A. 448
Ostrach, S. 424
Ouattara, Z. 137, 138
Oukhlef, A. 287
Ovarlez, G. 18, 19, 27, 131, 230, 384
Owens, R. G. 141, 144, 148, 149, 150
Oyeneyin, M. B. 72, 75
Ozahi, E. 246, 265
Özdemir, M. 244
Ozeki, E. 485
Ozoe, H. 424, 425
Ozogul, H. 132, 136, 137
Oztekin, A. 151, 164, 166

Padding, J. T. 284, 304, 305, 307, 315, 319, 320
Padhy, S. 163, 281, 283, 302
Paik, S. 461
Pain, J.-P. 83
Paine, P. L. 473
Pakdel, P. 152
Pakdemirli, M. 395
Pal, E. 50, 51, 54
Pal, R. 42, 175, 230
Paliwal, B. 377
Palle, S. 261
Pallinti, J. 505
Pan, D. 91, 101
Pan, L.-F. 3
Pan, S. 201, 221
Pan, T. W. 169, 170
Panda, S. 384
Pandit, A. B. 216, 505
Panfilov, M. 244, 245, 246
Pang, M. 200
Pang, Z. 50, 51
Pangarkar, V. G. 326, 369, 394
Panizza, M. 90
Pantangi, U. S. 426
Pantokratoras, A. 93, 485
Papadopoulos, K. 41
Papale, P. 175
Papanastasiou, A. C. 187
Papanastasiou, T. C. 119, 120, 185
Papathanasiou, T. D. 274
Papavassiliou, D. V. 253
Pareek, V. 110, 112
Parise, J. A. R. 263
Park, B. J. 3, 36, 37
Park, B. O. 3, 36, 37
Park, C.-W. 265
Park, H. C. 261, 281, 295, 296, 306, 311
Park, J. D. 110, 112
Park, K. H. 41, 129
Park, N. A. 41, 164, 524, 526, 527, 528
Parker, A. 282, 322
Parker, H. W. 281, 320
Parker, J. A. 281
Parks, C. F. 68, 78, 83

Parmaj, N. B. 378
Parnas, R. S. 239, 268
Parnell, T. 11
Parrot, J. F. 242
Partal, P. 3, 175
Parthasarathy, M. 18
Parthasarathy, R. N. 185
Parvazinia, M. 304
Pasari, S. N. 280, 320, 323
Pascal, F. 281, 304
Pascal, H. 276, 281, 303
Pascal, J. P. 91, 92, 281, 303
Paschedag, A. R. 188, 200, 215
Pashias, N. 42
Pasquino, R. 110, 145, 168, 169
Patankar, N. A. 170
Patel, K. 278
Patel, O. P. 136, 138, 139, 393, 465
Patel, S. A. 11, 56, 87, 90, 93, 132, 136, 138, 139, 363, 366, 367, 383, 384, 390, 393, 421, 422, 424, 426, 430, 440, 441, 442, 448, 449, 450, 455, 456, 458, 465, 473, 483, 484, 485, 494, 496, 499, 500, 502, 519
Patel, V. C. 49, 51, 54
Paterson, A. 281, 320
Paterson, M. S. 175
Pathak, M. 198, 202, 225
Pati, S. 444
Patil, R. C. 382
Patnaik, B. S. V. 444
Patnana, V. K. 300, 93, 96, 367, 377, 378, 381
Patrick, H. V. L. 161
Patruyo, L. G. 324
Patton, T. 3
Patwardhan, A. W. 50, 51, 54
Patwari, A. N. 410
Patzold, R. 168, 358
Paul, I. 444
Pavlidis, M. 196, 197
Pavlov, K. B. 394
Payatakes, A. C. 258
Payne, L. W. 281, 320
Pazwash, H. 108, 123, 124, 125, 129, 138
Pearlstein, A. J. 505
Pearson, G. 187
Pearson, J. R. A. 49, 239, 303, 320
Peddieson . 395
Peden, J. M. 59, 66, 68, 78, 91
Peel, L. C. 72, 75
Peev, G. 406
Pei, D. C. T. 431
Pelagagge, P. M. 239
Pellerin, F. 258
Peng, C. 22
Peng, J. 121
Peng, S. 166
Penny, G. S. 78
Pereira, F. A. R. 68, 476, 489, 499
Pereira, L. A. A. 188, 215
Pereira, R. G. 86
Perez, A. T. 18, 19

Perez-Martin, O. 288, 324
Perez-Patricio, M. 510
Peri, S. S. 78
Perrault, R. 510, 516
Perrin, C. L. 299
Perrine, R. L. 245
Perry, W. L. 366, 374
Persson, S. 185
Petera, J. 141
Peters, E. A. J. F. 144, 160, 284, 304, 305, 307, 315, 319, 320
Peters, G. W. M. 144, 152, 165
Peters, I. R. 23
Petersen, E. E. 98, 365, 366, 368, 369, 372, 373, 374, 375, 376, 377, 378, 386, 394, 422, 433
Petford, N. 3, 7
Petipas, C. 482
Petit, L. 482
Petrou, M. F. 107
Petsalis, S. 425
Petsch, R. 244
Petty, C. A. 307, 313, 319
Peysson, Y. 110, 285, 286, 358, 359
Pfeffer, R. 185, 407
Pfeiffer, F. 249
Pfennig, A. 474
Phair, J. W. 3
Phan, R. T. 260
Phan-Thien, N. 56, 57, 65, 66, 68, 90, 141, 143, 144, 149, 150, 153, 155, 157, 160, 165, 166, 258, 307, 473, 482, 527, 528
Phelan, Jr. F. R. 239, 274
Phelipot-Mardele, A. 196, 228
Philipp, A. 187
Philippoff, W. 160, 192, 203
Phillips, B. J. 369, 374, 377
Phillips, R. J. 168, 170, 273, 358, 359
Phillips, S. D. 186
Phillips, T. N. 141, 144, 143, 148, 149, 150, 167
Piau, J.-M. 162, 164, 191, 197, 228, 367, 434
Picaro, T. 239, 320
Pickett, J. 334
Picologlou, B. F. 3
Picon- Feliciano, R. 249
Picot, J. 185, 234
Pieransky, P. 510
Pierce, F. T. 239
Piermarini, G. J. 528
Pierre, A. 196, 228
Pilate, G. 142, 151, 164
Pilitsis, S. 258, 307
Pillai, K. K. 262
Pillapakkam, S. B. 194, 223, 224
Pilz, C. 201, 206, 209, 210, 211
Pimenta, F. 367, 395
Pincus, P. A. 14
Pinder, K. L. 248, 276, 295, 296, 317
Pinelli, D. 72, 75
Pinho, C. 265
Pinho, F. T. 51, 93, 97, 101, 148, 165, 283, 315
Pinto, A. M. F. R. 189, 195, 198, 227, 236

Pinto, F. 22
Pinton, J.-F. 198
Pipe, C. J. 165
Pipes, R. B. 315, 316
Piri, M. 259
Pisani, L. 242, 243
Pitchumani, R. 239
Pitois, O. 230
Pitt, M. J. 88
Pittman, J. F. T. 367, 369, 373, 375, 427
Plichta, C. 141
Pliskin, I. 86
Plog, J. P. 185
Plumb, O. A. 262, 264
Podczeck, F. 88
Podgorski, T. 170, 171
Podladchikov, Y. 175
Podolsak, A. K. 307
Pohlhausen, E. 424
Poirier, D. R. 239, 272
Polesskii, E. P. 161, 162
Poletto, M. 170, 358
Polishchuk, A. M. 283
Pollard, T. 517, 520, 524
Pollard, T. D. 517, 520
Pollert, J. 161
Polyanin, A. D. 186, 363
Ponche, A. 169
Poncin, S. 232
Ponter, A. B. 180
Poole, R. J. 19, 31, 40, 110, 121, 123, 127, 132, 135, 147, 364, 367, 382, 384, 390, 391, 392
Pop, I. 433, **463**
Pope, D. 88
Pope, G. A. 279, 283, 286, 295, 308, 311, 320, 322
Pordesimo, L. O. 83
Porteous, K. C. 148
Porter, J. E. 363
Poschl, C. 221
Potapov, A. 198, 217, 228
Potter, J. M. 448
Potucek, F. 406
Pouligny, B. 509
Poulikakos, D. 259, 303
Pouliquen, O. 27
Powell, R. L. 359, 482, 510
Power, G. 56, 142
Powley, M. B. 41
Pradipasena, P. 27
Prakash, O. 274, 313, 318, 393, 412
Prakash, R. 313, 411, 412
Prakash, S. 59, 67, 73, 75, 78, 84
Pralat, K. 364
Pralits, J. O. 97
Prandtl, L. 7
Prasad, B. 281
Prasad, D. V. N. 313, 315, 316, 318
Prashant 118
Pravesh, R. 289, 412
Premlata, A. R. 190, 199
Prentice, J. H. 3

Prhashanna, A. 384, 422, 434, 435, 438, 439, 440, 455, 457, 458
Price, A. H. 145, 146, 151, 155, 157, 160, **498**
Pricl, S. 3, 29
Prieler, R. 88
Prieto, J. L. 196, 233
Prieve, D. C. 168
Prilutski, G. 10, 156
Primicerio, M. 357
Prince, R. G. H. 181
Princen, H. M. 41
Prins, J. A. 434, 448
Pritchard, D. 163
Prodanović, M. 283
Prokunin, A. N. 83, 101
Proski, T. 107
Protopapas, A. L. 295, 299
Proudman, I. 49
Prud'homme, R. K. 41, 230, 240, 320
Pruess, K. 296
Pruitt, G. T. 161
Prunet-Foch, B. 175
Pruppacher, H. R. 49
Puget, F. P. 215
Pui, D. Y. H. 265
Pulletikurthi, V. 444
Pulley, J. W. 87, 497
Pullum, L. 3, 105
Puncochar, M. 151, 252
Puri, V. M. 83
Puris, B. I. 161, 162, 369, 374, 376, 377
Purohit, N. K. 109, 128
Putz, A. M. V. 117
Pye, D. J. 309, 320
Pye, K. W. 364
Pyle, D. L. 175

Qi, F. 163
Qi, H. 91
Qi, M. 485, 493
Qi, T. 225, 233
Qiang, S. 181, 231
Qu, J. 72, 78
Quach, A. 516
Quadir, J. A. 78
Quan, G. 280, 286, 297
Quan, S. 505
Quilliet, C. 4, 139
Quinn, B. W. 282, 312
Quinn, M. J. 231
Quintana, G. C. 194, 222, 223, 227
Quintans-Riveiro, L. C. 175, 206
Quintard, M. 278, 284, 304, 309
Quintard, M. Y. 276, 304
Qureshi, Z. H. 435, 444

Rabenjafimanantsoa, H. A. 155, 163
Rabiger, N. 181, 182, 225
Rackemann, D. W. 197
Radilla, G. 244, 246, 284, 286, 295, 304, 320
Radl, S. 175, 206, 227

Radulovic, P. T. 365, 373, 394
Rae, D. 108, 113
Raghuraman, J. 192, 217, 222, 281, 300
Raghavan, V. 449
Rahaman, K. 224, 504
Rahli, O. 268, 269, 273
Rai, S. R. 186
Raichura, R. C. 260
Rainey, P. G. 162, 395
Raja, A. H. 136, 138, 139, 367, 393, 421, 426, 430
Rajagopal, K. R. 3, 40, 147, 395
Rajagopalan, D. R. 144, 145, 147, 150, 151, 152, 155, 158, 160
Rajeswari, G. K. 394
Rajitha, P. 88, 89, 90, 91, 101
Raju, B. H. S. 463
Rajvanshi, S. C. 142, 147
Rallison, J. M. 143, 149, 151, 158, 160, 167, 234, 496, 497
Ramaccini, F. 27
Ramchandra, V. 411
Ramagopal, A. 396
Ramamurthy, G. 426
Raman, J. R. 276, 300, 301, 304
Ramarao, B. V. 359
Ramaswamy, H. S. 3, 25, 377, 423
Ramaswamy, S. 161, 171, 172, 235
Ramesh, K. V. 368, 393
Rameshwaran, P. 144, 151
Ramey, Jr. H. J. 279, 304
Ramge, P. 107
Rami, K. 91, 98, 99, 101, 497, 499
Ramírez, N. 286
Ramirez, N. E. 324
Ramkissoon, H. 143, 160, 170, 194, 224, 504
Ramkrishna, D. 233, 234
Ramteke, R. R. 367, 393
Ramya, K. A. 27
Ranade, V. V. 256
Ranganathan, S. 266
Rangel, C. 276, 295, 313, 320
Rangel, N. 265
Rangel, R. H. 145, 155, 168, 170
Rankin, P. J. 18, 111
Rao, A. R. K. 280, 323
Rao, B. K. 369, 374, 377, 378, 409
Rao, E. V. L. N. 182
Rao, J. H. 367, 375
Rao, M. K. 96, 382
Rao, M. A. 3
Rao, P. K. 93, 96, 97, 382
Rao, P. T. 281, 286, 312
Raper, J. 88, 101
Rapier, A. C. 248
Rashidi, M. 433
Rashidov, U. 239
Rasmussen, H. K. 144, 146, 151, 160
Rasul, M. G. 197
Rathna, S. L. 57, 147, 394
Ratner, B. D. 266
Ratulowski, J. 131
Raufaste, C. 4, 139

Raval, V. K. 184
Ray, H. S. 528
Ray, S. 443
Raymond, F. 215
Raynaud, P. G. 242
Raynor, P. C. 239, 269, 273
Reddy, C. R. 367, 388, 393, 485, 499, 500
Reddy, J. N. 141, 149
Reddy, R. K. 51, 256
Ree, T. 15
Reed, J. C. 290
Reed, J. S. 239
Rees, D. A. S. 284, 287
Reichelt, W. 260, 261
Reilly, C. D. 175
Reilly, I. G. 423, 427, 430
Reimers, C. E. 90
Reimert, R. 242
Reiner, M. 40
Reiss, L. P. 186, 187, 400
Reizes, J. A. 73, 78, 109, 128, 129, 188, 190, 215, 502
Ren, S. 19, 21
Ren, S. R. 73
Ren, X. H. 244
Renard, P. 255
Renardy, M. 150, 165, 186, 397, 398
Renardy, Y. Y. 234
Renaud, M. 72, 73, 74, 75, 243, 251, 252, 257, 263, 264, 267, 277, 293, 295, 303, 312, 336, 347, 405
Renksizbulut, M. 216
Repetti, R. V. 82
Resendiz-Tolentino, O. 172
Reverdy-Bruas, N. R. 240
Reynolds, G. K. 88, 101
Reynolds, P. A. 59, 66, 68, 75, 91
Rha, C. 27
Riba, J. P. 334
Ribeiro, F. S. 313, 315
Ribeiro, G. S. 86
Riberio, V. M. 51
Richardson, J. F. 2, 14, 42, 50, 51, 91, 99, 101, 248, 266, 290, 333, 335, 341, 342, 357, 363, 367, 369, 372, 373, 375, 409, 427, 434, **481, 497, 499**
Richet, P. 175
Richter, D. 145, 165
Riddle, M. J. 54, 155, 168, 170, 358
Ridgway, K. 262
Riethmuller, M. L. 189, 195, 198, 227, 236
Riley, D. S. 186
Riley, E. E. 164
Riley, N. 448, 449
Rindt, C. C. M. 443
Ringhofer, M. 510
Rios-Rojas, C. 510
Ripepe, M. 175
Risso, F. 176, 188
Rivero, D. 22
Rivkind, V. Y. 215
Rivlin, R. S. 39
Rizvi, S. S. H. 3
Roberts, B. 155, 172

Robertson, J. M. 108, 123, 124, 125, 129, 138
Robisson, A. 2
Roblee, L. H. S. 262
Rochelle, S. G. 395
Rockefeller, H. A. 91, 101
Rockwell, D. 164, 166
Rode, S. 244
Rodin, G. J. 56, 58, 61, 62, 79, 91, 92
Rodrigue, D. 56, 58, 65, 91, 191, 194, 195, 202, 208, 206, 209, 210, 211, 215, 221, 224, 225, 227
Rodriguez de Castro, A. 284, 285, 286, 287, 295, 304, 320
Rodriguez, M. 163
Rodriguez, S. 288, 322, 324
Rodriguez-Rodriguez, J. 1
Rodts, S. 284
Roger, R. P. 87, 499
Rogers, S. M. 239, 279, 324
Roghair, I. 196, 226
Rohani, S. 260, 358
Roig, A. 54
Rojas, M. R. 286
Roldan-Garcia, C. 26
Romero, C. 288, 322, 324
Roodhart, L. P. 67, 78, 83
Roquet, N. 135
Rosant, J.-M. 215
Rose, H. E. 248
Rosen, M. 281, 320
Rosen, S. L. 168
Rosentrater, K. A. 3
Roshani, S. 260, 263, 265
Rossen, W. 287
Rossen, W. R. 3, 175, 287, 304, 315, 319
Rossi, E. 112
Rosso, A. 285, 286, 304
Rosso, F. 357
Rosti, M. E. 358, 360
Rotem, Z. 368
Roth, N. 236
Rothstein, J. P. 155, 165, 172, 239, 308, 314, 319, 323, 327
Rouboa, A. 443, 446
Round, G. F. 162
Rouse, Jr. P. E. 310
Roussel, N. 42, 312, 314
Roustaei, A. 284, 287
Roux, S. 107
Roversi-M, D. 324
Roy, C. 3
Roy, R. A. 175, 195
Roy, S. 41, 367, 373
Royer, P. 239
Roza Oshaghi, M. 181
Rozanski, J. 283, 336, 337, 339, 349, 350
Rozant, O. 239
Rozhkov, A. 175
Rozhkov, A. N. 163
Rubatat, L. 112
Rubin, A. L. 268
Rubio-Hernandez, F. J. 22, 23, 24
Ruckenstein, E. 324, 396, 401, 429, 434
Rudolph, M. L. 3, 199, 228

Author Index

Rühs, P. A. 285
Rumpf, H. 247, 248
Rumscheidt, F. D. 234
Russel, W. B. 357
Ruszczycky, M. A. 161
Ryan, M. E. 10, 156, 175, 195, 215, 234
Ryan, N. W. 15
Ryskin, G. M. 215
Ryu, B. H. 3, 36, 37
Ryu, S. H. 168, 169, 170

Sa, J.-Y. 444
Saad, E. I. 255, 256
Saak, A. W. 107
Saar, M. O. 3
Sabiri, N. E. 88, 89, 90, 91, 101, 244, 245, 277, 280, 281, 286, 288, 293, 294, 295, 296, 297, 303, 311, 312, 334, 336, 337, 338, 347, 349, 369
Saboni, A. 215, 401
Sacheti, N. C. 142, 148, 513
Sacilik, K. 239
Sada, E. 175
Sadhal, S. S. 236
Sadiq, T. A. K. 268
Sadowski, T. J. 41, 282, 286, 287, 290, 291, 292, 293, 304, 306, 307, 309, 320
Sáez, A. E. 22, 275, 286, 288, 295, 314, 319, 322, 324
Saez, A. Z. 22
Saffman, P. G. 168
Sagert, N. H. 231
Saha, A. K. 87, 91
Saha, G. 109, 128
Saha, S. C. 444
Sahimi, M. 243, 252, 285, 295, 347
Sahin, C. 145, 162, 165
Sahraoui, M. 272
Sahu, A. K. 93, 96, 97, 367, 375, 382, 384, 429
Sahu, K. C. 190, 196, 199, 217, 221
Sai, P. S. T. 326
Saint- Michel, B. 186
Sairamu, M. 421, 422, 440, 457
Saitoh, T. 435
Sajiki, T. 435
Sakaguchi, T. 216
Sakai, K. 68
Salami Hosseini, M. 285
Salami, E. 505
Salamon, T. R. 141, 319
Saleh, S. 244
Salman, A. D. 88
Salt, D. L. 15
Saltiel, C. 269
Samano, D. 182, 200, 206, 227, 229, 233
Sameen, A. 444
Samson, G. 196, 228
Sanchez-Rivera, D. 283
Sandeep, K. P. 83, 282
Sanders, J. V. 161, 162
Sanders, R. S. 88
Sandiford, B. B. 309, 320, 321
Sandkuhler, P. 111

Sangani, A. S. 228, 230, 255, 256, 270, 271, 272, 273, 411
Sanjuji, B. 175, 206
Sano, Y. 162
Santini, R. 268, 269, 273
Santos, A. 265
Santos, T. G. M. 107
Saradhy, Y. P. 185
Saramito, P. 120, 122, 135, 141
Sarasua, L. G. 155, 168, 170
Sargenti, M. L. 288, 322, 324
Sarkar, K. 197, 202, 225, 235
Sarma, G. K. 443
Saroa, M. S. 395
Sarpkaya, T. 162, 395
Sasmal, C. 97, 101, 365, 366, 367, 382, 393, 395, 396, 419, 421, 422, 440, 448, 457
Sastry, S. K. 82, 368, 369, 377, 391, 392, 423, 438
Satheesh, V. K. 274, 312, 313, 315, 411
Satish, M. G. 300, 405
Sato, M. 524
Sato, T. 66, 67, 484, 514
Satrape, J. V. 144, 151
Saunders, M. G. 444
Savari, S. 45
Savicki, D. L. 239
Savin, V. G. 108, 113
Savins, J. G. 57, 59, 62, 83, 288, 523
Savorani, L. 3
Savreux, F. 137
Sawant, S. B. 216, 505
Sayegh, N. N. 240
Saylor, B. Z. 253
Sayre, R. M. 115, 150, 473, 504
Schaink, H. M. 169
Schatzmann, M. 524
Schaub, K. 516
Schechter, R. S. 56, 210
Scheele, G. F. 233
Scheid, C. M. 215, 499
Scheidegger, A. E. 240, 245, 252, 253
Scherr, P. 473
Scherzer, O. 221
Schieber, J. D. 314, 319
Schiller, V. L. 76
Schimanski, H. 41
Schlichting, H. 370, 372, 376
Schmid, R. 239, 245
Schmidl, W. D. 216
Schmidt, W. 312, 314
Schmiedel, J. **497**
Schnitzlein, K. 260, 261
Schowalter, W. R. 14, 23, 187, 307, 321, 323, 367, 370, 373, 374, 394
Schramm, L. 3, 42, 175, 230, 239, 240
Schugerl, K. 175, 206, 230, 231, 331, 403
Schuler, B. O. G. 176, 177, 178, 180, 181
Schulte, G. 175
Schumpe, A. 206, 230, 403, 410
Schurz, J. 18, 107, 109, 520, 523, 524
Schwarz, W. H. 142, 147, 396, 503, 511, 512, 524
Scirocco, R. 358

Scotford, I. M. 3
Scott Blair, G. W. 2, 523
Scott, K. J. 358
Scott, T. C. 505
Scriven, L. E. 187, 244
Scrivener, O. 161
Secchi, E. 285, 299
Sechet, P. 176
Sedahmed, G. H. 408
Sedahmed, H. G. 368, 394
Sederman, A. J. 244, 263
Sedran, T. 21
See, H. 18
Seeling, C. H. 202, 206
Seetharamu, K. N. 444
Segre, G. 82, 169
Seguin, D. 244, 245, 277, 288, 405
Seidl, B. 286
Sek, J. 288, 289, 290, 295, 296, 297, 298, 305
Selby, R. J. 279
Selen, S. H. A. 144, 165
Selerland, T. 312, 314, 315, 317
Sellers, H. S. 524
Sellgren, A. 2
Sellin, R. H. J. 161, 278, 307, 309, 369, 396
Semenov, A. N. 510
Semin, B. 499
Sen, S. 109, 124, 126, 129
Seo, J. -H. 96
Seperhrnoori, K. 283, 286, 308, 311, 320
Serpemen, Y. 403
Serra-Tosio, J.-M. 240
Serth, R. W. 367, 374, 376, 395
Seshadri, V. 101
Sestak, J. 522, 523
Sethi, R. 284, 298
Settles, G. S. 175
Settles, J. 121
Sevilla, A. 22, 23, 24
Seybold, H. J. 283, 285, 286, 299
Seyed-Ahmadi, A. 91
Seyer, F. A. 295, 306, 307, 304
Sezen, M. C. 523
Sgreva, S. 110, 112, 130, 131
Shadid, J. N. 376
Shah, C. B. 259
Shah, M. J. 365, 366, 368, 369, 372, 373, 374, 375, 376,
 377, 378, 386, 394, 422, 433
Shah, S. P. 107
Shah, S. N. 67, 72, 75, 78, 83, 168
Shah, Y. 4
Shah, Y. T. 206, 230, 403
Shahcheraghi, N. 499
Shahsavari, S. 286, 314, 316, 317
Shail, R. 87, 497, 500
Shair, F. H. 98
Shaker Shiran, B. 239
Shamakova, L. M. 57
Shang, C. 176, 186
Shankar Subramanian, R. 109, 126, 130, 131, 175, 233, 236
Shao, X. 91, 101

Shaokun, J. 196, 225, 226
Shapira, M. 269, 274
Shapiro, M. 269, 274
Shapley, N. T. 144, 145, 147, 150, 151, 152, 158
Shaqfeh, E. S. G. 145, 158, 163, 165, 283, 302, 311, 359
Sharaf, D. M. 190
Sharif, A. O. 367, 369, 373, 375, 473
Sharma, A. 93, 94, 377, 403
Sharma, H. G. 147
Sharma, K. K. 423, 430
Sharma, M. G. 83
Sharma, M. K. 91, 101, 276, 312, 335, 336, 342, 499
Sharma, M. M. 146
Sharma, O. P. 385, 396
Sharma, R. B. 186
Sharma, S. K. 191, 195, 206, 225, 228
Shattuck, M. D. 244
Shavit, U. 122, 201, 233
Shaw, M. T. 305, 510, 511
She, K. 88
Shearing, P. R. 243
Shedid, M. 423
Sheffield, R. E. 252, 280, 295
Sheikh, N. A. 284
Shen, J. 188, 199, 203
Shende, T. 283, 285, 286, 304
Shenoi, R. A. 269
Shenoy, A. R. 87, 91
Shenoy, A. V. 259, 278, 303, 363, 367, 386, 387, 394, 417,
 422, 425, 427, 428, 429, 433, 434, 466
Sheppard, S. 244
Sherman, P. 3, 522, 524
Sherrard, C. P. 367, 369, 373, 375, 427
Sherwood, A. A. 157
Sherwood, J. D. 505
Shevchuk, M. S. 268
Shew, W. L. 198
Shi, Y.-D. 3
Shiang, A. H. 164, 166
Shiau, T. 166, 167, 168
Shibata, H. 162, 206
Shibata, J. 225
Shibata, T. 162
Shibu, S. 274, 312, 314, 315, 318
Shie, C. F. 499
Shifang, H. 160
Shigekane, Y. 191, 195, 202, 206
Shih, A. 285, 309
Shih, T. M. 435
Shih, Y.-P. 366, 374, 375, 433
Shilov, D. J. 108
Shilov, V. N. 255
Shilton, N. C. 331
Shima, A. 186, 187
Shimizu, K. 234
Shimokawa, S. 66, 67, 484, 514
Shinnar, R. 185
Shinohara, M. 482
Shiona, T. 1
Shiono, F. 1
Shiraishi, Y. 62

Shirotsuka, T. 185, 186, 191, 193, 223, 401
Shmakov, Y. I. 57
Shmulevich, I. 239
Shoemaker, C. F. 3
Shokri, N. 284
Shook, C. A. 2, 56
Shook, J. 510, 516
Shook, J. W. 516
Shore, H. J. 184
Shosho, C. E. 195, 215
Shu, C. 91, 101
Shu, S. 449
Shukla, R. 405, 406, 407, 408
Shuler, M. L. 231, 331
Shulman, Z. P. 186, 187, 369, 374, 376, 377, 422, 429, 430
Shumway, G. 393
Shvets, Yu I. 433
Shvetsov, I. A. 282, 295, 320
Shyam, R. 97, 101, 367, 384, 393, 419, 422
Shyy, W. 269
Sica, L. U. R. 21
Siddiqa, S. **457**
Sigli, D. 62, 91, 142, 143, 151, 153, 154, 157, 170, 239
Sikorski, D. 191, 196, 203, 221, 228
Silberberg, A. 82, 169
Silman, R. W. 523
Siman, R. R. 499
Simon, M. 485
Simpkins, P. G. 175
Simpson, B. K. 377, 423
Singh, A. 461
Singh, B. 366, 369, 374, 375
Singh, D. 281, 297, 298, 335, 337, 338, 343
Singh, J. P. 198, 217, 233, 283, 302
Singh, P. 169, 170, 273, 313, 319
Singh, R. 239, 308, 323, 327
Singh, R. K. 107
Singh, S. N. 449
Singh, S. P. 520
Singh, T. 90, 91, 499, 500
Singh, U. K. 315
Sinha, N. K. 88
Siquier, S. 286
Sirignano, W. A. 165, 236, 255, 257
Sirkar, K. K. 405
Sisavath, S. 258
Siska, B. 65, 78, 101, 277, 296, 308, 336, 350, 351, 358, 359
Siskovic, N. 278, 304, 306
Sittoni, L. 105
Sivakumar, M. 351
Sivakumar, P. 93, 94, 96, 97, 378
Sivakumar, S. 448
Skadsem, H. J. 3
Skartsis, L. 267, 269, 272, 312, 313, 315, 319
Skauge, A. 239
Skauge, T. 239
Skelland, A. H. P. 184, 235, 290, 366, 368, 375
Skjetne, E. 244, 246
Skobeleva, A. A. 163
Skoog, J. 192, 206, 209, 211
Slater, M. J. 216, 505

Slatter, P. 3, 27
Slattery, J. C. 2, 41, 55, 56, 57, 59, 60, 62, 63, 64, 66, 67, 68, 70, 192, 206, 222, 241, 259, 365, 367, 385, 386, 484, 518
Slichter, C. S. 248
Slobodov, E. B. 355
Sloot, P. 273
Slot, J. M. 169
Smagin, I. 198, 202, 225
Smit, G. J. F. 259, 282, 303, 363
Smith, D. E. 151
Smith, D. H. 253
Smith, D. W. 249
Smith, F. W. 281, 282, 309, 322
Smith, G. S. 525
Smith, J. R. 175, 239, 506
Smith, K. A. 244
Smith, M. D. 150, 165
Smith, P. A. 3
Smith, T. 216
Smith, T. N. 107, 108, 111, 113, 115, 124, 126, 129, 131
Smolinski, J. M. 185
Smolka, L. B. 175
Snabre, P. 216
Snijkers, F. 145, 163, 168, 169
Soares, A. A. 82, 91, 93, 160, 368, 378, 394, 411, 422, 443, 446
Soares, E. J. 123, 127
Sobczak, R. 510, 514, 516
Sobey, I. J. 45
Sobolik, V. 369, 397
Sobti, A. 149, 153, 155, 284, 304, 306
Sochi, T. 246, 259, 283, 286, 287, 288, 295, 299, 308
Sodre, J. R. 263
Sofra, F. 3
Sohde, T. 268
Sohn, S.-M. 169
Sojka, P. E. 185
Sokovnin, O. M. 222
Solomon, M. J. 149, 155, 158, 159
Som, A. 422, 429
Soman, A. M. 326
Sommerfeld, M. 236
Son, J. E. 280, 295, 320, 323
Son, S. Y. 185
Song, D. 56, 61, 62, 72, 79, 80, 83, 388, 483, 484, 485, 486, 487, 488, 500, 501, 502
Song, K.-W. 91, 168, 169
Song, T. 109, 111
Song, X. 50, 51
Sontti, S. G. 199, 228, 236, 189
Sood, A. 161, 171, 172
Sood, A. K. 161, 171, 172
Soong, T. T. 273
Sorbie, K. S. 239, 282, 286, 299, 321, 322
Sorour, M. M. 444
Sostarecz, M. C. 197, 202, 224, 225
Soucemarianadin, A. 67, 70
Soundalgekar, V. M. 426, 466
Sousa, R. G. 189, 195, 198, 227, 236
Souvaliotis, A. 314, 319

Spaid, M. A. A. 274
Spalding, D. B. 275, 411
Spanjaards, M. M. A. 168
Sparks, R. S. J. 186
Sparrow, E. M. 270, 422, 430, 442
Spelt, P. D. M. 312, 314, 315, 317
Spence, A. 49, 481
Spencer, A. J. M. 39
Spicer, P. 107, 110, 111, 112
Spielman, L. 273
Spivak, R. 198, 217, 228
Spratt, K. S. 4
Springer, D. S. 239
Squires, Jr. W. 497, 498
Squires, L. 497, 498
Sreenivasulu, B. 368, 39
Sridhar, T. 10, 33, 41, 151, 154, 156, 157, 160, 295, 307, 494
Srinivas, A. T. 422, 443, 446, 447, 448
Srinivas, B. 368, 393
Srinivas, B. K. 260, 261, 276, 286, 294, 295, 303, 311, 312, 335, 337, 339, 340, 342, 343, 346, 350
Srinivas, K. V. 326
Srinivasan, R. 27
Srivastava, A. 384
Srivastava, A. C. 395
Srivastava, A. K. 520
Srivastava, R. C. 366, 373
Srivastava, V. K. 194, 230
St. Pierre, C. 425
Staley, L. M. 3
Stalnaker, J. F. 499
Stamatoudis, M. 233, 476
Stamboltzis, G. A. 88
Standish, N. 320
Stapf, S. 215, 243, 244
Stary, Z. 128
Stastna, J. 151, 522, 523
Stavland, A. 304, 309
Steenhoven, A. A. V. 443
Stefani, C. 107
Steffe, J. F. 3, 22, 29, 290
Steg, I. 26, 27
Steger, R. 278, 286, 304, 308
Stein, H. N. 22
Stein, S. 194, 221
Steinberg, V. 41, 165
Stejskal, J. 406
Stelter, M. 185
Stenzel, K. H. 268
Stephenson, J. L. 244
Stevenson, J. F.,516
Stevenson, P. 249
Stewart, C. S. 231
Stewart, P. 3
Stewart, W. E. 46, 244, 248, 249, 289, 291
Stokes, G. G. 48, 49, 60, 471
Stokes, J. R. 2, 3, 35
Stone, H. A. 144, 169, 234, 235
Stoop, N. 299, 302, 315
Storck, A. 244

Stow, F. S. 162
Street, J. R. 186, 187, 400
Strnadel, J. 485
Stroeve, P. 18, 234, **482**
Strom, J. R. 505
Strong, A. B. 216
Struble, L. J. 16, 25
Stuff, R. 239, 245
Su, X. 146, 172
Subbaraman, V. 511, 513, 514
Subramanayam, N. V. 91
Subramaniam, G. 67, 83
Suekane, T. 244
Sueyasu, Y. 192
Suga, T. 175
Sugeng, F. 143
Sugiyama, K. 1, 221
Sugiyama, S. 279, 307
Suh, I.-S. 403, 410, 411
Suhubi, E. S. 395
Sukanek, P. C. 443
Sukhatme, S. P. 443
Sulaymon, A. H. 358
Sullivan, F. W. 268
Sullivan, R. R. 268, 269
Sullivan, S. P. 299
Sumer, B. M. 51
Summers, J. 42
Sun, A. 42
Sun, B. 198, 201, 221, 226
Sun, D. 220
Sun, J. 143, 165
Sun, Q. 120, 132, 135, 137
Sun, Q.-C. 107, 110, 111, 120, 121, 123, 132, 135, 137
Sun, W. 198, 233
Sun, X. 72, 78, 201, 217, 220
Sun, Y. 22, 199, 225, 233
Sun, Z. 242, 261, 263
Sun, Z.-H. 84, 285
Sunada, D. K. 244, 260
Sundaram, M. 260
Sundararajakumar, R. R. **510**
Sundararajan, T. 56, 58, 61, 62, 71, 72, 74, 75, 79, 82, 87, 90, 91, 95, 101, 229, 257, 275, 279, 300, 301, 334, 335, 336, 337, 338, 343, 344, 345, 346, 347
Suo, Z. 196
Surati, A. I. 180
Suresh Kumar, R. 150
Suri, P. 56, 87, 90, 449, 450, 455, 456, 458, 510, 514, 518, 519
Sury, A. 485
Sussman, M. 191, 195, 202, 203, 206, 223
Sutterby, J. L. 15, 57, 65, 67, 154, 192, 472, 476, 509, 513, 514, 515
Suzaki, K. 435
Suzucki, T. 268
Suzuki, H. 3, 35
Suzuki, M. 1
Suzuki, T. 449
Svarovsky, L. 357
Svoboda, K. 88

Swift, G. W. 525, 526, 527
Swire, H. W. 444
Syrakos, A. 132
Szabo, M. T. 282, 321
Szabo, P. 160
Szady, M. J. 141, 150, 319
Szembek-Stoeger, M. 406, 407
Szesni, N. 249, 265
Szewczyk, A. A. 368, 373, 376

Taamneh, Y. 91
Tabak, I. 239
Tabak, S. 239
Tabakova, S. 49, 54
Tabata, M. 49
Tabuteau, H. 107, 191, 196, 203, 221, 228
Tachibana, M. 59, 66, 67, 72, 153, 154
Tadaki, T. 190, 204
Tadrist, L. 268, 269, 273
Taghi Esfidani, M. 181
Tahir, M. I. 270, 271, 272
Takada, M. 514, 517
Takagi, S. 215
Takahashi, K. 369, 376, 377
Takahashi, M. 369, 376, 377
Takahashi, T. 181, 186, 202, 203, 206, 352, 354, 355, 356
Takamatsu, H. 449
Takayama, S. 199, 228
Takemura, F. 215
Takeuchi, M. 435
Taksermani, U. 278, 307, 308, 309, 311
Tal (Thau), R. 255, 257
Taliaferro, S. 367, 375
Talini, L. 110, 168, 358, 359, 360
Talmon, A. M. 105
Talon, L. 284, 285, 286, 287, 299, 304
Talwar, K. K. 314, 319
Tam, C. K. W. 254, 299
Tam, K. C. 10
Tamaddon-Jahromi, H. R. 145, 172
Tamayol, A. 271
Tan, S. H. 480
Tanaka, H. 421, 423, 426
Tanaka, T. 216
Tanasawa, I.,187
Taneda, S. 51
Tang, G. H. 243, 262, 284, 286, 298
Tang, L. 463
Tang, P. 88, 101
Tang, X. D. 358
Tanguy, P. A. 3
Taniguchi, T. 141, 146
Taniyama, I. 66, 67, 484, 514
Tanner, R. I. 2, 18, 27, 29, 39, 42, 55, 56, 61, 62, 91, 92, 93, 119, 121, 131, 141, 142, 143, 144, 147, 148, 149, 150, 153, 154, 157, 158, 160, 163, 258, 307, 342, 363, 397, 482, 484, 485, 493, 496, 499, 509, 521
Tarbuck, K. J. 262
Tardy, P. M. J. 239, 299, 303, 320
Tarlet, D. 107
Tassart, M. 276, 295, 304, 335, 342, 343, 350

Tate, R. W. 185
Tatersall, G. H. 3
Tatham, J. P. 286, 324
Tatum, J. A. 152
Taud, H. 242
Tavener, S. J. 49, 481
Tavlarides, L. L. 233
Taylor, G. I. 191, 230
Taylor, M. 27
Taylor, M. A. 88
Taylor, S. E. 23
Tazukov, F. Kh. 425
te Bokkel, D. W. 194, 206, 208, 225
Teamah, M. A. 444
Teeuw, D. 308
Tehrani, M. A. 169
Teichman, R. 101
Teixeira, J. A. 242
Teke, I. 435
Teke, K. 45
Ten Bosch, B. I. M. 43, 49
Ten Cate, A. 51
Teng, S. 499
Teo, C. J. 91, 101
Teodovair, P. 262
Terasaka, K. 180, 181, 182, 206
Termonia, Y. 239
Terrones, G. 3, 175
Terzaghi, K. 248
Teske, M. N. 185
Thakur, P. 384
Thauvin, F. 266
Thelen, H.-J. 436
Theodoropoulou, M. 239
Thete, S. S. 231
Thinakaran, R. 248, 260
Thirriot, E. C. 244
Thirriot, F. 258
Thistle, H. W. 185
Thizon, P. 206
Thomas, A. D. 105, 109
Thomas, D. 351
Thomas, N. H. 175, **506**
Thomas, R. H. 142, 160
Thomas, S. 282
Thompson, J. L. 281
Thompson, K. E. 253, 299
Thompson, M. C. 372
Thompson, R. L. 21, 27, 123, 127
Thompson, T. L. 88
Thovert, J. F. 244
Thumati, V. T. 483, 484, 485, 486, 487, 494, 495, 497, 501, 502
Thyrion, F. C. 326
Tian, S. 483
Tian, X.-W. 84, 285
Tiefenbruck, G. 142, 148, 154, 168, 169, 192, 193, 223, 226, 400
Tien, C. 59, 192, 221, 258, 422, 423, 425, 426, 427, 428, 429, 430, 438, 451, 452, 453, 454, 456, 459, 466
Tien, C. L. 448

Tierney, J. W. 262
Tigoiu, V. M. 527
Time, R. W. 155, 163
Timol, M. G. 395
Timonen, J. 273
Ting, R. Y. 187
Tirrell, M. 276, 322
Tirtaatmadja, V. 33, 154, 157, 160, 494
Tiu, C. 10, 11, 17, 22, 41, 59, 66, 67, 109, 111, 143, 144, 146, 153, 154, 157, 261, 280, 282, 288, 289, 290, 292, 293, 295, 296, 304, 306, 307, 309, 311, 312, 320, 323, 483, 484, 485, 489, 492, 493, 514, 517, 518, 519, 529
Tiwari, A. K. 132, 138, 139, 366, 382, 384, 440, 442, 448
Tiwari, N. 384
Tiwari, S. S. 50, 51, 54
Tjaden, B. 243
Tobis, J. 266
Todorova, D. 406
Toften, T. H. 365, 373
Tohidi, M. 196, 228, 236
Tokpavi, D. L. 105, 117, 121, 132, 136
Toll, S. 271
Tomazic, P. 88
Tomboulides, A. G. 54
Tomita, Y. 57, 59, 60, 67, 83, 385, 396
Tomiyama, A. 216
Tong, P. 4
Tonini, R. D. 331, 335, 342, 409
Tonmukayakul, N. 163
Toor, A. P. 149, 153, 155
Toose, E. M. 234
Torobin, L. B. 51, 85
Torres, M. F. 295
Torrest, R. S. 91, 514
Torzecki, J. 279
Tosco, T. 284, 298
Tosun, I. 263
Toussaint, F. 3
Townsend, P. 56, 142, 143, 144, 149, 151, 165, 166
Toxopeus, S. 121
Tozeren, H. 482
Trahan, J. F. 497, 498
Trambouze, P. 231
Tran Son Tay, R. 166, 510, 524, 526, 527
Tran, A. 3, 199, 228
Tran, Q.-K. 109
Tran-Cong, S. 90
Trappe, V. 111
Travkin, V. S. 241, 259
Traynis, V. V. 105, 108, 111
Treybal, R. E. 182, 228
Tribollet, B. 164, 396
Trim, L. 88
Trinh, D. T. 109
Tripathi, A. 56, 58, 61, 62, 71, 72, 74, 75, 79, 82, 84, 87, 90, 91, 95, 101, 223, 225, 229, 312, 314, 315, 347
Tripathi, G. 368, 394
Tripathi, M. K. 190, 196, 199, 217, 221
Tripathi, P. K. 369, 391, 393
Trivedi, M. 366, 393, 403

Trnka, O. 88
Trouton, F. T. 33
Trull, T. W. 175
Tryggvason, G. 175, 227
Tsakalakis, K. G. 88
Tsakiroglou, C. D. 239, 259, 303
Tsamopoulos, J. 27, 107, 109, 110, 111, 113, 117, 118, 120, 121, 122, 186, 196, 197, 201, 299
Tsay, S.-Y. 394
Tseng, S. 91, 499
Tsotsas, E. 261, 263
Tsou, F.-K. 91, 164
Tsuei, H.-S. 422, 425, 426
Tsuge, H. 180, 181, **505**
Tsuji, Y. 236
Tsujino, T. 186, 187
Tsukada, T. 194, 208
Tucker, G. S. 83
Tuinier, R. 141, 146
Tullock, D. 482
Tullock, D. L. 90, 473, 482
Tuminello, W. H. 516
Tunan, N. F. 288, 295, 304, 307, 312
Tung, M. A. 3
Tuoc, T. K. 368, 369, 394
Turcotte, G. 41
Turian, R. M. 61, 66, 67, 70, 161, 258, 484, 485, 489, 513, 516, 522 Turki, S. 500
Turney, M. A. 359
Tyabin, N. V. 108, 112
Tytus, R. P. 185
Tzounakos, A. 191, 195, 221, 227, 491

Uchida, K. 410, 411
Uchida, Y. 524
Udaykumar, H. S. 101
Uddin, M. S. 200, 203, 226
Uehara, H. 435
Ugwu, J. O. 239, 298, 308, 323, 327
Uhlherr, P. H. T. 10, 17, 20, 25, 26, 33, 41, 52, 57–59, 62, 63, 64, 65, 66, 67, 68, 70, 75, 76, 91, 92, 98, 107, 109, 111, 113, 114, 115, 116, 117, 118, 122, 123, 125, 126, 129, 130, 143, 144, 146, 153, 154, 157, 160, 162, 233, 282, 312, 351, 358, 359, 475, 476, 479, 480, 481, 483, 484, 485, 489, 492, 493, 494, 511, 513, 514, 517, 518, 519, 520, 521, 528, 529
Ui, T. J. 87, 499, 526, 527
Ulaganathan, N. 351
Ulbrecht, J. 56–61, 65, 66, 67, 72, 73, 75, 91, 143, 154, 156, 180, 181, 182, 185, 191, 193, 204, 205, 206, 210, 217, 219, 223, 225, 227, 228, 231, 233, 279, 296, 300, 302, 334, 335, 337, 338, 343, 352, 354, 355, 356, 358, 385, 387, 394, 397, 401, 405, 406, 409, 410, 411, 426, 427, 428, 429, 482, 484, 510, 523
Ulbrichova, I. 72, 75, 109, 128, 343, 348
Ultman, J. S. 5, 83, 142, 148, 153, 154, 164, 396
Um, M. K. 295
Unnikrishnan Nair, V. R. 357, 359
Unnikrishnan, A. 91, 101, 357, 359, 499

Unny, T. E. 366
Uno, S. 505
Unsal, E. 277, 286, 320
Upadhyay, S. N. 175, 274, 280, 297, 312, 318, 327, 335, 337, 338, 369, 374, 376, 377, 378, 381, 391, 392, 393, 394, 406, 407, 409, 411
Urciuolo, M. 239
Urseanu, M. I. 229, 505
Usera, G. 155, 168, 170
Usman, K. 132
Usui, H. 3, 35, 162, 366, 369, 376, 377, 378, 397
Utracki, L. A. 234

Vaezi G. F. 88
Vafai, K. 240
Vaidya, A. 66
Valente, J. S. 175
Valentik, L. 108, 109, 113, 123, 124, 125, 128, 129
Valvatne, P. H. 259, 299
Valverde, J. M. 18, 19
Vamerzani, B. Z. 197, 223
Van Atta, C. W. 368, 373, 394
van Baten, J. M. 216, 229, **505**
Van Bokhorst, E. 283, 315
Van Brakel, J. 241
van Buren, V. 242
van de Ven, T. G. M. 239, 320, 482
van de Vosse, J. F. 233
Van den Akker, H. E. A. 51
van den Bogaard, H. A. 215
van den Brule, B. H. A. A. 15, 39, 130, 144, 155, 160, 161, 163, 166, 168, 170, 171
van Den Ende, D.,234
van der Merwe, D. F. 244
van der Westhuizen, J. 259
van der Zee, S. E. A. T. M. 242
Van Dyke, M. D. 49
Van Hecke, M. 22
van Heel, A. P. G. 161, 166
van Kesteren, W. G. 105
van Krevelen, D. W. 180
Van Os, R. G. M. 167
van Pollen, H. K. 304
van Santvoort, J. F. M. 307, 315
van Schijndel, S. J. G. 196, 226
van Sint Annaland, M. 196, 226
van Vliet, B. M. 88, 264
van Wazer, J. R. 41, 510
van Wijngaarden, L. 206, 235
van Wunnik, J. 284, 286, 305
Vanamala, U. M. 426
Vanderby, Jr. R. 240, 266, 312, 313, 315
Vandu, C. O. 175
Varanasi, P. P. 234
Varchanis, S. 27
Varela, S. 155, 168, 170
Vargas, A. S. 86
Varley, J. 195
Varma, Y. B. G. 326
Varshney, A. 165
Vasil'chenko, S. V. 198, 228

Vasilic, K. 312, 314
Vasukiran, M. 80
Vatani, A. 265
Vazquez-Quesada, A. 112
Velan, M. 351
Velazquez-Navarro, J. F. 23
Veldhuis, C. H. J. 85
Velez, C. 4
Velez-Cordero, J. R. 182, 200, 206, 227, 229, 233
Venerus, D. C. 186, 187
Vengadesan, S. 444
Venkataraman, P. 265
Venkatesh, M. 378
Venkateswarlu, D. 192, 217, 225
Veno, K. 369, 376, 377, 378
Venumadhav, G. 88, 91, 101, 499
Verboven P. 239, 263
Verhelst, J. M. 166
Verma, A. 90
Verma, N. 499
Verma, P. D. 142, 148, 513
Verma, R. L. 395
Vermant, J. 145, 163, 168, 169, 358
Vernet, A. 155, 168, 170
Verrelli, D. I. 91
Vickers, G. T. 88, 101
Vidal, V. 175
Vidyasagar, T. 390, 473, 483, 484, 485, 500
Vieira, S. M. J. 45
Vignes, A. **505**
Vignes-Adler, M. 175
Vijaya Lakshmi, A. C. 351
Vijaysri, M. 274, 312, 313, 314, 315, 318
Vikhansky, A. 430, 441
Villimpoc, V. 443
Virto, L. 279
Vishnevskiy, V. K. 433
Viswanadham, R. 268
Viswanathan, K. 194, 206, 210, 225, 230
Viturro, R. E. 18, 19
Vlassopoulos, D. 3, 22, 244, 284, 288, 299, 320
Vogelpohl, A. 181, 182, 225
Voigt, J. 175, 231
Volarovich, M. P. 108, 113
von der Lieth, J. 175
von Garnier, A. 242
von Seckendorff, J. 241, 242, 246, 248, 249, 262, 264, 265
Vortmeyer, D. 262
Vorwerk, J. 278, 281, 282, 286, 304, 305, 307, 308, 309, 324, 350
Vossers, G. 206, 235
Vossoughi, S. 295, 306, 307, 304
Vradis, G. C. 295, 299
Vryzas, Z. 19
Vyaz'min, A. V. 363

Wacholder, E. 505
Wachs, A. 91, 105, 119, 120, 430, 441, 485, 493
Wagner, M. G. 192, 206, 222
Wagner, N. J. 22, 26, 27, 28, 39, 42, 82
Waheed, A. 474

Wahyudi, I. 251
Wairegi, T. 190
Waisbord, N. 299, 302, 315
Waitukaitis, S. R. 22
Wajc, S. J. 262
Wakabayashi, H. 1
Wakiya, S. 473
Waldram, S. P. 242, 252, 257
Walitza, E. 3
Walkama, D. M. 299, 302, 315
Walker, L. M. 185
Wallbaum, U. 31
Wallick, G. C. 57, 59, 62, 83
Walrond, K. W. 276, 309
Walters, H. G. 45
Walters, K. 1, 2, 16, 18, 29, 33, 41, 42, 56, 61, 62, 78, 141,
    142, 145, 146, 147, 148, 149, 151, 153, 155, 156,
    157, 158, 160, 166, 170, 295, 309, 319, 394, 496,
    497, 499, 521, 523
Wampler, F. C. 278
Wan, B. 107
Wan, Z. 105, 109, 111
Wanat, R. L. 394
Wanchoo, R. K. 149, 153, 155, 191, 195, 206, 225, 228,
    284, 304, 306
Wang Q. 242, 244, 245, 312
Wang, C. 405
Wang, C. Y. 87, 269, 273, 498, 500
Wang, F. H. L. 277, 286, 320
Wang, J. 91, 170, 198, 226
Wang, K. K. 516
Wang, L. 22, 91, 101, 166, 188, 215, 358
Wang, N. 198, 226
Wang, P. 82, 141, 169, 435
Wang, Q. 242, 244, 261, 263
Wang, S. 172
Wang, S.-Q. 3
Wang, T.-Y. 376, 394, 422, 424, 433, 444, 445, 446,
    463, 464
Wang, W. , 411
Wang, W.-H. 181
Wang, X. 200, 233, 266
Wang, Y. 405
Wang, Z. 146, 172, 198, 201, 221, 226, 262, 264
Wang, Z. J. 146, 443
Wapperom, P. 165, 398
Warburton, F. L. 269
Warichet, V. 144, 150
Warnica, D. 216
Warren, B. C. H. 185
Warren, P. B. 244
Warshay, M. 191, 192, 203, 208, 225
Warson, H. 27
Wasan, D. T. 273
Waslo, S. 229
Wasserman, M. L. 56, 57, 59, 60, 62, 67, 385, 518
Watamura, K. 1, 146, 161, 221
Watamura, T. 1
Watanabe, K. 162

Waters, N. D. 142, 160
Watson, E. L. 3
Watson, J. S. 354
Weaire, D. 175
Weber, M. E. 50, 51, 54, 72, 86, 87, 88, 95, 160, 185, 188,
    190, 214, 226, 425, 471, 472, 473, 496, 502,
    503, 505
Webster, M. F. 144, 145, 149, 151, 165, 166
Wegener, M. 188, 200, 215
Wegner, T. H. 244, 245
Wehbeh, E. G. 497, 526, 527
Wei, W. 199, 400
Wei, Y. 27, 28
Weidman, P. D. 87, 90, 155, 172, 368, 394
Weinberg, R. F. 175
Weisenborn, A. J. 43, 49
Weissenberg, K. 30
Weitzenbock, J. R. 269
Wellek, R. M. 401, 402, 403
Wells, Jr. C. S. 368, 373
Wen, C. Y. 282, 283, 298, 312, 320, 335, 336, 337, 338, 350
Wenyuan, F. 196, 225, 226
Wesfreid, J. E. 91
Wesselingh, J. A. 215
West, T. J. 239, 279, 324
Westerberg, K. W. 385
Whalley, D. C. 141
Wham, R. M. **473, 474, 478, 504, 505**
Wheat, J. A. 268
Wheaton, F. W. 368, 377
Whitaker, S. 196, 241, 259, 308, 390
White, A. 161, 162, 282, 304, 395
White, D. A. 161
White, D. B. 175, **506**
White, J. L. 2, 40, 394, 395
White, J. W. 423, 466
White, M. L. 268
White, R. R. 326
Whitelaw, D. S. 185
Whitelaw, J. H. 185
Whiting, P. 358
Whitlock, M. 22
Whitmore, R. L. 107, 108, 109, 111, 113, 123, 124, 125,
    128, 129, 138, 368, 434, 475, 476, 478, 479, 481
Whitney, M. J. 56, 58, 61, 62, 79, 91, 92
Whorlow, R. W. 41, 169
Wiest, J. M. 39
Wiggins, E. J. 268, 269
Wild, G. 326
Wilkes, E. D. 186
Wilkes, J. O. 185
Wilkinson, D. 255
Wilkinson, W. L. 191, 192, 235, 276, 286, 296, 297, 298,
    335, 337, 338, 339, 342, 348, 350
Wilks, G. 435
Willets, W. R. 156
Willhite, G. P. 277, 304, 309, 321, 322
Williams, D. 83
Williams, D. J. 141

Author Index

Williams, E. M. 88
Williams, J. G. 266
Williams, M. C. 513, 515, 516
Williams, M. J. 443
Williams, P. R. 33, 175, 187
Williams, R. W. 157
Williamson, C. H. K. 85
Williamson, R. S. 72, 75
Wilsey, E. F. 248
Willson, C. S. 253
Wilson, A. S. 411
Wilson, C. A. M. E. 358
Wilson, K. 27
Wilson, K. C. 2, 73, 78, 128, 129, 170, 233
Wilson, P. S. 4
Wilson, P. A. 269
Windhab, E. J. 185
Wineman, A. S. 395
Winkler, B. 510
Winterberg, M. 263
Wise, G..239
Wissler, E. H. 282, 306
Withers, P. M. 83
Witherspoon, P. A. 296
Witten, T. A. 14
Wittler, R. 231
Wodie, J.-C. 246
Wolf, C. J. 368, 373, 376
Wollersheim, D. E. 421, 423, 426, 437, 438
Won, D. 168, 170
Wong, K. L. 461
Wong, P. 91, 101
Woods, J. K. 312, 314, 315
Woudberg, S. 259
Wreath, D. 283, 286, 308, 311, 320
Wronski, S. 406, 407
Wu, B. Y. 166, 396
Wu, C. C. 166, 396
Wu, D. 109
Wu, D. C. 278, 300
Wu, G. H. 166, 396, 397
Wu, J. 162, 242, 244, 261, 263, 298, 372
Wu, S. 84, 146
Wu, Y.-S. 259, 296, 304, 321, 327
Wubulikasimu, Y. 19
Wunderlich, A. 244, 307
Wunsch, O. 109, 200, 520
Wyart, M. 23
Wyllie, M. R. J. 265

Xiaoli, G. 306
Xie, C. 315, 319
Xiong, J. 161
Xiong, Y.-L. 141, 145, 166, 170
Xu, C.-X. 91
Xu, D. 23
Xu, F. 199, 206, 400
Xu, L. 84, 172, 285, 311
Xu, M. 109
Xu, Q. 474, 499

Xu, S.-M. 84, 285, 311
Xu, X. 199, 400
Xu, X. -Y. 160, 177
Xu, Y. 22, 358
Xu, Z. 50, 51, 146, 200
Xue, L. 240, 241, 242, 266

Yabe, A. 215
Yadav, S. 91
Yaginuma, T. 91, 101
Yahia, A. 3
Yala, N. 186, 187
Yamada, K. 3
Yamaguchi, S. 195
Yamamoto, H. 527
Yamamoto, K. 71, 79
Yamamoto, T. 175
Yamamoto, Y. 1
Yamanaka, A. 57, 65, 67, 70, 153, 154, 192, 193, 227, 391, 393, 423, 463
Yamanaka, S. 194, 204, 206, 208, 225
Yamashita, S. 215, 251
Yamayoshi, W. 268
Yan, X. 188, 215
Yang, B. 144, 151
Yang, C. 51, 97, 199, 201, 226
Yang, D. 141, 145, 166
Yang, F.-K. 3
Yang, G. 352
Yang, H. 199, 233
Yang, H. T. 281, 304
Yang, J. 3, 242, 244, 245, 261, 263, 312
Yang, J. H. 272
Yang, K. 225
Yang, L. M. 91, 101
Yang, M. 90, **499**
Yang, S. 449
Yang, S.-M. 235
Yang, W.-C. 331, 332, 352
Yang, W.-J. 187, 193, 206
Yang, X. 198, 226
Yao, C. 273, 411
Yao, J. 191, 202, 211
Yao, L.-S. 429, 434, 444
Yap, S. D. 3
Yaparpalvi, R. 185
Yaremko, Z. M. 90
Yarmola, E. G. 4
Yaron, I. 3
Yarusso, B. J. 18, 19
Yashchenko, A. 239, 276
Yasser Ibrahim, A. A. 351
Yasuda, K. 16
Yates, W. E. 185
Ybert, C. 215
Ye, X. 4
Yeh, F. 516
Yeh, H.-C. 425, 434
Yelshin, A. I. 242
Yen, H. H. 444

Yen, W. S. 320, 324
Yentov, V. M. 283
Yeow, A. Y. 312, 314, 315, 317
Yeow, Y. L. 202, 206
Yerushalmi, J. 185
Yildirim, O. E. 233
Yildiz, S. 424
Yim, J. 282, 320
Yip, R. 167, 244, 284, 287, 288, 304, 307, 314, 315, 319, 320
Yliruusi, J. 3
Yokouchi, Y. 244
Yoo, H. J. 187,
Yortsos, Y. C. 259, 287
Yoshida, Y. 191, 195, 202, 203, 206, 217, 222, 223, 228
Yoshimura, A. S. 41
Yoshioka, N. 59, 56, 57, 59, 60, 66, 67, 68, 71, 79, 107, 108, 113, 118, 131, 132, 133, 135
Yotsumoto, Y. 1
You, C. 91
You, R. 193, 223
Youguang, M. 196, 225, 226
Younes, E. 107, 128
Young, B. D. 88, 264
Young, M. J. 262, 264
Yovanovich, M. M.,366, 376, 448, 449
Yow, H. N. 88
Yu, A. B. 264
Yu, B. 298
Yu, C. P. 273
Yu, H. S. 186
Yu, K. L. 234
Yu, M. 485, 493
Yu, P. 166, 170
Yu, S.-C. 51
Yu, Y. H. 283, 298, 312, 335, 336, 337, 338
Yu, Z. 82, 141, 169, 485, 493
Yuan, H. 449
Yue, P. 180, 188, 199, 200, 203, 227, 233
Yue, X. 3
Yuki, T. 153, 154, 227, 391, 393, 423, 463
Yun, M. 298
Yurun, F. 144, 150
Yurusoy, M. 395

Zagoskin, S. N. 222
Zagoskina, N. V. 222
Zahed, A. H. 169
Zaidi, A. 352, 410, 411
Zaitsev, V. F. 186
Zakhem, R. 90
Zaki, W. N. 335, 341, 342, 357, 409
Zaltsgendler, E. A. 422, 429, 430
Zamani, N. 239
Zamankhan, P. 199, 228
Zami-Pierre, F. 284, 309
Zana, E. 148, 154, 168, 169, 187, 193, 202, 209, 211, 234, 400
Zandbergen, P. J. 234

Zapas, L. 39
Zapparov, F. I. 425
Zapryanov, Z. 49, 54
Zarei, M. 22
Zareifard, M. R. 377
Zarraa, M. A. 408
Zdravkovich, M. M. 51
Zenit, R. 1, 82, 141, 145, 149, 151, 153, 155, 158, 159, 163, 182, 188, 195, 200, 202, 206, 207, 210, 226, 227, 229, 233
Zeppenfeld, R. 107, 111
Zerai, B. 253
Zhadanovich, N. V. 369, 374, 376, 377
Zhang, C. -H. 107, 110, 111, 121, 123, 132, 135
Zhang, C. S. 145, 153, 161, 172
Zhang, G. 176, 186, 485, 493
Zhang, H. 22, 449
Zhang, J. 231, 200
Zhang, J.-D. 91
Zhang, J.-J. 121
Zhang, J.-P. 352
Zhang, K. 72, 78
Zhang, L.,199, 226
Zhang, M. 285
Zhang, N. 242, 261, 263
Zhang, S. 3
Zhang, X. 184, 221, 358
Zhang, X. X. 375
Zhang, X. Z. 23
Zhang, Y. 161, 216, 358
Zhang, Z. 22, 84, 285, 311
Zhao, B. 463
Zhao, C. 188
Zhao, C.-J. 449
Zhao, J. 285
Zhao, X. 121, 201, 221
Zheng, G.-H. 482
Zheng, K. 188, 215
Zheng, L. C. 375
Zheng, R. 90, 143, 150, 153, 160, 258, 307, 527, 528
Zhizhin, G. V. 368, 373
Zholkovskiy, E. K. 255
Zhong, H. 91
Zhong, H.-J. 91
Zhong, W. H. 315, 319
Zhou, C. 188, 199, 203
Zhou, J. Z. Q. 288, 295, 304, 307, 312
Zhou, L. 242, 261, 263
Zhou, Y. 188
Zhu, C. 181, 199, 200, 233, 358
Zhu, J. 220, 283, 300, 301, 352, 405
Zhu, K.-Q. 121
Zhu, L. 41, 42, 128, 521
Zhu, P. 225
Zhu, Q. 358
Zhu, S. 22
Zhu, Z. 50, 51
Zick, A. A. 255
Ziegenhagen, A. J. 57, 59, 60

Zierenberg, J. R. 384
Zijm, W. H. M. 1, 175
Zimm, B. H. 278, 309, 322
Zimmels, Y. 299
Zimmerman, R. W. 239, 258
Zimmerman, W. B. 497, 498
Ziolkowska, I. 263
Ziolkowski, D. 263
Zisis, Th. 133, 134, 135
Ziskind, G. 463

Zitha, P. L. 304, 315, 319
Zitoun, K. B. 369
Zlokarnik, M. 40
Zonfrilli, F. 110
Zoric, J. 263
Zou, R. P. 264
Zukauskas, A. 412
Zun, I. 216
Zuritz, C. A. 67, 82, 83, 282

Subject Index

accelerating motion 82, 142, 160, 172, 476
Acrivos equation 451
adsorption of polymers 161, 276, 279, 281, 287, 309, 320
aggregation of particles 4, 5, 29, 82, 170, 358, 360
agriculture waste slurry 2, 3, 7, 185, 239, 331
aiding-buoyancy mixed convection 421, 422, 432, 434, 441, 443, 444, 446, 448, 461, 463, 465
alignment of particles 267
anisotropy 5, 141, 199, 253, 266, 274, 287, 295
anomalous effects in porous media 6, 286, 320, 338
apparent viscosity 2, 12, 14, 18, 21, 30, 55, 83, 311
    rheopexy 24, 26, 27
    thixotropy 21, 28, 109, 110, 112, 122, 132, 198, 283, 301
Archimedes number 50, 77, 190, 196, 221, 347
arrays of cylinders 244, 273, 307, 314, 316, 319
    parallel flow 269, 270
    transverse flow 268, 271, 313, 316
arrays of sphere 256, 275, 301, 307
aspect ratio of 90, 165, 188, 440, 498
    bubbles 188, 204
    cylinder 165, 166, 266, 440, 498
    spheroids 90
average shear rate for a sphere 75, 222, 226, 514, 518, 519, 523, 527, 530
axial dispersion in packed beds 278, 320, 335, 350
azimuthal velocity 62, 463

Basset term 54
bead-spring models 38
bed expansion behavior of 6, 331, 337
    three-phase fluidized beds 352, 355
    two-phase fluidized beds 331, 334, 337, 343, 351
benchmark problem 141, 147, 149, 164
Best number 50
Bingham model 19, 27, 110, 119, 129
Bingham number 115, 125, 131, 196, 221, 365, 382, 391, 403, 420, 430, 434, 441, 442, 457, 459, 461, 467, 472, 497, 502, 503, 506
Bingham plastic fluids 137, 185, 418, 420, 421, 422, 430, 431, 434, 440–443, 448, 457–461, 465, 467, 469, 472, 483, 485, 492, 494, 495, 502, 506, 523, 527
    bubbles in 197, 201, 220, 229
    cylinders in 133
    in porous media 299
    sphere in 118, 123, 448, 457, 465, 472, 483, 485, 492, 494, 495
Bingham yield stress 19
Blake model 248, 250, 293
Blake-Kozeny model 248, 250, 276, 289, 335
blood viscosity 510, 516
Boger fluids 11, 143, 149, 153, 156, 160, 166, 182, 233, 314, 350, 493, 494
bond number 190, 221

boundary conditions 46, 255
    at cell surface 255
    on stress 56
boundary element method 234
boundary layer flow 165, 214, 302, 424, 425, 429, 434, 438, 439, 446, 448, 451, 461, 467
    over a cylinder 365, 375, 396, 434, 435, 438, 439
    laminar 363, 422
    over a plate 365, 370, 371, 373, 374, 424, 425
    of power-law fluids 363, 365, 376, 425
    over a sphere 62, 150, 161, 365, 385, 446, 451, 461
    thickness of 214, 302
    visco-elastic 394, 466
    visco-plastic 390, 434
boundary wall effects 6, 54, 57, 66, 87, 93, 107, 143, 154, 189, 244, 260, 311, 471
Boussinesq-Basset expression 54
breakage of drops 175, 234
break-up of jets and sheet 185
Brinkman equation 254, 260
Brinkman number 385, 395, 413, 418, 421, 467
bubble chains 182, 195
bubble columns 175, 363, 403
bubble ensembles 5, 176, 217, 220, 228, 230, 403
bubble swarms 4, 176, 217, 220, 228, 230, 403
bubbles
    in boiling polymer solutions 394
    coalescence of 230
    collapse of 176, 180, 400
    in confined flows 235
    creeping motion of 219, 401
    critical radius of 206, 210
    deformation of 194, 228, 235
    dilational characteristics of dispersions 230
    final volume of 178
    formation of 178
    free motion of 188
    growth of 186
    hydrodynamics of 206
    in low gravity fields 175
    mass transfer to 186, 192, 398
    in microgravity 175
    in Newtonian fluids, shapes of 188
    in non-Newtonian fluids, shapes of 29, 191
    non-spherical 188
    oscillating 221
    rupturing of 410
    swarms of 176, 217, 220, 228, 230, 403
    terminal velocity-volume behavior of 206
bulk porosity of beds 241, 260, 264, 269, 283, 333
bulk rheological characteristics of 14, 83, 112, 286, 321
    polymer solutions 14, 83, 112, 286, 321
Burgers visco-elastic fluid model 37
Burke-Plummer equation 249

701

capillary bundle model 246, 249, 257, 259
capillary models 246, 249, 257, 259, 308
    deficiencies of 252, 253, 287
    of fluidized beds 334, 336, 343, 347
    of laminar flow 289, 308
    of porous media 246, 249, 257, 297, 308, 323, 405
    of pressure loss 246, 249, 289, 297, 308, 323
capillary number 209, 211, 223, 234
capillary viscometry 509, 517
Carman-Kozeny equation 248
Carreau model fluids 16, 57, 58, 67, 93, 97, 143, 483, 499
    bubbles in 194, 201, 210, 227, 403
    drops in 194, 401
    fluidization with 276, 335
    mass transfer in 376, 403
    packed bed flows of 279, 282, 286, 295
    porous media flow of 279, 282, 286, 295
    sphere in 57, 58, 67, 76, 499
Carreau number 64, 65, 68, 230, 493
Carreau viscosity equation 16, 68, 217, 284, 295, 314, 401
Carreau-Yasuda equation 198
Casson model 19, 109, 118, 430
Casson yield stress 20, 129
Caswell method 511, 513
Caswell-Schwarz method 512
cavitation 175
cell models 228, 229, 255, 257, 271
    cylindrical 255, 257, 271
    free surface 228, 256, 257, 299, 300, 302, 313, 343, 358, 403
    zero vorticity 229, 256, 257, 273, 300, 302, 314, 315, 316
chains of bubbles 182
chains of spheres 90, 91, 98, 170, 499
channeling in beds 249, 260, 261, 267, 281, 332
characteristic time of fluids 39, 40, 156, 227, 310, 521
clay suspensions 22, 26, 67, 108, 117, 131, 138, 161, 368
coalescence of bubbles 175, 199, 230, 233
coil-stretch transition 311
Colburn factor 378, 381, 383, 391
coleman-noll model 39, 142, 192
collapse of bubbles 176, 186, 400
composite shape factor 99
concrete rheology 3, 28, 29, 41
conductance of porous media 242, 259
conduit models 246, 249, 257, 259
confined flows 235, 363
confining walls 160, 188, 235, 242, 248, 260, 263, 376, 471, 472, 495, 500, 502
    effect on drag of non-spherical particles 496
    effect on fluid sphere drag 502
    effect on sphere drag 472, 491, 493, 495
    effect on sphere heat transfer 500
    packed bed flows 260, 311
conglomerates of spheres 91, 499, 500
consolidated porous medium 260, 275
constant flow conditions 180, 183, 282, 320
constant pressure conditions 181, 282, 320
constant relaxation time 494
constant shear viscosity 156, 166, 308

constant viscosity 10, 12, 156, 233, 511
continuity equation 46, 55, 186, 370, 398, 418
continuous phase 175, 183, 191, 204, 214, 216, 222, 401, 503, 505
convection
    forced 363, 366, 370, 375, 384
    free 417, 421, 423, 430, 433, 448
    mixed 418, 419, 421, 422, 423, 430, 432, 433, 434, 443, 444, 446, 448, 461, 463, 465
    natural 417, 423, 430, 431, 434, 452
converging-diverging flow 253, 283
cooking oil 8
copper spheres 451, 452
corn syrup 8, 9, 156, 160
cosmetics 3, 7, 175
countercurrent gas-liquid flow 326
creeping flow 49, 54, 244, 271, 473
    approximation of 49, 54
    cessation of 84, 122, 475
creeping flow region 49, 54, 153, 244, 320
    drag coefficients 49, 54, 55, 78, 118, 123, 131, 153
    experimental results 66, 78, 123, 225
    mass transfer in 369, 384, 401, 406
    visco-elastic fluids in 146, 153, 319
creeping motion 208, 214, 271
    of bubbles 208, 214, 216, 225
    of drops 214, 401
    of spheres 49, 55, 78, 118, 153
critical Reynolds number
    cylinder 51, 78
    sphere 51, 78
cross viscosity equation 15, 16
cross viscosity model 15, 16
cross-buoyancy mixed convection 432
cusped bubbles 188, 191, 195, 203
Cygan and Caswell method 513
cylindrical boundaries wall effects 260, 311, 471

dairy waste systems 7
Darcian flow 244, 259
Darcy flow regime 244, 259
Darcy's equation 242
Davidson-Schuler model of bubble formation 177
Deborah number 40, 147, 211
    critical 223, 304, 306, 314, 319, 395, 397
    definition 40, 147, 211
    porous media and 288, 304, 306, 314, 395
deformation 504, 511, 520
    of bubble 194, 215, 228, 235
    of drops 197, 223, 233, 234, 235
    elastic 28, 112
    plastic 113
    rate of 4, 33
deformation rate tensor 47, 48, 55, 143
degradation, mechanical 161, 276, 295, 308, 319, 324
deviatoric normal stresses 10
die swell 29, 164, 306
diffusion 23, 130, 186, 240, 323, 397, 398, 423, 455, 500, 521
diffusive heat transfer 8, 382, 393, 424, 455, 500

Subject Index

dilatant fluids 12, 21, 83, 84, 186, 277, 298, 407, 438; *see also* shear-thickening fluids
dilute/semi-dilute drag reducing polymers 15, 22, 153, 161, 186, 287, 308
dimensional analysis 5, 39, 54, 63, 87, 126, 146, 185, 246
disintegration of jets and sheets 176, 185
dispersion 230, 257, 288, 320, 350
    axial 278, 281, 320, 335, 350
    gas-liquid 230, 231
    in packed beds 257, 278, 281, 288, 320, 335, 350
dissipation 84, 217, 247, 311, 370, 385, 418
diverging-converging flow 253, 283
drag coefficient; *see also* drag correction factor for
    bubbles 190, 206, 210, 213, 216, 220, 222
    for creeping motion of spheres 55, 90
    cylinder 73, 92, 99, 131, 164
    disk 91, 95, 97
    drops 191, 213, 216, 222
    ellipsoids 91
    form 49, 71 (*see also* pressure drag)
    free falling spheres 73, 77, 108, 146, 161
    friction 48, 71
    isometric particles 87, 90, 137, 138
    lower bound on 64
    multiparticle assemblages 253, 256, 279, 300
    in Newtonian fluids 47, 50, 87, 206, 213, 299
    non-spherical particles 87, 90, 91, 93, 137, 138
    numerical predictions of 55, 71, 118, 299, 300
    numerical values of 49, 55, 58, 61, 62, 71, 90, 118, 299, 300
    oblates 91
    of particle assemblages 253, 256, 279, 300
    in porous media 253
    pressure 48
    prolates 91
    rising light spheres 84
    rods 91
    Schiller-Naumann drag equation 76
    sphere 47, 49, 50, 55, 58, 61, 62, 90, 118, 146, 162
    Stokes 49, 118, 146
    upper bound on 63
    in visco-elastic fluids 146, 161, 164, 168, 222
drag correction factor of
    bubbles 206, 210
    a cylinder 73, 92, 99, 131, 164
    non-spherical particles 73, 92, 95, 137, 138
    a sphere 58, 61, 62, 65, 118, 146, 256
drag force in
    shear-thinning fluids 55, 90, 191, 213, 216, 279, 300, 312
    visco-elastic fluids 146, 168, 222, 319
    visco-plastic fluids 118, 131, 164, 221
drag reducing fluids 161, 308
drag reduction in external flows 141, 142, 161, 308
drops
    breakage of 234, 235
    coalescence of 230, 233
    in confined flows 235
    creeping motion of 213
    deformation of 233, 235
    ensembles of 228
    formation of 182
    free falling 209, 213, 225
    free motion of 213, 225, 228
    hydrodynamics of 215
    in Newtonian fluids, shapes of 188, 209
    in non-Newtonian fluids, shapes of 206, 222, 223, 224, 234
    in power-law fluids 216
    terminal velocity-volume behavior of, size of 206, 209
    in visco-elastic fluids, size of 222, 234, 235
    volume of, size of 182, 185
Dupuit equation 250
dynamic parameter 124, 126

eccentricity of bubbles 190, 204
Einstein's expression for viscosity 230
elastic deformation 28, 112
elastic forces 40, 156
elastic turbulence 41
elastic stress 161
elasticity number 40, 164, 395, 397
electro-rheological fluids 18, 283, 297, 299
ellipsoidal bubbles 146, 188, 191, 199, 215, 216, 472, 499
Ellis model
    bubbles and 187, 192, 210, 230
    packed bed and 280, 295, 299
    porous media and 280, 295, 299
    sphere settling in 57, 62, 70, 75
Ellis number 63, 70, 75
Ellis viscosity model 16, 63
elongation viscosity 32, 33, 37, 39, 186, 211, 278
elongational flow 32, 37, 165, 186, 211, 234, 277
emulsion viscosity 230, 234
Eotvos number 190, 196, 206, 211, 226
equal volume sphere diameter 99, 204, 221, 263, 393, 406, 504
equilibrium
    bubbles 110
    rigid particles 105, 110, 138
    in visco-plastic media 105, 110, 138, 521, 526, 527
Ergun equation 249, 251, 260, 265, 274
excess pressure drop 257, 278, 304, 313
excess pressure loss 257, 280, 304, 313
extensional viscosity 32, 33, 40, 186, 211, 234, 278
extrapolation methods 511, 513
extremum principles 56
Eyring model 15, 57, 186, 376

falling ball method 45, 54, 82, 476, 509, 516, 517
falling ball viscometry 45, 54, 82, 476, 509, 510
    Newtonian fluids 45, 54, 476, 509, 511
    shear dependent viscosity and 82, 511, 517
    visco-elastic fluids and 509, 521
    yield stress 523
    zero shear viscosity and 510, 511, 516
falling cylinder viscometer 524, 527
falling needle viscometer 526
fibrous beds 312
fibrous media 312

fixed beds 240, 263
flow
    boundary layer 363, 413
    confined 471
    diverging-converging 253, 258
    in fibrous beds 312
    in periodically constricted tubes 253, 258
    in porous media 246, 249, 257, 259
    over a cylinder 86, 91, 131, 161, 164
    over a disk 86, 91, 95, 137
    over a sphere 45, 107, 146, 161
    over spheroidal particles 91, 95, 138
flow curves 12
flow induced degradation polymers 161, 287, 295, 308, 319, 324
flow regimes in porous media 244
flow visualization 89, 115, 155, 193, 227, 288, 360
fluidization 331
    bed expansion behavior and 337, 352
    gas-liquid-solid 352
    incipient 332
    liquid-solid 331
    with Newtonian fluids 332
    with power-law fluids 334, 341
    with visco-elastic fluids 350
fluidization velocity 333
fluidized beds 331, 337
    expansion behavior of 337, 352
    heat transfer in 405, 409
    incipiently 333
    mass transfer in 405, 409
    three-phase 352
fluid-like zones 113, 117, 382, 457
forced convection 363
    cone 393
    cylinder 375
    disk 393
    hemisphere 393
    plate 370
    sphere 384
    spheroid 393
Forchheimer equation 260, 284, 298
form drag 48, 71; *see also* pressure drag
formation
    of bubbles 178
    of drops 233, 235
free convection 417
free fall terminal velocity 45, 50, 78
free falling drops 209, 213, 224
free falling spheres 73, 77, 118, 146, 161
free settling velocity 45, 50, 77
free surface cell model 228, 257, 299, 300, 313, 343, 358, 403
friction; *see also* drag
    in packed beds 289, 308, 323
    pressure loss-throughput relationship 246, 249, 295, 308, 323
frictional pressure drop 246, 249, 298, 308, 323
frictional pressure gradient 246, 249, 297, 308, 323
Froude number 40, 190, 200

Galileo number 50, 190, 333
gas hold-up 354
gas-liquid flow 352
gas-liquid-solid fluidization 352
gel formation 6, 276, 282, 309
generalized Newtonian fluid (GNF) 11, 56, 62, 186, 216, 288, 299, 304, 312
    bubbles in 176, 191, 206, 213
    spheres in 55, 107
    non-spherical particles in 86, 90, 131, 138
Giesekus fluid model 144, 147, 197, 200, 313, 397
Giesekus method 510
Grashof number 419, 420, 425, 430, 434, 438, 440, 444, 448, 449, 450, 455, 456, 457, 458, 461, 467
gravity-yield parameter 107
growth of bubbles 186

Hadamard-Rybczynski solution 214, 223
Hagen-Poiseuille equation 250, 253, 258
Happel's cell model 228, 255, 313, 343, 358, 405
heat transfer 378, 382
    in bubble 398, 401, 403
    convection 378
    free convection 417
    in drop 401, 403
    in fluidized beds 409
    mixed convection 417
    in packed beds 405, 409
    in porous media 393, 409
    in power-law fluids 384, 405, 409
    in three phase fluidized beds 405, 410
    in tube bundles 411
Hedstrom number 122, 430
Herschel-Bulkley model 21, 28, 108, 118, 122, 196, 221, 228
    porous media and 276, 286, 297, 314, 317
hexagonal array 270, 274, 315, 317
hindered settling 357; *see also* sedimentation
hold-up in two-phase flow 354
hydraulic diameter 412
hydraulic radius 250

incipient fluidization 331, 333, 336
incipiently fluidized beds 331, 333, 336
inertial flow regime 245, 249, 259
inflation of fluid cavity 186
inhomogeneity in
    hindered settling 358, 360
    fluidization 350
in situ rheological characteristics 282, 315, 320
integral rheological models 37
interlocking of particles 333
interstitial space in porous medium 240, 249
interstitial velocity 250
inviscid flow 12, 177, 179, 180, 247
isotropic porous media 240, 246, 247, 250, 260, 266, 284, 293
isotropic pressure 10

kaolin slurries 14, 109, 279, 286
K-BKZ fluid model 143, 144

Kelvin-Voigt model 37, 38
Kozeny constant 246, 268, 269, 273, 293
Kozeny-Carman equation 250, 253, 280, 289, 293, 296
Kumar-Kuloor model of
    bubble formation 178
    drop formation 182

laminar boundary layers 165, 214, 364
laminar flow 246, 249, 257, 259
    capillary model of 246, 249, 257, 259
    porous media and 246, 249, 257, 259
linear flow curves 8, 17
Lockhart-Martinelli parameter 326
low molecular weight liquids 8
lubrication flow approximation 273, 304

Mach number 173
magneto-rheology 18, 111, 510
magneto-sphere viscometer 510
Marangoni number 211, 401
mass transfer
    in bubbles 398, 403
    convective 363
    cylinder 375, 434
    in drops 401, 403
    enhancement in 403, 500
    in fluidized beds 409
    forced convection, from spheres 384
    free convection, from spheres 448
    gas-liquid 4, 201, 231, 398, 400, 410
    mixed convection, from spheres 461
    in packed beds 405, 409
    particle-liquid 374, 384
    in porous media 393
    in power-law fluids 374
    rates of 363
    from solid surfaces 374
    from sphere ensembles 403, 406, 410
    in three-phase fluidized beds 410
    in tube bundles 411
    in visco-elastic fluids 394
    visco-elasticity and 394
mathematical models of
    pseudoplastic behavior 14
    thixotropic behavior 28
    visco-elastic behavior 37
    visco-plastic behavior 19
Maxwell model 142, 164, 172, 210, 223, 319, 397, 400
    porous media and 281
Maxwellian relaxation time 39, 210, 211
meat extract rheology 7, 17
memory effects 29, 55, 235, 258
meter model 283, 285, 295
migration of particles 5, 29, 82, 131, 141, 168, 169, 235, 358, 401
minimum fluidization velocity for
    two-phase systems 332, 334, 350, 351
    three-phase systems 352
mixed convection heat transfer 430, 443, 461
mixing in packed beds 287, 320, 410
mobility in porous media 242, 277, 286, 314, 316

molecular network theories 16
momentum equations 46, 49, 56, 217, 415
morton number 190, 204, 211, 237
multi-phase mixtures 2, 4, 7, 41, 176, 363, 471
multi-particle interactions in
    visco-elastic fluids 319, 350, 359
    visco-plastic fluids 299, 316, 360

natural convection 417
Navier-Stokes equations 10, 49, 253, 259, 473
negative thixotropy 26
negative wake 29, 145, 152, 165, 173, 202, 210, 227, 236
Newtonian fluids
    bubbles in 188, 214
    definition of 8, 10
    drops falling in 215
    falling ball viscometry of 45, 54, 476, 509
    generalized 11, 56, 62, 186, 216, 288, 299, 304, 312
    in porous media 244
    sedimentation in 357
    settling in 48, 87
Newtonian viscosity 7
Newton's law of viscosity 8, 10
nonlinear flow curves 12
non-Newtonian fluids 11
    characterization of 11, 12
    definition 11, 12
    types of 11, 12
non-shear-thinning visco-elastic fluids 11, 149, 153, 156, 159, 166, 233, 319, 493, 496
non-spherical particle shapes 86, 102, 138, 264, 294
normal stress
    continuity of 213
    deviatoric 10
    primary 30, 31
    secondary 30, 31
    in steady shearing flows 29, 31
    unequal 30
no-shear condition 209, 212, 213
no-slip boundary condition 46, 118, 136, 206, 371
Nusselt number for a 371, 373, 375, 379, 388, 412, 424, 432, 436, 439, 449, 452, 466
    cylinder in Bingham fluids 382, 440
    cylinder in power-law fluids 375, 436
    plate in Bingham fluids 429
    plate in power-law fluids 370, 425
    sphere in Bingham fluids 390, 448
    sphere in power-law fluids 384, 444

Oldroyd fluid model 142, 147
    falling ball viscometry and 147, 509
    heat transfer and 395
    mass transfer and 400, 401
opposing-buoyancy mixed convection 432
oscillating bubble 221
oscillating sphere 155, 520, 524
oscillatory behaviour of settling spheres 145, 153, 161, 171
oscillatory measurements 34
oscillatory wake 154, 202
Oseen drag expression 49
overshoot in velocity 144, 160

packed beds 239
    Blake-Kozeny model of 248, 289
    dilute/semi dilute drag reducing polymers in 308
    dispersion in 320
    fibrous systems 266, 312
    friction in 246, 288
    heat transfer in 405
    mass transfer in 405
    miscellaneous effects 320
    mixing in 320
    particle shape effects 264, 312
    rheological parameters from flow in 320
    slip effects 323
    visco-elastic fluids in 304, 319
    visco-plastic fluids in 295, 317
    wall effects 260, 311
passage time distribution of particles 83
Peclet number 23, 186, 320, 382, 385, 398, 400, 402, 412
penetrometer for yield-stress measurement 528
periodically constricted tubes (PCT), flow in 257
permeability 6
    of fibrous beds 269, 275
    increase in 267
    of packed beds 242, 246, 261
    of porous media 6, 242, 246
    reduction in 267, 273, 277, 279, 308
Phan-Thien-Tanner fluid model 143, 155, 165, 200, 283, 315, 395
PIV measurements 51, 110, 132, 152, 155, 167, 195, 210, 227, 232, 244, 307, 319
plastic deformation 113
plastic viscosity 17, 129, 382
polymers
    adsorption of molecules of 321
    dilute/semi dilute drag reducing 161, 308
    extensional viscosity of 32, 33, 37, 165, 186, 211, 278
    in porous media 288, 308, 311, 320
    retention of 321
    solvent interactions with viscoelastic 149, 159
pore friction factor 252
pore Reynolds number 252
pores
    blind (dead end) 242
    blockage of 278, 309, 322
    geometry of 242
    non-uniformities in 242
    plugging of 278, 309, 322
    size distribution of 240, 242
    volume of 241
porosity
    bed 241
    bulk 241
    fluidization and 334
    local 242
    mean 241
    radial distribution 242
porous media 240
    anomalous effects in 320
    capillary model of 248
    classification of 240
    compressible 239, 264

conductance of 242, 259
consolidated 240
Darcy flow regime 244
description of 240, 241
dispersion in 320
drag in 246, 253, 288, 299
fibrous 266, 312
field equations for flow through 257, 259, 303
granular 240
heat transfer in 405
homogeneous 240, 250, 267, 288, 293, 303, 327
increase in permeability in 267
inertial flow regime 274
mass transfer in 405
Newtonian fluids in 246, 264
non-Newtonian fluids in 288, 312, 317, 319
permeability 241
visco-elastic fluids in 304, 319
visco-plastic fluids in 287, 289, 316
Powell-Eyring model 15, 57, 186, 376
power-law fluids
    boundary layer flow of 363, 375, 384, 398, 417, 425, 433, 444
    bubbles in 184, 191, 209, 216, 225, 228
    convection in 363, 375, 384, 398
    creeping sphere motion in 58
    Darcy's law for 303
    deficiencies of 14, 59
    drops in 184, 191, 216, 228
    falling ball viscometry 509
    in fixed beds 289, 312
    fluidization with 334, 341, 358
    heat transfer in 363, 375, 384, 398, 418, 425, 433, 444
    inadequacy of 14, 59
    liquid jets in 185
    mass transfer in 363, 375, 384, 398, 418, 425, 433, 444
    non-spherical particles in 86, 90
    in porous media 288, 312, 317, 319
    sedimentation in 358
    spherical particles in 58, 66, 71, 77, 83, 84
Prandtl number 366, 372, 395, 405, 409, 412, 419, 420, 421, 424, 425, 426, 427, 430, 432, 433, 434, 435, 436, 437, 438, 440, 444, 445, 446, 447, 448, 451, 452, 453, 455, 456, 457, 461, 463, 464, 465, 468, 500, 502, 507
pressure distribution on the surface 79, 196
pressure drag 48, 71
pressure drag coefficient 48, 71
pressure drop
    due to a settling particle 85
    excess 257, 278, 304, 313
    fluidization and 331, 337
    frictional 288, 312, 317, 319
    particle shape and 264, 312
    two-phase flow 326
    wall effects on 260, 311
pressure loss
    capillary model of 248
    excess 257, 278, 304, 313
    for generalized Newtonian fluids 288
    particle shape and 264, 312

Subject Index

pressure loss-throughput relationship 246
    capillary models 249
    conduit models of 249
    drag theories and 253, 288, 299
    empirical correlations for 247
    field equations for 257, 259, 303
    submerged object models of 246, 253, 288, 299
primary normal stress difference 30, 31
pseudoplastic fluids
    drops in 184, 191, 216, 228
    falling ball viscometry and 510
    heat transfer in 363, 375, 384, 398, 418, 425, 433, 444
    mass transfer in 363, 375, 384, 398, 418, 425, 433, 444
    mathematical models of 14
    wall effects in 311, 482
pulsating flow
    over a cylinder 384
    over a sphere 393
purely viscous fluids
    flow field with yield stress 112
    static equilibrium with yield stress 107
    wall effects with yield stress 491
    wall effects without yield stress 482
    with yield stress 105
    without yield stress 45
    yield stress values with yield stress 128

Rabinowitsch-Mooney equation 277, 280, 292
radial porosity distribution in a bed 260, 262
Rayleigh-Taylor instability 182
Reiner-Rivlin model 57, 58, 91, 194, 200, 224
relaxation times 111, 120, 127, 156, 396, 521
    characteristic 37, 185, 234, 396, 521
    Maxwellian 39, 172, 210, 396
Reynolds number for Ellis model fluids 63, 291
Reynolds number for power-law fluids 59, 218, 290, 291, 298, 318
rheograms 8, 311
rheological behaviour of 1
    agro-slurries 3
    cosmetics 3
    dairy waste slurries 3
    egg albumen 3
    muds 3, 26
    salvia 3
    semen 3
    toiletries 3
rheological equation of state 5, 11, 39, 42, 47, 149, 154, 510
rheological models 5, 11, 39, 42, 47, 149, 154, 172, 186, 510
rheometry 41, 509
rheopectic behaviour 24, 26
rheopexy 24, 26
Richardson number 417, 419, 420, 425, 430, 433, 434, 443, 445, 446, 448, 463, 464, 465, 468
rising spheres 84
rod bundles, flow and heat transfer 266, 312, 411
rod climbing effect 29
rolling ball viscometry for 509, 521
    Newtonian fluids 521
    non-Newtonian fluids 522
rotating sphere viscometer 523

Schiller-Naumann drag formula 76
Schmidt number 370, 375, 387, 398, 400, 425, 455, 468
second order fluid 142, 144, 148, 164, 170, 194, 394
secondary normal stress difference 30, 31
sedimentation in 357
    Newtonian fluids 357
    non-Newtonian fluids 358
    power-law fluids 358
    slurries 357
segregation in
    fluidization 338, 358, 359
    sedimentation 358, 359
settling under
    dynamic conditions 162
    static conditions 50, 55, 118, 170
settling velocity
    time-dependent 129, 160, 171
shapes of bubbles
    in Newtonian fluids 188
    in non-Newtonian fluids 191
shapes of drops
    in Newtonian fluids 188
    in non-Newtonian fluids 191
shear rate
    average 75, 286
    characteristic 286
    effective 286
    falling ball viscometry, and 517
    surface average 286, 517
    at wall 286
shear rate dependent viscosity 11, 517
shear stress
    average 517
    at cell surface 228, 256, 259
shear-thickening fluids 12, 21, 83, 84, 187, 277, 334, 358, 377
shear-thinning fluids 11, 12, 55, 84, 222, 288, 326, 334, 358, 370, 375, 384, 426, 429, 438, 446, 453, 465, 482, 510
shear wave velocity 148, 164, 210, 211
Sherwood number 374, 385, 398, 401, 403, 405, 406, 454, 455, 468
Sisko model 57, 58
slip effects in beds 323
slip line theory 131
"soup-bowl" effect 29
specific surface area 243, 251
spheres
    bulk porosity for 241
    cessation of motion of 107, 146
    creeping motion of 49, 55
    in dilatant fluids 83
    drag reducing fluids and 161
    free fall velocity for 50, 77
    free settling of 50, 77
    in generalized Newtonian fluids 55, 83, 84
    interactions between 168
    mass transfer in 384, 394, 448, 450, 457
    Nusselt number 384, 397, 448, 450, 457
    regular arrays of 253, 257
    settling of 50, 77

spheres (*cont.*)
    shear rate around 75, 514
    steady motion of 49, 55
    terminal velocity of 50, 78
    transient motion of 82, 159
    in visco-elastic fluids 146
    in visco-plastic fluids 105
sphere-to-tube diameter ratio 472
sphericity 88, 99, 264, 265, 312
Spriggs model 154, 281
stagnation point 51, 62, 79, 121, 152, 314, 378, 388, 394, 437, 461
standard drag curve 50, 56, 68, 72, 76, 84, 105, 122, 253
static equilibrium 107, 118, 138
strain-hardening 33, 149, 185, 200, 311
stress relaxation 37, 129, 157
submerged objects models 253, 299
surface average particle diameter 264, 312
surface roughness 264, 312
surface tension 18, 177, 180, 185, 187, 209, 221, 223, 235, 353, 401
Sutterby fluid model 15, 57, 65, 193, 421, 423
Sutterby number 65, 70
synovial fluid 3, 7, 28

tangential stress 7, 11, 14, 18, 213
tangential velocity 213
terminal velocity
    of cylinders 86, 138
    effect of orientation 86, 139
    effect of shape 86, 138
    of non-spherical particles 86, 138
    of spheres 50, 77, 162
terminal velocity-volume behaviour of bubbles 206
thixotropy 24, 109, 131, 198, 283, 301
three-phase fluidized bed (TPFB) systems 352
time-dependent fluids 24
time-independent fluids 11
tortuosity factor 242, 246, 252, 265, 277, 286, 295, 316, 347
transient flow 54, 120, 141, 151, 160, 161, 172, 173, 195, 276, 279, 319
transitional flow 297
transverse flow over a
    circular cylinder 91, 99, 131, 164
    elliptical cylinder 93, 132, 142, 151, 269, 314, 384, 440, 471
    plate 137
    semi-circular cylinder 384, 442
    square cross-section bar 93, 269, 384, 440, 471
Trouton ratio 29, 33, 39, 310
tube bundles 266, 312, 411, 357
tubeless siphon 29
turbulence, elastic 41
turbulent flow 297
two-phase flow in packed beds 326

unconsolidated porous media 5, 240, 247, 249, 266, 288, 293, 320
unsteady motion 54, 82, 159, 160, 173, 195

Vane method for yield stress 41, 129, 521, 528
variational principles 55, 57, 63, 108, 118, 217, 223, 366, 385, 403
velocity
    discontinuity in 206
    fluidization 332, 333, 352
    free fall 50, 77, 509
    free rise 206
    free settling 50, 77, 509
    minimum fluidization 332, 333, 352
    over-shoot in 160
    sedimentation 357, 358
    shear wave 148, 164, 210, 211
    superficial 40, 238, 250, 305, 316, 332, 337, 351
    terminal 50, 78, 206, 509
    time-dependent 129, 159, 160, 171
velocity profile 7, 61, 114, 145, 167, 288, 319, 366, 402, 434, 473, 481
velocity-volume behaviour of bubbles 206
visco-elastic fluids
    bubbles in 188, 222
    characteristic time for 37, 120, 127, 185, 210, 234, 396, 523
    convection and 394, 466
    drag coefficient and 146, 162, 164
    drag reduction in 147, 161
    drops in 188, 222, 308
    fluidization with 350
    mass transfer in 394, 466
    in porous media 304, 319, 320
    sedimentation in 358
    wall effects in 147, 149, 493
viscometry
    capillary 510, 517
    falling ball 509
    falling cylinder 524
    falling needle 526
    rolling ball 521
    rotational sphere 523
visco-plastic fluids
    bubbles in 220, 221
    cylinders in 131
    drag force in 118, 131, 137, 138
    interaction between two spheres 110, 120, 131
    porous media flow 295, 316
    spheres in 107
    thixotropic behaviour of 109, 110, 111, 112
    wall effects in 491
viscosity
    at elevated pressure 15, 510
    at elevated temperature 15, 510
    Carreau equation for 16
    Ellis equation for 16
    elongational 32
    falling ball method for 509
    falling cylinder method for 524
    falling needle method for 526
    Newtonian 8
    Newton's law of 8
    plastic 17, 129, 382
    rolling ball method 521

temperature dependent 19, 459, 510
zero shear 16, 511 (*see also* zero shear viscosity)
viscosity values for common substances 9
viscous dissipation 84, 217, 247, 311, 370, 385, 418
void volume 241, 334
voidage 241, 334; *see also* porosity
Voigt model 37, 38
volume averaging of equations 257, 259, 303
vortex shedding 54, 93, 96, 101, 132, 136, 145, 162, 165, 168, 367, 378, 444, 481

wake characteristics 29, 49, 79, 94, 96, 136, 143, 146, 150, 151, 168
wall correction factor 504
wall effects 260, 311, 472
wall factor 260, 311, 472
Walters fluid model 142, 466
Weber number 190, 204, 215, 238, 398, 401

Weissenberg number 40, 141, 149, 158, 163, 168, 223, 319, 394, 397, 494
Williamson model 57, 71
worm-like micellar solutions 145, 155, 165, 171, 186, 195, 275, 314

yield stress 12, 17, 18, 519, 528
    Bingham 19, 21
    Casson 19
    falling ball viscometry 519
    measurement of 41, 129, 521, 528
yield-gravity parameter 107
Young's modulus 28, 29

zero shear viscosity 16, 510, 511, 516
    determination of 510, 512, 516
    extrapolation methods for evaluation of 511, 516
    falling ball viscometry 510, 511, 516
zero vorticity cell model 229, 256, 257, 273, 300, 302, 314, 315, 316